Internal Friction and Ultrasonic Attenuation in Crystalline Solids

Proceedings of the Fifth International Conference on
Internal Friction and Ultrasonic Attenuation in Crystalline Solids
August 27-30, 1973, Aachen, Germany

Edited by D. Lenz and K. Lücke

Volume II

Springer-Verlag
Berlin Heidelberg GmbH 1975

Dr.-Ing. Dieter Lenz
Oberingenieur am Institut für Allgemeine Metallkunde und Metallphysik, Rhein.-Westf. Technische Hochschule Aachen

Prof. Dr. rer. nat. Kurt Lücke
Direktor des Instituts für Allgemeine Metallkunde und Metallphysik, Rhein.-Westf. Technische Hochschule Aachen

ISBN 978-3-540-07436-6   ISBN 978-3-642-95275-3 (eBook)
DOI 10.1007/978-3-642-95275-3

CONTENTS

V O L U M E II
================

AMPLITUDE DEPENDENT INTERNAL FRICTION

CONTENTS VOLUME I

THEORY OF THE INTERACTION OF DISLOCATIONS
WITH ELECTRONS AND PHONONS[†]

A. D. Brailsford
Theoretical Physics Division, A.E.R.E., Harwell, England and
Scientific Research Staff, Ford Motor Company, Dearborn, Michigan 48121[*]

ABSTRACT

A formal theory of the dynamics of a dislocation in a visco-elastic medium is outlined. It is shown that when the interaction with thermal phonons is ignored, the dislocation experiences a drag force arising from certain generalized viscosities. There are two of these in an elastically isotropic body. Contributions to the bulk component from electron and phonon scattering processes are derived from Boltzmann equations determining the particle distributions among stationary states in phase space. This scheme leads to a unified theory of thermo-elastic damping, electron and phonon "viscosity", electron drag for large mean free path, the non-linearity mechanism associated with lattice anharmonicity and re-radiation damping. The influence of thermal phonons in producing a drag force (the flutter mechanism) is not discussed in any detail although its origin within the present formalism is also elucidated.

[†]Presented at the International Conference on Internal Friction and Ultrasonic Attenuation in Crystalline Solids held at Aachen, W. Germany on August 27-30, 1973.

[*]Present address.

1. INTRODUCTION

The possible physical origins of the drag force acting upon a dislocation moving in a real crystal have been appreciated for some time. The main problem in this field is not so much to discover new phenomena as to assess the relative importance of different mechanisms in the known variety of solid state types (e.g. insulators and metals--either alkali, noble or transition metals). Even aside from the direct effect of the discrete lattice structure on the dislocation motion (through the Peierls' force) this is still a complex matter involving at least six different phenomena, the uniqueness of which has been a matter of some dispute in the past. In this situation it is advantageous to have available a unified formalism incorporating all the physical ingredients which have been separately invoked in earlier work. One may hope thereby to assess the relative importance of these phenomena and also perceive the improvements which are necessary in application to specific materials.

It is clear that to achieve this aim one first needs a formalism of dislocation dynamics that takes into account the dissipative nature of the host medium, a major problem which to our knowledge has not been adequately explored in the past. Accordingly we begin by outlining an approach to this topic in the following section. This highlights the micro-scopic constitutive parameters of interest, here called the generalized viscosities. It is then shown in Section 3 how these may be obtained from the detailed dynamics of the elementary particles of the system (electrons and phonons, respectively), while Section 4 contains some detailed results of the formalism for the dislocation drag force.

2. OUTLINE OF DISLOCATION DYNAMICS

Suppose the perfect body is an elastically isotropic continuum with shear modulus G and Poisson's ration ν. Expanding the displacement field $\underline{u}(\underline{r})$ in the Fourier series

$$\underline{u}(\underline{r}) = \sum_{\underline{k},\lambda} \underline{e}_{\underline{k}\lambda} \, u_{\underline{k}\lambda} \, \exp(i\,\underline{k}\cdot\underline{r}) \quad , \tag{1}$$

where $\underline{e}_{\underline{k}\lambda}$ is the polarization vector for the mode λ (one longitudinal and two transverse) the potential and kinetic energies are[1]

$$U = \tfrac{1}{2} V \sum_{\underline{k},\lambda} k^2 C_\lambda \, u^*_{\underline{k}\lambda} \, u_{\underline{k}\lambda} \quad , \tag{2}$$

and

$$T = \tfrac{1}{2} \rho V \sum_{\underline{k},\lambda} \dot{u}^*_{\underline{k}\lambda} \, u_{\underline{k}\lambda} \quad , \tag{3}$$

where ρ is the density and the moduli C_λ are

$$C_1 = 2G(1-\nu)/(1-2\nu) \quad ; \quad \text{(longitudinal)} \quad , \tag{4}$$

and

$$C_2 = C_3 = G \quad ; \quad \text{(transverse)} \quad . \tag{5}$$

We account for all dissipative processes in the body by introducing a dissipative function Ψ analogous to the treatment of Landau and Lifshitz[2]:

$$\Psi = \int \left\{ \eta(\dot{\epsilon}_{\alpha\beta} - \tfrac{1}{3}\delta_{\alpha\beta}\,\dot{\epsilon}_{\gamma\gamma})^2 + \tfrac{1}{2}\zeta\,\dot{\epsilon}_{\gamma\gamma}^{\,2} \right\} dV \quad , \tag{6}$$

where $\dot{\epsilon}_{\alpha\beta}$ are the time derivatives of the usual strain components, $\delta_{\alpha\beta}$ the Kronecker delta symbol and η and ζ are two viscosity coefficients.

However, as will eventually become clear (see the Appendix) we intend these viscosities to represent all losses in the medium (including thermoelastic effects) and thus their physical meaning is different from that in Ref. 2. They will be called generalized viscosities to emphasize the distinction. In terms of the $u_{\underline{k}\lambda}$ we find

$$\Psi = \tfrac{1}{2} V \sum_{\underline{k},\lambda} k^2 \zeta_{k\lambda} \, \dot{u}_{\underline{k}\lambda}^{\;*} \, \dot{u}_{\underline{k}\lambda} \qquad (7)$$

where

$$\zeta_{k1} = \tfrac{4}{3} \eta + \zeta \quad ; \qquad \text{(longitudinal)} \qquad (8)$$

and

$$\zeta_{k2} = \zeta_{k3} = \eta \quad ; \qquad \text{(transverse)} \quad . \qquad (9)$$

The possible dependence[3] of the generalized viscosities upon wave vector, which follows from the analysis of later sections, has been anticipated, although it of course does not follow from (6).

The inclusion of a dislocation in the body is described by supplementing the potential energy with a term, U_1, representing the effect of the fictitious body forces which induce its strain field,[4] namely

$$U_1 = \tfrac{1}{2} \sum_{k\,\lambda} A_{\underline{k}\lambda} \, u_{\underline{k}\lambda} \quad . \qquad (10)$$

Here

$$A_{k\lambda} = G \left\{ (\underline{k} \cdot \underline{n})(\underline{e}_{\underline{k}\lambda} \cdot \underline{b}) + (\underline{k} \cdot \underline{b})(\underline{e}_{\underline{k}\lambda} \cdot \underline{n}) \right\} f_{\underline{k}}, \quad (11)$$

$\underline{b}$ is the Burgers' vector, $\underline{n}$ is a unit vector normal to the slip plane, $f_{\underline{k}}$ is

$$f_{\underline{k}} = i \int_C \exp(i \, \underline{k} \cdot \underline{r}) \, dS \quad , \qquad (12)$$

and the integration is over the area in the slip plane enclosed by the contour C bounding the slipped region.

The potential and kinetic energies of the system, together with the dissipative function contain all the information necessary to describe the internal dynamics of the body. In particular, it must be possible to derive a theory of dislocation dynamics from them. Below we will indicate the scheme which leads to the required result although it will be pointed out that the analysis is still incomplete in one aspect, namely treatment of the flutter mechanism.

The general idea is contained in considering motion which to a first approximation represents uniform translation of the dislocation at constant velocity. Since there is no change of shape, the factor $f_{\underline{k}}$ may be written as

$$f_{\underline{k}} = f_{\underline{k}}(0) \exp(i \, \underline{k} \cdot \underline{r}_o) \quad , \qquad (13)$$

where $\underline{r}_o$ represents some identifiable point on the dislocation and $f_{\underline{k}}(0)$ represents the integration in (12) with this point taken as origin. Thus $f_{\underline{k}}(0)$ depends only upon the shape of the dislocation, and may be called the form factor, whereas $\underline{r}_o$ describes its position in space (e.g. for a kink[1] in a dislocation, $\underline{r}_o$ describes the position of the kink, whereas for a long straight dislocation it describes the position in space of some point fixed within the dislocation line). The assumption of essentially uniform velocity implies therefore $\underline{r}_o \simeq \underline{r}_o(0) + \underline{v}_D t$ where $\underline{r}_o(0)$ is the initial position, henceforth taken as zero. The essential point is that $f_{\underline{k}}$ is time dependent because of the dislocation movement.

With the $u_{\underline{k}\lambda}$ as generalized coordinates, Lagrange's equation of motion, supplemented by the frictional force derived from Ψ, is of the form

$$\rho V \ddot{u}_{\underline{k}\lambda} + V k^2 C_\lambda u_{\underline{k}\lambda} + V k^2 \zeta_{k\lambda} \dot{u}_{k\lambda} - A_{-\underline{k}\lambda} = 0 \qquad (14)$$

where $A_{-\underline{k}\lambda} = A_{\underline{k}\lambda}^*$. Thus, taking Fourier transforms with time[5]:

$$u_{\underline{k}\lambda} = \sum_\omega u_{\underline{k}\lambda,\omega} \exp(-i\omega t) \quad , \qquad (15)$$

we find

$$u_{\underline{k}\lambda,\omega} = A_{\underline{k}\lambda,\omega}^* \Big/ \rho V \left\{ \omega_{k\lambda}^2 - \omega^2 - (i\omega\zeta_{k\lambda}k^2/\rho) \right\} \qquad (16)$$

where $A_{\underline{k}\lambda,\omega}$ is given by

$$A_{\underline{k}\lambda} = \sum_\omega A_{\underline{k}\lambda,\omega} \exp(i\omega t) \qquad (17)$$

and $\omega_{k\lambda} = k(C_\lambda/\rho)^{\frac{1}{2}}$ are the sound wave frequencies. This, together with the normal sound waves

$$u_{\underline{k}\lambda} = \alpha_{\underline{k}\lambda} \exp(i\nu_{k\lambda}^{(+)}t) + \beta_{\underline{k}\lambda} \exp(i\nu_{\underline{k}\lambda}^{(-)}t) \qquad (18)$$

where

$$\nu_{\underline{k}\lambda}^{(\pm)} = \pm \left\{ \omega_{k\lambda}^2 - (\zeta_{k\lambda}k^2/2\rho)^2 \right\}^{\frac{1}{2}} + i(\zeta_{k\lambda}k^2/2\rho) \quad , \qquad (19)$$

defines the complete solution of (15), and enables us to evaluate the energy and dissipation in the solid for any prescribed dislocation motion.

In particular, for uniform translation at a velocity much less than the sound velocities, c_1 and c_2, this procedure has previously[1] been followed in a derivation of the dislocation effective mass from the kinetic energy of the body. In detail, since then $A_{\underline{k}\lambda,\omega} = A_{\underline{k}\lambda}(0)\, \delta_{\omega,\underline{k}\cdot\underline{v}_D}$, where $A_{\underline{k}\lambda}(0)$ is obtained from (11) upon replacing $f_{\underline{k}}$ by $f_{\underline{k}}(0)$, we have shown that

$$T = T_{TH} + \tfrac{1}{2} m^{*} v_D^{2} \quad , \tag{20}$$

where T_{TH} is the contribution to the kinetic energy from thermal excitations (i.e., from (18) when there are no external stresses) and m^{*} is the dislocation effective mass. Similarly, ignoring for the moment the thermal waves, evaluation of the potential energy has been interpreted in terms of the dislocation strain-energy.[6] Here, we have also derived by the above analysis, a contribution to the power dissipation, P, since this is equal to twice the dissipative function[2]:

$$\tfrac{1}{2} P = \Psi = \tfrac{1}{2} B_o v_D^{2} \quad , \quad \text{say} \quad , \tag{21}$$

where

$$B_o = \frac{1}{V} \sum_{\underline{k}\lambda} \zeta_{k\lambda} \left| \frac{k\underline{v}\cdot\underline{k}\,A_{\underline{k}\lambda}(0)}{\rho\omega_{k\lambda}^{2}} \right|^{2} \quad , \quad v_D \ll c \quad , \tag{22}$$

(and $\underline{v}$ is a unit vector along the direction of motion) as may be verified by substitution of (15) in (7).

A detailed investigation of the longitudinal mode contribution to (22) will be the major concern of the following sections. First, however, we establish the connection with a formal theory of dislocation dynamics. Again ignoring thermal vibrations for the present, suppose in addition there is a spatially uniform constant external stress acting in the slip plane in the slip direction. In the present context this implies an additional potential energy term, U_3, where

$$U_3 = \int_C n_i\, \sigma_{ij}\, b_j\, dS \tag{23}$$

7

which, for the long straight dislocation moving with velocity $\underline{v}_D$ along the direction $\hat{x}$ is equivalent to

$$U_3 = -\sigma b L x \quad , \tag{24}$$

where x is the displacement. Thus if we take the expressions (20), (21) and (24) and formally apply the Lagrange technique again, with now x and v_D as coordinate and velocity, we obtain the well-known equation

$$m^* \dot{v}_D = \sigma b L - B_o v_D \quad , \tag{25}$$

the drag force, $-B_o v_D$, being given by[2] $-(\partial\Psi/\partial v_D)$. This much appears to be nothing new until it is noted that it supplies an important check of self-consistency. Namely that the assumption of uniform motion, which is the basis of its derivation, is compatible with its solution, at least at times $t > (m^*/B)$.

Suppose now we restore the thermal excitations[7] (18) into the complete solution for $u_{\underline{k}\lambda}$. These, when substituted back into the potential energy terms lead to cross product terms between the thermal stress and dislocation component. Hence the equation of motion replacing (25) contains driving stresses corresponding to all frequencies and wave vectors of the excitation spectrum. The assumption of precisely uniform drift is thus never completely consistent since the dislocation "flutters" during its average translation and one must revert to (13) and assume $\underline{r}_o(t)$ is composed of both uniform drift and flutter. This implies a further change in the ultimate equation of motion to replace (26), the solution of which must be consistent with the assumed fluttering motion underlying its derivation. While we have not carried through this program to date, a treatment of this flutter mechanism of dislocation drag clearly appears within the scope of the present framework. Fortunately a thorough treatment of this topic for a kink in a dislocation, using a somewhat different approach (but implying a Lagrangian description of the kink motion), has already been given by Eshelby[8] in an investigation of some aspects of an earlier derivation by Leibfried.[9] The main point to be made here is that this fluttering is a dynamic effect resulting from the response of the dislocation in its motion to the thermal excitations of the medium in which it resides, whereas the drag force which occurs in (25) arises only from the loss processes inherent in the excitations themselves, the dynamics of the dislocation playing no role beyond the assumed motion per se.

In conclusion we might mention that while the above Lagrangian scheme has been discussed here within the context of motion at a uniform velocity (on the average), it can also be applied to specific cases where changes in dislocation shape are well defined, such as the problem of a vibrating dislocation loop. We obtain thereby a fundamental derivation of the Koehler-Granato-Lücke vibrating string model in terms of the microscopic parameters of the host medium. This particular application of the general technique will be discussed elsewhere.[10]

3. THE GENERALIZED VISCOSITIES

For other than the flutter mechanism we have shown that the drag coefficient B is determined from (22) once the generalized viscosities $\zeta_{\underline{k}\lambda}$ are known. In this section the method of obtaining these from the microscopic quantum states of the system is presented.

The line of reasoning is essentially due to Maris.[11] Suppose in the deformed body at steady state the energies of the elementary particles of the system (electrons or phonons) are denoted by E_q where the suffix q labels the energy level, E_q^o, from which E_q derives in the absence of deformation. For example, if these levels are sensitive only to the local dilatation, Δ, then

$$E_q = \hbar\omega_{q\lambda} (1 - \gamma\Delta) \quad , \quad (\text{phonons}) \quad , \quad (26)$$

or

$$E_q = \frac{\hbar^2 q^2}{2m} + E_1 \Delta \quad , \quad (\text{free electrons}) \quad , \quad (27)$$

where γ is Gruneisen's constant, m the electron effective mass and E_1 the deformation potential.[12] Since in general, therefore, the stress is given by $\sigma_{\alpha\beta} = V^{-1}(\partial\mathcal{E}/\partial\epsilon_{\alpha\beta})_S$ where $\mathcal{E}$ is the total internal energy of the body, S its entropy and V the volume

$$\sigma_{\alpha\beta} = \frac{1}{V}\left[\frac{\partial}{\partial\epsilon_{\alpha\beta}} \left(\sum_q N_q E_q \right)\right]_S \quad , \quad (28)$$

where N_q is the occupation probability of the state q. As the entropy is a functional of only[13] the N_q, the derivative in (28) is to be taken with these variables held fixed. Thus, for the relations (26) and (27),

$$\sigma_{\alpha\beta} = - \delta_{\alpha\beta}(\gamma/V)\sum_q N_q E_q^o \quad , \quad (\text{phonons}) \quad , \quad (29)$$

or

$$\sigma_{\alpha\beta} = \delta_{\alpha\beta}(1/V)\sum_q N_q E_1 \quad , \quad (\text{electrons}) \quad , \quad (30)$$

$$= \delta_{\alpha\beta} \, n \, E_1 \quad\quad\quad\quad\quad (31)$$

where $n = (\sum_q N_q/V) = (N/V)$ is the electron density.[14]

These stresses refer only to contributions from modifications of the spectrum of the elementary excitations and do not account for stresses arising from the static potential energy of deformation. These can be incorporated in the phonon case quite simply by measuring the stress (29) relative to its value for an adiabatic change (when N_q has its equilibrium value in the deformed state, which by definition must be the same as in the undeformed state, i.e., N_q^e). Thus to first order in the strain, including now terms from the static deformation we have

$$\sigma_{\alpha\beta} = C_{\alpha\beta\gamma\delta} \, \epsilon_{\gamma\delta} - \delta_{\alpha\beta}(\gamma/V) \sum_q (N_q - N_q^e) E_q^o \quad , \qquad (32)$$

where $C_{\alpha\beta\gamma\delta}$ are the adiabatic elastic constants.

Evidently more general distortions of the energy spectra than (26) and (27), involving shear strains, will produce additional contributions to the shear stresses. For nearly free electron metals these are known to be small but one has no assurance that this is generally true. Such generalizations could be included by extending the analysis to be given below. However for simplicity they will here be ignored.

Equation (32) therefore represents the stress-strain law for the model electrical insulator whose phonon energies change with deformation according to (26). Comparing this with the relation of the type

$$\sigma_{\alpha\beta} = \sigma_{\alpha\beta} + \sigma'_{\alpha\beta} \quad , \qquad (33)$$

where $\sigma_{\alpha\beta}$ is the elastic stress and $\sigma'_{\alpha\beta}$ the dissipative stress obtained from Ψ, Eq. (6),

$$\sigma'_{\alpha\beta} = \delta_{\alpha\beta} \, \zeta \, \dot{\Delta} \quad , \qquad (34)$$

for the same deformation (the present model has zero shear viscosity, $\eta = 0$) one sees that evaluation of (32) (and (31) for metals) will lead to an expression for the generalized bulk viscosity ζ_k, and hence the drag force B_o.

A. Phonon Contribution to ζ_k

The steady state phonon occupation numbers required for the evaluation of (32) are obtained from the Boltzmann equation in the form first given by Woodruff and Ehrenreich[15] in their study of sound attenuation. The only generalization necessary is that one must consider deformation waves for arbitrary frequency and wave vector, as in (1) and (15).

The equation satisfied by N_q is then

$$\frac{\partial N_q}{\partial t} + \underline{v}_{-q} \cdot \frac{\partial N_q}{\partial \underline{r}} - \frac{\partial E_q}{\partial \underline{r}} \cdot \frac{1}{\hbar}\frac{\partial N_q}{\partial q} = \left[\frac{\partial N_q}{\partial t}\right]_{coll} \quad , \qquad (35)$$

where $\hbar$ is Planck's constant and, as before, q is an abbreviation for the wave vector $\underline{q}$ and polarization index λ of the mode. Collisions are assumed to relax the distribution to the Planck distribution characteristic of the local deformation and local temperature $T' = T + \delta T$, the latter being incorporated to allow for thermoelastic effects. Thus in the single relaxation time approximation[16]

$$\left[\frac{\partial N_q}{\partial t}\right]_{coll} = \frac{1}{\tau}\left\{N_q - N_q^o(T')\right\} \quad , \qquad (36)$$

where

$$N_q^o(T') = \left\{\exp\left(E_q/KT'\right) - 1\right\}^{-1} \quad , \qquad (37)$$

K being Boltzmann's constant.

To lowest order (37) yields a linear equation for $n_q^{(1)}$, where

$$N_q \equiv N_q^e + n_q^{(1)} \quad , \tag{38}$$

and N_q^e is the Planck distribution at the temperature T for the undeformed energies E_q^o ($\equiv \hbar\omega_{q\lambda}$), as a function of the dilatation Δ and temperature variation δT. One finds[17] that if

$$\Delta = \sum_{\underline{k},\omega} \Delta_{\underline{k}\omega} \exp i(\underline{k} \cdot \underline{r} - \omega t) \quad , \tag{39}$$

and

$$\delta T = \sum_{\underline{k},\omega} \delta T_{\underline{k},\omega} \exp i(\underline{k} \cdot \underline{r} - \omega t) \quad , \tag{40}$$

then

$$n_q^{(1)} = \sum_{k,\omega} \frac{C_q}{\hbar\omega_q} \left[\frac{\gamma T(1 + i\underline{k} \cdot \underline{v}_q \tau) \Delta_{\underline{k},\omega} + \delta T_{\underline{k},\omega}}{(1 - i\omega\tau + i\underline{k} \cdot \underline{v}_q \tau)} \right] \exp i(\underline{k}\cdot\underline{r} - \omega t), \tag{41}$$

where $C_q = \hbar\omega_q(\partial N_q^e/\partial T)$ is the contribution of the mode q to the specific heat. The interpretation of this result, as it reflects in the stress-strain law (32) for arbitrary Δ and δT, is discussed in the Appendix. In fact, δT is not a free variable, but is here constrained by the fact that collision processes must conserve the zero order energy[18],

$$\sum_q \hbar\omega_q \left[\frac{\partial N_q}{\partial t} \right]_{coll} = 0 \quad , \tag{42}$$

which one can show is equivalent to the constraint that the equation of heat conduction is obeyed. This determines δT, but not in a very transparent manner, unless it is assumed that 1) the distinction between longitudinal and transverse sound velocities may be ignored and 2) that the q dependence of τ may be ignored. Neither is particularly realistic and the ensuing results have to be treated with caution for this reason. Nevertheless, adopting these approximations yields

$$\delta T_{\underline{k},\omega} \simeq \frac{i\omega\tau a}{(k\ell - a)} \gamma T \triangle_{\underline{k}\omega} \tag{43}$$

where $a = \tan^{-1}[k\ell/(1-i\omega\tau)]$, and a stress amplitude $\sigma_{\alpha\beta:\underline{k}\omega}$ accompanying each Fourier component of

$$\sigma_{\alpha\beta:\underline{k}\omega} \simeq C_{\alpha\beta\gamma\delta} \; \varepsilon_{\gamma\delta:\underline{k}\omega} - \delta_{\alpha\beta}\left\{ 1 + [i\omega\tau a/(k\ell - a)]\right\} \gamma^2 C T\triangle_{\underline{k},\omega} \quad , \tag{44}$$

which follows from (32), (here C is the total specific heat).

Comparing (44) and (34), since ζ_k is defined in terms of the total out-of-phase component of the stress, we obtain

$$\zeta_{k1} \simeq \gamma^2 C\,T\,\tau\,\mathrm{Re}\left\{a/(k\ell - a)\right\} \quad , \tag{45}$$

where Re denotes the real part. This has the following limiting behavior when $\omega\tau < 1$ (i.e., low dislocation velocity)

$$\zeta_{k1} \simeq \gamma^2 C(3\tau/k^2\ell^2)/\left\{ 1 + \omega^2(3\tau/k^2\ell^2)^2\right\} \quad , \qquad k\ell < 1 \quad , \tag{46}$$

and

$$\zeta_{k1} \simeq (\pi\gamma^2 T/2k)(C/c) \quad , \qquad k\ell > 1 \quad , \tag{47}$$

where c is the assumed common sound velocity. Thus for deformation waves of wavelength less than the phonon mean free path, ζ_{k1} is independent of frequency. Presently we will show that this arises from phonons propagating essentially parallel to the wave fronts. When $k\ell < 1$, however, ζ_{k1} is frequency dependent. The generalized viscosity then results from thermoelastic effects, since with the model considered $3\tau/k^2\ell^2$ is the time necessary for heat to diffuse a distance of the order of a half wavelength.

Various refinements of these results are suggested by a more detailed investigation of the loss mechanisms. It can be shown, for example, that the large $k\ell$ regime corresponds to the region where the Golden Rule perturbation theory of quantum mechanics is valid. Evaluation[17] of the power dissipation with a model interaction Hamiltonian consistent with (26) gives a result analogous to (47) but with the additional feature that $\hbar\underline{k}$ has the significance of a momentum transfer to the phonon during a scattering process. Since this is only possible for phonons of wave vector $q > \frac{1}{2}k$ we find that the total specific heat C should really be replaced by $\overline{C}_k$ where

$$\overline{C}_k = \sum_{q > \frac{1}{2}k} C_q \quad . \tag{48}$$

Additionally we can assess the effect of differing sound velocities in this region, for the physical reasoning[17] underlying (47) suggests that we should then replace C/c by $\sum_\lambda C_\lambda/c_\lambda$ where c_λ is the sound velocity for the modes of polarization λ and C_λ their specific heat. A more accurate version for the generalized bulk viscosity is therefore

$$\zeta_{k1} \simeq (\pi\gamma^2 T/2k) \sum_\lambda \overline{C}_{k\lambda}/c_\lambda \quad , \qquad (k\ell > 1) \quad , \tag{49}$$

to incorporate both effects discussed above.

B. Electron Contribution to ζ

It is seen from (31) that the electronic contribution is determined by the change in density and in each electron energy with the deformation. These changes arise because of the electronic charge, the local variations in ionic charge and temperature inducing density changes in the electron gas which in turn tend to screen the effect of the initial deformation. $E_1 \triangle/e$, where e is the electronic charge, is the total resulting potential at zero frequency. The above may also be found from the solution of a Boltzmann equation.

This equation is again of the form (35) if E_q is now given by (27). The only difference resides in the collision term, for which we here adopt the Ansatz that the effect of collisions is to relax the distribution to a Fermi distribution characteristic of the local particle energies, the local temperature T', and a Fermi energy ϵ_F corresponding to the undeformed state.[19] Thus we take

$$\left(\frac{\partial N_q}{\partial t}\right)_{coll} = -\frac{1}{\tau}\left\{N_q - f_o(E_q, T')\right\} \quad , \tag{50}$$

where

$$f_o(E_q, T') = \left\{1 + \exp\left[(E_q - \epsilon_F(T))/KT'\right]\right\}^{-1} \quad , \tag{51}$$

is the Fermi function.

As before, the Boltzmann equation is linearized with respect to $\triangle$ and δT to yield

$$N_q = f_o(\epsilon_q, T) + \sum_q \left\{\frac{(1 + i\underline{k} \cdot \underline{v}_q \tau)E_1 \triangle_{\underline{k}\omega} - (\epsilon_q - \epsilon_F)\, \delta\, T_{\underline{k}\omega}/T}{(1 - i\omega\tau + i\underline{k} \cdot \underline{v}_q \tau)}\right\} \frac{\partial f_o}{\partial \epsilon_q}$$
$$\times \exp i(\underline{k} \cdot \underline{r} - \omega t) \quad , \tag{52}$$

where $\epsilon_q = \hbar^2 q^2/2m$ and $\underline{v}_q = \hbar \underline{q}/m$. From this we obtain the amplitude of the $(\underline{k}, \omega)$ Fourier component of the number density. I.e., if this density is $n + \delta n(\underline{r}, t)$, where n is the average value, and

$$\delta n(\underline{r}, t) = \sum_{\underline{k}, \omega} \delta n_{\underline{k}\omega}\, \exp i(\underline{k} \cdot \underline{r} - \omega t) \quad , \tag{53}$$

then $\quad \delta n_{\underline{k}, \omega} \simeq - (3n/2\epsilon_F) \left\{ \left(1 + \frac{i\omega\tau a}{k\ell} \right) E_1 \triangle_{\underline{k}\omega} + \frac{a}{k\ell} \frac{\partial\epsilon_F}{\partial T} \delta T_{\underline{k}\omega} \right\} \quad . \quad$ (54)

This results from integrating (52) over all states. Here $\ell = v_F \tau$ is the electron mean free path,

$$\frac{\partial\epsilon_F}{\partial T} = - \frac{\pi^2}{6} \frac{K^2 T}{\epsilon_F} \quad , \quad\quad\quad\quad (55)$$

and (again) $a = \tan^{-1}\{k\ell/(1 - i\omega\tau)\}$. The variable charge distribution gives rise to a potential, E_s/e, say, satisfying

$$\nabla^2 E_s = - 4\pi e^2 \delta n \quad , \quad\quad\quad\quad (56)$$

which together with the potential, E_i/e, resulting from the positive charge displacement

$$E_i = ne^2 \int \left| \underline{r} - \underline{r}' \right|^{-1} \triangle(\underline{r}') \, d\underline{r} \quad , \quad\quad\quad (57)$$

defines the total energy change $E_1 \triangle = E_s + E_i$. Thus, from the preceding we find

$$E_1 \triangle_{\underline{k}\omega} \simeq \frac{\frac{2}{3} \epsilon_F \triangle_{\underline{k}\omega} - \frac{a}{k\ell} \frac{\partial\epsilon_F}{\partial T} \delta T_{\underline{k}\omega}}{\left\{ 1 + \frac{k^2}{q_{TF}^2} + \frac{i\omega\tau a}{k\ell} \right\}} \quad , \quad\quad\quad (58)$$

where $q_{TF} = (6\pi ne^2/\epsilon_F)^{\frac{1}{2}}$ is the inverse of the Thomas-Fermi screening length. In addition it is required that the electron density charge arrives from the drift terms in the Boltzmann equation only, since scattering is a local event. Thus it follows that, for electrons,

$$\sum_q \left(\frac{\partial N_q}{\partial t} \right)_{coll} = 0 \quad , \quad\quad\quad\quad (59)$$

or

$$\frac{i\omega\tau a}{k\ell} E_1 \Delta_{\underline{k}\omega} - \left(1 - \frac{a}{k\ell}\right) \frac{\partial \epsilon_F}{\partial T} \delta T_{\underline{k}\omega} \simeq 0 \quad , \tag{60}$$

a result derived from (50) and (52). Consequently solution of (58) and (60) for E_1 together with (31) and (54) leads to the electronic contribution to the generalized viscosity:

$$\zeta_{k1} = (3nE_1^2 \tau/2\epsilon_F) \, \text{Re} \, [a/(k\ell - a)] \quad , \tag{61}$$

which has the limiting forms for low frequencies ($\omega\tau < 1$)

$$\zeta_{k1} \simeq \tfrac{2}{3} n\epsilon_F \left(\frac{3\tau}{k^2\ell^2}\right) \Big/ \left\{ 1 + \left(\frac{3\omega\tau}{k^2\ell^2}\right)^2 \right\} \left\{ 1 + \frac{k^2}{q_{TF}^2} \right\}^2 \quad , \quad k\ell < 1 \quad , \tag{62}$$

and

$$\zeta_{k1} \simeq \pi n\epsilon_F/3v_F k \left\{ 1 + \frac{k^2}{q_{TF}^2} \right\}^2 \quad , \qquad k\ell > 1 \quad . \tag{63}$$

Thus, as with the phonon contribution, for long wavelengths the viscosity depends upon the heat diffusion time whereas for short wavelengths it also is independent of relaxation time. Additionally, the short wavelength limit can be obtained from quantum mechanical perturbation theory[20,21] through a calculation of the power dissipation.

4. EVALUATION OF THE DRAG FORCE

It was shown in Section 2 that once the dislocation shape is specified (so that the form factor in $A_{\underline{k}\lambda}(0)$ is known) the drag force constant B_o can be evaluated from a knowledge of the generalized viscosities $\zeta_{k\lambda}$. In this Section we summarize the results for a long straight dislocation. Other geometries, such as the motion of a kink along a dislocation, will only be referred to briefly, details may be found elsewhere.[17]

With the (incomplete) models of (26) and (27) only ζ_{k1} is non-zero so that longitudinal deformations only contribute to the sum in (22). The resulting expression for B_O (or more precisely the drag force per unit length, $B = B_O/L$) has been evaluated[17] for ζ_{k1} given by (45) to yield*, for an edge dislocation,

$$ B_{ph} = B_{TE;ph} + B_{s,ph} \quad , \tag{64} $$

where

$$ B_{TE,ph} \simeq \frac{3\,C\,T}{4\pi c\,\ell_{ph}}\,(\gamma\,s^2\,b)^2\,\ell n\left(\frac{c}{v}\right) \quad , \tag{65} $$

and

$$ B_{s,ph} \simeq \frac{3T}{2c}\left(\frac{\gamma s^2 b}{2\pi}\right)^2 \int_0^{q_D} q^3\,C_q\,dq \quad . \tag{66} $$

Here s is the ratio of transverse to longitudinal sound velocities, q_D the Debye wave vector and ℓ_{ph}, the phonon mean free path, is the same as the ℓ parameter of Section 3(A), the suffix being added for clarity. $B_{TE,ph}$ is the drag associated with thermoelastic losses[22] and originates from the long wavelength components of the deformation (see (46)) while $B_{s,ph}$, usually called the phonon scattering[23] or non-linearity contribution[24], derives from the short wavelength components satisfying (47). These identifications are discussed fully in Ref. 17.

_____

*These are obtained from (45) with $\overline{C}_k$ replacing C and a cut-off in k at $2q_D$, the maximum wave vector transfer. The further refinement using (49) for the short wavelength deformation components will not be discussed here.

Numerical evaluation of the integral in (66) leads to the form

$B_{s,ph} = (45\, q_D\, b\, \gamma^2\, s^4/16)(n\, K\, \theta\, b/c)\, \beta\, (T/\theta)$ where the temperature dependence

contained in β is illustrated in Fig. 1. From this we find, with

$\gamma = 1.5$, $s = \frac{1}{2}$, $c = 5 \times 10^5$ cm/sec, $n = 10^{23}/cm^3$ and $q_D = 1.76 \times 10^8/cm$

a room temperature value $B_{s,ph} \simeq 10^{-4}$ dyne. sec/cm^2. This drag component

is seen to be linear at high temperatures ($T \geq \theta$), whereas when $T \ll \theta$,

$B_{s,ph} \propto T^5$. The thermoelastic contribution is relatively small (for

example, with $\ell \sim 40b$ at room temperature, $B_{TE,ph} \sim \frac{1}{13}\, B_{s,ph}$) and is

generally ignored in analyses of experimental data.

Electronic contributions to the drag force may be derived with

the aid of (61), and are found to be

$$B_{TE,e\ell} \simeq \frac{n\epsilon_F\, s^4\, b^2}{4\pi v_F \ell_e}\, \ell n\, \left(\frac{v_F}{v_D}\right) \quad, \tag{67}$$

and

$$B_{s,e\ell} = \frac{n\epsilon_F\, s^4\, b^2\, k_F}{6\, v_F}\, \Phi\, \left(\frac{2k_F}{q_{TF}}\right) \quad, \tag{68}$$

where

$$\Phi(x) = \frac{1}{2}\left[(1+x^2)^{-1} + x^{-1}\tan^{-1}x\right] \quad, \tag{69}$$

and the electron mean free path is now designated by ℓ_e. The result (68),

which we shall call the electron scattering contribution, is the term

derived from perturbation theory by Holstein[20], apart from minor detail,

and differs from our earlier version[21] only through the choice of a cut-

off of $2k_F$ (where $k_F = (3\pi^2 n)^{\frac{1}{3}}$ is the Fermi radius) instead of q_D in the

range of deformation waves considered. As with the Grüneisen model con-

sidered earlier, the scattering contribution (68) to the drag is the

larger, yielding $B_{s,e\ell} \simeq 10^{-5}$ dyne. sec/cm^2 for $\epsilon_F = 5$ eV. In this instance, however, because of the electron gas degeneracy, the electronic component is independent of temperature.

With the models discussed here the four contributions listed represent the dominant mechanisms of drag on a dislocation in uniform motion. There are no additional effects associated with the kinetic theory forms of the isothermal viscosities of the electron and phonon systems.[25,26] These are already contained in the analysis outlined above, as we explain in more detail in the Appendix. The essential reason is that for the long wavelength components of the deformation the dominant source of loss is through heat conduction, it is only the short wavelength components for which isothermal conditions prevail. But then, simultaneously, for these, the effective force $-(\partial E_q / \partial \underline{r})$ exerted on the particles of the system tends to average to zero over distances over the order of the mean free path. It is only a fraction, $\sim (2/k\ell)$ of the total number, moving parallel to the wave fronts of the deformation that extract energy from the field. Thus the effective isothermal viscosity is, for the electron gas,

$$\zeta_e' \sim \frac{2}{k\ell_e} \cdot n m \cdot \tfrac{1}{3} v_F \ell_e \quad , \qquad (70)$$

and for the phonon system

$$\zeta_{ph}' \sim \frac{2}{k\ell_{ph}} \cdot \frac{\mathcal{E}_T}{c^2} \cdot \tfrac{1}{3} c \, \ell_{ph} \quad , \qquad (71)$$

where $\mathcal{E}_T / c^2$, with $\mathcal{E}_T$ the thermal energy density, has been identified with the mass density of kinetic theory. It will be noted that (70) and (71) are almost identical with (63) and (47). Consequently $B_{s,ph}$ and $B_{s,e\ell}$ might equally well be called phonon and electron viscosity contributions since the scattering approach, or the viscosity approach, have now been shown to be identical. However it is essential in the latter to bear in mind the ineffectiveness of most particle of the system, exemplified in (70) and (71), in determining its response to imposed forces.

So far we have concentrated on uniform motion. For the oscillatory motion of interest in internal friction studies one must go back to the theory of Section 2 and, as long as one ignores "flutter", insert an oscillatory form for $\underline{r}_0(t)$, say $\underline{r}_0(t) = \underline{R} \cos(\Omega t + \chi)$, in $f_{\underline{k}}$. To terms linear in R therefore, $A_{\underline{k}\lambda}$ has frequency components at $\pm \Omega$. The remaining analysis then follows through as before except that now one should retain the complete denominator in (16) to obtain[17]

$$B_0(\Omega) = \frac{1}{V} \sum_{\underline{k}\lambda} \zeta_{k\lambda} \left| \frac{k(\underline{v} \cdot \underline{k}) A_{\underline{k}\lambda}(0)}{\rho \left\{ \omega_k^2 - \Omega^2 - i\Omega \zeta_{k\lambda} k^2/\rho \right\}} \right|^2 \quad . \tag{72}$$

The resonance absorption which is seen to contribute to $B_0(\Omega)$ is usually called "re-radiation damping" since it may also be derived[8] by considering the power radiated from a vibrating dislocation segment in a non-viscous medium. The reason such an approach is possible is because although individual terms in the k sum (integral) are large near the resonance at Ω/c the width on the wave vector scale is very narrow. The net result from this phenomenon, for an edge dislocation[17]

$$B_R(\Omega) = \frac{1}{8} (1 + s^4) \rho b^2 \Omega \quad , \tag{73}$$

is consequently independent of the viscosity. At normal operating frequencies the above re-radiation term is quite small ($B_R \sim 3 \times 10^{-8}$ dynes. sec/cm^2 at 10 MHz) and in metals can be neglected compared with the electronic component. This mechanical resonance effect is an additive contribution to the drag force. The thermoelastic and scattering contributions still remain. While the latter may be shown[17] to be unmodified by the oscillatory motion, the thermoelastic drag is reduced by a factor of two and the argument of the logarithm in (65) becomes $(\Omega\tau)^{-1}$ instead of (c/v).

Finally, to complete the list of commonly invoked mechanisms in simple metals or insulators we quote without derivation an adaptation of the result of Eshelby, for the drag force on a kink in uniform motion (on the average), due to the flutter mechanism[8,22,27]:

$$B_{F\ell} \simeq F\left(\frac{w}{a}\right) \cdot \frac{1}{n_o a} \cdot \frac{w^2}{10ac}\, \ell_{TH} \quad . \tag{74}$$

This result incorporates Lothe's correction for the finite width w of the kink through the factor F,

$$F(x) = \left\{ \int_o^{\frac{w\theta}{aT}} \frac{z^3}{(e^z - 1)}\, dz \right\} \Big/ \left\{ \int_o^{\frac{\theta}{T}} \frac{z^3\, dz}{(e^z - 1)} \right\} \quad , \tag{75}$$

where a is the distance between neighboring close-packed rows in the slip plane, and a dependence upon the kink density, n_o, in order to relate lateral motion of the kink to forward displacement of the dislocation. At high temperatures $B_{F\ell}$ and $B_{s,ph}$ are roughly of the same order of magnitude if $w \simeq a$ and one uses (66). For a more detailed comparison, valid for all temperatures, it is necessary to evaluate the analogue of $B_{s,ph}$ for a kink, not a straight dislocation, since there are effects associated with the dislocation shape[17] on the damping in (22). These may be systematized and may be quite important for vibrating dislocation loops but such detail will not be discussed here.

5. SUMMARY

In this review we have attempted to give the formal structure of the theory of the dislocation drag force, showing how it is related through the generalized viscosities to the excitation spectra and transport properties of the host medium. This program has been carried to completion for two simple idealized models, a free electron metal and a Debye insulator.

The emphasis throughout has been placed more upon establishing physical mechanisms than upon obtaining detailed quantitative predictions for particular materials, since the question of duplication of processes by different formulations has been the source of some confusion in the past. Hopefully this is now resolved by the present work, if not, we hope that at least it furnishes a useful background for further discussion.

Granted the premise, however, it is evident that much more needs to be done on the subject. Clearly one must confirm that the flutter mechanism follows from the method outlined earlier and consider more complete energy spectra for the deformed state so that a non-zero shear viscosity will ensue. Similarly the complications of anisotropic elasticity must be confronted and all manifestations of the discreteness of the lattice investigated, such as phonon dispersion and the role of normal and umklapp processes. In metals these questions are augmented by similar problems in the electronic properties, together with the added complexities of the nature of the electron-lattice coupling in the transition and rare earth metals for example. Indeed, the subject of the dislocation drag force so intimately weds the static and dynamic properties of both the electron and phonon systems that one may anticipate with confidence many interesting new developments in the future.

APPENDIX

In this section we outline the connection between the ~~treatment~~ given in the text and the standard method of analysis using conventional viscosities and thermoelastic parameters. To this end we write the general relation for the stress accompanying the dilatation in the insulator as

$$\sigma^{ph}_{k\omega} = \varkappa_{ad} \triangle_{k\omega} - \gamma C \left\{ \left(1 + \frac{i\omega\tau a}{k\ell} \right) \gamma T \triangle_{\underline{k}\omega} + \frac{a}{k\ell} \delta T_{\underline{k}\omega} \right\} \quad , \tag{A1}$$

which follows from (32) and (41). We define the wave vector and frequency dependent volume expansion coefficient by

$$\alpha_{k\omega} = \gamma C (a/k\ell)(\varkappa_{ad} - \gamma^2 C T) \quad , \tag{A2}$$

and a bulk viscosity $\zeta'_{\underline{k}\omega}$ by

$$\zeta'_{k\omega} = \gamma^2 C T \tau a/k\ell \quad . \tag{A3}$$

Thus (A1) is also

$$\sigma^{ph}_{\underline{k}\omega} = \varkappa \triangle_{\underline{k}\omega} + \zeta'_{k\omega} \dot{\triangle}_{\underline{k}\omega} - \varkappa \alpha_{k\omega} \delta T_{k\omega} \quad , \tag{A4}$$

where
$$\varkappa = \varkappa_{ad} - \gamma^2 C T \quad . \tag{A5}$$

So defined, $\alpha_{k\omega} \delta T_{k\omega}$ is the volume expansion amplitude accompanying a temperature wave of amplitude $\delta T_{\underline{k}\omega}$ when the viscosity $\zeta'_{\underline{k}\omega}$ is neglected, whereas $\varkappa$ is the isothermal bulk modulus for a Gruneisen solid. Finally $\zeta'_{k\omega} \dot{\triangle}_{\underline{k}\omega}$ is the dissipative stress when $\delta T_{\underline{k}\omega} = 0$. These conform with the standard definitions for $\omega\tau < 1$ and $k\ell \ll 1$ and are therefore permissible generalizations to finite frequency and wave vector. To confirm this we take the standard result[2] for the entropy production

$$\dot{S} = -\int \frac{\text{div } \underline{J}}{T} \, d\underline{r} + \frac{1}{T} \int \sigma'_{\alpha\beta} \, \dot{\epsilon}_{\alpha\beta} \, d\underline{r} \tag{A6}$$

where $\underline{J}$ is the heat flux, and the equation of heat conduction for a thermal conductivity, χ,

$$C \frac{\partial T}{\partial t} - \chi \nabla^2 T = -\varkappa \alpha T \frac{\partial \Delta}{\partial t} \quad , \tag{A7}$$

to find the power dissipation $P = T\dot{S}$. There results (considering the bulk viscosity only)

$$P = \int \left(\zeta' \frac{\partial \Delta}{\partial t} - \varkappa \alpha \, \delta T \right) \frac{\partial \Delta}{\partial t} \, d\underline{r} + \frac{C}{2T} \int \frac{\partial}{\partial t} (\delta T)^2 \, d\underline{r} \quad , \tag{A8}$$

which has a time average

$$\langle P \rangle = V \sum_{\underline{k}, \omega} \omega^2 \left\{ \zeta' + \text{Im} \, \frac{\varkappa \alpha \, \delta T_{k\omega}}{\omega \Delta_{k\omega}} \right\} |\Delta_{\underline{k}\omega}|^2 \quad . \tag{A9}$$

Consider now our definition of the generalized viscosity as the total out-of-phase component of $\sigma_{\underline{k}\omega}/\Delta_{\underline{k}\omega}$. It follows that

$$\zeta_{k1} = \text{Re} \, \zeta'_{k\omega} + \text{Im} \, \frac{\varkappa \alpha_{k\omega} \, \delta T_{\underline{k}\omega}}{\omega \Delta_{\underline{k}\omega}} \tag{A10}$$

whilst the power dissipation is easily shown by extension of the analysis leading to (22) to be

$$P = V \sum_{\underline{k}, \omega} \omega^2 \zeta_{k1} |\Delta_{\underline{k}\omega}|^2 \quad . \tag{A11}$$

Thus (A10) and (A11) are together further consistent with standard results valid at long wavelengths and low frequencies. They show also that the generalized bulk viscosity incorporates both thermoelastic and viscous damping in the conventional sense.

With the above partitioning, consider now the contributions to each physical mechanism separately. From (A3) one finds (for $\omega\tau < 1$)

$$\text{Re } \zeta'_{k\omega} \simeq \gamma^2 C T \tau \quad , \quad k\ell < 1 \qquad (A12)$$

$$\simeq \gamma^2 C T \pi/2c \quad , \quad k\ell > 1 \quad . \qquad (A13)$$

(A12) is the analogue of the kinetic theory, being of the form of the product of the "mass density" $(\mathcal{E}_T/c^2)$ and the "diffusivity" $(\chi/C \equiv \frac{1}{3}c^2\tau)$ for the phonon system. (A13) is the same result modified by the ineffectiveness concept, as we have explained in the argument leading to (71). Similarly, from (A3) and (43),

$$\text{Im}\left\{\frac{\varkappa\alpha_{k\omega}\delta T_{k\omega}}{\omega\Delta_{k\omega}}\right\} \simeq \gamma^2 C T \tau \frac{\left[(3/k^2\ell^2) - \omega^2(3\tau/k^2\ell^2)^2\right]}{1 + \omega^2(3\tau/k^2\ell^2)^2} \quad , \quad k\ell < 1 \quad , \quad (A14)$$

$$\simeq \gamma^2 C T \tau (\pi/2k\ell)^2 \quad , \qquad\qquad k\ell > 1 \quad . \quad (A15)$$

Comparison of these results shows that for long wavelength deformations at low frequencies the normal viscous processes are completely dominated by thermoelastic effects, the former assume the major role for short wavelengths when the kinetic theory form is no longer valid. For <u>transverse</u> deformation components there is of course no thermoelastic contribution but analogous reasoning to the above suggests that here again the final resulting drag force contribution will still be independent of mean free path since the fraction of deformation modes for which the kinetic theory result is valid is only $\sim (q_D\ell)^{-3}$ of the total number.

A similar analysis may be carried through for the electron gas contributions. In particular one finds

27

$$\zeta'_{k\omega} = \frac{2}{3} \frac{n \epsilon_F \tau}{\{1 + k^2/q_{TF}^2\}^2} \frac{a}{k\ell} \quad , \quad \text{(electrons)} , \quad \text{(A16)}$$

showing the same essential variation of the viscosity with wave vector as the phonon contribution.

REFERENCES

1. A. D. Brailsford, Phys. Rev. 142, 388 (1966).

2. L. D. Landau and E. M. Lifshitz, Theory of Elasticity (Pergamon Press Ltd., 2nd Edn., 1970) p. 153.

3. The viscosities derived in Sections 3 and 4 for a single wave of frequency ω and wave vector k depend upon both k and ω. This implies that the viscosity is both non-local in space and retarded in time in general. However to avoid over complicating the analysis this sophistication is ignored in this outline of the theory.

4. A. D. Brailsford, Phys. Rev. 142, 383 (1966) or Ref. 2, p. 126 for example.

5. To simplify the notation in subsequent sections it is convenient to regard the motion as cyclic over some very large interval T so that the allowed ω values are discrete. This of course is actually the case for uniform motion if spatial periodic boundary conditions are also used.

6. A. D. Brailsford, J. Appl. Phys. 37, 2842 (1966).

7. The finite lifetime of these excitations, as implied by the complex character of the frequencies $\nu_{k\lambda}$ should henceforth be ignored as a first approximation.

8. J. D. Eshelby, Proc. Roy. Soc. (London) A266, 222 (1962).

9. G. Leibfried, Z. Physik 127, 344 (1950).

10. M. P. Read and A. D. Brailsford, in preparation.

11. H. J. Maris, Phil. Mag. 16, 331 (1967); Phys. Rev. 188, 1303 (1969); 188, 1308 (1969).

12. Some authors use the term deformation potential to describe the effect of inhomogeneous deformation without the accompanying screening charge. Here E_1 denotes the total self-consistent change in electron energy.

13. L. D. Landau and E. M. Lifshitz, *Statistical Physics* (Pergamon Press Ltd., Oxford, England, 1969) p. 146.

14. In the electron case the derivative with respect to strain is also with the total number of particles fixed.

15. T. O. Woodruff and H. Ehrenreich, Phys. Rev. 123, 1553 (1961).

16. The relaxation processes derive from anharmonic terms in the lattice potential energy and in a discrete model lead to normal and umklapp processes. These lead to relaxations to different distributions since the normal processes conserve wave vector. However these two relaxation times are not readily obtained from an analysis of experimental data and the one time τ we use should be interpreted as the thermal conductivity relaxation time in insulators.

17. A. D. Brailsford, J. Appl. Phys. 43, 1380 (1972).

18. A. Akhieser, J. Phys. (USSR) 1, 277 (1939).

19. It will be noted that this Ansatz differs from that used earlier by us in J. Appl. Phys. 186, 959 (1969), in which the relaxation was to a distribution characterized by the unperturbed energies, a Fermi energy determined by the local density, drifting with the local lattice velocity. When $\delta T = 0$ these two approaches are equivalent and, although the earlier treatment is the more fundamental we have adopted that given in the text for greater continuity with the phonon contribution.

20. T. Holstein in an Appendix of B. R. Tittmann and H. E. Bömmel, Phys. Rev. $\underline{151}$, 178 (1966).

21. A. D. Brailsford, Phys. Rev. $\underline{186}$, 959 (1969).

22. See for example J. Lothe, J. Appl. Phys. $\underline{33}$, 2116 (1962).

23. A. Hikata, R. A. Johnson and C. Elbaum, Phys. Rev. $\underline{B2}$, 4856 (1970).

24. A. Seeger and H. Engelke, Dislocation Dynamics, edited by A. R. Rosenfield, G. T. Hahn, A. L. Bement, Jr. and R. I. Jaffee (McGraw-Hill Book Company, New York, 1968) p. 623.

25. W. P. Mason, J. Appl. Phys. $\underline{35}$, 2779 (1964).

26. W. P. Mason, Phys. Rev. $\underline{97}$, 557 (1955).

27. A. D. Brailsford, J. Appl. Phys. $\underline{41}$, 4439 (1970).

Fig. 1   The temperature variation of the phonon scattering contribution

to the drag force.

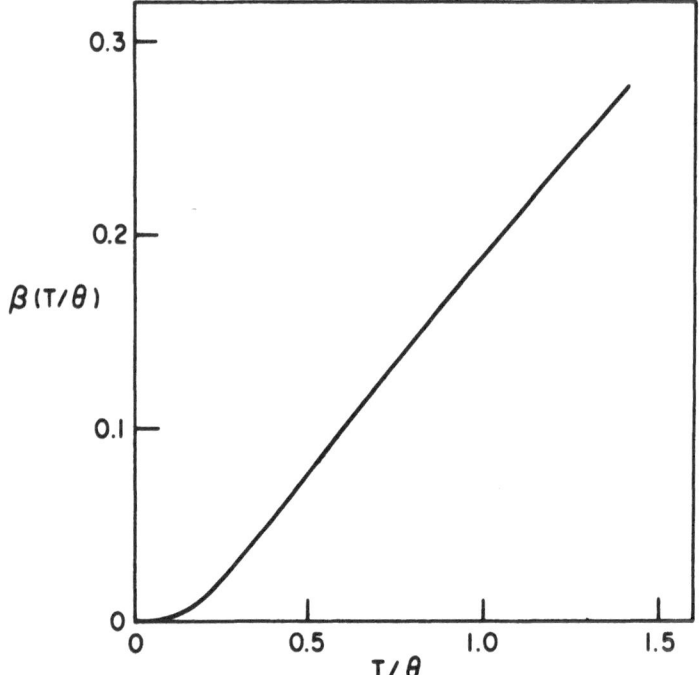

EFFECTS OF DISLOCATION INTERACTIONS WITH ELECTRONS AND PHONONS[†]

A. V. Granato
Department of Physics and Materials Research Laboratory
University of Illinois at Urbana-Champaign, Urbana, Illinois 61801

ABSTRACT

The effects of dislocation interactions with electrons and phonons can be observed in measurements of ultrasonic attenuation, thermal conductivity, etch pit displacements and macroscopic plastic flow in superconductors. Phonon reradiation scattering predominates over strain field scattering at low temperatures. The same string model used in ultrasonic attenuation studies can be used to describe the strength losses of metals entering the superconducting state.

I. Introduction

Dislocation interactions with electrons and phonons can be observed in measurements of ultrasonic attenuation, thermal conductivity, etch pit displacements and macroscopic plastic flow in superconductors. In Section II we review briefly the theoretically expected effects. In Section III, a limited selection of experimental results are examined semi-quantitatively to determine the extent to which the theoretical expectations are realized. A summary is given in Section IV.

II. Dislocation Drag Mechanisms

Known agents providing dislocation drag are: (1) phonons, (2) electrons, (3) radiation and (4) point defects. Phonon drag is thought to be of importance in all materials at high temperatures. At low temperatures in metals when the phonons are removed, electron drag remains. When the electrons are removed, as in superconductors or insulators, radiation drag remains for oscillating dislocations. Dislocation motion under stress in a crystal at finite temperature is analogous to Brownian motion in a force field. The phonons exert a viscous drag and also lead to fluctuations in displacements. The latter give rise to thermally assisted breakaway from pinning point effects, but these are not drag effects and will not be discussed here. Thermal fluctuations allow diffusion of pinning points, however, and this produces drag effects for dislocations moving slowly enough to carry along the pinning points.

A. Phonon Drag

The earliest estimate of phonon drag by Leibfried (1) in 1950 remains one of the simplest of, and about as accurate as, any of the many succeeding more sophisticated calculations. Leibfried supposed that for a plane lattice sound wave incident on a dislocation, all of the

[†]Work supported by the National Science Foundation.

energy in a strip of width of the order of atomic dimensions would be scattered. The scattering cross section per unit length σ is then $\sigma = b$ where b is the Burgers vector. The force per unit length acting on the dislocation is then given by

$$F \sim \sigma E \tag{1}$$

where E is the phonon energy density. E has the dimensions of energy/vol. or force/area which when multiplied by σ gives the force/length F acting on the dislocation. For a dislocation in an isotropic flux of phonons, there is no net force. However, if it moves through this flux with speed v, a fraction v/c of the waves now appear in the forward direction rather than the backward so that the net force/length is given by

$$F = g\sigma Ev/C = b\sigma = Bv \tag{2}$$

where g is a geometrical factor arising from averages which must be taken over a distribution of angles between incident phonons and the dislocation. Leibfried estimated this factor to be $g \sim 1/10$. Eq. 2 defines the drag coefficient B to be

$$B = g\sigma E/C \quad . \tag{3}$$

1. Radiation Scattering Cross Section. The first mechanism for which a detailed calculation of the scattering cross section σ was given was for reradiation scattering by Eshelby (2) (1949). If a phonon incident on a dislocation has a shear stress component on the slip plane in the slip direction, then the phonon drives the dislocation into forced oscillation. The oscillating dislocation radiates a cylindrical wave. The calculation is similar to that for an electron in an electromagnetic field. Eshelby found that the scattering cross section σ_R for this process (sometimes known as the flutter process) goes as

$$\sigma_R \sim d^2 \omega^3 \tag{4}$$

where d is the dislocation oscillation amplitude and ω is the frequency. For high frequencies, the dislocation motion is inertia-limited and

$$d \sim \left[M(\omega) \omega^2 \right]^{-1} \tag{5}$$

In Eq. 5 the dislocation mass depends logarithmically on frequency. The resulting frequency dependence from Eqs. 4 and 5 is approximately given by (3,4)

$$\sigma_R \simeq 2b(\omega_D/\omega)^{1/2} \quad , \tag{6}$$

where ω_D is the Debye frequency. Eshelby's calculation is a continuum elasticity calculation which should be valid for $\omega \ll \omega_D$. The reradiation scattering σ_R as given in Eq. 6 increases with decreasing ω. This increase continues until the dislocation resonant frequency

$$\omega_o \simeq \pi C/L \quad , \tag{7}$$

where L is a dislocation segment length between pinning points, is reached. Below this, the dislocation displacement d in Eq. 4 is tension limited, and $\sigma_R \sim \omega^3$. The resonance is sharp, and limited by radiation damping (5). For a random distribution of loop lengths, the resonance is broadened out somewhat and shifted to lower frequencies. For screw dislocations above the resonant frequency, an additional scattering is expected for phonons incident on dislocations at angles for which the phonon phase matches that of traveling waves on the dislocation (6).

2. Strain-field Scattering Cross Section. The second known scattering mechanism arises from non-linear elastic effects. The material close to the dislocation line is finitely strained, and scatters sound waves because of changes in the elastic constants and density in that region. This mechanism was first discussed by Nabarro (1951) (7) and has subsequently been considered further by many authors. This effect is reviewed in detail by Brailsford at this conference. The resulting scattering cross-section is given by

$$\sigma_S \simeq \frac{\gamma^2}{8} \frac{\omega}{\omega_D} b \qquad (8)$$

From Eqs. 6 and 8, for $\omega = \omega_D$ and $\gamma \simeq 2$, one finds $\sigma_R/\sigma_S \simeq 4$. However Nabarro (4) estimates that $B_R/B_S \simeq 1/3$ because of the different way that the geometrical factor g enters for the two effects.

In summary, phonons are expected to produce drag by two mechanisms, reradiation and strain field scattering. The reradiation cross section has an underdamped resonance at ω_o, falling as ω^3 for lower frequencies and as $\omega^{-1/2}$ for higher frequencies. The strain field scattering is linear in frequency, and both are expected to be comparable at $\omega = \omega_D$ or $T = \theta$, where θ is the Debye temperature. The resulting drag should be linear in velocity with a drag coefficient $B \sim 10^{-4}$ in cgs units at $T = \theta$.

3. Expected Temperature Dependence for Experimental Results. Drag effects can be observed through the measurement of the drag coefficient B in ultrasonic attenuation measurements, etch pit velocity measurements at high velocities, and cross sections σ may be determined from thermal resistance measurements.

The temperature dependence of B arises through that for the factors $\sigma(T)$ and $E(T)$ in Eq. 3. For high temperatures, $\sigma = \sigma(\omega_D)$ and $E = 3NkT/V$ where V is the volume per atom, so that $B \sim T$.

For low temperatures, $E \sim T^4$ and $\sigma_S \sim \omega$ for the strain field scattering. Defining the dominant phonons at temperature T to have a frequency $\nu = 3.8kT/h \simeq 10^{11}T(Hz/K)$, $\sigma_S \sim T$. For reradiation scattering $\sigma_R \sim \omega^{-1/2} \sim T^{-1/2}$. Thus, one expects the low temperature dependences of B for reradiation scattering (B_R) and strain-field scattering (B_S) to be

$$B_R \sim T^{7/2} \quad \text{and}$$
$$B_S \sim T^5, \text{ for } T \ll \theta \quad . \qquad (9)$$

Pinning point drag effects would be expected to carry the exponential temperature dependence of the diffusion constant of the pinning point.

Thermal conductivity measurements should offer a means for measuring $\sigma(\omega)$, as such measurements can be made with relatively high precision. Using the kinetic theory relation

$$K \sim \frac{1}{3} Cv\ell \quad , \qquad (10)$$

where K is the thermal conductivity, C the specific heat, v the velocity of sound and ℓ the mean free path, together with

$$\ell^{-1} = \Lambda\tau \qquad (11)$$

where Λ is the dislocation density, the frequency dependence may be inferred from the temperature

35

dependence of K. Since $C \sim T^3$ at low temperatures, one then expects

$$K_R \sim T^{7/2} \quad , \text{ and}$$

$$K_S \sim T^2 \quad , \text{ for } T \ll \theta \tag{12}$$

for the temperature dependence if the thermal conductivity is limited by the dislocation resistivity.

B. Electron Drag

Electrons are also scattered by dislocations. The interaction between the displacement field and the electrons is well known from ultrasonic attenuation studies (8).

A number of theories for this effect have been given. Brailsford (9) has recently reviewed these theories and finds results which are essentially in agreement with those of Holstein (10) and Kravchenko (12) from a theory based on electron scattering produced by the deformation potential of the strain field of a moving dislocation. This theory predicts that B_e is independent of both conductivity and temperature, and proportional to the electronic density N_e, i.e.

$$B_e \sim N_e \tag{13}$$

The magnitudes of B_e expected are typically of order 10^{-5} in cgs units, or about one order of magnitude smaller than phonon drags at room temperature.

III. Experimental Evidence

A. Phonon Drag Measurements

1. Ultrasonic Measurements. Ultrasonic measurements are analyzed in terms of the vibrating string model introduced by Koehler (13) and developed further by Granato and Lücke (14). Under the influence of an oscillating shear stress at low frequency, the dislocations oscillate as strings with the maximum displacement limited by the dislocation tension and the segment length L, as illustrated in Fig. 1. At high enough frequencies, a dislocation viscous drag is felt and the displacement is limited by the drag force instead. If a pinning point is added the low frequency displacement is greatly reduced, while the high frequency displacement is affected only in the neighborhood of the pinning points, which are widely separated.

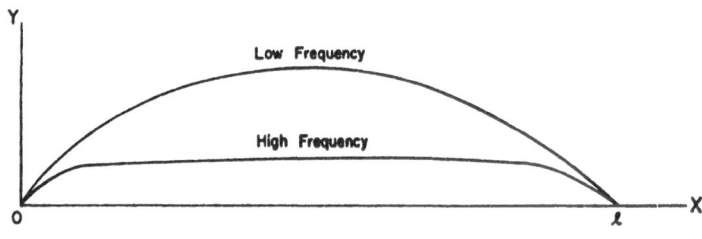

FIG. 1

Schematic dislocation displacement y(x) as a function of coordinate x for low frequencies and high frequencies. At low frequencies the displacement is limited by tension forces. At high frequencies, the displacement is limited by viscous forces.

The decrement expected as a function of frequency is shown in Fig. 2. For large damping, the behavior is like that of a relaxation effect. In the absence of phonon and electron drag

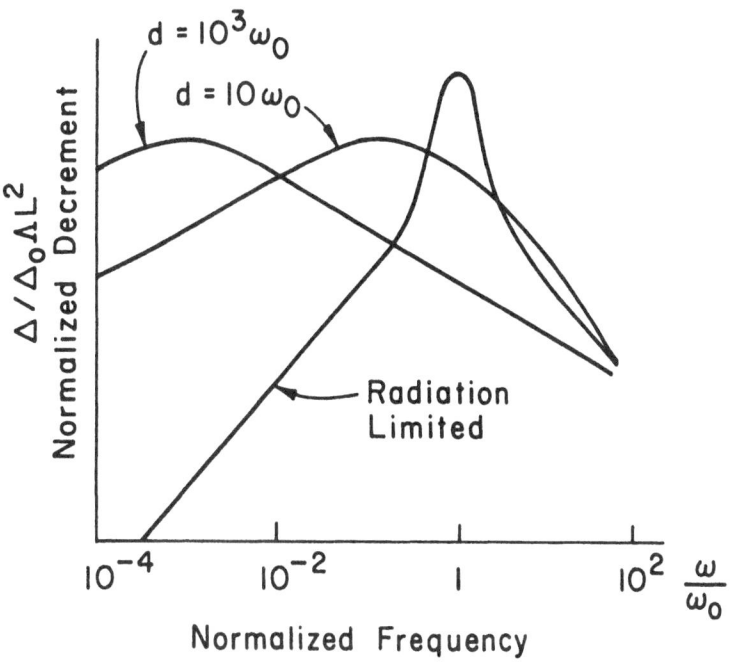

FIG. 2

Normalized decrement vs. normalized frequency for different damping constants.

reradiation drag remains. This minimum drag has two effects. The first is to limit the maximum of the loss at the resonant frequency, and the second is to change the frequency dependence of the decrement since the radiation drag is itself frequency dependent (5):

$$B_R = \rho b^2 \omega/8 \tag{14}$$

The curves shown in Fig. 2 are for a given loop length L. For a distribution of loop lengths, the maxima are broadened out somewhat (5).

An example of the effect of adding pinning points is shown in Fig. 3. These are measurements of the ultrasonic damping in Cu (15) as it is affected by pinning points introduced by cobalt gamma irradiation. One sees that the low frequency damping is strongly reduced, but the high frequency damping is hardly affected by the pinning points. At high frequencies, the attenuation is simply proportional to Λ/B, and is independent of the difficult to determine dislocation segment lengths and dislocation tensions and their distribution functions. By estimating the dislocation density Λ from etch pit counts, the drag constant was found from these measurements to be in reasonable agreement with Leibfried's estimate. In this technique the accuracy is limited by the measurement of Λ and the need to separate the dislocation component of attenuation from the total attenuation. The latter is done by assuming that light irradiations affect only the dislocation component.

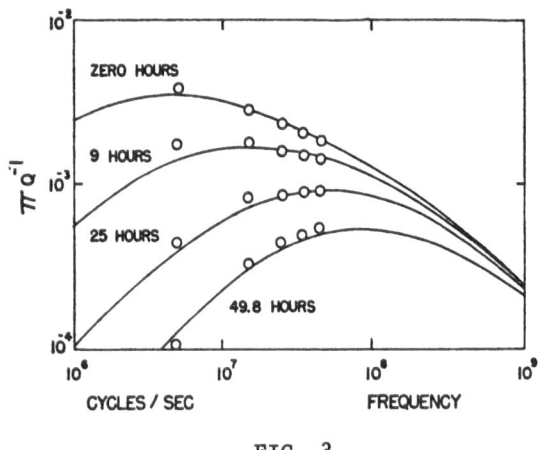

FIG. 3

Decrement as a function of frequency for several times
during cobalt gamma irradiation in a 6000 Curie source.
The solid curves are theoretical (after Stern and Granato.)

Ultrasonic measurements of attenuation and velocity in copper by Alers and Thompson (16)
showed that B is linear in temperature between liquid nitrogen and room temperature. This is
again in agreement with the Leibfried theory and shows furthermore that $B_p \gg B_e$, where B_p and
B_e are the phonon and electron drag, respectively.

The accuracy in these early measurements was not high (estimated at about a factor of four
by Stern and Granato). In addition, subsequent measurements by different workers on the same
materials showed disagreements of about one order of magnitude.

2. Etch-pit measurements. Dislocation velocities as a function of stress have been measured in many materials by measuring etch pit displacements as a function of time of application
of external stress. It is normally found that in relatively pure materials the velocity depends
strongly on stress at low velocities, but becomes linear in velocity at relatively high velocities which are still below the shear wave velocity. If it is assumed that the only force of
importance in the linear range is the viscous drag force, then Eq. 2 applies, and values for the
drag constant can be obtained. The accuracy of the earliest such measurements also was not high,
but more recent efforts to eliminate extraneous forces (recently reviewed by Vreeland (17)) has
led to values of the drag constant as a function of temperature with accuracies which are probably better than to within a factor of two.

3. Comparison of drag constants from ultrasonic and etch pit measurements. A strong
attempt to improve the accuracy of ultrasonic measurements was made by Fanti, et. al. (18) for
NaCl and LiF. Many systematic errors were identified and accounted for. Ultrasonic attenuation
and velocity were measured at two frequencies as a function of cobalt gamma irradiation. Measurements of attenuation and velocity at one frequency are equivalent to measurements of attenuation as a function of frequency if the segment length distribution is known. Measurements at
two frequencies provide a check of the assumed random segment length distribution function. The
check showed that the actual distribution is not the commonly assumed random distribution but
was close enough so that the estimated accuracy was to better than a factor of two. The

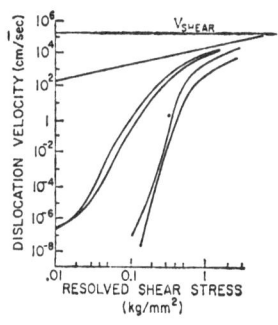

FIG. 4

Dislocation velocities as a function of applied
stress in NaCl. The straight line is a linear
extrapolation of ultrasonic velocities to higher
stresses. The curved lines are the etch-pit
measurements of Gutmanas et. al. (After Fanti, et. al.)

resulting drag limited velocities in NaCl as determined by Eq. 2 are shown in Fig. 4 as a straight
line. The line is extrapolated to higher stresses to compare with the velocities as determined
in etch pit measurements by Gutmanas, et. al. (19). A number of conclusions may be reached from
these results.

a. Since velocities determined ultrasonically are orders of magnitude higher than
velocities as deduced from etch pit measurements at low stresses, the etch pit measurements do not measure instantaneous velocities in this stress regime, but only
average velocities. The dislocations are free to oscillate at relatively high speeds
between pinning points. This was first noted by Baker (20).

b. Since the deduced drag constant is of the order predicted by Leibfried, the
scattering cross section is of order of atomic dimensions.

c. The viscous drag determined ultrasonically is strictly linear in velocity. If
this were not so, then the observed dependence on radiation dose, frequency and
stress amplitude would not be obtained.

d. At the yield stress of soft crystals at room temperatures, relativistic velocities (relative to sound speeds) cannot be achieved due to phonon drag. This answers
an old question in dislocation dynamics concerning the possibility of inertial
effects in plastic flow.

e. The same mechanism responsible for ultrasonic attenuation also limits dislocation
etch pit velocities at high speeds.

f. Different mechanisms operate in different ranges. At low stresses, dislocation
etch pit velocities are determined by interactions with obstacles. In an intermediate range, viscous drag limits the speeds. For high stresses, relativistic
effects limit speeds to sound velocities. Since the mechanisms are different, there
is no hope of finding a single relation which describes the results over the entire
stress range.

From these room temperature measurements we are not able to infer which mechanism of scattering is the more effective. Also, the accuracy of ultrasonic and etch pit measurements is not yet sufficiently precise to distinguish between the low temperature dependences predicted in Eq. 9.

4. Thermal conductivity measurements. We should expect to be able to distinguish reradiation scattering from strain-field scattering in thermal conductivity measurements, according to Eq. 12. Systematic measurements of the effect of dislocations on the thermal conductivity of LiF down to about 2°K were made by Sproull, et. al. (21). They found that the thermal resistivity was approximately proportional to the dislocation density. The temperature dependence was approximately T^2 as expected for static strain-field scattering, but the magnitude of the effect was two to three orders of magnitude greater than that predicted for strain field scattering. This magnitude however is that expected for reradiation scattering; the ratio of the two from Eqs. 6 and 8 being of the order of 10^3 for this temperature range.

It seems to have been implicitly assumed by most that the reradiation mechanism was somehow inoperative in this temperature range, and these measurements inspired heroic efforts to find ways of increasing the estimated magnitude of the strain field scattering (for detailed references, see reference 9). If, in fact, Eq. 8 did underestimate strain-field scattering by 2 - 3 orders of magnitude, then the resulting estimated drag constant for ultrasonic attenuation and etch pit measurements would be 2 - 3 orders of magnitude greater than those observed, which are in agreement with Leibfried's estimate. This fact was not noticed.

5. Direct observation of reradiation scattering at 10 - 100 MHZ. Reradiation scattering was directly measured (22) in LiF in the 10 - 100 MHZ range. Thin walls of dislocations were put in LiF crystals by shear and signals from the dislocation walls were observed. These are seen in Fig. 5 as the small echoes between the large echoes from the specimen and faces. The dislocations in the wall are each forced into oscillation by the impressed MHZ sound wave. Each dislocation sends out a cylindrical sound wave, but they are all in phase and combine to yield by

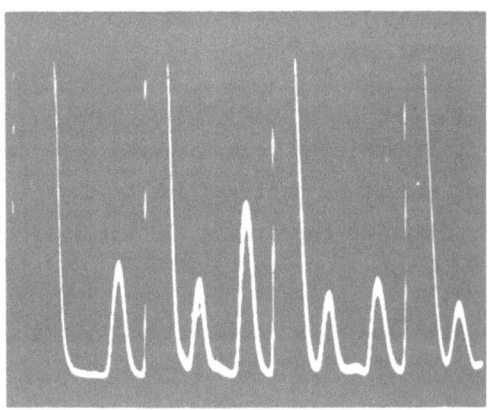

FIG. 5

Photograph taken on a Matec ultrasonic comparator screen which shows the rectified reradiated wall echoes for a deformed LiF crystal versus display time. (After Schwenker and Granato)

Huygen's principle a macroscopic plane sound wave which is then detected by a quartz transducer. The proof that these are reradiated waves and not strain field scattered waves is obtained from orientation and radiation studies. The signal is only obtained when there is a resolved shear stress in the dislocation slip planes. Also the signal disappears when the specimen is lightly irradiated so as to immobilize the dislocations. The frequency dependence of the radiated wave was found to agree with that of Eq. 4 when account was taken of the effect of damping on the dislocation displacement amplitude.

6. Thermal conductivity of LiF in the .03 to 1°K range. Using the same dislocation wall specimen configuration, Anderson and Malinowski (23) showed that reradiation scattering is far more important than strain-field scattering for the thermal conductivity of LiF. Their results for one specimen are shown in Fig. 6 as a function of cobalt gamma irradiation. The conductivity results have been normalized to the values obtained for the part of the specimen containing no dislocation wall where the conductivity was limited by specimen boundary scattering. When a dislocation wall is introduced, the conductivity drops to about half its original value. Immobilizing the dislocations by pinning restores the conductivity at low temperatures or frequencies. As the dislocation radiation is increased, the cross-over temperature moves to higher frequencies. The pinned dislocations remain in the specimen, but produce no detectable resistivity, showing that the reradiation scattering is far greater in magnitude than the strain field scattering. These measurements cover the range from 4×10^9 HZ to 10^{11} HZ, and show that

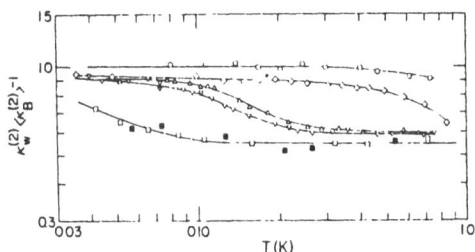

FIG. 6

Reduced thermal conductivity for the wall region of sample 2, $K_w{}^{(2)} \langle K_B{}^{(2)} \rangle^{-1}$, vs temperature: □ - deformed sample; ■ - deformed sample, remeasured; ▽ - 800-R total irradiation; △ - 1000-R total irradiation; ◊ - 26 000-R total irradiation; 0 - 136 000-R total irradiation. (After Anderson and Malinowski)

reradiation scattering is still effective at 1°K. The remaining conductivity is presumably due to that carried by the phonon modes having no resolved shear stress on the slip planes. Indeed, separate ballistic phonon measurements which resolve the different phonon modes by time-of-flight analysis showed that only those modes with resolved shear stress components in the slip plane were reduced by the dislocation wall.

7. Combined pinning point and phonon drag. Pinning point drag is reviewed by others at this conference. The reader is referred to these reviews and the literature (24-29) for details. We wish here merely to mention the type of effects observed when both the pinning point drag (B_D) and the phonon drag (B_P) are large.

It has been shown by Lücke and Schlipf (28) and by Simpson and Sosin (29) that the effect of pinning point drag can be described in a certain rather good approximation within the formalism of the string model simply by smearing out the drag force provided by the pinning points, and then using the usual string relations. The drag force per unit length is

$$F_D = \rho_o v/\mu \quad , \tag{15}$$

which defines the drag damping constant as

$$B_D = \rho_o kT/D \quad , \tag{16}$$

where ρ_o is the density of pinning points, μ is the mobility, and D is the diffusion constant of the pinning point in the core of the dislocation.

A relaxation is obtained when $\omega\tau \sim 1$, where

$$\tau = \frac{BL^2}{\pi^3 C} \tag{17}$$

and C is the dislocation tension. The relaxation may be produced by changing ω, T, or ρ_o. Simpson and Soxin (29), who assume that the drag is athermal, have emphasized the dependence of τ on ρ_o. They point out that if mobile pinning points are added to a dislocation line, for example by irradiation, then it is possible to obtain a peak in the damping as a junction of irradiation time in contrast to the more familiar pinning laws. That is, for mobile irradiation induced pinning points, B changes instead of L.

For a fixed number of mobile pinning points, a relaxation should be obtained as a function of temperature. Wire and Granato (30) found a relaxation in the elastic constant of LiF at 10 MHz near 100°K. In this case, the characteristic L^2, L^4 string laws were found for gamma irradiation, indicating that the radiation induced defects are not mobile. However, the characteristic loop length L at low temperature was found to be 5-6 times smaller than that at high temperature. That is, the loop length seems to suddenly decrease around 100°K. Otherwise the behavior is normal. If this is a drag effect, then a peak would be expected in the attenuation at the same temperature. However, no peak was found in the attenuation, rather the attenuation showed a step-like behavior, increasing suddenly in the neighborhood of 100°K.

This behavior can be explained if it is supposed that B in Eq. 17 is the sum of a pinning point drag B_D and a phonon drag B_P. At high frequencies, the phonon drag is large enough to arrest the peak, so that the total B does not decrease as the temperature is raised. The peak should then be observable at lower frequencies, where the phonon drag is smaller. Indeed, the relaxation appears to be the same as that observed to give a peak at kHz frequencies by Taylor (31,32) and by Gibbons and Chirba (33).

B. Electron Drag Measurements

1. Ultrasonic measurements. As noted earlier, electron drag is expected to lead to a drag coefficient of order of 10^{-5} independent of temperature. Ultrasonic measurements on aluminum by Hikata et. al. (34) are shown in Fig. 7. The results can be analyzed as the sum of a temperature dependent phonon component which is linear in temperature at high temperature, and a temperature independent electron component $\sim 10^{-5}$. The method used by Hikata, et. al., is to measure the

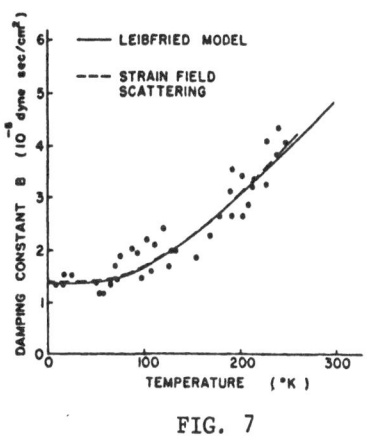

FIG. 7

Damping parameter B as a function of temperature. (After Hikata, Johnson and Elbaum)

change in attenuation with an applied bias stress, assuming that the dislocation segment length distribution is random. This method requires no independent measurement of the dislocation density.

Since the electronic drag depends only on electronic density, it is difficult to test Eq. 13. One possible method becomes available using superconductors, however, since the number of electrons which can be scattered by dislocations is reduced in the superconducting state by reducing the temperature. No direct ultrasonic measurements of B_e have yet been made. The electronic drag can be measured however by an indirect method using the amplitude-dependent component of the attenuation. Amplitude dependence arises from dislocation breakaway from pinning points with increasing stress (14). It was found by Tittman and Bommel (11) that a larger stress is required to produce the same degree of breakaway in the normal state than in the superconducting state. This occurs because the dislocations are unable to exert as large a depinning force when their displacement is drag-limited rather than tension-limited. Hikata and Elbaum (35) used an explicit relation (14) for the drag-dependent depinning force to analyze their amplitude dependent measurements to obtain a value of B_e in the normal state in agreement with the calculation of Holstein and Kravchenko.

2. Plastic deformation of superconductors. A possibility for determining the electronic density dependence of B_e arises from analysis of the temperature dependence of the flow stress changes which occur when superconducting materials are switched between the normal and superconducting states. For these macroscopic flow measurements, it has been found that the drag coefficient plays a decisive role. For superconductors in the normal state, the phonon drag is negligible and only electron drag and radiation drag remain. In the superconducting state at low temperatures, only radiation drag remains, and it has been shown (36) that this drag is weak enough so that all dislocation segments in superconductors are underdamped. This permits inertial effects to operate, allowing dislocations to overcome obstacles at lower stresses than for the normal state.

The decrease in strength of superconductors was discovered recently by Pustovalov, Startsev and Fomenko (37-40) and by Kojima and Suzuki (41). For constant strain-rate tests, the stress required to maintain the deformation rate drops. For creep measurements at constant stress,

Soldatov, Startsev, and Vainblatt (42) found that the strain rate increases by factors of up to 250. For stress-relaxation experiments at constant strain, Suenaga and Galligan found that the stress drops suddenly (43,44). The stress-change effects observed are typically of the order of 0.1% to 10%, but effects as large as 53% have been reported (40). Since the first measurements, the dependence of the effect on crystal structure, purity, type of superconductor, temperature, strain hardening, magnetic field, strain rate, alloying, and crystal orientation have been explored. These effects have recently been reviewed by Kostorz (45).

Ths same string model used to discuss damping measurements can be used to calculate the strength charge of superconductors entering the superconducting state. The model is entirely analogous to that of a loaded spring in a viscous medium and is easily understood with the help of Fig. 8. A dislocation line moving toward obstacles in position 1 of Fig. 8(a) meets the obstacles with a velocity v_o at position 2. The static equilibrium position under an applied stress is position 3. If the viscous damping is larger than a critical value, the dislocation line approaches the static equilibrium position as in the solid line of Fig. 8(b). If the damping is less than critical, the dislocation line overshoots to position 4 and oscillates about the static-equilibrium line. In position 4, the force exerted by the dislocation line on the obstacle is greater than in the static-equilibrium case. Alternatively, one may say that a smaller stress is needed in the low-damping case to produce the same force on the obstacle.

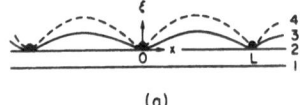

(a)

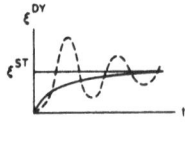

(b)

FIG. 8

(a) Schematic of the motion of the dislocation line. 1, dislocation line approaches pinning points; 2, dislocation line just touches pinning points; 3, static equilibrium position of dislocation line; 4, overshoot position of underdamped dislocation. (b) Displacement as a function of time for an underdamped dislocation (dashed curve) and for an overdamped dislocation (solid line).

This model predicts that the stress change should be proportional to the density of superconducting electrons. Measurements illustrating this effect are shown in Fig. 9. The solid line is data for lead by Suenaga and Galligan (44), the dashed line is data for indium by Alers, et. al. (46) (closed squares) and by Hutchinson and Pawlowicz (47) (closed circles). The temperature dependence of the density of superconducting electrons is different for indium than for lead since the energy gap is different.

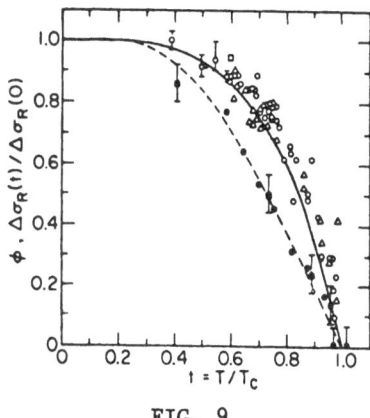

FIG. 9

The temperature dependence of $\Delta\sigma$ for Pb (open symbols) according to Ref. 44 and for In (closed symbols) according to Ref. 46 (squares with error bars) and to Ref. 47 (circles). The dashed line is Mühlschlegel's calculation of the superconducting electron density according to the BCS theory for weak superconductors, and the solid line is $1 - t^4$.

IV. Summary

It has been established from thermal conductivity studies that reradiation scattering of phonons by dislocations is larger than strain field scattering at low temperatures.

Values of the viscous drag constant determined ultrasonically appear to be in reasonable agreement with those found from etch pit studies.

The electronic drag is typically about an order of magnitude smaller than phonon drag near the Debye temperature.

There are two kinds of pinning point drag effects observed during irradiation of specimens, corresponding to mobile and immobile irradiation induced pinning points. In the former case, a peak may be obtained in the damping, while in the latter, the usual pinning laws are obtained. Also, the peak may be arrested at high frequencies by phonon drag.

The same model used to describe ultrasonic effects is also useful in describing the plastic properties of superconductors. Since these fields have developed rather independently, this represents a partial unification of these two fields at least over a certain region.

The string model has proven capable of providing understanding over a wide range of phenomena in terms of a relatively simple and easy to visualize model. It has also proven to be a fruitful model, with recent noteworthy extensions being its usefulness in describing pinning point drag effects and plastic flow in superconductors.

References

1. G. Leibfried, Z. Phys. 127, 344 (1950).

2. J. D. Eshelby, Proc. R. Soc. A197, 396 (1949).

3. A. V. Granato, Phys. Rev. 111, 740 (1958).

4. F. R. N. Nabarro, Theory of Crystal Dislocations, p. 505. Oxford U. P., Oxford, England (1967).

5. J. A. Garber and A. V. Granato, J. Phys. Chem. Solids 31, 1863 (1970); in Fundamental Aspects of Dislocation Theory (Special publication 317) J. A. Simmons, R. deWit and R. Bullough, eds., p. 419. National Bureau of Standards, U.S.A. (1970).

6. T. Ninomiya, in Fundamental Aspects of Dislocation Theory (Special publication 317) J. A. Simmons, R. deWit and R. Bullough, eds., p. 315. National Bureau of Standards, U.S.A. (1970).

7. F. R. N. Nabarro, Proc. R. Soc. A209, 278 (1951).

8. A. B. Pippard, Phil. Mag. 46, 1104 (1955).

9. A. D. Brailsford, Phys. Rev. 186, 959 (1969).

10. T. Holstein, in appendix in ref. 11.

11. B. R. Tittmann and H. E. Bommel, Phys. Rev. 151, 178 (1966).

12. V. Ya Kravchenko, Fiz. Tverd. Tela 8, 927 (1966) [Sov. Phys. Solid State 8, 740 (1966)].

13. J. S. Koehler, in Imperfections in Nearly Perfect Crystals, W. Shockley, J. H. Holloman, R. Maurer and F. Seitz, eds., p. 197. Wiley, N. Y. (1952).

14. A. V. Granato and K. Lücke, J. Appl. Phys. 27, 583 (1956).

15. R. M. Stern and A. V. Granato, Acta Met. 10, 358 (1962).

16. G. A. Alers and D. O. Thompson, J. Appl. Phys. 32, 283 (1961).

17. T. Vreeland, in Metallurgical Effects at High Strain Rates, to be published.

18. F. Fanti, J. Holder, and A. V. Granato, J. Acoust. Soc. Amer. 45, 1356 (1969).

19. E. Y. Gutmanas, E. M. Nadgornyi and A. V. Stepanov, Sov. Phys. Solid State 5, 743 (1963).

20. G. S. Baker, J. Appl. Phys. 33, 1730 (1962).

21. R. L. Sproull, M. Moss and H. Weinstock, J. Appl. Phys. 30, 334 (1959).

22. R. O. Schwenker and A. V. Granato, Phys. Rev. Letters 23, 918 (1969); J. Phys. Chem. Solids 31, 1869 (1970).

23. A. C. Anderson and M. E. Malinowski, Phys. Rev. 5B, 3199 (1972).

24. R. Kamel, Acta. Met. 9, 65 (1961).

25. G. Schoeck, Acta. Met. 6, 617 (1963).

26. A. A. Blistanov and M. Shaskol'skaya, Fizika Tverdogo Tela 6, 728 (1964). (Translation in Sov. Phys. - Solid State 6, 508 (1964).

27. P. Schiller, Phys. Stat. Sol. 5, 391 (1964).

28. K. Lücke and J. Schlipf, in The Interactions Between Dislocations and Point Defects, B. Eyre, ed., p. 118. H. M. Stationery Office, London (1968).

29. H. M. Simpson and A. Sosin, Phys. Rev. 5, 1382 (1972).

30. G. Wire and A. V. Granato, to be published.

31. A. Taylor, J. Appl. Phys. 32, 1799 (1961).

32. A. Taylor, Acta. Met. 10, 489 (1962).

33. D. F. Gibbons and V. G. Chirba, Acta. Met. 10, 484 (1962).

34. A. Hikata, R. A. Johnson and C. Elbaum, Phys. Rev. Letters 24, 215 (1970); Phys. Rev. B2, 4856 (1970).

35. A. Hikata and C. Elbaum, Phys. Rev. Letters 18, 750 (1967).

36. A. V. Granato, Phys. Rev. Letters 27, 660 (1971); Phys. Rev. B4, 2196 (1971).

37. V. V. Pustovalov, V. I. Startsev, D. A. Didenko and V. S. Fomenko, Fiz. Metal. Metalloved. 23, 312 (1967).

38. V. I. Startsev, V. V. Pustovalov and V. S. Fomenko, Trans. Jap. Inst. Metals, Suppl. 9, 843 (1968).

39. V. V. Pustovalov, V. I. Startsev and V. S. Fomenko, Physico-Technical Institute of Low Temperatures, Academy of Sciences of the Ukrainian SSR, Kharkov, Report, 1968 (unpublished) and Fiz. Tverd. Tela 11, 1382 (1969) [Sov. Phys. Solid State 11, 1119 (1969)].

40. V. V. Pustovalov, V. I. Startsev and V. S. Fomenko, Phys. Status Solidi 37, 413 (1970).

41. H. Kojima and T. Suzuki, Phys. Rev. Letters 21, 896 (1968).

42. V. P. Soldatov, V. I. Startsev and T. I. Vainblat, Phys. Status Solidi 37, 47 (1970).

43. M. Suenaga and J. M. Galligan, Scr. Met. 4, 697 (1970).

44. M. Suenaga and J. M. Galligan, Scr. Met. 5, 63 (1971).

45. G. Kostorz, Phys. Stat. Sol. 58, 61.9 (1973).

46. G. A. Alers, O. Buck and B. R. Tittman, Phys. Rev. Letters 23, 290 (1969).

47. T. S. Hutchison and A. T. Pawlowicz, Phys. Rev. Letters 25, 1272 (1970).

DISLOCATION RESONANCE DAMPING AND THE NATURE OF PINNING

D. Lenz and K. Lücke

Institut für Allgemeine Metallkunde und Metallphysik der
RWTH Aachen, Germany

I. Introduction

Many important aspects of the behaviour of metals and alloys are determined by interactions between point defects and dislocations. In order to
investigate these interactions each property which depends on the
dynamic behaviour of dislocations can be chosen to be studied. The
present paper reviews studies of this interaction by internal friction methods. Since these methods are extremely sensitive and yield
particularly detailed information, a rather large literature including some excellent survey articles (e.g. /1/ to /7/) has accumulated
in this area. In the present paper, mainly the amplitude independent
internal friction at small strain amplitudes and work on high-purity
metals (mostly fcc. metals) will be considered, covering the frequency
range from 1 Hz to 100 MHz. Even for this limited field, the paper does
not intend to give a complete coverage but restricts itself to the more
important results with emphasis on those of the Aachen School.

Internal friction - the property which reflects a materials ability
to convert mechanical energy into thermal energy of the lattice - is
caused in metals under a wide range of experimental conditions largely by dislocations (e.g. /1/ to /7/). Dislocation internal friction
comes about because the dislocation-strain-response to an externally
applied shear stress is retarded. It has first been quantitatively
described by the theory of dislocation resonance damping by Granato
and Lücke /8,9/ based on the vibrating string model of Koehler /10/.
In this theory the particular mechanism for energy dissipation remains unspecified and different mechanisms may be applied under
different circumstances. In the present paper, it is the hindering
of dislocation motion by point defects which will mainly be considered. Such an effect which will be denoted as dislocation pinning
was first demonstrated by the irradiation experiments by Thompson
and Holmes /11/ who also introduced the idea of point defects acting
as firm pinning points at the dislocations.

Since then, most of the results on dislocation pinning in the literature have been interpreted with help of the dislocation resonance theory using mostly the concept of firm pinning points and phonon drag as dissipating mechanism. This approach led to much valuable information about dislocations as well as about point defects.

On the other hand, low frequency results (e.g. in the 1 Hz range) are mostly excluded from applying the resonance theory. Also in interpreting the high-frequency results some difficulties were piling up recently (e.g. /12/). Furthermore, different point defect interaction processes (some of them not yet sufficiently quantitatively formulated) were included into the discussion of dislocation motion, e.g. breakaway from /5, 13 to 17/ and diffusion of point defects /5,18,19, 20/ or reorientation of axial defects in the dislocation stress field /21 to 24/. In some papers (e.g. /18/) such concepts were considered as being quite different from or even in contradiction to those of the old dislocation resonance theory.

For this reason the present paper intends
(i) to contribute to the clarification of the theoretical situation by showing that practically all these concepts can be treated in special cases in the frame of the dislocation resonance theory. Different dislocation point defect interaction concepts lead to different contributions to dislocation restoring force and/or to the dissipational force and thus do not change the general predictions of resonance theory.
(ii) to contribute to the clarification of the experimental situation by checking to which extent the results can be interpreted by the dislocation resonance theory, which of the restoring and dissipating mechanisms apply in each case and, most of all, which observations (including those in the 1 Hz-range) cannot be explained by the present concepts.

In cases not treated before or in order to stress similarities, derivations of the final formulas are given. Here simple physical arguments are preferred to rigorous mathematical treatments.

Sec. II presents a qualitative survey on the different possibilities of pinning and of investigating dislocation point defect interaction. Sec. III gives a survey on the situation of dislocation resonance damping. First, an extended version of the Granato-Lücke theory is presented (Sec. IIIa); then different dissipational mechanisms,

particularly those due to dislocation/point defect interaction, are
quantitatively described (Sec. IIIb) and finally, the resulting
frequency and temperature dependence of damping is compared with
experiments (Sec. IIIc). In Sec. IV the quantitative aspects of
pinning are discussed. First, the theoretical predictions following
for damping and modulus defect from the different pinning mechanisms
(Sec. IVa) are compared with experimental results (Sec. IVb), then
the theoretical (Sec. IVc) and experimental situation (Sec. IVd)
regarding kinetics of pinning is described. In Sec. V, finally,
the conclusions regarding the nature of pinning are summarized.

II. Qualitative Aspects of Dislocation Pinning

a)   Examples of internal friction phenomena involving
     dislocation/point defect interaction.

The statements appear to be generally accepted: (i) Dislocations are the
origin of internal friction; (ii) Point defects influence dislocation
internal friction. The first statement may be illustrated e.g. by considering the influence of plastic deformation, the second e.g. by
considering the influence of point defects introduced by alloying,
quenching or irradiation upon internal friction.

Fig. 1 gives an example for both effects in a single crystal of high
purity copper by showing the dependence of ultrasonic attenuation α
on plastic compressional strain ε /25/. Curve 1 shows that the deformation results in a considerable increase in damping at low strain
values. This is considered to be caused by introduction of new dislocations by deformation. (The peaked $\alpha(\varepsilon)$ dependence with increasing
strain indicates the development of special dislocation structures).
Curve 2 shows that during annealing at $110^{\circ}C$ a mayor part of this
increase recovers and curve 3 that by irradiation the damping is further
decreased even to values below the value at small strains α_o. These
effects are considered to be caused by pinning of these dislocations
by deformation induced (curve 2) or irradiation induced (curve 3)
defects.

Fig. 2 (curve 1) gives an example of impurity pinning /26/: The attenuation of a set of well annealed Ge doped Cu single crystals decreases
with increasing Ge-content up to about 300 ppm Ge (curve 1). After heavy
γ-irradiation final damping values are obtained (curve 2) which, for low

impurity contents (< 300 ppm Ge), may be attributed to the non-dislocation damping ("background") of the samples. (The increase of α at higher concentrations is connected with changes in substructure). Fig. 3 /27/ shows an example of pinning during isothermal irradiation of copper; a saturation of the pinning with time and an increase of the pinning rate with temperature is observed. Fig. 4 /28/ shows results on room temperature γ-irradiated high purity Pb; here again irradiation pinning takes place but depinning occurs if the irradiation is stopped.

Fig. 5 gives an example for the Hz-range /29/. During warm-up of copper deformed at 78 K the Hasiguti damping peaks appear (curve 2); if the warm-up is carried through to sufficiently high temperatures (about 300 K), the peaks anneal out and are no more observed during subsequent cooling (curve 3). The damping in curve 2 is attributed to deformation-induced point defects having migrated to the dislocations and probably being dragged along by the moving dislocation. The annealing out of the peaks (curve 3) is attributed to additional point defects migrating to the dislocations and immobilizing there the dragged point defects. Only when new defects come to the dislocation at low temperatures, e.g. after low temperature irradiation, damping peaks reappear (curve 4). (These interpretations are confirmed by simultaneous modul measurements /29/).

In addition to these few examples there is a great number of experiments which also give clear evidence for dislocation pinning by point defects. Before discussing further results possible mechanism by which dislocation/point defect interaction influences internal friction will be considered.

b) Different interaction effects.

The interaction of point defects with dislocations leads to a formation of a Cottrell atmosphere, i.e. to an enrichment of defects near the dislocation. It is customary (e.g. /30/) to separate this interaction into a short range part, i.e. the interaction with the dislocation core, and a long range part which is of elastic nature. For the latter mostly (except for vacancies) para-elastic interaction has been assumed which is characterized by

$$H(r) = Ab/r; \qquad c(r) = c_o \exp (H(r)/kT) \qquad (1)$$

Here is H(r) the enthalpy of interaction, A an interaction constant, r the distance from the dislocation and c(r) the equilibrium concentration of defects.

In internal friction work the situation has mostly been simplified in such a way that the short range interaction is assumed to be very strong (and attractive) and the long range interaction to be very weak. This leads to the concept of discrete dislocation pinning points in which the atmosphere outside the core is neglected. This assumption is justified not only by the smaller interaction forces exerted by the more distant defects upon the moving dislocation, but also by the much larger density of the atmosphere near the dislocation core.

Because of its simplicity the discrete pinning point concept has been widely applied to describe the results of dislocation internal friction experiments, and also in the present paper the extented Cottrell atmosphere will be disregarded. On the other hand, also long range interaction - although small for a single defect - might become effective, if a great number of defects is involved. With the interaction force $-dH/dr=Ab/r^2$ (Equ. (1)) the influence of a zylindrical shell of the volume $2\pi r dr$ decreases only with $1/r$, so that, for a concentration c independent of r, the integrated influence would even increase (logarithmically) with increasing r.

This is important for defects with axial symmetry (e.g. for foreign interstitials in b.c.c. or split-interstitials in f.c.c. lattices). In this case the interaction energy (here being comparatively large) depends upon the orientation of the dipol axes of these defects with respect to the stress field of the dislocation. Hence, the defects are able to lower their energy also by a reorientation and thus tend to form around the dislocation a "Snoek atmosphere" characterized by preferred orientations of the dipol axes /21,22/. The equilibrium atmosphere is attained within the meantime of a single atomic jump for each defect and extends over the radius of a cylinder given by $H(r)=Ab/r\approx kT$.

Fig. 6 shows a synopsis of some dislocation/point defect configurations which are presumed to be realistic and give rise to different internal friction effects /31/:

1. Only the dislocation can move and the point defects are immobile.

If the point defects are situated at the dislocation line forming pinning points, the dislocation is able to move at low stresses only between the anchoring points. Under the periodic shear stress applied in an internal friction experiment the dislocation may thus be looked at as a vibrating string fixed at its ends. This is the concept of the Köhler-Granato-Lücke model /8,10/ which has frequently been used to interpret experimental results. On the other hand, a reduction of the dislocation vibrations can also be obtained by long range interaction, i.e. by a Cottrell or Snoek atmosphere. With increasing stress, break-away of the dislocation from the pins will occur /8/. Since here mainly short range interaction is involved, this process is expected to be thermally activated (Teutonico, Granato, Lücke /13/).

2. Also the point defects can move, but only in times large compared to the dislocation vibration time.

In this case, dislocation pinning and thus also the internal friction effects become time dependent. There are several possibilities for the motion of the point defects: (i) Randomly distributed point defects might migrate to the dislocations to form there a Cottrell atmosphere or, upon arrival at the dislocation, to convert into dislocation pinning points (Thompson and Holmes /11/, Granato and Lücke /32/). (ii) Point defects with axial symmetry might cause pinning effects by reorientation with respect to the stress field of the dislocation (Hornung /24/. (iii) Since dislocations present a path of increased diffusivity, pinning points might move along the dislocation thus changing the adjacent loop length (Bauer and Yamafuji /18/). If the defects encounter efficient traps somewhere at the dislocation (e.g. at nodal points, jogs or other point defects) they also might form clusters there.

3. The point defects can move in times comparable with the dislocation vibration time.

Here the vibration of the dislocation induces vibrational motions of interacting point defects, which, in turn, cause additional contributions to damping and modulus change. Again the different possibilities of motion already listed in the preceding paragraph have to be considered (c.f. also Sec. III b): (i) The pinning points move together with the oscillating dislocation loop, i.e. about perpendicular to the original dislocation line (Lücke and Schlipf /5/, Simpson and Sosin /19/, Winkler-Gniewek, Schindlmayr, Schlipf /20/). (ii) Defects with axial symmetry carry out periodic rotations in the oscillating stress field of the vibrating dislocation (induced Snoek-effect, Schoeck /21/, Schoeck and Seeger /22/, Sokolowski and Lücke /33/). (iii) The pinning points move parallel to the dislocation thus changing the lengths of the adjacent oscillating loops (Lücke and Schlipf /5/, Winkler-Gniewek, Schindlmayr, Schlipf /20/).

This considerable variety of possible dislocation/point defect inter-action phenomena represents the basic material for the construction of theoretical damping models. Additionally, combinations of the different phenomena appear to be conceivable. Finally, dislocation properties such as stacking faults, kinks and jogs may be incorporated in the models.

In the case of short-range interaction between dislocation and pinning point, it is mainly the applied stress amplitude σ_o and the temperature T of measurement that determine which of the above listed mechanisms really takes place. Fig. 7 shows a damping mechanism "phase diagram" according to Schlipf and Lücke /5/: (i) at low T and σ_o the dislocation is firmly pinned during all of the stress cycle; this is the vibrating string configuration giving rise to dislocation resonance damping. (ii) with increasing σ_o dislocation breakaway occurs; because of thermal activation the minimum amplitude for breakaway damping decreases with increasing temperature. (iii) At elevated temperatures, finally, the effects connected with pinning points diffusing with or along the dislocation are taking place.

c) <u>Direct study of breakaway.</u>

The most direct way for studying dislocation/point defect interaction is to measure the force at which the dislocation breaks away from an immobile pinning point. Up till now, mainly micro-strain experiments or measurements of the amplitude dependence of internal friction have been used to derive such information. The evaluation of the first type of measurements requires some knowledge of structural parameters and that of the second type some additional theoretical assumptions (i.e. a theory of amplitude dependent damping /8,13/). Therefore, here a more direct experiment /34/ will be considered in which the stress is slowly varied as in micro-strain experiments, but instead of the strain ε the MHz-damping α is measured as function of stress τ (Fig. 8a).

The idea behind this experiment is to induce dislocation breakaway by an increasing external stress and simultaneously use the MHz attenuation to get information about the instantaneous value of the dislocation loop length. As shown schematically in Fig. 8b for the double loop model, a sudden increase in loop length (from L to 2L) and thus a rather sharp increase in α is expected when τ reaches the breakaway stress τ^*. In reality the increase of the $\alpha(\tau)$-curve should be more gradual because of the distribution in loop length (Fig. 8c).

Fig. 9 shows a set of $\alpha(\tau)$-curves measured on high purity Cu irradiated with different γ-doses at room temperature. Without irradiation the curve has the predicted S-shape (c.f. Fig. 8c). With increasing dose, $\alpha(0)$ is decreased (due to the decrease in loop length caused by accumulation of radiation induced pinning points) and the onset of breakaway is shifted to higher stresses (because decreasing loop length leads to increasing breakaway stress). Furthermore, the $\alpha(\tau)$-curves exhibit a hysteresis which increases with increasing irradiation dose. This can be explained by the assumption that after breakaway from the severely pinned initial position the moving dislocation experiences a (less severe) interaction with irradiation induced glide obstacles. As indicated in Fig. 10 /35/ such glide obstacles might balance the line tension restoring force and prevent the dislocation from moving back into its initial position after unloading. This means that the strongly pinned initial position and the corresponding minimum attenuation is attained only at negative stress values.

Fig. 11 shows a complete tension/compression cycle /35/. After passing through the first loading curve (dashed), a closed stable "butterfly" shaped attenuation/stress hysteresis loop is obtained. This loop corresponds to a stress/dislocation-microstrain hysteresis loop and can be considered as a mechanical analog to the magnetic hysteresis. The two minima indicate the stresses for which the dislocations travel through the glide plane position with highest pinning point density. The separation of the minima along the stress axes is a measure of the frictional stress τ_F due to the glide obstacles on the path of the moving dislocations. The asymmetry of the curve indicates the existence of a net internal stress acting upon the dislocations under consideration.

Another example /36/ shows that this kind of attenuation/stress experiment provides an interesting tool for the study of dislocation point defect interaction. Fig. 12 shows schematically a sequence of loading and annealing treatments of a sample of Elmore-copper irradiated under zero stress at a temperature between two pinning stages (c.f. Fig.28). During irradiation at 295 K pinning points from the lower stage (III) are accumulated at the dislocation. The sample is then subjected to a stress τ_A and, with this stress applied, annealed at a temperature within the higher pinning stage (IV) where stage IV-pins are formed at the bowed out dislocation. Upon further loading breakaway from these pins occurs. During subsequent unloading on its way back first repinning of the dislocation by the stage (IV)-pins, then breakaway from these pins and finally repinning by the stage III and the ingrown pins at its initial position takes place. During the following stress cycle, a minimum in the $\alpha(\tau)$ curve is expected when the dislocation passes through the position of the stage (IV) pinning points. Fig. 13 shows the $\alpha(\tau)$ dependence for such an Cu sample successively annealed under 3 stresses within stage IV after irradiation within stage III /36/.

For a quantitative evaluation of $\alpha(\tau)$-experiments, use can be made of the theoretical treatment of the double-loop breakaway (Lücke, Granato and Teutonico /14/). In this model (Fig. 14) a change of stress leads to a change of the depth of 2 potential wells (characterizing the pinned and the unpinned state) and of a potential maximum inbetween (characterizing the activated state). At low temperatures breakaway occurs only if the maximum is sufficiently reduced by the stress in order to allow thermal activation to take place; i.e. the breakaway stress depends strongly upon temperature. At high temperatures, however, the exchange between the wells is so fast that always equilibrium is attained. Then the stress for which unpinning is observed is given by the stress at which the 2 potential wells have an equal depth so that the probabilities for being occupied are equal. One obtains /34/

$$\tau^* = \sqrt{16\ H_o G/L^3} \tag{2}$$

not depending upon temperature with G = shear modulus, H_o = binding energy, L = total length of the double loop. Such a temperature independent breakaway stress has indeed been observed /34/. With the help of Equ. (2) and under proper consideration of the loop length distribution the interaction energy has been calculated and a reasonable

magnitude (0.2 eV in copper) has been obtained /34/. Furthermore
it has been found that in irradiated Cu by different annealing treatments pinning points of different interaction energies can be produced
/80/. This clearly indicates the formation of pinning point clusters
at the dislocation (c.f. Sec. IV d).

III. Theoretical Aspects of Dislocation Resonance Damping.

a) The generalized vibrating string model.

The most simple mathematical model for the movement of a dislocation
of Burgers vector b subjected to an oscillating shear-stress $\sigma = \sigma_0 \cos \omega t$
can be written as

$$Ky + B\dot{y} = b\sigma_0 \cos \omega t \tag{3}$$

Here it is assumed (Fig. 15a)
(i) that the displacement y of the dislocation from its zero-stress
position y = 0 is constant along the dislocation line, e.g. that an
initially straight dislocation moves like a rigid rod. (In more realistic
cases y must be considered as a kind of average value.)
(ii) that the restoring force -Ky pulling back the dislocation to its
zero-stress position can be described by a spring constant K;
(iii) that the moving dislocation experiences a Newtonian type frictional
force ('drag') -B$\dot{y}$;
(iv) that this frictional force is so large that dislocation inertial
effects play no role. (up to now, there are no internal friction results
being in contradiction to this assumption.)

Solving Equ. (3) for y(t) and calculating the logarithmic decrement
and the modulus defect $\Delta M/M$ in the usual way leads to expressions
having the forms of a simple Debye relaxation (Fig. 16):

$$\frac{\delta}{\pi} = Q^{-1} = \Delta_R \frac{\omega \tau_R}{1 + \omega^2 \tau_R^2} \quad ; \quad \frac{\Delta M}{M} = \Delta_R \frac{1}{1 + \omega^2 \tau_R^2} \tag{4}$$

with relaxation strength Δ_R and relaxation time τ_R obtained to be

$$\Delta_R = \Lambda G b^2 / K \quad ; \quad \tau_R = B/K \tag{5}$$

(Λ = dislocation density, G = shear modulus). One recognizes that only

τ_R and thus only the frequency of the damping maximum depends upon B (i.e. upon the drag mechanism), but that Δ_R and thus the height of the maximum is independent of it. Furthermore, one recognizes that from measurements of this type of internal friction for known Gb^2 only the values for the parameter combinations Λ/K, B/K and B/Λ can be derived.

In general, there are two kinds of restoring forces to be considered: (i) Forces due to the internal stress fields. They are caused by other dislocations and/or by interacting point defects contained in the surrounding (Cottrell- or Snoek- atmosphere).
(ii) Forces due to the line tension of the vibrating dislocation. In the original Köhler-Granato-Lücke treatment, the dislocation is considered to be fixed at its ends (loop length L) and restored by its own line-tension C (Fig. 15b). In this case, where K is a function of x (=coordinate along the dislocation line) and is given by $(C/y)\,\partial^2 y/\partial x^2$ Equ. (3) takes the form /10,8/

$$ Cy' + B\dot{y} = b\sigma_0 \cos \omega t \tag{6} $$

The exact expressions for δ and Δ M/M following from Equ. (6) /37, 38,3,19,16/ are analytically difficult to handle. According to Granato and Lücke /8/, however, they can be approximated very well by the Debye-relaxation formulas Equ. (4) with

$$ \Delta_R = \frac{1}{\varkappa}\,\frac{Gb^2}{2}\,\frac{\Lambda L^2}{C} \simeq \frac{\Lambda L^2}{\varkappa} \; ; \qquad \tau_R = \frac{BL^2}{\gamma C} \tag{7} $$

and $\varkappa = \pi^4/16 \simeq 6.09$ and $\gamma = \pi^2 \simeq 9.87$. These values have been repeatedly reconfirmed by other approximations. E.g. for low frequencies /38,3/ the values $\varkappa = 6$ and $\gamma = 10$ and for the region near the damping maximum /16/ $\varkappa = 6.073$[+)] and $\gamma = 9.9016$ have been obtained [++)]. Other values given in literature (e.g. $\gamma = 12$; /7/) are less correct. Only

---

[+)] By a misprint, the value H = 1/0.988 was listed.

[++)] For the sake of clarity it shall be emphasized that, in contrast to statements in literature /19,61/, the values $\varkappa = 6$ and $\gamma = 10$ are not more rigorous than the old values $\pi^4/16$ and π^2. Both sets are approximations, the former for low frequencies, the latter for frequencies near the maximum.

for very large frequencies ($\omega > 10/\tau_R$) the term ($\omega\tau_R$)^2 in the expression for the modulus must be replaced by ($\omega\tau_R$)$^{3/2}$/38/. At these frequencies, however, $\Delta M/M$ is already very small ($\approx 0.01\ \Delta_R$) so that nearly in the whole frequency range in which essential changes in $\Delta M/M$ occur (particularly at both sides of the damping peak) it is more correct to apply ($\omega\tau_R$)^2 than ($\omega\tau_R$)$^{3/2}$ (c.f. Fig. 16).

A comparison of Equ. (7) with Equ. (5) shows that K corresponds to 12.2 C/L^2 in the expression for Δ_R and to 9.9 C/L^2 in the expression for τ_R. Since, however, the other constants in these equations are only approximately known, it is very difficult to differentiate whether an investigated dislocation motion corresponds more to a rod vibrating in a potential field or to a vibrating string with fixed ends. Thus in the following,

$$K = 12\ C/L^2 \tag{7a}$$

will be used.

In general, measured $\delta(\omega)$-curves exhibit broader peaks than predicted by Equ. (4). This is explained by a distribution of relaxation times τ_R, i.e. of K- and/or B-values (Equ. (5)). Especially in the original Granato-Lücke treatment /8/ an exponential loop length distribution function (where L means then the mean free loop length) is assumed. This leads to the theoretical curve shown in Fig. 16 which exhibits a broadening of the peak by about a factor of 2. The different parts of this curve show again the L-dependence predicted by Equ. (7), only with different factors κ and γ, if for L the average loop length is introduced. In case of the <u>exponential distribution</u>, one obtains /8,38,3/ for the damping maximum

$$\left(\frac{\delta}{\pi}\right)_{MAX} = 0.36\ \frac{Gb^2}{\lambda}\ \frac{\Lambda L^2}{C}\quad ; \quad \omega_{MAX} = \frac{0.71\ C}{B L^2} \tag{8}$$

for the low frequency asymptotes ($\omega \ll \omega_{max}$)

$$\left(\frac{\delta}{\pi}\right)_{Low} = \frac{0.93\ Gb^2\ \Lambda L^4 B}{C^2}\ \omega \quad ; \left(\frac{\Delta M}{M}\right)_{Low} = \frac{0.49\ Gb^2\ \Lambda L^2}{C} \tag{9}$$

and for high frequencies ($\omega \gg \omega_{max}$)

$$\left(\frac{\delta}{\pi}\right)_{HIGH} = \frac{0.81\, Gb^2 \Lambda}{B} \cdot \frac{1}{\omega} \quad ; \quad \left(\frac{\Delta M}{M}\right)_{HIGH} \approx 0 \qquad (9a)$$

If the distribution of Λ over different glide systems is to be taken into account, an orientation factor Ω must be introduced into the expressions for δ and $\Delta M/M$ /8,39/.

It shall be noted that, according to Seeger and Schiller /40/, it might be physically more correct to describe dislocation motion as motion of kinks than as motion of a continuous elastic string. It has been shown /2,41/, however, that Equ. (6) essentially holds for the kink-model and that only the numerical constants come out slightly different. Also without leading to principal changes most of the following considerations can be reinterpreted in terms of kink-motion.

b) Possible mechanisms for the drag on moving dislocations

As stated above there is quite a number of possible mechanisms leading to frictional forces upon moving dislocations. Some of them as well as the resulting drag constants B will now be discussed. Special emphasis is given to the cases 2 to 5 where the drag is caused by dislocation/point defect interaction. Here as well as in case 6 the drag constant exhibits a strong (exponential) temperature dependence so that the corresponding damping effects can only be observed in a limited temperature or frequency range. As will be shown, these mechanisms can all be treated as stress aided thermally activated diffusion of dislocations and thus be quantitatively described by the same formalism. It must be emphasized that by determining the value of B in Equ. (3) all these mechanisms - as well as the not thermally activated mechanisms (e.g. phonon drag) - lead to the very same type of frequency dependence (namely that of Equ. (4)). Recognizing the common underlaying formalism it appears somewhat misleading to strictly distinguish between dislocation resonance and dislocation relaxation as sometimes done in literature.

1. Physical drag mechanisms

The interaction of moving dislocations with phonons and electrons as well as thermoelastic losses during dislocation motion lead to small values of B which become observable in internal friction experiments at high frequencies (mainly in the MHz-range). In the temperature range of primary interest here (T > 77K) the dislocation-phonon interaction predominates (e.g. /42/). For this case the appropriate theories yield (e.g. /42,43/)

$$B_{Ph} = const. \ T \tag{10}$$

2. Breakaway drag

Internal friction phenomena due to thermally activated breakaway processes of dislocations moving across a field of immobile point defects have been considered by Granato and Lücke /8/, Weertman /37/, Friedel /15/, Teutonico, Granato and Lücke /13,14/, and Schlipf and Schindlmayr /16/. In case of attractive interaction, random distribution of the point defects and applied stresses smaller than the mechanical breakaway stress, the dislocation assumes a zigzag-form /44,15,45/ (Fig. 17a). Its motion then occurs by stress aided thermally activated breakaway from the interacting point defect and glide of the freed segment until it is re-anchored by the next point defect. If H is the activation enthalpy for breakaway (i.e. the interaction enthalpy), V the activation volume, ν the effective attack frequency and λ the average displacement of the segment freed by a single activation event, the average velocity of the dislocation under the stress σ is

$$\bar{\dot{y}} = \lambda \nu \left\{ exp\left(- \frac{H - V\sigma}{kT}\right) - exp\left(- \frac{H + V\sigma}{kT}\right) \right\} \tag{11}$$

The first term gives the number of jumps in the direction of stress and the second one opposite to it. For $V\sigma \ll kT$, an expansion of $exp\ (\pm\ V\sigma/kT)$ leads to $\bar{\dot{y}} \sim \sigma$ and thus to the drag constant

$$B = \frac{b\sigma}{\bar{\dot{y}}} = \frac{bkT}{\lambda V \nu} exp\left(\frac{H}{kT}\right) \tag{12}$$

This expression has a rather general form. For the special case of breakaway of a zigzag dislocation, the zigzag length ℓ and the atomic fraction of defects c are introduced and $V \simeq \ell b^2$ and $c = b^3/2b \cdot 2\ell\lambda$ obtained. The latter expression follows from the assumptions that $\lambda \ll \ell$ (as valid for dilute solution where $\lambda/\ell \approx (2c)^{1/3}(H_B/2Gb)^{2/3}$ that only defects situated in the 2 atomic planes just above and below

61

the glide plane act as anchoring points and that the area $2\lambda\ell$ swept
by the dislocation by a single activation event contains in the average
just one pinning point (steady state condition for the zigzag-length).
Introducing these expressions into Equ. (12) leads to the breakaway
drag constant

$$B_B = \frac{4gc\,kT}{b^3\,\nu_B}\, exp\left(H_B/kT\right) \qquad (13)$$

with $g = 1$. In the more thorough treatment by Schlipf and Schindlmayr
/16/ g is a factor of magnitude 1 accounting for dislocation line-tension effects and being weakly dependent upon the applied stress.

Damping and modulus change resulting from this breakaway drag are obtained by introducing B_B into Equs. (5) and (7). However this result
is correct only if the number of point defects in contact with the loop
length L is large compared to 1 (i.e. $L \gg \ell$). The largest deviations
are obtained for only one pinning point per loop, (i.e. $L \approx 2\ell$,
Fig. 14). In this case the fractions N_1 and N_2 of the total number of
loops being in the state 1 (pinned) or 2 (unpinned) as function of time
are determined by the differential equation

$$\frac{dN_2}{dt} = N_1\nu_B\, exp\left(-\frac{H_B - V\sigma}{kT}\right) - N_2\nu_B\, exp\left(-\frac{H_B + V\sigma}{kT}\right) \qquad (14a)$$

$$\approx \left\{\frac{1}{2} - N_2 + \frac{V\sigma_0}{2kT}\cos\omega t\right\} 2\nu_B\, exp\left(-\frac{H_B}{kT}\right) \qquad (14b)$$

From the solution $N_2(t)$ one obtains in the usual way (with $V \approx \ell b^2$)

$$\Delta_R = \frac{1b G\Lambda V}{2kT} \qquad ; \qquad \tau_R^{-1} = 2\nu_B\, exp\left(-H_B/kT\right) \qquad (14c)$$

The comparison with Equ. (5) and (13) shows that the damping values following from both models agree for $\omega\tau_R \gg 1$. This is expected since only
in this limit   the motion of the loop $L \gg \ell$ consists of breakaway
jumps of isolated segments 2ℓ instead of successive ("catastrophic")
breakaway along the whole loop. It is to be noted that the expressions
Equ. (14c) deviate from those given in literature for a similar model
/17/.

3. Point defect drag

Due to the interaction between point defects and dislocations, mobile
point defects are carried along by a moving dislocation (Fig. 17b) and, in
turn, exert a retarding force on the dislocation. Originally, long range
interaction, i.e. the motion of a Cottrell atmosphere /46/ has been considered (e.g. /47/). The drag due to short range interaction, i.e. the
motion of pinning points, and the resulting internal friction effects/48/
have been quantitatively described by Lücke and Schlipf /5/. They derived the two   limiting   cases (one pinning point and very many
mobile pinning points per loop) to be considered here and also discussed
by Simpson and Sosin /19/. Since during the motion of the point defect
the dislocation remains alsways in contact with the defect, it is assumed that the defect experiences the highly increased diffusivity
attributed to the dislocation core region. This enhanced diffusion
strongly facilitates the occurance of point defect drag.

The resulting drag can be simply calculated from Equ. (12) which is
valid, since here the mean dislocation velocity again is determined
by stress-aided thermally activated events, namely diffusional steps
of the dragged point defect. Thus here the values $\lambda = $ b and
$V = b^2 \ell = b^2/n$ (n = number of point defects per unit length of the
dislocation) must be introduced into Equ. (12) leading to the point
defect drag constant

$$ B_D = \frac{n\,kT}{b^2\,\nu_D}\, exp\left(H_D/kT\right) \tag{15a} $$

with H_D being the activation energy of diffusion of the defects near
the core and ν_D the attack frequency. The point defect drag constant
B_D does not depend upon the interaction strength, since the distance
between point defect and dislocation adjusts itself just to that value
for which the total force exerted on the dislocation by the dragged
point defects compensates the force exerted on the dislocation by the
external stress. Only the limiting stress at which the dislocation
breaks away from the defects is determined by the interaction strength.

In order to demonstrate the physical nature of the damping due to point
defect dragging, Fig. 18 qualitatively shows the displacement of a dislocation loop (with nL = 3 point defects in contact) and the resulting
damping as function of temperature and frequency, respectively. At low

temperatures (or at high frequencies) the point defects are immobile and thus do not contribute to damping. At high temperatures (low frequencies) their mobility is so large that they move completely in phase with the oscillating dislocation loop and give rise to maximum dislocation strain but cause no damping. Only at medium temperatures (medium frequencies) limited point defect mobility results in an additional dislocation strain (as compared with low temperatures) as well as in a phase shift, i.e. in damping. Hence, if point defect drag damping is measured as function of frequency, the observed relaxation maximum shifts to higher frequencies with increasing temperature with its height being constant (Equ. (8) and (9)).

It is expected that Equ. (15) inserted into Equs. (5) and (7) yields a good approximation for the resulting internal friction effects, if nL (i.e. the number of point defects on the loop of length L) is large and if in Equ. (15) n is replaced by n_{eff} = (n + 1/L) in order to account for the influence of the end points of the loops L upon the length of the freely moving segments. The largest deviation occurs if the loop contains only a single pinning point. For this case, the exact calculation /5,19/[+)] shows that κ = 8 and

$$\tau_R = (\kappa TL/4 Cb^2 \nu_D) \exp(H_D/kT) \tag{15b}$$

i.e. n_{eff} = 2,48/L has to be used in Equs. (7) and (15) instead of κ = 6,1 and n_{eff} = 2/L. Thus, even for the one defect case, the correct values for Δ_R and τ_R differ only by 30% from those derived from the continuous drag model.

4. Superposition of phonon drag and point defect drag

In order to treat phonon drag and point defect drag simultaneously, one must consider that the phonon drag acts continuously upon the dislocation line, whereas the defect drag acts only at discrete points.

[+)] Owing to a misprint, a factor 2 is missing for κ in /5/. Another treatment given in literature/49/ differs from this one by limiting the diffusion path of the point defect to only 1 atomic distance in each direction. This is correct only for $\omega \tau_R \gg$ 1 whereas for $\omega \tau_R \leq$ 1 serious deviations occur.

Simpson and Sosin/19/ have analytically investigated this superposition. However, since their general results are very complex, here more physical arguments will be presented which explain the different cases in a very simple manner.

At high temperatures, as mentioned above, point defect drag can be neglected and one obtains the phonon resonance damping maximum defined by Equ. (7) and (8) with $B = B_{Ph}$ and loop length L (c.f. broken line in Fig. 19). At low temperatures, however, one obtains the superposition of two maxima: at high frequencies (where the point defects may be taken as immobile compared with the high dislocation velocity) the resonance damping maximum resulting from $B = B_{Ph}$ and loop length $\ell = L/(n+1)$ and at low frequencies (where the phonon drag can be neglected) the point defect drag maximum determined by $B = B_D$ and $\ell = L$. If T is raised, B_D decreases and the defect drag maximum simply shifts to higher frequencies without changing its height (Δ_R is independent of B, c.f. Equ. (5)).

Thus it appears appropriate to describe the combined drag either as a single relaxation process (at high temperatures) or as a superposition of two such processes (at low temperatures). For the two maxima being seperated but close together, the situation becomes more complex; also the analytical solution /19/ gives no simple answer for this case. Wire and Granato /50/ simply superimposed the drag constants instead of the relaxation processes by setting

$$B = B_{Ph} + B_D \tag{16}$$

This gives a single relaxation effect and would be only correct for a continuous distribution of the point defects along the dislocation. Otherwise, it represents an approximation which leads to the correct high temperature behaviour and to the defect drag peak at low temperatures, i.e. it neglects the high frequency peak determined by $B = B_{Ph}$, $\ell = L/(n+1)$.

It shall be noted that a superposition of the above described breakaway drag and the phonon drag lead to the same type of effects.

5. Diffusion along the dislocation

Yamafuji and Bauer /18/ showed that the oscillating dislocation loop induces diffusion of point defects along the dislocation; Lücke and Schlipf /5/ pointed out, that this motion gives rise to an additional damping peak. The exact calculations by Winkler-Gniewek, Schindlmayr and Schlipf /20/ revealed, however, that the effect comes out to be strongly amplitude dependent.

6. Reorientation drag

Defects with axial symmetry reorientate if a moving dislocation comes close to them and loose this preferred orientation again, after the dislocation has removed itself far enough (Fig. 17c). According to Schöck and Seeger /21,22/, this dragging of a Snoek atmosphere causes a retarding force on the dislocation and internal friction effects. The resulting drag constant comes out to be /33/

$$ B_s = \frac{12 \pi c \, k T}{b^3 \, \nu_s} \, exp\left(H_s / kT\right) \tag{17} $$

As to be expected this expression has again the form of Equ. (12), only the rate determining step is here the diffusion jump with an activation energy H_s leading to a rotation of the dipol axis. A quantitative discussion of the internal friction effects has been given by Sokolowski and Lücke /33/. It shall be emphasized that the relaxation time is not given here by the reorientation time
$\tau_s = (1/\gamma_s) exp\,(H_s/kT)$ as is discussed in literature /22/, but by the vibrating string expression Equ. (5)/33/.

7. Bordoni drag

Seeger /2,51,62/ and other authors/63,64/ interpreted the Bordoni-type damping peaks by the transition of dislocations well nearly parallel to a Peierls-valley into the neighbouring valley by means of double-kink formation. Although the thorough treatment of this process is more complex, according to Engelke /65/ and Schlipf and Schindlmayr /66/ under simplified circumstances an expression

$$B_K \simeq \frac{5kT}{b^2 L \nu_K} \exp\left(H_K / kT\right) \qquad (18)$$

can be derived for use in Equ. (7). Here (c.f. Fig. 17 d) is $\lambda = b$ and as to be seen by comparison with Equ. (12) $V = b^2 L/5$. Here the value L/5 instead of twice the kink width 2 w arises since the thermally activated event consists of the formation of two kinks each of width w but additionally of the diffusion of the kinks towards the ends of the loop. Consequently the "attack frequency" ν_K is also dependent on the viscosity of kink motion. The kink width is given by $w \simeq Gb^4/2W_K$ and the enthalpy of double kink formation by $H_K = 2W_K$ /2,6,7/.

8. Non-Newtonian drag

There are many cases where the frictional force is not proportional to the velocity of the dislocation, i.e. B is not constant. This makes the basic differential equation for the motion of the dislocation (Equ. (3) or (6)) nonlinear and leads in general to amplitude dependent damping. An example is the loss due to sound radiation of the vibrating dislocation /67/ (c.f. / 52, 42/). The nonlinear behaviour is obtained for all mechanisms, if the force F acting at the dislocation becomes large. For example, in case of breakaway drag the factor $g/\lambda V$ in Equ. (12) and (13) depends slightly on stress /16/. Another nonlinearity is introduced if $\sigma V \gg kT$, since then the second term in Equ. (11) (i.e. the back-jumprate), becomes small and an exponential stress dependence is obtained. In case of point defect drag or Snoek drag nonlinear effects occur already at comparatively small stresses at which the dislocation breaks away from the atmosphere. The breakaway from diffusing point defects takes place, if the dragging force per defect exceeds the maximum interaction force between point defect and dislocation, i.e. $\sigma b/n > f_{max}$ /5/. In the second case /33/ no retarding Snoek-atmosphere can be formed if σ and thus the dislocation velocity $\dot{y}$ becomes so large that the time a defect stays in the kT-circle (radius R) is smaller than the reorientation time τ_j (R/$\dot{y}\tau$ = RB$_s$/b$\sigma\tau \approx 12 \pi cA/\sigma b^3 < 1$).

If the point defects are distributed nonrandomly but, for instance, are segregated along the dislocation, the zig-zag configuration with its symmetry for forth- and back jumping is lost. In this case a strongly amplitude dependent breakaway damping takes place with the second term

in Equ. (11) mostly negligeably small /13, 14/. The breakaway as well
as the Bordoni-typ damping become frequency-independent at T = O K
(static hysteresis). Weertmann /37/ and Naundorf and Lücke /53/ proposed models leading to frequency-independent and, at the same time,
amplitude-independent damping. Instead of the Newtonian force in
Equ. (3) these models make use of a "frictional" force opposing dislocation motion which is proportional to the amount of the dislocation
displacement y:

$$F = - const \cdot \frac{\dot{y}}{|\dot{y}|} \cdot y \qquad (19)$$

In this way a quasi-linearity of the differential equation (3) is
retained.

c) Comparison with Experiments

In most MHz-experiments the damping is measured as function of frequency at low strain amplitudes ε_0 where no amplitude dependent
effects are observed. In most KHz- and Hz-experiments, however, damping
and modulus are measured at fixed frequency as function of amplitude.
In many experiments additionally the temperature T is varied. The
present paper considers only low amplitude measurements, i.e. ampli-
tudeindependent internal friction effects. Furthermore in the present
section the structural parameters, i.e. density and arrangement of
dislocations as well as point defects, will be taken as constant and
the dependence of internal friction on variations of ω and T will be
discussed. Experiments involving changes of the point defect concentration
(e.g. irradiation experiments) will be discussed in Sec. IV.

1. Non-dislocation background

In a considerable number of cases experimental results can be interpreted with reference to dislocation damping or modulus defect only
to a very limited extent. The reason is that damping and modulus resulting from simultaneously acting non-dislocation mechanism are not
known. It is believed that the best method to obtain these "background"
values consists of an intense irradiation with neutrons, or electrons or
γ- rays /3/.

For higher frequencies (MHz- and KHz-range) and not too high temperatures, this seems to lead to a complete immobilization of the dislocations due to pinning by irradiation induced defects (e.g. Fig.
(1) and (2)). In the Hz-region, however, complete pinning by irradiation
seems to be difficult to obtain, in fact even irradiation induced increases
of damping have been observed /19, 29, 61/.

2. Frequency dependence

In many experiments in the MHz-range a $\delta(\omega)$-dependence with a maximum
and a shape predicted for an exponential loop length distribution has
been observed (e.g. Fig. 21 and /54 to 60, 70, 71/. Evaluation of such
results according to Equ. (8) yield the parameter combinations $\Lambda L^2/C$
and B/Λ. By taking Λ-values from independent measurements (etch pit
counts), one obtained, e.g. for Cu at room temperature, $B = 6 \cdot 10^{-4}$ dyn
sec/cm^2 which is in an order of magnitude agreement with theoretical
predictions for the phonon drag /42/. Furthermore, with $C = Gb^2/2$ also
very reasonable L-values (10^{-5} to 10^{-3} cm for well annealed Cu-crystals)
have been calculated /e.g. 89, 59/. Similar results have been obtained
in the KHz-range by evaluation of damping and modulus measurements with
aid of Equ. (9) (e.g. /3/).

The more serious question to which extent MHz-damping values are compatible
with KHz-results with respect to the theory of dislocation resonance
damping has   received different answers for many years /69,70, 71, 12/.
Recently, Naundorf and Lücke /53/ investigated the internal friction of
one and the same sample simultaneously in the KHz- and MHz-frequency-
range. They convincingly proved what had been deduced before by comparing results for different samples investigated in this frequency
range, namely that the KHz-damping is about a factor 10^2 higher than
the value obtained by extrapolation of the damping observed at MHz-fre-
quencies (Fig. 20 a). The detailed analysis of the data yielded about
the same loop length L for both frequency ranges, but a drag constant
B to be larger by a factor of about 30 for the KHz-range.

This result can be most easily understood by assuming that phonon drag
determines MHz-damping whilst in the KHz-range a different mechanism
dominates. There are mainly two observations which cast some light on
the KHz-mechanism:

(i) the loop length dependences of modulus defect and damping in the
KHz-range is controlled by the L^2/L^4-law of Equ. (9 a) (e.g. /3, 53/;
(ii) the damping is found almost frequency independent from KHz-range
down to the Hz-range /72, 73/. Observation (i), which will be discussed in Sec. IV b, strongly favours a relaxation mechanism given
by Equs. (3) or (6) and (7). Observation (ii), however, is in severe
contradiction to it and requires a variation of the underlying differential equation (3) or (6). Withour closer specification of the underlaying atomistic mechanism Naundorf and Lücke /53/ proposed to replace the drag term $B\dot{y}$ by the expression given by Equ. (19). At the
moment this seems to be the only way which reasonably allows for the
observations (i) and (ii) and, at the same time, yields amplitude independent damping.

3. Temperature_dependence

Many authors have investigated the temperature dependence of MHz-dis-
location damping in order to derive B(T) /54-56, 74-77/. If one assumes
B to increase almost linearly with temperature as is expected for
the phonon drag (Equ. (10)), the decrement δ(T) at constant ω should
exhibit a maximum characterized by $B(T_{max}) = \gamma C/L^2 \omega$
according to Equ. (7). Several authors observed a damping maximum as
function of temperature at fixed frequency and consequently tried to
derive the B(T)-dependence from the measured δ(T)-curve /e.g. 54, 76/.
Kaufmann, Lenz and Lücke /60/, however, obtained evidence, that these
δ (T)-maxima might be caused not simply by the temperature dependence
of the dislocation resonance damping but by an unwanted pulse-echo specific
side effect, namely the deformation of the sample due to the difference
in thermal expansion of sample and quartz /56, 59/. On samples where
this quartz-sample deformation had been avoided, no maximum was observed in their δ(T)-curves /60/.

For further clarification, Kaufmann et al. /60/ varied both temperature
and frequency in their measurements on high-purity copper (Fig. 21) in
order to derive the temperature dependence of the damping maximum, Equ.
(8), obtained from $\delta(\omega)_T$-plots. They carefully avoided quartz-sample
deformation effects but found a rather disturbing behaviour: samples
of different purity and/or pretreatment lead to quite different B(T)-
curves. Thus, they concluded that also these curves did not reveal the
B(T) function, but are falsified by uncontrolled impurities effects

(c.f. Sec. IV b). One arrives at a similar conclusion by comparing results of different authors which often show not even a qualitative agreement /6/.

Up to now, also for the KHz-range no clear results leading to an understanding of the temperature dependence of damping exist /12/. Similar effects as in the MHz-range may play a role. Thus it must be concluded that, until further clarification, only the order of magnitude but not the temperature dependence of B can be derived from high frequency internal friction measurements of metals[+).

In the Hz-range experiments, usually the temperature dependence of damping and modulus is measured. Frequently relaxation maxima caused by dislocation motion were observed, e.g. the Hasiguti peaks (e.g. /78/ c.f. Fig. 5) and Bordoni peaks (e.g. /79/). Such peaks as function of temperature indicate that the dragging force is determined by thermally activated processes. There is strong evidence that the Hasiguti peaks are caused by interaction between dislocations and point defects. However, the real nature of the responsible point defects and the underlaying interaction processes (e.g. break-away, dragging or re-orientation) has not been clarified yet.

[+) As an interesting example for non-metallic materials the work of Wire and Granato /50/ shall be mentioned in which a temperature dependence of the modulus defect described by Equ. (16) is re-proted.

IV. Quantitative Aspects of Dislocation Pinning

a) Quantitative predictions following from the different pinning mechanisms

Dislocation pinning e.g. by irradiation or plastic deformation, may be caused by influencing each of the 3 parameters K, B, Λ of Equ. (4) and (5).

1. Influence upon K

According to Equ. (5), an increase in K reduces τ_R as well as Δ_R and thus shifts the maximum of $\delta(\omega)$, (Equ. (4)), to higher frequencies and lower decrement values without changing its high frequency asymptote. Quantitatively one obtains for the maximum (Equ. (8)).

$$\delta_{MAX} \sim 1/\omega_{MAX} \sim 1/K \tag{20}$$

and for low frequencies $\omega \ll \omega_{MAX}$ (Equ. (9))

$$\delta_{Low} \sim (\Delta M/M)^2_{Low} \sim 1/K^2 \tag{21}$$

and for high frequencies $\omega \gg \omega_{MAX}$ (Equ. (9 a))

$$\delta_{HIGH} = Const \quad ; \quad (\Delta M/M)_{HIGH} = Const \approx 0 \tag{22}$$

The right side of Fig. 22 shows this behaviour schematically.

Concerning the physical reasons for an increase in K, e.g. by irradiation, one has to realize that a random distribution of point defects has only little influence upon small displacements of a dislocation. Only after the point defects have moved within the stress field of the dislocation. i.e. into positions of lower energy, the dislocation experiences an additional restoring force K; in order to move the dislocation this energy gain would have to be reinvested as external work. Such a dislocation can be considered to be situated in a potential well. According to the different types of interaction (Sec. III b), several possibilities must be considered.

(i) Pinning by discrete pinning points /11/. Here the irradiation induced point defects are assumed to migrate towards the dislocations where they create additional pinning points reducing the free dislocation loop length L. Since K is proportional to $1/L^2$ (c.f. Equs. (5) and (7)), Equ. (20) resembles the L^2/L^{-2}-law for height and frequency of the damping maximum and Equ. (21) the frequently discussed "L^2/L^4-law" for the ratio of modulus defect and damping at low frequencies. If n is the number of irradiation induced pinning points per cm and p the corresponding number per initial loop length L_o, the resulting value of K is given by

$$\frac{K}{K_o} = \frac{L_o^2}{L^2} = L_o^2 \left(\frac{1}{L_o} + n\right)^2 = (1+p)^2 \tag{23}$$

(K_o = initial value)

(ii) Pinning by a Cottrell atmosphere. Here the point defects have to diffuse to the vicinity of the dislocation. As pointed out above, their effect can mostly be neglected in comparison to the effect of those reaching the dislocation and forming fixed pinning points. Hornung /24/, however, pointed out that the Cotrell atmosphere effect should be observable at the beginning of pinning, when the number of pinning point is still small, but the defects further away have already carried out some jumps towards the dislocation.

(iii) Pinning by a Snoek atmosphere. In contrast to the case (i) where long range diffusion is required, here the reorientation of the axial defects needs only a single diffusional jump. Hornung /24/ calculated the restoring force constant K resulting from such a Snoek-atmosphere. For a resting dislocation[+)] and euquilibrium one obtains

$$K_s = 6 G^2 (\lambda_1 - \lambda_2)^2 c / b^3 kT \tag{24}$$

[+)] In /24/ the atmosphere formed around a dislocation vibrating with an amplitude $\hat{y}$ has been calculated. For its formation, however, the vibrations had to keep going for times larger that the reorientation time τ_s, whereas in many cases the times allowed are much shorter and only long enough to take the measurements. For this reason, here a resting dislocation has been considered by setting $\hat{y}$= b. For considering the influence of the dislocation amplitude on K_s one has to add the factor $(b/\hat{y})^2$.

with c being the concentration and λ_1 and λ_2 the dipole moments of the axial defects. If, on the other hand, the effect of Snoek pinning is attributed to additional discrete pinning points, one derives from Equs. (7 a) and (23)

$$K_p = \frac{12C}{L^2} - \frac{12C}{L_0^2} = \frac{24C}{L_0^2} p \left(1 + \frac{p}{2} \right)$$

(25)

Setting equal Equ (24) and (25) one obtains the apparent pinning point number, p_{app} i.e. the number that would lead to the same restoring force as the Snoek atmosphere:

$$p_{app} = \frac{G^2 (\lambda_1 - \lambda_2)^2 c L_0^2}{4 b^3 C kT} = 4 \cdot 10^{-2} \frac{Gb^3}{kT} \frac{L_0^2}{b^2}$$

(26)

Here is assumed that $p \ll 1$, $C = Gb^2/2$ and $\lambda_1 - \lambda_2 = o,27b^3$ as calculated for the dumbbell interstitial in copper /24/.

As result one sees that direct changes of the loop length as well as changes of the Cottrell or Snoek atmospheres influence internal friction only by changing the parameter K. Since internal friction measurements yield only the parameter K, it is difficult to distinguish by such type of measurements between these different pinning mechanisms. In particular, all these mechanisms lead to the observed shift of the $\delta(\omega)$ -maximum and to the $\delta/(\Delta M/M)^2$-law.

2. Influence_upon_B

Since, according to Equs. (4) and (5), only τ_R and not Δ_R depends upon B, an increase in B results in a shift of the $\delta(\omega)$-curve towards lower frequencies with the height of the maximum unchanged (Equ. (8), left side of Fig. 22)

$$\delta_{max} = const; \quad \omega_{max} \sim 1/B$$

(27)

On the low frequency side of the peak the damping increases with B (Equ. (9))

74

$$\delta_{LOW} \sim B; \quad (\Delta M/M)_{LOW} = \text{const} \tag{28}$$

At the high frequency side of the maximum the increase in B results in a decrease of damping and modulus defect Equ. (9 a); (right side of left peak in Fig. 22)

$$\delta^2_{HiGH} \sim (\Delta M/M)_{HiGH} \sim 1/B^2 \tag{29}$$

(only for very high frequencies one has $(\Delta M/M) \sim B^{-3/2}$).

One recognizes that such an increase in B has qualitatively the same effects on the high frequency side as on the low frequency side an increase in K (Equ. (21)), namely an increase of modulus and a decrease of damping. Thus it is not possible to conclude from an observed (e.g. irradiation induced) increase of modulus and decrease of damping at fixed frequency wether K (e.g. the loop length) or B has changed. This can be decided only by quantitative comparison, since in the first case $\delta/ (\Delta M/M)^2$ (Equ. (21)) and in the second case $\delta^2/ (\Delta M/M)$ (Equ. (29)) is a constant. One recognizes further (Fig. 22) that at intermediate frequencies irradiation causes no damping changes but only modulus increases.

Physical reason for an increase of B by irradiation can be found in the mechanism 2) to 6) of Sec. III b. If the irradiation induced point defects are immobile at the measuring frequence or temperature B might increase because of the more difficult breakaway; if they are mobile, a pinning by point defect drag might be obtained; if they are of axial symmetry, an induced Snoek-effect might occur. All these mechanisms depend exponentially upon temperature so that the log ω -abszissa in Fig. 18 can be replaced by a 1/T-abszissa without any changes of the curves. It is not possible to distinguish between these mechanisms by internal friction measurements except by discussion of the measured values of the relaxation times.

If damping is caused by point defect drag and if the number of such point defects per dislocation loop is small, only those loops contribute to damping at which a defect happens to be situated. An increase of the concentration of such defects, e.g. by irradiation, is expected to cause an increase of the number of contributing loops, i.e. of the "effective" dislocation density Λ_{eff} (left side of Fig. 22). In this case

$$\delta_{MAX} \sim \Lambda_{eff} \; ; \quad \omega_{MAX} = const \tag{30}$$

The corresponding behaviour of the modulus can also be seen on the left side of Fig. 22: one recognizes that an increase in Λ_{eff} leads to an increase in modulus. In contrast, an increase of the true dislocation density, i.e. a creation of new dislocations (with dragging defects attached) would lead to a decrease of the modulus.

b) Comparison with Experiments

Irradiation experiments are of particular value for the interpretation of damping studies. They allow in a structure-conserving and controllable way the introduction of point defects which interact with moving dis-locations, finally rendering them immobile. The available experimental re-sults will now be checked wether it is K, B or Λ_{eff} which is influenced by the irradiation. Also some recovery measurements after plastic de-formation will be considered.

1. MHz-range

Here a great number of irradiation experiments, many of them by the Aachen group /69, 56, 58, 59, 60, 34/, have yielded a $\delta(\omega)$-behaviour which can quantitatively be interpreted in terms of changing K. In particular, the occurance of the predicted damping maximum, its shift according to Equ. (20) at constant high-frequency asymptote (Equ. (22)) has been observed in many cases. This behaviour has been found for pinning during isothermal irradiation (e.g. Fig. 23 a) and for pinning during annealing after low temperature irradiation or deformation where irradiation or deformation produced point defects are assumed to migrate to the dislocations, and also for annealing at elevated temperatures where those pinning point-defects anneal out and depinning occurs (e.g. Fig. 23 b).

Results of this kind have always been interpreted with help of the Gra-nato-Lücke theory in the sense of loop length changes. In order to check the question whether or not reorientation effects contribute to the changes in K, the pinning point numbers p_{exp} determined according to Equ.

(8) and (23) from the measured shift of the damping maxima during irradiation have been compared with the apparent point numbers p_{app} calculated according to Equ. (26). With the quantities L_o being determined from the frequency profile before irradiation and c from the applied irradiation dose (using for copper a cross section for defect production σ_d = 15 barn for 3 McV-electrons), p_{app} values smaller than p_{exp} by several orders of magnitude are obtained. This means that the effect of reorientation is too small to account for the observed pinning and that the discrete pinning point picture is justified. Similar results were obtained for the kHz-range /53/. For the Hz-range no reliable p_{exp}-values are available. For deformed specimens the estimates of p_{app} are not accurate enough for strict conclusions /29/.

2. KHz-range

Also most KHz-experiments exhibit a decrease of damping and an increase of modulus during irradiation or during annealing after low temperature irradiation or deformation. Some of them show a very close fullfillment of the $\delta / (\Delta M/M)^2$ = const (i.e. the L^4/L^2)-relation (c.f. Fig. 20 b /53/). Particularly convincing are the results given in /53/, since there, by high dose irradiation, the true background values for damping and modulus had been determined and no free parameters for fitting (as in some other work) were left. According to Equ. (21), the L^4/L^2-relation means that also in this frequency range only K is changed by the irradiation. Such irradiation-experiments in the KHz-range as well as those in the MHz-range are considered to be the strongest support for the Granato-Lücke theory, since they show exactly the results predicted by this theory for a change in loop length. The fact that the dislocation drag mechanism acting in the MHz- and the KHz-range , i.e. the B values in Equ. (3) or (6) seem to be different (Sec. III c 2) does not influence this conclusion.

On the other hand, there are also measurements for which the relation $\delta \sim (\Delta M/M)^2$ is not fullfilled, i.e. for which the pinning point numbers p calculated either from modulus or from damping with help of Equ. (9) come out to be different. Simpson and Sosin /19/ tried to explain this behaviour by assuming that the irradiation induced point defects at the dislocation give rise to an additional point defect drag instead of becoming additional firm pinning points. This would mean that by irradiations not K but B is increased. According to Equ. (29), also this would result in a decrease of damping, if it is further assumed that the KHz measuring frequencies are on the high frequency side of the point defect drag damping maximum, (c.f. Fig. 22). There are some arguments concerned with this assumption:

(i) The observed deviations from the $\delta \sim (\Delta M/M)^2$-law are indeed in the direction that the exponent of $(\Delta M/M)$ is smaller than 2 even down to 1, i.e. $\delta \sim (\Delta M/M)$ /19/. On the other side, in contrast to the firm pinning exponent 2, in no case exactly the exponent 1/2 (or 2/3) predicted for point defect drag (Equ. (29)) has been reported /61/.

(ii) In some of their experiments Simpson and Sosin and coworkers observed that with increasing irradiation time the damping first increases and then decreases after passing through a maximum ("peaking effect" / 61 /). Such an effect is expected, if the measuring frequency ω is initially smaller than the maximum frequency of the point defect drag peak and B increases with irradiation time causing the damping maximum to move through the frequency ω (Fig. 22). This interpretation of peaking, however, has not been proved by evaluation of both damping and modulus as function of irradiation time as would be necessary for firm conclusions. Instead there are other experiments exhibiting "peaking" not being attributable to point defect drag: a recent KHz-investigation has yielded a peaking effect in damping, it is true, but no corresponding modulus effect /53/. In this case the peaking could be retraced to an irradiation induced apparatus effect since it vanished when damping was measured directly by decay curves instead of amplification factor. Furthermore, in the amplitude range as high as $\varepsilon_o = 10^{-6}$ as used mostly by Simpson and Sosin /19/ peaking might be caused by amplitude dependent effects. E.g. the measurements of Lücke et al (Fig. 6 in ref. /81/) give a small damping peak if the amplitude $\varepsilon_o = 10^{-6}$ is chosen, whereas at $\varepsilon_o = 10^{-7}$ the damping decreases monotonically with time.

(iii) The assumption of point defect drag lead to a strong temperature dependence of the peaking curves (Fig. 22). This, however, has not been observed and has been explained by a non-thermal diffusion mechanism the nature of which remained unspecified /19, 61/.

Summarizing, it is concluded that the interpretation of the KHz-damping irradiation effects by point defect drag (i.e. by an influence upon B) still needs confirmation and is certainly not of general validity. The firm pinning point interpretation (i.e. the irradiation influence upon K) appears to be proved in a number of cases.

3. Hz-range

The very few irradiation experiments carried out in the Hz-range (for electron irradiation /68, 29/) have shown indeed an increase of damping by irradiation (e.g. Fig. 5 and 24). This can happen only if B (or Λ_{eff}) is increased by the irradiation. So far, no detailed analysis exists for this frequency range. In particular it is uncertain which of the drag mechanism described in Sec. III b is active here (e.g. dragging or breakaway). The same is true for the numerous pinning measurements after coldwork where several pronounced relaxation peaks (e.g. Hasiguti peaks) are observed /6, 7/.

c) Theoretical predictions for the kinetics of pinning

For studies of pinning 3 types of experiments have successfully been used:

(i) isothermal annealing: irradiation (or plastic deformation) at low temperatures where the produced defects are immobile is followed by isothermal annealing at elevated temperatures where the defects are able to migrate to the dislocations.

(ii) isothermal irradiation (or deformation) of the sample at temperatures where the produced defects are mobile.

(iii) constant heating rate (or isochronal warm-up) after low-temperature irradiation (or deformation) of the sample.

Best defined are experiments where irradiation is used for defect production (e-, γ-, n-irradiation) /3/. In order to describe the results of such experiments, several types of quantitative models have been developed. In all of them, only a single kind of migrating point defect is considered to influence only the restoring force (K in Equ. (3) or L in Equ. (7)) but not the drag constant B of the dislocation.

1. Pinning as first order reaction

Here it is assumed that the transition of the migrating defect from the lattice to the dislocation can be described simply by a first order reaction (e.g. /82/):

$$\frac{dN_L}{dt} = -\frac{N_L}{\tau_{LD}} + \gamma \qquad (31)$$

$$\frac{dN_D}{dt} = \frac{\Lambda}{\varphi L}\frac{dp}{dt} = \frac{N_L}{\tau_{LD}} \qquad (32)$$

Here is $N_L(t)$ the number of defects per cm^3 being in the lattice, $N_D(t)$ the number having reached the dislocation up to the time t, τ_{LD} the time constant for the transition lattice-dislocation and γ the (constant) production rate of defects. It is further assumed that only a certain fraction φ of the defects having reached the dislocation create a pinning point. The pinner number p(t) is then obtained by solving Equ. (31) for the proper boundary conditions, by introducing the resulting N_L-value into Equ. (32) and solving this equation for p. The results for the above listed types of experiments are the following (c.f. Fig. 25):

(i) Isothermal annealing. Here is $\gamma = 0$ and one obtains

$$p(t) = p_0 \left(1 - exp(-t/\tau)\right) \qquad (33)$$

The asymptotic behaviour is given by $p = p_0 t/\tau$ for $t \ll \tau$ and $p = p_0$ for $t \gg \tau$.
Correspondingly, the concentration N_L decreases exponentially from p_0 to 0.

(ii) Isothermal irradiation. Here one obtains

$$p(t) = \varphi\gamma \left\{t - \tau + \tau\, exp(-t/\tau)\right\} \qquad (34)$$

For $t \ll \tau$ this function shows an initial region with $p = \varphi\gamma\, t^2/2\tau$ and for $t \gg \tau$ a steady-state region with $p = \varphi\gamma\, (t - \tau)$.

In the first region N_L increases linearly starting from $N_L = 0$, in the second N_L stays constant and the pinning rate is then given by the production rate.

(iii) Constant heating rate. With the heating rate $\beta = dT/dt$ one obtains

$$p(T) = p_0 \left\{1 - exp\left[-\frac{kT^2}{\beta H_m \tau_0}\, exp(-H_m/kT)\right]\right\} \qquad (35)$$

Here it is assumed that

$$\tau = \tau_0 \exp (H_m/kT) \qquad\qquad (35\text{ a})$$

and that the activation enthalpy of defect migration $H_m \gg kT$. The function p(T) going from p = 0 at low to p = p_0 at high remperatures is S-shaped the inflection point and half width being given by /33/ (Fig. 25 c)

$$T_0 = \frac{H_m}{k(x - 2\ln x)} \quad \text{and} \quad \Delta T = \frac{2,32\, T_0}{x - 2\ln x} \qquad\qquad (35\text{ b})$$

with x = ln $(H_m/k\beta\tau_0)$.

2. Pinning as diffusion problem

Here the number of defects reaching the dislocation is determined by the equation for diffusion of the defects in the stress field of the dislocation and the boundary condition $(dN_L/dr)_{r=R} = 0$ with R = $(\pi\Lambda)^{-1/2}$ being half the dislocation distance (Fig. 26) /82, 83/. If the angular dependence of the point defect dislocation interaction is neglected and the interaction potential taken from Equ. (1) one obtains

$$\frac{\partial N_L}{\partial t} = D\left[\frac{\partial^2 N_L}{\partial r^2} + \frac{1}{r}\frac{\partial N_L}{\partial r} + \frac{Ab}{kT}\left(\frac{1}{r^2}\frac{\partial N_L}{\partial r^2} - \frac{N_L}{r^3}\right)\right] \qquad\qquad (36)$$

According to Ham /84/, the solution of Equ. (36) is approximately given by the solution of this equation without drift term (i.e. for A = 0), but with the boundary condition c(r_A) = 0 with r_A/b = A/kT (c.f. Fig. 26). This means the dislocation can be considered as a zylindrical unsaturable sink of the radius r_A with the pinning rate given by the point defect into this zylinder.

The solution of the diffusion equation without drift is covered up to 97% by an asymptotic solution which leads to a pinning rate given by Equ. (33) and (35 a) with

$$\tau = \frac{R^2}{2D}\left(\ln\frac{R}{r_A} - \frac{3}{5}\right) ; \quad \tau_0 \approx \frac{R^2}{2b^2 \nu_{m0}} \ln\frac{R}{r_A} \qquad\qquad (37)$$

This means, the diffusion treatment justifies the treatment of pinning as first order reaction and determines the value of the time constant τ. Strong deviations from the reaction rate occur only for the initial behaviour ($t \ll \tau$). Instead of the above t^1- and t^2-time-laws for the experiments (i) and (ii), here the relationships

$$p(t) = 4 \, \varphi R N_{Lo} \, (\pi D)^{-1/2} \, t^{1/2} \tag{38 a}$$

and by integration

$$p(t) = \frac{8}{3} \, \varphi R N_{Lo} \, (\pi D)^{1/2} \, t^{2/3} \tag{38 b}$$

are obtained. That also in the case of drift assisted diffusion such a $t^{1/2}$-law is valid and not the $t^{2/3}$-law proposed by Cottrell and Bilby /85/ has been shown by Seeger /86/ and Schindlmayr and Lücke /87/. The $t^{2/3}$-law would follow if in Equ. (36) only the drift terms are considered and the random diffusion terms neglected /83/; for the initial behaviour, however, obviously just the opposite is correct.

In contrast to this long range diffusion Hornung /24/ treated the short range diffusion of axial defects in the stress field of a dislocation. He obtained as short time approximation two terms both obeying an exponential time law, Equ. (33) (i.e., both treatable again by the simple rate theory). The first term describes the reorientation of the defects, i.e. the formation of a Snoek atmosphere. Here τ is given by the elementary jump time τ_S (c.f. Section III b 6) and p_o by p_{app} of Equ. (26). The second term having the time constant $2 \tau_s$ characterizes the first few migration steps towards the dislocation. The p_o-value of the second term might have the same magnitude as the first or be even larger.

3. Pinning derived from a system of coupled reactions

There consists considerable evidence that the kinetics of pinning is determined not only by the migration of defects to the dislocation but also by some other simultaneously occurring processes. Some of them are now discussed.

(i) The point defects may move along the dislocation line to the dislocation nodes where they are assumed to have a lower energy than at the line /88/. Thus an enrichment of defects at the nodes will be found.

With increasing temperature this equilibrium enrichment will decrease and, correspondingly, the number of defects on the line will increase. Since the defects situated at the nodes are not expected to influence dislocation mobility the resulting damping will decrease with increasing temperature.

(ii) Because of accumulation of point defects at the dislocations defect clusters may form even for rather small concentrations of defects produced in the lattice.

(iii) Since the concentration of dislocated atoms (about 10^{-8} for a dislocation density of $\Lambda = 10^7$ cm^{-2}) is much smaller than the concentration of impurity atoms in high-purity metals (about 10^{-5}) most of the defects on their way towards the dislocation encounter impurity traps and eventually become captured. Only the small number of defects produced very close to the dislocations within radius r_F in Fig. 26 happen to avoid trapping and reach the dislocation with a diffusion rate corresponding to that in the pure lattice /88, 89/. [It is only at elevated temperatures (or in case of saturable traps at rather high doses) that most irradiation produced defects are evaporated from (or not captured by) the traps and reach the dislocations]. In this range apparent activation energies different from that for simple defect migration should be pertinent. In a warm-up experiment, the direct diffusion from within the radius r_F would result in a first pinning stage and the untrapping in an additional pinning stage at a higher temperature T_p. Different species i of impurity atoms would lead to different pinning stages the temperatures T_{pi} of which would be given by the relation $kT_{pi} \approx H_{Bi} + H_m$ (binding energy + migration energy)/ 90, 27/.

(iv) At comparatively high temperatures point defects might disappear from the dislocation, e.g. by reevaporation into the lattice or by processes leading to climbing of the dislocation /88, 28, 58, 89/.

These 4 processes ((i) to (iv)) altogether lead to the commonly observed phenomenon that, in the end, the number of pinning points is smaller than the number of point defects produced.

In the present case of several coupled processes a quantitative understanding of the pinning kinetics can be obtained by application of chemical reaction rate theory. Here one assumes the defects to be distributed among different "reservoirs" (e.g. lattice, dislocation line, node, trap etc.)

and transitions between these different reservoirs to be controlled by approximate time constants $\tau_{\nu\mu}$. A reaction scheme applicable in the present case is shown in Fig.27. The wanted dislocation pinning rate is then determined by a system of coupled differential equations /88, 27/.

d) Comparison with Experiments

Hitherto, it has not been definitely clarified yet which of the reactions ((i) - (iv)), described above take place under the different experimental situations. In the following some results relevant to this question will be discussed. In most of the papers, dislocation pinning has been attributed to a single process, namely to the diffusion of one type of defect to the dislocation where it converts into a pinning point. The property of interest is the number of additional pinning points p as function of the time t of irradiation, recovery or heating (c.f. Fig. 25). In the MHz-range p is derived mostly from attenuation and in the KHz- and Hz-range from modulus measurements.

1. Pinning_results_described_by_a_single_diffusion_process

Warm-up (type (iii)-) experiments have been carried out in all 3 (MHz-, KHz- and Hz-) frequency ranges /3, 6/. In Fig. 28 where p and dp/dT is plotted as function of T an example is given for a copper crystal show-ing 3 pinning stages /60/. Such survey experiments have mostly been per-formed to qualitatively attribute the observed pinning stages to different diffusion processes, but seldom to derive quantitative data. E.g. acti-vation energies were not determined since mostly the heating rate was not systematically varied and the peak width ΔT (Equ. (35 b)) must be assumed to be not the natural one, but broadened due to unknown distri-bution effects.

Recovery (type (i)) experiments after irradiation as well as after cold work have been repeatedly carried out (e.g. /3, 6/). Many of them have been successfully evaluated by assuming the pinning point number p pro-portional to t^{α} . In earlier work, mostly the Cottrell-Bilby exponent $\alpha =$ 2/3 (e.g. /91/), more recently the value $\alpha = 1/2$ predicted by Equ. (38 a) (e.g. /97/) and sometimes also $\alpha = 1$ /60, 97/ has been reported. Often the accuracy of measurement is hardly sufficient to distinguish between exponents in the range $(1/2 \leqslant \alpha \leqslant 1)$. In all cases where p was followed to large enough values, a saturation value p_o has been observed which was mostly attributed to defect depletion in the lattice. Often exponential behaviour as predicted by Equ. (33) is obtained, sometimes in a form in which the exponent t/τ is replaced by $(t/\tau)^\alpha$ with $\alpha < 1$ (e.g. /89/). This might be an interpolation between Equ. (38 a) for the initial and Equ. (33) for the final behaviour.

In isothermal irradiation experiments (type (ii)), often a linear (steady-state) pinning behaviour, sometimes with an initial transition region, has been observed. Such behaviour is predicted by Equ. (34) and (38 b) respectively. For large times, however, the p(t) curve bends over to smaller p-values (c.f. curve for 343 K, Fig. 3). For small times , however, the quantitative evaluation is often difficult because of the effect of "spontaneous" pinning: If, for example, the temperature of irradiation lies between two pinning stages (e.g. 310 K, Fig. 28) the pinning points of the stages of lower temperature (II and III, 225 K and 260 K resp.) reach the dislocations almost instantaneously ("spontaneously") causing linear pinning. In order to obtain the contribution due to the higher stage (Fig. 28, stage IV at 360 K) this spontaneous pinning which might be larger than that to be investigated must first be subtracted.

Most pinning measurements intend to derive the activation energies from the temperature dependence and use it to draw conclusions about the nature of the migrating defect. Indeed, many good looking Arrhenius plots were obtained, particularly from measurements of τ in recovery experiments. Because of the following difficulties, however, one cannot generally interpret such activation energies as those for lattice migration of the defect:

(i) Height and temperature of the pinning stages reported in different papers exhibit considerable discrepancies /6, 92/. Fig. 29 shows an example of the influence of background impurity level. Also the activation energies show such differences and sometimes do not agree with values from other types of investigation.

(ii) In some cases these Arrhenius-plots yielded very large activation energies /89/ and, correspondingly, very small τ_0 values (Equ. 37). This leads to the result that also $m = \tau_0 \nu_m$, the number of diffusion steps during the time τ_0 , becomes very small, sometimes even < 1. This unreasonable result seems to indicate a superposition of several processes /93, 90, 89/.

(iii) Mostly the saturation values p_o in recovery experiments are much
smaller than the number of irradiation produced defects, and the steady-state
pinning rate $\dot{p}(t \gg \tau)$ in irradiation experiments is much smaller than
the defect production rate γ, ($\dot{p}$ even decreases at large p-values). Furthermore p_o and p increase with increasing temperature up to a certain temperature, then decrease. Since according to the most simple diffusion model
the total number of pinning points created and the number of defects
produced should be identical, one has to conclude that the defects undergo
still other reactions rather than only diffusion to the dislocation.

2. Pinning results described by a set of coupled processes

For these reasons it appears necessary to additionally include some of
the processes discussed in Sec. IV c and depicted in Fig. 27 into the
equations determining the pinning kinetics. Thompson et al. /88/ were
the first who tried to explain their results on Cu γ-irradiated in the
range from 333 K to 393 K in such a way. They used a model consisting of the
reservoirs "lattice", "dislocation" and "nodes". Neglecting the diffusion
back from the dislocation into the lattice they were able to describe
each of their KHz-pinning-curves by the 3 time-constants τ_{LD}, τ_{DN} and
τ_{ND}. In particular, the temperature dependence of these time constants
was found to give good linear Arrhenius-plots and quite reasonable values
for the corresponding activation energies.

Despite of this success, some criticisms have been put foreward against details of this model:

(i) According to this model, a small decrease in temperature of an irradiated sample should lead to decrease in pinning (and vice versa) since
it would cause point defects on the dislocation to condensate at the nodes
where they are supposed to have a lower energy. In MHz-measurements by
Winterhager, John and Lücke /94/ (and recently also in KHz-measurements
/53/), however, no such change in pinning connected with temperature
changes has been observed. The latter results show that, once they are
formed, pinning points are quite stable. This leads to the conclusion that a
pinning point does not consist of a single point defect but of a defect
cluster which is not mobile along the dislocation line. This point of
view is supported by direct measurements of the pinning point strength
/80/.

(ii) In their model Thompson et al. /88/ do not account for the interaction between irradiation induced defects and the impurities in the lattice although they discuss this process. That it is this interaction which plays an important role can be easily recognized from the occurance of quite a variety of different pinning stages. They can be caused either by different types of migrating defects or by different types of traps (sec. IV c 3). That the latter explanation seems to be conclusive is demonstrated by Fig. 29 which shows warm-up curves for samples prepared under identical conditions from high purity copper obtained from different suppliers. In each case several pinning stages are observed, but their number, temperature and height are different for the different brands of copper. This can only be explained by assuming that each stage is caused by a single species of impurity and that the concentrations of these impurities are different in the different types of copper. Since measurements of irradiation damage rate by means of electrical resistivity /95/ showed that at T > 140 K no interaction between normal metallic impurities and migrating defects takes place, it is further concluded that the impurities influencing pinning at T > 200 K must be present at rather small concentrations and exhibit especially strong interaction with defects(e.g. non-metallic impurities). Due to their small number their influence is not observable by present resistivity measurements, but only by extremely sensitive internal friction measurements.

Recently, Schindlmayr, John et al. /87, 27/ included the process of point defect trapping and pinning point clustering into their quantitative interpretation of pinning curves. As stated above, both processes contribute to the observation that the total number of measured pinning points is mostly considerably smaller than the number of irradiation produced defects. The detrapping is considered to be the reason for the increase of pinning rate with temperature and the cluster formation to result in the slowing down of the pinning rate at high doses. In the framework of the model of Thompson et al. /88/ all these effects would to be explained by the temperature dependent condensation of defects in the nodes.

It is commonly observed that during further warm-up the pinning disappears and the initial state before pinning is reestablished /e.g. 11, 88, 59, 97, 98, 96/. As found e.g. for Cu above 400 K /93, 58, 92/, also depinning occurs in several stages, but also these depinning stages are not in agreement with each other for different types of high purity copper. This indicates that not only the kinetics of pinning, but also the kinetics of depinning is impurity influenced. If a sample is irradiated in the temperature range of depinning, a dynamical equilibrium between the pinning- and depinning-reaction is obtained (Fig. 4). As has been shown for Pb irradiated near room temperature /28/, a temperature and also dose rate dependent saturation value for the pinning point number has been found. It is not yet clarified which mechanism controls depinning (c.f. Sec. IV c 3).

V. General Discussion and Conclusions

Because of the highly complex nature of internal friction phenomena due to dislocation point defect interaction their interpretation is often difficult. If point defects generated in the lattice (e.g. during an irradiation) diffuse with respect to the dislocation the dislocation response to the stress applied in the internal friction experiment might be changed. It has become custom to call such a change "pinning" if it results in an increase of the modulus and "depinning" if the contrary is the case. The present paper intended to show that it is oversimplified, however, to think of pinning as accumulation of point defects interacting with the dislocating ("pinning points"). It is still more difficult to make firm conclusions from damping changes, since, depending on the pinning mechanism, an accumulation of point defects may result not only in a decrease, but also in an increase of damping. Thus in pinning experiments always questions like the following arise: What is the nature of the pinning mechanism? What is the nature of the pinning stages? What is the nature of the pinning point defect? What is the nature of the depinning process? In the following the answers to these questions given (or left open) in the present paper will be summarized.

1. The nature of the pinning mechanism

As has been shown above, all hitherto in literature discussed concepts concerning the influence of point defects on low amplitude dislocation damping can be incorporated into dislocation resonance theory. There are two groups of mechanisms (c.f. Equ. (3)):
(i) K-mechanisms: i.e. mechanisms influencing the restoring force $(-Ky)$; here we have the formation of firm pinning points, of a Cottrell atmosphere, of a Snoek atmosphere, as well as dislocation/dislocation interaction effects (expected after cold work /99/ and not treated here).
(ii) B-mechanisms: i.e. mechanisms influencing the dissipational force $(-B\dot{y})$; here we have the breakaway of dislocations from point defects, the diffu-

sion of point defects with and along the moving dislocation and the reorientation of axial defects in the stress field of the moving dislocation
(induced Snoek effect). The B-processes are thermally activated in contrast
to the other dissipational mechanisms listed in Sec. III b which (except
for the Bordoni drag) have a much weaker temperature dependence.

Wether pinning is caused by a K- or a B-process can be checked directly
either by measuring the shift of the frequency profile or, at fixed frequency, by comparing the changes in damping to those of the modulus. It
turned out that pinning as observed in the MHz-range and at least in some
KHz-experiments is definitely caused by K-processes and that in the Hz-
range both types of processes contribute to internal friction effects.
The assumption made by some authors that pinning effects in the KHz-range are caused by a B-mechanism (point defect drag) has not been strictly
proved yet by quantitative comparison of damping and modulus results,and
even arguments against this interpretation have been forwarded.

Qualitatively the different mechanisms within one and the same group
(either B- or K-) lead to the same damping or modulus behaviour in the
cause of pinning so that it is difficult to further distinguish between
them. There are mainly two ways for such a differentiation: a) the discussion of the observed numerical values of relaxation strengths, relaxation times and activation energies and b) the close inspection of the pinning kinetics with respect to its time and temperature dependence.

Regarding the K-mechanisms, it has been shown in the case of Cu, the
most thoroughly investigated metal, that in the MHz- and KHz-ranges pinning
after irradiation seems to be caused by firm pinning points. The number
of irradiation induced point defects seems to be too small to account for
the observed effect on the basis of a Snoek atmosphere of split interstitials. Also the often observed $t^{1/2}$-time law is in agreement with the
firm pinning point concept. After cold work, the concentration of mobile
defects is not known accurately enough for definite conclusions. Some authors /23/concluded here from the exponential time law and from the activation energy (0,67eV) observed in Hz-measurements of isothermal annealing
that in cold-worked copper near 230 K Snoek-pinning takes place, but
others forwarded arguments against it /33/.

Regarding the B-mechanisms very few evaluations have been carried out. For
example, there is agreement that the Hasiguti peaks are caused by B-processes, but it is still open which of these processes is responsible for

which peak. In cold-worked Cu (Hz-range) the 230 K peak with an activation energy of 0,7eV has been attributed to the induced Snoek effect /23/
but again not uncontradicted /33/. The situation in the KHz-range is not
better. Even if for some results the interpretation as B-type pinning
/19, 61 / would be correct, there is no definite proof yet that it is
just point defect drag and no other B-mechanism.

2. The nature of the pinning stages

As has been shown mainly by warm-up experiments, pinning after irradiation
or cold-work occurs mostly in discrete pinning stages in definite temperature ranges. This can be explained in different ways:
(i) Direct migration of single species of defects (vacancies, interstitials) to the dislocations. Since the crystal contains many impurity traps
at low temperatures only those defects produced in a cylindrical volume
around the dislocation of a radius given by the average distance between traps can arrive at the dislocation before becoming trapped.
(ii) Migration of defects to the dislocations after having been trapped.
This takes place only at temperatures high enough to evaporate the defects from the traps. In this case defects from all over the crystal can
arrive at the dislocations.
(iii) Transport of impurity atoms to the dislocations by the irradiation
induced defects (vacancies). Such a process is to be expected /98/ and
its mechanism has been treated theoretically in another context e.g. in
/100/.
(iv) In cold-worked or extremely strong irradiated samples migration of
multiple defects (e.g. double vacancies) instead of single defects.
(v) Processes at the dislocations themselves, e.g. evaporation of defects
from the nodes back onto the dislocation lines or dissolution of pinning
point clusters.
(vi) Formation of a Snoek atmosphere of axial defects in the stress field
of the dislocations.

Evidently such pinning stages cannot be directly compared to the recovery
stages of electrical resistivity. Resistivity recovery occurs mainly by
mutual annihilation of interstitials and vacancies and by clustering in
the lattice.   In both cases the diffusion path is much
shorter than that for defect migration to the dislocations. Furthermore,
typical resistivity measurements require defect concentrations of $> 10^{-2}$
ppm, whereas pinning experiments need only $\simeq 10^{-6}$ ppm. At these low concen-

trations, mutual defect annihilation as well as clustering can be neglected. Only after plastic deformation, where defects of the same type are created close to one another, multiple defects and defect clusters in the lattice are to be considered.

Which of the above processes (i)-(vi) takes place within a given pinning stage and what is the nature of the migrating defect is not known definitely and this question shall not be discussed here in detail. The differences in opinion regarding the interpretation of the resitivity recovery stages appear also with respect to the interpretation of pinning stages. In the case of Cu some authors (e.g./101/) assume the existence of 2 types of interstitials, split interstitial and crowdion. The latter existing only at low temperatures (< 160 K) and the migration of the split interstitial in stage III and IV (i.e. up to 400 K). Recently /102/, evidence seems to accumulate that only one type of interstitial, the split interstitial, exists which migrates in stages I and II, that free vacancies migrate in stage III and that stage IV is characterized by vacancy/impurity interactions. As demonstated e.g. by Fig. 29 difficulties in interpretation of pinning stages are mainly due to the extremely strong influence of smallest amounts of impurities. The differences in the temperatures of the pinning stages for different types of high purity copper strongly suggest that many of these pinning stages are due to point defect/impurity interactions.

3. The nature of a pinning point

This question is closely connected with the question about the nature of pinning stages. Thus there are again several possible answers:
(i) A pinning point is formed by a single defect (vacancy, interstitial). This is supported by the linear increase of the pinning point number frequently observed in isothermal irradiation experiments (steady state pinning). However, there is the difficulty to understand why these defects do not annihilate at jogs and make the dislocation climb.
(ii) A pinning point is formed by a point defect cluster. This possibility explains in a very good way the bending over in the pinning curves at large irradiation times and even the time law observed then ($t^{1/4}$, c.f. Sec. IV c,/87/). In contrast to single defects such a cluster is expected to be immobile along the dislocation line, thus explaining the stability of pinning (Sec. IV d).

(iii) Pinning points are formed by foreign atoms (single atoms or clusters) transported to the dislocations by the irradiation induced defects. This possibility would explain the above mentioned impurity dependence of the pinning stages as well as the difficulty mentioned in (i) concerning dislocation climb.

As a result, it cannot be decided at present, whether the point defects themselves or foreign atoms transported by them are responsible for the pinning. Evidently the linear pinning (i) as well as the slowing down of the pinning rate at long times (ii) can be explained by both assumptions. It seems to be proved, however, that several types of pinning points (e.g. single type defects and clusters) exist. As discussed in Sec. II c, this can be concluded from breakaway measurements /34, 80/ which showed that the pinning strength per single point increased with increasing irradiation dose.

4. The nature of depinning

As long as the nature of the pinning points is not clarified yet, even less can be said about the nature of the depinning. In particular, it is not clear, wether the depinning at high temperatures, in the course of which the preirradiation state of the sample is re-established, is only due to increasing clustering of pinning points or wether other processes like re-evaporation of pinning defects into the lattice or dislocation climb are involved. Moreover, it must be recognized that in case of thermally activated drag (Sec. III c) a modulus decrease (characteristic of depinning) with increasing temperature is obtained without loss of pinning points merely by going through the corresponding relaxation maximum (c.f. Fig. 22) since pinning points, firm below the relaxation temperature, are mobile above this temperature. However, this is only an "apparent" depinning, since it does not lead to a change in the number of pinning points. Instead, the effect is strictly reversible with respect to temperature.

5. Concluding remark

It has been shown that the formal predictions of the theories of dislocation resonance and of dislocation pinning have been verified by the experiments in many instances. On the other hand, despite the large amount of work accumulated in this area, only few of the physical details of

dislocation pinning are really clear. It appears, however, that by the
more recent work important aspects and promising lines of approach have
been revealed. Thus, further progress also in the detailed understanding
of pinning, and more frequent applications of dislocation damping for
investigating other solid state problems can be expected.

Acknowledgement

Most of all, the authors want to thank their colleagues of the damping-
group and of the irradiation group of the Aachen institute. Much of
their still unpublished work has been mentioned in this article and
numerous discussions with them contributed to the views presented here.
The authors like to express their gratitude to the Deutsche Forschungsgemeinschaft which supported most of the internal friction and pinning
work of the Aachen institute reported in this article.

References

1   A.V. Granato, K. Lücke, in "Physical Acoustics" (W.P. Mason, Ed.),
        Vol. IV a, p. 266 ff., Academic Press, (New York 1966).

2   A. Seeger, P. Schiller, in "Physical Acoustics" (W.P. Mason, Ed.),
        Vol. III a, Chap. 8, Academic Press, (New York 1966).

3   D.O. Thompson, V. Paré, in "Physical Acoustics" (W.P. Mason, Ed.),
        Vol. III a, Chap. 7, Academic Press, (New York 1966).

4   D.H. Niblett, in "Physical Acoustics" (.P. Mason, Ed.), Vol. III a,
        Chap. 3, Academic Press (New York 1966).

5   K. Lücke, J. Schlipf, in "The Interactions between Dislocations and
        Point Defects" (B.L. Eyre, Ed.) Vol. 1, p. 118, AERE-Rep.-
        5944, Harwell (1968).

6   R. De Batist, "Internal Friction of Structural Defects in Crystalline
        Solids", North Holland Publ., Amsterdam (1972).

7   A.S. Nowick, B.S. Berry, "Anelastic Relaxation in Crystalline Solids"
        Academic Press, New York (1972).

8   A.V. Granato, K. Lücke, J. Appl. Phys. 27, 583, (1956).

9   A.V. Granato, K. Lücke, J. Appl. Phys. 27, 789, (1956).

10 J.S. Koehler, in "Imperfections in Nearly Perfect Crystals" (W.
        Shockley et al., Eds), p. 197 ff., Wiley, New York (1952).

11 D.O. Thompson, D.K. Holmes, J. Appl. Phys., 27, 713 (1956).

12 V.K. Paré, H.D. Guberman, J. Appl. Phys., 44, 32, (1973)

13 L.J. Teutonico, A.V. Granato, K. Lücke, J. Appl. Phys., 35, 220, (1964).

14 K. Lücke, A.V. Granato, L.J. Teutonico, J. Appl. Phys., 39, 5181 (1968).

15 J. Friedel, in Nat. Phys. Lab. Symp. on "The Relation between the
        Structure and Mechanical Properties of Metals" Vol. I, Her
        Majesty's Stationery Office, London (1963).
    J. Friedel,"Dislocations", Pergamon Press, Oxford (1964).

16 J. Schlipf, R. Schindlmayr, to be published.

17 R.R. Hasiguti, Phys. Stat. Sol., 9, 157, (1965).

18 K. Yamafuji, C.L. Bauer., J. Appl. Phys. 36, 3288, (1965).
    E.G. Oren, C.L. Bauer, Act Met. 15, 773, (1967).

19 H.M. Simpson, A. Sosin, Phys. Rev. B 5, 1382, (1972).

20 V. Winkler-Gniewek, R. Schindlmayr, J. Schlipf, to be published
    V. Winkler-Gniewek, Thesis, TH Aachen (1973)

21 G. Schoeck, Phys. Rev. 102, 1458 (1958).

22 G. Schoeck, A. Seeger, Acta Met. 7, 469, (1959).

23 F.J. Wagner, phys. stat. sol. (b) 51, 589, (1972).
    F.J. Wagner, phys. stat. sol. (b) 54, 135, (1972).

24 W. Hornung, phys. stat. sol. (b) 54, 341, (1972).

25 H. Rosinger, A. Polaković, D. Lenz, K. Lücke, to be published.

26 F. Fraikin, D. Lenz, K. Lücke, to be published.

27 R. John, D. Lent, K. Lücke, to be published.

28 D. Lenz, K. Lücke, in "The Interactions between Dislocations and Point
        Defects" (B.L. Eyre, Ed.) Vol. I, p. 239, AERE-Rep. 5944,
        Harwell (1968).

29 H. Brumme, H. Ebener, G. Sokolowski, J. Physique Suppl. C2/32,
        C2-147, (1971).
    A. Schnell, G. Sokolowski, K. Lücke, to be published.

30 E. Kröner, "Kontinuumstheorie der Versetzungen und Eigenspannungen"
        Springer, Berlin (1958).
    F. Friedel in "The Interactions between Dislocations and Point Defects"
        (B.L. Eyre, Ed.) Vol. I, p. 1, AERE-Rep. 5944, Harwell (1968).
    R. Siems, "Wechselwirkungen zwischen Defekten in Kristallen" Berichte
        der Kernforschungsanlage Jülich, Jül-545-FN (1968).

31 K. Lücke, J. de Physique, 7 Suppl. C2, C2-145, (1971).

32 A.V. Granato, A. Hikata, K. Lücke, Acta Met 6, 470, (1957).

33 G. Sokolowski, K. Lücke, to be published.
    G. Sokolowski, Thesis, TH Aachen (1973).

34 A. Ostermann, D. Lenz, K. Lücke, to be published.
    A. Ostermann, Thesis, TH Aachen, (1970).

35 H. Börger, D. Lenz, K. Lücke, to be published.

36 D. Lenz, B. Edendorfer, K. Lücke, Scripta Met., 5, 387, (1971).

37 J. Weertmann, J. Appl. Phys. 26, 202, (1955).

38 O.S. Oen, D.K. Holmes, M.T. Robinson, U.S. Atom Energy Comm. Report
        ORNL-3017, 3,(1960).

39 E.G. Henneke, R.E. Green jr., Trans. AIME, 242, 1071,(1968).

40 A. Seeger, P. Schiller, Acta Met. 10, 348,(1962).

41 T. Suzuki, C. Elbaum, J. Appl. Phys. 35, 1539, (1964).

42 A.D. Brailsford, J. Appl. Phys. 43, 1380, (1972).
    A.D. Brailsford, in "Internal Friction and Ultrasonic Attenuation
        in Crystalline Solids" (D. Lenz, K.Lücke Eds.) Vol. II,
        Springer Berlin-Heidelberg-New York (1974).

43 G. Leibfried, Z. Physik 127, 344, (1950).

44 N.F. Mott, in "Imperfections in Nearly Perfect Crystals" (W. Shockley
        et al. Eds.), 173, Wiley, New York (1952).

45 R. Schindlmayr, J. Schlipf, to be published.

46 A.H. Cottrell, M.A. Jaswon, Proc. Roy. Soc. London, A 199, 104, (1949).

47 G. Schoeck, Acta Met. 11, 617, (1963).

48 A.A. Blistanov, M.P. Shaskolskaya, Sov. Phys. Sol. State 6, 573, (1964).

49 P. Schiller, phys. stat. sol. 5, 391, (1964).

50 G. Wire, Thesis, Urbana (1972).
    G. Wire, A.V. Granato, to be published.

51 A. Seeger, in "Encyclopedia of Physics" Vol. 7/I, 383, Springer
        Berlin (1955).
    A. Seeger, Phil. Mag. 1, 651 (1956).

52 A. Hikata, C. Elbaum, in "Internal Friction and Ultrasonic Attenu-
        ation in Crystalline Solids" (D. Lenz, K. Lücke Eds.) Vol. II
        Springer, Berlin-Heidelberg-New York (1974).

53 V. Naundorf, K. Lücke, to be published.
    V. Naundorf, K. Lücke, in "Internal Friction and Ultrasonic Solids"
        (D. Lenz, K. Lücke Eds.) Vol. II, Springer Berlin-Heidelberg-
        New York (1974).

54 G.A. Alers, D.O. Thompson, J. Appl. Phys. 32, 283, (1961).

55 R.M. Stern, A.V. Granato, Acta Met. 10, 358, (1962).

56 D. Lenz, K. Lücke, Z. Metallkunde 60, 375, (1969).

57 T. Suzuki, A. Ikushima, M. Aoki, Acta Met. 12, 1231, (1964).

58 H. Inagaki, F. Hultgren, K. Lücke, Acta Met. 18, 713 (1970)

59 P. Winterhager, K. Lücke, J. Appl. Phys. 44, 4855, (1973).

60 H.R. Kaufmann, D. Lenz, K. Lücke, to be published.
    H.R. Kaufmann, Thesis, TH Aachen (1973).

61 A. Sosin, in "Internal Friction and Ultrasonic Attenuation in Crysta-
        lline Solids" (D. Lenz, K. Lücke Eds.) Vol. II, Springer-
        Berlin-Heidelberg-New York (1974).

62 A. Seeger, H. Donth, F. Pfaff, Disc. Faraday Soc. 23, 19, (1957).

63 V.K. Paré, J. Appl. Phys. 32, 332, (1961).

64 A.D. Brailsford, Phys. Rev. 137, A 1562, (1965).

65 H. Engelke, phys. stat. sol. 36 , 231, 245, (1969).

66 R. Schindlmayr, J. Schlipf, to be published.

67 F.D. Eshelby, Proc. Roy. Soc. London, A 197, 396, (1949).

68 D. Keefer, R. Vitt, Acta Met. 15, 1501, (1967).

69 A.V. Granato, R.M. Stern, J. Appl. Phys. 33, 2880, (1962).

70 C.R. Heiple, H.K. Birnbaum, J. Appl. Phys. 38, 3294, (1967).

71 H. Akita, N.F. Fiore, J. Appl. Phys. 42, 2203,(1971).

72 R. den Buurmann, D. Weiner, Scripta Met. 5, 573,(1971).

73 K. Beißner, E. Biller, Scripta Met. 7, 535, (1973).

74 W.P. Mason, A. Rosenberg, J. Appl. Phys. 38, 1929,(1967).

75 A. Ikushima, T. Kaneda, Trans.J. Inst. Met. 9, 38,(1968).

76 W.A. Fate, J. Appl. Phys. 43, 835,(1972).

77 A. Hikata, R.A. Johnson, C. Elbaum, Phys. Rev. Letters 24, 215,(1970).

78 R.R. Hasiguti, N. Igata, G. Kamoshita, Acta Met. 10, 442,(1962).
    G. Sokolowski, H. Ebener, K. Lücke phys. stat. sol (a), 19, 493,(1973).

79 D.H. Niblett, J. Wilks, Advan. Phys. 9, 1,(1960).
    J. Völkl, W. Schilling, phys. d. Kond. Mat. 1, 296,(1963).

80 A. Ostermann, D. Lenz, K. Lücke, J. Physique 7/32, C2-149,(1971).

81 K. Lücke, G. Roth, G. Sokolowski, Acta Met. 21, 237,(1973).

82 G.J. Dienes, G.H. Vineyard, "Radiation Effects in Solids", Inter-
        science Publ., New York (1957).
    A.C. Damask, G.J. Dienes, "Point Defects in Metals", Gordon and
        Breach, New York-London (1963).

83 R. Bullough in "The Interactions between Dislocations and Point
        Defects" (B.L. Eyre Ed.), Vol. I,22, AERE-R 5944 Report
        Harwell (1968).
    R. Bullough, R.C. Newman, Rep.Prog. Phys. 33, 101,(1970).

84 F.S. Ham, J. Appl. Phys. 30, 915,(1959).

85 A.H. Cottrell, B.A. Bilby, Proc. Phys. Soc. 62, 49,(1949).

86 A. Seeger, private communication.

87 R. Schindlmayr, K. Lücke, in "Internal Friction and Ultrasonic
        Attenuation in Crystalline Solids" (D. Lenz, K. Lücke, Eds.)
        Vol. II, Springer Berlin-Heidelber-New York (1974)

88 D.O. Thompson, O. Buck, R.S. Barnes, H.B. Huntington, J. Appl. Phys. 38
        3051 (Part I), 3057 (II), 3068 (III), (1967).

89 P. Winterhager, K. Lücke, in "The Interactions between Dislocations
        and Point Defects" (B.L. Eyre, Ed.) Vol. I, 214 AERE-R 5944
        Report Harwell (1968).

90 R. John, D. Lenz, K. Lücke, in "Internal Friction and Ultrasonic
        Attenuation in Crystalline Solids" (D. Lenz, K. Lücke, Eds.)
        Vol. II, Springer Berlin-Heidelberg-New York (1974).

91 A.V. Granato, K. Lücke, R.M. Stern, Métaux-Corrosion-Industries 433,
        3,(1961).

92 V. Naundorf, G. Roth, K. Lücke, Cryst. Latt. Def. 2, 205,(1971).

93 D. Keefer, A. Sosin, Acta Met. 12, 1041,(1964).

94 P. Winterhager, G. Roth, R. John, K.Lücke, J. de Physique C2, 32/7,
        C2-151 (1971).

95 R. Lennartz, F. Dworschak, H. Wollenberger, Verhandl. DPG (VI) 8,
        363,(1973).

96 A. Sosin, Acta. Met. 10, 390,(1962).
    D. Keefer, J.C. Robinson, A. Sosin, Acta. Met. 13, 1135,(1965).

97 D. Lenz, K. Lücke, A. Ostermann, W.A. Sibley, J. Acoust. So . Am.
        $\underline{45}$, 1374, (1969).

98 D. Lenz, K. Lücke, Cryst. Lattice Def. $\underline{1}$, 297, (1970).

99 I. Ino, T. Sugeno, Acta Met. $\underline{15}$, 1197, (1967).

100 T.R. Anthony, Acta Met. $\underline{17}$, 603, (1969); J. Appl. Phys. $\underline{41}$, 3969, (1970).

101 A. Seeger, in "Internal Friction and Ultrasonic Attenuation in
        Crystalline Solids" (D. Lenz, K. Lücke, Eds.) Vol. II,
        Springer Berlin-Heidelberg-New York (1974).

102 P. Erhart, H.G. Haubold, W. Schilling, Festkörperprobleme XIV,
        Adv. in Sol. State Phys., 87, (1974).

103 U. Förster, H.R. Kaufmann, D. Lenz, V. Naundorf, K. Lücke, to be
        published.

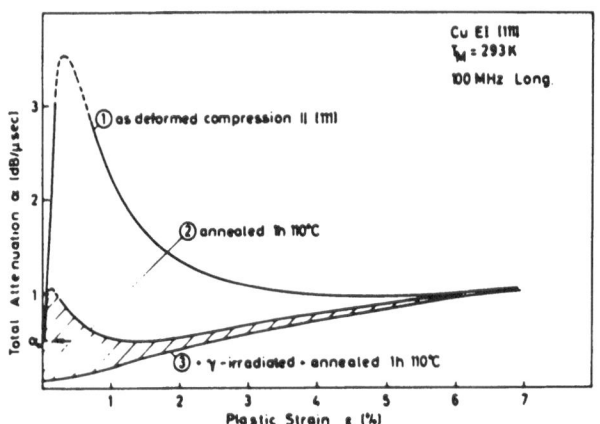

Fig.1   Attenuation α at 10 MHz as func-
         tion of plastic strain (compression
         along ⟨111⟩)in high purity Cu
         single crystals, (Curve 1) and
         after different treatments of the
         deformed samples (curves 2 and 3)
         /25/.

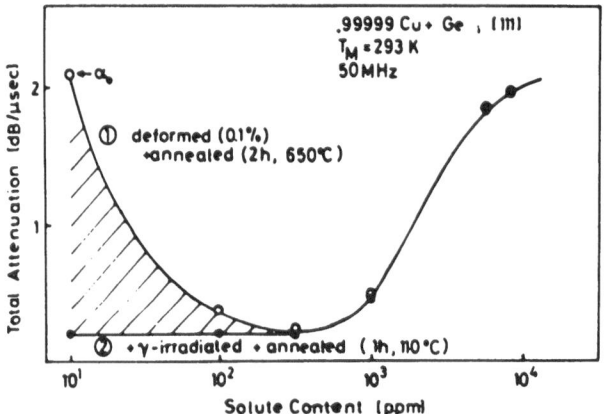

Fig.2   Attenuation α at 50 MHz in Cu-Ge
         single crystals as function of im-
         purity content. Open circles:
         standard treated crystals.
         Solid circles: same after heavy 3
         MeV γ-irradiation. Points at 10
         ppm correspondent to the "pure"
         Cu-sample /26/.

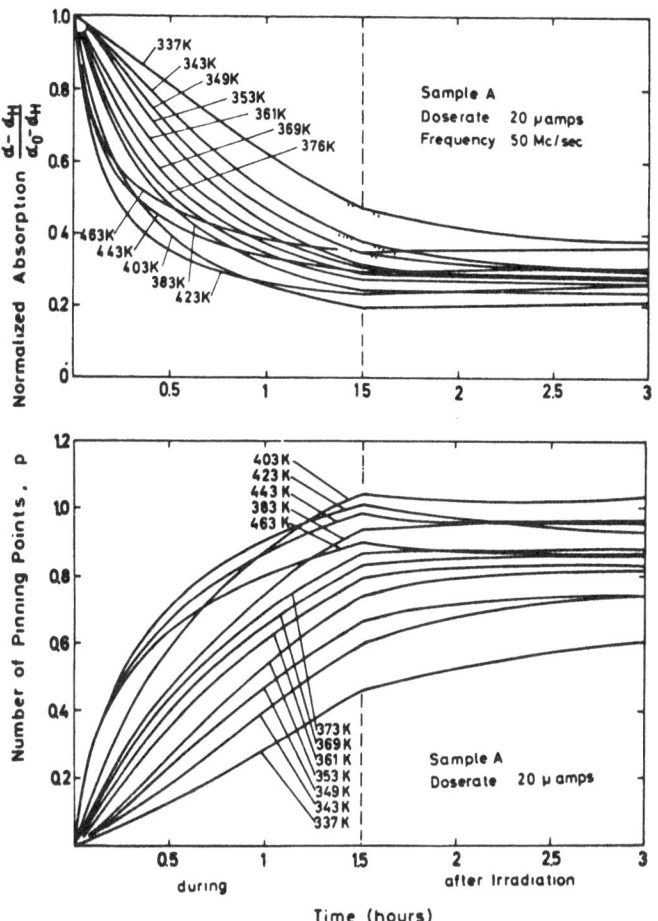

Fig.3 Normalized dislocation attenuation
(α_o initial value, α_H non-dislocation
background) and number of pinning points p
during (0 to 1.5 hrs) and after 3 MeV
γ-irradiation /27/.

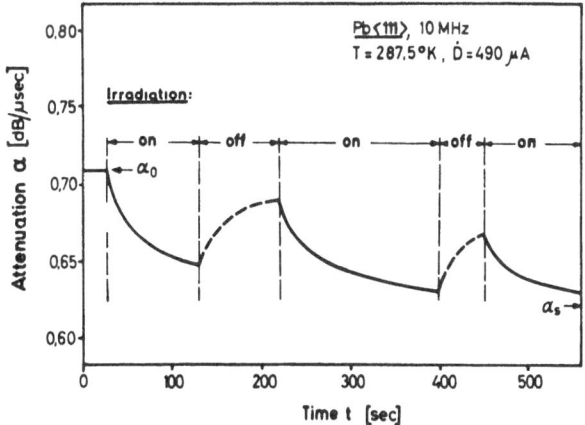

Fig.4 Influence of 3 MeV γ-irradiation upon
attenuation α in high purity Pb single
crystal at 288 K. During irradiation-
"on" pinning, during -"off" depinning
takes place /28/.

99

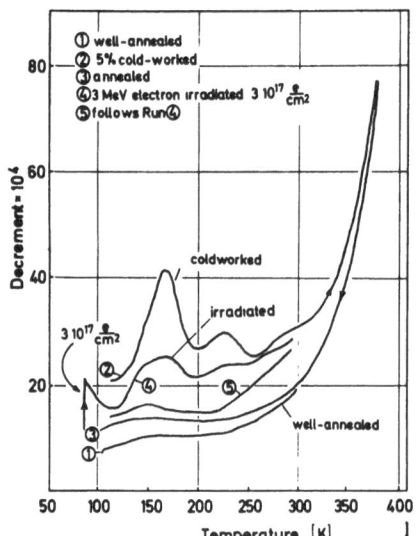

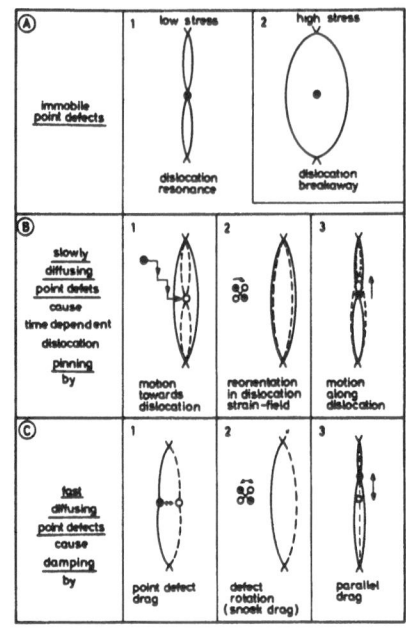

Fig. 5 Damping behaviour of
high purity Cu at 1 Hz
after cold-work and ir-
radiation resp./29/.

Fig. 6 Synopsis of point defect/
dislocation interaction
mechanisms. Sequence (a) to
(c) corresponds to increasing temperature at fixed frequency or to decreasing frequency at fixed temperature
(at low stress amplitudes).
At high stress amplitudes
breakaway effects may be
superimposed.

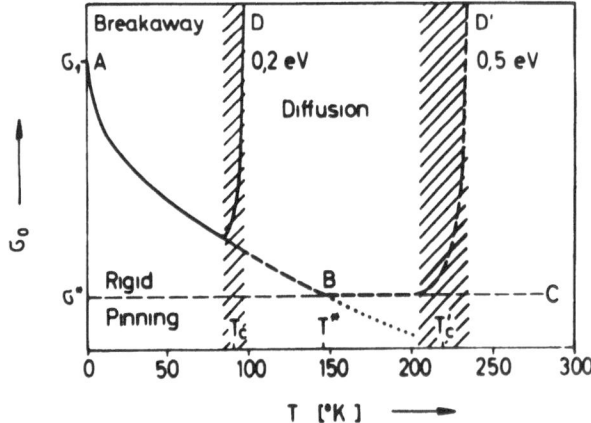

Fig. 7 "Phase diagram" according to /5/ showing
stress-amplitude/temperature domains for
dislocation damping mechanisms (A1), (A2)
and (C) depicted in Fig. 6.

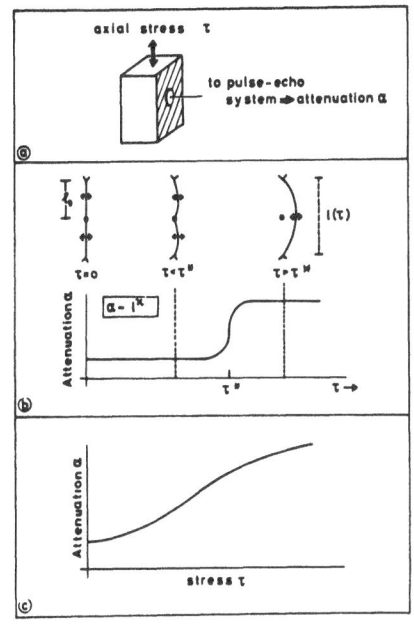

Fig. 8

(a) Schematic graph of pulse-echo-sample (with quartz attached) under axial bias-stress τ.

(b) above: pulse-echo dislocation resonance motion (small arrows) of dislocation double loop breaking away under increasing τ.

below: expected change of attenuation α.

(c) distributions of network length and pinning points result in a range of break-away stresses τ^* and cause gradual increase of α with increasing τ.

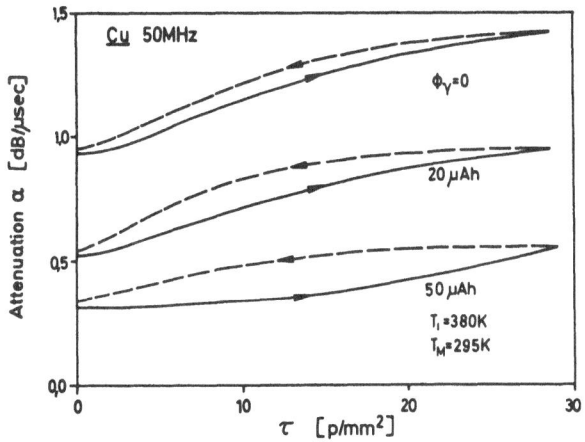

Fig. 9

Attenuation as function of bias-stress τ measured at 295 K on a Cu single crystal after different 3 MeV γ-doses applied at 380 K /34/.

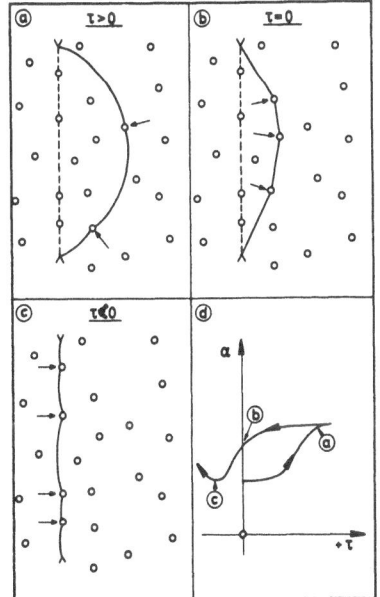

Fig. 10

(a) to (c): Schematic picture of dislocation loop on glide plane with obstacles under different bias stress values τ.
(d): dependence of dislocation attenuation on τ (dislocation positions (a) to (c) indicated) /35/.

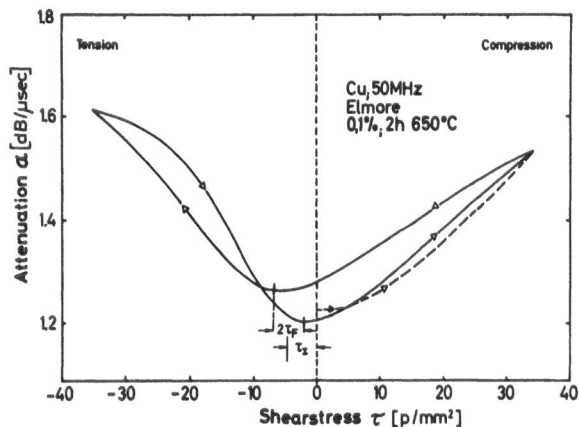

Fig. 11

Attenuation as function of cyclic bias stress. Broken curve: first loading /35/.

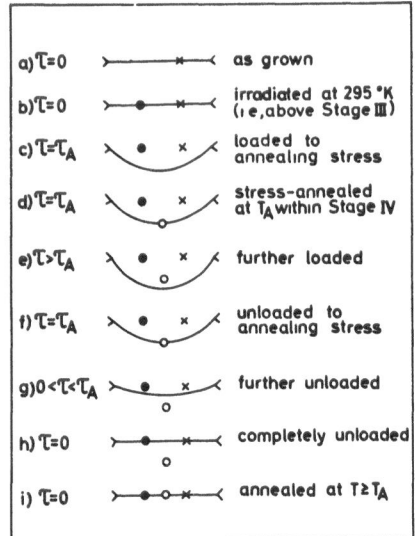

Fig. 12

Schematic sequence of an $\alpha(\tau)$-experiment: sample γ-irradiated above stage III (295 K); then annealed under constant stress τ_A within stage IV; then subjected to cyclic bias stress (c.f. Fig. 13) /36/.

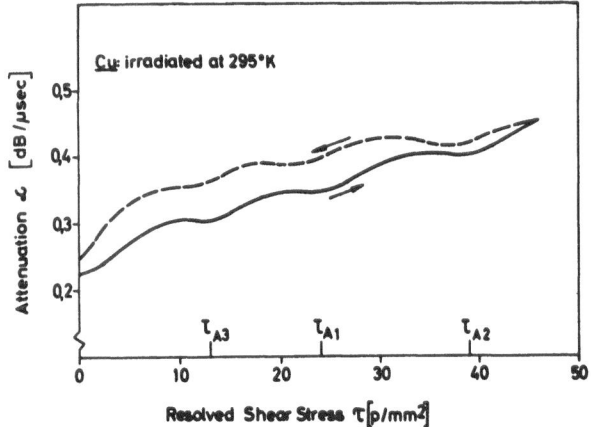

Fig. 13

Attenuation α as function of bias stress τ for γ-irradiated sample after annealing at 3 stress levels τ_{A1} to τ_{A3} according to the scheme in Fig. 12.

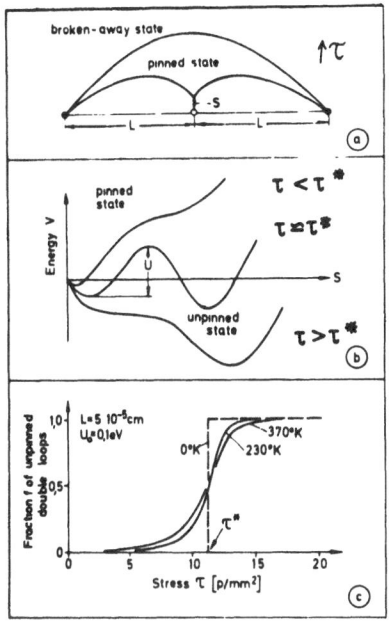

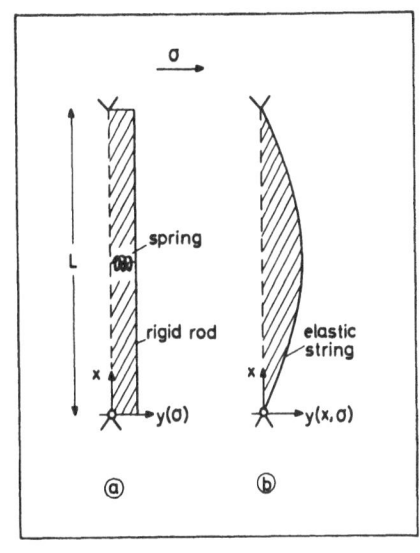

Fig. 14

(a) Breakaway of double loop under stress τ.
(b) Energy of double loop as function of antipin distance s for three stress levels.
(c) Fraction of broken-away double-loops as function of τ. τ^* is the stress at which the two energy minima in (b) are of equal depth.

Fig. 15

Simplified dislocation models (the hatched area represents dislocation strain due to stress σ):
(a) rigid rod of "loop"-length L attached to initial position by means of an elastic spring representing restoring forces.
(b) dislocation as elastic string.

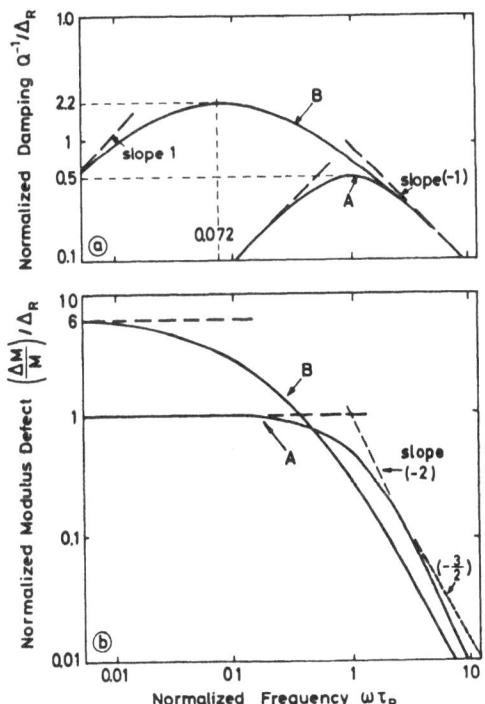

Fig. 16

Frequency dependence of damping and modulus defect for overdamped dislocation resonance. Curves A: delta-looplength distribution; Curves B: Exponential looplength distribution (Δ_R, τ_R are defined by Equ.(7)).

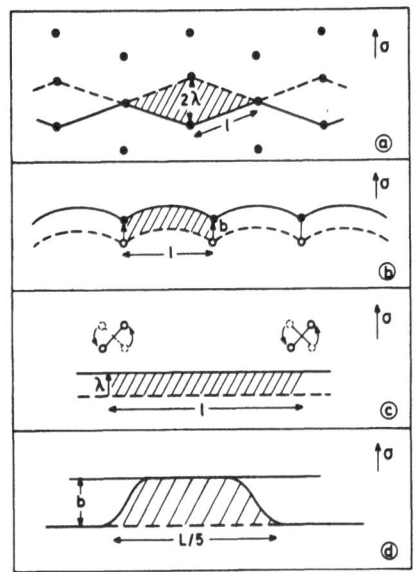

Fig. 17

Schematic representation of stressaided thermally activated dislocation motion mechanism (hatched area corresponds to dislocation strain per activation event):
(a) Zig-zag-dislocation moving through a field of immobile point defects.
(b) dislocation dragging mobile point defects.
(c) dislocation reorientating point defects with axial strain field.
(d) dislocation forming double kink.

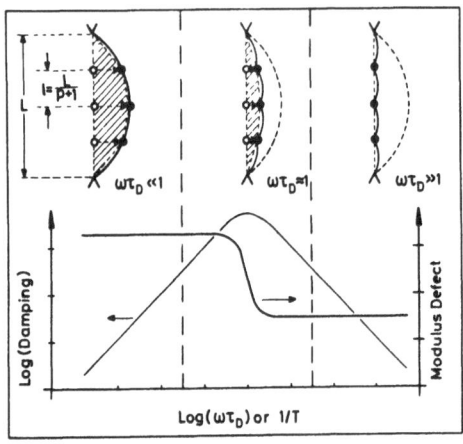

Fig. 18

(upper part): Dislocation loop of length L dragging point defects in three frequency ranges at constant temperature (or corresponding temperature ranges at constant frequency). The hatched area represents the modulus defect due to point defect drag)
(lower part): Schematic plot of corresponding damping and modulus defect as function of normalized frequency (or reciprocal temperature).

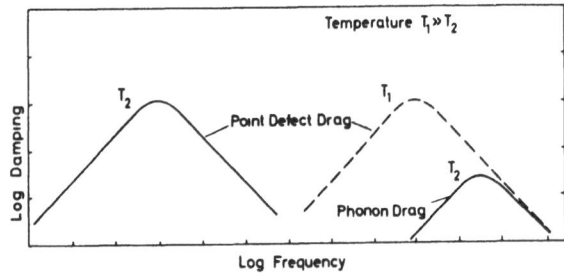

Fig. 19

Schematic frequency dependence of damping for the superposition of point defect- and phonon-drag at two temperatures $T_1 \gg T_2$. The small temperature dependence of phonon drag resulting in a minor shift of the T_2- phonon-drag maximum to a lower frequency has been neglected.

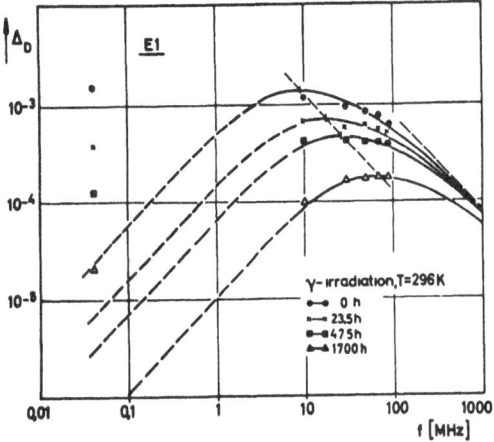

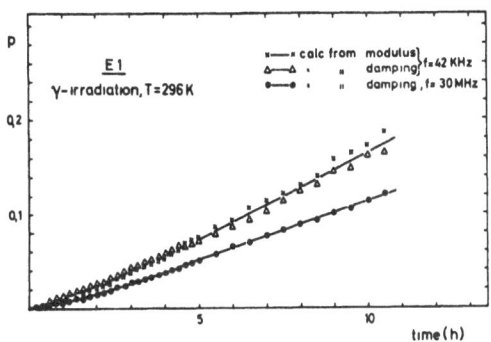

Fig. 2o

above: Frequency dependence of dislocation decrement Δ_D measured on
one and the same Cu crystal at 42
KHz and in the MHz range. Curves
represent theoretical $\Delta_D(f)$-depen-
dence /53/.
below: Comparison of pinning point
numbers p derived from the above
KHz- and MHz- measurements resp.
/53/.

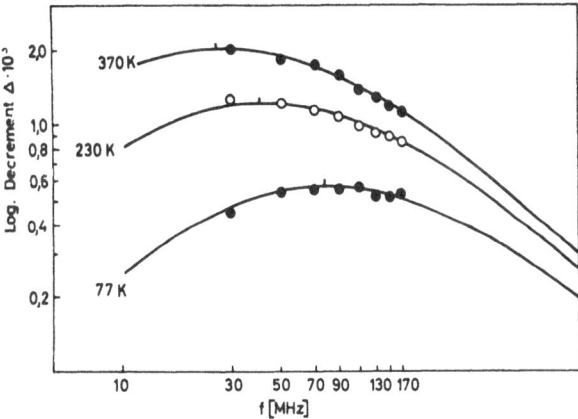

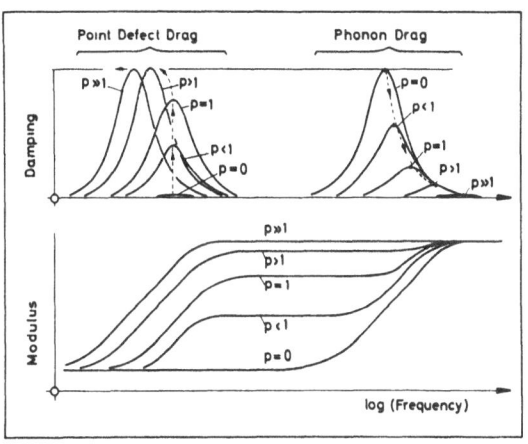

Fig. 21

Frequency dependence of dislocation decrement in Cu at three tem-
peratures/6o/. Solid curves are
theoretical dislocation resonance
master curves for exponential
looplength distribution.

Fig. 22
Influence of pinning point number
on the frequency dependence of
dislocation damping and modulus
for the superposition of dislocation phonon drag and point defect
drag.

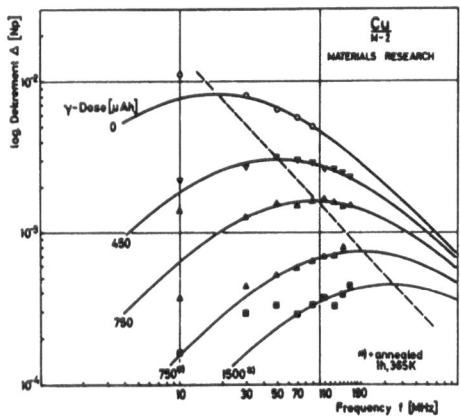

Fig. 23

(a) Frequency dependence of dislocation damping in Cu showing the influence of dislocation pinning due to different γ-doses applied at 31oK /1o3/.

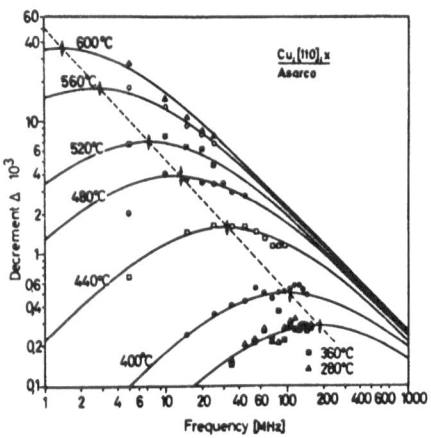

(b) Frequency dependence of dislocation damping in room temperature γ-irradiated Cu showing dislocation depinning in the course of an isochronal annealing treatment up to 6oo°C /58/.

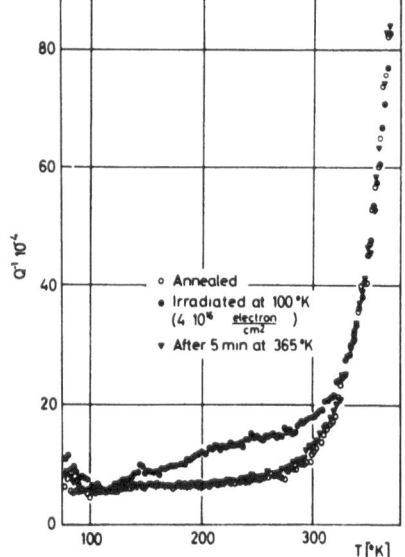

Fig. 24

Temperature dependence of Hz-damping in Cu before and after electron irradiation at 1ooK. The irradiation induced damping increase anneals out after 5 min. at 365K /29/.

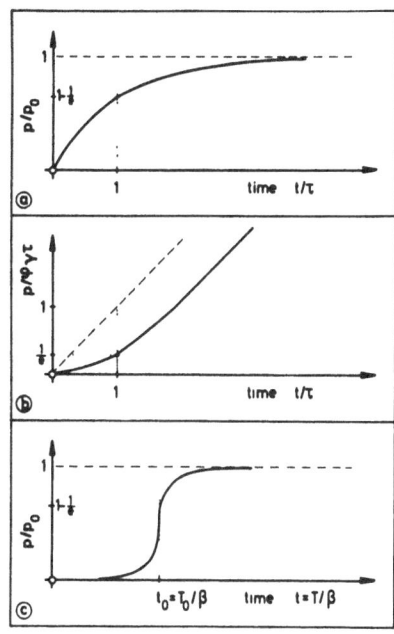

Fig. 25

Schematic plots of pinning point number as function of time.

(a) during annealing after low temperature point defect production.

(b) during isothermal irradiation.

(c) during warm-up (isochronal annealing) with β = heating rate.

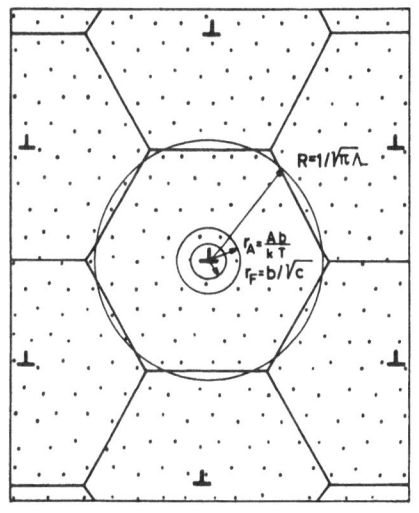

Fig. 26

Schematic picture of dislocations with appropriate diffusion volumina of radius R in a lattice containing impurities (dots); r_A is the radius of dislocation/point defect (drift) interaction (= sink radius) and r_F depicts a radius equal to mean impurity/impurity distance.

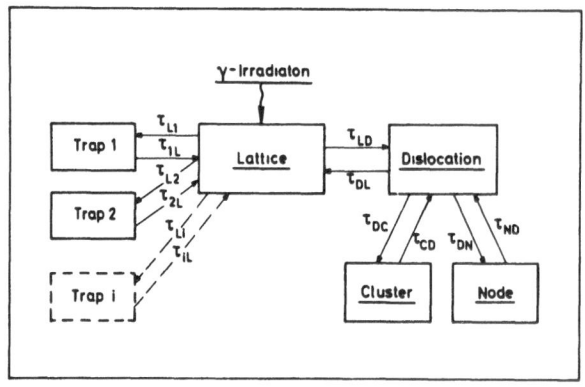

Fig. 27

Model for rate theory treatment of pinning
kinetics: Lattice containing dislocations
(providing nodes and defect clusters as sinks
for γ-irradiation produced point defects)
and different species of point defect traps.

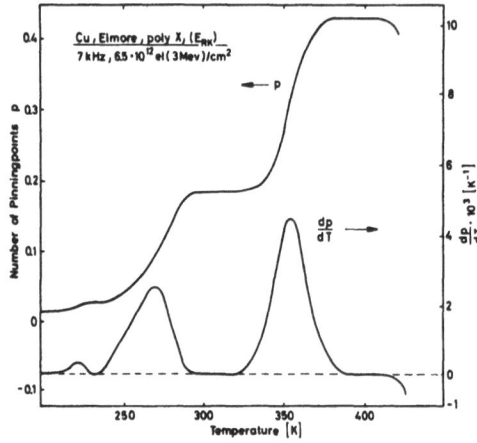

Fig. 28

Number of pinning points and pinning
rate as function of temperature during warm-up of Cu after 200 K
electron irradiation /60/.

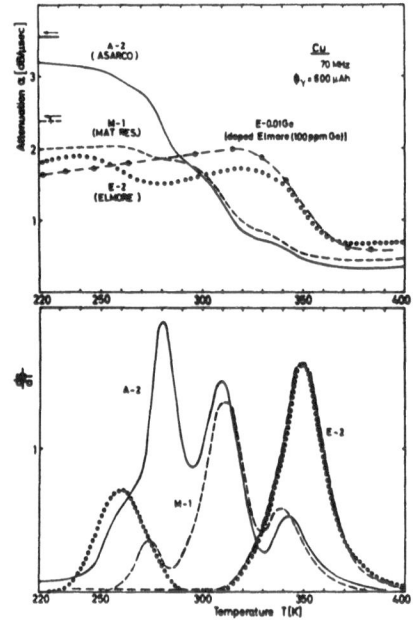

Fig. 29

Ultrasonic attenuation and its derivative with respect to temperature
after 220 K γ-irradiation of different
brands of high purity Cu and Cu-100 Ge
/103/.

DEFECT DRAGGING BY DISLICATIONS. THEORY AND EXPERIMENT IN THE AMPLITUDE INDEPENDENT AND DEPENDENT REGIONS

A. Sosin
University of Utah, Salt Lake City, Utah 84112 (U.S.A.)

ABSTRACT This paper reviews the theory of "defect dragging". In this theory, account is made for the contributions to internal friction (decrement) and to elastic moduli by the dragging of point defects attached to dislocation lines as the dislocation segments are bowed by an applied stress. Pertinent experiments are also reviewed. The dragging model accounts for the observed frequency dependence of decrement in the low (few kHz and below) frequency range. The model also predicts the initial increase in decrement (peaking effect) found on bombardment in copper, aluminum, and probably tungsten. The peaking effect, a function of temperature of irradiation and pre-irradiation sample history, is vividly present in copper, concomitant with a rapid, monotonic decrease in the modulus defect--as predicted by the dragging model. In aluminum, the increase is substantially slower and, at cyrogenic temperatures above 78° K, the modulus defect shows very little variation. These observations in aluminum are also consistent with the dragging model. In fact, copper and aluminum samples that have been studied provide support for the dragging model in different value ranges of a universal parameter which includes frequency of measurement, damping constants for dragged point defects and dislocation segments, dislocation loop length, and dislocation line tension.

Dislocation damping has been analyzed based on the vibrating-string model of Koehler (1) and of Granato and Lücke (2) for a score of years. The success of this model, particularly in the amplitude-independent region, over such a long period of time speaks to the success of the model. In fact, at high frequencies--in the high kiloHertz region and above --the KGL model leaves little to be desired.

At the same time, there has been recognized difficulties at lower frequencies. No mathematical developments of which I am aware have resolved these difficulties until the recent publication (3) of a "defect dragging" model for dislocation damping. It is tempting to present the defect dragging model, hereafter called the SS model, as totally different in orientation from the KGL model. Whereas the KGL model, as applied by essentially all investigators, speaks of the shortening (or lengthening) of dislocation segments by the addition (or removal) or point defects, in the SS model there is no change in dislocation loop lengths. The dislocation network in a metal is endowed with its characteristic distribution of loop lengths and maintains these lengths whether point defects are added or subtracted. In the SS model, the effect of defect additions is in the number of damping centers and, to a lesser extent at lower frequencies, to the damping of the portions of the dislocation segments free of point defects. By contrast, the KGL model employs a constant value of damping--the "line damping constant"-- throughout; the SS model centers on an effective damping constant which increases with increasing numbers of point defects on dislocation lines.

With this apparently radical change in emphasis, one might anticipate that there should be little common ground between the KGL and SS models. However, the established successes of the KGL theory and apparent success of the SS theory should prepare one for a confluence of the models, in the spirit of the correspondence principle. In fact, the dragging model is a more general formulation, with the KGL model as the high frequency limit.* The reason for this is, in retrospect, easy to discern. The SS model concentrates on the motion of pinning points dragged forth and back by the dislocation line responding to an external stress. In the high frequency limit, these defects, whatever their nature, are unable to respond to the applied stresses and, in the ultimate limit, remain stationary. But stationary dragging points are just the pinning points of the KGL theory.

A rather detailed presentation of the dragging model has already been presented elsewhere. Here I shall present only the more succinct developments and attempt to raise some questions which remain to be examined. To set this stage, I start with a review of some of the difficulties that have surrounded the KGL model for some considerable period. I restrict attention, until later, to the amplitude-independent region of dislocation damping.

The first difficulty concerns a very fundamental consideration. At low frequencies, the magnitude of damping that is observed experimentally exceeds the prediction of the KGL model. To see this, consider the two fundamental expressions of the KGL model.

$$\delta = a_\delta (\pi E b^2/C^2)_\omega B \Lambda \ell^4 \tag{1}$$

and

$$\Delta E/E = a_E (E b^2/2C) \Lambda \ell^2 \tag{2}$$

where E is Young's modulus (4×10^{11} dyne cm^{-2}), ΔE is the "modulus defect", the difference between the true elastic modulus and the measured modulus, b is the magnitude of the Burger's vector (2.5×10^{-8} cm), C is the dislocation line tension (4×10^{-4} dyne), ω is the angular frequency of measurement, B is the line damping ($\sim 4 \times 10^{-4}$ dyne cm^{-2} sec), Λ is the dislocation segment average length (typically found to be $\sim 5 \times 10^{-4}$ cm in annealed, unirradiated samples, and a_δ and a_E are numerical constants whose values depend on the nature of distribution

---

*The SS model has been formulated without account of dislocation line inertial effects. It, therefore, becomes inapplicable at high frequencies, as currently formulated. Extension of the SS model to include inertial effects is formally straightforward.

of lengths of dislocation segments.** If all dislocation segments are assumed to be of equal lengths--a "δ-function distribution"--$a_\delta = (5!)^{-1}$ and $a_E = (3!)^{-1}$. If an "exponential distribution" is assumed, $a_\delta = a_E = 1$.***

Now it is commonly observed that $\Delta E/E \cong 2 \times 10^{-2}$. Combining Eqs. (1) and (2), we can compute the expected decrement:

$$\delta = (a_\delta/a_E) \ 2\pi \ (_\omega B \ell^2/C)(\Delta E/E) \sim 2 \times 10^{-4} \qquad (3)$$

where $_\omega \sim 6$ kHz is assumed and the constants characteristic of the exponential distribution have been used. The calculated value of δ is lower by another factor of 20 if the constants for the δ-function distrubution are used. In any case, the value of the expected decrement is low--substantially lower·than observed in experiments at the assumed 1 kHz frequency.

It may be argued that the constants in Eq. (3) are imprecisely known, as some indeed are (e.g., B and ℓ). But the point is that decrement values of 5×10^{-3} are found at frequencies below 1 kHz as well. In fact, the observations indicate that the linear dependence on frequency predicted by Eq. (1) breaks down below a few kHz in copper. The precise frequency dependence is not yet known--and is a matter of urgent concern currently. There is some evidence (6) that the decrement increases as the frequency is decreased; at a minimum, it is suspected to remain constant with frequency at low frequencies. This matter has been investigated and summarized in detail recently by Heiple and Birnbaum (7).

The disparity in frequency dependence between the KGL theory and low frequency experiments has been a matter of concern for a long time. Sometimes the matter is coupled with another limitation of the original KGL model, a limitation explored by Granato, Lücke, and coworkers (8); namely, the KGL theory is a 0° K theory. The effects of temperature are neglected. (The SS model is ostensibly a 0° K theory, as well.) Even today there is no simple guide toward a temperature correction. It is by no means evident that the inclusion of temperature could alter the frequency problem.

Another problem area for the KGL theory concerns the length dependences shown in Eqs. (1) and (2). It is the exception, not the rule, to find the famous $\ell^2 - \ell^4$ dependence predicted there. It is, therefore, important to appreciate the conditions under which this "law" is valid. Suppose that Eqs. (1) and (2) are fully valid and that point defects behave as

---

**An orientation factor $\sim$ unity is omitted here. The numerical values of the constants are typical for copper.

***The Eqs. (1) and (2) have been given in a number of places (3,4,5) but, more generally, different numerical constants are used, these corresponding to the work of Granato and Lücke. In the frequency range in which these equations apply (e.g., where the dependence of δ on $_\omega^1$ applies), the equations written here are more rigorous, representing an exact solution of the string equation, with the neglect of inertial effects. The original GL expressions are leading terms of expansions. The discrepancies are small. The numerical factor in the modulus defect given by GL is $8 \cdot (3!)\pi^{-4} \approx 1/2$, the SS factor. In the decrement GL give $\approx 8 \cdot (5!)\pi^{-5} \approx \pi$, the SS factor.

as firm pinning points--no dragging. If the distribution of dislocation loop lengths follows the most frequently assumed "exponential distribution" (1),

$$f(\ell) = \Lambda(\bar{\ell})^{-2} \ e^{-\ell/\bar{\ell}} \tag{4}$$

where $\bar{\ell}$ is the average loop length, then it does follow that

$$\ell = \ell_o/(1 + n), \tag{5}$$

as is assumed by almost all investigators. Here n is the average number of pinning points on length ℓ_o. But Eq. (5) is only correct for the exponential distribution. This was demonstrated by Thompson and Holmes (9) as they showed that the distribution, Eq. (4), remained invariant in form with the addition of pinning points. With any other distribution function, Eq. (5) does not follow. It was shown by Rosenstock (10) that, if one starts with a delta-function distribution of loop lengths, the appropriate statistics are Poisson and the resulting variations of decrement and modulus defect with n are (3)

$$\delta = \pi\Lambda E b^2 {}_\omega B \ell_o^4 C^{-2} \ [e^{-n}(\frac{1}{6n^2} + \frac{1}{n^3} + \frac{3}{n^4} + \frac{4}{n^5}) - \frac{4}{n^5} + \frac{1}{n^4}] \tag{6}$$

$$\frac{\Delta E}{E} = \Lambda E b^2 \ell_o^2 C^{-1} \ [\frac{1}{2n^2} \ (1 + e^{-n}) + \frac{1}{n^3} \ (e^{-n} - 1)]. \tag{7}$$

It is clear that a plot of δ vs $(\Delta E/E)^2$ will not give a straight line in this case. However, it has been shown by several investigators (11), including myself, that agreement in such a plot is better approached if one postulates that we are dealing with two different families of dislocations, each with its own distribution, each such function assumed either explicity or implicity to be exponential in character. It is also important to recall that high frequency experiments (12) (near 1 MHz) fully display the dislocation resonance predicted by the GL theory if an exponential distribution is assumed. A delta-function distribution predicts too narrow a frequency dependence of decrement.

It was in such a background that Dr. Simpson and I started a new investigation into dislocation damping, using apparatus with very markedly improved sensitivity over those I had previously used. The sensitivity is most important, since it allowed measurements to be made at substantially lower point defect densities than formerly. Furthermore, the present system is essentially continuous in recording and regenerative. The first observations that we made, in quite impure copper bombarded by electrons at temperatures between 17° C and 101° C, see Fig. 1, was that the "Granato-Lücke plot", δ vs $(\Delta E/E)^2$, was clearly not linear. In fact, we found (13,14) that

$$\delta \propto (\Delta E/E)^1 \tag{8}$$

to a surprisingly good accuracy over the electron fluence ranged used, except for short sections near the beginning of the irradiation.

In all candor, at this point we were quite at a loss to explain these observations and were inclined to lay the blame entirely at the "excessive" amounts of impurities in the 99.9% copper being used. So we sought refuge in that material I have extensively used in the past-- 99.999% purity copper. To our amazement, we found the results to be considerably more complex than in the prior material and more complex than I had ever previously observed in the same material.

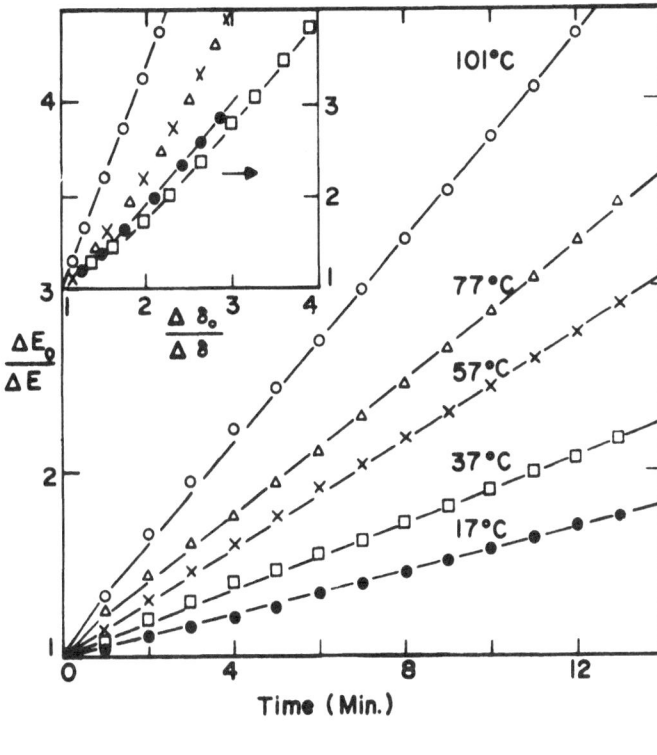

FIG. 1

The normalized inverse modulus defect plotted as a function of time during electron irradiation at the indicated temperatures. The sample material was 99.9% purity copper; the production rate of interstitials and vacancies was 2.5×10^{10} $sec^{-1}cm^{-3}$. The inset shows a plot of the normalized inverse modulus defect vs. the normalized inverse decrement. See Ref. 13.

It is almost axiomatic that an investigator considers his own previous observations as near-gospel. I'm afraid that I submitted to this weakness so that we spent an extended period seeking to find the purported experimental difficulties which were giving rise to these "anomalous" results. Needless to say, such experimental difficulties were not forthcoming. And, in this search, we made a number of observations which I believe are among the most fascinating ones in the field of internal friction.

The prime observation we made (13,14,15,16) with pure copper was that the decrement increased initially upon irradiation. Subsequently the decrement, in copper, always decreased. This observation has become known as the "peaking effect". The nature of the peaking effect is complex.

Some of the characteristics of the peaking effect and related effects are shown in Figs. 2 through 4 (14).

Fig. 2 shows the peaking effect during 1 MeV electron irradiation and simultaneous measurements at 57° C, as a function of sample history. The pre-irradiated annealing treatments for the several runs (on one sample throughout) are: E-2, 500° C, 30 min.; E-3, 500°C, 10 min.; E-4, 600° C, 30 min.; E-5, 700° C, 30 min.; E-6, 750° C, 30 min.; E-7, 500° C, 10 min.; E-8, 750° C, 30 min. The behaviors of the decrement for E-7 and E-8 are shown in Fig. 3. It is clear that annealing above 500° C tends to erase the peaking effect; 750° C is sufficient to erase it completely. Run E-7 shows the peaking effect cannot be reproduced in a sample by further annealing; the implication is that further plastic deformation would re-introduce the peaking effect. Here lies part of the answer why previous investigators have generally missed the peaking effect since it is rather common to anneal a sample at more elevated temperatures before beginning an irradiation study.

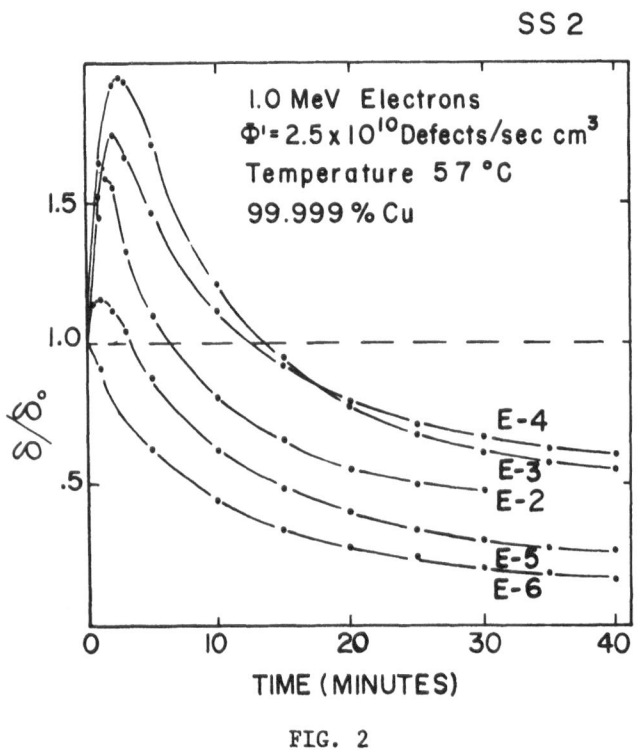

FIG. 2

The decrement normalized to its preirradiation value and plotted as a function of irradiation. Preannealing temperatures were E-2, 500°C; E-3, 500°C; E-4, 600°C; E-5, 700°C; E-6, 750°C. See Ref. 14.

The peaking effect is the most obvious discrepancy between observation and the GL theory as previously applied. The GL model predicts monotonic decreases in decrement as pinning points are added to dislocation segments. Clearly what was needed is a mechanism whereby more energy is dissipated with the addition of point defects to dislocation segments. However, prior to searching for such a mechanism, it was important to investigate the previous failure of the $\ell^2 - \ell^4$ law, but in pure copper. Accordingly, a sample was annealed at 750° C to erase any peaking effect, then irradiated at a number of temperatures. The results are included in Fig. 4. Obviously, the first power dependence between decrement and modulus is reproduced.

With these main observations, we turned our attemtion to mechanisms and the defect dragging model emerged. The model is described rather fully in Reference 3 and will be outlined below. The main feature is that we endow the point defects on dislocations with the ability to follow dislocation motion to a greater or lesser extent. The extent depends on several parameters, also discussed below. The fact that a point defect is dragged along by an oscillating dislocation segment must contribute to energy dissipation. The peaking effect is a compromise between two effects. If the dislocation loop length is not too long initially, the additional damping of the dragging of the first point defect causes more energy dissipation than the reduction due to decreased loop oscillation. If a second point defect is added to the segment, it too contributes to the energy dissipation but it inevitably decreases the motion of the first point defect (as well as the whole dislocation segment), thereby decreasing the original energy dissipation contribution of the first point defect. This logic leads to a maximum in the decrement when the additional dissipation due to the (n + 1)-st defect is not sufficient to offset the reduction in dissipation of the previous n point defects. Subsequent point defects act to diminish the decrement.

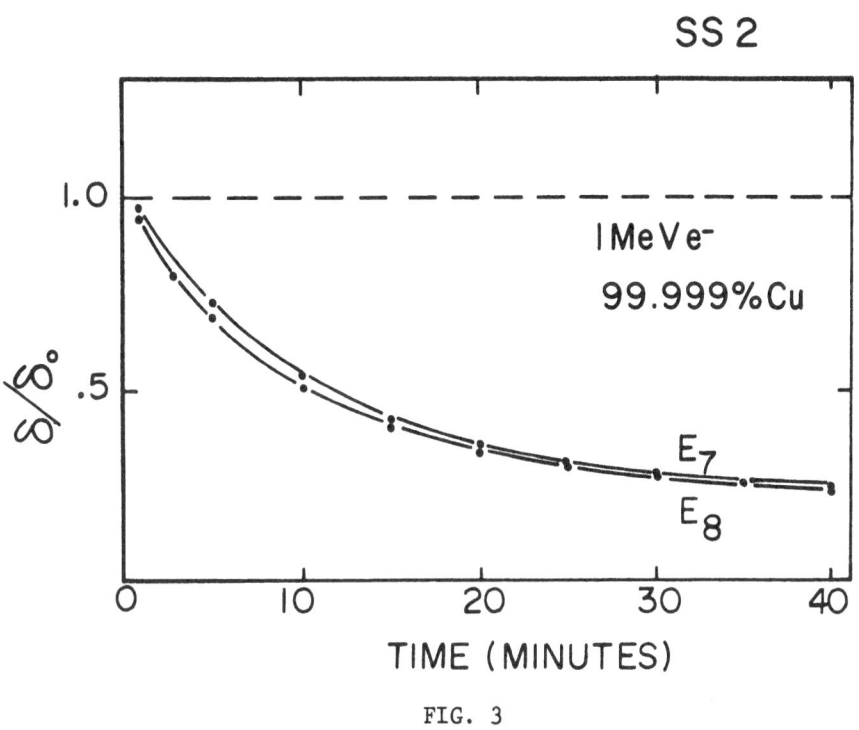

FIG. 3

Same as Fig. 2. Annealing temperatures were E-7, 500°C; E-8, 750°C.

As we made the observations described above and developed the dragging model, we worked in an almost completely isolated environment. That is to say, we believed that the peaking observations were entirely new and developed the dragging model without recourse to previous work. But this is an appropriate place to record the credits of others who, in one way or another, have contributed to this area.

On the experimental side, the earliest related findings are probably those of Routbort and Sack(6) in their 1966 report that the decrement appeared to increase in copper with decreasing frequency, below 40 kHz. This observation is consistent with the dragging model, although the high frequency is somewhat surprising. Nielsen (17) reported that the decrement initially increased in copper bombarded with protons at 20° K but failed to explore the effect since his interest centered on the "bulk effect" which occurs at much higher defect densities. Besides its historical value, Nielsen's observations are most important in its temperature--as low as 20° K.

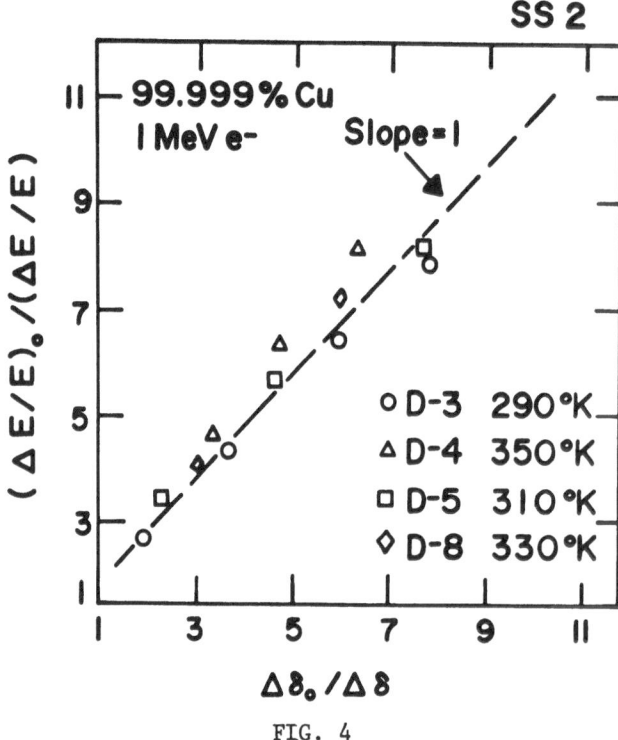

FIG. 4

Inverse normalized modulus defect vs. inverse normalized decrement of 99.999% purity copper irradiated with electrons at the indicated temperatures. See Ref. 14.

On the theoretical side, Kamel (18), following some his observation on quenched samples,
suggested that oscillating dislocations drag vacancies, but he failed to present any analysis
for this suggestion. Yamafuji and Bauer (19) and others have discussed the motion of point
defects along dislocation lines, rather than perpendicular to it, as does the dragging model.
It remains as unanswered question why lateral motion can be effectively ignored, as the success
of the dragging model would indicate. Lücke and Schlipf (20), in the 1968 Harwell Conference,
explored the character of dislocation damping when pinning points _diffuse_ perpendicular and
parallel to dislocation segments, under oscillating stresses. They found, for the case of small
bow-out and many pinning points, a relaxation peak in decrement given by

$$\delta = \frac{\pi}{6} \Lambda \ell^2 \; \frac{\omega_y \, \tau_y}{1 + \omega^2 \, \tau_y^{\,2}} \tag{9}$$

where

$$\tau_y = \frac{\gamma(n+1) \; \ell kT}{8 \; Gb^2} \; [\nu_D^{-1} \, \exp(\frac{U_c}{kT})], \tag{10}$$

where $\gamma \approx 2/3$, n is the number of pinning points on length ℓ of dislocation, k is Boltzmann's
constant, T is temperature, G is the shear modulus, ν_D is an atomic attack frequency and U_c is
the activation energy for such diffusion. Lücke and Schlipf note the earlier work of Schiller
(21) on the same problem but point out that Schiller's result incorrectly omits the unbracketed
pre-exponential factor which may be large.

The temperature dependence in Eq. (10) comes, of course, from the assumed diffusion.
The work of Simpson and Sosin (3), discussed below, shows that dragging, closely related to the

perpendicular motion studied by Lücke and Schlipf, occurs over the entire range from 4° K to over 400° K in copper. Apparently, defect dragging cannot be wholly ascribed to simple diffusion. We will return to this point later, also.

I now turn to the SS model calculations. The starting point is the Newtonian equation of motion originally introduced by Koehler (1). Neglecting inertial effects, we have

$$B(x)\rho(x) \frac{\partial y(x,t)}{\partial t} - C \frac{\partial^2 y(x,t)}{\partial x^2} = b\sigma_o e^{-i\omega t} \tag{11}$$

where $\rho(x)$ is the density of dragging points (or the line itself) and $B(x)$ is the viscous drag constant for the defect which is at a distance x along the dislocation segment from one end. C is the dislocation line tension, y is the displacement of the dislocation at position x, b is Burger's vector, ℓ is the length of the dislocation segment and $\sigma_o e^{-i\omega t}$ is the applied oscillating stress. The boundary conditions are that the displacement is zero at the two ends of the segment: $x = \ell$ and $x = 0$.

In the solutions to these equations, we take

$$B(x)\rho(x) \rightarrow B_{\ell} + \sum_{i=1}^{n} B_d \delta(x - x_i) \tag{12}$$

where B_{ℓ} is the frictional contribution of the dislocation line itself--the damping constant that has received a great amount of attention in the past--and B_d is the viscous drag constant associated with each of n dragging point--a constant which has received no adequate attention to date.

The problem has been treated at two levels. The first case is the case in which the point defects are considered to be spread out uniformly along the dislocation segment. We call this the "loaded-line case." Clearly this case is equivalent to a dislocation with no point defects on it, but whose damping constant varies as the concentration of point defects, in fact, increases. The point is that the solution to this problem has been given earlier, although we were not aware of this either during our development. Oen, Holmes, and Robinson (4), for example, calculated the results in this case for a constant value of B--that is, a value of B which did not depend on x or t. (The SS model solution is precisely the same, since we have taken B to be independent of time also. In fact, there is a slow time dependence since the number of dragging points may change with time but the characteristic time for point defect accumulation is very long compared with a period of oscillation in the dislocation loop). Their results are presented in Fig. 5. Note that the abscissa is angular frequency, ω, the parameter which was of interest to Oen et al. In the dragging model, a "universal parameter" μ replaces ω but there is no other change required. This plot of Oen et al. is particularly interesting, as noted further below, since their plot for the solution of what we would call the loaded-line case is a doubly logarithmic plot and since they have calculated the decrement and modulus defects for both the delta function and exponential loop length distributions. In Reference 3, the graphical solution is presented only for the delta-function distribution on a purely linear plot: Fig. 6.

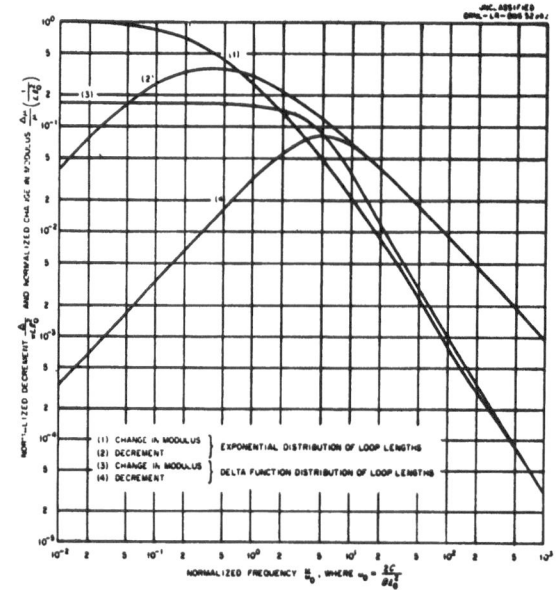

Normalized Decrement $\Delta/\pi L l_0^2$ and Normalized Change in Modulus $(\Delta\mu/\mu)(1/L l_0^2)$.

Asymptotic Forms for the Change in Modulus and the Decrement

$$\omega_0 = 2C/B l_0^2$$

Distribution of Loop Lengths	$\Delta/\pi L l_0^2$		$(\Delta\mu/\mu)(1/L l_0^2)$	
	Low Frequencies	High Frequencies	Low Frequencies	High Frequencies
Exponential	$4\dfrac{\omega}{\omega_0}$	$\dfrac{\omega_0}{\omega}$	1	$\left(\dfrac{\omega_0}{\omega}\right)^{3/2}$
Delta function	$\dfrac{4}{120}\dfrac{\omega}{\omega_0}$	$\dfrac{\omega_0}{\omega}$	$\dfrac{1}{6}$	$\left(\dfrac{\omega_0}{\omega}\right)^{3/2}$

FIG. 5

Theoretical plot of decrement and modulus for exponential and delta function distributions of dislocation loop lengths. The abscissa is angular frequency, as calculated by Oen at al. The same plot applies to the defect dragging model, with angular frequency replaced by the universal parameter, μ , and a change in abscissa scale. See Ref. 4.

For the delta-function distribution, the solutions are

$$\delta = \frac{\pi E b^2 \ell^2 \Lambda}{2 \, C\mu^2} \left[1 - \frac{1}{\mu}\left(\frac{\sinh\mu + \sin\mu}{\cosh\mu + \cos\mu}\right)\right] \tag{13}$$

and

$$\frac{\Delta E}{E} = \frac{E b^2 \Lambda \ell^2}{2C\mu^3} \left[\frac{\sinh\mu - \sin\mu}{\cosh\mu + \cos\mu}\right] \tag{14}$$

with

$$\mu^2 = \frac{\omega B \ell^2}{2C} . \tag{15}$$

In the limit $\mu \to 0$, Eqs. (13) and (14) reduce to Eqs. (1) and (2). This limit corresponds to several possibilities, particularly to $\omega \to 0$, the low frequency limit, and to $B \to 0$, the zero damping limit. Consider the latter first. We let

$$B = B_o + mB_d , \tag{16}$$

where B_o is the viscous drag constant for defects which produce the pre-irradiation damping, m is the number of defects added per unit length of dislocation, and mB_d is the additional drag due to these defects. As μ increases from zero, the decrement and modulus defect behave as shown on a linear plot in Fig. 6, or on the logarithmic plot in Fig. 5. The most obvious feature is the initial decrement increase--the peaking effect. The logarithmic plot shows up the frequency dependence strikingly. On the high frequency side (actually the high μ side, in the spirit of the defect dragging model), the decrement goes as ω^{-1}. It is this frequency dependence which, according to the defect dragging model, conflicts with the ω^{1}-dependence of the GL pinning model, presumably giving rise to a soft frequency dependence. It is well to recall here that Routbort and Sack (6) reported their apparent increase in decrement with decreasing frequency in experiments performed at rather low frequencies--1.5 kHz (as well as 40 kHz). Combining observations from Figs. 5 and 6 with experimental observations, and recalling that frequency is only one factor in the universal parameter, μ, we conclude that, as general guide lines in copper, the frequency range below somewhere about 0.5 kHz should show an inverse dependence on frequency, the range roughly from 0.5 kHz to somewhere about 20 kHz should be a transition region where the frequency dependence moves away from an inverse dependence characteristic of the upper frequency range of the dragging model to the lower frequency range of the pinning model, and above this the proportionality to frequency should dominate, as predicted by GL, up to the low megaHertz region. Presumably, at very low frequencies--perhaps about 1 Hz or less--the decrement should once again take on a linear dependence on frequency; experimental evidence for this would be very valuable.

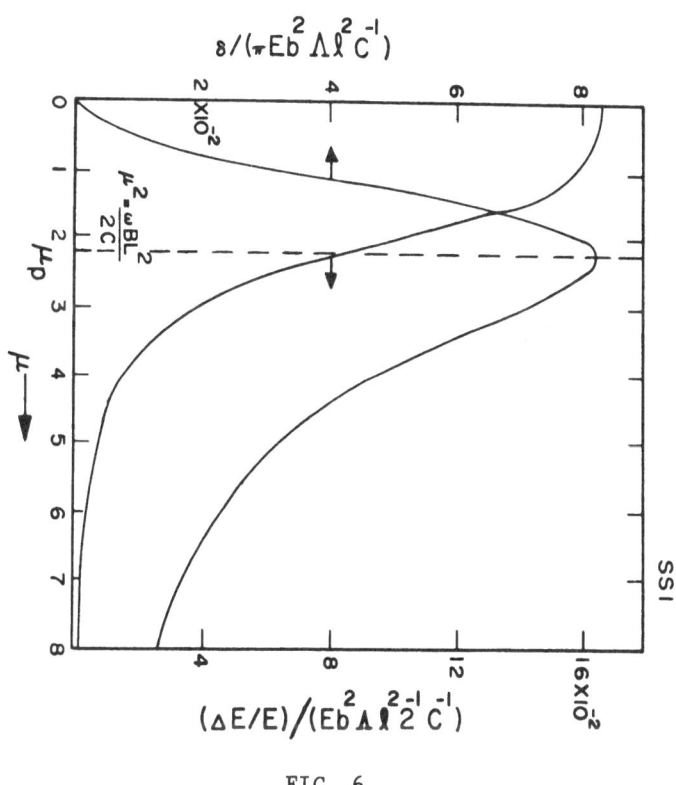

FIG. 6

The same theoretical curves of Fig. 5 for the delta-function distribution, plotted on a linear scale. See Ref. 3.

119

Fig. 6 shows another interesting prediction. At high values of μ, reached in our experiments by increasing B_d through the addition of point defects, the modulus defect should vary more rapidly with defect density (or frequency!) than the modulus; the modulus defect should vary as $\mu^{-3/2}$. This is entirely reverse compared with the GL model, where the faster variation is always associated with the decrement. In fact, we (14) have found this faster variation of the modulus defect. However, the approach to a -3/2 dependence was never completly reached. It was found, however, that a better fit to the large μ limits was found by introducing an additional component to the losses, in particular one that might represent modulus effects due to dislocations that are indeed pinned. In other words, we once again have evidence, of a not particularly firm nature, that there are two families of dislocations.

Fits to the full loaded-line model are shown in Figs. 7 and 8 (14). (No account of a second dislocation family is included here). The fits involve the arbitrary choice of μ_o, the initial value of μ (i.e., at the beginning of irradiation), and a parameter which includes the rate of defect production and diffusion during the irradiation. In the cases where a peak is shown, E-3, E-4, and E-5, the values of μ_o^2 of 1.1, 1.3, and 1.5 were used; in the case where the peak is absent, E-6, $\mu_o^2 = 10.0$. According to the load line analysis, $\mu_p = 2.2$ at the peak.

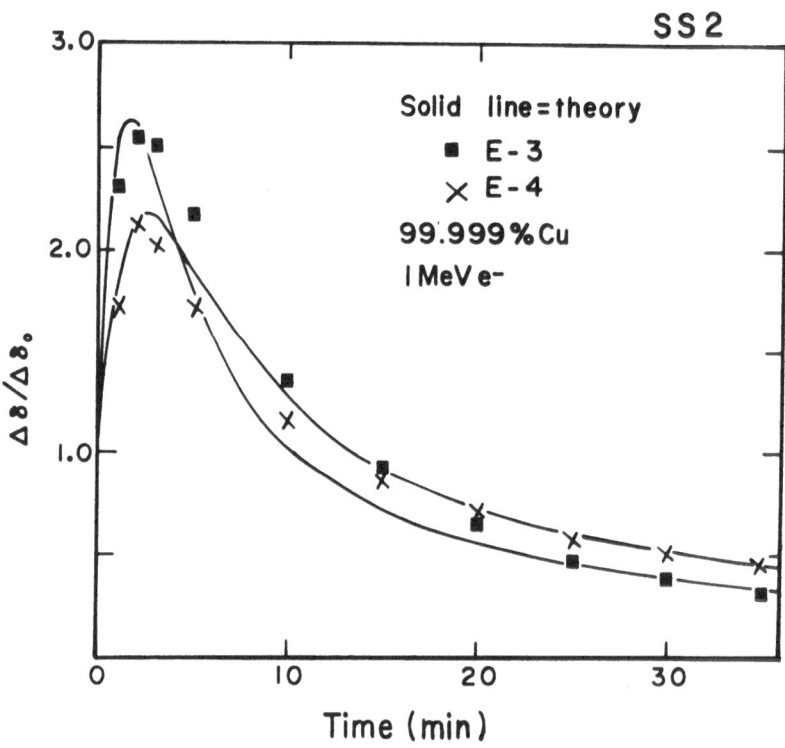

FIG. 7

Normalized decrement plotted as a function of irradiation time and compared with theory (solid lines). See Ref. 14.

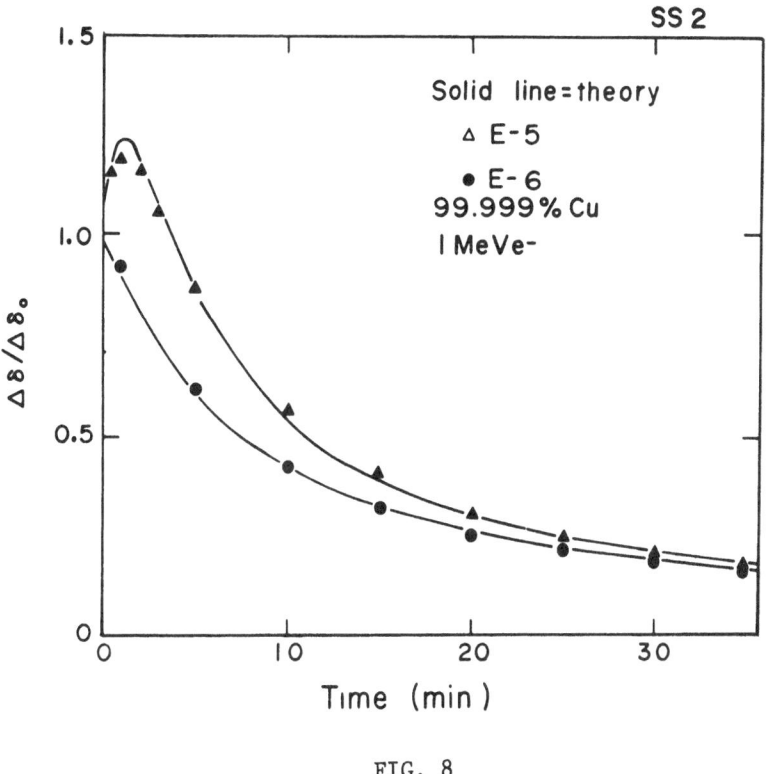

FIG. 8

Same as Fig. 5.

The remark in the previous paragraph, that $\mu_p = 2.2$, applies for a delta-function distribution of dislocation loop lengths. This could be taken as evidence that the delta-function distribution is appropriate. Since another distribution has not as yet been examined, such a conclusion is premature. Oen et al. have integrated over the exponential distribution to arrive at the results in Fig. 6, so that a check against this should be rather simple.

At this point it is appropriate to inquire into the origin of the proportionality of the decrement to the first-power of the modulus defect, Eq. (8). The theory does not show such a behavior precisely in any range of μ. However, a reasonable proportionality can be found (14) for a limited range, for $n \approx 5\text{-}10$. The linearity is very good, over a range from 2-20, roughly, if a second dislocation family is introduced, as just mentioned. Considering the non-monotonic behavior of the decrement, it is apparently too much to expect full linearity but the use of this pseudo-linearity is good enough to serve as a guide for experiments in which the dragging point density must be deduced from modulus and decrement measurements.

All of the above analysis comparing the defect dragging model with experiment can be criticized on still another basis. How realistic is the loaded line extreme in the case of one or a few point defects added to a dislocation segment? The answer is provided, in part, by the second level at which the dragging model has been pursued: cases of discrete pinning. The full solution for the case of one pinning point on a segment, with line damping also, is given in Reference 3. In addition, the exact solutions for discrete pinning have been calculated for cases of 1, 2, ..., 10 dragging points, without the inclusion of line damping. The latter bank of solutions has been assembled into graphical solutions for the decrement and the modulus defect as shown in Figs. 9 and 10 (14). Notice that the parameter for the families of plots is closely related to the universal parameter of the loaded-line case. It is $\mu_d = \mu^{1/2}/n$, with $B_o = 0$, and n becomes the abscissa variable.

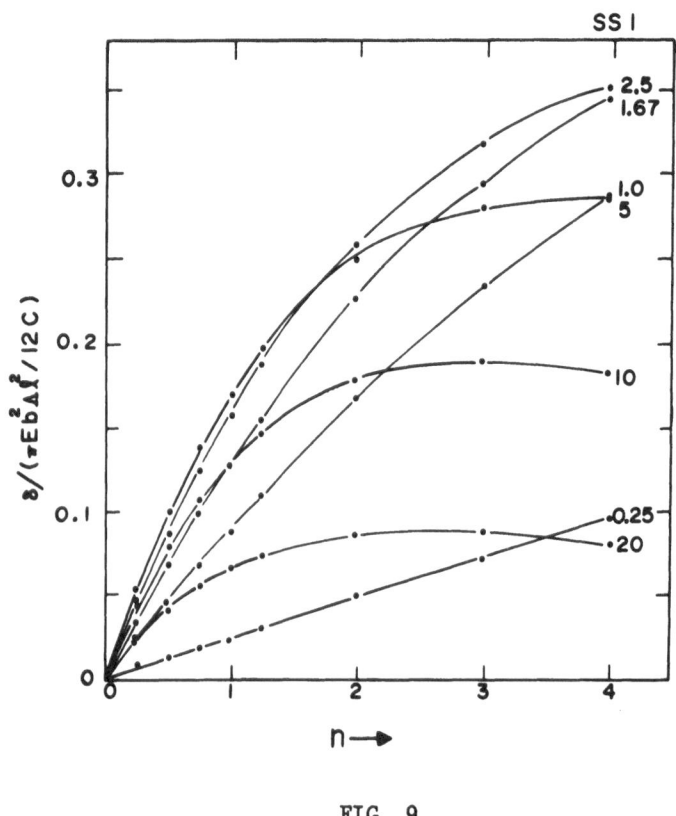

FIG. 9

Decrement plotted as a function of the average number, n, of defects per dislocation segment for a delta-function distribution of dislocation loop lengths, for the case of discrete dragging points. Values of $\mu_d = B_d \ell/C$ are noted beside each curve. See Ref. 3.

The curves of Fig. 2, particularly those for high values of μ_d, show the peak in decrement and subsequent decrease, as n increases. The plots are limited by the amount of calculation required to generate each curve. The abscissae in the two figures are limited by $n = 4$, or 5, since it requires the weighted use of cases for n sufficiently greater than the abscissa value to assure minimum error. Poisson weighting has been used; weighting according to an exponential distribution remains to be done.

The question is, then, how well do the calculations compare with the discrete cases?

The answer is: surprisingly well.

The extent of agreement can be judged in part by Fig. 11, where the cases calculated in Fig. 9 are compared with values of decrement calculated for the loaded-line case via Eq. (13). As one would anticipate, the largest discrepancy is found for the highest values of B_d, since loading the entire line with a "smoothed-out" defect distribution is more extreme in its consequences than placing a single obdurate point defect somewhere on a dislocation segment, then averaging over its random positions.

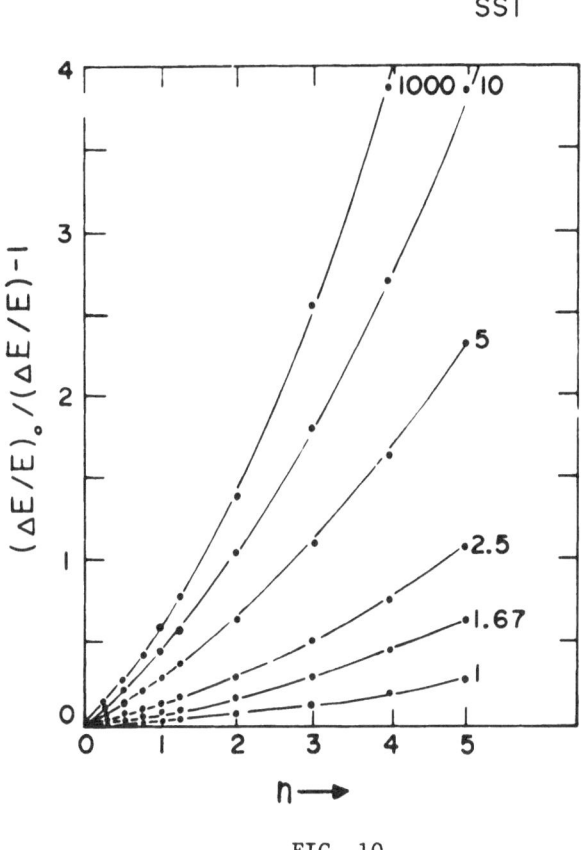

FIG. 10

Normalized inverse modulus defect plotted as in Fig. 9. See Ref. 3.

Another way in which the agreement between the loaded-line analysis and discrete analysis can be appreciated is to compare analytically the values given for particular values of n, with no recourse to averaging methods. We have done this for n = 1, once again with B_o = 0. Letting the point defect assume any position along the length of the dislocation loop, and averaging over all cases, the expressions, for n = 1, for the decrement and modulus defects are:

$$\delta = \frac{\pi \Lambda E b^2 \ell}{4 \omega B_d} \int_o^1 \frac{u^2 (1 - u)^2 \, du}{u^2 (1 - u)^2 + \mu_d^{-2}} \tag{17}$$

and

$$\frac{\Delta E}{E} = \frac{\Lambda E b^2 \ell^2}{12 C} \left\{ 1 - 3 \int_o^1 \frac{u^3 (1-u)^3 \, du}{\mu_d^{-2} + u^2 (1-u)^2} \right\}. \tag{18}$$

A comparison with Eqs. (13) and (14) shows that the pre-exponential factor of Eq. (17) is 1/4 the corresponding factor in Eq. (13) and that the multiplying constants are identical in Eqs. (18) and (14), recalling that B_d = ℓB when B_o = 0.

A full comparison will be presented elsewhere, in a compilation of statistics of the dragging model. However, it is interesting to compare the loaded line analysis with this discrete case in the limits. In the limit $\mu \to 0$, the agreement is complete. In the limit $\mu \to \infty$, δ and $\Delta E/E$ go to zero in all cases, which is agreement of a sort, also. However, a better measure of the agreement is the rate at which these go to zero. The ratio of the modulus defects calculated in the two cases for the limit $\mu \to \infty$ is unity, full agreement in this limit, also. The ratio of the decrement for the loaded-line case to the discrete (single dragging point) case, is four.

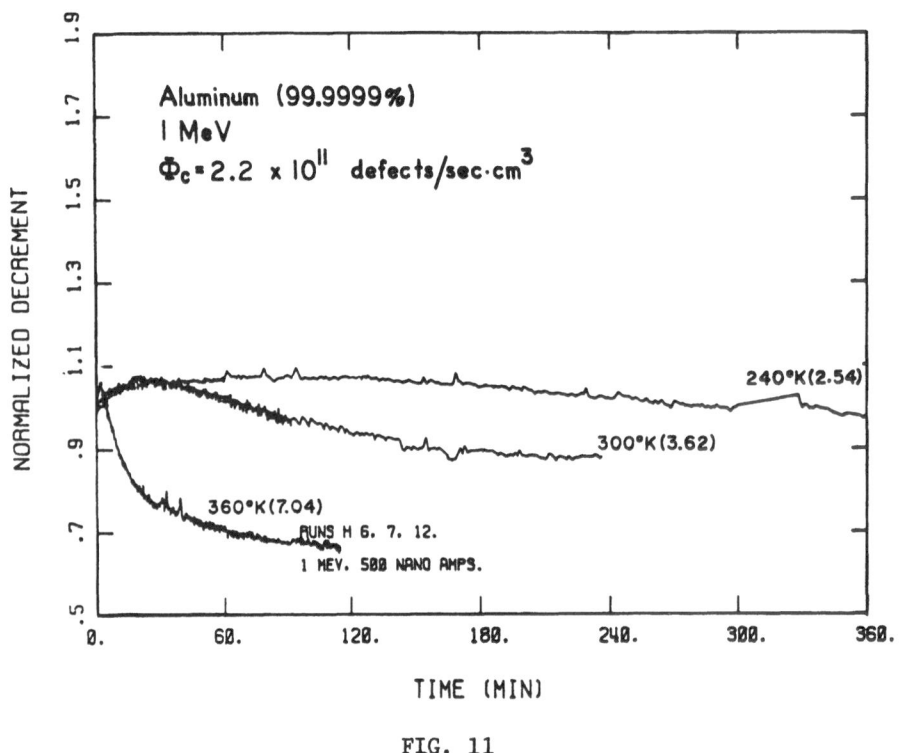

FIG. 11

Normalized decrement plotted vs. time of irradiation for aluminum. The temperatures of irradiation are noted next to each curve. See Ref. 23.

The conclusion from comparisons between the loaded-line case analysis with the discrete-dragger analysis is that the loaded-line analysis provides a convenient and relatively accurate basis for comparing experiment with theory. The conclusion from comparisons between the loaded-line analysis and experiments in copper is that the defect dragging theory adequately explains all of the observations we have made in copper in the amplitude-independent region of internal friction.

There is nothing in the formulation of the SS model which restricts the model to copper; it is formulated as a general theory. Does it apply to other metals? There is one small piece of evidence that it does apply to tungsten. Muss and Townsend (22) irradiated tungsten with deuterons. Their data show that the pre-irradiated value of decrement is below the first point taken after exposure to protons. All subsequent decrement values are lower. Muss chose to draw a monotonically decreasing curve through these data, obscuring the peaking effect which, I believe, he was in fact observing.

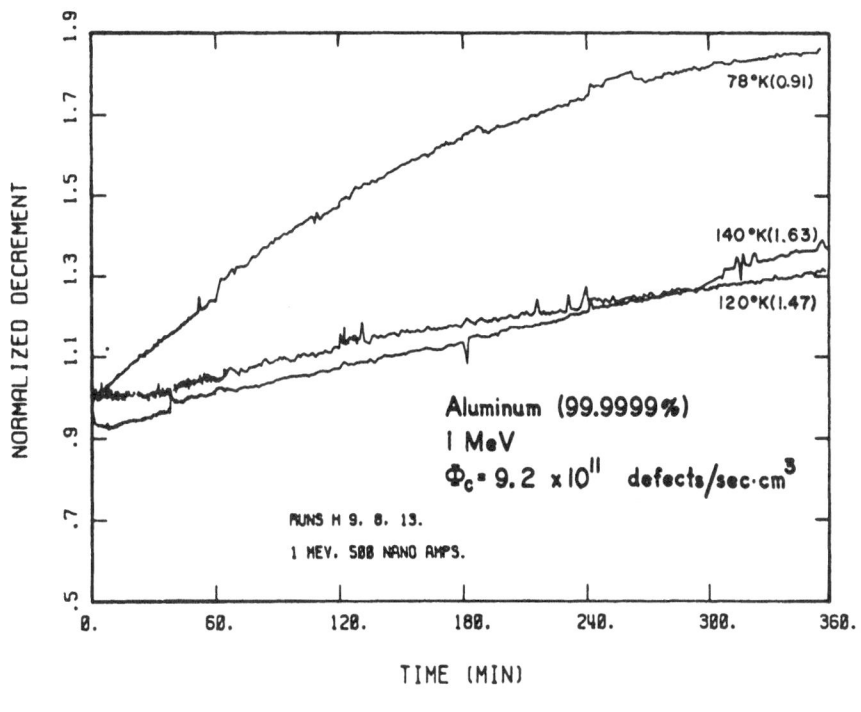

FIG. 12

Same as in Fig. 11.

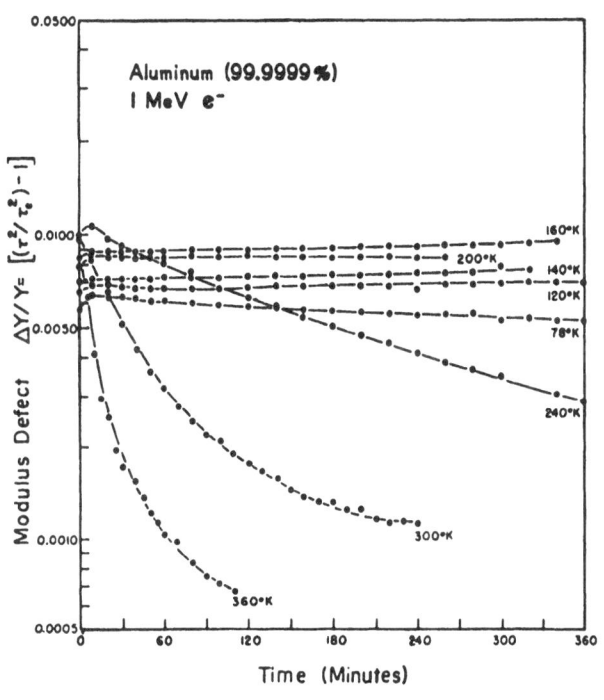

FIG. 13

Modulus defect plotted vs. time of irradiation for aluminum. The temperatures of irradiation are noted next to each curve. See Ref. 23.

In order to test the universality of the dragging model, we have examined aluminum also. This study is still in progress but some of the results have been published (23). A portion of these results are presented in Figs. 11-14.

Figs. 11 and 12 show that the decrement increases upon irradiation (for particular thermal history) and that the nature of this peaking effect depends on the temperature of irradiation; the effect of irradiation temperature is described more fully below. It turns out that the peaking effects in copper and aluminum are quite similar, but the rate of approach to maximum decrement and beyond in aluminum is distinctly more sluggish than in copper. Fig. 13 presents the most interesting data, although in a deceptive manner. The initial increases in modulus defect should be ignored; these represent small temperature increases in the sample brought about by the higher beam current needed in aluminum than in copper. At elevated temperatures--240° K and above--the modulus defect decreases in a simple fashion, as observed in copper. But below 240° K, the modulus defect shows only the slightest variation during the entire length of our irradiation exposure. This behavior has puzzled us for some time and, in our publication of these results, we have speculated concerning interstitial-vacancy recombination on dislocation lines. Very recently we have realized that the flat behavior of the modulus defect can be explained by the dragging model, without further complications. This is best appreciated by inspection of Fig. 5. Note that for relatively low values of μ (ω, on the plot of Fig. 5), the modulus defect is expected to show a flat response to increases in μ, exactly as observed in Fig. 13. Furthermore, one would anticipate that the flat behavior occurs while the decrement climbs toward its peak and that the approach to the peak is extended. This, too, is consistent with the experimental results in aluminum. Thus, it appears that our data on copper have established the dragging model for the region of the universal curve, Fig. 6, in which the modulus defect undergoes its fastest variations and the decrement may go through a maximum; the experiments on aluminum have provided confirmation of the universal curve for the earlier portions.

The effects of thermal history are similar in aluminum and copper, occuring at lower temperatures in aluminum. Fig. 14 demonstrates this. Runs K-13 and K-16 were performed following annealing at 300° C; runs K-17 through K-19, following anneals at 350° C. Clearly the peaking effect shows signs of eradication at 300° C, and is completely erased at 350° C.

The effect of temperature of irradiation on aluminum are presented in part in Figs. 11-14. Fig. 15 shows the effects on decrement for copper. The effect is a curious one. The time-to-peak, t_p, decreases with increasing temperature, as one would expect based on considerations of enhanced defect diffusion in the lattice at elevated temperatures. But this "normal" behavior disappears below about 150° K and an "abnormal" behavior is exhibited at lower temperatures. There t_p decreases with decreasing irradiation temperature. In addition to the behavior of t_p, there is a concomitant analogous behavior of the normalized decrement (the actual decrement divided by the initial value of the decrement at the given temperature before irradiation).

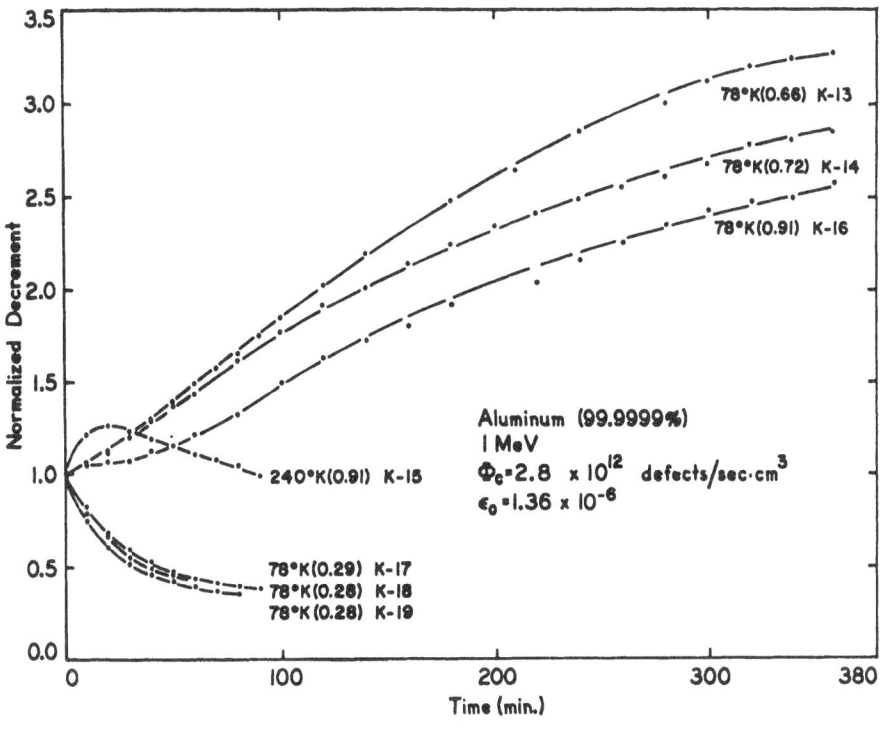

FIG. 14

Normalized decrement plotted vs. time of irradiation for aluminum in successive irradiations. The temperatures of irradiations are noted next to each curve. The values in parentheses are preirradiation decrement values, multiplied by 10^3. Preirradiation annealing temperatures were 300°C for K-13 to 15; 350°C for K-16; 400°C for K-17 to 19.

The full explanation of the behavior of t_p and peak normalized decrement as a function of irradiation temperature remains open. A potential complication lies in the presence of the Bordoni peaks which have their maxima at about 80° K in copper and 120° K in aluminum. Certainly an accounting of the behavior of the Bordoni peak must be made at the same time. (The sensitivity of the Bordoni peak to irradiation is not simple, either. We found (24) a high level of sensitivity of this peak to irradiation. In contrast, Thompson (25) reported only a slight sensitivity. It appears that the purity and history of the material is very important.) However, the behaviors of t_p and peak normalized decrement are quite similar in the cases of copper and aluminum which we have explored, even though the Bordoni peak was fully present in the copper samples and had been eliminated by previous thermal history in our aluminum samples. Apparently the Bordoni peak is a complication, but not an overriding one.

It is clear that the enhanced diffusivity of point defects with increasing temperature must play a notable role. I feel very confident that the normal behavior of t_p lies mainly in this enhanced diffusivity of a point defect--whether it be a vacancy or interstitial. Notice also, that the peak normalized decrement is relatively mildly dependent on irradiation temperature in this range. This would be consistent with the temperature dependence of B_d, leading to different peak values when the analysis is based on discrete dragging point cases. The more troublesome aspects lie at low temperatures.

Our explanation of the abnormal behavior of t_p has been based on the nature of the point defects which arrive at dislocations. We have formulated a model for this behavior which analytically describes the observed behavior of the initial "pinning" rates observed by Thompson and Buck and by ourselves based on the conversion of self-interstitial defects from one crystallographic configuration to another. Reference 27 should be consulted for the full details since this matter is only of peripheral interest in this discussion. That same formulation can account for the behaviors of t_p, as observed by Thompson and Buck (26) and by us (27). However, the matter is currently in an unsatisfactory condition based on two facts. First, we have not grappled with the behavior of the peak normalized decrement. The arguments for an increase in B_d with increasing temperature would not seem, at first glance, appropriate at low temperatures. Secondly, Lücke and co-workers (28,29,30) have shown that the minimum in the initial "pinning" rates are quite different when one compares decrement observations with modulus observations and, more important, when one makes the observations at one base temperature rather than at the temperature of irradiation, as Thompson and Buck and we have done. They have also shown the sensitivity of results to strain amplitude.

It may be important to notice that the experiments of Thompson and Buck were performed at 11 kHz; of Lücke et al., at 5 kHz; ours were at about 0.5 kHz. Thus all of these experiments, particularly the latter ones, were in the frequency range where defect dragging becomes or is important, compared with simple dislocation pinning. In a speculative manner, I propose that these experiments may, after all, not be related to interstitial conversion but to the nature of the dragging coefficient, B_d. At high temperatures, the diffusive aspects of defects attempting to follow dislocations, as discussed by Lücke and Schlipf (20), would lead to a normal dragging temperature dependence. At low temperatures, the motion of defects following dislocations should more accurately be described as dragging. Such dragging might well become more difficult as the temperature is decreased, leading to an inverse dependence of B_d on temperature.

The discussion of the last paragraph is heavily speculative but does emphasize the need for a fuller description of defect dragging. Clearly, the dragging of a defect is not to be interpreted to literally mean that an interstitial atom, for example, is carried along with the oscillating dislocation. Rather, a set of atoms alternate in assuming the interstitial role as the dislocation moves through the lattice. Similar remarks hold for a vacancy. In this regard, it is worthwhile to consider the extent of the motion we envision for dragged defects. From Reference 3, it can be shown that the displacement of a single dragging point at the center of a dislocation segment of length ℓ, ignoring line damping, is

$$y(1/2\,\ell) \approx \frac{b\sigma_o\ell^2}{2C\mu_d} \quad .$$

(19)

At the peak of the decrement curve, Fig. 6, $\mu_d^2 \simeq 5$. Taking this value and other values for copper, the displacement is calculated to be

$$y\,(1/2\,\ell) = 30 \times 10^6 \; \epsilon_o \; (\text{Å}) .$$

(20)

For a strain amplitude of $\varepsilon_o = 10^{-6}$ the displacement is some 30 Angstroms, not an inconsiderable amount.

The entire question of viscous damping in dislocation motion has been examined by Paré and Guberman (5) very recently, including both line damping, ala the GL model, and defect dragging. The examination is based heavily on their experiments with copper neutron-irradiated to 10^{17} neutrons/cm^2. The samples were examined from 20° K to 330° K before and after room temperature irradiation. The samples were resonated between 12 and 17 kHz. Their results are in apparently complete contradiction of expectation based on the GL model. That is, they find typically only about one "pinning point" added per dislocation segment during this rather extended bombardment and after raising the sample temperature to 100° C--a treatment known to allow large numbers of point defects to reach dislocations. The discrepancy is at least one of an order of magnitude, more likely two orders. Concomitantly, their deduced value of line damping is 15 to 65 times the normally accepted value. They failed to discuss the first problem, concentrating on the meaning of the high damping constant they find. Their interpretation is that the GL model is apparently insufficient or incorrect in some manner. In an attempt to resolve this, they consider defect dragging and, based on calculations using their own data, rule against the acceptability of defect dragging, as well.

The resolution of this matter is not yet established. The small deduced pinning point numbers reported by Paré and Guberman cast suspicion on their experiments or analysis since all related experiments, whether analyzed on the GL or SS models, amply demonstrate the large number of point defects which reach dislocation segments.

Nevertheless, it may be profitable to examine their criticism of the dragging model. They calculate a typical dragging point displacement of about 20 Å, essentially in agreement with the estimate given above. They then argue that the energy loss associated with one dragging point, per cycle, is ~ 0.01 eV (I would estimate a somewhat higher figure). They then state: "At room temperature, far more than this is available, with high probability per unit time, from thermal activation. Thus the dragging would have to be looked upon as a type of stress-assisted diffusion. The implication would then be that measurements were being made on the high-temperature side of a thermally activated relaxation peak. No such peak is visible...."

The conclusion of Paré and Guberman is basically correct. In fact, we contend that such a peak, of sorts, does exist. In fact, our overall picture of dislocation damping, neglecting Bordoni and related peaks, is one that shows two peaks of the variety presented by Oen et al. in Fig. 5, one at lower frequencies or values of μ arising from defect dragging, the second at higher frequencies being the well established GL dislocation relaxation peak. But it is important to appreciate that the peaks are not ideal relaxation peaks due to the dependence of μ on dragging point concentration, the unknown dependence of B_d on temperature, etc.

It is evident that the physics of dragging remains to be explored.

I close with the consideration of some miscellaneous topics which we have looked at

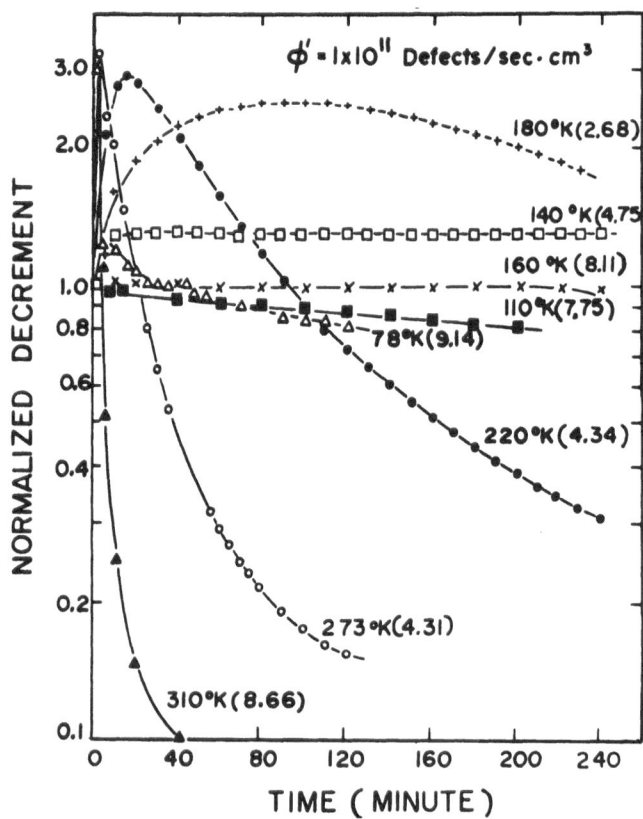

Normalized decrement of copper plotted vs. time of irradiation at the indicated temperatures. The temperatures of irradiations are noted next to each curve. The values in parentheses are preirradiation decrement values, multiplied by 10^3.

One might well ask why the peaking effect has been so obscure until now. There are several reasons which we can now understand. First, it is clear that the previous sample history is of crucial importance, as mentioned before. Many investigators have purposely pre-annealed their samples at elevated temperatures to provide an initial characterized state for their samples. In this regard, it is evident that our samples are less fully characterized and we are attempting to obtain better characterized samples in future investigations. Still another factor that has obscured the peaking effect is the frequency: some investigators have worked at higher frequencies. A third factor concerns the existence, or apparent existence, of two (or more) dislocation families. If true, this is a legitimate complicating factor; if not true, it may turn out to be the analytic complication introduced artifically by experimentors to explain their results. For example, D. W. Keefer (31) has re-analyzed data reported in gold for electron irradiations at 20° K and finds that the modulus defect and decrement were directly proportional over the range studied. At the time of this investigation, we fitted the data with a two-dislocation fit based on the GL model. The fit on such a basis was then considered reasonable, but certainly not perfect. The dragging point fit is much better, and involves considerably fewer adjustable parameters.

There is still another reason for the failure to note the peaking effect earlier. The first data taken in the earlier experiments at Atomics International were taken after some considerable avount of irradiation. As a result, measurements were made intermittently, with the accelerator shut down. Currently the measurements are made almost continuously. It is easy to see, using Fig. 6, how we could have gone over the peak before making the first measurement and felt that the decrement was decreasing monotonically. With hindsight, this is the presumed reason that we then felt that our decrement measurements were not generally trust-worthy and relied almost solely on modulus measurements.

At the completion of much of the work already reported for copper and the development of much of the dragging model, we felt that the frequency dependence is a critical factor in verifying the dragging model. At this time, I feel this is still basically correct but I anticipate that we will find no clear cut ω^1 or ω^{-1} behavior in our frequency range. In fact, our experiments to date (32) on frequency dependence seem to confirm this: there appears to be little frequency dependence in the near 1 kHz range. Nevertheless, a full frequency study remains one of our goals.

There is also a matter of amplitude dependence. Our early data (32) show that the amplitude dependence has a relatively minor effect on the dragging model. We have also made an early analysis (33), using computer techniques, to analyze amplitude dependence. Our results based on this approximate analysis, are very similar to those of Heiple and Birnbaum (7) based on the GL model. However, the matter of amplitude dependence is too little developed to make firm conclusions. Unfortunately the analysis for amplitude dependence in the dragging model is more difficult than in the GL pinning model. The reason is the following. In the pinning model, breakaway from any pinner on a dislocation segment is catastrophic, leading to breakaway from all the pinners on that segment. This is not true in the dragging model, although such a catastrophy occurs rather frequently.

Finally, I would like to venture an opinion as to the importance of the dragging model, assuming, of course, its credibility. The dragging model obviously resolves many of the difficulties previously blamed on the pinning model; this point has been stressed in this article. Perhaps of equal importance is that the dragging model may open up avenues for more complete understanding of the nature of interaction between dislocations and point defects. Certainly the pinning model has made notable contributions here. But, in retrospect, the pinning model may have been too successful. In the frequency range in which it has served best, it has treated the point defect as a completely inert specie. This approach diminishes some of the potential for exploring the details of interaction. The details of interaction are of major importance in defect dragging.

ACKNOWLEDGMENTS It is a pleasure to acknowledge the contributions made in our work by a team of enthusiastic individuals including W. Even, G. R. Edwards, D. F. Johnson, W. Pederson, S. S. Liu, R. Wang, W. Huff, and J. McIlwain. The contributions of S. L. Seiffert in almost all phases of the work, particularly in the work on aluminum in his thesis, deserve special acknowledgments. And, the many splendid contributions to all of the theory and experiments of H. M. Simpson cannot be over emphasized. The continuing experimental support of J. Corey is always appreciated. The financial support of the U.S. Atomic Energy Commision was vital in this work, as is the present support of the U.S. National Science Foundation. The calculational support of L. Thompson is appreciated as is the permission of M. Robinson to publish Fig. 5.

REFERENCES

1. J. S. Koehler, Imperfections in Nearly Perfect Crystals, p. 197 (W. Shockley et al, editors), Wiley (1952).

2. A. Granato and K. Lücke, J. Appl. Phys. $\underline{27}$, 582, 789 (1956).

3. H. M. Simpson and A. Sosin, Phys. Rev. $\underline{B5}$, 1382 (1972).

4. O. S. Oen, D. K. Homes and M. T. Robinson, U.S.A.E.C. Report No. ORNL-3017, p. 3 (1960) unpublished.

5. V. K. Paré and H. D. Guberman, J. Appl. Phys. $\underline{44}$, 32 (1973).

6. J. L. Routbort and H. S. Sack, J. Appl. Phys. $\underline{37}$, 4803 (1966).

7. C. R. Heiple and H. K. Birnbaum, J. Appl. Phys. $\underline{38}$, 3294 (1967).

8. L. Teutonico, A. Granato and K. Lücke, J. Appl. Phys. $\underline{35}$, 220 (1964).

9. D. O. Thompson and D. K. Holmes, J. Appl. Phys. $\underline{27}$, 713 (1956).

10. H. Rosenstock, Phys. Rev. $\underline{B5}$, 1402 (1972).

11. D. O. Thompson and V. K. Pare, J. Appl. Phys. $\underline{31}$, 528 (1960); D. W. Keefer, J. C. Robinson, Acta Met. $\underline{13}$, 1135 (1965); A. Sosin, Acta Met. $\underline{10}$, 390 (1962).

12. R. M. Stern and A. Granato, Acta Met. $\underline{10}$, 358 (1962).

13. H. M. Simpson, A. Sosin, G. R. Edwards and S. L. Seiffert, Phys. Rev. Letters $\underline{26}$, 897 (1971).

14. H. M. Simpson, A. Sosin and D. F. Johnson, Phys. Rev. $\underline{B5}$, 1393 (1972).

15. H. M. Simpson, A. Sosin and S. L. Seiffert, J. Appl. Phys $\underline{42}$, 3977 (1971).

16. H. M. Simpson, A. Sosin and S. L. Seiffert, unpublished.

17. R. L. Nielsen, Ph.D. dissertation, University of Pittsburgh (1968) unpublished.

18. R. Kamel, Acta Met. $\underline{9}$, 65 (1961).

19. K. Yamafuji and C. L. Bauer, J. Appl. Phys. $\underline{36}$, 3288 (1965).

20. K. Lücke and J. Schlipf, Proceedings of the Conference on the Interactions between Dislocations and Point Defects, Harwell (1968).

21. P. Schiller, Phys. Stat. Sol. $\underline{5}$, 391 (1964).

22. D. R. Muss and J. R. Townsend, J. Appl. Phys. $\underline{33}$, 1804 (1962).

23. S. L. Seiffert, H. M. Simpson and A. Sosin, J. Appl. Phys., to be published.

24. To be published.

25. D. O. Thompson and D. K. Holmes, J. Appl. Phys. $\underline{30}$, 525 (1969).

26. D. O. Thompson and O. Buck, Phys. Stat. Sol. $\underline{37}$, 53 (1970).

27. H. M. Simpson, A. Sosin and S. L. Seiffert, J. Appl. Phys. $\underline{42}$, 3977 (1971).

28. G. Roth, G. Sokolowski and K. Lücke, Phys. Stat. Sol. <u>40</u>, K77 (1970).

29. G. Roth, G. Sokolowski and K. Lücke, J. de Physique <u>32</u>, C2-145 (1971).

30. K. Lücke, G. Roth and G. Sokolowski,

31. See Reference 13, Footnote 5. For the original work, see D. W. Keefer, J. C. Robinson and A. Sosin, Acta Met. <u>14</u>, 1409 (1966).

32. To be published.

33. S. S. Liu, M.S. dissertation, University of Utah (1972) unpublished.

PLASTIC DEFORMATION AND INTERNAL FRICTION

D.N. Beshers and R.J. Gottschall
Henry Krumb School of Mines
Columbia University
New York, New York 10027   USA

Recent advances in our understanding of both plastic deformation and internal friction make a review of their connection most timely. Limitations of space have forced a restriction of this review to a small selection of topics within that range: the Bordoni peak, some of the Hasiguti peaks, and a brief look at some recent work at very high amplitudes of oscillation. The emphasis throughout will be on plastic deformation as a variable of internal friction experiments, and on what internal friction tells us about plastic deformation.

For brevity and clarity, a particular point of view will be adopted throughout, and the evidence compared with it. Not all the evidence is perfectly in accord with the point of view adopted, but much is, and the insights gained by consistent application of a single point of view may help to atone for its shortcomings.

The point of view adopted here is firstly that the Bordoni peak tends to increase with increasing plastic deformation, unless some process intervenes to change the course of plastic deformation. This thesis is in contradiction to the saturation behavior which is widely reported (1).

Secondly, and more particularly, the Seeger-Paré theory of the Bordoni peaks will be adopted. According to Seeger (2,3), the Bordoni peak is attributable to the thermal generation of double-kinks along dislocations. An essential point of Seeger's treatment is the dependence of the activation energy ΔH^* on σ, the shear stress acting on the dislocation. Such calculations have been extended since by Dorn and Rajnak (4,5) and by Kocks, Argon, and Ashby (6). All these calculations emphasize the dependence of ΔH^* on σ; Kocks et al. show that the whole structure of the saddle point changes as σ increases.

Paré (7) pointed out that considerations of thermal equilibrium place conditions upon

Final State ━ ━ ━ ━

Initial State ━━━━━━

FIG. 1

A dislocation line which has bowed under the effect
of a shear stress, represented in the extreme kink
model. Nucleation of a double-kink anywhere along
the long solid segment in the middle leads to the
final state shown as dashed, nucleation of a reversed pair of double-kinks along the long dashed
section leads to the final state shown as solid.

the circumstances in which this process can give rise to observable internal friction. The
essential features of Paré's model are shown in Figs. 1, 2, and 3. Fig. 1 shows a dislocation
line bowed out under the influence of a shear stress, of either internal or external source,
acting on the slip plane. The Peierls stress is large enough for the whole configuration to
be described by kinks, but the pre-existing kinks might well be described by a continuum
approximation (3,8). If a double-kink is generated along the long segment, the shear stress
will tend to drive the kinks apart to the ends of that segment, where they will interact with
the pre-existing kinks, so that the whole line effectively moves to a new position. The shear
stress σ does work $b\sigma \cdot \Delta A$ over the area ΔA swept out during the process. If ξ denotes the
separation of the newly created kinks, ΔA is greater than $b\xi$ by the area swept out by the pre-
existing kinks on either side. A simple line tension argument shows that if the center of a
dislocation loop advances by a step b, then $\Delta A \gtrsim \frac{1}{2}b\ell$ where ℓ is the length of dislocation line
between end points.

The Gibbs free energy ΔG of the system is shown schematically as a function of ξ in

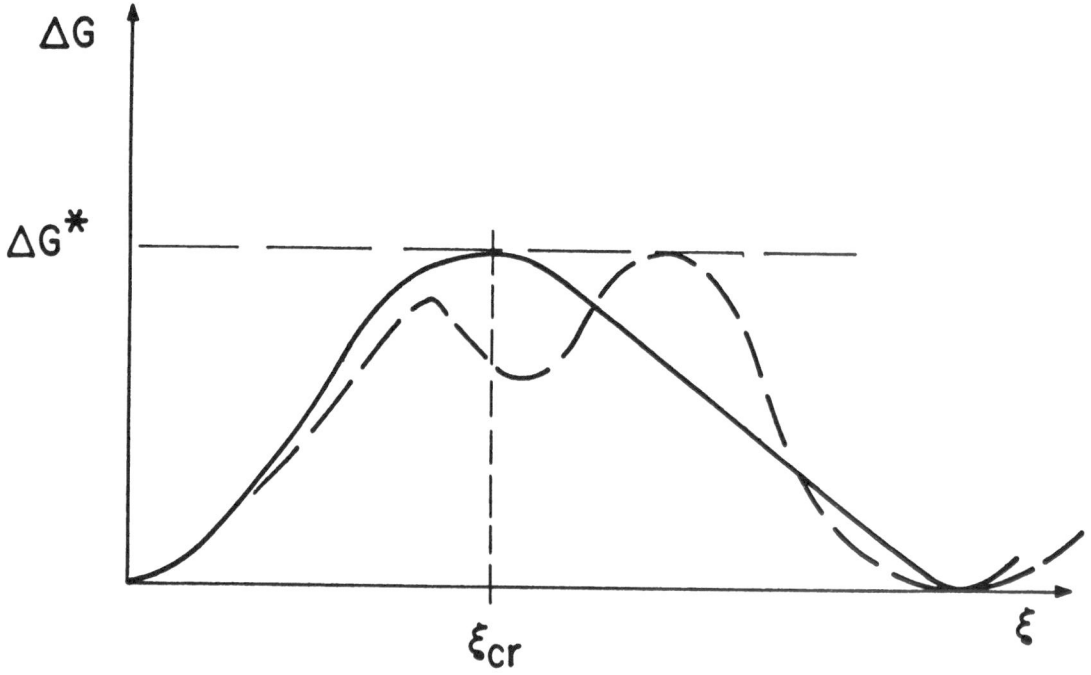

FIG. 2

Two hypothetical curves of free energy ΔG versus kink separation ξ, for $\eta = b$, with σ_i satisfying equation (2). The internal friction is expected to depend on the difference in strain between the initial state ($\xi = 0$) and the final state, and on ΔG^*, so these curves should give nearly the same internal friction.

Fig. 2 and of η, the direction normal to the line, in Fig. 3. It must be emphasized that detailed calculations (2,4,6) give a more complicated picture. In Fig. 2, ΔG decreases at large ξ linearly with ξ as a result of the work term $-b\sigma\Delta A$. At smaller values of ξ, kink-kink interaction cannot be neglected, and there is a maximum in ΔG as a result. When viewed as a function of η in Fig. 3, ΔG goes through a maximum around $\eta = b/2$ and then reaches a minimum around $\eta = b$, due to the action of the Peierls stress. However, the depth of the minimum around $\eta = b$ depends on the kink spacing ξ and the stress σ acting on the line through the term $b\sigma\Delta A$, so that $\Delta G = \Delta G(\sigma)$. Both the maximum ΔG^* and the second minimum ΔG_f, measured relative to the initial configuration, depend on σ. The distribution of dislocations between the two configurations depends on the Boltzmann factor $\exp(-\Delta G_f/kT)$. The relaxation strength for internal friction contains the factor (7) $\exp(-|\Delta G_f|/kT)$. Accordingly, the internal friction is a maximum when $\Delta G_f = 0$, that is, when the dislocation is a bistable element with equal

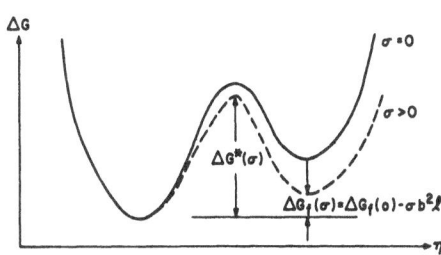

FIG. 3

Schematic curves of ΔG versus η. In the upper (solid)
curve, $\xi \approx \xi_{cr}$, while in the lower (dashed)
curve, $\xi > \xi_{cr}$.

probability of being in either configuration. If we take $\Delta G_f(0)$ to be $2W_k$ (3,6), then

$\Delta G_f(\sigma) = 2W_k - b\sigma\Delta A$, and the condition

$$\Delta G_f \approx 0 \qquad\qquad\qquad [1]$$

becomes $\sigma\Delta A \approx 2W_k/b$ and if we write $\Delta A = b\ell$, the criterion for an observable internal friction

peak derived by Paré is

$$\sigma\ell \approx 2W_k/b^2 \qquad\qquad\qquad [2]$$

The Paré criterion is a central feature of the Seeger-Paré model. Paré's ideas received

strong confirmation from Alefeld's (9) observation that the Bordoni peak can be brought out in

an annealed sample by applying a static bias stress, or by using a very large vibratory stress.

However, when the vibratory stress is small, and there is no bias stress, the only way

in which Paré's criterion [2] can be fulfilled is through the action of a long-range internal

shear stress σ_i on the dislocation segment. In the relatively pure materials in which the

Bordoni peak is usually studied, σ_i has its origin primarily in other dislocations, including those at grain boundaries. Dislocation segments subject to an internal stress large enough to satisfy Paré's criterion should be bowed out perceptibly, and indeed electron microscope pictures often show such bowout. Fig. 1 is then sketched correctly, except that the labels initial state and final state are arbitrary, since the two states are approximately equally populated. Indeed, if Fig. 1 should be turned upside-down, corresponding to a change of sign of σ_i, it would still represent a situation which would contribute to a Bordoni peak.

Which features of this model are significant for experimental results? The contribution of each dislocation segment to the relaxation strength is governed by $(\Delta\varepsilon)^2$, where $\Delta\varepsilon$ is the change in strain accompanying a transition from one state to the other. The strain $\Delta\varepsilon$ depends only on the two end points, the bottoms of the wells in Fig. 2, and not at all on the intermediate configurations. For example, the dotted line in Fig. 2 coincides with the solid line at the end points, so the two curves correspond to the same $\Delta\varepsilon$. Similarly, the relaxation time depends primarily on ΔG^* so in this respect as well there will be little difference between the solid and dotted curves in Fig. 2. Note, however, that the situation shown in Fig. 2 implies that the internal stress fulfills Paré's criterion (the two states have equal free energy), and therefore the observed ΔG^* is $\Delta G^*(\sigma_i)$ which is smaller than $\Delta G^*(\sigma)$ as shown in Fig. 3. Now in Paré's criterion [2], σ_i depends on ℓ, a quantity expected to show a statistical distribution within any one specimen, the distribution varying from specimen to specimen Therefore, specimens for which the average value of ℓ differs will be expected to show different average values of ΔG^* about the mean. These changes and distributions of ΔG^* will lead to corresponding changes and distributions of relaxation time. In sum then we expect a relaxation strength governed by the dislocation geometry, and a distribution of relaxation times which will change with specimen conditions.

It is possible to derive an expression for $N(\Delta\varepsilon)^2$ from a simple network model of dislocations, where N is the number of dislocation segments per unit volume. If the density of dislocations is ρ, and if they are all arranged as the edges of cubes of length ℓ, then N is ρ/ℓ. The area swept out in each event, ΔA, we have already taken as $b\ell$, and this determines $\Delta\varepsilon$, so $N(\Delta\varepsilon)^2$ is proportional to $(\rho/\ell) \times (b\ell)^2$ or to $\rho\ell$, a result derived by Paré. Other theories have arrived at different powers of ℓ (2,3,8).

The network model may be better than it seems at first sight. Remember first that the Bordoni peak is usually measured on a deformed specimen, <u>after</u> the deforming stress has been removed. The mobile dislocations will have retreated somewhat from the barriers which re-strained them, and will then thread their way through the peaks and valleys of the long-range internal stress distribution around them. Sometimes this may lead to the formation of attrac-tive junctions, but often it may not. What determines ℓ then is the average distance over which the dislocation has one sign of curvature, that is, the average distance between inflec-tion points. This distance will be essentially the same as the spacing between the sources of internal stress, which sources are just the dislocations running perpendicular to the slip plane, one-third of the total number. The spacing of these sources is given by the square root of the number of dislocations running perpendicular to the slip plane; i.e. $\ell = (\rho/3)^{\frac{1}{2}}$. Thus the "network" formula holds whether there is a real network, or just an assemblage of dislocations interacting through their elastic stress fields.

The next step is to relate ρ and ℓ to measurable parameters of plastic deformation. It is now very well established (10,11) that the flow stress σ_f, in tension, is related to the dislocation density, in pure single-phase materials, by the expression

$$\sigma_f = \sigma_o + \alpha\mu b\sqrt{\rho} \qquad [3]$$

where σ_o represents all other sources of hardening, together with thermal effects, α is a numerical constant of the order of $\frac{1}{2}$, and μ is the shear modulus of the material. We will neglect σ_o, so that σ_f is taken proportional to $\sqrt{\rho}$, or ρ proportional to σ_f^2. The network model also gives $\rho = 3/\ell^2$, so the product $\rho\ell$ in the relaxation strength is $\sqrt{3\rho}$ which is pro-portional to σ_f. Writing the above is symbols, using Δ_M for the relaxation strength, we have

$$\Delta_M \propto N(\Delta\epsilon)^2 \propto \rho\ell \propto \sqrt{\rho} \propto \sigma_f \qquad [4]$$

Thus Paré's theory has been converted (12) to a statement that the Bordoni peak height should be proportional to the flow stress, provided that the internal stress field in the unloaded, but work-hardened, specimen is large enough so that σ_i satisfies [2] for all, or most of the dislocation segments in the specimen.

Paré carried his theory one step further by supposing that the values of σ_i at the several segments are statistically distributed, and assumed a Gaussian distribution of σ_i. This assumption is a very reasonable one, since it gives the average value of σ_i as zero, which is correct, and it also drops off very rapidly at high values of σ_i, as is surely the

139

case physically. With this assumption, Paré estimated that the temperature of the Bordoni peak T_B might shift as much as 27% ($\Delta T/T_B$) during the course of work-hardening, and that the peak width might increase by a factor of 2. In arriving at these conclusions, he used an expression for the dependence of activation energy on stress which is in doubt by a multiplicative constant, but which in functional form is not too far from the band of later results (6).

Since some theories (3,8) suggest that Δ_B should be proportional to $\rho \ell^2$, it is noteworthy that the network model predicts that $\rho \ell^2$ is a constant. This prediction is borne out in the case of Schoeck's theory (13) of the cold-work peak: for constant nitrogen concentrations the peak height is independent of the degree of cold-work, as long as the cold-work is greater than the saturation value for that particular concentration of nitrogen (14,15). However, the ℓ^2 dependence does appear in a modified form in pinning experiments, discussed below.

It cannot be emphasized too strongly that the vast preponderance of the evidence in the field of plastic deformation is that the single parameter which best describes the internal state of a specimen is σ_f. Another example may be found in the many successes of the assumption that the thermally-activated strain rate is a function only of stress and temperature (16). By comparison, the strain ε is rather a complicated integral over the deformation history of the specimen, and the work hardening coefficient $d\sigma/d\varepsilon$ (or $d\tau/d\gamma$) has a more direct physical significance (17).

Comparison of Seeger-Paré Model with Experiment

The most extensive data suitable for comparison with the predictions of the Seeger-Paré model are data on Ni (12,18,19). More observations have been made on other materials, particularly Cu, but most authors characterize their specimens by ε rather than by σ_f. The situation in Ni is complicated because two peaks occur (Fig. 4). The upper one (Peak X) has been identified as the Bordoni peak, the lower (Peak Y) we shall discuss later. The peak heights Δ_X and Δ_Y for a single crystal, with background subtracted, are plotted versus τ, the Schmid resolved shear stress for the primary slip systems, in Fig. 5. The general behavior is seen clearly: Δ_X rises linearly with τ in Stage II and then falls sharply in Stage III, where Δ_Y rises. Polycrystalline specimens show the same general behavior (18,19), but the peaks are up to an order of magnitude larger. In each of the three studies the largest value of Δ_Y

CODE	CRYSTAL NO.	$\tau\,kg/mm^2$	ν_C
A	13	6.0	.5
B	13	7.3	0

log dec x 10⁴ ... $\log dec \times 10^4$

$(\nu_r - \nu_C)$ kHz

A

B

ν_r

B

ν_r

A

28.0

27.5

27.0

26.5

25

20

15

-160 -80 0 40

TEMPERATURE (°C)

FIG. 4

Damping versus temperature for deformed Ni (12) in Stage III. Peak Y is at -130°C, Peak X near -40°C. Note the changes in temperature and width of Peak X.

FIG. 5

Variation of peaks X and Y with τ (12).

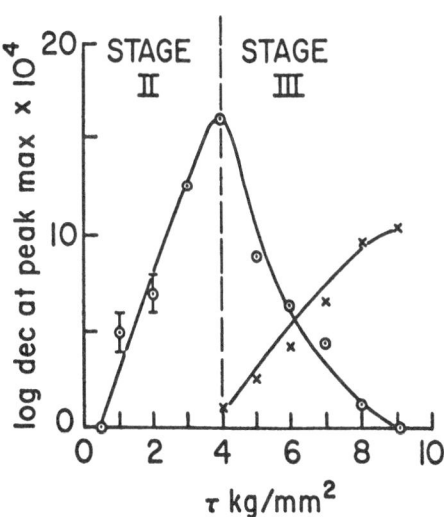

CRYSTAL 14

⊙ PEAK X

× PEAK Y

log dec at peak max x 10⁴

STAGE II   STAGE III

20

10

0

0   2   4   6   8   10

$\tau\ kg/mm^2$

observed is rather smaller (1/3 to 1/2) than the largest value of Δ_X, this rough proportion being observed over a range of 10 or so in Δ_X. The primary glide dislocations appear to be responsible for Peak X, because, as shown in Fig. 6, the orientation dependence in Stage II is all accounted for by this hypothesis. The linear rise of Δ_X is a striking confirmation of the Seeger-Pare model.

One may inquire as to why this correlation between peak height and σ_f has been overlooked in other materials. Three reasons seem of importance: first, a failure to use σ_f as the variable describing the experiment; second, many experiments have not treated the deformation variable cleanly, a point we will deal with below; and third, Ni is favorably situated with respect to room temperature deformation.

Taking up the last point, it is well-known (20,21) that the Bordoni peak height is affected by each of the various stages of migration of point defects, particularly Stage III. For Ni, room temperature falls well below Stage III, while for the other common fcc metals, particularly Cu, room temperature is just a little above Stage III. For these other metals one therefore expects the height of the Bordoni peak to be more sensitive than for Ni to details of treatment and thermal history. It is only for identical specimens identically treated that one should hope to find consistent behavior as to peak height. Since the variou annealing processes of point defects are known to be sensitive to small amounts of impurities (22,23), it may be expected that variations in the trace impurities may have a distinct effect on the Bordoni peak. We should compare only data on the same specimen, successively deformed if possible, and certainly only on similar specimens deformed at the same temperature, with the same subsequent thermal history.

There are four studies of polycrystalline Cu deformed at room temperature, which approximate to these conditions (19,24-26). In that of deFouquet et al. (19) shown in Fig. 7 particular attention was given to controlling the microstructure and thermal history. Unfortunately, the deformation was described by strain ε in these studies, but we have converted to σ_f by taking account of the grain size of the specimens, and using published (27) curves of σ_f vs. ε, with grain size as a parameter. The results are shown in Fig. 8, where it is apparent that there is a linear rise of peak height, followed by a decrease at higher flow stresses. The scatter of the data is surprisingly small in view of our previous comments. Thus, we may say that Cu, like Ni, obeys the Seeger-Paré prediction of a linear

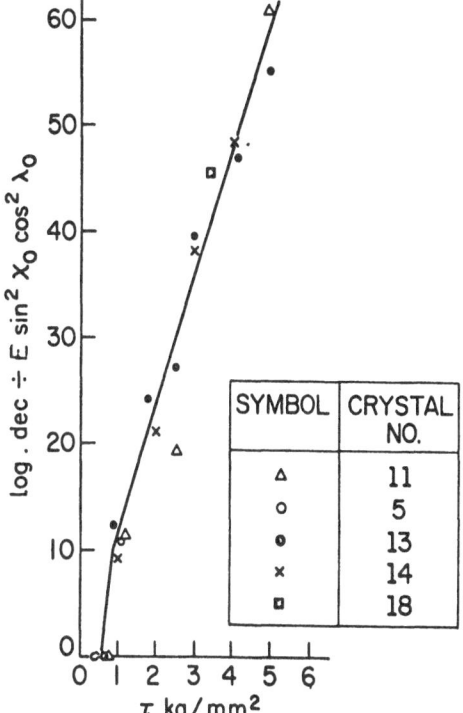

SYMBOL	CRYSTAL NO.
△	11
○	5
◑	13
×	14
⊡	18

FIG. 6

Height of Peak X, corrected for orientation, on hypothesis that primary dislocations are responsible, for all crystals of ref. (12).

FIG. 7

Height of Bordoni Peak (P_O) and Hasiguti P_1, for Cu versus prestrain after de Fouquet et al. (19).

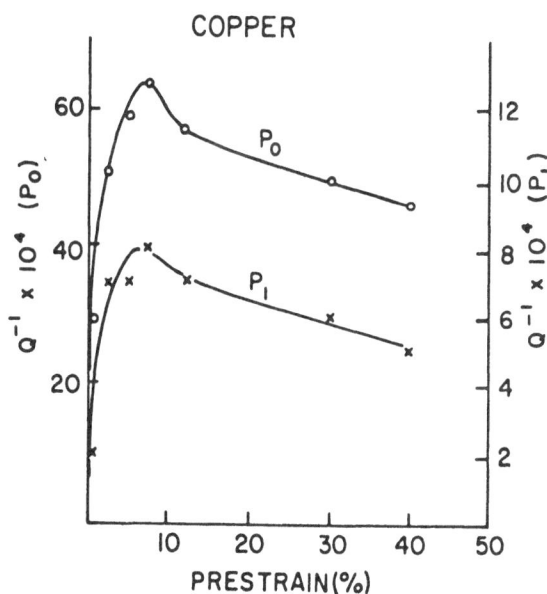

rise of peak height with σ_f, but again there is a fall in peak height at higher values of σ_f.

The data for other fcc metals are sparse. Those of Mecs and Nowick (28) for Ag are plotted versus flow stress and are not inconsistent with our picture, if one point is ignored in their Fig. 8. They also show an extreme case of a single crystal deformed to a large strain, but with a low flow stress, and a correspondingly low Bordoni peak. The data of Grandchamps (29) on polycrystalline Au show a monotonic increase of peak height with ϵ, to 25% strain. Lacking data for σ_f, we can only say that the peak increases with deformation instead of saturating. We may take it then that the normal behavior for fcc metals is for the Bordoni peak to increase monotonically, often linearly, with the flow stress, at least up to the onset of Stage III of deformation.

For the bcc metals, Gibala et al. (30), when studying the α peak in Mo and Nb, found the peak height to increase as $(\tau - \tau_0)^2$, where τ_0 is the resolved shear stress for yielding. There is no indication of a deviation from this law, even at a strain of 50%, corresponding to $\tau = 8 \text{kg/mm}$. Chambers and Schultz (31) report a fall-off of the α-peak at strains of 20 - 30% in fine-grained Mo, which probably correspond to much higher stresses; see also (19). Takita and Sakamoto (32) report a monotonic increase with deformation of a peak they found around 50°K and 100kHz in α-Fe. In hcp Zr, de Fouquet et al. (19,33) report a peak which increases monotonically, perhaps parabolically, with strain up to 10%. It appears that the behavior exhibited by Ni may be typical of the metals.

Data which are apparently contradictory with the Seeger-Pare model have been reported by Thompson (34) and by Grandchamp (29). They carried out pinning experiments, on Cu and Au respectively, which indicated that the damping is proportional to ℓ^2. The discrepancy may be removed by noticing that in the deviation of the expression for $N(\Delta\epsilon)^2$, we took ℓ to represent both the network length and the length free to move. When additional pinning points are introduced, these lengths are no longer equal. If we represent the network length by L, N is ρ/L, while $\Delta\epsilon$ is proportional to ℓ, now the length free to move as a whole, so that instead of [4] we have

$$\Delta_M \propto N(\Delta\epsilon)^2 \propto \frac{\rho}{L} \ell^2 \qquad [5]$$

In experiments at constant ρ and L, such as those of Thompson and Grandchamp, a dependence of Δ_M on ℓ^2 is to be interpreted as evidence that the number of segments remains constant, but

144

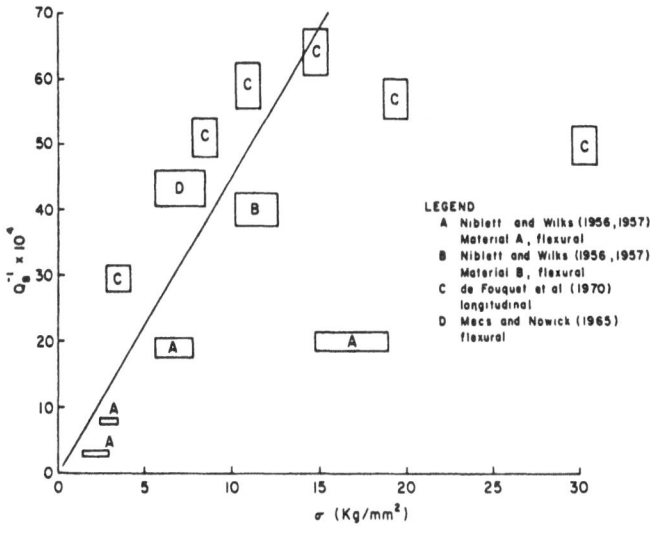

FIG. 8

Bordoni Peak versus σ for Cu. σ was
calculated from original data on ε
following (27). Refs. A, B (24,25),
C (19), D (26).

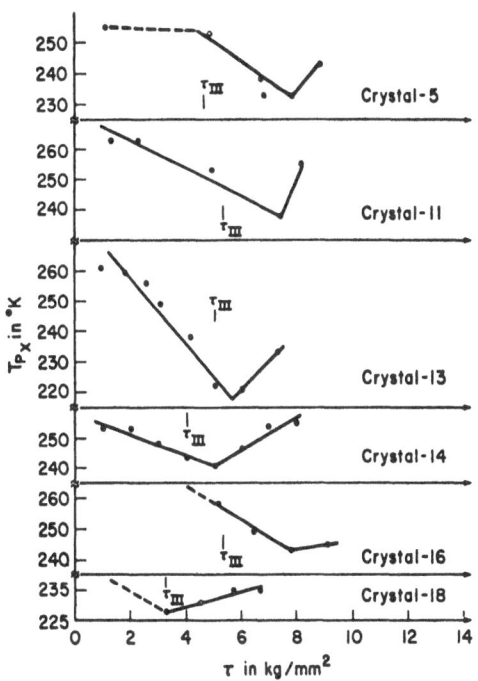

FIG. 9

T_x versus τ, plotted from tabulated
data of ref. (12).

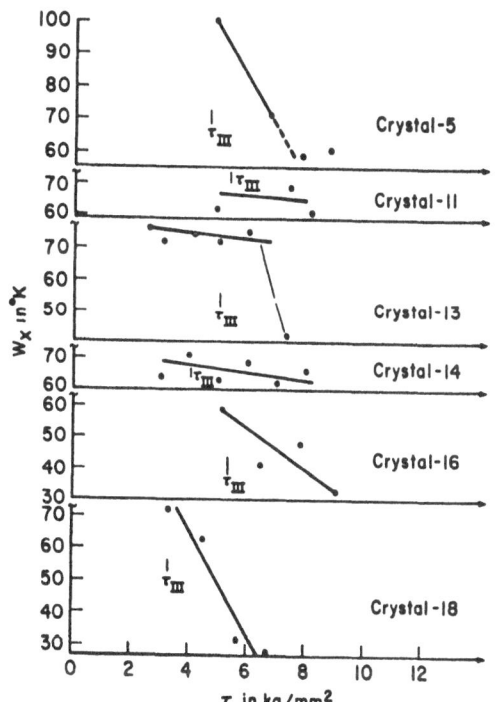

FIG. 10

W_x versus τ, plotted from tabulated
data of ref. (12).

the length free to move in the Bordoni process has been reduced by the pinning points. As far as the predictions represented by [4] and [5], the Seeger-Paré model is very well confirmed.

The model also predicts that the peak temperature and width will change during the course of plastic deformation. This prediction has been verified for polycrystalline Ni (18); the changes are smaller for single crystals (12). Cu exhibits considerably smaller shifts of T_B, but the absolute value of T_B is also smaller, so that $\Delta T/T_B$ is appreciable. Brown and Niblett (35) report a range of 80° to 98° at 20kHz, giving $\Delta T/T_B \sim 20\%$, in single crystals. Fig. 18 of (1), which exhibits all the data for Cu up to 1961, shows a variation of almost 30% in the reported values of T_B at 30kHz. A distinct tendency for the α-peak in Nb and Ta to shift with microstructure is shown by Gibala et al. The peak widths observed in Cu are wider than a single Debye peak, according to Niblett (1), by a factor of 3 to 5, which is considerably greater than predicted by Paré (1) or found for Ni (18). The explanation for these large widths may lie in the suggestion of Thompson and Holmes (36) that the Bordoni peak in Cu is a superposition of several peaks.

This leads to an important point about the peak magnitudes. Thompson and Holmes drew their conclusion from observations on the modulus defect of a single crystal of Cu, extended (probably) only into Stage I. The accompanying internal friction data showed only a small Bordoni peak on a rising background, hardly recognizable, but using the modulus defect data, and assuming that to each step in the modulus there corresponded a Debye internal friction peak according to the standard anelastic solid (37), they calculated an internal friction curve which was in fairly good agreement with a peak observed in a polycrystalline specimen. That is, the single crystal showed much the same dislocation strain as the polycrystalline specimen, but there was much less accompanying internal friction in the single crystal. Similarly, Ni crystals (12) show much smaller peaks than polycrystals, (18,19), as do Ag (28) and Al (38). The explanation is that in the single crystals σ_i fails to fulfill the Paré criterion [2]. Since the measurements are made in the unloaded condition, two specimens may have the same dislocation density, and thus the same flow stress, but relax to different dislocation arrangements, with different σ_i, on unloading. Presumably, the grain boundaries, and the concurrent geometric dislocations (27,39), which may be pileups, inhibit the relaxation of σ_i on unloading. In confirmation of this view, the peak heights of polycrystalline Ni specimens are accounted for by the Seeger-Paré model (12), so it is the single crystals that are

146

deficient in damping, just as they were for Thompson and Holmes (36).

Attempts have been made (28,38) to draw quantitative conclusions concerning the Bordoni peak by introducing a known density of dislocations by bending single crystals to a radius R. The geometrical dislocation densities ρ_g (ρ_g = 1/bR) achieved with R of 1 to 2 cm, correspond, using equation [3], to flow stresses of the order of 10^{-1}kg/mm^2, which is much less than the usual critical resolved shear stress, and so almost negligible with respect to the statistical dislocations introduced by work-hardening during the bending. The point may be made even more strongly. The only requirement of bending is that a Burgers circuit around the crystal have a net closure failure corresponding to ρ_g, a condition which imposes little restriction on the dislocations when $\rho \gg \rho_g$.

So far, we have accounted for the rise of the Bordoni peak. The fall at higher deformation must be associated with the onset of another process. At least for Ni, the fall of the Bordoni peak (Peak X) is accompanied by a tendency for change of sign of ΔT, i.e. an increase, and a reduction of width. The data of Venkatesan and Beshers (12) have been plotted in Figs. 9 and 10, to show this tendency. The fall of the peak, the positive ΔT, and the narrowing are all indicative of a decrease in σ_i, on the Seeger-Paré model. Such a decrease is consistent with our knowledge of Stages III. Stage III of plastic deformation has a lower rate of work-hardening than Stage II, and there is much evidence that the process responsible for the lower rate is cross-slip of screw dislocations. It is entirely plausible that the introduction of a new mode of dislocation motion should lead to greater relaxation on unloading and thus to lower values of σ_i.

One of the objections to the previous picture is that Peak X occurs at too high a temperature, and therefore too high an activation energy, to be a Bordoni peak. An answer to this objection is now at hand. Harrison (40) has recently suggested that the dislocations in Ni and Al are not split into partials, but undergo a change of structure in the core to lower their energy, keeping b equal to a lattice translation vector. It is also true that the Bordoni peak in Al also occurs at higher temperatures than might be expected, above those of Cu, Ag, and Au. (See Table 1) Now $2W_k$ varies approximately as μb^3 (5,6), and so a change from the partial to the whole dislocation in fcc ought to increase the activation energy by $3^{3/2} \approx 5.2$. However, if both partials must form double kinks, making a total of four kinks, the energy increase is only $3^{3/2}/2$ or 2.6. The activation energy for Peak X in Ni is about

TABLE 1

Values for $2W_k$, the Activation Energy, in eV.
(The values are from (1) except for Ni (41).)

Metal	Ag	Cu	Au	Pt	Al	Pd	Ni
$2W_k$	0.12	0.14	.16	.19	.25	.26	.4

0.4 eV, so a reduction of 2.6 would give an energy of about 0.16 eV, intermediate between, Au and Pt. The corresponding reduction for Al would be from 0.25 eV to about 0.1 eV, where it would fit very nicely below Ag. With the suggested correction the sequence of increasing $2W_k$ corresponds, with two exceptions (Cu, Au; Pt, Pd), to the sequence of melting temperatures. The activation energies are not sufficiently well-determined to say more. It appears then that Harrison's suggestion may explain the anomalously high values of $2W_k$ for Ni and Al. A more complete, but unpublished, correlation of $2W_k$ with μb^3 has been given by Chambers (42).

We may push this line of thought one step further: the suggested activation energy for Ni split into partials, 0.16 eV, is in the range found by Sommer (41) for Peak Y (0.16 to 0.22 eV). It seems entirely possible that Peak Y is a Bordoni peak for split dislocations of Ni, these split dislocations being formed under special conditions in Stage III. This would explain why Peak Y grows as Peak X falls. A peak somewhat similar to Peak Y also occurs in Al (12).

Hasiguti Peaks

The Hasiguti peaks are relatively unstable peaks which occur at temperatures rather above the Bordoni peak. Numerous studies agree that these peaks are attributable to dislocation- (point imperfection) interactions. A number of models along these lines, put forward by Hasiguti and others have been reviewed by Hasiguti (43).

The reason for mentioning Hasiguti peaks in this paper is that Fig. 7 shows a very close correlation between the Bordoni peak and P_1 in their dependence on ϵ, suggesting that these two peaks arise from the same dislocation segments undergoing assentially the same motions, but with different rate-limiting processes. Indeed, it has been shown (43) that there is an inverse relationship between the two peaks, in that when one grows, the other decays, in the course of annealing at constant dislocation structure. This inverse relationship is interpreted as meaning that the addition of a point defect converts a particular

dislocation segment from participating in the Bordoni peak to participating in P_1 (or one of
the other peaks), the subsequent removal of the point defect restoring participation in the
Bordoni peak. A similar picture has been found for Au (44,45), and other materials. These
observations suggest that the Seeger-Paré model for the Bordoni peak may apply, with appropriate modifications, to the Hasiguti peaks as well.

Fig. 11 shows one way in which Fig. 2 may be modified to give a Hasiguti peak instead
of a Bordoni peak. The presence of a point defect prevents a newly-formed double-kink from
spreading to the ends of the line. Only when the double-kink is thermally activated past the
point defect, can it sweep out sufficient area to satisfy Paré's criterion [2]. Otherwise
the final state of the process is that in which the double-kink is held up at a point defect,
with ΔG_f considerably greater than kT leading to greatly reduced internal friction according
to the discussion preceding equation [1]. A point defect situated as either of those shown
in Fig. 11 will act to change the activation energy and so the relaxation time, but not the
relaxation strength, which depends only on the end points. Let us assume further that for
each of the specimens of Fig. 7 the proportion of dislocation segments with a suitable point
defect was the same. Then this modified Seeger-Paré model explains the close relationship
between P_1 and the Bordoni peak shown in Fig. 7.

Further evidence in support of this view may be found in the work of Koiwa and
Hasiguti (46). They reported curves of the height of P_1 versus ε which show a maximum like
those of Figs. 7 and 8; in particular, their 1965 paper gives a curve very similar to that of
de Fouquet et al. in Fig. 7. These data then help to substantiate the contention, based on
Fig. 8, that the Bordoni peak in Cu does fall off at higher values of ε and σ_f.

A slightly different kind of support may be found in the work of Tung and Sommer (47)
on Ti. They report a peak of Hasiguti type with a hydrogen pair (H_2) as the point imperfection. They obtained an excellent straight line on plotting the height of their peak versus
$\varepsilon^{\frac{1}{2}}$. They interpret $\varepsilon^{\frac{1}{2}}$ as proportional to $\rho^{\frac{1}{2}}$, which in turn is proportional to ρ on the network model, in agreement with the prediction of the Seeger-Paré model.

Turning to the relaxation times, the Hasiguti peaks differ from the Bordoni peaks in
having a relatively well-defined peak temperature, although some sensitivity to microstructure
and history is left, particularly for P_3. This is shown in (43), and also in de Batist's book
(48).

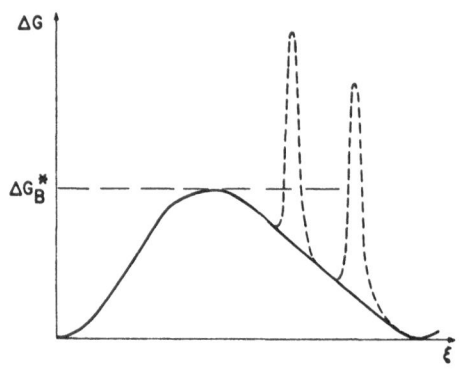

FIG. 11

Modification of Fig. 2 to give
a model of Hasiguti Peak

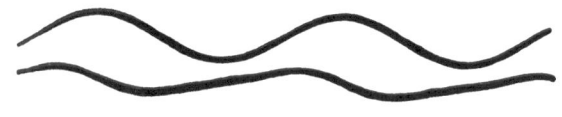

FIG. 12

Driving signal (sinusoidal) and
pick-up (distorted) with fundamental
balanced out (approximately) for a
Ti specimen.

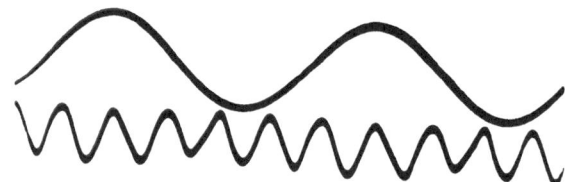

FIG. 13

Driving signal (sinusoidal; long period)
and pick-up (distorted; short period)
with fundamental balanced out for a
brass specimen.

150

There are several possible explanations. First, the basic argument about the shift of the Bordoni peak rests on the dependence of the activation energy on σ as shown in Fig. 3. That argument affects the Bordoni peak barrier shown in Figs. 2 and 11, but not the Hasiguti peak barrier directly. There is, however, a stress dependence of ΔG represented by the long linear fall of ΔG vs. $\dot{\eta}$ to the right of the peak in Fig. 3, in the region of the point defects in Fig. 11. Since the activation energy must be measured from the base line to the peak, we do expect some variation of the activation energy. There will be considerable variation within one sample, if the defects are distributed over the whole of the segment, leading to a rather broad peak, but between samples there should be rather less variation in ΔG^* for the Hasiguti peaks than for the Bordoni peaks. Since the Hasiguti peaks also have higher values of ΔG^* than the Bordoni peaks, the relative variation $\delta \Delta G^*/\Delta G^*$, should be considerably less than for the Bordoni peaks, leading to smaller values of $(\Delta T_H/T_H)$.

Many details of the model, such as the placement of the point imperfection and the nature of its interaction with a kink, need quantitative formulation before comparing the predictions of the model to experiment. The presence of the imperfection may even modify the process of double-kink generation.

The modified Seeger-Paré model just presented is related to two previous models. Okuda and Hasiguti (49) suggested that thermal unpinning might occur along a dislocation running in a close-packed direction with double-kink formation modified by the presence of a point imperfection. Hasiguti (50) later suggested the diffusion of a kink between pinning points, ignoring the Peierls stress; this theory was used by Tung and Sommer (47) to fit their data on Ti-H. The model presented above is a combination of these two in that double-kink generation is an essential first step, which is followed by a diffusive motion of one of the kinks, either pre-existing or newly formed, past a point imperfection.

However, the model suggested here is not one involving thermal unpinning, in Hasiguti's phrase. There is no need to consider motion of the point imperfection. The model is still one of a dislocation segment, which has become a bistable element under the influence of the Peierls potential and σ_i (Fig. 1). The point imperfection only changes the barrier between the two stable states, not the states themselves, at least in the first approximation. For the element to be bistable, Paré's condition [2] must be satisfied. The point imperfection will itself contribute to σ_i, and so may possibly make a contribution towards fulfilling [2].

This contribution is probably not large because of the rapid fall off of the stresses around a point imperfection.

Criticisms of Experimental Method

We turn now to a brief review of some serious shortcomings in the treatment of plastic deformation as an experimental variable. As previously, the Seeger-Pare model will be adopted as a point of view.

We have already indicated that σ_f is a much better indicator than ϵ of the internal state of a deformed specimen. Many investigators report only ϵ, and not even a grain size, so it becomes quite difficult to use their data in comparison with theory. Thermal history may be quite important; even the temperature of anneal before deformation may be significant. The temperature of deformation, and the specimen's subsequent history, are variables insufficiently explored for many materials.

With respect to mode of deformation and of measurement, serious shortcomings are noticeable. It would be most desirable to have the original deformation a uniform one, with a known orientation if the specimen is a single crystal. It is equally desirable that the measuring stress be uniform, with a known orientation. The only lower frequency investigation that meets these criteria is that of Venkatesan and Beshers (12), who used a simple tensile deformation and a vibrational specimen configuration which is essentially a "longitudinal pendulum" in which the vibratory stress is nearly uniform over the length of the specimen. These criteria may also be met easily in pulse echo measurements above 1 MHz (35).

Many practical reasons, such as the achievement of a certain frequency range, have led investigators to use other methods with a non-uniform vibratory stress. If the sample was deformed homogeneously to start with, this may present no problem. If the sample was deformed inhomogeneously, as in torsion, or bending, and the measuring stress wave was not of the same type, then the results may refer only to some particular part of the sample which is uncharacteristic of it. For example, if a cylinder were deformed in torsion and then examined by longitudinal waves, the lightly deformed central region would be equally weighted in the results with the more heavily deformed outer region.

This criticism of inhomogeneous deformation becomes stronger when we remember the importance of σ_i in the Seeger-Pare theory. An inhomogeneously deformed specimen may be expected to have variations in σ_i, leading to variations in ΔG^* across the sample.

Compressive deformation is often used, and taken to be equivalent to tensile deformation, but it is not really so (51) for plastic deformation, and barreling, which is inhomogeneous, will occur unless precautions are taken.

A number of the specimens used at Cornell by Paré (7) and others were deformed by rolling. The still more serious criticism here is that rolling leads to a complex distribution of σ_i on a macroscopic scale (52), which surely affects the value at a dislocation segment as well.

High Amplitude Vibrations; Acoustic Emission

Internal friction observations at high amplitudes of oscillation may be expected to have a particularly close connection with plastic deformation. Indeed, at sufficiently large amplitudes plastic deformation does occur in the course of the vibratory experiment (53), leading eventually to fatigue. Before fatigue is reached, a number of phenomena are observed. We give a brief, non-critical review of this area, followed by a brief report of some recent experiments.

The Granato-Lücke (54) damping, attributable to the breaking away of dislocations from pinning points, is well known. The original theory was valid only for $0°$ K. It has been extended to finite temperatures by Teutonico, Granato, and Lücke (55), together with Schlipf (56). An extensive and detailed working-out of the consequences has been given in a series of papers by Blair, Hutchison and Rogers (57) and by Blair (58). These papers are concerned with a regime in which the vibratory stress is large enough so that the dislocations can break away from pinning points, but not from network nodes; that is the stress is still small enough that dislocation sources do not operate. A result confirmed by the analysis of Blair et al. (57) is that the Granato-Lücke damping goes through a maximum as the stress increases. A fall in damping followed by a further rise at higher amplitudes has often been observed (53).

It is to this higher stress regime, in which relatively little work has been done, that we wish to draw attention. Mason (53) reviewed much of the older work in which it was first suggested that the damping beyond the Granato-Lücke range is attributable to the operation of Frank-Read (59). A more recent effort along these lines is that of Peguin, Perez, and Gobin (60), who, working with Al at frequencies near 2 Hz, decomposed the damping into two parts, Δ_H and Δ_P. Δ_H they associated with thermally assisted unpinning, their calculation along the lines of (55) preceding (57). Δ_P was attributed to plastic strain and analyzed

153

assuming that the plastic strain rate is thermally activated, the activation energy being reduced by the effective stress acting on the dislocations. The coefficient which relates activation energy and stress is called the activation volume V^* (it would be better to express it as bA^* where A^* is then called the activation area). Peguin, Perez, and Gobin showed that their data could be fitted by their theoretical expressions. They called attention particularly to the overlap between the regions of Δ_H and Δ_P which sometimes occurs. A good separation was found in an alloy of Al-3%Zn, annealed, furnace-cooled, and strained 2.5%.

This point was pursued by Burdett (61), using polycrystalline Fe near 1 Hz. He confirmed the separation into Δ_H and Δ_P, deriving from his results a value of V^* in agreement with one obtained from more conventional testing methods. Other authors whose work contributed importantly, Gelli (62), Bratina (63), and Baker and Carpenter (64), are mentioned in these papers (60,61).

Mason has pioneered the use of high amplitudes at 20 kHz, working out the operation of a practical acoustical transformer in great detail (65). This apparatus was applied by Mason and Mac Donald (66) to study the changes in damping and modules defect which precede ultrasonic fatigue in brass. They also made metallographic observations and succeeded in correlating the two sets of observations. The initial rise in damping was found to coincide with the appearance of slip bands, further changes being associated with the production of microcracks from slip bands, and so on.

Some very recent experiments at Columbia carried out by W.P. Mason, M.C. Jon, J.T. Kuo, and one of the authors (DNB), have attempted to use these high amplitude oscillations for the study of acoustic emission. The original thought was that emission of sound from events of plastic deformation, such as the operation of a slip band, would occur just at the peak of the oscillating stress, for stress waves of a suitable amplitude, just large enough to cause the emitting event. It would then be possible to look only for those events which occur near the stress peaks, and particularly those which occur repetively. Such a scheme supposes that any emission will take place at a much higher frequency than the driving frequency of 20 kHz, but there is ample basis for believing that such is the case.

We have actually made some observations of this sort, which will be reported elsewhere, but in the course of these experiments we have observed another phenomenon, the strong generation of high harmonics of the driving wave in the high amplitude region.

154

Harmonic generation is of course expected in non-linear behavior, and a dependence of damping on amplitude is a manifestation of non-linear behavior. Harmonic generation by non-linear bowing motion of dislocations has been discussed theoretically, and compared with lattice anharmonicity, by Hikata and Elbaum (67). They found that the second harmonic was generated about equally by the two mechanisms, while dislocations generated the third harmonic much more strongly. Hikata, Sewell, and Elbaum (68) applied a static bias stress to several crystals and observed the resulting changes in the amplitude of the 3rd harmonic and the attenuation of the fundamental. The specimens differed in purity, orientation, and, eventually, in degree of plastic deformation. The changes observed were in qualitative agreement with the theory (67). The stress amplitude of vibration in these experiments was an order of magnitude less than the yield stress, corresponding to strains of the order of 10^{-5}.

The recent work at Columbia has been done at much higher strains, with a much richer generation of harmonics. Fig. 12 shows the driving signal and the detected signal from a sample of Ti at a vibratory strain of 2×10^{-4} which showed no amplitude dependence. The detected signal was passed through a filter to reduce the fundamental, to make harmonics more visible, but there is not much to see. By comparison, Fig. 13 shows similar signals from brass at strain amplitude of 4×10^{-5}, in which the 5th harmonic is plainly dominant. At higher amplitudes, it is not necessary to filter out the fundamental in order to see the harmonics, although it helps for the analysis.

We have been able to show to date some correlation of harmonic generation with processes of deformation. For example, at the onset of twinning in Sn, the harmonic generation increases extremely rapidly. This technique seems to offer more promise than just measurements of damping and modulus defect in identifying and characterizing processes of deformation.

Summary

By focusing on plastic deformation as a variable, we are able to correlate and explain a great many of the observations on the Bordoni peak in fcc metals, and apparently related peaks in bcc and hcp metals. The characteristics of these peaks are: (1) the peak increases monotonically, with flow stress, at least through Stage II of plastic deformation; this increase is found to be linear where the data are sufficient to establish a law; (2) the peak temperature depends somewhat (up to 20 to 30%) on σ_f; and (3) the peak width also varies with

σ_f. These characteristics were explained using the Seeger-Paré model, in which the linear increase of peak height arises from the factor $\rho\ell$ in the relaxation strength, while the variations in peak temperature and width arise through the dependence of activation energy on stress, and the identification of σ_i as the important component of stress. These conclusions are reinforced by the discussion on Hasiguti peaks.

In this review, to a large extent, we have used information obtained from studies of plastic deformation to help understand the Bordoni peak. The question naturally arises, what does all this knowledge of the Bordoni peak tell us about plastic deformation? We first notice that the actual dislocation motion involved in the Bordoni peak is quite small, and may possibly not be closely connected with the thermally activated processes which occur during plastic deformation. For example, when the critical step is surpassing a localized obstacle, the thermal activation occurs at a place where both dislocation curvature and the obstacle combine to give an internal stress which is varying quite rapidly in the vicinity (69). There may be as much of the Hasiguti peak as of the Bordoni peak in such a process. Similarly, the thermally activated cutting of a forest dislocation may or may not be representable in terms of kink generation and mobility. The important connection between the Bordoni peak and the large scale plastic deformation arises through Seeger-Paré model in the relation of the factor $\rho\ell$ in the relaxation strength to the flow stress, and through the variation with σ_i of the peak temperature and width. The successes of the Seeger-Paré model, outlined above, must give us considerable confidence in the basic picture of elementary plastic deformation. There are mobile dislocations arranged in a network with average length L, and in the absence of pinning points the mobile length ℓ approaches L. There is an internal stress field σ_i, and we get some limited information concerning it. A more detailed theory, combined with careful experiments, might lead to even more information about this very important subject.

The high amplitude phenomena described in the last part involve elementary processes of plastic deformation, so we expect to build towards a more detailed understanding of them on the foundations of the Bordoni peak and Hasiguti peak studies.

Acknowledgements

This work was supported by the Army Research Office, Durham, North Carolina.

References

1. D.H. Niblett, Physical Acoustics, Vol. III A, Ed. W.P. Mason, Academic Press, N.Y. (1966).

2. A. Seeger, Phil. Mag. 1, 651 (1956).

3. A. Seeger, J. de Phys. 32, C2-193 (1971).

4. J.E. Dorn and S. Rajnak, Trans. AIME 230, 1052 (1964).

5. P. Guyot and J.E. Dorn, Can. J. Phys. 45, 983 (1967).

6. U.F. Kocks, A.S. Argon and M.F. Ashby, Thermodynamics and Kinetics of Slip, to be published.

7. V.K. Paré, J. Appl. Phys. 32, 332 (1961).

8. A.D. Brailsford, Phys. Rev. 122, 778 (1961); 128, 1033 (1962); 137A, 1562 (1965); 139A, 1813 (1965); J. Appl. Phys. 36, 3941 (1965).

9. G. Alefeld, J. Filloux, and H. Harper, in Dislocation Dynamics, Ed. A.R. Rosenfield, McGraw-Hill Book Company, New York (1968).

10. F.R.N. Nabarro, Z.S. Basinski and D.B. Holt, Adv. in Phys. 13, 193 (1964).

11. H. Conrad, K. Okasaki, V. Gadgil and M. Jon, in Electron Microscopy and Structure of Materials, Ed. G. Thomas, Berkeley, University of California Press, (1972).

12. P.S. Venkatesan and D.N. Beshers, J. Appl. Phys. 41, 42 (1970).

13. G. Schoeck, Acta Met. 11, 617 (1963).

14. D.P. Petarra and D.N. Beshers, Acta Met. 15, 791 (1967).

15. W. Koester, L. Bangert and R. Hahn, Arch. Eisenhütt. Wes. 25, 569 (1961).

16. J.C.M. Li, Can. J. Phys. 45, 493 (1967).

17. H. Mecking and K. Lücke, Z. Metallk. 60, 185 (1969)
    H. Mecking, Acta Met. 7, 279 (1969).

18. A.W. Sommer and D.N. Beshers, J. Appl. Phys. 37, 4603 (1966).

19. J. deFouquet, P. Boch, J. Petit and G. Rieu, J. Phys. Chem. Solids 31, 1901 (1970).

20. L.J. Bruner and B.M. Mecs, Phys. Rev. 129, 1525 (1963).

21. S. Okuda, Phys. Rev. 34, 3107 (1963); J. Phys. Soc. Japan 18, SI, 187 (1962?).

22. O. Mercier, A. Isoré and W. Benoit, Scripta Met. 6, 961 (1970); Helv. Phys. Acta 45, 858 (1972).

23. J.W. Corbett, Solid State Physics, Suppl. 7, Ed. F. Seitz, D. Turnbull, Academic Press (1966).

24. D.H. Niblett and J. Wilks, Phil. Mag. 1, 415 (1956).

25. D.H. Niblett and J. Wilks, Phil. Mag. 2, 1427 (1957).

26. B. Mecs and A.S. Nowick, Acta Met. 13, 771 (1965).

27. A.W. Thompson, M.I. Baskes and W.F. Flanagan, Acta Met. 21, 1017 (1973).

28. B. Mecs and A.S. Nowick, Phil. Mag. 17, 509 (1968).

29. P.A. Grandchamp, J. de Phys. 32, C2-229 (1971).

30. R. Gibala, M.K. Korenko, M.F. Amateau and T.E. Mitchell, J. Phys. Chem. Solids 31, 1889 (1970).

31. R.H. Chambers and J. Schultz, Acta Met. 10, 466 (1962).

32. K. Takita and K. Sakamoto, Scripta Met. 4, 403 (1970).

33. J. Petit, M. Quintard, R. Soulet and J. deFouquet, J. de Phys. 32, C2-215 (1971).

34. D.O. Thompson, in Reinstoffprobleme, Ed. E. Rexer, Akademie-Verlag, Berlin (1957).

35. G.R. Brown and D.H. Niblett, J. Phys. D6, 809 (1973).

36. D.O. Thompson and D.K. Holmes, J. Appl. Phys. 30, 525 (1959).

37. A.S. Nowick and B.S. Berry, Anelastic Relaxation in Crystalline Solids, New York, Academic Press (1972).

38. A.F. Mayadas, Ph.D. Thesis, Cornell University, (1966).

39. M.F. Ashby, Phil. Mag. 21, 399 (1970).

40. E.A. Harrison, Acta Met. 21, 1111 (1973).

41. A.W. Sommer, Ph.D. Thesis, Columbia University (1965).

42. R.H. Chambers, Bull. Am. Phys. Soc. 11, 217 (1966); T. Trozera, T.E. Firle and R.H. Chambers, AEC Report GA-7978 from General Dynamics (July 29, 1967).

43. R.R. Hasiguti, 3. Intern. Symposium, Reinstoffe in Wissenschaft and Technik, Dresden, 1970; J. Less-Common Metals 28, 249 (1972).

44. B. Bays, W. Benoit and P.A. Grandchamp, J. de Phys. 32, C2-153 (1971).

45. W. Benoit, B. Bays, P.A. Grandchamp, B. Vittoz, G. Fantozzi, J. Perez and P. Gobin, J. Phys. Chem. Solids 31, 1907 (1971).

46. M. Koiwa and R.R. Hasiguti, Acta Met. 11, 1215 (1963); 13, 1219 (1965).

47. P.P. Tung and A.W. Sommer, NA-72-305 Tech. Report of North American Rockwell Corp. to Office of Naval Research (1972).

48. R. de Batist, Internal Friction of Structural Defects in Crystalline Solids, Amsterdam, North Holland (1972).

49. S. Okuda and R.R. Hasiguti, Acta Met. 11, 257 (1963).

50. R.R. Hasiguti, Phys. Stat. Sol. 9, 157 (1965).

51. E. Schmid and W. Boas, Plasticity of Crystals, London, F.A. Hughes (1950): translated from Kristallplastizität, Springer, Berlin (1935).

52. G.E. Dieter, Mechanical Metallurgy, New York, McGraw-Hill (1961).

53. W.P. Mason, in Resonance and Relaxation in Metals, Plenum Press, New York (1964).

54. A.V. Granato and K. Lücke, J. Appl. Phys. 27, 583 (1956).

55. L.J. Teutonico, A.V. Granato and K. Lücke, J. Appl. Phys. 35, 220 (1964).

56. A.V. Granato, K. Lücke, J. Schlipf and L.J. Teutonico, J. Appl. Phys. 35, 2732 (1964).

57. D.G. Blair, T.S. Hutchinson and D.H. Rogers, Can. J. Phys. 49, 633 (1971).

58. D.G. Blair, J. Appl. Phys. 43, 37 (1972).

59. D.N. Beshers, J. Appl. Phys. 30, 252 (1958).

60. P. Peguin, J. Perez and P. Gobin, Trans. AIME 239, 438 (1967).

61. C.F. Burdett, Phil. Mag. 24, 1459 (1971); J. Phys. D4, 2017 (1971).

62. D. Gelli, Phys. Stat. Sol. 12, 829 (1965).

63. W.J. Bratina in Physical Acoustics IIIA, Ed. W.P. Mason, Academic Press, New York (1966).

64. G.S. Baker and S.H. Carpenter, J. Appl. Phys. 38, 3557 (1967).

65. W.P. Mason in Microplasticity, Ed. C.J. McMahon, Jr., Interscience Publishers, New York, (1968).

66. W.P. Mason and D.E. MacDonald, J. Acoustical Soc. Am. 51, 894 (1972).

67. A. Hikata and C. Elbaum, Phys. Rev. 144, 469 (1966).

68. A. Hikata, F.A. Sewell and C. Elbaum, Phys. Rev. 151, 442 (1966).

69. D.N. Beshers, Scripta Met. 5, 469 (1972).

INTERACTIONS OF DISLOCATIONS

WITH

PHONONS AND ELECTRONS

PHONON AND ELECTRON DRAG OF DISLOCATIONS

V.I.Alshits and V.L.Indenbom

A.V.Shubnikov's Institute of Crystallography
Academy of Sciences of the USSR, Moscow

Introduction

In processes of the internal friction as well as in the plastic deformation of crystals the dislocation motion can be limited by two qualitatively different phenomena: the thermofluctuation overcoming the barriers and the dynamic dislocation drag caused by the energy transfer to different elementary excitations (phonons, electrons, excitons etc.). The first process is promoted with the temperature increase, for the fluctuation waiting time decreases the second one is impeded because of the increase of the density of the elementary excitations gas. The dynamic drag manifests itself not only at the dislocation motion between and over the barriers which is characteristic of the amplitude-independent internal friction and of the fast dislocation mobility but also at small-scale thermal oscillations during the overcoming the barriers (e.g. in case of amplitude-dependent internal friction).

Only several years ago the questions of the existence, role and major mechanisms of the dynamic drag were under discussion. At present the progress of the theory permits already to answer reliably the former questions about the drag mechanisms and to explane the experimental data available. The theory predicts also a number of new effects not yet observed experimentally which could be of significant physical interest. In the report the present state of the problem of the phonon and electron dislocation drag is analysed taking into account the recent works by the authors which provide a general approach to the problem.

A. Phonon Mechanisms of Dislocation Drag

Let us express the elastic field of a dislocation moving with the velocity v small as compared to the sound velocity c as a packet of plane waves

$$\mathcal{E}(\bar{r}, t) = \mathcal{E}(\bar{r} - \bar{v}t) = \int \frac{d\bar{q}}{(2\pi)^3} \, \mathcal{E}_{\bar{q}} \, e^{i(\bar{q}\bar{r} - \Omega_{\bar{q}}t)} \quad . \tag{A.I}$$

Here $\mathcal{E}_{\bar{q}}$ is the Fourier-transform of the static dislocation strain field, $\Omega_{\bar{q}} = \bar{q}\bar{v}$. Evaluating the energy dissipation per unit length of the dislocation as the sum of damping of separate waves of the packet (A.I) one has

$$D = \int \frac{d\bar{q}}{(2\pi)^3} \, \Omega_{\bar{q}}^2 \, \eta(\bar{q}, \Omega_{\bar{q}}) \, \mathcal{E}_{\bar{q}} \, \mathcal{E}_{\bar{q}}^* \tag{A.2}$$

where $\eta(\bar{\kappa}, \omega)$ is the effective viscosity (the imaginary part of the elastic modulus) for the wave with the wave vector $\bar{\kappa}$ and frequency ω. To the dissipation (A.2) corresponds the dynamic drag coefficient measurable directly in experiment:

$$B = D / v^2 \quad . \tag{A.3}$$

The aim of the theory thus consists in the estimation of the viscosity taking into account the spatial and time dispersion.

A.I. Phonon Viscosity

The concept of the phonon viscosity was introduced in the works on the ultrasound attenuation in crystals. When re-establishing the equilibrium of the phonon gas perturbed by the sound wave the phonons behave like a gas with an effective viscosity $\eta_{ph} \sim E\tau_{ph}$ (E is the thermal energy density, τ_{ph} is the phonon relaxation time). Mason [I,2] used the concept of the phonon viscosity to evaluate the phonon drag of dislocations putting

$$\eta(\bar{q}, \Omega_{\bar{q}}) = \eta_{ph} = const, \qquad D = \eta_{ph} \int \frac{d\bar{q}}{(2\pi)^3} \, \Omega_{\bar{q}}^2 \, \mathcal{E}_{\bar{q}} \, \mathcal{E}_{\bar{q}}^* \quad . \tag{A.I.I.}$$

Since near the dislocation the deformations increase inversely proportionally to the distance the integral (A.I.I.) diverges quadratically at the upper limit and has to be truncated at some value $q_m \sim 1/r_0$. In his calculations Mason proceeded from the fact that r_0 has the meaning of the dislocation core radius and should therefore be of the order of the lattice parameter a. It is easy to prove however that Mason's estimate is incorrect in principle and is invalid even as an order of magnitude. Actually, the integral (A.I.I) is determined at the upper limit while namely for large q it is inadmissible to neglect in (A.2) the spatial dispersion of the phonon viscosity, for it leads to incorrect results.

The concept of the phonon viscosity without dispersion has the sense only for adiabatic perturbing fields with characteristic scales of time and spatial inhomogeneities large compared to the relaxation time τ_{ph} and the mean free path $l_{ph} = c\tau_{ph}$ of phonons, respectively. For a sound wave with the wave vector $\tilde{\kappa}$ and frequency $\omega = c\kappa$ this restrictions

$$\chi\kappa^2 \ll \omega, \qquad \kappa l_{ph} \ll 1, \qquad \omega\tau_{ph} \ll 1 \qquad (A.I.2)$$

are equivalent (here $\chi \approx \frac{1}{3}c l_{ph}$ is the temperature conductivity). As applied to partial waves from the packet (A.I) these inequalities however are not equivalent. In this case this is the adiabaticity condition $\chi q^2 \ll \Omega_q$ that is the strongest restriction. It necessitates to exclude from consideration the most essential region $q \gtrsim \frac{v}{c} l_{ph}^{-1}$ which corresponds to the cut-off radius $r_0 \sim \frac{c}{v} l_{ph}$. Thus, formula (A.I.I) may be used to evaluate the damping of the longest-wave part of the packet (A.I) whose contribution to the dislocation drag is negligible - Mason's estimate should be reduced by at least a factor of $(c l_{ph}/va)^2$!

A.2. Thermoelastic Dissipation

The thermoelastic dissipation can be considered as a particular case of the manifestation of the phonon viscosity when the deformations are reduced. to the dilatation, the change of the phonon spectrum to the local heating and cooling of the crystal and the equilibrium reestablishing in the phonon subsystem - to heat fluxes. The well-known calculations by Eshelby [3] were modified by Lothe [4] who pointed out correctly the macroscopic character of the thermoelastic mechamism and its inapplicability to the region $R < l_{ph}$. Dynamic heating and cooling of the crystal during the dislocation motion were evaluated by Lothe with the aid of the Grüneisen constant which gave the underestimated value of the energy dissipation. In Sec.A.5 a comparison of this mechanism with others based on a more general treatment will be given.

A.3. Phonon Scattering

The simplest description of processes occuring arbitrarily close to the moving dislocation was suggested by Leibfried [5] who considered the phonon scattering at the dislocation.

As a result of the simple semiqualitative treatment Leibfried obtained the following estimate of the dislocation drag coefficient

$$B = \frac{aE}{10\,c} \; . \qquad\qquad (A.3.1)$$

Leibfried did not calculate the cross-section of the phonon scattering at the dislocation assuming that its radius is of order of the lattice parameter. In this sense formula (A.3.1) is relevant for any scattering mechanism with the cross-section radius of order a . As Nabarro [6] has shown two mechanisms have to be distinguished: the phonon wind (the nonlinearity mechanism) caused by the anharmonicity of the crystal and the flutter mechanism connected with phonon re-radiation by the dislocation vibrating in the thermal field of the lattice. According to Lothe [7] at the room temperature the two effect are of the same order and can be estimated using formula (A.3.1). However this conclusion is valid only if one evaluates the anharmonicity through the Grüneisen constant which underestimates the result (formula (A.3.1) corresponds to values of B which are approximately by a factor of ten smaller than those observed experimentally).

A quantum-mechanical calculation of the phonon wind taking into account the crystal anharmonicity through the Murnaghan moduli was first performed in the work by Alshits [8] . According to [8]

$$B_w \approx \left[4 + \left(\frac{|n|}{\mu} - 6 \right)^2 \right] \frac{\hbar}{a^3} \left(\frac{\kappa_D a}{2\pi} \right)^5 f(T/\theta) \; . \qquad (A.3.2)$$

Here

$$f(x) = x^5 \int_0^{1/x} \frac{e^t \, t^5 \, dt}{(e^t - 1)^2} \; , \qquad\qquad (A.3.3)$$

n is the Murnaghan modulus, μ the shear modulus, κ_D the Debye boundary of the phonon spectrum, θ the Debye temperature.

The calculation of the flutter mechanism made in a recent paper by Alshits and Sandler [9] allowed first to compare the two scattering mechanisms considering their temperature run. According to [9]

$$B_{fl} = \hbar \kappa_D^3 f_1(T/\theta), \qquad f_1(x) = x^3 \int_0^{1/x} \frac{e^t t^3}{(e^t - 1)^2} \frac{dt}{\widehat{n}^2 + \ln^2[(xt)^{-2} - 1]} . \qquad (A.3.4)$$

From formulae (A.3.2) and (A.3.4) it is seen that at high temperatures both effects are linear in T . At low temperatures: $B_w \sim T^5$ and $B_{fl} \sim T^3$. Correspondingly the ratio of the drag coefficients in these temperatures ranges should tend asymptotically to the values

$$\frac{B_{fl}}{B_w} = \left(\frac{T_o}{T}\right)^2 \quad (T \ll \theta), \qquad \frac{B_{fl}}{B_w} = d = const \quad (T \gg \theta) . \qquad (A.3.5)$$

A numerical estimation shows that, for instance, for copper $T_o \approx 15°K$, $d \approx 1/15$. Hence, at low temperatures ($T \ll \theta/10$) the flutter mechanism should prevail while at high temperatures the phonon wind dominates.

A.4. Role of the Dislocation Core

At high temperatures formulae (A.3.2) and (A.3.4) are much less reliable for with lowering the average wave length of phonons the problem becomes more and more sensitive to both the deviation of the phonon spectrum from the Debye model and the strain field structure in the dislocation core. The relevant corrections are not of great importance for the flutter mechanism describing the low temperature dislocation drag. But the estimate of the phonon wind dominating at high temperatures has to be corrected. In the next section we show that the consideration of the deviation of the phonon spectrum from the Debye model leads to a new channel of dissipation of "slow" phonons. Here we confine ourselves to a simple illustration showing how the consideration of the existence of the dislocation core affects the temperature dependance $B_w(T)$.

Adopting the model of Lothe [4] as a simplest dislocation core model we introduce a smooth truncation of the field of the dislocation at small distances:

$$\mathcal{E}(\vec{r}) = \mathcal{E}^0(\vec{r})(1 - e^{-r/\Lambda}) . \qquad (A.4.I)$$

Correspondingly, formula (A.3.3) which describes the temperature dependence of the phonon wind should be replaced by the expression

$$f_2(x) = x^5 \int_0^{1/x} \frac{dt \, e^t t^5}{(e^t - 1)^2} \frac{arctg \, \beta x t}{\beta x t} \qquad (A.4.2)$$

where $\beta = 2\kappa_D \Lambda$ (for typical values of κ_D and $\Lambda = 3a$: $\beta \approx 30$). It is seen from (A.4.2) that at low temperatures ($\beta x \ll 1$) the function $f_2(x)$ practically does not depend on β and coincides with $f(x)$. With rising the temperature the function $f_2(x)$ tends rather rapidly to the linear dependence: $f_2(x) \approx x/2\beta$.

A.5. Role of the Phonon Relaxation

The scattering mechanisms considered above were evaluated neglecting the phonon relaxation. This is permissible only in case $\Omega_{\vec{q}} \tau_{ph} \gg 1$, i.e. for the short wave part of the packet (A.I) corresponding to wave vectors $q \gg \frac{c}{v} \ell_{ph}^{-1}$. The estimates obtained are thus valid for sufficiently fast dislocations and low temperatures since the major contribution to the dissipation is due to short waves but they are inapplicable for slow dislocations and high temperatures when $\frac{c}{v} \ell_{ph}^{-1} > \kappa_D$. The evaluation of the contribution of the region $\frac{v}{c} \ell_{ph} < R < \frac{c}{v} \ell_{ph}$ intermediate with respect to the regions of the phonon scattering and phonon viscosity necessitates to solve a kinetic problem for phonons in the field of a moving dislocation. A consequent calculation of the phonon scattering at the dislocation taking into account the phonon relaxation and including the phonon wind, the phonon viscosity and the thermoelastic losses as particular cases was made in a recent paper by Brailsford [IO] and independently in a more general formulation in the work by Alshits and Malshukov [II] .

The generality of the treatment permitted the authors [IO,II] to clear the relative role of the phonon viscosity, the thermoelastic losses and the phonon wind. According to [IO,II] the phonon viscosity contributes negligibly to the dissipation compared to the thermoelastic losses which, in their turn, contribute less than the phonon wind. However, the establishing of such a hierarchy of mechanisms still not allow the authors of [IO,II] to state that did the relaxation processes are not essential compared to scattering processes, for the Debye approximation they used did not permit to analyze one more effect: the relaxation of "slow" phonons.

Due to the departure of the phonon dispersion law $\omega(\kappa)$ from the linearity in the short wave part of the spectrum with increasing the temperature the fraction of the "slow" phonons with low group velocities $V_{gr} = \frac{\partial \omega}{\partial \kappa}$ and, respectively, with short free pathes $V_{gr} \tau_{ph} K_D < 1$ increases. The contribution of these phonons to the dissipation not taken into account in the works [IO,II] differs qualitatively from that of the "Debye phonons" and determines the relaxational component of the dislocation drag at high temperatures. One can prove that the damping in the system of "slow" phonons can be characterized by the effective viscosity which "freeze out" with lowering the temperature according to the lowering of the "slow" phonon density. The total contribution to the dissipation due to the phonon wind and the "slow" phonon relaxation - the two mechanisms determining the dislocation drag at not too low velocities - results in the drag coefficient

$$B = \left[4 + \left(\frac{|n|}{4} - 6\right)^2\right] \frac{\hbar}{\theta a^3} \left(\frac{K_D a}{2\pi}\right)^5 \left[\varphi(T/\theta) + \lambda \, \psi(T/\theta)\right]. \qquad (A.5.I)$$

Here

$$\psi(x) = \frac{1}{x^2} \frac{e^{1/x}}{\left(e^{1/x} - 1\right)^2} \quad , \qquad (A.5.2)$$

$\varphi(x) = \beta f_2(x)$ is the function that is practically independent of β in the range $\beta x \gg 1$ we are interested in for at lower temperatures the flutter mechanism prevails; λ is a dimensionless phenomenological parameter of order unity which is a model characteristic of the real phonon spectrum to be determined from experiment. On fig.I the functions $\varphi(x)$ and $\psi(x)$ are presented and on fig.2 the dimensionless drag coefficient

$$B(T)/B(\theta) = \frac{\varphi(T/\theta)}{\varphi(1)} (1 - \Delta) + \Delta \cdot \psi(T/\theta) \qquad (A.5.3)$$

is plotted as a function of temperature the dependence being compared to the experimental data. The geometric sense of the parameter $\Delta = \frac{1}{\varphi(1)} \cdot \frac{\lambda}{1 + \lambda}$ is shown on fig.2.

A.6. Optic Phonon Contribution to the Energy Dissipation

So far in the analysis of the dynamic dislocation drag only the acoustic branches of the phonon spectrum were considered. For crystals whose phonon srectra contain also optic modes it is necessary to clear the role of the damping in the optic system. Usually optic modes are characterized

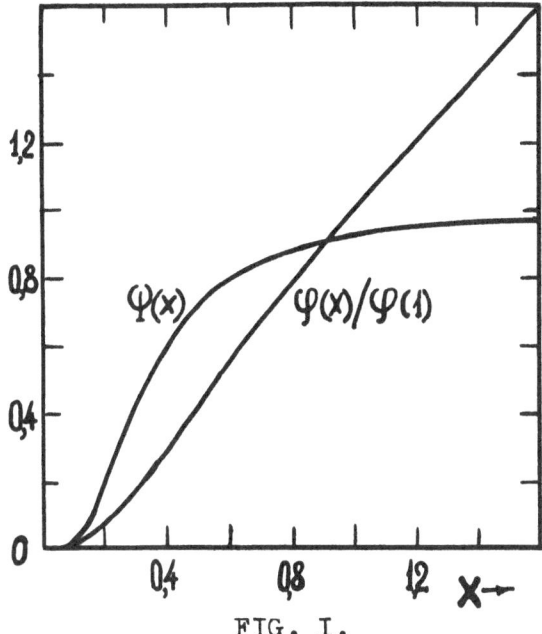

FIG. I.

The functions $\Psi(x)$ and $\varphi(x)/\varphi(1)$ describing the temperature dependence of the "slow" phonon relaxation and phonon wind.

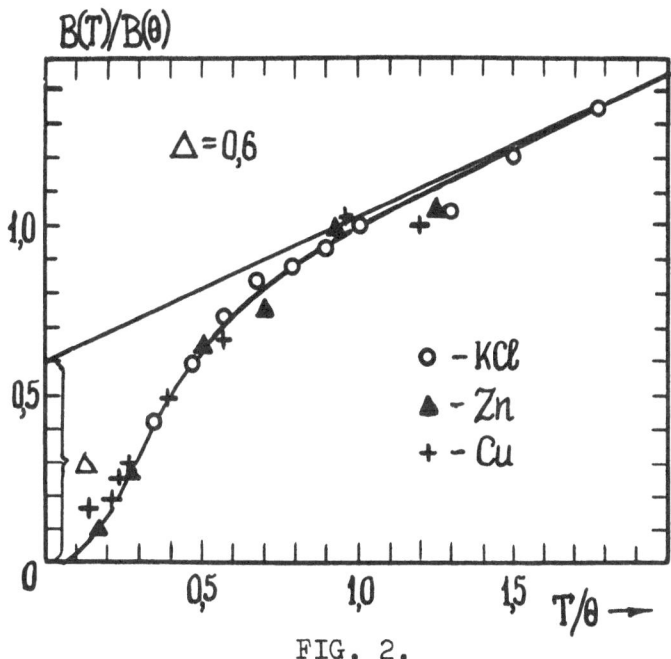

FIG. 2.

Comparison of the theory and experiment. Solid line is the theoretical curve. Experimental data correspond to measurements of fast dislocation mobility [33] in Zn and Cu crystals and of internal friction in KCl crystal (Startsev et al, to be published). Absolute values of B corresponding to (A.5.I) are in acodence with experimental data too, for example: $B_{Cu}^{theor}(\theta) = 1,6 \cdot 10^{-4} \frac{dyn \cdot sec}{cm^2}$ and $B_{Cu}^{exp}(\theta) = 2,1 \cdot 10^{-4} \frac{dyn \cdot sec}{cm^2}$.

by the dispersion law $\omega(\kappa)$ with the range of variation (ω_0, $\omega_0 + \Delta\omega$) in a narrow band $\Delta\omega \ll \omega_0$.

Analyzing dissipative processes in the system of optic phonons it is necessary, as before, to bear in mind the qualitatively different character of the interaction with the dislocation of "fast" ($V_{gr}\tau_{ph}K_D > 1$) and "slow" ($V_{gr}\tau_{ph}K_D < 1$) phonons. While the interaction of the "fast" phonons with the dislocation has the nature of scattering the "slow" phonons relax in its field like a viscous gas.

The calculation shows that at low temperatures ($T \ll \theta_0 = \hbar\omega_0/k_B$) the contribution of the optic phonons to the dissipation is exponentially small. With rising the temperature the function $B_{op}(T)$ approaches a linear function of the type of $AT + B$. The contribution of the "fast" phonons constitutes a fracture of the acoustic component of the phonon wind (A.3.2) of order $10^{-1}\left(\frac{\theta}{\theta_0}\right)^4 \frac{\omega_0}{\Delta\omega}$ and that of the "slow" phonons differs from the corresponding term in formula (A.5.I) by the factor $10^{-1}\left(\frac{\theta}{\theta_0}\right)^4$.

Usually the magnitude of θ_0 is appreciably higher than the Debye temperature θ and the dissipative processes in the acoustic phonon system should be dominated over the damping in optic branches. An exception are crystals having "soft" optic modes with a small dispersion in the phonon spectrum (e.g. under the conditions of phase transformations).

A.7. Raman Scattering of Phonons

When moving in the Peierls relief the dislocation periodically changes the core configuration and its velocity ocsillates about the average value V . Accordingly configurational and dynamical oscillations of the elastic field of the dislocation at the ground frequency $\Omega = 2\pi V/a$ and higher harmonics $\Omega_n = \Omega n$ arise. This causes the inelastic (Raman) scattering of phonons with the energy change by $\Delta E = \pm \hbar\Omega_n$. The prevailing of the Stokes component of scattering ($\Delta E > 0$) over the anti-Stokes one ($\Delta E < 0$) determines the energy dissipation and the dislocation drag. This mechanism was considered first by Alshits [I2,I3] .

The calculation shows that the dynamical oscillations give a small contribution to the dissipation compared to the configurational oscilla-

tions, the latter leading to the viscous dislocation drag with the temperature dependence which coincides with the temperature dependance of the
phonon wind. The effect can be noticable as compared to the phonon wind
only in crystals with sufficiently high Peierls relief: $\sigma_p/\mu \geqslant 10^{-3}$
(σ_p is the Peierls stress).

A.8. Radiation Damping

With configurational and dynamical dislocation field oscillation is
connected one more channel of dissipation - it is the radiation damping
arising due to the emission of the elastic waves by the dislocation. The
radiation damping caused by the periodical variation of the core form was
investigated by a number of authors. In the recent papers [I4,I5] it was
shown that this effect can be essential only at subsonic velocities. At
low dislocation velocities more essential for radiation losses proved to
be dynamical oscillations. The influence of the non-uniformity of the dislocation motion in the Peierls relief on the level of the radiation damping was studied by Hart [I6] under the assumption of the single mode oscillation at the first harmonic. The non-self-consistent approach by Hart led
to erroneous conclusions for low dislocation velocities. In fact, with
lowering the velocity the radiation assumes more and more many-modal character and for a correct solution of the problem a self-consistent determination of the dislocation motion in the periodic potential field is needed taking into account the reaction of the radiation.

Such approach was first developed in the works by the authors [I2,I7]
where the stationary motion of the dislocation in the Peierls relief is
investigated in the continuum approximation.

According to [I2,I7], at high velocities when the kinetic energy of the
dislocation exceeds by far the Peierls energy the presence of the relief
immaterially perturbes the dislocation motion, the radiation occurs mainly
at the first harmonic and the radiation damping diminishes with the velocity
proportionally to v^{-2}. As the velocity decreases the degree of the non-uniformity of the dislocation motion increases and hence increase the radia-

tion damping, the radiation at higher harmonics becoming more essential.

A stationary motion with the average velocity below a certain criti-cal value $V_c \sim C \sqrt{\mathfrak{S}_p/\mu}$ is impossible. The minimum possible average velo-city V_c corresponds to the motion when at the top of the relief the dis-location has zero kinetic energy. The phenomenon of the critical velocity still takes place under the conditions of viscous dissipation when an ad-ditional viscous drag force $\bar{F} = -B\dot{x}$ acts on the dislocation. The critical velocity decreases with the viscosity B increase and vanished starting from some threshold value B_c. At $B > B_c$ the stationary motion realizes for any velocity V. In this case an effect of the type of "dry friction" should occur: when lowering the velocity the radiation stress approaches to the static Peierls stress $\mathfrak{S}_p$ not going to zero when the velocity tends to zero.

An analogeous problem was solved [18] also for a tangential motion of the kink along the dislocation with the consideration of the secondary Peierls relief. All above mentioned qualitative peculiarities hold for the kink too.

B. Electron Drag of Dislocations

The pumping of energy to the electronic subsystem of the crystal be-comes noticeable when compared to the phonon drag at low temperatures when the phonon gas is frozen out. The general expressions (A.2), (A.3) for the energy dissipation and dislocation drag coefficient are still valid and the problem consists in the calculation of the effective viscosity caused by the interaction of the elastic waves of the packet (A.I) with electrons. The effects arising here can be considered by analogy with the correspond-ing phonon drag mechanism.

B.I. Electron Viscosity

By analogy with (A.I.I) a formula for the electron drag connected with the viscisity of the electron gas $\eta_e \sim N_e \mathcal{E}_F \tau_e$ can be written. Here N_e is the density of conduction electrons, $\mathcal{E}_F$ is the Fermi energy, τ_e is the electron relaxation time. As in the case of the phonon viscosity one should point out Mason's [2] mistake who used the unjustified concept of the electron viscosity without spatial dispersion to calculate the energy dissipation near the dislocation core. It corresponds to the continuation of the integration in the formula of type (A.I.I) to $q_m \sim 1/a$. The inpermissibility of this procedure follows directly from the theory of the ultrasound damping [19] according to which the spatial dispersion of the electron viscosity can be neglected only for the long-wave part of the packet (A.I) $q l_e \ll 1$. Kravchenko [20] was the first to point out the necessity of the truncation at $q_m \sim l_e^{-1}$ in the integration. Then, independently, Tittman and Bömmel [21] noted it but in the works [22,23] the above mentioned mistake repeated.

B.2. Electron Wind

The scattering of electrons at the deformation potential of the moving dislocation can be considered on the analogy with the calculation of the phonon wind. According to Holstein [24] in the region approaching the dislocation axis the effect is eqivalent to the electron wind resuling in the temperature independent drag

$$B_e = \alpha \, \frac{a \, N_e \, \mathcal{E}_F}{V_F} \tag{B.2.I}$$

where α is the numerical factor of order 10^{-I}. As compared with expression (A.3.I), here the electron velocity V_F at the Fermi surface stands instead of the sound velocity c and the energy density of Fermi electrons $N_e \mathcal{E}_F$ stands instead of the thermal energy density E .

The calculation of the electron wind does not take into account the relaxation of electrons and is valid only provided $\Omega_q \tau_e \gg 1$. For fast enough dislocations this limitation is immaterial, for the major contribution to the dissipation is that of the short wave length part of the packet (A.I). However for slow dislocations with $v \lesssim a/\tau_e$ for all q : $\Omega_q \tau_e \lesssim 1$ and the problem of the electron drag of dislocations should be solved taking into account the relaxation of the electron gas in the process of scattering.

B.3. Electron Relaxation

The electron drag of slow dislocations ($\Omega_q \tau_e < 1$) was investigated

by Kravchenko [20] in the framework of the kinetic approach. The calculation of the damping of waves from the packet (A.I) was performed using the
theory of electronic ultrasound attennuation [I9] considering the spatial
dispersion of the electron viscosity. According to [20] the electron drag
of dislocations is practically independent of temperature and is described
by the expression of type (B.2.I) with inessential relaxation additions.
An effort to revise Kravchenko's theory undertaken by Huffmann and Louat
[25] turned out to be incorrect which was pointed out by Brailsford an[26]
and Kravchenko [27] .

B.4. Electron Drag of Dislocations in Superconductors

The decrease of the normal electron density at the superconducting
transition of the metal produces a jump-like enhancement of the dislocation
mobility that influenced strikingly the macroscopic plastic properties of
the crystal. Usually the temperature dependence of the dislocation drag,
ultrasound attennuation, creep rate, etc. is given directly by the temperature dependence of the density of normal electrons [28] . However for fast
dislocations a new dissipative process [29,30] could be expected: at $\hbar \Omega_q > 2\Delta(T)$
($\Delta(T)$ is the energy gap of the superconductor)
the electron-holes paines should be generated, the intensity of the process
being independent of temperature. In the analysis of macroscopic experiments
one has also to take into account quasi-static effects [3I] arising at the
superconducting transition of an inhomogeneously deformed crystal.

B.5. Electron Drag and Dynamic Conductivity

In the work by the authors [32] a single description of the interaction
of electrons with both fast and slow dislocations is given. This permits
to ascertain the relationship between the electron drag and the dynamic
electric conductivity of the metal. Both quantities were expressed in terms
of the two-particle Green's function of the electron. Only the asymptotics
of this function for short waves and small frequences proves to be essential which enables one to calculate in a general case the electron drag of
dislocations, kinks, crowdions, as well as short-wave length ultrasound
attennuation and to point out the relation of these effects to the anoma-

lous skin effect and other manifestations of the dynamic electronic conductivity.

Conclusion

The present state of the theory of the dynamic dislocation drag enables
one to solve the problem of the comparative role of macroscopic and microscopic mechanisms. One can state that namely the short wave length processes
which do not permit a macroscopic description and are determined by the
short wave length asymptotics of the dynamic elastic moduli and of the
electron conductivity give the major contribution to the phonon and electron
dislocation drag.

The theory provides already an opportunity of direct quantitative comparison with experiment, describes correctly the drag intensity and its
temperature dependence, predicts a number of new effects to be proved experimentally and also provides the ways of using the dynamic drag for the
investigation of some fundamental questions of the theory of fluctuation
motion, theory of phonon spectra and electronic theory of metals.

References

I. W.P.Mason. Journ.Acoust.Soc.Amer., 32, 458, (I960).

2. W.P.Mason. Journ.appl.Phys., 35, 2779 (I964).

3. J.D.Eshelby. Proc.Roy.Soc.Lond., AI97, 396 (I957).

4. J.Lothe. Journ.appl.Phys., 33, 2II6 (I962).

5. G.Leibfried. Z.Phys., I27, 344 (I950).

6. F.R.N.Nabarro. Proc.Roy.Soc., A209, 278, (I95I).

7. J.Lothe. Phys.Rev., II7, 704 (I960).

8. V.I.Alshits. Fiz.tverd.Tela, II, 2405 (I969).

9. V.I.Alshits, J.M.Sandler. to be published .

IO. A.D.Brailsford. Journ.appl.Phys., 43, I380 (I972).

II. V.I.Alshits, A.G.Malshukov. Zh.eksper.teor.Fiz., 63, I849 (I972).

I2. V.I.Alshits. In "Dinamika Dislokatsii", Phys.-Techn.Inst.Low Temp.
              of Ac.Sci.Ukr.SSR, Kharkov, I968, p.52.

I3. V.I.Alshits. Fiz.tverd.Tela, II, I336 (I969).

14. N.Flytzanis, V.Celli. Journ.appl.Phys., $\underline{43}$, 3301 (1972).

15. S.Jshioka, Journ.Phys.Soc.Jap., $\underline{34}$, 462 (1973).

16. E.W.Hart. Phys.Rev., $\underline{98}$, 1775 (1955).

17. V.I.Alshits, V.L.Indenbom, A.A.Shtolberg. Zh.eksper.teor.Fiz.,
    $\underline{60}$, 2308 (1971).

18. V.I.Alshits, V.L.Indenbom, A.A.Shtolberg. Phys.Stat.Sol.,$\underline{50}$,59 (1972).

19. A.I.Ahiezer, M.I.Kaganov,G.Ja.Lubarskii. Zh.eksper.teor.Fiz.,
    $\underline{32}$,837 (1957).

20. V.Ja.Kravchenko. Fiz.tverd.Tela, $\underline{8}$, 927 (1966).

21. B.R.Tittman, H.E.Bömmel. Phys.Rev., $\underline{151}$, 178 (1966).

22. W.P.Mason, A.Rosenberg. Journ.appl.Phys., $\underline{38}$, 1929 (1967).

23. W.P.Mason, D.E.McDonald. J.appl.Phys., $\underline{42}$, 1836 (1971).

24. T.Holstein. Phys.Rev., $\underline{151}$, 187 (1966).

25. G.P.Huffman, N.P.Louat. Phys.Rev., $\underline{176}$, 773 (1968).

26. A.D.Brailsford. Phys.Rev., $\underline{186}$, 959, (1956).

27. V.Ja.Kravchenko. Fh.eksper.teor.Fiz.,Pisma, $\underline{12}$, 551 (1970).

28. W.P.Mason. Appl.Phys.Letters, $\underline{6}$, 111 (1956).

29. M.I.Kaganov, V.D.Natsik. Zh.eksper.teor.Fiz.,Pisma, $\underline{11}$, 550 (1970).

30. G.Huffman, N.Louat. Phys.Rev.Letters, $\underline{24}$, 1055 (1970).

31. V.L.Indenbom, J.Z.Estrin. Zh.eksper.teor.Fiz.,Pisma, $\underline{17}$, 675 (1973).

32. V.I.Alshits, V.L.Indenbom. Fh.eksper.teor.Fiz., $\underline{64}$, 1808 (1973).

33. K.M.Jassby, T.Vreeland, Jr. Scripta Met., $\underline{5}$, 1007 (1971).

On the Temperature Dependence of the Dislocation Drag Constant as derived by Ultrasonic Attenuation Measurements.

H.R. Kaufmann, D. Lenz, K. Lücke

Institut für Allgemeine Metallkunde und Metallphysik
der Technischen Hochschule Aachen, W. Germany.

The Granato-Lücke-theory of dislocation /1/ damping predicts among other things that the decrement measured as a function of frequency has a maximum normally situated in the MHz region. For the decrement Δ_m and the frequency f_m of this maximum the following expressions are derived:

$$\Delta_m = K_1 \Lambda L^2 \qquad (1)$$

$$\Delta_m \cdot f_m = K_2 G \Lambda / B \qquad (2)$$

Here K_1 and K_2 are constants, G is the shear modulus, Λ the dislocation density, L = the mean dislocation loop length and B the dislocation drag constant. In many experiments the existence of such a maximum has been verified /2-8/, and by evaluating the maximum with the help of Eq. (1) and (2), informations about the parameters Λ, L and B have been obtained.

In the present paper, measurements of the maximum as a function of temperature will be described. From such measurements rather direct information is expected about the temperature dependence of the drag constant B since the temperature dependence of the remaining quantities in Eq. (2) is small. The pulse echo attenuation measurements were carried out on copper single crystals.

Investigation of the Quartz-Sample-Deformation.

Similar ultrasonic measurements of the temperature dependence of B in Cu have been published by other authors /9,10/. In such measurements, however, the true temperature dependence of B may easily be masked by an effect which is typical for the pulse echo technique and which experimentally is difficult to avoid. This effect is due to the difference in thermal expansion of sample and quartz /5/.

If the bonding material to be used as couplant between quartz and
sample is fluid or viscous the different changes in dimension of
sample and quartz taking place at temperature changes are taken up
by the bond. At lower temperatures, however, when the bond is solid
the different dimension changes of sample and quartz give rise to
a slight deformation of the sample which, in turn, cause an increase
in the measured attenuation.

The effect of this quartz-sample deformation is shown in Fig. 1. Here
the attenuation measured at room temperature is plotted versus frequency. Nonaq was used as bonding material. The crosses show α(f)
after the crystal was slightly deformed (about 0.1%). It was then
annealed for 4 h at 650°C (solid circles). The α(f) dependence shown
by open squares was measured after the sample, including quartz and
bond, had been cooled down to liquid nitrogen temperature and reheated
to room temperature. It is to be seen that the attenuation has markedly
increased by this treatment. An annealing for 1 h at 130°C reduced
the attenuation increase to the open circles and after a further
annealing at 600°C the open triangles are obtained. This shows that
the increase in attenuation caused by the quartz-sample deformation
during temperature cycling completely recovers during a high temperature anneal and that in this way, the sample can be restored to its
initial state.

Fig. 2 shows the attenuation of such a well annealed sample as a
function of temperature during a slow (1K/min) thermal cycle between
room temperature and 77 K. The Nonaq bond (solid line) solidifies
at about 220 K   below which temperature a marked attenuation
increase is observed. The dashed and dashed-dotted lines refer to
Plexol as bonding material. According to the lower solidifying temperature of Plexol (about 170 K) the increase of the attenuation starts
at this lower temperature. The details of these curves as well as
of the warming up curves will be discussed in a later paper /11/.

The quartz-sample deformation effects were only observed, if the
samples had been annealed at higher temperatures (in Figs. 1 and 2
at 650°C). This is to be seen in Fig. 3. Curves a and b refer to a
sample which after cutting from the bulk single crystal was only
lapped to its final shape without subsequent deformation and
annealing. Curve a was obtained by adjusting manually the exponential
curve of the Matec pulse echo comparator to the echo pattern on its

CRT-screen and curve b was obtained from the Matec automatic attenuation recorder. These curves agree within the experimental accuracy and exhibit no maxima or minima. However, after the sample was slightly γ-irradiated (curve c) or subsequently annealed 4 h at 650°C (curve d) the previously described effect of quartz-sample deformation is observed.

Investigation of the True Temperature Dependence of Attenuation.

The above results illustrate that the interpretation of the temperature-dependence/pulse-echo-attenuation-experiments needs considerable caution. By measuring frequency profiles it could be clearly shown that the attenuation maxima of the above curves result from dislocation loop length changes, i.e. they are caused by unpinning of the dislocations under stresses due to the differential thermal expansion of sample and quartz and that they are not a result of the temperature dependence of B. Additional measurements in which the present authors tried to reproduce the circumstances of the above mentioned published experiments /9,10/ lead to the opinion, that the maxima as function of temperature described in these papers are also due to the quartz-sample deformation and cannot be interpreted by the temperature dependence of B.

Therefore here it will be attempted to deduce the temperature dependence of B from α(T) curves obtained with samples which exhibited no quartz-sample deformation effect (e.g. curve a,b in Fig. 3). For these measurements, five high-purity (99.999%) copper samples of three different producers have been used. The capital letter A refers to Asarco, M to Materials Research and E to Elmore copper. Four of these samples were subjected to the s.c. standard treatment /8/ including 4 h annealing treatment at 650°C. Since they were expected to show at low temperatures the quartz-sample-deformation effect(to be seen in Fig. 3) they were only measured in the temperature range from 170 to 370 K with Plexol bond or from 230 to 370 K with Nonaq bond. The sample E_1 had not been annealed and thus could be measured down to liquid nitrogen temperature without deviation from linearity.

By measuring for each the temperature dependence of the total attenuation α and that of the background attenuation α_B and by substracting these two quantities from another, the temperature dependence of the dislocation damping $\alpha_d = \alpha - \alpha_B$ has been obtained. As an example, Figs. 4 and 5 show the temperature dependence of the two quantities α, α_B for sample E_1. In both cases at all frequencies investigated in the range from 10 to 170 MHz, a linear increase of the attenuation with temperature is observed. Fig. 6 shows as an example the frequency dependence of the decrement Δ for 77, 230 and 370 K derived from the data of Figs. 4 and 5. In Fig. 7 and 8, finally, the frequency and the height of the decrement/frequency-maximum is plotted as function of temperature for all samples investigated.

It is evidently impossible to interpret Figs. 7 and 8 under the assumption that only one of the 3 parameters, Λ, L or B, in Eq. (1), (2) changes with temperature. For example, Fig. 9 shows the quantity B/Λ obtained from Figs. 7 and 8 by means of Eq. (2). If Λ is assumed to be constant, these curves would directly show the temperature dependence of B. However, at the same time, the maximum decrement Δ_m (see Fig. 8), i.e. the product ΛL^2 (see Eq. (1)) changes with temperature. Thus it is questionable which two of the 3 quantities change with T or whether all 3 do. Furthermore, none of the temperature dependences obtainable under simple assumptions appear to be very reasonable: If one keeps Λ constant then one obtains for 4 samples a decrease of B with temperature (but with different slopes) and for one an increase. If one allows Λ and L to vary in order to obtain a reasonable, i.e. linear B(T)-dependence, then the functions Λ(T) and L(T) come out very different for the different specimens and, moreover, do not look plausible at all.

The present results show, that, even without quartz-sample deformation, there are considerably difficulties in deducing B(T) from ultrasonic measurements. Since the behaviour of samples with different impurities and pre-treatment is different, it is concluded that, besides phonon drag, interactions of point defects with dislocations contribute to the temperature dependence of dislocation damping. Only the magnitude of the drag constant comes out to be of the right order. This can be seen from Fig. 10, where the absolute values for B obtained from Fig. 9 by setting $\Lambda = 10^7 \text{cm}^{-2}$ are compared with the curves predicted by the theories of Leibfried /12/ and Brailsford /13/. Also the directly measured values of B /14/ are of this magnitude.

Acknowledgement

The authors thankfully acknowledge a financial support by the
Deutsche Forschungsgemeinschaft.

References

1   A.V. Granato, K. Lücke; J. Appl. Phys. 27, 583 (1956)

2   R.M. Stern, A.V. Granato; Acta Met 10, 358 (1962)

3   T. Suzuki, A. Ikushima, M. Aoki; Acta Met 12, 1231 (1964)

4   C.R. Heiple, H.K. Birnbaum; J. Appl. Phys. 38, 3294 (1967)

5   D. Lenz, K. Lücke; Z. Metallkde 60, 375 (1969)

6   H. Inagaki, F. Hultgren, K. Lücke; Acta Met 18, 713 (1970)

7   H. Akita, N.F. Fiore; J. Appl. Phys. 42, 2203 (1971)

8   P. Winterhager, K. Lücke; J. Appl. Phys. 44, 4855 (1973)

9   G.A. Alers, D.O. Thompson; J. Appl. Phys. 32, 283 (1961)

10 W.A. Fate; J. Appl. Phys. 43, 835 (1972)

11 H.R. Kaufmann, D. Lenz, K. Lücke; to be published

12 G. Leibfried; Z. Phys. 127, 344 (1950)

13 A.D. Brailsford; J. Appl. Phys. 43, 1380 (1972)

14 K.M. Jassby, T. Vreeland; Phil. Mag. 21, 1147 (1970)

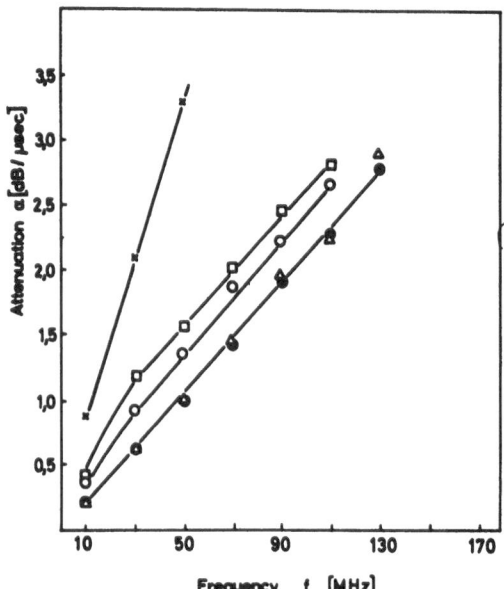

Fig. 1   Attenuation vs. frequency measured at room temperature
with Nonaq bond. The sample was 0.1% deformed in compression (x), then annealed 4 h at 650°C (●), then
with ultrasonic quartz bonded to its surface cooled
to 77 K (◻), then annealed 1 h at 130°C (○), then
annealed 4 h at 600°C (△).

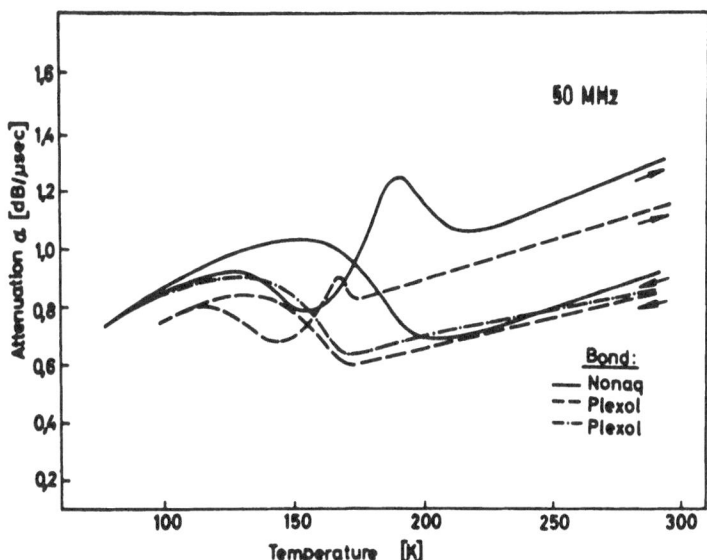

Fig. 2   Attenuation at 50 MHz in an well annealed sample during
slow thermal cycles between RT and 77 K using Nonaq and
Plexol bonds.

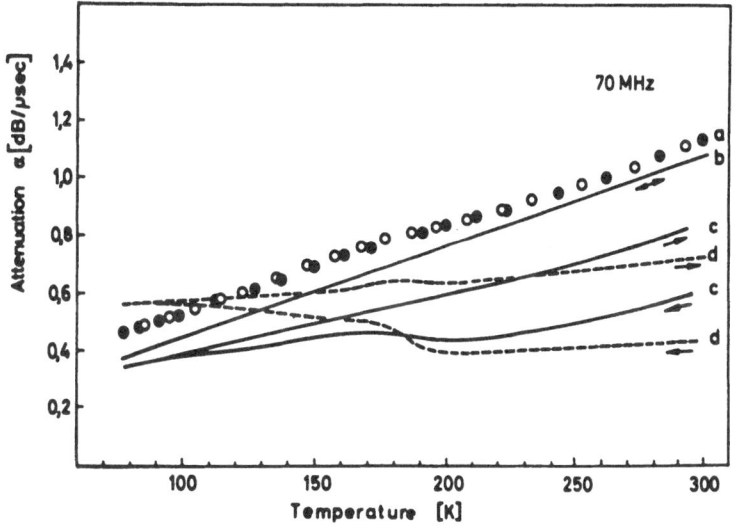

Fig. 3   Attenuation at 70 MHz during slow thermal cycles.
Curve a: measured by comparison of pulse train with
calibrated exponential curve. Curve b: automatically
recorded. Curve c: after 40 uAh γ-irradiation.
Curve d: after 4 h 650°C additional annealing.

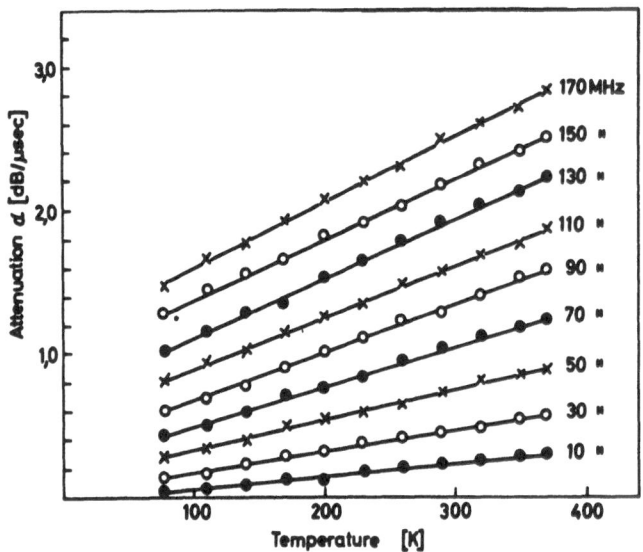

Fig. 4   Temperature dependence of the total attenuation
of sample E_1.

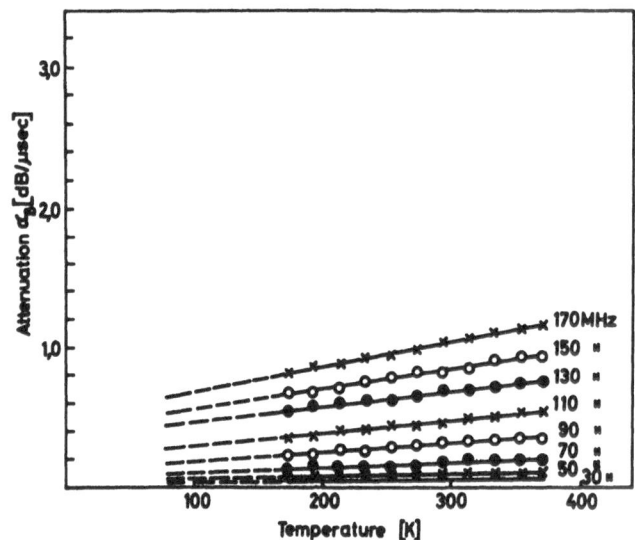

Fig. 5   Temperature dependence of the "background"-attenuation α_B
after heavy γ-irradiation of sample E_1.

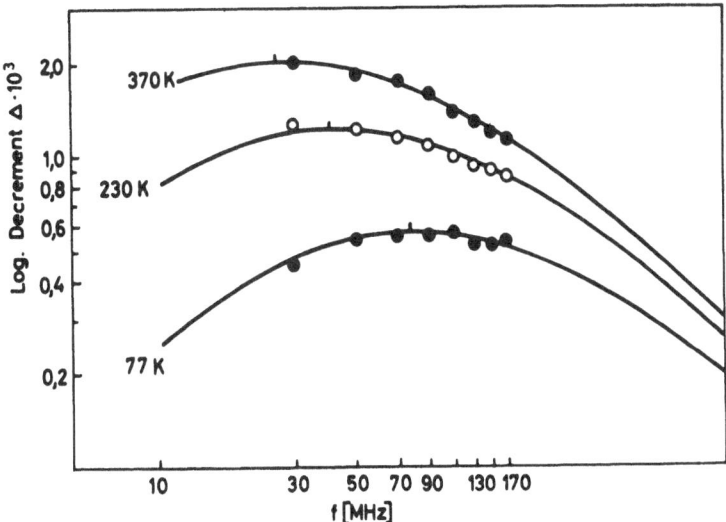

Fig. 6   Dislocation decrement vs. frequency of sample E_1 obtained
from Figs. 4 and 5. Solid curves: Granato-Lücke /1/
theoretical dependence for exponential loop length
distribution.

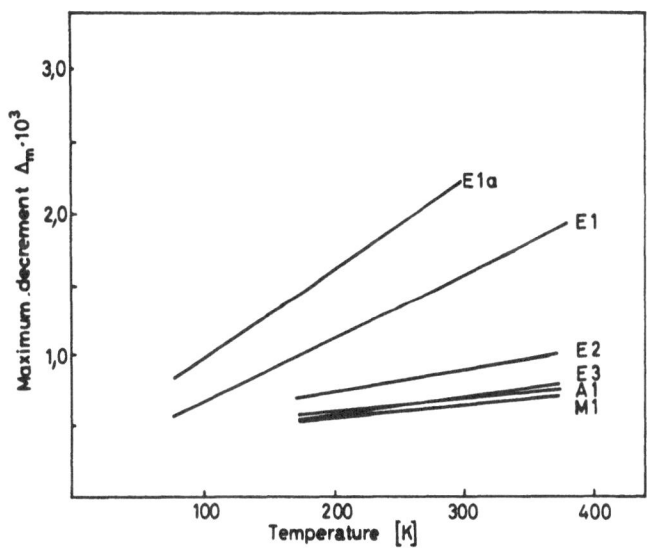

Fig. 7   Temperature dependence of the maximum decrement Δ_m
for the samples investigated.

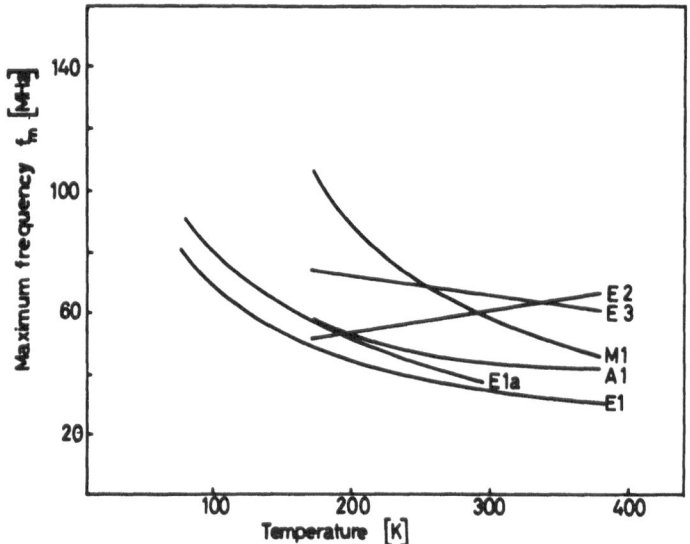

Fig. 8   Temperature dependence of the maximum frequency f_m.

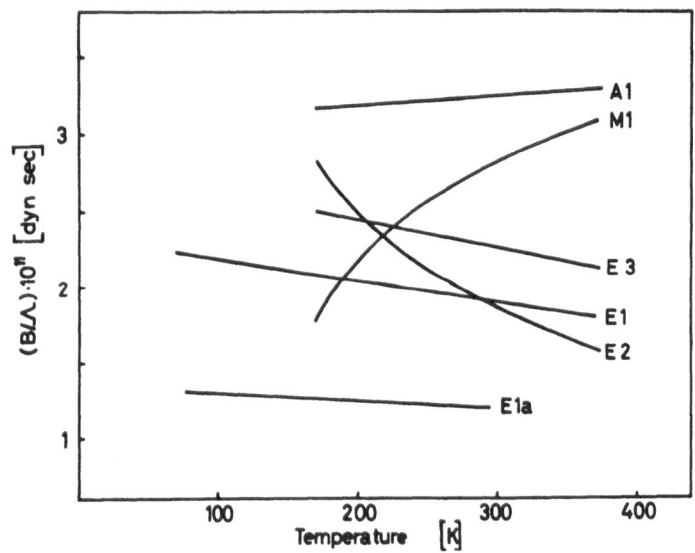

Fig. 9   Temperature dependence of the quantity B/Λ
         c.f. Eq. (2).

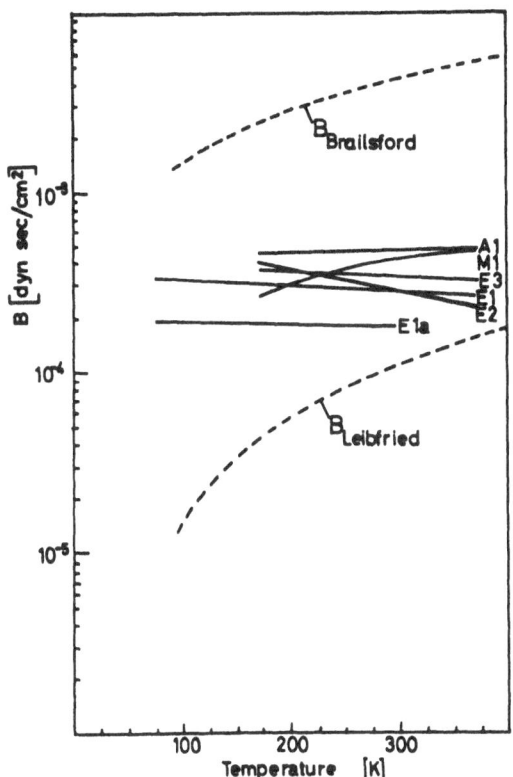

Fig. 10 Temperature dependence of the drag constant B derived from Fig. 9 with $\Lambda = 10^7 \text{cm}^{-2}$ compared with theoretical B(T)-dependences /12,13/.

RADIATION DAMPING OF DISLOCATIONS IN SODIUM CHLORIDE AT LOW TEMPERATURES

A. Hikata and C. Elbaum
Brown University, Providence, Rhode Island 02912, U.S.A.

Introduction

In a previous article (hereafter call I (1)), we reported experimental results on the temperature dependence of the dislocation drag coefficient, B, in Sodium Chloride in the temperature range 70 - 300°K. These results for B and its temperature dependence agreed reasonably well with the theoretical predictions based on the thermal phonon scattering by moving dislocations, i.e., the resistive force is identified with a viscous type damping. In the same article, however, we also mentioned that in the temperature range 2°K to 70°K contributions from mechanisms other than the viscous damping seem to become appreciable, and that it may no longer be justifiable to carry out the analysis with the expression containing viscous damping alone as a resistive force. It is this point that we treat here in some detail.

Experimental Technique

The experiments consist of measuring concurrently changes in attenuation, $\Delta\alpha$, and in velocity, $\Delta(\frac{\Delta V}{V})$, as a function of an applied, dynamic bias stress. The types of samples used and the technique for determining $\Delta\alpha$ were described in previous publications (1,2). The changes in velocity $\Delta(\frac{\Delta V}{V})$ were determined by means of an interferometric technique, which is a modified version of a method described by Blume (3). Particular attention was directed to eliminating spurious indications of velocity changes associated with changes in echo amplitude. Details of this technique will be published separately (4).

Results and Discussion

Examples of the $\Delta\alpha$ as a function of frequency ν obtained at 4.2°K are shown in Fig. 1. Here, in contrast to the behavior of $\Delta\alpha$ - ν relation above 70°K (1), $\Delta\alpha$ increases with frequency essentially linearly up to a frequency in excess of 100MHz (the exact value of this frequency varies from sample to sample), goes through a "hump," then levels off and becomes independent of frequency; in some cases the "hump" does not appear.

As mentioned in the Introduction, we concluded that at low temperatures (below 70°K) contributions from mechanisms other than viscous damping become important.

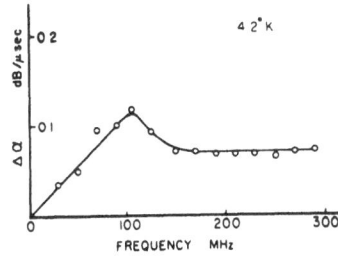 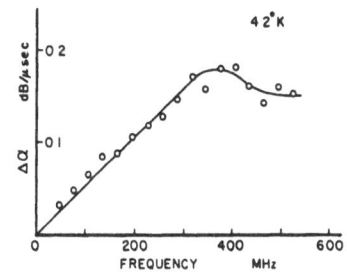

FIG. 1

Two examples of attenuation change, $\Delta\alpha$, as a function of frequency, due to application of bias stress, T=4.2°K.

We have examined, therefore, other loss mechanisms, and concentrated on radiation damping as the most likely source of dissipation, in terms of consistency with the experimental results. From an analysis of energies for motion of dislocations (5,6,7) and of the stress amplitude of the ultrasonic wave used, it is concluded that the dislocation contributions to the ultrasonic attenuation measured in this study originate from the motion of built-in kinks (geometrical kinks). The following analysis is based on this assumption.

Eshelby (8) considered radiation loss from moving kinks, but the attenuation derived on the basis of his calculation is not consistent with the present experimental results. His treatment is based on a continiuum model and does not take into account the effect of energy barriers for kink motion. When the kink barriers are included, the kink motion may not be smooth; instead it may be accelerated and decelerated in crossing the barrier, even if the average velocity v is kept constant. Such a change in velocity also is a source of energy radiation. This problem has been treated by several investigators including Hart (9) and Nabarro (10). H. Suzuki (11) also considered this problem from the analogy between charged particles in an electric field and dislocations in a stress field. According to his calculation, the stress required to move dislocations with a certain velocity v over an energy barrier W is given approximately by

$$\tau = (\pi \rho b \ell / 8 a^2 c_t)(W/M)^2 \{M/(\tfrac{1}{2} Mv^2 + \tfrac{1}{2} W)\}^{1/2} \tag{1}$$

where ℓ is the length of the dislocation, ρ is the density, c_t is the velocity of elastic shear waves and M is the mass of the dislocation. If one applies this formula to the case of kink motion by replacing ℓ with a (kink hight), W by $\sigma_K b^2/4$, and with $M = 2\mu a^2 b^2/\pi w c_t^2$ (w being the kink width and σ_k is the stress required to move a kink without thermal activation), the following expression is obtained:

$$\tau = \sigma_k (\pi/16)(b/a)(\pi w/4a)(\sigma_k/\mu)^{1/2}(1 + 8\mu a v^2/\pi \sigma_k w c_t^2)^{-1/2} . \tag{2}$$

When the velocity is small so that the kinetic energy is neglected against the potential energy, the above expression becomes independent of velocity,

$$\tau = \sigma_k(\pi/16)(b/a)(\pi w/4a)(\sigma_k/\mu)^{1/2} \equiv \tau_{dpk}. \tag{3}$$

This stress may be identified as the dynamic Peierls stress (12) for kinks, τ_{dpk}. If the applied stress is larger than this stress, kinks can move continuously. As the velocity increases the resistive force decreases, and in the extreme cases where the potential energy is neglected against the kinetic energy, the resistive force becomes inversely proportional to the velocity;

$$\tau = \sigma_k(\sqrt{2}\,\pi^2/8^3)(b^2/a^2)(w^2/ab)(\sigma_k/\mu)(c_t/v). \tag{4}$$

Recently Alshits et al (13) made a detailed calculation of this problem and arrived at essentially the same conclusion. In fact the above expressions for the two extreme cases (equations 3 and 4) agree with the corresponding expressions of Alshit within a numerical factor of order unity.

Thus, both Suzuki's and Alshits et al's analysis predict that at low velocities (below critical velocity v_c), the resistive force for kink motion is independent of the kink velocity and given by τ_{dpk}; and at high velocities, the resistive force is inversely proportional to the kink velocity. If it is assumed that the critical velocity v_c is determined by the condition

$$8\mu a v_c^2/\pi\sigma_k w c_t^2 = 1,$$

then for values of $\sigma_k/\mu = 10^{-6}$, and w=5a, v_c becomes $4.5 \times 10^{-3} c_t$. The distance a kink travels in the quarter cycle of the applied oscillatory stress is thought to be in the order of 50 interatomic spacings. Then the critical velocity v_c can be achieved at a frequency of 125 MHz, which is well in our experimental frequency range.

According to this model, the kink will accelerate indefinitely (in the high velocity regime) if applied stress is the only stress acting on the kink. In order to prevent this from occurring, Alshits et al introduced a large viscous damping without identifying its origin. It is difficult, however, to find such a large viscous damping at the low temperatures under discussion. We consider here, instead, the case of geometrical kink chains pinned at both ends. In this case, the interaction between kinks as well as the pinning points will prevent the divergence from occurring.

In the following the attenuation and the modulus defect (in terms of the velocity change) caused by the radiation mechanisms mentioned above are presented. Only the two limiting cases, low and high velocity regimes, are considered.

1) Low velocity regime.

$F = \tau_{dpk}ab$: the resistive force is independent of velocity.

In order to incorporate the velocity independent resistive force (dynamic) into the equation of motion, we use the "equivalent viscous damping method," which postulates the equivalence of work done by the real force $\tau_{dpk}a$ b ($\equiv B_1$) and by the equivalent viscous damping force $b_1\dot{x}_m$ at the end of each cycle;

$$4 \int_0^{\frac{\pi}{2\omega}} B_1 (\frac{dX_m}{dt}) dt = 4 \int_0^{\frac{\pi}{2\omega}} b_1 (\frac{dX_m}{dt})^2 dt,$$

or $b_1 = (4/\pi)(B_1/\omega X_{mO})$. It should be noted that the equivalent damping coefficient b_1 depends on the amplitude of the mth kink X_{mo} and no longer is a material constant. By substituting b_1 for B in the viscous damping, and taking only the first term of a series expansion, one obtains for the amplitude of the oscillation X_{mO},

$$X_{mO} = \{K_m^2 - (4B_1/\pi M)^2\}^{1/2} / (\omega_1^2 - \omega^2),$$

where

$$K_m = (\frac{2}{n+1}) \cot (\frac{\pi}{2} \frac{1}{n+1}) \frac{\sigma_o \alpha b}{M} \sin (\frac{\pi m}{n+1}),$$

$$\omega_1 = (4c/M)^{1/2} \sin (\frac{\pi}{2} \frac{1}{n+1}),$$

n is the number of kinks between pins, σ_o is the amplitude of the applied stress and c is the interaction constant between kinks. The displacement becomes

$$X_m = X_{mO} \cos (\omega t - \phi_m)$$

where

$$\tan \phi_m = (4/\pi)(B_1/M)\{K_m^2 - (4B_1/\pi M)^2\}^{-1/2}.$$

From these quantities, one obtains

$$\alpha = \frac{1}{2c_t} \frac{1}{n+1} \frac{N \alpha b}{\sigma_o e^{-\alpha y}} \frac{\mu}{\omega} \frac{\omega}{\omega_1^2 - \omega^2} \sum_{m=1}^{n} \frac{4}{\pi} \frac{B_1}{M} \{1 - (\frac{\frac{4}{\pi} \frac{B_1}{M}}{K_m e^{-\alpha y}})^2\}^{1/2} \qquad (5)$$

$$(\frac{\Delta V}{V}) = \frac{1}{2} \frac{1}{n+1} \frac{N \alpha b}{\sigma_o e^{-\alpha y}} \frac{\mu}{\omega_1^2 - \omega^2} \sum_{m=1}^{n} K_m e^{-\alpha y} \{1 - (\frac{\frac{4}{\pi} \frac{B_1}{M}}{K_m e^{-\alpha y}})^2\}. \qquad (6)$$

The above expression for α loses meaning unless

$$(\frac{4}{\pi} \frac{B_1}{M} / K_m e^{-\alpha y})^2 < 1.$$

The relative magnitudes of K_m^2 and $(4B_1/\pi M)^2$ are practically determined by $\{\sigma_o \sin(\frac{\pi m}{n+1})\}^2$ and $(\tau_{dPk})^2$ (the factor $\frac{2}{n+1} \cot (\frac{\pi}{2} \frac{1}{n+1}$ in K_m changes from 1 to $4/\pi$ as n increases from 1 to ∞). The amplitude of the applied measuring wave σ_o is in the order of 10^4 dynes/cm^2 and τ_{dPk} is estimated to be 10^3 dynes/cm^2 ($\tau_{dPk}/\sigma_{Pk} = 10^{-2}$, $\sigma_{Pk}/\mu = 10^{-6}$). If one sets a criterion

$$\{\tau_{dPk}/\sigma_o \sin (\frac{\pi m}{n+1})\}^2 < 0.1$$

for $(4B_1/\pi M)^2$ to be neglected against K_m^2, 20 kinks out of 99 kinks of the chain (the first 10 and the last 10 kinks) fail to meet this criterion. For a kink chain containing less than 9 kinks, all the kinks meet this criterion. Within this approximation, we may discard the term $(4B_1/\pi M)^2$ against K_m^2.

As mentioned previously, the quantities we measure are not the attenuation α nor $(\frac{\Delta V}{V})$ themselves, but the change of these quantities $\Delta\alpha$ and $\Delta(\frac{\Delta V}{V})$, due to the dynamic bias stress. The role of the bias stress is thought to be a slight increase of loop lengths ℓ, on the average, by depinning dislocations from weak pinning points, as described in I. This means that the effect of bias stress is to increase n slightly in the above expressions. The factors n/n+1 and

$$\frac{2}{(n + 1)^2} \cot^2 (\frac{\pi}{2} \frac{1}{n + 1})$$

do not change appreciably for small changes in n when n is sufficiently large. Therefore the contributions to $\Delta\alpha$ and $\Delta(\frac{\Delta V}{V})$ should come mainly from the change in the factor ω_1. In this case and together with the approximation mentioned above, the $\Delta\alpha$ and $\Delta(\frac{\Delta V}{V})$ are calculated to be

$$\Delta\alpha = \frac{\partial}{\partial\omega_1}(\alpha)\,\Delta\omega_1 = \frac{1}{c_t}\frac{1}{n+1}\frac{4}{\pi}\frac{N\alpha^2 b^2\mu\tau_d Pk\omega_1\Delta\omega_1}{M\sigma_o e^{-\alpha y}}\frac{\omega}{(\omega_1^2 - \omega^2)^2} \tag{7}$$

$$\Delta(\frac{\Delta V}{V}) = \frac{\partial}{\partial\omega_1}(\frac{\Delta V}{V})\Delta\omega_1 = \frac{2}{(n+1)^2}\cot^2(\frac{\pi}{2}\frac{1}{n+1})\frac{N\alpha^2 b^2\mu\omega_1\Delta\omega_1}{M}\frac{1}{(\omega_1^2 - \omega^2)^2}\;. \tag{8}$$

One can see from these expressions that $\Delta\alpha$ has an "inverse" amplitude dependence and increases linearly with frequency when $\omega_1^2 \gg \omega^2$, while $\Delta(\frac{\Delta V}{V})$ is independent of frequency as well as of stress amplitude.

Fig. 2 shows the results of concurrent measurements of $\Delta\alpha$ and $\Delta(\frac{\Delta V}{V})$ taken at frequencies of

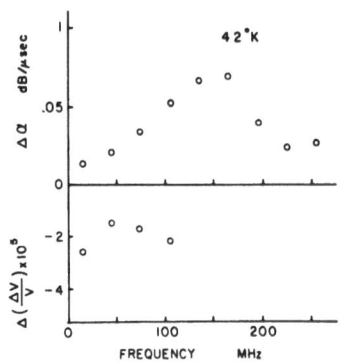

FIG. 2

Attenuation change, $\Delta\alpha$, and velocity change, $\Delta(\frac{\Delta V}{V})$, as a function of frequency, due to application of bias stress. T=4.2°K.

15, 45, 75, and 105 MHz and at a temperature of 4.2°K. In accordance with the predictions of the expressions (7) and (8), $\Delta\alpha$ increases linearly with frequency, while $\Delta(\frac{\Delta V}{V})$ appears to be independent of frequency, though the experimental points are somewhat scattered. Using a pair of the experimental values of $\Delta\alpha$ and $\Delta(\frac{\Delta V}{V})$ (for example, $\Delta\alpha$ = 0.03 dB/μsec, and $\Delta(\frac{\Delta V}{V})$ = 2 x 10^{-5} at 75 MHz), one can calculate the ratio τ_{dPk}/σ_o from the expressions (7) and (8):

$$\frac{\Delta\alpha}{\Delta(\frac{\Delta V}{V})} \simeq \frac{\omega}{c_t} \frac{\tau_{dPk}}{\sigma_o} \frac{\pi}{2}$$

or $\tau_{dPk}/\sigma_o \simeq 1.2$ x 10^{-1}. Since the stress amplitude of the measuring wave σ_o is in the order of 10^4 dynes/cm^2 or $10^{-7}\mu$, τ_{dPk} should be in the order of $10^{-8}\mu$, which agrees with the theoretical estimate $\tau_{dPk}/\sigma_K = 10^{-2}$ by Weiner (14), combined with $\sigma_k/\mu = 10^{-6}$ by Schottky.[8]

Fig. 3 shows the effect of the measuring wave amplitude on $\Delta\alpha$. Though the effect is not

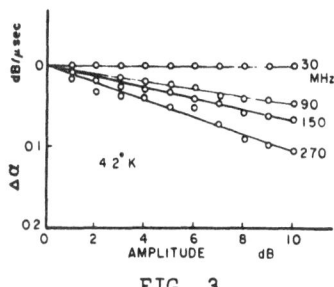

FIG. 3

Attenuation change as a function of measuring wave amplitude, at frequencies of 30, 90, 150 and 270 MHz, in the presence of a bias stress. T=4.2°K.

large, it clearly indicates the "inverse" amplitude dependence, i.e., the attenuation decreases as the amplitude of the measuring wave increases. The effect also increases with increasing frequency, as expected.[*] This inverse amplitude dependence stems from the assumed nonlinearity in the resistive force. The same effect, however, could result from the nonlinearity in the static interaction force between kinks as proposed by Suzuki and Elbaum (14) and by Alefeld (15). Such a mechanism with the viscous damping as a resistive force, however, fails to explain the frequency dependence of $\Delta\alpha$ and $\Delta(\frac{\Delta V}{V})$ observed in these experiments.

2) High velocity regime.

$F = \frac{Q\mathbf{a}b}{v}$ where

$$Q = \sigma_k (\frac{\sqrt{2}}{8^3} \frac{\pi^2}{\mathbf{a}^2} \frac{b^2}{})(\frac{w^2}{\mathbf{a}b})(\frac{\sigma_k}{\mu}) c_t \quad .$$

Here again the equivalent viscous damping method is used. The equivalent damping coefficient q becomes

$$q = \frac{2Q\mathbf{a}b}{x_{mO}^2 \omega^2}$$

---

[*]The $1/\sigma_o$ dependence of $\Delta\alpha$ given above is the result of the approximation $(4B_1/\pi M)^2 \ll K_m^2$. As σ_o becomes smaller, this approximation becomes less valid. If this term is included, the amplitude dependence of $\Delta\alpha$ becomes less significant. The data given in Fig. 3 show this trend.

The amplitude X_{m0} and phase ϕ_m are given by

$$X_{m0} = \frac{K_m + [1 - \{2(\omega_1^2 - \omega^2)(2Q\,b)/K_m^2\,M\omega\}]^{1/2}}{\sqrt{2}\,(\omega_1^2 - \omega^2)}$$

$$\tan \phi_m = \frac{2Q\mathbf{a}\,b}{M\omega X_{m0}^2}/(\omega_1^2 - \omega^2) \quad ,$$

which leads to

$$\alpha = \frac{1}{2c_t}\,\frac{N\mathbf{a}bQ\mu}{(\sigma_0 e^{-\alpha y})^2}\,\tan\,(\frac{\pi}{2}\,\frac{1}{n+1})\,\sum_{m=1}^{n}\,\frac{1}{\sin(\frac{\pi m}{n+1})} \quad ,$$

$$(\frac{\Delta V}{V}) = \frac{1}{2}\,\frac{N\mathbf{a}^2 b^2 \mu}{M}\,\frac{2}{(n+1)^2}\,\cot^2\,(\frac{\pi}{2}\,\frac{1}{n+1})\,\frac{1}{\omega_1^2 - \omega^2} \quad .$$

As can be seen, the attenuation α is independent of the frequency ω, which is consistent with the experimental observation at high frequencies. However, α is also independent of ω_1, the resonant frequency. This fact raises some question concerning the role of the bias stress. The effect of the bias stress has been interpreted as a decrease of the resonant frequency ω_1 through unpinning of weak pinning points. Since the above expression for α does not contain ω_1, the attenuation increase caused by the bias stress cannot be interpreted in this way. The net effect of the factors containing n is to increase both α and $\frac{\Delta V}{V}$, but slowly, especially at large n. Therefore the only effect the bias stress can produce is in N. In the above analysis, N is considered to be independent of bias stress. As the frequency of the applied stress is increased, however, not all the moving kinks shift from the low frequency, i.e., low velocity regime, to the high frequency, i.e., high velocity regime. Because of the limited traveling distances available, the kinks situated near the pinning points remain in the low frequency regime. The effect of the bias stress is then thought to shift, through the unpinning, those kinks in the low frequency regime to the high frequency regime, effectively increasing N. It should be noted that α is proportional to $1/\sigma_0^2$, i.e., should have strong "inverse" amplitude dependence. The experimental verification of this matter was not possible because of the insufficient dynamic range of our attenuation measurement instrument, when operated at high frequencies.

Another feature of the experimental results shown in Fig. 1 is a small "hump" in the $\Delta\alpha - \nu$ relation in the transition region between the frequency dependent and frequency independent regions. This hump is thought to arise from the effect illustrated schematically in Fig. 4. In this figure, α_1 portrays the frequency dependence of the attenuation in the absence of a bias stress, according to equs. (3) and (4), with emphasis on the linear dependence of α on ν and the ν independent regions. The application of a bias stress causes α to change from α_1 to α_2. It is noted that a shift of the transition between the two regions would appear at lower frequencies, when a bias stress is applied, because of the unpinning effect. Thus the measured, incremental attenuation, $\Delta\alpha = \alpha_2 - \alpha_1$, displays all the qualitative features, including the hump, as

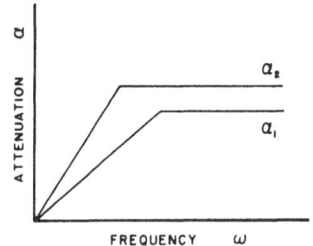

FIG. 4

Schematic representation of attenuation as a function of frequency, α_1 without bias stress, α_2 in the presence of a bias stress.

seen in Fig. 1.

The velocity change $\Delta(\frac{\Delta V}{V})$, on the other hand, contains a factor $(\omega_1^2 - \omega^2)$, and does not depend on σ_0. Therefore, there should be no amplitude dependence, and at high frequencies a resonance should be observed. Unfortunately, our velocity measurement technique is limited to the frequency of up to 110 MHz at present. Therefore, full confirmation of the validity of the model used is still subject to verifying the above predictions.

Finally we examine the connection between the low temperature mechanisms discussed above and the viscous damping assumed to prevail at higher temperatures (i.e., $T > 70°K$). Fig. 5 shows

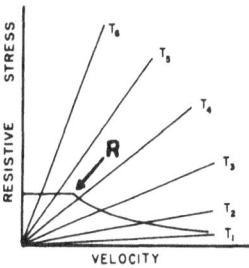

FIG. 5

Proposed (schematic) dependence of resistive force acting on dislocation (or kink), as a function of dislocation (or kink) velocity. Straight lines labeled T_1 to T_6 represent the case of viscous damping for increasing temperature ($T_1 < T_2 ... < T_6$). Curve labeled R represents the case of radiation damping.

schematically the resistive force dependence on velocity, for various temperatures. The straight lines labeled T_1 to T_6 are assumed to represent the resistance due to viscous damping as the temperature increases from T_1 to T_6 (the slope of the straight line is the damping coefficient B). The curve R composed of a velocity independent region at low velocities, and a part proportional to the reciprocal of the velocity at high velocities, represents the resistance due to radiation damping, as discussed above (it is assumed that to a first approximation this curve is independent of temperature). The transition from high to low temperature behavior is viewed as follows. At any set of conditions, the largest value of the resistive force dominates; thus at high temperatures (say T_6) and all but the smallest velocities, viscous damping applies. As the temperature is lowered, the viscous damping becomes less and less important for a given velocity. At the lowest temperatures viscous damping becomes negligible for all velocities and radiation damping dominates throughout. This behavior could also account for the fact that dislocation damping measured at low frequencies (KHz region and below) generally displays a frequency independent decrement and is therefore not consistent (16,17,18) with the predictions of the Granato-Lücke theory (19), which is based on viscous damping only. Indeed, at low frequencies the dislocation velocities are generally small and could be in the region to the left of the viscous damping line that corresponds to the temperature of the experiment. Under these conditions the damping would be governed by the radiation loss depicted by the curve R, and would display the feature mentioned above.

Research supported by the National Science Foundation, the Advanced Research Projects Agency and the Office of Naval Research.

References

1. A. Hikata, J. Deputat and C. Elbaum, Phys. Rev. B 6, 4008 (1972).

2. A. Hikata, R. A. Johnson and C. Elbaum, Phys. Rev. B 2, 4856 (1970); 4, 674 (E) (1971).

3. R. J. Blume, Rev. Sci. Instr. 34, 1400 (1963).

4. B. B. Chick, to be published.

5. A. Seeger, Phil. Mag. 1, 651 (1956).

6. G. Schottky, Phys. Stat. Sol. 5, 697 (1964).

7. W. T. Sanders, J. Appl. Phys., 36, 2822 (1965).

8. A. Seeger and P. Schiller, Acta. Met. 10, 348 (1962).

9. E. W. Hart, Phys. Rev. 98, 1775 (1955).

10. F. R. N. Nabarro, "Theory of Crystal Dislocation," p. 511 (Clarendon Press, 1967).

11. H. Suzuki, "Introduction to Dislocation Theory" (in Japanese) (AGNE Publishing Co., Tokyo, 1967).

12. J. H. Weiner, Phys. Rev. 136, A863 (1964).

13. V. I. Alshits, V. L. Indenbom and A. A. Shtolberg, Phys. Stat. Sol. B 50, 59 (1972).

14. T. Suzuki and C. Elbaum, J. Appl. Phys., 35, 1539 (1964).

15. G. Alefeld, J. Appl. Phys., 36, 2642 (1965); G. Alefeld, J. Filloux and H. Harper, "Dislocation Dynamics," A. R. Rosenfield et al, ed. p. 191, (McGraw Hill, New York, 1968).

16. R. den Buurman and D. Weiner, Scripta Met. 5, 573 (1971).

17. K. Lücke and G. Roth, Scripta Met. 5, 757 (1971).

18. V. K. Paré and H. D. Guberman, J. Appl. Phys., 44, 32 (1973).

19. A. Granato and K. Lücke, J. Appl. Phys., 27, 583 (1956).

PHONON DISLOCATION DRAG AND THE FORM OF
THE PHONON SPECTRUM

J.W. Martin and R. Paetsch
Department of Structural Properties of Materials
The Technical University of Denmark
DK-2800 Lyngby, Denmark

In a recent paper (1) Brailsford has given a general theory of the phonon dislocation drag constant B_{ph}. A central part of this theory involves the evaluation of a generalized viscosity ζ, which depends on both the frequency Ω and wave vector $\underline{s}$ of the strain wave, and which in the general case is expressed as a complicated thermodynamic average over the phonon modes. In Brailsford's paper ζ and B_{ph} were found on the basis of a simple Debye model. Here we use other models for the phonon spectrum and investigate their effect on the form of B_{ph}, and further we calculate values of B_{ph} for a series of metals using the Debye model. It is to be noted that our results are only applicable at temperatures near the Debye temperature or above.

Our purpose is to investigate the assumptions made in the theory about how the relaxation time τ, the phonon frequency ω and the Gruneisen parameter γ depend on the phonon mode involved (labelled by $\underline{Q}$). In the simplest case, it was assumed that τ and γ are constants, independent of $\underline{Q}$, and that the phonon spectrum is of the simple Debye type. Here we consider how refining this approximate model affects the values of B_{ph}.

The relaxation time τ is a measure of the phonon-phonon interaction. Rather than having a constant single value of τ, we could choose to take different values for each branch of the phonon spectrum. Alternatively we could calculate $\tau(\underline{Q})$ separately for each mode. Experimental curves of $\tau(\underline{Q})$ are not available: we must be content with an average value.

The phonon spectrum gives us the richest choice, both in the number of models available, and in the existence of experimentally determined spectra. Among the simpler models may be noted the Born model, which has a Debye like spectrum in each acoustic branch. In the alkali halides the optical branches are often approximated by an Einstein model.

The Gruneisen parameter γ is a measure of the strain wave-phonon interaction. In the alkali halides there are a number of calculations of γ for the different branches, and even for γ as a function of $\underline{Q}$. Experimental data on long wavelength γ values comes from the third order elastic constants, and it may be noted that it is possible to derive a tensorial γ which gives for example the different responses to edge and screw dislocations.

With these possibilities in mind, let us turn to the basic features of the theory of B_{ph}. We know that the theory includes a thermoelastic effect, described by a coefficient A:

$$\frac{\Delta T}{T} = - A\eta \qquad (1)$$

where ΔT is the thermoelastic temperature change, η is Lagrangian strain and A, like ζ, depends on the strain wave frequency Ω and wave vector s. A is proportional to γ, or rather a thermodynamic average <γ> over the modes Q. This is of significance in the case of screw dislocations in cubic metals; for shear strains <γ> vanishes. Indeed we know that A is zero since there is then no thermoelastic effect.

The general expression for ζ contains two terms. One of these arises from the thermo-elastic effect and is proportional to $<\gamma>^2$. As we have seen, this term vanishes under certain conditions. The other term is proportional to $<\gamma^2>$; only in the simplest model, when γ is assumed constant, will this be the same as $<\gamma>^2$; in the general case, it will be somewhat greater.

We now consider a refinement of Brailsford's model, which we refer to as the separate branch model. The assumptions of the model are that in each branch of the phonon spectrum (labelled by p) τ and γ have constant values τ_p, γ_p. The phonon spectrum is taken as Debye type in each acoustic branch, with a sound velocity c_p, and of the Einstein type in the optical branches in alkali halides.

We consider metals first. If the values of τ_p in the different branches are not too differ-end from each other, then we find that the major contribution to B_{ph} arises from the pho-non scattering region, and at high temperatures (T ≳ θ_D) has a similar form to Eqn (53) of (1)

$$B_{ph} = \frac{kT}{32\pi^2} q_D^4 s^4 b^2 \sum_p \gamma_p^2 / c_p \qquad (2)$$

using similar notation. Values of γ_p for this model can be obtained from the third order elastic constants (2), and using published values for Al (3) increases B_{ph} by a factor 1.08 over the single γ value. Another point to notice here is the dependence on q_D^4 due to a weighting of the averages to large wave vectors. This indicates the clear necessity for an accurate density of states function, if B_{ph} is to be evaluated accurately.

The effect of the form of the phonon spectrum is most clearly seen in the case of the alkali halides. We take a Debye model for the acoustic branches, an Einstein model for the optical branches. Thus the group velocity in the optical branches is zero, and the evaluation of B follows a different course. Basically we find that while for a Debye model we obtain a factor

$$\sum_p \gamma_p^2 / c_p$$

in the Debye-Einstein model this must be replaced by

$$\sum_p \gamma_p^2 \tau_p \, q_D$$

This means that B_{ph} is changed by a factor of approximately $q_D \tau / c$ or about 10-40 times at the Debye temperature. We emphasize that this is the result for a particular model. From experiment we know that B_{ph} is linear in T for the alkali halides at these temperatures. Clearly the assumption of a zero group velocity throughout the branch is too drastic.

TABLE 1

Values of B_{ph} at the Debye Temperature for Metallic Elements in Units of $10^{-5} \, \mathrm{Nm}^{-2}$ s

3	Li	0.59	37	Rb	0.56	77	Ir	8.24
4	Be	4.59	38	Sr	0.28	78	Pz	8.68
6	C	0.14	39	Y	0.75	79	Au	8.48
			40	Zr	0.48	80	Hg	5.41
11	Na	0.74	41	Nb	2.45	81	Tl	2.64
12	Mg	2.10	42	Mo	2.48	82	Pb	3.80
13	Al	4.80	43	Tc	9.41	83	Bi	0.80
14	Si	0.46	44	Ru	14.26	84	Po	0.93
			45	Rh	7.95			
19	K	0.35	46	Pd	8.37	89	Ac	0.52
20	Ca	0.44	47	Ag	5.39	90	Th	1.01
21	Sc	1.03	48	Cd	3.98			
22	Ti	1.38	49	In	3.41	58	Ce	0.73
23	V	2.45	50	Sn	1.58	59	Pr	0.24
24	Cr	4.96	51	Sb	0.56	60	Ne	0.35
25	Mn	1.79	52	Te	0.64	61	Pm	0.36
26	Fe	4.72				62	Sm	0.25
27	Co	6.37	55	Cs	0.29	63	Eu	1.00
28	Ni	6.57	56	Ba	0.21	64	Gd	0.18
29	Cu	5.25	57	La	0.27	65	Tb	0.50
30	Zn	4.89	72	Hf	0.93	66	Dy	0.47
31	Ga	1.40	73	Ta	2.99	67	Ho	0.95
32	Ge	0.84	74	W	3.34	68	Er	0.89
33	As	0.08	75	Re	8.66	69	Tm	1.25
34	Se	0.46	76	Os	5.37	70	Yb	0.46
						71	Lu	0.37

Finally we give some results for calculated values of B_{ph} based on the simple Debye model, across the periodic table. The data for these calculations, the polycrystalline elastic moduli and values of γ and θ_D are taken from (4). Since B_{ph} is proportional to T at higher temperatures we give the value of the coefficient B_0 in

$$B_{ph} = B_0 \left(\frac{T}{\theta_D}\right) \tag{3}$$

that is, the value of B_{ph} at the Debye temperature.

It is seen that the values of B_0 at either end of the table are relatively small, and that largest values of B_0 occur among the transition elements. Comparison of these values with some recent experimental values for Al and Cu and Zn (5) show reasonable order of magnitude agreement, within a factor two or so.

We should like to draw some general conclusions, considering the high temperature region only. Firstly, for metals the value of B_{ph} is sensitive to the form of $\gamma(\underline{Q})$ and $\omega(\underline{Q})$, but not to assumptions about the relaxation time τ. Secondly, in alkali halides the value of B is sensitive to all three of γ, ω and τ. Further the temperature dependence, even in the high temperature region, is affected by these factors

1.   A.D. Brailsford, J.Appl.Phys. 43, 1380 (1972)
2.   K. Brugger and T.C. Fritz, Phys.Rev. 157, 524 (1967)
3.   V.P.N. Sarma and P.J. Reddy, Phys.Stat.Sol. (a) 10, 563 (1972)
4.   K.A. Gscheidner, Solid State Physics Vol.16, Academic Press, N.Y. (1964)
5.   T. Vreeland and K.M. Jassby, Mat.Sci.Eng. 7, 95 (1971)

ULTRASONIC ATTENUATION IN CRYSTALS UNDER HIGH PRESSURE

Yosio Hiki and Tadashi Maruyama

Tokyo Institute of Technology, Oh-okayama, Meguro-ku, Tokyo, Japan

Introduction

A number of experimental results on dislocation damping have been accumulated, and many data have been analyzed successfully on the basis of the string model of vibrating dislocation proposed by Granato and Lücke (1). It is important that one can obtain from these experiments knowlege about the resistive forces against dislocation motion, the magnitude of which is represented by the damping constant, B, defined as the force per unit length of moving dislocation per unit velocity of the dislocation. There are many theories concerning the resistive forces (2), and the applicability of these theories to real damping phenomena may be examined by means of damping experiments. There is, however, difficulty in comparing the theoretical values of the damping constant with those of experiment, because typical ultrasonic experiments can only provide the ratio of the damping constant to the density of mobile dislocations (3), and the latter quantity cannot be determined so accurately. Another way of testing the theories is to study the dependence of the damping constant on some parameters, for example, on temperature (4,5). We intended to study the effect of hydrostatic pressure on the damping constant, expecting to obtain some information on the damping mechanism of dislocations. Copper crystals were chosen as specimens, because there appeared many examples of adaptation of the string-model analysis in the case of that material (3, 6–10).

Experimental Method

The attenuation experiment was carried out using a high-pressure vessel as shown in Fig. 1. The pressure of hydraulic fluid produced by a hand-operated single-stage oil pump was intensified by a couple of pistons having an area ratio of 17.4. The hydraulic fluid employed was Idemitsu' Daphne Oil, a kind of lubricant oil. Steady working pressures up to 8000 kg/cm^2 were produced in the high-pressure cylinder in which a specimen and a pressure gauge were inserted. The gauge was a 2 m length of 0.12 mm-diameter manganin wire, of about 120-ohms resistance, wound around a small bakelite cylinder. The change in resistance of the wire, ΔR, due to the pressure, P, was measured with a potentiometer. The sensitivity of the gauge, $(1/R)(\Delta R/ \Delta P)$, was $2.41 \times 10^{-6}/kg \cdot cm^{-2}$. The gauge was calibrated by making use of the phase transition of ammomium fluoride (NH_4F) at 3653 kg/cm^2 at $25^{o}C$.

The accuracy of the pressure measurement was ± 12 kg/cm^2. Electrical leads to the manganin gauge and a lead to the ultrasonic transducer were introduced into the high-pressure vessel using insulating conical pyrophyllite sleeves.

The attenuation of ultrasound in the specimen was measured by the conventional pulse-echo technique using a commercial attenuation comparator with usual accuracy of about 0.05 dB/μ sec. A 6.5-mm diameter X-cut quartz transducer with the fundamental frequency of 10 MHz or 20 MHz was bonded to an end face of the specimen with Nonaq stopcock grease. The grease sustained good bonding even in the high-pressure hydraulic oil. The high-voltage electrode was a stainless steel disk placed on the transducer and pressed slightly with a spring, and the specimen itself was grounded.

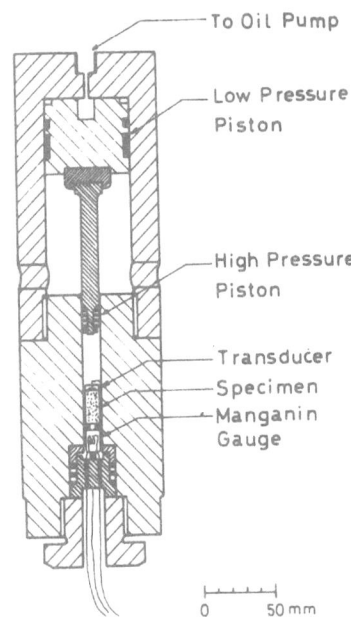

To Oil Pump

Low Pressure Piston

High Pressure Piston

Transducer
Specimen
Manganin Gauge

0   50 mm

Fig. 1. High-pressure vessel.

The specimens were copper single crystals of cylindrical shape, being 9 mm in diameter and 15 mm in length. Raw materials for the specimen were electrolytic copper plates of 99.99+% purity, which were zone refined five times in a CO gas atmosphere. Three cylindrical crystals with <100>, <110>, and <111> axial orientations were then grown by the Czochralski method. The orientations were checked by x rays, and found out to be accurate within 1°. The crystals were cut with an acid cutter, and the end faces were formed flat and parallel to within 2 μ by hand polishing. The specimens were annealed at 1000°C for 24 h in vacuum, and the parallelism of the end faces was again corrected. The second annealing at 800°C for 7 h was done before the attenuation measurement.

Results and Analysis

An example of the pressure dependence of the ultrasonic attenuation is shown in Fig. 2. The attenuation increases gradually with pressure, and the change is reversible when the pressure is lower than about 4000 kg/cm^2. A remarkable increase in attenuation occurs when the pressure exceeds that value, and the change is not reversible. It can be considered that new dislocations are introduced in the specimen when the stress due to the difference between the compressibilities of the specimen and the quartz transducer is increased to a definite value. Usual measurements were made under pressures not higher than 4000 kg/cm^2.

The attenuations of three annealed specimens with different orientations

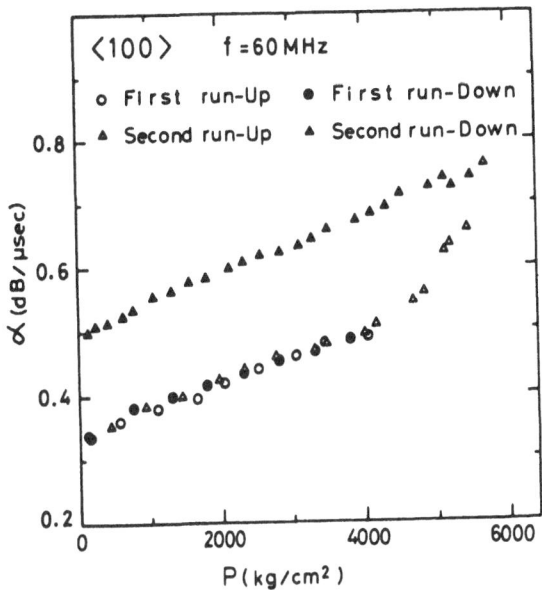

<figure>
<100>   f = 60 MHz

o First run-Up   • First run-Down
▲ Second run-Up   ▲ Second run-Down

α (dB/μsec)

0.8

0.6

0.4

0.2

0   2000   4000   6000
P(kg/cm²)
</figure>

Fig. 2. Reversible and irreversible changes of attenuation with pressure.

were measured as a function of frequency while the hydrostatic pressure was held constant, and the pressure was changed in steps of 500 kg/cm² from zero to 4000 kg/cm². Some of the typical data are shown in Fig. 3 with full lines. After the attenuation measurement, the specimens were irradiated with Co^{60} γ rays to a total dose of about 10^7 r. The values of attenuation decreased after irradiation, and further irradiation produced no appreciable change. The results on the irradiated crystals are shown with dotted lines in the same figures. It is concluded that the annealed specimens contain ample dislocations which are introduced by handling the specimens, and that the total dose of the γ-irradiation is enough to pin down the mobile dislocations.

The values of the attenuation measured in hydraulic oil are larger than those measured in air, as shown in Fig. 3. The reason is that the losses of acoustic energy from the specimen surfaces into the surrounding are larger in the case of oil because of good acoustic matching between the two materials. The apparent change of the attenuation under pressure may also be partly due to the increase of acoustic matching with increasing pressure. Further, the effect of the surrounding oil on the apparent attenuation is largest at a frequency of 10 MHz. This is reasonable, because the spread of the ultrasonic beam due to diffraction is larger at lower frequencies and therefore the energy loss from the side face of specimen is largest at 10 MHz. Because of the small inner volume of the high-pressure vessel, specimens of rather small diameter were used in the experiment, and the effect mentioned above was inevitable. We could obtain, however, the true values of the dislocation damping by finding the difference in the attenuation before and after the γ-irradiation.

The difference in the attenuation is converted into the decrement, Δ, and several examples of the frequency dependence of the decrement are shown in Fig. 4. According to the theory of Granato and Lücke, the decrement of a specimen containing dislocations is written in the form (3)

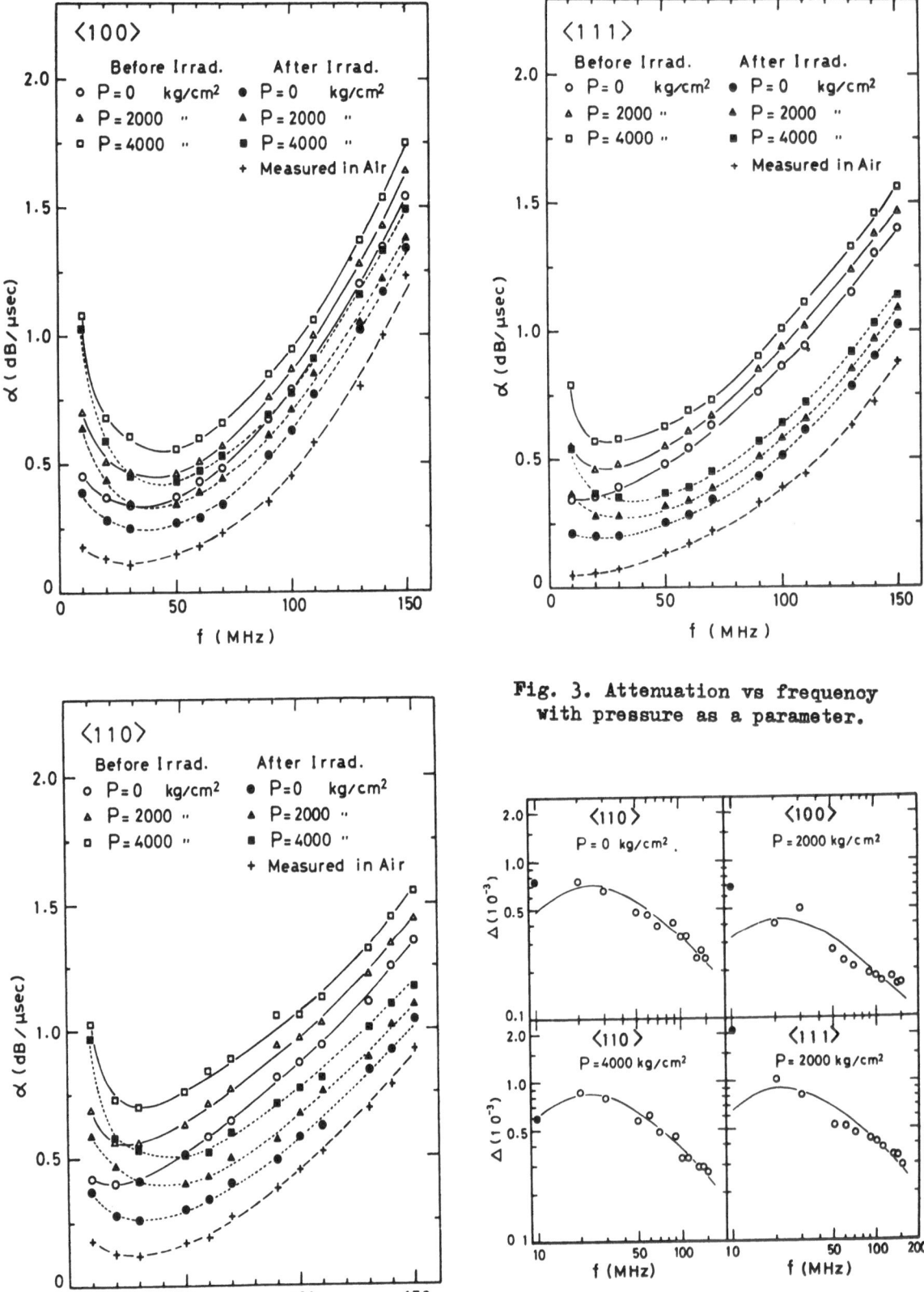

Fig. 3. Attenuation vs frequency
with pressure as a parameter.

Fig. 4. Decrement vs frequency.

$$\Delta = A\omega\tau/(1 + \omega^2\tau^2), \qquad (1)$$

$$A = n\,\Omega Gb^2 \Lambda\, L^2/\pi^3 C, \qquad (2)$$

$$\tau = n'BL^2/\pi^2 C. \qquad (3)$$

In the above formulas, the symbols have
the following meanings: ω is the angular
frequency of the sound, Ω the orientation
factor of the specimen, G the shear modulus, b the Burgers vector, Λ the dislocation density, L the pinning length, C
the effective dislocation tension, and B
the damping constant. The constants n and
n' take the values of n = 1, n' = 1 or
n = 4.4, n' = 11.9 when the distribution
of the pinning points is of delta-function
type or exponential type, respectively.
A computer calculation was carried out to
obtain the most probable values of two
parameters A and τ to fit Eq. (1) to the
experimental values of the decrement. The
10 MHz values were omitted from the computation because they were rather scattered,
probably due to the diffraction effect.
The fitted curves are drawn in Fig. 4,
which show reasonable agreement between
theory and experiment. The fitted values
of A and τ are plotted in Fig. 5. They
vary irregularly when the pressure is
changed, but it seems that the changes of
the two parameters are parallel with each
other. The ratio of τ and A is found to
decrease linearly with pressure, as shown
in the same figures.

These results can be explained as follows: As can be seen from Eqs. (2) and (3),
both expressions for A and τ contain the
pinning length L, while the ratio of two

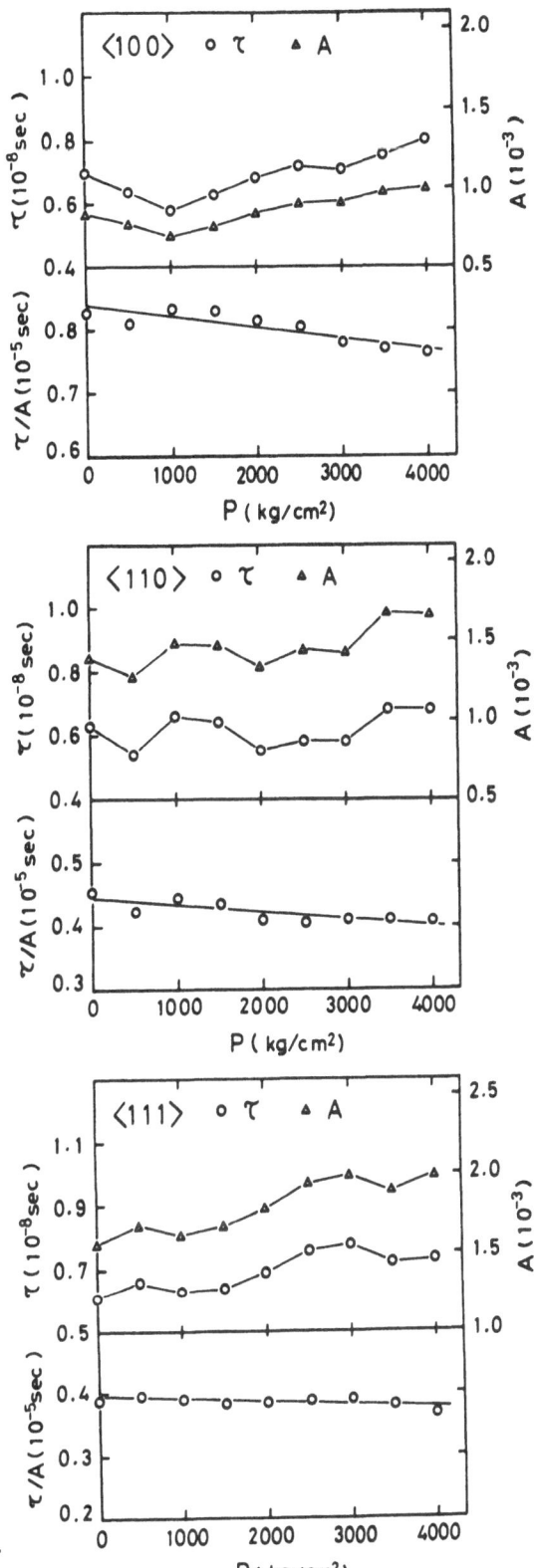

Fig. 5. τ, A, and τ/A vs pressure.

quantities is independent of the pinning length;

$$\tau/A = n'B/n\Omega Gb^2 \Lambda .$$ (4)

It is considered that the variation of A and τ is partly due to change in the pinning length. The origin of the change may be small stresses produced in the specimen during the pressure-increasing process. The hydraulic oil becomes very viscous at high pressures, and a non-hydrostatic force may be applied to the specimen in the transient state. The effect of the change of the pinning length is eliminated in the ratio τ/A, and its pressure dependence is considered as coming from the quantities contained on the right hand side of Eq. (4). The change of Gb^2 with pressure can be easily evaluated, and is found to be very small. If the dislocation density is not changed by pressure, which is a reasonable assumption for the present pressure range, and if the type of the distribution for the pinning points is always the same, one can determine the pressure dependence of the damping constant from the experimental (τ/A)-vs-P relation. The values of $(1/B) \times (dB/dP)$ thus determined for three crystals are given in Table I, where the values and the probable errors are obtained by least-squares fits of the experimental data.

Table I

Experimental values of the pressure dependence of the damping constant.

	<100>	<110>	<111>
$(1/B)(dB/dP)$ in 10^{-12} cm^2/dyne	-22 ± 3	-23 ± 5	-10 ± 3

Discussion

Various mechanisms have been proposed for the origin of the resistance to the motion of dislocations (2). Electronic effects may not be dominant at room temperature. Among vibrational effects, the contribution of radiative damping and thermoelastic damping may not be large when the velocity of the dislocation is not so high as in the case of the attenuation experiments. The phonon scattering and the phonon viscosity, both involving nonlinearity mechanisms, will be considered here as to whether they are suitable for explaining the present experimental results.

The damping constant for the resistive force caused by the scattering of thermal phonons by a moving dislocation can be expressed as (2)

$$B = \alpha \langle \varepsilon \sigma/c \rangle_{Av},$$ (5)

where ε is the energy density of phonons, σ the scattering cross section, c the velocity of sound, and α is a constant. The average is taken over all phonon modes. The energy density is given by the formula

$$\varepsilon = \frac{3k^4 T^4}{2\pi^2 c^3 \hbar^3} \int_0^{x_D} \frac{x^3}{e^x - 1}\, dx,$$

$$x = \hbar\omega/kT, \quad x_D = \Theta/T, \quad \Theta = (\hbar c/kL)(6\pi^2 N)^{1/3}. \tag{6}$$

Here ω is the angular frequency of phonons, Θ the Debye temperature, L the dimension of the crystal, N the total number of atoms, and the other symbols have their usual meaning. When the phonons are scattered by the strain field of a dislocation, the scattering cross section is, by Klemens (11)

$$\sigma = \alpha' \gamma^2 b^2 \omega/c, \tag{7}$$

where α' is a constant, b the Burgers vector, and γ the Grüneisen parameter which represents the anharmonicity of the crystal lattice. In the above expressions, the quantities which depend upon the hydrostatic pressure are considered to be L, b, c, Θ, and γ. The changes of L and b with pressure can easily be related to the bulk modulus of the material. The pressure dependence of the sound velocity, c, is calculated from the changes of the density and the second-order elastic constants of the material. The second-order constants of cubic crystals under pressure, P, are shown to be (12,13)

$$c_{11}(P) = c_{11} - (P/3K)(2c_{11} + 2c_{12} + C_{111} + 2C_{112}),$$

$$c_{12}(P) = c_{12} - (P/3K)(-c_{11} - c_{12} + 2C_{112} + C_{123}), \tag{8}$$

$$c_{44}(P) = c_{44} - (P/3K)(c_{11} + 2c_{12} + c_{44} + C_{144} + 2C_{166}),$$

where $K = (c_{11} + 2c_{12})/3$ is the bulk modulus, and C_{IJK}'s are the third-order elastic constants, the experimental values of which are available for copper (14). The pressure dependence of Θ is also calculated from the changes of c and L. Finally, the change of the Grüneisen parameter, γ, with pressure can be calculated as follows. It can be shown that the thermodynamically-defined Grüneisen parameter $\gamma = \beta V/K C_v$ (β is the volume thermal expansion coefficient, V the volume, and C_v the heat capacity at constant volume) is related to the mode Grüneisen parameter, $_{ik}\gamma_j^T$, as (15)

$$\gamma = (1/3)_{ii}\gamma_j^T = {_{11}}\gamma_j^T; \quad {_{ik}}\gamma_j^T = -\left[\partial \ln \omega_j / \partial \eta_{ik}\right]_{T,\eta=0}, \tag{9}$$

where ω_j is the angular frequency of the j-th phonon mode, and η_{ik} is the component of the Lagrangian strain tensor. The mode Grüneisen parameter is expressed as a combination of the second- and third-order elastic constants as follows (16):

$$_{ik}\gamma^T_j = -(1/2w)\left[c^T_{abik} + c^S_{aubk}U_uU_i + c^S_{aubi}U_uU_k + c^{ST}_{aubvik}U_uU_v\right]N_aN_b,$$

$$w = c^S_{aubv}U_uU_vN_aN_b,$$

(10)

where N_j and U_j are the components of the propagation and polarization vectors of the j-th phonon mode, and the superscript T or S expresses the isothermal or adiabatic constant. The pressure dependence of the mode Grüneisen parameter, $_{11}\gamma^T_j$, can be calculated as

$$\partial_{11}\gamma^T_j/\partial P = (\partial_{11}\gamma^T_j/\partial\eta_{ik})(\partial\eta_{ik}/\partial P) = -s^T_{ikss}(\partial_{11}\gamma^T_j/\partial\eta_{ik}),$$

(11)

where s^T_{ikss} is the elastic compliance constant. When the higher order mode Grüneisen parameter defined as (16)

$$_{iklm}\Gamma^T_j = -\left[\partial^2 \ln w_j/\partial\eta_{ik}\partial\eta_{lm}\right]_{T,\eta=0}$$

(12)

is used, Eq. (11) is modified as

$$\partial_{11}\gamma^T_j/\partial P = -(1/3K)(_{1111}\Gamma^T_j + 2_{1122}\Gamma^T_j).$$

(13)

The higher order mode Grüneisen parameter for any phonon mode can be calculated on the basis of the quasi-harmonic approximation (16), and can be expressed as a combination of the second-, the third-, and the fourth-order elastic constants:

$$_{iklm}\Gamma^T_j = 2_{ik}\gamma^T_j{}_{lm}\gamma^T_j - (1/2w)\left[c^{TT}_{abiklm} + c^{ST}_{aubklm}U_uU_i + c^{ST}_{aubilm}U_uU_k\right.$$

$$\left. + c^{ST}_{aubmik}U_uU_l + c^{ST}_{aublik}U_uU_m + c^{STT}_{aubviklm}U_uU_v\right]N_aN_b.$$

(14)

We have evaluated the pressure dependence of damping constant, $(1/B)(dB/dP)$, using appropriate average values of the sound velocity and the Grüneisen parameters. Houston's method is conveniently used for averaging, this asserting that the average value of any quantity in a cubic crystal is approximated well by a combination of its values for several crystallographic directions of high symmetry (16,17). In the present case, all longitudinal and transverse phonon modes with [100], [110], and [111] propagation directions were used for the averaging. Unfortunately, there is no data on the fourth-order elastic constants of copper. The values estimated from the temperature changes of the second-order elastic constants (18) were used for the computation. The final result was: $(1/B)(dB/dP) = -4.3 \times 10^{-12}$ cm^2/dyne.

Leibfried considered that the scattering cross section of phonons by a dislocation is a constant quantity of the order of the Burgers vector (19). In that case, the pressure dependence of the damping constant becomes smaller than the value obtained above, because the contribution of the change of the Grüneisen parameter is missing. Brailsford treated the problem of the dislocation drag very extensively (20), and derived an expression for the damping constant due to phonon scattering as follows:

$$B = \frac{3T}{2c} \left(\frac{\gamma s^2 b}{2\pi}\right)^2 \int_0^{q_D} q^3 C_q \, dq, \tag{15}$$

where s is the ratio of transverse to longitudinal velocities, C_q the contribution of the phonon with wave vector q to the specific heat, and q_D the Debye wave number. Here again the Grüneisen parameter, γ, appears. After a calculation similar to those described above, the pressure dependence of the damping constant was found to be close to the value obtained previously.

According to Mason (21,22), the damping constant originating from phonon viscosity can be expressed as

$$B = \alpha'' \Delta c \, \kappa/c^2 C_v, \qquad \Delta c = \sum E_j (_{ik}\gamma_j^T)^2 - \gamma^2 \rho C_v T, \tag{16}$$

where α'' is a constant, E_j the contribution of the j-th mode phonon to the thermal energy, ρ the density and κ the thermal conductivity of the material. Near room temperature the thermal conduction is mainly restricted by the umklapp process. In this case, the thermal conductivity is approximated by the expression (23)

$$\kappa = \frac{12}{5} \frac{4^{1/3}}{\gamma^2} \left(\frac{k}{h}\right)^3 \frac{Ma\,\Theta^3}{T}, \tag{17}$$

where M and a are the atomic mass and spacing. In these expressions, the pressure-dependent quantities are c, $_{ik}\gamma_j^T$, γ, ρ, a, and Θ, and also C_v and E_j, because the Debye temperature changes with pressure. The procedure in obtaining the pressure dependence of the damping constant is the same as those described previously. Because of the large positive contribution of the change of thermal conductivity, one obtains a positive pressure dependence of the damping constant. The computed results are summarized in Table II.

Table II

Calculated values of the pressure dependence of the damping constant.

	Klemens	Leibfried	Brailsford	Mason
$(1/B)(dB/dP)$ in 10^{-12} cm^2/dyne	-4.3	-1.8	-5.1	+4.6

When the calculated values are compared with the experimental results of Table I, it is noticed that the mechanism of phonon scattering due to a strain field of dislocation presents the pressure dependence of the correct sign but is somewhat small in magnitude. In the above calculation, the Grüneisen parameter was used as the quantity representing the anharmonicity, which might result in an underestimation of the scattering cross section. Furthermore, the effect of crystal anisotropy was smoothed out in the calculation. The experimental results show different pressure dependence for different directions of sound propagation. We are now performing an exact treatment of the problem of the scattering of phonons by a strain field in an anisotropic crystal on the basis of finite elasticity theory, and the calculation of the pressure dependence of the damping constant will appear in the near future.

References

(1) A. V. Granato and K. Lücke, J. Appl. Phys. 27, 583, 789 (1956).

(2) F. R. N. Nabarro, Theory of Crystal Dislocations, p. 505. Clarendon Press, Oxford (1967).

(3) R. M. Stern and A. V. Granato, Acta Met. 10, 358 (1962).

(4) A. Hikata, R. A. Johnson and C. Elbaum, Phys. Rev. B 2, 4856 (1970).

(5) A. Hikata, J. Deputat and C. Elbaum, Phys. Rev. B 6, 4008 (1972).

(6) G. A. Alers and D. O. Thompson, J. Appl. Phys. 32, 283 (1961).

(7) T. Suzuki, A. Ikushima and M. Aoki, Acta Met. 12, 1231 (1964).

(8) T. Kaneda, J. Phys. Soc. Japan 28, 1205 (1970).

(9) H. Inagaki, F. Hultgren and K. Lücke, Acta Met. 18, 713 (1970).

(10) H. Akita and N. Fiore, J. Appl. Phys. 44, 20 (1973).

(11) P. G. Klemens, Proc. Phys. Soc. A68, 1113 (1955).

(12) F. W. Sheard, Phil. Mag. 3, 1381 (1958).

(13) T. Bateman, W. P. Mason and H. J. McSkimin, J. Appl. Phys. 32, 928 (1961).

(14) Y. Hiki and A. V. Granato, Phys. Rev. 144, 411 (1966).

(15) K. Brugger, Phys. Rev. 137, A1826 (1965).

(16) Y. Hiki, T. Maruyama and Y. Kogure, J. Phys. Soc. Japan 34, 725 (1973).

(17) W. V. Houston, Rev. Mod. Phys. 20, 161 (1948).

(18) Y. Hiki, J. F. Thomas, Jr. and A. V. Granato, Phys. Rev. 153, 764 (1967).

(19) G. Leibfried, Z. Physik 127, 344 (1950).

(20) A. D. Brailsford, J. Appl. Phys. 43, 1380 (1972).

(21) W. P. Mason, J. Acoust. Soc. Amer. 32, 458 (1960).

(22) W. P. Mason and T. B. Bateman, J. Acoust. Soc. Amer. 36, 644 (1964).

(23) H. M. Rosenberg, Low Temperature Solid State Physics, p. 53. Clarendon Press, Oxford (1963).

INTERNAL FRICTION STUDY OF GOLD SINGLE CRYSTALS
IN THE MEGACYCLE RANGE[1]

K. Akune, M. Mondino[2], B. Vittoz

Laboratoire de Génie Atomique de l'Ecole Polytechnique Fédérale de Lausanne

(Switzerland)

INTRODUCTION

It is well known that the ultrasonic attenuation is a useful tool for the study of metals. Particularly, parameters such as dislocation length and dislocation density can be determined with the aid of Granato and Lücke model (1).

This work, undertaken with the aim of understanding the evolution of the dislocation pattern in slightly deformed gold, is part of a program of study of the dislocation recovery in f. c. c. crystals (2), (3).

Preliminary results in gold (4) have shown the existence of two internal friction maxima during isochronal annealing. In order to understand these phenomena we have studied the orientation dependences and the influence of impurities as well as irradiation effects.

EXPERIMENTAL PROCEDURE

The samples investigated in this experiment are gold single crystals with the three orientations ($\langle 100 \rangle$, $\langle 110 \rangle$, $\langle 111 \rangle$) and of purity 99, 999 %. A single crystal was grown by the Bridgemann method from an oriented germ of 99, 9999% purity in a high purity graphite crucible. After the examination of orientation by the usual Laue method, the single crystal was spark cut into several specimens and then polished by a diamond-headed milling cutter in order to obtain a good parallelism and smoothness of the two surfaces. Finally all the specimens were annealed for 4 hours at 900°C in vacuum before undergoing further treatments (e. g. deformation, irradiation[3]).

The geometric characteristics of the specimens are as follows :

> form : cylindrical (10 mm in diameter, 15 - 20 mm in length)
>
> precision of orientation : 2 - 3°
>
> parallelism between two surfaces : better than 30"

[1] This work was supported by the "Fonds National Suisse de la Recherche Scientifique", subsidy no. 2.776.72

[2] Centro Atomico Bariloche (CNEA), Argentina

[3] The treatments described above shall be referred to as standard treatments henceforth

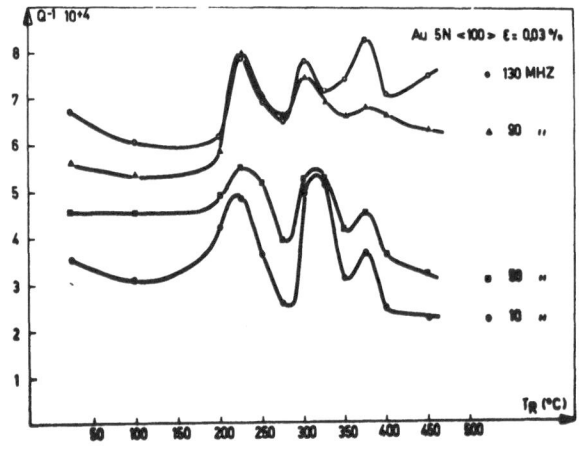

Fig. 1

Internal friction of a gold single crystal (99,999%, ⟨100⟩) as a function of annealing temperature. Each curve corresponds to a different measuring frequency

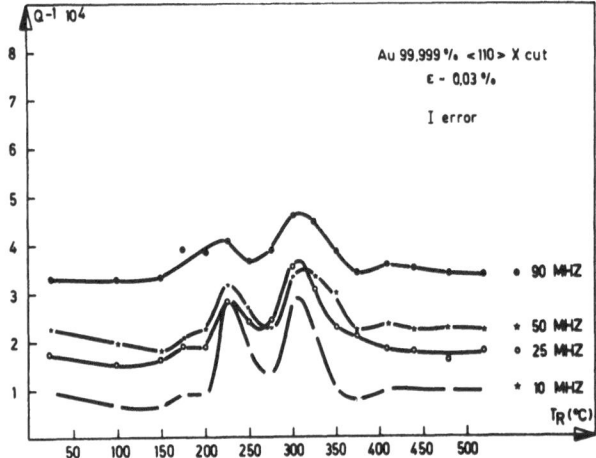

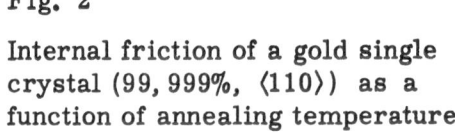

Fig. 2

Internal friction of a gold single crystal (99,999%, ⟨110⟩) as a function of annealing temperature

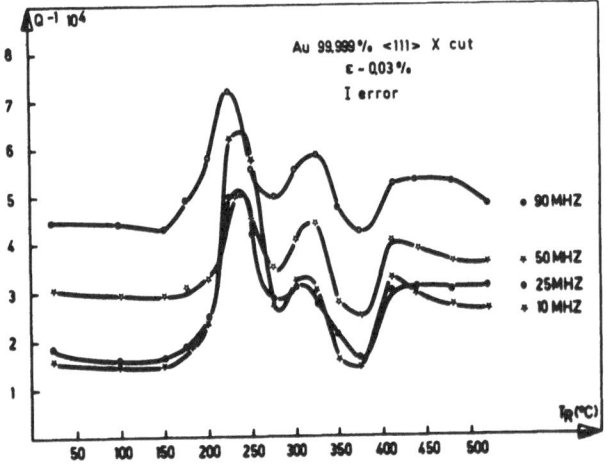

Fig. 3

Internal friction of a gold single crystal (99,999%, ⟨111⟩) as a function of annealing temperature

The specimens have also undergone one or both of the following treatments :

1) very slight deformation by compression at room temperature (deformation ratio = 0,03 - 0,06%)

2) neutron irradiation in water (integrated flux = 10^{14} or 10^{15} nvt).

The specimens thus prepared were annealed isochronally for 1hour at increasing temperatures up to 500°C. After each annealing, a quartz transducer (x cut, 5mm in diameter, fundamental frequency = 5 MHz or 10 MHz) was bonded on one face of the specimen with Nonaq Stopcock grease and the internal friction was determined from the measurement of ultrasonic attenuation at room temperature in the frequency range of 5 to 150 MHz. The attenuation was measured by the pulse echo technique using a commercial attenuator (Matec, Inc.).

EXPERIMENTAL RESULTS AND ANALYSIS

The specimens having undergone the standard treatments, were deformed very slightly (deformation ratio = 0,03 - 0,06%) by compression at room temperature. After each isochronal annealing of 1 hour, the attenuation was measured at room temperature as a function of frequency. The figures 1, 2 and 3 show the internal friction as a function of annealing temperature for three different orientations (⟨100⟩, ⟨110⟩, ⟨111⟩). For all the orientations, two internal friction maxima are observed ; one located at about 210°C and the other at 320°C, their characteristics being slightly different for each orientation.

Some specimens were neutron irradiated with an integrated flux of 10^{14} nvt without being deformed. In order to prevent any increase of the temperature of the specimens, the irradiation was done in the water of the swimming pool reactor SAPHIR at Würenlingen, Switzerland . Two months after the irradiation the same measurements were carried out on these samples. The figure 4 shows the internal friction versus annealing temperature curves obtained for a sample with ⟨110⟩ orientation. The results are completely different from those obtained by Inagaki et al.(5) on a γ-irradiated copper single crystal.

The following comments may be made from this figure :

1) The internal friction diminishes after the irradiation and it recovers the initial value at higher annealing temperatures.

2) The two internal friction maxima observed on the slightly deformed samples have disappeared. Instead, a broad maximum appears at higher annealing temperatures.

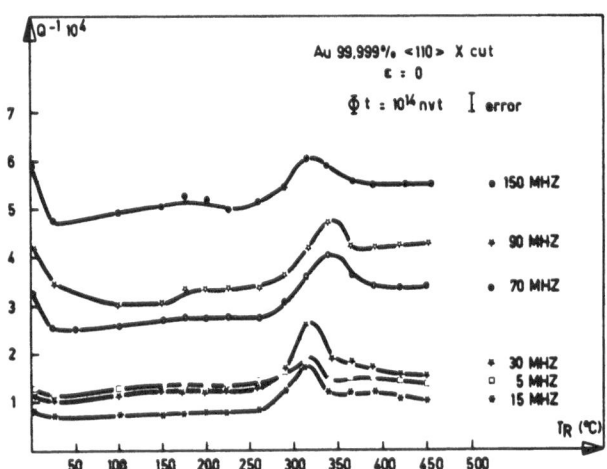

Fig. 4   Internal friction of a gold single crystal (99,999%, ⟨110⟩) as a function of
         annealing temperature. The sample was neutron irradiated without defor-
         mation (integrated flux = 10^{14} nvt)

The same measurements were also carried out on a gold single crystal with ⟨100⟩ orientation
containing 100 ppm silicon as impurity. The single crystal was grown by the same method as
the pure one, then homogenized for four days at $950^{\circ}C$ in vacuum and finally cut and polished.
The specimen, after undergoing the standard treatment, was slightly deformed (deformation
ratio = 0,05%) by compression at room temperature.

The figure 5 shows the internal friction versus annealing temperature curves obtained on this
sample. The results are rather dispersed. Nevertheless one can distinguish the following
characteristics for this specimen :

1)   The internal friction level is higher than that of pure
      single crystals.

2)   No internal friction maximum appears. The internal friction
      increases at higher annealing temperatures.

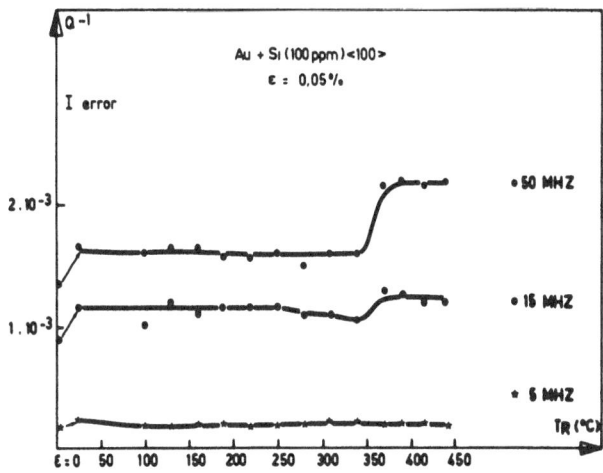

Fig. 5 Internal friction as a function of annealing temperature for a gold single crystal
($\langle 100 \rangle$) containing 100 ppm silicon

Although it is a little difficult to make comparison with the preceding results - the purity of
this sample being of 99, 99 % - the same experimental results for a single crystal ($\langle 100 \rangle$) ,
neutron irradiated (integrated flux = 10^{14} nvt) after being deformed by compression (deformation ratio = 0, 05 %) , are shown in figure 6. For this sample one cannot distinguish clearly the two maxima.

In the experimentally determined internal friction are included various kinds of losses due to
other causes than the movement of dislocations (6). The exact determination of this back -
ground is difficult both experimentally and theoretically. Supposing here that the movement
of most of the dislocations is blocked when the internal friction takes the minimum value, we
shall consider this minimum as an approximate background. For this purpose, we have
shown in figure 7 a serie of minimum values of the internal friction as a function of measuring frequency, observed in three different samples with $\langle 110 \rangle$ orientation, having undergone different treatments, i. e.   :

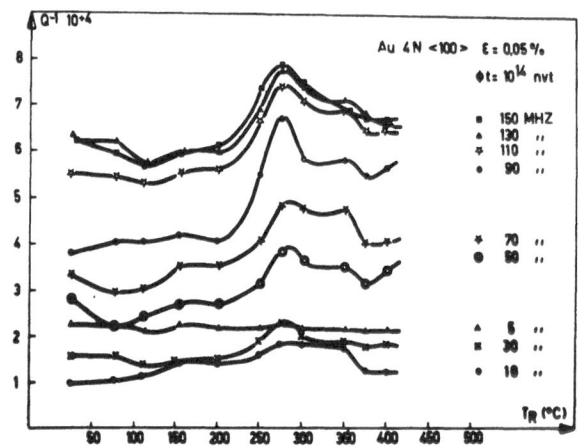

Fig. 6   Internal friction of gold single crystal (99, 99%, ⟨100⟩) as a function of annealing temperature. The sample was neutron irradiated (integrated flux = 10^{14}nvt) after being deformed by compression (deformation ratio = 0, 05%) at room temperature

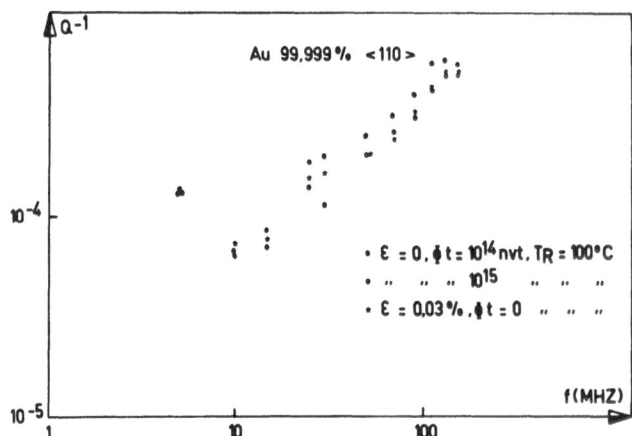

Fig. 7   Background of gold single crystals (99, 999%, ⟨110⟩). The samples have undergone the following treatments :
1)   neutron irradiated without deformation (integrated flux = 10^{14} nvt)
2)   neutron irradiated without deformation (integrated flux = 10^{15} nvt)
3)   deformed by compression at room temperature (deformation ratio = 0, 03%)
These values were obtained after having annealed the samples for 1 hour at 100°C.

1) Neutron irradiated without deformation ; integrated flux = 10^{14} nvt

2) Neutron irradiated without deformation ; integrated flux = 10^{15} nvt

3) Deformed by compression at room temperature ; deformation ratio = 0,03%

These minima were obtained after having annealed the samples for 1 hour at 100°C.

One observes that (1) and (3) give almost the same values, and (2) slightly higher values. The increase in background at lower frequencies is considered to be principally due to diffraction (7), and that at higher frequencies to thermoelastic effects (8)[*].

Considering (1) as a background, we substracted it from the experimentally determined internal friction. Some examples are shown in figure 8.

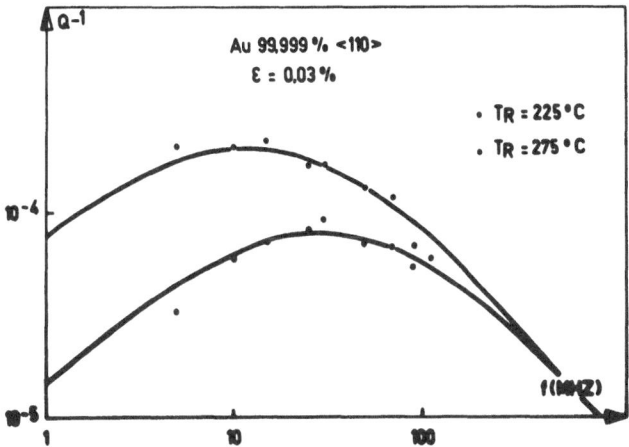

Fig. 8   Internal friction of a gold single crystal (99,999%, ⟨110⟩) deformed 0,03% by compression. Here the background is substracted. The solid curves correspond to the theoretical curves of the Granato-Lücke model

The internal friction-frequency relationship thus obtained is pretty well represented by the theoretical curve of the Granato-Lücke model (1) corresponding to an exponential distribution of average dislocation length, i.e. :

---

[*]   Concerning a concrete calculation of these effects
      for gold, see (4)

$$N(l) \, dl = \frac{\Lambda}{L^2} \exp\left(-\frac{l}{L}\right) dl \qquad (1)$$

$$\Delta = \int_0^\infty \Omega \, \Delta_0 \, l^3 \, \frac{\omega\tau}{1+\omega^2\tau^2} \, N(l) \, dl \qquad (2)$$

where the usual notations are used.

The same theory allows calculation of the density, and the average length of dislocations (Λ and L) if one knows the resonant frequency ω_m and the corresponding maximum decrement Δ_m^*.

$$\Delta_m = 2,2 \; \Delta_0 \, \Lambda \, L^3 \qquad (3)$$

$$\omega_m = 0,084 \; \frac{\pi^2 C}{L^2 B} \qquad (4)$$

Here again the usual notations are used[**].

The exact determination of Δ_m and ω_m is difficult. We can however calculate approximate values of Λ and L as a function of annealing temperature by making use of the formulae (3) and (4).

The results of such a calculation are shown in the figures (9) and (10).

These results are very dispersed. It seems, however, that the dislocation-density remains constant while the average dislocation-length varies during isochronal annealing.

DISCUSSION

In figure 10 we remark that the average dislocation-density is about $2.10^{10} \, m^{-2}$. This result is consistent with that obtained by Ramsteiner, by electron microscopy for gold single crystals at stage 1. At the same time, the above author indicates that the dislocations existing in gold single crystals are of edge type (9), (10).

On the other hand, the analysis made in the preceding section seems to imply that the dislocation density remains constant while the average length varies during annealing.

We may therefore reasonably assume that the increases in the internal friction are caused by a liberation of edge type dislocations (depinning), the latter rearranging themselves by different kinds of movement, i.e. glide (at lower temperature) and climb (at higher temperature).

---

[*] In this work we have used the internal friction Q^{-1} instead of the decrement Δ. They are related by $Q^{-1} = \Delta/\pi$

[**] The constant used in this work are $B_{100} = 2,0.10^{-5}$; $B_{110} = 3,4.10^{-5}$; $B_{111} = 2,7.10^{-5}$; $C = 2,6.10^{-9}$ (in MKS units). For the details of the calculation see (4).

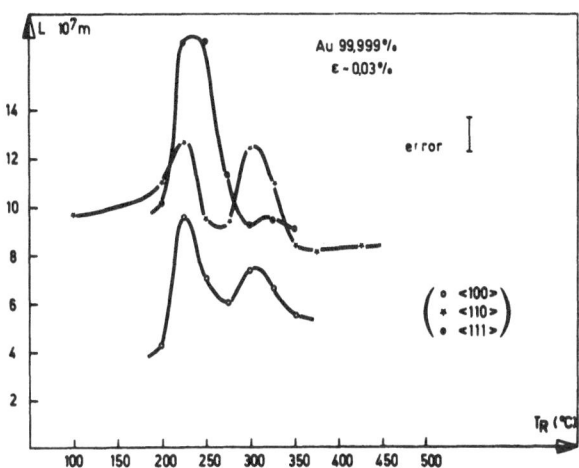

Fig. 9   Variation of the average length of dislocations L as a function of annealing temperature

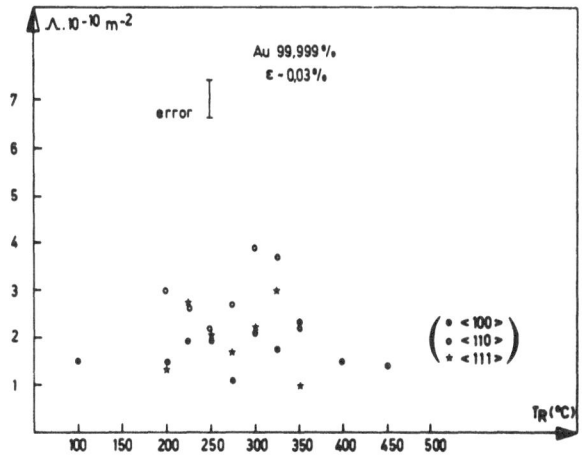

Fig. 10   Variation of the dislocation density Λ with annealing temperature

The decreases, on the other hand, are perhaps due to the pinning of dislocations by point defects. The stacking fault tetrahedron might also play a role in these processes (11).

It is interesting to compare the results shown in figure 4 with the results obtained by Inagaki et al. (5) for γ- irradiated copper single crystals. By performing the same kind of experiments, they found two distinct depinning stages, although the nature of the point defects contributing to the depinning process is not clearly mentioned. Our results show only one distinct depinning stage at about $320^{o}C$, followed by a repinning stage.

The role of stacking fault energy is not clear. In fact, we performed the same experiments for aluminium and silver single crystals, the former having stacking fault energy higher than gold and the latter lower (12). We could not, however, establish a distinct correlation between these metals due to the difference in stacking fault energy. The results for aluminium single crystals obtained by Kopetskiy et al (13) are also significant ; as well as a slight increase in the internal friction at lower annealing temperature, which we also observed for the slightly deformed samples, their results show complicated features for highly deformed samples.

In order to get a better understanding of the phenomena, the aid of other experimental methods would be useful. A study by electron microscopy of slightly deformed gold single crystals after annealings at different temperatures, is under way in our Laboratory. We hope it will throw light on our understanding in near future.

REFERENCES

(1) A. V. Granato and K. Lücke, J. Appl. Phys., 27, 583 (1956)

(2) K. Akune, A. Isoré and W. Benoit, Helv. Phys. Acta, 46, 22 (1973)

(3) K. Akune, M. Mondino and B. Vittoz, to be published in Helv. Phys. Acta

(4) K. Akune, W. Benoit and B. Vittoz, to be published in Mém. Sci. Rev. Mét.

(5) H. Inagaki, F. Hultgen and K. Lücke, Acta Met., 18, 713 (1970)

(6) See for example R. Truell, C. Elbaum, B. B. Chick, Ultrasonic Methods in Solid State Physics, Academic Press

(7) H. Seki, A. V. Granato and T. Truell, J. Acoustics. Soc. Am., 28, 230 (1956)

(8) K. Lücke, J. Appl. Phys., 27, 1433 (1956)

(9) F. Ramsteiner, Mater. Sci. Eng., 1, 206 (1966)

(10) F. Ramsteiner, Mater. Sci. Eng., 1, 281 (1966)

(11) B. Escaig, Cryst. Latt. Defects, 1, 211 (1970)

(12) J. Hirth and J. Lothe, Theory of Dislocations, Mc Graw Hill

(13) Ch. V. Kopetskiy, L. L. Rokhlin and V. S. Shkirov, Russian Metallurgy, 6, 107 (1971)

SOLUTE EFFECTS ON THE DISLOCATION DAMPING
BEHAVIOR OF CU-BASED ALLOYS

S.K. Banerji and J.C. Bilello
Department of Materials Science
State University of New York
Stony Brook, New York 11790

ABSTRACT: The dislocation damping phenomena in single crystals of copper and copper-nickel
solid solution alloys have been studied through high frequency ultrasonic attenuation
experiments. The attention is focussed here on the solute effects and the influence
of dislocation multiplication on the dynamics of dislocation motion in solid solution alloys. Quantitative characterization of the microstructure by etch-pitting
technique has been correlated with the attenuation characteristics and micromechani-
cal parameters in these alloys to gain more insight into the nature of dislocation
interactions with random localized obstacle fields.

Introduction

The ultrasonic attenuation studies have proved to be a very successful nondestructive way
of studying the dynamics of dislocation motion which is so crucial to the understanding of the
mechanical behavior of solids. Johnston (1) has pointed out that the mechanical properties of
materials should generally be reconsidered from the standpoint of dynamical behavior of dislocation motion. The frictional force on a moving dislocation is one of the fundamental dynamic
properties of a dislocation. If $\vec{F}$ is the externally applied force acting on a unit length of
dislocation, and $\vec{V}$ is the velocity of dislocation, one may define a damping constant "B" by the
following relation:

$$\vec{F} = B \cdot \vec{V} \qquad\qquad (1)$$

Hence, the measurement of the damping constant B and the effect of various parameters such as
impurities, and dislocation density etc. on this quantity are of great significance to the
understanding of the dynamics of dislocation motion. Experimentally, B can be measured in two
ways. One method is to measure the velocity of dislocations as a function of stress (2). This
method has been discussed in detail by Mason (3). The second method is based on the measurement
of amplitude independent high frequency internal friction as a function of frequency (4,5) and
Mason (3) has shown that these two independent methods give the same physical parameter represented by the damping constant B. The second method is often preferred because it is nondestructive in nature and the fact that only the dynamic behavior of dislocation motion can be studied
by this method since immobile dislocations play no part in dislocation damping and internal

friction. Under the present experimental conditions of high frequency (megacycle)-low amplitude ($\varepsilon \backsim 10^{-7}$) ultrasonic wave propagating within the solid, there is strong evidence that the damping is caused by dislocation resonance (6), unless the pinning length is unusually short. Koehler (7) first suggested a theoretical treatment of this problem by making an analogy between the motion of a pinned dislocation segment oscillating under the influence of an externally applied sinusoidal stress, and the forced oscillation of a vibrating string. Granato and Lucke (4) made significant improvements in the physical and analytical details of the model presented by Koehler which remains to be the best one available thus far. The damping constant B is derived from the asymptotic behavior of the log decrement (Δ) at higher frequencies in accordance with the theory of Granato and Lucke. The decrement is related to the attenuation by the following relation: $\Delta = \alpha /8.686 f$, where α is the attenuation in db/μsec. and f is the frequency in Mc/sec. This has been discussed in good detail by Stern and Granato (5) and a repetition of these ideas will be redundant.

The origin of the mechanism impeding the dislocation mobility and the nature of frictional forces acting on a moving dislocation depends strongly on the velocity of dislocations. In ultrasonic experiments such as those employed in the present study, dislocations are considered to move at high speed close to the velocity of sound in the material. Under such conditions the damping in pure metals is considered to arise from the interaction of moving dislocations with the lattice vibrations (3,8,9). There are several experimental evidences supporting this phonon interaction mechanism (5,10). However, very little is known about the effect of impurity atoms on the damping constant. Takamura and Morimoto (11), and Ookawa and Yazu (12) have treated this problem briefly on the basis of dynamic interaction of moving dislocations with an impurity atom. A very limited number of experimental results are available dealing with impurity effects on the damping constant (13-16), and they do not quite agree with either of these theories.

Experimental

The present investigation was undertaken to shed some new light into the understanding of solute effects and the influence of dislocation multiplication on the damping constant of Cu and Cu-Ni alloys. Ultrasonic attenuation experiments using a pulse-echo technique (17,18) have been carried out in the frequency range 5-200 Mc/sec on single crystals of Cu and Cu-Ni alloys of varying Ni concentrations. The crystals were prepared in the laboratory and cut into a rectangular parallelopiped geometry using a spark erosion technique. The damage caused by spark cutting was etched away in a suitable etchant. The samples were approximately 12 mm long and 7 mm X 7 mm in cross-section with the cross-sectional plane being a (110) plane and one of the side faces being a (111) plane. Quartz transducers of fundamental frequency 5 and 10 Mc/sec were bonded to the samples with Salol which gave satisfactory results in all cases. Higher frequencies were generated by driving the transducers at their odd harmonics. The attenuation of longitudinal waves propagating in the [110] direction was measured using a Matec Model 9000 Attenuation Comparator and related accessories. The results were correlated with a quantitative characterization of the dislocation structure using an etch-pitting technique. Dislocation densities were measured on a (111) side face parallel to the propagation direction of the ultrasonic wave. Livingston's etch (19) was able to satisfactorily reveal dislocations in each case. The dislocation densities were altered by uniaxial compression in an Instron machine, the compression axis being coincident with the direction of propagation of ultrsonic wave. Thus, the attenuation could be measured as a function of changing dislocation structure, temperature, and solute content.

Results and Discussion

Figures 1-5 show the room temperature (298°K) frequency dependence of differential decrement (Δ) for Cu, Cu-0.01 at.% Ni, Cu-0.1 at.% Ni, Cu-1 at.% Ni, and Cu-10 at.% Ni, respectively. The differential decrement is defined as the log decrement of any sample at a given frequency in a "lightly deformed" state (resolved shear strain, $\gamma \sim 0.4$-0.5%) when the decrement at the same frequency in an undeformed state is subtracted out. This is an useful quantity because it differentially measures the decrement between two levels of dislocation density, and because of its differential nature it minimizes the errors arising from the measurement system, and sample geometry etc. This quantity has been used earlier in the measurement of damping constant B for other materials (20,21). The decrement is calculated from the measured attenuation value by the formula mentioned earlier. All attenuation measurements were observed to be independent of the amplitude of the ultrasonic pulse. This was ascertained by observing the decrement or the attenuation as a function of amplitude using a stepped attenuator in between the pulsed-oscillator and the sample. The attenuation values at a given frequency remained unchanged upto the maximum available amplitude of the system. Hence, the measurements are definitely indicative of a strain amplitude-independent type of damping. Figures 1-5 all show a decrement maximum (Δ_m) at a certain frequency (f_m). These decrement maxima are not related to any relaxation type of mechanism because the positions of these maxima do not tend to shift with temperature. So it is fair to consider that these decrement maxima are caused by an over-damped resonance of dislocations.

The data points in figures 1-5 are the actual measured values whereas the solid lines are drawn on the basis of the following theoretical equation as derived from the Granato-Lucke theory (4):

$$\Delta = \Omega \, \Delta_o \Lambda \, L^2 \, \frac{\omega \, \tau}{1 + \omega^2 \tau^2} \qquad (2)$$

The various symbols in equation (2) have the following meaning: Ω is the orientation factor to take into account the fact that the resolved shear stresses on the slip systems are smaller than the externally applied stress; Λ is the dislocation density; L is the average loop length of the dislocation segment; $\Delta_o = 8 \, G \, b^2/\pi^3 C$, where G is the shear modulus, b is burgers vector, C is the dislocation line tension; $\tau = B \, L^2/\pi^2 \, C$, where B is the damping constant; and $\omega = 2\pi \cdot f$, where f is the frequency of the alternating applied stress. The data points in all these figures seem to lie in good agreement with the theoretical curves, thus justifying an interpretation and analysis of these results on the basis of this well accepted theory. The dashed lines in figures 1-5 represent the asymptotic behavior of equation (2) at high frquencies. From equation (2), the maximum decrement (Δ_m) and the frequency f_m at which this occurs are given by the following relations:

223

$$\Delta_m = \Omega \ \Delta_o \Lambda \ L^2/\ 2$$

$$f_m = 1\ /\ 2\pi\cdot\tau = \pi\ C/\ 2\ B\ L^2 \quad \Big\} \quad \text{for } \delta\text{-function distribution of } L \quad (3)$$

or,

$$\Delta_m = 2.2\Omega \ \Delta_o \Lambda \ L^2$$

$$f_m = 0.084\ \pi\ C/\ 2\ B\ L^2 \quad \Big\} \quad \text{for exponential distribution of } L \quad (4)$$

The results of figures 1-5 tend to indicate an agreement with delta-function distribution of loop lengths L. At frequencies $f \gg f_m$, equation (2) reduces to:

$$\Delta \to \Delta_\infty = \frac{4\ \Omega\ G\ b^2\ \Lambda}{\pi^2\ B\ f} \quad (5)$$

when the asymptotic value of log decrement (Δ_∞) at higher frequencies (f) is known from equation (2), one can immediately use equation (5) to obtain the value of B provided Λ is known. So it is absolutely necessary to have a precise quantitative knowledge of the dislocation content of the same samples that are being tested. This was done using an etch-pitting technique on the (111) side faces of the samples. Thus, one could determine dislocation densities after successive stages of compression e.g., undeformed, lightly deformed, and "heavily deformed" states ($\gamma \approx 3\%$), until the dislocation density is not too high ($\approx 10^8$); and then make a quantitative correlation with the observed damping characteristics. Table I summarizes the results of etch-pit analyses on various crystals after successive steps of deformation.

TABLE I

Average Dislocation Densities of Various Crystals After Successive Steps of Deformation.

(1) Sample	(2) State of Deformation	(3) Shear Stress τ^* (gms/mm^2)	(4) Shear Strain γ (%)	(5) Average Dislocation Density Λ (#/cm^2)
Cu	Undeformed Lightly Deformed Heavily Deformed	0 411.7 943.9	0.00 0.36 2.74	1.1×10^6 7.7×10^7 1.1×10^8
Cu-0.01 at.% Ni	Undeformed Lightly Deformed Heavily Deformed	0 458.1 991.5	0.00 0.42 2.69	4.5×10^6 1.8×10^7 7.0×10^7
Cu-0.10 at.% Ni	Undeformed Lightly Deformed Heavily Deformed	0 389.4 1014.6	0.00 0.48 3.05	1.7×10^6 2.0×10^7 4.9×10^7
Cu-1.00 at.% Ni	Undeformed Lightly Deformed Heavily Deformed	0 405.0 847.3	0.00 0.49 3.15	2.0×10^6 3.5×10^7 1.2×10^8
Cu-10.0 at.% Ni	Undeformed Lightly Deformed Heavily Deformed	0 984.4 1554.5	0.00 0.49 3.42	2.5×10^6 7.7×10^6 9.5×10^7

Now returning to a discussion of figures 1-5, one observes in each case a decrement maximum (Δ_m) at a frequency f_m. The maximum value of the decrement decreases consistently with increasing Ni concentration and the position of the maximum (f_m) shifts towards higher frequencies. Stern and Granato (5) observed the same effect in Cu as induced by irradiation. The decrement maximum decreased in magnitude and moved towards higher frequencies with increasing irradiation time. Hasiguti et al (21) have also observed the same effect in lightly deformed iron (0.2% compression) when aged at 20°C for various lengths of time. So, it is clear from these observations that any parameter which tends to reduce the effective pinning length L, will cause the decrement maxima to decrease in magnitude and shift it towards higher frequencies. From the data in figures 1-5, one can now calculate the damping constant B and other parameters such as L^2/C etc on the basis of the above analysis. The results are tabulated in Table II. The columns (2) and (3) of Table II clearly show the decrease in the magnitude of Δ_m and shift of f_m to higher frequencies as a function of increasing Ni concentration. The decrease in Δ_m is rather significant in the early stages of Ni addition, and beyond a certain small concentration of Ni, Δ_m decreases only gradually with further addition of Ni atoms.

TABLE II

Experimental Results of Copper and Copper-Nickel Alloys.

(1) Sample	(2) Δ_m	(3) f_m , Mc/sec	(4) Δ_{∞} at $f=10^9$ Mc/s	(5) $(L^2/C) \cdot \Lambda$ dynes^{-1}	(6) L^2/C cm^2/dyne	(7) B/Λ dynes-sec	(8) B d-sec/cm^2
Cu	1.38×10^{-2}	9.5	2.64×10^{-4}	4.46×10^3	4.2×10^{-3}	3.7×10^{-11}	3.9×10^{-5}
Cu-0.01% Ni	8.35×10^{-3}	19.7	3.28×10^{-4}	2.69×10^3	6.0×10^{-4}	3.0×10^{-11}	1.3×10^{-4}
Cu-0.10% Ni	2.91×10^{-3}	27.7	1.62×10^{-4}	9.40×10^2	5.4×10^{-4}	6.0×10^{-11}	1.0×10^{-4}
Cu-1.00% Ni	2.52×10^{-3}	29.1	1.47×10^{-4}	8.09×10^2	4.0×10^{-4}	6.7×10^{-11}	1.3×10^{-4}
Cu-10.0% Ni	2.07×10^{-3}	46.6	1.93×10^{-4}	6.24×10^2	2.5×10^{-4}	5.4×10^{-11}	1.3×10^{-4}

The B coefficient for pure copper (see col.8, Table II) as calculated from these data using equation (5) is considerably lower than the values reported by some other workers (5,10,22); but it is in reasonably good agreement with the values reported by Suzuki et al (13). Suzuki and his co-workers have hinted that this discrepancy may be due to the lack of a precise knowledge of the dislocation densities in crystals used by Stern and Granato (5) and Alers and Thompson (10), since their work was published before the development of Livingston's etchant (19). Suzuki et al (13) have used this reliable etch-pitting technique, as employed in the present study, to suitably reveal dislocations on the (111) faces of their crystals. This enables a precise and quantitative characterization of the dislocation density of the same samples that are being tested. When this is done, one observes (see figure 6) that the damping constant (B) is quite sensitive to the inherent dislocation content of the material. In figure 6 the value of the damping constant B for pure copper as reported by various workers, including the present work, is plotted against the dislocation density as measured or estimated by the respective authors. A wide range of scatter in the data points is observed, but a pattern does seem to emerge as shown by the dashed line which is tentatively drawn through the data points of the present work. In view of the limited available data at the present time, the authors do not necessarily claim a linear relationship as shown by the dashed line; but the thrust of this discussion is to emphasize that there exists a "certain" relationship between B and Λ. The question as to what that specific relationship is, remains unsettled at this point in time. Physically it seems quite obvious that there should exist such a relationship. Whatever be the mechanism of damping or source of B, it does arise only due to some kind of interaction between the moving dislocations and the lattice vibrations (phonons) and/or impurities. If the mobile dislocation density is high, the number of such interactions would be large, and one would expect a large frictional force resisting the motion of dislocations and hence, a large value of the damping constant. Consequently, as the dislocation density decreases, the number of such interactions also decreases and one would expect a smaller value of B, approaching zero in an ideal dislocation-free crystal.

The B values for various Cu-Ni alloys are also tabulated in column (8) of Table II. These values are again considerably lower than those of Kaneda (16) which is the only available data on the B coefficients of Cu-Ni alloys thus far. Kaneda (16) has worked in a limited range of Ni concentration (0-0.3 at.%) and has made no attempt to characterize and correlate his results with the actual dislocation structure of his crystals. The difference between his B values for Cu-Ni alloys and those of the present investigation may well be due to the difference in dislocation densities of the crystals used in the two different studies, as pointed out in the preceding paragraph. Suzuki (23) has cited an unpublished value of 2.8×10^{-4} cgs for the B coefficient of Cu-0.1 at.% Ni which is in better agreement with the present value of B for that composition.

Figure 7 shows the concentration dependence of B (lower curve) for Cu-Ni alloys at room temperature. The error bars indicate the limiting values depending upon the scatter in the dislocation densities of various crystals. It is observed that the very first addition of Ni in Cu increases the value of B quite significantly, almost by a factor of 3, but subsequent additions of Ni has no substantial effect on B. Ikushima and Kaneda (14,15) have also observed that the B coefficient of Cu increases appreciably by the addition of small amounts of Pt or Pd. Their data also seem to indicate that the B values, after the initial significant rise, do tend to level off with further solute additions. However, when one plots the quantity B/Λ, which is normalized with respect to the dislocation density, a linear relationship is obtained (upper straight line, fig.7) as a function of Ni concentration. This suggests that either of the two or both models, namely, dislocation-phonon interaction or dynamic interaction of moving dislocations with the random localized obstacle field, could be operative. For the former mechanism to govern it is necessary that the thermal energy density should change in proportion to the Ni concentration, that is, the phonon distribution of Cu must change as a function of the impurity content, causing an increase in the interaction between the moving dislocations and lattice vibrations. On the other hand, the theory of dynamic interaction of moving dislocations with the impurity atoms predicts (11,12) that the frictional force resisting the motion of dislocations is directly proportional to N_0, the number if impurities in a given volume. This could also explain the observed concentration dependence in fig.7. However, the question as to which of the two mechanisms is operative, remains unsettled at this time. It is also possible that both the mechanisms make a contribution to the observed behavior. The work is still in progress and it is hoped that this problem will be resolved in near future.

Acknowledgement

The authors wish to gratefully acknowledge the support of U.S.Atomic Energy Commission for this research project.

References

1. W.G.Johnston, Jour.Appl. Phy., 33, 2716, (1962)

2. W.G.Johnston,and J.J.Gilman,Jour. Appl. Phy., 30, 129, (1959)

3. W.P.Mason, Jour. Acoust. Soc. Amer., 32, 458, (1960)

4. A.Granato and K.Lucke, Jour. Appl. Phy., 27, 583 & 789, (1956)

5. R.M.Stern and A.V.Granato, Acta Met., 10, 358, (1962)

6. A.V.Granato, "Dislocation Dynamics", ed. A.R.Rosenfield, G.T.Hahn, A.L.Bement,Jr., and R.I. Jaffee, McGraw-Hill, New York, p.117, (1968)

7. J.S.Koehler, "Imperfections in Nearly Perfect Crystals", ed. W.Shockley, J.H.Hollomon, R. Maurer, and F.Seitz, John Wiley & Sons, New York, p.197, (1959)

8. J.D.Eshelby, Proc. Roy. Soc., A197, 396, (1949)

9. G.Leibfried, Z. Physik, 127, 344, (1950)

10. G.A.Alers and D.O.Thompson, Jour. Appl. Phy., 32, 283, (1961)

11. J.Takamura and T.Morimoto, Jour. Phys. Soc. Japan, 18, Suppl. I, 28, (1963)

12. A.Ookawa and K.Yazu, Jour. Phys. Soc. Japan, 18, Suppl. I, 36, (1963)

13. T.Suzuki, A.Ikushima and M.Aoki, Acta Met., 12, 1231, (1964)

14. A.Ikushima and T.Kaneda, Scripta Met., 2, 89, (1968)

15. A.Ikushima and T.Kaneda, Trans. Japan Inst. Metals, 9 (Suppl.), 38, (1968)

16. T. Kaneda, Jour. Phys. Soc. Japan, 28, 1205, (1970)

17. R.L.Roderick and Rohn Truell, Jour. Appl. Phy., 23, 267, (1952)

18. W.P.Mason and H.J.McSkimin, Jour. Acoust. Soc. Amer., 19, 464, (1947)

19. J.D.Livingston, Acta Met., 10, 229, (1962)

20. N.Igata, R.R.Hasiguti and K.Domoto, Proc. First Int. Conf.on Fracture, 2, 883, (1965)

21. R.R.Hasiguti, N.Igata and M.Shimotomai, Trans. Japan Inst.of Metals, 9 (Suppl.), 42, (1968)

22. T.Vreeland, Jr., "Dislocation Dynamics", ed. A.R.Rosenfield, G.T.Hahn, A.L.Bement, Jr., and R.I.Jaffee, McGraw-Hill, New York, p.529, (1968)

23. T.Suzuki, "Dislocation Dynamics", ed. A.R.Rosenfield, G.T.Hahn, A.L.Bement, Jr., and R.I. Jaffee, McGraw-Hill, New York, p.551, (1968)

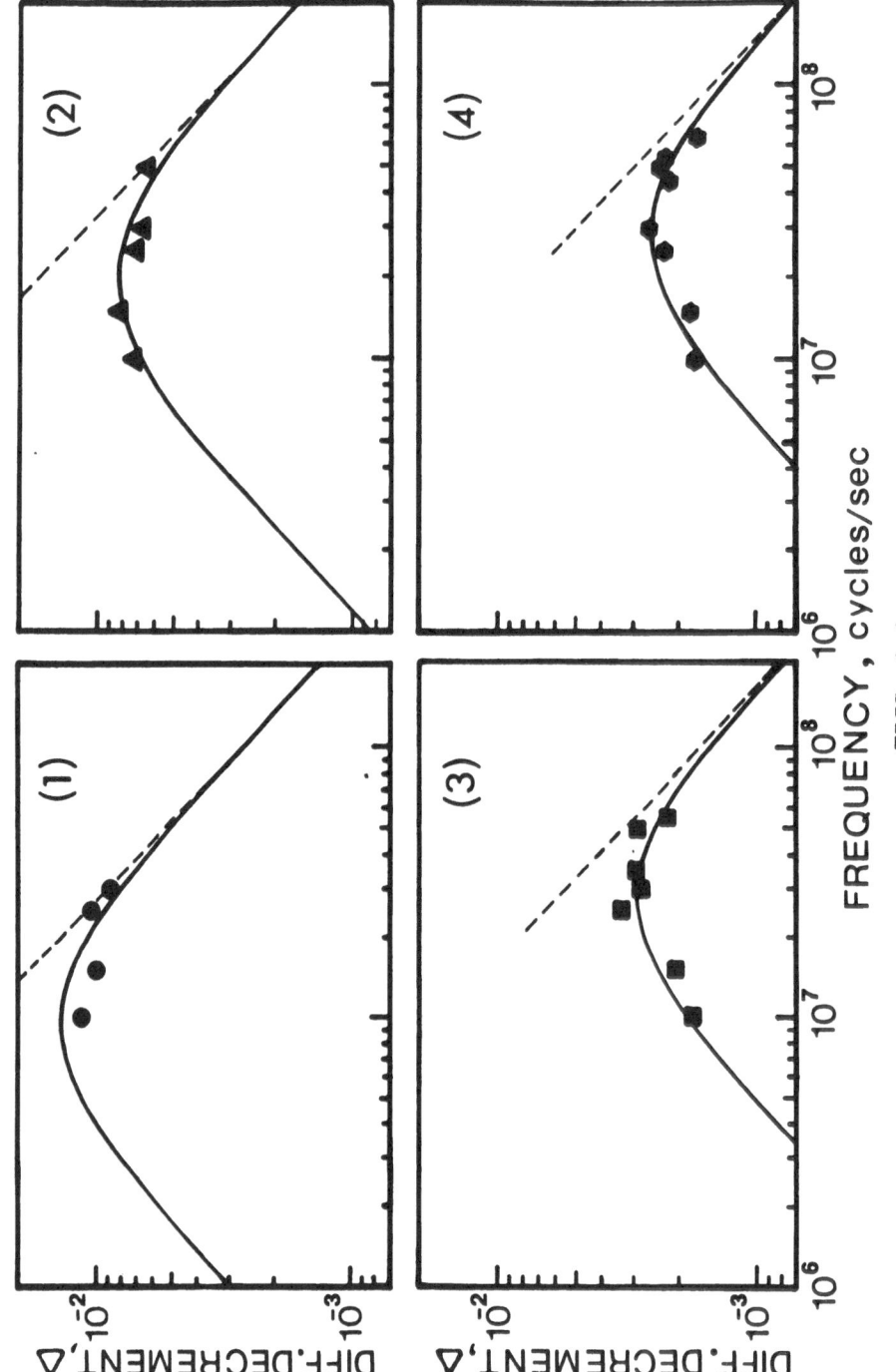

FIGS. 1-4

Frequency dependence of differential decrement at room temperature for (1) Pure Cu, (2) Cu-0.01 at.% Ni, (3) Cu-0.1 at.% Ni, and (4) Cu-1 at.% Ni

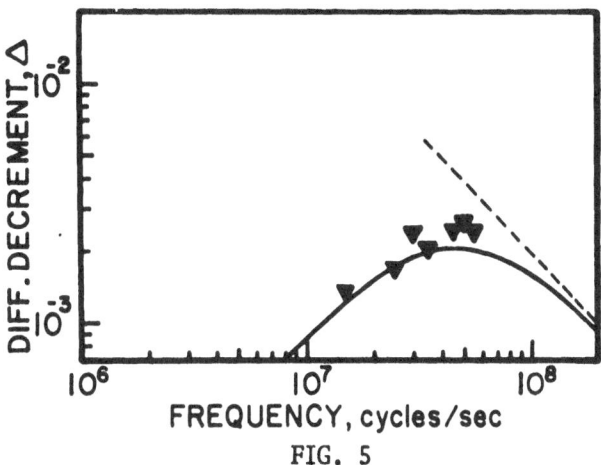

FIG. 5

Frequency dependence of differential decrement at room temperature for Cu-10% Ni

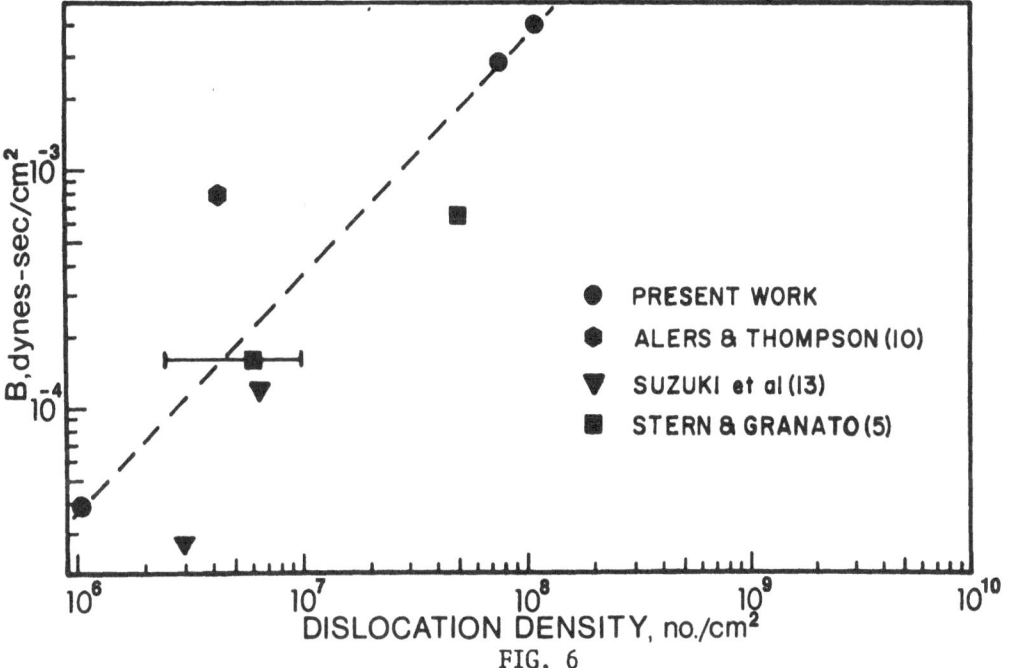

FIG. 6

The room temperature damping constant (B) for pure Cu vs. dislocation density as reported by various workers.

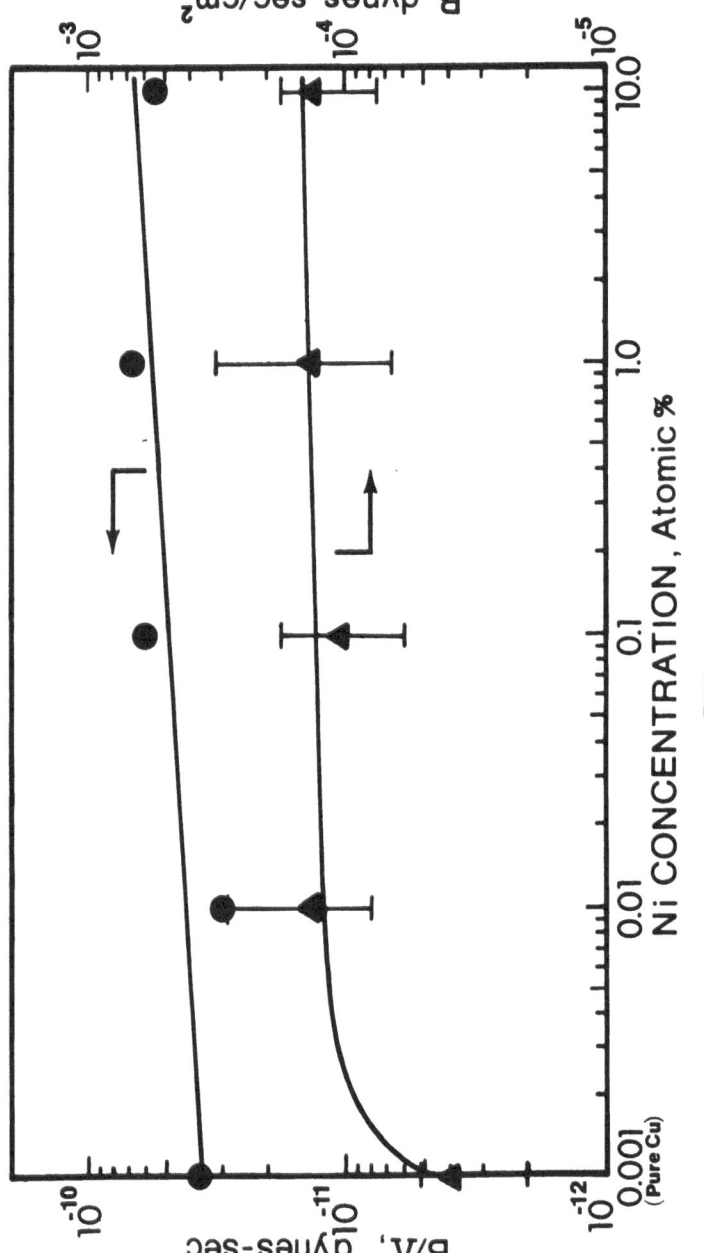

FIG. 7

The room temperature damping constant B (lower curve), and dislocation density normalized damping constant B/Λ (upper line) for Cu-Ni alloys as as a function of Ni concentration.

DISLOCATION RESONANCE

AND

IRRADIATION

THE NATURE OF "DISLOCATION PINNING" BY RADIATION INDUCED DEFECTS, WITH

PARTICULAR REFERENCE TO THE SIMPSON-SOSIN PEAK.

A. Seeger

Max-Planck-Institut für Metallforschung, Institut für Physik,
and Institut für theoretische und angewandte Physik der Universität Stuttgart.

Abstract: The paper develops a physical picture of the interaction between point defects and dislocations as studied by measurements of the internal friction and the elastic modulus defect. The case of a defect with <100> - symmetry that is capable of both rotational and migrational jumps is used to illustrate the general ideas. Particular emphasis is laid on the effects of the long-range interaction between point defects and dislocations. These are contrasted with the usual dislocation-pinning description. The theoretical conclusions are compared with experiments on cold-worked and irradiated nickel and copper. On the basis of internal friction, modulus defect and relaxation experiments it is shown that in these metals the self-interstitial atoms may give rise to a Snoek effect, exist in a <100> - dumbbell configuration, and migrate with an activation energy corresponding to that of the well-known annealing stage III. An earlier interpretation of the recovery stages of f.c.c. metals is thus confirmed in an independent way. It is shown that the present results lead to a natural explanation of the Simpson-Sosin peak in irradiated metals including its strong temperature dependence.

1. Introduction

Beginning with the work of Thompson and Holmes (1) and of Dieckamp and Sosin (2) the effect of irradiation on the internal friction associated with the movement of dislocations in metals has been used for a long time to study point defects generated by radiation. An early review has been given by Holmes (3); for later summaries the reader is referred to Thompson and Paré (4), Sosin and Keefer (5), and Sosin (6).

Most of the experimental work concerned with the study of point-defect-dislocation interactions by means of internal friction or modulus measurements is interpreted in terms of the "vibrating-string model" (3,7,8). Within the framework of this model it is usually assumed that the effect of point defects on the dislocation motion may be described as "pinning" of the dislocation, i.e., as a shortening of the dislocation lengths associated with the arrival of point defects at the dislocation lines. Strong experimental support for the pinning model came very early from the work of Thompson and Holmes (1), who demonstrated that the change in Young's modulus and in logarithmic decrement during neutron irradiation of copper somewhat above room temperature obeyed quite well the dose dependence predicted by the pinning model.

Although the pinning model has since been used extensively in the interpretation of radiation damage and other experiments, it is nevertheless unsatisfactory from a theoretical viewpoint. Through their long-range strain fields dislocations interact with point defects far outside the dislocation core. It is therefore not very likely that the effect of point defects on the dislocation motion is an "all - or - nothing effect", i.e., that the point defects leave the dislocation motion unaffected until they reach the dislocation core but that as soon as they do so, they pin the dislocation completely. Rather, one expects that the influence of a point defect on the dislocation movement and hence on the internal friction and modulus defect associated with it increases gradually as the defect approaches a dislocation line.

The theoretical treatment of the long-range interaction is considerably more complicated than that of the simple pinning model. It requires appropriate mathematical tools, which have been developed in an accompanying paper (9). The present paper is concerned with outlining the general physical ideas which emerge from such an approach, using the <100> - dumbbell self-interstitials in f.c.c. metals as an example, and with discussing a number of relevant experiments.

2. The Dumbbell Self-interstitial in FCC Metals.

In 1960, Seeger, Schiller, and Kronmüller (10) pointed out that the <100> - dumbbell, which according to calculations based on atomistic models (11,12) is the most stable configuration of self-interstitials in copper and presumably in the majority of the f.c.c. metals, should give rise to a relaxation effect analogous to the Snoek effect of carbon atoms in α - Fe (which have the same symmetry). As discussed by these authors (10),the <100> - dumbbell self-interstitials are capable of two basic jump motions, namely a rotation of the dumbbell axis without displacement of the "centre of gravity" of the dumbbell, and a migrational jump, which is automatically accompanied by a change in the orientation of the dumbbell axis. If we denote the frequency of a jump over one of the saddle points for "rotation" by ν^R and the analogous quantity for "migration" by ν^M, the usual treatment of the relaxation effect in a weak homogeneous field (13,14) leads to the relaxation times

$$\tau_{1,2} = \frac{1}{6(\nu^R+2\nu^M)} \qquad (1a)$$

$$\tau_3 = \infty \qquad . \qquad (1b)$$

The diffusion coefficient of the interstitial atoms is given by (15,16)

$$D_I = \frac{2}{3} a^2 \nu^M \leq \frac{a^2}{18} \frac{1}{\tau_{1,2}} \qquad , \qquad (2)$$

where a is the length of the elementary cube edge.

The existence of the Snoek effect associated with <100> - dumbbell interstitials in f.c.c. metals has been demonstrated in Ni (in which case the experiments gave $\nu^R \gg \nu^M$) (10,16-21) and in Cu ($\nu^M \gg \nu^R$, ref. 20,21). For a theoretical treatment of the relaxation strength and its orientation dependence see Seeger, Mann and v.Jan (12). In the following sections we shall use the example of the dumbbell self-interstitials in f.c.c. metals to illustrate the interaction of point-defects giving rise to a Snoek effect with dislocations.

3. The Impedence of Dislocation Movement by the Long-range Interaction with Point Defects.

For definiteness, we consider a uniform concentration of <100> - dumbbell interstitials in an f.c.c. metal with the dumbbell axis distributed equally over the x-,y-, and z-directions. At time t=0 we switch on the (elastic) interaction with a long straight dislocation line. This interaction changes the equilibrium distribution of the dumbbell axes over the three possible crystallographic orientations in such a way that the Gibbs free energy of the system "dislocation line plus dumbbells" is lowered. Since the interaction energy increases as the reciprocal distance from the dislocation line, a potential well forms around the dislocation line. It approaches its equilibrium shape according to a time-law $(1 - \exp[- t/\tau_{1,2}])$. If the dislocation vibrates under the influence of an oscillating external stress of constant amplitude with a frequency $f \gg \tau_{1,2}^{-1}$, the amplitude of the dislocation motion and hence the modulus defect and the internal friction due to the dislocation motion decreases gradually as the potential well develops. These effects may be observed as <u>mechanical after-effects</u> (22); they can thus be used to determine $\tau_{1,2}$ experimentally. Measurements of $\tau_{1,2}$ by the mechanical after-effects are of considerable practical importance, since they permit the determination of relaxation times which are several powers of ten larger than those obtainable from internal friction measurements with a torsion pendulum. Examples will be given later (Sect.6).

In the preceding paragraph we considered only the redistribution of the dumbbell axes under the influence of the interaction with the dislocation and disregarded the fact that the free enthalpy of the system dislocation - dumbbells may be lowered further if the dumbbells respond to the inhomogeneity of the dislocation stress field, i.e., if they take up a non-uniform spatial distribution by means of migrational jumps. By arguments analogous to those given above this process can be measured as after-effect of internal friction or modulus, too. The rate of development of the dislocation potential well and hence the elastic after-effect will thus be proportional to the migrational frequency ν^M. The following line of reasoning shows that to a good approximation the initial time dependence should obey a $([1 - \exp(-t/\tau_3)]$ - law with a relaxation time τ_3 differing from $\tau_{1,2}$: We may write down a set of rate equations for the probabilities of finding a dumbbell of one of the three possible orientations at any one of the interstitial sites of the crystal. Since for $f \gg \nu^M$ the dislocation-dumbbell interaction may be considered as effectively time-independent, these rate equations constitute a set of 3N coupled differential equations (N = number of atoms in the crystal) of first order in time with time-independent coefficients. The solutions of such equations are exponential functions in time with time constants ("relaxation times") following from the solution of the secular equation. The three relaxation times τ_i (i=1,2,3) associated with spatially uniform distributions have been given in (1a,b). We have now to ask for the relaxation times associated with the approach to the equilibrium distribution determined by the space-dependent interaction with the dislocation line. The appropriate way to handle this problem is through a spatial Fourier decomposition of the dumbbell distribution. As in the analogous problem of lattice vibrations in crystals one is led to relaxation times $\tau_i(\underline{k})$ which are functions of the wave-vector $\underline{k}$. The experimentally observed relaxation times depend thus on the Fourier decomposition of the interaction energy, i.e., in the case of elastic interactions with dislocations on the Fourier transforms of the stress fields. The relaxation times $\tau_i(k)$ may be considered to develop, with increasing k, from the relaxation times τ_i of (1a,b) (i=1,2,3), to which they reduce for k→0. In this way it can be seen that in the presence of dislocations one should observe one relaxation time very close to (1a) (this is the justification for the statements in the first paragraph of this section) and another one, τ_3, obeying

$$\frac{1}{8\nu^M} \leq \tau_3 \leq \infty \quad . \tag{3}$$

Hornung (23) has shown that for a screw-dislocation in a <110> - direction in an elastically isotropic crystal one finds

234    $\tau_3 = \dfrac{1}{7\nu^M} \quad .$ (4)

Eq.(4) is fairly close to the lower limit of (3); the upper limit follows from the fact that (3) must include the special case of (1b).

The preceding argument considers essentially one migrational jump and is therefore valid only for times $t \ll 1/\nu^M$. If a substantial fraction of the dumbbells carries out several migrational jumps during the time of observation, a diffusion description becomes appropriate. In the accompanying paper (9) it is shown that by introducing partial diffusion coefficients $\underline{D}^{ij}$, which describe not only spatial diffusion but also the changes of the dumbbell axes from orientation i to orientation j, it is possible to give a continuum approximation to that regime. It is found that for diffusion and reorientation in the stress field of straight dislocation lines, the initially exponential time-dependence goes gradually over into a t^{-1} - dependence (9,23).

The t^{-1} - dependence can hold only as long as the presence of other dislocations may be neglected, i.e., as long as the competition of different dislocations for the point defects diffusing towards them does not yet make itself felt. Taking into account this competition by introducing appropriate external boundary conditions, the time dependence for long times is changed into an exponential one with a reciprocal time-constant equal to (apart from numerical factors of order unity) the product of defect diffusion coefficient D_I (Eq. 2) and dislocation density ρ.

The various effects discussed in this section may be treated from a unified point of view by the method of Seeger and Hornung (9), provided the interaction energy between a dislocation line and an individual point defect is smaller than kT, i.e., if the main effect on the dislocation motion is caused by point defects that are outside the dislocation core. If the number of point defects interacting with a dislocation line is so small that measurable effects are observed only when some of the defects reach the dislocation core, the pinning picture to be discussed in Sect. 4 is more appropriate. Fig. 1 summarizes the main results of preceding discussion.

4. Pinning of Dislocations by Point Defects.

As mentioned in Sect. 1, the "pinning model" of dislocation damping makes the extreme assumption that point defects in the neighbourhood of a dislocation line do not interfere with its movement until they have reached the dislocation core, where they are supposed to pin the dislocation line completely and thus to reduce the dislocation loop length appearing in the "vibrating string model" of dislocation damping. It is clear that for point defects possessing an appreciable long-range interaction with dislocations, such as the dumbbell self-interstitials in f.c.c.metals, this picture can be a good approximation only if the concentration of the point defects involved is sufficiently small, so that the total effect of the long-range interaction remains small compared with the pinning effects.

The important quantity in the pinning picture is the number of point defects arriving per unit length of the dislocations due to diffusion and drift in the field of force of the dislocations. Most workers in the field use the Cottrell-Bilby law (24,25), which states that, starting at t=0 from a uniform concentration C_0 of point defects interacting with a straight edge dislocation according to a potential energy

$$U(\underline{r}) = \frac{A \cos \phi}{r} \tag{5}$$

$(r, \phi$ = polar coordinates), the number of point defects having arrived on a unit length of a dislocation line after time t is given by

$$n(t) = 3.49 \ C_o \ (\frac{A \ D \ t}{kT})^{2/3} \ . \tag{6}$$

However, the Cottrell-Bilby result (6) is incorrect because of inadmissible simplifications in the treatment of the diffusion equation. The correct result reads (26)

$$n(t) = 2 \ C_o \ \frac{A}{kT} \ (\pi \ D \ t)^{1/2} \ , \tag{7}$$

provided we are justified in disregarding the competition of neighbouring dislocations for the point defects. If this is taken into account, for very long times the time-law (7) goes over into an exponential one similar to that discussed at the end of Sect.3.

The long-range interaction picture of Sect.3 and the pinning picture of the present section constitute two limiting cases. In a given experiment both aspects may play a rôle. The development of a general theory which includes both aspects as special cases remains a task for the future.

Experimentally the pinning picture and the long-range-interaction picture may be distinguished by comparing the damping and the modulus defect due to dislocations. Since in the pinning picture the dislocation damping varies as the fourth power of the distance L between the pinning points (dislocation loop length), whereas the modulus defect varies as L^2, one expects a proportionality of the modulus defect to the square of the damping. This has indeed been observed in irradiation experiments involving very small numbers of point defects (1,27,28), i.e., under conditions where the pinning picture is most likely to apply. By contrast, the long-range interaction picture predicts a relationship between damping and modulus defect which is closer to a first power, a result borne out by more recent experiments.

5. The Enhancement of Internal Friction by Dislocation - Point-Defect Interactions.

In Sect. 3 and 4 we have been concerned with the reduction of the internal friction associated with the motion of dislocations by either long-range interaction with or pinning by point defects. In the present section we shall discuss the enhancement of internal friction by the interaction of point defects with oscillating dislocations. We illustrate this for the example of the dislocation-enhanced Snoek effect discussed earlier (16,29,30):

As discussed in the first paragraph of Sect. 3, the redistribution of dumbbell axes around a fixed or very rapidly oscillating dislocation (which may be referred to as the development of a "Snoek cloud") takes place in times of the order of magnitude of $\tau_{1,2}$. If under the influence of the external stress the dislocation oscillates with a frequency f approximately equal to $1/2\pi\tau_{1,2}$, the distribution of the dumbbell axes and hence the strain associated with it is out of phase from the applied stress by approximately 90°. Hence a large energy dissipation results. This effect has been termed "dislocation-enhanced Snoek" effect since the oscillating applied stress giving rise to the Snoek distribution of the dumbbell axes is amplified by the oscillating stresses resulting from the dislocation movement. As long as the dislocation amplitude is small compared with the extension of the "Snoek cloud" (i.e., the width of the potential well

236

established by the redistribution of dumbbell axes), the enhancement is amplitude independent and is easy to treat theoretically. It leads to a relaxation peak of essentially the same position and width as the unenhanced Snoek peak. If the dislocation amplitudes are comparable with or larger than the extension of the Snoek cloud, the effect becomes amplitude dependent and the relaxation peak broadens and shifts. This regime is very difficult to treat quantitatively.

The existence of the dislocation-enhanced Snoek effect means that if one wants to study the Snoek relaxation in a quantitative way in the presence of mobile dislocations, he first has to reduce the dislocation amplitudes until the amplitude independent regime is reached. This may be done by suitable annealing treatments, making use of the impedence of dislocation motion by the point defects giving rise to the Snoek effect as discussed in Sects. 4 and 5. However, since during such anneals the concentration of point defects close to the dislocations is enhanced, they may result in an initial increase of the damping before the amplitude dependent contribution is eliminated.

6. Experimental Evidence for the Internal Friction Effects Associated with Dumbbell Self-Interstitials in FCC Metals.

Let us first consider a situation in which the jump frequency for the rotation of <100> - dumbbells, ν^R, is very much larger than ν^M, the jump frequency for migrational steps. Then the "rotation effects", i.e., Snoek effect and dislocation-enhanced Snoek effect, are clearly separated in temperature from the "migration effects" on internal friction and damping. The activation energy associated with the latter is the enthalpy of migration of self-interstitials, H_I^M. From the agreement or disagreement of H_I^M with the activation energies measured in the various recovery stages conclusions about the assignment of self-interstitial migration to one of the recovery processes may be drawn.

The condition $\nu^R \gg \nu^M$ is satisfied for Ni. Fig. 2 shows the $\tau_{1,2}$ values obtained from the mechanical after-effect of internal friction or modulus defect of single or polycrystals plastically deformed in tension (21), from the temperature dependence of Young's modulus after low-temperature tensile deformation of polycrystals (31), from the magnetic after-effect of the initial susceptibility of plastically deformed or electron irradiated nickel (18) and from the temperature dependence of the internal friction of plastically deformed or electron irradiated single or polycrystals (2o). The data extend over five and a half powers of ten and give

$$\tau_{1,2} = 2 \cdot 10^{-14} \text{ s } \exp(H_I^R/kT) \tag{8}$$

with

$$H_I^R = (0.87 \pm 0.03) \text{ eV} \tag{8a}$$

and corresponding error limits on the preexponential factor.

Fig. 2 contains further relaxation times obtained by Friedrich (32) from measurements of the socalled resistivity susceptibility χ_R(33) on polycrystalline nickel deformed 8% in tension at 77 K. Within experimental error they give the same activation energy as the measurements on which (8a) was based. The absolute values of the relaxation times are somewhat lower than those found in the after-effect measurements of Wagner (21), but may still be compatible with these within experimental error.

Wagner (16,20) demonstrated that after elimination of the dislocation - enhancement effect the internal friction experiments showed the orientation dependence that is theoretically predicted (12) for a defect with tetragonal symmetry. He showed further (20) that magnetic fields high enough to eliminate or at least to change completely the ferromagnetic domain structures leave both the internal friction and the mechanical after-effect unaffected. This result is significant in two respects: (i) It shows that the interpretation in terms of mechanical energy losses (16) is justified and that the effect is not due to magnetostrictive effects. (ii) The defects giving rise to the effect must have <100> - symmetry, since because of the <111> - directions of the spontaneous magnetization inside the ferromagnetic domains of nickel, this is the only symmetry that should not give rise to large and easily detectable energy losses associated with the motion of the domain walls under the influence of the applied stress.

In addition to the relaxation time $\tau_{1,2}$ just discussed, the mechanical after-effect measurements on plastically deformed nickel exhibit a second, considerably larger, relaxation time. If this after-effect is analyzed with the help of Hornung's theory (23) of the relaxation due to the first migrational step of a <100> - dumbbell towards a screw dislocation the τ_3 - values showed in Fig.2 are obtained. Because of recovery effects at higher temperatures the available data cover only two and a half powers of ten ; they may be described by

$$\tau_3 = 2 \cdot 10^{-15} \text{ s } \exp(H_I^M/kT) \tag{9}$$

with

$$H_I^M = (0.99 \pm 0.05) \text{ eV} \tag{9a}$$

and corresponding error limits on the preexponential factor.

We have designated the activation energy in (9) as the migration enthalpy H_I^M of self-interstitials, (i) since it agrees within experimental error with the activation energy with which the $\tau_{1,2}$ - effect anneals out, (ii) since the $\tau_{1,2}$ - and τ_3 - after-effects are always coupled and possess relative intensities in accord with the theoretical predictions (23) and, (iii) since the experimental evidence on the entire effect is strongly in favour of attributing it to the <100> - dumbbell self-interstitials. (See also the discussion on Cu at the end of this section.)

The τ_3-after-effect is followed by a slowly decreasing after-effect, which may be described by a $t^{1/2}$-law and which gives an activation energy of (0.95 ± 0.1) eV (21) coinciding with H_I^M within experimental error, in agreement with the predictions of Sect. 3 and 4.

In contrast to nickel, mechanical after-effect measurements on plastically deformed copper single and polycrystals do not show two exponential contributions with widely separated relaxation times $\tau_{1,2}$ and τ_3. If these relaxation times are determined by fitting time-laws of the functional form used for nickel to the Cu data, one obtains a temperature-independent ratio $\tau_3/\tau_{1,2}$ of about 2. This corresponds to the theoretical expectations (Sect. 3 and 4) if ν^R is smaller than ν^M.

The result $\nu^R < \nu^M$ means that in Cu the jumps giving rise to the Snoek effect of dumbbell self-interstitials are also responsible for the annealing out of the defects. The number of cycles that can be used to investigate the Snoek effect is therefore very limited and experimental studies on Cu are much more difficult than those on Ni. In spite of these difficulties Wagner (20) has succeeded in measuring the Snoek-effect of <100> - dumbbell self-interstitials on plastically deformed copper in an inverted torsion pendulum. The relaxation times obtained in this way are shown in Fig. 3. Also included are the relaxation times that are found by analysing an isotherm of Young's modulus measured by D.Keefer (34,35) on electron irradiated copper and those obtained by Bischoff (31) from the temperature dependence of Young's modulus of polycrystalline copper deformed at 77 K either cyclically or in tension.

It is seen that the relaxation times identified with $\tau_{1,2}$ fall very well on one and the same Arrhenius line

$$\tau_{1,2} = 2\cdot10^{-15}s \; exp(H_I^M/kT) \tag{10}$$

$$H_I^M = (0.67 \pm 0.03) \; eV \tag{10a}$$

with a corresponding uncertainty in the preexponential factor. In Fig. 3 the torsion-pendulum measurements of Walz (36) and of Völkl and Schilling (37) have been left out, since they appear to be representative of a strongly dislocation-enhanced Snoek maximum (though a smaller relaxation peak appearing in Völkl's and Schilling's work after light anneal falls on the Arrhenius line of Fig. 3 and is presumably due to the Snoek effect proper of dumbbell self-interstitials).

In Cu the after-effect described by the relaxation times $\tau_{1,2}$ and τ_3 is followed by a $t^{1/2}$-dependence with an activation energy of (0.70 ± 0.05) eV (21). Thus the entire mechanical after-effect of cold-worked copper in the temperature range 175 K to 200 K (with the exception of a fast initial process presumably associated with dumbbells located very close to dislocation cores, which for simplicity are left out of our discussion) is characterized by the same activation energy. This is in contrast to the situation obtaining in nickel, and constitutes, within the frame-work of the present interpretation, clear evidence that the rotational movement of dumbbell interstitials in copper proceeds more slowly than the elementary migrational step.

Whereas on nickel it has been demonstrated extensively that the $\tau_{1,2}$-relaxation occurs not only in cold-worked but also in electron-irradiated samples (comp. Fig. 2), the experiments on copper sofar presented all pertain to cold-work, Keefer's data (34,35) being the only exception. Since self-interstitials are one of the two primary defects created in radiation-damage experiments, it is important to ask the question whether any other of the phenomena attributed to self-interstitials in Cu have also been observed in irradiation experiments. In view of the difficulties, described above, inherent in the experimental study of the Snoek effect of self-interstitials in copper we shall concentrate on the phenomena associated with long-range self-interstitial migration. These may be observed by measuring internal friction and elastic modulus either during irradiation (at-temperature experiments), or during or after a warm-up following low-temperature irradiation (isochronal or isothermal annealing experiments).

A series of important experiments have been carried out by Thompson and Buck (27,38,39). They measured Young's modulus and internal friction (at a frequency of 11 kHz) of copper single crystals during γ-irradiation at various temperatures between 333 K and 393 K. Among other information they deduce time-constants τ_{LE} characterizing the diffusion of irradiation-produced point defects to dislocations (Fig. 3). We see that these τ_{LE}-values possess the same temperature dependence as $\tau_{1,2}$ but are about nine orders of magnitude larger than these. This supports strongly the interpretation of τ_{LE} as describing the competition of neighbouring dislocation lines for self-interstitials generated in the lattice (39). The displacement parallel to τ-axis required to bring the Thompson-Buck values onto the $\tau_{1,2}$ Arrhenius plot (Fig.3) is approximately of the order of magnitude as but somewhat larger than expected from estimated dislocation densities.

Winterhager (40) has carried out both at-temperature and annealing experiments on γ-irradiated copper in the megacycle frequency range. Fig. 3 includes the time-constants obtained by isothermal anneals after 220 K - irradiations. They give the same activation energy as the $\tau_{1,2}$ - and τ_3 - determinations. The preexponential factor lies between that of $\tau_{1,2}$ and the Thompson-Buck data. At-temperature experiments by Winterhager (40) gave similar results. The natural interpretation of these findings is that Winterhager has observed the long-range migration of the defects giving rise to the Snoek effect, i.e., of dumbbell self-interstitials, to dislocations in specimens with a considerably smaller dislocation density than those employed by Thompson and Buck.

We finally mention measurements of Young's modulus at about 5 kHz on high-purity copper after 3 MeV-electron irradiation at 78 K and different annealing treatments by Naundorf, Roth and Lücke (41). These authors find that during anneals at temperatures between 270 K and 310 K the modulus defect is reduced with an activation energy of about 0.7 eV. From their data it is not possible to deduce time-constants or the time-law of annealing, but on the basis of an estimated number of jumps the same explanation as that given above for Winterhager's observations appears to be natural.

The evidence presented sofar indicates strongly that all the internal-friction and modulus-defect effects described in this section are due to the same type of defect. About the nature of this defect we may state the following:

(i) It must be one of the two elementary intrinsic point defects produced in radiation damage experiments, i.e., a monovacancy or a single self-interstitial. This follows from the facts that under the conditions of the Thompson-Buck experiment these two defects are the only ones generated in detectable numbers, and that on account of the rather high temperatures at which the experiments were performed the instantaneous bulk concentrations of these defects are so low that the formation of divacancies or di-interstitials by defect diffusion is completely negligible.

(ii) The defect must be able to give rise to a Snoek effect, i.e., must have lower than cubic symmetry.

(iii) The defect must possess tetragonal symmetry, i.e., a preferred <100> - axis. This follows from both the orientation dependence of the Snoek effect and the absence of a magnetic-field effect in Ni.

(iv) Impurity effects, i.e., interactions between immobile impurity atoms and the migrating intrinsic defects, are unimportant for a number of reasons: (a) As convincingly argued by Thompson and Buck (39), the large number of jumps observed by them precludes any appreciable impurity effects. (b) The same activation energy is observed over an extremely wide range of ratios between the concentrations of impurities and intrinsic defects. (c) The same activation energy is observed for short-range diffusion (essentially one migrational jump) and for long-range diffusion. (d) A dependence of the activation energy on the origin and pre-history of the specimens has not been found.

Since it does appear very improbable that monovacancies in close-packed metals such as Cu and Ni show deviations from cubic symmetry, let alone of the required strength and symmetry, we must conclude that the observed effects are entirely due to self-interstitials. This conclusion is confirmed by the fact that the symmetry following from the experiments as well as the relaxation strength is in full agreement with the theoretical predictions (12) for self-interstitials in Cu. Another gratifying feature is that the possibilities $\nu^R \gg \nu^M$ and $\nu^M \gg \nu^R$, which are quite characteristic for the model used, have been borne out by the two metals investigated in detail, nickel and copper. We may therefore state that the assumptions made at the beginning of the present section have been fully verified and that the migration enthalpies of <100> - dumbbell self-interstitials in Cu and Ni are correctly given by Eqs. (9a) and (10a).

7. Interpretation of Recovery Stages.

The results of Sect. 6 have considerable significance for the interpretation of the recovery stages in f.c.c. metals, a subject which is still controversal (42). The migration enthalpies obtained for self-interstitials in nickel and copper agree within experimental error with the activation energies of stage III recovery (43-47). The results of Sect. 2 form thus a proof, outlined earlier (48), for the assignment of stage III to self-interstitial migration. We emphasize that this proof is independent of the other experimental proofs, e.g., the "historical" one, based mainly on the comparison of quenching, cold-work and irradiation data (49), the proof based on the absence of a low-temperature free-migration stage in Au and Pb (50,51) and the proof based on the quantitative analysis of high-temperature equilibrium and self-diffusion data (52).

The identification of self-interstitial migration with recovery stage III allows us to test the assumption, implicit in the comparison of the magnitude of the Snoek-effect of single crystals of different orientations, that comparable concentrations of self-interstitials are involved. K.Maier (53) has deformed copper single crystals of <111> and <100> orientations at temperatures of 4.2 K, 77 K, 173 K, and 188 K in tension by amounts comparable with those used by Wagner (20). He found that the isochronal recovery curves of the residual electrical resistivity had the same shape for the two crystallographic orientations. Stage III obeyed approximately second order kinetics with an activation energy of about 0.7 eV. The absolute value of the resistivity recovery of the <100>-orientated crystals was about 15% higher than that of the <111>-orientated crystals. This result demonstrates that the absence of the Snoek-effect in <100>-crystals cannot be attributed to the absence of the stage-III defects in these samples.

8. The Simpson-Sosin Peak

In this section we discuss the superposition of the reduction of the internal friction and of the modulus defect resulting from long-range and pinning interactions between dislocations and point defects, treated in Sect.3 and 4, and of the enhancement of the internal friction by dislocation-point-defect interactions, treated in Sect.5. For definiteness we consider an isothermal irradiation experiments of the type carried out by Thompson and Buck (27,28,39) at temperatures which are low enough for the relaxation time $\tau_{1,2}$ of the self-interstitials to be larger than the reciprocal measuring frequency, f^{-1}, but small enough for the rotational or migrational mobility of the defects to be appreciable during the irradiation time. If the self-interstitials are deposited homogeneously in the crystal, the dislocation-enhanced Snoek effect increases linearly with irradiation time as long as the dislocation motion is essentially un-impeded. Diffusion and drift during irradiation causes the interstitial to migrate closer to the dislocations and thus to increase the enhancement effect. By these arguments one expects, at the beginning of an isothermal irradiation experiments performed under the above-mentioned conditions, an increase in the internal friction. However, with growing point-defect concen-tration and increasing drift towards the dislocations the gradual reduction of the dislocation motion makes itself felt, and eventually the internal friction will decrease until the dislo-cation contribution is completely eliminated.

From the preceding discussion it emerges that in experiments of the Thompson-Buck type one expects the long-time decrease of the internal friction and the modulus defect discussed in Sect.6 to be preceded by a maximum of these quantities as a function of irradiation time. In retrospect it may be stated that Thompson and Buck (27) did not observe such a maximum because it occured at irradiation times at which, in their particular set-up, the temperature was not yet sufficiently stable. In similar experiments, in which, however, the initial heating problems were much less serious, Simpson, Sosin and coworkers observed maxima of the nature discussed above on both copper (54,55) and aluminium (56) during electron irradiation. We do not wish to enter here into a discussion of the more detailed experiments (57) or the explana-tions advanced in the literature (see the contribution of A.Sosin to this conference) but em-phasize the particular features predicted by the present interpretation of the Simpson-Sosin peak.

The outstanding feature of the mechanism outlined above is that it relates the height of the maximum to the dislocation-enhanced Snoek effect of the self-interstitials. This means that for $2\pi f \cdot \tau_{1,2} \gg 1$ the effect should decrease rapidly with decreasing temperature, a prediction that is clearly borne out by the experiments but remains unaccounted by Simpson's and Sosin's dragging model (58). Another prediction, not yet subjected to a quantitative experimental test, is that in the regime $2\pi f \cdot \tau_{1,2} \gg 1$ the initial slope of the internal-friction - vs. - time curve should be inversely proportional to the measuring frequency, since it represents simply the low-temperature side of the dislocation-enhanced self-interstitial Snoek relaxation maximum.

Acknowledgment

The author acknowledges gratefully discussions with Privatdozent Dr. W.Frank, Dr. W. Hornung, Dr. D.Keefer, Prof. H.Kronmüller, Prof. A.Sosin, Dr. F.-J.Wagner, and Dipl.-Phys. D.Werneth.

References

1. D.O.Thompson and D.K.Holmes, J.Appl.Phys. <u>27</u>, 713 (1956)

2. H.Dieckamp and A.Sosin, J.Appl.Phys. <u>27</u>, 1416 (1956)

3. D.K.Holmes, in: Radiation Damage in Solids (D.S.Billington, ed.),p.777, Academic Press, New York and London 1962

4. D.O.Thompson and V.K.Paré, in: Physical Acoustics, Vol.IIIA (W.P.Mason, ed.) p.294, Academic Press, New York and London 1966

5. A.Sosin and D.W.Keefer, in: Advances in Materials Research, Vol.2 (H.Herman, ed.), p.159, Interscience Publishers, New York etc. 1968

6. A.Sosin, in: Vacancies and Interstitials in Metals (A.Seeger, D.Schumacher, W.Schilling, J.Diehl, eds.), p.729, North-Holland Publ.Comp., Amsterdam 1970

7. J.S.Koehler, in: Imperfections in Nearly Perfect Crystals (W.Shockley, J.H.Hollomon, R.Maurer, and F.Seitz, eds.), p.197, John Wiley + Sons, New York 1952

8. A.V.Granato and K.Lücke, J.Appl.Phys. <u>27</u>, 583, 789 (1956)

9. A.Seeger and W.Hornung, this volume

10. A.Seeger, P.Schiller and H.Kronmüller, Phil.Mag. <u>5</u>, 853 (1960)

11. J.B.Gibson, A.N.Goland, M.Milgram and G.H.Vineyard, Phys.Rev. <u>120</u>, 1229 (1960)

12. A.Seeger, E.Mann and R.v.Jan, J.Phys.Chem.Solids <u>23</u>, 639 (1962)

13. D.Polder, Philips Res.Rep. <u>1</u>, 1 (1945)

14. C.Zener, Suppl.Nuovo Cimento VII [X], 544 (1958)

15. F.Ramsteiner, W.Schüle and A.Seeger, phys.stat.sol. <u>7</u>, 937 (1964)

16. A.Seeger and F.J.Wagner, phys.stat.sol. <u>9</u>, 583 (1965)

17. H.Kronmüller, A.Seeger and P.Schiller, Z.Naturforschg. <u>15 a</u>, 740 (1960)

18. H.Kronmüller, H.-E.Schaefer and H.Rieger, phys.stat.sol. <u>9</u>, 863 (1965)

19. A.Seeger, in: Magnetismus - Struktur und Eigenschaften magnetischer Festkörper (H.Ringpfeil, ed.), p. 160, VEB Deutscher Verlag für Grundstoffindustrie, Leipzig 1967

20. F.J.Wagner, phys.stat.sol.(b) <u>51</u>, 589 (1972)

21. F.J.Wagner, phys.stat.sol.(b) <u>54</u>, 135 (1972)

22. F.J.Wagner and A.Seeger, Phys.Letters (Netherlands) <u>30 A</u>, 274 (1969)

23. W.Hornung, phys.stat.sol. <u>54</u>, 341, 441 (1972)

24. A.H.Cottrell and B.A.Bilby, Proc.Phys.Soc. <u>A 62</u>, 29 (1949)

25. R.Bullough and R.C.Newman, Acta Met. <u>10</u>, 971 (1962)

26. A.Seeger, to be published

27. D.O.Thompson, O.Buck, R.S.Barnes and H.B.Huntington, J.Appl.Phys. <u>38</u>, 3051 (1967)

28. D.O.Thompson and O.Buck, phys.stat.sol. <u>37</u>, 53 (1970)

29. G.Schoeck, Phys.Rev. <u>102</u>, 1458 (1966)

30. G.Schoeck and A.Seeger, Acta Met. <u>7</u>, 469 (1959)

31. A.Bischoff, Dr.rer.nat.thesis, Universität Stuttgart 1973

32. E.Friedrich, Dr.rer.nat.thesis, Universität Münster 1973

33. W.Hellenthal and U.Lotter, Z.angew.Physik <u>26</u>, 84 (1969)

34. D.Keefer, J.C.Robinson and A.Sosin, Acta Met. <u>13</u>, 1135 (1965)

35. D.Keefer, private communication

36. E.Walz, phys.stat.sol. <u>7</u>, 953 (1964)

37. J.Völkl and W.Schilling, Phys.Kondens.Materie <u>1</u>, 296 (1963)

38. D.O.Thompson, O.Buck, H.B.Huntington,and P.S.Barnes, J.Appl.Phys. 38, 3057 (1967)

39. D.O.Thompson and O.Buck, J.Appl.Phys. 38, 3068 (1967)

40. P.Winterhager, Dr.Ing.thesis, T.H.Aachen 1968

41. V.Naundorf, G.Roth, and K.Lücke, Cryst.Lattice Defects 2, 205 (1971)

42. W.Schilling and K.Sonneberg, J.Phys.F : Metal Phys. 3, 322 (1973)

43. W.Schilling, G.Burgers, K.Isebeck, and H.Wenzl, in: Vacancies and Interstitials in Metals (A.Seeger, D.Schumacher, W.Schilling, and J.Diehl, eds), p.255, North-Holland Publishing Comp., Amsterdam 1970

44. H.Kronmüller, A.Seeger, H.Jäger, and H.Rieger, phys.stat.sol. 2, K 105 (1962)

45. F.Dworschak and J.Koehler, Phys.Rev. 140, A 941 (1965)

46. I.A.Gindin, M.B.Lazareva, V.M.Matsevityy,and Ya.D.Starodubov, Fiz.metal.metalloved. 28, 466 (1969)

47. J.Polák, phys.stat.sol. 40, 677 (1970)

48. A.Seeger, in: Vacancies and Interstitials in Metals (A.Seeger, D.Schumacher, W.Schilling, and J.Diehl, eds.), p. 999, North-Holland Publishing Comp., Amsterdam 1970

49. A.Seeger, Handbuch der Physik, Vol. VII/1 (S.Flügge, ed.), p. 383, Springer-Verlag, Berlin-Göttingen-Heidelberg 1955

50. A.Seeger, J.Phys.Soc.Japan 18, Suppl. III, 260 (1963)

51. A.Seeger, Phys.Letters (Netherlands) 35A, 135 (1971)

52. A.Seeger, J.Phys.F.: Metal Phys. 3, 248 (1973)

53. K.Maier, Diplomarbeit,Universität Stuttgart 1971

54. H.M.Simpson, A.Sosin, G.R.Edwards and S.L.Seiffert, Phys.Rev.Letters 26, 897 (1971)

55. H.M.Simpson, A.Sosin, and S.Seiffert, J.Appl.Phys. 42, 3977 (1971)

56. S.L. Seiffert, H.M.Simpson, and A.Sosin, J.Appl.Phys. 44, 3404 (1973)

57. H.M.Simpson, A.Sosin, and D.F.Johnson, Phys.Rev. 5, 1391 (1972)

58. H.M.Simpson and A.Sosin, Phys.Rev. 5, 1382 (1972)

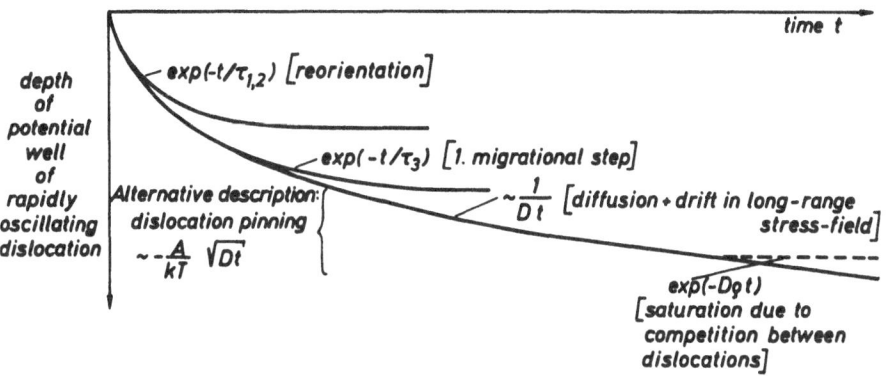

Fig. 1:

Reduction of dislocation motion due to the interaction with a
uniform initial distribution of dumbbell interstitials

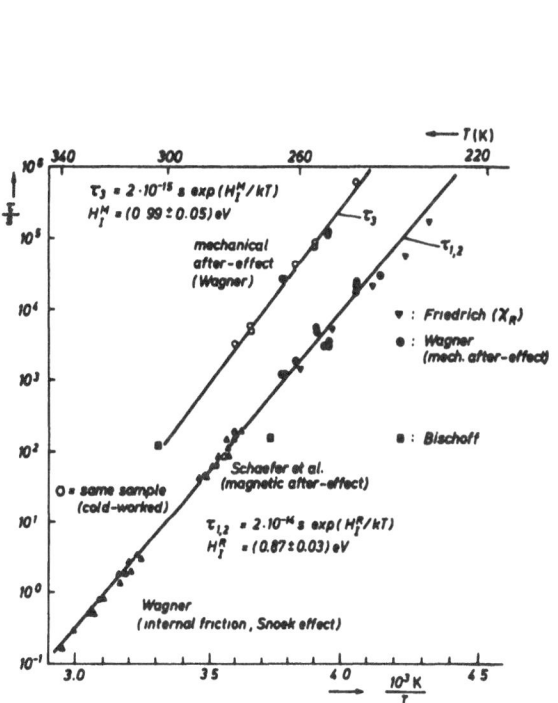

Fig. 2:

Arrhenius plots of the relaxation times measured on cold-
worked or electron-irradiated
nickel

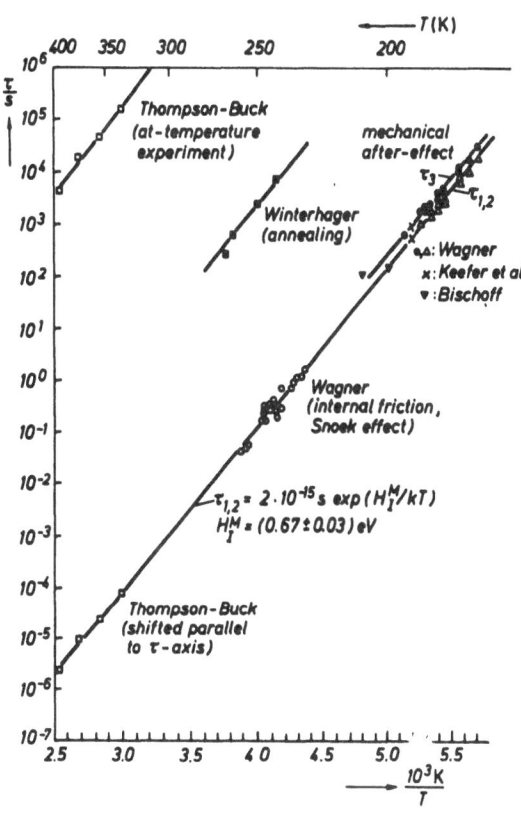

Fig. 3:

Arrhenius plots of the relaxation
times measured on cold-worked or
electron (γ)-irradiated copper

DISLOCATION DAMPING DUE TO MOBILE PINNING POINTS

Wladis Winkler-Gniewek, J. Schlipf and R. Schindlmayr
Institut für Allgemeine Metallkunde und Metallphysik,
Technische Hochschule Aachen, W.-Germany

The damping of cold-worked metals shows several relaxation maxima /1/ which
are due to the interaction of dislocations with intrinsic or extrinsic point
defects. The present authors feel that at least some of these peaks are due
to the stressinduced motion of pinning points on dislocations.

The drag exerted by mobile pinning points (dragging points) on a dislocation
has been the subject of theoretical investigations by many authors /2-7/.
However, the majority of these treatments have been incomplete in that they
allowed only for a special way of motion of the dragging points, i.e. either
in a direction parallel to the dislocation line (longitudinal motion) or in
a direction parallel to the direction of motion of the dislocation (transversal motion). The situation is illustrated in fig. 1. The total force
exerted by the dislocation on a dragging point is $F^{(\sigma)}$. It can be decomposed into a transversal force $F_T^{(\sigma)}$ and a longitudinal force $F_L^{(\sigma)}$. If the
dragging point possesses transversal mobility only, F_T causes a relaxational
motion. The relaxation time is given by $\tau_T = (\bar{l}^2-q^2)/2\mu m_T\bar{l}$, where m_T is
transversal mobility and μ is the line tension. It is independent of the
applied stress, but depends strongly on the position q of the dragging point.
The transversal motion, therefore, will be greatly influenced by the presence
of longitudinal motion.

In the other extreme, where we have longitudinal mobility only, the force
$F_L^{(\sigma)}$ does not lead to a relaxational motion, since no mechanical restoring
force is present in this case. The entropy force $F_L^{(V)}$ /8,9,10/ due to the
vibration of the free dislocation segments does not give rise to a restoring force either. However, if the statistical character of the motion is
taken into account, we are led to define a distribution function /8/, which

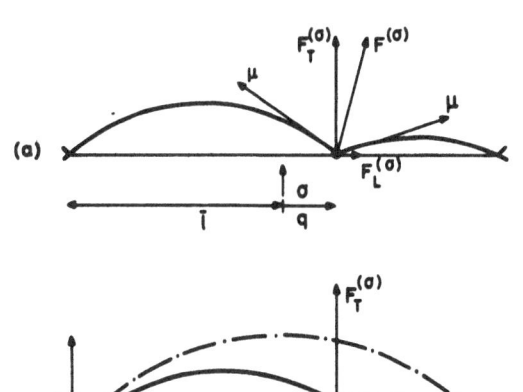

$$F_T^{(\sigma)} = \sigma\,b\bar{l}$$

$$F_T^{(\mu)} = \frac{2\mu la}{\bar{l}^2-q^2}$$

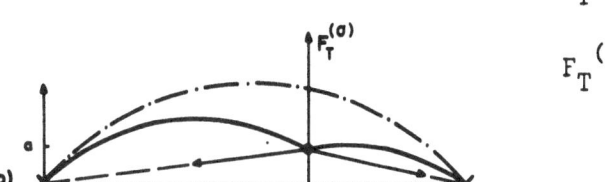

$$F_L^{(\sigma)} = \frac{\sigma^2 b^2}{2\mu}\,\bar{l}q$$

$$F_L^{(V)} = \frac{kT}{\bar{l}^2-q^2}\,q$$

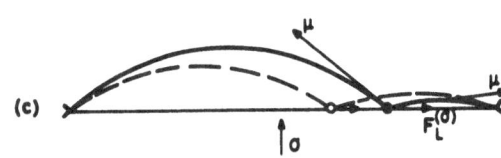

FIG. 1

Stress induced migration of a mobile pinning point on a dislocation line. (a) Mechanical forces acting on the pinning point. (b) Transversal motion under the combined action of the external force $F_T^{(\sigma)}$ and the line tension force $F_T^{(\mu)}$. The broken line indicates the equilibrium position of the dislocation. (c) Longitudinal motion under the combined action the external force $F_L^{(\sigma)}$ and the entropy force $F_L^{(V)}$.

is the probability density of finding the pinner at q. The gradient of the probability density represents a thermodynamic restoring force, giving a relaxation time

$$\tau_L = \bar{l}^2/m_L kT(\pi^2+\alpha_o^2). \tag{1}$$

m_L is longitudinal mobility and

$$\alpha_o^2 = \frac{1}{2}\,(\sigma_o/G)^2(Gb^3/kT)\,(\bar{l}/b)^3 \tag{2}$$

defines a normalized stress amplitude. τ_L depends on stress amplitude, but is independent of the position of the individual dragging point. As can be seen from fig. 1b, the longitudinal force component $F_L = F_L^{(\sigma)} + F_L^{(V)}$ may become very small, if transversal motion is superimposed on the longitudinal motion. In the limit of high transversal mobility, longitudinal motion may be completely suppressed because the pinner follows any glide motion of the dislocation and thus the longitudinal force component becomes zero. This again is strong evidence that both modes of motion have to be considered simultaneously.

Even confining ourselves to the case of one dragging point per dislocation, the problem in its full generality is too involved to be presented here /11/. There are two limiting cases, however, for which the distribution function and the corresponding damping and modulus effect can be obtained analytically. They are characterized by the inqualities $\tau_L \gg \tau_T$ and $\tau_T \gg \tau_L$ respectively. Since nothing is known experimentally about the relative magnitude of the two mobilities involved, we have to consider both possibilities. It will be shown below how the two cases can be distinguished experimentally.

If $\tau_L \gg \tau_T$, the longitudinal motion taking place during a stress period is negligibly small and we may replace the periodic longitudinal force by its time average. Then after a time of the order of τ_L the longitudinal equilibrium distribution will be established. The resulting distribution function has been calculated in /11/:

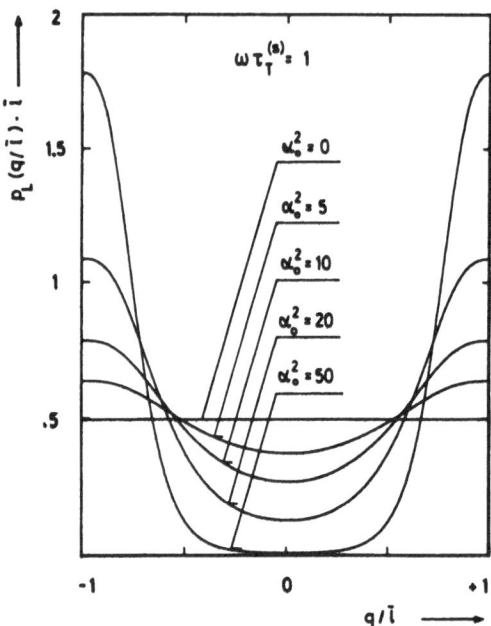

FIG. 2

Longitudinal distribution function for a single pinning point with both longitudinal and transversal mobility, in the limiting case $\omega \tau_L \gg 1$ and under a periodic applied stress. α_0 is normalized stress amplitude as defined in the text and $\tau_T^{(S)}$ denotes the relaxation time of a pinner at a symmetric position in the middle of the dislocation line ($q = 0$).

$$p_L(q) = C \exp\left(-\frac{\alpha_0^2}{2}\left[1 - q^2/\bar{l}^2\right]\left[1 - \frac{arc\,tan\,\omega\tau_T}{\omega\tau_T}\right]\right) \qquad (3)$$

where $\omega\tau_T = \omega\tau_T^{(S)}(1-q^2/\bar{l}^2)$. A plot of this function is shown in fig.2 for different stress amplitudes. When the stress is zero, there is no preferred position of the pinner along the dislocation /6/. With increasing stress the probability density increases more and more towards the ends of the dislocation, and decreases in the middle. The knowledge of the distribution along the dislocation line enables us to calculate the damping due to the transversal motion of the dragging points /12/. The result is

$$\delta_T = \Lambda\bar{l}^2\,\frac{\pi}{2}\,\frac{Gb^2}{2\mu}\int_{-\bar{l}}^{\bar{l}}\frac{\omega\tau_T}{1+\omega^2\tau_T^2}\,p_L(q)\,dq \qquad (4)$$

and is shown in fig. 3 for different strain amplitudes. The peak height decreases with increasing strain amplitude. At the same time the peak maximum is shifted to higher temperatures. Since τ_T is a function of the position of the pinner along the dislocation q, we have used $\tau_T^{(S)} = \tau_T(q = 0)$ as the temperature dependent scaling factor on the abscissa.

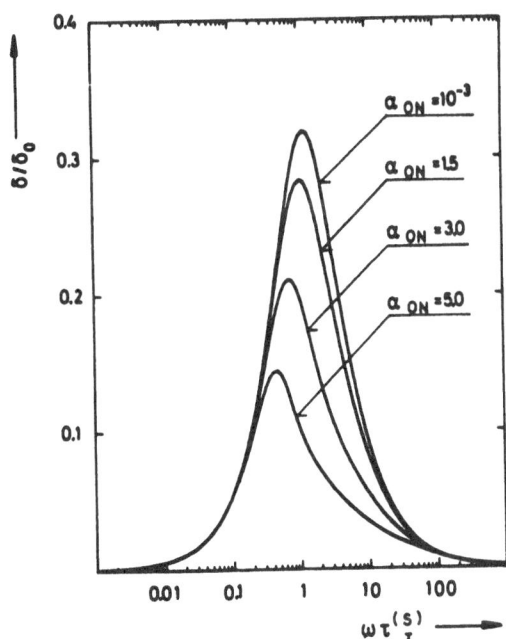

FIG. 3

The transversal damping peak appearing when the longitudinal mobility is very small ($\omega\tau_L \gg 1$). The subscript N of α_{0N} indicates that the temperature T in α_0 has been normalized to a fixed reference temperature T_N = 3oo K. δ_0 = (π /2) $\Lambda\bar{l}^2$; Λ = dislocation density.

Similarly, the modulus defect associated with this damping peak has been
obtained:

$$\left(\frac{\Delta G}{G}\right)_T = \frac{1}{2} \Lambda \bar{l}^2 \frac{Gb^2}{2M} \int_{-\bar{l}}^{\bar{l}} \frac{1 - q^2/\bar{l}^2}{1 + \omega^2 \tau_T^2} \, p_L(q) \, dq \qquad (5)$$

and is shown in fig. 4. Contrary to the damping, however, the modulus defect
is not very sensitive to changes of the strain amplitude.

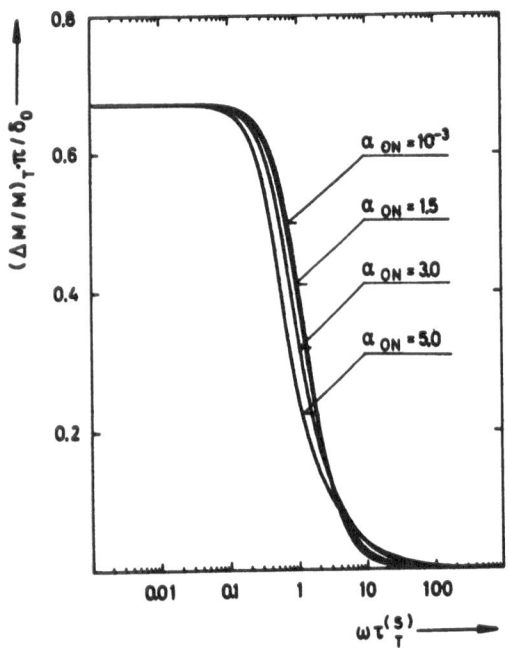

FIG. 4

Transversal modulus defect when the
longitudinal mobility is very small
$(\omega \tau_L \gg 1)$. $\delta_o = (\pi/2) \Lambda \bar{l}^2$.

The other limiting case, characterized by $\tau_T \gg \tau_L$, is treated by a similar
procedure /12/. Obviously the transversal motion during a stress period can
now be neglected while the longitudinal force component causes periodic
changes of the longitudinal distribution, with frequency 2ω. This in turn
produces a phase lag of the dislocation strain. The corresponding damping
curves in a first approximation are simple Debye peaks:

$$\delta_L = \frac{\pi}{2} \left(\frac{\Delta G_{max}}{G}\right)_L \frac{2\omega \tau_L}{1 + 4\omega^2 \tau_L^2} \qquad (6)$$

$$\left(\frac{\Delta G}{G}\right)_L = \frac{1}{2}\left(\frac{\Delta G_{max}}{G}\right)_L \frac{1}{1+4\omega^2\tau_L^2} \tag{7}$$

where τ_L is a function of stress as given by equation (1) and the relaxation strength at small strain amplitudes is given by:

$$\left(\frac{\Delta G_{max}}{G}\right)_L = \frac{1}{2}\Lambda\bar{l}^2\frac{Gb^2}{2\mu}\left(\frac{\sqrt{2}}{\alpha_0}e^{\alpha_0^2/2}\left[\int_0^{\alpha_0/\sqrt{2}}exp(\xi^2)d\xi\right]^{-1}-\frac{1}{\alpha_0^2}-\frac{1}{3}\right) \tag{8}$$

The longitudinal damping and modulus defect are shown in fig. 5 and 6, respectively, for several strain amplitudes. The behavior of the damping peak as a function of strain amplitude is contrary to the behavior of the transversal peak above: The peak height is zero at very small stress amplitudes, increases with increasing stress and asymptotically reaches a final value. The peak maximum at the same time shifts to lower temperatures.

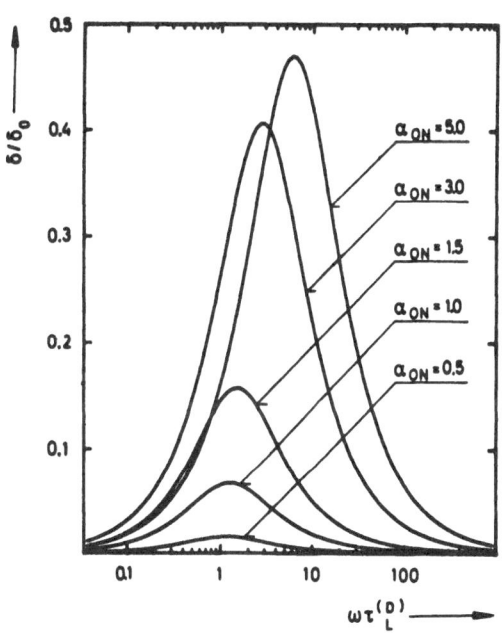

FIG. 5

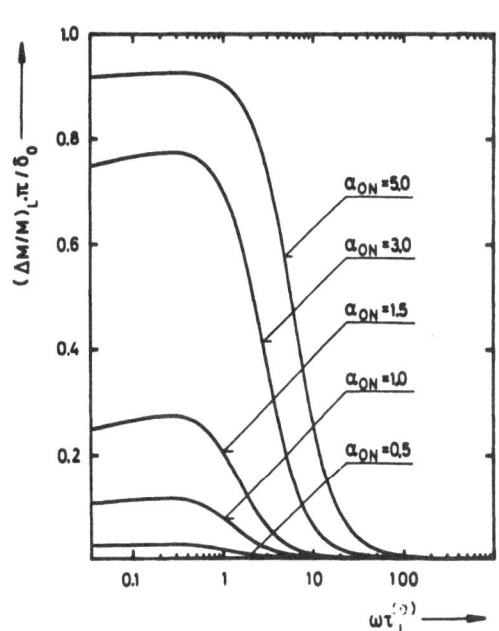

FIG. 6

Fig. 5: The longitudinal damping peak appearing when the transversal mobility is very small ($\omega\tau_T \gg 1$). $\tau_L^{(D)}$ is the longitudinal relaxation time when $\alpha_0 = 0$. $\delta_0 = (\pi/6)\Lambda\bar{l}^2$.

Fig. 6: Longitudinal modulus defect when the transversal mobility is very small ($\omega\tau_T \gg 1$). $\delta_0 = (\pi/6)\Lambda\bar{l}^2$.

The different behavior of the two peaks is due to the fact that the transversal force $F_T^{(\sigma)} \sim \sigma_0$ whereas the longitudinal force $F_L^{(\sigma)} \sim \sigma_0^2$. In the limit of small stress, therefore, the energy dissipated during a cycle is in the first case $\Delta W \sim \sigma_0^2$, and in the second case $\Delta W \sim \sigma_0^4$. Dividing by the elastic energy $W \sim \sigma_0^2$ we obtain for the transversal peak $(\Delta W/W)_T = $ const. and for the longitudinal peak $(\Delta W/W)_L \sim \sigma_0^2$. At higher stresses of course these simple considerations are no longer valid /11/.

The modulus defect in the present case exhibits the same stress dependence as the damping does. This is contrary to the behavior observed when the transversal motion is more important. Therefore, the stress dependence of the modulus defect may also be used for identifying the mode of motion underlying the observed process.

The fact that the longitudinal force component is also very sensitive to transversal mobility (cf. fig. 1b) provides additional information in order to assert which of the two limiting cases is realized in a given experiment. Let us first suppose $\tau_L \gg \tau_T$. Then we expect the transversal peak of fig. 3 and the modulus change fig. 4 to appear at a low temperature where $\omega \tau_T = 1$. As we increase the temperature we will reach a point where $\omega \tau_L = 1$. But at this temperature $\omega \tau_T \ll 1$, and the transversal motion is so fast that no longitudinal force es exerted any more. Thus if $\tau_L \gg \tau_T$, i.e. if the longitudinal mobility is smaller than the transversal mobility, the longitudinal peak will be absent and there will be a transversal peak only.

On the other hand, if $\tau_T \gg \tau_L$, we have at low temperature $\omega \tau_L = 1$, and the longitudinal peak of fig. 5 appears there. Upon increasing the temperature we reach a point where $\omega \tau_T = 1$, which leads to a transversal peak (fig.7). Its behavior is different from the previous one, however, because now $\omega \tau_L \ll 1$ and the dragging point easily follows any change of the longitudinal force /11/. The peak height decreases in a different way with increasing stress, and the position of the maximum is practically constant. Again, the modulus defect is almost independent of stress (fig.8). Thus, two peaks should be present, if the longitudinal mobility is larger than the transversal mobility: The low temperature peak being due to longitudinal motion, the high temperature peak being due to transversal motion.

As can be seen from fig. 3, the longitudinal peak shows the characteristics of Hasiguti's peak P_1 /1/. Accordingly, the transversal peak, fig. 7, should be present at some higher temperature. We suggest that peak P_3 accompanying P_1 is the transversal peak. A detailed comparison will be given elsewhere /12/.

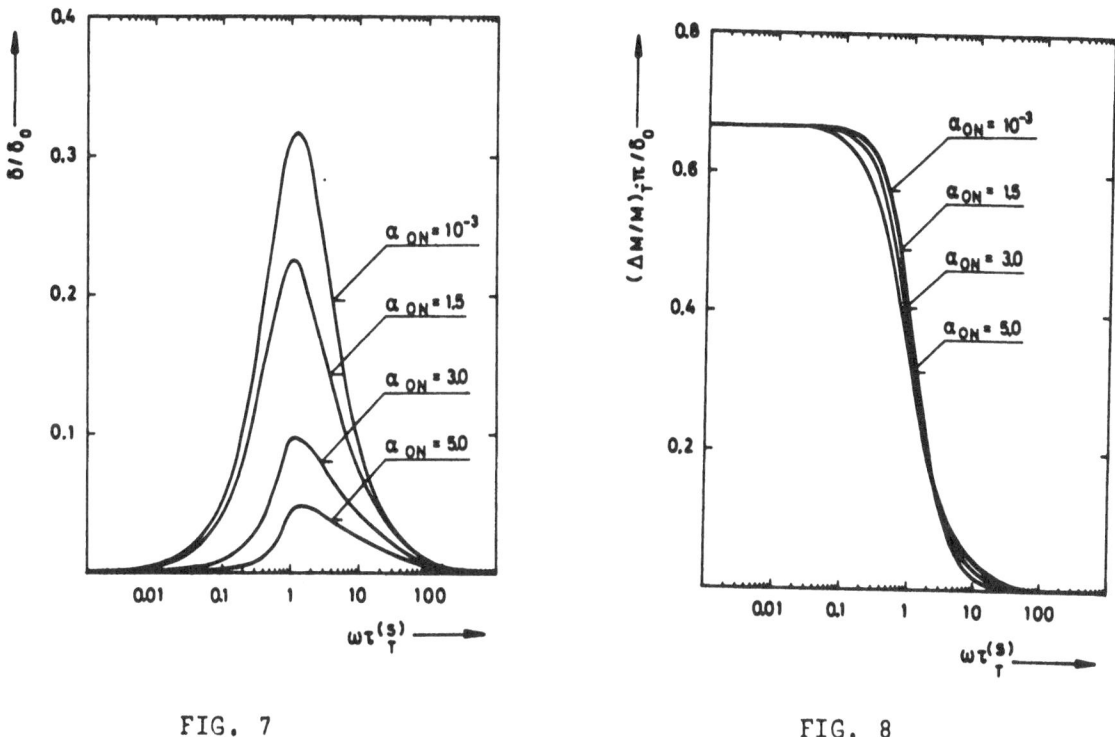

FIG. 7   FIG. 8

Fig.7: The transversal damping peak appearing when the longitudinal mobility is large ($\omega \tau_L \ll 1$). $\delta_o = (\pi/2) \wedge \bar{I}^2$.

Fig.8: Transversal modulus defect when the longitudinal mobility is large ($\omega \tau_L \ll 1$). $\delta_o = (\pi/2) \wedge \bar{I}^2$.

References

1) M. Koiwa and R.R. Hasiguti, Acta met. $\underline{11}$, 1215 (1963)

2) A.H. Cottrell and M.A. Jaswon, Proc. Roy. Soc. A $\underline{199}$, 104 (1949)

3) P. Schiller, phys. stat. sol. $\underline{5}$, 391 (1964)

4) S.K. Turkov and T.D. Shermergor, Soviet Physics-Solid State,
    $\underline{6}$, 2807 (1965)

5) P.D. Southgate and K.S. Mendelson,
    J. appl. phys. $\underline{36}$, 2685 (1965)

6) K. Lücke and J. Schlipf, Proc. Symp. Harwell $\underline{1}$, 118 (1968)

7) H.M. Simpson and A. Sosin, Phys. Rev. B $\underline{5}$, 1393 (1972)

8) G. Alefeld, Phil. Mag. $\underline{11}$, 809 (1965)

9) C.L. Bauer, Phil. Mag. $\underline{11}$, 827 (1965)

10) E. Bode, Phil. Mag. $\underline{13}$, 275 (1966)

11) For details see: W. Winkler-Gniewek, Thesis, Aachen 1973

12) W. Winkler-Gniewek and J. Schlipf, to be published.

On the Analysis of Dislocation Pinning

R. Schindlmayr, K. Lücke

Institut für Allgemeine Metallkunde und Metallphysik
der Technischen Hochschule Aachen, W.-Germany

Form of the pinning curve

It is well known that electron or gamma irradiation of metals causes
a pinning of the dislocation. This is mostly assumed to be due to a
reduction of the free loop length of the dislocations by irradiation
induced point defects diffusing to the dislocations and acting there
as pinning points. This decrease of loop length causes a decrease
of dislocation damping or a corresponding increase of modulus from
which the number of radiation induced pinning points can be calculu-
lated /1, 2, 3/.

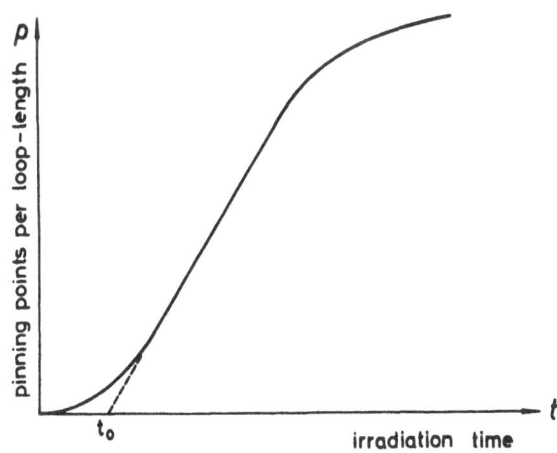

Fig. 1: Number of pinning points per loop length vs. irradiation
time.

Fig. 1 shows schematically a pinning curve obtained in this way in which the number of pinning points p (per unit volume) is plotted versus irradiation time t. In such curves one recognizes three more or less pronounced parts: an initial transient part, a linear part and a part with a decreasing slope. In the present paper a model for the interpretation of the three parts of the pinning curve will be given. This model differs from that given by Thompson et al, /4/.

The transient part

The dislocation is assumed to be a cylindrical sink of radius R_1 and infinite strength. Then the concentration of diffusing particles is always zero at $r = R_1$. The influence of neighbouring dislocations is taken into account by setting the concentration gradient zero at $r = R_2$ with $R_2 = 1/\sqrt{\pi \Lambda}$ being the mean distance between neighbouring dislocations and Λ the dislocation density. The initial concentration of defects is assumed to be zero and the defect production rate $\emptyset$ to be homogeneous and independent of time. Then the concentration c(r, t) of defects is obtained as solution of the diffusion equation:

$$\frac{\partial c}{\partial t} = D\left(\frac{\partial^2 c}{\partial r^2} + \frac{1}{r}\frac{\partial c}{\partial r}\right) + \emptyset \tag{1}$$

for the boundary conditions

$$c(R_1, t) = 0 \; ; \quad \left.\frac{\partial c}{\partial r}\right|_{R_2} = 0 \; ; \quad c(r, 0) = 0 \tag{2}$$

and the number of pinning points p as function of time by

$$p = -2\pi R_1 D \int_0^t \left.\frac{\partial c}{\partial r}\right|_{R_1} dt' \tag{3}$$

As a result of these calculations one find for small times

$$p = 4\sqrt{\pi} \; \emptyset \, R_1 \sqrt{D} \; t^{3/2} \tag{4}$$

256

In order to understand this result one must realize that during the time t a shell of the thickness $\approx 2\sqrt{Dt}$ is depleted. If one assumes for a moment that instead of Equ. (2) the condition $c(r, 0) = const = c_o$ and $\emptyset = 0$ applies, one thus obtains directly $p'(t) = 4\pi R_1 \sqrt{Dt} \, c_o$, i.e. $p' \propto t^{1/2}$. The case of the original boundary condition Equ. (2) can then directly be obtained by the integration

$$p = \int_0^t p'(t-t') \emptyset \, dt'$$

leading to $p \propto t^{3/2}$ as stated in Equ. (4).

In reality, however, the dislocation is no cylindrical sink but a stress field giving rise to an interaction potential between dislocation and point defect mostly assumed to be $W = Ab/r$, if the azimutal dependence is neglected (A is the interaction constant). For this reason a drift term $(D/kT)\nabla$ (c.gradW) has to be added to the right side of Equ. (1). If one considers only this term and neglects the original diffusion term in Equ. (1), the solution of this equation leads in the case of constant initial c_o to the well known Cottrell-Bilby-law with $p \propto t^{2/3}$ accordingly, and , in the case of irradiation with constant $\emptyset$ to a $t^{5/3}$-law. It has not yet been possible to solve the diffusion equation containing both terms /5/. For small times, however, it can directly be seen that the pinning following from the diffusion term ($p \propto t^{3/2}$) is always larger than that following from the drift term ($p \propto t^{5/3}$). Thus one can conclude that for small times the drift term can be neglected and the diffusion term must be used, i.e. that also in the case of drift Equ. (4) is obtained.

According to Ham /6/ the radius R_1 of the cylinder must be chosen as $R_1/b \approx A/kT$. This can be seen if one simplifying assumes that inside R_1 the interaction is so strong that there the defects reach the dislocation directly by drift, and that outside R_1 the interaction can be neglected so that there only Fick-type diffusion takes place. Thus one recognizes that with this temperature dependent effective capture radius R_1 Equ. (4), i.e. $p \propto t^{3/2}$, gives an approximate description for the initial defect flux to the dislocation in which the interaction between point defect and dislocation is taken into account although the drift term itself is neglected.

It is now proposed by the present authors to interprete the transient part of the pinning curve by such a diffusion law

$$p = 4 \sqrt{\pi} \frac{A}{kT} b \phi \sqrt{\mathcal{D}} \, t^{3/2} \qquad (5)$$

This seems to agree with the experiments where exponents 3/2 and 1/2, respectivly , are more often reported than 2/3 and 5/3. This result is in contrast to that of Thompson et al. /4/ where an initial t^2-law has been reported.

The linear part

For large times, Equs. (1) to (3) give a linear behaviour p = ϕ (t-t_o), where t_o is the intercept of the extrapolated linear part with the time axis. This means that in this range the rate of diffusion has obtained such large values that the pinning rate $\dot{p}$ becomes equal to the production rate of defects ϕ. Such behaviour, however, does not agree with the experiments. It is found, instead, that the pinning rate is always smaller than the production rate and that it increases with temperature /4, 7, 8/. Thompson et al. /4/ tried to explain these effects by the assumption of a thermal equilibrium between point defects situated on normal places of the dislocation and on dislocation nodes where they do not act as pinning points. Against this interpretation, however, arguments have been forwarded: (i) such equilibrium does not occur /9/, and (ii) the pinning behaviour is strongly influenced by small amounts of impurities /10/. For these reasons it will here be concluded that the different slopes of the linear part of the pinning curve are largely determined by the presence of impurity atoms.

The irradiation induced defects are known to have an attractive interaction to foreign atoms. Thus, on their way to the dislocation, many of the point defects will be trapped at impurity atoms and the pinning rate will decrease. This can be described by adding to the diffusion equation (1) an absorbtion term which represents the trapping rate of the defects:

$$\frac{\partial c}{\partial t} = \mathcal{D} \left(\frac{\partial^2 c}{\partial r^2} + \frac{1}{r} \frac{\partial c}{\partial r} \right) + \phi - \alpha c \qquad (6)$$

258

c is now the concentration of free point defects in the lattice, i.e. the concentration of those not trapped by impurities on dislocations. Assuming that the impurity atoms are spherical sinks of radius R_o and infinite strength it can be shown that the absorption constant α is proportional to the (volume) concentration c_i of impurity atoms and given by

$$\alpha = 4\pi D R_o c_i \qquad (7)$$

By solving Equ. (6) for the boundary conditions (2) and introducing the solution into Equ. (3), one obtains for large times again a constant pinning rate $\dot{p}$. This, however, is much smaller than the pinning rate $\dot{p} = \emptyset$ obtained without absorption:

$$\frac{\dot{p}}{\emptyset} = \frac{1/R_a^2}{R_o c_i} \approx \frac{\Lambda}{R_o c_i} \qquad (8)$$

i.e. the ratio of pinning rates is approximately equal to the ratio of the numbers of dislocated atoms and impurity atoms. Such a factor has to be introduced also into Equ. (5) in order to consider the presence of impurity atoms also for the transient part of the pinning curve.

For dislocation densities of $\Lambda \approx 10^7$, impurity concentrations of $c_i \approx 10^{-5}$ and an interaction radius R_o of three lattice constants, the ratio $\dot{p}/\emptyset \approx 10^{-3}$ is obtained, i.e. 99,9% of the produced defects are trapped by the impurity atoms. Since the traps loose their effectivity when the thermal energy kT surpasses the binding energy between defect and trap, the absorption α decreases with increasing temperature from the value given by Equ. (7) at very low temperatures. Then an increase of the pinning rate ratio $\dot{p}/\emptyset$ with increasing temperature is not to be expected.

As an experimental example, Fig. 2 gives the temperature dependence of the pinning rate in copper for electron irradiation as obtained from kilocycle damping measurements /11/. One recognizes that from 80 to about 150 K the pinning rate is temperature independent, but that between 150 and 400 K an increase by about three orders of magnitude occurs. This increase is assumed to be due to the decrease

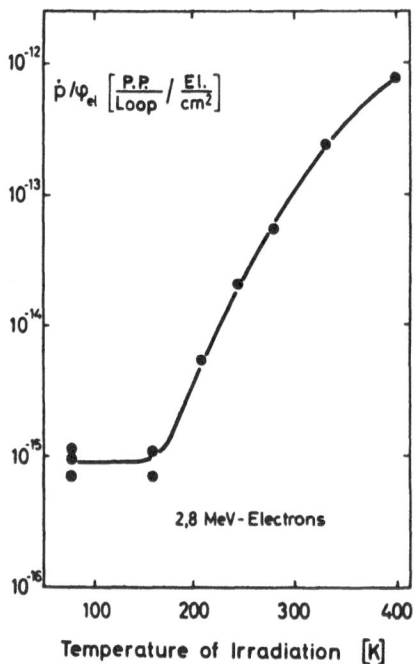

Fig. 2: Pinning rates vs. irradiation temperature for copper
        irradiated with 2,8 MeV electrons

of absorption with temperature. Since it streches over a rather large
temperature range, it is further assumed that different kinds of impurity atoms are present which, due to the different binding energies,
loose their effectivity as traps at different temperatures.

This assumption is verified in Fig. 3 /lo/ which shows the pinning
rate during warm up after a 200 K irradiation as obtained from MHz-
damping measurements. One recognizes different discrete maxima which
are assumed to correspond to different binding energies and thus
to different kinds of impurity atoms. These maxima are reproducible
but different for the three different investigated types of high-
purity copper. This agrees with the results of mass spectroscopic analysis
which also gave pronounced differences in kind and amount of impurities.

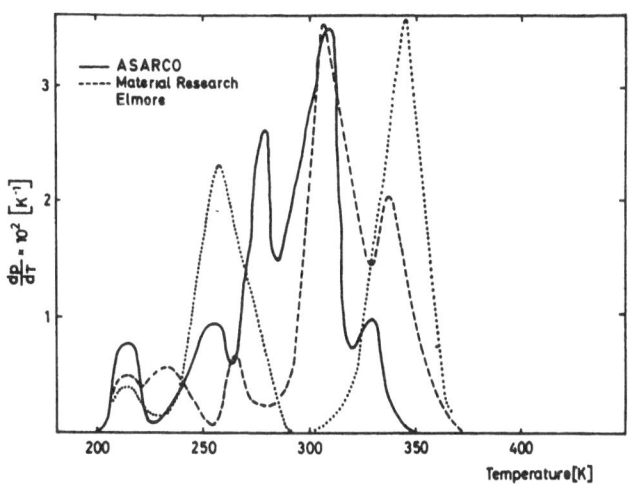

Fig. 3: Differentiated pinning curves obtained from MHz-damping
measurements during warm up after a 200 K - γ - irradiation
for 99,999% copper of three different producers.

Part with decreasing slope

The decrease of the slope following the linear part means that not all
defects arriving at the dislocation act as pinning points. Therefore
it is here assumed that this is mainly due to the coagulation of the
single defects at the dislocation i.e. due to the formation of point
defect clusters on the dislocations. One has to imagine that a point
defect having reached and diffusing along a dislocation either meets
a cluster already present with which it combines or it meets another
single point defect with which it combines thus forming a new cluster
It is further assumed that also clusters act as pinning points, but
that they are immobile (in contrast to single defects) and can dissolve
only at much higher temperatures. Fig. 4 gives a schematic representation of this model.

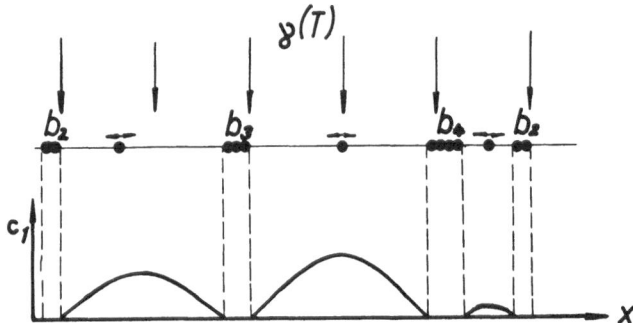

Fig. 4: Schematical plot of clustering on dislocations. In the
lower part the concentration profile of single defects
(c_1) along the dislocation is shown.

Under these suppositions, the density of single defects (c_1) and of
clusters (c_c) can be described by a simple system of rate equations:

$$\dot{c}_1 = \gamma - c_1 F_1(c_1) - c_c F(c_1)$$ (9)

$$\dot{c}_c = c_1 F(c_1)$$ (10)

Here is γ the arrival rate of defects at the dislocation, $F(c_1)$ the rate
of combination of single defects at the dislocations with the already
present clusters and $F_1(c_1)$ the combination rate of two single defects with one another forming a new cluster. The function $F(c_1)$ represents the current of the single defects into the clusters. It has
been calculated by solving an one-dimensional diffusion equation,
supposing a constant arrival rate γ (i.e. neglecting the transient
part) and assuming the clusters being sinks with infinite strength
(i.e. with zero concentration on their surfaces). In a similar way
also the function $F_1(c_1)$ has been determined. It resulted that $F_1(c_1)$
and $F(c_1)$ are proportional to $\sqrt{c_1}$.

If one assumes that the total pinning point number $p = c_1 + c_c$ the function $p(t)$ can be obtained by solving Equs. (9) and (10). It turns out that for small t the number of single defects and for large t the number of clusters dominates in p and one obtains

$$\gamma t \sqrt{Dt} \ll 1 \; : \; p \approx c_1 = \gamma t$$

$$\gamma t / c_1 \gg 1 \; : \; p \approx c_0 = \left(\frac{\gamma b_1^2}{D}\right)^{1/4} (\gamma t)^{1/4} \tag{11}$$

The linear increase of p given in Equ. (11) corresponds to the linear part of the pinning curve. The increase with $t^{1/4}$ as given by Equ. (11) is in contrast to the predictions of Thompson et al. /4/. Since these authors assumed for long times a thermal equilibrium between point defect at dislocations and at nodes, they obtained a linear increase with reduced slope. A recent comparison with experimental observations /8/ seems to favour the present $t^{1/4}$-law. Additionally, temperature change experiments exhibit great stability of the pinning point numbers with respect to small temperature changes. Also this contradicts the assumption of thermal equilibrium, but is in agreement with the assumption of cluster formation.

References

/1/   A. Granato and K. Lücke, Physical Acoustics, Vol 4 Part A
       edited by W.P. Mason, Academic Press (1966)

/2/   D.O. Thompson, V.K. Paré, Physical Acoustics, Vol. 3 Part A,
       edited by P.W. Mason, Academic Press (1966)

/3/   V.K. Paré, H.D. Gubermann, J. Appl. Phys. **44**, (1972), 32

/4/   D.O. Thompson, O. Buck, H.B. Huntington and R.S. Barnes,
       J. Appl. Phys. 1967, **38** , 3051

/5/   A. Seeger, phys. stat. sol. 41, 509, (1970)

/6/   F.S. Ham, J. Appl. Phys. **30**, (1959) 915

/7/   P. Winterhager, K. Lücke, J. Appl. Phys. **44**, (1973), 4855

/8/   R. John, D. Lenz, K. Lücke this issue

/9/   P. Winterhager, G. Roth, R. John, K. Lücke, J. de Physique, **32** (1971) C2-151

/10/   Förster, Kaufmann, D. Lenz, K. Lücke, Naundorf to be published

/11/   Zeckau, G. Roth, Wollenberger to be published

Analysis of Dislocation Pinning Experiments
in Copper

R. John, D. Lenz and K. Lücke

Institut für Allgemeine Metallkunde und Metallphysik
der Technischen Hochschule Aachen, W.-Germany.

Extension of the Model of Thompson et al.

The interpretation of the time laws of dislocation pinning during isothermal irradiation is of great interest since it would allow to
conclude from pinning curves about possible reactions of the irradiation induced point defects (PD) during their diffusion in the lattice
and along the dislocations. The most sophisticated model in this
respect is given by Thompson et al. /1,2/. It implies an diffusion
controlled exchange of irradiation induced point defects between three
reservoirs: lattice, dislocations, nodal points (L,D,N in Fig. 1); i.e.
point defects produced e.g. by γ-irradiation in the lattice can
migrate to the dislocations and along the dislocations to nodal points
where their energy is further decreased. Only while at the dislocations, the point defects are assumed to act as pinning points.

Thompson et al. used this model to interprete their measurements of
the changes of damping and modulus in the kHz range during γ-irra-
diation of copper above room temperature. They found a good agreement
between the predictions of their model and measured curves and were
able to deduce from them the activation energies of the point defect
diffusion in the lattice and along the dislocations as well as the
interaction energies of the PD with dislocations and nodal points.

The present own pinning experiments on copper, however, cannot be
interpreted by the above model, even not qualitatively. Our experiments make necessary introduction of at least two additional reservoirs (indicated by dashed lines in Fig. 1) which means a twofold
extension of the Thompson model:

(i) Warm-up experiments after low temperature irradiation led for
different brands of high-purity copper to drastically different
pinning curves both for MHz- and kHz-measurements /3/. This indicates
that small amounts of impurities strongly influence the observed

pinning rate so that the interaction between PD and the impurity atoms cannot be neglected in the reaction scheme. For this reason, the additional reservoir I for PD trapped by impurities has been added in Fig. 1. Also Thompson et al. discussed such possibility, but disregarded it.

(ii) The pinning points obtained after irradiation below 430°C are always found to be very stable with respect to small temperature changes. Fig. 2 shows an example: After an Co^{60} γ -irradiation at 353 K (10.5 h irradiation at a dose rate of 1.8×10^4 r/h) temperature changes of $\pm 10°$ have been carried out. These led to only small attenuation changes (solid curve) which are probably due to the temperature dependence of the constants entering the Granato-Lücke theory of dislocation resonance damping /4/. The model of Thompson et al., however, predicts a strong unpinning (i.e. attenuation increase) at a temperature decrease since the PD on the dislocations should then prefer the nodal point positions (because of the lower energy) where they do not act any more as pinning points. Correspondingly, at a temperature increase the PD should move back to the dislocations causing an increase in pinning (i.e. an attenuation decrease). The dashed curves in Fig. 2 are those predicted by the model of Thompson et al.[+], with the parameters given in the paper by Thompson et al. Because of the strong discrepancy with the measured curves, it is concluded that no such thermal equilibrium of pinning points with sinks on the dislocations can take place. It is assumed, instead, that, to a major extent, the observed pinning points are made up by point defect clusters which dissolve only at much higher temperatures as observed by Inagaki et al. /5/ and thus are stable with respect to the small temperature changes applied in Fig. 2. Thus in Fig. 1 an reservoir C for PD in clusters has been added.

Own experiments.

As first shown by Winterhager /6/, the Elmore copper (purity 99.999%) used in the present experiments exhibits two clearly defined pinning stages above 220 K, stage C near 265 K and D near 360 K. This can be

---

[+] Thompson et al. did not check their model by experiments of the present type.

recognized in Fig. 3 where the pinning point number p measured during warm-up after irradiation at 220 K is shown. In the following isothermal MHz-attenuation experiments during irradiation in the temperature range of stage D will be described.

All experiments were carried out on one and the same single crystal sample (dimensions 1x1x1 cm) which was standard treated according to Winterhager /7/ and then repeatedly irradiated. Before each irradiation the sample has been reconverted to the state after deformation by an anneal at 650 K. The irradiation was 3 MeV-Bremsstrahlung produced by a Van de Graaff accelerator; its electron beam current (20 µA) was equivalent to a flux of $\approx 10^{11}$ photons $cm^{-2} sec^{-1}$ at the sample surface. The results are given in Fig. 4a, which shows the normalized attenuation at 50 MHz (sound propagation in ⟨111⟩-direction) vs. time t, measured at 12 temperatures between 337 and 463 K during ($0 < t < 1,5h$) and after irradiation.

In Fig. 4b the number of pinning points per dislocation segment calculated from the attenuation curves of Fig. 4a by means of the Granato-Lücke theory /4/ is given. These curves show, more or less pronounced, an initial transient with slopes increasing with time, then temporarily a straight part and finally slopes decreasing with time. With increasing temperatures the straight part slopes of all the curves (for T < 420 K) increase. After switching off the irradiation all the curves first show a discontinuous decrease of the slope and subsequently, depending on temperature, some further pinning, almost constant values, or depinning with time finally leading to constant pinning point values at prolonged times. Fig. 5 shows a magnified plot of the initial part of the curves of Fig. 4b. One recognizes that even at the lowest temperatures the curves start with a final slope, i.e. a linear ("spontaneous") pinning is superimposed to the transient which alone would start with zero slope.

The isothermal pinning curves measured by Thompson et al. /1/ in the same temperature range using kHz measurements and a different type of copper (ASARCO, purity 99.999%) show features similar to those presented in Figs. 4b and 5; e.g. the increase of the slopes of the linear parts with increasing temperature and the decrease of the slopes with time after sufficiently large doses. Other features of the kHz curves, however, e.g. the very long straight parts ("steady state" behaviour) and positive intercepts of these lines when extrapolated back to zero time, are not observed in the MHz experiments.

For a quantitative evaluation of stage D pinning, the spontaneous pinning has to be subtracted from the measured pinning curves. In Fig. 6 the remaining stage D contribution has been presented in a log-log plot. According to Schindlmayr and Lücke /8/, the diffusion processes from lattice to dislocation, i.e. the transient part of the pinning curves, should follow a $t^{3/2}$-law whereas for the clustering processes, i.e. for the long-time parts of pinning curves a $t^{1/4}$-law should apply. One recognizes from Fig. 6 that these two predictions about the slopes of the pinning curves for very small and very large times are approximately fullfilled. One recognizes further that no pronounced region with the slope 1 appears. This means that the straight parts in Fig. 4b do not correspond to steady-state behaviour, but are merely inflection regions due to the transition from the initial transient behaviour (exponent > 1) to the clustering behaviour (exponent < 1). This means that the clustering process starts, before the true steady-state pinning takes place.

For this reasons, a further evaluation of the data seems to be useful only in the two regions where only one process occurs, i.e. at very small and very long times. For small times a quantity A, for long times a quantity B has been derived from the measurements according to

$$\dot{p} = \dot{p}_{spon} + (At)^{1/2} \quad \text{(small t)}$$

$$\dot{p} = \text{const} \cdot B^{1/4} t^{-3/4} \quad \text{(large t)}$$

where $\dot{p}$ is the total pinning rate measured, $\dot{p}_{spon}$ is the spontaneous pinning rate and t is the irradiation time.

An Arrhenius plot of A and B yields satisfactory straight lines (Fig. 7). From the one derived from A (the transient), an activation energy of $U_G = 1.1$ eV is obtained. Since during its diffusion from lattice to the dislocation, the PD has to overcome many impurity traps, this activation energy must be a combined quantity determined both by its activation energy for migration in the pure lattice and by its binding energy to the impurity traps. The activation energy corresponding to the long-time parts of the pinning curves comes out to be about $U_v = 0.2$ eV. This small value is not unreasonable, if one takes into account that clustering occurs by pipe diffusion of the PD along the dislocation lines.

Interpretation of the results.

The following qualitative interpretation of the different features of the pinning curves seems to be in close agreement with the observed facts (see also Schindlmayr and Lücke /8/).

(i) The initial transient is attributed to the building up of a concentration gradient of the point defects in the lattice after the beginning of irradiation. Only when the magnitude of this gradient is so that the rate of PD diffusion from the lattice to the dislocations equals the rate of PD production in the lattice by the irradiation a steady state behaviour, i.e. a linear increase of the pinning point number with time, is reached.

(ii) The increase of the slope of the linear part of the pinning curves (i.e. near the inflection points) with increasing temperature is interpreted by the assumption that not all the PD produced in the lattice reach the dislocations but that some of them are trapped by impurities. With increasing temperature the rate of emission of PD from the traps and thus their rate of arrival at the dislocations increases.

(iii) The decrease of the slope of the pinning curves with time for times beyond the inflection point is attributed to the formation of PD clusters on the dislocations. The fact that at different temperature the same irradiation dose yields different pinning point numbers (Fig. 4b) can then be explained by clusters of different sizes. The depinning after switching off the irradiation then indicates the dissolution of clusters. This effect is observed only at higher temperatures (> 390 K) thus explaining the stability of pinning point numbers at lower temperatures (see Fig. 2).

(IV) The spontaneous pinning observed at the beginning of the curves is attributed to those PD which start to move in stage C and which at the much higher temperatures of stage D considered here migrate so fast that no transient but only the linear (steady-state) behaviour can be observed. The discontineous decrease in slope after switching off the irradiation (Fig. 4b) is then to be explained by the instantaneous termination of spontaneous pinning. Such a well defined spontaneous pinning can only be observed if the two pinning stages are clearly separated as in the present case (Fig. 3).

Further evidence for the formation of clusters in the flat part of the pinning curve is obtained from the observed dose rate dependence of pinning. Fig. 8 shows pinning numbers vs. dose at 353 K for three different dose rates. One recognizes that with decreasing dose rate the curves bend down from the straight part at smaller dose values. Since a lower dose rate leads to a lower concentration of single PD at the dislocations, also the probability of nucleating clusters, i.e. stable pinning points, decreases with decreasing dose rate. Thus the result of Fig. 8 means that for lower dose rates, the chance of a PD to arrive at a cluster already present instead of forming a new one is increased. Thus with lower dose rates less but larger clusters are produced.

Acknowledgements

The authors have benefitted from many discussions with Drs. P. Winterhager and R. Schindlmayr. They thankfully acknowledge a financial support by the Deutsche Forschungsgemeinschaft.

References

1   D.O. Thompson, O. Buck, R.S. Barnes and H.B. Huntington;
    J. Appl. Phys. 38, 3051 (1967).

2   D.O. Thompson, O. Buck, R.S. Barnes and H.B. Huntington;
    J. Appl. Phys. 38, 3057 (1967).

3   U. Förster, H.R. Kaufmann, D. Lenz, K. Lücke, V. Naundorf, G. Roth;
    to be published
    see also the paper by
    D. Lenz, K. Lücke in this volume.

4   A. Granato, K. Lücke;
    J. Appl. Phys. 27, 583 (1956).

5   H. Inagaki, F. Hultgren, K. Lücke;
    Acta Met 18, 713 (1970).

6   P. Winterhager; Thesis Aachen 1968
    P. Winterhager, K. Lücke;
    Proceedings of the Symposium on "The Interaction Point Defects",
    Harwell U.K., Vol. I, p. 214 (1968).

7   P. Winterhager, K. Lücke;
    J. Appl. Phys. 44, 4855 (1973).

8   R. Schindlmayr, K. Lücke this volume.

<u>Figures</u>

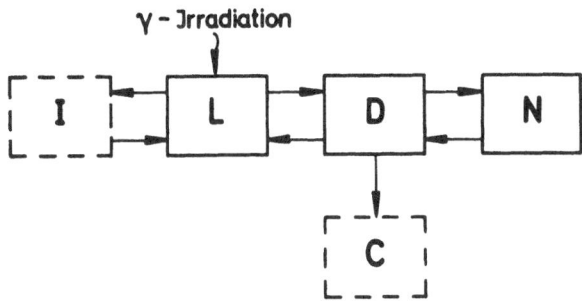

Fig. 1 Models for calculating the pinning rate due to γ-irra-
        diation. Full lines : Model of Thompson et al. /1,2/;
        dashed lines : proposed extensions. (L = lattice,
        D = dislocation, N = nodal points, I = impurity traps,
        C = clusters).

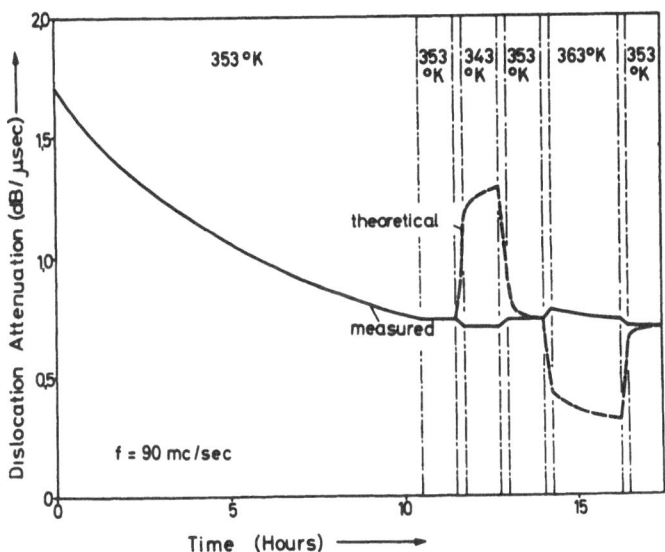

Fig. 2 Dislocation damping during (0-10.5 h) and after (>10.5 h)
        γ-irradiation as a function of time at various temperatu-
        res (Dose rate: 1.8 x 10^4r/h).

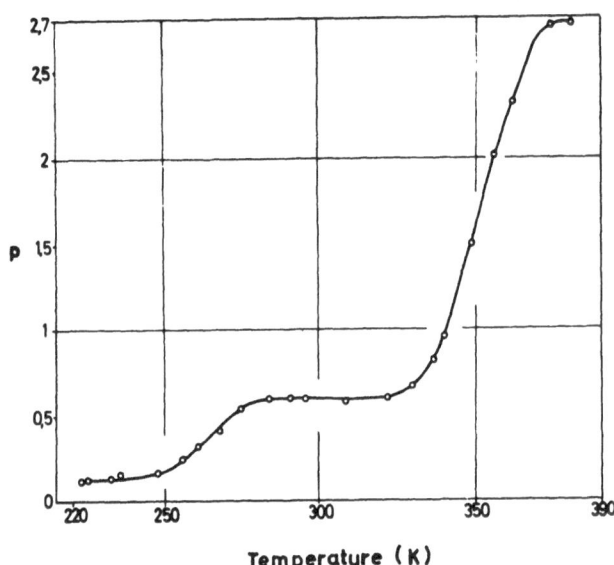

Fig. 3   Pinning point number p as a function of temperature
during isochronal annealing after γ-irradiation
at 220 K /6/.

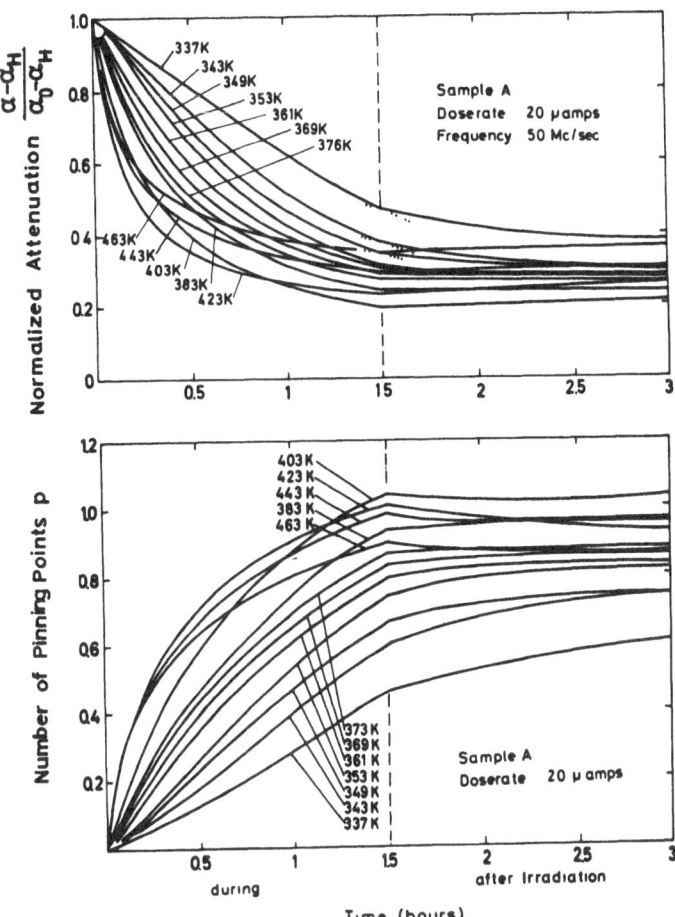

Fig. 4 (a) Normalized absorption ($\triangleq$ attenuation) at 50 MHz and
(b) pinning point number p vs. time during (0-1.5 h) and
after (> 1.5 h) γ-irradiation at different temperatures. (Flux: 10^{11} photons $cm^{-2}sec^{-1}$; α_0: attenuation at zero time, α_H: nondislocation ("background")
attenuation).

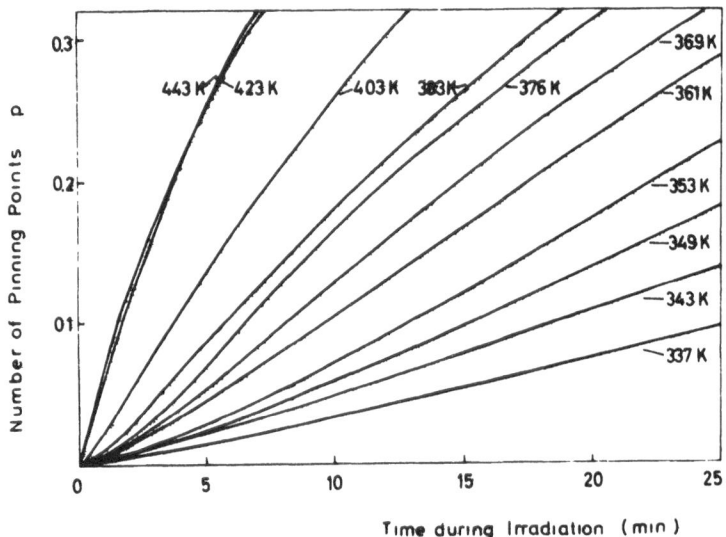

Fig. 5   Pinning point number vs. irradiation time
          (Initial part of Fig.4b)

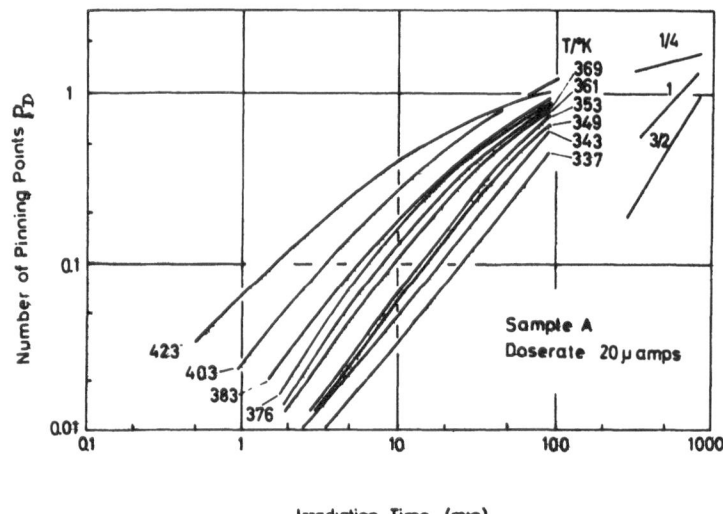

Fig. 6   Stage D pinning point number vs. irradiation time.
          (the straight lines at the right hand side indicate
          different slopes i.e. time-exponents).

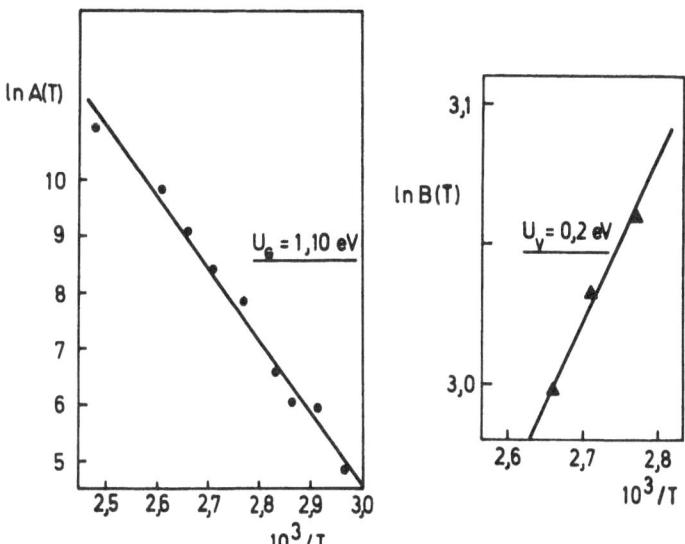

Fig. 7   Arrhenius plots of the pinning rate constants A and B
         evaluated from theoretical fits to the initial part
         (left) and to the long-time part (right) of the pinning
         curves of Fig. 4.

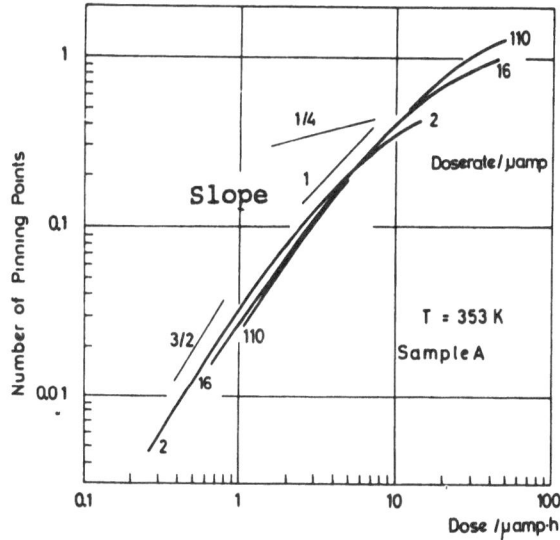

Fig. 8   Pinning point number vs. irradiation dose at different
         dose rates of γ-irradiation at 353 K.

DISLOCATION PINNING IN ELECTRON-IRRADIATED COPPER AT DIFFERENT ELECTRON ENERGIES AND IRRADIATION TEMPERATURES

G. Roth, H. Wollenberger, Ch. Zeckau and K. Lücke
Institut für Allgemeine Metallkunde und Metallphysik der RWTH
Aachen und Van de Graaff-Labor, Aachen der Kernforschungsanlage
Jülich

Different experimental results of dislocation pinning in γ-irradiated /1/ and electron irradiated /2/ copper have been explained by the production of different types of defects (on-line and off-line crowdions) by the different kinds of irradiation /3/. Since both kinds of experiments were performed in different laboratories with completely different methods, the above conclusion did not seem to be very convincing. In order to check the experimental basis, a movable gold target has been installed into the beam handling system of the Aachen Van de Graaff irradiation facility. Samples mounted onto the internal friction measuring device (described in /4/) could thus be irradiated directly with electrons, if the target was removed, or by γ-rays produced in the gold target, if this was brought into the beam. Hence, the type of irradiation could be changed without any handling of the sample. The pinning rates have been determined by means of modulus measurements and evaluated according to Lücke and Granato /5/. The measurement temperature was equal to the irradiation temperature. After annealing at 4oo K the modulus change was measured at 78 K.

In a first series of experiments the fractional increase of the number of pinning points p per loop caused by a 4oo K anneal after a 78 K irradiation has been determined. In Fig. 1 the ratio of Δp (4oo K) obtained after the 4oo K anneal devided by Δp (78 K) determined after the 78 K irradiation is plotted versus Δp (78 K) for both kinds of irradiation. There is an average increase of the number of pinning points by about three orders of magnitude. This result can be interpreted by a large fractional defect trapping at impurity atoms at the 78 K irradiation temperature and a detrapping during the 4oo K anneal /6/. Obviously, the fractional increase of Δp by annealing at 4oo K is not influenced by the type of irradiation. The above mentioned discrepancies between different experimental results from different laboratories must be due to other effects.

In a second series of experiments pinning rates were measured at 78 K and at 4oo K irradiation temperature. The results are collected in Table I. The ratio of pinning rates amounts to the same order of magnitude as the fractional increase of the number of pinning points in Fig. 1. This equivalence must be expected within the framework of the above given interpretation since the defects become no longer trapped during the 4oo K irradiation and reach in total the dislocations. An interstitial vacancy recombination and interstitial clustering can be excluded during annealing because of the very low radiation induced defect concentrations ($c < 10^{-9}$).

Furthermore, Table I shows pinning rate ratios being larger by a factor of about 2 for the γ-irradiation than for the electron irradiation. Since no difference was found for the annealing experiments the difference observed here must be due to some change in the defect production process at 4oo K when compared with that at 78 K. The primary difference of both kinds of irradiations is the average electron energy of the Compton electrons of about o.5 MeV in case of the γ-irradiation and 2.8 MeV for the electron irradiation. Hence, the different ratios in Table I could be caused by the different electron energies. In order to check this point irradiations have been performed using electrons of different energies between o.2 MeV and 2.8 MeV. As a result the pinning rates normalized to 2.8 MeV are obtained as function of electron energy for both irradiation temperatures as shown in Fig. 2. As can be seen, the ratio r defined in Table I increases with decreasing electron energy. Because of the large experimental error of the data in Fig. 2 of about $\pm$ 3o% it cannot be decided exactly at which electron energy the ratio r reaches two when it is taken to be unity at 2.8 MeV. But this value is certainly reached somewhere between o.4 MeV and o.7 MeV. Since the average energy of the compton electrons produced by 2.8 MeV-Bremsstrahlung (thick target) is expected to ly within this energy range /7/ the $r_\gamma / r_{e\ell}$ values in Table I are easily explained by this effect.

The curves in Fig. 2 also show that the threshold energy for pinning point production is smaller for 4oo K than for 78 K irradiation temperature. A detailed discussion of this result in terms of the underlying mechanism of defect production, however, requires a careful evaluation of the energy dependent displacement probability from the data in Fig. 2. Since the Mott scattering cross section increases less than linearly with $\mathcal{E}_{max}$ the nearly exponentially increasing pinning rate indicates an exponentially increasing displacement probability. Furthermore, the five orders of magnitude change found for 4oo K shows that displacements caused by recoil energies below 15 eV really occur with negligible probability. A detailed evaluation of the data in Fig. 2 with respect to the influence of temperature on the displacement process is given elsewhere /8/.

References.

/1/ D.O. Thompson and O. Buck, phys. stat. sol 37, 53 (1969)

/2/ G. Roth, G. Sokolowski and K. Lücke, Journ. d. Physique 32, C 2-145 (1971)

/3/ W. Frank and A. Seeger, Crystal Lattice Defects 5, 141 (1974)

/4/ G. Roth and K.F. Rittinghaus, Z. Angew. Phys. 32, 331 (1972)

/5/ A. Granato and K. Lücke, J. Appl. Phys. 27, 583, 789 (1956)

/6/ R. Lennartz, F. Dworschak and H. Wollenberger, J. Phys. (F)

/7/ O.S. Oen and D.K. Holmes, J. Appl. Phys. 3o, 1289 (1959)

/8/ G. Roth, H. Wollenberger, Ch. Zeckau and K. Lücke, to appear in Radiation Effects 1975

Table I

sample number	1	2	3	average value
r_γ	$2.1 \; 10^3$	$2.6 \; 10^3$	$1.5 \; 10^3$	$2.1 \; 10^3$
r_{el}	$9.1 \; 11^2$	$9.4 \; 10^2$	$9.2 \; 10^2$	$9.2 \; 10^2$
r_γ/r_{el}	2.3	2.8	1.7	2.3

Ratio r of the pinning rates at 4oo K and 78 K, measured at three samples 1 - 3. r_γ relates to γ-irradiations, r_{el} to 2.8 MeV electron irradiations.

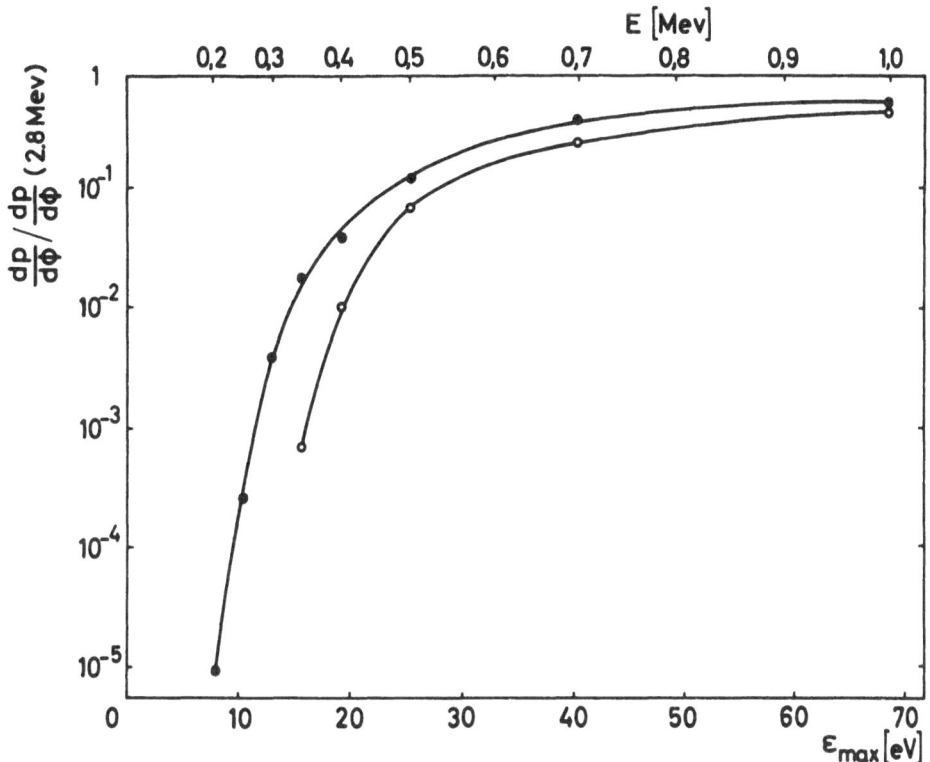

Fig. 2

Pinning rate normalized to 2.8 MeV electron energy vs.
maximum transferred energy ε_{max} and electron energy E
for 78 K (O) and 400 K (●) irradiation temperature.

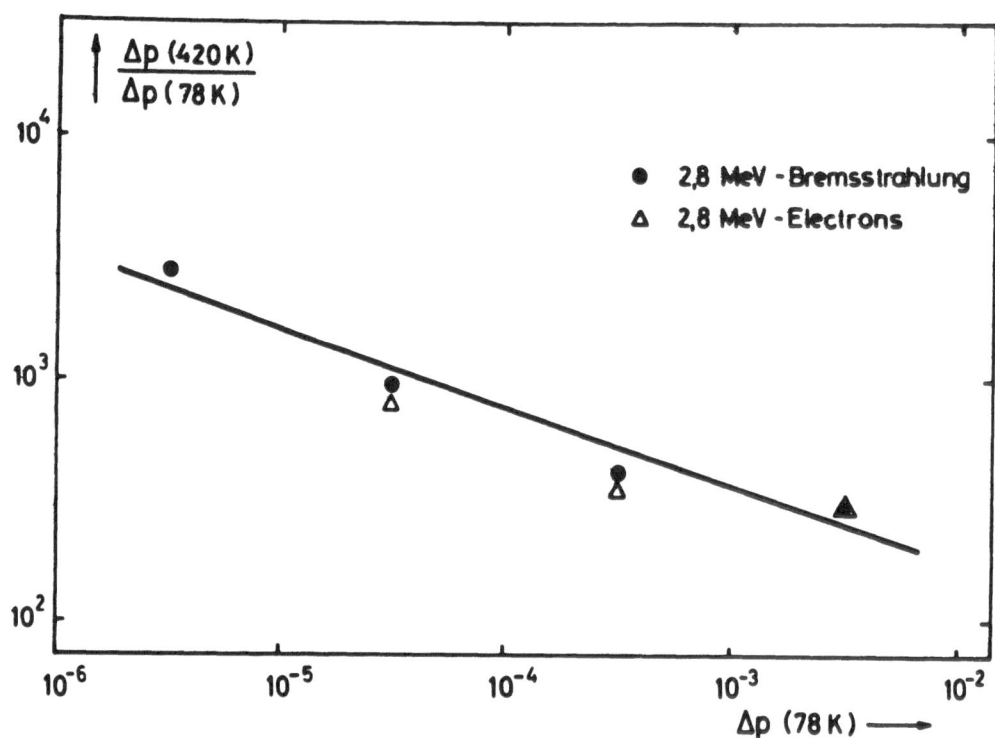

Fig. 1

Relative gain of pinning points: $\Delta p(400 \text{ K})/\Delta p(78 \text{ K})$ by
annealing treatment at 400 K after irradiation at 78 K.
($\Delta p(78 \text{ K})$ is the number of pinning points before,
$\Delta p(400 \text{ K})$ after annealing.) This gain was found to be
dependent on irradiation dose, ($\Delta p(78 \text{ K})$ is a measure of
dose), but was found to be independent of the type of
irradiation.

Dislocation Damping in Cu at kHz- and MHz-Frequencies

V. Naundorf and K. Lücke

Institut für Allgemeine Metallkunde und Metallphysik
der Technischen Hochschule Aachen, W.-Germany.

For evaluation of dislocation damping data, mostly the Granato-Lücke- (G.-L.-) theory /1,2/ is used. It allows to derive from the damping data information about the dislocation density , the free loop length L and the damping constant B. The model was especially successful in interpreting the influence of irradiation on internal friction phenomena. Under the additional assumption that by irradiation only the loop length L is changed, it predicts for the kHz-range the so-called L^2-L^4-dependence for modulus defect and damping, respectively /2,3/, and for the MHz-damping as function of frequency a maximum with a hight and a frequency proportional to L^4 and $1/L^2$, respectively. In both frequency ranges these predictions have repeatedly been verified with good accuracy.

Comparing measurements in the MHz- and kHz-region, however, there remained the question whether or not, from the point of the G.-L.-theory, the data obtained in both frequency regions were compatible also with each other. For example, Heiple and Birnbaum /4/ and Akita and Fiore /5/ showed that the damping data over the whole frequency region could sufficiently be matched by the predicted frequency dependence of the decrement using the same damping constant B for both ranges. On the other hand, the experiments of Paré and Gubermann /6/ indicated for the kHz-region a damping constant B of 15 - 65 times of that usually accepted in MHz-work.

Most of the attempts to connect kHz- and MHz-damping data suffer from the disadvantage that the experiments in the different frequency ranges are performed on different samples. For this reason, an apparatus has been built which allows to measure, on one and the same sample and simultaneously, MHz-damping by the impuls-echo method (compressional waves along the sample axis) and kHz-damping and the related modulus defect by a standing wave (longitudinal vibration of the sample). The sample was a Cu-single crystal (99.999% purity) of about 55 mm length and 10 mm diameter in $\langle 111 \rangle$ -orientation.

Fig. I shows a substantial part of the experimental results. Damping is plotted versus frequency for several states of γ -irradiation. From the data the background damping determined as the damping after extensive irradiation (which leads to complete dislocation pinning) was removed. The appropriate orientation factor of 0.5 /7/ has been applied to the kHz-data to account for the different types of wave propagation in the two methods of measurement. As one can see, the MHz-data are fully accounted for by the G.-L.-theory the predictions of which are given by the full curves. But one can see, too, that the kHz-damping is about a factor of 70 higher if compared with the extrapolation of the MHz-data.

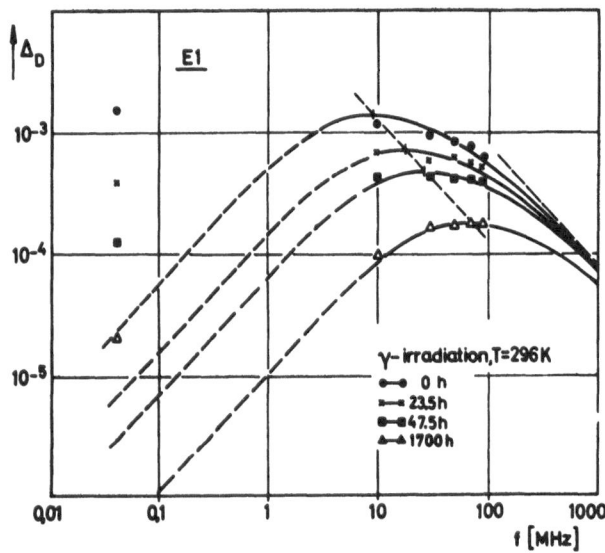

Fig. I: Frequency dependence of the decrement Δ_D, after different amounts of γ -irradiation. The inserted curves are those predicted by the Granato-Lücke-theory.

This discrepancy cannot be explained by an additional amplitude dependent damping in the kHz-range. As to be seen in Fig. II for the sample being in the same irradiation state as in Fig. I, the damping does not depend upon the amplitude of vibration in the range used in the present experiments ($\mathcal{E} = 3 \times 10^{-8}$).

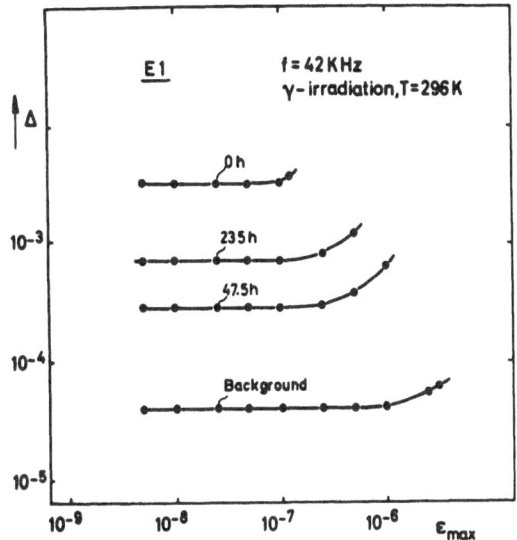

Fig. II: Amplitude dependence of Δ at 42 kHz after different
           amounts of γ-irradiation.

As the MHz-data alone are well described by the G.-L.-theory,
they have been evaluated on the basis of this theory by deducing
the dislocation parameters Λ/B, $\Lambda L^2/C$, and BL^2/C (C = line
tension). These are shown in Table 1 for the unirradiated crystal.
Assuming the values of B $\approx$ 6 x 10^{-4} dyn sec/cm^2 and C = 0.6 x 10^{-4} dyn,
one would obtain Λ = 5 x 10^6 cm^{-2} and L $\approx$ 5 x 10^{-5} cm. Evaluating
also the kHz-data in the usual way by the G.-L.-theory, the param-
eters listed in the first row of Table 1 are obtained. With the
assumption that Λ is the same in both experiments, the comparison
of the two parameter sets leads to the conclusion that
$B_{kHz} \approx 10 B_{MHz}$ and $L_{kHz} \approx 1,5 L_{MHz}$. This means that the loop
lengths derived in the two frequency ranges come out to be
approximately equal, but that there is a strong discrepancy
concerning B.

Fig. III shows the results of an irradiation run of 10 hours at
room temperature. Measured quantities are modulus and damping
at 42 kHz and the damping at 30 MHz as function of irradiation
time. The data were normalized with respect to the initial values
after subtraction of the background values. The nearly linear

time dependence of these quantities is expected for this irradiation temperature because it lies directly above a pinning
stage. Both sets of experiments were evaluated again in the
usual way with the help of the G.-L.-theory, in order to get
the number p of irradiation induced pinning points per initial
loop length L.

TAB.1- DATA OF UNIRRADIATED CRYSTAL E1

	disloc. damping	disloc. modul. defect	dislocation parameters (calc. from G.-L.-theory)		
			Λ/B (dyn·sec)$^{-1}$	$\Lambda L^2/C$ dyn^{-1}	BL^2/C sec
long. stand. wave f=42 KHz	$3.05 \cdot 10^{-3}$	$5.8 \cdot 10^{-3}$	$7.3 \cdot 10^8$	$2.5 \cdot 10^2$	$3.3 \ 10^{-7}$
long.compress. wave(imp.-echo) f_{max}=9.0 MHz	$1.4 \cdot 10^{-3}$	—	$7.3 \cdot 10^9$	$1.1 \cdot 10^2$	$1.5 \cdot 10^{-8}$

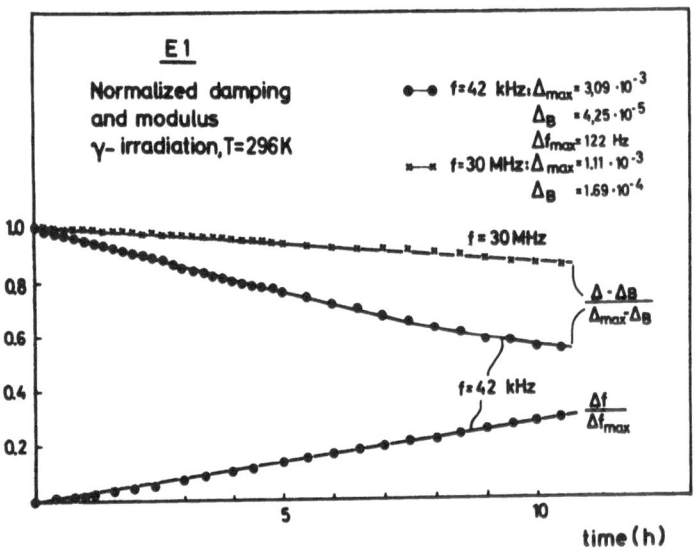

Fig. III: Normalized damping and modulus defect as function of
           irradiation time for 42 kHz and 30 MHz.

Fig. IV and V show the results. As to be seen in Fig. IV the
pinning point numbers calculated from the modulus change (p_M)
are equal to that calculated from the change in kHz-damping (p_Δ)
over about two powers of ten. This clearly indicates the validity
of the L^2-L^4-dependence in the kHz-region. Comparing these pinning
point numbers with that derived from the MHz-data, Fig. V shows
a difference by about 30%. This difference, however, can be fully
reduced to the difference in loop length obtained for the two
frequency ranges. Considering the factor 1.5 in loop length and
calculating the number of pinning points per unit length instead
per initial loop length L, the numbers derived from kHz-data and
MHz-data come out to be about equal.

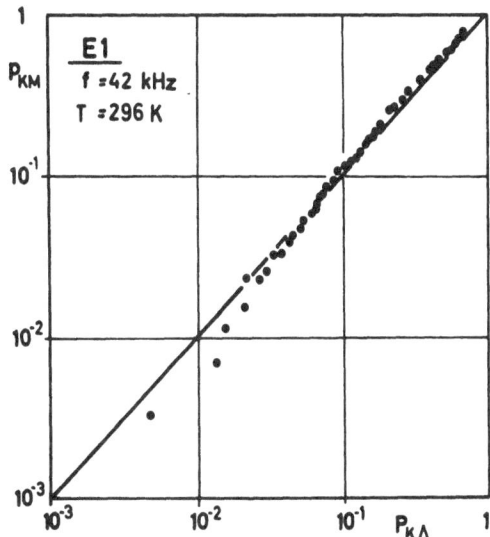

Fig. IV: Pinning points per initial loop length, evaluated
          from damping (p_Δ) and modulus defect (p_M) at 42 kHz.

The main results of these experiments can be summarized by the
following four points:

1. In the MHz-region, the damping before and as function of
   irradiation, can well be explained by the G.-L.-theory,
   especially with regard to the frequency and loop length
   dependence. This is in agreement with the literature /2,3/.

2. In the kHz-region, damping and modulus change before and as
   function of irradiation, can well be explained by the G.-L.-
   theory, especially with regard to the loop length dependence.
   This is in agreement with the literature /2,3/.

3. The value of the damping constant B derived from kHz-
   measurements is about a factor of 10 larger than that
   derived from MHz-damping. This difference which is not
   due to an amplitude dependence is in agreement with /6/ ,
   but in contradiction to /4/ and /5/ .

4. The loop lengths derived from the kHz- and from the MHz-
   data come out to be different by a factor of 1.5, but the
   number of irradiation induced pinning points per unit length
   come out to be the same in both frequency ranges. Such a
   comparison could not yet be achieved before.

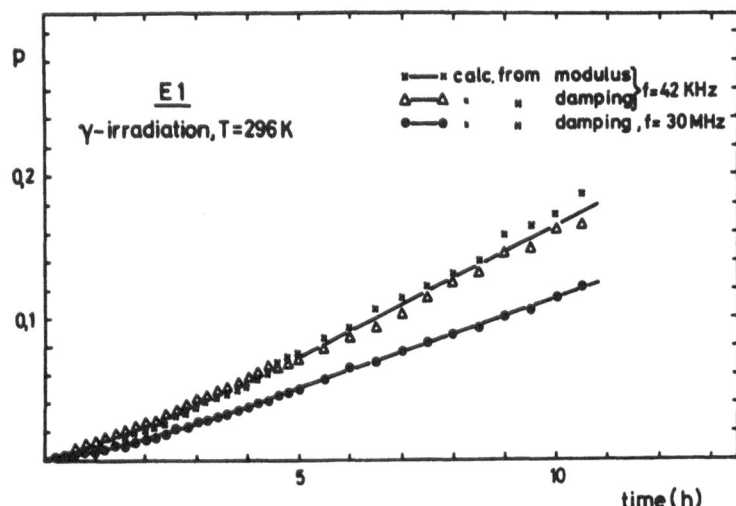

Fig. V: Number of pinning points per loop length vs. irra-
         diation time as evaluated from kHz- and MHz-data
         (Fig. III).

These results mean that the G.-L.-theory describes rather properly the features connected with the loop length dependence in
the whole frequency range and with the frequency dependence in
the MHz-range, but that it fails to describe, in its present
form, the frequency dependence in the kHz-region. This conclusion
is in agreement with measurements in the lower frequency range
which indicate that there, in contrast to the G.-L.-theory, the
dislocation damping is nearly independent of frequency /8,9/.
As will be shown at another place /10/, this discrepancy cannot
be solved by assuming a superposition of dislocation components

with different parameters or by assuming impurity drag instead
of phonon drag as the source of energy dissipation. At the
moment it seems to be most probable to the authors that the
solution of the problem lies in a friction term being independent of frequency but proportional to the strain amplitude.

References:

1 A. Granato and K. Lücke; J. appl. phys. $\underline{27}$ (1956) 583

2 A.V. Granato and K. Lücke; in Physical Acoustics, IV,
   Part A, p. 225, ed. W.P. Mason, New York (1966)

3 D.O. Thompson and V.K. Paré; in Physical Acoustics III,
   Part A, p. 293, ed. W.P. Mason, New York (1966)

4 C.R. Heiple and H.K. Birnbaum; J. appl. phys. $\underline{38}$ (1967) 3394

5 H. Akita and N.F. Fiore; J. appl. phys. $\underline{42}$ (1971) 2203

6 V.K. Paré and H.D. Gubermann; J. appl. phys. $\underline{44}$ (1973) 32

7 E.G. Henneke II and R.E. Green Jr., Trans. AIME $\underline{242}$ (1968)1071

8 R. den Buurman and D. Weiner; Scripta Met. $\underline{5}$ (1971) 573

9 K. Beißner and E. Biller; Scripta Met. $\underline{7}$ (1973) 535

10 V. Naundorf and K. Lücke; to be published;
   V. Naundorf; Thesis, Aachen, 1974

DISLOCATION PINNING OF MOLYBDENUM AFTER LOW TEMPERATURE NEUTRON IRRADIATION

S. Okuda and H. Mizubayashi

Japan Atomic Energy Research Institute
Tokai-mura, Naka-gun, Ibaraki-ken, Japan

ABSTRACT

The dislocation pinning by the stage I defects in Mo single crystals is studied after fast neutron irradiation at liquid helium temperature. There are the two largest substages centered at about 28 and 42°K in the stage I recovery of Mo. The defects of these two substages have a maximum strain field along <110> crystalline direction (probably <110> split interstitials). It was found that the pronounced increase in elastic modulus occurred in the temperature range centered at about 30°K during warm up after low temperature irradiation. This modulus increase was very small in the less pure specimens with the resistivity ratio of even about 2000. The structure of the differential modulus change curves ($\Delta E/\Delta T$) seemed to suggest that both defects of the above two substages contributed to the pinning of dislocations. Therefore, the defects of the two types appeared to perform a long range migration during their recovery. This point is discussed in detail.

1. Introduction

The dislocation pinning due to an arrival of point defects to dislocations has been used effectively as a sensitive tool to study the migration of point defects. These experimental results have been explained mostly by the vibrating-string model after Koehler-Granato-Lücke (K-G-L model) (1). Qualitatively, the dislocation pinning causes a shortening of dislocation loops, thus decreases both internal friction and modulus defect. Whereas the nature of pinning is not well understood at present, the firm pinning assumed in the K-G-L model is a rough approximation. Therefore, some secondary effects which complicate the behavior of internal friction appear, sometimes (2,3). The modulus defect does, however, decrease always by pinning, except the very special case such as overdamped case.

In the present experiments, this pinning experiments were applied for the study of the stage I defects in Mo. Since the recent studies on the stage I recovery in Mo have revealed the complicated substage structures in this stage (4), the determination of a long range migration stage by the dislocation pinning would be desirable to clearify the problem. In the previous measurements using polycrystalline Mo, the dislocation pinning was found to occur from about 30°K (5). These measurements were extended to single crystals with higher purity in the present work.

2. Experimental Procedures

Specimens of single crystal were cut into the shape of about $0.2 \times 3 \times 20$ mm^3 from the single crystal rods (nominal purity 99.992%) purchased from Material Research Corporation. One end of specimens is thicker to ensure a rigid gripping in the specimen holder. After etching to remove the strained surface, the specimens were annealed at $1600 \sim 1800°C$ in vacuum of better than 1×10^{-8} mmH$_g$. Their resistivity ratios (R.R.) after the annealing varied from 2200 to 4300 for those from different single crystal rods. The procedures of irradiation and post-irradiation measurements were similar as those described previously (5,6).

Specimens were irradiated using the in-pile irradiation facility (LHTL in JAERI) at about 5°K with neutron flux of nearly fission spectrum of 1×10^{12} n/cm^2.sec (>0.1 MeV)*, then transferred to the measuring cryostat and the internal friction and dynamic modulus were measured during warm-ups (warm-up rate $\sim 0.5 \sim 1°K$/min). The flexural vibration of the specimen (~ 500 cps) was both excited and detected by an electrostatic method. The internal friction (Q^{-1}) was obtained from both a drive force at constant amplitude of vibration and a free decay. The change in Young's modulus ($\Delta E/E \approx -2\Delta P/P$) was measured from a period of vibration at resonance (P). These measurements were repeated every 30 sec. The maximum strain amplitude was of the order of 10^{-6}, in a little amplitude dependent region.

---

* During the course of the present experiments, the mode of reactor operation (nominal 10MW) was a little changed. This was indicated in the figures by 10MW $(1+\alpha)$ and 10MW $(1+\alpha')$, the increases in the flux in these operations were less than $\sim 20\%$.

3. Experimental Results

Typical examples for the <100> specimens (the specimens with a stress axis of vibration parallel with <100> crystalline direction) with R.R.=4000 are shown in Figs. 1 and 2. The pronounced hardening occuring around 30°K during warm-up after neutron irradiation is seen in the figures. Fig.3 shows derivatives of modulus change $(\Delta E_T/E)/\Delta T$ during the initial warm-up of the above two specimens and also of the polycrystalline specimen of low purity, together with a recovery spectrum of electrical resistivity $(\Delta \rho_T/\Delta \rho_0)/\Delta T$ for polycrystalline specimen (data from Ref. (7)). The stage I recovery in electrical resistivity gives rise to two pronounced substages at about 28°K and 42°K. It seems that the modulus defect recovery for 2hr irradiation corresponds to the 28°K substage and that for 0.5hr irradiation being smaller in the amount of modulus recovery, extends over both 28°K and 42°K substages.

Results on the <100> specimens of lower purity (R.R.=2200) are shown in Fig.4. For the less pure specimen, the modulus recovery or pinning is remarkably reduced. On the other hand, the specimen with the same purity (from the same rod) but deformed 0.5% at room temperature before irradiation shows again a pronounced pinning, but the largest pinning stage shifts to ∿ 40°K. Furthermore, in the deformed specimen, a pinning occurs also around 50 ∿ 70°K where the detrapping of impurity trapped interstitials is suggested (6). A similar tendency is also seen in the polycrstal specimen as shown in Fig.3. Fig.5 shows results on the <111> specimen. In general, the modulus recovery is less in the <111> than in the <100> specimens even if the purities are similar. The temperature coefficients of Young's modulus are also smaller in the <111> specimens. These difference might possibly come from the different dislocation structures in the two types of specimens.. In Fig.5, a pinning after the first irradiation of 1hr is far from complete, disappears after warmed up to 308°K and a further pinning proceeds after the second irradiation. After the second irradiation, the modulus curve becomes a little complicated because of the overlapping of the modulus effect associated with two relaxation peaks.

The derivatives of modulus change after low temperature irradiation of the <100> and <111> specimens with various treatments are compared in Fig.6. The above mentioned effects of deformation are seen better in the figure for the specimens of both orientations. Namely, due to the introduction of probably fresh dislocations, the pinning becomes more pronounced, the main pinning stage moves from ∿ 30°K to ∿ 40°K and also the pinning around 60°K or above appears. The results up to room temperature are shown in Fig.7.

As it can be seen in the figure, there is no other large pinning stage up to room temperature. In Fig.7, corresponding decreases in the internal friction are also seen in the above pinning stages.

In Fig.8, the relative modulus changes (measured at 10°K) referred to as-irradiated moduli after each warm-up runs are shown for various irradiation times and purities. In Fig.9, the internal frictions (measured at 20°K) are shown as a function of warm-up temperatures. Whereas the stage I pinning is clearly seen in these figures, the behaviors for higher temperatures are not simple and further study is needed.

4. Discussion

Before entering into discussion, it is noted that the bulk effect of Frenkel pairs on Young's modulus is negligibly small at least for irradiations of less than 2 hr. Preliminary measurements give $\Delta E/E \leq -3\%/at.\%$ of the Frenkel defect, as the bulk effect.

It is generally observed that the stage I recovery in Mo can be divided into at least five substages, namely I_1 ($\sim 15°K$), I_2 ($\sim 28°K$), I_3 ($\sim 35°K$), I_4 ($\sim 42°K$) and I_5 ($\sim 60°K$) (4,8). Among these substages, I_2 and I_4 are the most pronounced ones. From the very detailed study of recovery kinetics, Afman (9,10) proposed that I_2 is due to a free migration of interstitials (correlated annihilation) and I_4 due to the defects of a new type, possibly close pairs dissociating along <111>. Later, Moser (11) modified this model and proposed an interstitial extending along <111> for I_4. Rizk et al. (12,13) and Bichon (14) have criticized a too much confidence on the Afman's kinetics analysis and considered that I_2 and I_4 are due to close pairs and correlated annihilation of free interstitials, respectively.

In our previous experiments (6), it was shown that there are two relaxation peaks which are considered to be associated with the stress induced rotation of the I_2 and I_4 defects, respectively, and both defects are considered to be of <110> symmetry or the <110> split interstitials with different activation parameters. Then, it might be natural to assume that the one is free interstitials and the other is interstitials in close pairs, and furthermore, I_2 is due to the latter and I_4 to the former. The close pair model seems, however, to be probable for neither I_2 nor I_4 as it will be shown below. By the way, the new types of defects by Afman and Moser do not seem reconcilable with the observed <110> symmetry of the defects.

In the present results, the effects of amount of dose and dislocation density on the pinning spectra suggest that the pinning in I_4 appears more pronounced when the pinning has not been completed in I_2. Therefore, the present results suggest that the two kinds of interstitials with <110> symmetry perform a long rang migration in both I_2 and I_4, respectively. To explain the two pinning stages, it may be considered that the <110> split interstitials perform a free migration in I_2 and some trapped types of <110> split interstitials release from traps in I_4. At this moment, however, it would be premature to make any further speculations without further study.

Acknowledgment

The authors with to express their gratitude to the members of LHTL group and JRR-3 for their invaluable help throughout the experiments.

References

1. D. O. Thompson and V. K. Paré, Physical Acoustics, Vol. 3A, (W. P. Mason ed.), p. 293, Academic Press, London/New York, (1966)

2. S. Okuda and R. R. Hasiguti, Acta Met. 11, 257 (1963)

3. H. M. Simpson, A. Sosin, G. R. Edwards and S. L. Seiffert, Phys. Rev. Letters, 26, 897 (1971)

4. R. de Batist, J. Nihoul and L. Stals eds., Defects in Refractory Metals, S. C. K. /C. E. N., Mol (1972)

5. S. Okuda and H. Mizubayashi, Phys. Stat. Sol. (a) 16, 355 (1973)

6. S. Okuda and H. Mizubayashi, Crystal Lattice Defects, in press.

7. R. Hanada, S. Takamura, S. Okuda and H. Kimura, Trans. Japan Inst. Metals, 11, 434 (1970)

8. J. Nihoul, Radiation Damage in Reactor Materials, Vol. 1, p. 3, IAEA, Vienna (1969)

9. H. B. Afman, Phys. Stat. Sol. (a) 4, 427 (1971)

10. H. B. Afman, ibid., 11, 705 (1972)

11. P. Moser, p. 59 in Ref. (4)

12. R. Rizk, P. Vajda, F. Maury, A. Lucasson and P. Lucasson, Phys. Stat. Sol. (a) 14, 135 (1972)

13. R. Rizk, P. Vajda, A. Lucasson and P. Lucasson, ibid, 15, K105 (1973)

14. P. Bichon, Thesis, CEA-R-4301 (1972)

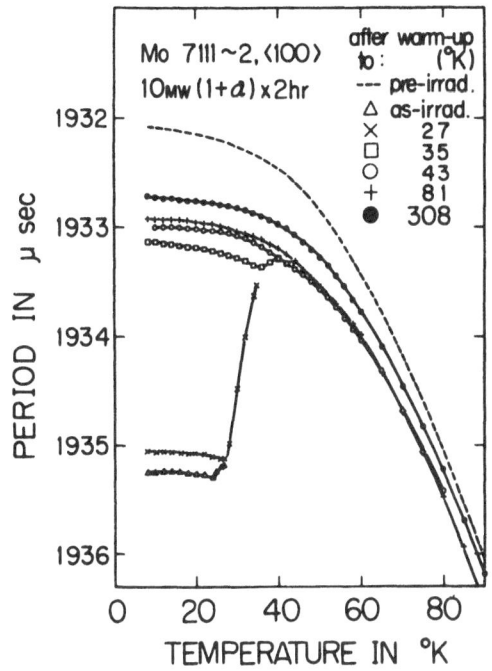

FIG. 1

Dynamic modulus of the <100> single
crystal Mo during warm-up after fast
neutron irradiation near liquid helium
temperature. (R.R.=4000)

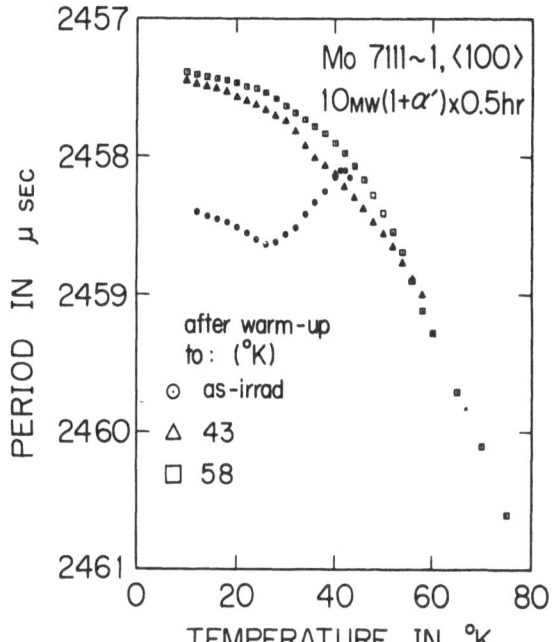

FIG. 2

Similar curves as Fig.1 for lower dose.

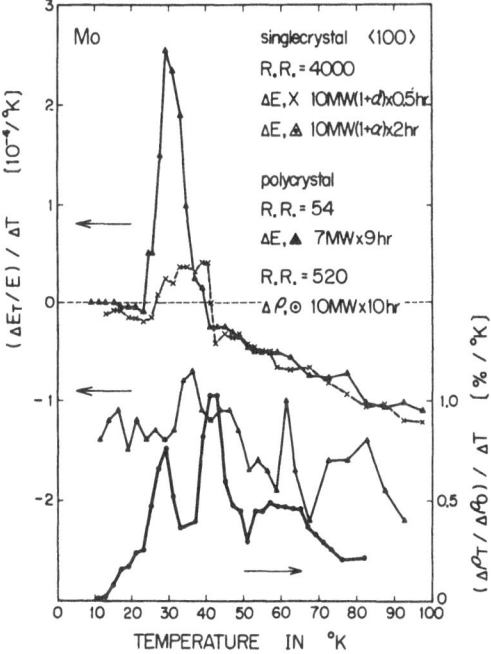

FIG. 3

Derivative curves for recovery of dynamic
modulus $(\Delta E_T/E)/\Delta T$ of the polycrystal and
<100> single crystal Mo during first warm-up
after fast neutron irradiation. Recovery
spectra of the electrical resistivity is also
shown for comparison purpose.

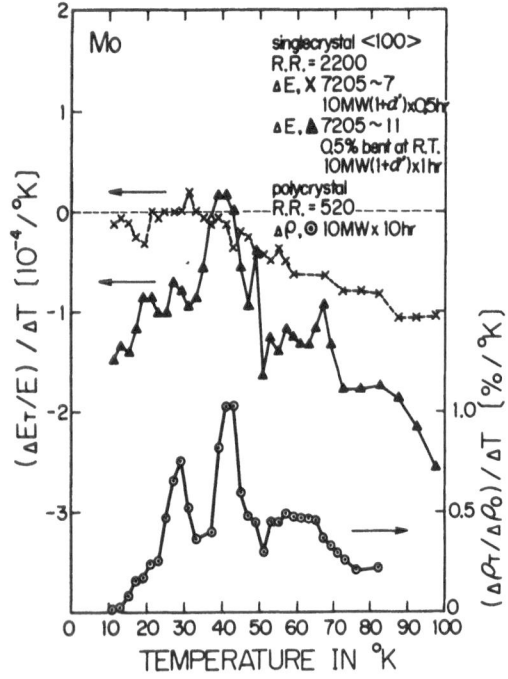

FIG. 4

Similar curves as Fig.3 for differently treated specimens.

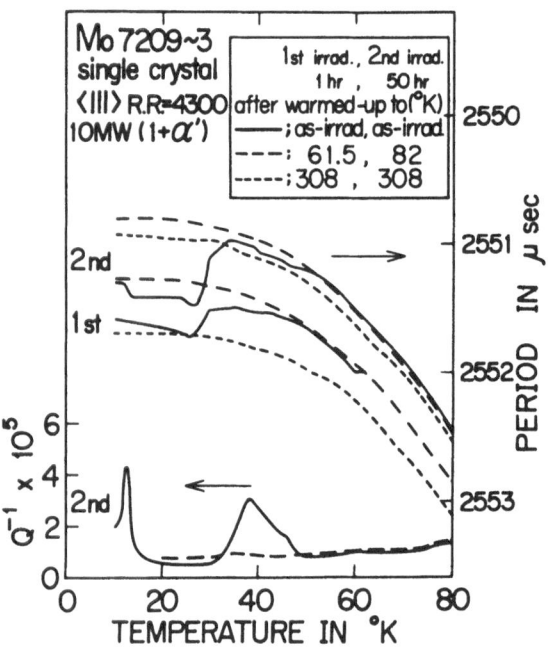

FIG. 5

Internal friction and dynamic modulus of the <111> single crystal Mo during warm-up after fast neutron irradiation. Irradiations were made twice. The second irradiation followed after warm-up to 308°K after the first irradiation.

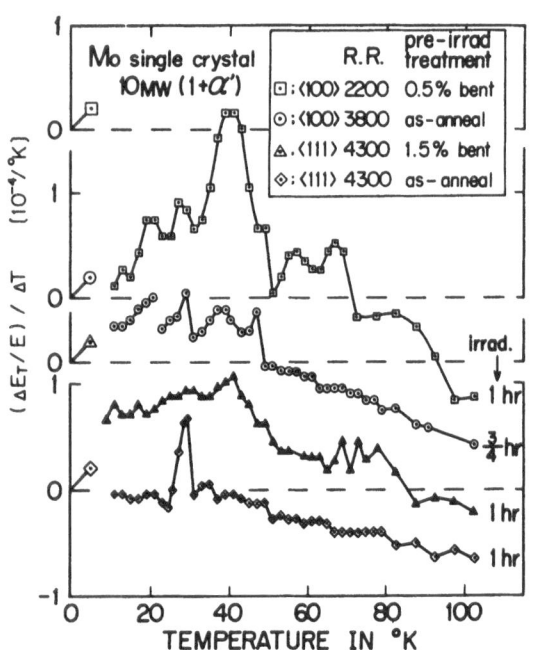

FIG. 6

Derivative curves for recovery of dynamic modulus $(\Delta E_T/E)/\Delta T$ during warm-up after fast neutron irradiation for the <100> and <111> single crystal Mo with various pre-irradiation treatments.

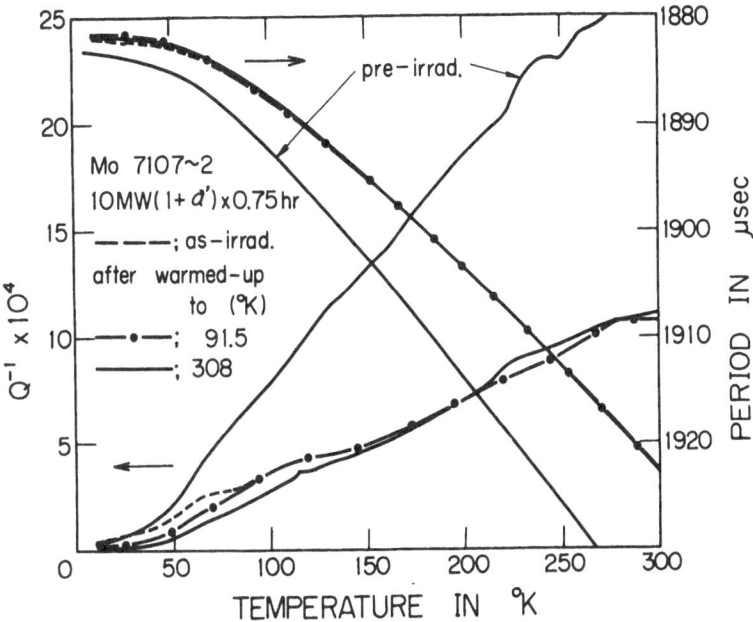

FIG. 7

Internal friction and dynamic modulus of the <100> single
crystal Mo during warm-up after fast neutron irradiation.
(R.R.=3800)

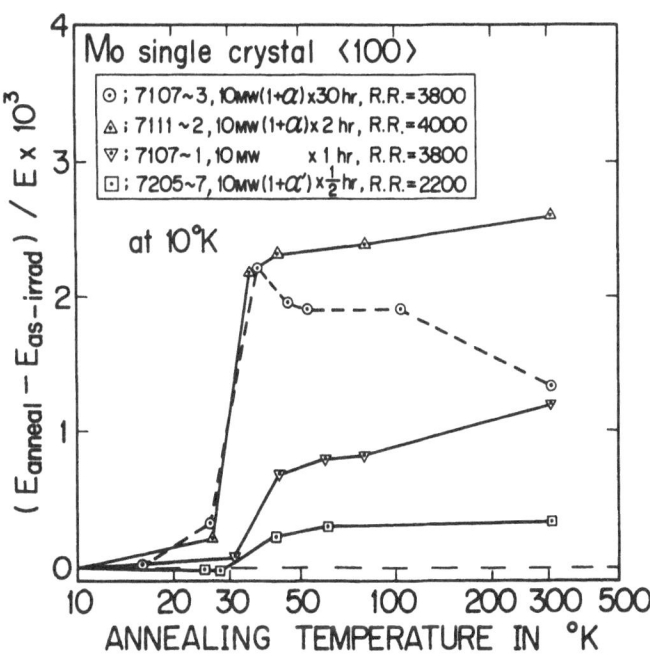

FIG. 8

Relative modulus changes referred to
as-irradiated moduli after each warm-
up runs after fast neutron irradiation. All measurements were made at
10°K.

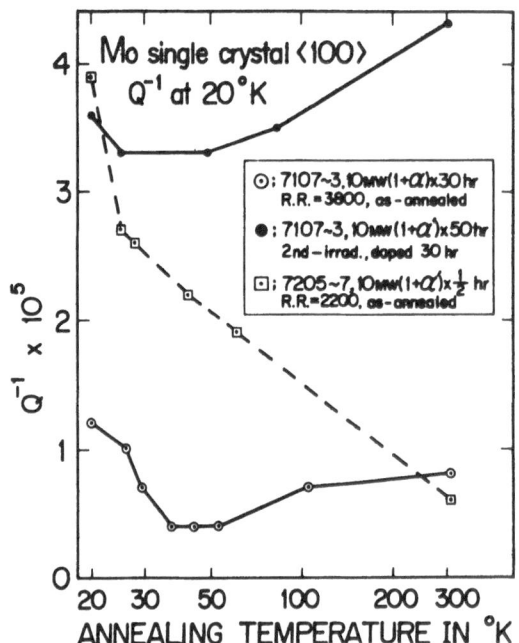

FIG. 9

Internal friction measured at 20°K as
a function of warm-up temperatures of
the ⟨100⟩ single crystal Mo after fast
neutron irradiation.

INTERNAL FRICTION OF NIOBIUM IRRADIATED AT LOW TEMPERATURE

N. Igata, F. Watari, A. Yamaguchi, K. Miyahara and S. Sato
Department of Metallurgy and Materials Science,
University of Tokyo, Tokyo, Japan

Introduction

Hitherto relaxation spectra of niobium have been investigated by many authors and the δ (1, 2), α' (3, 4, 5), α (3, 5 - 19) and β peaks (7, 10, 11, 12, 14, 15, 16, 17, 20, 21) have been observed. Internal friction of neutron irradiated niobium has been also reported. Stanley et al. (14) showed that the α peak decreased (in annealed specimens, increased then decreased) and the β peak increased after irradiation above 10^{18} nvt (> 1 MeV) at 65°C. The objective of this study is to clarify the role of point defects induced by low temperature irradiation in the α and β peaks. Since the α and β peaks are sensitive for interstitial impurity atoms, the materials of various contents of interstitial impurity are used.

Experimental Procedure

The contents of interstitial impurity atoms, metallic purities and treatments are shown in TABLE 1. The specimen sizes were 0.5 - 2.0 mm in thickness, 5 - 8 mm in width and 50 - 100 mm in length.

TABLE 1
The Purities and Treatments of Specimens

Specimen Number	Metallic Purity (wt %)	Interstitial Impurity (wt ppm)				O+N (at ppm)	Cold Work (%)
		C	N	O	H		
A	99.99	30	20	30	2	306	0 - 10
B	99.8	< 100	10	80	2	531	0 - 10
C	99.9	< 50	30	90	0.1-0.4	722	0
D	99.8	< 100	20	110	2	772	0
E	99.9	10	60	179	4	1450	0 - 12
F	electron beam melted	< 50	50	270	2	1900	0 - 12
G	electron beam melted	< 50	60	410	0.5	2700	3

The irradiation was performed at the liquid helium loop in KUR I (Kyoto University Reactor). The total dose was between $3 \times 10^{15} - 6.8 \times 10^{16}$ nvt (> 1 MeV) and irradiation temperatures were below 15°K. Only specimen A was irradiated by 4×10^{16} nvt (> 1 MeV) at 120°C. The irradiated specimens were further annealed between 150°C and 450°C.

Internal friction was measured by transverse vibration method from liquid helium or liquid nitrogen temperature to room temperature. The frequency range was between 3×10^{2}Hz and 2×10^{3}Hz. The measurement was performed in situ in loop or after taking out from the liquid helium loop into liquid helium or liquid nitrogen baths.

Experimental Results and Discussions

A typical experimental result is shown in FIG.1. This is the result for specimen E cold worked 10%, irradiated 4×10^{16} nvt (> 1 MeV) below 15°K and measured in situ in loop from liquid helium temperature after irradiation. After "stage I annealing" at 100°K the α' and α peaks were suppressed and the β_1 and β_2 peaks became larger. Then after Stage III annealing at 152°C for 1 hr the α' , α , β_1 and β_2 peaks decayed. In the case of the annealed specimen the α peak was very small or below detection limit both before irradiation and after irradiation, and the β_1 , β_2 and sometimes β_3 became larger. But those peaks also decayed after Stage III annealing.

The α peak

FIG.2 shows the ratio of the α peak height after irradiation to that before irradiation as the function of total dose. The ratio decreased with the irradiation dose. The change was more sensitive compared with that reported by Stanley et al. (14). This shows that low temperature irradiation is more effective than room temperature irradiation because of the difference of the annealing behavior of simple defects such as free interstitials which are movable below Stage III. Above results after irradiation is compared with the decreasing tendency of the α peak due to interstitial impurity atoms. FIG.3 shows the relation between the α peak height and interstitial atoms content. The reason why O+N contents are taken as interstitial atoms is because carbon atoms are often below detection limit and the role of hydrogen atoms seems to be different from O or N atoms (22). From the comparison of FIG.2 and FIG.3, it was revealed that irradiation induced point defects pinned dislocation in the same way as interstitial impurity atoms.

On the other hand, the maximum of the α peak height versus strain was observed for unirradiated specimens. This suggests that the α peak is not only the function of dislocation density but also that of dislocation configurations or pinning by point defects produced by cold work. In our previous works (5), the α peak was attributed to be the relaxation due to overcoming intrinsic barriers of non-screw dislocations and interpreted by double kink mechanism. The decrease of the α peak was interpreted by dislocation pinning by point defects. Assuming that the distribution of dislocation loop length is exponential function (23), that the double kink crosses over only one Peierls barrier or the displacement is one atomic distance in this temperature range, and that double kink formation (24) is impossible when loop length l is less than critical length d_{cr} , the next equation was tentatively derived by integrating the displacement of dislocation loops which are longer than d_{cr}.

$$Q_{max}^{-1} = \frac{G\, b^4\, d_{cr}}{k\, T_p} (1 + \frac{d_{cr}}{L_c}) \exp(-\frac{d_{cr}}{L_c}) \tag{1}$$

where Q_{max}^{-1} is the α peak height, G is shear modulus, b is Burgers vector, d_{cr} is the critical separation of two kinks for overcoming Peierls potential, L_c is the average dislocation loop length, and Λ is effective dislocation density. The relation shown in the equation (1) can interpret FIG.2 and FIG.3. When we assume L_c = a/c (a is the atomic distance and c is the concentration of pinning atoms), d_{cr} can be calculated from FIG.3 to be 740 a. Then Λ is estimated to be $2 \times 10^6/cm^2$ from the intersect at C = 0. The dislocation density observed by electron microscope was $3 \times 10^9/cm^2$. This suggests that only small fraction ($\sim 10^{-3}$) of observed dislocation density is contributive to the α peak, but its fraction seems to be closely related with dislocation configuration or interstitial impurity contents.

The β peak

FIG.4 shows the frequency dependence of the β_1 and the β_2 peaks (3, 5, 14, 21). The frequency factors and activation energies are $10^{12.1}$/sec. and 0.40eV for the β_1 peak and $10^{11.9}$/sec. and 0.45eV for the β_2 peak. From that the frequency factors of the β_1 and β_2 peaks are below Debye frequency, it is suggested that those peaks are not due to the relaxation of point defects.

The effect of neutron irradiation on the β_1 and β_2 peaks are shown in FIG.5. The plotted lines show that the β_1 and β_2 peaks increase with irradiation dose in annealed specimens while there is the maximum versus neutron dose in cold worked specimens. This tendency can be also observed in the dependence of the β_1 and β_2 peaks on the contents of interstitial impurity atoms as shown in FIG.6. In both FIG.5 and FIG.6, the β_1 and β_2 peaks show maxima at appropriate pinning point densities. This shows that β peaks are not attributed to the relaxation of simple point defects but the other mechanisms. In our previous report (21), there was also the maximum in the correlation between the α and the β peak (in this case the β_2 peak) and the tentative model was proposed, in which the β peak is due to the catastrophic break away process which is initiated by thermal unpinning. It was because if the β peak is the relaxation of point defects, as the β peak increases the α peak must be suppressed, the results were not so. The conditions that adjacent dislocation loops l_1 and l_2 are thermally unpinned and initiate the catastrophic break away (25) is given by $2L_b > l_1 + l_2 > L_b$, where L_b is the break away length under the given stress amplitude. The swept area of dislocation loop L_N , dislocation network length, which is produced by catastrophic break away was assumed (26), and the β peak height was obtained from the product of the above swept area and the probability which is given by the integration of the distribution function of dislocation loops between $2L_b > l_1 + l_2 > L_b$.

$$Q_{max}^{-1} = \frac{L_N^2\, \Lambda}{24} (\exp(-\frac{L_b}{L_c}) - \exp(-\frac{2\,L_b}{L_c})) \tag{2}$$

where Q_{max}^{-1} is the β peak height, and Λ is effective dislocation density.

From the above relation it can be understood that there is the maxima in the β_1 and β_2 peaks versus neutron dose and interstitial impurity contents. As an example at the β_2 peak maximum versus interstitial impurity content, the peak height is 3×10^{-4}, $L_N^2 \Lambda \simeq 1$ and the brancket term becomes 0.24 from the condition for the maximum of the β peak. Then this results suggest that small fraction ($\sim10^{-2}$) of observed dislocation density contributed to the β_2 peak. Both the β_1 peak and β_2 peak seem to be closely related to the point defects (including interstitial impurity atoms) which can migrate and pin dislocations between Stage I and Stage III.

Summary and Acknowledgements

Summarizing the above experimental results and discussions, the followings are concluded.

(1) The α peak is suppressed by neutron irradiation below $6.8\times10^{16}n/cm^2$ (> 1 MeV). This comes from pinning effect of free migrating point defects which are produced by low temperature irradiation. The α peak changes more sensitively in the case of low temperature irradiation than in the case of room temperature irradiation. It suggests that the annealing of the simple point defects is more effective in the latter than in the former.

(2) The β_1 and β_2 peaks are also sensitive for irradiation doses. The maxima of the β_1 and β_2 peaks were observed versus irradiation dose and interstitial impurity contents.

(3) The α peak could be interpreted by double kink formation mechanisms.

(4) The β_1 and β_2 peaks would be due to the similar mechanism. The β_1 and β_2 peaks would not be due to the relaxation of the simple defects but due to the interaction between dislocations and point defects.

The authors appreciate Dr. H. Yoshida and Dr. M. Nakagawa and their colleague for executing irragiation experiments with our groups. The authors also thank Prof. R. R. Hasiguti and Prof. S. Yoshida for their kindly discussions.

Reference

1. F. M. Mazzolai and M. Nuovo, Solid State Com. 71, 103 (1969)

2. C. Wert, J. Phys. Chem. Solids 31, 1793 (1970)

3. R. H. Chambers, Physical Acoustics, Vol.III A, edited by W. P. Mason, p. 123, Academic Press, New York (1966)

4. T. A. Trozera and T. E. Firle, General Atomics, GA-7661 (1967)

5. N. Igata, A. Kohyama, K. Miyahara, S. Sato and H. Yoshida, Mechanical Behavior of Materials, Vo. 1, p. 58, The Society of Materials Science, Japan (1972)

6. R. H. Chambers and J. Schultz, Acta Met. 8, 585 (1960)

7. L. J. Bruner, Phys. Rev. 118, 399 (1960)

8. R. H. Chambers and J. Schultz, Phys. Rev. Letters 6, 273 (1961)

9. R. G. Bordoni, M. Nuovo and L. Verdini, Phys. Rev. 123, 1204 (1961)

10. R. H. Chambers and J. Schultz, Acta Met. 10, 466 (1962)

11. R. De Batist, Phys. Stat. Sol. 2, 661 (1962)

12. R. De Batist, Phys. Stat. Sol. 3, 1475 (1963)

13. M. F. Amateau, R. Gibala and T. E. Mitchell, Scripta Met. 2, 123 (1967)

14. M. W. Stanley and Z. C. Szkopiak, J. Nucl. Mat. 23, 163 (1967)

15. M. W. Stanley and Z. C. Szkopiak, J. Mat. Science 2, 559 (1967)

16. M. J. Kramer and C. I. Bauer, Phys. Rev. 163, 407 (1967)

17. M. F. Amateau, T. E. Mitchell and R. Gibala, Phys. Stat. Sol. 36, 407 (1969)

18. R. Gibala, M. K. Korenko, M. F. Amateau and T. E. Mitchell, J. Phys. Chem. Solids 31, 1889 (1970)

19. B. Escaig, Scripta Met. 5, 199 (1971)

20. M. W. Stanley and Z. C. Szkopiak, J. Mat. Science 3, 610 (1968)

21. N. Igata, F. Watari, K. Miyahara, S. Sato and H. Yoshida, in press in Crystal Lattice Defects, Gordon and Brench, Science Publishers Ltd. (1973)

22. H. Y. Chang and C. A. Wert, International Meeting on Hydrogen in Metals at Kernforschungsanlage, Jülich, Germany, p. 558 (1972)

23. J. S. Koehler, Imperfections in Nearly Perfect Crystals, p. 197 (1952)

24. A. Seeger, Phil. Mag. 1, 651 (1956)

25. A. Granato and K. Lücke, J. Appl. Phys. 27, 583 (1956)

26. K. Lücke and J. Schlipf, The Interaction between Dislocations and Point Defects, AERE - R5944, Vol. 1, p. 118 (1968)

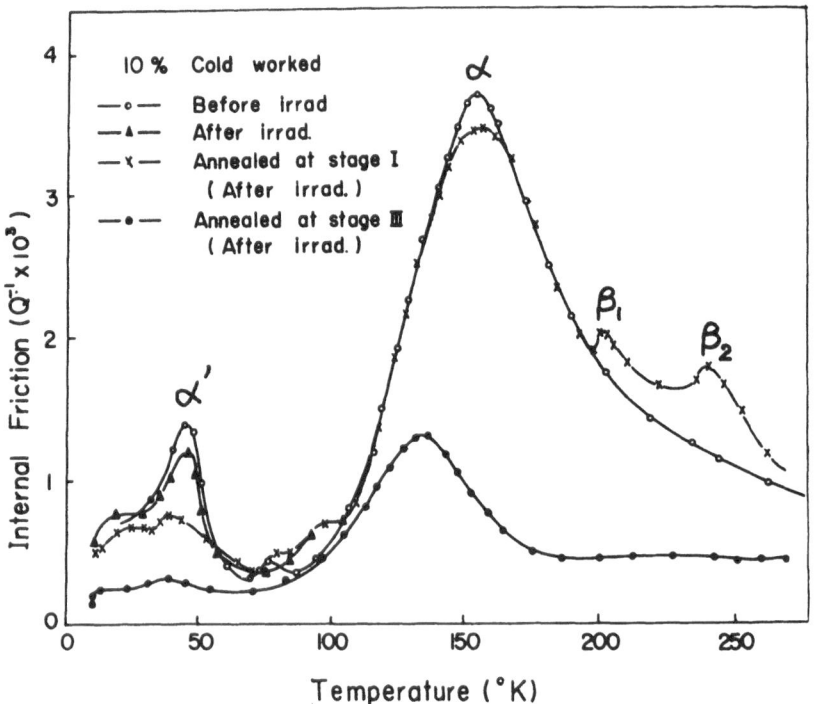

FIG. 1

Relaxation Spectra of Niobium of Specimen E Cold Worked 10%, and Irradiated 4×10^{16} nvt (> 1 MeV) below 15°K

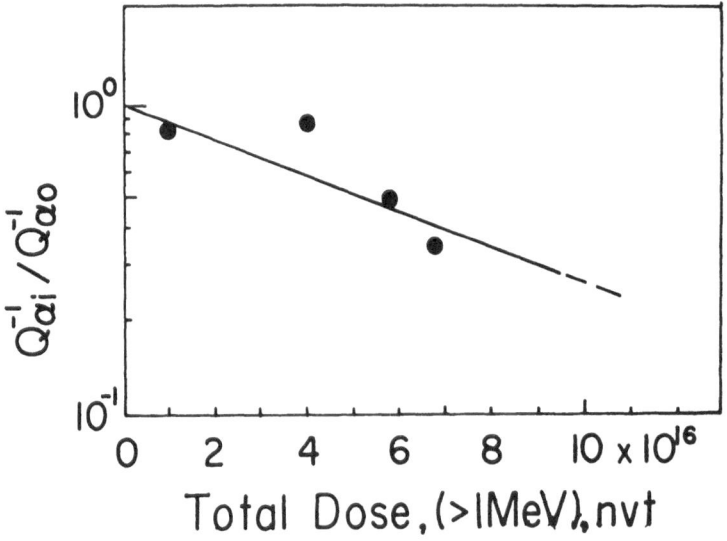

FIG. 2

The Ratio of the α Peak Height after Irradiation to that before Irradiation as the Function of Total Irradiation Dose

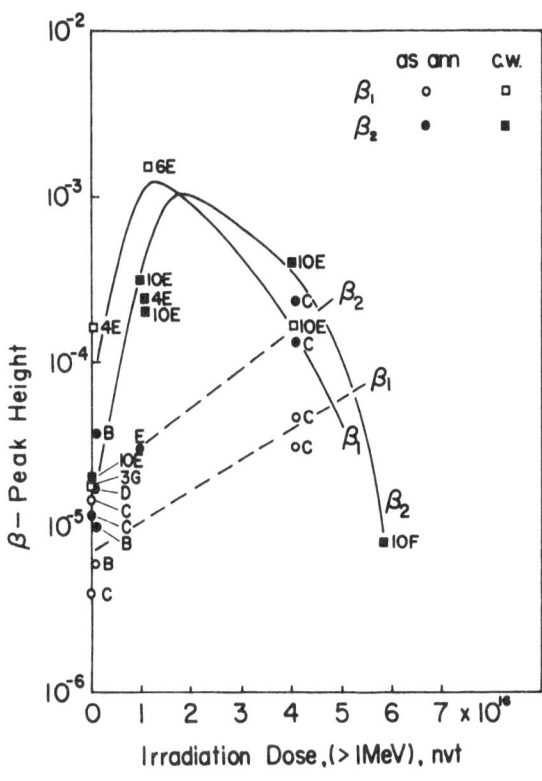

FIG. 5

The Effect of Irradiation on the β_1 and β_2 Peak Heights

(A ~ E show specimen group and the numbers show cold work degree.)

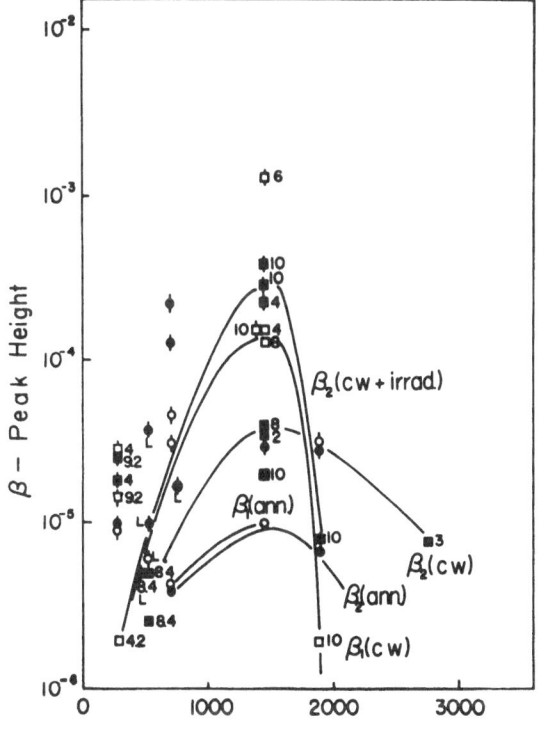

FIG. 6

The Effect of the Interstitial Impurity Contents on the β_1 and β_2 Peak Heights

(A ~ E show specimen group and the numbers show cold work degree.)

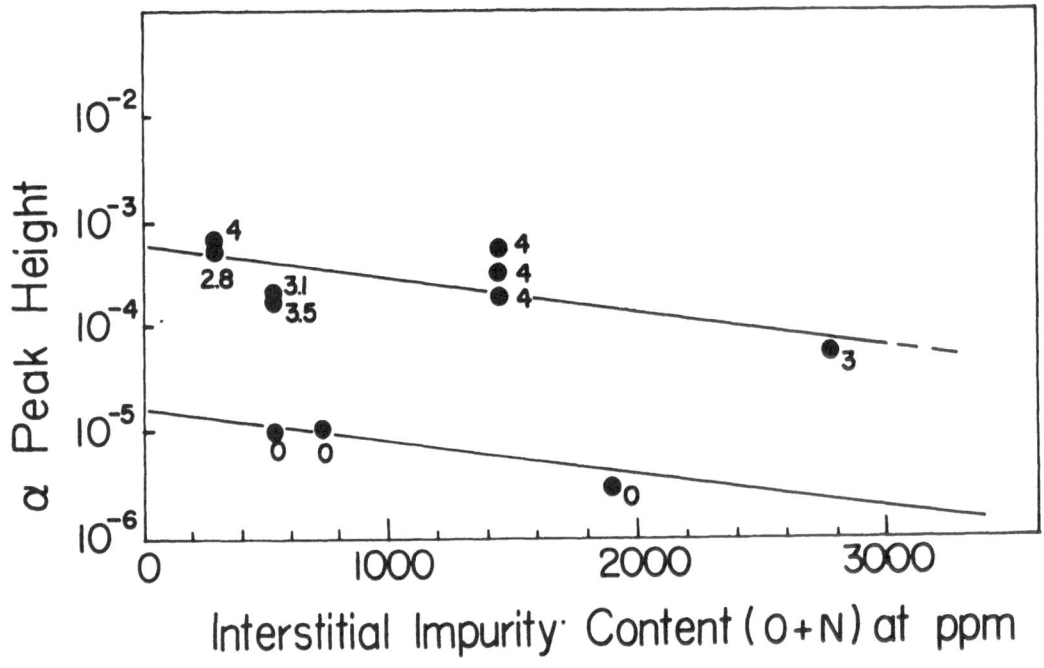

FIG. 3

The Relation between the ∝ Peak Height and Interstitial Impurity Contents
(The numbers show cold work degree.)

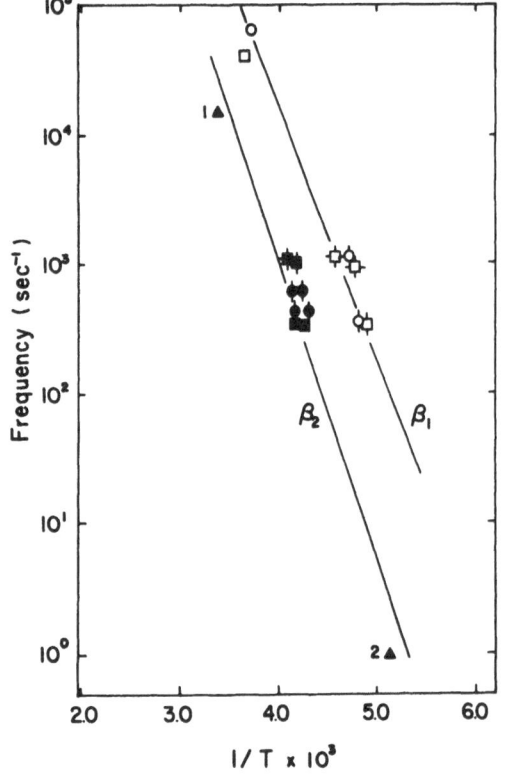

FIG. 4

The Frequency Dependence of the β_1 and β_2
(1 ▲ : Ref.8., 2 ▲ : Ref.14.)

	as ann.	ann. irrad.	ann. irrad. ann.	c.w.	c.w. irrad.	c.w. irrad. ann.
β_1	○	⬙	◈	□	⊡	⊞
β_2	●	⬩	✦	■	⬛	✚

COLD WORK EFFECTS

THE BORDONI PEAK IN COPPER AND SILVER SINGLE CRYSTALS

D. H. Niblett, G. R. Brown and M. Zein
Physics Laboratory, The University,
Canterbury, Kent, England.

Introduction

One of the most interesting internal friction phenomena is the relaxation peak which was first observed by Bordoni (1) in cold-worked face-centred cubic metals. This Bordoni peak has been the subject of a considerable amount of research, which has been reviewed, for example, by Niblett (2), Seeger (3), Nowick and Berry (4) and De Batist (5).

The principal features of the experimental results have been satisfactorily explained by a theory, originated by Seeger (6), that the peak is caused by a thermally-activated relaxation process involving dislocations lying parallel to a close-packed direction in the crystal lattice. Seeger's theory has been modified by Paré (7), Seeger and Schiller (8,9), Alefeld (10) and Engelke (11,12) in attempts to account for the secondary features of the peak: it is considerably broader than a single relaxation peak and its temperature depends to some extent on the amount of prior deformation and on the impurity content of the specimen.

The early work, on both polycrystalline material and single crystals, suggested that, for a given metal at a particular frequency, the Bordoni peak always occurred at the same temperature, apart from variations of a few degrees due to different amounts of prior deformation or impurity content. However, this was not the case for some ultrasonic attenuation measurements on aluminium, copper, silver and gold by Mongy et al (13,14,15) and Mongy (16). For each metal they deformed a single crystal by compression in a <110> direction, and measured the attenuation in the <111>, <100> and <110> directions at frequencies between 5 and 50 MHz. In each case there were considerable differences between the temperatures of the peak for different orientations; for example, for copper at 10 MHz the peak occurred at 134, 156 and 191 K respectively. From the frequency dependence of the temperature of the peak, they estimated the respective activation energies to be 0.040, 0.079 and 0.186 eV for the three orientations in copper, compared with 0.116 eV for polycrystalline copper. Their results for silver were similar to those for copper, except that the lowest activation energy occurred for the <100> direction.

The Bordoni peak in silver single crystals was studied at kilohertz frequencies by Mecs and Nowick (17). They carefully controlled the dislocation distribution, by using specimens of special orientation and by deforming crystals into different stages of the stress-strain curve, but they did not measure a crystal of <110> orientation, the orientation which the results of Mongy et al (15) suggest should show the biggest displacement of the peak from its temperature in polycrystalline silver.

Kilohertz Frequency Measurements

Internal friction measurements were made on single crystals of high purity copper and silver at frequencies between 10 and 25 kHz in order to see whether the orientation dependence observed by Mongy et al (14,15) at megahertz frequencies was also present in the kilohertz frequency range. Peak temperatures ranging from about 50 to 120 K would be expected for different orientations by extrapolating the frequency dependence of the megahertz frequency results.

The crystals were in the form of cylindrical rods of 6 mm diameter or bars with dimensions 10cm x 5mm x 5mm. The rods were deformed by longitudinal extension at room temperature in a Hounsfield tensometer; the ends were then cut off, leaving a specimen about 10cm long. The bars were deformed in stages by applying a compressive force normal to a pair of 10cm x 5mm faces. The orientation of each crystal was checked after each deformation by the Laue back-reflection X-ray method, and each specimen was left at room temperature for several days after deformation before its internal friction was measured.

The internal friction of the crystals was measured for longitudinal vibrations at temperatures between 20 and 200 K, as described by Brown and Niblett (18). The maximum strain amplitudes used were less than 10^{-7}, and no amplitude dependence of the internal friction was observed.

The results for copper have been described in detail by Brown and Niblett (18), and are summarised in Table 1.

Three copper rods, with their axes in <111>, <100> and <110> directions, were each extended by 5%. The results for the crystals with axes in the <111> and <100> directions were very similar to those obtained for similarly deformed polycrystalline copper at the corresponding frequencies. However, the peak was very much larger for the crystal of <110> orientation, and was displaced to a higher temperature. The peak for this crystal was asymmetric, the damping being relatively higher on the high temperature side. A second copper crystal of <110> orientation was extended 7%; in this case the displacement of the temperature of the peak was rather less.

Two single crystal bars of copper were deformed by compression in a <110> direction, and their internal friction was measured in a direction perpendicular to the axis of prior deformation. For one crystal this measurement axis was a <111> direction, and for the other a <110> direction. For both orientations the peak was displaced to a higher temperature than that for cold-worked polycrystalline copper, while the magnitude of the displacement decreased with increasing prior deformation.

TABLE 1

Summary of Principal Results for Copper at Kilohertz Frequencies.

Orientation	Axis of prior deformation	Prior deformation	Resolved shear stress (N mm^{-2})	Frequency (kHz)	Temperature of peak (K)	Maximum damping (Q^{-1}x10^4)
<111>	<111>	5% extension	42	23.0	81	16
<100>	<100>	5% extension	38	12.0	78	14
<110>	<110>	5% extension	13	18.3	98	107
<110>	<110>	7% extension	17	19.0	91	62
<111>	<110>	3.5% compression	16	23.3	90	8
<111>	<110>	5% compression	21	23.3	85	7
<110>	<110>	1.5% compression	8	19.6	98	5
<110>	<110>	3.3% compression	16	19.6	82	3

TABLE 2

Summary of Results for <110> Copper Single Crystal at 10.2 MHz.

Compression	Resolved shear stress (N mm^{-2})	Temperature at peak (K)
As received	–	137
0.1%	3	137
0.4%	6	136
1.4%	11	135
2.2%	15	134
2.8%	18	130
3.3%	21	128
3.9%	24	127
4.6%	28	127
5.1%	34	127
6.9%	42	129
8.6%	50	130

Similar measurements on silver single crystals have been described by Brown and Niblett (19). Figure 1 shows the internal friction of three silver rods, with orientations as shown in the stereographic triangle on the figure, after extensions of 5%. Specimens B and C, which had been subjected to resolved shear stresses of 7 N mm^{-2} compared with 30 N mm^{-2} for specimen A, showed asymmetric Bordoni peaks at a temperature about four degrees higher than that for specimen A.

A silver bar was compressed in a <110> direction and its internal friction was measured in a <111> direction; figure 2 shows the damping after various amounts of deformation. After the application of resolved shear stresses of 1 and 4 N mm^{-2} the internal friction rises monotonically with increasing temperature. As the deformation is increased the background internal friction falls and the Bordoni peak becomes visible; a resolved shear stress of 7 N mm^{-2} produces a peak similar in shape to that for the rods stretched with similar stresses. Further increases in prior deformation cause the peak to become more symmetrical.

There is thus no evidence from these measurements at kilohertz frequencies for an orientation dependence of the temperature of the Bordoni peak for either copper or silver. Grandchamp (20) has also found, in the case of gold, that the Bordoni peak at kilohertz frequencies occurs at nearly the same temperature for crystals of different orientations.

The shifts of the temperature of the peak observed in our crystals deformed in a <110> direction are believed to be due to the small resolved shear stresses used. A symmetrical Bordoni peak only occurs after the application of a suitably high resolved shear stress; smaller stresses produce a broad, asymmetric peak at a somewhat higher temperature, the damping being relatively higher on the high temperature side. The magnitude of the temperature shift appears to be greater in copper than in silver.

It is interesting to note that the behaviour of the Bordoni peak for copper single crystals deformed in a <110> direction is very similar to that for polycrystalline copper. Figure 3 shows our measurements, at about 20 kHz, on specimens of polycrystalline copper extended by 0.7%, 3%, 10% and 21%. As in the single crystals, the broad, asymmetric peak displaced to a higher temperature is clearly seen for the smaller deformations.

Megahertz Frequency Measurements

As our results at kilohertz frequencies do not agree with the temperatures of the peak obtained by extrapolating the results of Mongy et al (14,15), we decided to make some further measurements at megahertz frequencies. To date we have only measured one copper single crystal, of 99.999% purity, in the form of a cylinder of 12mm diameter and 10mm length with its axis in a <110> direction. The attenuation of 10.2 MHz longitudinal waves was measured by a conventional ultrasonic pulse technique after deforming the crystal in stages by compression along its axis. The temperature of the peak after each deformation is shown in Table 2, and some of the graphs of attenuation against temperature are shown in Figure 4.

It is seen that, as at kilohertz frequencies, the temperature of the peak first decreases as the deformation is increased. This decrease is most marked for resolved shear stresses in

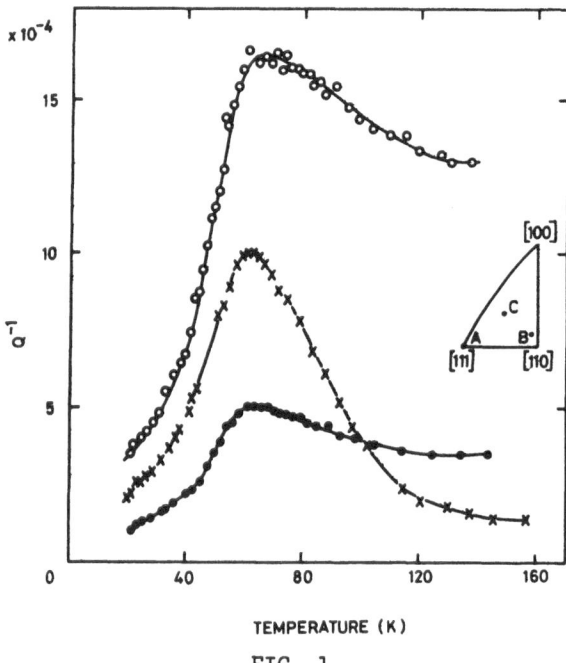

FIG. 1

The internal friction of silver single crystals, extended 5%.
  × specimen A   o specimen B   ● specimen C

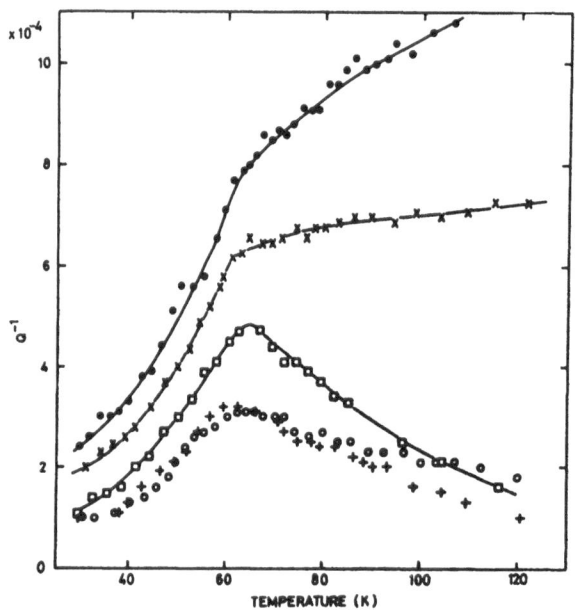

FIG. 2

The internal friction of a silver single crystal of <111>
orientation, after deformation in a <110> direction.
  ● 1 N mm^{-2} × 4 N mm^{-2} o 7 N mm^{-2}
    + 11 N mm^{-2} ◘ 16 N mm^{-2}

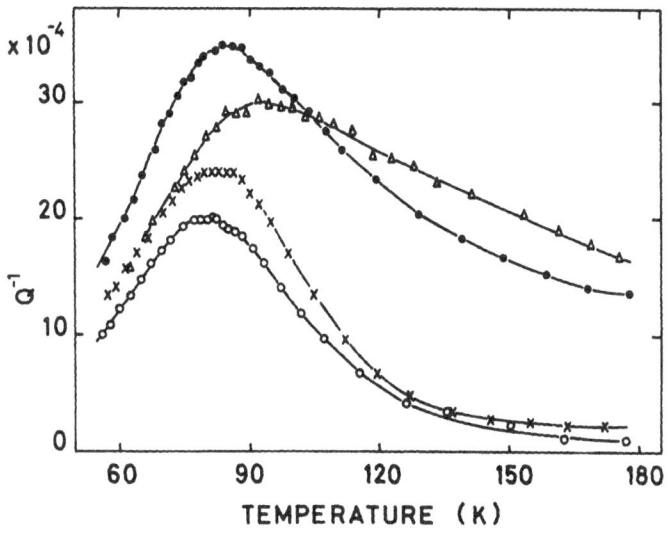

FIG. 3

The internal friction of polycrystalline copper.

Δ extended 0.7%   ● extended 3%   o extended 10%   × extended 21%

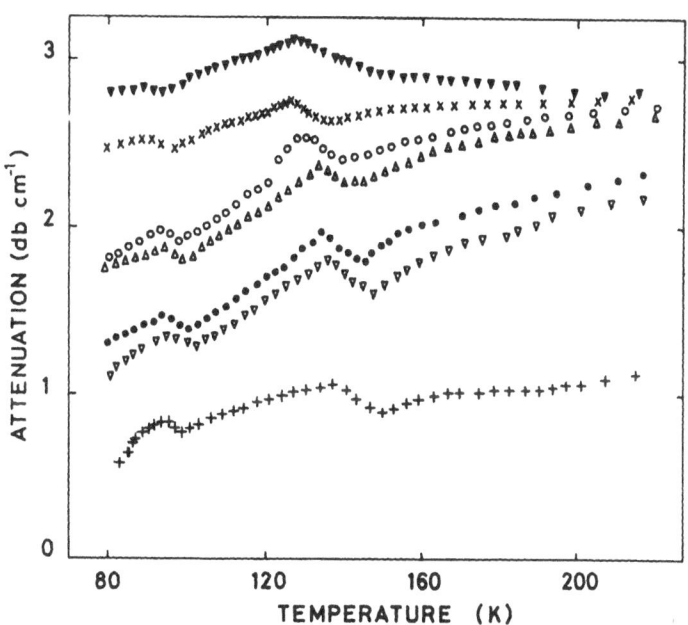

FIG. 4

The ultrasonic attenuation of a copper single crystal of <110> orientation.

+ as received   ∇ 6 N mm^{-2} ● 11 N mm^{-2} Δ 15 N mm^{-2}

o 18 N mm^{-2} × 24 N mm^{-2} ▼ 28 N mm^{-2}

the range from 15 to 21 N mm^{-2}. After a compression of 5% we found the peak at 127 K, whereas Mongy et al (14) observed it at 191 K for a similar crystal. For higher deformations there appears to be a slight increase in the temperature of the peak, as was also found for poly-crystalline copper at kilohertz frequencies (Figure 3). Our values of the temperature of the peak at 10.2 MHz are close to the value (135 K) obtained by Mongy et al (14) for polycrystal-line copper; we have also found the peak in polycrystalline copper at about 130 K. Thus, at these frequencies too, we are unable to confirm the orientation dependence of the temperature of the Bordoni peak.

Discussion

Our measurements at kilohertz frequencies, together with those of Mecs and Nowick (17) and Grandchamp (20), show that the temperature of the Bordoni peak in copper, silver and gold is substantially independent of orientation in the kilohertz frequency range. The fact that our results for the temperature of the attenuation peak at 10.2 MHz for a crystal of <110> orientation are close to that for polycrystalline copper means that we are also unable to con-firm the orientation dependence at megahertz frequencies previously reported by Mongy et al (14).

We attribute the observed shifts in the temperature of the peak to the different stresses used to deform the specimens. The temperature initially decreases with increasing stress, while for large stresses there appears to be a slight increase. This type of behaviour is similar to that reported for silver by Mecs and Nowick (17) and for nickel by Sommer and Beshers (21) and Venkatesan and Beshers (22), but the temperature shifts are generally smaller in mag-nitude than those for nickel. The displacement of the peak to higher temperatures for the lightly-deformed specimens can be explained by the fact that these specimens are likely to contain an appreciable number of relatively long dislocation loops. Engelke (11,12) has shown that the temperature of the peak should increase with increasing loop length. Thus the long loops in the lightly-deformed specimens should produce additional damping at the higher temperatures, so that a broad asymmetric peak is observed at a relatively high temperature. Our experiments confirm that the parameter which determines the shape of the graph of internal friction against temperature is the resolved shear stress used for the prior deformation.

References

1. P. G. Bordoni, Ric. Sci. 19, 851 (1949).

2. D. H. Niblett, Physical Acoustics 3A, ed. W.P. Mason, p.77. Academic Press, New York (1966).

3. A. Seeger, J. Phys. (Paris, Suppl.) 32, C2, 193 (1971).

4. A. S. Nowick and B. S. Berry, Anelastic Relaxation in Crystalline Solids. Academic Press, New York (1972).

5. R. De Batist, Internal Friction of Structural Defects in Crystalline Solids. North-Holland, Amsterdam (1972).

6. A. Seeger, Phil. Mag. 1, 651 (1956).

7.   V. K. Paré, J. Appl. Phys. 28, 332 (1961).

8.   A. Seeger and P. Schiller, Acta Met. 10, 348 (1962).

9.   A. Seeger and P. Schiller, Physical Acoustics 3A, ed. W. P. Mason, p.361. Academic Press, New York (1966).

10. G. Alefeld, Lattice Defects and their Interactions, ed. R. R. Hasiguti, p.407. Gordon and Breach, New York (1967).

11. H. Engelke, Phys. Stat. Solidi, 36, 231 (1969).

12. H. Engelke, Phys. Stat. Solidi, 36, 245 (1969).

13. M. Mongy, K. Salama and O. Beckman, Solid State Comm. 1, 234 (1963).

14. M. Mongy, K. Salama and O. Beckman, Nuovo Cim. 34, 869 (1964).

15. M. Mongy, K. Salama and O. Beckman, Nuovo Cim. 36, 10 (1965).

16. M. Mongy, Ark. Fys. 29, 343 (1965).

17. B. M. Mecs and A. S. Nowick, Phil. Mag. 17, 509 (1968).

18. G. R. Brown and D. H. Niblett, J. Phys. D: Appl. Phys. 6, 809 (1973).

19. G. R. Brown and D. H. Niblett, J. Phys. D: Appl. Phys. (to be published).

20. P. A. Grandchamp, J. Phys. (Paris, Suppl.) 32, C2, 229 (1971).

21. A. W. Sommer and D. N. Beshers, J. Appl. Phys. 37, 4603 (1966).

22. P. S. Venkatesan and D. N. Beshers, J. Appl. Phys. 41, 42 (1970).

CONTRIBUTION TO THE STUDY OF RELAXATION PEAKS
IN COLD-WORKED SILVER AND NICKEL

J.L BESSON and P.BOCH

Laboratoire de "Physique des Vibrations". LIMOGES. FRANCE

ABSTRACT. The relaxation effects due to dislocations and point defects at low temperature in room temperature cold-worked f.c.c metals, have been studied in silver and nickel. Internal friction and Young's modulus measurements have been performed by using a resonant bar method in the kilocycle range and completed in the megacycle range by using an ultrasonic technique. Silver was found to exhibit a classical behaviour, very similar to copper whereas nickel showed a peculiar one. Precise numerical data are given for the activation energies and the frequency factors of the peaks in silver. An effect which may be attributed to the entropy term of the Alefeld theory is examined.

Introduction

The samples are polycrystalline cylindrical rods, 100 mm long and 7 mm in diameter for 99.999 % pure silver and 150 mm long and 5 mm in diameter for 99.999 % pure nickel. For ultrasonic measurements, samples 10 mm long are cut from the previous rods. The initial state is heavily cold-worked. Annealing in vacuo (10^{-5} Torr) for 4 hours at 970°K for silver and 3 hours at 870°K for nickel leads to a fully annealed state, chosen as a standard state. The average grain size is about 0.1 mm for silver and 0.05 mm for nickel.

Measurements can be done from liquid helium temperature up to room temperature in the kilocycle range and from liquid nitrogen temperature up to room temperature in the megacycle range.

The evolution of absorption spectra versus plastic strain ε_p is studied after deformation at room temperature. The influence of isochronal thermal treatments on the annealing of the peaks is investigated.

Experimental

The kilocycle range apparatus (1) works between 10 and 100 kc/s, using electrostatic drive and detection for longitudinal vibrations of the ends of resonant bars supported vertically by three needle points in the nodal plane. The specimens are enclosed in an evacuated container (5.10^{-5} Torr). The strain amplitude of vibration is 10^{-8} to 10^{-7}. The Young's modulus is calculated from the resonant frequency of the sample measured within 1 c/s. The reproducibility is better than 5.10^{-3}. The internal friction is obtained automatically by an electronic device from the decay of the amplitude of the free vibrations of the specimen. The internal friction is evaluated with a better accuracy than 5.10^{-2}.

The ultrasonic apparatus (2) uses 10 Mc/s longitudinal pulses. The attenuation between two succeeding echoes, obtained with a reproducibility better than 5.10^{-2}, allows to calculate the internal friction. The interest of longitudinal ultrasonic waves lies in the fact that they provide the same mode of deformation as the kilocycle range apparatus : i. e. longitudinal vibrations.

Effect of Cold Work

Silver.

The standardized samples are plastically deformed in compression on a 5 T - Zwick machine at a rate of 0.1 mm/min. Buckling is avoided by holding the rod in position in a V-shaped groove inside a brass block. The greater permanent deformation is ϵ_p = 5 %. In addition, two samples have been deformed in tension at a rate of 1mm/min. at ϵ_p = 5 % and 14.5 %.

After a 14.5 % deformation at room temperature it appears on the graph of internal friction against temperature three peaks similar to those observed in copper under the same conditions (Fig.1) and they have been identified as the Niblett and Wilks peak, P_{NW}, at 67°K, the Bordoni peak, P_o, at 76°K and the Hasiguti peak P_1 at 200°K (3).

Background internal friction being substrated, the Bordoni peak height increases quickly with plastic strain up to ϵ_p = 5 %. For the same plastic deformation (ϵ_p = 5 %) the peak observed after extension is about one third smaller than the peak produced by compression. For the greater deformations the peak height remains fairly constant (Fig. 2).

When the plastic strain increases from 1.25 % to 15 %, the position of the peak shifts 11°K towards lower temperature whilst the peak width at half height increases from $5.2.10^{-3}$°K^{-1} to $7.2.10^{-3}$°K^{-1} which is consistent with Paré's previsions (4).

The Hasiguti peak height seems to undergo a similar evolution though it is difficult to give absolute data for the peak height because of the influence of the high temperature side of the Bordoni peak whose contribution to the whole internal friction is not yet negligeable at the temperature at which the Hasiguti peak occurs.

Nickel.

The standardized samples are cold worked in compression at a rate of 0.08mm/min. Annealed aluminium plaquettes, placed between the rods and the guide, have allowed to reach a 10 % permanent deformation.

After applying a stress, equal to the macroscopic elastic limit (σ = 3.56kgf/mm2), which doesn't introduce any permanent deformation (ϵ_p = 0), it appears no peak on the Q^{-1} vs T curve ; the Young's modulus is very similar to the modulus of the standard state.

For slight deformations (ϵ_p = 0.25 %) the P_x peak appears at 220°K ; in heavily cold worked states (ϵ_p = 9.5 %) the P_y peak appears at a lower temperature (140°K) ; for intermediate deformations, the two peaks are both present (Fig.3). After a 1.85 % plastic strain, their heights are similar. The two peaks have a complex strusture. At each peak is associated a modulus defect (Fig.4). The modulus defect, $\Delta E/E$, is always much more than

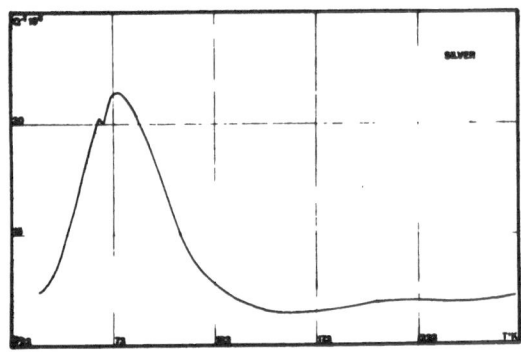

FIG.1

Internal Friction versus Temperature. 14.5 % cold-worked silver.

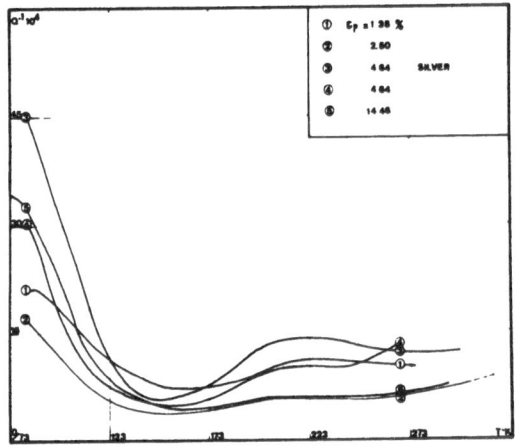

FIG.2

Internal Friction versus Temperature in cold worked silver
for different amounts of plastic deformation in compression
and extension.

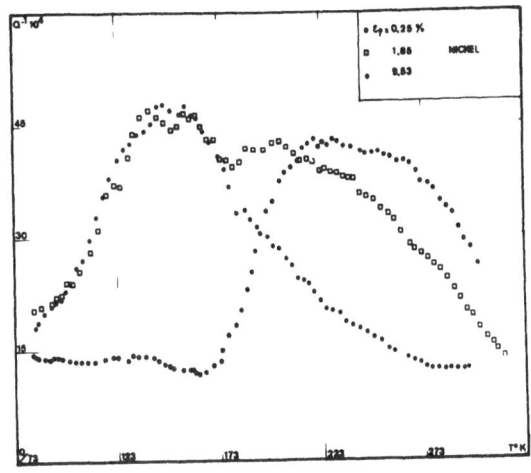

FIG.3

Internal Friction versus Temperature in cold worked nickel
for different amounts of cold-work.

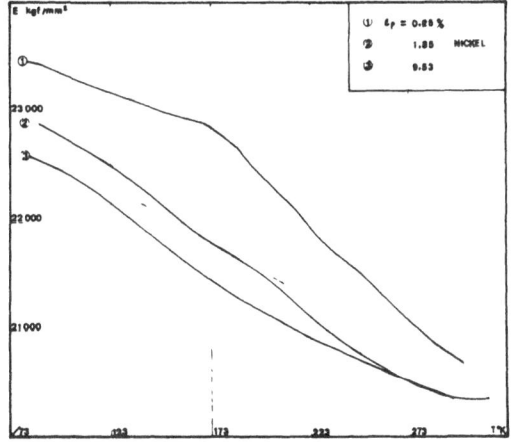

FIG.4

Young's modulus versus Temperature in cold worked nickel
for different amounts of cold-work.

twice the peak height, which shows that they don't correspond to a simple relaxation process.

These results are consistent with Sommer and Beshers'ones (5) and we have used their nomenclature.

Effect of Annealing

Silver.

The Hasiguti peak P_1 disappears after annealing at rather low temperatures. For a 1.25 % plastic deformation, the peak height decreases of one third after 3 months' ageing at room temperature. A subsequent annealing of 4 hours at 350°K reduces by half the peak height. The Hasiguti peak has completely annealed out after 4 hours at 575°K whereas the Bordoni peak is very high (Fig.5).

The Bordoni peak drops very abruptly. The higher the deformation is, the lower the anneal-out temperature (Table 1). It has been verified that the recrystallisation occurs at the same time.

TABLE 1.

Anneal-out Temperatures

ε_p	1.25	4.64	14.46
T °K	725	575	475

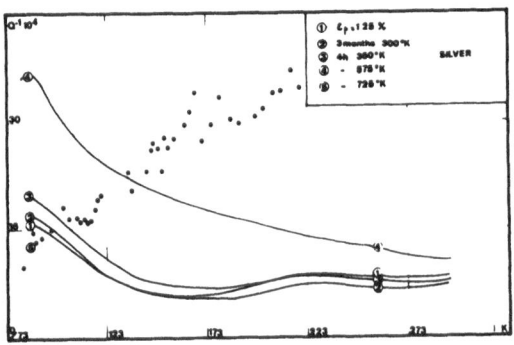

FIG.5

Internal Friction versus Temperature in 1.25 % cold worked Silver
after different annealing treatments.

Nickel

For slight and moderate deformations, a rapid decrease of the P_x and P_y peaks is observed after annealings at relatively low temperatures (Fig.6). After 3 hours at 475°K, the peaks are no longer observed. The internal friction is very low and doesn't change a lot with the measurement temperature. At the decrease of the internal damping at room temperature, corresponds an increase of the Young's modulus.

For the great amounts of cold work (ε_p = 9.53 %), where only P_y appears (Fig 7), the peak height decreases with the increasing annealing temperatures up to 3 hours at 475°K, then it increases between 525°K and 575°K and decreases again to vanish when the recrystallisation takes place. The high temperature components of the peak anneal first, leading to a shift of the position of the peak of 30°K towards lower temperatures.

On the initial state (Fig.8), heavily cold-worked and aged at room temperature, which shows a relatively small P_y peak, 3 hours at 475°K achieves only a small decrease of the peak ; the following annealing treatments, up to 575°K, make the peak increase. After 3 hours at 650°K, P_y is practically removed and it appears a broad irregular bump in the temperature range corresponding to P_x. After 3 hours at 725°K the internal friction is characteristic of a fully annealed state.

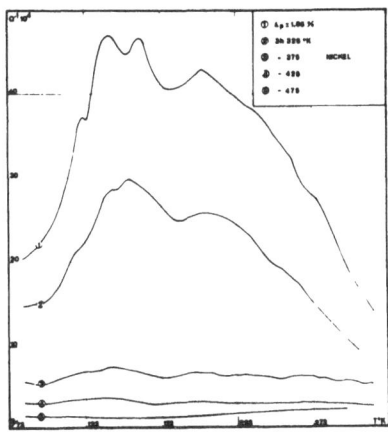

FIG.6

Internal Friction versus Temperature in 1.85% cold worked nickel
after cumulative annealing treatments

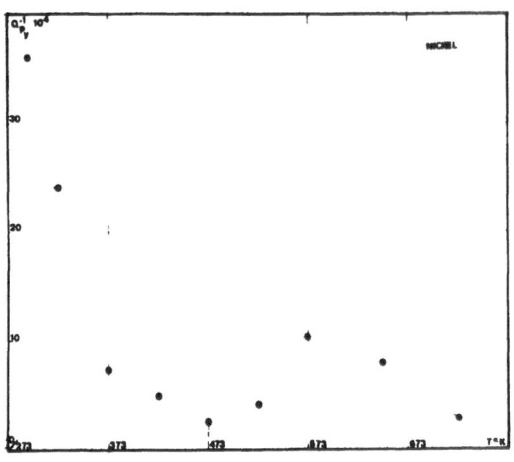

FIG.7

P_y peak Height versus Annealing Temperature
in 9.53 % cold worked nickel

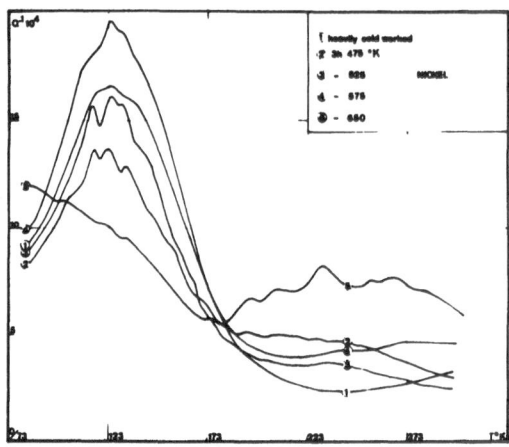

FIG.8

Internal Friction versus Temperature in heavily cold worked nickel
after cumulative annealing treatments.

Activation Energies and Frequency Factors of the Peaks in Silver.

The activation energies and the frequency factors have been determined from the frequency shift of the peak temperature measured in the kilocycle range for a rod vibrating on the first and the third overtone for P_{NW} and P_o and the fifth overtone for P_1 (Fig.9). The temperature of the Bordoni peak has been determined at 10 Mc/s which extends to upper frequencies the range explored by Bordoni and coworkers (6). In the case of P_1, the apparent temperature was corrected by substracting to the total damping the contributions due to the background internal friction and the Bordoni peak (7).

The results, given in Table 2, are supported by the temperatures of the peaks observed at 50 c/s (8).

TABLE 2

Activation Energies and Frequency Factor for Silver

	P_{NW}	P_o	P_1
E ev	0.09 ± 0.01	0.130 ± 0.008	0.37 ± 0.02
$\tau_o^{-1} s^{-1}$	$5.10^{11} \pm 0,75$	$35.10^{12} \pm 0.5$	$17.10^{13} \pm 0.5$

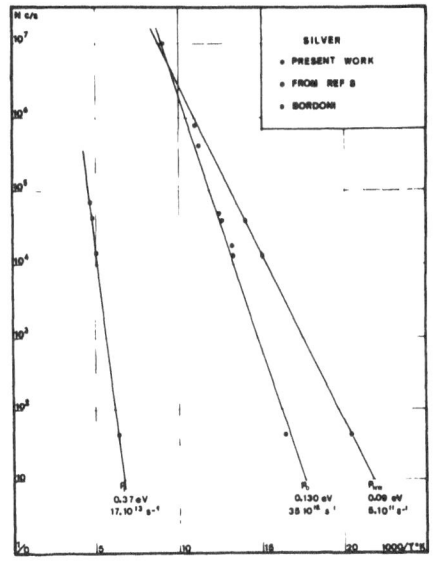

FIG.9

Logarithm of the vibration frequency against the reciprocal of the temperature.

* our measurements
● Bordoni and coworkers
✳ from Vogl data.

An Entropy Effect.

We have found a noticeable difference between the Bordoni peak heights measured at 15 kc/s and 10 Mc/s which could be interpreted as an effect of the entropy.

Let us consider the Alefeld modification of the Paré condition (9) for a Bordoni peak to appear within the Seeger theory (10).

$$abL\ \sigma \geqslant 2\ W_k - T\Delta S.$$

where a is the lattice constant, b the Burgers vector of a dislocation, L its length, σ the external or internal shear stress applied to the dislocation, W_k the kink energy, T the absolute temperature and ΔS the entropy difference between the states of the dislocation with and without a double kink.

For Silver we get (11) :

$$abL\ \sigma \geqslant 1.27\ W_k \text{ for } T = 70°K$$
$$abL\ \sigma \geqslant 0.86\ W_k \text{ for } T = 110°K$$

That warrants that, for a given distribution of internal stresses, far enough from saturation, the Bordoni peak height found by ultrasonic measurements should be higher than that measured in the kilocycle range.

Conclusion

The relaxation peaks observed at low temperature in room temperature cold worked silver have features and a behaviour similar to those observed in copper and the other f.c.c. metals, but nickel.

Our results on nickel show the peaks observed by Sommer and Beshers and corroborate the fact that if P_y appears only for high dislocation densities, P_x on the contrary is not present after heavy cold work. They also show that the dislocation pinnings and reorderings resulting from annealing treatments on moderate deformed states make P_y disappear by reduction of the number of the active dislocations ; the process is more complex (12) for the high dislocation densities where there is a competition between several mecanisms. The recrystallisation is then necessary to cancel the peak, whose height goes through a minimum for a state corresponding to the more efficient pinning.

References

1. J.C. GLANDUS and P. BOCH, C.R. Acad. Sci. Serie A. 275, 1119 (1972)

2. C.GAULT, Thèse de 3ème Cycle, POITIERS (1971)

3. D.H. NIBLETT, Physical Acoustics III Part. A, p 77, ed. by W.P. MASON, Academic Press New-York (1966)

4. V.K. PARE, J. Appl, Phys. 32, 332 (1961)

5. W. SOMMER and D.N. BESHERS, J. Appl. Phys. 37, 4603 (1966)

6. P.G. BORDONI, M. NUOVO and L. VERDINI, Suppl. Nuovo Cimento. 18, 55 (1960)

7. J.L. BESSON, J.C. GLANDUS and P. BOCH. Mém. Sci. Rev. Mét. 69, 299 (1972)

8. K. CHOUNTAS, W. DONITZ, K. PAPATHANASSOPOULOS and G. VOGL. Phys. Stat. Sol. (B) 53, 219 (1972) and G. VOGL private communication.

9. G. ALEFELD. Lattice Defects and Their Interactions, p 409, ed. by R.R. HASIGUTI. Gordon Br. Sc. Pub. (1967) and Thesis Jülich (1968).

10. A. SEEGER. Phil. Mag. 1,651 (1956)

11. J.L. BESSON, C.GAULT and P. BOCH. C.R. Acad. Sci. Série B, 273, 789 (1971)

12. B. DUBOIS, O. DIMITROV. Mém. Sci. Rev. Mét. 64, 641 (1967).

- MODIFICATION OF INTERNAL FRICTION PEAKS OF COLD WORKED ALUMINIUM AFTER FATIGUE TEST -

J.L. CHEVALIER, P. PEGUIN

CSTB/CEN-G - BP 85 - Centre de Tri - 38041 GRENOBLE-CEDEX
FRANCE

G. FANTOZZI, C. ESNOUF, J. PEREZ, P.F. GOBIN

INSA de Lyon - 69621 VILLEURBANNE
FRANCE

ooo O ooo

1 - INTRODUCTION -

The internal friction and modulus anomaly of pure aluminium fatigued at 78 K by torsion have been measured against temperature. Results were compared to those obtained after cold-working in the same conditions[1]. This type of experiment shows the existence, at low temperature, of

   a) the B_1 and B_2 Bordoni peaks

   b) the P_A and P_B interaction peaks

In this way, curve 1 in figure 1 shows $\delta = f(T)$ spectrum of the aluminium deformed by torsion of 0,5 %.

Refering to the interpretation for the whole of the spectrum[2-3-4] :

   a) B_1 and B_2 peaks are a result of intrinsic properties of dislocations ; the models based on kink formation and diffusion describe the results in a fairly satisfactory manner.

   b) P_A and P_B peaks can be interpreted in terms of interaction mechanism between intrinsic point defects and dislocations. This type of study has been carried out on the same metal which had been neutron irradiated at low temperature before deformation[5-6] : it shows that, on the one hand, the Bordoni relaxation is reduced to a very narrow B_2 peak and that, on the other hand, the characteristics of P_A and P_B interaction peaks are modified.

The role played by the intrinsic defects at least in the case of P_B peak, has been shown beyond any doubt : electron irradiation after deformation and annealing at 220 K, during which the interaction peaks disappear, brings about the reappearance of the P_B peak [7].

The present investigation deal with the effect of fatigue caused by torsion at 78 K on the internal friction spectrum of aluminium referred to above. The defects created during the fatigue test lead to the formation of a striking peak P_B and to a modification of the characteristics of the Bordoni relaxation.

A comparison can be made between the effects of irradiation and fatigue.

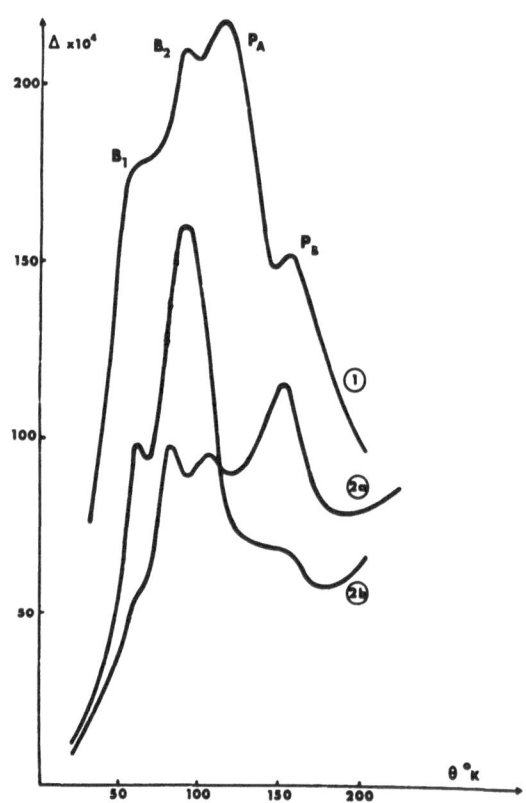

Figure 1 - Internal friction after 0,5 % deformation at 10 K (curve 1), after fatigue test : 1.000 cycles at 78 K, ε_f = 0,5 % (curve 2a) and the same fatigue test followed by an annealing at 230 K (curve 2b) -

We have measured the internal friction as a function of temperature in pure aluminium (99,999 %) subjected to fatigue by torsion at 78 K[1]. The experiment starts at a temperature of 78 K with the specimen, not having been subjected to previous heating. The internal friction is measured with a torsional pendulum oscillating at a frequency of approximately one hertz. The experiment was carried out under identical conditions as those of the experiments dealing with deformation*, which allows us to compare the anelastic effects due to each type of treatment. We have subjected the specimen to a sufficiently strenuous treatment by fatigue with regard to its length of life which gives a typical effect. Figure 1 shows the effect of 1.000 cycles at a fatigue amplitude ε_f = 0,5 %.

The effect due to fatigue has two essential differences with regard to that caused by cold-work only : on the one hand, the characteristics of the Bordoni peaks B_1 and B_2 are altered, on the other hand, the P_B peak situated around 150 K is noticeably accentuated while the P_A peak is almost non existent[7].

a - *Bordoni relaxation* -

Figure 2 shows the effects of the number of cycles on the Bordoni relaxation after annealing at 230 K. The width half way up the curve decreases from 70 K after one cycle to 35 K after 10.000 cycles.

We also see that the B_1 peak height decreases in function of the number of cycles while that of B_2 changes hardly at all. It is stressed that this reduction of B_1 height is more noticeable as the B_2 peak becomes more narrow. After fatigue, the B_2 peak appears at a slightly lower temperature than after deformation only (see figure 2). During successive heatings, the change of the B_2

---

* The measurements have been carried out on the same pendulum with a strain amplitude of vibration close to 3.10^{-5} and during a linear heating at 60K/h speed.

The B_2 peak height increases after treatment carried out between 90 K and 230 K, the growth being more rapid above 150 K. Then after a slight drop between 230 and 250 K, no change is observed till 320 K. Beyond this temperature there is a rapid dropping off of the peak until it disappears after 470 K.

b – *P_B peak* –

The P_B peak obtained after fatigue can be compared to those obtained after other types of treatments :

- its height is striking : it is much greater than that observed after deformation (figure 1) or even after irradiation and deformation[5]

- its temperature remains constant : 150 K

- a series of isochronal annealings shows that the subsequent P_B peaks are smaller than the first. The P_B peak recurs up to an elevated annealing temperature (280 K : figure 3).

After eliminating the peak by maintaining the temperature at either 180 K or 230 K, small additional deformations were applied in order to show precisely the nature of the defects which cause the phenomenon[8]. The results shown in figure 4 show that the peak reappears only after the annealing at 180 K. The measurements of osciliation period of the pendulum confirm this fact : thus the period defect situated around 150 K and linked to the P_B peak only reappears after the treatment at 180 K.

.../...

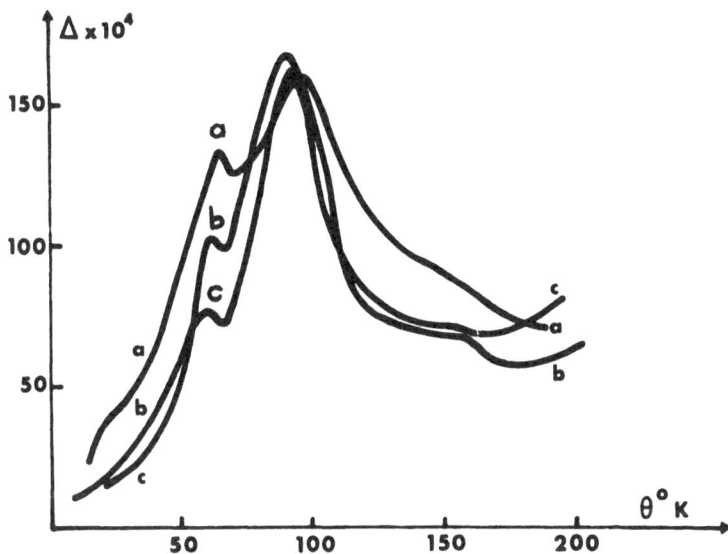

Figure 2 - *Effect of number of cycles (ε$_f$ = 0,5 %) on the Bordoni relaxation after fatigue test and a subsequent annealing at 230 K ; curve a : 1 cycle - curve b : 1.000 cycles - curve c : 10.000 cycles -*

3 - <u>DISCUSSION</u> -

We will consider the above results in terms of, first, the effect of fatigue on the Bordoni relaxation and, then, on the P$_B$ peak. We will thus try to formulate a global interpretation of the behaviour of aluminium after fatigue.

a -<u>Effect of fatigue on the Bordoni relaxation</u> -

The fundamental and secondary properties of the Bordoni relaxation can be correctly interpreted by the relaxation theory due to the thermally activated formation of double kinks, the kink diffusion explaining in a satisfactory way the secondary features[9-14]. In particular, the relaxation time seems to be very sensitive to loop lengths[15] and in consequence the peak width and temperature depend on the distribution of the dislocation lengths : for the large loops, the peak is situated at a higher temperature and it is much larger.

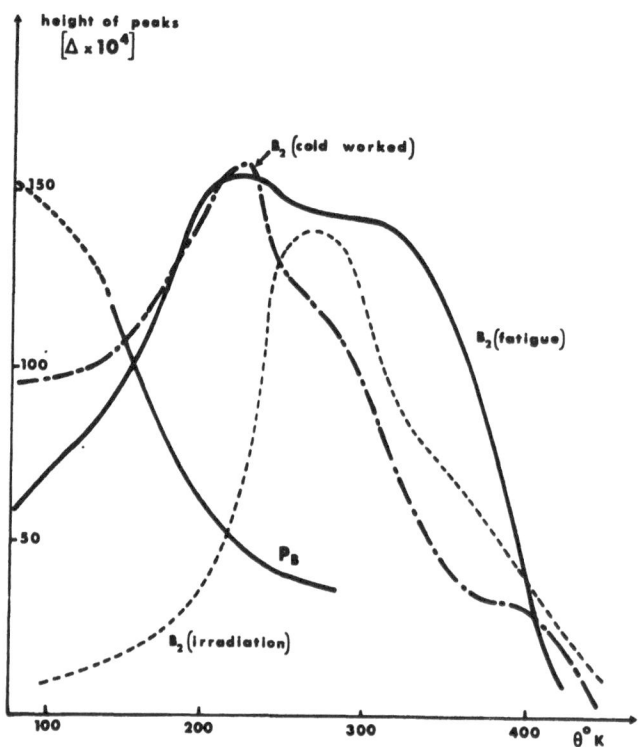

Figure 3 - *Variation of the height of the P_B peak after fatigue test and of the height of the B_2 peak after fatigue, deformation or neutron irradiation plus deformation with annealing temperature.*

Thus, we see that the B_2 peak width and temperature are smaller after fatigue than after deformation (figure 2) : this result can be attributed to the reduction of the number of long dislocation segments with the number of cycles and this is confirmed by electronic microscopy. In the same way, the reduction of the B_1 peak when the number of fatigue cycles increases can be perhaps attributed to a more effective pinning of the dislocations which bring about this B_1 component. The same feature was noted after neutron irradiation[5]. Since the Bordoni peak height is proportional to Λl^2 (Λ = active dislocation density, l = length of dislocation) its change is therefore related to the variations in Λ and l. For low annealing temperatures, the dislocation density is almost constant and as a result the change of B_2 height is essentially due to the variation of the number of pinning points on the dislocations. We thought it worth while to compare this change with those obtained after deformation only or after neutron irradiation and deformation (figure 3). Up to 230 K the change in B_2 is comparable : the number of pinning points on the dislocations situated in the Peierls valleys diminishes. This effect is much successively more noticeable as we pass from deformation, to fatigue, to irradiation. This increase in the B_2 peak height coincides with the disappearance of the P_B peak : we therefore think that the reduction of the number of pinning points is caused by either annihilation or condensation of single defects. From 220 K, the B_2 peak reduces in height, this reduction which takes place in 3 stages after coldwork, takes place in only 2 stages after fatigue (or irradiation) : 220-260 k and 310-420 K. Electronic microscopic examination shows, in the case of fatigued (or irradiated) aluminium, the formation of several clusters and dislocation loops. In the case of coldworked only aluminium, this formation is not observed. It is likely that the nucleation of these loops caused by the super saturation of point defects is more probable near impureties[16]. However, once the loops are formed, they trap the majority of the impureties. Thus, the decrease of the amplitude of the B_2 peak around 320 K observed after deformation is no longer observed so neatly after fatigue or irradiation this decrease being attributed to the migration of impurities to the dislocations. Above 350 k, the impurities could be released and furthermore the dislocation network changes.

b - *Effect of fatigue on the P_B Peak* -

The P_B peak is generally attributed to the interaction between dislo-cations and intrinsic point defects[3-4-7]. Therefore it is necessary to take into consideration point defects created during the fatigue ; these defects, when they come near the dislocations, can form the pinning agents responsible for the relaxation mechanism of the P_B peak[*].
The exact process of energy dissipation is not yet clearly established but the hypothesis of a thermally activated depinning is now reasonable. If we take into account the temperature range, these defects are likely to be of interstitial nature.

---

[*] The defect configuration hardly allows the appearance of the P_A peak.

From 150 K to 180 K at least, their diffusion along the dislocations to form clusters can explain the fast annealing of the P_B peak (figure 3) ; Therefore, these clusters constitute pinning points which are harder but less numerous :

- below 150 K (P_B peak temperature), the length of free dislocation is larger and therefore the B_2 peak height as well as the period increases.

- above 150 K, the depinning no longer takes place and the period is smaller than in the case where the peak P_B was evident.

A small additional deformation could destroy these clusters and redistribute point defects along the dislocations. The dislocation depinning from more simple pinning defects is possible again and thus the P_B peak reappears (figure 4). After annealing at 230 K, the clusters are either annihilated by the vacancy defects migration or eliminated on the dislocations. In this case, a small additional deformation does not bring about the reappearance of the original relaxing elements.

The same effect due to additional deformation was obtained in the case of fatigued aluminium which was purer (99,9999 %), which confirms the role of intrinsic defects in the relaxation process related to the P_B peak.

.../...

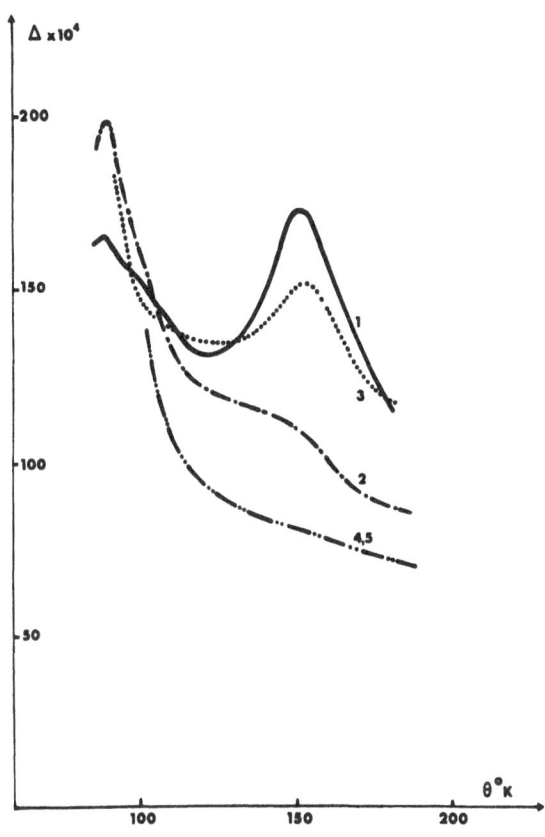

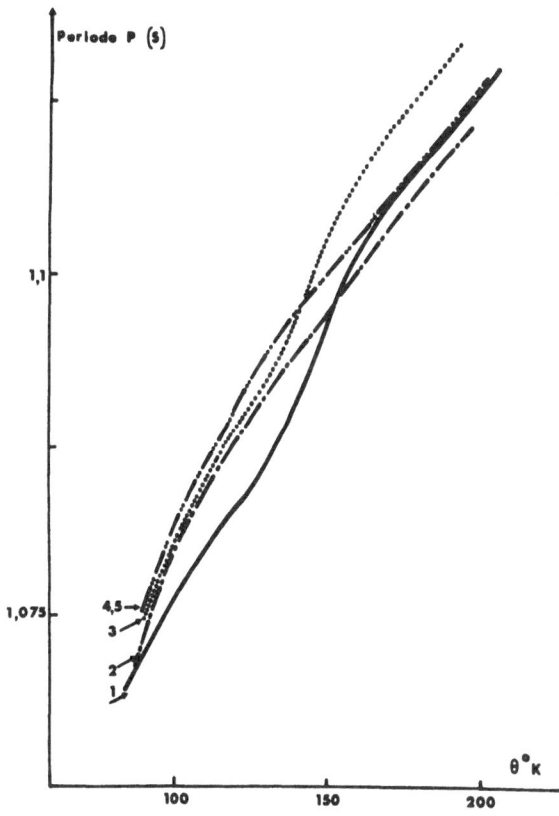

Figure 4 - Internal friction and period of the pendulum against temperature after fatigue test : 10.000 cycles, $\varepsilon_f = 0,5$ % (curve 1) and after an annealing at 180 K (curve 2), after a subsequent deformation of 0,12 % (curve 3) and an annealing at 230 K (curve 4) and, then, after an additionnal deformation of 0,12 % (curve 5) -

4 - *CONCLUSION* -

In this work, it has been shown that fatigue changes the characteristics of the Bordoni relaxation and gives a very striking P_B peak.

The reappearance of the P_B peak caused by a small additional deformation allows us to show the existence of two stages in the elimination of the defects responsible for the relaxation : for an annealing temperature of the order of 180 K, the defects rearrange themselves but in the case of an annealing temperature of 230 K, they are probably eliminated.

Analogies (in particular for the Bordoni relaxation) appear in the behaviour of fatigued or irradiated and strained aluminium. This analogy has been aeready reported[17]. This stresses the important role of point defects in fatigue phenomena.

Finnaly, it would appear to us that the observation of anelastic properties is a useful tool in the theorical study of the fatigue mechanism.

5 - *REFERENCES* -

1 - J.L. CHEVALIER, thesis, Grenoble (1972)

2 - J.L. ROUTBORT, thesis, Cornell University (1966)

3 - J. VOLKL, W. WEINLANDER and J. CARSTEN, Phys.St.Sol., 10, 739 (1965)

4 - G. FANTOZZI, J. PEREZ, P. PEGUIN and P.F. GOBIN, Mém.Sci.Rev.Mét., 66, 2, 185 (1969)

5 - G. GUENIN, J. PEREZ and P.F. GOBIN, Rad.Effects, 10, 17 (1971)

6 - G. FANTOZZI, F. FOUQUET and P.F. GOBIN, Mém.Sci.Rev.Mét., 9, 641 (1972)

7 - O. MERCIER, W. BENOIT, P. MOSER, G. FANTOZZI, J. PEREZ and P.F. GOBIN, V int. Conf. on Internal Friction Aachen (1973)

8 - B. BAYS, thesis, E.P.F. Lausanne (1970)

9 - A. SEEGER, J. Phys., 32, C2, 195 (1971)

10 - A. SEEGER, Phil. Mag., 1, 651 (1956)

11 - V.K. PARE, J. Ap. Phys., 3, 332 (1961)

12 - A. SEEGER and P. SCHILLER, J. Phys. Soc. Jap., Sup 1, 178 (1963)

13 - G. ALEFELD, "Lattice defects and their interactions", Gord,Br.Sc.Pub. (1967)

14 - H. ENGELKE, Phys.St.Sol., 36, 231 and 245 (1969)

15 - G. FANTOZZI, thesis, Lyon (1971)

16 - M.J. MAKIN, "Electron microscopy in material science", Ac. Press, New York, 389 (1971)

17 - J.L. CHEVALIER, D.F. GIBBONS and L. LEONARD, J.Ap.Phys., 43, 1, 73 (1972)

IMPURITIES EFFECTS ON RELAXATION PEAKS IN COLD-WORKED GOLD[*]

O. Mercier, A. Isoré and W. Benoit[+]
Laboratoire de Génie Atomique de l'Ecole Polytechnique Fédérale de Lausanne
(Switzerland)

INTRODUCTION

After plastic deformation at low temperature of f.c.c. metals, several internal friction peaks are observed, they are called the Bordoni peak [1] and Hasiguti peaks [2]. The results obtained on gold by Benoit [3], Bays and al. [4],[5], Grandchamp and al. [6],[7],[8], show strong variations of the internal friction spectrum depending on the origin of the gold. Mercier and al.[9] have studied this problem trying to understand the role played by impurities. Generaly speaking trace impurities seem to influence the shape and the amplitude of peak P_2, (the peak is narrower and smaller in the presence of impurities) and also to change the recovery of the Bordoni peak and of the modulus defect (depinning of dislocations, observed at about 200 K, disappears when the samples are not annealed under atmosphere). The role of impurities is not obvious, they can migrate towards the dislocations and pin them ; they can also act as traps for the point defects (either interstitials or vacancies), thus reducing the flux of defects to the dislocations. By this work we have tried to ascertain the precise influence played by the species and location of trace impurities on the spectrum of Hasiguti peaks and we have described the main characteristics of these peaks via a theoretical model.

EXPERIMENTAL PROCEDURE AND SPECIMENS

The apparatus used in this work is an inverted torsional pendulum [10], frequency range : 0,3 - 3 Hz, temperature range 77 - 700 K, amplitude of deformation during measurement $\varepsilon_m = 5 . 10^{-7}$.

The sample materials used in this experiment were : polycristalline gold from Semi-Alloys (USA) and from Métaux Précieux SA (Switzerland). The dilute alloys came from Métaux Précieux SA (Switzerland). The amount of impurities was measured by Johnson-Matthey, England ; and the residual resistivity was measured at our Laboratory (Table I).

* This work was supported by the "Fonds National Suisse de la Recherche Scientifique", subsidy no. 2.776.72

+ Now Visiting Professor at the "Institut für Theoretische und Angewandte Physik, Universität Stuttgart"

TABLE I

Origin	Purity	Ag Fe Cu Mg Pa C (ppm wt)	Pre-annealing	$\dfrac{\rho\,300K}{\rho\,4K}$
Semi-Alloys	99.998%	10 13 2 <1 1	4h 800°C high vacuum	200
Métaux precieux SA Switzerland	99.999%	2 1 <1 <1 – <5	4h 800°C High vacuum 4h 800°C air 24h 900°C air	120 200 400
idem	99.99%	40 1 <1 <1 –	4h 800°C High vacuum	300
idem	99.99%	2 25 <1 <1 –	4h 800°C High vacuum	40

The specimens were wires of 100 mm in length and 1 mm in diameter. After machining, the samples are annealed either under vacuum or under atmosphere for 4 hours at 800°C, or for 24 hours at 900°C under atmosphere.

EXPERIMENTAL RESULTS

Measurements on 99,998 wt% gold samples

In fig. 1 are reported the results obtained on 99,998 wt% gold samples, from Semi-Alloys, deformed 6% in torsion at 77 K. The various curves represent the evolution of a maximum of internal friction according to a program of successive linear increments of temperature, as a function of time. This maximum is the Hasiguti peak P_2 (P_3).

The activation energy of the peak ($E_R = 0,33$ eV) and the frequency factor ($f_o = 10^8$–10^9 s^{-1}) can be obtained by plotting the log. of the frequency of measurement versus the inverse of the temperature of the maximum [5]. The broadening factor α of a relaxation peak is given by three different methods provided that the peak's activation energy is known (fig. 2) [11] :

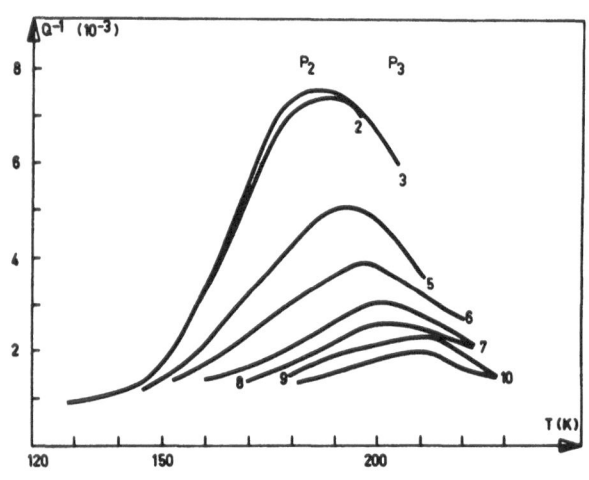

Fig. 1   Variation of the internal friction as a function of successive linear increments of temperature ; increment no. 1, final temperature = 198 K ; no. 2, 203 K; no. 3, 208 K; no. 4, 213 K; no. 5, 218 K; no. 6, 223 K; no. 7, 233 K; no. 8, 243 K; no. 9, 253 K; no. 10, 263 K. Samples were 99,998% wt gold wires from Semi-Alloys, deformed 6% in torsion at 77 K.

(i)   the width of the peak at half height

(ii)   the asymptotic slope, in a diagram where log. of internal friction of the peak (after subtraction of the background) is plotted as a function of the inverse of temperature

(iii) the ratio between the modulus defect connected with the peak and the height of this peak

334

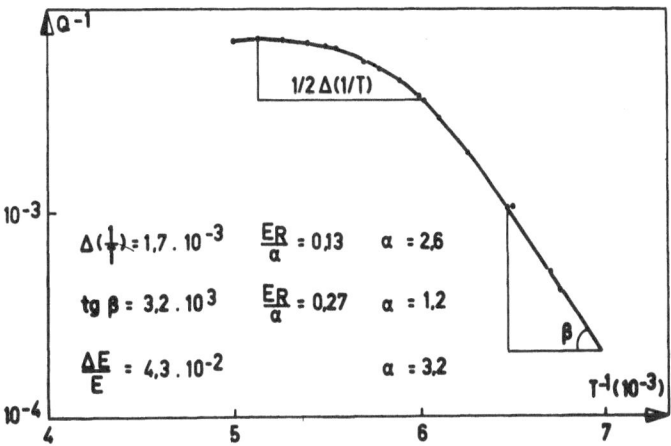

Fig. 2 Log of the internal friction Q^{-1} versus the inverse of the temperature of the peak P_2. Measurements of the broadening factor, α, taken from the half height peak width and the asymptotic slope

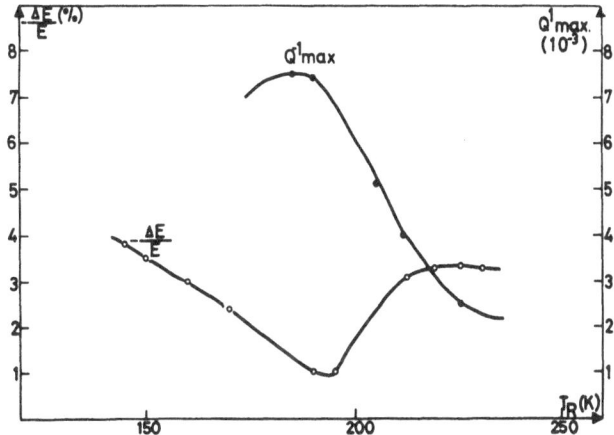

Fig. 3 Variation of peak height Q^{-1}_{max} and the modulus defect $\Delta E/E$ as a function of the annealing temperature T_R; for 99,998% wt gold samples deformed 6% in torsion at 77 K

The broadening factors of P_2 are 2,6 ; 1,2 ; 3,2 for each method respectively. As these three values are markedly different, the peak P_2 is not only a broadened peak, but it must result from the superposition of several peaks.

In fig. 3 is reported the evolution of the peak's height and the evolution of the modulus defect (measured at 140 K) as a function of annealing temperature. These two parameters evolve in the opposite direction. The maximum of the peak's height and the minimum of the modulus defect are observed after a similar annealing at 190 K. During the peak's recovery, the modulus defect increases.

The curves in fig. 4 represent the evolution of the characteristic parameters of the peak (broadening factor (width at half height), peak height, temperature of the maximum) as a function of the annealing temperatures. The peak's height (Q^{-1}_{max}) decreases, and the temperature of the maximum moves towards the high temperatures whereas the broadening

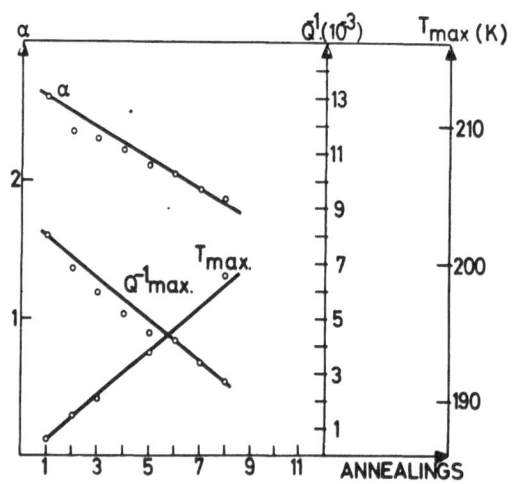

Fig. 4   Evolution of the characteristic parameters of the peak P_2 (broadening factor α, (width at half height), peak height Q^{-1}_{max}, and the temperature of the peak T_{max}) as a function of the annealing temperature, in a 99,998% wt gold sample

factor decreases, as the annealing temperature increases.

Measurements of gold samples containing small amounts of impurities

The results for internal friction are depicted in fig. 5 ; measurements were made on 99,999 wt % gold samples from Métaux Précieux SA, which were deformed 6% in torsion at 77 K. All samples were subjected to the same experimental conditions except for pre-annealing : 4 hours at 800°C under high vacuum (curve (a)), 4 hours at 800°C under atmosphere (curve (b)), 24 hours at 900°C under atmosphere (curve (c)).

The parameters which characterize the two peaks are shown in Table IIa. It can be noted :

. The presence of a peak at about 190 K ; this is peak $P_2(P_3)$. Its height ranges between 5,5 - 8,5. 10^{-3} ; its broadening factor fluctuates greatly (between 1,3 and 2,5).

The presence of another peak, situated at about 224 K ; this is peak P_4 [12]. Its height is maximum for a pre-annealing of 4 hours at 800^oC under atmosphere. It disappears completely after a long anneal under atmosphere (24 hours at 900^oC). It is nearly a perfect Debye peak ($\alpha = 1, 1$)

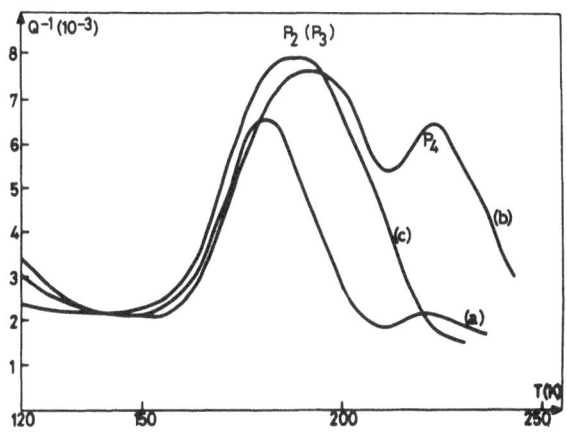

Fig. 5

Measurements of internal friction Q^{-1} on 99, 999% wt gold samples, from Métaux Précieux SA, which were deformed 6% in torsion at 77K and pre-annealed for :
(a) 4 h at 800^oC under high vacuum;
(b) 4 h at 800^oC under atmosphere ;
(c) 24 h at 900^oC under atmosphere

Similar measurements have been made on gold samples containing silver or iron as impurities, deformed 6% in torsion at 77 K. Table IIb gives the characteristics of peak P_2 for these measurements ; it is higher and its broadening factor, measured from the half height peak width, is smaller than for gold samples pre-annealed for 24 hours at 900^oC. Peak P_4 does not appear.

TABLE IIa

Metal	Pre-annealing	$\frac{\rho\,300K}{\rho\,4K}$	Temperature (K)		Height 10^{-3}	α
Au 99.999%	4 h 800°C under high vacuum (10^{-8}T)	120	P_2	187	5,7	1,3
			P_4	224	0,2	-
Au 99.999%	4 h 800°C under air	200	P_2	195	5,5	1,7
			P_4 -	224	4,7	1,1
Au 99.999%	24h 900°C under air	400	P_2	195	8,5	2,5
			P_4	-	-	-

TABLE IIb

Metal	Pre-annealing	$\frac{\rho\,300K}{\rho\,4K}$	Temperature (K)		Height 10^{-3}	α
Au + 40ppm Ag	4h 800°C under vacuum	300	P_2	197	11	1,8
Au + 25ppm Fe	4h 800°C under vacuum	50	P_2	198	10	1,8

337

In fig. 6 are shown curves depicting the evolution of the modulus defect under the same conditions and for identical samples as those indicated in fig. 5.

. Sample (a) shows only one stage of diminution of modulus defect, at about 200 K ; during this stage peak P_2 goes through a maximum.

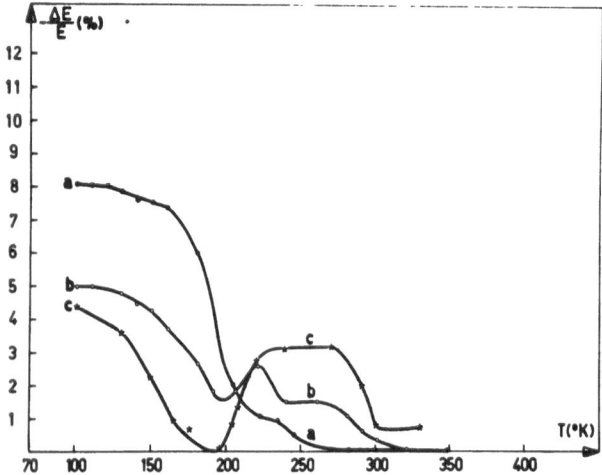

. Sample (b) shows a more complex evolution. A first pinning stage occurs together with the growth of P_2. The subsequent depinning occurs simultaneously with the disappearance of P_2. The growth of P_4 is accompanied by a new pinning stage.

. Sample (c) shows one stage of diminution of modulus defect followed by a stage of increase; during the first stage P_2 increases, and during the second stage, it disappears.

Fig. 6   Variations of the modulus defect $\Delta E/E$ as a function of the annealing temperature T_R measured on 99, 999% wt gold samples, from Métaux Précieux SA, which were deformed 6% in torsion at 77 K and pre-annealed for : (a) 4h at 800°C under high vacuum ; (b) 4h at 800°C under atmosphere ; (c) 24h at 900°C under atmosphere

For sample (c) the evolution of the peak's parameters is similar to that described in fig. 4. However, for sample (a), this evolution is different, as shown in fig. 7. During the peak's recovery, the temperature of the maximum decreases, and so does the broadening factor ; at the same time, the modulus defect goes on decreasing.

In summary, the Hasiguti peaks show the following properties :

. The broadening factor of the peak P_2 fluctuates greatly (1, 3-2, 5), dependent upon the metal purity. The most high and narrow peak is observed for the least pure samples. The peak P_2 is not only a broadened peak, it can be decomposed into several subsidiary peaks.

. The modulus defect decreases when the peak increases. Morover, the activation energy of the recovery of the modulus defect (0, 3eV) is approximatively equal to the activation energy of the peak; therefore mobile point defects probably account for the peak.

A supplementary peak, P_4, can also appear. An interstitial impurity (carbon or hydrogen), accounts for this peak. This impurity causes a pinning of the dislocations which seems to be present until recrystallization occurs. It can be eliminated by pre-annealing under atmosphere at high temperature.

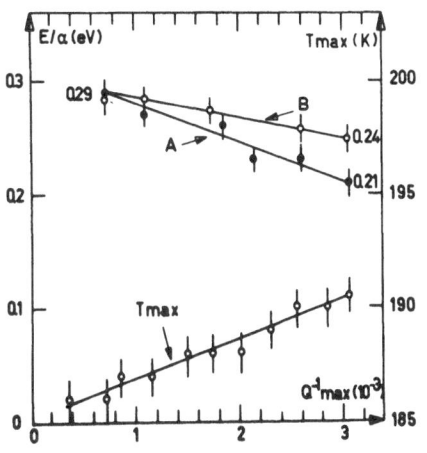

Fig. 7

Evolution of the characteristic parameters of the peak P_2 (broadening factor measured from (A), half height peak width, (B) asymptotic slope ; and the temperature of the peak T_{max} as functions of the peak height Q_{max}^{-1} ,on a 99, 999% wt gold sample pre-annealed for 4h at 800°C under high vacuum (a)

THEORIES

The basic hypothesis assumes that the peaks can be explained by the interaction between mobile point defects, and dislocations. This model is developed from the theories of Granato-Lücke [13] and Schoeck [14].

The dislocation which is subjected to an external alternating stress (σ) applies a force (F) on the point defect. The latter, under the influence of this force (F), will move with a speed (v).

$$v = \frac{D \cdot F}{k_0 T}$$

$D = D_0 \exp(-E_M/k_0 T)$; D is the diffusion coefficient of the point defect ; E_M is the activation energy for diffusion ; k_0 is the Boltzmann constant ; T is the absolute temperature.

The force (F) is given as a function of the external stresses that act on the dislocation.

$$F = \frac{1}{n} (-m \ddot{\xi} - B_L \dot{\xi} - k\xi + \sigma b) \qquad [15]$$

ξ is the average value of the dislocation displacement from equilibrium over the length l ; m is the effective mass of the dislocation per unit length ; B_L is the damping constant (interaction dislocation - lattice) ; k is the average elastic constant of the dislocation ; b is the magnitude of the Burgers vector ; l is the loop length ; n is the number of mobile point defects on the dislocation.

The calculation of the internal friction and the modulus defect of a dislocation loop of length l, subjected to an alternating stress $\sigma = \sigma_0 \cos \omega t$ gives the following results :

$$q^{-1} = \Delta l^3 \frac{\omega \tau}{1 + (\omega \tau)^2} \; ; \quad \frac{\delta e}{e} = \Delta l^3 \frac{1}{1 + (\omega \tau)^2} \; ; \quad \text{with} \quad \tau = B/k = \frac{B l^2}{C \pi^2} \; ; \; B = B_L + \frac{k_0 T}{D} \cdot \frac{n}{l} \; ;$$

$\Delta = \frac{8 b^2 G}{\pi^4 C} g^2$; q^{-1} is the internal friction due to one loop per unit volume ; $\frac{\delta e}{e}$ is the modulus defect due to one loop per unit volume ; C is the line tension ; G is the shear modulus ; g^2 is an orientation factor.

The equations describe exactly a relaxation peak, according to the theory of the ideal linear solid, provided that the background B_L can be neglected.

Random distribution of mobile point defects

The number of dislocations per unit volume, the length of which ranges between (l) and $(l) + d(l)$, is given by [16] :

$$\rho (l) \, dl = \frac{\Lambda}{L^2} e^{-l/L} \, dl$$

where L is the average length of the loops ; Λ is the density of dislocations.

For the total internal friction and modulus defect we have :

$$Q^{-1} = \sum_{n=0}^{1/b} \int_o^\infty dl \, \rho (l) \cdot P (l, n) \, q^{-1} (n, l) \; ; \quad \frac{\Delta E}{E} = \sum_{n=0}^{1/b} \int_o^\infty dl \, \rho (l) \cdot P (l, n) \frac{\delta e}{e} (n, l)$$

where $P (l, n)$ is the probability that n mobile point defects be present on a dislocation loop of length (l).

After simplification we have obtained the following computed qualitative evolution of the internal friction peaks (fig. 8).

8a   When the defect concentration (c) is increased, an increment of the peak is observed, together with an increment of the temperature of the maximum and a broadening of the peak .

8b   Evolution of the peak when some defects disappear from the loops by migration along the dislocation and by forming clusters.

8c   Evolution of the peak when the number of strong pinning points (impurities) increases, thus causing the length (l) to decrease.

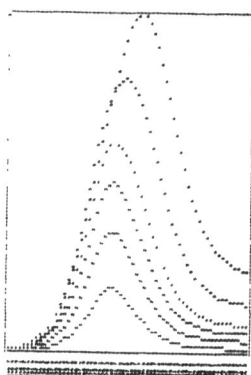

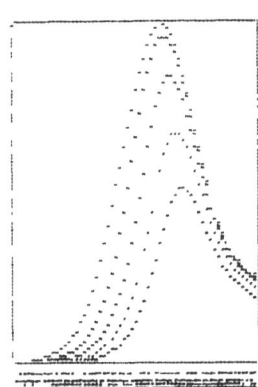

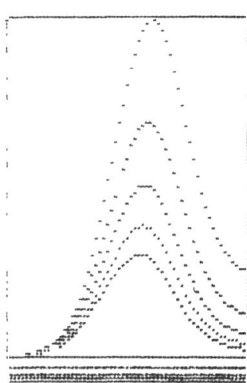

Fig. 8a

Computed evolution of the internal friction peaks when the defect concentration (c) is increased

Fig. 8b

Computed evolution of the internal friction peaks when some defects disappear from the loops

Fig. 8c

Computed evolution of the internal friction peaks when the number of strong pinning points increases

CONCLUSION

The Hasiguti peaks can be explained by a mechanism of interaction between dislocations and mobile point defects ; the variable broadness of these peaks can be accounted for by different concentrations of the defects on the dislocation loops. If the lattice contains substitutional impurities, which trap the intrinsic point defects, the number of these defects on the dislocation loops will be smaller, therefore the peak will be narrower. That is exactly what we observed after doping our sample with iron or silver (Table IIb).

The peak's growth can be explained by a flux of interstitial point defects towards the dislocations, probably as small agglomerates (temperature of pinning stage (180K) which corresponds to the end of the stage II of recovery ; energy of pinning : 0, 3eV, [4]). When the concentration of point defects on the dislocation loops increases, the peak grows, simultaneously it broadens and moves towards high temperatures (as in fig. 8a) .

Two processes of peak diminution can be observed :

. A migration of the interstitial point defects along the dislocation to its nodes, kinks or jogs ; as a consequence, the peak moves to higher temperatures and its width decreases (as in fig. 8b) ; the modulus defect increases.

. A pinning of the dislocations by an impurity (carbon or hydrogen) ; the number of strong pinning points increases ; as a consequence, the maximum moves to lower temperatures and its width decreases (as in fig. 8c) ; the modulus defect decreases.

REFERENCES

[1]   P. G. Bordoni, J. Acoust. , Soc. Am. 26, 495 (1954)

[2]   S. Okuda and R. R. Hasiguti, Acta Met. , 11, 257 (1963)

[3]   W. Benoit, Mém. Sci. Rev. Mét. , 66, 763-778 (1969)

[4]   B. Bays, W. Benoit, P. -A. Grandchamp, J. de Phys. , Coll. C2, 32, C2-152-157 (1971)

[5]   B. Bays, Mém. Sci. Rev. Mét. , 68, 629-638 (1971)

[6]   P. -A. Grandchamp, B. Bays, W. Benoit, Helv. Phys. Acta, 43, 754-756 (1970)

[7]   P. -A. Grandchamp, J. de Phys. , Coll. C2, 32, C2-229-241 (1971)

[8]   P. -A. Grandchamp, B. Bays, W. Benoit, Mém. Sci. Rev. Mét. , 68, 617-627 (1971)

[9]   O. Mercier, A. Isoré, W. Benoit, Scripta Met. , 6, 961-964 (1972)

[10]   B. Bays, P. -A. Grandchamp, Rev. Phys. Appl. , 5, 327-332 (1970)

[11]   A. S. Nowick and B. S. Berry, Anelastic Relaxation in Crystalline Solids, Acad. Press (1972)

[12]   O. Mercier, A. Isoré, W. Benoit, Helv. Phys. Acta, 46, 18-21 (1973)

[13]   A. Granato and K. Lücke, J. Appl. Phys. , 27, 583 (1956)

[14]   G. Schoeck, Acta Met. , 11, 617, (1963)

[15]   B. Vittoz, Fifth Int. Conf. on Internal Friction and Ultrasonic Attenuation in Crystalline Solids, Aachen 1973

[16]   J. S. Koehler, Imperfections in nearly perfect Crystals (John Wiley and Sons, Inc. New York, 1962)

INTERNAL FRICTION PEAKS OBSERVED IN

W SINGLE CRYSTALS AFTER SMALL DEFORMATION

Georges E. Rieu
Ecole Nationale Supérieure de Mécanique et d'Aérotechnique
University of Poitiers, France

Visiting National Research Council Associate at
National Aeronautics and Space Administration
Lewis Research Center, Cleveland, Ohio U.S.A.

Introduction

Internal friction peaks observed in deformed BCC metals are strongly influenced by extrinsic material properties. For instance, impurity interactions with dislocations and impurity pinning contribute to the internal friction spectrum and make difficult any interpretation of intrinsic effects. Also, deformation behavior is highly orientation dependent, particularly in W (1).

In the present work, we tried to control these two major experimental conditions by using W single crystals preoriented and free of impurities in solution. In order to differentiate between edge and screw dislocation we deformed the W specimens only a few percent at room temperature to enhance the proportion of screw dislocations.

Materials and Measurements

We used tungsten single crystal rods .25 inches in diameter and 2 inches in length, oriented $[100]$, $[110]$, $[111]$ along the axis, within $\pm 2^{\circ}$ of the specific orientation. All the specimens were zone refined and had a resistivity ratio greater than 50,000 which corresponds to a total point defect concentration less than 1 atomic ppm (2). A polycrystalline rod of 99.995 percent tungsten was studied for comparison.

The internal friction measurements were made at about 45 kHz between 30-800°K, using longitudinal vibration and capacitive drive and detection. The sample was held at the middle by three sharp tungsten screws. The strain amplitude was lower than 10^{-7}. The instrumental background decrement was about 10^{-6}.

The samples were deformed in compression at room temperature using an Instron machine at a strain rate of 0.01 inch/min. Guides of annealed copper prevented bending of the sample without introducing any significant radial stresses.

Experimental Results

Annealed Samples

To eliminate any residual stresses due to polishing or manipulating, we annealed all of the samples 2 hours at $1900^{o}K$ under a vacuum of about $2x10^{-9}$ torr.

The internal friction spectrum observed after annealing is shown in figure 1. As with polycrystalline molybdenum (3), our polycrystalline tungsten shows a very low damping over the entire temperature range $(Q^{-1}<4x10^{-6})$. It is associated with a linear variation of modulus with temperature (Fig. 2). In contrast, the high purity single crystal specimens show damping (or peaks) as high as $3x10^{-4}$, and exhibit several modulus defects.

The [111] crystals show two small peaks at low temperature at around $170^{o}K$, and a broad composite peak over the temperature range $300-700^{o}K$. The corresponding modulus variation is perturbated between $450-600^{o}K$ and decreases rapidly above $600^{o}K$. This faster decrease corresponds exactly to the high temperature fall-off of the broad peak.

Similar high temperature behavior is shown by the [100] crystal but smaller in amplitude; however, no peak is detected at low temperature.

Besides the two small low temperature peaks, the [110] crystal shows two distinct peaks at $320^{o}K$ and $460^{o}K$. The $320^{o}K$ peak is slightly affected by temperature. Cooling to liquid nitrogen temperature makes it higher (x1.15) and shifts it about $15^{o}K$ toward lower temperatures. Heating to $475^{o}K$ makes it lower (x0.8) and shifts it about $12^{o}K$ toward higher temperatures. After aging by heating to $730^{o}K$ at a rate of $0.5^{o}K$/min, it almost disappears (x0.08). The $460^{o}K$ "peak" is too narrow to be a relaxation peak and looks more like a "phase transformation" spike. It does not correspond to any significant change in the modulus.

After 2.5% deformation and 2 hours annealing at $1800^{o}K$ under a $2x10^{-9}$ torr vacuum, these two peaks disappeared and a broad, temperature stable peak appeared. The disappearance of the $320^{o}K$ peak and the linear variation of the corresponding modulus suggest that this peak may be a more stable configuration of the $320^{o}K$ peak.

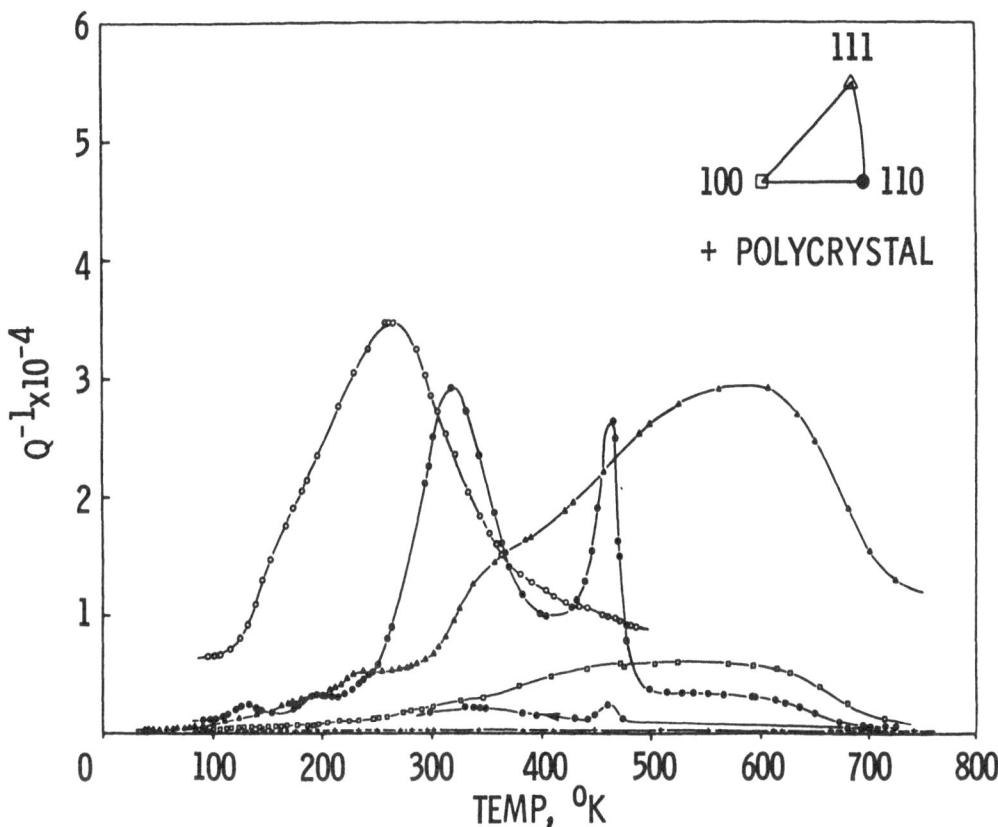

FIG. 1 - Internal friction curves for tungsten single crystals oriented on the three
corners of the stereographic triangle and for polycrystalline tungsten after
2 hours annealing at 1900°K. Frequency 45 kHz.
o - [110] crystal, after 2.5% deformation at 300°K and 2 hours annealing at
1800°K

FIG. 2 - Variations of the resonant
frequency corresponding to
the internal friction curves
shown in fig. 1.

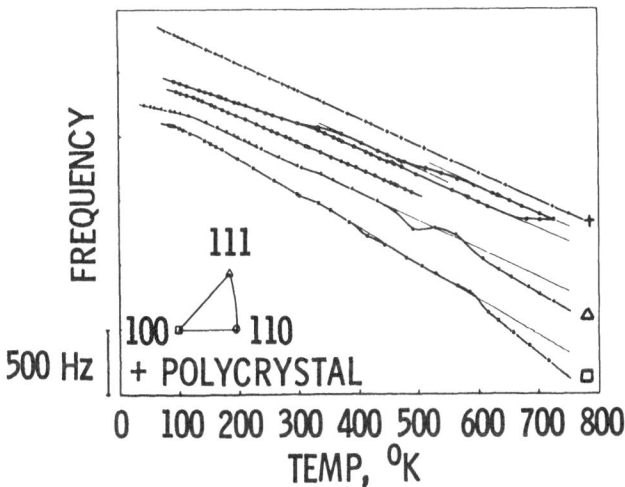

Deformed Samples

The samples were deformed in compression at room temperature. The corresponding stress-strain curves for the "corner" orientations and for the polycrystalline material are given in figure 3. The [100] and [111] crystals exhibited a very low proportional limit which we

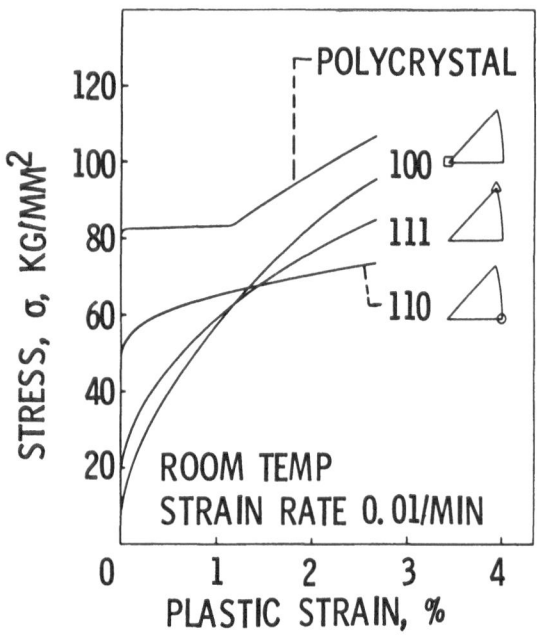

FIG. 3 - Representative Stress-strain curves for tungsten single crystal and tungsten polycrystal deformed in compression at room temperature

associate with the high purity of the samples. They yielded smoothly and showed significant strain hardening. The [110] crystals showed sharper yields at higher stress and exhibited less work hardening. Similar stress-strain behavior has been observed previously in tension for less pure single crystals (1)(4). In contrast, polycrystalline samples yielded very sharply at higher stresses, showing no strain hardening until more than 1 percent deformation.

The internal friction curves obtained after 2.5 percent compression at room temperature (Fig. 4) show several distinct peaks: a small low temperature peak, around 70°K, a large double peak at about 170° and 220°K, a broad complex peak, particularly strong in the [110] crystal between 200°K and 500°K, and a high temperature peak around 800°K. All of these peaks appear for all the samples, except the high temperature peak for the polycrystalline specimen.

The 70°K peak which appears clearly for every corner orientation and the polycrystal can perhaps be associated with the α' peak observed in Nb by Chambers (5). The corresponding modulus defect (Fig. 5) is about twice the maximum damping height (subtracting the 170°K peak contribution). We thus assume that it corresponds to a single relaxation process.

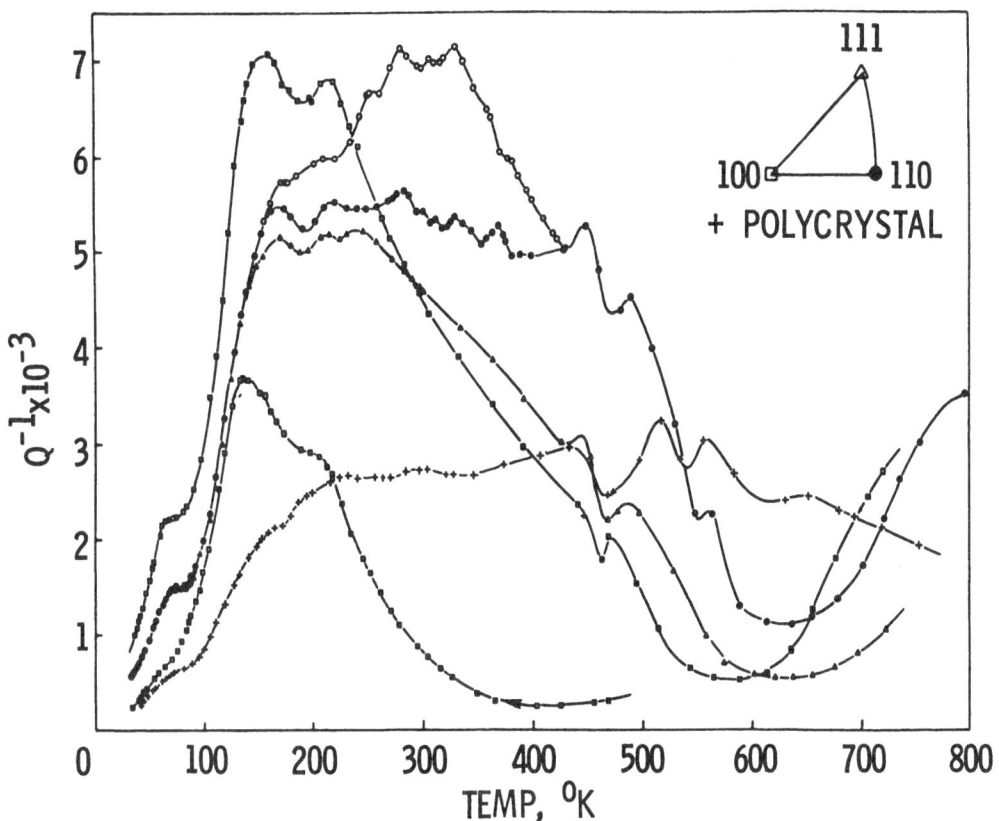

FIG. 4 - Internal friction curves for tungsten single crystal oriented on the three
corner of the stereographic triangle and for polycrystalline tungsten after
2.5% deformation in compression at room temperature. Measurement made at about
45 kHz.
o - [110] crystal aged by heating to 400°K at 0.5°K/min.

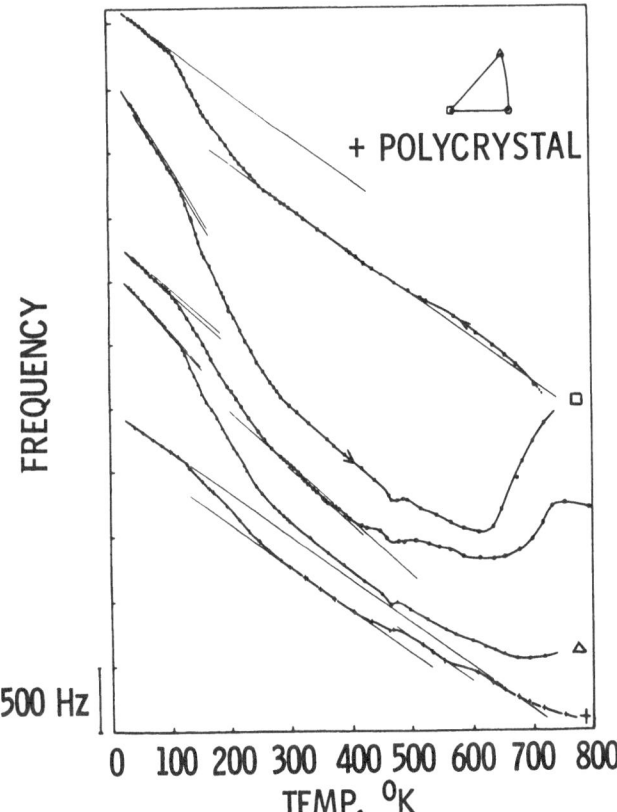

FIG. 5 - Variations of the resonant
frequency corresponding to
the internal friction curves
shown in fig. 4.

From the temperature at which it appears we estimate it to have an activation energy of about 0.1 eV.

The 170°K and 200°K peaks correspond respectively to the α and β peaks reported by Chambers (6) and Secretan (7), and to the α composite peak reported by Muss and Towsend (8). We cannot separate the respective contribution of each one of these peaks to the modulus defect (Fig. 5), but the magnitude of this effect indicates a complex relaxation process. Their activation energies are 0.21 eV and 0.66 eV, respectively.

The high temperature side of this double peak is obscured by the 200°-500°K broad complex peak, particularly for the [110] crystal. This complex peak is affected by temperature. It increases and shows a more defined maximum around 320°K after heating to 400°K at 0.5°K/min ([110] crystal). It is almost eliminated by heating to 720°K ([100] crystal) revealing the 170°-200°K double peak and the corresponding modulus defect.

The 800°K peak is presumed to be the same peak observed by Chambers (9) (γ peak), Martinet (10) and Schnitzel (11). In fact, its temperature for this frequency is remarkably consistent with Martinet and Schnitzel's results. This temperature corresponds to an activation energy of 1.5 eV. An important inverse and irreversible modulus defect occurs at the same time which shows a saturation 50 degrees lower than the maximum of the peak. We notice also in figure 4 a small perturbation occurring in the internal friction and modulus for all the samples at the temperature of the spike.

Discussion

These two series of peaks observed after annealing and deformation correspond to different structures; after annealing, we can assume that the crystals contain some dislocations and vacancies formed by thermal stresses during the cooling, and by the local plastic deformation occurring around the support points of the sample. The existence of active dislocations is confirmed by the presence of the 170°K and 220°K peaks on the [111] crystal.

The peaks observed at high temperature after annealing have the shape of a "dislocation background" with a decrease of internal friction occurring between 600°K and 700°K. This temperature region corresponds to stage III recovery of W. The smaller damping observed during the cooling suggests a pinning of the active dislocations. However, the more rapid decrease of the modulus for this temperature range rejects this interpretation.

In contrast, this pinning effect is suggested by the the modulus variation after deformation, and one can think of a vacancy migration toward the dislocations. The 800^oK peak would then be associated in this case with a vacancy-dislocation interaction occurring during the pinning. Martinet (10) studied the recovery of this peak and concluded the occurrence of this kind of interaction. However, the 100^oK difference between the migration temperature of these defects (600-700°K) and the temperature of the maximum of this peak (800^oK), and also the saturation observed in the variation of the modulus, make vacancy-dislocation interaction arguments unacceptable.

In fact, electron microscopy observation after a few percent deformation (2%) at room temperature (12) shows clearly that the dislocation substructures are only long screw dislocations parallel to the [111] direction. This suggests their active participation in the process responsible for these peaks; we will suggest therefore an interpretation in agreement with Seeger-Sěsták theorie (13). In essence then, the 70^oK peak must be the expected peak corresponding to the formation of double kinks in non-screw dislocations, and the 170^oK peak corresponds to the motion of kinks in screw dislocations.

For the [110] crystal, Rose et al (1) pointed out the fact that the only two operating slip directions lie in a plane whose normal is perpendicular to the deformation axis. Therefore, dislocation motion will require high stress. Accordingly, we can expect this sample to have more discontinuities on the screw dislocations and large internal residual stresses. But Stephens (12), and Arsenault and Lawley (14) noticed that the dislocation mobility is sensitive to the internal stresses. We can interpret the broad 200-500°K peak by the motion of kinks under stress fields that can include different defects (jogs, vacancies, etc) acting over several atomic diameters. These effects are less important on [111] and [100] crystals because for these orientations more slip directions are active. The dislocation motion will be easier during deformation and the crystal will contain, therefore, less internal residual stress.

The 800^oK peak must be due to the formation of double kinks in screw dislocations. In this case, we interpret the modulus defect by a possible readjustment during the double kink formation toward a more stable geometric kink configuration.

These interpretations of the 200-500°K and 800°K peaks explain particularly:

(1) the increase of the 200-500°K peak and 170°-220°K peaks after heating, because the corresponding stress release allows an easier motion of kinks.

(2) the diminution of the $170-220^{\circ}$K peak after heating to 720°K because of the decrease in the number of geometric kinks.

(3) the $200-500^{\circ}$K peak diminution or disappearence after heating to 720°K for the preceding reason and also because of the large stress release.

(4) the saturation of the modulus defect after reaching a stable geometric kink configuration.

(5) the correspondance between the height of the $170-220^{\circ}$K peak and the magnitude of the inverse modulus defect. Both are directly related to the number of geometric kinks.

For the polycrystalline samples we do not see the 800°C peak. This is understandable because impurity pinning would dominate here and mask the double kink process.

In conclusion, the high purity single crystalline W shows an important sensitivity to any small deformation. The orientation affects the internal friction spectra when this orientation is favorable to the creation of internal residual stresses. Clearly, further experiments are needed to confirm this interpretation. These are currently in progress.

Acknowledgements

The author gratefully acknowledges an associateship from the National Research Council, and the National Aeronautics and Space Administration for making available this research. I wish to thank Dr. H. H. Grimes head of the Department of Solid State Physics at NASA for his interest in this work.

References

1. R. M. Rose, D. P. Ferriss and J. Wulff, Transactions of the Metallurgical Society of AIME, Volume 224, Oct 1962 - 981.

2. H. Schultz, Acta Metallurgica, Vol. 12, 1964 - 649.

3. G. Rieu, J. de Fouquet, J. de Physique, Colloque C2, supplement 7, tome 32, 1971, page C2-221.

4. R. G. Garlick, NASA E-2649 - 1966.

5. R. H. Chambers, T. E. Firle, T. Trozera and G. Buzzelli, Final Summary Report, General Dynamics, G.A. 7978, 1965.

6. R. H. Chambers, T. E. Firle, Technical Report No. AFML-TR-65-28, General Atomic Division of General Dynamic Corporation - 1965.

7. B. Secretan, These E.P.U.L. Lausanne 1964.

8. D. R. Muss and J. R. Townsend, J. Appl. Phys. 33, 1962, 1804.

9. R. H. Chambers, Physical Acoustics, Vol. IIIA, Chap. 4 (W. P. Mason, ed.) Academic Press, New York (1966)

10. B. Martinet, These E.P.U.L. Lausanne 1964.

11. R. H. Schnitzel, Transactions of the Metal. Society of AIME, Vol. 233, 1965, 186.

12. J. R. Stephens, Metallurgical Transactions, Vol. 1, May 1970, 1293.

13. A. Seeger and B. Šesták, Acta Met. Vol. 5, 1971, 875.

14. R. J. Arsenault and A. Lawley, Phil. Mag., vol. 15, 1967, 549.

INTERACTION BETWEEN DISLOCATIONS AND POINT DEFECTS AND ULTRASONIC PROPERTIES OF SLIGHTLY COLD WORKED ALUMINIUM

A. VINCENT, J. PEREZ and P.F. GOBIN

Department of Acoustics and Department of Metallurgy - I.N.S.A. - 69621 VILLEURBANNE, FRANCE

Introduction

Measurements of ultrasonic attenuation and velocity during compressional or tensile tests have been done by several workers (1,2,3). This method has appeared very sensitive to study the movement of dislocations during plastic deformation of different materials. More recently it has been shown that unpinning of dislocations during the application of external quasi static compressional (4) or tensile (5) stress could be also studied by this method.

For instance on figure 1, it can be seen in the case of aluminium attenuation $\Delta\alpha$ and velocity $\frac{\Delta v}{v}$ changes against tensile quasi static stress : two stages are observed :

(i) in stage I there is an increase in attenuation and a decrease in velocity while deformation   remains in macroscopic elastic range*; this stage has been attributed to dislocation breakaway from point defects .

(ii) above a stress called σ_p deformation ε increases rapidly and there is some evidence to associate corresponding attenuation and velocity changes to the start of plastic deformation.

The purpose of this paper is to deal with a quantitative analysis of observations   done in stage I.

Experimental Method

The material used throughout this expriment was polycristalline aluminium 99,999% pure (the mean diameter of the grains was about 2mm). The dimensions of the specimen were 10mm x 10mm x 70mm. After annealing at 500°C, it was prepared and glued as usually (1) to specially designed grips of an INSTRON tensile machine. The tensile stress speed was about 1g/mm^2/s in the macroscopic elastic range. The temperature of the specimen could be regulated between -110°C and +70°C.

---

*It has been verified otherway that stress-strain loops show an hysteretic effect less than 10^{-6}.

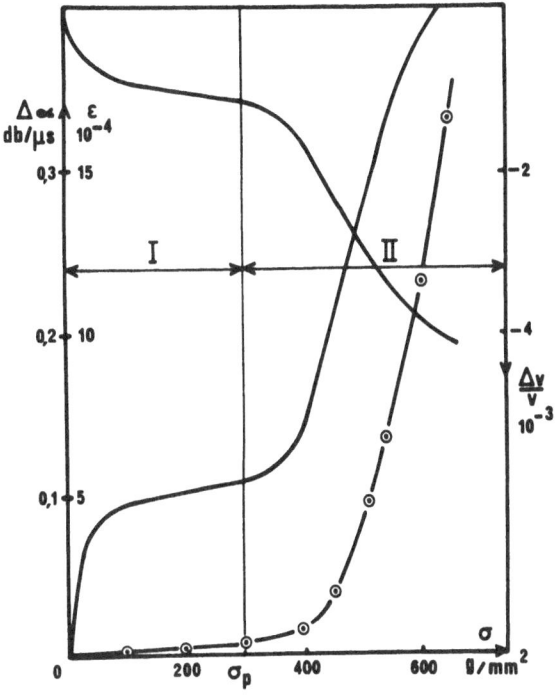

FIG.1

Attenuation changes Δα , velocity changes $\frac{\Delta v}{v}$ and strain ε against stress at 27°C. Prestrain 0,1 x 10^{-2}, and recovery 72h in situ.

The attenuation and velocity changes were measured along the axis of tensile stress by a pulsed method at a frequency of 4 MHz.

Experimental Results

In order to study the microstructure effect (dislocation density and number of pinning point defects) from attenuation and velocity changes observed in stage I we did two kinds of experiments : on one hand we looked for the recovery time effect, after the specimen was prestrained about 0,26 x 10^{-2} ; the figure 2 shows attenuation and velocity changes against stress for recovery times of 1mn30s at 27°C (curves A and A'), 20h at 27°C (B and B') and 72h at 50°C (C and C'). For each test tensile stress was limited to small values in order to keep the same dislocation network between different experiments.

On the other hand the prestrain effect at constant recovery time of 20h at 27°C has been observed. The figure 3 shows of 0,1 x 10^{-2} (curves A and A'), 0,26 10^{-2} (B and B') 0,75 x 10^{-2} (C and C') and 3,5 x 10^{-2} (D and D').

Discussion of Results

Unpinning of dislocations from point defects has been wideley studied last years. In order to build a model fitting to the preceding results observed in stage I we have

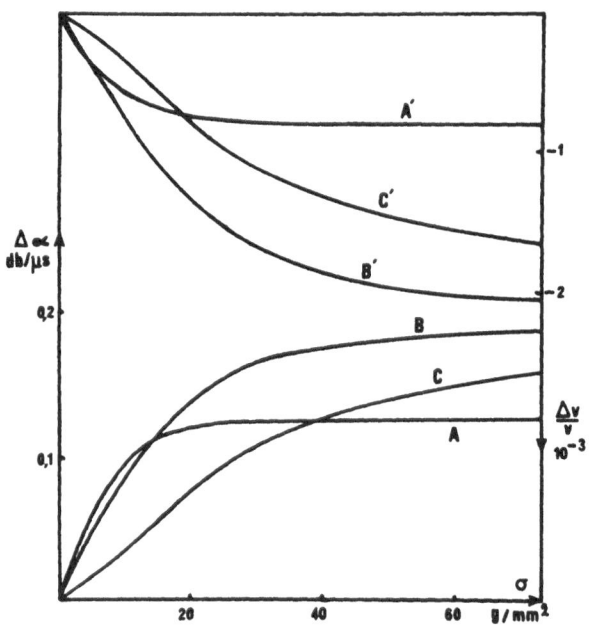

FIG 2

Attenuation Δα and velocity changes $\frac{\Delta v}{v}$ against stress σ at 27°C. Recovery 1mn30s at 27°C (A and A'), 20h at 27°C (B and B') and 72h at 50°C (C and C').

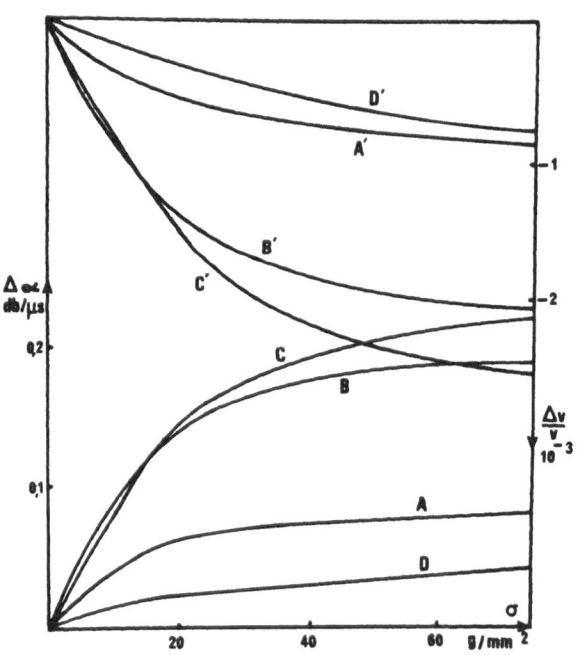

FIG 3

Attenuation Δα and velocity changes $\frac{\Delta v}{v}$ against stress σ at 27°C. Prestrain $0,1 \times 10^{-2}$ (A and A'), $0,26 \times 10^{-2}$ (B and B'), $0,75 \times 10^{-2}$ (C and C') and $3,5 \times 10^{-2}$ (D and D').

successively considered three basic hypothesis associated to available theories.

(i) At first, the simplest theory was proposed by GRANATO and LUCKE (6) who assumed that dislocations pinned by point defects can mechanically beakaway under the action of a stress ; this hypothesis was considered in a previous paper (7) : it was found that it applied rather well to the exprimental results but the value of binding energy between dislocations and point defects obtained from exprimental data appeared too low (less than 10^{-2} eV).

(ii) Then the thermally assisted unpinning of dislocation can be considered. Hence, the pinned linear defects have a combined static hysteresis-relaxation behavior (8). In this case, attenuation and velocity changes due to the thermomechanical unpinning of dislocations should be strongly temperature dependent. Actually, experimental results do not show such a dependance (except for that due to damping constant change) (9).

(iii) Finally, this feature leads us to consider the fraction $f(\sigma)$ of broken away dislocations as a function of stress at thermal equilibrium. As shown by LUCKE et al (8), $f(\sigma)$ presents only a slight variation with temperature.

The most general model considering the thermal breakaway of dislocation loops pinned by randomly distributed point defects leads to sophisticated calculations. So, we do the following assumptions :

(i) There is in the crystal, a population of major loops which possess two configurations of stable équilibrium : a completely pinned and a completely unpinned configuration.

(ii) The probability of breakaway from the pinned state and the probability of repinning from the unpinned state are assumed to be similar to that of a double loop of length 2l pinned at its middle ; in other words it is considered that the unpinning and repinning events are essentially governed by the unpinning from and repinning to the pinning point lying between the largest segments of the major loop.

(iii) There is a distribution in the length 2l. We assume the following exponential distribution

$$N(l)\ dl = \frac{\Lambda}{L_N l_o^2}\ l \exp\left(-\frac{l}{l_o}\right)\ dl$$

in accordance with the normalization condition

$$\int_0^\infty N(l)\ dl = \frac{\Lambda}{L_N}$$

where Λ is the dislocation density, L_N is the mean value of length of major loops and l_o is the most probable value of l.

This kind of distribution was already used by LI (10) and is supported by the physical fact that there are neither loop (governing the breakaway) of lenght zero, nor lopp of infinite lenght, and that N(l) has a maximum value at lo which depends upon the total number of point defects pinning the dislocations.

(iiii) The contribution of dN free dislocations loops (of length L_N) to the ultrasonic attenuation and velocity effects are in the low megahertz range (6):

$$d\alpha = K_\alpha . L_N^5 . dN$$

$$\frac{dv}{v} = - K_v . L_N^3 . dN$$

K_α and K_v being constants which are about 4×10^6 and 4×10^{-3} with our exprimental conditions. Thus the probability for a major loop to be in the unpinned configuration is given by :

$$f(\sigma) = \frac{1}{1 + \frac{\nu_2}{\nu_1} \exp \frac{U_1 - U_2}{kT}}$$

ν_1, ν_2 frequency factors for unpinning and pinning ;

U_1, U_2 activation energies for unpinning and pinning;

T temperature, K Boltzmann constant.

Linearizing the COTTRELL force between the dislocation and a point defect LÜCKE et al analysed the situation of the double loop pinned at the middle (8). The values obtained for U_1 and U_2 lead to :

$$U_1 - U_2 \simeq U_0 (1 - \frac{\sigma^2 f^2 l^3}{2U_0 G})$$

G shear modulus; f orientation factor which average value will be assumed to be about 0,2.

Moreover GRANATO et al calculated the frequency factors ν_1 and ν_2 (11) ; from their formulas it can be deduced

$$\frac{\nu_2}{\nu_1} \simeq 10^{-9}/ (U_0 .l)^{1/2}$$

U_0 interaction energy between dislocation and point defect.

Total ultrasonic attenuation (and velocity) changes produced by the unpinning of dislocations is given by the attenuation in the unpinned configuration minus the attenuation in the pinned configuration. By neglecting this second term*the derived ultrasonic changes will be :

---

*It has been shown in a previous paper (7) that this correcting term may have some experimental evidence for $\frac{\Delta v}{v}$ (because of the L_N^3 dependance instead of the L_N^5 for $\Delta\alpha$) but it is not of fundamental importance for the present work.

$$\Delta\alpha \simeq K_\alpha L_N^5 \int_0^\infty f(\sigma) N(l) \cdot dl$$

$$\frac{\Delta v}{v} \simeq - K_v L_N^3 \int_0^\infty f(\sigma) N(l) \cdot dl$$

thus one finds that

$$\Delta\alpha = K_\alpha \Lambda L_N^4 \cdot I$$

$$\frac{\Delta v}{v} = - K_v \Lambda L_N^2 \cdot I$$

$$I = \int_0^\infty \frac{\dfrac{1}{l_0^2} \exp(-\dfrac{1}{l_0})}{1 + \dfrac{10^{-9}}{(U_0 l)^{1/2}} \exp\left[\dfrac{U_0}{kT}\left(1 - \dfrac{\sigma^2 f^2 l^3}{2 U_0 G}\right)\right]} dl$$

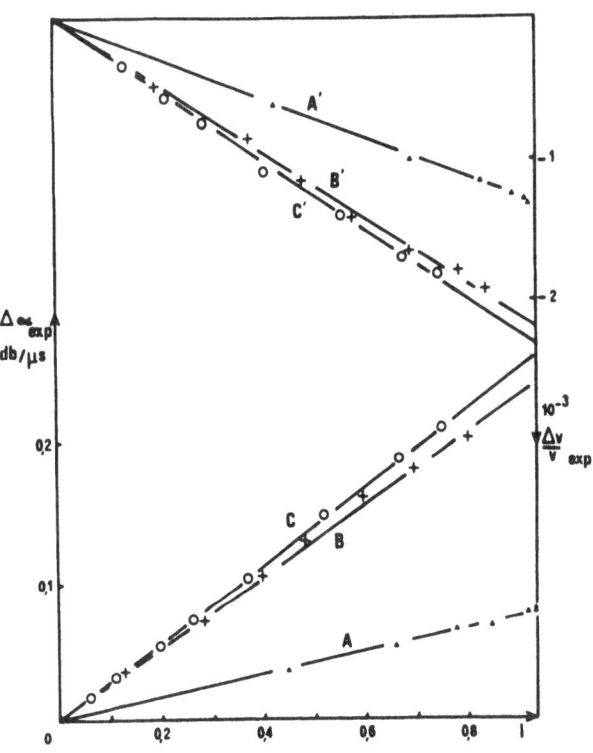

FIG.4

Attenuation $\Delta\alpha_{exp}$ and velocity $\dfrac{\Delta v}{v_{exp}}$ changes plotted with analytical expression I. Curves[*] corresponding to that of the recovery study of FIG.2

[*]For curves A and A' this good fit has been obtained with a numerical integration from length l_0 instead of lenght zero. This can be explained by the fact at short recovery time there are fewer short loops l.

357

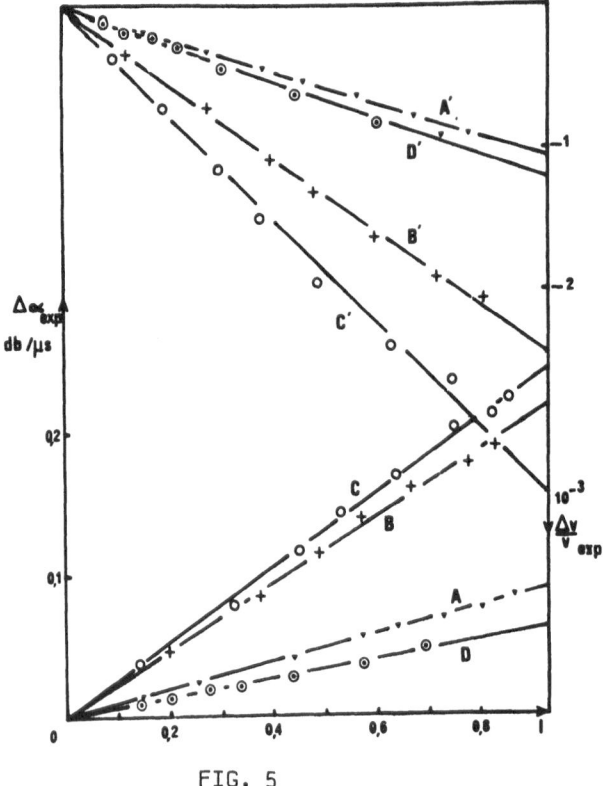

FIG. 5

Attenuation $\Delta\alpha_{exp}$ and velocity $\frac{\Delta v}{v}_{exp}$ changes plotted with analytical expression I. Curves corresponding to that of the prestrain study of FIG.3.

From this expression it appears that the function dependance of $\Delta\alpha$ (or $\frac{\Delta v}{v}$) against σ, will be mainly governed by the two parameters U_o, l_o. So we programmed a computer to search for the couple (U_o, l_o) leading to the best fit between this analytical expression of $\Delta\alpha$ and the experimental dependance against stress σ.

On figure 4 and figure 5 it has been plotted experimental changes in attenuation $\Delta\alpha_{exp}$ (and in velocity $\frac{\Delta v}{v}$ corresponding to each curve of figure 2 and figure 3. It can be seen that the agreement between the preceding theoritical model and exprimental data is rather good in the whole range of stress involved in stage I.

Then, by taking the limit values of $\Delta\alpha$ and $\frac{\Delta v}{v}$ (so called respectively $(\Delta\alpha)_1$ and $(\frac{\Delta v}{v})_1$ corresponding to an infinite values of σ, it can be deduced numerical values of L_N and Λ.

$$L_N \simeq 10^{-5} \frac{10(\Delta\alpha)_1^{1/2}}{(\frac{\Delta v}{v})_1} \quad cm$$

$$\Lambda \simeq 2,5 \times 10^{11} \frac{(\frac{\Delta v}{v})_1^2}{(\Delta\alpha)_1} \quad cm/cm^3$$

All results are put together in table I for recovery time tests, and table II for different prestrain ratio.

TABLE I

Physical Parameters Deduced From Recovery Time experiments

Recovery time	U_o from $\Delta\alpha$ (eV)	U_o from $\frac{\Delta v}{v}$ (eV)	l_o from $\Delta\alpha$ (cm)	l_o from $\frac{\Delta v}{v}$ (cm)	Λ (cm/cm3)	L_N (cm)
1mn 30s at 27°C	0,15	0,10	$10,0 \times 10^{-5}$	$7,1 \times 10^{-5}$	$1,2 \times 10^6$	$4,1 \times 10^{-4}$
20h at 27°C	0,13	0,15	$5,3 \times 10^{-5}$	$4,8 \times 10^{-5}$	$6,8 \times 10^6$	$3,0 \times 10^{-4}$
72h at 50°C	0,13	0,13	$2,9 \times 10^{-5}$	$2,6 \times 10^{-5}$	$7,4 \times 10^6$	$3,0 \times 10^{-4}$

TABLE II

Physical Parameters Deduced From Prestrain Experiments

Prestrain	U_o from $\Delta\alpha$ (eV)	U_o from $\frac{\Delta v}{v}$ (eV)	l_o from $\Delta\alpha$ (cm)	l_o from $\frac{\Delta v}{v}$ (cm)	Λ (cm/cm3)	L_N (cm)
$0,1 \times 10^{-2}$	0,15	0,11	$6,7 \times 10^{-5}$	$3,7 \times 10^{-5}$	$3,1 \times 10^6$	$2,9 \times 10^{-4}$
$0,26 \times 10^{-2}$	0,13	0,15	$5,3 \times 10^{-5}$	$4,8 \times 10^{-5}$	$6,8 \times 10^6$	$3,0 \times 10^{-4}$
$0,7 \times 10^{-2}$	0,18	0,13	$5,8 \times 10^{-5}$	$3,4 \times 10^{-5}$	$11,8 \times 10^6$	$2,7 \times 10^{-4}$
$2,8 \times 10^{-2}$	0,10	0,11	$2,4 \times 10^{-5}$	$2,1 \times 10^{-5}$	$2,7 \times 10^6$	$2,2 \times 10^{-4}$

The main following remarks can be done :

a - the value of the interaction energy U_o between dislocations and point defects appears to be a constant, about 0,13eV, whatever the exprimental conditions are. This result agrees very well with that obtained from low frequency internal friction experiments done on the same material (12).

b - lo decreases with increasing recovery time (table I). This result, as for the recovery of attenuation after deformation of aluminium observed by several authors (1,13), can be attributed to the diffusion of point defects towards the new dislocations created during prestrain. The value of binding energy and the conditions of migration allows us to identify these point defects to impurity atoms.

c - The density of dislocations seems to increase with recovery time (table I). This can be explained if we assume that just after prestrain all major loops have not received at less one point defect ; thus the major loops free of pinning agent do not contribute to the unpinning process and subsequent attenuation and velocity changes. From a more general point of view it can be said that the measured density is that of dislocation lines able to be unpinned at low stress.

d - The apparent density of dislocations increases with increasing prestrain (table II) in agreement with the idea of the generation of dislocations during plastic deformation, then seems to reach a maximum and decreases ; an equivalent result was already obtained by HIKATA at al (1) who measured the attenuation and velocity changes during plastic deformation. Althought in disagreement with the conclusion of Mc DONALD at al (14), this feature may be attributed to dislocations-dislocations interactions which reduce loops freedom of vibration.

Conclusion

In order to demonstrate completely the validity of this theoritical treatment of our data further experiments have to be made : on one hand to identify more surely the point defects, we intend to do the same study on purer aluminium ; on the other hand to verify the hypothesis of equilibrium conditions, higher speed of stressing and lower temperature must be used. Nevertheless the theoritical treatment proposed here leads to numerical values of physical parameters such as density of unpinnable dislocations, mean lenght of major loops and binding energy between dislocation line and point defect.

References

1 - A. HIKATA, R. TRUELL, A. GRANATO, B. CHICK and K. LÜCKE, J. Appl. Phys., 27, 4, 396 (1956)
   A. HIKATA, B. CHICK, C. ELBAUM and R. TRUELL, Acta. Met., 10, 4, (1962)

2 - W.F. CHIAO and R.B. GORDON, Trans. AIME, 233, 1164, (1965)

3 - W. SACHSE and R.E. GREEN, J. Phys. Chem. Solids, 31, 1955 (1970)

4 - D. LENZ, B. EDENHOFER and K. LÜCKE, Scripta. Met., 5, 5, 387 (1971)

5 - A. VINCENT, J. PEREZ and P.F. GOBIN, J. Phys., 32, 651, (1971)

6 - A.V. GRANATO and K. LÜCKE, J. Appl. Phys. 27, 583, (1956)

7 - A. VINCENT, J. PEREZ and P.F. GOBIN, J. Phys. Sup., 11-12, 33, 170, (1972)

8 - K. LÜCKE, A.V. GRANATO, L.J. TEUTONICO, J. Appl. Phys. 39, 11, 5181, (1968)

9 - A. VINCENT, Thèse LYON (1973)

10- J.M.C. LI - Physics of Strength and Plasticity, p.245, Edited by Ali S. ARGON (1969)

11- A.V. GRANATO, K. LÜCKE, J. SCHLIPF and L.J. TEUTONICO, J. Appl. Phys., 35, 9, 2732 (1964)

12- J. PEREZ, P. PEGUIN and P.F. GOBIN, J. Phys., sup. 7, 32, 127 (1971)

13- I. HOLWECH, J. Appl. Phys.,31, 5, 928 (1960)

14- S.G. Mc DONALD and N.F. FIORE, Scripta. Met. 4, 135 (1970)

EFFECT OF MANGANESE AND CHROMIUM ON COLD WORK PEAKS IN IRON

L.E. Buchanan and R. Kennedy
Metallurgy Department, University of Strathclyde
Glasgow, Scotland.

ABSTRACT

The characteristics of cold work peaks due to nitrogen and/or carbon in iron containing up to 2% Mn or 1% Cr have been studied. The isolated peaks gave a good fit to a lognormal distribution of relaxation times and showed consistent trends in the half width parameter β, which was found to increase with decreasing nitrogen content, higher quenching temperatures and with increasing substitutional solute content. The latter effect was particularly marked with manganese which also resulted in a lowering of peak height. The substitutional solutes produced a significant lowering of peak temperature. In all cases the peaks produced by carbon alone were of almost negligible height. The results have been interpreted in terms of a modified Schoeck model.

The Koster or Cold Work Peak (CWP) has been found, after deformation, in many body centred cubic metals containing interstitial impurities such as N, C, O and H. Numerous systems, including Nb-N (1), Nb-O (2), Ta-O (3) and Fe-H (4) have been investigated but the most extensively studied systems have been the Fe-N and Fe-C (5-10). In these the peak occurs at a temperature of about 200°C for a frequency of 1Hz with an activation energy in the region of 150 KJ/mol. One feature of the published results, however, is the wide range of peak characteristics which have been reported and the large number of variables found to affect the peak. Despite the fact that many models have been proposed to explain the effect, there is still doubt as to the correct theoretical interpretation of the peak. The present work was intended primarily to determine the effect of the substitutional solutes manganese and chromium on the height, breadth and activation energy of the CWP in Fe containing N and/or C.

Experimental Procedure

Using Swedish iron (.025% C, .001% Si, .011% S, .002%P, .105% O, others <.01%) as a base material, various alloys containing up to 2% Mn and up to 1% Cr, together with a "pure iron" cast were prepared by vacuum melting employing carbon deoxidation. After forging and rolling the materials were swaged and drawn to .75 mm diameter wire. A 20 minute 950°C recrystallisation treatment was found to remove texture and produce a consistent grain size. Particular care was taken to remove residual C and N by wet hydrogen treatment, to a level undetected by Snoek peak

measurement, before nitriding and/or carburising with ammonia/hydrogen or n-heptane/
hydrogen mixtures respectively. Interstitial levels were determined by Snoek peak
(11) and/or chemical analysis. The wire specimens were quenched from a suitable
soaking temperature and cold drawn by 24% reduction of area immediately prior to
CWP determination.

All of the internal friction measurements were carried out on a Ke type
torsion pendulum. The free decay was measured by an optical lever system and a
'photodyne' light spot follower. The latter was automated to record the amplitude
of successive vibrations on punched tape, allowing direct input of the damping data,
together with temperature and frequency measurements, to a computer for calculation
and analysis of results. The frequency range used was from 0.5 to 3.5 Hz.

EXPERIMENTAL RESULTS

A typical experimental damping v. temperature profile for an Fe/N alloy
is shown in Fig. 1. This emphasises one difficulty in the study of the CWP, since
it is superimposed on a rapidly rising background related to the grain boundary peak.
This latter effect is itself affected by deformation and by both interstitial and
substitutional solute content (12), so that it is impossible to measure the background contribution in isolation. Various graphical techniques have been suggested
(13, 14) for background removal, but in the present work the boundary damping contribution was estimated by assuming that it could be described mathematically as
the low temperature side of a Debye curve which was computed to fit the experimental
points above and below the temperature range of the CWP, making allowance for the
apparatus and background dislocation damping levels. Fig. 1 also shows this
computed boundary damping and the remaining cold work peak (which still includes
the other background contributions).

Various series of experimental tests were carried out on the pure iron,
manganese and chromium alloys with varying percentages of nitrogen and carbon
present, the results being analysed as indicated above. Examples of the cold work
peaks obtained with manganese alloys containing .022% N are shown in Fig. 3. This
shows clearly the three main effects of manganese on the peak characteristics,
namely, a progressive reduction in peak height, a broadening of the peak and a
reduction in peak temperature with increasing manganese content. A similar trend
was noted with the two chromium alloys studied. This is indicated in Fig. 4 for
alloys of a somewhat higher nitrogen content (.03%).

One feature of the experimental results was that the cold work peaks
developed in all of the alloys which contained carbon as the only interstitial
were of almost negligible height, even after ageing for considerable times at peak
temperature. This confirms the observations of certain (9), though not all (6)
previous workers.

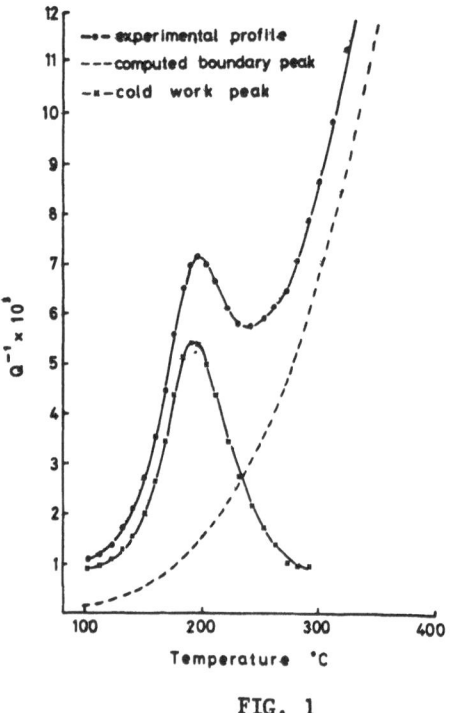

FIG. 1

Experimental damping profile showing
boundary damping removal and CWP.

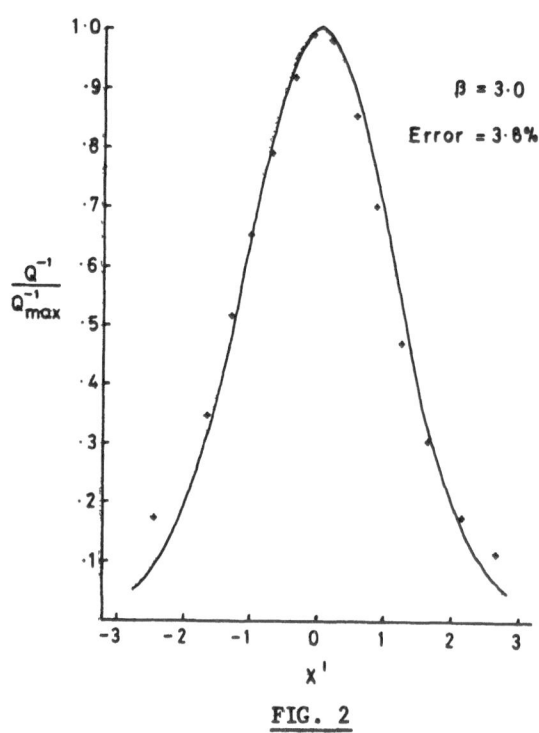

FIG. 2

Fit of normalised damping values to
theoretical lognormal distribution.
(Fe .02% N)

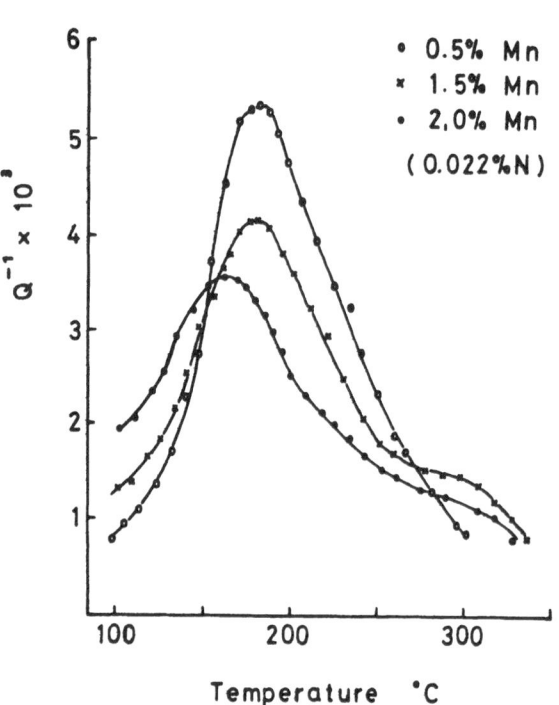

FIG. 3

Isolated cold work peaks for Mn alloys
quenched from 850°C.

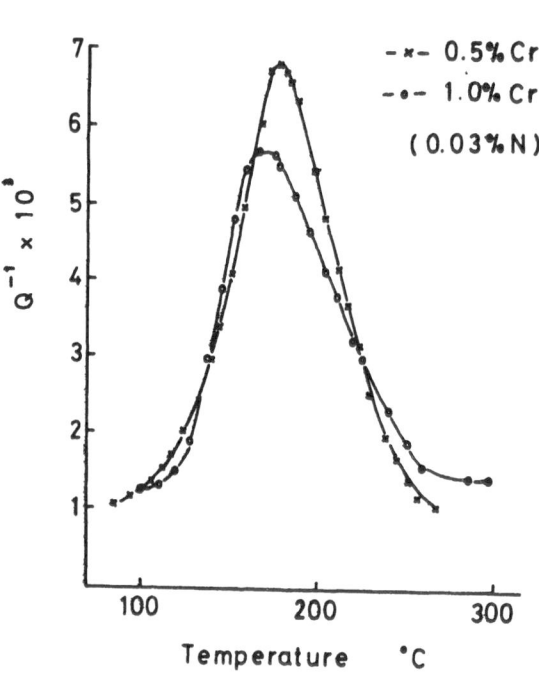

FIG. 4

Isolated cold work peaks for Cr
alloys quenched from 950°C.

The CWP is much broader than the theoretical Debye curve corresponding to a single relaxation time and it has been suggested that this may be explained in terms of a lognormal distribution of the relaxation times contributing to the peak. Nowick and Berry (15) have shown that the extent of peak broadening can be characterised by the distribution parameter β ($\beta = 0$ being equivalent to a single relaxation process). Following their analysis, the CWP's obtained were replotted (after background removal) in terms of the normalised damping Q^{-1}/Q^{-1} max against the temperature variable $X' = H/2.303R$ $(1/T - 1/T_m)$, where H is the most probable activation energy (obtained from peak shift data) and T_m is the peak temperature. The best β value for each curve was then determined by computerised interpolation of the tabulated data provided by the above authors, the root mean square error between experimental and theoretical values being adopted as the criterion for assessing the degree of fit. The results of this analysis for a selection of the experimental runs are given in Table 1.

TABLE 1

β PARAMETERS FOR COLD WORK PEAKS

	No.	Alloy Fe +	Quench Temp.	β	R.M.S. Error %
	1	.015%N	590°C	3.0	7.6
a	2	.026%N	"	2.75	5.1
	3	.041%N	"	2.25	5.2
b	4	.02%N	590°C	2.9	3.8
	5	.02%N	850°C	3.2	3.6
	6	.12%Mn/.026%N	850°C	3.0	7.5
	7	.5%Mn "	"	3.3	8.7
c	8	.7%Mn "	"	3.5	4.8
	9	1.5%Mn "	"	4.0	3.4
	10	2.0%Mn "	"	5.0	4.1
d	11	.5%Cr/.03%N	950°C	3.0	6.0
	12	1.0%Cr "	"	3.5	7.8

It was estimated that a root mean square error of 5% was within the limits imposed by the accuracy of temperature and damping measurements, background removal and the interval between the β values evaluated. In view of this it would appear that a reasonable fit is obtained in most cases to the lognormal distribution. An example of the relationship between experimental points and theoretical distribution giving an RMS error of 3.8% is shown in Fig. 2.

As indicated in Table 1, the β values show trends with (a) nitrogen content, (b) quenching temperature, (c) manganese, and (d) chromium content. These are discussed below.

The activation energy of the CWP was estimated by measurement of peak shift with frequency. Unfortunately the small range of frequency available with the torsion pendulum and the broadened nature of the peak itself make accurate determination difficult and the values obtained are subject to an error óf about ± 10%. Such factors may be partially responsible for the wide range of activation energies reported in the literature. The results obtained in the present work, together with the relevant peak temperature data are compared in Table 2 with other reported data for cold drawn specimens.

TABLE 2

PEAK TEMPERATURES AND ACTIVATION ENERGIES

Alloy	Temp. C (for 1 Hz)	Energy KJ/mol	R. of A. %	Author
Fe N	–	150	5 to 80	Koster et al (5)
Fe .01%C	215	138	25	Kamber et al (6)
Fe .01%N	207	138	25	"
Fe .016%N	200	159	15	Petarra et al (9)
Fe .025%N	195	160±10%	24	Macdonald (11)
Fe .014%C+N	215	185±12%	24	Present Work
Fe 1.5%Mn N	170	145±10%	24	" "
Fe 1.0%Cr N	180	136±10%	24	" "

It will be noted that in the present work there is an indication of lower activation energies in the manganese and chromium alloys. In spite of the limited accuracy of the values it is felt that this trend may be significant in view of the pronounced effect shown by both Mn and Cr on the peak temperature, which is substantially reduced, and the general relationship existing between peak temperatures and activation energies (16).

Discussion

Various mechanisms have been put forward to explain the CWP. Notably that of Schoeck (7) has been held by many authors to give reasonable agreement with experiment. Schoeck suggests that the anelastic strain is produced by the movement of dislocations dragging their interstitial solute atmospheres and that the activation energy of the peak H, is the sum of the activation energy for diffusion of the interstitial Ho, and the binding energy between the interstitial and the dislocation Hd. The relaxation time $\tau = C\alpha L^2/B$ where C is a constant, α is the damping force per unit length of dislocation, L is the average free dislocation length and B is the line tension of the dislocation. α itself depends on the concentration of interstitials at the dislocation line C_d, and the diffusion coefficient of the interstitial atom D, being given by the relationship

$$\alpha = \frac{C' \, R^2 \, C_d \, kT}{Db^3}$$ where k and T have their usual significance,

R is the radial distance over which the interstitials contributing to the peak are included, b is the Burgers vector of the dislocation and C' is a constant.

From the results in Table 1 it would appear that the breadth of the CWP can be explained in terms of a lognormal distribution of relaxation times, which, since τ is proportional to L^2 can be interpreted as a lognormal distribution in the squares of the dislocation loop lengths. This has been suggested by previous authors (4, 13). The decrease in β (Table 1a) with increasing nitrogen content can thus be explained by a reduction in the spread of loop lengths as interstitial pinning increases. The increase in β noted with higher quenching temperature (Table 1b), although small, was consistent and is in agreement with a higher dislocation density producing a larger distribution in L values. This is also in accord with the observation of Gibala (4) that β increased substantially with increasing deformation in Fe-H alloys.

It is difficult, however, to see how the peak broadening observed with Mn and Cr (Table 1c,d) can be explained simply by changes (an increase) in the distribution of loop lengths. An alternative view is that substitutional solutes increase the range of α values through their effect on the range of both C_d and D. This is probable since Mn, for example, has been shown to effect both the distribution and precipitation kinetics of nitrogen in iron (17). The lognormal distribution in τ would depend therefore on both α and L^2.

The observed lowering of the peak temperature and the apparent decrease in activation energy due to manganese and chromium are considered to be interrelated effects. A lower peak temperature and hence a reduced value of τ suggests that the principle factor in determining α is the interstitial concentration at the dislocations C_d, which would decrease with increasing concentration of Mn and Cr. The expected decrease in D due to increase in these substitutional solutes would lead to an increased peak temperature and cannot therefore be the overriding factor.

It is also difficult to explain the indicated decrease in activation energy in terms of the basic Schoeck model in which $H = H_o + H_d$. Again manganese would increase H_o and the observed effect would then require a substantial reduction in H_d the binding energy between the interstitial atom and the dislocation. Ino and Sugeno (10) however, have suggested that the activation energy should take account of interaction between clustered interstitial atoms and postulate that $H = H_o + H_d + nH_s$, where n is the number of interacting impurity atoms in a cluster and H_s is the binding energy between two such atoms. On this basis the effect of

Mn and Cr could be explained in terms of their ability to reduce the degree of clustering n, which is in agreement with the previous suggestion that they reduce Cd.

The reduction in peak height with increasing Mn or Cr content may be explained in part by the progressive peak broadening which occurs, since this leads to an overall decrease in the peak height compared to that for a single relaxation process or peak of lower β value (15). The dependence of peak height on interstitial content is not well explained by the Schoeck model which gives Q_m^{-1} proportional to ΛL^2, where Λ is the dislocation density. However, it is suggested that variation in the degree of drag α, via Cd, will alter the range of loop lengths which contribute strongly to the peak for a given test frequency. The peak height would then depend on the relationship between α, L and the overall dislocation loop length distribution.

Acknowledgements

The authors are indebted to Dr. D.M. Macdonald who was responsible for some of the initial experimental work.

References

1. N. Dahlstrom, C.C. Dollins and C. Wert, Acta Met. 19, 955, (1971).

2. E. De Lamotte and C. Wert, J. Phys. Soc. Japan, 19, 1560, (1964).

3. G. Schoeck and M. Mondino, J. Phys. Soc. Japan, 18, Suppl. I, 149, (1963).

4. R. Gibala, Trans. Met. Soc. AIME, 239, 1574, (1967).

5. W. Koster, L. Bangert and R. Hahn, Arch. Eisenh. 25, 569, (1954).

6. K. Kamber, D. Keefer and C. Wert, Acta Met. 9, 403, (1961).

7. G. Schoeck, Acta Met., 11, 617, (1963).

8. P. Barrand and G.M. Leak, Acta Met. 12, 1147, (1964).

9. D.P. Petarra and D.N. Beshers, Acta Met. 15, 791, (1967).

10. H. Ino and T. Sugeno, Acta Met. 15, 1197, (1967).

11. G.J. Couper and R. Kennedy, J.I.S.I., 205, 642, (1967).

12. D.M. Macdonald, Ph.D. Thesis, University of Strathclyde, (1969).

13. J. McGrath and R. Rawlings, Acta Met. 14, 1, (1966).

14. V.W. Dickenscheid and M. Peehs, Arch. Eisenh. 40, 251, (1969).

15. A.S. Nowick and B.S. Berry, IBM Journal 5, 297, (1961).

16. E.T. Stephenson, Trans. A.I.M.E., 233, 1183, (1965).

17. J.F. Enrietto, Trans. Met. Soc. AIME, 224, 43, (1962).

INTERNAL FRICTION IN POLYCRYSTALLINE COPPER AND α-BRASSES
IN THE MICROPLASTIC REGION AT 90-300K

R. Threader[o] and P. Feltham[*]

[o]The Middlesex Polytechnic, Enfield, England
[*]Brunel University, Uxbridge, London

ABSTRACT   The energy loss per cycle of tension/compression was studied in Cu/Zn specimens
containing 0-30 at% of zinc, at amplitudes not exceeding the macroscopic flow-stress,
over the range 0.05 to 0.50Hz, at 90-300K. Its dependence on amplitude, frequency,
temperature, zinc content and grain size is shown to be consistent with a model in
which both, the hysteric and frequency-dependent contributions to the loss, are
ascribed to the work expended by migrating dislocations in passing through the
heterogeneous intragranular stress-fields.

Introduction

Although the problem of the passage of dislocations through the "statistical" stress-
field of a crystal was studied theoretically already over 25 years ago (1), it lay almost
dormant until relatively recently, when modern experimental and mathematical techniques promised
to facilitate fruitful, new, approaches. Resurgence of interest has also been catalysed by the
growing realisation that viable theories of the plastic response of crystalline materials must
take into account, at least statistically, the role of collective and cooperative processes in
the kinetics of actual dislocation ensembles (2,3).

These considerations prompted us to attempt to study dislocation kinetics by internal
friction: a potentially informative method. We aimed at conditions of deformation in which
significant dislocation movement would occur, yet where, at the same time, complications associated with gross plastic flow are avoided. More specifically, by using several variables, such as
amplitude, frequency, temperature, alloy content and grain size, we hoped to obtain a fairly
"organic" representation of the energy loss associated with restricted dislocation movement; the
latter was to be induced by subjecting the materials to tension/compression cycles at amplitudes
within the microplastic region, i.e. below the level of stress at which structural changes are
induced through flow.

Cylindrical tensile specimens with shoulders of conventional shape were machined from oxygen-free high-conductivity copper bars of 99.995% purity, and from α-brasses having nominal zinc-contents of 10, 15, 20 and 30 at%. The gauge length was 2.2cm, the diameter 0.64cm. The principal impurities were small amounts of iron ($<$10 ppm), tin and bismuth. Specimens, wrapped in brass foil to minimise de-zincification, were heated at 1000K in argon for periods of the order of 15 minutes until a grain size of 35 μm had developed. Grain-sizes were checked by means of standard ASTM charts. The flow stress of the recrystallised copper at room temperature was 263kp/cm^2; corresponding values for the brasses were 855 (90/10), 855 (85/15), 930 (80/20), and 915kp/cm^2 (70/30).

In addition several 90/10 specimens were prepared with mean grain-sizes ranging from 12 to 200 μm. They were used only for studying the effect of grain dimensions on the internal friction.

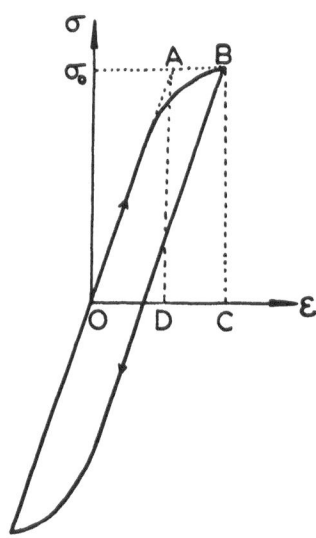

Fig. 1. Schematic representation of a hysteresis loop. The maximum "elastic" energy stored per unit volume is represented by the triangle OAD, the total work of deformation per half-cycle by the area OBC. The area within the whole loop represents the loss per cycle, ΔW.

Deformation in tension/compression was applied in a "hard", motor-driven "Mayes Universal Tensile Machine", model "DM U10". In any given experiment the absolute value of the strain rate was very nearly constant, and the related stress/time relation was, to a good approximation, of symmetrical saw-tooth shape. Within the preplastic region investigated, the hysteresis loops for consecutive cycles were closed, and similar in shape and size, indicating that no significant cumulative changes of structure occurred in the material as a result of repeated stress-cycling. All loops had inversion symmetry, as is also indicated in the schematic diagram in Fig.1.

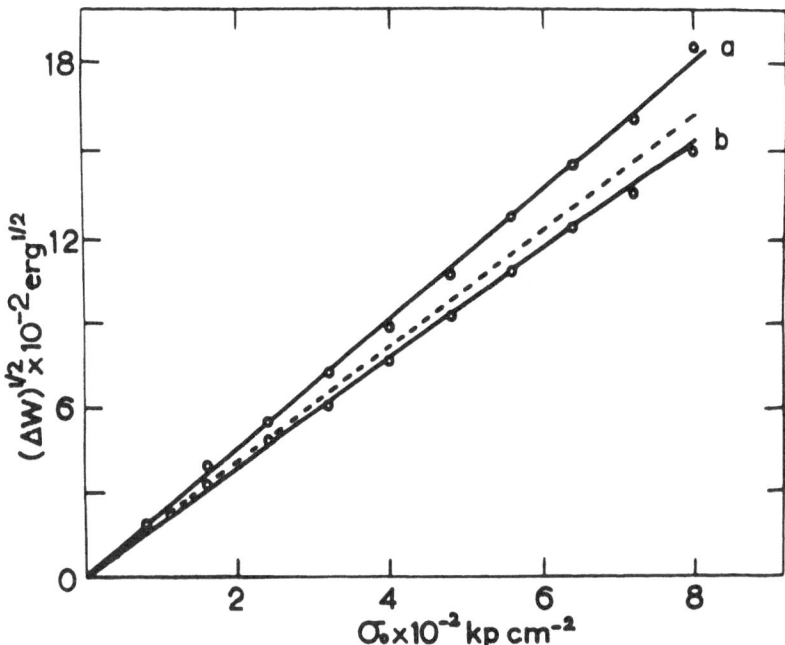

Fig.2. The relation between the energy dissipated per cycle
per unit volume and the stress amplitude. The lines denoted
by "a" and "b" refer to 80/20 and 70/30 brasses respectively.
Results for copper and the 90/10 and 85/15 brasses fall close
to the "dashed" line. T=290K, ω = 0.10 Hz.

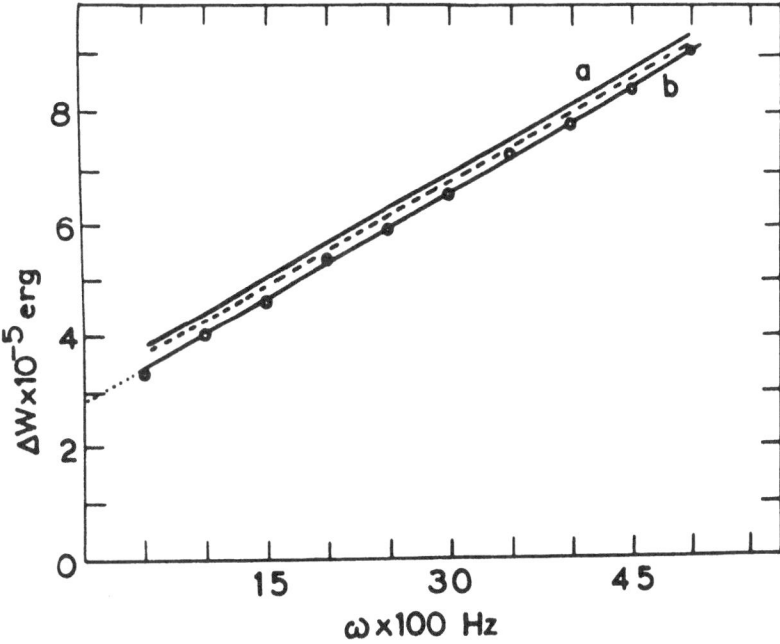

Fig.3. The energy dissipated per cycle per unit volume as
function of the "saw-tooth" frequency. Symbols as in Fig.2.
T=290K, σ_0 = 320 kp/cm^2.

Tests with a hard-steel "blank" showed that contributions to the measured loss per cycle, from the machine or accessories, was negligible under all conditions in the experiments. The highest tensile strain never exceeded about 10^{-3}; extension was measured with a sensitive linear transducer.

The energy, ΔW, dissipated per cycle per unit volume of the material, was derived from the area of the hysteresis loop. A simple cryostat and conventional techniques were used to cool the specimen on "cycling" below room temperature. Results of the measurements made are shown in Fig.2-4.

Discussion and Conclusions

The experimental correlations show that ΔW has the functional form

$$\Delta W = \sigma_0^2 (A + B\omega T)\left[1 + \phi(c)\right],\tag{1}$$

where A and B are constants for a specimen of given grain size L, and represent the hysteric and relaxational loss respectively. Fig. 4, and corresponding results for the other grain sizes referred to, show that A, but not B depends on L.

The function $\phi(c)$ increases uniformly with the zinc concentration, from zero at c=0, to about 0.15 at c=25at% of zinc, and then decreases to about -0.10 at c=30% The influence of alloying on the loss is thus not pronounced; the occurrence of a maximum in $\phi(c)$ close to a concentration corresponding to Cu_3Zn suggests that ordering, known to occur in α -brasses (4), somewhat enhances the effective friction acting on moving dislocations.

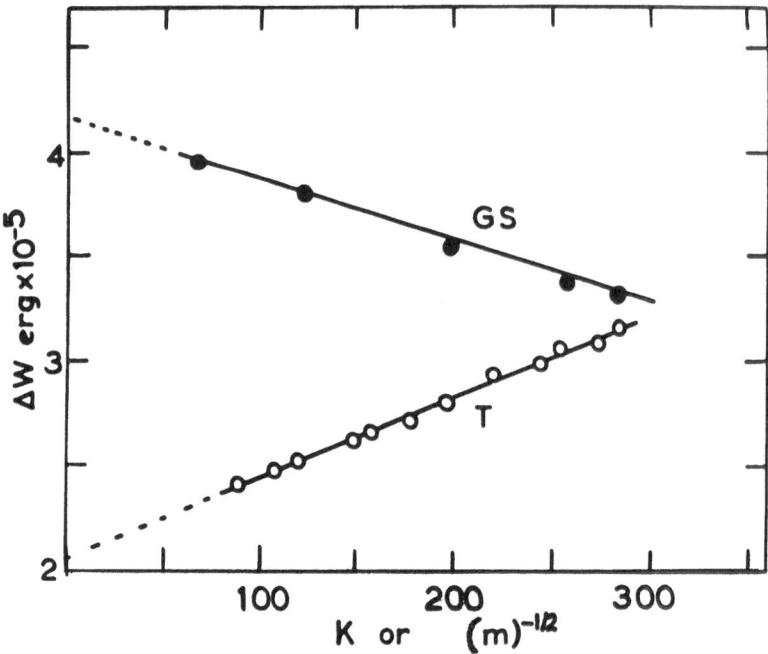

Fig.4. Dependence of ΔW on temperature and grain size. Upper curve: 90/10 brass, T=290K, $\sigma_0 = 256kp/cm^2$ and ω =0.146Hz. Lower curve: 85/15 brass, $\sigma_0 = 320kp/cm^2$ and $\omega = 0.025Hz$.

The effect of grain size on the hysteric component of the loss ΔW deduced from Fig.4 and similar data for other grain sizes, which fall on lines parallel to the one shown, can be represented in eq.(1) by writing

$$A = A_0 \left(1 - \alpha L^{-\frac{1}{2}}\right) \approx A_0 \Big/ \left(1 + \alpha L^{-\frac{1}{2}}\right), \tag{2}$$

where A_0 and α are constants independent of L. The approximate formulation of eq.(2), which is also given, and which will be referred to below, is readily justified on the basis of the weak dependence of ΔW on L.

A dimensionless measure of the internal friction may be obtained either by using the ratio $\Delta W/W_e$, where $W_e = \sigma_0^2/E$ and E is Young's modulus, with W_e denoting twice the "elastic" energy represented by the area OAD or, more conventionally, in terms of the ratio $\Delta W/W$, where $\frac{1}{2}W$, corresponding to the area OBC in Fig.1, is the total energy expended per unit volume in deforming the material on loading up to σ_0, i.e. along the path OB. Within the limits of the attainable accuracy W, like W_e, is also proportional to σ_0^2, so that both of the dimensionless measures of the loss per cycle are independent of the stress amplitude. Values of $\Delta W/W$ obtained at room temperature, at a frequency of 0.10Hz, with σ_0 = 320 kp/cm^2 were, with zinc contents in at% given in brackets, 0.185(0), 0.184(10), 0.188(15), 0.214(20) and 0.187(30). At a frequency of 0.5Hz the corresponding values are very nearly twice as high; this is in accord with the trend apparent in Fig.3.

The preservation of the loop shape and size, on repeated cycling to a given amplitude below the flow stress, suggested a process of dislocation movement in which the grain structure remained invariant. Further, the presence of a hysteric, frequency-independent contribution to the loss per cycle, could then be explained if the dissipative mechanism consisted, at least in part, in the stress-assisted penetration of dislocations through the internal stress-fields. Thus a moving dislocation or segment, may become trapped at a barrier, e.g. by entering into a dipole configuration with another dislocation. It could be released, and new segments could be induced to move, on increasing the stress. Such a drift between extended barriers would lead to a hysteric loss on "cycling".

Already a considerable time ago Weertman and Salkovitz (5) showed that internal friction due to the migration of dislocations through a "statistical" distribution of barriers would yield an internal friction given by

$$\frac{\Delta W}{W} \approx N \lambda b \, G/\sigma_y , \tag{3}$$

where N is the total density of dislocations in the crystal, λ the mean distance a dislocation, or segment, would jump on being activated, b is the Burgers vector, G the shear modulus, and $\sigma_y > \sigma_0$ is the flow stress of the crystal. As in the present results, the amplitude σ_0 does not appear explicitly in eq.(3).

The model, being of the "exhaustion" type, is open to certain criticisms (3,6); nevertheless eq.(3) has the merit of showing the hysteric, frequency-independent mode of internal friction implied by it. Further, as the well-known Hall-Petch relation of the grain-size dependence of the flow stress appears to be applicable to polycrystalline brasses in the microplastic region (7), the grain-size dependence of the internal friction implied by eq.(2) is readily expressed in terms of that of the flow stress σ_y appearing in eq.(3).

A somewhat more general interpretation of the origin of the frequency-independent part of the loss can be based on the following consideration. Models of the type considered by Weertman and Salkovitz imply a logarithmic form of low-temperature creep (1). This is also the case with similar, "stochastic" models not relying on the exhaustion hypothesis (3,6). Now Ross Macdonald has shown (8) that if the creep strain at low stresses is written

$$\varepsilon \propto \sigma_0 \, ln \, (1 + \frac{t}{t_0}) , \qquad (4)$$

where t_0 is a characteristic "retardation time" then the corresponding internal friction would be essentially frequency independent at "low" frequencies, i.e. for which ωt_0 is less than about 1. A typical value of t_0 for metals at room temperature is about one second (8), so that with the range of frequencies used in the present work the criterion for the occurrence of hysteretic damping is likely to be satisfied.

An interesting inference from such an interpretation, also considered from a different standpoint by Asano (9) in relation to hysteresis loops of the type shown in Fig.1, is the need to associate the functional form of the loss not simply with the character of the frictional force opposing the movement of an individual dislocation, but to see it also as a reflection of the stochastic and cooperative features of the kinetics of the process.

Concerning the relaxation term in eq.(1), we suggest the following interpretation. The grain-size dependence of ΔW shows that grain-boundaries may be effective obstacles to the drift of dislocations, so that a certain "piling-up" or "sedimentation" (10) may occur at them; this would induce local relaxation as the external stress is reversed. An additional source of relaxational loss would arise from the "vibrating-string" behaviour of dislocation segments temporarily or permanently pinned at localised barriers.

By an appropriate modification of the "vibrating-string model"(11) the form of the term in eq.(1) depending linearly on frequency can be explained. However, instead of associating the friction with the "phonon viscosity" - which can make but a negligible contribution to ΔW at the low frequencies considered - the friction has again to be sought in the heterogeneous internal stress-field which the loops, expanding over relatively large areas, must penetrate. This process, and the "hysteretic" one referred to above, would therefore have, essentially, the same origin: the resistance to glide due to the dislocation-trapping action of the intragranular stress-fields, including extended and localised barriers.

We believe that temperature appears in eq.(1) because an increasing number of barriers to dislocation motion become "transparent" on raising the temperature, resulting in a parallel increase in the modulus defect and, hence, in the internal friction. A consistent interpretation of the observations is thus obtained.

The structure sensitivity of $\triangle$W, implied by the assumed, dominant role of the internal stress fields, as well as its dependence on grain size, should render further work along the present lines most rewarding in relation to current efforts to obtain a deeper insight into the role of the microstructure in the kinetics of dislocation ensembles.

References

1. N.F. Mott and F.R.N. Nabarro, Report of a Conference on Strength of Solids, p.1, The Physical Society, London (1948).

2. V.L. Indenbom and A.N. Orlov, Proc.Second Internat.Conf. on the Strength of Metals and Alloys, Asilomar, 2, 385 (1970).

3. P. Feltham, Reviews on the Deformation Behaviour of Materials, 1, - (1973). In the press.

4. C.J. Spears, This Conference.

5. J. Wertman abd E.L. Salkovitz, Acta Met. 3, 1 (1955).

6. P. Feltham, J.Phys. (London) D, (1973). In the press.

7. W.L. Phillips and R.W. Armstrong, Metals Trans. 3, 2571 (1972).

8. J. Ross Macdonald, J. Appl.Phys. 32, 2385 (1961).

9. S. Asano, J. Phys.Soc. Japan 29, 952 (1970).

10. A.I. Landau, Phys.Stat.Sol.(a) 15, 343 (1973).

11. A. Granato and K. Lücke, J.Appl.Phys. 27, 583 (1956).

INTERNAL FRICTION AND YOUNG'S MODULUS STUDIES IN COPPER
AFTER HIGH FREQUENCY FATIGUE (20 kHz)

H. Müllner, P. Bajons, H. Kousek and B. Weiss
II. Physikalisches Institut der Universität Wien
(University of Vienna, Austria)

Many observations indicate, that dislocation multiplication and point defect production play an important role in metal fatigue. Much has been reported about changes in dislocation structure after cyclic deformation, the reader is referred to the review article (1). Recently resistivity measurements were also performed, which yield additional information about point defect arrangements (2,3). However at present only a small amount of data is available about interaction processes between dislocations and point defect during metal fatigue (4,5,6). This may partly be attributed to the fact that low frequency fatigue experiments require considerable time to accumulate for a large number of test cycles. Therefore the present paper deals with internal friction and Young's modulus measurements after high frequency (ultrasonic) fatigue. This high frequency method (7) permits to obtain results after a high number of cycles in a reasonable time.

Experimental Details

Annealed polycrystalline copper samples with the purities of 99,98% and 99,999% were employed for this investigation. After annealing in vacuum (1.10^{-5} torr) for 3 hours at a temperature of 650°C the resulting average grain size was about 100μm. The samples were used in form of bars with dimensions of 100mm length and a cross section of 3×3 mm^2.

The high frequency equipment consisted of an ultrasonic frequency generator (Branson Sonic Power Co., Danbury, Conn.) which served as a power supply for a ceramic transducer. The transducer, amplifier and sample were parts of a resonant system, which operated at a frequency of 20 kHz. The length of the sample was chosen to allow the build up of standing longitudinal waves, which caused tension – compression fatigue about mean zero strain. The strain distribution along the length of the sample approximated a sinusoidal curve. The total strain amplitude ε_o (where ε_o is the sum of the elastic and plastic parts of the strain amplitude) was computed from the measured displacement amplitude. In order to prevent heating caused by the high frequency, cyclic deformation was performed in pulsed operation (8) and in a non corrosive coolant (water, freon). After fatigue exposure at room temperature the samples were subjected to a temperature treatment of $60^\circ C$ for 20 minutes in order to obtain reproducible conditions in regard to point defect diffusion.

For the measurement of the internal friction and Young's modulus a resonant bar method was used with a very sensitive capacitive excitation and detection system (9,10). The samples were excited in their fundamental mode of transverse vibration at a resonant frequency of about 1 kHz. The modulus was measured in the self exciting mode and the internal friction was determined by the decrement of the free decay of the vibration amplitude. The ability to vary the vibration amplitude (characterized by the maximum strain amplitude at the surface of the sample) down to less than 1.10^{-8} permitted measurements in the amplitude dependent as well as independent range. The error in measurement of the relative change of the resonant frequency was about $0,1\%$ (therefore that of the relative change of the modulus about $0,2\%$) and that of the internal friction (Q^{-1}) values was less than 1%.

Experimental Results

The dependence of the internal friction of fatigued samples on the amount of total strain amplitude ε_o for a constant number of cycles N is illustrated in fig. 1 for the 99,98% copper. Similar results were obtained for the 99,999% copper. Whereas the undeformed state is characterized by an amplitude dependence in some cases down to a vibration amplitude of less than 1.10^{-8}, the internal friction of fatigued samples shows a different behaviour depending on the amount of ε_o. In the case of very small deformation amplitudes (eg. $\varepsilon_o = 2,5.10^{-5}$) the independent part of the internal friction (Q_I^{-1}) is increased, whereas the dependence on vibration amplitude remains similar to that of the annealed sample. At higher deformation amplitudes (eg. $\varepsilon_o = 3,5.10^{-4}$ and $\varepsilon_o = 7,4.10^{-4}$) the same numbers of cycles ($N = 5.10^5$) caused a decrease in both the independent and dependent part of Q^{-1}. For relative high values of ε_o (eg. $\varepsilon_o = 7,4.10^{-4}$) the decrease of Q_I^{-1} was less pronounced. However with increasing N a dependence on vibration amplitude was again observed. This is shown in fig.2 for a sample fatigued with total strain amplitude $\varepsilon_o = 7,4.10^{-4}$ after different numbers of cycles.

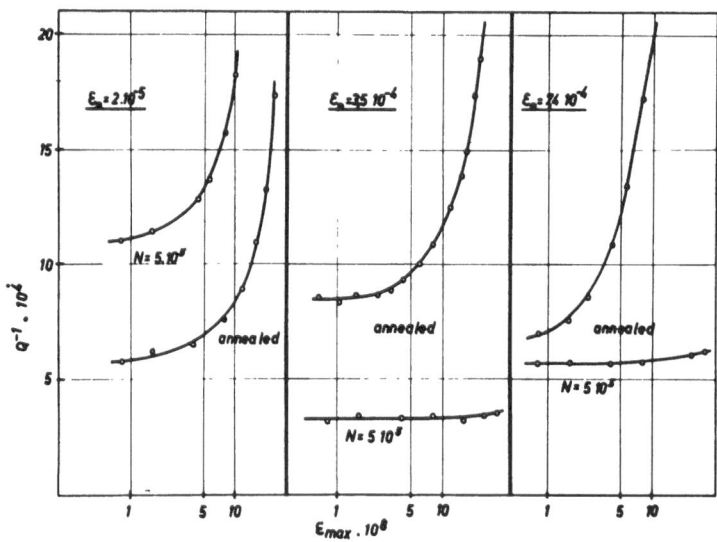

FIG. 1

Plot of Q^{-1} versus $\mathcal{E}_{max}$ (vibration amplitude) after high frequency
fatigue with different total strain amplitudes $\mathcal{E}_0$; number of cyclesN

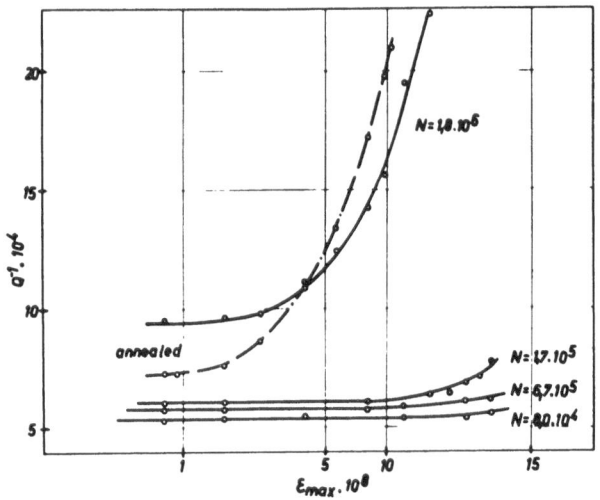

FIG. 2

Plot of Q^{-1} versus $\mathcal{E}_{max}$ after high frequency fatigue with total strain
amplitude $\mathcal{E}_0 = 7,4.10^{-4}$ at different number of cycles.

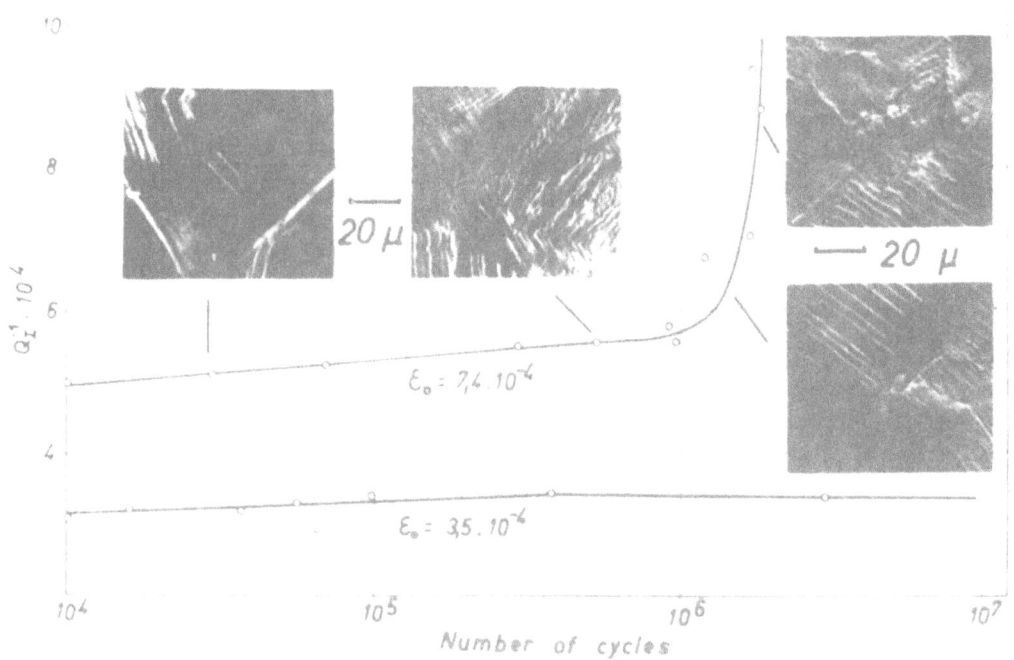

FIG. 3
Amplitude independent internal friction (Q_I^{-1}) versus number of cycles,
ε_o... total strain amplitude; SEM micrographs, see text.

In fig. 3 a plot of Q_I^{-1} versus N for two different ε_o is shown. In the case of
$\varepsilon_o = 3,5.10^{-4}$ there was only a slight increase with increasing N, while for $\varepsilon_o = 7,4.10^{-4}$
a sharp rise in the Q_I^{-1} values was observed. To obtain information on the surface structure,
samples fatigued with the higher amplitude were studied in the scanning electron microscope.
The micrograph obtained after 3.10^5 cycles reveals several fatigue slip bands. With increasing N the density increased considerably, whereas the internal friction values showed only
a slight increase. The sharp rise in the Q_I^{-1} values could be attributed to the occurence
of microcracks. Only a few numbers of cycles (about 10^4) were sufficient for further growth
of the crack. Measurements of the Young's modulus of fatigued samples indicated a similar
effect as the internal friction studies. The relative decrease of the modulus ($-\Delta E/E$) as
a function of N at constant ε_o is illustrated in fig. 4. The calculation of the ($-\Delta E/E$)
values was based on the modulus of the annealed samples. At the lower strain amplitudes

the relative decrease of the modulus saturated as a function of N, while the sharp rise
in the ($-\Delta E/E$) values for higher ε_o could also be explained by the occurence of microcracks.

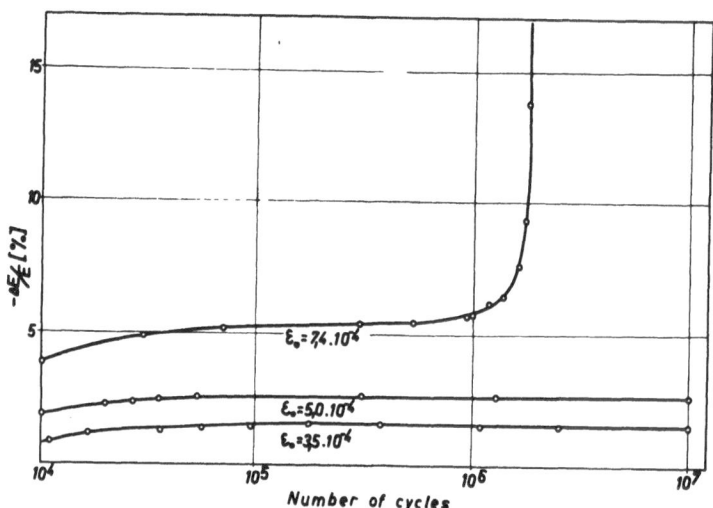

FIG. 4

Relative decrease of Young's modulus of Cu versus number of cycles
at different total strain amplitudes ε_o

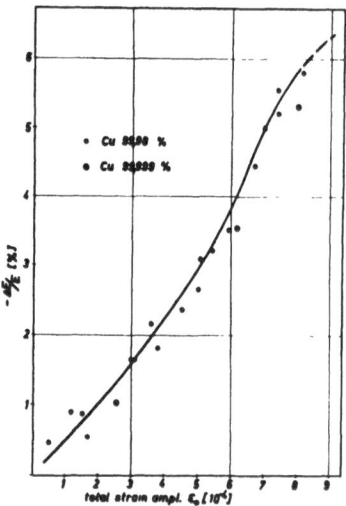

FIG. 5

Relative decrease of Young's modulus of Cu versus total strain amplitude ε_o
at the constant number of cycles $N = 4.10^5$

The decrease of the modulus as a function of ε_0 is illustrated in fig. 5. The ($-\Delta E/E$) values were taken from the region of saturation (fig. 4). As a function of ε_0 the ($-\Delta E/E$) values increased and seemed to saturate at higher total strain amplitudes. In addition to the values measured an 99,98% copper samples, values of the 99,999% copper are also added in fig. 5. The relative large scatter of the values is due to the fact, that each ($-\Delta E/E$) value was obtained from a different specimen. In the case of the pure material the observed deviation of approximatly 0,5% could be attributed to the larger modulus defect of the undeformed samples of this material. However for deformation studies of this type, the purity of the material seemed to be of minor importance.

Discussion

In order to analyse the experimental results reported in this paper, it seems to be necessary to differentiate between two areas, characterized by certain ε_0 and N values: one region in which no visible crack formation occurs and the other, where cracks form.

To obtain information about changes in the dislocation density and in the average dislocation line length, the vibrating string model which was developped by Koehler (11) and Granato and Lücke (12), has been successfully employed in several deformation studies (5,6,13). According to this theory (GL theory) a dislocation network which is characterized by a certain density Λ and average line length l, is responsible for a decrease in modulus and for an amplitude independent part of the internal friction:

$$\Delta M = (E_i - E)/ E = k_1 \Lambda l^2 \quad \text{and} \quad Q_I^{-1} = k_2 \Lambda l^4$$

where ΔM is modulus defect, E_i, E is ideal, measured modulus respectively, and k_1, k_2 are constants.

The dependence of Q_I^{-1} on the frequency of measurement ω, as predicted by the theory could be neglected in this work because ω changed so slightly that it was included into the constant k_2. From simultaneous measurements of ΔM and Q_I^{-1}, changes in the dislocation density and average free line length during fatigue can be obtained. One difficulty in applying this theory is, that even in fully annealed samples, the dislocation network causes a modulus defect depending on the purity of the material under consideration. Recovery experiments (14) revealed that in the case of the 99,98% pure copper ΔM of the undeformed samples was negligible within the mentioned experimental error of 0,2%. For this material the ($-\Delta E/E$) values of the fig. 4 could be used without correction. The GL-theory can only be applied for that part of ε_0 and N values where no crack formation can be observed. Restricted to that part, the following expressions $(\Delta M)^2/Q_I^{-1}$ and $(\Delta M/Q_I^{-1})^{1/2}$ were plotted versus N in fig. 6. As predicted by the GL- theory, the first expression can be assumed to be proportional to Λ, the second one to be proportional to the average pinning point density (proportional to 1/l).

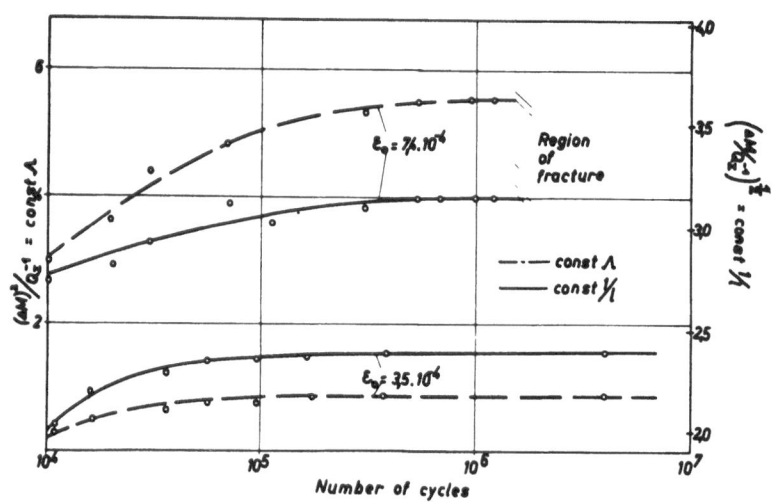

FIG. 6

Plot of dislocation density (const. Λ) and average pinning point density (const. 1/l) versus number of cycles; ε_0...total strain amplitude

It is seen from fig. 6 that Λ as well as 1/l saturate with increasing N and are dependent on ε_0. This dependence of Λ and 1/l on ε_0 and N is qualitatively in good agreement with results obtained from resistivity measurements (3) and transmission electron microscopy studies (7). Nevertheless the applicability of the GL-theory for the relative complex dislocation structure after fatigue at this high amplitudes seems to be surprising. It is known, that this dislocation structure consists of bundles of fragmented dislocation dipoles and loops separated by areas relatively free of dislocations (15). Recent electron microscopy studies indicate that the regions between the bundles contain a certain number of long dislocations (15,16). Because of the strong dependence of internal friction and modulus defect on the average free line length it is assumed, that the applicability of GL-theory is mainly restricted to these long dislocations.

For that part of N and $\mathcal{E}_o$ where cracks are to be expected, a relation between crack initiation, modulus defect, and internal friction was observed. A similar change of the modulus defect due to microcracks after low frequency fatigue at the temperature of $78^{\circ}K$ was reported by den Buurman and Snoep (5). The absence of such an effect in their internal friction results was attributed to the independence of Q^{-1} on dimensional changes. The observed rise in the $(-\Delta E/E)$ values they associated with dimensional changes only. In contradiction to their results the fig. 3 and fig. 4 presented in this paper showed a sharp rise of both quantities. For this reason it is assumed that regardless of dimensional effects which are obviously involved, the internal defect structure which changed considerably during fatigue plays an important role.

Furthermore, it was observed that in that part of $\mathcal{E}_o$ and N where fracture can be expected, the internal friction again becomes amplitude dependent. It may be speculated that this experimental fact could be a more sensitive indicator for crack initiation at a very early stage than the change of resonant frequency. For further clarification simultaneous measurements of modulus, internal friction, and direct observation of surface fatigue damage in the scanning electron microscope are planned.

References

1.   J. C. Grosskreutz, Phys. Stat. Sol., 47, 11 (1971)

2.   j. Polak, Czech. J. Phys., B 19, 315 (1969)

3.   W. Kromp and B. Weiss, Scripta Met., 5, 499 (1971)

4.   P. Bajons and B. Weiss, Scripta Met., 5, 511 (1971)

5.   R. den Buurman and A. P. Snoep, Acta Met., 20, 407 (1972)

6.   R. den Buurman, Scripta Met., 6, 975 (1972)

7.   B. Weiss, Aluminium, 48, 741 (1972)

8.   W. Kromp, K. Kromp, H. Bitt, H. Langer and B. Weiss, Proc. of Ultrasonics Int.,London (1973)

9.   R. Kaniak and H. Müllner, Zeitschr. f. Metallk., 66, 724 (1972)

10.. H. Kousek and H. Müllner, to be published

11.   J. S. Koehler, Imperfections in Nearly Perfect Crystals, 197 Wiley, New York (1952)

12.   A. Granato and K. Lücke, J. Appl. Phys. 27, 583, 789, (1956)

13.   K. Boissner, E. Biller and D. Lübbers, Scripta Met. 7, 529, 535 (1973)

14.   H. Müllner and P. Bajons, to be published

15.   J. R. Hancock and J.C. Grosskreutz, Acta Met., 17, 77 (1969)

16.   H. Mughrabi, Vortrag Frühjahrstagung d. DPG u. DGM, Münster (1973)

ULTRASONIC INTERACTION WITH SINGLE DISLOCATIONS AS
OBSERVED WITH THE ELECTRON MICROSCOPE

I. Hansson and A. Thölén
Laboratory of Applied Physics I
Technical University of Denmark
Building 307 - 2800 Lyngby, Denmark

ABSTRACT - Stainless steel and aluminium specimens have been subjected to ultrasound and rearrangement of single dislocations or dislocation elements have
been observed in an electron microscope before and after the ultrasonic treatment.

Introduction

As an ultrasonic wave passes through a perfect crystal, energy is
absorbed due to the interaction between the wave and the lattice. Attenuation
of the wave is caused by the exchange of energy between the wave with phonons
and electrons. In a crystal which contains defects (for example dislocations,
grain boundaries, point defects), the number of possible attenuation mechanisms
increases.

Ultrasonic waves interact very strongly with dislocations, the actual
process depending on a number of parameters which include strain amplitude,
frequency, temperature, the material and its prehistory. For small strain
amplitudes (10^{-8} - 10^{-6}) in fcc materials at subzero temperatures, the dominating process is the passage of the Peierl's barriers in going from one
equilibrium position to another [1,2]. Larger strain amplitudes cause the
dislocations to swing between pinning points and, with increasing amplitude,
cause them to break away from these. The Granato-Lücke model (for a review
see |3|) has been extensively used in an effort to describe the attenuation
process.

Under the influence of yet higher strain amplitudes ($>10^{-4}$), the material
starts to deform plastically with the occurrence of large scale dislocation
rearrangement and multiplication. Electron microscopy has been used to compare
the overall structure before and after the ultrasonic treatment and Langenecker et al have used this technique [4,5,6]. His test specimens were subjected
to the combined action of <u>uniaxial</u> stress and superposed ultrasonic strain of
varying amplitude. It was found that increasing the strain amplitude effectively

lowers the critical shear stress for continued plastic deformation much in the same way as an increase in temperature would do. The one important difference between the effect of increasing the temperature and superimposing ultrasonic waves is that the attenuation of ultrasonic waves caused by lattice defects is a much more selective process. After exposure to macrosound in aluminium a cell structure is developed. The acoustic stress level alone can hardly account for the high stresses needed for the breakaway of dislocations from pinning points. However, Langenecker suggests that a local rise in temperature also occurs which would ease dislocation motion. Strain amplitudes less than 10^{-4} give rise to reversible changes in the dislocation structure but higher strain amplitudes yield visible changes in the whole dislocation pattern.

Wood et al [7-10] have investigated fatigue and fatigue cracking in fcc, bcc and hcp materials by microsonic methods and found major differences between low and high frequency (17 kHz) fatigue. The critical factor seems to be the limited time for diffusion processes at the higher frequency.

Some Russian workers [11] have compared the dislocation structure in nickel fatigued at low and high frequency (20 kHz). The internal dislocation structure as judged by electron microscopy is however the same in the two cases. The structure depending only on the number of loading cycled and the temperature.

A large number of results have been obtained on the interaction between ultrasonic waves and dislocations out of which only a minor fraction have been mentioned here. Common to these investigations is that either the information about the dislocation behaviour has been obtained in an indirect way or one has observed the result of large scale deformations in the electron microscope. The missing link seems to be the actual observation of single dislocation during exposure to ultrasound. This might be done by equipping an electron microscope with an ultrasonic stage and observing dislocation motion as a function of frequency etc. Such a stage is presently under construction. As a preliminary investigation changes in the arrangement of individual dislocations that occurred in metal specimens which were treated by ultrasound outside the electron microscope were recorded.

Experimental

Two materials were chosen for this investigation - a stainless steel and a commercial aluminium. The steel was an austenitic 18/8 steel obtained in sheet form with a grain size of 0.01 mm and it was well annealed and contained relatively few dislocations. The aluminium was a commercial grade (SM 5050-14) and was taken from specimens which had been subjected to creep deformation at 146°C $(0.45 \cdot T_m)$ With a stress of 54 N/mm² until fracture. The fracture occurred after a strain of 10% in roughly six hours. These specimens were used primarily for investigations of grain boundary behaviour during creep. They were used here because they represent a material with quite different mechanical properties than the steel and they were also known to contain a specific dislocation structure.

The specimens used were of disc type with a diameter of 3 mm and a thickness of 0.15-0.20 mm. They were then electropolished using a jet method (TENUPOL) until perforation occurred and subsequently observed with an accelerating voltage of 100 kV in a Philips 300 electron microscope.

The general procedure adopted was to photograph areas of interest both before and after the ultrasonic treatment. Some difficulties were encountered as it proved quite difficult to be sure one had the exact same area and to take both sets of pictures under the same diffraction conditions. Many pictures were thus taken from neighbouring areas and with slightly different diffraction conditions and afterwards it was usually found possible to find matching areas with the same diffraction conditions.

The exposure to ultrasound was made in a special specimen holder (Fig. 1) made of aluminium and consisting of two main parts (Fig. 1:2, 3). In the lower part (Fig. 1:3) there is a cavity (Fig. 1:5) with the same dimensions as the specimen and in this cavity the specimen sits. The parts are then screwed together to ensure good acoustic contact between the holder and the specimen. The screws (Fig. 1:1) goes through the aluminium parts and are screwed into a steel plate (Fig. 1:4) to be able to press the two parts very firmly together. This contact between the specimen and the holder is further improved by first dipping the specimen into oil before placing it into the cavity. A number of oils were tried in the connection with ultrasound velocity and attenuation measurements with Apiezon C, an oil for diffusion pumps being found the most suitable.

The ultrasonic treatment was performed in a cleaner of the water tank type. The cleaner has an effective average power of 250 W and the tank volume is about 10 dm^3. The three driving crystals are made of lead zirconate titanate ceramic and have a reasonance frequency of 45 kHz. The specimen holder was lowered into a plastic beaker (1000 ml) with about 2 mm wallthickness and filled with water 1 cm above the surface of the holder, the beaker itself then standing in the water tank for a couple of minutes. The reason for using the beaker was to decrease the ultrasonic wave amplitude (and the resulting strain amplitude) because if an electropolished folie sample is simply placed in the tank itself, the thinned part will be destroyed due to the formation of cavities in the cleaning liquid.

We have tried to estimate the strain amplitude which should depend on input intensity, attenuation in the beaker and the specimen holder, reflection at various interfaces and the resonance frequency of the holder itself (50 kHz), a value of the strain amplitude lying between 10^{-4}-10^{-3} was obtained. We are now trying to find some way of directly measuring the strain amplitude in the specimen.

After the treatment the specimens were cleaned in xylene and methanol to remove traces of the contact oil and then dried by putting them on a sheet of blotting-paper before again putting them into the specimen holder of the electronmicroscope. Great care was exerted in the handling of the specimens in order to avoid unnecessary damage. The damage due to the handling of the specimens was checked by observing the same area in the microscope before and after a dummy treatment in the ultrasonic specimen holder and was found to be negligible. The single tilt rather than the double tilt and the rotation holders of the Philips electron microscope was used since the last two have a much more complicated clamping system. The use of the double tilt holder would also have made it even more difficult to reproduce the original diffraction conditions.

Results

Figs 2b and 2c show an area of a stainless steel specimen before and after the ultrasonic treatment. Although the pictures were not taken under exactly the same diffraction conditions, it is obvious that some major changes have occurred. New dislocations have been nucleated and redistribution of the ones which were present from the beginning has occurred. As a comparison, Fig. 2a containing the same area just before a test with an ultrasonic flaw tester (it is difficult to calculate the strain amplitude, but it is estimated to be less than 10^{-6}). The end result is shown in Fig. 2b and very little change has thus occurred between 2a and 2b. This also gives confidence in the other results as it is seen that careful handling of the specimens causes negligible structure changes.

Figs 3a and 3b show another area from the same specimen and the rearrangement of dislocations in the active slip plane is quite remarkable. The appearance of new slip traces clearly indicates that motion of dislocations has occurred during the process and that the grain boundary has acted as a source for new dislocations. In this low stacking fault energy alloy, whole dislocations split up into partials giving rise to stacking faults which can be seen in the figure.

Figs 4a and 4b show a twin boundary in the same steel specimen before and after the treatment. The twin boundary shows line contrast which stems from one or more lattice dislocations encountering the boundary along slip planes and finally getting stuck. The number of lines in Fig. 4b is larger than in Fig. 4a showing that dislocation motion has taken place. In this connection, it is of interest to consider the influence of ultrasound on grain boundaries.

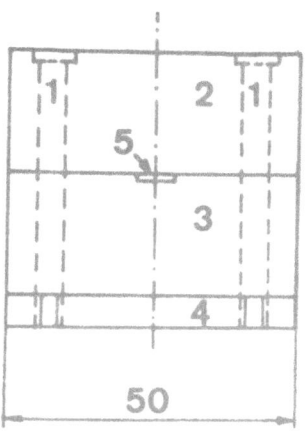

FIG. 1

Specimen holder

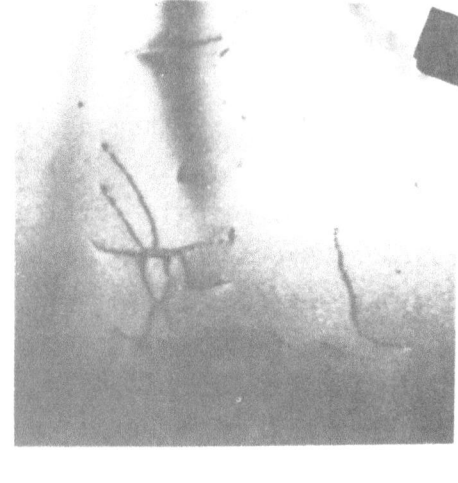

FIG. 2a

Stainless steel before the treatment

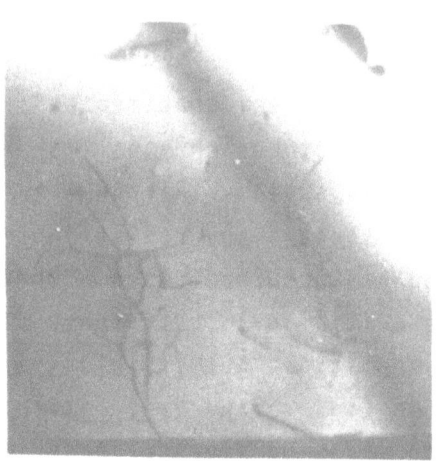

FIG. 2b

Treated 1 min. with an ultrasonic flaw tester

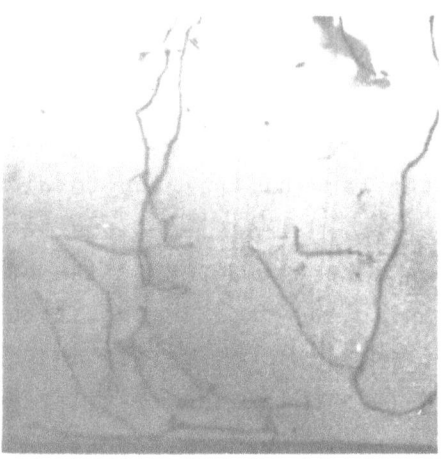

FIG. 2c

Treated 3 min. in the ultrasonic cleaner

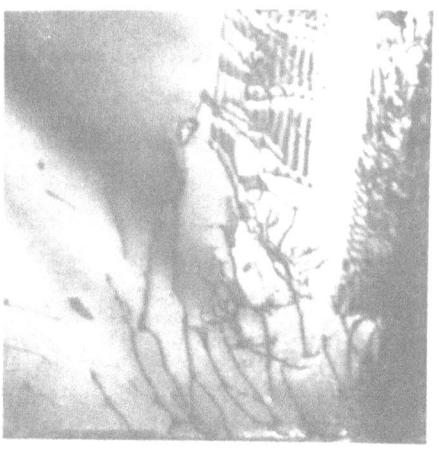

FIG. 3a

Slipbands in a stainless steel specimen. Before the treatment

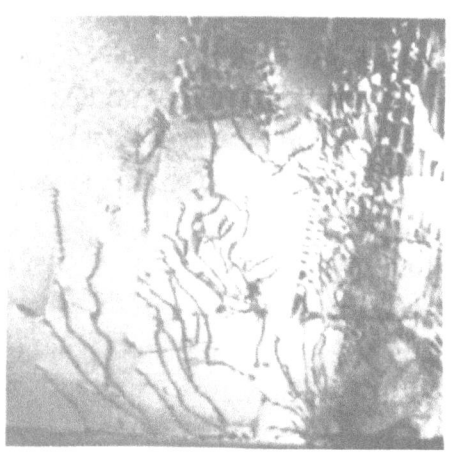

FIG. 3b

Treated 3 min. in the ultrasonic cleaner

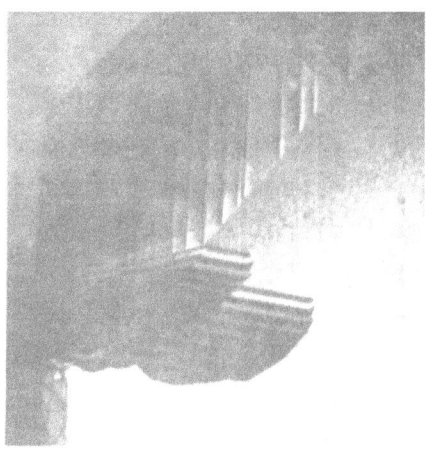

FIG. 4a

Twin boundary in stainless
steel. Before treatment

FIG. 4b

Treated 3 min. in the
ultrasonic cleaner

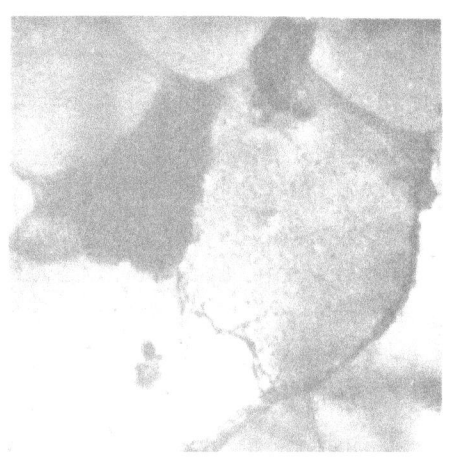

FIG. 5a

Creep-tested Al. Before
treatment

FIG. 5b

Treated 3 min. in the
ultrasonic cleaner

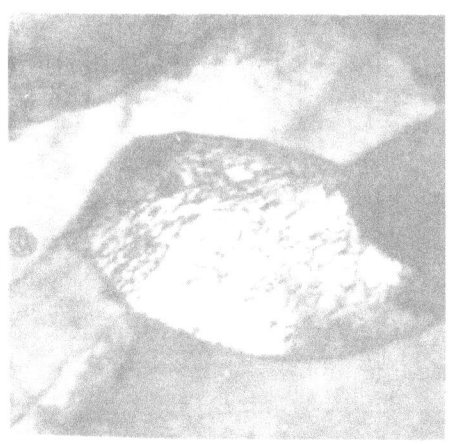
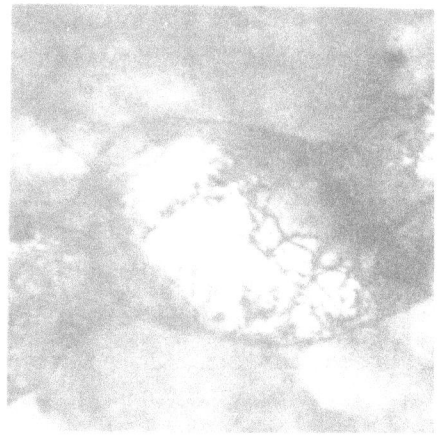

FIG. 6a

Creep-tested Al. Before
treatment

FIG. 6b

Treated 3 min. in the
ultrasonic cleaner

A polycrystalline specimen causes attenuation of an ultrasonic wave in a number of ways: thermoelastic processes, scattering at the boundaries, diffraction effects and effects due to viscous grain boundary behaviour. The most interesting for the present purposes is the direct interaction with the grain boundary. Grain boundaries assume different characters depending on the angle between the two grains. In the low angle limit, they can be discribed as arrays of dislocations. For larger misorientations, this is still valid in a mathematical sense although a more atomistic description would be appropriate here. It is clear that a boundary can cause attenuation of an ultrasonic wave by letting its "dislocations" move either sideways (migration) or along the bounddary, (sliding). The important factors governing these processes are the applied strain amplitude, the character of the boundary, temperature, and the diffusivity of vacancies.

One prerequisite for easy grain boundary sliding is that the plane of the boundary not be a symmetry plane. The contrary is the case in a coherent twin boundary where "sliding" only can occur through the movement of lattice dislocations along the common (111)-plane, a process which, as seen in Fig. 4, does not occur.

A very interesting feature is seen at the edges of the stacking faults in Fig. 4. These are bounded by smoothly curved partial dislocations before the treatment. During the application of ultrasound, these partial dislocations seem to be pinned in one or two places and the segments between then possibly swing forwards and backwards thus absorbing energy and giving rise to a highly irregular structure.

In the aluminium, on the other hand, the situation is quite different. In Figs 5a and 5b the same grains before and after the testing are shown and the changes which have occurred are quite apparent. The number of dislocations has decreased drastically which is due to the fact that the critical shear stress is exceeded.

Figs 6a and 6b show that the dislocations that were originally present tend to straighten out and lie along common crystallographic direction. This was often observed.

Conclusion

It is seen that quite large changes in different materials occur under the influece of ultrasound. In the stainless steel dislocation multiplication took place and single dislocation or parts of them have moved.

In the creep-tested aluminium specimen, however, a general decrease in the number of dislocations was observed and a straightening out of existing dislocations took place. The remaining dislocations tend to move so as to form a low angle boundary. This tendency to alter the dislocation density and arrangement should be borne in mind when considering the attenuation of ultrasound as a function of dislocation density. These preliminary results indicate that there is much to be gained by introducing a stage containing an ultrasonic transmitter into the electron microscope and observing the in situ movement of the dislocations.

References

1. Physical Acoustics Vol III part A, ed W.P. Mason, Academic Press, 3, 77 (1966)

2. Physical Acoustics Vol III Part A, ed W.P. Mason, Academic Press, 3, 361 (1966)

3. Physical Acoustics Vol IVa Part A, ed W.P. Mason Academic Press, 4, 225 (1966)

4. K.H. Westmacott and B. Langenecker, Phys. Rev. 14, 221 (1965)

5. B. Langenecker, IEEE Transactions on Sonics and Ultrasonics SU-13, 1 (1966)

6. B. Langenecker, Phys. Rev. 145, 487 (1966)

7. W.P. Mason and W.A. Wood, J. Appl. Phys. 39, 5581 (1968)

8. W.P. Mason and W.A. Wood, J. Appl. Phys. 40, 4514 (1969)

9. D.E. Mac Donald and W.A. Wood, J. Appl. Phys. 42, 5531 (1971)

10. C.M. Gilmore, D.E. Mac Donald and W.A. Wood, Acta Met. 20, 953 (1972)

11. I.A. Gindin, I.M. Neklyudov, M.P. Starolat, G.N. Malik and O.I. Volchok, Soviet Physics - Solid State 12, 1964 (1971)

Relation between Hasiguti Peaks and Pinning Stages

in Cold-Worked and Electron-Irradiated Copper

G. Sokolowski and K. Lücke

Institut für Allgemeine Metallkunde und Metallphysik

der Technischen Hochschule Aachen, W.-Germany

In the low frequency internal friction studies of the annealing of cold-worked metals mainly two groups of experiments have been reported: In one group the authors have concentrated on modulus measurements and discussed the results in terms of dislocation pinning; in the other group the authors have concentrated on damping measurements in order to study the Hasiguti peaks. It is the purpose of the present experiments to measure in the 1 Hz-region both quantities simultaneously and to find out how they are related.

For this purpose, wire shaped samples of 99.999% ASARCO-copper have been in situ tensile-deformed at 78 K and then warmed up to 360 K. Modulus and damping have been measured at about 1 Hz by a torsion pendulum apparatus which is described in /1/. Additionally to the cold-work the effect of electron-irradiation at 78 K has been investigated.

Fig. 1 gives an example of the experimental results after cold-work. In the upper part (solid line) the modulus defect during warm up is plotted vs. temperature. One recognizes three regions of this curve, at about 135, 190 and 300 K, where a decrease of the modulus defect occurs and which have to be interpreted as pinning stages. At about 170 and 220 K, however, an increase of modulus defect is observed. The damping curve in the lower part of the figure gives damping peaks just at these temperatures: the Hasiguti peaks P_2 at 167 K and P_3 at 227 K. Because of this correspondence, it will be concluded that the increases of modulus defect are attributed to the relaxation peaks in damping.

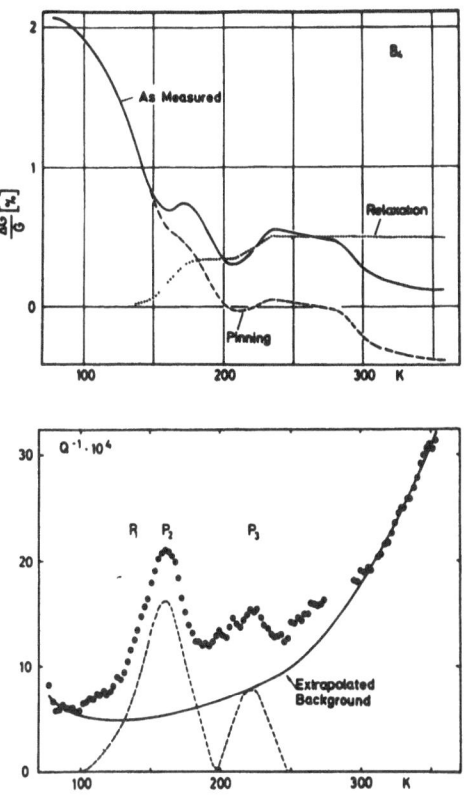

Fig. I: Modulus defect annealing up to 360 K (upper part)
          and decrement (lower part) vs. temperature after
          7.4% tensile deformation at 78 K.

An exact quantitative comparison of both curves is not possible, since
for both modulus and damping, the superimposed pinning and relaxation
effects cannot be separated unequivocally. Therefore, first the relaxation peaks (broken lines in the lower part of Fig. 1) were isolated
by subtracting an extrapolated background (full line) from the measured
damping curve. Then, from these, the corresponding modulus defect (dotted line in the upper part) was calculated assuming for the damping
simple Debye relaxation.

By substracting this curve from the measured one, the dashed curve is
obtained. In case of linear superposition of relaxation and pinning,
this curve would describe the pure pinning effect. One recognizes that
this curve, in contrast to the original one, exhibits a stepwise but
monotonic decrease, i.e. only pinning. Furthermore, the pinning stages
at 135 and 190 K appear now much less separated than in the measured

curve. Because of above mentioned uncertainties one cannot decide
whether or not the small remaining separation is a result of an inac-
curate correction and, therefore, whether there exists between 100
and 200 K only a single pinning stage or whether this separation reflects two pinning stages.

At any case, the present experiments indicate clearly that it is not
possible - as often done - to analyse alone modulus curves without
considering damping or to interprete modulus curves alone from the
point of pinning. In particular, one cannot derive alone from modulus
curves the exact position and height of the pinning stages and conclude that an increase of modulus defect is caused by a true depinning
(in contrast to relaxation effects). On the other hand, it is also
not possible - as often done - to conclude alone from damping measurements that an observed maximum is an relaxation peak. The decrease of
the damping at the high temperature side of the maximum can also be
caused by pinning which is more clearly observable by modulus measurements. (Examples for this behaviour will be given in /2/.)

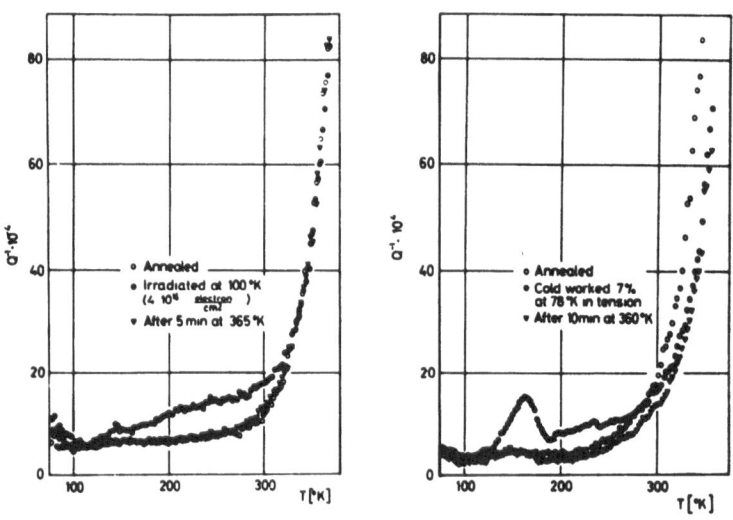

Fig. II: Decrement vs. temperature after electron irradiation
          (left part) and cold-work (right part).

In order to compare the damping behaviour of cold-worked and irradiated copper, Fig. 2a (full circles) gives the temperature dependence of damping during annealing after 7% deformation at 78 K and Fig. 2b (full circles) after electron-irradiation at 100 K. In both experiments apparatus and sample material were the same. As the main difference, one recognizes that the pronounced peak P_2 at about 160 K only appears after cold-work and not after irradiation. Fig. 3 gives the corresponding differentiated modulus curves for which
$g = -(1/G \times dG/dT - 1/G_o \times dG_o/dT)$ - the subscript zero indicates

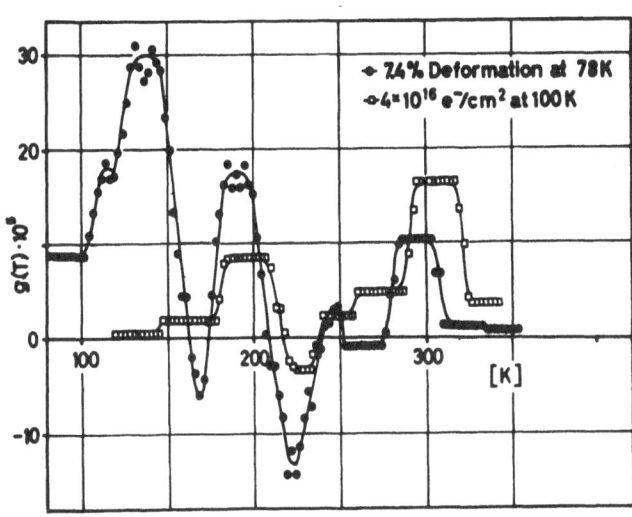

Fig. III: Differentiated modulus defect after cold-work and electron irradiation.

the undeformed resp. unirradiated state - is plotted and which show the pinning stages as minima (filled circles after cold-work, open squares after irradiation). The temperature position of the pinning stages agrees relatively well for the two treatments. The main difference is that the pinning stage at 135 K is completely missing after irradiation.

It is concluded from this that in this pinning stage (stage II_A) a point defect is involved which only appears after cold-work and not after irradiation and which - because of the reproducibility of this stage - is of intrinsic nature. It appears reasonable to assume that this defect is the di-interstitial From the additional observation that the peak temperature of stage II_A is shifted with the dislocation density, it can be concluded further that the rate-determining process is

not the dissociation but really the migration of the di-interstitials
to the dislocation. Since the damping peak P_2 occurs also only after
cold-work it appears reasonable to assume, that P_2 is caused by the
same point defects which cause also the pinning stage II_A, i.e. also
by the di-interstitial. Also the fact that the temperature of stage
II_A (where the defects migrate to the dislocations) is below the
temperature of peak P_2 (where they interact with the dislocations) is
in agreement with this interpretation.

In Fig. 4 the temperatures of the pinning stages and damping peaks
are indicated schematically. The coordination of the pinning stage II_A
to the damping peak P_2 is represented by a solid connecting line. The
lowest row indicates the di-interstitials proposed,

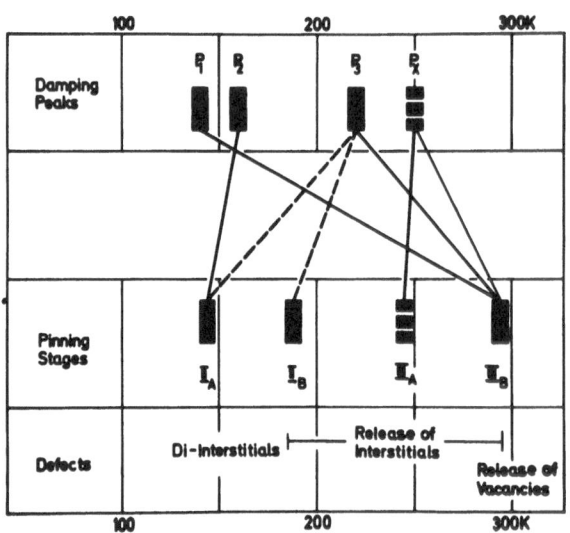

Fig. IV: Schematic diagram of damping peaks and pinning
stages after cold-work.

but it gives only a scheme coming out from these measurements. For
stage II_B at about 190 K the release of interstitials from traps is
assumed here. Proceeding on the assumption that also P_3 only exists after cold-work, it is obvious to consider also P_3 as to be caused by
the defects of stage II_A. But it cannot be excluded that the defect of
stage II_B is responsible. Therefore, broken lines connect P_3 with II_A
and II_B. A solid line goes to stage III_B, since according to Hasiguti
and co-workers an anneal at about 290 K, that is in the range of

stage III_B, further increases the peak P_3. To that one has to assume that also in stage III_B interstitials are released from traps.

The height of stage III_A, superimposed to the beginning of stage III_B, is found to be very different. It clearly appears only in the measurements of Völkl and Schilling /3/ and Wagner /4/ [+). In the majority of the experiments after cold-work this stage is, as in the present work, very small respectively cannot be observed /5-7/. Moreover, in those experiments with a pronounced pinning in the range of stage III_A an additional damping peak near 250 K has been reported and attributed to the relaxation of split-interstitials /3,4/. As to be seen in the lower part of Fig. 1, this peak does not appear in the present work, although with its reported height of $Q_{max}^{-1} = 6 \times 10^{-4}$ it should be clearly recognizable. This seems to indicate that the occurance of this damping peak, which will be named P_x here, and this pinning stage are connected with each other. Therefore, stage III_A and the damping peak P_x are marked in a dotted way, but connected with a solid line. A further solid line goes from P_x to stage III_B, because from the experimental results existing till now it cannot be excluded that this stage may be involved. Because of this irreproducibility it seems impossible to regard stage III_A (or possibly stage III_B) and peak P_x as intrinsic properties of cold-worked copper, as it is done in the arguments for the two-interstitial recovery model /8/.

Because after deformation at 78 K the peak P_1 is growing only by an annealing at about 300 K /9/, it should be caused by point defects migrating in stage III_B at about 300 K. According to Hasiguti P_1 is increasing strongly with the oxygen content of the copper /7/. This is in agreement with the behaviour of stage III_B observed in the present work /1/. This stage increases at higher temperatures of pre-deformation annealing, to which corresponds a higher content of oxygen in solution. One can assume here that the oxygen atoms act as traps for point defects, so that with increasing oxygen content more defects responsible for P_1 arrive at the dislocations. Since interstitials

---

+) However, in these experiments the measurements were only extended to 240 K /4/ resp. 270 K /3/, that is only to the beginning of stage III. Therefore, it cannot be decided, whether the pronounced pinning at about 250 K is caused by an increased stage III_A or an increased stage III_B.

migrate also at lower temperatures and P_1 is built up not before the temperatures of stage III_B it can be assumed obviously, that P_1 is caused by vacancies. Therefore, stage III_B represents also the migration of vacancies trapped at oxygen atoms besides that of trapped interstitials.

So some information about the point defects involved in the Hasiguti peaks is achieved by the comparison of modulus and damping measurements. Also the growth of the peaks observed in certain temperature ranges becomes more understandable.

References:

1   G. Sokolowski, H. Ebener, and K. Lücke; phys. stat. sol.(a)19, 493, (1973)

2   G. Sokolowski and K. Lücke, to be published

3   J. Völkl and W. Schilling; Phys. der kond. Mat. 1, 296 (1963)

4   F. J. Wagner; phys. stat. sol.(b) 51, 589 (1972)

5   W. Lems; Physica 28, 445 (1962), 30, 445 (1964), Thesis Delft, 1963

6   A. J. Brouwner and C. Groenenboom-Eygelaar; Acta Met. 15, 1597 (1967)

7   G. Fantozzi, D. Boulanger, and P. Gobin; J. Inst. Metals 96, 236 (1968)

8   A. Seeger, in Vacancies and Interstitials in Metals, Ed. A. Seeger, D. Schuhmacher, W. Schilling, and J. Diehl, North-Holland Publishing Company, Amsterdam 1970

9   M. Koiwa and R. R. Hasiguti; Acta Met. 11, 1215 (1963)

STUDY OF DISLOCATION PINNING KINETICS IN COLD WORKED
COPPER AND NICKEL BY ELASTIC MODULUS MEASUREMENTS

A. Bischoff and G. Kralik

Institut für Sondermetalle/Metallkunde am Max-Planck-
Institut für Metallforschung, Stuttgart, and Institut
für Metallkunde der Universität Stuttgart, Germany.

The thermal recovery behavior of lattice defects produced by cold
working two fcc metals, copper and nickel, has been investigated by
elastic modulus measurements which yield information about the interaction between point defects and dislocations in these materials.

Experimental Details

The modulus measurements were made on polycrystalline tensile samples
(with a gauge length of 15 mm, diameter of 2 mm and with shoulders of
5 mm diameter) using the Förster-Elastomat to excite the longitudinal
vibration mode of the sample at a resonant frequency of about 20 kilo-
cycles per second. The maximum strain amplitude produced by this means
was less than 10^{-6}. The modulus was measured continuously either
during heating at constant rate from 78 to 540 K (called modulus isochrone) or at constant temperature (modulus isotherm).

Results and Discussion

The upper part of Fig. 1 shows the recovery of the elastic modulus of
copper during heating at two different constant rates. Within the two
main recovery stages (designated as stages II and III), several
experimentally reproducible substages can be seen, especially in the
temperature derivative of the modulus isochrone shown in the lower part
of Fig. 1. This fine structure is due to the pinning stages which are
associated with the stress induced diffusion of point defects to dislocations. This phenomenon was investigated by means of isothermal

recovery measurements in the appropriate temperature regions. The $t^{1/2}$-pinning law proposed by Seeger et al (1,2) for cubic and tetragonal defects has been confirmed experimentally for times not too large (2,3,4,5). This relationship is

$$n(t) \quad \frac{\sqrt{D(T,U)}}{T} \, t^{1/2} \tag{1}$$

where $n(t)$ is the number of pinning points at time t, U is the migration energy, and D in the diffusion coefficient for the defect.

From a knowledge of the kinetic behavior, it is possible to determine the activation energy for migration of the point defects. In principle this can be obtained directly from the slopes of the isotherms (2,3,4); however, in this case such an analysis was not possible because of the extensive overlapping of the individual processes, especially in the regions of stages III and IV. Rather, the activation energy for migration was evaluated from the horizontal shift of the two modulus recovery curves shown in Fig. 1. Since these curves were determined from samples having identical thermal and mechanical histories, one can assume that for equal values of the modulus the pinning point concentrations are also equal. The calculation of the pinning point concentration at a given temperature is carried out according to the rather simple model proposed by Friedel (6). According to this model, the point defects move to dislocations in a cylindrically symmetric strain field with the drift velocity $v = \frac{F(r)D(T)}{kT}$. The temperature in this Einstein-Nernst relation is considered to be a linear function of time (i.e., the heating rate $dT/dt = \mu = $ constant). The driving force is

$$F = - \text{grad } W = \frac{n \cdot W_o \cdot b^n}{r^{n+1}} \quad , \tag{2}$$

where W is the interaction potential and n is a constant having the value 1 for interstitials and 2 for vacancies. During the time interval $t \ldots t + dt$, those defects originally lying within a cylindrical shell, $r(t) \ldots r(t) + dr$, migrate into the cylindrical core region with radius b around the dislocation. The time rate of change of the point defect concentration per unit length of dislocation line is

$$\frac{dp(r,t)}{dt} = \frac{2\pi r(t) C_o \, dr/dt}{b^3} \quad , \tag{3}$$

where b^3 means the atomic volume and C_o is the bulk concentration of point defects per unit volume. Since all point defects lying within the shell between r(t) and r(t) + dr of the dislocation line can move to within the cylindrical element with radius b in time t, r(t) can be evaluated from

$$\int_{r(t)}^{b} dr' = - \int_{o}^{t} v \, dt \, . \qquad (4)$$

From this informations one can obtain the increase in defect concentration per unit length of dislocation core as a function of temperature during heating at a constant rate μ from

$$p(T) = \int_{o}^{T} \left[\frac{2}{n+2} \frac{\pi C_o}{b^2} \left(\frac{A}{\mu}\right)^{\frac{2}{n+2}} \frac{e^{-U/kT}}{T} \left(E_1 \left(\frac{U}{kT}\right)^{-\frac{n}{n+2}}\right) \right] dT \, , \qquad (5)$$

where A is a known function of D_o, W_o, b and n. The function $E_1(\chi)$ is defined by

$$E_1(\chi) = \int_{\chi}^{\infty} \frac{e^{-\chi'}}{\chi'} \, d\chi' \, . \qquad (6)$$

Thus, using Eq. (5) the activation energy for migration of the point defects can be obtained from the shift in the modulus recovery curve when a different constant heating rate is used as is shown in Fig. 1. Based upon the assumptions that C_o = const and that the pinning point concentration is equal for equal values of the modulus, one finds from Eq. (5)

$$\mu_1^{-1} E_1 \left(\frac{U}{kT_1}\right) = \mu_2^{-1} E_1 \left(\frac{U}{kT_2}\right) \, . \qquad (7)$$

This transcendental equation has been solved by an iterative procedure, yielding the activation energy U corresponding to the current state of annealing. It is worth noting that this relation becomes independent of n, i.e., independent whether the defects are interstitials or vacancies.

The activation energies obtained in this manner and shown in Fig. 1 are those corresponding to the dashed modulus curve. Clearly, in the temperature regions of the various substages, plateaus are found which indicate uniform activation processes. From the magnitude of the energy of the various substages and by comparison with other experiments (5) the defect associated with the different substages can be identified. These pinning defects are assumed to be dumbbell-interstitials (stage

III_a), divacancies (stage III_b), single vacancies (stage IV_a), and trapped point defects (stage IV_b).

Knowing the pinning point concentration p as a function of temperature (see Eq. 5), it is possible to calculate a theoretical modulus defect curve due to that defect species. This is obtained from the equidistant dislocation pinning relation proposed by Granato and Lücke (7,8):

$$Y = \frac{\Delta E}{E_e} = \alpha \Lambda \left(\frac{l_o}{1+p(T)l_o}\right)^2 , \qquad (8)$$

where α is a constant (≈ 0.2), Λ is the dislocation density, l_o is the average free dislocation length, E_e is the defect free elastic modulus, and $\Delta E = E - E_e$ is the resulting modulus defect. However, the special case for the dumbbell-interstitial is more realistic, if one takes into account the additional decrease of the interstitial concentration (C_o in Eq. 5) caused by vacancy-interstitial (V - I) annihilation. The temperature dependent contribution to C_o can be estimated from measurement of the isochronal resistance recovery (5) and application of a theory proposed by Waite (9).

In Fig. 2 are plotted calculated modulus curves with and without consideration of the V - I annihilation for two average free dislocation lengths. Using apparently reasonable values of 2×10^{-6} for the point defect concentration and 3×10^{-6} cm for the average free dislocation length, the calculated result is in agreement with the present experimental value at 275 K, which corresponds to stage III_a.

According to Hornung and Wagner (3,10), the local dislocation pinning by dumbbell-interstitials is preceded by long range interactions between interstitials and dislocations resulting in restriction of dislocation movement and, correspondingly, a decrease in the modulus defect. Here, two distinct processes occur: reorientation of the dumbbell-interstitials and diffusion of these defects in the direction of the dislocations. These two processes result in what is called the orientation after-effect and diffusion after-effect (3,10), respectively. Each process approximately obeys an exponential time law, and the resulting modulus defect is

$$- \frac{\Delta E}{E_e} = B_i \, e^{-t/\tau_i} \qquad (9)$$

where τ_i is the relaxation time for a migration step (i = 1) or for a reorientation step by rotation (i = 2) of the dumbbell-interstitial. The after-effect amplitude B_i can be regarded as constant. The relaxation times are associated with the activation energies U_i of the respective processes according to

$$\tau_i = \tau_{io}\, e^{U_i/kT} \quad . \tag{10}$$

These relations have been confirmed experimentally by Wagner (3) by isothermal after-effect measurements on Cu and Ni. However, a constant heating rate has been used in the present experiments, and hence Eqs. (9) and (10) must be modified accordingly if this model is to be applied. This can be done by considering $-\dfrac{\Delta E}{E_e}$ as a differential magnitude

$$d\left(-\frac{\Delta E}{E_e}\right) \propto e^{-T/\mu\tau_i}\left[\frac{T}{\tau_i^2}\frac{\partial \tau_i}{\partial T} - \frac{1}{\tau_i}\right] dT \quad . \tag{11}$$

By integrating this relationship for specified values of the parameters U_i, τ_{io} and μ, the modulus defect for the present case can be determined. Using the values from Wagner's isothermal measurements (3), calculations of the peak temperature corresponding to the present experimental conditions have been carried out. Reasonable agreement between experiment and theory has been found as can be seen in table I.

TABLE I

Material studied	Orientation after-effect				Diffusion after-effect			
	τ_{20}(s)	U_2(eV)	$T_{theor.}$	$T_{exp.}$	τ_{10}(s)	U_2(eV)	$T_{theor.}$	$T_{exp.}$
Cu (99.999%)	$2\cdot10^{-15}$	0.67	200 K	198 K	$6\cdot10^{-15}$	0.67	207 K	208 K
Ni (99.995%)	$2\cdot10^{-14}$	0.86	272 K	268 K	$2\cdot10^{-15}$	1.01	300 K	302 K

μ = 2.48 K/min ; ε = 5 %

Tabulated here are the peak temperatures of the orientation and diffusion after-effect stages for Cu and Ni for both theory and experiment. Because of the agreement of the calculated positions of the stages with those obtained experimentally, it is considered reasonable for copper to attribute stage II_b at about 198 K to the orientation after-effect and stage II_c at about 208 K to the diffusion after-effect.

In the case of Ni, corresponding stages are exhibited (Fig. 3). They are situated partially within the strong modulus defect rise between 175 and 260 K (presumably caused by the magnetostrictive 'ΔE-effect'). Here stage II_b at 268 K is attributed to the orientation after-effect, stage II_c at 302 K to the diffusion after-effect, and III_a at 346 K to the pinning after-effect produced by dumbbell-interstitials.

Summary and Conclusions

By means of simple models relating the interaction of point defects with dislocations, we have attempted to correlate the various substages found in the modulus recovery curve with the relevant defect interactions.
i) With the assumption of stress induced point defect diffusion, taking into account also vacancy-interstitial annihilation, the theoretical curve for dumbbell-interstitial pinning could be fitted with the sub-stage III_a.
ii) Using activation energies and frequency factors from F.J. Wagner's work on the isothermal recovery of the shear modulus the two substages II_b and II_c could be attributed to the orientation and the diffusion after-effect produced by the dumbbell-interstitials.

Acknowledgement

The authors wish to acknowledge helpful discussions with Prof. V. Gerold and Dr. F.J. Wagner and to thank Dr. E. Epperson for correcting the English style. The work has been supported by the Deutsche Forschungs-gemeinschaft.

References

1. A. Seeger, Phys. Letters 31 A, 93 (1970).

2. D. Werneth, Diplomarbeit, Universität Stuttgart (1970).

3. F.J. Wagner, phys.stat.sol. (b) 54, 135 (1972).

4. G. Fantozzi, D. Boulanger and P. Gobin, J. of the Inst. of Metals 96, 236 (1968).

5. A. Bischoff, G. Kralik, to be published.

6. J. Friedel, in "Dislocations", p. 405 Oxford (Pergamon Press).

7. A. Granato, K. Lücke, J.Appl.Phys. 27, 583, 789 (1956).

8. N.F. Mott, Phil.Mag. 43, 1151 (1952).

9. T.R. Waite, Phys.Rev. 107, 463 (1957).

10. W. Hornung, phys.stat.sol. (b) 54, 441 (1972).

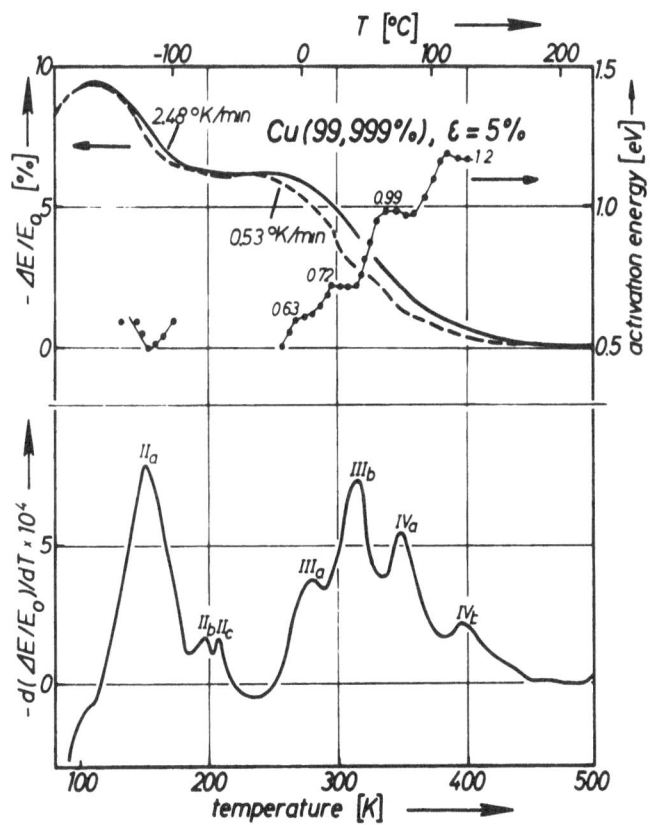

FIG. 1

Isochronal Recovery of the Elastic Modulus in Copper

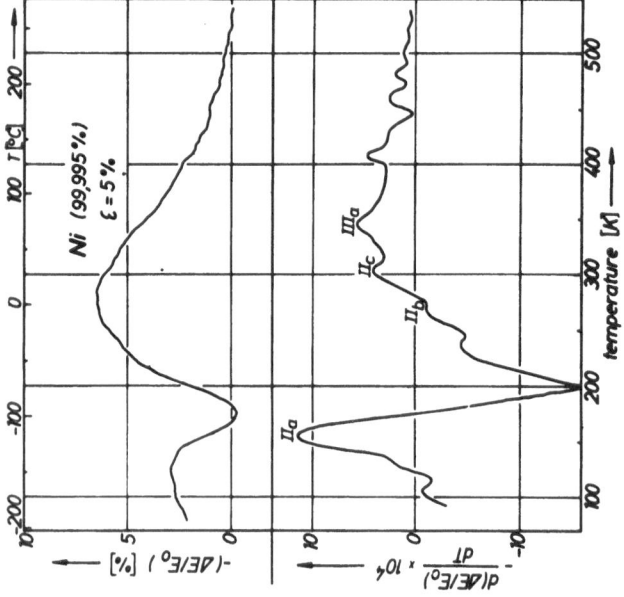

FIG. 3

Isochronal Recovery of the Elastic Modulus
in Nickel

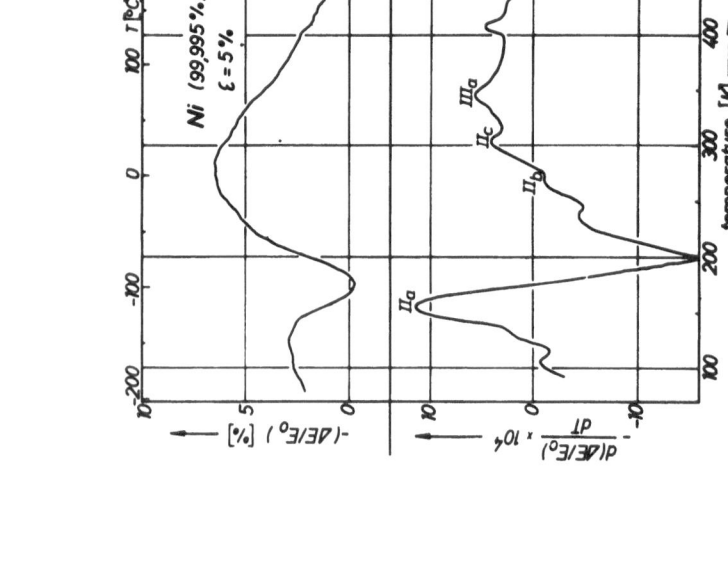

FIG. 2

Theoretical and Experimental Recovery
of the Elastic Modulus in Copper

ANELASTIC EFFECTS DUE TO DISLOCATIONS IN LOW TEMPERATURE COLD-WORKED TITANIUM AND ZIRCONIUM

J. PETIT, M. QUINTARD, P. MAZOT, J. de FOUQUET
Laboratoire de Mécanique et de Physique des Matériaux
E.R.A. au C.N.R.S. n° 123
Université de Poitiers
E.N.S.M.A., rue Guillaume VII - 86034 - POITIERS

INTRODUCTION

The present investigation deals with the internal friction peaks observed in cold-worked Zirconium and Titanium, in the low temperature range. Earlier
work (1) revealed the existence in the two metals of similar relaxation peaks
after straining at room temperature ; the damping measurements were then performed in longitudinal vibrations, at about 20 KHz (2).

The purpose of this paper is to present experiments made recently in
flexural vibrations, at about 400 Hz between 77°K and 300°K, in specimens just
strained by tension at 77°K, or, at room temperature ; it will chiefly be seen
that in contrast with Titanium, the relaxation spectra obtained in Zirconium
are very dependent on the deformation temperature, and on the other hand, that
for the two metals a large range of frequency must be investigated to characterize the broad observed relaxation peaks.

INTERNAL FRICTION SPECTRA IN ZIRCONIUM AND TITANIUM
AS A FONCTION OF THE DEFORMATION TEMPERATURE

a) Zirconium.

In figure (1) are plotted the damping Q^{-1} and the resonnance frequency /
N, versus absolute temperature T, in M.R.C. 99,99% Zirconium specimen (main impurities are listed in table I) after the following treatments :

     1 - annealed 3h at 800°C (N is not reported in this case) ;

     2 - annealed 3h at 800°C, then strained by ten per cent at room tempe-
        rature ;

     3 - annealed 3h at 800°C, then strained by ten per cent at 77°K ;

     4 - annealed 3h at 800°C, then strained by ten per cent at 77°K and
        subsequently aged for 1h at 20°C.

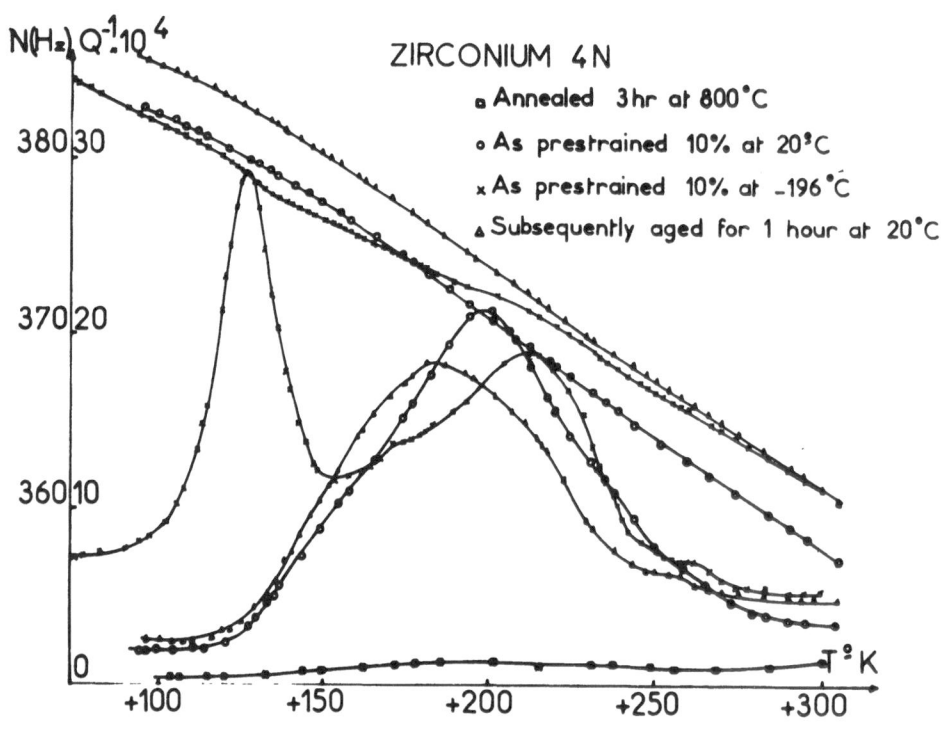

FIG. 1

Temperature dependence of the internal friction Q^{-1} and resonant frequency N of 99,99% Zirconium after annealing 3h at 800°C - after prestraining at room temperature and at 77°K - and after aging at 293°K.

The most remarkable features are :

- the absence of peak in the fully annealed material ;
- the analogy between the curves obtained after treatments 2 and 4, revealing in each case the existence of a broad peak at slightly different temperatures, which will be labelled P_2 ;
- the existence after straining at 77°K of a narrow and well developed peak at 126°K called P_1, and of an other peak at about 220°K which will be seen distinct from P_2, called P_3 ;
- the important variation of frequency during the first heating run after straining at 77°K, especially at about 200°K.

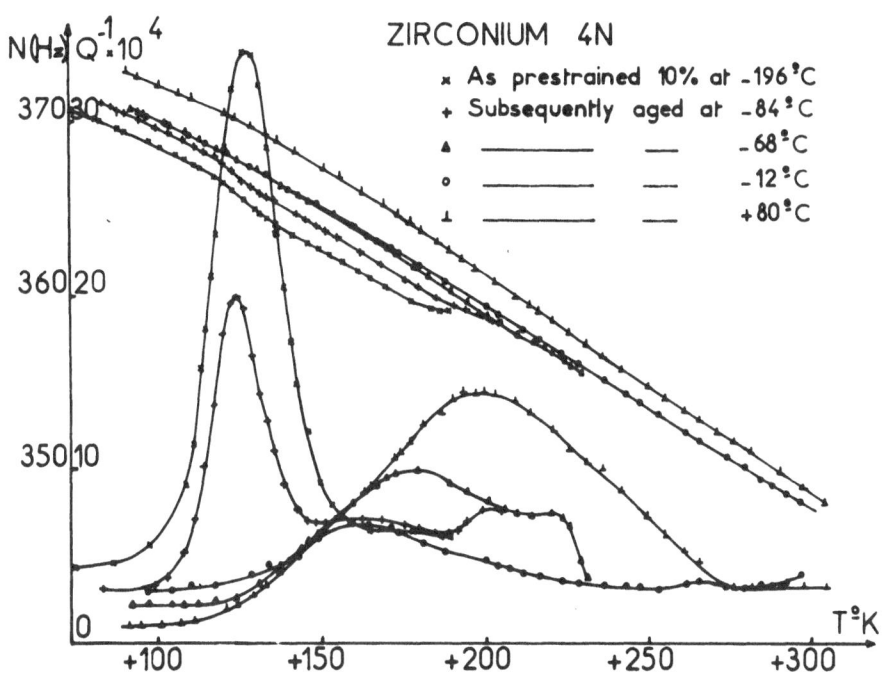

FIG. 2

99,99% Zirconium pre-stained 10% at 77°K. Internal friction Q^{-1} and
resonant frequency N vs temperature T°K after aging at rising temperatures.

Figure (2) shows internal friction and frequency variations after straining at 77°K during successive runs at rising temperatures. It may be noted that the increasing of the frequency at about 200°K coincides with the desappearing of P_1. These results are in very good agreement with thoses of SAVINO (3) obtained at lower frequency :

- P_1 looks like a DEBYE peak and desappears between 170°K and 200°K at any measurement vibration frequency ;

- the height and the shape of P_3 are very sensitive to the previous treatments and to the rate of heating ; the peak temperature is independent of the frequency ; consequently, and in contrast with P_2, P_3 is not a true relaxation peak ;

- after heating at room temperature a single peak close to P_2 is obtained only after heating at 350°K.

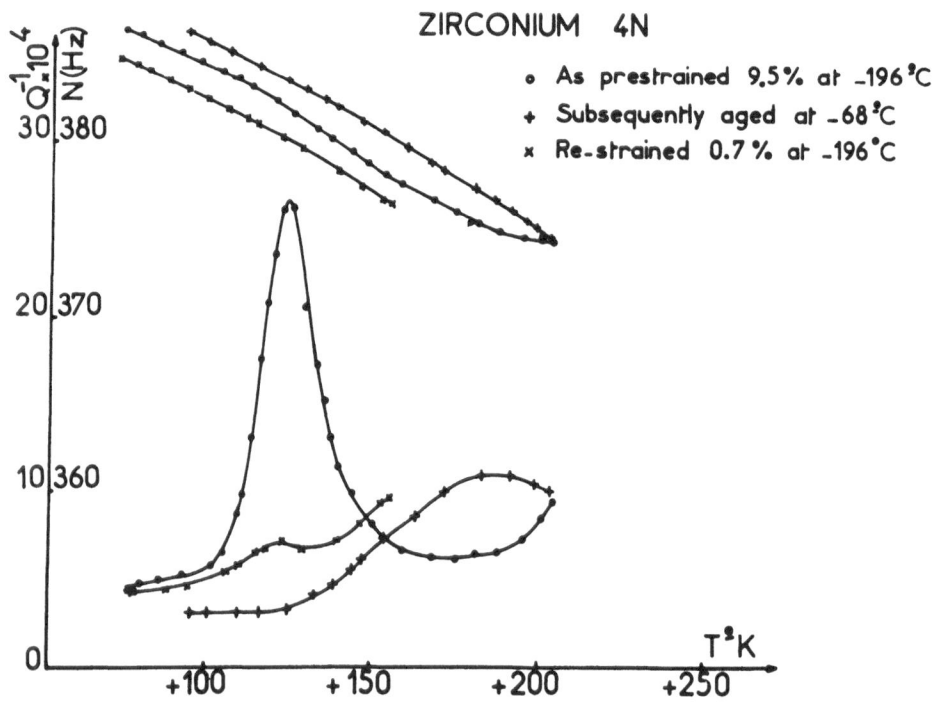

FIG. 3

The effect of subsequent 0,7% plastic deformation on a 99,99%
Zirconium sample pre-stained 9,5% at 77°K and annealed at 205°K.

The frequency increase during the desappearing of P_1 suggested that the
decrease and the annealing of P_1 observed at relatively low temperature could
be associated with a strong pinning of dislocations by interstitial impurities
or by self-interstitial defects. The test depicted in figure (3) has been performed to check that assumption : a specimen prestrained by 9,5% at 77°K was
annealed at 205°K, then strained again by 0.7% at 77°K ; the corresponding peak
height after re-straining shows that the annealing at 205°K is defenitive ;
therefore one must conclude that the defects or the configuration of defects
involved in the relaxation mechanism for P_1 are completely eliminated at about
200°K.

b) Titanium.

Similar tests were performed in 99,97% Titanium specimens (main impurities are given in table II) for nearly the same frequencies and the same deformation conditions.

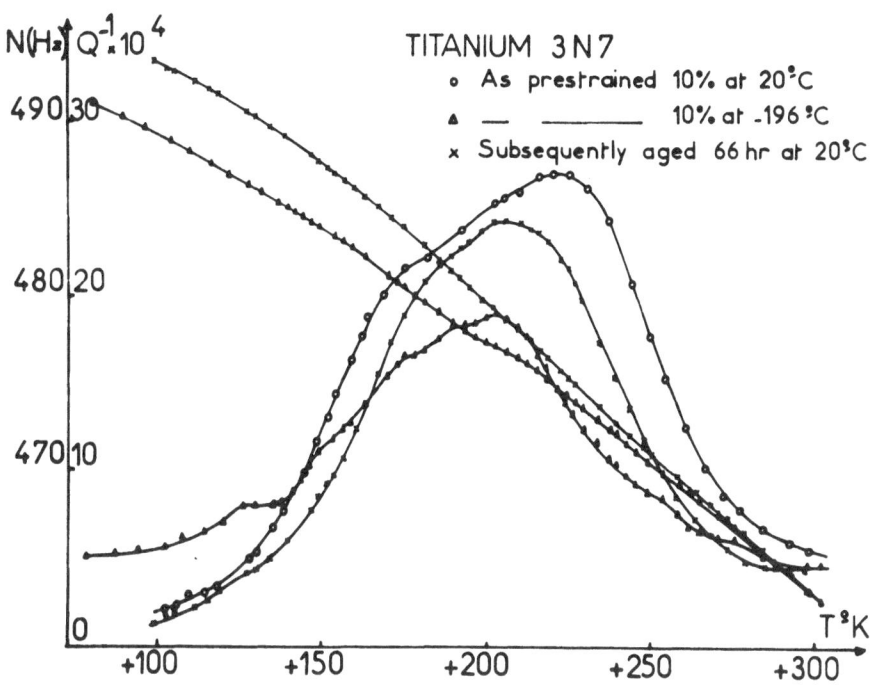

FIG. 4

Temperature dependence of the internal friction Q^{-1} and resonant frequency N of 99,97% Titanium after prestraining at room temperature, at 77°K and after aging at 293°K.

Table II

Main impurities in 99,97% Ti specimen

	ppm
H	3
O	125
N	2
C	78
Fe	30
Hf	12

Figure (4) shows the temperature dependence of both the internal friction and resonant frequency of specimen prestrained 10% in tension : at room temperature - at 77°K - and after heating at room temperature. The corresponding curves show the existence in each case of a broad peak quite similar to P_2 in Zirconium, in the 130 - 270°K range ; no peak corresponding to P_1 has been obtained in the specimen prestrained at 77°K, in the temperature and frequency range covered in this investigation. However, as in Zirconium, it must be noted a fast frequency increase at about 190°K, after straining at 77°K, and the important damping decrease

Table III

Principal characteristics of P_1 and P_2 in 99,99 Zirconium
and in 99,97% Titanium strained by 10% in tension.

		H(ev)	τ_0	$T_M°K$ at 400Hz	
Zr	P_1	$0,15 \pm 0,015$	$10^{-9 \pm 1}$	126	DEBYE PEAK
Zr	P_2	$0,38 \pm 0,05$	$10^{-14 \pm 1,5}$	200	Q_M^{-1} 0,65 N/N $\beta>3(4)$
Ti	P_2	$0,48 \pm 0,05$	$10^{-14 \pm 1,5}$	225	Q_M^{-1} 0,45 N/N $\beta>5(4)$

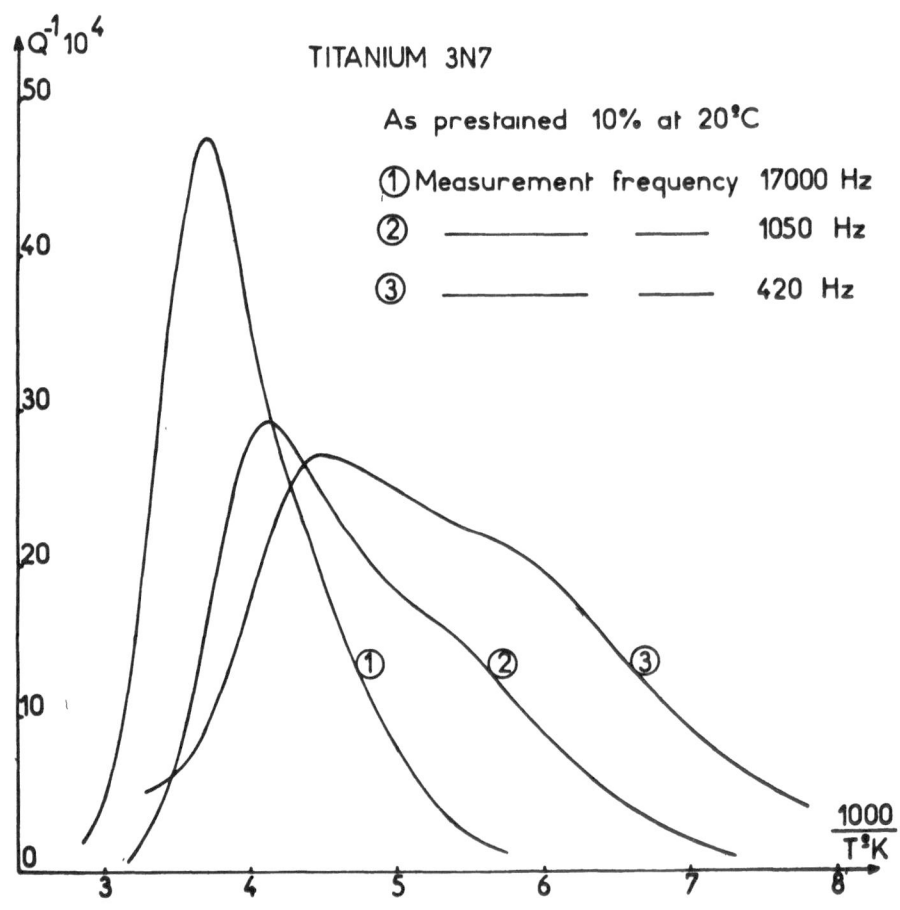

FIG. V

Internal friction peak P_2 in 99,97% Titanium pre-strained 10%
at room temperature for three different frequencies.

associated with the frequency increase at low temperature, after aging at room temperature.

DISCUSSION

The question arises as to what mechanisms govern the P_1 and P_2 peaks in Zirconium and Titanium. The principal characteristics of these two peaks, for ten per cent pre-strained materials, from our measurements and the SAVINO's results,are listed in table III.

The activation energies H for P_2 peaks, have been obtained from the temperature shift with frequency. However, measurement in Titanium,.at different frequenciesbetween 400 and 16,000Hz (see Fig. 5), show that it is difficult to explain the broadening of the peak when frequency decreases, whithout assume the existence of two or three peaks with activation energies from 0.2 to 0.55 for Titanium and 0.2 to 0.4 for Zirconium ; the corresponding attempt frequencies are respectively from 10^{10} up to 10^{14} in the two cases.

SAVINO and BISOGNI (3) suggest that P_1 peak in Zirconium corresponds to the BORDONI peak (5,6), and that P_2 peak is related to a dislocation de-pinning process involving an Hydrogen effect. The activation energy and the low value obtained for the frequency factor are in favour of such an interpre-tation for P_1 ; nevertheless it is necessary to explain the desappearing of the peak at relatively low temperature, and also the fact that it doesn't at least partially reappear after unpinning ; moreover, the agreement between P_1 and a DEBYE peak is remarkable. Therefore, another possibility is that P_1 peak is related to point defects created by straining at 77°K (interstitial for example) in Zirconium, defects which desappear between 170°K and 200°K ; the relaxation effect could then arise from the jump of the above point defects in the field of the moving dislocations ; such a process agrees with a low attempt frequency ; the absence of P_1 peak in Titanium would be due to some difference between the point defects induced by deformation in the two metals.

On the other hand, in early work (1) P_2 was considered to be a BORDONI peak or a HASIGUTI peak (7,8). Characteristics of P_2 peak are effectively in good agreement with those reported for HASIGUTI peaks. However, in view of the broad attempt frequency and activation energy distribution obtained for these peaks, and the large variation previously reported with cold-work and purity(1), it doesn't seem that peak P_2 may be attributed to only one relaxation process as well in Zirconium as in Titanium. The idea advanced by SAVINO (3) of a SCHOEK peak (9) due to Hydrogen may be not rejected ; nevertheless, measure-ments made in Zirconium specimen containing various amounts of Hydrogen and Oxygen were not decisive on this point. One another intriguing fact is the ma-gnitude of the activation energy and frequency factor obtained for the low temperature component of P_2, and the relative insensivity to the previous treatment and purity ; as a consequence, it is not excluded that P_2 peak is constitued by a HASIGUTI (or a SNOEK-KOSTER) peak on the high temperature side and by a BORDONI peak on the low temperature one. On the basis of these results, it appears that it is first necessary to characterize better the various relaxation phenomena observed in the two metals, and the conditions generating them before to support either well defined mechanism, as well for P_1 peak as for P_2.

ACKNOWLEDGMENT

It is a pleasure to thank E.J. SAVINO and E. BISOGNI for communication of their results prior publication, and valuable discussions with one of us.

REFERENCES

(1) J. Petit, M. Quintard, R. Soulet and J. de Fouquet, Journal de Phys. C2-32-21 (1971).

(2) P. Boch, These Doc. Sc. Poitiers (1968).

(3) E.J. Savino, These Doc. Sc. Buenos Aires (1971).

(4) Nowick A.S, Berry B.S. IBM Journal 297 (1961).

(5) P.G. Bordoni, J. Acoustic Soc. Am. 26-495 (1954).

(6) A. Seeger, P. Shiller, Acta. Met. 10-348 (1962).

(7) R.R. Hasiguti, N. Igata, G. Kamoshita, Acta. Met. 10-442 (1962).

(8) R.R. Hasiguti, M. Koiwa, Acta. Met. 13-1219 (1965).

(9) G. Schoek, Phys, Stat. Sol. 8-499 (1965).

IRRADIATION EFFECTS ON HASIGUTI PEAKS IN GOLD AND ALUMINIUM

O. Mercier[*], W. Benoit[**]
P. Moser[***]
G. Fantozzi, J. Perez, P. Gobin[****]

INTRODUCTION

The Hasiguti peaks are observed on cold-worked gold and aluminium samples [1]. These peaks are usually explained in terms of interaction between dislocations and point defécts. This can be demonstrated by introducing irradiation point defects into a sample containing convenient dislocations [2]. Bays [3] has already shown that only the measurement of samples containing both dislocations and point defects showed Hasiguti peaks.

Similar measurements have been conducted on gold and aluminium samples, irradiated either with electrons at 20 K, or with neutrons at 77 K.

EXPERIMENTAL PROCEDURE

The experiments of irradiation with electrons have been conducted "in situ" on an inverted pendulum in line with the accelerator. The lower part of this pendulum is immersed in the liquid hydrogen of the irradiation cryostat. After irradiation, this liquid is evacuated. During the entire experiment, the specimen does not undergo any handling. The pendulum (previously described) [4] oscillates at about 1 c/s and the maximum strain during measurements was 10^{-5}. Cold work was produced by 5% tensile strain at 20 K.

Irradiation with neutrons has been done in a reactor at 77 K[+]. After irradiation the samples are kept at the temperature of 77 K so as to enable their de-activation, then they are mounted without deformation by means of mechanical arrangement in an inverted pendulum [5] : frequency range : 0, 3 to 3 Hz ; temperature range : 77 - 700 K ; maximum strain during measurement : 5.10^{-7}. Cold work was produced by 6% torsion strain at 77 K.

The aluminium specimens (99, 995%) are preannealed one hour at 650 K and have the shape of a cylindrical rod (ϕ 3mm ; L = 150mm) thinned by a spark machine in their central region (e = 0, 5mm ; l = 30mm) where they are irradiated.

[*]   Laboratoire de Génie Atomique, EPF-Lausanne, Switzerland
[**]   Now at the "Institut für Theoretische und Angewandte Physik", Univ. Stuttgart, Germany
[***]   DRF-G/Physique du Solide, CEN-Grenoble, France
[****] Laboratoire Physique des Matériaux, INSA-Lyon, France
[+] Cold circuitry ESKIMO of the Saphir Reactor at the EIR, Würenlingen, Switzerland

The gold specimens (99, 998 wt%) are preannealed 24 hours at 900°C under atmosphere; the electron irradiated samples have the form of foil (0, 3 mm in thickness, 3 mm in width and 40 mm in length). The neutrons irradiated samples have the form of wire (1 mm in diameter, 100 mm in length).

In order to obtain a network of dislocations that is suited to these experiments :

. the Hasiguti peaks must have disappeared

. the loops of dislocations must be large, i.e. the modulus defect must be high.

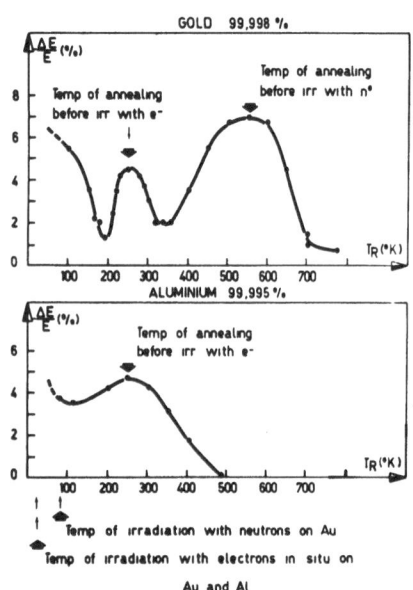

Fig, 1   Evolution of the modulus defect $\Delta E/E$ on gold and aluminium samples deformed plastically at low temperature as a function of annealing temperature

In fig. 1 the modulus defects of gold and aluminium samples, deformed plastically at low temperature are shown as a function of the annealing temperature. Two maxima of modulus defect are observed after annealing at 250 and 570 K for gold, and one maximum after annealing at 250 K for aluminium. Therefore, before irradiation with electrons, the gold samples and the aluminium samples are annealed at the temperature of the first maximum, after plastic deformation (5% by tensile strain). The Hasiguti peaks P_2 and P_B have disappeared at this temperature.

Before irradiation with neutrons, the gold samples are plastically deformed at 77 K (6% by torsion) and then annealed at about 570 K, which is the temperature of the second maximum of the modulus defect. All the point defects have disappeared at this temperature, which is just less than recrystallization temperature.

IRRADIATION OF ALUMINIUM SAMPLES WITH ELECTRONS

In fig. 2 the curves obtained for aluminium samples irradiated with electrons at 20 K are shown [6]. The curve denoted by (a) shows the measurements of internal friction obtained after plastic deformation, that by (a') after annealing at 220 K, curves (b), (c), (d) after irradiations at different doses (0, 16.10^{19}; 1, 6.10^{19}; 3, 6.10^{19} e/cm² respectively). One can see (Table I) :

(i)   The height of the Bordoni peak decreases when the dose increases.

(ii)   The peak P_A does not appear again.

(iii) The peak P_B appears again, its height first increases then decreases.

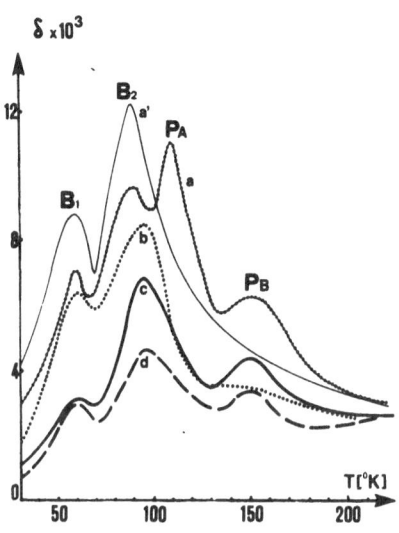

Assuming that the height of the Bordoni peak after irradiation is proportional to ΛL^2 (Λ : density of dislocations, constant during the experiment ; L : length of the loop) and that the height of the Hasiguti peak (P_B) is proportional to a first approximation to $\varkappa \Lambda L^2$ ($\varkappa$: fraction of loops pinned by the defects that account for the peak), the ratio of the heights of the two peaks will give the fraction of the pinned loop. This ratio increases with irradiation dose, i.e. as a function of the number of defects created.

Fig. 2   Evolution of the internal friction on an aluminium sample :

a)   after plastic deformation of 5% by traction at 20 K
a')   after annealing for one hour at 220 K
b) c) d) after an irradiation with 2 MeV electrons at 20 K (doses = 0, 16.10^{19}; 1, 6.10^{19}; 3, 6.10^{19} e/cm² respectively)

IRRADIATION OF GOLD SAMPLES WITH ELECTRONS

In fig. 3 the curves obtained for gold samples irradiated with 2, 2 MeV electrons at 20 K are shown. As before, curves (a) and (a') represent the internal friction measured after plastic deformation and after annealing at 270 K respectively. Curves (b), (c) and (d) represent the internal friction after irradiation at different doses (1, 12.10^{16}; 1, 12.10^{17}; 1, 12.10^{18} e/cm²) respectively. One can see (Table II) :

(i)   the height of the Bordoni peak decreases as the dose increases

(ii)   peak P_2 reappears ; its height first increases and then decreases. The peak's height is much smaller than after plastic deformation

(iii) it is difficult to distinguish between P_2 and P_3.

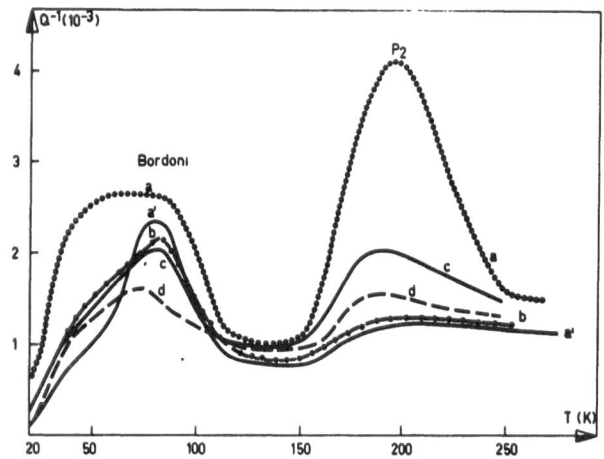

Fig. 3

Evolution of the internal friction on a gold sample :

a) after plastic deformation of 6% by traction at 20 K

a') after annealing for one hour at 270 K

b)c)d) after irradiation with 2,2 MeV electrons at 20 K (doses = 1, 12.10^{16}; 1, 12.10^{17}; 1, 12.10^{18} e/cm² respectively)

TABLE I

dose e/cm² \ height	0	1,6 . 10^{18}	1,6 . 10^{19}	3,6 . 10^{19}
B_2 . 10^3	10 ± 0,2	7,3 ± 0,2	5,7 ± 0,2	3,4 ± 0,2
P_B . 10^3	0	0,25 ± 0,1	1,1 ± 0,1	0,95 ± 0,1
$\varkappa = \dfrac{P_B}{B_2}$	0	0,03 ± 0,01	0,19 ± 0,04	0,28 ± 0,05

TABLE II

dose e/cm² \ height	0	1,12 10^{16}	1,12 . 10^{17}	1,12 .10^{18}
B . 10^3	2,2 ± 0,1	2,0 ± 0,1	1,9 ± 0,1	1,5 ± 0,1
P_2 .10^3	0	0,3 ± 0,1	0,8 ± 0,1	0,7 ± 0,1
$\varkappa = \dfrac{P_2}{B}$	0	0,15 ± 0,05	0,4 ± 0,1	0,5 ± 0,1

IRRADIATION OF GOLD SAMPLES WITH NEUTRONS

The measurements on gold samples irradiated with neutrons at 77 K are depicted in fig. 4. The curve denoted by (a) indicates the measurement of internal friction on a cold-worked sample as a function of the temperature. Curve (a') represents the measurement after annealing for one hour at 570 K and curves (b), (c), (d) after irradiation doses of 10^{13}, 10^{14}, 10^{15} neutrons/cm² respectively. The peak's height as a function of the dose shows a maximum, as for the measurements made on samples, irradiated with electrons. Moreover the peak as a function of the dose first widens, then moves towards higher temperatures.

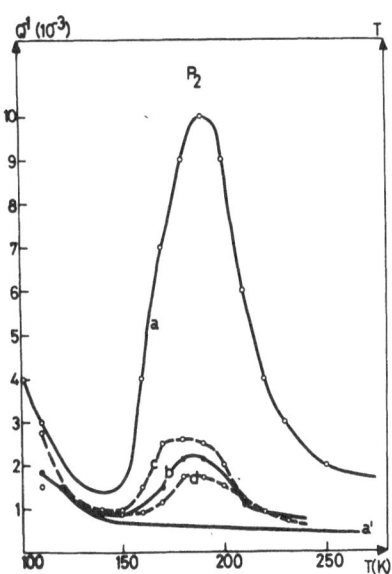

The curves in fig. 5 show the evolution of the peak P_2 in an irradiated sample (dose : 10^{14} neutrons/cm^2) as a function of successive linear increments of temperature.

The curves 2 to 5 show the increment and the broadening of peak P_2 during a first stage of recovery ($T_P \cong 175\,K$). The curves 6 to 10 show the second stage of recovery. The temperature of the peak's maximum increases by about 20°, whereas, at the same time, its height decreases. Measurements do not allow to distinguish between P_2 and P_3. P_3 seems to be a consequence of P_2.

Fig. 4   Evolution of the internal friction on a gold sample :

a)   after plastic deformation of 6% by torsion at 77 K
a')   after annealing for one hour at 570 K
b)c)d)   after an irradiation with neutrons at 77 K (doses = 10^{13}, 10^{14}, 10^{15} n/cm^2 respectively)

Fig. 5

Measurements of the internal friction on a gold sample irradiated with neutrons (dose = 10^{14} n/cm^2) according to a program of successive linear increments of temperature as a function of time. Increment no. 1, final temperature 150 K; no. 2, 165 K; no. 3, 180 K; no. 4, 195 K; no. 5, 203 K; no. 6, 212 K; no. 7, 220 K; no. 8, 230 K; no. 9, 240 K; no. 10, 250 K.

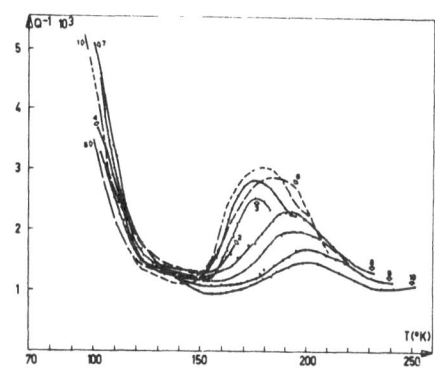

The curves in fig. 6 represent the evolution of the peak height and of the modulus defect, measured at 140 K as a function of annealing temperature. These results show that :

(i) the height of the peak is approximatively proportional to the importance of dislocation pinning

(ii) this pinning is observed simultaneously with the increase of the peak ; an energy of pinning of about $0,3eV$ [3] is measured. This energy can be compared with the activation energy of the peak $(0,33eV)$.

The two observations above infer that the peak is the result of an interaction between mobile point defects and dislocations.

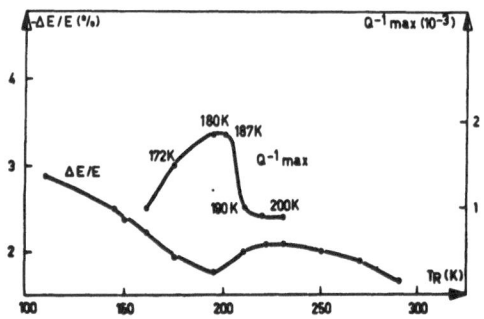

Fig. 6

Evolution of the modulus defect measured at 140 K and the height of the Hasiguti peak as a function of annealing temperature in an irradiated gold sample (dose = $10^{14} n/cm^2$)

an irradiated gold sample (dose 10^{14} neutrons/cm²)

The curves denoted by (a) and (a') in fig. 7 indicate the measurement of internal friction before and after annealing at 250 K in that was maintained for 20 days at 77 K.

The curve (b) represents the measurement on a sample subjected to the same conditions, but maintained for 80 days at 77K. In this case, peak P_2 hardly appears and during the first increment of temperature peaks P_1 and P_3 are observed. This means that at 77K a recovery of the defects accounting for peak P_2 is possible; this recovery involves the apparition of peaks P_1 and P_3. It can be explained assuming that at 77K the small agglomerates of point defects

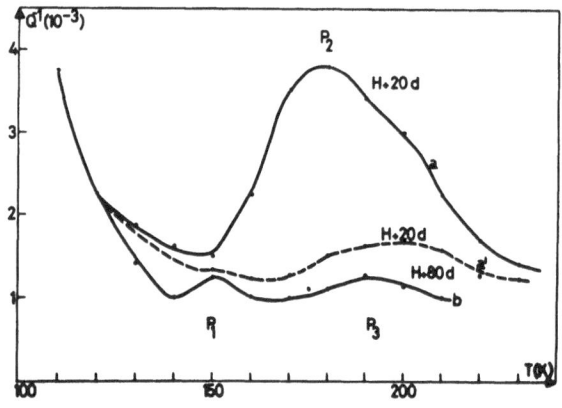

Fig. 7   Evolution of the internal friction in a gold
sample :
a) after an irradiation with neutrons (10^{14}n/cm²) and annealing for 20 days at 77K
a') after annealing at 250K
b) after an irradiation with neutrons (10^{14}n/cm²) and annealing for 80 days at 77K

created by irradiation slowly grow into larger agglomerates. The critical stability radius decreases with temperature . This process can no longer take place at higher temperature; the small agglomerates will remain and will be able to produce peak P_2 .

DISCUSSION

In summary, the following properties of Hasiguti peaks can be presented :

. The Hasiguti peaks (P_B, P_1, P_2, P_3) appear after plastic deformation at low temperature, and after irradiation with electrons and neutrons, provided that a dislocation network is already present. These peaks appear in a similar way, after irradiation, both in gold and aluminium.

. The peak P_A for aluminium does not reappear after irradiation.

. The height of peak P_2 or P_B increases and then decreases as a function of the dose whereas the Bordoni peak only decreases.

. As a function of dose or of annealing, peak P_2 first increases and widens, then migrates towards high temperatures.

. The maximum of the peak is observed simultaneously with the minimum of the modulus defect.

. At the temperature of liquid nitrogen (77K) a slow recovery occurs, further observation of P_2 is therefore impossible.

Considering these properties one can explain the peaks by the interaction between mobile point defects and dislocations. These point defects pin the dislocations before and during the increase of the peaks ; they are mobile while the peaks are observed. The model described in a preceding paper [7] can quite well explain the results of this communication.

Our experiments support the following model of recovery for aluminium :

. At the beginning of stage II of recovery, small agglomerates of interstitial point defects slowly grow into larger agglomerates. The critical stability radius decreases with temperature , this process can no longer take place at higher temperature. The small agglomerates will remain and migrate towards the dislocations ; their interaction with these dislocations creates peak P_B in aluminium.

Then through migration along the dislocations, they will form wider agglomerates either on the nodes of dislocations or on the geometrical kinks. P_B decreases .

At the temperature of stage III of recovery the vacancies arrive at the dislocations where they cause either the disappearance of interstitial agglomerates, or new pinning points by forming clusters. On average, the modulus defect will only decrease at the end of stage III of recovery when the second process will become predominant.

In gold our experiments demonstrate a very similar behaviour. We can conclude that the same model could be convenient for gold.

REFERENCES

[1]   S. Okuda and R. R. Hasiguti, Acta Met., 11, 257 (1963)
        G. Fantozzi, J. Perez, P. Peguin, P. F. Gobin, Mém. Sci. Rev. Mét., 66, 151 (1969)
        W. Benoit and al., J. Phys. Chem. Solids, vol. 31, 1909-1912 (1970)

[2]   D. Keefer and R. Vitt, Acta Met., 15, 1501 (1967)

[3]   B. Bays, W. Benoit, P.-A. Grandchamp, J. de Phys., Coll. C2, 32, C2-55-56 (1971)

[4]   R. Pichon, F. Vanoni, P. Bichon, G. de Keating-Hart, P. Moser, Rev. Phys. Appl.
        5, 427 (1970)

[5]   B. Bays, P.-A. Grandchamp, Rev. Phys. Appl., 5, 327-332 (1970)

[6]   G. Fantozzi, J. Perez, W. Benoit, P. Moser, Z. Kabsch, Radiation Effects
        (to be published)

[7]   O. Mercier, A. Isoré, W. Benoit, Fifth Int. Conf. on Internal Friction and Ultra-
        sonic Attenuation in Crystalline Solids, Aachen, 1973

EFFECTS OF STRAIN AMPLITUDE AND BIAS STRESS ON
DISLOCATION INTERNAL FRICTION IN MOLYBDENUM SINGLE CRYSTALS

T.E. Mitchell, E.A. Kenik, P.S. Sklad and R. Gibala
Department of Metallurgy and Materials Science
Case Western Reserve University
Cleveland, Ohio 44106

1. INTRODUCTION

Internal friction peaks associated with dislocation relaxations exist in many materials[1]. In fcc metals, there are the Bordoni and Hasiguti peaks, while in bcc metals, the α, β and γ peaks are thought to involve dislocations. The Bordoni and α peaks have a number of characteristics in common, characteristics determined largely by experiments on the amplitude independent damping of single crystals deformed systematically to various strains at different temperatures[2,3]. However, since additional information can be obtained from amplitude dependent damping and bias stressing[4], these types of experiments have been extended in the present study to include these variables.

2. EXPERIMENTAL PROCEDURE

Molybdenum single crystals oriented near the center of the stereographic triangle were grown from 3 mm diameter polycrystalline rods by electron beam zone melting at pressures $\sim 10^{-6}$ torr. Tensile specimens were cut from these crystals, ground, electropolished and outgassed at a temperature $\sim 1800°$K and a vacuum $\sim 10^{-8}$ torr. Specimens to be used for resonant bar measurements at ~ 100 kHz had diameters ~ 2.5 mm and gage-lengths ~ 35 mm, while torsional pendulum specimens had diameters ~ 0.07 mm and gage-lengths ~ 70 mm. Final resistivity ratios between room temperature and 4.2°K were ~ 2000.

Outgassed specimens were deformed in tension on a floor model Instron testing machine. The two testing temperatures used (450 and 156°K) were obtained using standard liquid baths. Specimens for internal friction studies were cut from the gage length of the tensile specimens. Resonant bar measurements were made with a two component piezoelectric system at $\sim$ 100 kHz longitudinal resonance with strain amplitudes from 5×10^{-7} to 5×10^{-5} using an impedance bridge[5]. Torsional pendulum measurements at $\sim$ 1 Hz were made using an inverted pendulum with strain amplitudes from 5×10^{-7} to 1.5×10^{-4}; tensile bias stresses up to 5 kg/mm^2 and torsional bias stresses up to 6 kg/mm^2 were possible[6]. Internal friction was measured at temperatures from 4.2 to 310°K.

3. RESULTS

3.1 Crystals Deformed at 450°K

(a) Effect of Deformation. Figure 1 shows curves of internal friction versus temperature as measured by the torsional pendulum for a specimen which was deformed in eleven successive steps at 450°K. The three-stage hardening curve is also shown in the figure. The results are very similar to those reported in Mo crystals by Korenko et al[7] for resonant bar measurements. In stage I, the damping is low and a step-like internal friction spectrum develops with increasing strain. During stage II and stage III deformation, a diffuse α peak becomes apparent, which gradually increases in height, sharpens and moves to lower temperatures, stabilizing at a peak temperature $\sim$ 90°K. At the highest strains, the high temperature background damping tends to decrease, as observed for resonant bar measurements.

(b) Effect of Strain Amplitude. Internal friction curves illustrating the effect of strain amplitude for crystals deformed various amounts at 450°K are shown in Fig. 2 (resonant bar) and Figs. 3 and 4 (torsional pendulum). Fig. 2(a) shows that, for a crystal deformed into stage I to produce a step-like damping spectrum, the effect of strain amplitude is to increase the level of damping until, at an amplitude of 5×10^{-5}, an α peak rises out of the background at $\sim$ 120°K. However, for a crystal deformed into the beginning of stage II, Fig. 2(b) shows that the damping is more sensitive to the

5×10^{-5}. The torsional pendulum specimen in Fig. 3 is deformed to about the same extent as the resonant bar specimen in Fig. 2(b); the behavior is similar except that the higher strain amplitude (1.4×10^{-4}) of the torsional pendulum is seen to be capable of producing a prominent α peak. In Fig. 4 the well-developed α peak of a specimen deformed to the end of stage II is enhanced considerably in height and moved to lower temperatures with increasing strain amplitude.

(c) <u>Effect of Bias Stress</u>. Fig. 5 shows the effect of a torsional bias stress on the internal friction of two crystals, one deformed to the end of stage I and the other to the end of stage II. It is seen that in the first case the overall damping level is decreased by the bias stress, whereas in the second case it is increased. The different behavior may be explained by the fact that the α peak is not properly developed in the crystal deformed to the end of stage I, so that the major effect of the bias stress is to decrease the level of the step-like background. For the crystal deformed to the end of stage II, the α peak is relatively well developed and the major effect of the bias stress is to enhance the height of the α peak.

3.2 <u>Crystals Deformed at 156°K</u>

(a) <u>Effect of Deformation</u>. The effect of deformation at 156°K on the internal friction spectrum of Mo is as described by Korenko et al[7]. The maximum damping levels produced are much lower than those resulting from deformation at 450°K. A step-like background is produced by low deformations at 156°K, but high deformations only give a broad maximum, which may be interpreted as a developing α peak.

(b) <u>Effect of Strain Amplitude</u>. Fig. 6 shows the effect of strain amplitude on the resonant bar internal friction of two crystals. Fig. 6(a) shows that, for a crystal deformed 1.2% at 156°K, the effect of increasing strain amplitude is first to enhance the step-like damping and then to induce a small α peak. By contrast, the more highly deformed crystal in Fig. 6(b) reveals only a broad higher temperature maximum at the highest strain amplitude (5×10^{-5}). These results are very similar to those illustrated in Fig. 2 for deformation at 450°K. Fig. 7 gives equivalent internal friction curves for a torsional pendulum specimen deformed 12% at 156°K; the results are similar to Fig. 6(b) except that, because of the higher strain amplitudes (up to 1×10^{-4}), the broad maximum becomes more obvious, although it does not approach an "equilibrium" α peak.

(c) <u>Effect of Bias Stress</u>. The effect of bias stress on crystals deformed at 156°K is similar to that shown in Fig. 5 for the lower deformation at 450°K. That is, since the α peak is not well developed and the major part of the internal friction spectrum is the step-like background after deformation at 156°K, the effect of bias stress is to reduce the level of damping slightly.

4. DISCUSSION

We have discussed previously[7] three possible types of mechanisms for the α peak in bcc metals: (a) nucleation of kink pairs on non-screw dislocations, (b) motion of abrupt kinks on screw dislocations, and (c) depinning of kinks trapped by point defects.

The depinning model is favored because it more readily explains (a) the influence of impurities on the α relaxation, (b) the step-like change in the background damping from below to above the α peak, and (c) the similarity of the Bordoni peak in fcc metals to the α peak in bcc metals. The α peak occurs by the depinning of kinks on dislocation segments under high internal stress; application of high strain amplitudes (external stresses) allows depinning of kinks to occur on dislocation segments under lower internal stresses, so that the α peak is enhanced. The step-like background observed is damping of the vibrating string or kink type[8], which occurs on those dislocation segments that can contribute to the α peak. The α peak and background damping are thus closely connected, although they will have a different dependence on the magnitude of internal and external stresses. The present results will now be discussed in the light of these ideas.

Figs. 2(a) and 6(a) show that, for crystals deformed to low strains at 156°K and 450°K, it is possible to generate a well-defined α peak by the application of a high strain amplitude. At intermediate strains (Figs. 2 (b),3,6(b) and 7), only a broad ill-developed α peak is generated, presumably because the combination of internal and external stress is insufficient. At higher strains, the internal stress of itself is sufficient to give a well-developed α peak at low strain amplitudes, and high strain amplitudes (external stresses) enhance the α peak even further, as expected. The low strain results in Figs. 2(a) and 6(a) present some difficulty in interpretation because of the entirely different dislocation structures in crystals deformed at 450°K (predominantly edge in character) and at 156°K (predominantly screw in character[9,10]). However, it should be recognized that the low temperature structure contains edge segments in the form of cusps, dipole trails and elongated loop ends, and that the higher temperature structure contains edge dislocations in the form of dipoles and multipoles; in both cases these non-screw segments may be in optimum configurations under a high enough local internal stress that an α peak can be generated at high strain amplitudes. Both low strain structures tend to be broken or tangled up at higher strains, so that an α peak is difficult to generate until the overall internal stress is large enough.

With regard to the effect of a bias stress, the increase in the α peak height (Fig. 5) is explained by the increased number of non-screw segments which may be depinned under the combined action of the internal and external (bias) stress. The reduction in the background with bias stress indicates that the average loop length is being reduced. Usually such a reduction is achieved by higher impurity contents, annealing and very high deformations

(2,7). In the present case, the bias stress must have the effect of bowing out dislocation segments with long loop lengths, leaving shorter loop lengths to contribute to the background damping.

5. SUMMARY

The internal friction spectrum of Mo single crystals deformed systematically in tension at 450°K and 156°K has been investigated as a function of temperature, deformation strain, strain amplitude and bias stress, giving the following results:

1. After deformation at 450°K, a step-like background damping first develops and then a broad α peak, which grows, sharpens and moves to lower temperatures with increasing strain hardening. After deformation at 156°K, the step-like background also develops but it is only possible to generate a smaller broad α peak.

2. After low deformations, an α peak can be produced out of the step-like background by the application of high strain amplitudes.

3. After higher deformations, the α peak existing at low strain amplitudes can be increased in height and moved to lower temperatures by the application of high strain amplitudes and bias stresses.

4. The behavior of the α peak can be consistently interpreted in terms of the influence of internal and external stresses on the depinning of kinks from point defects. The step-like background is damping of the vibrating string or kink type.

ACKNOWLEDGMENT

This research was partly supported by the National Science Foundation, Grant No. GK4154.

REFERENCES

1. A.S. Nowick and B.S. Berry, Anelastic Relaxation in Crystalline Solids, Academic Press, p. 371 (1972).

2. R. Gibala, M.K. Korenko, M.F. Amateau and T.E. Mitchell, J. Phys. Chem. Solids 31, 1889 (1969).

3. P.S. Venkatesan and D.N. Beshers, J. Appl. Phys. 41, 42 (1970).

4. G. Alefeld, J. Filloux and H. Harper, Dislocation Dynamics, A.R. Rosenfield et al., eds., McGraw-Hill, p. 191 (1967).

5. P.S. Sklad, M.S. Thesis, Case Western Reserve University, 1972.

6. E.A. Kenik, M.S. Thesis, Case Western Reserve University, 1972.

7. M.K. Korenko, T.E. Mitchell and R. Gibala, Phil. Mag., to be published.

8. A.V. Granato and K. Lücke, Physical Acoustics, Vol. IV, Part A, W.P. Mason, ed., Academic Press, p. 225 (1966).

9. H.W. Loesch and F.R. Brotzen, J. Less Common Metals 13, 565 (1967).

10. D. Vesely, Phil. Mag. 27, 607 (1973).

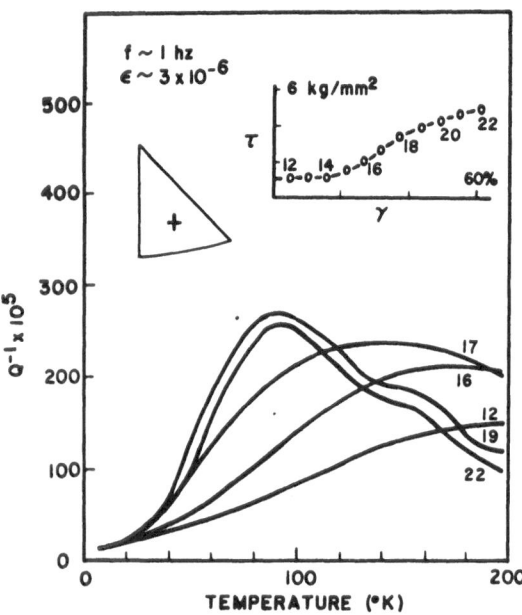

Figure 1: Torsional pendulum internal friction as a function of prestrain for "easy glide" molybdenum single crystal deformed at tension at 450 K.

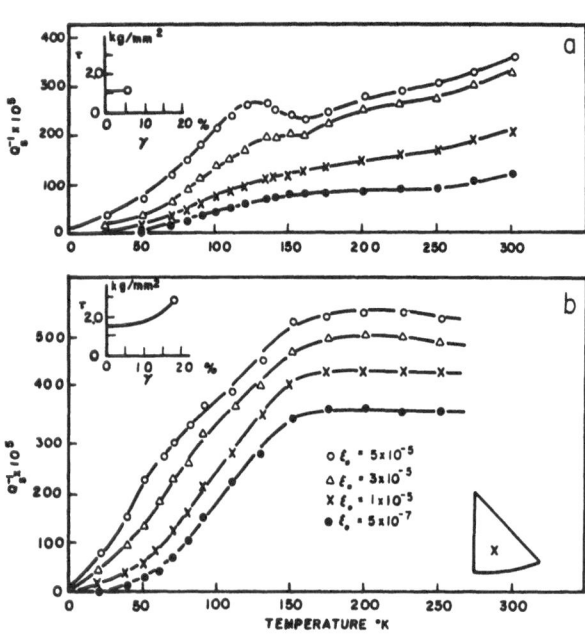

Figure 2: Resonant bar internal friction as a function of strain amplitude for "easy glide" molybdenum crystals deformed in tension at 450 K. a.) $\gamma = 5.6\%$ b.) $\gamma = 18.2\%$

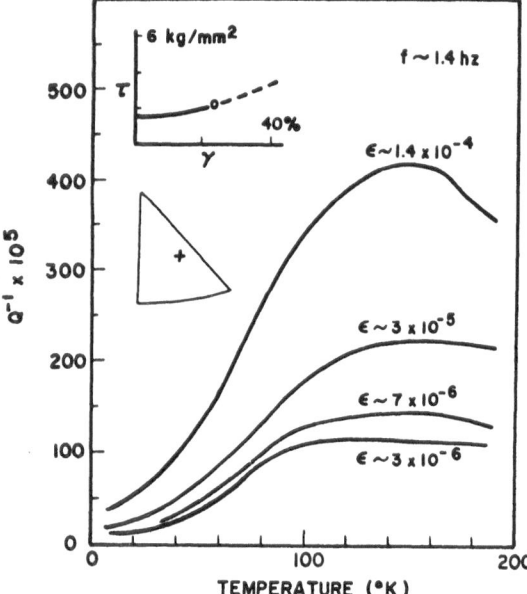

Figure 3: Torsional pendulum internal friction as a function of strain amplitude for "easy glide" molybdenum crystal deformed 23% at 450 K.

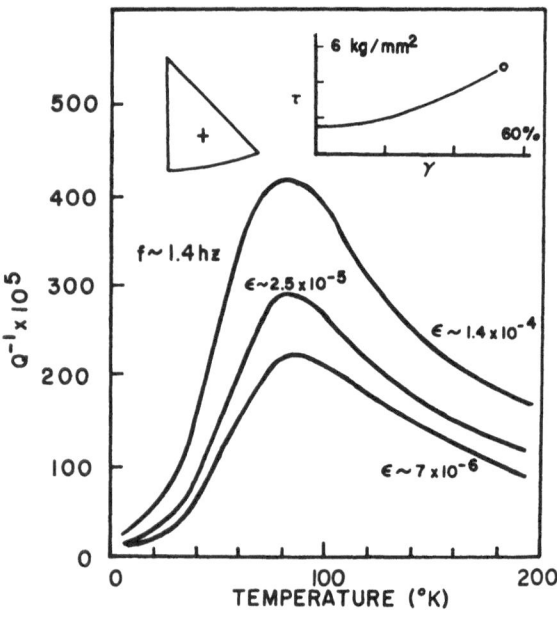

Figure 4: Torsional pendulum internal friction as a function of strain amplitude for "easy glide" molybdenum crystal deformed 55% at 450 K.

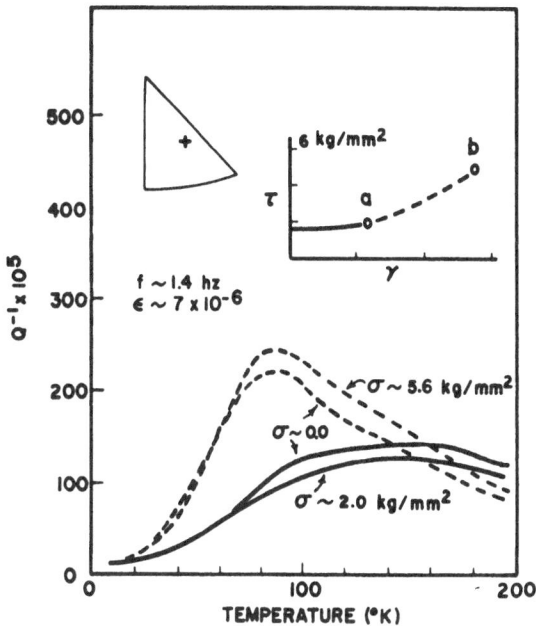

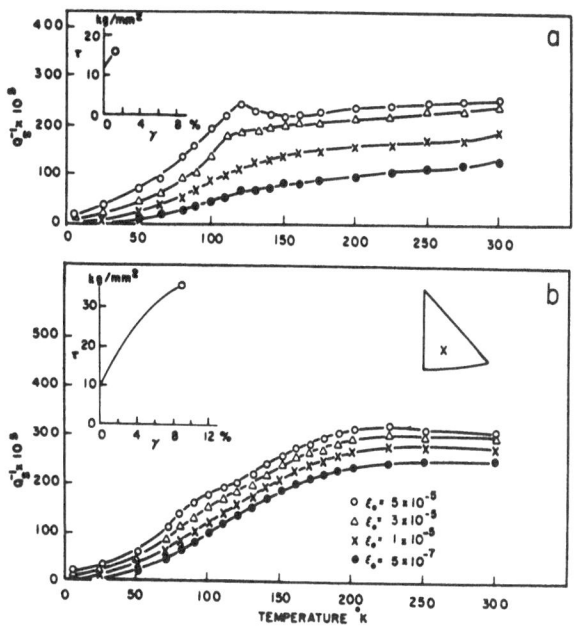

Figure 5: Torsional pendulum internal friction as a function of bias stress for "easy glide" molybdenum crystal deformed at 450 K. a.) γ = 23% b.) γ = 55%.

Figure 6: Resonant bar internal friction as a function of strain amplitude for "easy glide" molybdenum crystals deformed in tension at 156 K. a.) γ = 1.2% b.) γ = 8.6%

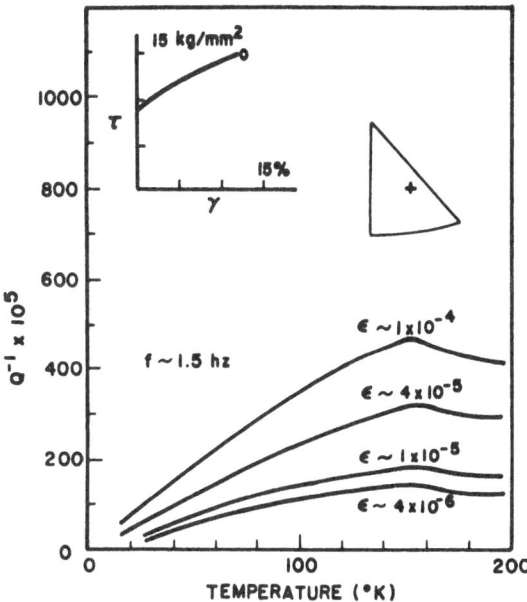

Figure 7: Torsional pendulum internal friction as a function of strain amplitude for "easy glide" molybdenum crystal deformed 12% at 156 K.

OBSERVATION OF EXTRINSIC INTERNAL FRICTION PEAKS IN DOPED ALUMINIUM NEUTRON-IRRADIATED OR COLD-WORKED AT LOW TEMPERATURES

R. GRYNSZPAN

Centre d'Etudes de Chimie Métallurgique. 15, rue G. Urbain, 94-Vitry. France

Introduction

In order to study the interactions which possibly occur between point defects and impurities, internal friction measurements have been made on diluted alloys, after fast neutron irradiation at low temperature, in the 1 cycle frequency range.

The results previously obtained on irradiated zone refined aluminium (1) have shown that no internal friction peaks appear below 200 K, although a sharp increase of the damping background occurs at higher temperatures (Fig. 1). It has been also found (1-4) that the presence of 50 to 200 atomic ppm of a foreign element in the matrix modifies strongly the spectrum of the pure metal, producing several internal friction maxima. The size, temperature and annealing behaviour of these maxima seem to be characteristic of the solute element.

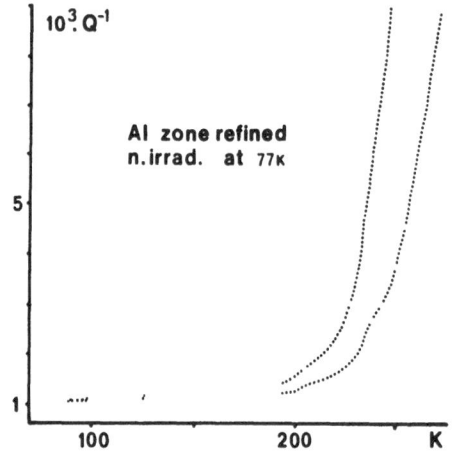

FIG. 1
Internal friction vs. temperature of two pure aluminium irradiated samples (dose $\sim 4.1 \times 10^{17}$ n.cm^{-2}). Strain amplitude and frequency of measurements : $\varepsilon \sim 1.7 \times 10^{-5}$, $\nu = 0.5$ Hz

As these modifications could either be related to irradiation effects only or could involve dislocations introduced by parasitic deformations, cold-work experiments have been made in addition to the study of the influence of the irradiation dose and the impurity content. Considering the large effect obtained with copper additions, we have centered our investigations on the aluminium-copper alloys.

Experimental

We have prepared the materials used for the present work by doping zone refined aluminium ($\sim$ 99.9997 %) with high purity copper, in a plasma furnace. The solute content varied from 10 to 200 atomic ppm, covering the whole range of concentrations of the previously studied alloys (2-4).

The samples were polycrystalline wires (diameter : 0.8 mm, length : 100 mm) obtained by drawing. They have been annealed in air at 773K during 10 hours and neutron irradiated at 20K in the C.N.R.S. loop in Fontenay-aux-Roses (Triton Reactor). The present report refers to the doses ϕ_1 (3.7×10^{16} n.cm^{-2}) and ϕ_2 (2.9×10^{17} n.cm^{-2}) measured for fast neutrons (Energy > 1 MeV). The samples were keeped several days in liquid nitrogen before study.

The solute content of each sample has been controlled by measuring, at 77K, the radioactivity of copper produced by the ^{63}Cu(n,γ)^{64}Cu reaction during the irradiation. The weight of the matrix was deduced from the sodium activity following the ^{27}Al(n,α)^{24}Na nuclear reaction.

The wires were fixed at 77K on an automaticaly drived torsion pendulum. The measurements were performed by using the free decay of the oscillations between two constant strain amplitudes corresponding to the mean value $\varepsilon \times 10^5$ = 1.1 $\pm$ 0.1. Our thermal treatment program, consisting in successive linear increases of the temperature, is schematicaly described in figure 2.

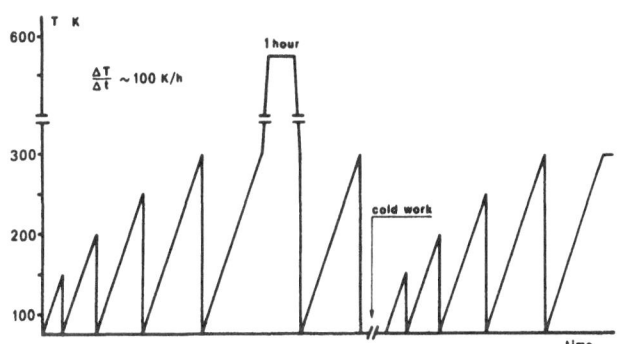

FIG. 2
Annealing program for an irradiated sample. Cold-work is made by torsion in situ

The spectrum obtained at 0.5 Hz for a sample with 150 atomic ppm of copper (Fig. 3) exhibits three groups of maxima located at 80-150K, 160-200K and 200-260K.

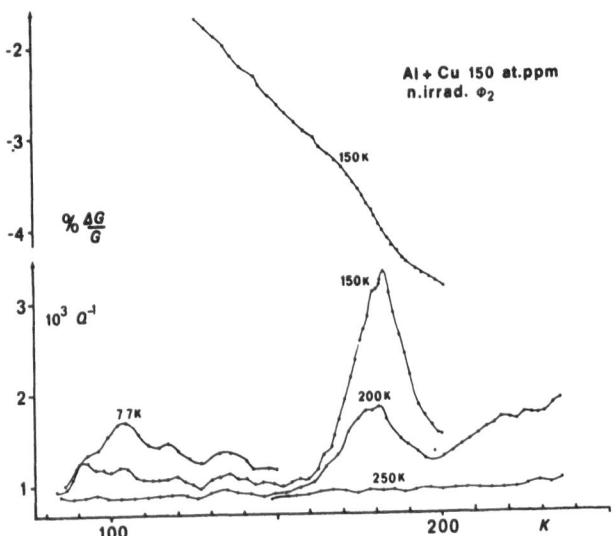

FIG. 3

Internal friction and modulus-defect of an aluminium copper alloy containing ∼150 atomic ppm irradiated at 20K. The sequence of the curves is obtained as described in figure 2. The temperature labeling each curve refer to the maximum temperature reached at the end of the preceding run.

Particular attention was paid to the maximum which occurs in the temperature range 160-200K, corresponding to the end of stage II and the beginning of stage III of the electrical resistivity recovery of an aluminium-copper alloy (5). This maximum is unstable ; it decreases sharply with the increase of the annealing temperature and has completly desappeared after heating up to 250K. Moreover, it presents a structure more complex than that showed by the initial results (1).

By increasing the frequency by a factor ∼ 2.4, the maximum shifts to a higher temperature. The shift corresponds to an apparent activation energy of 0.7 eV ± 0.1 eV. A shape analysis, using this value, indicated that the maximum does not correspond to a single Debye peak. It can be decomposed into two or several components. The main peak, noted π_1 (4), should be located at 182 ± 2K. A subsidiary peak occurs around 165K. Its presence may be confirmed by the observation of the spectrum of an alloy containing 50 atomic ppm (Fig. 4) which clearly shows a maximum at about 162K. This temperature can be related to that of the P_B Hasiguti peak found at low frequency by Perez (6) after cold-work.

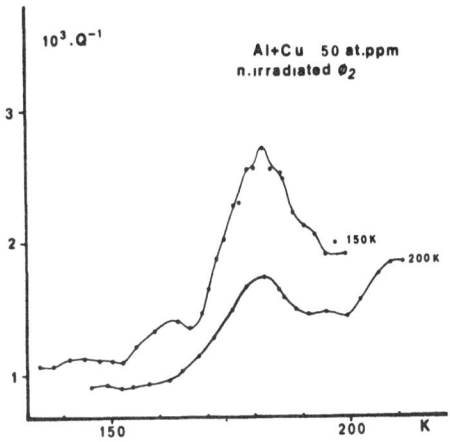

FIG. 4

Structure and annealing behaviour of the internal friction peak π_1 (at $\sim$ 182K) for an aluminium alloy containing $\sim$ 50 atomic ppm of copper, irradiated at 20K

The influence of solute content and of the neutron irradiation dose on the height of peak π_1 are reported on figure 5 for the doses ϕ_1 and ϕ_2. No peak appear for the irradiated pure metal. For the ϕ_2 dose and for the run following the annealing at 150K, there appears a rather linear relationship between the peak size and the impurity content up to 100 atomic ppm. At higher contents, a saturation is observed, also apparent for the lower dose ϕ_1, or after heating up to 200K. On the other hand, the maximum peak height is approximately proportional to the dose.

These results indicate that peak π_1 is an extrinsic peak due to the presence of the impurity. We do not observe it in the internal friction spectrum of a cold-worked pure aluminium sample (Fig. 6) which reaches its lowest damping value between 170 and 200K. Nevertheless, for an irradiated Al-Cu alloy, annealed in situ at high temperature to eliminate the point defects, a cold-work at 77K gives rise to the intrinsic Bordoni peak B_2 and the Hasiguti peaks P_A and P_B (6,7,8). It also produces the extrinsic peak π_1 and the P_c peak of Perez (called π_2 in the present work) which does not appear with a significant height in the pure metal.

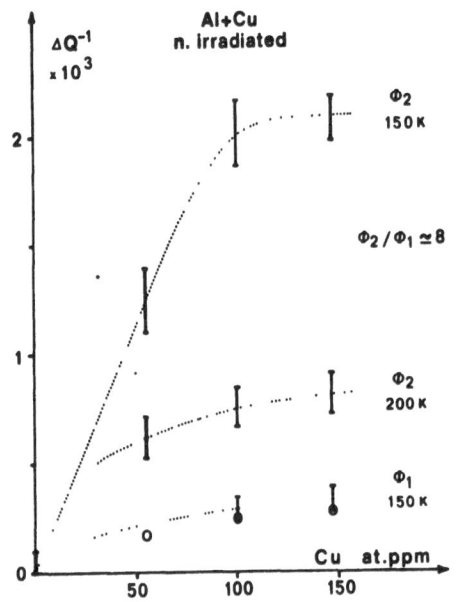

FIG. 5
Height of the internal friction peak π_1 vs. solute content for two irradiation doses $\emptyset_1$ and $\emptyset_2$. The temperature labelling each curve refers to the maximum temperature reached before measurement. The circles ploted on the lower dose curve $\emptyset_1$ are deduced from the heights found for the dose $\emptyset_2$, assuming they are proportional to the dose for a given impurity content

Discussion

Impurities are able to trap point defects and lead to the formation of point defect-impurity complexes as has been concluded from the resistivity recovery measurements (5,9). We have considered that these complexes could give rise to relaxation peaks. Possible mechanisms are :

1° a Snoek type relaxation of low symetry complexes, either under the action of the applied stress only (10) or in the field stress of the oscillating dislocations as proposed by Seeger and Wagner (11) for the relaxation of a dumbbell interstitial.

2° a relaxation involving parasitic dislocations in a Hasiguti-type mechanism, as Hasiguti and coworkers (12) have found in certain copper bases alloys.

The point defects involved in the impurity complexes could be interstitials or vacancies. Interstitial-copper complexes should be stable up to about 220K, enhancing the first part of stage III of the resistivity recovery as proposed by C. and D. Dimitrov (5). Destruction of these complexes could occur by detrapping of the interstitials or by annihilation with the antagonist defects, vacancies when the latter become mobile. On the other hand, vacancy-impurity complexes possibly formed under the neutron flux may have an anisotropic structure producing a relaxation phenomenon. If the vacancies

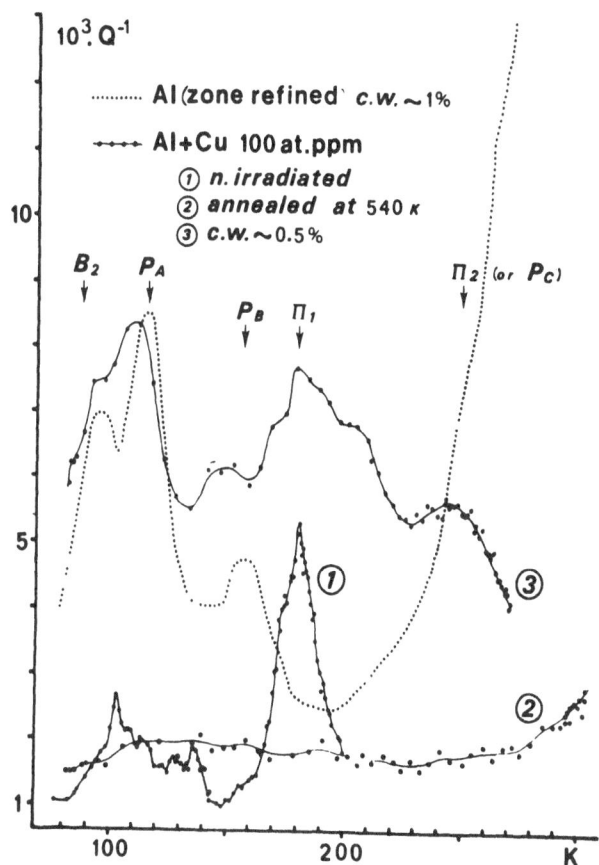

FIG. 6

Comparison of the internal friction spectra of cold-worked or irradiated
$(3.3 \times 10^{17} \text{ n.cm}^{-2})$ materials at 77K. Curve 1 is obtained using a mean strain
amplitude of $\varepsilon = 1.1 \times 10^{-5}$. All other measurements shown in this diagram
have been made with a strain amplitude of $\varepsilon = 1.7 \times 10^{-5}$

are mobile in stage III, they could migrate towards the complexes and increase their size. The so formed clusters would not be able to relax but would strongly pin dislocations giving a very low damping background as shown in figure 3 after annealing at 250K.

Since the internal friction background and the shape of the maxima adjacent to π_1 are not easy to determine for a cold-worked alloy, it is difficult to find out how the peak size of π_1 and the amount of cold-work are related. Therefore it is not yet possible to definitively choose between the proposed main mechanisms 1 and 2.

Acknowledgements

The author is grateful to Dr M. Fédoroff from the C.E.C.M. Vitry for helpful advice in activation analysis and calculator programming.

References

1. R. Grynszpan, C. Dimitrov and O. Dimitrov, C.R. Acad. Sci. Paris, 271 C, 261 (1970)

2. R. Grynszpan, Discussion-meeting on Internal Friction of Metals, Gemund, W. Germany (5-8 oct. 1970)

3. R. Grynszpan, Thèse de 3ème cycle Paris (june 1971)

4. R. Grynszpan and O. Dimitrov, Communication aux Journées d'Automne de la Soc. Française de Métallurgie, Massy-Palaiseau (oct. 1972) to be published in Mém. Sci. Rev. Mét.

5. C. Dimitrov-Frois and O. Dimitrov, Mém. Sci. Rev. Mét. 65, 425 (1968)

6. J.P. Perez, P. Peguin and P. Gobin, Rev. Phys. Appl. 4 (4) 437 (1969)

7. J.L. Routbort and H.S. Sack, Phys. Stat. Sol. 22, 203 (1967)

8. R.R. Hasiguti, N. Igata and G. Kamoshita, Acta Met. 10, 442 (1962)

9. C. Dimitrov and O. Dimitrov, Phys. Stat. Sol. 34, 545 (1969)

10. A.S. Nowick, Adv. Physics 16, 1 (1967)

11. A. Seeger and F.J. Wagner, Phys. Stat. Sol. 9, 583 (1965)

12. R.R. Hasiguti, C.R. "3. Internationales Symposium Reinststoffe in Wissenschaft und Technik" Dresden (4-8 may 1970)

THEORY OF DISLOCATION DAMPING IN DILUTE ALLOYS

J. Schlipf and R. Schindlmayr

Institut für Allgemeine Metallkunde und Metallphysik
Technische Hochschule Aachen, W.-Germany

In dilute alloys, the interaction of impurity atoms with dislocations may give rise to three damping mechanisms: (i) the breakaway mechanism due to breakaway of the dislocations from segregated atoms; (ii) the impurity drag mechanism, which is due to the drag exerted on the dislocation, if the pinning points are mobile. (iii) A third mechanism comes into play, if a free dislocation moves through an array of immobile point obstacles. It is this mechanism that will be considered in some detail here. As will be shown it also exerts a drag on the dislocation, and we therefore may call it the interaction drag mechanism.

It is well known [1,2,3] that in the presence of randomly distributed point obstacles the dislocation assumes a zigzag configuration (fig. 1), characterized by an average amplitude z and a mean loop length l. Both quantities depend on concentration and on whether the interaction is repulsive [1], attractive [2], or of mixed character [3]. The most interesting case seems to be that of a mixed population of an equal number of attractive and repulsive obstacles. Detailed calculations, therefore, have been made for this special case [4].

We consider a dislocation line, which is kept fixed at its end points and bows out under an applied stress σ. The overall shape of the dislocation is determined by the line tension (fig. 2). However, if point obstacles are present, the zigzagging will be superimposed on a small scale, and will thus controll the kinetics of motion. Nevertheless, as long as the total length of the dislocation L $\gg$ l, the dislocation can be approximated by a smooth string, which moves continuously.

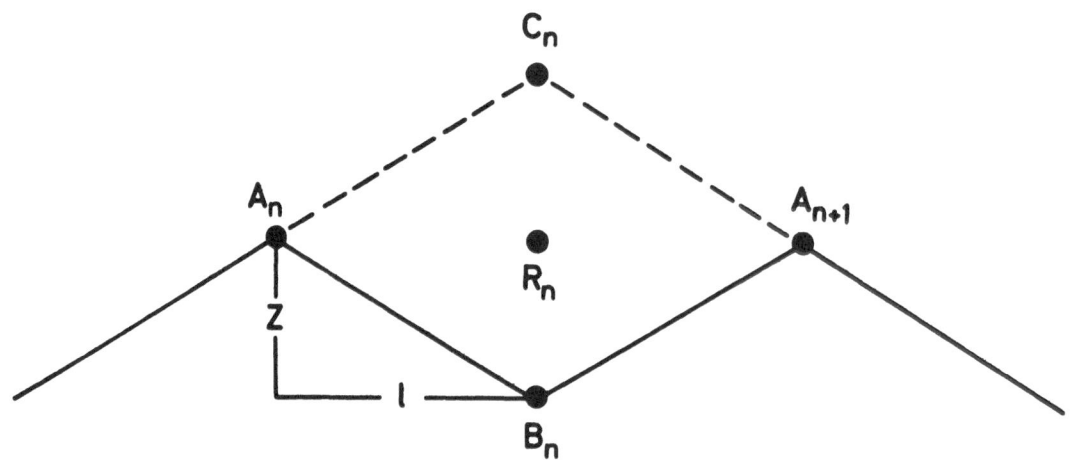

Fig. 1

Element of a zigzag dislocation showing zigzag length l and amplitude z.
A_n, B_n, A_{n+1}, C_n denote attractive obstacles, R_n = repulsive obstacle.

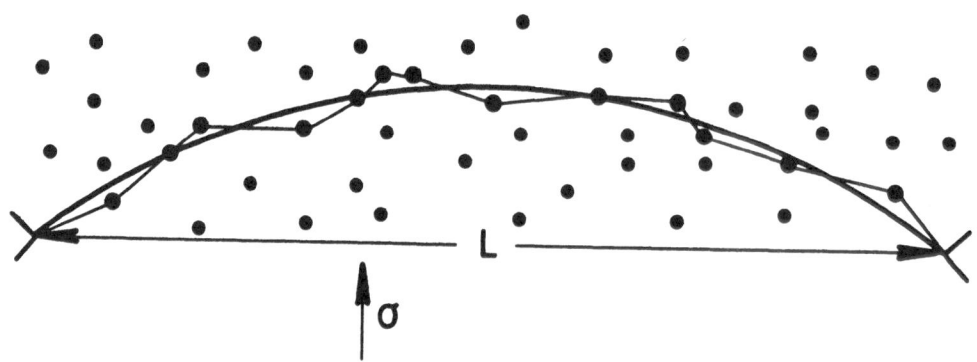

Fig. 2

A dislocation segment L bows out under an applied stress σ. The
average course of the dislocation is a parabolic arc. The small scale
zigzagging due to the interaction with the glide obstacles is indicated.

As has been shown elsewhere [4] a mobility m can be assigned to such a
dislocation, which may be defined by setting the velocity

$$v = mb^2 \sigma_{eff}.\qquad(1)$$

σ_{eff} is an effective stress which in the most simple case is composed
of the external stress σ and a back stress $\frac{\mu}{b} d^2y/dx^2$ due to the line
tension μ:

$$\sigma_{eff} = \sigma + \mu y''/b.$$

The solution to the corresponding equation of motion

$$\dot{y} = mb (b\sigma + \mu y'')\qquad(2)$$

in the case of constant mobility has been published by several
authors [5,6,7]. The damping and modulus defect can be obtained in
closed form and are given by

$$\delta \propto \Lambda L^2 \frac{Gb^2}{2\mu} \frac{1}{\omega\tau} \left[1 - \frac{\sinh\sqrt{\omega\tau} + \sin\sqrt{\omega\tau}}{\sqrt{\omega\tau}\left(\cosh\sqrt{\omega\tau} + \cos\sqrt{\omega\tau}\right)} \right]\qquad(3)$$

and

$$\frac{\Delta G}{G} = \Lambda L^2 \frac{Gb^2}{2\mu} \frac{1}{(\omega\tau)^{3/2}} \frac{\sinh\sqrt{\omega\tau} - \sin\sqrt{\omega\tau}}{\cosh\sqrt{\omega\tau} + \cos\sqrt{\omega\tau}}\qquad(4)$$

respectively. Λ = dislocation density. The corresponding curves are
shown in fig. 3.

$$\tau = \frac{L^2}{2\,m\,\mu b}\qquad(5)$$

and represents the relaxation time of a dislocation string of mobility m.

The closed form solution is not very practical in evaluating experiments. Therefore, an approximation has been worked out which gives both
simple and accurate results. It is well known (see for instance [4,8])
that the displacement y(x) of a smooth dislocation can be excellently
approximated by a cosine function.

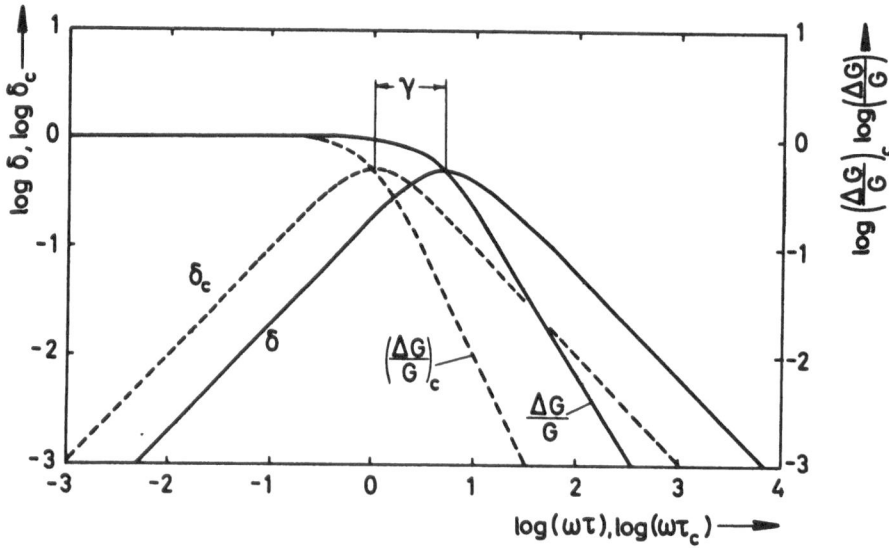

Fig. 3

Exact damping peak δ and modulus defect $\Delta G/G$ for the linear drag model, eq. (2) (solid lines). δ_c and $(\Delta G/G)_c$ give the cosine approximation discussed in the text (dotted lines).

We then have $y'' \sim - y$ and eq. (2) leads to a simple Debye peak of the damping and to a modulus defect given by

$$\delta_c = \frac{16}{\pi^4} \Lambda L^2 \frac{Gb^2}{2\mu} \frac{\omega \tau_c}{1 + \omega^2 \tau_c^2} \qquad (6)$$

$$\left(\frac{\Delta G}{G}\right)_c = \frac{16}{\pi^4} \frac{Gb^2}{2\mu} \frac{1}{1 + \omega^2 \tau_c^2} \qquad (7)$$

respectively.

$$\tau_c = \frac{L^2}{\pi^2 \, m \, \mu b} \qquad (8)$$

is the relaxation time associated with the cosine approximation.

The corresponding curves are shown in fig. 3. While the peak δ_c is located at $\omega \tau_c = 1$, δ has its maximum at $\omega \tau = y$, where $y \approx 4.951$. However, the two sets of curves can be made to almost coincide, if they are shifted by log 4.951 along the log $\omega\tau$ axis. The exact solution, therefore, can be excellently approximated by an empirical Debye peak

$$\delta_E = K_E \Lambda L^2 \frac{Gb^2}{2\mu} \frac{\omega \tau_E}{1 + \omega^2 \tau_E^2} \tag{9}$$

and a modulus defect

$$\left(\frac{\Delta G}{G}\right)_E = K_E \Lambda L^2 \frac{Gb^2}{2\mu} \frac{1}{1 + \omega^2 \tau_E^2} \tag{10}$$

where now

$$\tau_E \equiv \frac{\tau}{\gamma} = \frac{L^2}{2\gamma m \mu b} \ . \tag{11}$$

τ_E and K_E differ by only o.3 % from τ_c as given by (8) and $K_c = 16/\pi^4$. This again is evidence that the cosine displacement of the dislocation is an excellent approximation in dynamical problems.

Now let us have a closer look to what is going on in the small zigzags controlling the mobility in the interaction drag model. The elementary step in the motion of a zigzag dislocation is the breakaway from a pinning point at a reentrant angle (B_n in fig. 1) and its recapture at the complementary obstacle (C_n) [2]. In order to describe the kinetics of this process the following model has been designed [3] (fig. 4).

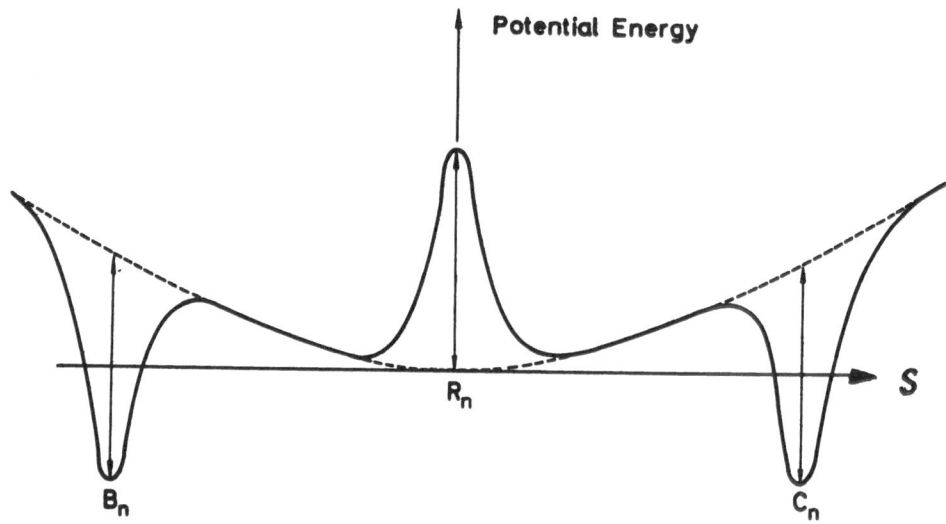

Fig. 4

Potential energy profile along the reaction path B_n, R_n, C_n for break-away and recapture. The reaction coordinate is denoted by S.

At low stresses the dislocation will interact with the attractive obstacles only, while the repulsive ones may be thought of as being located at the center (R_n) of each of the diamond-shaped average zigzag units. As the dislocation moves from B_n through R_n to C_n it sees a potential which is depicted schematically in fig. 4. There are minima at the attractive obstacles B_n and C_n and a maximum at the repulsive obstacle R_n.

Denoting by ω_0 the jump frequency from B_n to C_n at zero stress we have shown elsewhere [4] the net transition rate to be

$$\omega_0 \sinh \left(\frac{b^3 \, \sigma_{eff}}{c \, k \, T} \right)$$

when a stress σ is applied. σ_{eff} is the effective stress defined above and c is the concentration of obstacles. The dislocation velocity is then given by

$$\dot{y} = z \, \omega_0 \, \sinh \left(\frac{b^3 \, \sigma_{eff}}{c \, k \, T} \right) \tag{12}$$

where z denotes the jump distance. For small σ_{eff} (i.e. $\sigma_0 < ckT/b^3$) this reduces to the linear equation (2) treated above and yields a mobility

$$m = \frac{zb\,\omega_0}{ckT} \qquad (13)$$

For this linear case we have shown above that the cosine approximation gives an excellent result. It is tempting, therefore, to use this approximation also for the case of higher stresses. In so doing we specify the spatial part of the displacement, and eq. (12) holds for the time dependent part only. If we write $y(x,t) = \eta(t)\cos\frac{\pi x}{L}$ we have

$$\sigma_{eff} = \sigma - \frac{\mu}{b}\frac{\pi^2}{L^2} \qquad \text{and}$$

$$\dot{\eta} = z\,\omega_0\,sinh\left(\alpha - \eta/\eta_0\right) \qquad (14)$$

where $\alpha = \dfrac{\sigma}{G} \cdot \dfrac{Gb^3}{ckT}$; $\qquad (15)$

$$\frac{\eta_0}{b} = \frac{2}{\pi^2}\cdot\frac{kT}{2\mu b}\left(\frac{L}{b}\right)^2 c \ . \qquad (16)$$

Even so, a solution in closed form can be obtained only for very special forms of the periodic stress. The results shown in fig. 5 have been obtained by assuming a linearly varying stress, as in cyclic deformation:

$$\sigma = \sigma_0\,\frac{2\omega}{\pi}\,t \ ; \qquad\qquad -\frac{\pi}{2\omega} \le t \le \frac{\pi}{2\omega}$$

$$\sigma = \sigma_0\,\frac{2\omega}{\pi}\left(\frac{\pi}{\omega} - t\right) ; \qquad \frac{\pi}{2\omega} \le t \le \frac{3\pi}{2\omega}$$

and so on. This gives us $\dot{\sigma} = \pm\,\sigma_0\,\frac{2\omega}{\pi} = $ const.

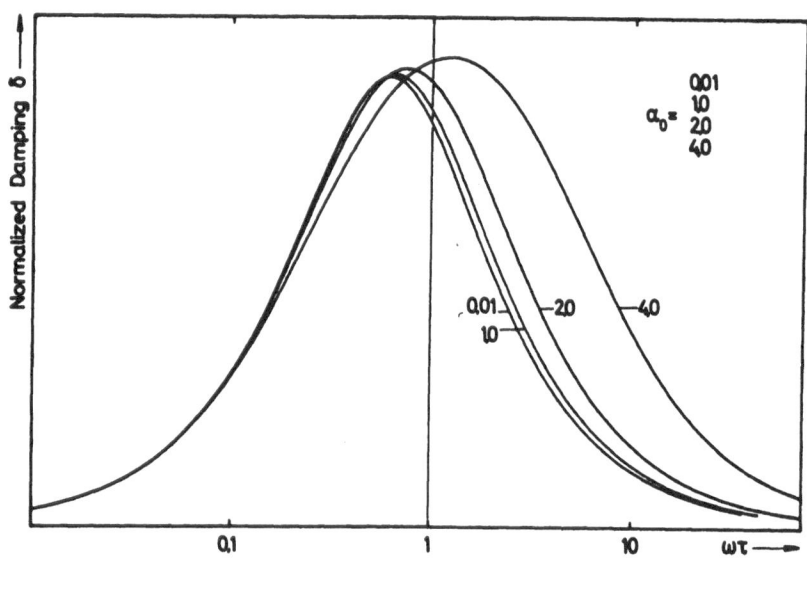

Fig. 5

Plot of the damping due to interaction drag as a function of log $\omega\tau$. α_o denotes the normalized stress amplitude.

Setting $u = \alpha - \eta/\eta_o$ eq. (14) for the quarter period of rising stress becomes

$$\dot{u} = \frac{2\omega}{\pi}\,\alpha_o - \frac{1}{\tau}\,\sinh u \qquad (17)$$

Upon integration of (17) we obtain $\eta(t)$:

$$\eta(t) = z\,\omega_o\,\tau\left[A + \frac{2\omega}{\pi}\alpha_o\,t - \ln\frac{1 - q_1\,e^{-\varkappa t}}{1 + q_2\,e^{-\varkappa t}}\right] \qquad (18)$$

where

$$A = \ln\frac{1 - p + \sqrt{p^2 + 1}}{1 + p - \sqrt{p^2 + 1}} \quad ; \quad \varkappa = \frac{2\omega}{\tau}\,\alpha_o\sqrt{p^2 + 1}\;.$$

446

$$q_1 = \frac{\sqrt{p^2+1} + p - 1}{\sqrt{p^2+1} + p + 1} \left\{ \left[1 + p^2 \cosh^2 \left(\frac{\pi \kappa}{2\omega} \right) \right]^{1/2} - p \cosh \left(\frac{\pi \kappa}{2\omega} \right) \right\}$$

$$q_2 = \frac{\sqrt{p^2+1} + p + 1}{\sqrt{p^2+1} + p - 1} \left\{ \left[1 + p^2 \cosh^2 \left(\frac{\pi \kappa}{2\omega} \right) \right]^{1/2} - p \cosh \left(\frac{\pi \kappa}{2\omega} \right) \right\}$$

$$p = \frac{\pi}{2} \frac{G b^2}{2\mu} \frac{1}{\alpha_0 \omega \tau} \quad .$$

From this the logarithmic decrement can be calculated in the usual way to give

$$\Delta = \left(\frac{2}{\pi} \right)^3 \Lambda L^2 \frac{G b^2}{2\mu} \left\{ \frac{2A}{\alpha_0} + \frac{2\omega}{\pi \kappa} \left[Li_2 \left(q_1 e^{-\pi \kappa/2\omega} \right) \right. \right.$$

$$\left. \left. - Li_2 \left(q_1 e^{\pi \kappa/2\omega} \right) + Li_2 \left(-q_2 e^{\pi \kappa/2\omega} \right) - Li_2 \left(-q_2 e^{-\pi \kappa/2\omega} \right) \right] \right\} \quad (19)$$

where $Li_2 (x) = - \int_0^x \frac{\ln (1 - \xi)}{\xi} d\xi$ denotes a function known as dilogarithm.

In fig. 5 $\delta = \Delta/\pi$ is plotted against $\omega \tau_c$ for several stress amplitudes σ_0. For very small σ_0 the simple Debye peak obtains, as it must be. With increasing σ_0 the high temperature ($\omega \tau_c < 1$) side of the peak remains unchanged. The low temperature ($\omega \tau_c > 1$) side is shifted towards higher $\omega \tau_c$ in such a way that the peak remains symmetric. At the same time the peak height changes slightly. This behavior is due to the fact that at low temperatures the mobility increases exponentially with stress.

On the other hand, at constant temperature the damping is amplitude dependent. At high temperatures ($\omega \tau_c < 1$) the damping decreases with increasing stress amplitude. At low temperatures ($\omega \tau_c > 1$) it first increases then decreases with increasing stress amplitude. Since the interaction drag model as presented here involves comparatively long dislocation segments interacting with random point obstacles it will apply to freshly deformed dilute alloys or materials of low purity.

References:

1) J. Friedel, Dislocations, p. 225, Pergamon Press (1964)

2) N.F. Mott, Imperfections in Nearly Perfect Crystals, Wiley (1950)

3) R. Schindlmayr and J. Schlipf, Phil. Mag. to be published

4) J. Schlipf and R. Schindlmayr, Phil. Mag. to be published

5) J. Weertman, J. appl. phys. 26, 202 (1955)

6) D.O. Thompson and V.K. Paré, Physical Acoustics

7) H.M. Simpson and A. Sosin, Phys. Rev. B, 5, 1382 (1972)

8) A. Granato and K. Lücke, J. appl. phys. 27, 583 (1956)

AMPLITUDE DEPENDENT

INTERNAL FRICTION

THEORY OF DAMPING DUE TO THERMALLY ACTIVATED BREAKAWAY
OF DISLOCATIONS FROM EQUALLY SPACED PINS**

D.H. Rogers*, D.G. Blair and T.S. Hutchison
Royal Military College
Kingston, Ontario
K7L 2W3, Canada

Abstract

The theory has been developed in several steps (1) The dynamics of the thermally assisted breakaway of a dislocation is studied using a new approximation: the "independent joint approximation". Above a stress level, σ_s, it is found that breakaway of the major loop is activated at a single pin, i.e., breakaway is catastrophic as in the Granto-Lücke theory, below σ_s more than one minor pin in a major loop length must be broken before major breakaway occurs. (2) Effective activation energy functions for the two regions are found. (3) The activation diagram is introduced to show on a temperature, stress-amplitude plane the variation in damping behaviour. (4) Three different mathematical approximations are used to derive expressions for the damping yielding five regions of different damping behaviour in the T, σ_s plane. Some of the experimental expectations resulting from the theory are discussed.

The Dynamics of Breakaway

We consider a dislocation of Burger's vector b which is pinned by a row of identical, breakable, pinning agents (minor pins) spaced a distance L_c apart. The dislocation is also pinned by unbreakable pins spaced a distance L_N apart where $L_N \gg L_c$. Each minor pin exerts a force $U'(y)$ on the dislocation, Figure 1 where y is the displacement of the "joint" from the pin. (By a "joint" we mean that point on the dislocation opposite·a pin, called an "anti-pin" by some authors). The pinning energy, $U(y)$ is assumed to rise monotonically from 0 at y=0 to U_o as y→∞ and to be a fixed function of y. We define an interaction range:

$$r \equiv U_o / U'_{max} \qquad [1]$$

We assume that the force $U'(y)$ is bounded, increases monotonically as y increases from 0 to y_1 and decreases monotonically thereafter, and that from $y=y_2$ to ∞ the force has the asymptotic behaviour

$$U'(y) = KU'_{max} \left(\frac{y_2}{y}\right)^{\gamma+1} \qquad [2]$$

where K, y_2 and γ are positive constants. The process of thermal activation involves the movement of a joint from a position $y_p < y_1$ to a position $y_s > y_2$ where the force exerted by the pin is the "same" (depending on some other

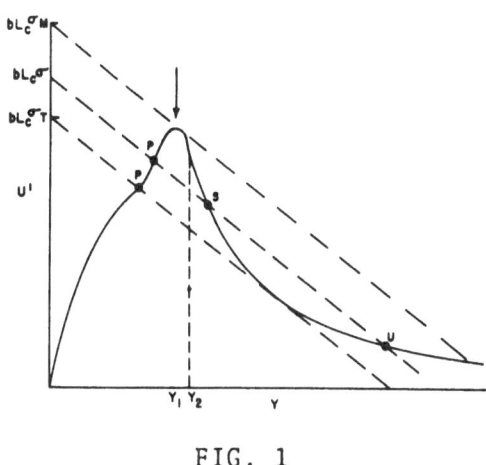

FIG. 1

The pinning force as a function of the distance of the "joint" from the pin.

factors) at both positions. Note that as the force exerted is changed, that y_p varies only alittle, y_s varies a lot, so that the energy difference between the two positions is substantially independent of $U(y_p)$.

Neglecting the Peierls' force and the viscous drag forces on the moving dislocation, we consider only the line tension force of coefficient C, and a force $b\sigma$ per unit length due to the applied shear stress σ. The energy of the dislocation is

$$E = \sum_i [U(y_i) - \tfrac{1}{2} b\sigma L_c (y_{i+1} - y_i) + \tfrac{1}{2} C (y_{i+1} - y_i)^2 L_c^{-1}$$
$$- b^2 \sigma^2 L_c^3 / 24C \qquad [3]$$

the four terms arising from the pins, the force from the applied shear stress, the stretching of the dislocation due to the displacement of the joints, and the bowing of the dislocation between the joints, respectively. When the whole dislocation is in equilibrium $\delta E/\delta y_i = 0$ so that the ith joint is acted on by three forces which may be written in dimensionless terms as

$$rU'(y_i/r)/U_0 - bL_c r\sigma/U_0$$
$$- \beta^{-1}\{(y_{i+1}/r) - (2y_i/r) + (y_{i-1}/r)\} = 0 \qquad [4]$$

where $\beta = L_c U_0 / Cr^2$ is a parameter first introduced by TGL[1] (except their r refers to another characteristic length in the triangle force law which they used.)

The strength of coupling between each joint and its neighbour is β^{-1}: where β is large the dislocation is flexible; where β is small neighbouring joints move together. We restrict ourselves to the case of large β in the following, i.e., long loop lengths.

In the independent joint approximation[2] we drop the term in equation (3) corresponding to the β^{-1} term in (4) since this term is small for long loop lengths. We assume also that broken pins exert no force on the dislocation and

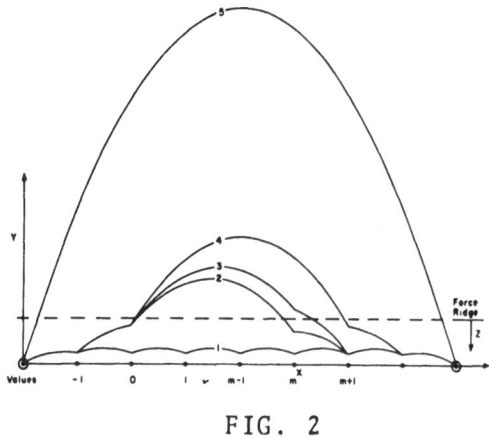

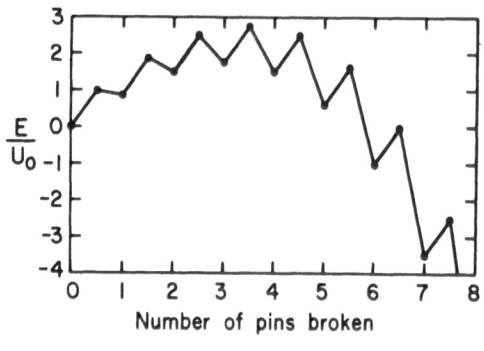

FIG. 2   FIG. 3

Figure 2:

The fully pinned configuration (1), the static configuration where m-1 pins are broken but the mth joint is pinned (2), the saddle state where the mth joint is on the other side of the force ridge (3), the stable state where the mth joint is unpinned (4), the configuration with all the minor pins broken.(5)

Figure 3:

The energies of stable and unstable states as consecutive pins are broken (by thermal activation) at a specific stress level $\sigma=0.35\sigma_s$. Breakaway of the whole major loop does not occur until four pins have been broken. The effective activation energy for major breakaway is approximately the maximum on this curve.

hence each contributes energy U_0 to it. Then, considering the equilibrium configuration in which the dislocation has broken away from m-1 consecutive pins, see Figure 2: configuration 2, 3 , we use equation 3 to write the difference between the energy of the configuration and the completely pinned configuration 1:

$$E_{ms} = U(y_m) + (m-1)U_0 - \tfrac{1}{2}(m-1)bL_c\sigma y_m - (m^3-m)b^2L_c^3\sigma^2/24C \qquad [5]$$

where we have also restricted our attention to low stress values so that we could set $U(y_p) \approx 0$ and $y_p \approx 0$ using the shape of the force law curve as indicated earlier. Equation (5) is the saddle state for the breaking of the mth pin; that is the mth joint is at "s" (Figure 1). The energy of configuration 2 of Figure 2: the stable state where the mth pin is still unbroken, is found by dropping the first and third terms from equation (5), i.e., replacing y_m by y_p

In this low stress region we can approximate equation (5) by

$$E_{ms} \approx mU_0 - (m^3-m)b^2L_c^3\sigma^2/24C \qquad [6]$$

which is approximately true for all pinning force law . If we assume the law of equation (2) a better approximation is

$$E_{ms} \approx mU_0 - (m^3-m)b^2L_c^3\sigma^2/24C$$
$$-[(\gamma+1)/\gamma]U_0\{\tfrac{1}{2}(m+1)y_2bL_c\sigma/U_0\}^{\gamma/(\gamma+1)} \qquad [7]$$

Where the additional term is small.

Studying Equation (6) we see that depending on the σ level, the energy of the saddle state associated with breaking the mth pin may either increase or decreases as m increases. Thus in Figure 3 the whole major loop, L_N, will not be broken away until four pins are broken one by one. Thus, unlike the mechanical breakaway of GL^3, thermally activated breakaway is not catastrophic, i.e, the breaking of the first pin does not cause the whole major loop to break away until a sufficiently high stress level is reached that E_{1S} is the highest point on the E vs m curve. By equating E_{1S} to E_{2S} in Equation (6) we find that the stress, σ_s, above which major breakaway is activated at a single pin is

$$\sigma_s = (4CU_0/b^2 L_c^3)^{\frac{1}{2}} \qquad [8]$$

As we move to lower stresses, major breakaway changes from being activated at m pins to m+1 pins at the stress

$$\sigma = \sigma_s/[m(m+1)]^{\frac{1}{2}} \qquad [9]$$

Thermal Activation

The activation energy for major breakaway at stresses above σ_s is clearly given by setting m=1 in Equation (6) or (7):

$$[10]$$
$$U_a(\sigma) = U_0[1 - (\gamma+1)\gamma^{-1}\{\gamma_2 b L_c \sigma/U_0\}^{\gamma/(\gamma+1)}]$$

By using Equations (9) and (6) we find that for $\sigma \ll \sigma_s$

$$E_{ms} = U_0 \sigma_s/\sigma \qquad [11]$$

for $\sigma < \sigma_s$ but approaching it in value, specific values of m may be found from Equation (9) and substituted into (6).

Ree and Eyring[4] have studied thermal activation over barriers in series and in parallel, and it is possible to show using their results that to a high precision the successive breakaway of pins by thermal activation when $\sigma < \sigma_s$ is

$$U_a(\sigma) = U_0 \sigma_s/\sigma \qquad [12]$$

The energy of the dislocation after breakaway of the whole loop, L_N, has occurred, may be calculated in the manner of Equation (3), and referred to the energy of the completely pinned state is

$$E_{BA} = -b^2 \sigma^2 L_N^3/24C \qquad [13]$$

Thus the activation energy for repinning is always this much greather than the activation energy for unpinning; and since $L_N \gg L_c$ this is usually a substantial amount.

There is, however, a stress σ_r below which the dislocation must be completely pinned; i.e. when the broken-away" dislocation bows out a distance less than the range of the pin and thus is not broken away. Thus when a sinusoidal stress

$$\sigma = \sigma_0 \sin \omega t \qquad [14]$$

is applied the major loops that breakaway during a half cycle will all be repinned by the time the stress reaches zero.

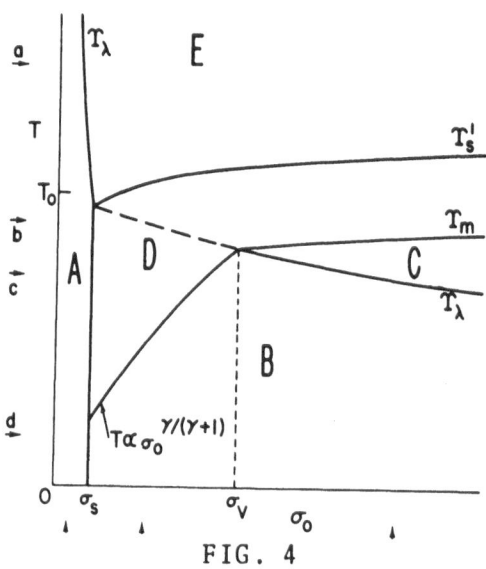
FIG. 4

Sketch of the stress-amplitude plane for very long loop lengths. (As L_c shortens region D disappears by moving $\sigma_v \to \sigma_s$). The activation region, labelledγ_λ, has a vertical width of almost $T/\ln(\nu_a/\omega)$.

The activation energies as derived above are shown in Figure 4. U_a reaches the σ axis at σ_M, the mechanical breakaway stress given by

$$\sigma_M = U_0/rbL_c \qquad [15]$$

We assume that thermal activation proceeds according to rate theory; thus if z is the fraction of the major loops that are not broken away then z changes according to:

$$dz/dt = -z\Gamma_a + (1-z)\Gamma_b \qquad [16]$$

where the Γ's are of the form

$$\Gamma_a = \nu_a \exp[-U_a(\sigma)/kT] \qquad [17]$$

Here ν_a is an attempt frequency which we assume is independent of σ, k is Boltzman's constant, T the absolute temperature. $U_a(\sigma)$ is the activation energy given by Equation (12) or (10) depending on whether σ is greater or less than σ_s.

The Activation Diagram

When a sinusoidal stress is applied to the specimen all or few of the major dislocation loops are broken away in a cycle depending on whether ω is smaller or greater than Γ_a. The boundary between the two situations, when $\omega \simeq \Gamma_a$, occurs at a temperature given by

$$T = U_a(\sigma)/k\ln(\nu_a/\omega) \qquad [18]$$

Thus a plot of U_a vs σ may be converted to a diagram of the T, σ_o plane (which we call an activation diagram[5]) by dividing the energy by $k \ln(\nu_a/\omega)$ - a constant for a given experiment. If an experiment is done with a stress amplitude and temperature such that the point T, σ_o is below or the left of the line of equation(18), then few major loops breakaway each cycle; if T, σ_o is above or to the right of the line, then all breakaway during each cycle; most of the breakaways occuring in a region of vertical width $\Delta U_a = KT$.

454

A bit of reflection shows us that this line determined by $\omega \approx \Gamma_a$ is really only a rough approximation; a better one would be $\Gamma_a \Delta t \approx 1$, where Δt is the time interval during which most of the breakaways occur, Δt is much shorter than $1/\omega$ of course. Taking this into account changes the boundaries to slightly higher temperatures and amplitudes: the most important change being the boundary between the E and C regions which becomes curved as shown in Figure 4 instead of being a straight line.

Calculation of the Decrement

Energy absorption from the cyclic stress (or sound damping) occurs when both: (a) some major loop lengths breakaway during the time of increasing stress, and (b) they repin before the next half cycle begins. The repinning is guaranteed by the disappearance of the broken away state at low enough stress. At stress levels above this repinning stress the activation energy for repinning is much greater than the activation energy for unpinning, so that the last term of Equation (16) can be dropped. The solution of the resulting equation is a very r rapidly changing function of the activation energy--enabling us to use two different mathematical approximations[7] for the regions A and B on the one hand, and, Regions C and E on the other.

Approximation 1: In regions A and B the number of broken away loops depends on how small $U_0(\sigma)$ becomes. It has its smallest value at $\sigma = \sigma_0$, thus we can assume that approximately all the loops that do breakaway do so when $\sigma = \sigma_0$. The expression for the decrement and modulus defect which result are:

Region A: $\qquad \Delta = \sqrt{2\pi} \Delta_0 (\sigma_0 T/H_1)^{\frac{1}{2}} (\nu_a/\omega) \exp[-H_1/\sigma_0 T]$ [20]

$$\text{where} \quad \Delta_0 = NGb^2 L_N{}^3/12C$$
$$\text{and} \quad H_1 = 4(2CU_0{}^3/L_c{}^3)^{\frac{1}{2}}/3bk$$

Here G is the shear modulus and N is the number of major loops per unit volume.
The modulus defect $\mu = \Delta/2$.

Region B: $\qquad \Delta = \sqrt{2\pi} \Delta_0 [T/H_2 \sigma_0{}^{\gamma/(\gamma+1)}]^{\frac{1}{2}} (\nu_a/\omega)$ [21]

$$\times \exp\left[-\frac{1}{T}\left\{\frac{U_0}{k} - \frac{(\gamma+1)}{\gamma} H_2 \sigma_0{}^{\gamma/(\gamma+1)}\right\}\right]$$

$$\text{where} \quad H_2 = (y_2 bL_c/U_0)^{\gamma/(\gamma+1)} U_0/k$$
$$\mu = \Delta/2$$

Approximation 2: In regions C and E the loops have all broken away during each stress half-cycle, absorbing a constant amount of energy. The decrement expressions are:

455

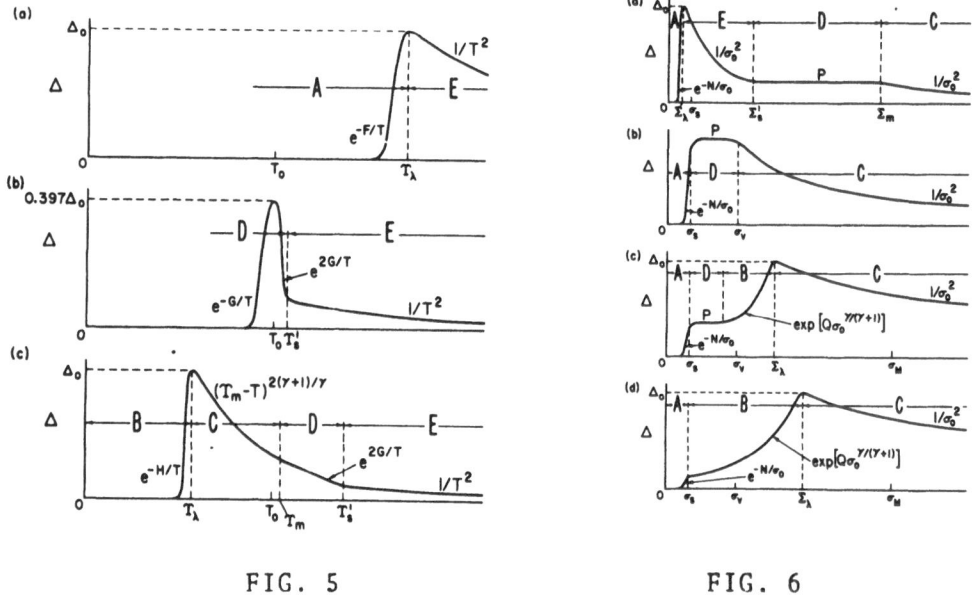

FIG. 5   FIG. 6

Figure 5: Schematic graphs of decrement as a function of temperature for very long loop lengths. The stress amplitude increases from (a) to (c). The dominant factors of the decrement as a funtion of temperature are shown where F, G, and H are positive constants. The specific temperatures and regions are indicated on Figure 4.

Figure 6: Schematic graphs of decrement as a function of stress amplitude for very long loop lengths. The temperature decreases from (a) to (d). The dominant factors of the decrement as a function of stress are shown where N and P are positive constants.

Region E:

$$\Delta = \Delta_0 (\Sigma_\lambda / \sigma_0)^2$$

where $\Sigma_\lambda = U_0 \sigma_s / kT \ln (\nu_a / \omega)$ to a first approximation.

[22]

Region C:

$$\Delta = \Delta_0 (\Sigma_\lambda / \sigma_0)^2$$

where $\Sigma_\lambda = [U_0 / y_2 bL_c]\{\gamma(\gamma+1)^{-1}[1-(kT/U_0) \ln(\nu_a/\omega)]\}^{(\gamma+1)/\gamma}$

Approximation 3: When the stress first rises above σ_s and single-pin activated breakaway begins, if the distance L_c is long enough the stress will rise to some value $\sigma(T)$ before the activation energy $U_a(\sigma)$ has changed by an amount kT. Since the width of the activation region is kT, this means that the stress reaches this value $\sigma(T)$ before there is any effective change in $U_a(\sigma)$. There is thus a region D where the activation energy is approximately constant at about U_0, which for long enough loop lengths is a triangular insertion in B,

456

and a corresponding strip between E and C. The transition probability becomes

$$\Gamma_o = \nu_a \, \exp(-U_o/kT) \tag{23}$$

The decrement becomes

$$\Delta = \tfrac{1}{2}\Delta_o [1 - \exp(-\pi\Gamma_o/\omega)]/[1+(\Gamma_o/2\omega)^2] \tag{24}$$

This represents a peak as a function of temperature, the maximum value of 0.397 Δ_o occuring at $\Gamma_o/\omega \approx 0.75$. The decrement is not a function of stress-amplitude. This result was obtained by Koiwa' and Hasiguti[8] and for temperatures below the peak by Saul and Bauer[9]. The maximum value of the $\sigma(T)$ boundary is at σ_V on the activation line, where

$$\sigma_V = \sigma_M/[\ln(\nu_a/\omega)]^{(\gamma+1)/\gamma} \tag{25}$$

which leads to the requirement that

$$\beta >> [\ln(\nu_a/\omega]^{2(\gamma+1)/\gamma} \tag{26}$$

for $\sigma_V >> \sigma_S$.

Discussion

Figures 5 and 6 show the way in which a series of decrement measurments on a specimen of fixed dislocation organization, all made at the same frequency should appear. As a function of T a peak should be seen at a temperature corresponding to the activation line γ_λ of Figure 4, asymmetrical except in region D (which may not exist). Th symmetrical peak of region D is only).397 the heights of the others. As a function of σ_o the peak between A, E regions seen at high temperatures, should be very sharp compared to that between B and C regions. Since the γ_λ line of Figure 4 is almost horizontal, the "half-width" of the peak between B and C should be as great as the width of region D if it is seen. The schematic curves of Figure 6 show this peak as being sharper than it will really look.

Acknowledgements

D.H.R. gratefully acknowledges the support of the Deutscher Akademischer Austauschdienst which enabled him to visit Dr. K. Lücke's Institut fur Metcl-lkunde where he had many fruitful discussions.

**The research was supported by the Defense Research Board of Canada.

References

1. Teutonico, L.J., Granato, A.V., and Lücke, K.; J. Appl. Phys. 35, 220 (1964)

2. Blair, D.G., Hutchison, T.S., and Rogers, D.H., J. Appl. Phys. 40,97 (1969)

3. Granato, A.V. and Lücke, K, J. Appl. Phys. 27, 583 (1956)

4. Ree, T. and Eyring, H. Rheology, Theory and Applications Einrich, F.R., Ed., Academic Press, N.Y. (1958) P. 102

5. Blair, D.G., Hutchison, T.S. and Rogers, D.H., Can. J. Phys. 48, 29 43 (1970)

6. Blair, D.G., Hutchison, T.S., and Rogers, D.H., Can. J. Phys. 49,633 (1071)
    ct eqn. 4.23
7. Blair, D.G., Hutchison, T.S. and Rogers, D.H., Can J. Phys. 48,2955 (1970)
8. Koiwa, M. and Hasiguti, R.R., Acta Met. 13,1219 (1965)
9. Saul,R.H. and Bauer, G.L. J. Appl. Phys. 39, 1469 (1968)

DISLOCATION UNPINNING IN SINGLE CRYSTALS OF CERAMIC OXIDES

I.G. Ritchie
Atomic Energy of Canada Limited
Whiteshell Nuclear Research Establishment
Pinawa, Manitoba ROE 1LO, Canada

Introduction

Strain amplitude dependent damping has been studied in single crystals of the ceramic oxides MgO and sapphire. The results have been interpreted using the theory developed by Blair, Hutchison and Rogers (1) for the thermally assisted unpinning of very long dislocation loops. This theory, referred to as the BHR theory in the following, has been outlined by Rogers (2) in the preceeding paper. For this reason only those theoretical results which have been used directly in the intrepetation of the experimental data will be repeated here.

In the BHR theory, a generalized pinning force has been incorporated into the Teutonico, Granato and Lücke model (3) for thermally activated breakaway. In addition, it has been shown that for calculation of the logarithmic decrement and modulus defect in the various regions of the temperature-amplitude plane, only the asymptotic behaviour of the pinning force, $U'(Y)$, need be specified. The asymptotic form of the pinning force can be expressed as

$$U'(Y) \rightarrow (U_0/\eta_2 r)(\eta_2 r/Y)^{\gamma+1} \text{ as } Y \rightarrow \infty \qquad [1]$$

where η_2 and γ are positive constants, U_0 is the binding energy between dislocation and pin, r is a range characteristic of the pinning force and Y is the displacement of the joint on the dislocation from the pin. It is assumed in the BHR theory that a dislocation network length, L_N, of Burgers vector, b, interacts with a row of isolated pins equally spaced at intervals L_c. From an experimental point of view, this assumption of equally spaced pins is restrictive. However, as shown in this paper, the assumption of equally spaced pins is probably valid for vibrationally conditioned ionic materials containing fresh dislocations. For the particular case when L_c is very long, the BHR theory predicts five regions of distinct damping behaviour due to thermally assisted unpinning. Of these, region D (which occurs only for very long loops) represents a portion of the temperature-amplitude plane where relaxation-type damping is observed. This region is of considerable interest, since it is probable that the damping versus temperature peak which occurs in this region is of the Hasiguti type. The condition defining very long loops and, therefore, the condition necessary to observe D-region damping is given by

$$\beta \gg [\ln(\nu/\omega)]^{2(\gamma+1)/\gamma} \qquad [2]$$

where ν is the effective attack frequency for unpinning and ω is the radian frequency of the applied stress. β is a dimensionless loop length parameter defined as

$$\beta = L_c U_0/Cr^2 \qquad [3]$$

where $C \simeq \frac{1}{2}Gb^2$ is the line tension of the dislocation and G the shear modulus of the material.

Experimental

The single crystals of MgO used in this study were obtained from the Norton Company. Specimens were cleaved with a <100> orientation. Some were irradiated in the WR-1 reactor to a total integrated neutron flux of 1.16×10^{20} n cm^{-2} (> 1 MeV) at about 450°C. Fresh dislocations were introduced into the MgO specimens by chemically polishing and sprinkling with a fine silicon carbide powder (Stokes et al [4]).

The single crystals of sapphire were obtained from Tycho Laboratories. These consisted of polished single crystal tapes of two orientations; one with the c-axis parallel to the longitudinal axis of the tape and the other with an a-axis parallel to the longitudinal axis of the tape. Fresh dislocations were introduced into these specimens by grinding the two large surfaces with a diamond wheel.

Strain amplitude dependent damping was observed at selected temperatures by monitoring the free decay of flexural oscillations of the specimen inserted in a counterbalanced reed pendulum. Raw data, in the form of amplitudes and periods of oscillation, were precisely logged by electronic instrumentation interfaced to a small computer. The counterbalanced reed pendulum and the electronic instrumentation used in this study have been described in detail by Ritchie and Sprungmann [5] and Ritchie et al [6]. Calculation of the logarithmic decrement as a function of strain amplitude, data processing and correction for the strain distribution in the specimen is also described in [6]. Temperature and temperature gradients in the specimen were controlled to within 1°C by the programmable temperature controller described by Sprungmann and Ritchie [7].

An important additional part of the equipment is a pendulum drive system; Ritchie et al [8]. This allows the logarithmic decrement, $\delta = \Delta w/2w$, to be calculated by measuring the energy input per cycle, Δw, to maintain a constant strain amplitude ε. $W = \frac{1}{2}E\varepsilon^2$ is the maximum energy of the maintained oscillations and E the elastic modulus. Thus the shape of a damping vs amplitude curve can be established by analysis of a free decay or by measurement of the energy input per cycle at discrete amplitudes. This is illustrated in Figure 1 where a damping versus amplitude curve at room temperature for a single crystal of MgO containing fresh dislocations has been established by both methods. As shown in this Figure, the results obtained from a free decay and the results obtained by reducing the amplitude in steps are in good agreement. However, there is a small deviation at the higher amplitudes when the drive system is stepped up in amplitude. Also shown in Figure 1 is the curve corrected for the strain distribution in the specimen. This has been interpreted as one of the family of curves from the BHR theory for very long loops and the regions of different damping behaviour described by Rogers [2] are indicated.

The drive system is also of importance in these experiments because it allows the specimen to be vibrationally conditioned by prolonged vibration at a precisely known constant amplitude. Many workers have discussed the production of long dynamically stable loop lengths by prolonged vibration (e.g. Bode [9] and Trott and Birnbaum [10]). Without stabilizing the distribution of pinning agents in this way, amplitude dependent damping is accompanied by time dependent damping which is probably associated with the dislocation core diffusion of pins discussed by Lücke and Schlipf [11].

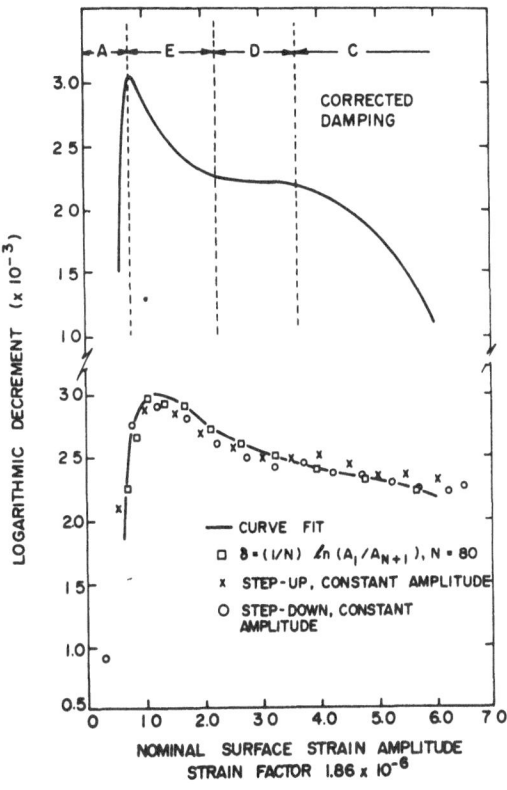

FIG. 1

Logarithmic decrement as a function of surface strain amplitude determined from free decay and from the energy input at constant amplitude. Specimen is a single crystal of neutron irradiated MgO. Temperature 22°C, frequency 4.5 Hz.

The lowest damping measured with our system is shown in Figure 2, where results are presented for a polished single crystal of sapphire at room temperature and 400°C. The rise in this essentially amplitude independent damping can be accounted for by thermoelastic relaxation. Results for a similar specimen containing fresh dislocations are also shown. Again, the corrected curve has been interpreted as one of the family of curves from the BHR theory for very long loops. These results are somewhat controversial since they imply the unpinning of nonbasal dislocations in sapphire at room temperature. However, nonbasal dislocations have been observed in transmission electron micrographs of diamond polished sapphire specimens by Hockey (12).

The results analysed in detail in this paper were obtained on a single crystal of neutron irradiated MgO. Vibration conditioning was carried out at a strain amplitude of 1.5×10^{-5} while the specimen was heated to 700°C, held for about 16 hours at 700°C and slowly cooled. The results shown in Figure 3 were obtained in the temperature range 150°C to 350°C upon reheating. Again, vibration conditioning at a strain amplitude of 1.5×10^{-5} for up to 15 minutes at each temperature was required before the shape of any damping vs strain amplitude curve was reproducible.

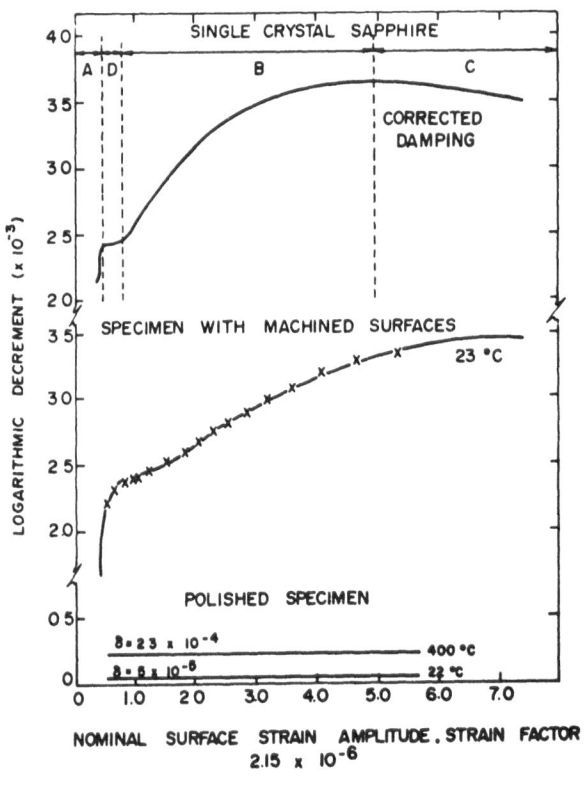

FIG. 2

Logarithmic decrement as a function of surface strain amplitude in single crystals of sapphire. Curves obtained as described in reference (6). X from $\delta = (1/N)\ln(A_1/A_{N+1})$ with N=160. Frequency $\simeq$ 6.9 Hz.

Results and Discussion

Figure 4 shows a selection of the curves of Figure 3 compared with the predicted shapes from the BHR theory. For the theoretical curves, Figure 4(b), temperature is increasing from the bottom curve upward. Five regions of different damping behaviour are delineated by the extrema of these curves. The experimental curves, Figures 3 and 4(a), show clearly the development of the knee at the AD boundary and the development of D-region damping. In addition, the movement of the BC peak as the temperature increases is from high amplitudes, through the experimental range, to low amplitudes, as predicted by the BHR theory.

Figure 5(a) shows the schematic curves determined from the BHR theory for the behaviour of damping as a function of temperature with strain amplitude decreasing from the bottom curve upward. The most interesting features of these theoretical curves are that the peak height drops by a factor of 2.52 in moving through region D and the peak position increases in temperature as the amplitude decreases.

These theoretical predictions are qualitatively confirmed as shown in Figure 5(b). As the strain amplitude increases, the peak height passes through a minimum and the peak position moves to higher temperatures. The quantitative prediction of a drop in peak height by a factor of 2.52

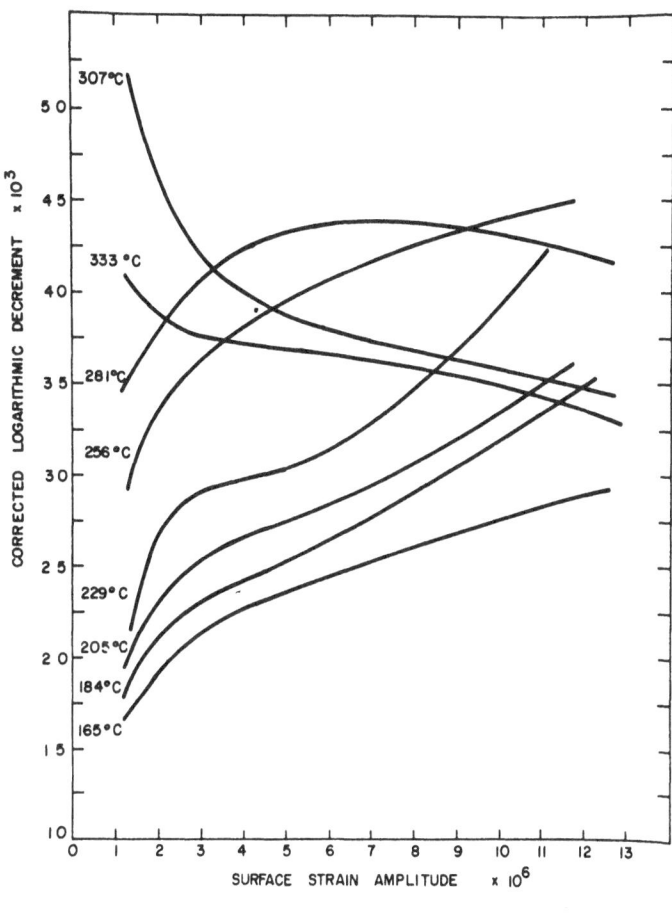

FIG. 3

Logarithmic decrement (corrected for the stain distribution in the specimen) as a function of
surface strain amplitude at various temperatures. Specimen is a single crystal of neutron
irradiated MgO with <100> orientation. Frequency 4.5 Hz.

will be difficult to check precisely since the experimental results undoubtedly contain contributions of damping due to other mechanisms. This leads to uncertainty in the background damping
to the various curves in Figures 3 and 5(b). However, not shown in Figure 5(b), are a second
set of amplitude dependent peaks at about 420^{0}C. Using this second set to estimate the background damping of the first set leads to an estimate of the drop in peak height of about 1.8 when
the strain amplitude is raised from 1.2×10^{-6} to 3×10^{-6}.

In the BHR theory, the complex damping behaviour, shown schematically in Figures 4(b) and
5(a), is conveniently summarized in a temperature-amplitude (T-ϵ) diagram. The main features
of the T-ϵ diagram for very long loops are shown in the inset to Figure 6. Experimentally,
the boundaries of the T-ϵ diagram can be estimated from the appropriate extrema of the damping
vs amplitude or damping vs temperature curves.

The different damping behaviour in the regions A-E is, according to the BHR theory, as
follows:

1. In regions A and E, unpinning is activated over a group of pins, while in regions B, C and
   D unpinning from the whole row of pinners is initiated at a single pin.

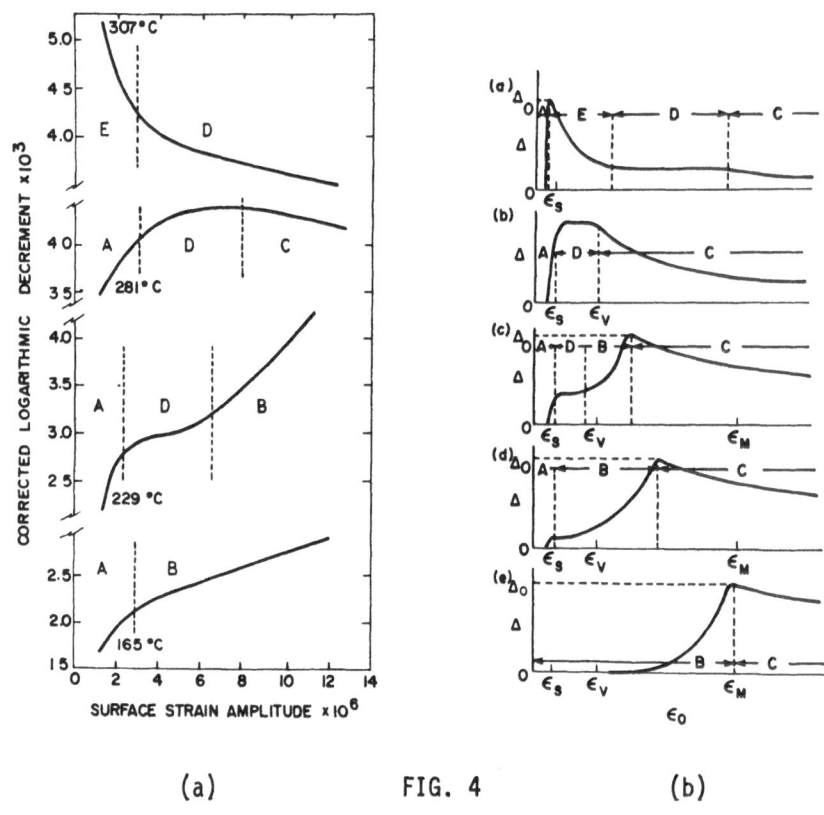

(a)   FIG. 4   (b)

(a) Selection of the experimentally determined damping vs strain amplitude curves at various temperatures.

(b) Behaviour of damping vs strain amplitude curves predicted by the BHR theory for thermally assisted unpinning of very long loops. Temperature is increasing from the bottom curve upward.

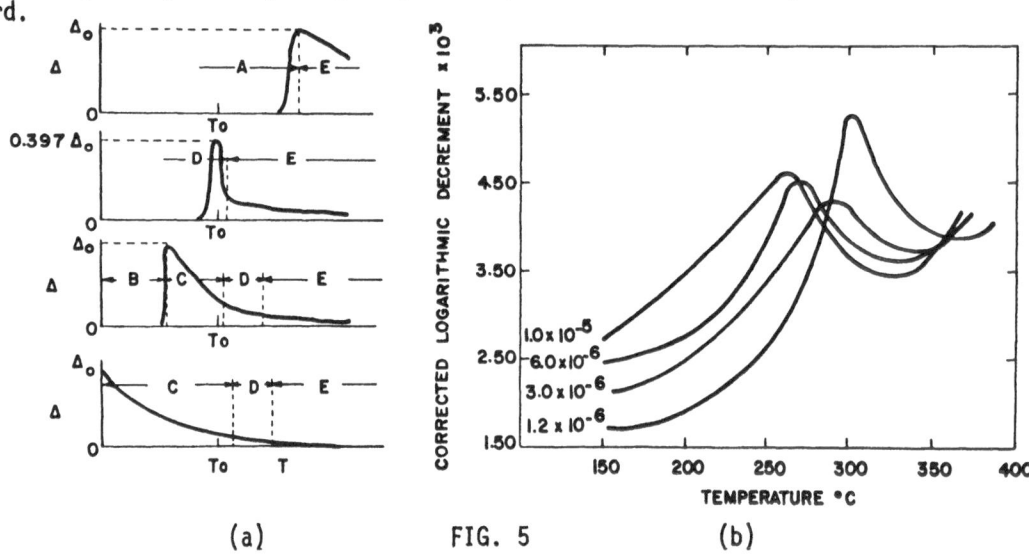

(a)   FIG. 5   (b)

(a) Damping behaviour as a function of temperature predicted by the BHR theory for very long loops. Amplitude is decreasing from the bottom curve upward.

(b) Selection of experimentally determined damping vs temperature curves at various surface strain amplitudes.

2. Regions A and B are characterized by few of the dislocations breaking away in a given cycle, while in regions C and E nearly all the dislocations unpin in a given cycle.

3. In region D, the activation energy for unpinning is effectively constant throughout the stress cycle resulting in relaxation-type behaviour.

Of most interest experimentally are the expressions for the critical amplitudes, ε_s, ε_v and (not shown in Figure 4) ε_M, the mechanical breakaway amplitude which occurs where the BC boundary intersects the zero temperature axis. For bending of a single crystal of MgO with <100> orientation, the active slip systems are {110} <1$\bar{1}$0> on which the dislocations experience an applied stress $\sigma = \frac{1}{2}E_{100}\dot{\varepsilon}$, where E_{100} is Young's modulus in the <100> direction and ε is the strain amplitude. Under these conditions, ε_s, ε_v and ε_M are given by the following three equations

$$\varepsilon_s = \frac{2}{E_{100}} \left[\frac{2\,G\,U_0}{L_c^3} \right]^{\frac{1}{2}} \qquad [4]$$

$$\varepsilon_v = \frac{2}{E_{100}} \cdot \frac{U_0}{br\,L_c} \cdot \frac{1}{[\ln(\nu/\omega)]^{(\gamma+1)/\gamma}} \qquad [5]$$

and

$$\varepsilon_M \simeq \frac{2}{E_{100}} \cdot \frac{U_0}{br\,L_c} \qquad [6]$$

In addition, the critical temperature, T_0, which occurs close to the intersection of regions A, D and E is given by

$$k\,T_0\,\ln(\nu/\omega) = U_0 \qquad [7]$$

Since T_0 can be determined quite accurately from the T-ε diagram, the value of U_0 can be calculated from equation [7]. Subsequently, L_c can be calculated from equation [4] and r estimated from equation [6]. Values of the dimensionless loop length parameter, β, and the pinning force constant, γ, can be estimated directly from the ratios of the experimentally determined critical strains, as shown in the following two equations:

$$\beta = 4\,(\varepsilon_M/\varepsilon_s)^2 \qquad [8]$$

and

$$\gamma = [\ln(\varepsilon_M/\varepsilon_v)/\ln \ln(\nu/\omega) - 1]^{-1} \qquad [9]$$

The experimental values and numerical constants used in the calculations presented here, together with the results of the calculations, are listed in Table 1.

The results in Table 1 indicate that the loop lengths obtained by vibrationally conditioning a single crystal of MgO containing fresh dislocations, are indeed very long ($L_c \sim 1 \times 10^{-4}$ cm or 3750 b). The binding energy of the pin to the dislocation is about 1 eV. The calculated value of the range characteristic of the pinning force, $r \simeq b/10$, is less than the estimated value of the core radius, $r_0 \simeq b/4$, for non-metals (Hirthe and Lothe (15)) and less than the displacement, $Y \simeq b/6$, at which Teutonico et al (3) estimate that the Cottrell type interaction force (16) reaches a maximum value. $\gamma \simeq 9$ is an interesting result indicating that the pinning force falls

465

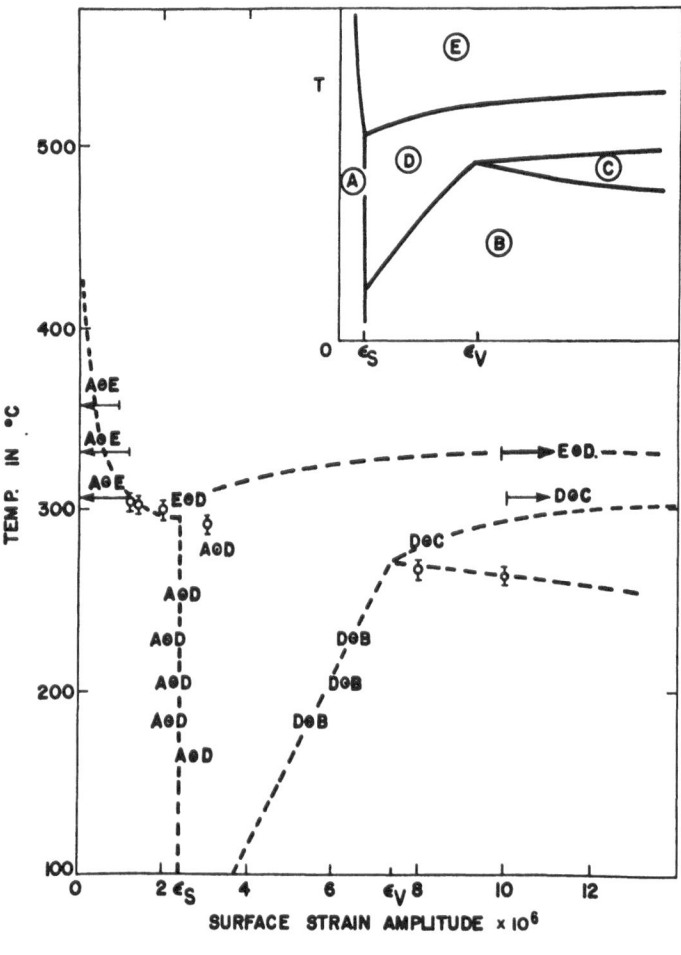

FIG. 6

Temperature-strain diagram for a single crystal of MgO containing fresh dislocations. Points indicated by φ are obtained from damping vs temperature curves such as those in Figure 5(b). Points indicated by θ (e.g. A θ D) are estimated from the curves in Figure 3 and Figure 4(a). Points labelled with an arrow, are boundaries off-scale on the experimental curves which have been estimated from their shapes.

The inset T-ε diagram contains the main features of the behaviour predicted by the BHR theory for very long dislocation loop lengths defined by condition [2].

off much more rapidly at large displacements than the elastic interactions derived by Cottrell (16) and Bullough and Newman (17). Both these latter interactions are characterized by $\gamma = 2$. Theoretical investigations of the interactions between dislocations and pinning agents in ionic materials have been limited to the investigation of the interactions between edge dislocations and point defects in NaCl. Bassani and Thomson (18) have calculated the interaction between an edge dislocation and a positive ion vacancy in NaCl. They found that the predominant contribution to the pinning interaction resulted from the repulsive or overlap part of the assumed Born-Mayer lattice potential. For MgO, a pinning force with the same characteristics as the overlap part of the Born-Mayer interaction would have $\gamma \sim 8$. It is interesting to note that positive ion vacancies in the form of v^- centres have been detected in specimens of MgO containing fresh dislocations by the method of electron spin resonance (Sargent (19)).

TABLE 1

Experimental Values, Numerical Constants and Derived Results

Experimental Results	Constants	Calculated Results
$T_o = 566^{\circ}K$	$E_{100} = 2.68 \times 10^{12}$ dyne cm^{-2}*	$U_o = 1.09$ eV
$\varepsilon_s = 1.4 \times 10^{-6}$	$G_{110} = 1.22 \times 10^{12}$ dyne cm^{-2}*	$L_c = 1.12 \times 10^{-4}$cm
$\varepsilon_v = 6.4 \times 10^{-6}$	$\nu = 1.3 \times 10^{11}$ sec^{-1}†	$r = 2.10 \times 10^{-9}$cm
$\varepsilon_M = 2.0 \times 10^{-4}$	$b = 2.979 \times 10^{-8}$cm	$\gamma = 9$
$\omega = 28.27$ sec^{-1}		$\beta = 81590$

* Chung et al (13)
† Calculated from equation (37) of Granato et al (14).

The value of $[\ln \nu/\omega]^{2(\gamma+1)/\gamma}$ is 11,042, 988 and 495 for $\gamma = 2$, $\gamma = 9$ and $\gamma \to \infty$ respectively. Thus the condition defining very long loops and, therefore, the condition necessary for the observation of D-region damping (i.e. $\beta \gg [\ln \nu/\omega]^{2(\gamma+1)/\gamma}$), is only reasonable for an inter-action characterized by a larger value of γ than is predicted by elastic interaction.

It is surprising that we observe qualitative agreement with a theory which assumes equidistant spacing of the pins. The production of loops of equal length by prolonged vibration would not normally be expected for longitudinal mobility of the pins in the core of the dis-location. In fact, Alefeld (20) and Lücke and Schlipf (11) have shown that for non-interacting pins, prolonged vibration should produce a bunching of the pins at ends of the network length. However, jogs and many of the possible point defects in MgO have an associated charge. This leads to long range Coulombic repulsive forces between pins of the same species and charge. Preliminary calculations of the distribution under these circumstances does predict equidistant spacing of the pins for reasonable assumptions of the parameters involved. These calculations lend some credance to our observed agreement with the BHR theory and will be published else-where.

Very similar results to those discussed in this paper have been observed in unirradiated single crystals of MgO containing fresh dislocations. Above the temperature range of the results reported here, there is a marked difference between the strain amplitude dependent damping in neutron irradiated and unirradiated specimens. These neutron irradiation effects will be reported elsewhere.

Summary

The Blair, Hutchison and Rogers theory of thermally assisted unpinning of very long dislocation loops has been qualitatively confirmed in many of its major predictions. Thus, it appears that vibration conditioning of fresh dislocations in MgO single crystals produces very long loops of approximately equal length. These fresh dislocations can be used as an effective probe for studying pinning agents such as those produced by irradiation.

Some preliminary tests indicate that fresh dislocations in sapphire single crystals give rise to similar results. The results indicate that for MgO the law of force between the dislocation and pinning agent is characterized by a more rapid fall off at large separations between the dislocation and pin than theoretically predicted for an elastic interaction. This is probably caused by a pinning force that is predominantly electrical in nature rather than elastic. Further corroborative studies will be required to determine the nature of the pinning agent.

Acknowledgements

The author wishes to acknowledge helpful discussions with Professors D.G. Blair, T.S. Hutchison and D.H. Rogers.

References

1. D.G. Blair, T.S. Hutchison and D.H. Rogers, Can. J. Phys., 49, 633 (1971).
2. D.H. Rogers, Preceeding Paper in this Conference.
3. L.J. Teutonico, A.V. Granato and K. Lücke, J. Appl. Phys., 35, 220 (1964).
4. R.J. Stokes, T.L. Johnston, and C.H. Li, Phil. Mag., 6, 9 (1961).
5. I.G. Ritchie and K.W. Sprungmann, J. Phys. E: Scientific Instruments 5, 1158 (1972).
6. I.G. Ritchie, J.R. Saltvold, H.K. Schmidt and K.W. Sprungmann, J. Phys. E: Scientific Instruments 6, 341 (1973).
7. K.W. Sprungmann and I.G. Ritchie, An Improved Reed Pendulum Apparatus and Techniques for the Study of Internal Friction of Ceramics Single Crystals, AECL-3794 (1971).
8. I.G. Ritchie, J.R. Saltvold, H.K. Schmidt and K.W. Sprungmann, to be published.
9. E. Bode, Phil. Mag., 13, 693 (1966).
10. B.D. Trott and H.K. Birnbaum, J. Appl. Phys., 41, 4434 (1970).
11. K. Lücke and J. Schlipf, The Interactions between Dislocations and Point Defects, AERE-R-5944, 1, 118 (1968).
12. B.J. Hockey, The Science of Ceramic Machining and Surface Finishing, NBS Special Publication 348, 333 (1972).
13. D.H. Chung, J.J. Swica and W.B. Crandall, J. Am. Ceram. Soc., 46, 452 (1963).
14. A.V. Granato, K. Lücke, J. Schlipf and L.J. Teutonico, J. Appl. Phys., 35, 2732 (1964).
15. J.P. Hirthe and J. Lothe, Theory of Dislocations, McGraw-Hill (1968) p. 212.
16. A.H. Cottrell, Report of a Conference on the Strength of Solids, University of Bristol, The Physical Soc., London, (1948) p. 30.
17. R. Bullough and R.C. Newman, Phil. Mag., 7, 529 (1962).
18. F. Bassani and R. Thomson, Phys. Rev. 102, 1264 (1956).
19. F.P. Sargent--private communication.
20. G. Alefeld, Phil. Mag., 11, 809 (1965).

MICROPLASTICITY AND HIGH-AMPLITUDE DAMPING IN TANTULUM SINGLE CRYSTALS

M.J. Cowling and D. J. Bacon
Department of Metallurgy and Materials Science
The University, P.O. Box 147, Liverpool.

1. Introduction

As a result of the considerable amount of research carried out over the past ten years on the mechanical properties of the b.c.c. transition metals, their yielding behaviour is now well-characterised (for review see (1)). The effects of temperature, interstitial content, crystallographic orientation and mode of testing are recognised to be due to the special nature of screw dislocations in the b.c.c. structure and the importance of dislocation-interstitial interactions. The core structure of screw dislocations is such that they are less mobile than non-screws, and macroscopic yielding is considered to be associated with their large-scale motion. The corresponding activation parameters are approximately 1.0eV and $50b^3$ for enthalpy and volume respectively. Additional information has been obtained from investigations of microyielding. These suggest that dislocations of several Burgers vectors contribute to slip in the microstrain region, but they are generally thought to be non-screw in character and to be more affected than screws by interstitials. Activation volumes tend to increase to several hundred b^3 in the microyield region, and measured activation enthalpies are generally in the range 0.1 - 0.4eV at 100K and 0.4 - 1.0eV at 300K.

In parallel with this work, several investigators have studied the thermally-activated damping peaks associated with dislocations in the b.c.c. metals (for review see (2-4)). In order of increasing peak temperature, the peaks, and the range of their activation energy measured in several metals, are δ (0.01 - 0.02eV), α (0.1 - 0.3eV), β (0.4 - 0.6eV) and γ (0.8 - 1.5eV) respectively. The peaks are reduced by anneals at a few hundred $^{\circ}$C, and, particularly in the case of β, increased by a decrease in interstitial content. Moderate prestrain at 295K enhances the α and β peaks, and prestraining at 77K is reported to enhance the α peak more than the β (2,5) and vice-versa (6). Damping shows a strong amplitude dependence at strain amplitudes $\approx$ $10^{-6}-10^{-5}$, and saturates at amplitudes $\gtrsim 5 \times 10^{-4}$. As the amplitude is increased in this range, the peak heights are increased, the peak temperatures are lowered, and the γ peak in particular becomes large and broad. Various interpretations of the peaks have been presented. The α peak has been assigned to dislocation breakaway from pinning points (7), double-kink relaxation on edge dislocations (6), and the motion of kinks on screws (8). The β peak has been interpreted as arising from point-defect complexes (9), double-kink relaxation (2), long-range motion of dislocations (4), double-kink relaxation on screws (6), and a dislocation-hydrogen interaction (8). Seeger and Šesták (8) associate the γ peak with double-kink relaxation on screws.

In arriving at their conclusions, Seeger and Šesták (8) made extensive use of both internal friction and microyield data. This seems a very fruitful line of approach, and the aim of the present work has been to further bridge the gap between internal friction and conventional mechanical testing by carrying out load-unload tests in the compression mode at strain sensitivities of ~10^{-6}. These microyield tests have been performed at temperatures in the range 77-295K on tantalum single crystals having a range of impurity concentrations and crystallographic orientations. Different dislocation distributions have been studied by pretraining the specimens at various temperatures. In the present paper we present our preliminary results on specimens of one orientation and two impurity concentrations.

2. Method

2.1 Material and testing equipment.

Single crystals of tantalum were prepared from 4mm diameter rods in an electron-beam furnace. In order to achieve different interstitial impurity levels (10), the rods were given either 2 passes or 4 passes of the molten zone. The 2 zone-pass (2 z.p.) material was analysed and found to have the following interstitial content in p.p.m. by weight : H<5, C≃20, N≈30, O<30. The crystallographic direction of the rod axis for the specimens discussed here was 50° from[111] and was such that the angle between the ($\bar{1}$10) reference plane and the plane of maximum resolved shearstress was + 24^O. For this orientation, the actual slip plane after macroyielding is close to ($\bar{2}$11) for temperatures in the range 77-295K (11).

As there has been some doubt as to the accuracy of several compression tests carried out in the microyield region (12,13), great care was taken to develop an accurate and reliable compression jig (14). It consists essentially of three high-modulus rods in parallel with the specimen and compression anvils. The rods are such that a 1 kN load on the jig produces a strain of only 10^{-4} in the specimen. This arrangement has the advantage that the jig can be preloaded before the specimen is stressed, thereby avoiding errors due to vibrations and misalignment effects at low machine loads, and, since the rods deform elastically throughout, a constant strain rate can be imposed on the specimen. In order to avoid errors due to specimen end effects, strain was measured by the use of two diametrically-opposed temperature-compensating resistance gauges on each specimen. The load on the specimen was determined by measuring the strain on one of the compression anvils with an annular parallel-plate capacitance transducer of high sensitivity. With strain and load plotted directly on an x-y recorder, a resolution of 1 kg of load and 2 x 10^{-6} strain per cm of chart was achieved. Accuracy of the system was such that the elastic modulus measured on the chart always fell within 5% of the dynamic modulus calculated from elastic constant data. Strain rate (cycle frequency) was determined directly by using a second pen on the recorder to plot strain vs. time. The jig was enclosed in a simple double-walled cryostat for temperatures in the range 77-295K.

2.2. Experimental tests

For the results presented here, specimens were tested at 77, 203 or 295K in either the as-grown condition or after a 1% prestrain at 77K. Prior to testing or prestraining, each specimen underwent a 1½ hour anneal at 500K in order to cure the cement used to bond the strain gauges. For one specimen, the cryostat was allowed to warm up slowly from 77K, and cycling tests were then carried out at intervals of approximately 20 degrees.

The majority of tests were simple load-unload tests carried out at stress levels below the microyield stress σ_{my}, which is defined as the stress required to produce a permanent plastic strain of 2×10^{-6}. (Due to the difficulty of associating dislocation motion with one slip system in the microyield region, all stresses and strains quoted here are axial). Measurements were made of the precision elastic limit σ_e, defined as the stress at which the loading and un-loading curves of a cycle are separated by a strain of 10^{-6}, and the variation of the fraction of energy dissipated per cycle $\Delta W/W$ with stress amplitude σ_m. (Here, ΔW is the area of the stress-strain loop and W is the area under the loading curve, and the ratio is therefore related to the quantities usually measured in damping experiments). Load-unload cycles were carried out at frequencies in the range 0.02 - 0.5Hz, and it was found that $\Delta W/W$ is almost independent of frequency in that range.

Although $\Delta W/W$ is almost independent of frequency in the range used here, plastic flow and relaxation effects are thermally activated (see Section 1). The activation volume v and activation enthalpy ΔH at various plastic strain levels ϵ_p in the microstrain region were there-fore determined from the compression-test cycles using the relations :
$$v \equiv kT \left(\delta \ln \epsilon_p / \delta \sigma \right)_T \text{ and } \Delta H \equiv -vT \left(\delta \sigma / \delta T \right) \dot{\epsilon}_p.$$ Volume v was obtained by performing load-unload cycles at stress amplitudes $\sigma_m < \sigma_{my}$ for several total strain rates in the range 8×10^{-6} to 4×10^{-4} s^{-1}; by measuring ϵ_p from the elastic line defined by stress cycles below σ_e, it was then possible to calculate the dependence of σ on $\dot{\epsilon}_p$ at a given ϵ_p. Similar tests at the three temperatures enabled plots to be made of σ vs. T at constant $\dot{\epsilon}_p$ for several ϵ_p values, and ΔH was calculated from these. It is recognised that this approach can introduce considerable error into the v and ΔH values, but more direct methods are not easy to achieve for the plastic strain range $10^{-6} \lesssim \epsilon_p \lesssim 10^{-5}$, and the results are believed to be sufficiently accurate for the interpretation required here.

3. Results

Results on the dependence of $\Delta W/W$ on temperature for the 4 z.p. zero prestrain specimen tested at temperatures in the range 77-295K are shown in Fig.1. Each datum point is the average of three measured values, and the two curves correspond to different constant stress amplitudes normalised with respect to the Young's modulus (σ_m/E). It can be seen that the damping decreases with increasing stress amplitude at each temperature, and that the specimen exhibits two broad damping peaks centred about 100K and 230K. From the activation energies and pre-exponential factors for tantalum given by Chambers (2), the peak temperature for the frequency used in Fig.1 should occur at approximately 107K, 190K and 470K for α, β and γ respectively. Although damping at small amplitudes generally tends to increase with increasing temperature in b.c.c. metals, the opposite trend in the plots of Fig.1 is not unlike that found by Chambers and co-workers (2) for high torsional amplitudes in tantalum.

Results for the variation of $\Delta W/W$ with stress amplitude σ_m at the three test temperatures for four starting conditions are shown in Fig.2. All the data points presented are the average of several measured values. The stress range over which measurements on $\Delta W/W$ could be made ($\sigma_e \lesssim \sigma_m < \sigma_{my}$) was fairly small, but it can be seen that $\Delta W/W$ initially decreases with

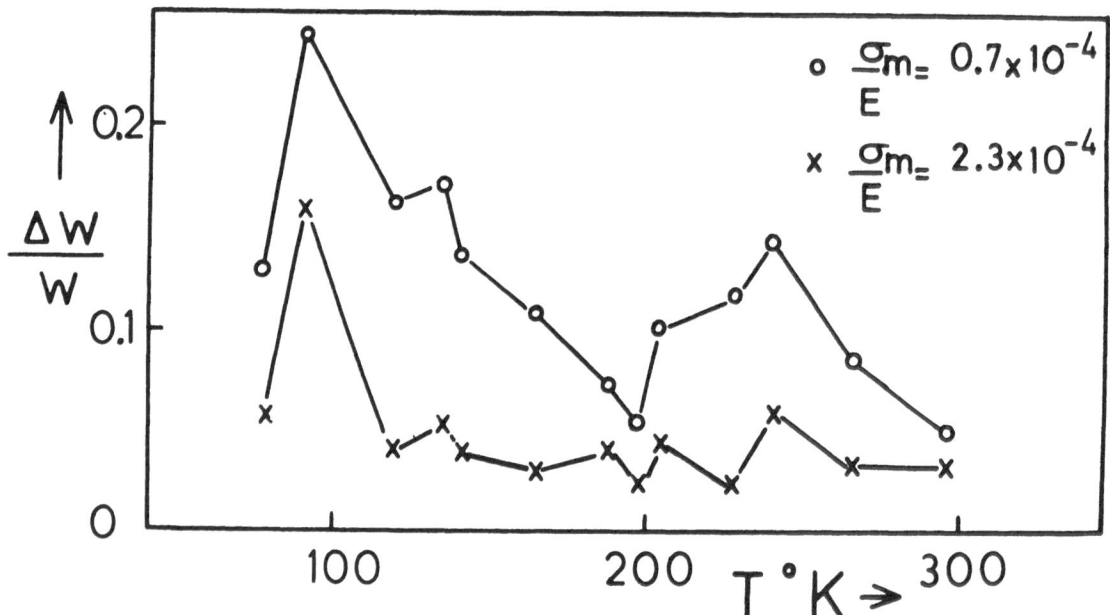

FIG. 1.

The variation of ΔW/W with temperature for an unprestrained 4 z.p. specimen at two stress amplitudes σ_m normalised with respect to the Young's modulus E. The cycle frequency was 0.5 Hz and 0.1 Hz for the low and high amplitudes respectively.

increasing σ_m for all specimen conditions. As noted in the Introduction, the torsion experiments of Chambers and co-workers (2,15) have shown the damping at small strains in b.c.c. metals to have an 'impurity-breakaway' form, i.e. damping increases with amplitude. It then decreases and finally becomes constant for strain amplitudes above about 5×10^{-4} however, and the results of Fig.2 are therefore thought to be equivalent to the high-amplitude torsion results. The ΔW/W values for the 2 z.p. specimens (Figs. 2(a) and (b)) show that the damping for a given σ_m tends to be larger at 77K than at 203K or 295K. The same feature is present in the results for the un-prestrained 4 z.p. specimens (Figs.1 and 2 (c)). The effect of the 1% prestrain at 77K on ΔW/W for the 2 z.p. material is not large at any of the three test temperatures, but its effect on ΔW/W for the 4 z.p. specimen at 295K is striking (Fig.2 (d)). This specimen had a lower value of σ_e, and exhibited reversible plastic strain to higher levels, than any other specimen tested.

The specimens had similar microyield characteristics to those found by other workers, i.e. the stress to produce a given value of ϵ_p was increased by additional impurities, and the temperature dependence of the stress was smaller for low ϵ_p values. In addition, the 1%

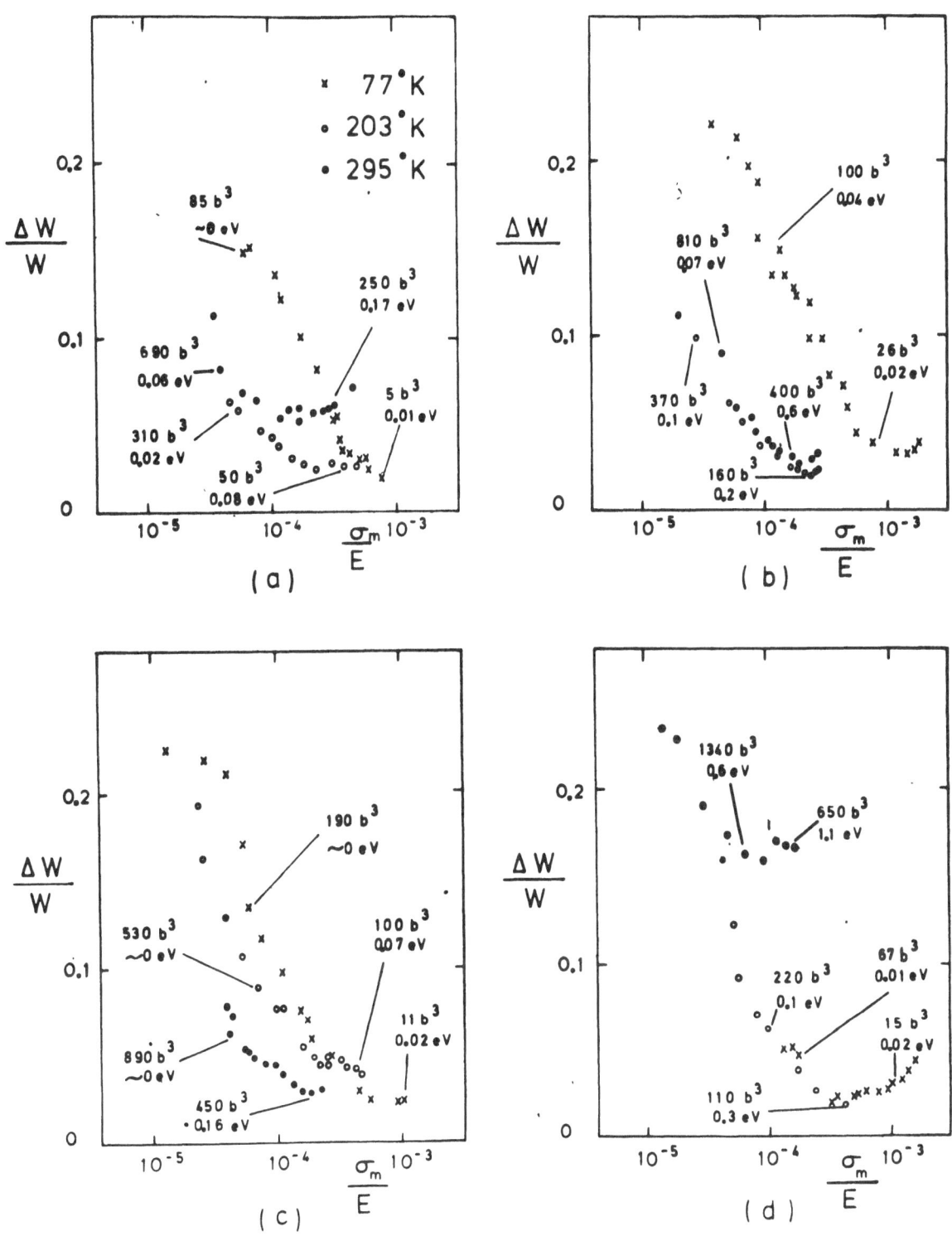

FIG.2.

The variation of $\Delta W/W$ with stress amplitude at three temperatures for the unprestrained and 77K - prestrained 2 z.p. samples ((a) and (b)), and the unprestrained and 77K - prestrained 4 z.p. samples ((c) and (d)). Also shown are v and ΔH values at specific amplitudes.

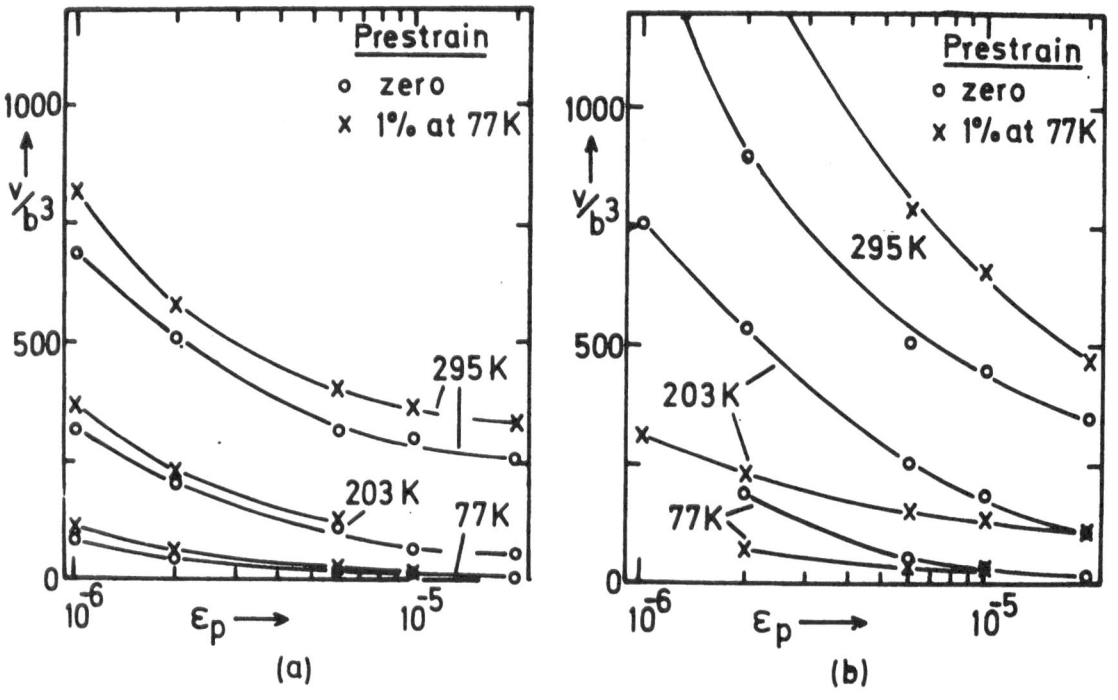

FIG. 3.

The variation of activation volume with plastic strain at three temperatures for (a) the 2 z.p. specimens and (b) the 4 z.p. specimens.

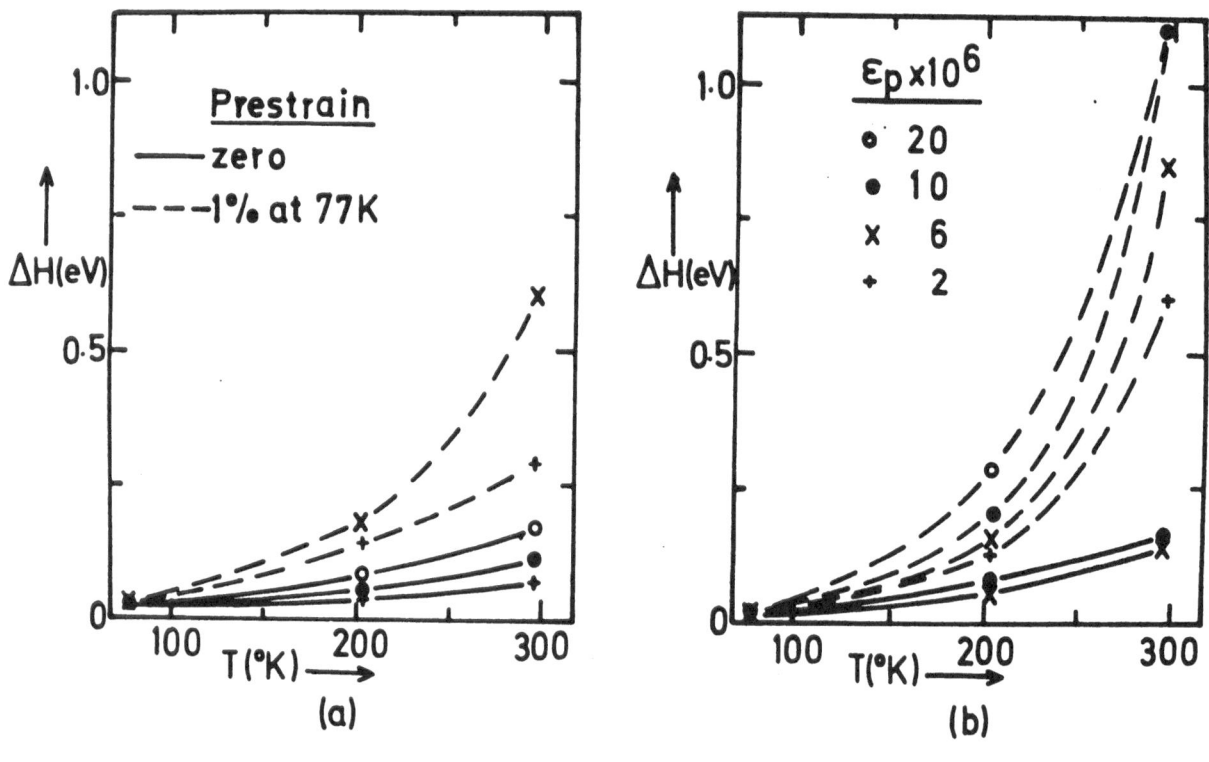

FIG. 4.

The variation of activation enthalpy with temperature at several plastic strain levels for (a) the 2 z.p. specimens and (b) the 4 z.p. specimens.

474

prestrain at 77K markedly increased the temperature dependence of the stresses for a given value of ϵ_p but lowered the value of the stress at 295K. From the experiments on the effect of temperature and strain rate on the stress at several ϵ_p levels, the activation volumes and enthalpies presented in Figs. 3 and 4 respectively were calculated. By the nature of the method used, errors in the ΔH values could be large, and are likely to be largest at 295K where they could be as high as 30%. The results show that v increases strongly for decreasing ϵ_p, in agreement with earlier observations on microyielding at rather higher strain levels (12,16,17). It can be seen that the effect of the prestrain at 77K on v for the less pure material (Fig.3(a)) is small, whereas the effect on the 4 z.p. crystal is to reduce v for small strains and low temperatures and to increase it for all strains at 295K. The prestrain produces v values at 77K and 203K in the higher purity material which are close to those for the 2 z.p. material. The effect of impurities in the annealed specimens is to reduce v at low strains. The results of Fig.4 show that ΔH has a non-linear dependence on T, and that it has values which are almost independent of impurity concentration in the annealed condition. It is increased by the prestrain, but is again almost independent of impurity content below about 200K. The effect of the prestrain is very marked at 295K, particularly in the higher purity material (Fig. 4 (b)).

4. Discussion

From the correlation between the temperature of the two broad peaks of Fig.1 and the temperatures predicted for the α and β peaks, it is tempting to assign the peaks of Fig.1 to the α and β relaxations. The physical interpretation of these processes by use of the activation parameters is unfortunately made difficult by the fact that the microyield tests could only be performed at three fixed temperatures. Furthermore, since v is very dependent on ϵ_p and ΔH varies non-linearly with T, it could be argued that processes occur in the microstrain region which invalidate the method used for obtaining v and ΔH. For example, changes in the mobile dislocation density would cause corrections to be made to the v values. These are difficult to estimate, however, and an attempt has been made to reduce their effect by carrying out all tests at stresses below σ_{my}. The form of the results in Figs. 3 and 4 therefore suggest that more than one process is rate-controlling in the microstrain region, as concluded by Arsenault et al (12).

In order to see how the variation of the activation parameters with ϵ_p is related to the variation of $\Delta W/W$ with σ_m at a given temperature, values of v and ΔH have been indicated at several points on the plots of Fig.2. It was suggested in the preceding section that the damping observed here is occurring at stresses above those for breakaway from interstitial impurities, and one interpretation of the curves of Fig.2 is that they are the high-amplitude remnant of an unpinning process. This is supported by the fact that the plots have the form expected for the high amplitude region of a thermal unpinning process (18). This is particularly true for the un-prestrained specimens and the prestrained samples tested at 295K, and it is only for these specimens at low strains that the activation volume is markedly impurity-dependent. Furthermore, the points on the strongly amplitude-dependent parts of the curves of Fig.2 give good Granato-Lücke plots.

A significant feature of the results for v and ΔH is that the activation enthalpy at 295K in the materials prestrained 1% at 77K is increased to relatively large values (~0.6 - 1.1eV for 4 z.p.), even for strain levels $\leq 10^{-5}$. These are close to the values found for macroscopic yielding in the b.c.c. metals. The purpose of the prestrain was to produce a dislocation structure predominantly screw in character, and this has been observed by transmission electron microscopy (19). We therefore conclude that the high damping observed in the prestrained 4 z.p. specimen at 295K arises as a result of screw dislocation motion, although the associated activation volume is somewhat larger than would be expected for double-kink relaxation (20). Despite the fact that the ΔH values at 295K for the pre-strained 2 z.p. specimen are fairly large (~0.3 - 0.6eV), the ΔW/W values are not. A possible explanation for this is that the damping peak for screw dislocation relaxation in this less pure material occurs at temperatures above 295K. Alternatively, the density of long, straight screw dislocations may be much lower than in the 4 z.p. prestrained specimen, and electron-microscope observations suggest that this is the case (19). The activation enthalpies we have measured for the temperatures where the α and β processes are believed to occur are considerably lower than the values given by internal friction experiments (2). This reflects the relatively high mean-stress levels used in the present work. Nevertheless, on the basis of the results for ΔH on the un-prestrained 4 z.p. specimen (Fig.4 (b)), we conclude that the 230K peak of Fig.1, which we believe to be the β peak, is not produced by the motion of screw dislocations. This agrees with the conclusions of Seeger and Šesták (8). Our results on the prestrained material also support their interpretation that screw relaxation is associated with the γ peak, although we have not yet observed the peak directly in a test of the kind which produced Fig.1.

The low values of ΔW/W at 77K in Fig.2(d) suggests that the height of the α peak may be small in the specimen containing long screw dislocations. This would appear to support the proposal of Seeger and Šesták (8) that the α relaxation arises from the lateral motion of kinks on screw dislocations. However, since our other results indicate that an impurity-breakaway process involving non-screw dislocations also occurs at low strain amplitudes, we are led to consider the possibility that the α peak is only significant once the unpinning process has occurred. The unpinning process could involve double-kink nucleation on non-screw dislocations for example. This would help to explain the difficulty many workers have met in attempting to separate the impurity and intrinsic-lattice effects involved in damping in the b.c.c. metals. Further study of these effects at large amplitudes using the technique employed here requires tests showing in detail how ΔW/W varies with T. This has been done for one specimen (Fig.1), and further tests are planned.

Acknowledgements

This work was supported by a research grant from the Science Research Council. One of us (M.J.C.) acknowledges the award of an S.R.C. research studentship.

References

1. J.W. Christian, Conf. on Strength of Metals and Alloys, vol.1, p.31, Amer. Soc. for Metals (1970).

2. R.H. Chambers, Physical Acoustics, vol. IIIA, chap.4 (W.P. Mason, ed.), Academic Press, New York (1966).

3. A.S. Nowick and B.S. Berry, Anelastic Relaxation in Crystalline Solids, Academic Press, New York (1972).

4. R. De Batiste, Internal Friction of Structural Defects in Crystalline Solids, North-Holland, Amsterdam (1972).

5. R. De Batiste, Phys. Stat. Sol. 2, 661, (1962).

6. B. Escaig, Scripta Met. 5, 199 (1971).

7. R. Gibala, M. K. Korenko, M.F. Amateau and T.E. Mitchell, J. Phys. Chem. Solids 31, 1889 (1970).

8. A. Seeger and B. Šesták, Scripta Met. 5, 875 (1971).

9. M.W. Stanley and Z.C. Szkopiak, J. Nucl. Mater. 23, 163 (1967).

10. B.L. Mordike, Z. Metallkde. 55, 304 (1964).

11. K.D. Rogausch and B.L. Mordike, Conf. on Strength of Metals and Alloys, vol.1, p.168 Amer. Soc. for Metals (1970).

12. R.J. Arsenault, C.R. Crowe and R.D. Carnahan, Reinststoffe in Wissenschaft und Technik, p. 345, Akademie-Verlag, Berlin (1972).

13. D.S. Tomalin, D.P. Pope and C.J. McMahon Jr., Met. Trans. 4, 1638 (1973).

14. M.J. Cowling and D.J. Bacon, J. Mater. Sci., in the press.

15. R.H. Chambers, T.E. Firle, T. Trozera and G. Buzzelli, General Atomic Division Report GA - 7978, General Dynamics, San Diego (1967).

16. R. Kossowsky, AIME Refractory Metals Conference, French Lick, Indiana, p. 47, Gordon and Breach, New York (1967).

17. J.D. Meakin, Can. J. Phys. 45, 1121 (1967).

18. K. Lücke, Interaction between Dislocations and Point Defects, Harwell Symposium, p.118, U.K.A.E.A. (1968).

19. M.J. Cowling, unpublished work.

20. A. Seeger and P. Schiller, Physical Acoustics, vol. IIIA, chap. 8 (W.P. Mason, ed.), Academic Press, New York (1966).

AMPLITUDE DEPENDENT INTERNAL FRICTION IN COPPER CRYSTALS

T.Nakamura* and K.Ishii

Department of Physics, Nagoya Institute of Technology

Shōwa-ku, Nagoya, Japan

Introduction

This paper reports on some experimental observations which are concerned with the amplitude dependent internal friction in copper crystals mainly at high strain amplitudes. The amplitude dependent internal friction in crystalline solids has been explained most satisfactorily by the theory of Granato and Lücke.[1] However, the theory is valid, as the authors predicted, within a small range of strain amplitudes above the breakaway. At higher amplitudes, the dislocations will move large distances in the lattice, and the multiplication would take place at sufficiently high amplitudes. In fact, some evidences have been obtained that the plastic deformation takes place during the measurements. For example, Mason[2] observed in lead crystals that the metal fatigued in a short time at high strain amplitudes. Whitworth[3] found in sodium chloride crystals that fresh dislocations were produced by the measurements at high strain amplitudes.

In the case of copper crystals,[4]-[6] recent studies on the early stages of plastic deformation have shown that the irreversible motion of dislocation occurrs at low stresses which are of the same order of magnitude as the stress amplitude of the measurements of internalf friction.

Based on these results and on the results of the amplitude dependent internal friction in white tin crystals by one of the authors,[7] it was expected that the irreversible motion of dislocations would take place during the measurements of internal friction in copper crystals under suitable conditions, and an attempt was made to observe the motion of dislocations directly by the etch pit technique which is now well established for copper crystals.

---

* Now at the Japan Patent Office.

Experimental Procedure

Copper crystals were grown from material nominally 99.999% pure. They were grown in a graphite boat from the seed crystal to length of 90 mm and rectangular shaped cross section of 6 x 2 mm^2, in a vacuum of 2 x 10^{-5} mm Hg. After growth, the specimens were acid sawed into 70 mm length.

The orientation of the crystals was arranged such that one pair of lateral surface was parallel to the {111} plane, and the specimen axis was parallel to the ⟨110⟩ direction. This orientation was chosen because equal stress will be applied on four slip systems under the transverse vibration, while no stress will be applied on the other eight systems. This would be a simple condition for the interpretation of results.

Some of the specimens were annealed for two days at 1060°C, followed by furnace cool. The specimens so prepared had the dislocation density of about 10^6 per cm^2. This density was too high for the observation of the motion of individual dislocations. Therefore, other specimens were subject to thermal cycling,[8] and the dislocation density was decreased to about 10^4 per cm^2. These specimens were further aged for 50 hours at 120°C, to observe the internal friction at high strain amplitudes.

The internal friction was measured by the transverse vibration. The etch pit observation was made in situ. Livingston etchant was used.[9] The displacement of dislocations after the specimen was vibrated was observed by repeated etching.

Results

1. Amplitude Dependence.

Figure 1, curve A shows the amplitude dependent internal friction taken with a specimen which has been subject to thermal cycling and then aged before the measurements. As shown in the figure, the internal friction has relatively low amplitude independent part and a high breakaway point. It was also observed that the amplitude dependence was irreversible when the measurements were made up to a high strain amplitude.

The curve B in the same figure shows an example of the amplitude dependent internal friction of a furnace cooled specimen, which was taken for comparison. In this case, the amplitude independent part is higher and the breakaway point is about an order of magnitude lower than before. The magnitude of internal friction of the furnace cooled specimen was not reproducible. It was sensitive to the condition of prior heat treatment, and probably to the possible disturbance due to handling.

2. Etch Pit Observation.

The etch pit observation was made in some detail for an aged specimen of which the amplitude dependent internal friction is shown in Fig.1. The result was that both the irreversible movement and the multiplication of dislocations occurred during the measurements. For the furnace cooled specimen, neither the irreversible movement nor the multiplication was observed within the range observed.

Figure 2 shows the substructure and dislocations revealed by etching in the middle part of the {111} surface. The dislocation density within subgrains averages about 2×10^4 per cm^2. The dislocations are more or less uniformly distributed, and at places regions of higher density are observed such as region A in the figure.

Subsequent etches were made after the measurements of internal friction at six points that are shown by the numbers (1) to (6) in Fig.1. Figure 3 shows the changes in the dislocation structure as the strain amplitude is increased, in the region A in Fig.2. The photographs (a), (b), (c) and (d) represent the strucures at the amplitudes (3), (4), (5) and (6), respectively. Among large sharp-bottomed pits which represent the sites of dislocations unmoved, a number of large flat-bottomed pits and small sharp-bottomed pits are observed. This shows, as in the case of steady loading test, that a number of dislocations have moved from the original sites. It will be seen in the figure, that the dislocations moved increase as the strain amplitude is increased. The movement was not observed at the strain amplitudes (1) and (2), in this region of high dislocation density. The multiplication is observed in the photograph (d), i.e., at point (6).

The movement of dislocation is observed more accurately in the region of low density. Figure 4 shows two examples of the successive movements of dislocations: One dislocation moved in the same sense, while the other returned to the original site. The successive movement of this kind was, however, observed only in a few cases, about 5% of all the dislocations moved. Most of the dislocations once moved from the original sites moved no more even when the amplitude was increased.

The number of dislocations moved is plotted against the strain amplitude in Fig.5. The dislocations generated is excluded because of difficulty of counting. The irreversible motion of dislocation was observed at point (1), where the stress amplitude was less than 10 gm per mm^2. The multiplication first occurred to a small extent at point (4). It was not observed at point (5), and at point (6), it occurred extensively. The stress amplitude at this point was about 50 gm per mm^2. Figure 6 shows another example of dislocation stucture in a region C in Fig.2, of low dislocation density, which was taken at point (6). It is observed that large numbers of dislocations are generated in the form of long slip bands. The number of dislocations moved from the original sites up to this point is about 20% of total dislocations observed before the measurements, as is plotted in Fig.5. This would mean that the multiplication occurred extensively when about half the dislocations in the active slip systems moved, since about 45% (= 4/9) of the dislocations observed should belong to the active slip systems.

Discussions

The results obtained shows that the amplitude dependent internal friction is reversible when the specimen was furnace cooled and measured at small strain amplitudes, and that the amplitude dependence is irreversible, when the specimen was aged before the measurements.

In the case of the furnace cooled specimen, the observed amplitude dependence will be consistent with the Granato-Lücke theory, in the sense that the irreversible motion was not observed. The 'Granato-Lücke plot' was not taken, because the amplitude independent part was not determined accurately.

In the case of the aged specimen, the etch pit observation showed that the dislocations moved increased as the strain amplitude was increased, and at higher strain amplitudes the multiplication occurred, in a way similar to the increase in the internal friction. This will be an indication that the observed amplitude dependent internal friction is directly related to the increase in these dislocations, say the mobile dislocations, The idea that the amplitude dependent internal friction would be caused by the increase in the dislocation density was proposed by Read.[10] In the present case, a dense cloud of impurities along the dislocation lines would be formed during the aging, and then the amplitude dependence of the Granato-Lücke type would be suppressed. As the strain amplitude is increased, the dislocations will breakaway from the original sites. Under relatively high stress amplitudes, the brokenaway dislocations will move large distances in the lattice, and will not return to the original sites, as observed, as a result of interactions with the defects such as forest dislocations. The amplitude dependence will be caused in this case, as well as in the case of generated dislocations, by the interaction of the mobile dislocations with the defects randomly distributed in the lattice, the number of the mobile dislocation increasing with increasing amplitude of excitation.

The stress amplitude for the initiation of the movement, less than 10 gm per mm^2, and the stress amplitude for the generation, about 50 gm per mm^2, appear to be of the same order of magnitude as those observed under steady loading tests, although the detailed comparison is not yet made.

Summary

(1) Amplitude dependent internal friction in copper crystals, 99.999% pure, and oriented such that one lateral surface was parallel to the {111} plane and the specimen axis parallel to the ⟨110⟩ direction, was measured by the transverse vibration, and the etch pit observation was made at the same time.

(2) When the specimen was furnace cooled from a high temperature, the amplitude dependence measured within 10^{-7} in strain was reversible, and the irreversible·motion of dislocation was not observed, being consistent with the Granato-Lücke theory.

(3) The etch pit observation was made in some detail with the specimen having the dislocation density of about 10^4 per cm^2, and aged for 50 hours at 120°C after the thermal cycling. In this case, the breakaway point in the internal friction, about 5×10^{-7}, was an order of magnitude higher than before, and the amplitude dependence was not reversible.

(4) The etch pit observation showed that the dislocations moved irreversibly in the amplitude dependent range. The number of dislocations moved increased as the strain amplitude was increased. At high strain amplitude, about 1×10^{-5} in strain, the multiplication occurred, when the internal friction increased steeply.

References

1. A.Granato and K.Lücke, J. appl. Phys. 27 (1956) 789.

2. W.P.Mason, J. acoust, Soc. Amer. 28 (1956) 1207.

3. R.W.Whitworth, Phil. Mag. 5 (1960) 425.

4. F.W.Young, J. appl. Phys. 33 (1962) 963.

5. K.Marukawa, J. phys. Soc. Japan 22 (1967) 499.

6. S.Kitajima, H.Honda and H.Kaieda, Proc. Int. Conf. on Strength of metals and Alloys, Supplement to Trans. JIM vol. 9 (1968) 740.

7. K.Ishii, J.Phys. Soc. Japan 28 (1970) 168, 1494.

8. S.Kitajima, M.Ohta and H.Kaieda, J. Japan Inst. Metals 32 (1968) 164.

9. J.D. Livingston, J. appl. Phys. 31 (1960) 1071.

10. T.A.Read, Phys. Rev. 58 (1940) 371.

Fig.1. Decrement vs. Strain amplitude.

Curve A, for a specimen with a dislocation density of about 10^4 per cm^2, and aged for 50 hrs at 120°C after thermal cycling. Curve B, for a specimen with a dislocation density of about 10^6 per cm^2, and furnace cooled after anneling at 1060°C.

Fig.2. Subboundaries and dislocations in the middle part of the (111) surface of a specimen for which the internal friction is shown in Fig.1, curve A.

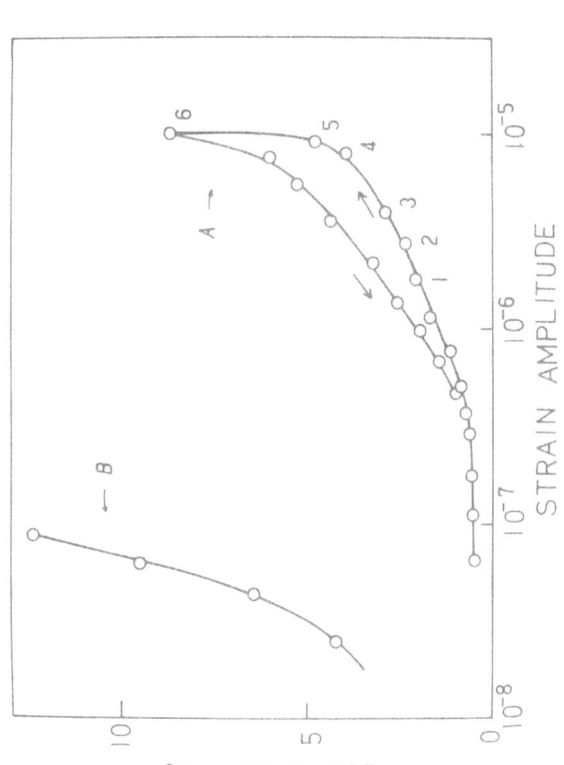

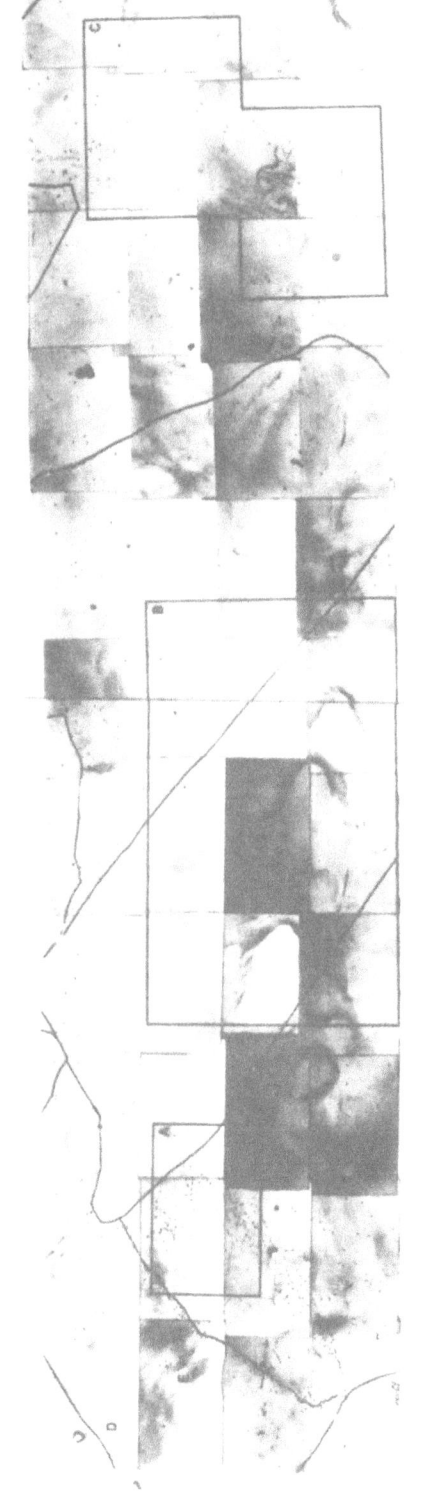

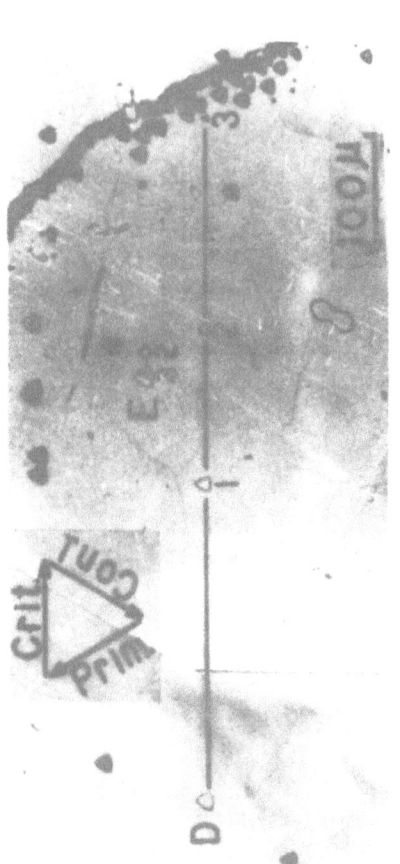

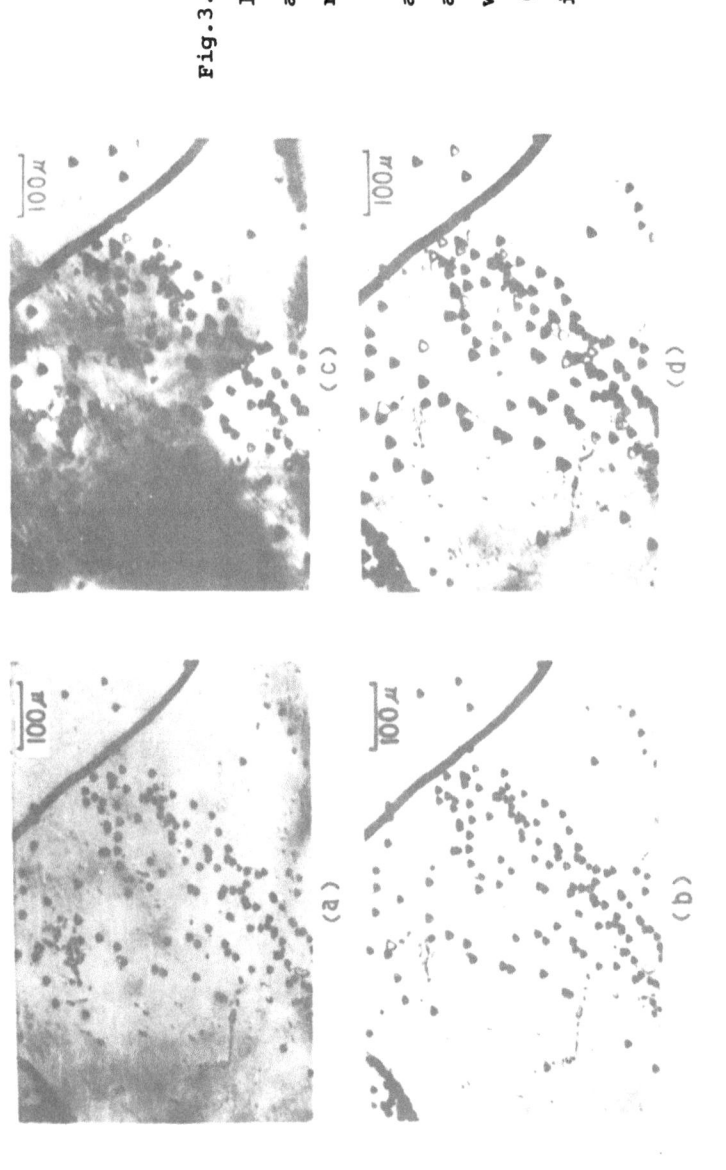

Fig.3. Showing the change in dislocation structure as the strain amplitude is increased, in the region A in Fig.2.

Photographs (a), (b), (c) and (d) show the structures after the internal friction was measured at points (3), (4), (5) and (6) respectively, in Fig.1.

(a)

(b)

(c)

(d)

Fig.4. Showing the successive movement of two dislocations as the strain amplitude is increased. The numbers in the photograph correspond to the numbers in Fig.1.

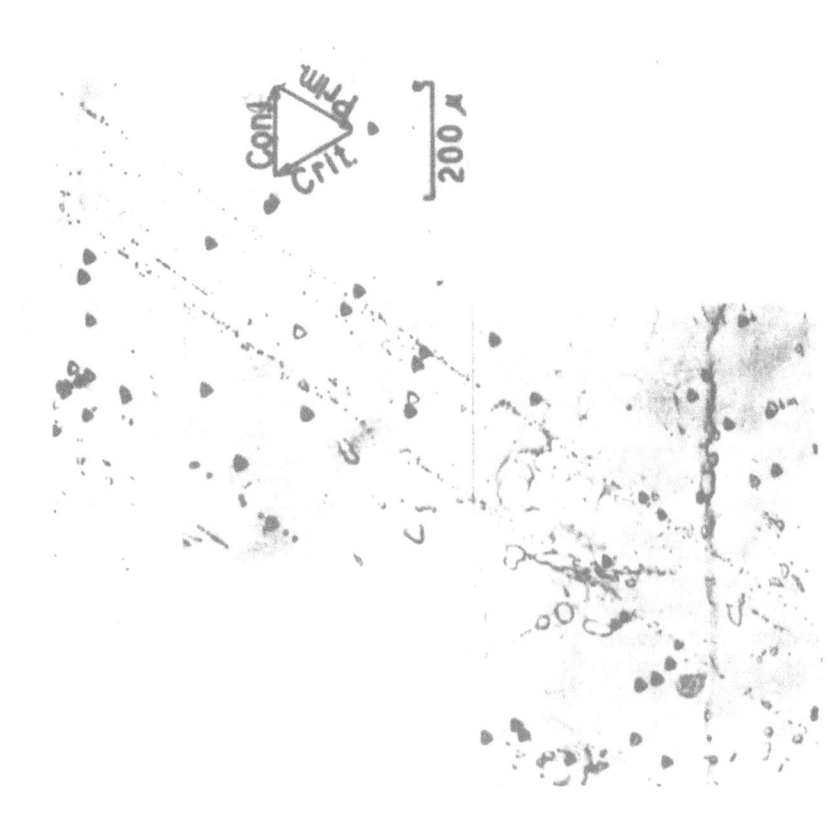

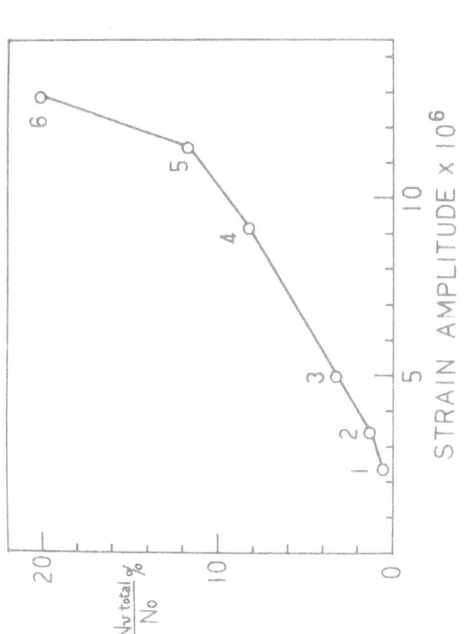

Fig.5. The ratio of the total number N_V total of dislocations moved to the total number No of dislocations observed (N_0 = 1945) is plotted against the strain amplitude of internal friction measurements shown in Fig.1.

Fig.6. Dislocation structure observed at the strain amplitude (6), in the region (C) in Fig.2.

AMPLITUDE-DEPENDENT INTERNAL FRICTION IN RHENIUM[*]

M. Callens-Raadschelders[**] and R. De Batist[***]
Solid State Physics Department, S.C.K./C.E.N., B-2400 MOL (Belgium)

ABSTRACT

The internal friction of a polycrystalline rhenium specimen after different deformation and ageing treatments has been investigated both as a function of temperature
and of strain amplitude. The mutual influence of these two parameters is interpreted in
terms of a double kink generation relaxation process and of double kink-stimulated mechanical break-away, according to the Seeger-Paré-Alefeld and Granato-Lücke theories.

1. Introduction

This paper reports dislocation damping processes in polycrystalline rhenium (hexagonal
close-packed), investigated through their effect on the temperature dependence (between 100 K
and 300 K) and on the strain amplitude dependence (between 4.10^{-6} and 7.10^{-5}) of the internal
friction, which is measured by means of the logarithmic decrement of freely decaying torsional
vibrations (frequency about 1 Hz). The material used is a (VP grade metal, 99.98 %) wire with a
diameter of 0.5 mm and a length of about 10 cm. All results, reported in this paper, are
measured with increasing temperature and decreasing amplitude and are obtained with one and the
same specimen, in which the amplitude dependence, after an annealing treatment of 1 hour at
2000 K in a vacuum of 10^{-8} Torr, was found to be particularly strong. Although a similar behaviour was observed in several other specimens, it was decided to select this one specimen for a
detailed study of the influence of different thermal and mechanical treatments. Indeed, the damping is found to be strongly influenced by changes in the internal state of the specimen (dislocation structure, internal stress distribution), caused by plastic deformation and/or ageing at
room temperature. We will discuss successively the different internal states of the specimen

---

[*] Work performed for the Association R.U.C.A. - S.C.K./C.E.N.

[**] Fellow of I.W.O.N.L.; Rijksuniversitair Centrum Antwerpen, Antwerpen (Belgium).

[***]Also at Rijksuniversitair Centrum Antwerpen, Antwerpen (Belgium).

and interpret in each case the internal friction behaviour in terms of double kink generation processes and their influence on dislocation relaxation and dislocation hysteresis.

Before starting a description of the experimental data, we want to emphasize that, for a better understanding of the results, one always has to consider three different temperature regions: one below 200 K, a second one at 200 - 220 K and a third one above 220 K. Furthermore, it will be necessary to consider separately the temperature dependence at a constant low strain amplitude and a constant high strain amplitude.

Table 1 gives the different treatments together with conjectures about the internal state, based upon the discussion given in the following sections.

TABLE 1.

Different Thermal and Mechanical Treatments applied to the Rhenium Specimen

Number of state	Subsequent applied treatments	Suggestions for internal state of material
1	Annealing 1 hour at 2000 K and 10^{-8} Torr	Long dislocation line segments lying in the Peierls valleys
2	0.5 % plastic deformation in torsion at room temperature	Creation of internal stresses
3	4 days ageing at room temperature	Dislocation rearrangement in the internal stress fields
4	0.4 % plastic deformation in torsion at room temperature	Creation of dislocation nodes
5	$7^1/_2$ months ageing at room temperature	Formation of a dislocation network
6	5 % plastic deformation in torsion at room temperature	Development of the dislocation network

2. Experimental Results and Interpretation of the Material in State 1

2.1. Amplitude Dependence

An example of the amplitude dependence of the internal friction of the rhenium specimen in state 1 at room temperature is shown in Fig. 1. In order to interpret this behaviour, we tried to fit the experimental points with the simple Granato-Lücke expression (1), giving the total damping as the sum of an amplitude-independent part Δ_I and an amplitude-dependent part

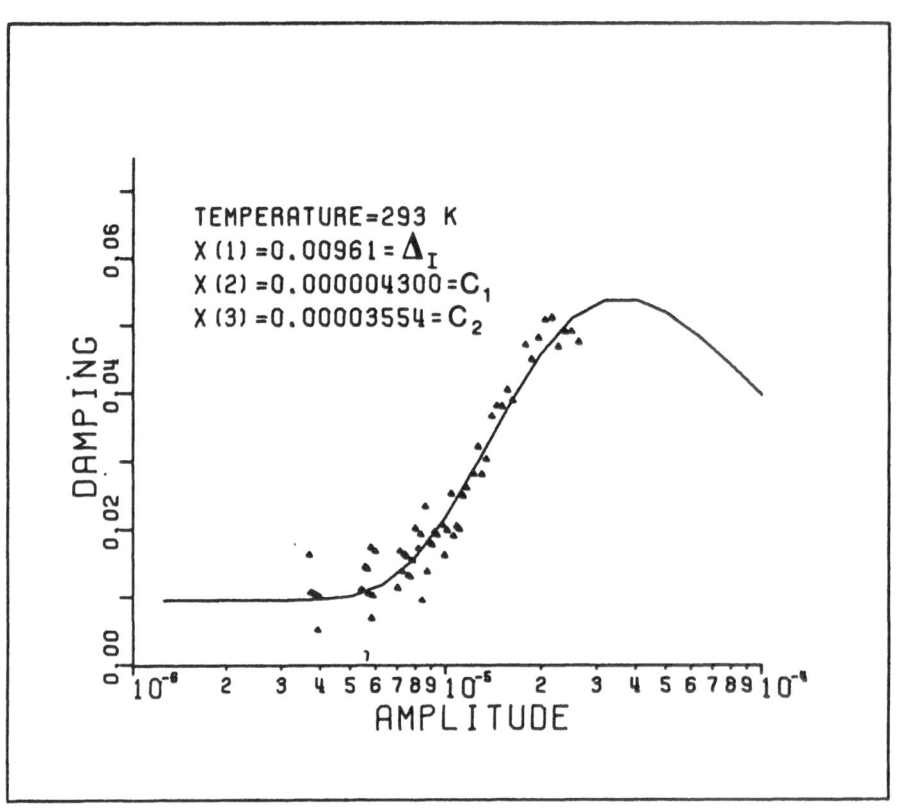

FIG. 1.

Granato-Lücke fit for the amplitude-dependent internal friction of rhenium in state 1 (see Table 1) at room temperature; ——— : theoretical curve.

Δ_H, with:

$$\Delta_H = \frac{C_1}{\varepsilon_o} \exp \left(-\frac{C_2}{\varepsilon_o}\right) \tag{1}$$

$$C_1 \sim \frac{\Lambda L_N^3}{L_C^2} \qquad \text{and} \qquad C_2 \sim \frac{1}{L_C}$$

where: ε_o = strain amplitude

       Λ = dislocation density

       L_N = average loop length between strong pinning points

       L_C = average loop length between weak pinning points.

Hence, C_1 is proportional to the dislocation density and to the square of the pinning point density and C_2 is proportional to the pinning point density.

Fitting is done by means of a computer program, in which trial values for C_1 and C_2 are estimated by applying the method of least squares to eq. (1) after taking the logarithm of both sides, followed by an iteration method (2) to correct the estimated values of Δ_I, C_1 and C_2 and to find the best fit with the exponential expression.

Although, in using this Granato-Lücke model (1), one neglects the influence of temperature (3), the influence of inhomogeneity in strain distribution in the wire (4, 5) and several
modifications proposed for the original Granato-Lücke model (e.g. 6 to 12), one obtains a theoretical curve which fits the experimental points rather well. The break-away occurs at a strain
of about 10^{-5}, whereas the maximum damping seems to occur at about 4.10^{-5}.

2.2. Temperature Dependence

The internal friction as a function of temperature is shown in Fig. 2 (lower part). At
low strain amplitudes (about 9.10^{-6}) the internal friction is nearly temperature-independent,
whereas at high strain amplitudes (about 2.10^{-5}) there is a stepwise increase at about 200 K.
The amplitude dependence at temperatures below and above 200 K is shown in the upper part of
Fig. 2. From the temperature dependence of the parameters C_1 and C_2, one derives a slight decrease of the pinning point density with increasing temperature, and a stepwise increase of the
density of dislocations, taking part in the break-away process.

2.3. Discussion

In earlier work (13, 14) we reported the presence of a relaxation peak in Re, occurring
at about 200 K for a frequency of 1 Hz and at 270 K for a frequency of about 300 Hz, leading to
an activation energy of about 0.38 eV and a τ_0 value of about 6.10^{-10} s. Based on these results
and on the behaviour of the relaxation strength after annealing or plastic deformation, this
peak was ascribed to a dislocation relaxation process. Further study of the internal friction
in several other rhenium wires has always shown a relaxation peak at about 200 - 210 K for
1 Hz. It turned out, however, that shape, height and temperature of the peak are strongly dependent on the prehistory of the specimen. Comparing with the present results we notice that
the relaxation peak occurs at the same temperature as the stepwise increase of Fig. 2. A similar behaviour has been reported e.g. by Alefeld et al. for f.c.c. aluminium (15) and by Korenko
for b.c.c. molybdenum (16). This confirms our hypothesis that we deal with a h.c.p. Bordoni
peak. The experimental results obtained for the other internal states of the specimen also
support this model.

The presence of a relaxation peak and of a stepwise increase in the amplitude dependence can be understood in the following way, according to the Seeger-Paré-Alefeld theory (17-22):
- the relaxation peak is due to the motion of dislocations, lying in a Peierls valley and jumping to another Peierls valley by the thermally-activated formation of pairs of kinks;
- at temperatures below the relaxation peak no thermal energy is available for the kinks to be
  formed and the small amplitude dependence comes from dislocations which are not lying in a
  Peierls valley;
- at temperatures above the peak temperature, thermal kinks are formed very easily and this

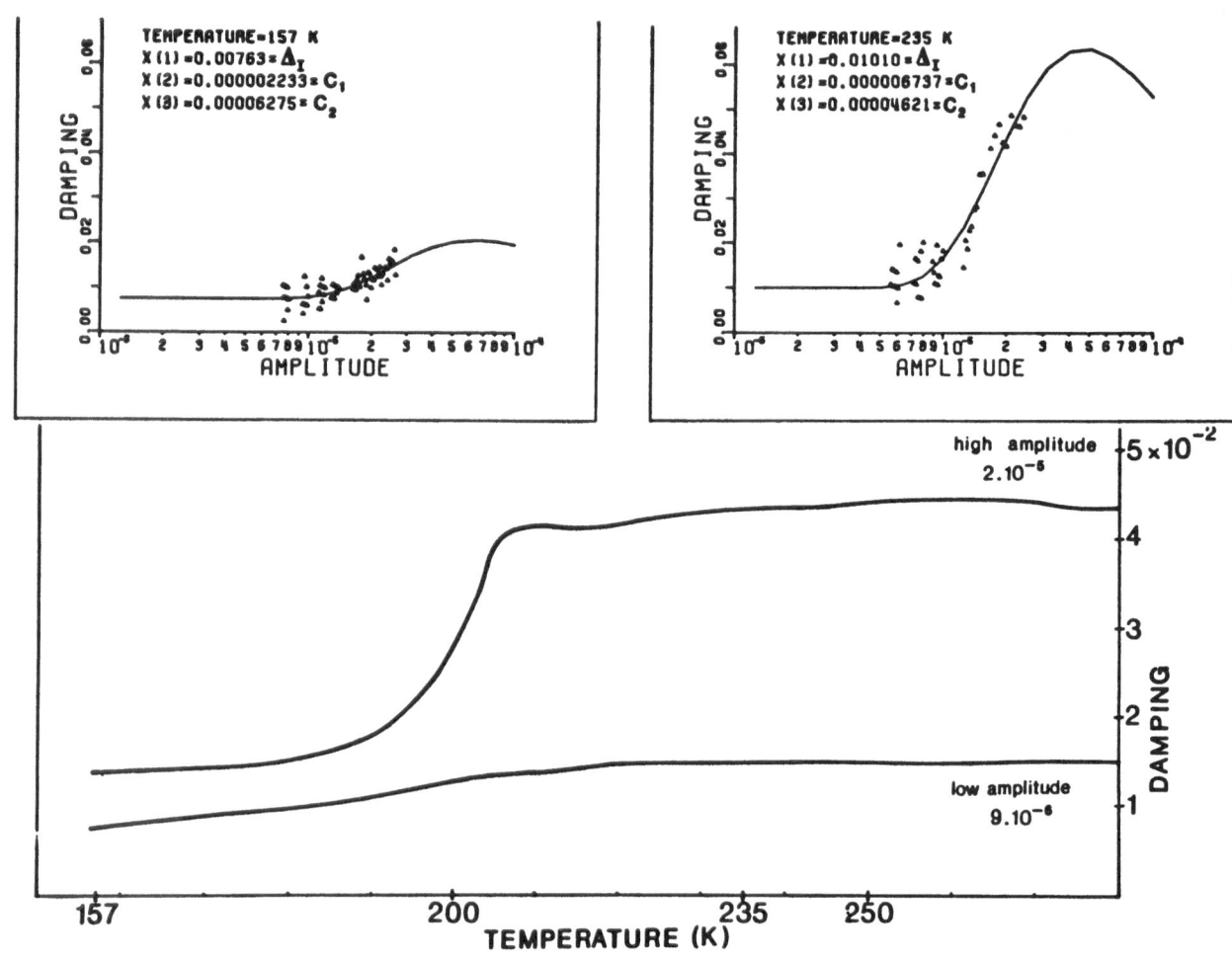

FIG. 2.

Lower part: internal friction vs. temperature for rhenium in state 1 (see
Table 1) at low and high amplitudes.
Upper part: Granato-Lücke fit of the amplitude dependence for T = 157 K
and T = 235 K.

kink formation will help the dislocations to break away from their weak pinning points. This
explains the stepwise increase in the hysteretic break-away damping;
- at small oscillation amplitudes, however, and in the absence of internal stress or external
bias stress, thermally-created double kinks are not separated far enough to be stable; so
no break-away will occur, and hence no step is observed in the amplitude dependence at small
amplitudes.

3. Experimental Results and Interpretation for Material in States 2, 3, 4, 5 and 6

3.1. State 2

Let us first consider the influence of the 0.5 % plastic deformation (cfr. states 1 and
2 in Fig. 3). In the temperature region below the Bordoni peak, only a slight increase of
damping appears, which indicates that only a small number of dislocations, non-parallel with a

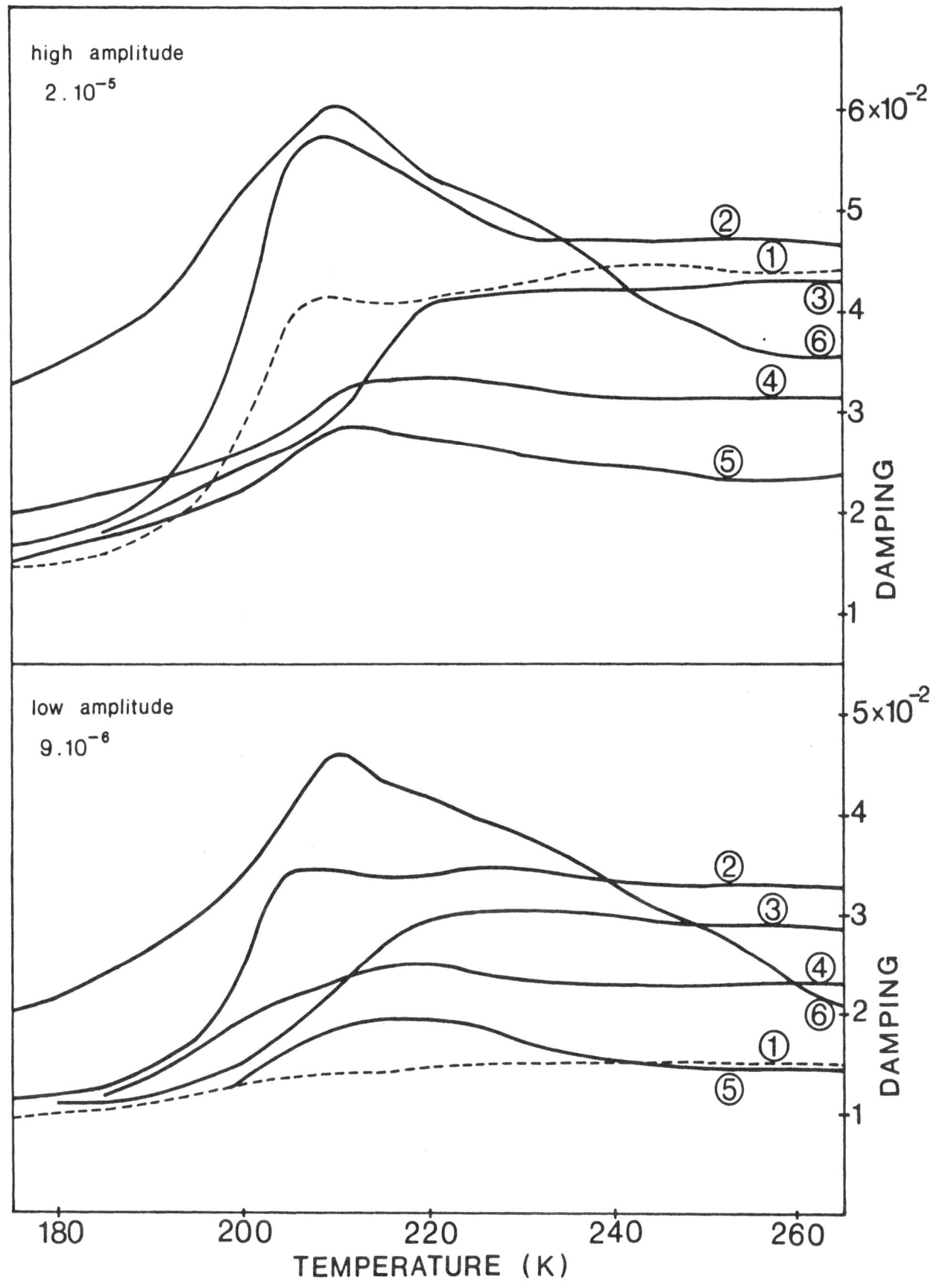

FIG. 3.

Temperature-dependent internal friction of rhenium in states 1 to 6 (see Table 1)

Peierls valley, have been created during the 0.5 % deformation. At temperatures above the Bordoni peak temperature, however, a large increase of damping, measured at low strain amplitudes appears, whereas the high amplitude damping only shows a small increase. In the region of the relaxation peak we find a stepwise increase for low strain amplitude measurements and a well-defined relaxation peak at high strain amplitudes (Fig. 3).

This behaviour, as well as the behaviour of the material in state 3, can again be interpreted in the frame-work of the Seeger-Paré-Alefeld theory. Indeed, this theory shows that, even above the double kink generation temperature, small oscillating stresses are not necessarily large enough to separate a double kink pair (see state 1 at small amplitudes). Extra energy, necessary to bulge the dislocation, has to come from high strain amplitudes (state 1 at high amplitudes) or from internal stresses (state 2 at small amplitudes) created by a small plastic deformation. According to Paré (20) and Alefeld (22), one can only have double kink generation if $ab\sigma L \geq 2 W_k - T\Delta S$, where: σ = internal and/or external stress

L = dislocation line length

W_k = kink energy

a = lattice parameter

b = Burgers vector

T = temperature

ΔS = entropy difference between a dislocation with and without a double kink.

Once the dislocation has been bulged by double kink formation, it can break away from the weak pinning points and amplitude-dependent damping occurs. Interpretation of the relaxation peak is based on the theoretical discussion given by Alefeld (22). He distinguishes between non-interacting kinks and interacting kinks, and derives in both cases expressions for the relaxation strength of the double kink generation peak. Alefeld predicts that, starting from low kink density (non-interacting kinks), the kink density will increase exponentially with internal stress until interaction will occur between kinks. In the case of non-interacting kinks, the relaxation strength is given by:

$$\Delta_M = \frac{\Lambda L^2}{4} \ \frac{Gb^2 a}{kT} \ F\ (\sigma,\ T) \tag{2}$$

with: Λ = dislocation density

L = average loop length

T = temperature

σ = internal and/or external stress

F (σ, T) = 0 for σ = 0 and increases exponentially with increasing stress. This is what we observe in the states 1 and 2. However, this exponential increase will not continue, but will be taken over by a "stringlike" relaxation strength, once the interactions between kinks become strong:

$$\Delta_M = 0 \quad \text{when} \quad abL\sigma \ll 2 \, W_k$$

$$\Delta_M = \gamma \, \frac{\Lambda L^2}{12} \, \frac{Gb^2}{E_L} \quad \text{when} \quad abL\sigma \gg 4 \, W_k \qquad (3)$$

with: W_k = kink energy

σ = internal and/or external stress

$0 \leq \gamma \leq 3/4$ depending on the position on the line where double kink generation occurs

E_L = line tension.

Further, Alefeld has shown that the maximum value of the relaxation strength for non-interacting kinks, obtained by inserting in eq. (2) the estimated value of the internal stress at which kink interaction becomes significant, is larger than the limiting value for interacting kinks (eq. (3)). This means that one can expect a maximum in the relaxation strength as a function of internal stress.

3.2. State 3

After 4 days ageing at room temperature, the peak height is strongly decreased whereas only a slight decrease is detected in the amplitude dependence above the peak (Fig. 3). Other measurements indicate that the material needs several hours ageing at room temperature to reach internal equilibrium (23, 24). Therefore, one expects that, following 4 days ageing at room temperature, the dislocations will have rearranged in accordance with the equilibrium internal stress state and, according to Alefeld's theory, will contain more kinks. Double kinks, created during an internal friction experiment, will interact with existing kinks and hence a relaxation strength of the form of eq. (3) will be observed. The newly created double kinks will still be able to help the unpinning mechanism and hence no significant change in amplitude dependence is expected.

3.3. State 4

The most striking observation for the treatment leading to state 4 (Fig. 3) is the drastic decrease of the damping at temperatures above the relaxation peak. No significant change appears at the temperature of the relaxation peak. This large reduction in amplitude-dependent damping after plastic deformation can only be due to the decrease of the average line length of the dislocations lying in the Peierls valley as a result of dislocation tangling. Indeed, the theory of Granato-Lücke predicts a ΛL^4 dependence for the amplitude-independent loss and a ΛL^3 dependence for the amplitude-dependent loss, whereas the relaxation strength of the double kinks is determined by ΛL^2 (22).

3.4. State 5

Comparing states 4 and 5 (Fig. 3), we notice a decrease of the background and a slight sharpening of the relaxation peak. This suggests that, during ageing, the dislocation network is ordered, so that more dislocations are lying in Peierls valleys. This behaviour of decreasing background and sharpening relaxation peak after ageing at room temperature has also been observed in some other rhenium specimen, deformed in torsion at liquid nitrogen temperature up to 23 % (23).

Looking at the amplitude dependence in stage 5 (Fig. 4), we notice that the functional relationship between damping and strain amplitude can no longer be described by a simple Granato-Lücke process. A second mechanism seems to be activated at higher amplitudes (23).

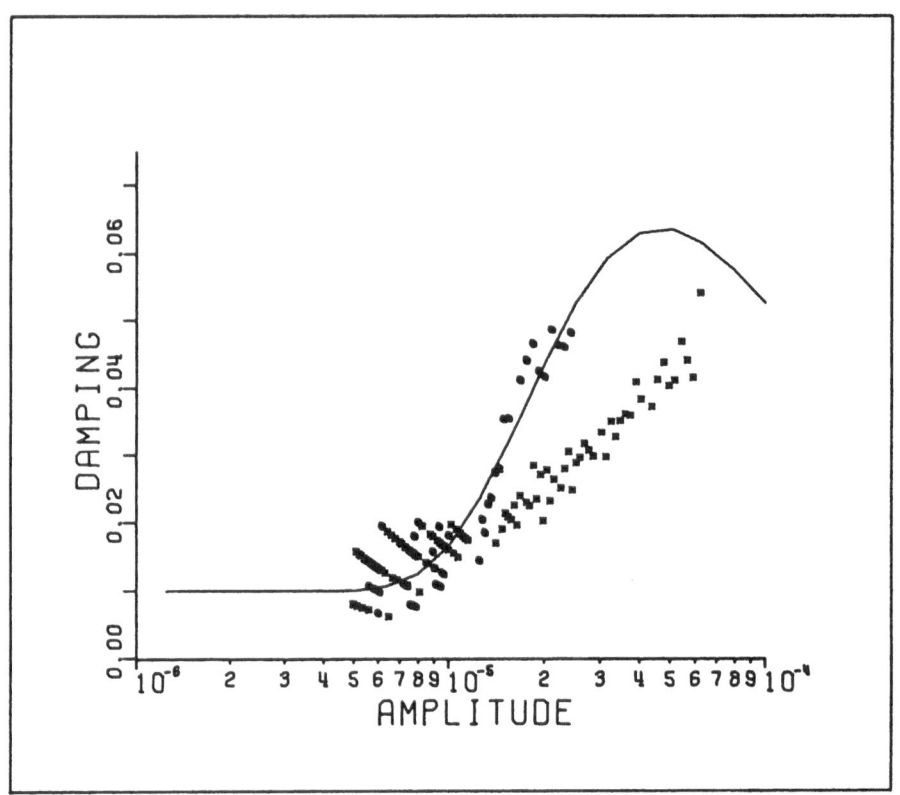

FIG. 4.

Amplitude-dependent internal friction of Re at 235 K in state 1 (◐) and state 5 (✱) (see Table 1); ———— : Granato-Lücke fit.

3.5. State 6

The material in state 6 (5 % plastic deformation) again shows a well-defined relaxation peak (Fig. 3), probably due to the presence of a dense dislocation network, containing large amounts of dislocations and internal stresses. Comparing with state 2 there are, however, several differences:
- the relaxation peak is much broader, suggesting that we have more than one relaxation process;

- the background below the peak temperature is higher, indicating that we have more dislocations non-parallel with the Peierls valleys;
- the background above the peak is lower, indicating that the strong unpinning process of state
  2 no longer exists;
- the amplitude dependence is independent of temperature: damping values at low and high strain
  amplitudes only differ by a constant amount. This shows that there is no longer a connection
  between the double kink generation and the amplitude dependence.

4. Discussion

The present results confirm that, also in h.c.p. material, double kink generation
strongly depends, not only on temperature, but also on the stress, acting on the dislocation
line (be it internal stresses or an alternating stress exerted during the internal friction
experiment). Comparing these measurements with the amplitude-dependent measurements of Alefeld
et al. (15) on polycrystalline aluminium (deformed 1 % and measured with increasing strain amplitude) on one side, and with the measurements of Korenko (16) on molybdenum single crystals
(measured as a function of increasing amount of deformation up to 50 %) on the other side, we
notice that our results of the material in state 1 and 2 are rather similar to Alefeld's measurements, whereas the results in state 5 and 6 are rather similar to Korenko's measurements.
The main difference between the results for states 2 and 6 in rhenium, and between the measurements on polycrystalline aluminium (15) and single crystal molybdenum (16) is in the behaviour of the background when a relaxation peak is present. Following a relatively small amount
of plastic deformation, such as in the state 2 rhenium and in the polycrystalline aluminium,
the background shows a stepwise increase as a function of temperature, whereas in the state 6
rhenium and the monocrystalline molybdenum, the background only shows a monotonic increase with
increasing temperature. These observations can be rationalized when taking into account the
differences in dislocation structure, expected after deformation of single crystals or polycrystalline material. Therefore, we can summarize as follows:
- for the appearance of a double kink generation peak we need dislocation lines lying in
  Peierls valleys, and we need internal or external stresses;
- in single crystals, deformed in stage I, we can have the required dislocation lines, but,
  due to the fact that only one glide system is activated, we do not have internal stresses;
- in stage II, internal stresses will build up due to tangling of dislocations; therefore, a
  relaxation peak will occur. Due to tangling, however, the motion of the dislocations will be
  restricted and the background above the peak will decrease;
- in polycrystalline specimens, a small deformation will create dislocations in different glide
  systems and internal stresses will occur even before a network starts to form; there will be
  no restriction for dislocation motion because tangling does not yet occur (strong amplitude-
  dependent background above the peak);
- large deformations in polycrystalline material will create a network; tangling will occur and
  hence, the background above the peak decreases and the relaxation peak increases.

References

1. A. Granato and K. Lücke, J. Appl. Phys. 27, 789 (1956).

2. L. Schotsmans, S.C.K./C.E.N. Mol, Belgium, note FO 4/7, Appl. Math. L.S./cj 621-75/70-184 (1970).

3. L. Teutonico, A. Granato and K. Lücke, J. Appl. Phys. 35, 220 (1964).

4. P. Peguin, J. Perez and P. Gobin, J. Sci. Instr. 42, 814 (1965).

5. F. Povolo and R. Gibala, Phil. Mag. 27, 1281 (1973).

6. J. Swartz and J. Weertman, J. Appl. Phys. 32, 1860 (1961).

7. J. Friedel, The Relation between the Structure and Mechanical Properties of Metals, p. 409. Her Majesty's Stationary Office, London (1963).

8. B. Trott and H. Birnbaum, J. Appl. Phys. 41, 4418 (1970).

9. D. Blair, T. Hutchison and D. Rogers, J. Appl. Phys. 40, 97 (1969).

10. D. Blair, T. Hutchison and D. Rogers, Can. J. Phys. 48, 2943 (1970).

11. D. Blair, T. Hutchison and D. Rogers, Can. J. Phys. 48, 2955 (1970).

12. D. Blair, T. Hutchison and D. Rogers, Can. J. Phys. 49, 633 (1971).

13. M. Raadschelders and R. De Batist, J. de Physique 32, 179 (1971).

14. M. Raadschelders and R. De Batist, Defects in Refractory Metals, Paper IV.4., p. 251. Eds. R. De Batist, J. Nihoul, L. Stals, S.C.K./C.E.N. Mol, Belgium (1972).

15. G. Alefeld, J. Filloux and H. Harper, Dislocation Dynamics, p. 191. Eds. A.R. Rosenfield, G.T. Hahn, A.L. Bement Jr., R.I. Jaffee, Mc. Graw-Hill Book Company (1967).

16. M.K. Korenko, Master of Science Thesis: Dislocation Relaxations in Molybdenum Single Crystals, Case Western Reserve University (1969).

17. A. Seeger, Phil. Mag. 1, 651 (1956).

18. A. Seeger, H. Donth and F. Pfaff, Disc. Faraday Soc. 23, 19 (1957).

19. V.K. Paré, Thesis, Cornell University AFOSR-TR-58-92, ASTIA AD 162133 (1958).

20. V.K. Paré, J. Appl. Phys. 32, 332 (1961).

21. G. Alefeld, R.H. Chambers and T.E. Firle, Phys. Rev. 140, A 178 (1965).

22. G. Alefeld, Lattice Defects and their Interactions, p. 407. Ed. R.R. Hasiguti, Gordon and Breach Science Publishers, New York (1967).

23. M. Callens-Raadschelders, to be published.

24. R. De Batist, to be published.

THE RELATION BETWEEN THE AMPLITUDE DEPENDENT
INTERNAL FRICTION AND THE TEMPERATURE PEAK OF
INTERNAL FRICTION DUE TO UNPINNING OF DISLOCATIONS

R. R. Hasiguti and M. Kobayashi*
Faculty of Engineering, University of Tokyo
Bunkyo-ku, Tokyo
and
Institute of Physical and Chemical Research
Wako-shi, Saitama-ken, Japan

Introduction

The unpinning of dislocations is manifested in the so-called Granato-Lücke amplitude dependent internal friction. If this kind of internal friction is measured at finite temperatures, the unpinning is assisted by thermal energy. On the other hand the thermal unpinning of dislocations is manifested in the so-called Hasiguti peak of internal friction.

Considerable efforts have hitherto been made from both sides in order to relate theoretically the above two manifestations of internal friction with each other. These have been successful qualitatively to some extent, but not successful enough quantitatively (1-4).

In this paper the present authors show experimentally rather than theoretically a relation between the Granato-Lücke amplitude dependent internal friction and the Hasiguti peak of internal friction making use of dislocations pinned by carbon atoms in iron.

Theory

The binding energy between a pinning point defect and a dislocation can be obtained, in principle, from either of the above two kinds of internal friction. If the binding energies obtained from both kinds of internal friction are identical with each other, then we can say that we have a definite

---

*Now at Mitsubishi Electric Corporation.

relation between the two kinds of internal friction.

In the case of a Hasiguti peak, the binding energy is directly related to the activation energy of the peak, or in simple models the binding energy is identical with the experimental activation energy of the peak (5).

On the other hand in the case of a Granato-Lücke internal friction, at least two factors must be considered in order to obtain the binding energy: namely the temperature and the dislocation density. Lücke and his associates (2) made detailed theoretical investigations of the effect of temperatures on the Granato-Lücke internal frictions, but their results are too complicated to be used in this experimental analyses. In the present research, therefore, the temperature effect was eliminated by extrapolating the experimental results to 0°K, where we have an original simple Granato-Lücke theory at 0°K (6).

The dislocation density brings forth the most difficult problem. Although we can obtain the total density of existing dislocations by means of various direct experimental methods, we do not know how many per cents of existing dislocations are contributing to the internal friction concerned. In other words we do not know the density of "effective" dislocations, which is considered to be a small fraction, say 0.1 to 10 %, of existing dislocations. It is impossible to obtain the exact value of the binding energy without an accurate knowledge of the "effective" dislocation density in the case of the Granato-Lücke theory (6).

Then the following procedure is taken in this paper. The binding energy obtained as the activation energy of a Hasiguti peak is put into the 0°K Granato-Lücke equation, which is solved with respect to the unknown effective dislocation density. If thus obtained effective dislocation density is reasonable in its magnitude, we would be able to consider that a relation between the two kinds of internal friction, for which we are searching, is established, or in other words we would be able to consider that we are observing two kinds of manifestation of dislocation unpinning from the same combination of dislocations and pinning point defects in the same specimen.

Experimental Procedures

The specimens should contain well identified pinning agents, so that the following iron specimens were prepared. The Johnson-Matthey iron specimens were annealed at 750°C for 5 days in wet hydrogen, zone-refined three times, and then doped with carbon which is the pinning agent to the content of 7 weight ppm. The content of carbon was measured by means of Snoek peak after quenching from 700°C. If the carbon content is above 15 ppm, this kind of experiments becomes difficult because of the over-pinning. The internal friction was measured by the conventional torsion pendulum in a magnetic field of about 110 oersteds in the temperature range from about 100°K to about 360°K.

Before starting the measurements, the specimens in the form of wire 0.8 mm in diameter were quenched from 700°C, cold worked about 10 % by twisting to increase the total dislocation density, and then set to the apparatus and cooled to the liquid nitrogen temperature. These procedures were taken as quick as possible in order to minimize the possible excessive pinning before starting the measurements.

Experimental Results and Discussions

Hasiguti Peak

Figure 1 shows an example set of measurements of internal friction as a function of temperatures. The measurements were performed during cooling runs to avoid a possible appearance of false peaks due to annealing effects. The vibrational strain amplitude used is 2.9×10^{-5}, and the vibrational frequencies used are shown in the figure. There appears a peak in each curve at about 190°K. The shape of the peak becomes clearer when its background is subtracted, as shown in Fig. 2 (a). The reciprocal absolute peak temperatures are plotted as a function of the logarithms of vibrational frequencies in Fig. 2 (b), from which an activation energy of 0.31 eV is obtained.

Many sets of similar experiments were performed changing the degree of aging of specimens at room temperature. The change of degree of aging corresponds to the change of degree of pinning by carbon atoms. The internal friction peaks and the background internal friction change in parallel with the change of degree of aging. For example, the background internal friction decreases as the aging proceeds; and the peak temperature shifts slightly to higher temperatures and the peak height decreases as the aging proceeds. When the aging proceeds sufficiently, the peak finally disappears. These behaviors of the peak and the background are understood, if the aging is considered to be the increase of carbon pinning points, which include to some extent carbon clusters in the later stage of aging.

From the above results it is concluded that the peak around 190°K is due to the thermal unpinning of dislocations from carbon pinning points, or a Hasiguti peak of dislocations pinned by carbon atoms. The activation energy of the peak averaged over seven sets of similar experiments of differently aged specimens is 0.34 eV. If a result of a specimen somewhat over-aged is omitted, the average is 0.33 eV, which should be recommended as the binding energy between a dislocation and a carbon atom.

Granato-Lücke Internal Friction

During the measurements shown in Fig. 1, the amplitude dependence

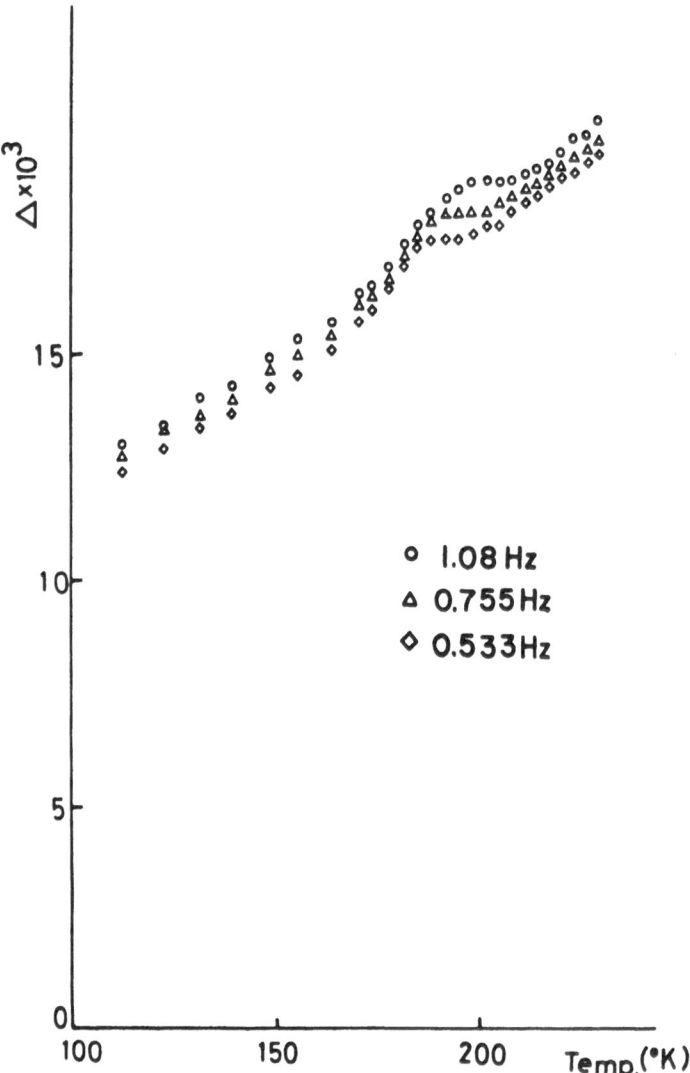

Fig. 1.
Decrements ($\Delta = \Delta W/2W$) as a function of
temperatures observed in 7 ppm carbon doped
iron in which dislocations are pinned by
carbon atoms.

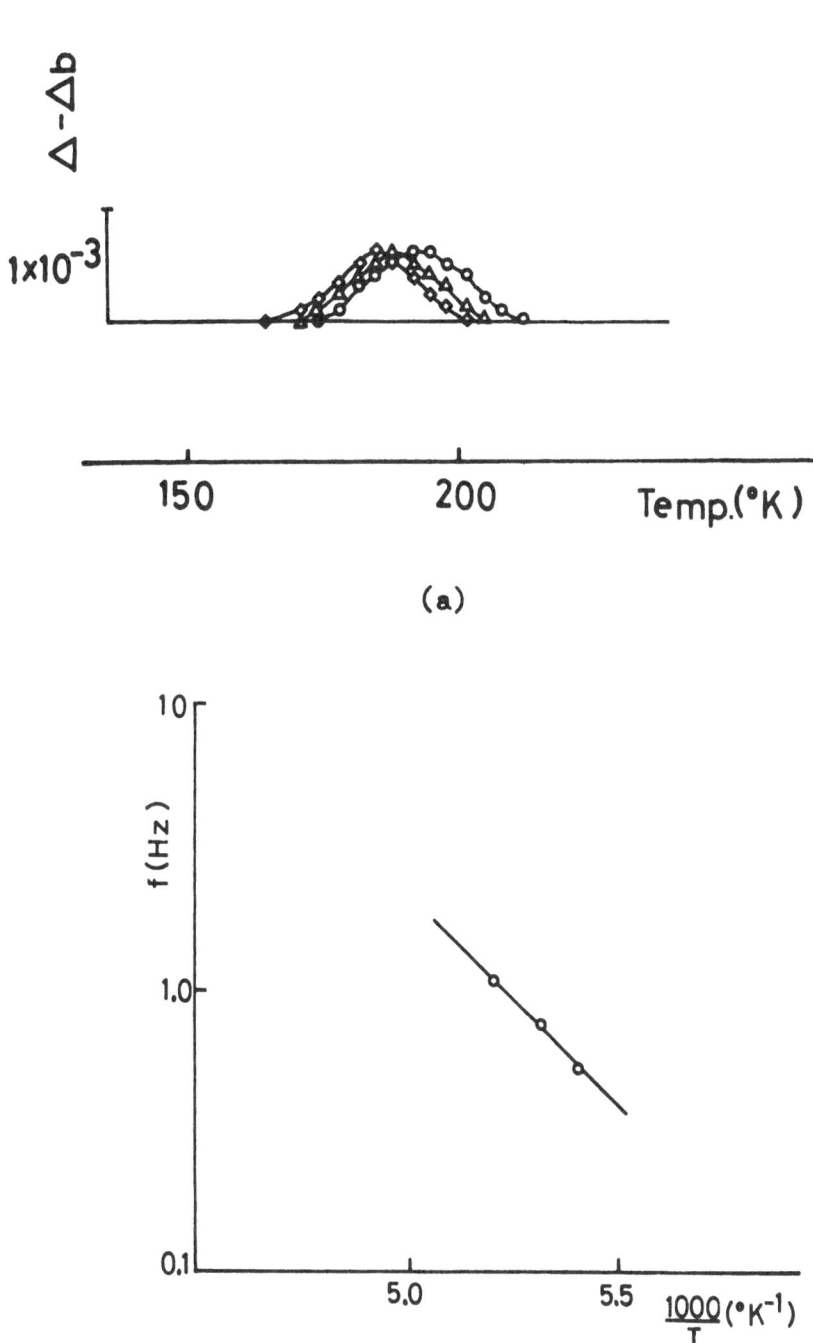

(a)

(b)

Fig. 2.
(a) Decrement peaks replotted from Fig. 1
after subtracting background decrements.
(b) Reciprocal absolute peak temperatures vs.
logarithms of vibrational frequencies.

of internal friction was measured at four constant temperatures shown in Fig. 3. The results are shown in the form of so-called Granato-Lücke plots in Fig. 3.

Two kinds of Granato-Lücke plots, which are different in the expression of the ordinate, are used in different literatures. The one uses $\log \Delta_H \mathcal{E}_0$ and the other uses $\log \Delta_H \mathcal{E}_0^{1/2}$. After examining many sets of experimental results making use of both kinds of plots, we found that the latter type or the square root amplitude type of plots is better, because the former type or the linear amplitude type gives often unreasonably small effective dislocation density. Therefore, only the square root amplitude type is used in this paper.

The Granato-Lücke straight lines at different temperatures are expressed by

$$\Delta_H \mathcal{E}_0^{1/2} = A(T)\exp(-B(T)/\mathcal{E}_0) \qquad (1)$$

where Δ_H is the decrement $(\Delta W/2W)$ and $\mathcal{E}_0$ is the vibrational strain amplitude. According to our results of ten sets of serial experiments between 100 and 300°K, the coefficients $A(T)$ and $B(T)$ change linearly with temperature T, namely

$$A(T)+aT = c_1 \qquad (2)$$
$$B(T)+bT = c_2. \qquad (3)$$

Here c_1 and c_2 are positive constants; b is a positive constant which shows that the slope of Granato-Lücke straight line always increases as the temperature is lowered as shown in Fig. 3; a can be either a positive or a negative constant, which shows that Granato-Lücke straight lines at different temperatures intersect with one another either in the positive region or the negative region of the abscissa values. In the case of Fig. 3 a is a negative constant. From Eqs. (2) and (3) we obtain $A(0°K)$ and $B(0°K)$, from which the 0°K Granato-Lücke straight line is obtained as shown by a broken line in Fig. 3.

In order to proceed with further numerical analyses according to the 0°K Granato-Lücke theory (6), we have to know the so-called breakaway stress or strain, which is the stress or strain amplitude, at which an observable deviation occurs from the horizontal part in the conventional amplitude dependent internal friction curve. Taking the accuracy of our experimental results into consideration, the breakaway strain is determined as the strain amplitude, at which the amplitude dependent component of internal friction deviates as much as 1×10^{-4} from the horizontal line. The most accurate and practical way of determining this breakaway strain will be shown elsewhere. The breakaway strain of the 0°K line in Fig. 3 was determined to be 7.5×10^{-6}.

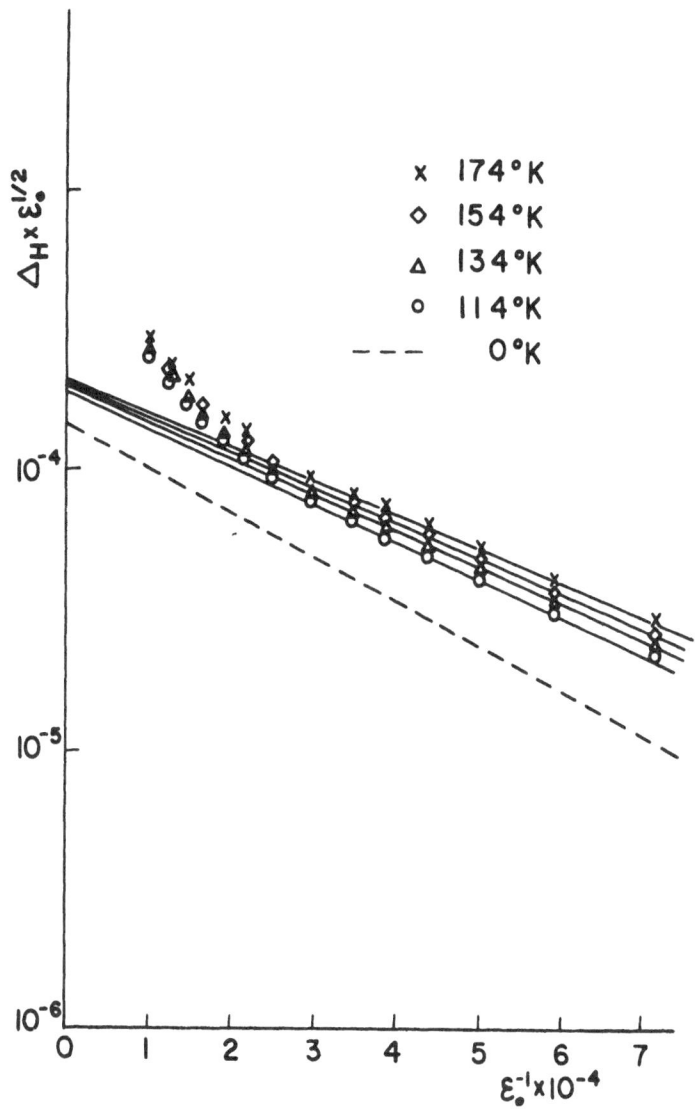

Fig. 3.

Granato-Lücke plots of amplitude dependent
decrements observed in 7 ppm carbon doped iron
in which dislocations are pinned by carbon
atoms. The 0°K line was obtained by means of
extrapolation.

Now we are ready to solve the 0°K Granato-Lücke equation or the 0°K
Granato-Lücke line shown in Fig. 3 according to the scheme shown in the section of "Theory". The results shown in Figs. 1, 2 and 3 were obtained from
the same specimen prepared under the same condition, so that we should use
the binding energy 0.31 eV obtained from Fig. 2 rather than the final average
value 0.33 eV in order to solve the 0°K line in Fig. 3. Using 0.31 eV, the
effective dislocation density was determined to be 1.7×10^7 cm^{-2}.

We made six sets of similar experiments, and the effective dislocation
density averaged over the six sets of results was found to be 2.1×10^7 cm^{-2}.
This effective dislocation density is quite reasonable, because this is of
the order of 1 % of the density of total existing dislocations, which is considered to be about 1×10^9 cm^{-2} in 10 % twisted iron wire. If this is accepted, it is concluded, according to the above scheme shown in the "Theory",
that the Granato-Lücke mechanical unpinning at 0°K and the Hasiguti peak
thermal unpinning are two different manifestations of unpinning processes of
the same dislocations from the same pinning points.

General Discussions

Discussions were already made to some extent in the preceding sections
of "Theory" and of "Experimental Results and Discussions". Here we make some
additional and more general discussions.

Effective Dislocation Density

To judge whether the obtained effective dislocation density is reasonable or not is an important problem in this paper. Before discussing this
problem the probable error of the effective dislocation density will be considered. It was shown in the preceding section that for the specimen of
Figs. 1, 2 and 3 the binding energy is 0.31 eV and the effective dislocation
density is 1.7×10^7 cm^{-2}. It is easily shown that the probable error of ±0.04
eV in the binding energy results in the change of the effective dislocation
density by a factor of $10^{\pm 0.6}$.

Now it is almost impossible at present to find the effective dislocation
density by any direct experimental method, while there are many indirect
methods by which we can estimate it. We make an indirect estimation in this
paper as follows. According to Koiwa and Hasiguti (7) the height of the
Hasiguti peak is given by

$$\Delta_h = \pi Q_h^{-1} = 0.4 \, \rho_e L_c^2 \qquad (4)$$

where Δ_h is the decrement peak height, Q_h^{-1} is the internal friction peak
height, ρ_e is the effective dislocation density, and L_c is the dislocation
loop length. L_c is obtained from the slope of Granato-Lücke straight line.
Now we put into Eq. (4) $\Delta_h = 6.3 \times 10^{-4}$ obtained from Fig. 2(a) and

$L_c = 5.3 \times 10^{-6}$ cm obtained from the 0°K line of Fig. 3. Then we get $\rho_e = 5.6 \times 10^7$ cm^{-2}, which is within the range of $1.7 \times 10^{7\pm0.6}$ cm^{-2}.

The above is an example of a crude estimation. More detailed and exact estimations of the effective dislocation density should be explored in future.

Remarks on the Nature of Unpinning

The Granato-Lücke unpinning is essentially a mechanical unpinning, although it is assisted by thermal energy at finite temperatures. Therefore, the binding energy appears through the mechanical breakaway stress, which depends on the loop length. This makes the Granato-Lücke unpinning a very complex phenomenon. On the other hand the Hasiguti unpinning is essentially a thermal unpinning, so that the binding energy appears directly as the activation energy, which does not depend on the loop length.

Summary

The unpinning of dislocations from pinning carbon atoms in high purity iron doped with 7 weight ppm carbon was observed by means of two kinds of internal friction, namely the 0°K Granato-Lücke amplitude dependent internal friction and the Hasiguti peak of internal friction. The 0°K condition in the Granato-Lücke internal friction was achieved by means of an extrapolation from higher temperatures. If the binding energies between a dislocation and a carbon atom obtained by the above two methods are identical with each other, we can say that the above two kinds of internal friction are two different manifestations of unpinning processes of the same dislocations from the same pinning points. It was found that this was the case, although somewhat roundabout discussions were necessary to reach the conclusion because of the lack of exact knowledge of the effective dislocation density. Finally, a recommended value of the binding energy between a dislocation and a carbon atom in iron is 0.33 eV.

References

1. K. Lücke and J. Schlipf, *Interaction between Dislocations and Point Defects*, ed. B. L. Eyre, H. M. Stationery Office, London, 1968, p.118.

2. K. Lücke, A. V. Granato and L. J. Teutonico, J. Appl. Phys. **39**, 5181 (1968).

3. R. R. Hasiguti, *Ann. Rev. Materials Sci.* Vol. 2, Annual Reviews, Inc. Palo Alto, Calif. 1972, p.69.

4. R. R. Hasiguti, J. Less-Common Metals, **28**, 249 (1972).

5. R. R. Hasiguti, *Proc. Third Internat. Symp. Reinststoffe Wissensch. Tech., Dresden*, Akademie-Verlag, Berlin, 1972, p.165.

6. A. Granato and K. Lücke, J. Appl. Phys. **27**, 583 (1956).

7. M. Koiwa and R. R. Hasiguti, Acta Met. **13**, 1219 (1965).

AMPLITUDE DEPENDENT INTERNAL FRICTION IN Zn MONOCRYSTALS.

A. Bortolotti; G. Martinelli, L. Passari
Istituto di Fisica dell'Università . Ferrara
G.N.S.M. Ferrara

1. Introduction.

In the last years several attempts have been made to explain the amplitude
dependent internal friction (1-8). One of the most complete models is that
developed by Granato and Lücke(2). This model is essentialy based upon the
mechanism proposed by Koehler (1).

In that model it is assumed that the dislocation network is characterized
by two different kind of pinning points, strong and weak pinning points.
The dislocations are supposed not to be able to break away from the first
type of pinning point (strong) when a stress is applied. These pinning points
are crossing of dislocations, cluster of defects and so on. Instead, the dis
location can break away from the second type of pinning point (impurity
atom, point defects,;,;.) when the strss is sufficiently high.
These two kinds of pinning poits are randomly distributed along the disloca
tions. An important parameter characterizing the distribution is the ave_
rage distance between the pinning points. Let be L_A the average distance
between the strong pinning points (network length) and L_C the average dis
tance between the weak pinning points (dislocation loop length).
Normally is $L_A > L_C$. If we apply an oscillating stress to the sample, the
dislocation loop under the action of the stress begin tooscillate like a
string. It is assumed that during its movement the dislocation looses ener
gy by a viscous damping.
When the maximum applied stress increases, the forces exerted by the dislo
cation line on the pinning points increase. These forces depend also on the
dislocation loop length. In this condition, for every applied stress and
for a certain dislocation-pinning point binding energy, a critical disloca

tion length exists. When the total length of two dislocation loops acting on
the same pinning point is the critical for the value of the stress applied,
the dislocation breaks away from the pinning point, and automatically the
condition for the break away is reached in all the nearest pinning points.
There is a catastrophic break away of all the loops in a network length,
limited by the strong pinning points.

Due to the random distribution of the pinning points this process occurs in
a range of stresses. When the stress decreases the dislocatin loops collapse
and become pinned again. The process repeats itself in the successive half
cycle. From this kind of process turns out a hysteresis loop in the stress-
strain diagramm.

Granato and Lücke computed the contribution due to these hysteretic losses
to the amplitude dependent internal friction. Between the experimental data
and the result of the Granato-Lücke theory there is only a qualitative
agreement limited to particular condition.

Improvments to this model have been proposed for ex. by Granato , Lücke and
alii (5,6) Taking in account the temperature effects.

An other improvement has been proposed by Roger (7) who pointed out that
when the dislocations break away the length distribution of the dislocation
loops changes. Because of the strong dependence of the internal friction
from the average dislocation loop length, this change in the distribution
length results in a second mechanism for the i. f. dependence   on the
amplitude.

Following this way Trott and Birnbaum (8) studied, with the help of a compu
ter, a model of a crystal containing a dislocation network and followed its
behaviour when the various parameters (L_C and L_A and the distribution law)
were changed and when various conditions were simulated (High or low tempera
ture). From this computer simulation turned out that the contribution to the
amplitude dependent internal friction of the hysteretic process and that due
to the change in the length distribution may be of the same order of magnitude.

Trott and Birnbaum have also extended their calculation to the case of random
orientation of dislocation and when there is a stress distribution.

The purpose of our work was to study the amplitude dependent internal fricti on of Zn monocrystals and to verify if it is possible to explain the experimental results using the existing theories.

2. Experimental.

Internal friction measurements have been carried out in vacuum (10^{-3}mm Hg) by means of a Bordoni apparatus (9-10). In this apparatus oscillations are drived and detected by an electrostatic method.

Resonant flexural modes of oscillation at about 2000 cps have been used. Q^{-1} has been determined either from the width of the resonance curve or from the logarithmic decrement of the oscillation.

The maximum strain amplitude has been controlled by the method described in a previous paper (11).

The monocrystalline samples were prepared by the Bridgmann method from high purity Zn. The crystals were prepared in shape of bars from which the samples were obtained cutting the extremities by an acid saw. The size of the samples were 50 x 5 x 1 mm.

The high purity Zn was analized by the spectroscopic method and the results are the following:

Pb: 0.0003 at %, Ni: 0.0003%,Cu:0.0003% Cd< 0.0001% , Fe, Mg, Sn, Al not visible.

The monocrystal were controlled and oriented by X rays.

Before the I. F. measurement the samples were annealed at 350°C for 12 hours and then slowly cooled in furnace. The samples were sensitive to the handling. For this reason all the measurements were carried out 24 hours after the samples were mounted in the apparatus.

The internal friction and the resonant frequency has been measured at room temperature and at the liquid nitrogen temperature.

The changes of I. F. and resonant frequency of one samples (n° 2) have been followed for 637 hours at room temperature after the low temperature measurements. The experimental results are reported on Fig. 1.

3. Discussion.

It is usual to write the total I.F. as the sum of two terms.

$$\Delta = \Delta_i + \Delta_a$$

where Δ_i is the low stress amplitude indipendent I.F. and Δ_a is the amplitude dependent part. According to Granato and Lücke we have:

$$\Delta_a = A \frac{\mathcal{E}_o}{\mathcal{E}} \exp\left(-\frac{\mathcal{E}_o}{\mathcal{E}}\right)$$

and if we plot $\log(\Delta_a \circ \mathcal{E})$ versus $1/\mathcal{E}$ we would obtain a straigth line.
How we may see in fig. 2 the experimental data does not agree with this
behaviour.

An analytical expression that fits very well our data is the following:

$$\Delta_a = \frac{1}{\mathcal{E}}\left[A_1 e^{-\frac{k_1}{\mathcal{E}}} + A_2 e^{-\frac{k_2}{\mathcal{E}}}\right]$$

In other words in the Granato-Lücke plot our experimental points are distributed on two straigth lines. In table I we give the value of the parameter
A_1, A_2, k_1, k_2 we used to fit the experimental points.

TABLE I

Values of the parameters A_1, A_2, k_1, k_2 for the various samples

SAMPLE	$A_1 10^9$	$A_2 10^9$	$K_1 10^5$	$K_2 10^5$	$\frac{A_1}{K_1} 10^4$	$\frac{A_2}{K_2} 10^4$	K_1/K_2	$\Delta_i 10^4$
1 T=300°K	0.316	50	0.077	2.39	4.12	20.9	0.032	2.40
T= 96°K	0.871	28.8	0.075	0.48	11.6	59.8	0.156	1.50
2a T=300°K	3.09	20.5	0.110	1.80	28.2	11.4	0.061	5.9
T= 96°K	6.1	5.2	0.159	0.83	32.2	6.2	0.191	3.-
2b T=300°K	18.5	25	0.197	1.175	93.8	21.2	0.168	3.6
T= 96°K	64	---	0.348	---	184.	---	---	3.8
3 T=300°K	4.33	28.8	0.143	2.44	30.4	11.8	0.058	6.7
T= 96°K	15.	11.7	0.297	1.65	50.	7.1	0.180	4.4
2b T=300°K								
15h 30'	25.2	100	0.198	3.29	126.5	30.3	0.060	3.5
283 h	1.78	14.15	0.099	1.53	18.	9.2	0.064	3.18
673 h	0.321	4.74	0.073	0.79	4.38	6.05	0.093	3.5

This kind of behaviour was already observed in aluminium monocrystals by
Hasiguti, Igata, Tanaka (12) and in Gold policrystals (13).
At this point we have to ask some questions. The first is,if this analytical
expression has any physical meaning. The second one is, if it is possible
using Trott and Birnbaum results to get at least qualitative explanation of
the observations.

We will begin with the second one. We can try to see it it is possible to distinguish the contribution to the amplitude dependent I.F. Δ_a due to the two different mechanisms: the hysteretic damping and that consequent to the change in dislocation loop length distribution (Trott and Birnbaum, stress amplitude dependent viscous damping).

To distinguish the two contributions we have used the ratio betweenn Δ_a and the correspondent modulus change $\Delta M/M$.

In fact according to Granato and Lücke for the hysteretic damping this ratio $\Delta_a / \frac{\Delta M}{M}$ is 1. For the viscous damping, from (14), in the low frequency limit (kilocycle range) this ratio result to be $k\frac{\omega}{\omega_0}$, where ω_0 is the resonance frequency of the dislocation loops, ω is the measurement frequency, k a numerical factor depending on what kind of distribution law has been supposed to be valid for the dislocation loop length (for example $k=5/\pi$ for a δ distribution).

Generally this ratio is very small because ω_0 is very large (it may depend on the amplitude through ω_0 that changes when the average dislocation length changes). Therefore the magnitude of this ratio can suggest the relative effective importance of the two mechanisms. In fig. 4 the values of $\Delta_a / \frac{\Delta M}{M}$ for the different samples and for differents temperatures are reported.

For the sample 2b the changes of the ratio at different time are also reported. In these figures we can observe that the hysteretic process is not the unique process and that it is important in a range of amplitude near the beginning of the amplitude dependence, then the relative importance of this process tend to diminish.

It does not seem that the change in temperature have a marked effect on the two processes. At low temperature we have increase or decrease of the hysteretic process depending on the samples.What seems evident is that the maximum is shifted towards higher amplitudes.

We have to point out that the effect of the low temperature is not simple. The thermic stress changes the distribution and density of dislocation. This is confirmed by the measurements at room temperature on the sample 2b after the low temperature measurement. The I.F. is completely different in respect to the previous measure at room temperature. The recovery of the sample is very slow.

These effects due to the low temperature measurements are very difficult to controll. The change in $\Delta_a / \frac{\Delta M}{M}$ during the recovery of the samples seems to indicate a decrease of the hysteretic process.

For a quantitative comparision with the results of Trott and Birnbaum it is necessary to normalize our data. But to do that we have to know many internal parameters of the material we do not know. Nevertheless we can try some comparisons.

If we observe the data of Trott and Birnbaum raprepresented in the Granato-Lücke plot, we notice a change of slope between the initial and the intermediate part. It is very easy to control that the ratio of these two slopes is linearly dependent on C_N (number of pinning points in a network length) at least in the range of values of C_N considered. The dependence is different for the viscous damping and for the hysteretic damping.

It seems useful to compare this ratio with that of the slope of the two straigth lines obtained in our plot. For our samples all the values for the room temperature measurements are very similar, about 0.06, and the same is true for the measurements carried out at liquid nitrogen temperature. In this case the ratios are near 0.18 (see table 1).

From the comparison with the ratios obtained by the data of Trott and Birnbaum we get $C_N \simeq 1-2$ at room temperature and $C_N \simeq 6$ at liquid nitrogen temperature. Consequently it seems that the effect of the temperature is to change the effective number of pinning points and not to change in an essential way the mechanism of the process.

If we study the variations of this ratio with time (sample 2b) we can deduce a small increase of C_N (from about 2 to about 3).Probably the dislocations created or depinned by thermal stresses are subsequentely pinned by diffusing impurity atoms. What we point out again is the extreme sensitivities of amplitude dependet I.F. to density and distibution of dislocations and to their smallest change.

For this reason it is very difficult to obtain the same results on different samples with different history and also in the same sample after different handlings.

In conclusion we can say that our analytical expression seems not to have physical meaning, but we may regard it as an approximation to the Trott and Birnbaum curves. The parameters present in this expression can be probably used to characterize the distribution and density of dislocations.

The data on modulus change seem be very important to confirm the presence of two kind of mechanisms depending on the stress amplitude and to suggest a way to determine the relative importance of the two processes.

References;

1. J.S. Koehler: Imprfections in nearly Perfect Crystals, p. 197- Wiley New York (1952)

2. A. Granato, K; Lücke: J. Appl. Phys. 27, 583 (1956)

3. "   "   "   "   "   "   "   27,789 (1956)

4. A. Granato, K. Lücke, L.J. Teutonico: J. Appl. Phys. 35, 220 (1964)

5. A. Granato,K. Lücke, J. Schlipf, L.J. Teutonico: J. Appl. Phys. 5,2732 (1964)

6. J. C. Swartz, J. Weertmann: J. Appl. Phys. 32 ,1860 (1961)

7. D.H. Rogers: J. Appl. Phys. 33, 781 (1962)

8. B.D. Trott, H.K. Birnbaum: J. Appl. Phys. 41, 4418 (1970)

9. P.G. Borboni: Nuovo Cimento, 4, 177 (1947)

10. M. Nuovo: Ric. Scientifica, 31 II A/1, 212 (1961)

11. A. Bortolotti, L. Passari: Nuovo Cimento, 35, 988 (1965)

12. R.R. Hasiguti, n. Igata, K. Tanaka: Acta Met. 13, 1083 (1965)

13. A. Bortolotti, G. Martinelli, L. Passari: Atti Accad. Sci. Ferrara 49, (1972)

14. O.S. Oen, D.K. Holmes, M.T. Robinson: ORNL 3017

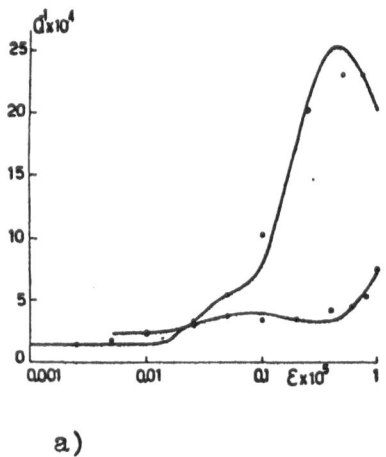

a)

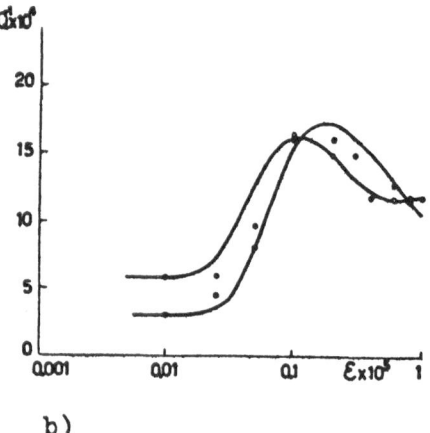

b)

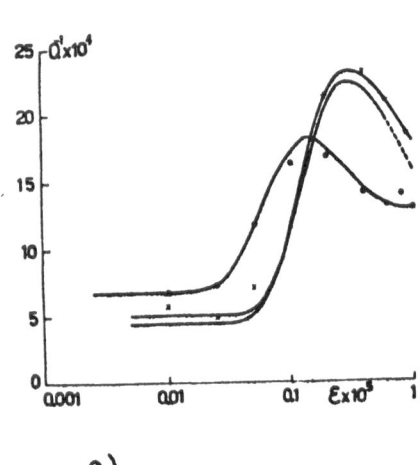

c)

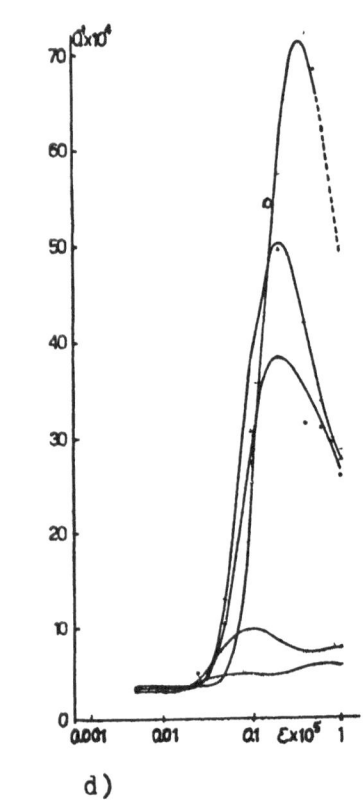

d)

Fig.1) Amplitude dependent I.F. of
various samples.

a) Sample 1 ● T= 300°K
           o T= 96°K

b) Sample 2a ● T= 300°K
            o T= 96°K

c) Sample 3 ● T= 300°K
            o T= 96°K

d) Sample 2b T= 300°K
      △ 18 h 30' after the liquid
      ✗ 283 h   Nitrogen measure
      ⊙ 673 H   ment.

514

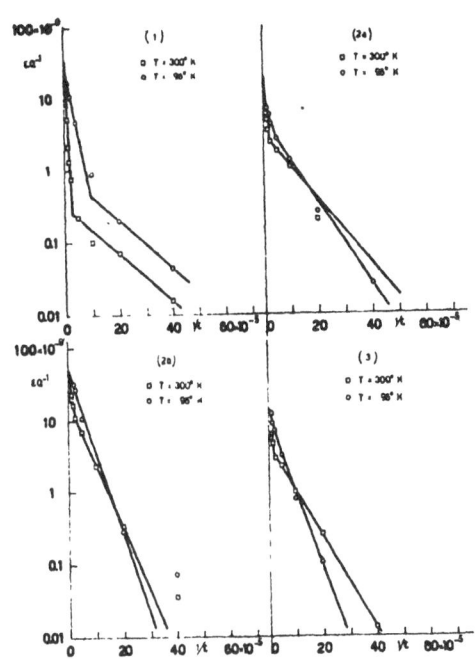

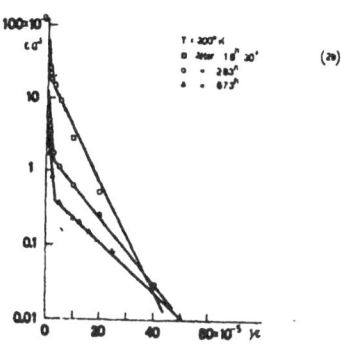

Fig.3

Fig.2

Fig. 2) Granato-Lücke plot of measurements at room and liquid
Nitrogen temperature.

Fig. 3) Granato-Lücke plot of measurements on sample 2b after
various Time.

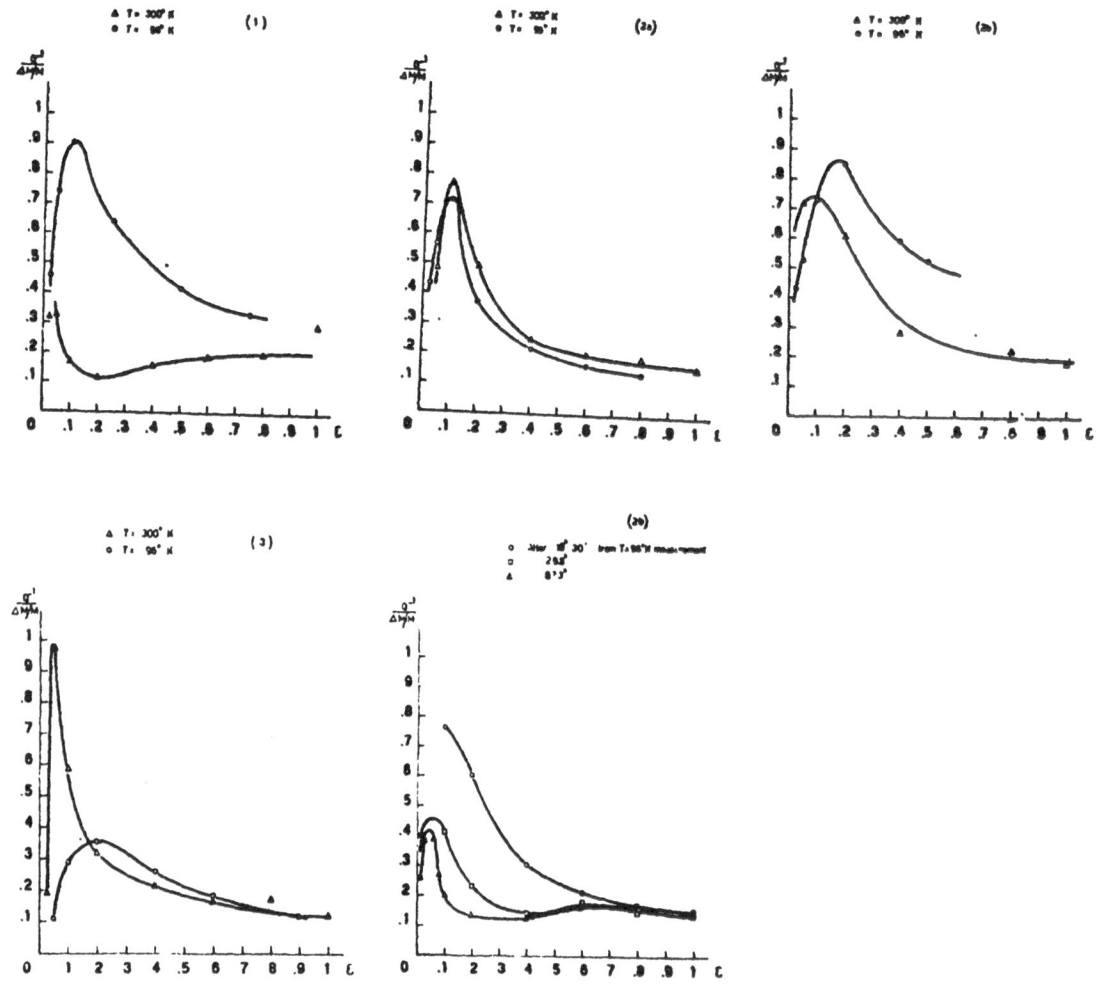

Fig;4) The ratio $\dfrac{Q^{-1}}{\Delta M/M}$ versus ε for various samples.

(ε in arbitrary unit)

THEORY OF DISLOCATION HYSTERESIS

V.L.Indenbom and V.M.Chernov

A.V.Shubnikov Institute of Crystallography
Academy of Sciences of the USSR, Moscow

Introduction

In many cases the main resistance to the dislocation glide
is due to point defects (both impurity and radiation-induced defects), their influence depending essentially upon the fine details of the real crystal structure arising in production, thermal treatment, radiation and plastic deformation. The main experimental techniques of the investigation of dislocations and
their interaction with different defects are the internal friction (variable external stress) and the mobility measurements
(constant external stress). Here the inverse problem has to be
solved, namely, the problem of determination of the crystal potential relief and of its connection with specific characteristics of the lattice and the defects. The dislocation in this
problem plays the role of a "measuring device".

At not too high temperatures when the diffusional motion
of dislocations and point defects can be neglected (it is just
the case that is considered below) the dislocation overcomes
the impeding defects by virtue of thermal fluctuations. The kinetics of the processes is completely determined by the elementary event of overcoming the impeding barrier by the dislocation
segment. The decisive characteristics of this elementary event

are the barrier height (activation energy) equal to the difference of segments energies in the stable and unstable configurations, the activation volume (or the activation length) and the effective frequency of segment oscillation in the direction of the barrier. Provided the force law of the dislocation interaction with the impeding defect is known, these parameters can be calculated (1, 2). If the point defects were situated equidistant along the dislocation line, the problem of relation of microscopic processes in the crystal to their macroscopic appearances would be very simple. Actually the point defects are distributed along the dislocation line in some arbitrary manner and each of them gives its individual contribution depending on the geometry of the position of neighbour defects. This circumstance considerably complicates the calculations and state the major problem arising when treating experimental data on the dislocation motion - the problem of separating out the influence of the geometric configuration statistics in finding out the energy characteristics of the dislocation-defect interaction. Since the type of defects is not known in advance the theory has not to be based on assumptions about concrete characteristics of point defects. The theory of this kind was developed by the present authors as applied to dislocation hysteresis problems (3-5) and to the dislocation mobility (6).

Dislocation Hysteresis

The magnitude of the internal friction is determined by the hysteretic losses in every event of the breakaway of the dislocation segment from the point defect. The problem consists in calculation of the segments broken away and their contribution to the internal friction at the given values of the oscillation amplitude τ and temperature T. It should be taken into account that an increase of the temperature results in an increase of the number of segments contributing to the hysteresis and in a decrease of the contribution of each segment.

Under rather general assumptions one succeeds in the expression of the vibration decrement $Q^{-1}(\tau, T)$ in terms of the only parameter $l_{min}(\tau, T)$ which determines the length of the shortest segment broken away at the given amplitude and temperature:

$$Q^{-1} = l_{min}^2 \cdot \int_{l \geq l_{min}} l\, n(l)\, dl \tag{1}$$

Here the function n(l) gives the length distribution of segments.

The theory developed enabled the authors to suggest a new method of the analysis of experimental data on the dislocation hysteresis. Up to now the variation of the internal friction magnitude with the variation of either the amplitude or the temperature was studied. But the change of the temperature changes both the individual contribution of the segment to the losses and the number of segments broken away. Therefore, this way is hopeless because the dislocation structure of the sample under investigation is unknown. We suggest the following new way: when changing, for instance, the temperature one has to change the amplitude of stresses in such a way that the magnitude of internal friction should be unchanged. To a fixed level of the internal friction corresponds, as seen from (1), a constant dislocation geometry. The fixation of the geometry permits to determine the characteristics of the dislocation-point defect interaction: it is just the relation between the temperature and vibration amplitude at the constant internal friction level that gives the dependence of the activation energy (in appropriate units) upon the stress.

The approximations used restrict the application of the developed theory to kilohertz frequency range and to not too high temperatures. The analysis of experimental data (4-5) obtained in sufficiently wide temperature and vibration amplitude intervals showed a good agreement with the theory developed.

Dislocation Mobility

The process of overcoming local barriers by dislocations at constant load depends essentially on the distribution of the barriers in the glide plane. Moving straightforward by virtue of breakaway in the areas with easily penetrable obstacles the dislocations can go round the heavy barriers. The statistics of the arrangement of barriers along the dislocation line essentially determines the dependence of the average dislocation velocity on stress. The qualitative character of the motion is determined by the relation between the time of the thermofluctuation overcoming the barrier and the time of the following tangential propagation of the segment broken away. As in case of dislocation hysteresis the condition $V(\tau, T) = $ const (V is the dislocation velocity, τ is the value of the constant external stress) provides an opportunity to separate out the characteristics of the dislocation segment interaction with defects at the same geometry. This condition fixing the geometry of the dislocation segment-defect interaction permits to derive from the dependence $T(\tau)$ at $V(\tau,T) = $ const the dependence of the activation energy on stress. Then the activation volume determination enables one to judge about the nature of the force law of the interaction of the dislocation with the impeding defect.

The analysis of the experimental data (6) on the dislocation mobility obtained in sufficiently wide ranges of temperatures and stresses showed a good agreement of the developed theory with experiment.

Determination of the Nature of the Defect

Thus, on the basis of the developed theory of dislocation hysteresis and dislocation mobility it is possible to eliminate effectively the influence of the geometric statistics when determining the energy characteristics of the dislocation-point defect interaction. The relation between the temperature and the amplitude of vibrations at the constant magnitude of the internal friction or the relation between the temperature and the stress at the constant dislocation velocity determines directly the dependence of the activation energy on stress. The extrapolation of the dependences $T(\tau)$ to the zero value of stress gives an estimate of the binding energy of the dislocation with the point defect. In all cases (for the internal friction as well as for the dislocation mobility) the experimental values of the binding energy are not less than 0.5 eV.

A more detailed information about the nature of the defects can be obtained from the dependence of the activation energy H on stress. The derivatives $\gamma(\tau) = -\dfrac{\partial H}{\partial \tau}$ and $d(F) = -\dfrac{\partial H}{\partial F}$ (where F is the magnitude of the external force acting on the defect from the dislocation segment) define the activation volume and activation length, respectively. These two values are very sensitive to the nature of the dislocation-defect interaction. For a dislocation segment of a length ℓ the external force is given by $F = \tau b \ell$ (b is the Burgers vector). In this case the activation volume and the activation length are connected by a simple relationship $\gamma = b \ell d$. The reconstruction of the force law from the dependence of the activation length on the external force F is valid only in the "binding energy approximation" (1) (the change of the segment length during the thermoactivation overcoming the barrier being neglected) when this law is immediately given by the reverse function F(d).

In Fig.1 the dependences of the activation length on the external force calculated (1,6) for different force laws are presented. Shown are also the values of the activation length derived from the experimental data on internal friction and mobility. The use of the log-log scale facilitates the comparison of theoretical data with experimental ones: the coincidence of the plots is achieved by their displacement along the coordinate axes.

As seen from Fig.1 experimental data both on the dislocation hysteresis and mobility are in the best agreement with the predictions of the theory for cases when the dislocation glide is limited by anisotropic centres producing tetragonal distortions in the lattice.

References

1. V.M.Chernov and V.L.Indenbom, Fiz.Tverd.Tela, 10, 3331 (1968).

2. V.M.Chernov, Fiz.Tverd.Tela 15, 323 (1973).

3. V.M.Chernov and V.L.Indenbom, Meeting Mechanisms of Internal Friction in Metallic Materials, Batumi 1968, in: Vnutrennee trenie v metallicheskikh materialakh, "Nauka", Moscow 1970, p.26.

4. V.L.Indenbom and V.M.Chernov, Conf. Mechanisms of Relaxation Phenomena in Solids, Moscow 1969, in: Mekhanizmy relaxatsionnykh yavleny v tverdykh telakh, "Nauka", Moscow 1972, p.87.

5. V.L.Indenbom and V.M.Chernov, Phys.Stat.Sol.(a) 14, 347 (1972).

6. V.M.Chernov, Fiz.Tverd.Tela 15, 1159 (1973).

7. R.H.Saul and C.L.Bauer, J. Appl.Phys. 39, 1469 (1968).

8. G.S.Baker and S.H.Carpenter, J. Appl.Phys. 38, 3557 (1967).

9. H.L.Caswell, Adv.Phys. 9, 38 (1960).

10. B.K.Kardashev, S.P.Nikanorov and O.A.Voinova, Phys. Stat.Sol.(a) 12, 375 (1972).

11. G.C.Das and P.L.Pratt, Second Intern.Conf. on the Strength of Metals and Alloys, Asilomar 1970, vol.1, p.105.

12. D.F.Stein and J.R.Low, J. Appl.Phys. 31, 362 (1960).

13. S.N.Valkovsky and E.M.Nadgornyi, Fiz.Tverd.Tela 12 2542 (1970).

14. G.A.Ermakov and E.M.Nadgornyi, Fiz.Tverd.Tela 14 3517 (1972).

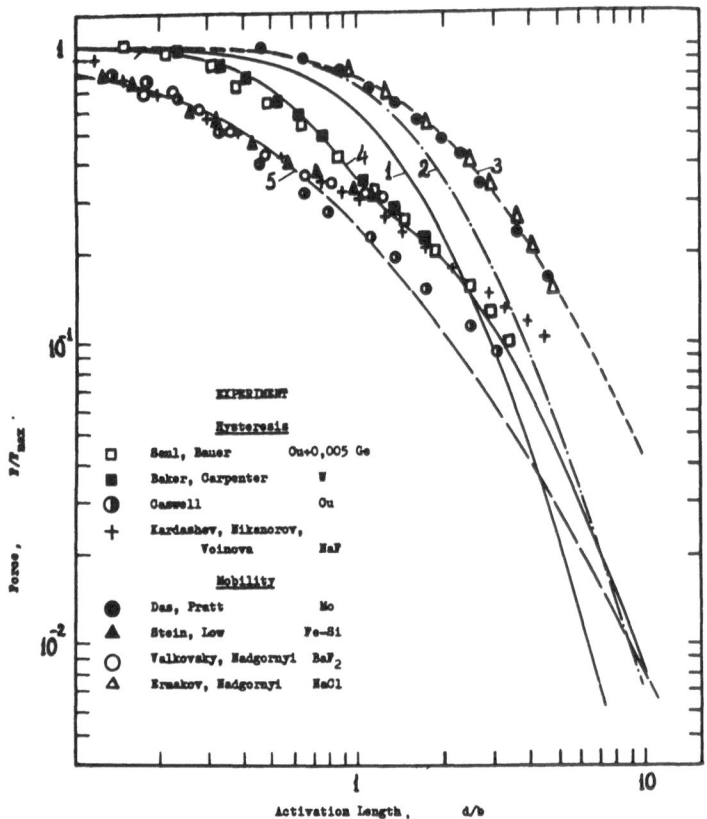

FIG.I

Comparison of the calculated (solid and dashed curves 1-5) and experimental (data points) force-distance curves. F_{max} defines the maximum value of an external force which may be counterbalanced by the point defect.

1 - edge dislocation and dilation centre.
2 - [211] edge dislocation with the [$\bar{1}$11] Burgers vector and ⟨100⟩ tetragonal defect.
3 - [10$\bar{1}$] screw dislocation and ⟨110⟩ tetragonal defect.
4 - [100] edge dislocation with [0$\bar{1}$1] Burgers vector and ⟨110⟩ tetragonal defect.
5 - Fleischer's approximation.

Fifth International Conference on
INTERNAL FRICTION AND ULTRASONIC ATTENUATION IN
CRYSTALLINE SOLIDS

RWTH AACHEN, August 27 - 30, 1973

SPONSORSHIP

International Union of Pure and Applied Physics (IUPAP), European
Physical Society (EPS), Arbeitsgemeinschaft Metallphysik der Deutschen
Physikalischen Gesellschaft (DPG), der Deutschen Gesellschaft für
Metallkunde (DGM), des Vereins Deutscher Eisenhüttenleute (VDEh).

FINANCIAL SUPPORT

Deutsche Forschungsgemeinschaft (DFG), Minister für Wissenschaft
und Forschung des Landes Nordrhein-Westfalen.

PREFACE

The methods of internal friction allow the study of numerous structural
features and processes in solids. In many cases these methods are
superior to other means of investigation or even the only ones able
to provide the desired information. For these reasons, and especially
since the appearance of Clarence Zener's "Elasticity and Anelasticity
of Metals" in 1948, the interest in the field of internal friction has
been increasing strongly. Both the depth of understanding of known phenomena and the area of application of internal friction to the investigation of new effects grew quickly.

As a consequence, worlwide international conferences on internal friction
were established. Until now five meetings took place, namely at Brown
University, Providence R.I. (1956 and 1969), at Cornell University,
Ithaca N.Y. (1961), at the University of Manchester (1965) and at the
Technische Hochschule of Aachen (1973). The sixth will be held in
Tokyo.

The present two volumes contain the refereed proceedings of the 5th
International Conference on Internal Friction and Ultrasonic Attenuation
in Crystalline Solids (ICIFUA). It was held from August 27 to 30, 1973
at the Institut für Allgemeine Metallkunde und Metallphysik of the
Rheinisch Westfälische Technische Hochschule Aachen. The 202 participating scientists (56 from Germany, 146 from 16 foreign countries) heard
9 invited and 100 contributed papers. Topics of the conference were all
important areas of current research interest: internal friction due to
electrons, phonons, magnetic effects, phase transformations, point de-

fects, grain boundaries and, most of all, dislocations. To the latter
the whole of the second volume is devoted.

This is the first time that the proceedings of an ICIFUA are published
in form of a book. The editors who were also responsible for the or-
ganisation of the meeting like to take the opportunity to thank all who
contributed to the success of the conference and to the publication
of the proceedings; especially they express their gratitude to all
their colleagues in the conference committee and the organizing committee,
to the sponsors of the conference, the Rektor and the officials of the
Technische Hochschule Aachen, the officials of the city of Aachen, the
members of the Institut für Allgemeine Metallkunde und Metallphysik and,
last but not least, to the referees of the printed version of these
proceedings.

The editors hope that these two volumes will not only provide a good
survey of the present state of the field of internal friction, but
also render new impulses to its further development.

Aachen 1974   D. Lenz, K. Lücke

CONTENTS

V O L U M E I
===============

POINT DEFECTS

GRAIN BOUNDARIES AND PHASE TRANSFORMATIONS

CONTENTS VOLUME II

VERY HIGH FREQUENCY PHONONS

J.K. Wigmore
Department of Physics, University of Lancaster, Lancaster, U.K.

ABSTRACT   The paper reviews the methods used for studying
phonons in the frequency range approximately
10-1000 GHz; this is roughly the region that is
too high for ultrasonic techniques and too low
for neutron diffraction. Techniques for these
phonons are mainly incoherent, and include the
use of heat pulses, superconducting and semiconducting bolometers, superconducting tunnel
junctions and fluorescence generators, optical
dichroism and Brillouin scattering. A brief
summary is given of the physical problems for
which very high frequency phonon techniques are
at present being used.

INTRODUCTION

The term "very high frequency phonon", for no definite
scientific reason, seems to have acquired the connotation of an
acoustic phonon between approximately 10 and 1000 GHz in
frequency. This is the difficult region of the phonon spectrum
of a solid, too high in frequency for conventional ultrasonic
techniques and in general too low for neutron diffraction. For
a long time, the only method of investigation was by thermal
conductivity. In recent years, however, several new techniques
have been developed for studying phonons in this regime: many
of these were described in detail at the Colloquium on Very High
Frequency Phonons, held at St. Maxime, France, in June 1972[1].
It now seems to be rather an opportune moment to be reviewing
the field, since Forkel et al[2] have only very recently reported
producing quasi-monochromatic phonons at 870 GHz by means of a
superconducting tunnel junction. Their technique, which will
be described later in this talk, is capable of even higher
frequencies and the great divide to the neutron diffraction

regime now has unquestionably been bridged. Almost all the studies to be described in this talk are very low temperature experiments, and unless otherwise stated it may always be taken that the temperature is below 4.2K.

The record (and likely to remain so) for the highest frequency of phonons both generated and detected by the coherent piezoelectric technique, is held by Ilukor and Jacobsen [3,4] at 114 GHz. Instead of the coaxial reentrant cavity, de rigueur at lower microwave frequencies, they used a simple cylindrical cavity completely filled with a single crystal of x-cut quartz with polished ends perpendicular to the electric field. They saw five pulse echoes, with a maximum signal to noise of ∿10 dB. However, very few ultrasonic experiments utilising coherent generation and detection have been carried out above about 10 GHz. One problem is that the piezoelectric coupling efficiency is inversely proportional to frequency[5], but much more serious are the rigorous requirements of crystal preparation. Unless both the transducer and the specimen, and the bond between them, are all flat and parallel to the accuracy of the ultrasonic wavelength, the detecting piezoelectric surface will not be driven in phase across its entire area, and cancellation will occur. At 10 GHz, the wavelength of a phonon is roughly the wavelength of light. Thus, much higher frequencies than this require impossibly precise polishing techniques.

INCOHERENT DETECTION OF ULTRASOUND

The most direct solution to the problem is clearly to use a method of detection that does not depend on the phase of the ultrasonic waves. Two types of bolometer have been tried, both at liquid helium temperatures. The superconducting bolometer[6,7] is a thin film of, for example, tin, at a bath temperature carefully stabilised on its superconducting transition. Its resistance at this point is a sharply varying function of temperature, and the bolometer responds rapidly (∿10 nanoseconds) to the small amount of heating produced by an incident phonon pulse. So far it has been used to detect only 10GHz ultrasonics, but this type of bolometer is also of great value in heat pulse experiments, to be described later[8]. The semiconducting avalanche bolometer utilises the rapid variation of current with temperature of a doped semiconductor biassed for impact ionisation. The particular advantages of this bolometer over the superconducting type are its independence of magnetic field and, over a range of several K, of temperature. It has been used with both 10 GHz[9] and 35 GHz[10] ultrasonics, and also in heat pulse experiments.

Some incoherent techniques involve direct interaction with the incident phonons without the intermediate thermalisation process that takes place in a bolometer. Several of the methods are based on the saturation by phonons of a paramagnetic spin system, but I shall describe only the method due to Anderson and Sabisky[11], since it has proved to be so versatile. The basic physical principle is that phonons equal in energy to the spin splittings are absorbed and change the relative populations of the states. The population differences can, of course, be measured by electron spin resonance but this is essentially a fixed frequency technique and does not allow the energy of the phonons being studied to be altered very greatly. Anderson and Sabisky showed that the population differences can also be monitored with great sensitivity by measuring the circular dichroism of an optical transition between the spin levels as ground state and an excited state. The splitting between the spin levels themselves can be varied with a magnetic field, thus varying the frequency of the phonons that are being detected but not the wavelength of the optical transition being observed (since the optical linewidth is much broader than the spin splittings). The system therefore acts as a monochromatic, variable frequency, incoherent phonon detector which the originators have used up to 24 GHz with ultrasonic phonons and as high as 340 GHz using incoherently generated phonons. The only drawback is that the detector is very slow and will not respond to acoustic pulses. Nevertheless, the technique has provided extremely valuable data in the study of liquid helium, most recently on the onset of superfluid flow in helium films[12].

Another optical technique for studying phonons is that of Brillouin scattering [13-15], in which

$$\text{photon } (\omega_1, \underline{k}_1) - \text{photon } (\omega_2, \underline{k}_2) \rightleftharpoons \text{phonon } (\omega^1, \underline{q})$$

The highest frequency phonon that can be studied has $|\underline{q}| = |\underline{k}_1| + |\underline{k}_2|$ and this limits the maximum frequency to less than ~ 100 GHz in a typical experiment in which a laser is used as the source. However, Sandercock[16] has shown that considerably higher frequencies can be monitored in a sample which has high optical absorption, and Carlson and Segmüller[17] have shown that it is feasible to use x-ray photons, so that $\underline{k}_1 + \underline{k}_2$ can pick out $\underline{q}$ anywhere in the Brillouin zone.

A very recent development in this area is the observation of phonon (or boson) echoes [18,19]. Joffrin and co-workers found that an unpolished piece of a non-centrosymmetric crystal such as CdS, when placed in the electric field of a microwave cavity and excited by two or more electromagnetic pulses, returned echo pulses whose timing depended on the separations of the driving pulses. There is an analogy - although not an exact one - between these experiments and the phenomenon of spin echoes. The first pulse of the sequence, at frequency ω, generates coherent phonon populations in the crystal at frequencies of both ω and $\omega/2$. The ω phonons are the result of the usual piezoelectric surface transduction. The $\omega/2$ phonons, however, are produced by a volume interaction

$$\text{photon } (\omega, 0) = \text{phonon } (\tfrac{\omega}{2}, \tfrac{q}{2}) + \text{phonon } (\tfrac{\omega}{2}, \tfrac{-q}{2}) \ .$$

The second pulse in the sequence, occurring after time τ, again of frequency ω, has the effect of time reversal; each phonon $(\tfrac{\omega}{2}, \tfrac{q}{2})$ is replaced by a phonon $(\tfrac{\omega}{2}, \tfrac{-q}{2})$ and vice versa and after a further time τ a photon $(\omega, 0)$ is created, the "echo". The variation of the amplitude of this echo as a function of τ measures the relaxation time T_2 of the coherent state. A three pulse sequence, analogously to spin echoes, yields further echoes from which it is possible to obtain also a "T_1" for the phonon system. What do T_1 and T_2 represent in the context of the phonons? The T_2 clearly measures the decay of coherence of phonons of a single frequency by elastic scattering, and therefore corresponds to the phonon lifetime measured at low temperatures by ultrasonic attenuation. T_1 on the other hand is a measure of the inelastic collisions of the phonons, and may be very long indeed (~ 0.1 seconds) at helium temperatures. The technique of phonon echoes provides a means of generating and detecting microwave phonons with unpolished specimens, and has been used between 1 and 36 GHz.

INCOHERENT GENERATION

In almost all the experiments that I have talked about so far, the phonons have been generated by conversion from microwaves, and are bounded therefore by the limits of microwave technology. Phonons of much higher frequencies can be studied by wholly incoherent techniques. The major disadvantage of these is the loss of monochromaticity, in the sense that a magnetron emits monochromatic radiation, although pulsed experiments do allow resolution of polarisation and of wavevector. Chronologically, the heat pulse technique came first, devised by von Gutfeld and Nethercot [20,8]. The phonon generator here is simply a thin metal film, of, for example, constantan excited by a fast ($\sim$0.1 microseconds) pulse of electrical current. The temperature of the film rises rapidly above that of the specimen, and the excess phonons are emitted from the heater into the specimen. Typically, a 1 watt excitation pulse raises the heater to about 5K at an ambient temperature of 1K. The dominant phonons being generated are thus about 100 GHz, although the exact frequency distribution of a heat pulse is still not quantitatively understood. On the assumption that the phonons in the heater film remain always in thermal equilibrium with the excited electrons, the phonon spectrum emitted should be the difference between the two Bose-Einstein distributions at the heater and specimen temperatures respectively[21]. It is difficult to obtain experimental data and such that there are cannot be regarded as conclusively supporting this picture [22-24]. The problem is that the assumption of thermal equilibrium between electrons and phonons should require a heater film thickness very much larger than the phonon mean free path due to electron-phonon scattering. In experiments, however, the former quantity is usually $\sim$10^{2}nm and the latter $\sim$10^{4}nm. The problem has been recently discussed by Maris[25] and by Perrin and Budd[26]. It follows that, at the present, when precise knowledge of frequencies is required, the heat pulse technique is best avoided.

The most powerful technique in this field is undoubtedly
that of phonon generation and detection by single particle
superconducting tunnelling, first used by Eisenmenger and Dayem[27].
Figure 1 illustrates the mechanism of generation in a
superconductor-insulator-(same) superconductor junction[28].

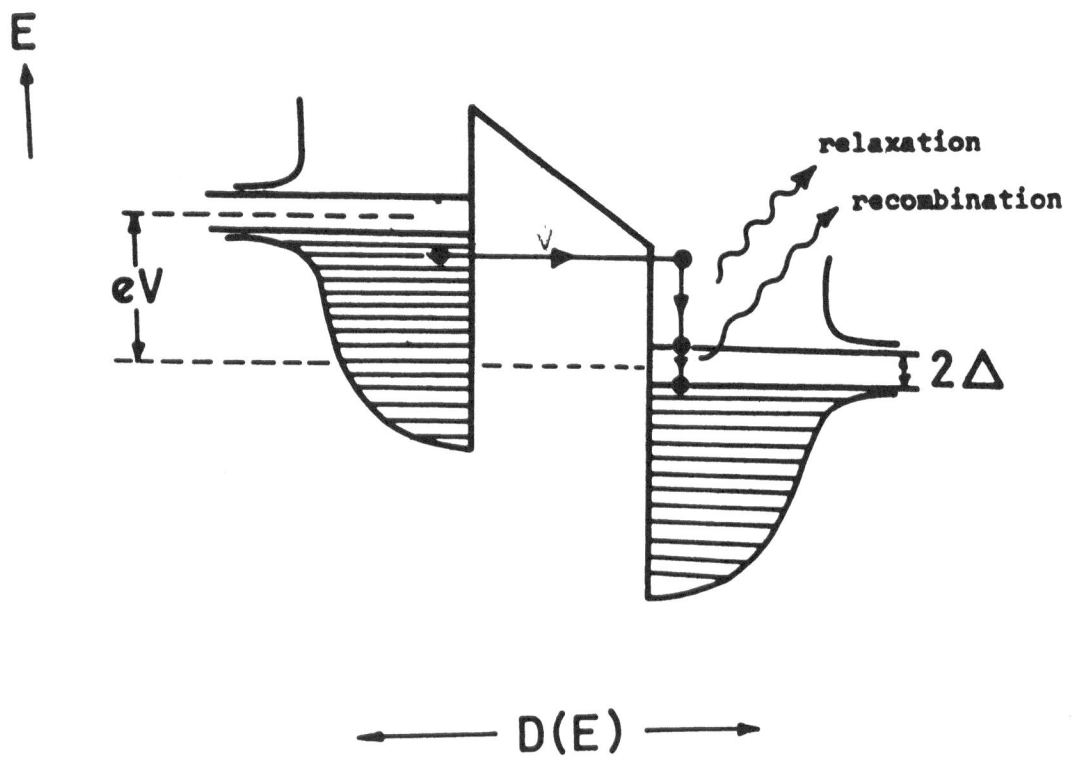

FIG. 1
Illustrates the generation of recombination and relaxation
phonons by a superconducting tunnel junction. Energy E;
density of states D(E).

The junction is biassed with a voltage V so that all states on
the left hand side are eV higher than the corresponding states
on the right. An excited quasiparticle on the left can tunnel
to an unfilled state of the same energy on the right. It is
now no longer in thermal equilibrium, and will decay to the
superconducting ground state by a two stage process. In the
first stage, known as 'relaxation', the quasiparticle drops down
from its excited state to the upper edge of the energy gap, with
the emission of a single phonon that may have energy anywhere

between 0 and eV-2Δ depending on the initial state. This process
is immediately followed by 'recombination', in which two
quasiparticles unite together in a Cooper pair with the emission
of a phonon of energy ∿2Δ. Figure 2 illustrates the frequency
distribution of the phonons generated by these two processes.

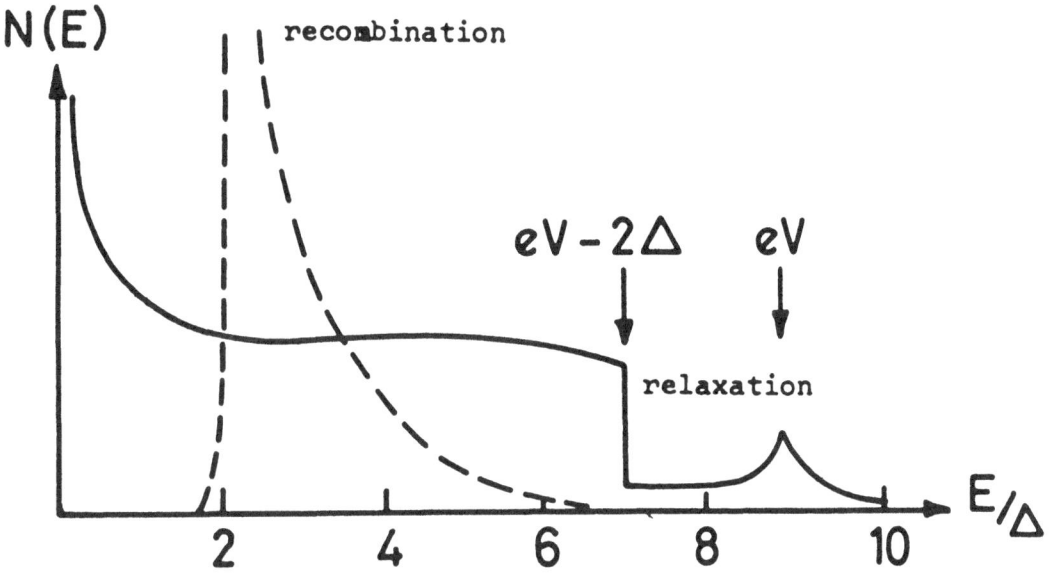

FIG. 2
The number of phonons N(E) emitted by the recombination and
relaxation processes for a junction biassed to V = 9Δ/e.

Detection of phonons takes place as the inverse of recombination;
a phonon with energy greater than 2Δ breaks a Cooper pair and
thus causes an increase in the tunnelling current. As a
detector, the junction is sufficiently fast and sensitive for
pulse experiments.

By varying the experimental procedure it is possible to
separate out the quasi-monochromatic recombination phonons from
the relaxation continuum. Dynes et al[29], using a "spectrometer"
based on resonant scattering in Ge:Sb that could be tuned by
applying uniaxial stress, showed that the phonons generated by
a Sn-SnO-Sn junction are almost all of the recombination type
with energy 2Δ. Why are there so few relaxation phonons with
energies up to eV-2Δ? It seems likely that these are rapidly

reabsorbed in the metal, so that a relaxation phonon is degraded into several 2Δ phonons and a single lower energy phonon that is not energetic enough to excite quasiparticles. Following on this idea, Narayanamurti and Dynes[30] showed that the relaxation phonons in this degrading process could also be supplied from an external source, for example, by a simple heat pulse generator of constantan evaporated on top of a superconducting film, with an insulating layer between. Such a superconducting fluorescence generator has the advantage that it is much easier to fabricate than a tunnel junction yet produces a quasi-monochromatic spectrum of phonons at energy 2Δ which is tuneable over a 50% range by applying magnetic field[30].

Other workers have concentrated on the relaxation phonons[31]. Although the relaxation spectrum is very broad, it does potentially allow the study of much higher frequency phonons than those emitted by recombination. The most outstanding feature of the spectrum is the sharp step at E = eV-2Δ. If a small modulation is superimposed on the bias voltage, V, and the detector signal is filtered at this modulation frequency, the major contribution is due to the phonons in a narrow band at the step energy. This technique was originated by Kinder et al in a CW mode[32], was extended to pulsed experiments by Kinder[33], and has resulted in the recent observation of phonon scattering at 870 GHz by Forkel et al[2], referred to at the beginning of this talk. In order to alleviate the degradation of these high energy phonons into 2Δ ones, a superconductor with a weaker electron-phonon interaction was sought. In the event, aluminium was used for which the 870 GHz is equivalent to 24Δ. The frequency was identified by resonant scattering from the 29 cm^{-1} transition of Si:O.

Thus far in the field of very high frequency phonons, the greatest ingenuity has gone into developing the methods, and relatively little effort into using them. Table 1 gives a summary of some of the areas in which the techniques described in this talk have found application, with a representative selection of references. It is clear that very high frequency phonon technology has relevance to a number of the current problems of interest in solid state and low temperature physics, and I believe that the field is still at the expansion stage of its development.

TABLE 1

Some physical problems studied using very high frequency phonons.

Liquid Helium (He^4):

Kapitza boundary resistance	11, 34, 35, 36
Dispersion curve	37, 38
Roton propagation	39
Helium films	11, 12

Second Sound in Solids:

Solid He^3 and He^4	40, 41
normal solids (NaF, Bi)	42, 43
phonon-phonon interactions	44, 45

Interaction of Phonons with other excitations:

electrons in metals	46, 47
electrons in semiconductors	17, 48
bound acceptors and donors	29, 49, 50
magnetic impurities	11, 22, 24, 51
dislocations	52

Elastic Wave Dispersion in Solids	53, 54
Solid-Solid Interfacial Thermal Conduction	23, 24, 55
Superconductivity	31, 49, 56

References

1. Proceedings published in J. Physique, $\underline{33}$ supplement $\underline{C4}$ (1972)

2. Forkel, W., Welte, M. and Eisenmenger, W., Phys. Rev. Letters $\underline{31}$, 215 (1973)

3. Ilukor, J. and Jacobsen, E.H., Physical Acoustics vol. 5 (Academic Press, New York) 221 (1968)

4. Jacobsen, E.H., J. Physique $\underline{33}$ supplement $\underline{C6}$, 25 (1972)

5. For example, see Tucker, J.W. and Rampton, V.W., "Microwave Ultrasonics in Solid State Physics" (North-Holland) 54 (1972)

6. Von Gutfeld, R.J., Proceedings of IEEE Symposium on Ultrasonics, Cleveland, Ohio (1966)

7. Andrews, J.M. and Strandberg, M.W.P., J. Appl. Phys. $\underline{38}$, 2660 (1967)

8. Von Gutfeld, R.J., Physical Acoustics vol. 5, (Academic Press, New York) 133 (1968)

9. Wigmore, J.K. and von Gutfeld, R.J., Ultrasonics $\underline{7}$, 117 (1969)

10. Dentschuk, M.B., Dobbs, E.R. and Wigmore, J.K. (to be published)

11. Anderson, C.H. and Sabisky, E.S., Physical Acoustics vol. 8, (Academic Press, New York) 1 (1972)

12. Anderson, C.H. and Sabisky, E.S., Phys. Rev. Letters $\underline{30}$, 1122 (1973)

13. Benedek, G.B. and Frisch, K., Phys. Rev. $\underline{149}$, 647 (1966)

14. Heinicke, W., Winterling, G. and Dransfeld, K., J. Acoust. Soc. Am. $\underline{49}$, 951 (1971)

15. Brya, W.J., Geschwind, S. and Devlin, G.E., Phys. Rev. $\underline{6B}$, 1924 (1972)

16. Sandercock, J.R., Phys. Rev. Letters $\underline{28}$, 237 (1972)

17. Carlson, D.G. and Segmüller, A., Phys. Rev. Letters $\underline{27}$, 195 (1971)

18. Joffrin, J. and Levelut, A., Phys. Rev. Letters $\underline{29}$, 1325 (1972)

19. Billman, A., Frenois, C., Joffrin, J., Levelut, A. and Ziolkiewicz, S., J. Physique $\underline{34}$, 453 (1973)

20. Von Gutfeld, R.J. and Nethercot Jr., A.H., Phys. Rev. Letters $\underline{12}$, 641 (1964)

21. Little, W.A., Can. J. Phys. $\underline{37}$, 334 (1959)

22. Narayanamurti, V., Physics Letters $\underline{30A}$, 521 (1969)

23. Weis, O., J. Physique, 33 supplement C4, 48 (1972)

24. Wigmore, J.K., Phys. Rev. 5B, 700 (1972)

25. Maris, H.J., J. Physique 33 supplement C4, 3 (1972)

26. Perrin, N. and Budd, H., Phys. Rev. Letters

27. Eisenmenger, W. and Dayem, A., Phys. Rev. Letters 18, 125 (1967)

28. Eisenmenger, W., "Tunnelling Phenomena in Solids" (edited by Burstein and Lundquist, Plenum Press, New York) 371 (1969)

29. Dynes, R.C., Narayanamurti, V. and Chin, M., Phys. Rev. Letters 26, 181 (1971)

30. Narayanamurti, V. and Dynes, R.C., Phys. Rev. Letters 27, 410 (1971)

31. Welte, M., Lassmann, K. and Eisenmenger, W., J. Physique 33 supplement C4, 25 (1972)

32. Kinder, H., Lassmann, K. and Eisenmenger, W., Physics Letters 31A, 475 (1970)

33. Kinder, H., Phys. Rev. Letters 28, 1564 (1972)

34. Guo, C.J. and Maris, H.J., Phys. Rev. Letters 29, 855 (1972)

35. Sherlock, R.A., Wyatt, A.F.G., Mills, N.G. and Lockerbie, N.A., Phys. Rev. Letters 29, 1299 (1972)

36. Pfeifer, C.D. and Luszczynski, K., Phys. Rev. 7A, 1055 (1973)

37. Anderson, C.H. and Sabisky, E.S., Phys. Rev. Letters 28, 80 (1972)

38. Narayanamurti, V., Andres, K. and Dynes, R.C., (to be published)

39. Dynes, R.C., Narayanamurti, V. and Andres, K., Phys. Rev. Letters 30, 1129 (1973)

40. Ackerman, C.C., Bertman, B., Fairbank, H.A. and Guyer, R.A., Phys. Rev. Letters 16, 789 (1966)

41. Ackerman, C.C. and Overton, W.C., Phys. Rev. Letters 22, 769 (1969)

42. Rogers, S.J., Phys. Rev. 3B, 1440 (1971)

43. Narayanamurti, V. and Dynes, R.C., Phys. Rev. Letters 28, 1461 (1972)

44. Ribbands, M.S. and Osborne, D.V., J. Physique 33 supplement C4, 119 (1972)

45. Narayanamurti, V. and Varma, C.M., Phys. Rev. Letters 25, 1105 (1970)

46. Von Gutfeld, R.J. and Nethercot, Jr. A.H., Phys. Rev. Letters $\underline{18}$, 855 (1967)

47. Long, A.R., J. Physique, $\underline{33}$ supplement C4, 73 (1972)

48. Maneval, J.P., Zylbersztejn, A. and Huet, D., Phys. Rev. Letters $\underline{27}$, 1375 (1971)

49. Dynes, R.C. and Narayanamurti, V., Phys. Rev. $\underline{6B}$, 143 (1972)

50. Fjeldly, T., Ishiguro, T. and Elbaum, C., Phys. Rev. $\underline{7B}$, 1392 (1973)

51. Wigmore, J.K., J. Physique, $\underline{33}$ supplement C4, 107 (1972)

52. Anderson, A.C. and Malinowski, M.E., Phys. Rev. $\underline{5B}$, 3199 (1972)

53. Andrews, J.M. and Strandberg, M.W.P., Phys. Rev. $\underline{172}$, 869 (1968)

54. Huet, D., Maneval, J.P. and Zylbersztejn, A., Phys. Rev. Letters $\underline{29}$, 1092 (1972)

55. Cheeke, J.D.N., Hebral, B. and Martinon, C., J. Physique $\underline{33}$, supplement C4, 57 (1972)

56. Dayem, A.H., Miller, B.I. and Wiegand, J.J., Phys. Rev. $\underline{3B}$, 2949 (1971)

ULTRASONIC ATTENUATION IN NORMAL METALS[*]

J. A. Rayne
Department of Physics, Carnegie-Mellon University
Pittsburgh, Pennsylvania 15213, U.S.A.

Abstract

This paper reviews recent experimental and theoretical
work on the electronic contribution to the attenuation of
sound in normal metals. Particular attention is paid to the
behaviour of shear waves, including electromagnetic breakdown effects and anomalies in the temperature dependence
of the shear wave attenuation in niobium and rhenium. New
magnetoacoustic data on caesium and copper are discussed,
as well as Doppler-shifted acoustic cyclotron resonance
experiments on aluminum and indium. An extensive bibliography of recent acoustic attenuation experiments at low
temperatures is provided.

---

[*]Work supported by National Science Foundation. This paper is based on
an invited talk at Fifth International Conference on Internal Friction
and Ultrasonic Attenuation in Crystalline Solids, Aachen, Germany
(August, 1973).

INTRODUCTION

The purpose of this paper is to give a summary of recent work on the attenuation of ultrasonic waves in normal metals. It will be convenient to divide the review into two sections dealing with effects in the absence and presence of a magnetic field, respectively.

I. ZERO-FIELD ATTENUATION

Theory

As is well known, the basic attenuation mechanism in a pure metal at low temperatures, apart from dislocation effects, is due to the interaction between the conduction electrons and the sound wave. Basically, the attenuation is due to the work done by electric fields, which maintain an electronic current screening the ionic motion in the presence of thermal phonons and impurities.

When the electronic mean free path ℓ is large compared to the sound wavelength i.e. $q\ell \gg 1$, q being the phonon wavenumber, the attenuation of a longitudinal wave is given by[1,2]

$$\alpha_\ell = \frac{\hbar q}{4\pi^2 \rho v_s} \oint R K_x^2 \, d\psi \qquad . \qquad (1)$$

Here ρ is the density and v_s is the relevant sound velocity in the metal. The integral is taken around the effective zone, for which the normal to the Fermi surface is perpendicular to the sound propagation direction, Ox. This simply reflects the fact that only electrons moving at right angles to q can interact with the sound wave. In the integrand R is the reciprocal Gaussian curvature (i.e. the product of the principal radii of curvature) at the appropriate point on the effective zone, while K_x is the so-called deformation parameter. The latter quantity measures how the Fermi surface (F.S.) distorts under a static strain field corresponding to the longitudinal elastic wave. From charge conservation, it may be shown that K_x satisfies the equation

$$\int_{F.S.} (K_x + k_x \cos\varphi) \, dS = 0 \qquad , \tag{2}$$

where φ is angle between the electron velocity $\underset{\sim}{v}$ and the propagation direction Ox.

Equation (1) may be written in the form

$$\alpha_\ell = Af \qquad (q\ell \gg 1) \qquad , \tag{3}$$

where the coefficient A depends on the geometry and deformation properties of the F.S. as well as the elastic properties of the metal, viz.,

$$A = \lim_{q\ell \to \infty} \left(\frac{\alpha}{f}\right) = \frac{\hbar}{2\pi \, \rho \, v_s^2} \oint R \, K_x^2 \, d\psi \qquad . \tag{4}$$

Clearly α_ℓ scales linearly with frequency and is independent of mean free path. For longitudinal waves, charge neutrality and this linear frequency dependence are preserved up to plasma frequencies of the order of 10^{13} cps. As we shall see in a moment, this is not true of shear waves.

When $q\ell \ll 1$ it may be shown that α_ℓ varies as $q^2\ell$, giving the familiar quadratic frequency behaviour. Between these limits, the dependence of α_ℓ on $q\ell$ is in general complex. However, for the free-electron model, it has a relatively simple form, viz.,

$$\alpha_\ell = A_{FE} \, f \, F(q\ell) \quad , \tag{5}$$

where

$$A_{FE} = \frac{\pi^2}{3} \frac{Nmv_F}{\rho v_s^2} \quad , \tag{6}$$

N being the electron density, m the electron mass and v_F the Fermi velocity. $F(q\ell)$ is the Pippard function defined by

$$F(a) = \frac{6}{\pi} \left[\frac{a}{3} \left(\frac{\tan^{-1}a}{a-\tan^{-1}a} \right) - \frac{1}{a} \right] \quad , \tag{7}$$

with $a = q\ell$. From Figure 1 it can be seen that $F(q\ell)$ varies as q for $q\ell \ll 1$ and saturates to unity for $q\ell \gg 1$, giving the correct limiting behaviour of α. It is to be noted from (5) that $\alpha/Af = F(q\ell)$, so that the graph also represents the dependence of α/f versus f with appropriate scaling.

For shear waves the attenuation in the limit $q\ell \gg 1$ can be written

$$\alpha_t = \frac{\hbar q}{4\pi^2 \rho v_s} \left[\oint R \, K_y^2 \, d\psi + \frac{1}{\pi^2} \frac{(\oint \mathscr{B} \tan\phi \, \cos\psi \, dS)^2}{\oint R \cos^2\psi \, d\psi} \right] \quad . \tag{8}$$

The first term is the so-called deformation contribution analogous to that discussed for longitudinal waves, while the second term is the electromagnetic contribution arising from the transverse screening current. The quantity $\mathcal{D}$ is the component of the deformation vector along the direction of particle motion Oy, while ψ is the angle between $\underset{\sim}{v}$ and Oy. Again we see that the attenuation scales linearly with frequency.

Equation (8) assumes complete screening, which is true only if the sound wavelength is large compared to the skin depth δ. Actually, the second term should be multiplied by a factor $\beta = [1 + q^4 \delta^4]^{-1}$, where the skin depth δ is given by

$$\delta = \frac{c}{(4\pi \omega \sigma_{EA})^{1/2}} \quad , \tag{9}$$

$\sigma_{EA} = \frac{3\pi}{4} \left(\frac{\sigma_o}{\ell} \right) \frac{1}{q}$ being the conductivity in the extreme anomalous limit,[2] The dependence of α_t on frequency will then be altered when $q\delta$ becomes appreciable. If $K_y \neq 0$ on the effective zone, the behaviour of α_t is as shown in Figure 2(b). Note that near $q\delta \sim 1$ there is a deviation from linearity, which is to be contrasted with the behaviour of α_ℓ shown in Figure 2(a). For propagation along a symmetry direction K_y may vanish on the effective zone and the first term in equation (8) is zero. The attenuation then exhibits a sharp drop at $q\delta \sim 1$ as shown in Figure 2(c). In this case α_t ultimately tends to a constant value, corresponding to the so-called collision-drag contribution to the attenuation. Normally the latter can be neglected for $q\ell \gg 1$, so that α_t is of the form

$$\alpha_t = \frac{Bq}{1+(q\delta)^4} \qquad , \qquad (10)$$

where the coefficient B again depends on the geometry and deformation properties of the F. S. in addition to the elastic properties of the metal. Breakdown effects are expected to become important for frequencies of the order
of 10^9 cps, although there are cases in which they can occur appreciably below this figure.

Longitudinal Wave Experiments

Peck and Dobbs[3] have reported detailed measurements of the attenuation
for longitudinal waves propagating along the principal symmetry directions
in high purity cadmium. Values of α_ℓ were obtained by measuring the difference $\alpha_n - \alpha_s$ down to 0.15K in a dilution refrigerator and extrapolating to
T = 0. Their data could be fitted to a modified free-electron formula

$$\alpha_\ell = A \ f \ F(q\ell) \qquad , \qquad (11)$$

where A is effective zone integral given by (4) and F is the Pippard function
discussed previously. The values of A corresponding to the limiting values
of α/f, together with the free-electron results, are shown in Table I. Also
shown for comparison are data for zinc obtained by Lea and Dobbs.[4]

For propagation in the basal plane, the observed anisotropy of A for both metals can be qualitatively understood in terms of the F.S. geometry with the assumption that K_x has the free-electron value $-\frac{1}{3} k_F$. The large values of A along [0001] are thought to be due to singularities in R on the effective zone, although it is possible that it is in part due to anisotropy in the deformation parameter. It is to be noted that the presence of a flat region on the effective zone would give an f^2 dependence for the attenuation and α/f would not then reach a true limiting value. Recent data on cadmium by Garfunkel et al.[5] at 9.3 GHz are not sufficiently precise to determine whether this behaviour is observed.

Berre and Vetleseter[6] have recently measured the attenuation of longitudinal waves propagating along the principal symmetry directions in ultra high-purity aluminum. Values of α_ℓ were again obtained from the difference $\alpha_n - \alpha_s$ extrapolated to T=0, the maximum $q\ell$ value being about 50 for propagation along [100] and about 30 along [110], [111]. Their data again appear to fit a modified free-electron formula quite well; the resulting values of A are summarised in Table 2, together with the free electron predictions. Also shown in the table are independent determinations of A obtained by Hepfer and Rayne.[7] These data were obtained by a different method, in which α_ℓ is computed from the difference between α_n at T=0 and at a temperature sufficiently high that the electronic attenuation is negligible. Clearly the results are in good agreement, so that there can be little doubt that the discrepancies between the experimental and free-electron values of A are significant.

Berre and Vetleseter have obtained theoretical estimates of A for each propagation direction, using the general formula for A given in equation (4). Values of R along the effective zone were obtained from the model Fermi surface of aluminum due to Ashcroft,[8] the integral being evaluated numerically with an assumed isotropic deformation parameter K_x. The explicit value of the latter can be obtained from the charge neutrality condition given by equation (2) and the known F.S. geometry. From Table 2, it can be seen that the calculated and experimental values of A are in good agreement.

This agreement is rather surprising. Hepfer and Rayne[7] have independently calculated the dependence of α on $q\ell$ for all three propagation directions, using the same model F.S. and the general form of the Pippard expression for α_ℓ. They find a limiting behaviour for an isotropic deformation parameter, which differs from that obtained by Berre and Vetleseter. Their work indicates that an <u>anisotropic</u> deformation parameter, again based on the Ashcroft model, gives satisfactory agreement with experiment. Figure 3 shows a comparison of their calculations with unpublished experimental data[9] on aluminum for [100] propagation. Note that the theoretical curve has interesting behaviour at high $q\ell$, again due to the presence of a flat spot on the effective zone for this propagation direction. As noted previously, this should lead to an effective f^2 frequency dependence. Unfortunately, the upturn is difficult to observe since it requires measurements at very high frequency on very pure, very accurately oriented samples.

Shear Wave Experiments

Figure 4 shows some recent data obtained by Page and Leibowitz[10] for shear waves propagating along [001] in tin. Although low frequency data are not indicated on the diagram, they lie within experimental error on the linear portion of the curve. The data certainly indicate that breakdown effects are occurring, but they really do not extend to sufficiently high frequencies to establish whether the behaviour is as shown in Figures 2(b) or (c), i.e. whether or not $K_y \neq 0$ on the effective zone. The curve shown in Figure 4 is given by equation (10), corresponding to the latter case; the maximum is at $q\delta = 3^{-1/4}$, which occurs at a frequency of 1 GHz. Figure 5 shows similar data obtained by Almond et al.[11] for shear waves propagating along [11$\bar{2}$0] in cadmium. Again the curves correspond to the form given by equation (10), with $q\delta = 1$ at about 470 MHz in contrast to 1.3 GHz for tin. This abnormally low value is thought to be due to the extremely low value of σ for the relevant effective zone in cadmium.

Rather controversial behaviour has been observed in the temperature dependence of α_t for both rhenium and niobium. Figure 6 shows data reported by Robinson and Levy[12] for shear waves propagating along [0001] in rhenium. There is a pronounced maximum in the temperature dependence of the attenuation at higher frequencies. Figure 7 shows in somewhat greater detail the data of Blessing and Leibowitz[13] for shear wave propagation along [100] in niobium. Again there is a well-defined maximum in the attenuation as a function of temperature, the $q\ell$ value at the maximum being essentially constant at $q\ell = 11.7 \pm 0.3$. The existence at these maxima is not consistent with equation (8), which gives a monotonic increase of α with decreasing T up to a saturation value at T=0.

A possible explanation of these results is that they are due to dislocation motion. At least in the case of niobium, the behaviour is not amplitude-dependent but there is still the possibility of a temperature dependence due to dislocations even for small amplitudes. Figure 8 shows attenuation data for longitudinal waves propagating along [100] in high purity aluminum. The behaviour is consistent with a background attenuation due to dislocations having the temperature dependence shown and it seems not a similar situation could hold in the case of rhenium and niobium, giving rise to a maximum in attenuation with decreasing temperature.

There are other explanations of the observed behaviour, one being that the peak is associated with the effects of electromagnetic breakdown. It has already been shown that α_t has a maximum as qδ passes through the critical value $3^{-1/4}$. Correspondingly α_t should have a maximum as a function of temperature, as ℓ the electron free path and hence δ the skin depth changes. In both cases the observed frequency dependence of the maximum does not appear to be consistent with this explanation. For niobium there is the additional complication that α seems to be accurately linear with frequency, so that breakdown effects are not present up to 400 MHz.

Another model for the maximum in α_t involves the presence of the collision-drag term in the attenuation. For the simple free-electron model with complete screening, the shear attenuation can be written

$$\alpha_t = \left(\frac{Nmv_F}{\rho v_s}\right) q \left(\frac{1-g}{ag}\right) [g + (1-g)] \qquad , \qquad (12)$$

where the function g is given by

$$g = \left[\frac{3}{2a^2}\left(\frac{1+a^2}{a}\right)\tan^{-1} a - 1\right] \qquad . \qquad (13)$$

The first term in parenthesis is the collision drag contribution, while the second is the electromagnetic contribution to α_t. Note the absence of a deformation effect in this case because of symmetry, i.e., $K_y = 0$ in the effective zone. The collision drag term has a maximum as a function of $q\ell$, as can be seen from curve B in Figure 9. However, from equation (12) we see that the electromagnetic term involves $(1-g)$, so that the overall dependence of α_t on $q\ell$ is a smooth function similar to that found for α_ℓ.

Robinson and Levy have considered the extension of the free-electron model to a simple two-band case, with different numbers of carriers, different effective masses and different relaxation times in each band. Interband transitions and deformation effects are neglected, although the latter restriction has been removed by Leibowitz.[13] The resulting formula for α_t is then modified to read

$$\alpha_t = \left(\frac{Nmv_F}{\rho v_s}\right) q \left(\frac{1-g}{g}\right)[A_1 g + A_2(1-g)] \qquad , \qquad (14)$$

where the coefficients A_1, A_2 are functions of the band parameters and, in the case of A_2, a term taking account of screening effects. The general behaviour of α_t with ℓ is shown in Figure 10, which shows the case with complete screening and equal numbers of carriers in both bands with $M = \dfrac{m_2}{m_1} = 0.405$. The q value for each curve increases from 1 to 10 in units of one-half the reciprocal electron mean-free-path unit; clearly the behaviour is suggestive of the experimental results. Indeed, Robinson and Levy find that with appropriate choice of the available parameters, viz. $M = 10$ and a screening parameter $\dfrac{\chi q^2}{ag} = 1.2 \times 10^8$ cm^{-2} corresponding to breakdown beginning at 600 MHz, they can get a fit to data. However, it is to be noted that the data in Figure 6 also show an effect for longitudinal waves and it is not clear how this behaviour fits into the model.

Finally, in connection with existence of maxima in the dependence of α on temperature, attention should be drawn to some unpublished theoretical work by Peverly.[14] He has been able to extend solutions for the attenuation in the free-electron model to arbitrary $q\ell$ and to <u>arbitrary</u> scattering functions $W(\theta)$, θ being the scattering angle between the final and initial electron states. The perturbed electron distribution function is expanded in a series of spherical harmonics, each of which has a different relaxation time and mean free path. Expressions for α_t and α_ℓ are then obtained in the form of continued fractions, viz.,

$$\alpha_t = \frac{Nmv_F}{3\rho v_{st}} \; q \; \left[\cfrac{1.3}{5/q\ell_2 + \cfrac{2.4}{7/q\ell_3 + \cfrac{3.5}{9/q\ell_4\cdots}}} \right] \qquad , \qquad (15)$$

and

$$\alpha_\ell = \frac{Nmv_F}{3\rho v_{s\ell}} \; q \; \left[\cfrac{2.2}{5/q\ell_2 + \cfrac{3.3}{7/q\ell_3 + \cfrac{4.4}{9/q\ell_4\cdots}}} \right] \qquad . \qquad (16)$$

In the isotropic scattering limit $\ell_2 = \ell_3\ldots = \ell$ and the original Pippard formulae for α_t and α_ℓ are obtained.

The effects of scattering with a strong forward lobe can be simulated by taking

$$
\begin{aligned}
W(\theta) &= R_1 & \theta &< \theta_1 \\
&= 0 & &\text{otherwise}
\end{aligned}
\qquad , \qquad (17)
$$

while the effects of a strong backward scattering can be described with

$$
\begin{aligned}
W(\theta) &= R_2 & \pi-\theta &< \theta_2 \\
&= 0 & &\text{otherwise}
\end{aligned}
\qquad . \qquad (18)
$$

Note that isotropic scattering in each case corresponds to θ_1, $\theta_2 = \pi$. Figure 10 shows α_t as a function of $q\ell_2$ for each of these assumptions, the solid curve representing isotropic scattering. Clearly strong backward scattering gives rise to a pronounced maximum in α_t which of course would produce a peak in the associated temperature dependence. Whether or not such a model could be applied to the observations in rhenium and niobium is not clear. A much more likely candidate for measurements is potassium, where Umklapp processes are likely to produce strong back-scattering.

II. ATTENUATION IN A MAGNETIC FIELD

Among the many effects that fall into this category are

1) Magnetoacoustic Effect

2) Doppled-shifted Acoustic Cyclotron Resonance (DSACR) and Acoustic Cyclotron Resonance

3) Quantum Oscillations in Attenuation including Giant Quantum Magnetoacoustic Effects

4) Direct Sound Generation and Helicon-Phonon Interaction.

Clearly a comprehensive summary of such a wide range of subjects is not possible in the available space. Accordingly this review will focus attention on just a few topics.

Magnetoacoustic Experiments

There is considerable interest in some recent measurements by Trivisanno et al.[15,16] on the magnetoacoustic effect in caesium. Data have been reported for both longitudinal and shear waves, the former being used principally to measure caliper dimensions of the Fermi surface. The resulting radial anisotropy agrees within experimental error with that obtained from dHvA experiments. For shear waves, there is a significant disagreement between the field dependence of the attenuation and that predicted by the Cohen, Harrison and Harrison[17] (CHH) theory.

Typical results for theory and experiment are shown in Figure 11, where it can be seen that the observed change in relative attenuation is _less_ than that predicted by theory. The difference between the zero and infinite field attenuation, which should be the electronic attenuation, scales sublinearly with frequency for qℓ large. Moreover, the saturation behaviour for $\underset{\sim}{H}$ parallel to $\underset{\sim}{\varepsilon}$ is different from that with $\underset{\sim}{H}$ perpendicular to $\underset{\sim}{\varepsilon}$, although theory predicts that the saturation effects are independent of the angle between $\underset{\sim}{H}$ and $\underset{\sim}{\varepsilon}$. Similar discrepancies with the CHH theory have been observed for longitudinal waves in potassium,[18] which has a very spherical Fermi surface. It thus appears that F.S. geometry alone cannot explain the caesium results. One interesting suggestion due to Peverley[14] is that the back-scattering effects discussed previously might be responsible for the sublinear frequency behaviour. Clearly the existence of a maximum in α_t as a function of qℓ could produce precisely the observed effect.

Some elegant experiments have been reported by Peverley and Khatri[19] on the use of the magnetoacoustic effect to obtain information about the electron-phonon relaxation time in copper. It can be shown[2,20] that the amplitude of the n-th oscillation associated with a given extremal orbit satisfies the equation (n >> 1)

$$\frac{A_n(T)}{A_n(0)} = \exp\left\{-\frac{\pi n m^*\lambda}{\hbar\Delta k}\left[\frac{1}{\tau(T)} - \frac{1}{\tau(0)}\right]\right\} \qquad . \qquad (19)$$

Here Δk is the extremal orbit dimension associated with the oscillatory behaviour, m^* is the corresponding effective mass and λ is the relevant sound wavelength. Since these quantities are known, it follows that the temperature dependence of the amplitude A_n can be used to obtain $r_n(T) - r_n(0) = \frac{1}{\tau(T)} - \frac{1}{\tau(0)}$. Data have been obtained for the central and non-central belly orbits in copper perpendicular to [100], using values of n up to 50. In the former case the relaxation rate is given by

$$r_n(T) - r_n(0) = (6.0 \pm 0.3) \times 10^6 \, T^3 \, sec^{-1} \quad , \qquad (20)$$

while in the latter it follows the relation

$$r_n(T) - r_n(0) = (2.9 \pm 0.15) \times 10^6 \, T^3 \, sec^{-1} \quad . \qquad (21)$$

The difference between the two rates is due to the anisotropy of the scattering over the F.S. of copper. Both results agree with other experimental determinations of these quantities[21,22,23] and with average relaxation rates computed from recent theoretical studies.[24] The magnetoacoustic method has the obvious advantage that it does not depend on surface preparation and is more representative of the bulk scattering effects.

It is rather interesting that at very low temperatures the relaxation rates are proportional to T^5 rather than T^3. This result strongly suggests that scattering out of the effective orbit width is then a diffusion type, multi-step process rather than a catastrophic event.

Peverley and Khatri have also measured the electron relaxation time for the [111] open orbit in copper by fitting the corresponding resonance line to a Lorentzian shape. In this case the electron-phonon relaxation rate exhibits a T^4 temperature dependence, intermediate between the extremes of diffusion and catastrophic scattering.

DSACR Experiments

As is well known, the resonance condition is given by the equation[2]

$$q \cdot \bar{v} - \omega - n\omega_c = 0 \qquad , \qquad (22)$$

where $\bar{v}$ is the mean electron velocity, ω is the frequency of the sound wave, ω_c is the relevant cyclotron frequency and n is an integer. It can be shown that this equation gives resonant or anti-resonant peaks in attenuation for fields B_n such that

$$B_n = \frac{qc\hbar}{2\pi ne} \left| \frac{\partial A}{\partial k_z} \right|_{ext} \qquad (\omega \ll \omega_c) \qquad , \qquad (23)$$

where the extremal F.S. area derivative $\left| \partial A / \partial k_z \right|_{ext}$ involves the component of the electron wave vector along the field.

Typical behaviour is shown in Figure 12, which shows recent data obtained by Hui and Rayne[25] on high-purity aluminum with longitudinal waves propagating along [001]. Two distinct series of attenuation of peaks A and B are observed, both being associated with orbit extrema on the second zone hole surface. The resulting extremal derivatives, calculated from equation (23), give better agreement with the predictions of the model F.S. due to Andersen and Lane[26] than that of Ashcroft.[8] No structure in the attenuation data appears to be associated with the third-zone electron surface. However, pronounced third-zone DSCAR peaks have been observed in indium,[27] which has a Fermi surface similar to that of aluminum. No reason for the difference in behaviour has as yet been adduced.

Where applicable, DSACR appears to provide a sensitive means of refining Fermi surface data. It has the advantage over helicon DSCR that non-local dispersive effects are absent.

ACKNOWLEDGEMENTS

Apologies are due to many authors, whose work has not been explicitly mentioned in this necessarily brief review. A complete bibliography of recent publications is given in the list of references at the end of this paper. Thanks are due to Drs. J. R. Leibowitz and J. R. Peverley for communicating the results of their work before publication.

REFERENCES

1. A. B. Pippard, Proc. Roy. Soc. A257, 165 (1960).

2. J. Mertsching, Phys. Stat. Sol. 37, 465 (1970).

3. D. R. Peck and E. R. Dobbs, Phys. Letters 35A, 196 (1971).

4. M. J. Lea and E. R. Dobbs, Phys. Letters 27A, 556 (1968).

5. M. P. Garfunkel, J. W. Lue and G. E. Pike, Phys. Rev. Letters 25, 1649 (1970).

6. B. Berre and A. Vetleseter, J. Low Temp. Phys. 7, 399 (1972).

7. K. Hepfer and J. Rayne, Phys. Rev. B4, 1050 (1971).

8. N. W. Ashcroft, Phil. Mag. 8, 2055 (1963).

9. J. A. Rayne, D. J. Meredith and E. R. Dobbs (unpublished).

10. E. Page and J. R. Leibowitz (private communication).

11. D. P. Almond, M. J. Lea and E. R. Dobbs, Phys. Letters 43A, 69 (1972).

12. D. A. Robinson and M. Levy, Aust. J. Phys. 24, 333 (1971).

13. G. V. Blessing and J. R. Leibowitz, Bull. Am. Phys. Soc. Ser. II, 18, 704 (1973).

14. J. R. Peverley (private communication).

15. J. Trivisanno and J. A. Murphy, Phys. Rev. B1, 3341 (1970).

16. B. Keramidas, J. Trivisanno and G. Kaltenbach, Phys. Rev. B6, 4412 (1972).

17. M. H. Cohen, M. J. Harrison and W. A. Harrison, Phys. Rev. 117, 937 (1960).

18. J. R. Peverley, Phys. Rev. 173, 689 (1968).

19. J. R. Peverley and D. S. Khatri, Bull. Am. Phys. Soc. 18, 366 (1973).

20. M. S. Phua and J. R. Peverley, Phys. Rev. B3, 3115 (1971).

21. R. E. Doezema and J. F. Koch, Phys. Rev. B6, 2071 (1972).

22. V. F. Gantmakher (private communication).

23. P. Haussler and S. J. Welles, Phys. Rev. 152, 675 (1966).

24. D. Nowak, Phys. Rev. B6, 3691 (1972).

25. S. W. Hui and J. A. Rayne, J. Low Temp. Phys. (to be published).

26. J. R. Andersen and S. S. Lane, Phys. Rev. B2, 298 (1970).

27. S. C. Hayden, Penn State University, State Center (Thesis, 1973), (unpublished).

ADDITIONAL REFERENCES

Normal State Electronic Attenuation

H. C. Huang, Wayne State University, Detroit (Thesis, 1970) - Hg

J. M. Perz and W. A. Roger, Can. J. Phys. 49, 296 (1971) - Ta

G. E. Pike and M. P. Garfunkel, Physica 55, 656 (1971) - Mo

Magnetoacoustic Effect

G. N. Kamm, Phys. Rev. B1, 554 (1970) - Cu

C. Guthmann, J. P. D'Haenens and A. Libchaber, Phys. Rev. B4, 1538 (1971) - Bi

R. V. Kollarits, J. Trivisanno and R. W. Stark, Phys. Rev. B2, 1508 (1972) - Mg

J. D. Gavenda and W. Royall Cox, Phys. Rev. B6, 4392 (1972) - Cu

C. Alquie and J. Lewiner, Phys. Rev. B6, 4490 (1972) - Ga

J. R. Leibowitz, E. Alexander, G. Blessing, T. Francavilla and J. R. Peverley, Proceedings of 13th International Conference on Low Temperature Physics (to be published) - Nb

C. Guthmann, J. P. D'Haenens and A. Libchaber, Phys. Rev. B8, 561 (1973) - Bi

J. R. Leibowitz, J. R. Peverley and E. Alexander, Phys. Letters 44A, 298 (1973).

Giant Quantum Oscillations and High Field Effects

S. Mase, Y. Matusumoto, T. Sakai, Y. Suido, Proc. 12th Int. Temp.
Conf. (Kyoto, Japan 1970) p. 575 - Bi, Sb

W. R. Cox and J. D. Gavenda, Phys. Rev. B3, 324 (1971) - Cu

R. W. Reed and F. G. Brickwedde, Phys. Rev. B3, 1081 (1971) - Mg

D. F. Snider and R. L. Thomas, Phys. Rev. B3, 1091 (1971) - Cr

S. Mase and T. Sakai, J. Phys. Soc. Jap. 31, 730 (1971) - Bi

H. Mori and S. Mase, J. Phys. Soc. Jap. 31, 738 (1971) - Bi

P. W. Murray and R. C. Young, Phys. Letters 37A, 217 (1971) - Sn

J. M. Perz and J. P. Kalejs, Phys. Can. 27, 63 (1971) - W

A. A. Galkin, E. P. Degtyar, S. E. Zhevago and A. I. Popovich,
Sov. Phys.-Dokl. 16, 382 (1971) - As

M. Mongy, J. Phys. Chem. Solids 33, 1355 (1972) - Au

G. Belessa, Phys. Rev. B7, 2400 (1973) - Hg

Doppler-Shifted Cyclotron Resonance and Acoustic Cyclotron Resonance

W. R. Cox and J. D. Gavenda, Phys. Rev. B3, 231 (1971) - Cu

S. W. Hui and J. A. Rayne, J. Phys. Chem. Solids 33, 611 (1972) - W

C. Alquie and J. Lewiner, Phys. Rev. B6, 4490 (1972) - Ga

Direct Sound Generation and Helicon-Phonon Interaction

E. R. Dobbs, D. J. Meredith and J. Young, Proc. 12th International
Low Temperature Conference (Kyoto, Japan 1970) p. 529 - Bi, Sb

K. O. Legg and D. J. Meredith, J. Phys. D3, 161 (1970).

I. I. Babkin, V. T. Dolgopolov, V. Ya. Kravchenko, JETP Letters 13,
402 (1971) - Bi

Yu. P. Gaidukov and A. P. Perov, Sov. Phys. Acoust. $\underline{17}$, 266 (1971) - Sn

D. K. Hsu, Wayne State University, Detroit (Thesis, 1971) - K, Bi

G. Turner, R. L. Thomas and D. Hsu, Phys. Rev. $\underline{B3}$, 3097 (1971) - K

K. R. Lyall and J. F. Cochran, Can. J. Phys. $\underline{49}$, 1075 (1971) - Ga

D. Hsu and R. L. Thomas, Phys. Rev. $\underline{B5}$, 4668 (1972) - Bi

K. C. Lee, A. M. de Graaf, D. Hsu and R. L. Thomas, Phys. Rev. $\underline{B8}$, 460 (1973)

Theory

T. M. Rice and L. J. Sham, Phys. Rev. $\underline{B1}$, 4546 (1970) - Electron-Phonon Interaction in Potassium

H. Fukuyama and T. Nagai, J. Phys. Soc. Jap. $\underline{31}$, 812 (1971) - Exitonic Instability in High Magnetic Fields

P. N. Trofimenkoff and J. W. Ekin, Phys. Rev. $\underline{B4}$, 2392 (1971) - Electron-Phonon Umklapp Scattering in Potassium

A. A. Slotskin and S. A. Sokolov, JETP Letters $\underline{14}$, 40 (1971) - Magnetic Breakdown Giant Oscillations of Sound Absorption

S. S. Ghatak and A. Ramakanth, Phys. Letters $\underline{35A}$, 371 (1971) - Ultrasonic Propagation in Kondo Systems

V. M. Gokhfel'd and S. S. Nederezov, JETP $\underline{61}$, 2041 (1971) - Quantum Magnetoacoustic Oscillations in Thin Films

R. Sandstrom, Ann. Phys. $\underline{70}$, 516 (1972) - Sound Propagation in an Anharmonic Metal

34

TABLE 1

Limiting values of α/f for longitudinal wave propagation in cadmium and
zinc along principal symmetry directions.

Metal	A(dB cm^{-1} MHz^{-1})[*]					
	[11$\bar{2}$0]		[10$\bar{1}$0]		[0001]	
	Expt	FE	Expt	FE	Expt	FE
Cadmium	0.048	0.156	0.165	0.156	0.99	0.356
Zinc	0.067	0.182	0.128	0.182	1.29	0.475

[*]Value of A obtained from fit of data to equation (11)

TABLE 2

Limiting values of α/f for longitudinal wave propagation in aluminum along principal symmetry directions.

Propagation Direction	$A(dB\ cm^{-1}MHz^{-1})$ [*]		
	B & V[†][a]	H & R[b]	F.E.
[100]	0.41 ± 0.01(0.41 ± 0.03)	0.38 ± 0.04	0.403
[110]	0.49 ± 0.01(0.48 ± 0.03)	0.50 ± 0.03	0.386
[111]	0.30 ± 0.01(0.30 ± 0.03)	0.28 ± 0.03	0.379

[*]Value of A obtained from fit of data to equation (11)

[†]Values in parentheses are calculated from equation (4) using isotropic deformation parameter.

[a] See ref. 6

[b] See ref. 7

Figure 1. Dependence of Pippard function F on $q\ell$.

Figure 2. Frequency dependence of the attenuation coefficient α for (a) longitudinal waves (b) shear waves with $Ky \neq 0$ on effective zone and (c) shear waves with $Ky = 0$ on effective zone.

Figure 3. Variation of α/f with scaled values of frequency for aluminum. The values of ℓ/ℓ_{ref} are computed from the observed residual resistivity ratios using the lowest purity sample as reference.

Figure 4. Frequency variation of electronic attenuation for shear waves propagating along [001] in tin. Data for lower frequencies corresponding to the linear part of the graph are omitted. The solid line is given by equation (10).

Figure 5. Frequency variation of electronic attenuation for shear waves propagating along [11$\bar{2}$0] in cadmium showing fit to equation (10).

Figure 6. Temperature dependence of attenuation of longitudinal and shear waves propagating along [0001] in rhenium.

Figure 7. Temperature dependence of shear wave attenuation in niobium for propagation along [100].

Figure 8. Temperature dependence of longitudinal wave attenuation in high-purity aluminum for propagation along [100] at 345 MHz. The dashed line is the assumed background attenuation due to dis-location motion.

Figure 9. Dependence of shear wave attenuation on $q\ell$ for free-electron model with $q\delta \ll 1$. Curve A is the total attenuation, while curve B is the collision-drag contribution.

Figure 10. Dependence of shear wave attenuation on electron mean free path ℓ for two band model with equal number of carriers in each band and with $M = m_2/m_1 = 0.405$. The q value for each curve increases from 1 to 10 in units of one-half the reciprocal electron mean free path unit. Electromagnetic breakdown effects are neglected, i.e., $q\delta \ll 1$.

Figure 11. Dependence of α_t/q on $q\ell_2$ for scattering function $W(\theta)$ given by equation (18). The values of θ_2 are indicated in the figure, the case of isotropic scattering corresponding to $\theta_2 = 180°$.

Figure 12. Dependence of relative attenuation on magnetic field for shear waves propagating in caesium with $q||[011]$ and $\underline{\epsilon}||[100]$ at 87 MHz. The full curve is the theoretical prediction of the CHH theory for $q\ell = 35$ while the dashed curve represents the experimental data. R is the cyclotron radius $\hbar k c/e B$.

Figure 13. Amplitude of transmitted pulse for longitudinal waves propagating in high-purity aluminum at 153 MHz with q, $\underline{B}$ parallel to [001]. The DSCR series A and B are shown with harmonic numbers marked accordingly. Sinusoidal oscillations at low fields are presumably geometric resonances.

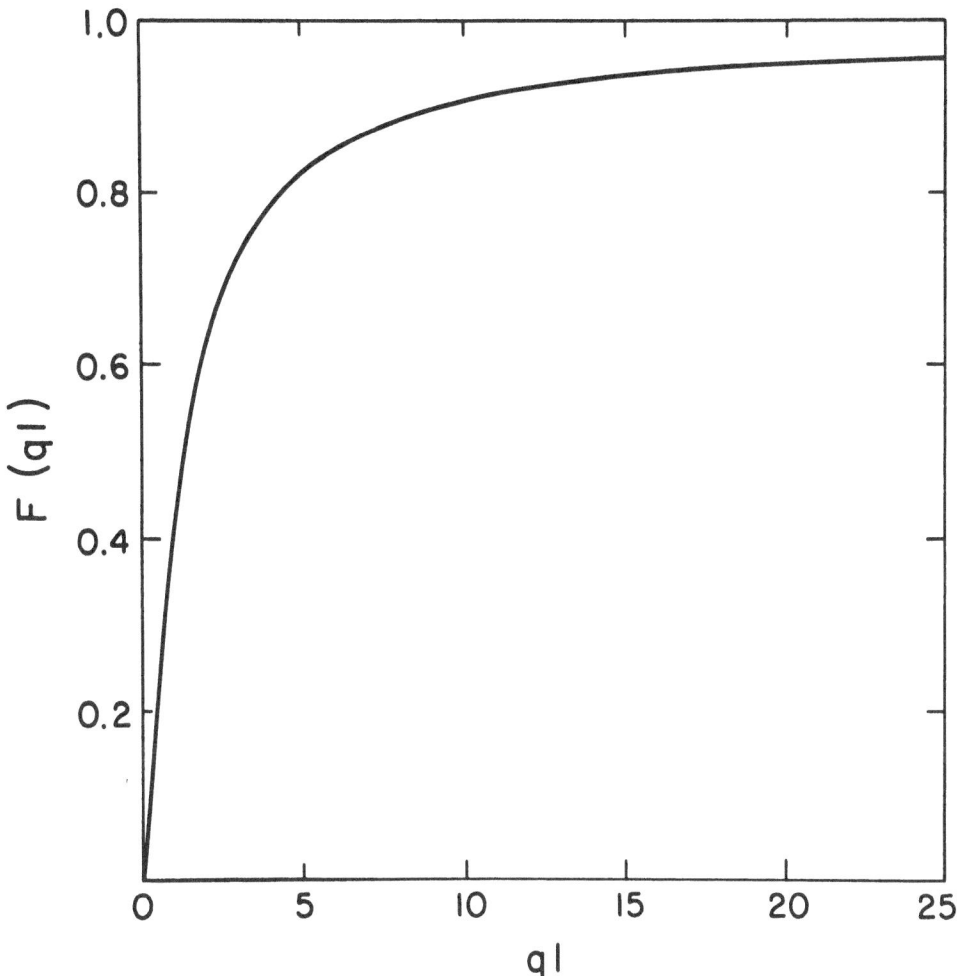

Figure 1

39

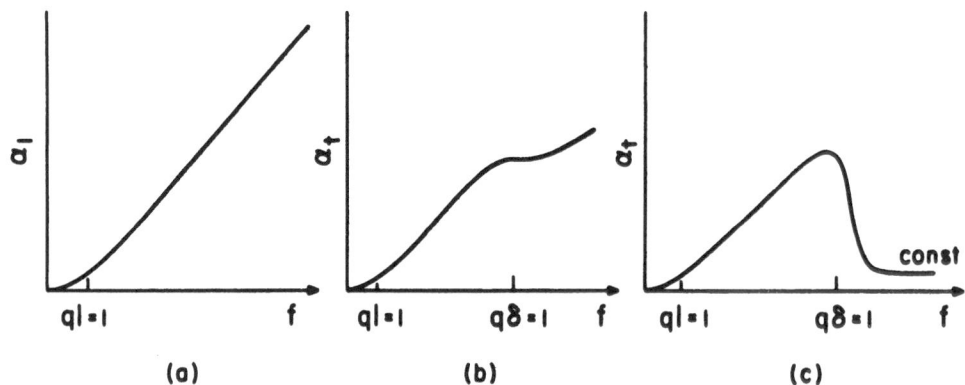

Figure 2

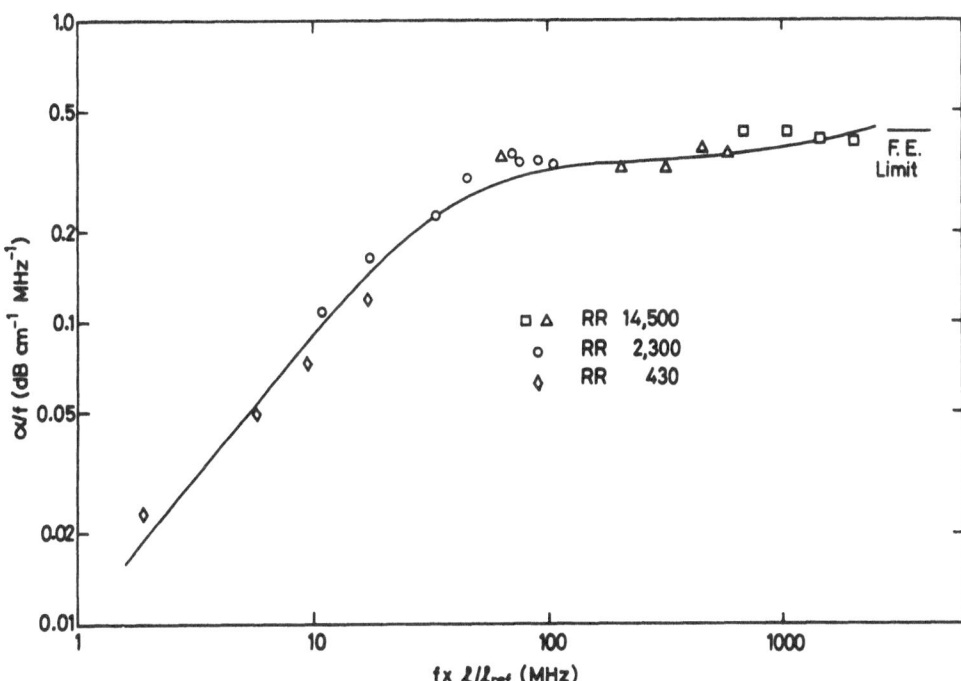

Figure 3

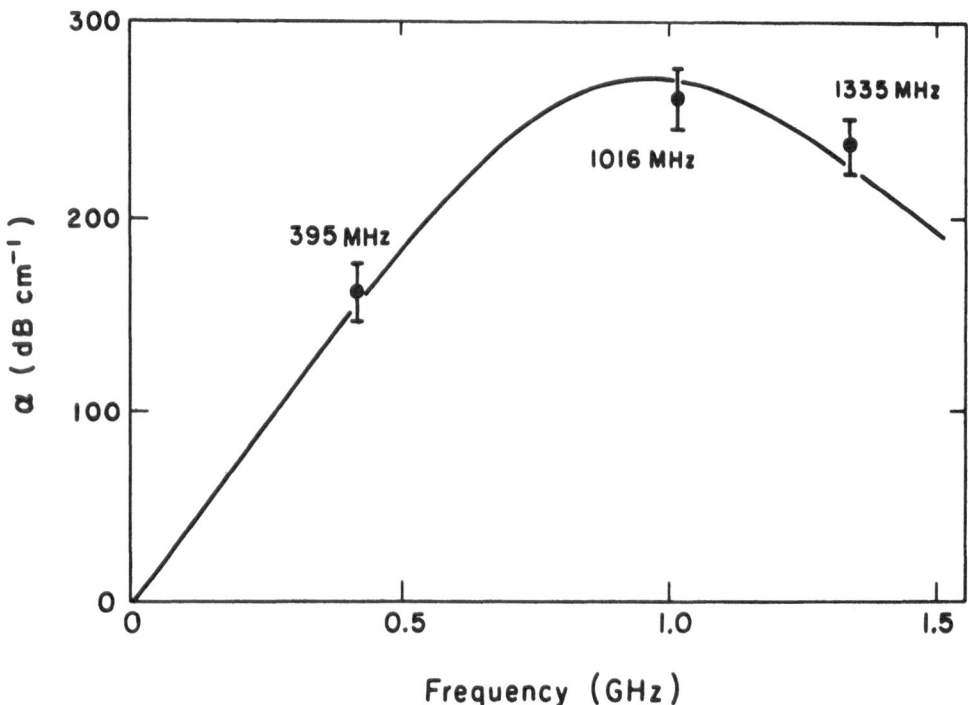

Figure 4

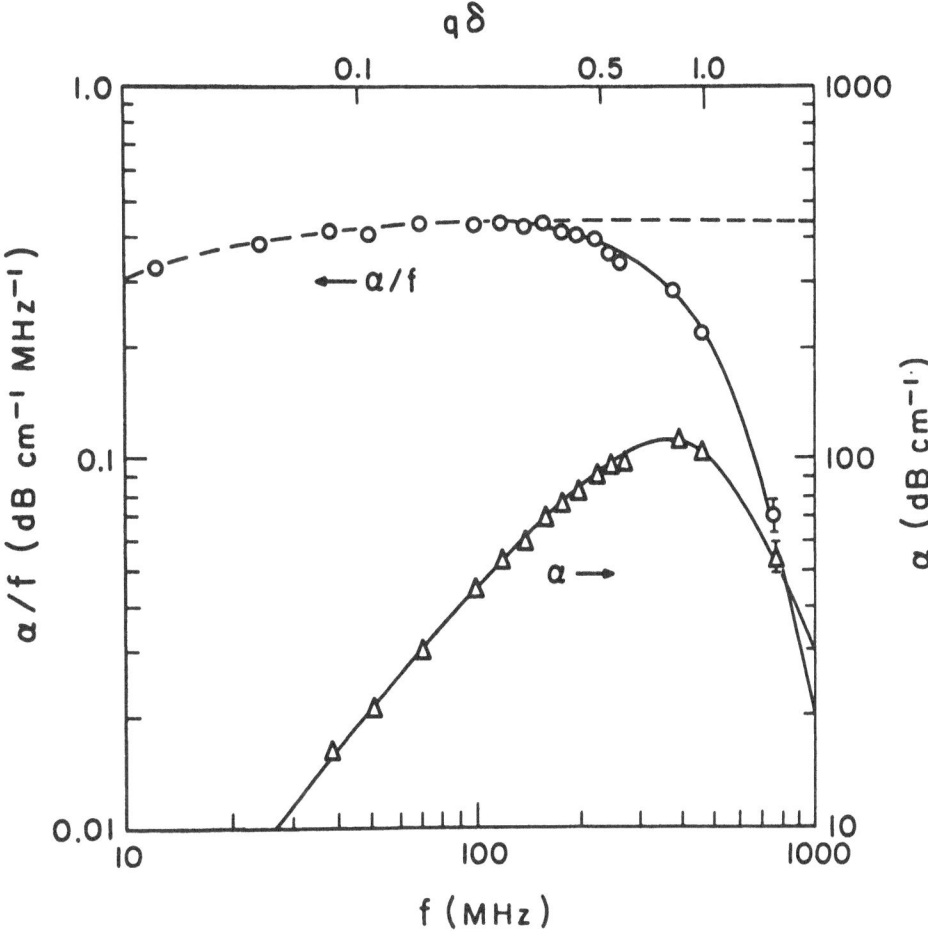

Figure 5

43

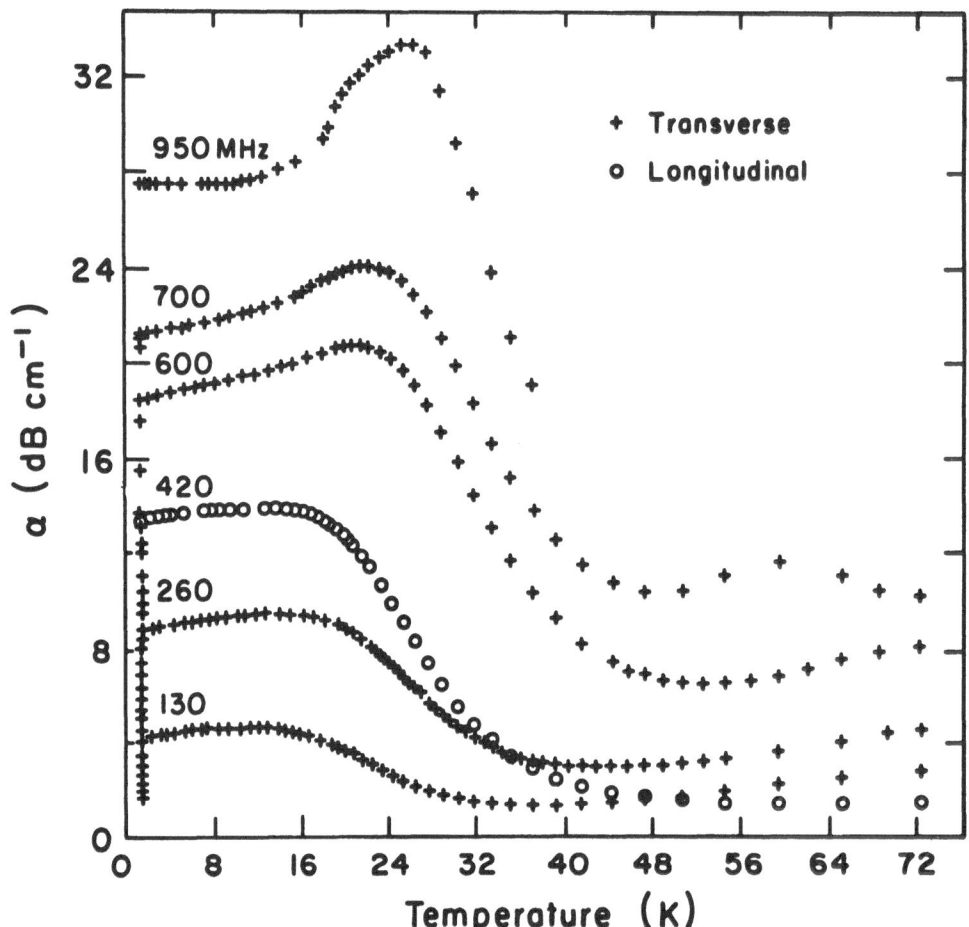

Figure 6

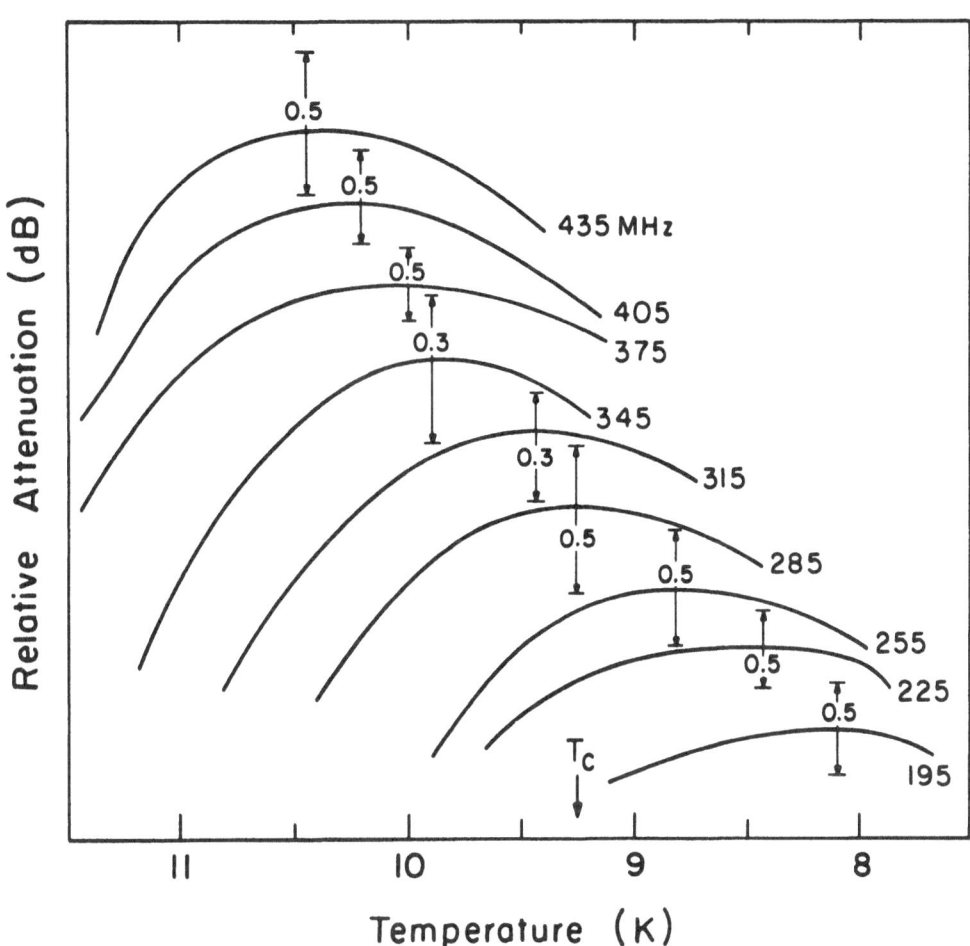

Figure 7

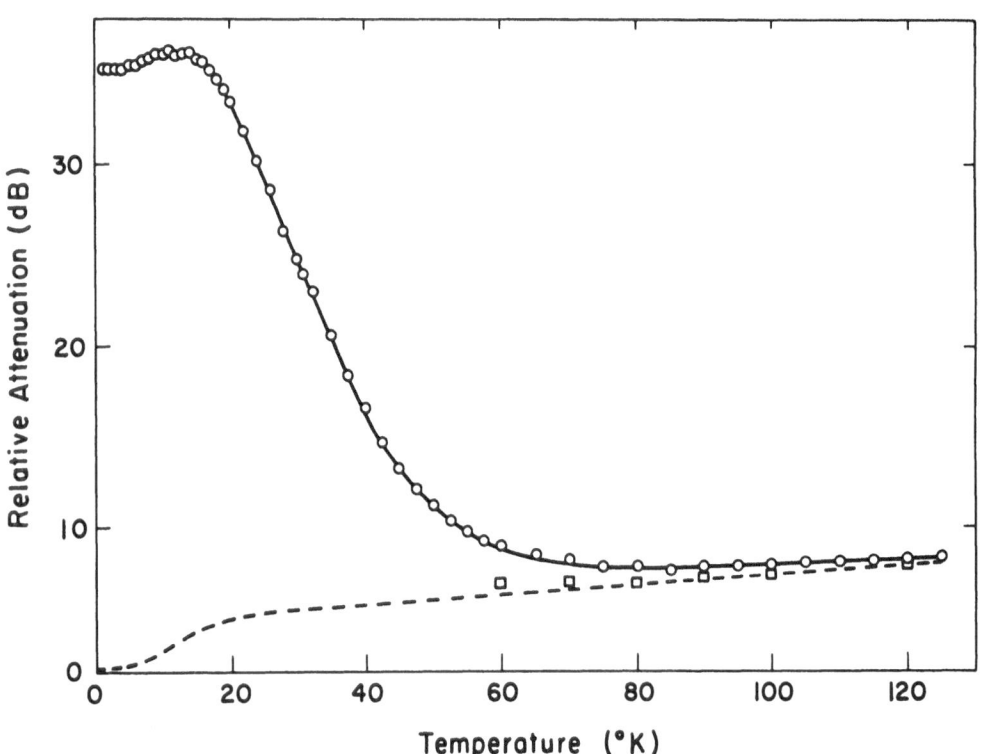

Figure 8

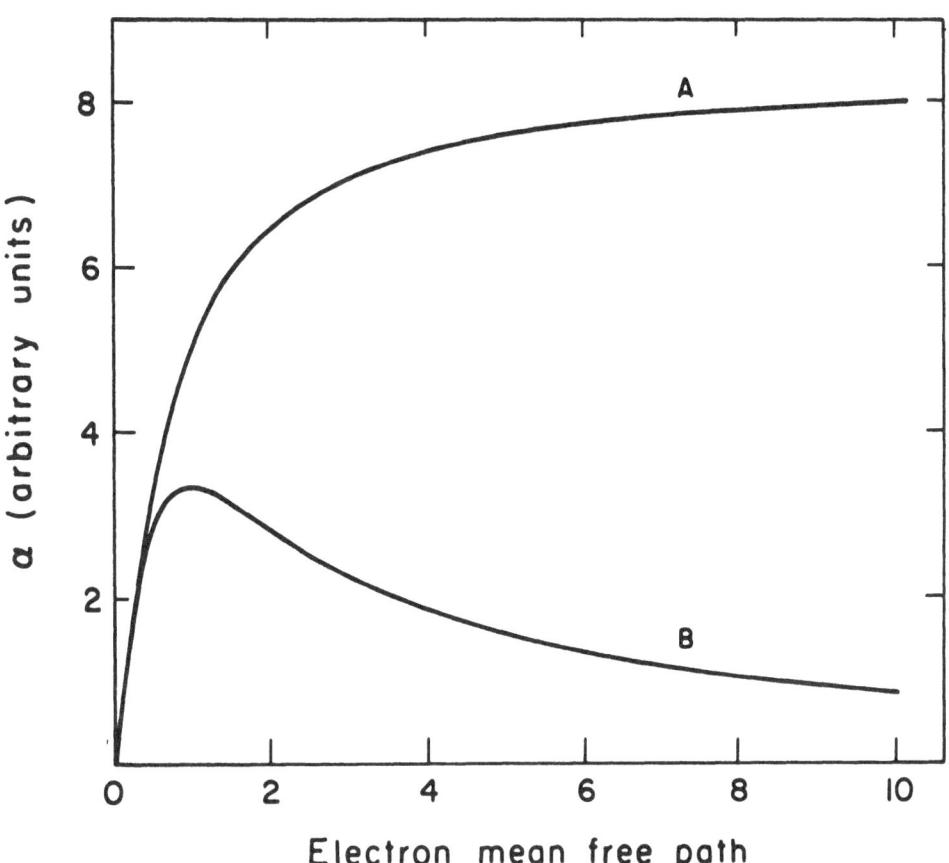

Figure 9

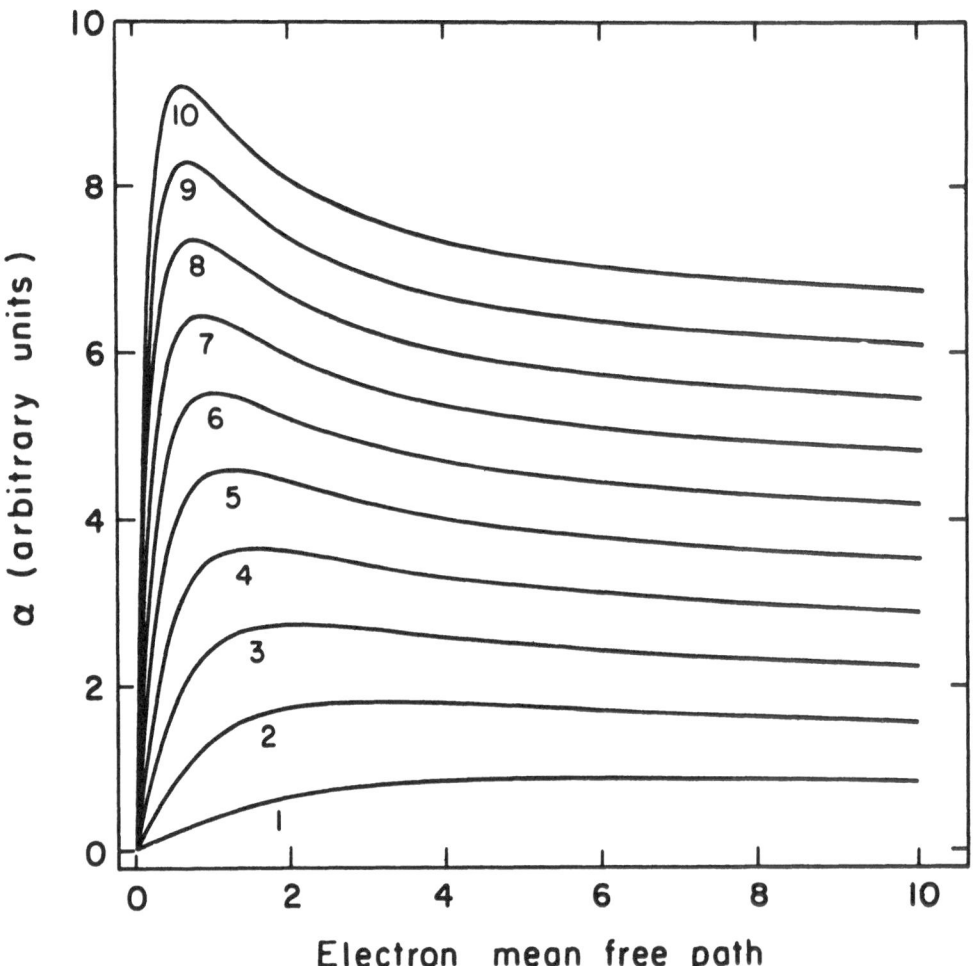

Figure 10

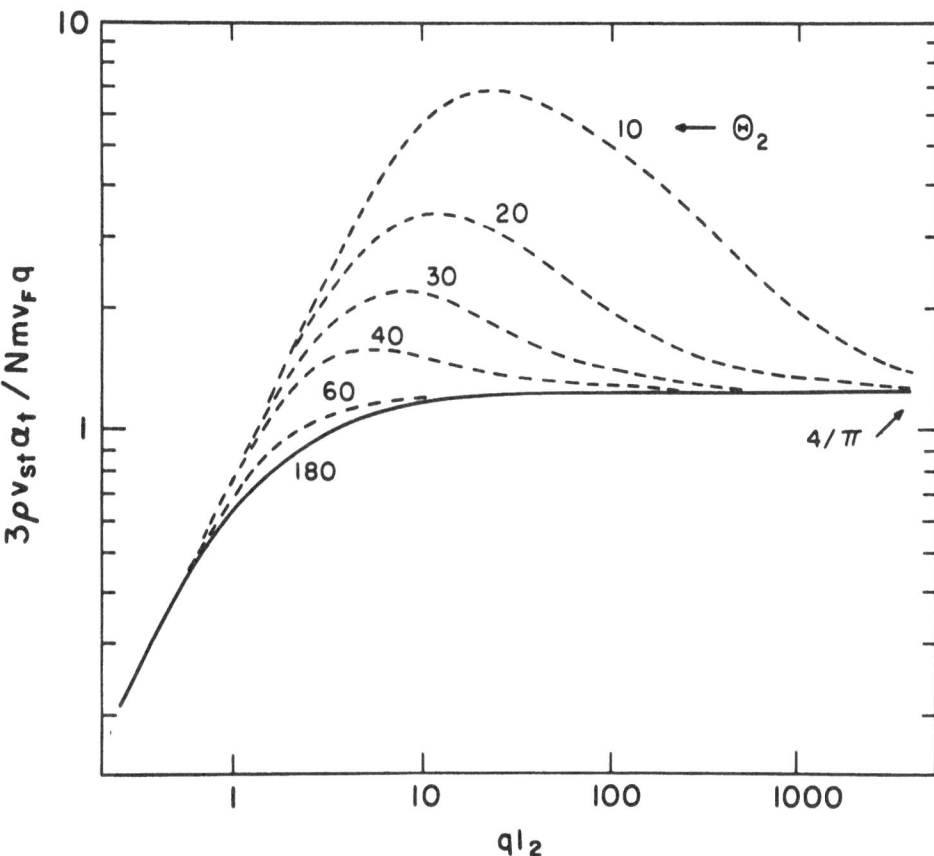

Figure 11

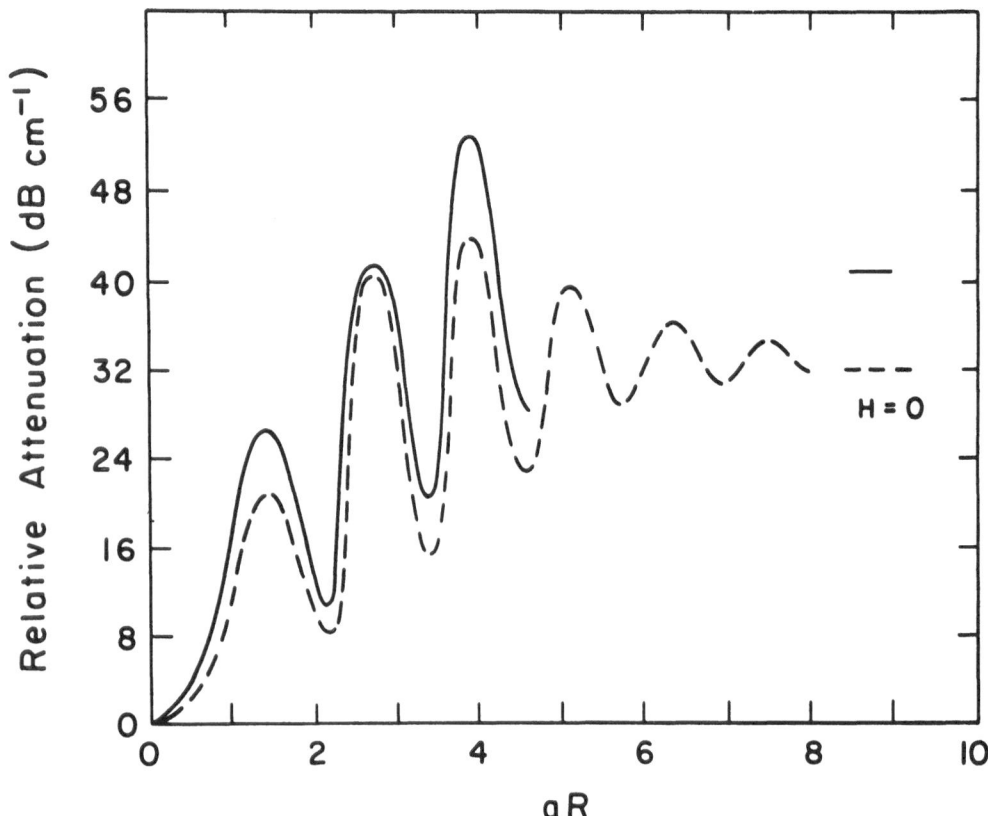

Figure 12

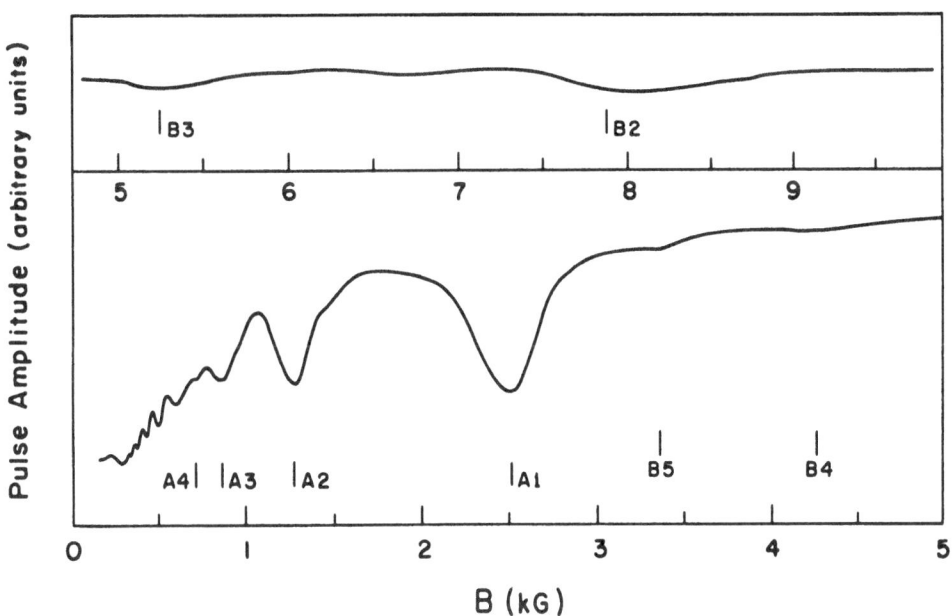

Figure 13

ULTRASONIC ATTENUATION IN SUPERCONDUCTORS

B. R. Tittmann
Science Center, Rockwell International
Thousand Oaks, California 91360

ABSTRACT The discovery, early development, and BCS theory of the ultrasonic attenuation in superconductors is reviewed. It is shown
that the BCS formulation containing the concept of a superconducting energy gap leads to a satisfactory explanation of
the temperature dependence of the attenuation for a wide range
of superconductors. Emphasis is placed on recent measurements
which exhibit drastic deviations from the simple behavior
described by BCS. Experiments are reviewed on the amplitude
dependence in attenuation associated with the motion of dislocations and on the influence of vortex lines on the magnetic
field dependence of the attenuation in type II superconductors.

Introduction

The field of ultrasonic attenuation studies in superconductors achieved its
first thrust with the work of Bömmel (1) and Mackinnon (2) who discovered independently that as they lowered the temperature of their samples into the liquid
helium range, the ultrasonic attenuation diminished abruptly with the onset of
superconductivity. Figure 1 displays Bömmel's original data on a single crystal
of Pb at 26.5 MHz. The plot, showing attenuation $\alpha(T)$, as a function of temperature T,
exhibits the characteristic increase in α as the temperature is lowered into the liquid
helium range, reflecting the increase in the electron mean free path, and consequently, the increase in the interaction between the sound wave and the conduction
electrons. At the transition temperature, T_C = 7.15°K, and in zero magnetic field, α is
seen to drop abruptly, then more gradually, and finally to level off as T is lowered further
onto T $\approx$ 0°K. In the presence of a magnetic field exceeding the critical field H_C for the
sample, $\alpha(T)$ continues smoothly along the curve begun above T_C and in all respects behaves as if
the sample were entirely normal.

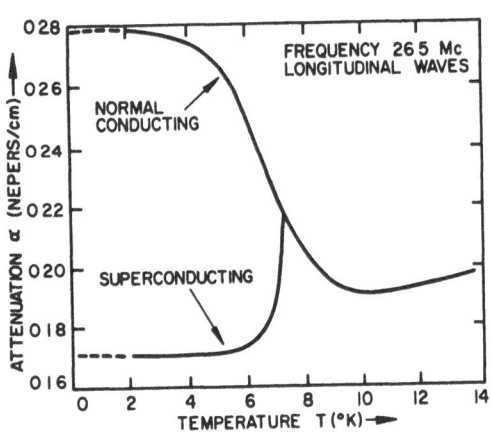

FIG. 1

Early measurements of the ultrasonic attenuation in lead by
Bömmel (1).

This behavior of the ultrasonic attenuation
was initially interpreted [Mason (3)] by imagining
that for T > T_C the "normal" conducting electrons
behave as a viscous medium, giving rise to viscous
damping of a sound wave propagating through the
crystal lattice. With the onset of superconductivity at T = T_C, "superconducting" electrons

were thought to appear suddenly being characterized by zero viscosity, and zero sound wave damping. With further decreases in temperature below T_c, the number of "superconducting" electrons were imagined to increase at the expense of the "normal" conducting electrons and the damping of the sound wave diminished in proportion to the volume fraction of "normal" electrons. At $T = 0$ K, all conduction electrons were thought to be "superconducting", so that the electronic contribution to the attenuation was zero. With the application of a magnetic field $H > H_c$, the "superconducting" electrons were thought to be destroyed and turned into "normal" conduction electrons, so that the electronic damping characteristic of the normal state was restored.

These interpretations led to models [for example, Mason (3)] which were partly successful in explaining many of the features observed, until Pippard (4) and Bardeen, Cooper, and Schrieffer BCS (5) produced their theories for the ultrasonic attenuation in normal and superconducting metals, respectively, which for the first time, explained most of the available data on a quantitative basis.

Theory of Ultrasonic Attenuation

A. Simple Physical Picture

During the passage of a plane longitudinal wave in a metal crystal, the lattice undergoes compressions and rarefactions alternately, so that an ion-density modulation is set up in the crystal. Accompanying this is a conduction electron density modulation which is imagined [after Pippard (4)] to be slightly out of step, so that a space charge is created leading to an electric field with the same space and time dependence as that of the sound wave. The difference in the ionic and electronic density modulations is very small compared to the individual modulations, a situation often referred to as "quasi-neutrality". Since the sound velocity is smaller by two orders of magnitude than the Fermi velocity for the conduction electrons, the sound wave, and therefore the accompanying electric field, may be visualized as "frozen-in" from the point of view of the conduction electrons. The electric field acting on the electrons is seen as the principal mechanism by which the sound wave imparts excess energy to the conduction electrons with a resultant perturbation of the electron velocity distribution. This distribution is imagined relaxing to the equilibrium distribution in a relaxation time τ, during which time, the conduction electrons give up their excess energy by random collisions with their environment. In this way, the damping of the sound wave is seen as a two-step process in which the energy first transmitted by the sound wave to the conduction electrons is ultimately lost as heat to the crystal lattice. The picture thus developed is useful in providing physical insight into the mechanism involved and the role played by some of the parameters. For example, in the regime in which the sound wavelength λ is large compared to the electronic mean free path ℓ, it is easy to see that an increase in ℓ lengthens the distance over which the electric field can accelerate the conduction electrons, and therefore increases the amount of energy transferred from the sound wave to the electrons, in agreement with the derived result that $\alpha \sim \ell$.

With the onset of superconductivity, an additional feature suddenly appears in the form of an attractive interaction between the conduction electrons which can be considered as arising in the following way: A negatively charged electron moving with the Fermi velocity of about 10^8 cm/sec in the periodic potential of the positively charged lattice ions exerts an attractive force on the nearby ions, causing a small displacement in their positions. Relaxation to their equilibrium positions is relatively slow for the ions, both due to their large mass and their slow speed (about 10^5 cm/sec). The combination of the electron with its positively charged "mantle" exerts an attractive force on any other electron in the vicinity within this time period, and hence gives rise to a retarded, attractive electron-electron interaction which is strongly dependent upon the dynamic properties of the lattice. For this reason it is commonly referred to as the electron-phonon interaction. The presence of a binding energy between pairing electrons places these electrons in a different category by separating them energetically from the unpaired conduction electrons. Thus, a small energy gap 2Δ is created between the pair "ground" state and the excited states of the conduction-electron energy spectrum. As the temperature is lowered below $T = T_c$, more and more electrons pair up with a resultant increase in 2Δ, until at $T = 0$, all the conduction electrons are paired, the gap is maximum, and according to BCS, $2\Delta(0) = 3.56 \, kT_c$ where k is Boltzmann's constant. The energy gap, except near T_c, is large, corresponding in frequency to about 10^{11} to 10^{12} cps which is much higher than that of the sound waves usually used in the laboratory. Thus the electric field associated with the sound wave is unable to break the binding between pair-electrons who therefore become energetically inaccessible to the energy transfer processes described above for the normal state. The attenuation diminishes with temperature in a way which reflects the temperature dependence of the energy gap. Raising the temperature above T_c or raising the magnetic field above H_c, breaks the pairing of the electrons, the energy gap vanishes, superconductivity is destroyed, and the normal ultrasonic attenuation is restored.

B. BCS Theory of Attenuation

The discussion below is an abbreviated derivation of the result originally obtained by BCS and follows one of several treatments discussed in the literature (6, 7, 8).

First a collision is considered between a normal conduction electron with energy E and wavevector k and a phonon with energy $\hbar\omega$ and wavevector q as shown below in the diagram.

The attenuation for a superconductor in the normal state is considered to behave in exactly the same way as in a normal metal and is

$$\alpha_n = \int C_{kk'} \left| M_{kk'} \right|^2 \overset{\text{absorption}}{\left\{ N(E)f(E) \, N(E') \, [1-f(E')] \right.}$$

$$\left. - N(E')f(E') \, N(E) \, [1-f(E)] \right\} \, dE \, ds \qquad (1)$$

where $N(E)$ and $N(E')$ give the density of states at energies E and E' respectively, and $f(E)$ and $[1-f(E')]$ are the probabilities of occupancy for the states with energy E and of vacancy of those of energy E' respectively. Both absorption and induced emission of phonons of frequency ω are taken into account. Both the matrix element $M_{kk'}$ between initial and final states and the constant factor $C_{kk'}$ involved in the transformation from a sum over k, k' into the above integral are assumed to be independent of E. Thus Eq. 1 can be separated in the form

$$\alpha_n = \int C_{kk'} \left| M_{kk'} \right|^2 ds \, x \int [f(E) - f(E+\hbar\omega)] \, N(E)N(E+\hbar\omega)dE \qquad (2)$$

where the integration of the first integral is over the Fermi surface. Taking the energy zero at the Fermi level, gives

$$\alpha_n = \hbar\omega\left[N(0)\right]^2 \int C_{kk'}\left|M_{kk'}\right|^2 ds \qquad (3)$$

where $N(0)$ is the total density of states.

For the superconducting state, BCS postulated that the energy per electron is

$$E = \left[\epsilon^2 + \Delta^2(T)\right]^{1/2}$$

where Δ is one-half the energy gap and ϵ is the Fermi energy. This statement leads directly to a new density of states.

$$N_s(E) = N(0)\frac{E}{(E^2-\Delta^2)^{1/2}}$$

and a new distribution function

$$f(E) = \left[1 + e^{E/kT}\right]^{-1}\quad .$$

Further, in the superconducting state, it is assumed that the factors $C_{kk'}$ and $M_{kk'}$ remain unchanged from those in the normal state. But the integral in Eq. 1 is changed because of the presence of coherence effects involved in the scattering of an electron from k to k', as is illustrated below.

In a normal metal, the transition from k, σ to k', σ' is independent from all other possible transition where σ designates the electron spin. BCS, however, postulated that the electrons most likely to form pairs were those with equal and opposite momenta and spins. The two processes illustrated in a) and b) are visualized as two events which can interfere with one another. In case (b) the pair k'↑, - k'↓ is occupied, as well as the single-particle state k↑; and scattering of a particle from - k'↑ to - k↓ gives a final state in which there is a single excited electron in k'↑ and a ground state pair k↑, - k↓. This scattering process is coherent with that in which the initial and final states only involve single-particle excitations k↑ and k'↑, respectively, as in a). Bardeen et al. (5) have shown that these processes can interfere constructively or destructively, so that the square of the matrix element $M_{kk'}$ must be modified by inclusion of a coherence factor $[1\pm(\Delta^2/EE')]$. For the case of electromagnetic absorption such as in the hyperfine interaction in NMR the interference is constructive and the plus sign is applicable. For ultrasonic absorption the processes interfere destructively and so for the superconducting state

$$\alpha_s = \int C_{kk'} \left| M_{kk'} \right|^2 ds \times \int [f(E) - f(E+\hbar\omega)] \cdot N(E) N(E') [1 - (\Delta^2/EE')] dE . \quad (4)$$

Combining Eqs. 3 and 4 gives

$$\frac{\alpha_s}{\alpha_n} = 2f(\Delta) = 2 \left[e^{\Delta(T)/kT} +1 \right]^{-1} \quad (5)$$

where $\Delta = \Delta(T)$ can be obtained from tables in the literature [for example, Mühlschlegel, (9)]. This simple expression relates the attenuation α_s/α_n and the energy gap Δ in a straight forward way and allows Δ to be easily calculated from ultrasonic data. Note that according to Eq. 5 at T_c, α_s/α_n drops with vertical tangent, because the gap is also changing with infinite slope in contrast to the drop for α_s/α_n as $(T/T_c)^4$, predicted by the simple two-fluid model described in the introduction.

The result in Eq. 5 was obtained under several simplifying assumptions but has since been shown to hold more generally or has been extended to include other conditions. For example, BCS derived Eq. 5 with the assumption that $q\ell > 1$. Since then, Tsuneto (10) has shown the result to hold for all $q\ell$. The assumption of $\Delta \ll \hbar\omega$ has been considered by Bobetic (11) who developed a numerical solution for arbitrary phonon energies and predicted discontinuities in the attenuation at temperatures for which $2\Delta = \hbar\omega$. This condition is likely to occur when magnetic impurities reduce the size of the energy gap. The theory has also been extended (12) to include effects due to a non-spherical Fermi surface. Anisotropy parameters for correcting the energy gap for anisotropy can now be calculated from data on the variation of transition temperature with residual resistivity. Holstein (13), Clairborne and Morse (14) and Leibovitz (15) have treated the case for transverse waves where deviations have been found in the regime $q\ell \geqslant 1$.

Aside from these and other extensions to the theory, the simple result of Eq. 5 has been found to hold remarkably well for many elements, compounds, and alloys. Figure 2 shows that even for a random solution alloy V-5at%Ta the ultrasonic data (16) gives an energy gap in good agreement with that predicted by BCS. Note that in this material the characteristic rise with decreasing T in the normal state attenuation is missing, because of the severe impurity limiting of the electronic mean free path.

Deviations from BCS

Soon after the advent of the BCS theory, pronounced deviations from predicted behavior were discovered in some systems. These have aroused considerable interest and are the subject of some of the excitement and controversy of the present state of the art of ultrasonic attenuation. The intent in this review is not to present a comprehensive review of all of these investigations, but rather to select two areas and treat these in greater depth.

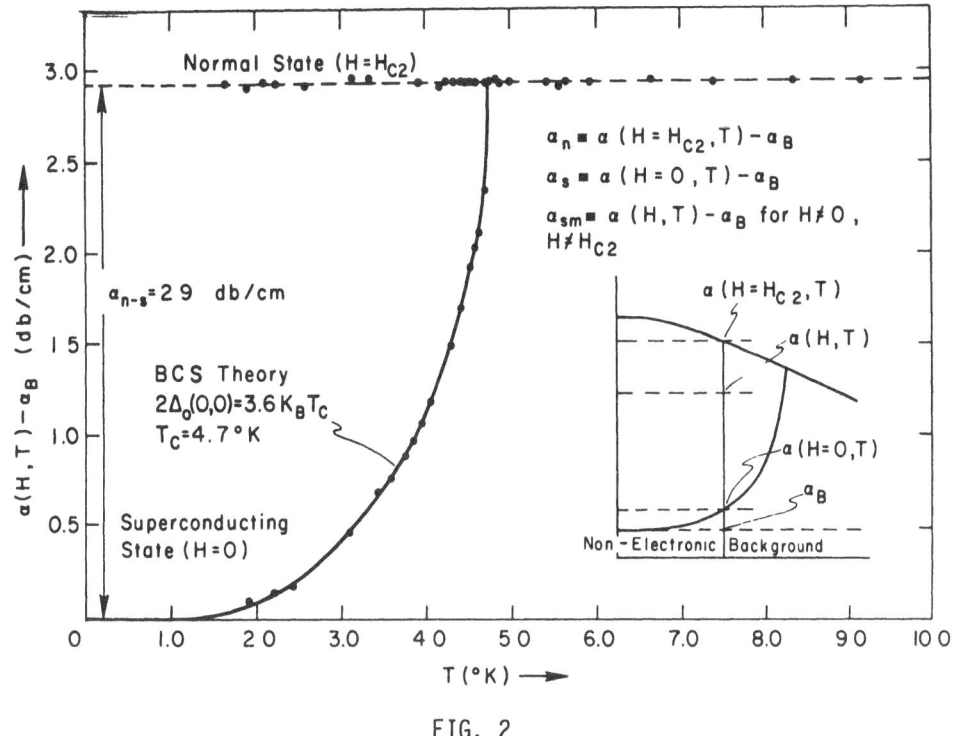

FIG. 2

Temperature dependence of the attenuation in the normal and superconducting state of a single crystal of the solid solution alloy
V-5.6at%Ta for longitudinal waves at 1.0 GHz (16)

A. Damping of Sound Waves by Dislocations

In experiments on pure Pb (17, 18, 19), In (18, 20) and Sn (18) at higher amplitudes
of the ultrasonic waves, investigators reported a significant amplitude dependent deviation from BCS behavior. This is illustrated for very pure Pb (18) at 50 MHz in Figs.
3 and 4. In Fig. 3, the normal-state attenuation curve has been corrected for the presence of the magneto-acoustic effect. This curve and the dashed curve (schematically)
for the superconducting state represent the typical behavior of most superconductors.
The experimental curve in the superconducting state shows the attenuation at an intermediate amplitude of the ultrasonic wave. When the power in the ultrasonic wave was
varied a corresponding amplitude dependence in the attenuation resulted. This amplitude
dependence was found to be weak above T_C but to increase suddenly with the onset of
superconductivity and to become very strong near 1K as seen in Fig. 4.

An explanation of these features was proposed by Tittmann and Bömmel (18) in terms
of a model centered upon an interaction between dislocations and conduction electrons.
According to this model, the conduction electrons interact with the stress field of a
moving dislocation in such a way as to cause a damping of its motion initiated by a
passing sound wave. With the onset of superconductivity and as the temperature is
lowered further, the number of "normal" electrons decrease in favor of the number of
Cooper pairs, which are assumed not to dampen the dislocation motion. Consequently,
the dislocations become increasingly free to move and can engage in a mechanism leading
to a strong amplitude dependence of the ultrasonic attenuation. Such a mechanism might,
for example, be the Granato-Lücke unpinning process in which high amplitude sound waves
are imagined to cause unpinning of dislocations from their impurity pinning points
thereby causing an additional absorption.

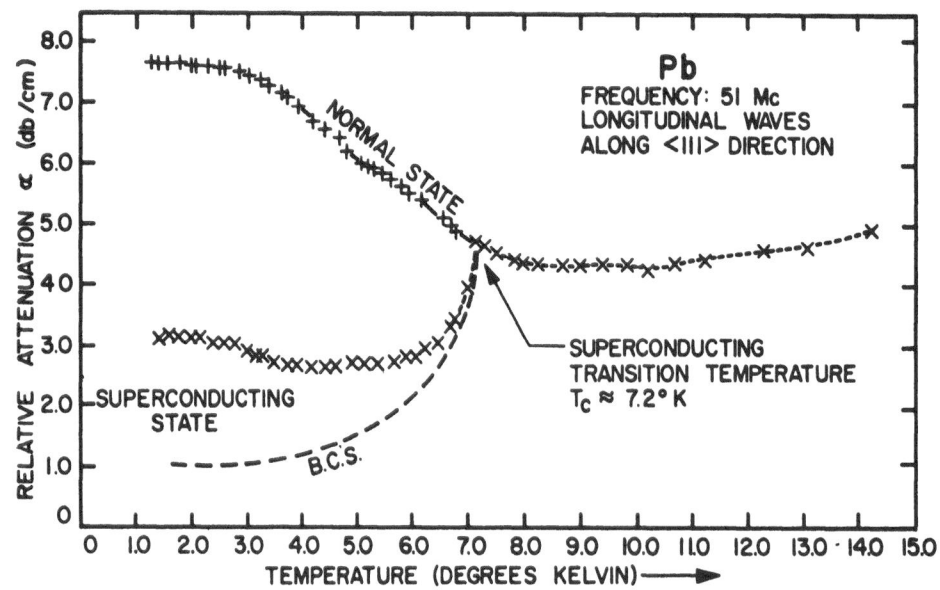

FIG. 3

Temperature dependence of ultrasonic attenuation
at medium amplitude in 99.999% pure single crystal
lead (18).

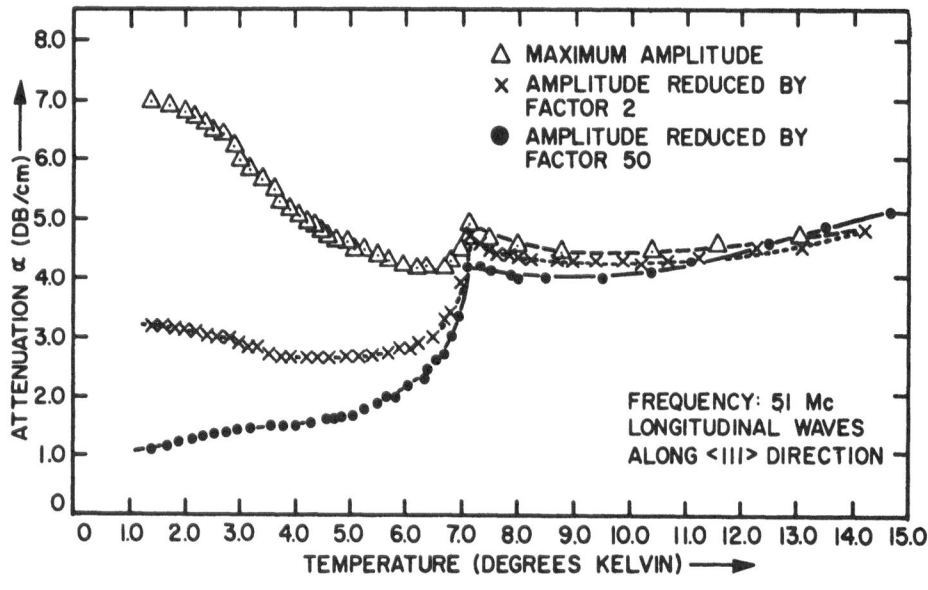

FIG. 4

Temperature dependence of attenuation in 99.999% pure
single crystal lead for three different transmitter
power levels (18).

B. Influence of Vortex Lines

The behavior of superconductors in an applied magnetic field may be characterized by the parameter $\kappa = \lambda/\sqrt{2}\,\xi$ where λ is the effective penetration depth of the magnetic field and ξ is the coherence distance between the two electrons in a Cooper pair. Superconductors with $\xi > \lambda$ or $\kappa < 1/\sqrt{2}$ are called type I superconductors in which flux penetration leads to the formation of macroscopic domains of alternating normal and superconducting regions. This condition is called the intermediate state and is strongly influenced by sample geometry and magnetic field direction. For $\xi < \lambda$ or $\kappa > 1\sqrt{2}$ the superconductor enters the mixed state characterized by vortex lines spanning the sample in an ordered array parallel to the applied magnetic field. This case is illustrated in Fig. 5 which shows schematically the configurations for the reduced magnetic field $h \equiv H/H_C$, and "superconducting electron" density n_s associated with a vortex line whose core is seen as nearly normal and has associated with it a quantity of magnetic flux equal to a single flux quantum $\phi_0 = 2.06 \times 10^{-7}$ Oe/cm^2.

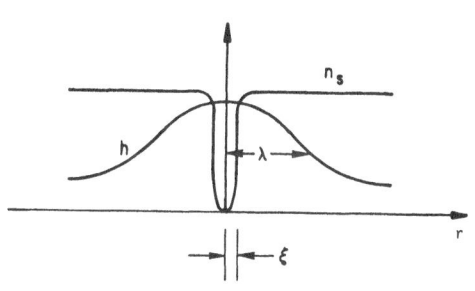

FIG. 5

Structure of one vortex line in a type II superconductor. The magnetic field h is maximum near the center of the line. Going outwards, h decreases because of the "screening" by annular currents in an "electromagnetic region" of radius $\sim \lambda$. The number of superconducting electrons per cm^3 n_s is reduced only in a small "core region" of radius ξ.

FIG. 6

Diagram representing equilibrium behavior in a magnetic field of superconductors with different values of κ. Solid lines represent various reduced critical fields.

A convenient way to represent the degree of type II behavior for a superconductor is in a κ-diagram such as shown in Fig. 6. Lines representing h_1, h_2, h_3, and $h = 1$ ($H = H_C$) have been drawn, separating the diagram into four regions in which the superconductor will have markedly different properties. As indicated, in one region it will be fully superconducting, in another the bulk will be normal but any part of its surface parallel to the field will have a superconducting sheath. Zero demagnetizing factor and bulk specimen dimensions (large compared to the magnetic field penetration depth) have been assumed. The behavior in the magnetic field of a

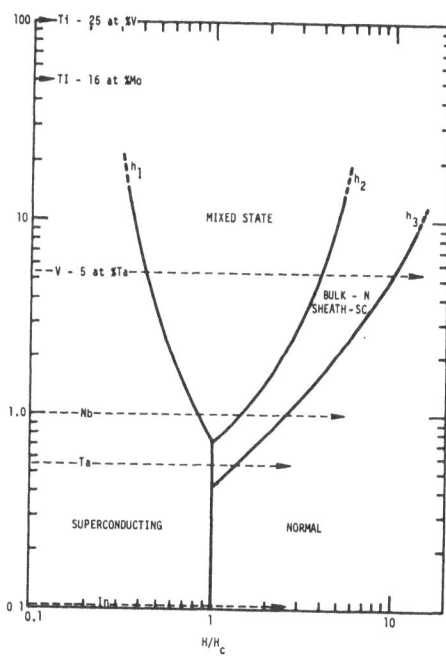

bulk superconductor of given κ can be traced by following a line parallel to the h-axis. The lines indicated are roughly representative of a few examples with different κ, of which Ti-16at%Mo and Ti-25at%V exhibit extreme type II behavior (27).

An example of a simple way in which vortex lines can influence the ultrasonic attenuation for low magnetic fields in the mixed state was given by Tittmann (16) in V-5at%Ta with κ ≈ 5. Figure 7 compares the ultrasonic data (16) for the field dependence of the attenuation for various temperatures by plotting reduced values of the field-dependent electronic attenuation as a function of the reduced field. A marked feature of this comparison is that as temperature is lowered from T_c, the slope of the curves near the field of first flux penetration changes.

At these low fields the vortex lines are few in number and only weakly interacting. In an effort to gain insight into the behavior of the attenuation, a model was constructed in which vortices are replaced by cylinders for which the energy gap Δ parameter is nearly zero. The matrix surrounding the cylinder acts with a BCS zero-field energy gap. If the normal core and matrix have nearly the same moduli, the total attenuation is given by

$$\alpha = (\alpha_s + \alpha_B)[1-V(H,T)] + (\alpha_n + \alpha_B)V(H,T)$$

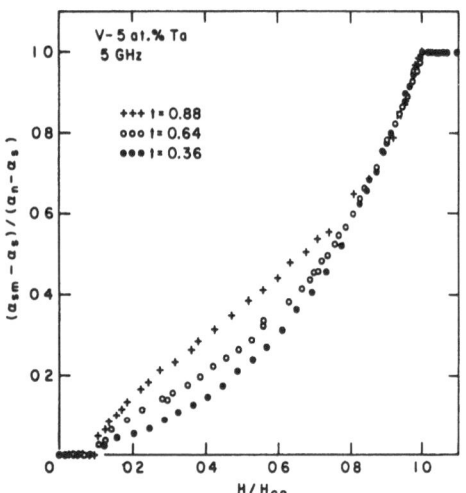

where α_B is the non-electronic background attenuation and $V(H,T)$ is the total volume of normal material assumed to be made up of N_V cores each of diameter D and unit length. Substituting the BCS result of Eq. 5, expressing V in terms of D, and letting $N_V = B/\Phi_0 \approx H/\Phi_0$ gives

$$\alpha \simeq (\alpha_s + \alpha_B) + \alpha_n \frac{\pi D^2}{4} \frac{H}{\Phi_0} \tanh \frac{\Delta}{2kT} \ .$$

The slopes are given by

$$\frac{\alpha - \alpha_s}{\alpha_n H} = \frac{\pi}{4} \frac{D^2}{\Phi_0} \tanh \frac{\Delta}{2kT} \ .$$

FIG. 7

Electronic attenuation as a function of magnetic field for several temperatures at the same frequency 0.54 GHz for V-5at%Ta (16).

Theoretical calculations (28) of an effective core diameter, gave $D \approx \pi \xi_G$ where $\xi_G \simeq 0.723\ \xi\ell/(1-T/T_c)$. The slope values calculated for $D = 2\xi_G$ are shown in Fig. 8 together with those obtained from the data of Fig. 7. Thus despite its simplicity, the model describes the observed magnetic field and temperature dependence and even predicts reasonably close magnitudes. In the model, the linear field dependence is a consequence of a linear increase in vortex density with increasing field. The rapid increase in the slope values with temperature reflects the swelling of the vortex core with temperature and consequently an increase in the volume of normal material at the expense of the volume of the superconducting matrix.

Experimental evidence presented by Tittmann and Bömmel was that 1) the amplitude dependence of the attenuation followed a semi-logarithmic dependence with inverse strain amplitude, commonly known as a Granato-Lücke plot, 2) deformation of the sample and subsequent annealing affected the amplitude dependence strongly, 3) the effect became weak in specimens of Pb grown with 0.1% Sn as impurity, 4) the effect is frequency independent over the range from 30 to 130 MHz in agreement with the Granato-Lücke theory (21) and 5) in very pure (six-nines) Pb at very high amplitudes time-dependent effects were observed similar to those thought to arise from temporary complete breakaway of dislocations from pinning points.

Theoretical evidence was presented independently by Holstein (22) and Kravchenko (23) who calculated an electronic damping factor for moving dislocations and predicted that it be independent of temperature in the normal state in agreement with later experiments by Platkov, Polunina, and Startsev (20). In these theories, the strain field of a dislocation is expressed as a Fourier sum of waves with wave vector q. The dominant q values are of a magnitude comparable with the reciprocal of a lattice spacing and hence the $q\ell$ parameter that governs the attenuation is always much greater than unity. Thus, simplified interpretations based on models involving electron viscosity concepts ($q\ell \ll 1$) are not applicable and more complex mathematical models must be used. The theoretical situation has been reviewed most recently by Brailsford (24).

The influence of the superconducting transition on dislocation mobility has recently received new attention in stress-strain experiments by for example, Alers, Buck and Tittmann (25) who interpretated changes in plasticity and creep rate during the transition into the superconducting state on the basis of a change in the number of mobile dislocations. Detailed interpretations of this phenomenon lead to considerable complexity, however, and the state of the art has most recently been reviewed by Granato (26).

On the basis of the accumulated evidence there is now little doubt that the amplitude dependence of the ultrasonic attenuation has as its source the damping of the moving dislocations by the conduction electrons which is severely reduced in the superconducting state. Furthermore, good agreement exists between experiment and the theories by Holstein and Kravchenko for the normal state. On the other hand, no detailed quantitative theory is as yet available for the superconducting state. In particular a simple application of the BCS formalism does not appear to explain the temperature dependence of the amplitude dependent attenuation in the superconducting state. As recently pointed out by Platkov, Polunina, and Startsev (20), the amplitude dependent attenuation measured in In at fixed amplitude in the superconducting state grows with decreasing temperature in the range from 3.4 to 1.8K and then drops again below 1.8K. This behavior already reported by Love and Shaw (17) has so far been only qualitatively and tentatively explained (20) as being due to a nonlinear velocity dependence of the dislocation electronic drag. Another point to be made is that so far no quantitative estimates of the damping factor B have been made from ultrasonic data in the superconducting state. Also no systematic studies have been made with controlled and known impurity concentrations. Such a study might give insight in the detailed nature of the pinning forces on dislocations. No detailed investigations have been reported for systems other than Pb and In. Thus considerably more work could be done to fully exploit this phenomenon leading to greater insight into the behavior of dislocations in a metal at low temperature.

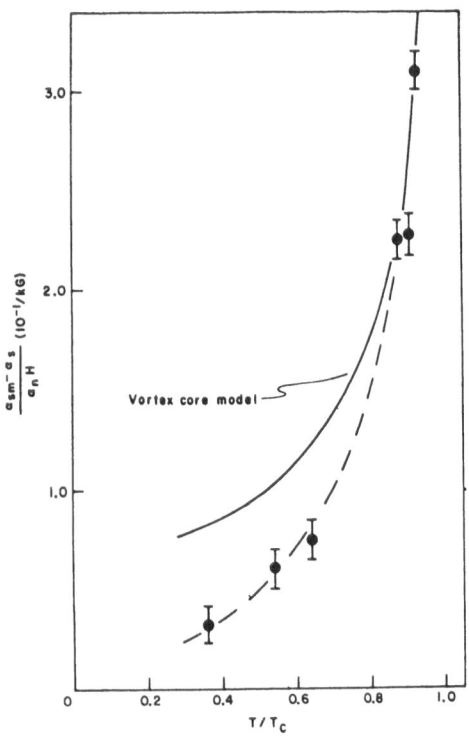

FIG. 8

Electronic attenuation per unit
magnetic field as a function of
temperature in the region near
the field of first flux penetration. The ultrasonic data are
compared with the results of a
vortex core model (16).

In the type II superconductors with low
κ and high purity (long electronic mean free path
compared with ξ) another interaction arises involving the vortices. As first reported by Forgan
and Gough (29) for Nb, a dip is observed in the
attenuation just above the field of first flux
penetration. As first suggested by Forgan and
Gough, and later supported by quantitative theoretical calculations by Cleary (30), the attenuation dip is caused by additional scattering from the
vortices of thermally excited unbound quasi-particles
(i.e., normal electrons in the superconducting
matrix of the mixed state). There are then two
competing mechanisms acting on the attenuation at
H_{c1}. First as vortices enter the specimen, the
attenuation increases because of absorption by the
normal electrons bound to these vortices as
described above for the case of V-5at%Ta. The
mechanism of scattering of the normal unbound electrons in the matrix by the vortices decreases their
mean free path which, as described earlier, decreases the attenuation. The effect of the vortex
scattering is estimated by a simple scattering
diameter a. The density of vortex lines threading
the specimen is $N = B/\Phi_0$ where B is the induction
in the specimen. Then the mean free path of the
electron in the presence of an induction B is

$$1/\ell(B) = 1/\ell(0) + (B/\Phi_0)a \quad .$$

In a pure superconductor, in which the attenuation
is mean-free-path limited, the BCS expression for
attenuation will then be modified as follows to
take account of vortex scattering.

$$\frac{\alpha_s}{\alpha_n} = \frac{2}{\exp(\Delta/kT) + 1} \quad \frac{\ell(B)}{\ell(0)} = \frac{2}{\exp(\Delta/kT) + 1} \quad \frac{1}{1 + [Ba\ell(0)/\Phi_0]}$$

Sinclair and Leibowitz (31) observed this effect also in pure V and found good
quantitative agreement between theory and experiment based on the above scattering
mechanism.

For the magnetic fields close to the upper critical field H_{c2} the vortex lattice
is so dense that it must be viewed as a collective ensemble of strongly interacting
elements. This regime has received considerable attention from both theoretical and
experimental viewpoints. One of the more successful attempts to explain this regime
has been the work by Carsey and Levy (32) who compared their data on pure Nb (resistivity ratio equal about 7000) to the theory of Houghton and Maki (33). Figure 9
shows the ultrasonic data for longitudinal waves parallel to the magnetic field. In
Figs. 9(b) and 9(d), data very near H_{c2} are shown in plots of the reduced attenuation
$(\alpha_n-\alpha_s)/\alpha_n$ versus $H_{c2} - H$ in units of gauss. In each case the straight lines are linear
fits to the data showing that there is a region very near H_{c2}, called the gapless region,
in which the attenuation depends linearly on H. Figures 9(a) and 9(b) show data for
$[(\alpha_n-\alpha_s)/\alpha_n]^2$ versus $H_{c2} - H$ showing that there is another region, called the BCS region,

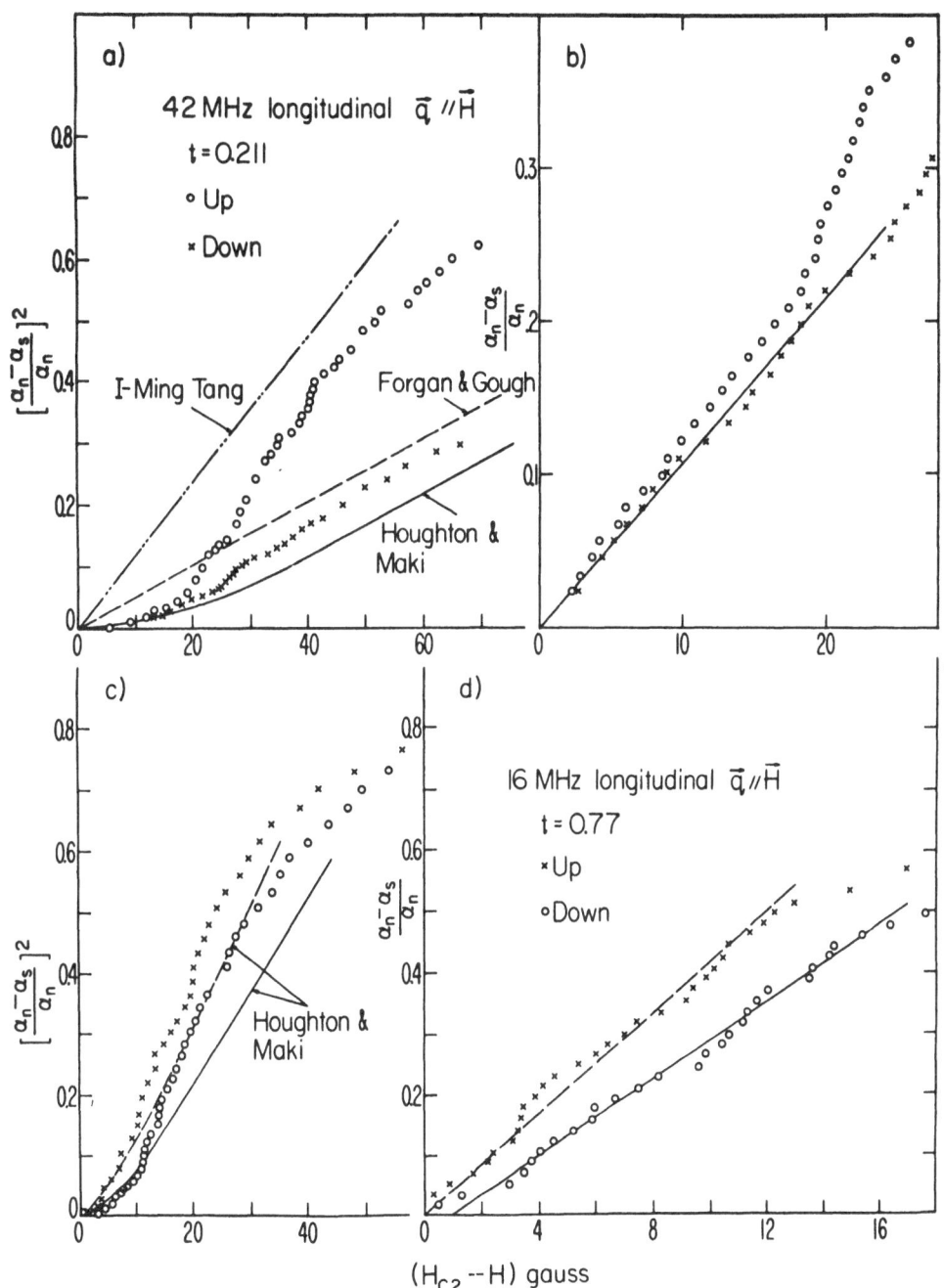

FIG. 9

Data obtained by Carsey and Levy (32) in Nb for longitudinal
waves compared with the predictions of the theories by Tang
(34), Forgan and Gough (35) and Houghton and Maki (33).

which is further away from H_{c2} and where the attenuation depends parabolically on the magnetic field. Figure 9(a) also shows the attenuation as predicted by Tang (34), the approximate fit to the data by Forgan and Gough (35) and the predicted fit to the data from Houghton and Maki (33). The experimentally determined slopes from the data at various temperatures are plotted in Fig. 10 as a function of temperature for the BCS region in 10(a) and the gapless region in 10(b). The solid lines are based on the theory with the constants adjusted to fit the data for low temperature. The results certainly go far in supporting the theory and leave little doubt regarding the predicted presence of two separate regimes in the magnet field dependence of the attenuation for the pure type II superconductor with $\xi/\ell \ll 1$.

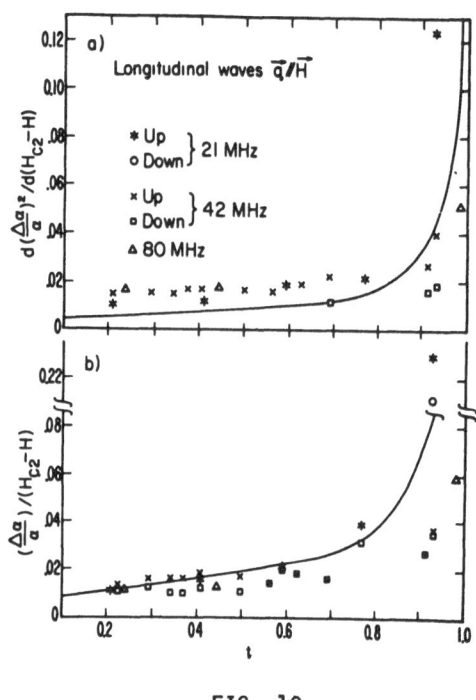

FIG. 10

A summary of slopes for the attenuation versus field data obtained by Carsey and Levy (32). The lines are drawn from the theory of Houghton and Maki (33).

At the present time there is no general theoretical treatment that treats superconductors with arbitrary ξ/ℓ nor even those with $\ell \approx \xi$. On the other hand, experimental data have been obtained on superconductors with $\xi \gtrsim \ell$ for example on V-5at%Ta with $\xi/\ell = 5$. As shown in Fig. 11 [Tittmann (16)] the data also exhibit well defined separate regions which appear gapless and BCS-like in behavior. Since this alloy is almost completely magnetically reversible, quantitative comparisons between experiment and theory could be carried out for the gapless regime. In this regime an expression derived by Maki (36) for the limit $\xi/\ell \gg 1$ was assumed to hold which is given as

$$1 - \frac{\alpha_s}{\alpha_n} = \frac{e\rho_n c}{8\pi^2 kT_c} \frac{H_{c2} - H}{(2\kappa_2^2 - 1)\beta} C_2(t)$$

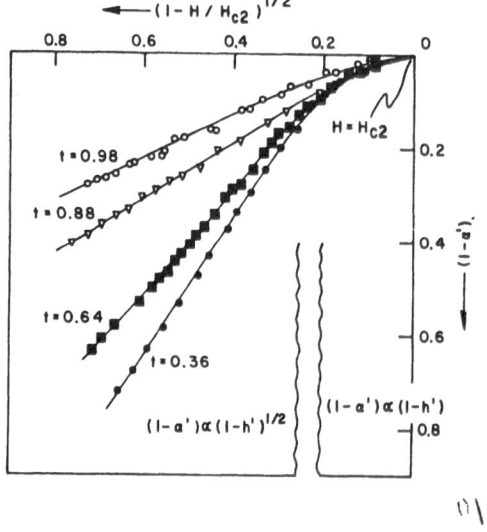

FIG. 11

Electronic attenuation $(1-\alpha')$ for V-5at%Ta as a function of field $(1-H/H_{c2})^2$ where $\alpha'=\alpha_s/\alpha_n$. Note that the data falls into two well-defined regions one "gapless" the other BCS-like in behavior (16).

where $C_2(t)$ is a universal function (36) of the reduced temperature $t \equiv T/T_c$, e is the electronic charge, k is Boltzmann's constant, c is the speed of light, $\beta \approx 1.16$, ρ_n is the normal state resistivity and κ_2 is the temperature dependent Abrikosov-Maki parameter. In order to test the validity of Maki's expression, κ_2 was calculated from the slopes of the ultrasonic data near H_{c2} in Fig. 11 and from complementary resistivity measurements. These values for κ_2 were compared with values obtained in previous calorimetric and magnetization studies and found to be in remarkably good agreement. Thus in spite of certain shortcomings much of the magnetic field dependent data near H_{c2} can be explained satisfactorily by the present theories.

More generally the results discussed here and other studies reviewed elsewhere (37, 38) show that although much work has been done, much more experimental and theoretical is necessary to improve the exact and detailed understanding of the attenuation in the superconducting state.

Potentials for the Future

The few examples discussed in this review serve as a reminder that the ultrasonic technique is a useful and sensitive tool for the investigation of the superconducting state, especially in view of the well-known advantage of the technique - that it can probe the bulk interior of the specimen. In spite of this advantage of bulk waves there are many experiments in which interesting effects could be observed and correlated with non-acoustic measurements if acoustic surface waves could be launched into the superconductor. These take on many forms: such as the Rayleigh wave traveling at a solid-air interface, the Love wave traveling in a surface layer, the Stonely wave traveling at the interface between two media, to name but a few. With the development of an extensive and sophisticated technology of surface wave excitation and detection the opportunity becomes increasingly practical to study the velocity and attenuation of surface waves in superconductors. One characteristic of surface waves which could be very useful in the investigations is that the penetration depth of the wave is directly proportional to the wavelength and therefore can be controlled easily by an adjustment of the frequency. Consequently, physical property gradients such as for example the superconducting sheath phase or purposely constructed metallurgical gradients involving perhaps magnetic impurities could be readily investigated. In another category are the superconducting thin films such as the amorphous superconductors which can only be grown as films and which show anomalous T_c values. With these, acoustic measurements are only feasible with the aid of surface waves. In short, a thrust into the field of surface waves could start an exciting set of investigations complementing the ever growing log of bulk wave studies.

Acknowledgments

It gives me much pleasure to thank Professor K. Lücke for inviting me to present this paper at the Technische Hochschule Aachen, West Germany. It also is a pleasure to thank Professor R. Dobbs for helpful discussions, and Dr. G. A. Alers for his critical review of the manuscript.

1. H. E. Bömmel, Phys. Rev. 96, 220 (1954).

2. L. Mackinnon, Phys. Rev. 98, 1181 (1955).

3. W. P. Mason, J. Acoust. Soc. Am. 27, 643 (1955).

4. A. B. Pippard, Phil. Mag. 46, 1104 (1955); 2, 1147 (1957).

5. J. Bardeen, L. N. Cooper, and J. R. Schrieffer, Phys. Rev. 108, 1175 (1957).

6. M. Tinkham, "Low Temp. Physics" Lectures delivered in 1961 at Les Houches at the University of Grenoble. Gordon and Breach, New York (1962).

7. J. A. Rayne and C. K. Jones, "Phys. Acoustics" (W. P. Mason and R. N. Thurston, ed.), Vol. VII. Academic Press, New York (1970).

8. G. Richayzen in "Superconductivity" (R. D. Parks, ed.) Vol. 1. Marcel Dekker, Inc., New York (1969).

9. B. Mühlschlegel, Z. Physik 155, 313 (1959).

10. T. Tsuneto, Phys. Rev. 121, 402 (1961).

11. V. M. Bobetic, Phys. Rev. 136, A1535 (1964).

12. D. Markovitz and L. P. Kadanoff, Phys. Rev. 131, 563 (1963).

13. T. Holstein, Res. Memo 60-94698-3M17. Westinghouse Res. Lab., Pittsburgh, Pennsylvania, unpublished (1956).

14. L. T. Clairborne and R. W. Morse, Phys. Rev. 136, A893 (1964).

15. J. R. Leibowitz, Phys. Rev. 136, A22 (1964).

16. B. R. Tittmann, Phys. Rev. B, 2, 625 (1970).

17. R. E. Love and R. W. Shaw, Rev. Mod. Phys. 34, 260 (1964).

18. B. R. Tittmann and H. E. Bömmel, Phys. Rev. Letters 14, 296 (1965); Phys. Rev. 151, 178 (1966).

19. P. A. Bezirglyi, V. D. Fil, and O. A. Shevenko, Zh. Eksp. Teor. Fiz. 49, 1715 (1965). [Sov. Phys. - JETP 22, 1172 (1966)].

20. V. Ya. Platkov, L. N. Polunina, and V. I. Startsev, Journal of Low Temp. Physics 10, 359 (1973).

21. A. Granato and K. Lücke, J. Appl. Phys. 27, 583 (1956); 27, 789 (1956).

22. T. Holstein, appendix of paper by B. R. Tittmann and H. E. Bömmel, Phys. Rev. 151, 178 (1966).

23. V. Ya. Kravchenko, Fiz. Tverd. Tela $\underline{8}$, 927 (1966)[Sov. Phys. - Solid State $\underline{8}$, 740 (1966)].

24. A. D. Brailsford, Proc. of Fifth Intern. Conf. on Int. Frict. and Ultras. Atten. in Cryst. Solids, RWTH, Aachen, Germany (1973), in press.

25. G. A. Alers, O. Buck and B. R. Tittmann, Phys. Rev. Letters $\underline{23}$, 290 (1969).

26. A Granato, Proc. of Fifth Intern. Conf. on Int. Frict. and Ultras. Atten. in Cryst. Solids, RWTH, Aachen, Germany (1973), in press.

27. R. R. Hake, Phys. Rev. $\underline{158}$, 356 (1967).

28. A. G. Van Vijfeijken, Philips Res. Rept. Suppl. $\underline{8}$ (1968).

29. E. M. Forgan and C. E. Gough, Phys. Letters $\underline{21}$, 133 (1966); $\underline{26A}$, 602 (1968).

30. R. M. Cleary, Phys. Rev. $\underline{175}$, 587 (1968).

31. A.C.E. Sinclair and J. R. Leibowitz, Phys. Rev. $\underline{175}$, 596 (1968).

32. F. Carsey and M. Levy, Phys. Rev. Letters $\underline{27}$, 853 (1971).

33. A. Houghton and K. Maki, Phys. Rev. $\underline{B4}$, 843 (1971).

34. I-M. Tang, Phys. Rev. $\underline{B2}$, 2581 (1970).

35. E. M. Forgan and C. E. Gough, Phys. Letters $\underline{26A}$, 602 (1968).

36. K. Maki, in "Superconductivity" (R. D. Parks, ed.) Vol. 2, Marcel Dekker, Inc., New York (1969).

37. M. Gottlieb, M. Garbuny, and C. K. Jones, "Phys. Acoustics" (W. P. Mason and R. N. Thurston. ed.) Vol. VII., Academic Press, New York (1970).

38. B. R. Tittmann, unpublished.

ULTRASONIC STUDIES OF ANTIFERROMAGNETIC RESONANCE

C. Elbaum
Department of Physics and Metals Research Laboratory
Brown University, Providence, Rhode Island 02912, U.S.A.

ABSTRACT

A brief review is given of the coupling mechanisms between elastic waves and spin waves, which allow one to study phonon-magnon interactions in general and antiferromagnetic resonance in particular. Selected experimental results are presented and discussed, with emphasis on the information they provide concerning the temperature and magnetic field dependence of various parameters that characterize antiferromagnets. Attention is given to the studies of these parameters near the Néel transition temperature, where the use of ultrasonic methods is particularly advantageous.

I. Introduction

Ultrasonic studies of antiferromagnetic resonance (AFMR), also referred to as antiferro-acoustic resonance - "AFAR" - have been developed quite recently (1-4), although theoretical considerations of resonant interaction between ultrasonic waves and spin waves in antiferromagnets go back some years (5). The main advantages of the method include the possibility of AFMR measurements at relatively low frequencies, which are not readily accessible with more conventional (electromagnetic) techniques, and the possibility of such measurements in bulk metals. The limitation in the use of electromagnetic methods at low frequencies, in general, is essentially one of insufficient coupling to antiferromagnetic spin waves, (6) which makes it difficult to detect resonant absorption. In the case of bulk metals, the difficulty with the use of electromagnetic waves for AFMR studies is the same as, for example, in nuclear magnetic resonance; namely, that their penetration is confined to the skin depth. The ultrasonic method, on the other hand, does not involve these limitations and constitutes, therefore, a very useful technique for AFMR studies, whenever samples of suitable dimensions and geometry are available.

In what follows a brief description is given in section II of the coupling or interaction mechanisms between ultrasonic waves (phonons) and antiferromagnetic spin waves (antiferromagnetic magnons). The resulting ultrasonic attenuation peaks, identified with resonant absorption, as well as their temperature and magnetic field dependence, are discussed in section III. Section IV gives a brief summary.

II. Phonon-Magnon Interactions

It is convenient to recall, first, that spin waves are oscillations in the relative orientations of spins on a lattice, while lattice vibrations, or ultrasonic waves are oscillations in the relative positions of atoms or ions on a lattice. Since the spins are associated either with all or with a specified sequence of ions, displacements of the latter, associated with an ultrasonic wave, will modulate the couplings among the spins. This modulation gives rise to phonon-magnon interactions. More formal statements of the problem are usually presented in terms of an appropriate Hamiltonian of the system. In general, however, the effects of lattice displacements on the various, possible parameters describing the coupling of spins are not well-known. It is customary, therefore, to express the phonon-magnon interactions in the Hamiltonian by phenomenological coefficients, usually called magnetoelastic coupling constants, the values of which can be determined from experiments.

The ultrasonic attenuation which results from the magnetoelastic coupling is characterized by different coupling constants, which depend on the elastic mode used and on the type of interaction involved. General symmetry arguments preclude coupling between spin and strain which is linear in spin variables. Among the allowed interactions, volume magnetostriction and single-ion magnetostriction are usually considered to be important.

The volume magnetostriction arises primarily from the strain modulation of the exchange interaction and affects only longitudinal elastic modes (there is no coupling, to first order, with transverse modes). The single-ion magnetostriction arises from spin-orbit and orbit-lattice interactions; in this case there is coupling of comparable magnitude to both longitudinal and transverse waves. It is on this basis that the experimentally observed resonance interaction (see section III) is attributed to single-ion magnetostriction. A brief outline of a formal approach to calculating the ultrasonic attenuation is, therefore, given below for this type of interaction (4).

To first order in strain the Hamiltonian has the form

$$H_{SI} = \sum_{\ell} \sum_{\alpha\beta\gamma\delta} G_{\alpha\beta\gamma\delta} \, S_\alpha^\ell \, S_\beta^\ell \, \varepsilon_{\gamma\delta} \tag{1}$$

Here G is the magnetoelastic coupling tensor, S are spin components, ε is the strain, and the summation index ℓ extends over the whole lattice. The interaction Hamiltonian depends only on single ion magnetostriction for ions (i and j) on both sublattices,

$$H_{Int} = H_{SI}^i + H_{SI}^j$$

In the particular case of cubic symmetry two identical terms are obtained for each sublattice, of the form:

$$H_{Int}^i = \sum_{i=1}^{N} \{\frac{1}{2} G_{11}[(3S_x^{i2} - S^2)\varepsilon_{xx} + (3S_y^{i2} - S^2)\varepsilon_{yy} +$$

$$+ (3S_z^{i2} - S^2)\epsilon_{zz}] + 2G_{44}[(S_x^i S_y^i + S_y^i S_x^i)\epsilon_{xy} +$$

$$+ (S_y^i S_z^i + S_z^i S_y^i)\epsilon_{yz} + (S_x^i S_2^i + S_z^i S_x^i)\epsilon_{xz}]\} \tag{2}$$

The ultrasonic attenuation coefficient, α, can then be calculated by use of the Golden rule,

$$\alpha = \frac{1}{v}\frac{2\pi}{\hbar^2}|< i|H_{Int}^{\cdot}|f >|^2 \delta(\omega) \tag{3}$$

where i and f denote initial and final states, v is the sound velocity, and $\delta(\omega)$ is a line width, usually defined in terms of spin wave lifetime. Equ. 10 can be used to evaluate the peak attenuation, α_M, when the magnetoelastic coupling constants and lifetimes are known. Conversely, it can be used to deduce the values of the coupling constants from measured values of α_M and of lifetimes estimated from resonance line widths.

Ultrasonic propagation changes associated with the magnetoelastic coupling have been studied extensively in a variety of contexts. Notable among these have been investigations near magnetic phase transitions, for example by Melcher and Bolef (7) and by Lüthi and his collaborators (8). These studies were concerned with the temperature and sometimes frequency dependence of ultrasonic attenuation and velocity in the "critical" region, i.e., with their behavior on approaching a phase transition. In contrast, the studies described here are directed at the temperature and magnetic field dependence of frequencies of the spin waves (magnons), as studied through their resonant interaction with ultrasonic waves. When these investigations are conducted near phase transitions, the main emphasis is on the critical behavior of the resonant frequency, and indirectly of the magnetization, rather than the ultrasonic attenuation itself.

III. AFMR Studies

The range of AFMR frequencies that can be studied by ultrasonic means is determined by the usual limitations in generating high frequency ultrasonic waves. The highest frequency that has been used so far is approximately 10 GHz (2)(4) and a number of experiments have been carried out in the range 30 to 700 MHz. Thus, even for the highest frequencies used, the wave vector $|k| \sim 10^5$ cm^{-1}. This is still a small value compared to the reciprocal lattice spacing. It follows that the wave vector dependence of the phonon-magnon interactions may usually be neglected, i.e., the problem is treated in the $\vec{k}=0$ approximation. This is equivalent to considering only the "uniform" spin wave mode.

The resonant interaction may be measured by any convenient ultrasonic method. Care must be taken, of course, to select appropriate elastic modes for which the phonon-magnon coupling is not expected to be zero. In this regard it is necessary to determine, either theoretically or experimentally, the selection rules which govern the phonon-magnon interaction in the solid under study.

Examples of ultrasonic attenuation peaks observed in single crystals of $RbMnF_3$ (1) just below the Néel temperature (T_N), in the absence of external magnetic fields, are shown in figures 1 and 2 (1). It may be noted that in the case of longitudinal waves (fig. 1) the resonance peaks are superimposed on large, "critical" attenuation, which makes it difficult to locate the peaks with accuracy. In the case of transverse waves, on the other hand, (fig. 2) the "critical" attenuation does not appear, and the resonance peaks are well defined.*

Attenuation peaks of a similar nature, but in the presence of a magnetic field, were observed in $GdAlO_3$ (2).

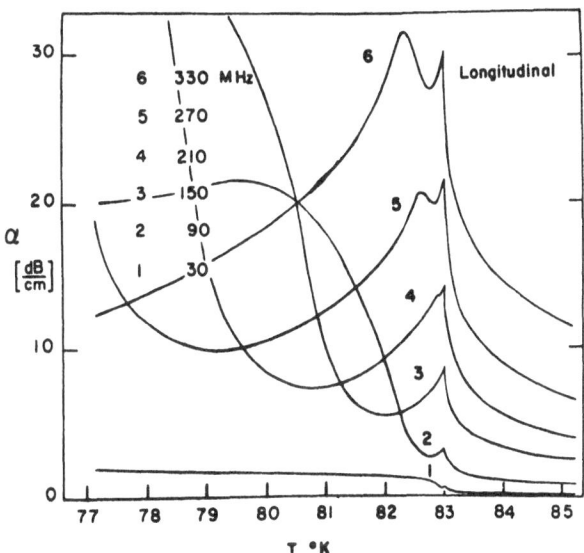

FIG. 1

Attenuation as a function of temperature, for longitudinal waves propagating along the [001] direction in $RbMnF_3$. Peaks below the Néel temperature T_N $(T_N \sim 83°K)$ are seen on curves 4, 5, 6. The "critical" attenuation on both sides of T_N is also seen. Frequencies are indicated on the figure (zero magnetic field).

These experimentally observed attenuation peaks have been attributed to antiferromagnetic resonance on the basis of a detailed analysis of their dependence on a number of variables, most notably on magnetic field. At the same time, when the dependence of the peak frequency and width on temperature, and of peak height on elastic mode is included in the analysis, various features of the AFMR, of the spontaneous magnetization, and the values of the magnetoelastic coupling constants can be deduced. Outlines of the main features of the observations and of the corresponding analyses follow.

In the case of transverse waves in $RbMnF_3$, when the peak frequency, ω_M, is plotted as a function of $T_N - T_M$, where T_M is the temperature at which the attenuation maximum occurs for a

---

*These observations allow one to attribute the critical and resonance absorptions, respectively, to volume magnetostriction and to single-ion magnetostriction; indeed, the critical attenuation appears only with longitudinal waves, whereas the resonance peaks appear both with longitudinal and with transverse waves (see section II).

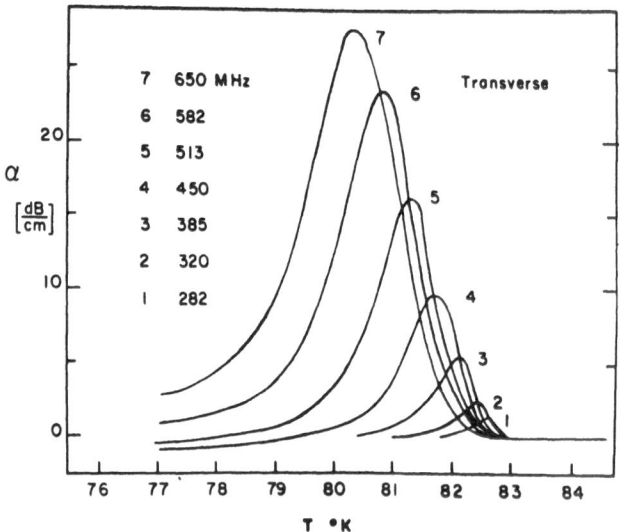

FIG. 2
Peaks in the attenuation of transverse waves in RbMnF$_3$ propagating along the [001] direction, polarized along the [100] direction. The peaks are shown from the background attenuation level measured above the Néel temperature T$_N$ (T$_N$ $\sim$ 83°K) for each frequency. Frequencies are indicated on the figure (zero external magnetic field).

given frequency, the following relation is found;

$$\omega_M \propto (T_N - T_M)^n \tag{4}$$

where n = 0.47 ± 0.05, and for T$_N$ - T$_M$ < 4°K. The maximum attenuation, α_M, for each peak, and the frequency ω_M are found to be related to each other and therefore to temperature by:

$$\alpha_M \propto \omega_M^{3.8} \propto (T_N - T_M)^{1.8} \tag{5}$$

The height of the attenuation peaks is found to depend on the elastic mode used. In particular, peaks associated with transverse waves propagating along the [110] direction and polarized along the [1$\bar{1}$0] direction are typically much higher than for [001] polarization and the same propagation direction.

The observed temperature dependence of ω_M (equ.4), is now compared with the temperature dependence of the AMFR frequency, ω_R, which can be deduced from the relation (9):

$$\omega_R \propto (\frac{K}{\chi})^{\frac{1}{2}} \tag{6}$$

where K is the anisotropy constant and χ is the perpendicular susceptibility. Near T$_N$ the perpendicular susceptibility is essentially temperature independent, therefore the dependence of K$^{1/2}$ on temperature must be examined; this is done as follows. The magnetoelastic coupling appears to originate from single ion anisotropy (see footnote, p.4), the theory of which (10) predicts the relation between K and the sublattice magnetization, M. Near T_N this relation is $K \propto M^4$, and

$$M \propto (T_N - T)^\beta. \tag{7}$$

Thus the exponent n of equation (4) would be

$$n = 2\beta \tag{8}$$

The value of β obtained from neutron scattering experiments (11) is 0.32, whereas scaling relations predict $\beta = 0.285$. Correspondingly n would be 0.64 or 0.57, which is to be compared with the experimental value of n = 0.47 ± 0.05. The origin of this discrepancy cannot be readily resolved. Since, as mentioned previously and discussed in more detail below, the identification of the peaks as being due to AFMR has been made separately on the basis of their magnetic field dependence, this result, i.e., $n \tilde{\sim} (2/3)\beta$, strongly suggests that $K \propto M^3$ near T_N.

The observed temperature dependence of α_M can be compared with theoretical estimates (5) of the absorption parameter for transverse waves at resonance expressed in the form:

$$\Gamma = \frac{Cb_2^2}{\lambda} \tag{9}$$

Here b_2 is the magnetoelastic coupling constant for the mode under consideration, λ is the relaxation rate of the AFMR modes and C is a substantially temperature independent parameter with appropriate dimensions. When the temperature dependence of b_2 and λ near T_N is taken into account (1), the relation for α_M becomes

$$\alpha_M \propto \Gamma \propto (T_N - T)^{1.92} \tag{10}$$

which is in fair agreement with equ. 5.

The relative heights of the attenuation peaks observed for different modes should scale with the corresponding magnetoelastic coupling constants. The coupling constants (designated by b_1 and b_2, respectively) for the transverse ultrasonic modes discussed above, namely, propagation along the [110] direction, polarization along the [1$\bar{1}$0] and [001] directions, have been studied previously in different contexts (5) (12). The results of these studies show that $b_1 \ll b_2$, which is in qualitative agreement with the observed peak heights.

The application of an external magnetic field generally splits the AFMR modes. In the simplest approximation the two AFMR frequencies for $\vec{k}=0$ are given by:

$$\omega_R = \gamma[(H_A^2 + 2H_A H_E)^{\frac{1}{2}} \pm H_o] \tag{11}$$

Where $\gamma = \frac{g\mu}{\hbar}$ is the magnetomechanical ratio, H_o is the external field, H_E and H_A are respectively the exchange and anisotropy fields. γ contains only fundamental constants (i.e., g, the spectroscopic splitting factor, μ, the Bohr magneton and $\hbar$, Planck's constant divided by 2π). It follows that the amount of splitting which results from the application of a known field (at least within the approximation of equ. 11) constitutes a direct method of checking whether the observed attenuation peaks are due to AFMR. Such checks have been carried out and they confirmed that the peaks were indeed associated with AFMR.

In addition to the checks mentioned, investigations of AFMR dependence on magnetic field have been performed for the purpose of comparing the experimental results with the behavior predicted by various models. Ince (6), for example, investigated both theoretically and experimentally AFMR in $RbMnF_3$ in the presence of a magnetic field. His calculation of the magnetic modes were based on the molecular field theory of an antiferromagnet. His experimental results of conventional AFMR studies, well below the Néel temperature, are accounted for by a four sublattice (two electronic and two nuclear) model. The interaction in the presence of a magnetic field between acoustic phonons and magnetic modes in a cubic antiferromagnet has been treated theoretically by Fedders (13). His work refers particularly to $RbMnF_3$. An experimental study on the magnetic field and temperature dependences of AFMR near the Néel temperature of $RbMnF_3$, using the ultrasonic method, was carried out by Jimbo and Elbaum (14). In order to compare their results with theoretical predictions, these authors followed an approach similar to that of Ince (6). Instead of using the results of the molecular field theory, however, which does not account well for the behavior of the sublattice magnetization near the Néel temperature, they adopted in their calculations of the magnetic modes the following relation:

$$B = \frac{M}{M_o} = A \left(1 - \frac{T}{T_N}\right)^{\frac{1}{3}} \tag{12}$$

where B is the ratio of sublattice magnetizations at $T°K(M)$ and at $0°K(M_o)$, and A is a coefficient deduced from experimental results. The results of the ultrasonically measured AFMR were found to be consistent with the values calculated on the basis of equ. 12. This provides further support for the form of the temperature dependence of the magnetization near T_N represented by equ. 12.

Ultrasonic investigations of AFMR are also useful in the study of the "magnetic phase diagram" of antiferromagnets, i.e., of the temperature-magnetic field relation which defines the phase boundary between the antiferromagnetic and paramagnetic states. The separation between the two phases occurs when the field applied to the antiferromagnet becomes approximately equal to twice the exchange field, and a second order phase transition to the paramagnetic state occurs; that is, the average orientation of the spins becomes parallel to the external field. At the same time one of the spin-wave mode frequencies goes to zero (15). An experimental study of this "exchange transition" has been carried out on the basis of the resonant interaction between ultrasonic waves and spin waves (2).

III. Summary

Resonant interactions between ultrasonic waves and spin waves in antiferromagnets, which were considered theoretically some years ago (5), have been observed experimentally (1-4). This method is very useful for studying antiferromagnetic resonance, especially at low frequencies. As is well known, the AFMR frequency has its maximum value at $0°K$, decreases with increasing temperature as a fractional power of T, and reaches zero at the Néel transition temperature, T_N. The capability of investigating AFMR by ultrasonic means at low frequencies (below 1 GHz) is especially important, therefore, in studies of the behavior of antiferromagnets very near T_N, where conventional (electromagnetic) techniques are usually inadequate.

Several studies of ultrasonic AFMR have been reported to-date, in particular on $RbMnF_3$ (1)(3)(14) and on $GdAlO_3$ (2)(4). These studies, in addition to demonstrating the existence of the resonant interaction, have provided new information on the temperature and magnetic field dependence of AFMR, near the Néel temperature. This information has been used to deduce the temperature dependence of the sublattice magnetization in $RbMnF_3$ near T_N.

References

1. T. Jimbo and C. Elbaum, Phys. Rev. Letters, 28, 1393 (1972).

2. P. Doussineau, B. Ferry, J. Joffrin and A. Levelut, Phys. Rev. Letters, 28, 1704 (1972).

3. T. Jimbo and C. Elbaum, Proc. of IEEE Ultrasonics Symp. Oct. 1972, p. 467.

4. M. Boiteux, P. Doussineau, B. Ferry and U. T. Höchli, Phys. Rev.

5. S. V. Peletminski, J. Exper. Theoret. Phys. U.S.S.R. (English Transl.) 37, 321 (1960).

6. W. J. Ince, Phys. Rev. 184, 574 (1969).

7. R. L. Melcher and D. I. Bolef, Phys. Rev. 186, 491 (1969).

8. B. Lüthi, T. J. Moran and R. J. Pollina, J. Phys. Chem. Solids, 31, 1741 (1970).

9. J. Kanamori and M. Tachiki, J. Phys. Soc. Jap. 17, 1384 (1962).

10. W. P. Wolf, Phys. Rev. 108, 1152 (1957).

11. A. Tucciarone, H. Y. Lau, L. M. Corliss, A. Delapalme, and J. M. Hastings, Phys. Rev. B4, 3206 (1971).

12. D. E. Eastman, Phys. Rev. 186, 645 (1967).

13. P. A. Fedders, Phys. Rev. B1, 375 (1970).

14. T. Jimbo and C. Elbaum (to be published).

15. F. Keffer, in "Handbuch der Physik," edited by S. Flügge (Springer, Berlin, 1966) vol. 18, Part 2.

PRINCIPLES OF POINT-DEFECT RELAXATIONS

A. S. Nowick
Henry Krumb School of Mines
Columbia University, New York, N.Y. 10027

I. Introduction and General Review

This paper will attempt to review some of the basic concepts involved in the analysis of anelastic relaxation phenomena due to point defects in crystals. Included will be those principles already well established in earlier papers, as well as some newer developments. In order to illustrate the general concepts in concrete terms, we will make considerable use of a widely studied example, namely the case of a substitutional-interstitial (s-i) impurity pair in bcc metals.

Point defects may be of the elementary type, which includes the substitutional impurity, s, the interstitial impurity (or self interstitial), i, or the vacancy, v. They may also consist of clusters of these elementary defects, the most important of which are pairs. If a given point defect possesses a lower symmetry than the point group of the crystal in which it resides, it will have more than one crystallographically equivalent orientation. Anelastic relaxation can then take place under an appropriate stress through statistical redistribution of defects among the several equivalent orientations. (1) A point defect can also undergo "reactions" which take it into other non-equivalent defect "species". The most common example of this is a defect pair which can exist in a series of shells of different separations, each shell having several equivalent orientations.

In the simplest cases of defect relaxations one obtains a single Debye peak of internal friction, tan ϕ, as a function of temperature, of the form

$$\tan \phi = (\delta s/s) \, [\omega\tau/(1 + \omega^2\tau^2)] \tag{1}$$

where the relaxation rate τ^{-1} obeys the Arrhenius equation

$$\tau^{-1} = \tau_0^{-1} \exp(-Q/kT) \tag{2}$$

In these equations, ϕ is the phase angle by which the strain lags behind the stress, s the appropriate compliance, δs is the relaxation magnitude (the difference between the relaxed and unrelaxed compliance), ω is the circular frequency, and Q the appropriate activation energy. This is the case of a "single-line relaxation spectrum".

More often one obtains a multiple-line spectrum consisting of two or more Debye peaks, either very close together or well separated (resolved). Various methods, usually involving computer fitting, have been used to obtain the separate Debye peaks from experimental data. These methods will not be discussed here.

The object of a basic theory of point defect relaxations is to interpret the relaxation rates, τ^{-1}, of the various Debye peaks in terms of various atomic jump frequencies and the magnitudes, δs, in terms of components of the λ-tensor, which describes the average distortion due to each of the various defect species. (1)

It may be helpful, in discussing the concepts which enter into the analysis of point defect relaxations, to begin by listing some <u>misconceptions</u> that have appeared in the literature of this subject. The following are noteworthy examples:

1) To each relaxation peak there corresponds a unique atom jump. (The converse statement also appears occasionally.)

2) To each relaxation peak there corresponds a single cluster composition (e.g., pairs for one peak, triplets for the next, and so on).

3) To each relaxation peak there corresponds a unique separation of a pair or larger cluster (e.g., nn pairs[*] for one peak, nnn pairs for the second, and so on).

4) If the nn configuration is strongly bound, and it is not possible to pass from one nn orientation to an equivalent orientation via the shortest atom jump, one must introduce higher-order (i.e. longer distance) atomic jumps in order to interpret the relaxation process.

As we proceed, we shall see that statement 1) above is often true, though not always, while the converse of 1) and statements 2), 3) and 4) are false, except for very special situations.

---

[*] nn ≡ nearest neighbor
nnn ≡ next-nearest neighbor

For proper interpretation of a relaxation spectrum due to point defects one must turn to the concept of relaxational normal modes, first introduced by Haven and van Santen (2), later explored by Wachtman (3), and more completely developed in a recent series of papers. (4-6) The approach begins from the kinetic equations for defect relaxation in the absence of stress, which take the form

$$dC_u/dt = \sum_v \nu_{uv} C_v \qquad (u = 1,2,\ldots n) \qquad (3)$$

in which each C represents a concentration (as a mole fraction), and the indices u and v run over both equivalent and non-equivalent defect configurations. (For a pair, equivalent configurations are the different orientations within a given shell or separation, while non-equivalent configurations are the various shells.) Finally, ν_{uv} is the probability per second for a defect to pass from configuration u to configuration v. The frequency ν_{uu} is defined, however as

$$\nu_{uu} = -\sum_{u \neq v} \nu_{uv} .$$

Now, it can be shown (5,6) that there exists a linear transformation of the set of concentrations C_u into another set, called the <u>normal coordinates</u>[*] C_i'' which obey the equations

$$\frac{dC_i''}{dt} = -\tau_i^{-1} C_i'' \qquad (i = 1,2,\ldots n) \qquad (4)$$

in which the quantities τ_i^{-1} are expressible in terms of the ν_{uv}'s.[**] The solutions to equations (4) are, of course simple exponentials with relaxation time equal to τ_i. When only one of these normal coordinates is present and the others are all zero, we say that we have a <u>normal relaxational mode</u>.[***] It can be shown that some of the normal modes can be excited by an appropriate applied stress; these are called the "mechanically active" modes. It should also be noted that not all the τ_i's are different; those normal modes which have the same τ_i are said to form a degenerate set (of two or, at most, three modes for cubic crystals).

---

[*] The symbol C' is reserved for an intermediate set of quantities called "symmetry coordinates".

[**] These ν_{uv}'s are, in turn, expressible in terms of the jump frequencies of defects between specific sites, denoted by w_1, w_2 ...

[***] The reader will note the close similarity to the case of vibrational normal modes, except that in the vibrational case the linear differential equations of motion are of second order. For further details on the concept of normal relaxational modes, particularly their diagramatic representation, see ref. 7.

In solving for the normal mode frequencies τ_i^{-1} (the "eigenvalues") and the normal modes themselves (the "eigenvectors") the methods of group representation theory are very helpful, if not essential. However, for the purposes of this paper we shall not require group theory since we will simply quote results as needed.

We begin with the "selection-rule table" for cubic crystals, as Table 1. This table gives the number of independent relaxation times (i.e., the number of Debye peaks) that can be excited by the two different types of shear stress, for each possible defect symmetry which is lower than cubic. The column "i.r." gives the group theoretical designation of the relaxational modes in question, and "deg." refers to the degeneracy (the number of modes which have the same τ_i). Next are given typical forms of the shear stresses which can excite these modes (or the corresponding uniaxial stresses $\sigma_{<100>}$ and $\sigma_{<111>}$).[*] Finally, under the headings of the various possible defect symmetries, there is given the number of Debye peaks that can be observed under that stress system. Clearly, the number of relaxational modes of each type is the product of this number and the degeneracy, listed in column 2. The appearance of a zero in the table means that no relaxation at all can take place under the shear stress indicated for the particular defect in question. These absences help to establish the symmetry of the defect involved. Thus, only for a tetragonal defect or a < 100 > orthorhombic defect is there a complete absence of relaxation for stress σ_{xy}, or what is equivalent, for uniaxial stress along a < 111 > direction of the crystal.

TABLE 1.

Selection Rule Table for Cubic Crystals
(The numbers in the table give the number of independent τ_i-values for each stress and defect symmetry.)

							DEFECT		
i.r.	deg.	stress	tetr.	trig	<110> ortho.	<100> ortho.	<100> mono.	<110> mono.	tricl.
E	2	$\sigma_{xx}-\sigma_{yy}$ (or $\sigma_{<100>}$)	1	0	1	2	2	1	2
T_2	3	σ_{xy} (or $\sigma_{<111>}$)	0	1	1	0	1	2	3

---

[*] The i.r. called A_1, which involves relaxational modes that can only be excited by hydrostatic stress, are omitted here because such modes have been only rarely studied in cubic crystals. For lower-than-cubic crystals, however, the A_1 modes may be more easily observed by employing an appropriate uniaxial stress. (8)

From the table it can be seen that for the lower symmetry defects there may be more than one Debye peak associated with a single defect species.

When more than one species is present simultaneously, the numbers of relaxation times for each i.r. is simply the sum of the numbers for the separate species. (6) The most important application is to defect pairs at a series of different separations (shells). Let us illustrate with the case of the s-i defect in the bcc lattice, as shown in Figure 1. Here the nn pair (first shell, or α defect species) has tetragonal symmetry; the nnn pair (second shell, or β species) is < 110 > orthorhombic; the tnn pair (third shell or η species) is < 100 > monoclinic. The number of τ_i-values is then dependent on how many shells we choose to take, and in fact, can be extremely large. For the moment, however, let us confine ourselves to the first two shells, α and β. From Table 1 and the symmetries of the α and β defect species, we see that such a situation would give rise to two relaxations of the E type and one of the T_2 type. Introducing the third shell (η species) would increase these numbers to 4 and 2, respectively.

Now, in this particular example, the isolated (dissociated) interstitial is itself tetragonal and gives rise to the well known Snoek relaxation. (9) In fact, the distortion is so large along the tetragonal axis of the i defect that even when it is associated with the s to form a pair, the other distortions are probably negligible by comparison. This means that for the s-i defect, the T_2 type relaxations are probably too small to be observed. (10) Accordingly, in further considerations of this example, we will deal only with the E type relaxations.

It should be noted that while the number of relaxation times when α and β defects are both present is simply the sum of the numbers due to each species taken separately, the expressions for these relaxation times will involve the rates of atom jumps which take α into β and of β into α, because of the coupling that occurs via the kinetic equations (3). The normal coordinates, C_i'', themselves will also involve linear combinations of the normal coordinates of the two species taken separately, as a consequence of this coupling. For the specific s-i example under consideration we obtain, for the two relaxation rates, the expression: (11)

$$\tau_{1,2}^{-1} = (w_3 + 2w_2) \pm [(w_3 - 2w_2)^2 + 2w_2w_3]^{\frac{1}{2}} \tag{5}$$

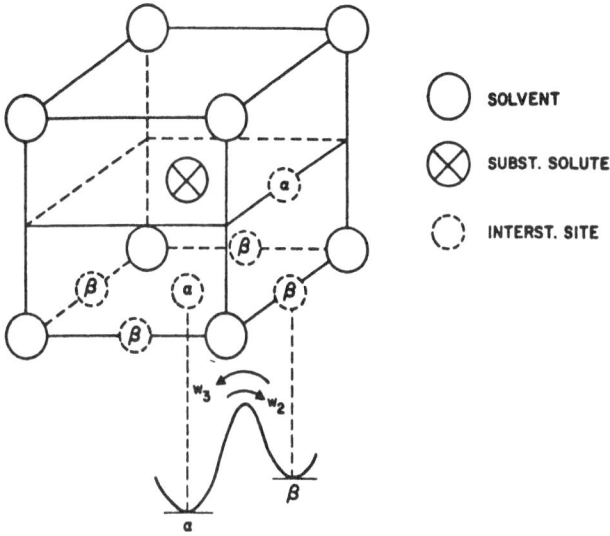

FIG. 1

Model of the bcc Lattice Showing s-i pairs in nn
(α) and nnn (β) Positions, and the Corresponding
Jump Frequencies.

where w_2 (defined in Figure 1) is the jump frequency (probability/sec) for passage from a β site to a specific α site, and w_3 is the reverse jump frequency. The symbol w_1 is reserved for the specific jump frequency of the dissociated interstitial (the "Snoek jump"), and indeed the well known expression for the Snoek relaxation rate is (9)

$$\tau_1^{-1} = 6w_1 \qquad (6)$$

An important approximation occurs when the nn (α) is more strongly bound than the nnn, so that $w_3/w_2 \gg 1$; equation (5) then simplifies to

$$\tau_1^{-1} = 2w_3 \quad ; \quad \tau_2^{-1} = 3w_2 \qquad (7)$$

It is reasonable to expect that, in such a situation, w_1 will lie between w_3 and w_2. Thus, the relaxation spectrum contributed by the α-β pairs will consist of two Debye peaks, one of which falls at temperature below the Snoek peak (the "fast relaxation", for which $\tau^{-1} = 2w_3$) and the other of which appears above the Snoek peak (the "slow relaxation", for which $\tau^{-1} = 3w_2$).

II. Partial Relaxation Magnitudes

We now turn from consideration of the relaxation rates, τ_i^{-1}, to the relaxation magnitudes, δs_i. Again, consideration of the above s-i defect in nn and nnn configurations only, will be helpful to illustrate the general concept. Specifically, in this case, we know the relaxation magnitude (of the E type relaxations) due to the separate defect species to be: (4)

$$\delta s_\alpha = (C_\alpha v_0/3kT) \ (\lambda_1^\alpha - \lambda_2^\alpha)^2$$
$$\delta s_\beta = (C_\beta v_0/3kT) \ [\lambda_3^\beta - (\lambda_1^\beta + \lambda_2^\beta)/2]^2 \tag{8}$$

where C_α and C_β are the total concentrations of the α and β species, λ_1, λ_2 and λ_3 are the principal values of the λ tensor for the two species α and β, and v_0 is the atomic volume. It follows that the total relaxation magnitude, i.e., the sum of the magnitudes of the two relaxations whose frequencies are given by equation (5), is

$$\delta s_t = \delta s_\alpha + \delta s_\beta \tag{9}$$

The question that remains is what are the magnitudes, δs_1 and δs_2, of the separate relaxations? In fact, we are led to wonder whether under some circumstances δs_1 will equal δs_α and $\delta s_2 = \delta s_\beta$ (or vice versa), in accordance with item 3) on the list of misconceptions presented earlier.

The quantities δs_1 and δs_2 are called partial relaxation magnitudes (as against the total value given by equation (9)), and the question of how to calculate these quantities has been worked out in detail in a recent paper. (12) What is involved is the calculation of the normal coordinates (the eigenvectors) that correspond to the relaxation rates (the eigenvalues), and the substitution of these normal coordinates into expressions for the relaxation magnitude. The partial relaxation magnitudes δs_i, so obtained, are not simply thermodynamic quantities, as is the total magnitude δs_t, for they depend on the kinetic equations (which determine the normal coordinates). Thus, they are mixed quantities in the sense that they depend both on kinetic and thermodynamic parameters. For our s-i example with only nn and nnn configurations allowed, the exact expressions for δs_i (i = 1,2) are:

$$\delta s_i = \frac{v_0}{3kT(1 + r_i^2)} \ \{\sqrt{C_\alpha}(\lambda_1^\alpha - \lambda_2^\alpha) + r_i\sqrt{C_\beta}[\lambda_3^\beta - (\lambda_1^\beta + \lambda_2^\beta)/2]\}^2 \tag{10}$$

$$(i = 1,2)$$

where

$$r_{\frac{1}{2}} = \frac{1}{(2w_2w_3)^{\frac{1}{2}}} \ \{(w_3 - 2w_2) \pm [(w_3 - 2w_2)^2 + 2w_2w_3]^{\frac{1}{2}}\} \tag{11}$$

Note also that C_α and C_β are interrelated through the jump frequencies, by

$$C_\beta/C_\alpha = 2w_2/w_3 \qquad (12)$$

(This last result may be obtained from the principle of detailed balance applied to the energy diagram of Figure 1, recalling that there are twice as many β sites as α sites.) For purpose of calculation we shall make the simplifying assumption that only the major (Snoek-type) distortion is of importance (as discussed earlier) and that this distortion is the same regardless of the location of the s atom. In that case $\lambda_1^\alpha \simeq \lambda_3^\beta$ (in accordance with the numbering convention used for tetragonal and < 110 > orthorhombic defects (1,5)) and all other λ coefficients are zero. Figure 2 shows the calculated ratio of $\delta s_1/\delta s_2$ as a function of w_3/w_2 on a log-log plot. At the same time, the ratio $\delta s_\beta/\delta s_\alpha$, obtained from equation (8) under the same assumption, is shown as a dashed line. It is clear from Figure 2, even in the limit of $w_3/w_2 \gg 1$ when the two relaxation rates as given by equation (7), that the ratio $\delta s_1/\delta s_2$ is not equal to $\delta s_\beta/\delta s_\alpha$; in fact, in the limit, the former quantity is 2.5 times the latter quantity. This is a clear refutation of misconception 3) which regards that we may assign one relaxation entirely to nn pairs and the other to nnn pairs.

The reason that $\delta s_1/\delta s_2 > \delta s_\beta/\delta s_\alpha$ even for large w_3/w_2 is that it is not possible to produce relaxation of β defects at low temperature solely through the agency of jumps out of β sites (of jump rate w_3). Rather, in order to reorient the β defects there must also be jumps from α sites into β sites. In spite of the fact that $w_2 \ll w_3$, an appreciable number of such jumps do take place as a part of the low-temperature relaxation because of the relatively large concentration of α defects, i.e., by virtue of equation (12). Thus, the low-temperature relaxational mode (with values τ_1^{-1} and δs_1) involves reorientation of both α and β defects and therefore has a larger magnitude than δs_β, which involves β alone. On the other hand, the high-temperature mode involves the major part of the α species' relaxation and is, of course, much larger than the low-temperature relaxation when $w_3 \gg w_2$, as a consequence of equation (12).

Figure 2 shows that as w_3/w_2 increases, $\delta s_1/\delta s_2$ decreases. Thus, for a sufficiently large value of w_3/w_2, the lower relaxation peak should be undetectable. From the figure, we note that for $w_3/w_2 \simeq 100$, $\delta s_1/\delta s_2 \sim 0.05$. This ratio may be taken as the limit of detectability of δs_1 for the usual measurement sensitivity and background damping.

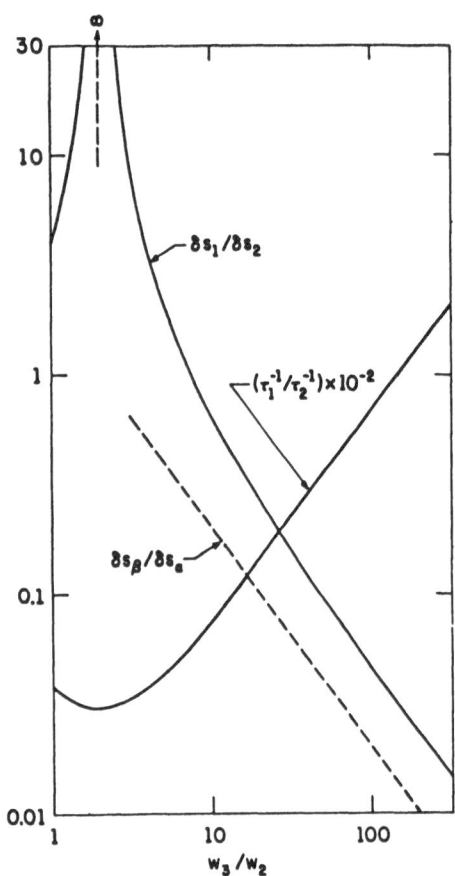

FIG. 2

Variation of relative relaxation rates $(\tau_1^{-1}/\tau_2^{-1})$ and relative relaxation magnitudes $(\delta s_1/\delta s_2)$ for the s-i defect in a bcc metal in the approximation of $\alpha \leftrightarrow \beta$ only. Also assumed is that $\lambda_1^\alpha = \lambda_3^\beta$, and that the other principal λ values are zero. The ratio $\delta s_\beta/\delta s_\alpha$ is shown for comparison.

We may also note that as w_3/w_2 increases, the s-i interaction becomes stronger. Correspondingly, the shift in temperature, δT, between the <u>high</u> temperature s-i peak and the Snoek peak should increase. A more convenient "rule of thumb" for the detectability of δs_1 may be obtained by translating the above limiting value of w_3/w_2 into a corresponding limiting value of δT. To do this, we make the admittedly crude assumption that

$$w_2 w_3 \simeq w_1^2 \tag{13}$$

i.e., that the Snoek jump frequency w_1 is the geometric average of the frequencies w_2 and w_3 defined in Figure 1. (This means that the Snoek activation energy, Q_1, is the arithmetic average of these two activation energies, Q_2 and Q_3, if we further assume that all pre-exponential factors w_0 are nearly the same.) It is then easily shown that corresponding to a ratio $w_3/w_2 < 100$, we must have a relative peak shift for the slow relaxation of $\delta T/T_1 < 0.075$ (where T_1 is the peak temperature of the Snoek peak). This means, for example, that for Fe-N ternary alloys where $T_1 \simeq 300°K$, the criterion for detectability of the low-temperature peak is that $\delta T < 23°K$, while for Ta-N and Nb-N where $T_1 \simeq 600°K$, δT must be $< 45°K$. For s atoms which give rise to shifts greater than these values, we expect the low temperature peak to be too small to be observable experimentally. Such s atoms may be considered as giving rise to strong s-i interactions, and in these cases only a single interaction peak due to s-i pairs should be observable, above the Snoek peak.

III. The s-i (bcc) Problem Including Dissociation

The limitation of the above discussion of the s-i problem in bcc crystals to only nn and nnn pairs, although very instructive, still gave a somewhat artificial view of this problem. The next alternative would be to extend the occupancy to any finite number, m, of shells. In this case, the formalism which begins with equation (3) and leads to the concepts of normal coordinates and relaxational normal modes can be applied, since there will be a finite system of equations. Yet in metal systems, where s-i binding energies are not usually more than ~ 0.1 eV (as against the situation in ionic crystals where larger Coulomb energies are involved in pair formation) the limitation to a finite number of shells is unrealistic.

The mathematical difficulties are more serious when one attempts to include dissociation, since strictly, the system of equations becomes infinite. The problem is to find a suitable method of approximation which serves to break the coupling between shells[*] that occurs through equations (3) without disallowing jumps which we know can actually occur. A general discussion of this problem will be presented in a later paper; here we shall consider it only for our example of the s-i defect in a bcc crystal.

In this example, the i atom at sufficient distance from the s atom (say at the m + 1 st shell) will behave as a Snoek defect, which is tetragonal in symmetry. Nevertheless, the occupation of the three equivalent orientations may not be the same as if the i atom were isolated because of the coupling to the m th shell. We therefore use the following method of approximation:

a) Treat the i-s pair in the m th shell as if it had tetragonal symmetry, i.e., label the sites as x, y and z types as in the Snoek defect, but allow the m th shell to be coupled both to the m - 1 st and the m + 1 st through equations (3).

b) Treat the m + 1 st shell in the same way as the m th, but assume that the concentrations in the 3 tetragonal orientations are identical to the corresponding ones in the m th shell. (If the m th shell is far out enough, the error inherent in this assumption should be

---

[*] i.e., the dependence of the concentrations in one shell on those of the neighboring shells.

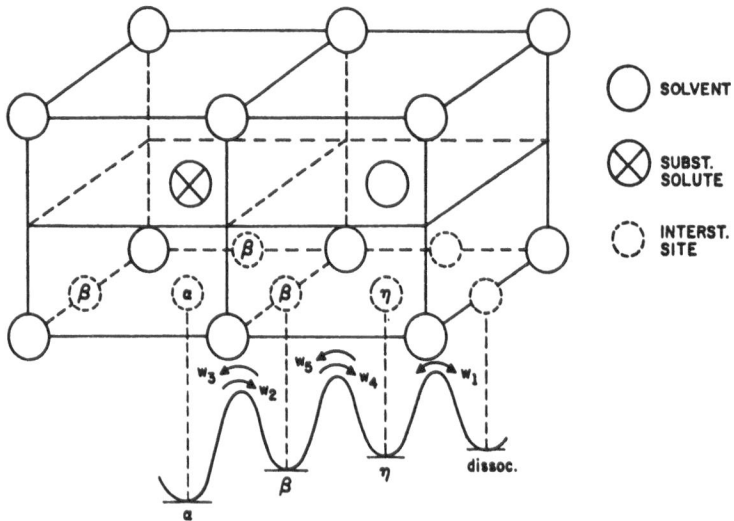

FIG. 3

The s-i Pair in a bcc Crystal Showing nn (α), nnn (β), tnn (η) and
Dissociated Positions. (Also shown is the free-energy curve and
the jump frequencies for the case in which the α pair is the most
strongly bound.)

small.

In this way the set of equations (3) become decoupled at the m th shell, but jumps
to further shells are included. One can let m be as high as one wishes, but for present purposes we will see that a good understanding of the problem may be obtained by stopping at
m = 3. The three shells are then labelled α, β and η, and the jump frequencies w_1 through w_5
are as shown in Figure 3.

As mentioned earlier, we will deal with the E type relaxations only, and since the η
shell is treated as tetragonal, its introduction leads to one more E relaxation than for the
α-β problem only. Straightforward calculation gives, for the secular determinant which determines the relaxation rates, τ^{-1}:

$$\begin{vmatrix} 4w_2 - \tau^{-1} & \sqrt{2}\,w_2 & 0 \\ \sqrt{2}\,w_3 & 2(w_3 + w_4) - \tau^{-1} & \sqrt{2}\,w_4 \\ 0 & \sqrt{2}\,w_5/4 & w_5 + \frac{9}{2}w_1 - \tau^{-1} \end{vmatrix} = 0 \qquad (14)$$

The three relaxation rates are then the solutions of this cubic equation in τ^{-1}. It is
interesting to examine two special cases, as follows.

<u>Case A</u> in which the α(nn) pair is relatively tightly bound: In this case, for which
the free energy curve is that shown in Figure 2, it is clear that w_3 is the largest jump rate,

w_2 the smallest, and all others may be considered as intermediate. Equation (13) may first be solved by keeping only large terms (treating τ^{-1} as a large quantity) to obtain, for the fastest relaxation rate:

$$\tau_1^{-1} \simeq 2w_3 \qquad (15)$$

In a similar way, the intermediate and slow relaxation rates are, respectively, found to be

$$\tau_2^{-1} \simeq w_5 + \frac{9}{2} w_1 \qquad (16)$$

$$\tau_3^{-1} \simeq 3w_2 \qquad (17)$$

Note that the fast and slow relaxations are the same as for the problem which considers only α and β pairs, equation (7), but that now there is an additional intermediate relaxation frequency, $w_5 + \frac{9}{2} w_1$. As before, one may expect the fast relaxation to appear as a peak at temperatures below the Snoek peak, and to involve relaxation of both the α and β defect species. The slow relaxation should appear above the Snoek peak, and, as before, involve mainly relaxation of the $\alpha(nn)$ pairs; this one is, therefore, the major peak. Finally, we now have an intermediate peak whose position depends on the relative magnitudes of w_5 and w_1. Thus, if the $\beta(nnn)$ pairs are only very weakly bound and $w_5 \simeq w_1$, τ_2^{-1} is practically the same as for the Snoek relaxation, equation (6); on the other hand, if β is more strongly bound, and $w_5 > w_1$, the intermediate peak may appear below the Snoek peak. The intermediate peak involves relaxation of both the β and η pairs.

It is quite possible that, in the majority of cases, the intermediate peak will be too close to the Snoek peak to be readily detected experimentally. Further, if the number of shells, m, were allowed to become greater than 3 in the method outlined above, there would be additional relaxations all clustered close to the Snoek peak and, therefore, difficult to detect.

Case B in which the $\beta(nnn)$ pair is most strongly bound: This situation may arise, for example, from strain-energy considerations, when the s atom is larger than the solvent atoms. (13) In this case, the free-energy curve of Figure 2 must be redrawn such that w_2 and w_5 are the large jump rates while w_3 and w_4 are small. Applying the same approximation methods to the problem of finding the roots of equation (14) as in Case A, we obtain

$$\tau_1^{-1} \simeq 4w_2 \; ; \quad \tau_2^{-1} \simeq w_5 \; ; \quad \tau_3^{-1} \simeq \frac{3}{2} (w_3 + w_4) \qquad (18)$$

where the first two are fast relaxations, expected to fall below the Snoek peak and the third is the slow relaxation which falls above the Snoek peak. In this case $C_\beta \gg C_\alpha$, C_η, and the slow relaxation will again be the major peak, this time due primarily to the relaxation of the β defect species.

Unfortunately, in both cases A and B we predict that the major relaxation peak occurs above the Snoek peak, thereby making it difficult to distinguish between these two cases. Let us now examine the experimental situation.

In spite of an extremely large number of experimental investigations of ternary bcc alloys containing both a substitutional and an interstitial impurity (for a comprehensive review see (14)), the situation remains quite confused. Some of the reasons for the complexities which are found and for disagreements among different investigators are as follows:

a) Various methods of analysis are used by different authors to obtain component Debye peaks from the experimental data.

b) When the concentration of substitutional impurity is relatively high (sometimes even at $\gtrsim$ 0.5%) triplet defects of the type s-i-s contribute to the relaxation spectrum.

c) When the concentration of interstitial impurity is high, i-i and s-i-i defects appear to contribute.

d) Some s atoms are strong scavengers for the i atoms and, even at very low concentrations, may give rise to higher clusters and even to precipitates.

Perhaps the most widely studied ternary is the Fe-Mn-N alloy. For this system the Snoek peak (at 1 Hz) appears at 23°C and the major "interaction peak" due to s-i pairs is at 35°C. Thus, for this system, $\delta T/T_1 = 0.04$, i.e. well within the rough limit of 0.075 given in our earlier calculation. Accordingly, the low-temperature peak should be detectable. There is, in fact, a low-temperature peak at 7°C in this system, as shown for example in Figure 4. However, several authors have shown that the height of the 7°C peak increases much faster than linearly with Mn content, and it has been suggested that the defect responsible for this peak involves two Mn atoms. (15) Nevertheless, the 7°C peak is present for alloys with as little as 0.1 - 0.2% Mn. This fact, plus other considerations, led Couper and Kennedy (16) to suggest that the 7°C peak is made up of two components, one due to simple Mn-N pairs and the other involving triplets consisting of two Mn and one N atom. It seems reasonable to regard that the peak for the 0.1% Mn alloy is due entirely to the former component. Based on the data of Enrietto (17) and of Couper and Kennedy (16) we then obtain $\delta s_1/\delta s_2 \sim 0.2 - 0.25$.

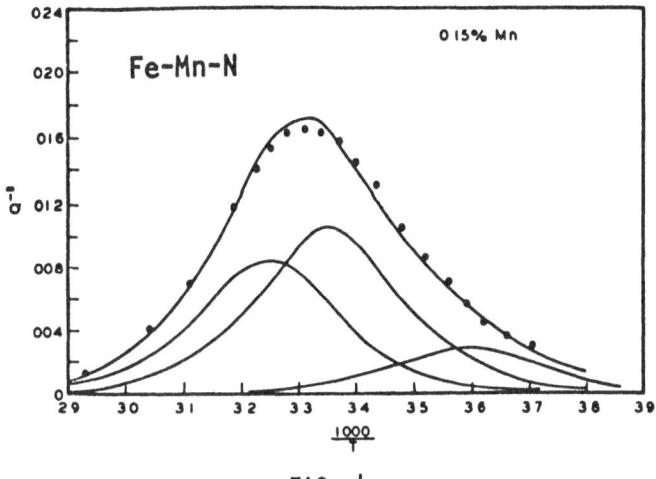

FIG. 4

Internal Friction vs. T^{-1} of an Fe-0.15%Mn-N Alloy
and its Resolution into Three Debye Peaks.
Frequency $\sim$ 1 Hz. From Enrietto (17).

From the separation of the peaks a ratio $w_3/w_2 = 25$ is obtained. (12) For this ratio, a theoretical value (of $\delta s_1/\delta s_2 = 0.2$) can be obtained from the calculation plotted as Figure 2. The agreement is quite satisfactory.

Ternary alloys of Nb and Ta containing O or N generally show just one interaction peak above the Snoek peak, and at high s concentrations develop a second high temperature peak attributed to s-i-s triplets. In these cases, the shift $\delta T/T_1$ is generally greater than 0.075. It is therefore not surprising that a low temperature peak (below the Snoek peak) has never been reported for these systems.

It is unfortunate that a more decisive demonstration of a low-temperature s-i peak has not been observed thus far. Also, the intermediate peak, which should be close to the Snoek peak has never been reported. It is hoped that the present analysis will aid the experimenter, first, in finding conditions in which only the s-i pair relaxations are obtained, and also, to focus attention on alloy systems in which the low-temperature and intermediate-temperature peaks can be observed.

IV. Concluding Remarks

We may utilize our discussion of the various concepts involved in point defect relaxations to re-examine the list of misconceptions presented at the beginning. Actually these misconceptions arose because many of them are sometimes true, and because they represent useful guides when they are valid. It is therefore more useful to indicate when each such concept is valid, or why it is not, than simply to say that it is wrong.

89

Turning to misconception 1), we must emphasize that anelastic peaks are due to relaxational normal modes associated with specific defects and not to atom jumps. Although the rates are controlled by atom jumps, it is first necessary to consider which defects are relaxing before discussing the jumps. In general, as illustrated by equation (5), a given relaxation does not involve only one atom jump. However, often one jump does dominate, and then, as in equation (7), one jump rate determines each τ^{-1}. The converse statement, that to each jump there corresponds a relaxation peak, is often not valid, as can be seen by the following argument. Whenever there are two (or more) alternative sets of jumps by which the same reorientations may be produced, both of the corresponding jump rates must enter into the expression for τ^{-1}. The two paths are then competing to produce the same relaxation; there will then not be a separate peak for each jump. The expression for τ_3^{-1} in equation (18) provides a particularly good illustration of this point with respect to the two jump rates w_3 and w_4.

In the case of misconception 2), examination of Table 1 shows that if pairs can exist in several shells, i.e., having several defect symmetries, the statement is clearly false. But if the binding energy in one particular shell is much stronger than for the others, and the symmetry of the pair in this configuration is high, we may come out with only one peak for the paired defects.

As for misconception 3), when several shells are occupied by a paired defect, we cannot in general assign one peak to each shell. We saw this for the example of the s-i pair in the case of nn $\leftrightarrow$ nnn ($\alpha \leftrightarrow \beta$) only. Here, if $w_3 \gg w_2$, the magnitude of the large peak, δs_2, was primarily due to the α defect, but the magnitude of the small peak, δs_1, had contributions from both α and β in comparable amounts. This was the reason why $\delta s_\beta / \delta s_\alpha$ did not equal $\delta s_1 / \delta s_2$ in Figure 2, even in the asymptotic approximation ($w_3/w_2 \gg 1$).

Finally, we turn to misconception 4). We have seen that when a nn pair is moderately strongly bound, its relaxation is controlled by dissociative-type jumps (nn $\rightarrow$ nnn), as for τ_2^{-1} in equation (7). This result is obtained even if the nnn site is too weakly occupied to give rise to a relaxation of its own. There is no need, therefore, to introduce a longer distant atom jump to account for the relaxation behavior. Of course, if the binding energy were very large, comparable to the difference in activation energy between the shortest jump and the next longer one, one might reach the point where the longer jump would become rate controlling. Thus far, however, there is no evidence from either diffusion or relaxation experiments to suggest that such jumps are taking place.

Acknowledgements

This work has been supported by a grant from the National Science Foundation. The author is grateful to Dr. A. Sagues for helpful discussions of the s-i (bcc) problem.

References

1. A.S. Nowick and B.S. Berry, Anelastic Relaxation in Crystalline Solids, Academic Press, New York, 1972, Chapter 8.

2. Y. Haven and J.H. van Santen, Nuovo Cim. Suppl. 7 (2), 605 (1958).

3. J.B. Wachtman, Jr., Phys. Rev. 131, 517 (1963).

4. A.S. Nowick and W.R. Heller, Adv. Phys. 14, 101 (1965).

5. A.S. Nowick, Adv. Phys. 16, 1 (1967).

6. A.S. Nowick, J. Phys. Chem. Solids 31, 1819 (1970).

7. A.S. Nowick, in Point Defects in Solids, Vol. 1, ed. J.H. Crawford, Jr. and L.M. Slifkin, Plenum Press, New York, 1972, Chapter 3.

8. A.S. Nowick, J. Chem. Phys. 53, 2066 (1970).

9. See, for example, reference 1, Chapter 9.

10. A.S. Nowick, Scripta Met. 7, 289 (1973).

11. R. Chang, J. Phys. Chem. Solids 25, 1081 (1964); see also, reference 1, p. 336.

12. A.S. Nowick, J. Phys. Chem. Solids 34, 1507 (1973).

13. J. Gouzou, J. Wegria and L. Habraken, J. Phys. 32, C2 - 25 (1971).

14. D.F. Hasson and R.J. Arsenault, in Treatise on Materials Science and Technology, Vol. 1, ed. H. Herman, Academic Press, New York, 1972.

15. M. Nacken and U. Kuhlmann, Arch. Einsenhütt. 37, 235 (1966).

16. G.J. Couper and R. Kennedy, J. Iron Steel Inst. 205, 642 (1967).

17. J.F. Enrietto, Trans. Met. Soc. AIME 224, 1119 (1962).

PHONONS

ELECTRONS

MAGNETIC EFFECTS

EFFECTS OF FERMI SURFACE ANISOTROPY ON THE ATTENUATION OF SHEAR WAVES IN CADMIUM AND ZINC

D.P. Almond, M.J. Lea and E.R. Dobbs

Department of Physics, University of Lancaster, Lancaster LA1 4YB, U.K.

We have made a detailed study of the electronic attenuation of ultrasonic shear waves in pure cadmium and zinc at low temperatures and present here some of our conclusions. The relationships of the geometries of the Fermi surfaces (1,2) to these measurements are particularly striking. For the interpretation of ultrasonic measurements these Fermi surfaces are closely approximated by free electron Fermi spheres with 12 segments absent (Fig.1). The segments represent the sheets of electron states, the butterflies and cigars in the 3rd and 4th zones, which were shown to be absent in cadmium and zinc by Stark and Falicov (2).

The electronic attenuation of shear waves in metals with complex Fermi surfaces has been described theoretically by Pippard (3). The complete expression for the attenuation is

$$\alpha = \frac{\hbar q}{4\pi^3 M v_s} \left\{ \int \frac{D^2 a \, dS}{1 + a^2 \cos^2\phi} + \left[\int \frac{D a^2 \sin\phi \, \cos\phi \, \cos\psi \, dS}{1 + a^2 \cos^2\phi} \right]^2 \middle/ \int \frac{a \sin^2\phi \, \cos^2\psi \, dS}{1 + a^2 \cos^2\phi} \right\} \qquad (1)$$

in which all the symbols have the same meaning as used by Pippard. Shear wave attenuation is composed of two components. The first, the deformation attenuation, results from changes in electron energy caused by the shear distortions of the Fermi surface. This is a basically local term rising to a peak at $q\ell \sim 2$ and falling off as $q\ell^{-1}$ at high $q\ell$ (q phonon wave number, ℓ electron mean free path). The second term, the electromagnetic attenuation, dominates at $q\ell \stackrel{>}{\sim} 2$. In this non-local region the electrons fail to follow instantaneously the ion cores and non-zero transverse currents flow in the metal. These currents set up electromagnetic fields which induce further electron currents in the metal. At frequencies of less than $\sim$ 200 MHz the induced currents exactly cancel the remaining currents. The second term in (1) is the attenuation caused by the flow of these additional currents. The integral in the numerator represents the induced current and the integral in the denominator the transverse conductivity. At high $q\ell$ the conductivity approximates to a line integral about the effective zone.

$$\sigma_q = \frac{e^2}{4\pi^3 q\hbar} \int R \cos^2\psi \, d\psi \qquad (2)$$

The attenuation of fast shear waves (polarized in the basal plane) along $[10\bar{1}0]$ and $[11\bar{2}0]$ in cadmium and zinc was computed over a range of a values from 0 to 100 ($a = q\ell$). These calculations and some measurements of total attenuation are presented as $\frac{\alpha}{\nu} \times q\ell$, in Fig.2. The measurements of electronic attenuation were made at $\sim$ 100mK by comparing ultrasonic pulse-echo heights in the normal and superconducting states. At this temperature ℓ is impurity limited and $q\ell$ varies linearly with phonon frequency. The anisotropy, a factor of $\sim$ 11 for these two directions, simply reflects the anisotropy in the line integral in σ_q. The absence of the butterflies greatly reduces the number of electrons in the $[11\bar{2}0]$ effective zone but has no effect on $[10\bar{1}0]$. In addition, as a result of reducing the effective area of the Fermi surface, the surface integrals in both the deformation term and the induced current terms are reduced from the free electron values. For $[10\bar{1}0]$ the deformation term becomes $\sim \frac{1}{3}$ the free electron value and the electromagnetic term $\sim \frac{1}{9}$. The peak at $q\ell \sim 3$, from the dominant deformation term, accounts for the unusual temperature dependence of the attenuation in this direction.

Measurements of the attenuation of fast shear waves along $[10\bar{1}0]$ cadmium in the temperature range $0 \rightarrow 30K$ are shown in Fig.3. As the temperature is raised above 100mK ℓ decreases and with it $q\ell$. At the higher frequencies decreasing $q\ell$ raises the effective $\frac{\alpha}{\nu}$ through a peak before it falls off at low $q\ell$.

The current neutrality condition, which was valid for all these lower frequency measurements, breaks down for $q\delta \overset{>}{\sim} 1$ (Pippard (4)). Here δ is the electromagnetic screening range defined by

$$\delta^2 = \frac{1}{\mu_0 \omega \sigma_q} \qquad (3) \quad (\omega = 2\pi\nu)$$

in which σ_q is the same transverse conductivity as equation (2). For cadmium $[11\bar{2}0]$ this conductivity is particularly small and hence δ is much larger than the equivalent free electron value. This property has made possible a detailed study of the attenuation where $q\delta \overset{>}{\sim} 1$ which would not have been possible in most other metals at frequencies below 1 GHz. Measurement of the attenuation in the frequency range $100 \rightarrow 400$ MHz are shown in Fig.4. At frequencies where the phonon wavelength becomes of the same order as δ, non-local effects reduce the effectiveness of electromagnetic coupling. The induced currents and hence the electromagnetic term in the attenuation fall off, vanishing altogether for $q\delta \gg 1$. From the measurements Fig.4, δ or $\sigma_{11\bar{2}0}$ can be deduced setting $q\delta = 1$ at 410 MHz which is in good agreement with a value of 375 MHz calculated from the simple model.

The measurements of shear wave attenuation in zinc show the same general features as found in cadmium and can be explained using a very similar Fermi surface model.

References

1. D.F. Gibbons and L.M. Falicov, Phil. Mag. 8, 177 (1963).
2. R.W. Stark and L.M. Falicov, Phys. Rev. Letters 19, 795 (1967).
3. A.B. Pippard, Proc. Roy Soc. A257, 165 (1960a).
4. A.B. Pippard, Phil. Mag. 46, 1104 (1955).

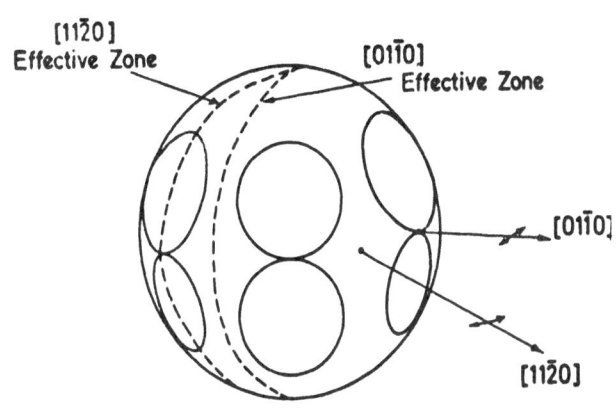

Fig.1   Model Fermi Surface for
Cadmium and Zinc

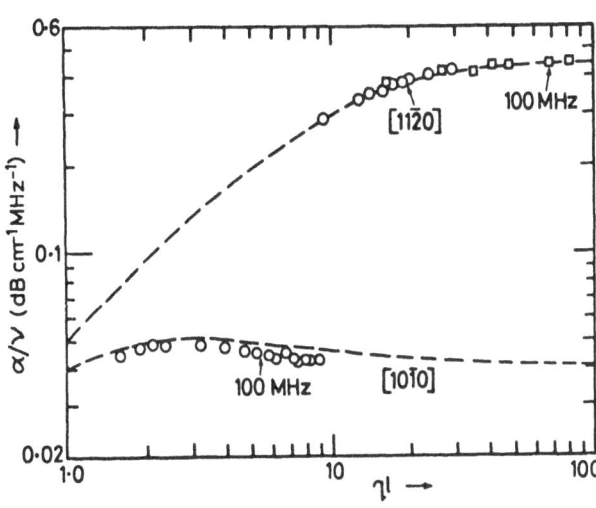

Fig.2   Electronic Attenuation in
Cadmium, ql from 1 to 100

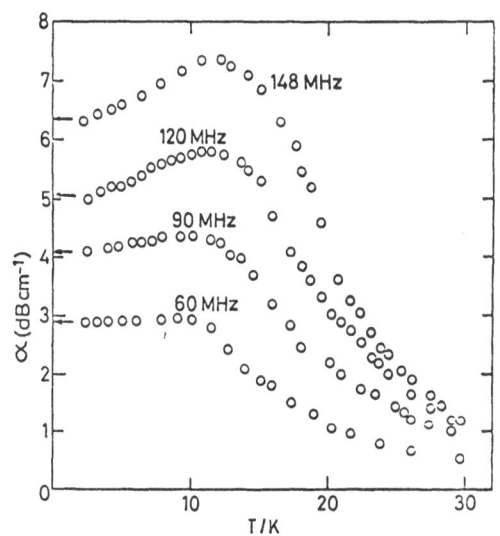

Fig.3   Temperature Dependent
Attenuation Peak,
Cadmium [10$\bar{1}$0]

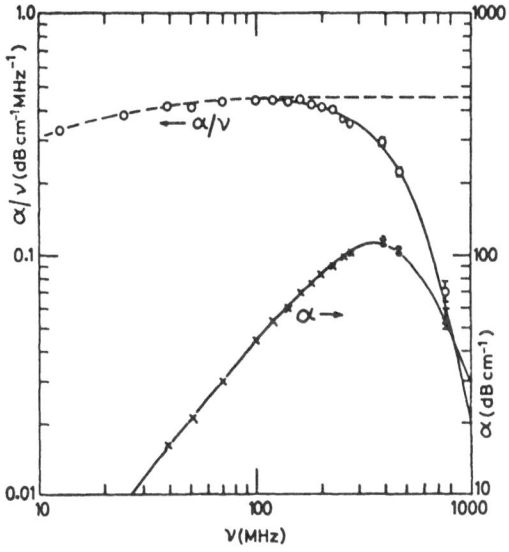

Fig.4   Screening Breakdown,
Cadmium [11$\bar{2}$0]

ULTRASONIC ATTENUATION IN n-GaP and p-InSb AT LOW TEMPERATURES

W. F. Boyle and R. J. Sladek
Dept. of Physics, Purdue University, West Lafayette, Ind., 47907, U.S.A.

ABSTRACT   Measurements of the attenuation, α, of 30MHz bulk ultrasonic waves have been made in single crystal samples of undoped and sulfur-doped n-GaP and of copper-doped p-InSb between T = 4.2K and 240K. Centered around some low T is a narrow maximum in α for piezoelectrically active waves. The maxima are due to the decreased screening of piezoelectric fields as charge carriers drop from band to impurity states with decreasing T. Although the experimental peaks cannot be accounted for quantitatively by means of theory using measured resistivity values, they can be fitted exactly by letting the electromechanical coupling factor and the electrical resistivity take suitable values. This implies that the samples are somewhat inhomogeneous in resistivity. We find that $|e_{14}| = 0.058 \pm 0.006$ C/m^2 for InSb and 0.084 ± 0.003 C/m^2 for GaP where e_{14} is the piezoelectric constant. No accurate value of $|e_{14}|$ for GaP has been published previously. The transfer of bond charge implied by the values of e_{14} for these and other zinc blende compounds is found to have a dependence on ionicity like that implied by recent theory.

Introduction

InSb and GaP normally crystallize in the cubic, zinc blende structure. (1) Since this structure has no center of symmetry, it is piezoelectric and, since the structure is cubic, the piezoelectric tensor has only one independent term, e_{14}. (2) The structure also implies that only certain pure acoustical propagation modes in the crystal are piezoelectrically active, i.e. accompanied by a longitudinal electric field. (3) This activity is expected to influence the attenuation, α, and velocity, v, of these types of ultrasonic waves which produce piezoelectric-ally active vibrations provided that the electrical conductivity, σ, of the specimen is not too large. Specific theoretical expressions have been developed which give the dependence of α and v on the ultrasonic frequency, ω, the conductivity relaxation frequency, ω_c (which equals σ/ϵ where ϵ is the dielectric permittivity), and the diffusion frequency, ω_D, of mobile electrons (or holes). (3)

Experiments on the II-VI compound CdS have indicated that modifications of the theory might be necessary to account for the measured attenuation if there are inhomogeneities in the electrical conductivity. (4,5,6) However, because of the difficulty of handling inhomogeneity effects theoreticallly and the lack of knowledge about the actual inhomogeneities present in a real sample, only limited efforts have been made to obtain a theoretical expression for α which includes the influence of inhomogeneities. (5,6) Other work on CdS has indicated that the electromechanical coupling factor, which provides a measure of the maximum influence that piezoelectric effects can have on the attenuation, may itself be reduced by the presence of inhomogeneities. (7) Experiments on CdS (6,8) also indicate that α may be affected by the trapping of the excess electrons or holes produced by the ultrasonic wave. An appropriate theoretical expression for α has been worked out by including a complex, temperature dependent trapping time in the theory. (9, 10)

Recently (11) attenuation and velocity versus temperature data were obtained on n-GaAs which, when compared with the theoretical expressions of Hutson and White (3), allowed apparently very accurate values for $|e_{14}|$ to be deduced which agreed with each other to within a few percent. Both values were also within experimental error of the values implied by the most accurate previous determinations of e_{14}. In order for it to be possible to determine values for $|e_{14}|$ from the measured attenuation and the Hutson and White theory, ω_c must change from being much larger than ω to being much smaller than ω in the temperature range covered. This requirement can be satisfied if, as the temperature of the sample is lowered, a large number of the conductivity determining charge carriers, i.e. electrons or holes, fall from high mobility band states into localized impurity states.

In this paper we shall report on the ultrasonic attenuation of two III-V semiconductors, InSb and GaP. The particular specimens used were selected so that the requirement mentioned above be satisfied. From our results we shall deduce values for $|e_{14}|$. For GaP this will be the first accurate determination of $|e_{14}|$. We shall show, in addition, that the values of e_{14} for InSb and GaP and those for other zinc blende structure semiconductors (both III-V's and II-VI's) are understandable in terms of a stress induced change in ionicity or transfer of electron bonding charge (12, 13) which has a functional dependence on the Phillips ioncity (14) like that derived recently by Hidaka. (15)

Experimental

An echo-voltage-ratio attenuation measuring system similar to that described elsewhere (16) was used to provide pulses of 30 MHz ultrasonic shear and longitudinal waves. Temperatures between 4.2K and 240K were achieved by means of a commercial, variable-temperature cryostat and temperature controller with GaAs diode sensor and were measured using commercial Ge or Pt resistance thermometers or a copper-constantan thermocouple.

The samples were X-ray oriented, large single crystals of copper-doped InSb (which was p-type in the extrinsic range) and of sulfur-doped, and nominally undoped, n-type GaP which were lapped flat to 0.05µ and parallel to 0.2 sec of arc.

Rectangular parallelopiped samples were obtained from material adjacent to that which provided the ultrasonic samples in order to allow electrical characterization measurements made by means of a potentiometric measuring system. A 4" Varian electromagnet was used to provide 4000 gauss for determining the Hall coefficient, R_H. Cryogenic liquids were used to lower the temperature of the electrical samples below 296K. Temperatures were determined with a copper-constantan thermocouple.

From our electrical measurements we found that our p-InSb(Cu) and n-GaP(S) samples had resistivities, ρ, ($= 1/\sigma$) such that $\sigma/\epsilon \gg \omega$ at high temperatures and $\sigma/\epsilon \ll \omega$ at low temperatures. The very large changes of ρ indicate that freeze out of band carriers into impurity levels must be occuring. Evidence for carrier freeze out in the n-GaP(S) sample was obtained directly from Hall data on that sample.

The attenuation of 30 MHz waves in our samples is given as a function of temperature in Fig. 1. Actual data points have been omitted to avoid clutter.

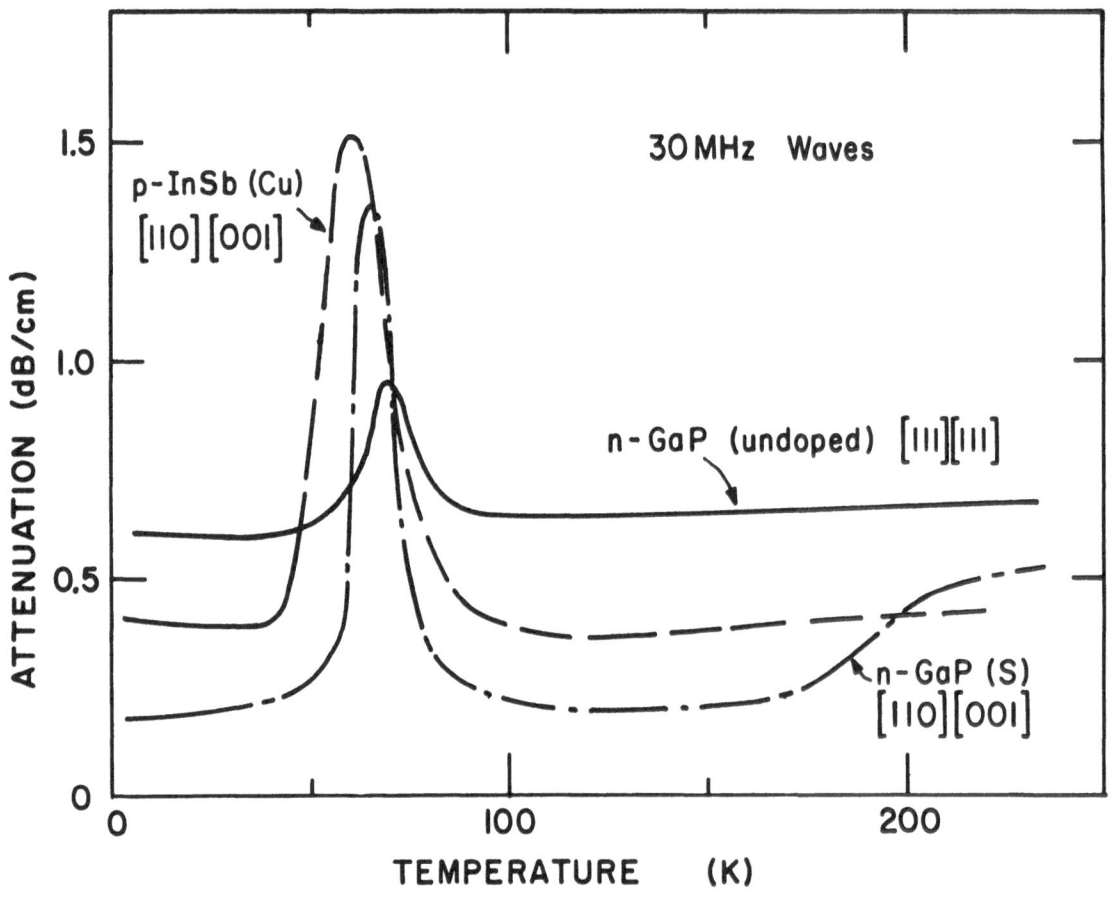

FIG. 1

Ultrasonic attenuation versus temperature. The directions of propagation and polarization of the waves are given by the first and second set of crystallographic indices, respectively.

From Fig. 1 it can be seen that there is a maximum in the attenuation whose location and height depend on the sample in question. Accompanying each maximum there is an extra dependence of the velocity on temperature. (16) Since these effects do not occur for nonpiezoelectrically active modes (16), the piezoelectric origin of the maxima is clear and we shall analyze them below. Before doing so, we note that outside the interval where it peaks, the attenuation is similar to that observed for dielectric crystals and arises, presumably, in the same way, i.e. partly from the experimental technique (17) and partly either from the relaxation of the ultrasonically produced disturbance by means of thermal phonon interactions at the higher temperatures (18,19) and from direct scattering of the ultrasonic phonons by (finite lifetime) thermal phonons at the lowest temperatures.(20, 21, 22)

In order to analyze the attenuation maxima we first note that for piezoelectric semiconductors which are homogeneous in conductivity, theory (3,9) gives for the attenuation due to piezoelectric coupling,

$$\alpha_{pz} = \frac{K^2 \omega}{2 v_o} \left[\frac{a\left(\dfrac{\omega^2}{\omega_c \omega_D'} + \dfrac{\omega}{\omega_c} a\right) + \dfrac{\omega}{\omega_c}}{\left(\dfrac{\omega}{\omega_c} - a\right)^2 + \left(1 + \dfrac{\omega^2}{\omega_c \omega_D'} + \dfrac{\omega}{\omega_c} a\right)^2} \right] \qquad (1)$$

where for [110] [001] waves: $K^2 = e_{14}^2 / C_{44} \epsilon$ and $v_o = \sqrt{C_{44}/d}$; while for [111] [111] waves: $K^2 = 4e_{14}^2/(C_{11}+2C_{12}+4C_{44})\epsilon$ and $v_o = \sqrt{(C_{11}+2C_{12}+4C_{44})/3d}$; d = density, $a = \omega\tau(1-f_o)/(f_o+\omega^2\tau^2)$, τ is the trapping time, f_o is the fraction of acoustically produced space charge which is mobile in the absence of trapping effects, $\omega_D' = \omega_D/b$, and $b = (\omega^2\tau^2+f_o^2)/f_o(\omega^2\tau^2+f_o)$. When the effects of carrier diffusion and finite trapping time can be neglected, i.e. ω_D is much larger than both ω and ω_c, $a = 0$, and $b = 1$, Eq. (1) reduces to the familiar relaxation form,

$$\alpha_{pz} = \frac{K^2 \omega}{2 v_o} \left[\frac{\omega/\omega_c}{1 + (\omega/\omega_c)^2} \right] \qquad (2)$$

which goes through a maximum with a value of $K^2\omega/4v_o$ at $\omega = \omega_c$ (i.e. at the temperature for which the resistivity becomes equal to $1/\omega\epsilon$). By equating $(\alpha_{pz})_{max}$ from Eq. (2) to $(\alpha_{expt})_{max}$ minus α_{back} where α_{back} is obtained from a smooth curve drawn to connect attenuation data outside the temperature interval where the maximum in α occurs, we were able to determine a value of $|e_{14}|$ for each of our samples. The results are given in Table 1.

TABLE 1

Values of $|e_{14}|$ Determined from the Maximum Attenuation of Waves Listed

p-InSb	0.058 ± 0.006 C/m^2	[110][001] Fast Shear
n-GaP(S)	0.085 ∓ 0.002 "	[110][001] Fast Shear
n-GaP (undoped)	0.082 ∓ 0.003 "	[111][111] Longitudinal

The values of $|e_{14}|$ listed in Table 1 agree within experimental error with those deduced from velocity measurements (16). The most accurate previous value of e_{14} for InSb is -0.071 ± 0.007 C/m^2. Our values of $|e_{14}|$ for GaP are the first accurate ones available, although a value of 0.1 C/m^2 of unspecified accuracy has been mentioned in the literature (23).

We turn now to the question of accounting for the shape and location of the whole attenuation peak observed for each sample. As a first attempt to do so we tried substituting the resistivity data for our p-InSb(Cu) and n-GaP(S) samples into Eq. (1) and (2) to calculate α_{pz} as a function of temperature. The results were disappointing. The calculated curves were somewhat different in shape and had their maximum values at somewhat different temperatures than did the respective $\alpha_{expt} - \alpha_{back}$ curves obtained from our ultrasonic measurements. Discrepancies persisted even when the trapping time and f_o in Eq. (1) were allowed to be temperature dependent. Since conductivity inhomogeneities have been found to be responsible for lack of agreement between calculated and experimental α's in CdS(6), inhomogeneities might be suspected as the cause of the discrepancies we have found. Unfortunately a direct quantitative check of this suspicion is not possible since no theoretical expression for α_{pz} is available which includes the effect of conductivity inhomogeneities, and we don't have any detailed knowledge about the homogeneity of our samples anyway.

Nevertheless, we have attempted to allow for inhomogeneities as indicated below. Macroscopic inhomogeneity, which causes the resistivity of an ultrasonic sample to differ from that of its partner electrical sample, we allow for by using an arbitrary resistivity of the form $\rho_o \exp (E/kT)$ instead of the measured resistivity of the electrical sample for determining the ω_c to be used in calculating α_{pz}. Other inhomogeneity effects we allow for by letting K be arbitrary rather than using the value of K obtained previously from the velocity data. (16) Values for the three quantities ρ_o, E, and K were obtained from the best fit between the α_{pz} versus T curve calculated using Eq. (2) and our experimental results.

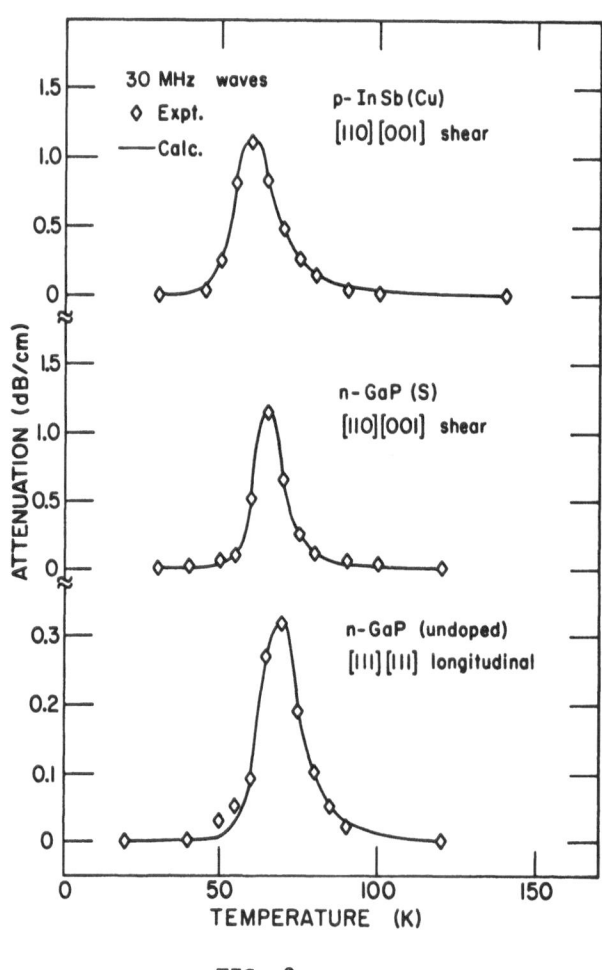

FIG. 2

Ultrasonic attenuation versus temperature. The curves were calculated using Eq. (2) and parameter values given in Table 2. Points give measured attenuation less background attenuation.

From Fig. 2 it can be seen that the calculated curves can reproduce both the shape and location of our experimental attenuation peaks. The values employed for K, ρ_o and E are given in Table 2., which also includes values of K obtained from velocity data and values of ρ_o and E from our measured resistivities.

TABLE 2

Values of the Coupling Constant K and of the Resisitivity Parameters.

	$K(10^{-2})$		$\rho_o(10^{-4}\Omega\text{-m})$		$E(10^{-2}eV)$	
	α fit	from Δv	αfit	Meas.	α fit	Meas.
p-InSb (Cu)	2.6	3.1	9.6	2.2	5.4	4.3
n-GaP (S)	3.3	3.3	0.036	0.032	9.27	9.25
n-GaP (undoped)	2.2	2.4	1.5	-----	7.5	----

The values of ρ_o and E deduced from fitting α are reasonable in terms of the electrical properties usually reported for similar materials. (24,25) From Table 2 it can also be seen that for GaP there is excellent agreement between the values of K obtained by both methods. In addition, for the n-GaP(S) sample, the values of ρ_o and E from α are quite close to the values deduced from the resistivity of our electrical sample. The results for n-GaP(S) indicate that when the amount of inhomogeneity is small, the experimental attenuation does approach that calculable from the Hutson and White theory. For our p-InSb(Cu) sample, however, the values of K, ρ_o, and E obtained from fitting α don't agree with those from velocity and resistivity data. These discrepancies may reflect nonuniformities in Cu doping.

Finally, it is appropriate to consider a microscopic interpretation of the piezoelectric constants. From previous work (13) has emerged a phenomenonological equation for the piezo-electric constant of zinc blende structure crystals. It is

$$e_{14} = (\zeta - \tfrac{1}{2}s) (e*/e) (16/a^2) \qquad (3)$$

where e_{14} is in units of C/m^2, ζ is the internal strain parameter which is calculable from elastic constant and optical vibrational frequencies (26), and represents the effect of displacement of ionic and (bonding) electronic sublattices relative to each other, s is the stress induced (electronic) charge redistribution index or change in ionicity, a is the lattice parameter in $\AA$, and e* is an effective charge which we believe (16) is well represented by the so called Szigeti charge. (27)

It was noticed some time ago (13) that s depends on f_i, the ionicity as defined by Phillips. (14) However only recently has theory been developed which yields an expression for se* as a function of f_i. (15) Because we believe it is more appropriate to do so (16), we shall use a similar functional form to represent the dependence of s on f_i, namely:

$$s = Bf_i/(1 + Cf_i^2) \qquad (4).$$

Hidaka (15) made an attempt to connect C with the force constant describing the interaction between electronic bonding charges on second nearest neighbor lattice ions. The connection is somewhat tenuous so that we regard both C and B as parameters to be obtained by making a best fit of Eq. (4) to all the values of s available for zinc blende structure compounds. Fig. 3 shows s plotted versus the ionicity and a curve calculated using Eq. (4). Included in Fig. 3 are the values of s deduced from our $|e_{14}|$ values for InSb and GaP. It was assumed that e_{14} has a negative sign as has been shown to be the case for a number of III-V's. (12)

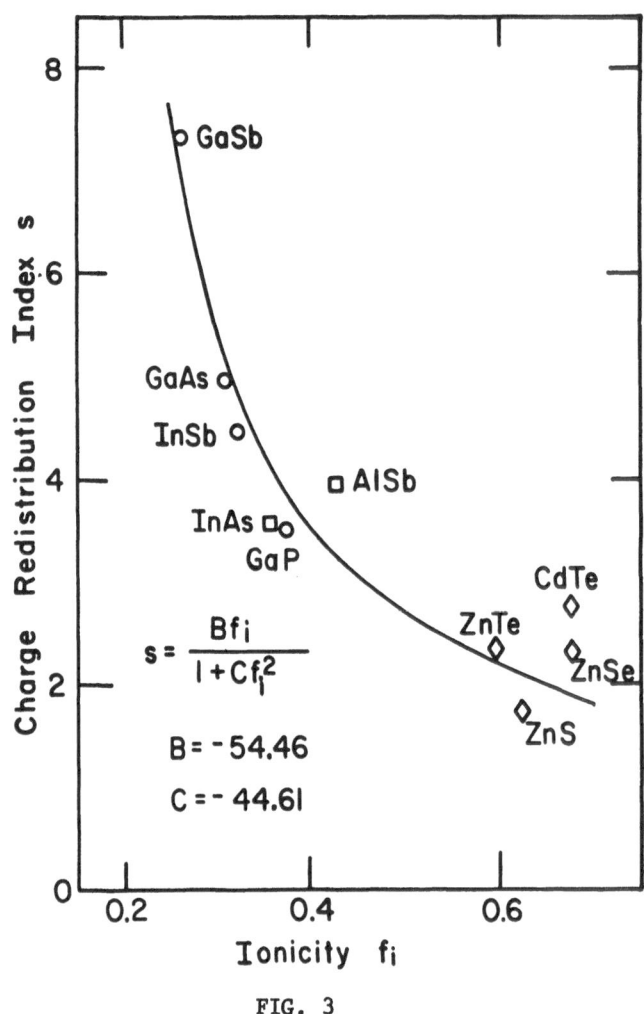

FIG. 3

Points give values of charge redistribution index deduced from e_{14} data using Eq. (3). Curve is a least squares fit of the equation shown to the data.

From Fig. 3 it can be seen that the calculated curve fits all the plotted points rather well. The fit is better than that achieved in previous attempts to relate s (13) or se* (15) to f_i. Part of the reason for our success is that contrary to the practice of some authors (13,15) we have used a Szigeti rather than a Callen effective charge. Martin (28) has shown that a transverse change is appropriate. However in discussing e_{14} he uses the Born charge. The Szigeti charge which we used does reduce to the Born charge in the limit of a zero polarization factor. Because of the partially ionic, partially covalent nature of the III-V's, we believe that the most appropriate charge would have a value somewhere between the Born and Szigeti charges.

Although we cannot give a detailed justification of the values of B and C which produce the best fit of Eq. (4) to our data, we should like to observe that the location of electronic bonding charge between atoms is involved in the theory leading to the expression on which Eq. (4) is based. The success of Eq. (4) thus implies support for the bond charge density model for bonding in solids. (13, 29)

Conclusions

1) The peak in the attenuation of piezoelectrically active ultrasonic waves which occurs as a function of temperature in each sample of InSb and GaP results from a change in the concentration of electrons (or holes) available to screen the piezoelectric interaction.

2) Quantitative interpretation of these peaks indicate the presence of conductivity inhomogeneities in the samples.

3) Values of the piezoelectric constant $|e_{14}|$ can be deduced from the size of these peaks relative to the background attenuation.

4) The correlation between the transfer of electronic bonding charge and ionicity can be significantly improved when a more appropriate value of the effective charge is used than has been used previously.

5) Interpretation of piezoelectric constants in terms of recent microscopic theory provides support for such theory and the dielectric theory of solids.

Acknowledgements

This work was supported at various times by ARPA-IDL grant DAHC-0213, USARO-D contract DA-31-124-ARO-D-17 and NSF-MRL grant GH 33574. Also we wish to thank Dr. Paul Norton, Syracuse University and Dr. Michael Foster, IBM for supplying the InSb and GaP (undoped) samples respectively.

References

1.   V. M. Goldschmidt, Trans. Faraday. Soc. 25, 253, (1929).

2.   W. G. Cady, Piezoelectricity, Vol. 1, p. 229, Dover Publications, N. Y. (1964).

3.   A. R. Hutson and D. L. White, J. Appl. Phys. 33, 40 (1962).

4.   E. Harnik and T. Yasar, J. Appl. Phys. 36, 2086 (1965).

5.   E. A. Davis and R. E. Drew, J. Appl. Phys. 38, 2663 (1967).

6.   V. E. Henrich and G. Weinreich, Phys. Rev. 178, 1204 (1969).

7.   R. B. Wilson, J. Appl. Phys. 37, 1932 (1966).

8.   I. A. Viktorov, Soviet Phys.-Dokl 12, 487 (1967).

9.   I. Uchida, T. Ishiguro, Y. Sasaki, and T. Suzuki, J. Phys. Soc. Japan 19, 674 (1964).

10. A. R. Moore and R. W. Smith, Phys. Rev. 138, A1250 (1965).

11. W. F. Boyle and R. J. Sladek, Solid State Commun. 12, 165 (1973).

12. G. Arlt and P. Quadflieg, Phys. Stat. Sol. 25, 323 (1968).

13. J. C. Phillips, and J. A. Van Vechten, Phys. Rev. Letters 23, 1115 (1969).

14. J. C. Phillips, Phys. Rev. Letters 22, 645 (1969).

15. T. Hidaka, Phys. Rev. B5, 4030 (1972).

16. W. F. Boyle, Ph.D. Thesis, (Purdue University), 1973 (unpublished).

17. M. J. Keck and R. J. Sladek, Phys. Rev. B2, 3135 (1970).

18. A. Akhieser, J. Phys. (USSR) 1, 277 (1939).

19. T. O. Woodruff and H. Ehrenreich, Phys. Rev. 123, 1553 (1961).

20. T. Landau and G. Rumer, Phys. Z. Sowjetunion 11, 18 (1937).

21. I. S. Ciccarello and K. Dransfeld, Phys. Rev. 134, A1517 (1964).

22. H. J. Maris, Proc. Phys. Soc. (London) 9, 901 (1964).

23. D. F. Nelson and E. H. Turner, J. Appl. Phys. 39, 3337 (1968).

24. T. Hara and I. Akasaki, J. Appl. Phys. 39, 285 (1968).

25. K. I. Vinogradova, D. N. Nasledov, Yu. G. Popov, and Yu. S. Smetannikova, Bull. Acad. Sci. USSR, Phys. Ser. (USA) 28, 863 (1964).

26. R. M. Martin, Phys. Rev. B1, 4005 (1970).

27. M. Hass, in Semiconductors and Semimetals, Vol. 3, p. 3, Edited by R. K. Willardson and A. C. Beer, Academic Press, N. Y. (1967).

28. R. M. Martin, Phys. Rev. B5, 1607 (1972).

29. M. L. Cohen, Science, 179, 1189 (1973).

Ultrasonic Attenuation in Thin Polycrystalline Tin Films

John U. Free, Jr.[*] and M.W.P. Strandberg
Massachusetts Institute of Technology
Cambridge, Massachusetts

This paper is a preliminary report on the study of
longitudinal ultrasonic waves reflected from a thin metallic
film at low-temperature. The experiments were performed on
evaporated tin films between 2.8-μ and 6.3-μ in thickness.
The measurements were made in the frequency range 8.9 GHz. to
9.4 GHz. The pulse echo technique, with a slight variation,
was used.   In Figure 1 we show a diagram of the experimental
apparatus.   The tin film was deposited on one end of a quartz
rod. The other end was excited by a conventional non-resonance
piezoelectric method (1) using an X-band re-entrant cavity.
Thus, amplitude measurements were made on the pulse echoes
reflected from the tin film. In order that relative measure-

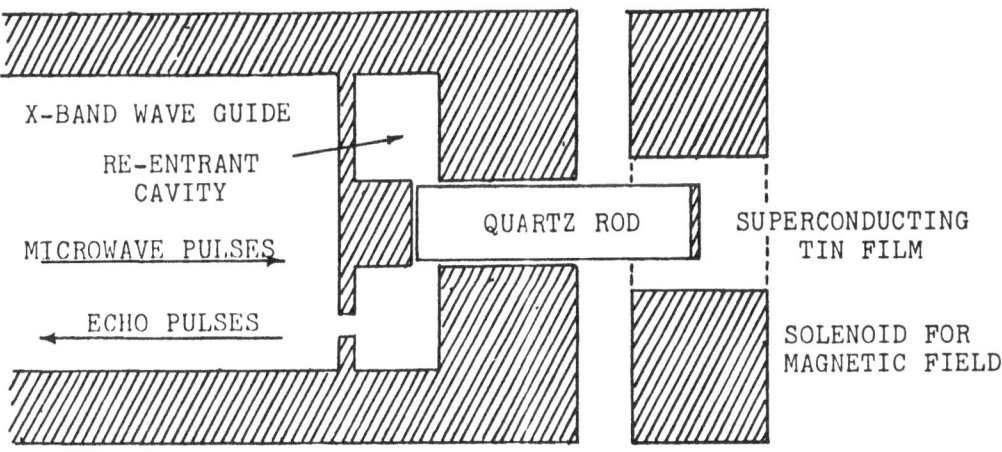

Figure 1. Apparatus for the observation of ultrasonic waves
reflected from a thin superconducting film.

---

[*] Author is now at Eastern Nazarene College, Quincy, Mass.

ments could be made, the difference between the ultrasonic wave
reflected from the film in the superconducting and normal state
was measured. The film was switched between the normal and
superconducting state by means of an external magnetic field.

When the pulse length is long compared with the film
thickness, the waves reflected back and forth in the film can
interfere. Thus, the amplitude of the wave is a superposition
of the waves returned from the film and the wave reflected from
the front surface. This is shown in Figure 2. We have assumed
that the front surface has some areas at which the tin and
quartz are not in contact. The two interfaces are characterized
by a complex amplitude reflection coefficient $r_{12}\,e^{i\phi_{12}}$ and
$r_{23}\,e^{i\phi_{23}}$ for the front and back surface respectively. If the
attenuation per wavelength is small, the losses in the tin film
can be characterized by $\alpha(\mathrm{cm}^{-1})$, the attenuation coefficient.

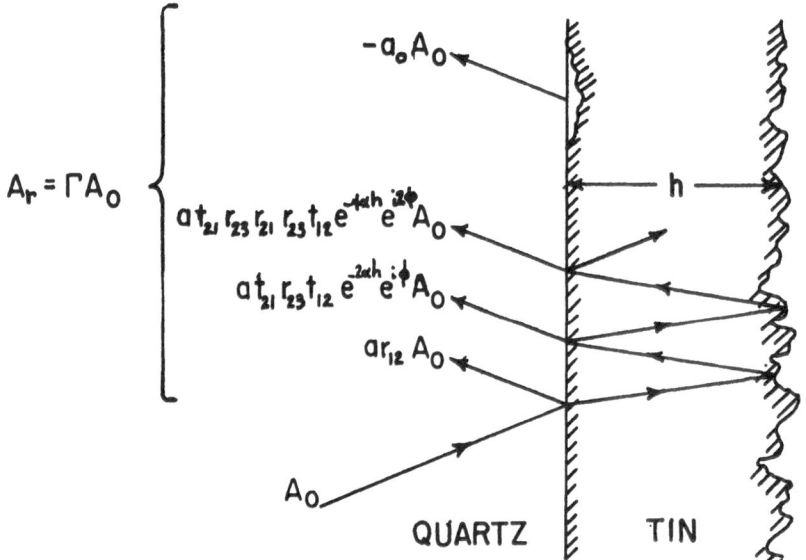

Figure 2. Multiple reflections of a wave in a thin film.

The total amplitude of the ultrasonic wave reflected from
the film, A_r, can be obtained by adding the sum of the geometric
series representing multiple reflections within the film to
the wave reflected from the areas not in contact (2). In the
case of normal incidence, the ratio of the total reflected
amplitude to the incident amplitude is

$$\Gamma = \frac{A_r}{A_o} = \left[\frac{r_{12}e^{i\phi_{12}} + r_{23}e^{i\phi_{23}} e^{-2\alpha h} e^{i\phi}}{1 + r_{12}e^{i\phi_{12}} r_{23}e^{i\phi_{23}} e^{-2\alpha h} e^{i\phi}} \right] a - a_o \tag{1}$$

where a and a_o are the total relative areas in contact and not
in contact, respectively. We have also substituted minus one
for the reflection coefficient at the areas not in contact.
The quantity ϕ is a measure of the phase retardation within
the film.

$$\phi = \frac{4\pi h}{\lambda} = \frac{4\pi h \nu}{V} \tag{2}$$

where λ (cm) is the ultrasonic wavelength in the tin film,
ν (sec^{-1}) is the frequency, and V (cm/sec) is the longitudinal
ultrasonic velocity. In principle, the experiment seems straight
forward. One would use the acoustical mismatch equation to
calculate the amplitude reflection coefficient, r, for two
two solids in perfect contact;

$$r_{ij} = \frac{\rho_j V_j - \rho_i V_i}{\rho_j V_j + \rho_i V_i} \tag{3}$$

where ρ_i (gm/cm^{-3}) is the density of the ith medium, and V_i
is the longitudinal ultrasonic velocity for the ith medium.

One would then use Equation 1 to calculate the ultrasonic
attenuation coefficient, and relative areas in contact and not
in contact. In practice, it is not this simple. For example,
the back surface may not be smooth compared to the ultrasonic
wavelength. Thus, only a fraction of the wave is specularly
reflected. The magnitude of the reflection coefficient, r_{23},
is therefore a fraction of the theoretical value. The front
surface may also be non-ideal. Due to impurities on the surface the bonds could be weak, and thus, the magnitude and
phase of the amplitude reflection coefficient for the front
surface, $r_{12} e^{i\phi_{12}}$, could deviate from the ideal reflection
coefficient.

One is also plagued with the normal problems associated
with ultrasonic experiments at 10 GHz. For example, if the
crystallographic axis and the cylinder axis of the quartz rod
are not parallel the ultrasonic wave will walk off axis as
the wave bounces back and forth in the quartz rod. This leads
to a loss of energy between successive echoes. A second problem
appears if the end faces of the quartz rod are not parallel.
In this case, the phase can vary over the end surface of the
rod. Since the signal detected depends on the phase, a fraction
of the energy in the wave is detected. The variation of the
phase angle on the interface causes other problems. The areas
in Equation 1 are the results of integrating over the end
surface of the quartz rod. The introduction of a variable phase
angle causes the areas to become effective areas, as in the

use of Fresnel zones in diffraction. This also causes the
two areas to be out of phase by an effective phase angle.
Thus, one must cope with a number of troublesome effects when
performing ultrasonic attenuation measurements at 10 GHz.

In the course of studying the thin films, we found the
area not in contact to vary from zero up to about 60 per cent
of the total area. In this paper we will consider just the
samples in which we believe the film to be in contact over the
total area. If the area not in contact is zero, Equation 1
becomes

$$\Gamma = \frac{A_r}{A_o} = \frac{r_{12} e^{i\phi_{12}} + r_{23} e^{i\phi_{23}} e^{-2\alpha h} e^{i\phi}}{1 + r_{12} e^{i\phi_{12}} r_{23} e^{i\phi_{23}} e^{-2\alpha h} e^{i\phi}} \tag{4}$$

The attenuation in the film can be conveniently broken
up into two terms: (i) a residual attenuation due to grain
boundaries, impurities, and other defects, and, (ii) an
attenuation due to the conduction electrons. Thus, the
attenuation coefficient can be expressed as

$$\alpha = \alpha_o + \alpha_{ele} \tag{5}$$

where α_o is the residual attenuation coefficient and α_{ele} is
the electronic attenuation coefficient. When the metal is a
superconductor the electronic attenuation goes to zero at absolute zero. Thus, the attenuation in tin in the superconducting
state is essentially the residual attenuation, α_o.

The difference between the amplitudes of the ultrasonic
waves reflected from the tin film in the superconduction state

and the normal state is expressed by

$$D_{S,N} = 10 \log \left| \frac{\Gamma_S}{\Gamma_N} \right|^2 \qquad (6)$$

$$= 10 \log \left[\frac{\left(\dfrac{r_{12}^2 + r_{23}^2 e^{-4\alpha_0 h} + 2 r_{12} r_{23} e^{-2\alpha_0 h} \cos(\phi - \phi_{12} + \phi_{23})}{1 + r_{12}^2 r_{23}^2 e^{-4\alpha_0 h} + 2 r_{12} r_{23} e^{-2\alpha_0 h} \cos(\phi + \phi_{12} + \phi_{23})} \right)}{\left(\dfrac{r_{12}^2 + r_{23}^2 e^{-4(\alpha_0 + \alpha_{ele})h} + 2 r_{12} r_{23} e^{-2(\alpha_0 + \alpha_{ele})h} \cos(\phi - \phi_{12} + \phi_{23})}{1 + r_{12}^2 r_{23}^2 e^{-4(\alpha_0 + \alpha_{ele})h} + 2 r_{12} r_{23} e^{-2(\alpha_0 + \alpha_{ele})h} \cos(\phi + \phi_{12} + \phi_{23})} \right)} \right]$$

In Figure 3 we have plotted $D_{S,N}$ as a function of film thickness, h. From the figure, one can see there are two distinct regions for $D_{S,N}$. The first region is characterized by $D_{S,N}$ always being positive. In this region the film is said to be over coupled because the wave returned from the film dominates the wave reflected from the front surface. In the second region, $D_{S,N}$ is both positive and negative. The film is under coupled in this region because the wave reflected from the front surface now dominates. If the parameters, r_{12}, $r_{23} e^{-2\alpha_0 h}$, or $e^{-2\alpha_{ele} h}$ in Equation 6 are changed, the form of $D_{S,N}$ stays the same, but the point where the wave reflected from the front surface equals the waves returned from the film, increases or decreases, depending on how the parameters are changed.

In Figures 4 and 5 we show the data for two typical films. The film in Figure 4 is in the over coupled region, while the film in Figure 5 is from the under coupled region. The solid curves were obtained from Equation 6.

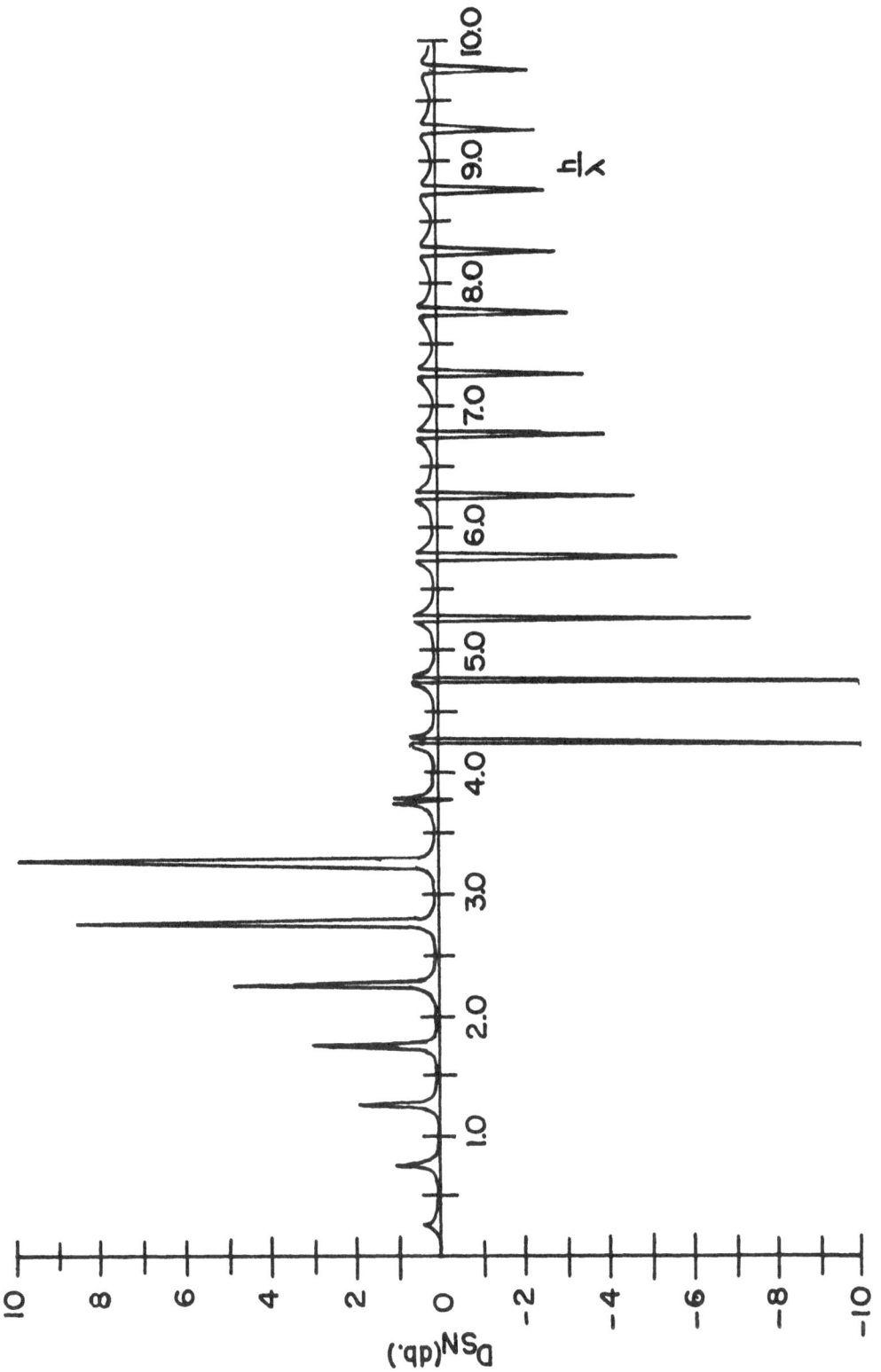

Figure 3.   A plot of the amplitude reflection coefficient for a thin film as a function of film thickness.

113

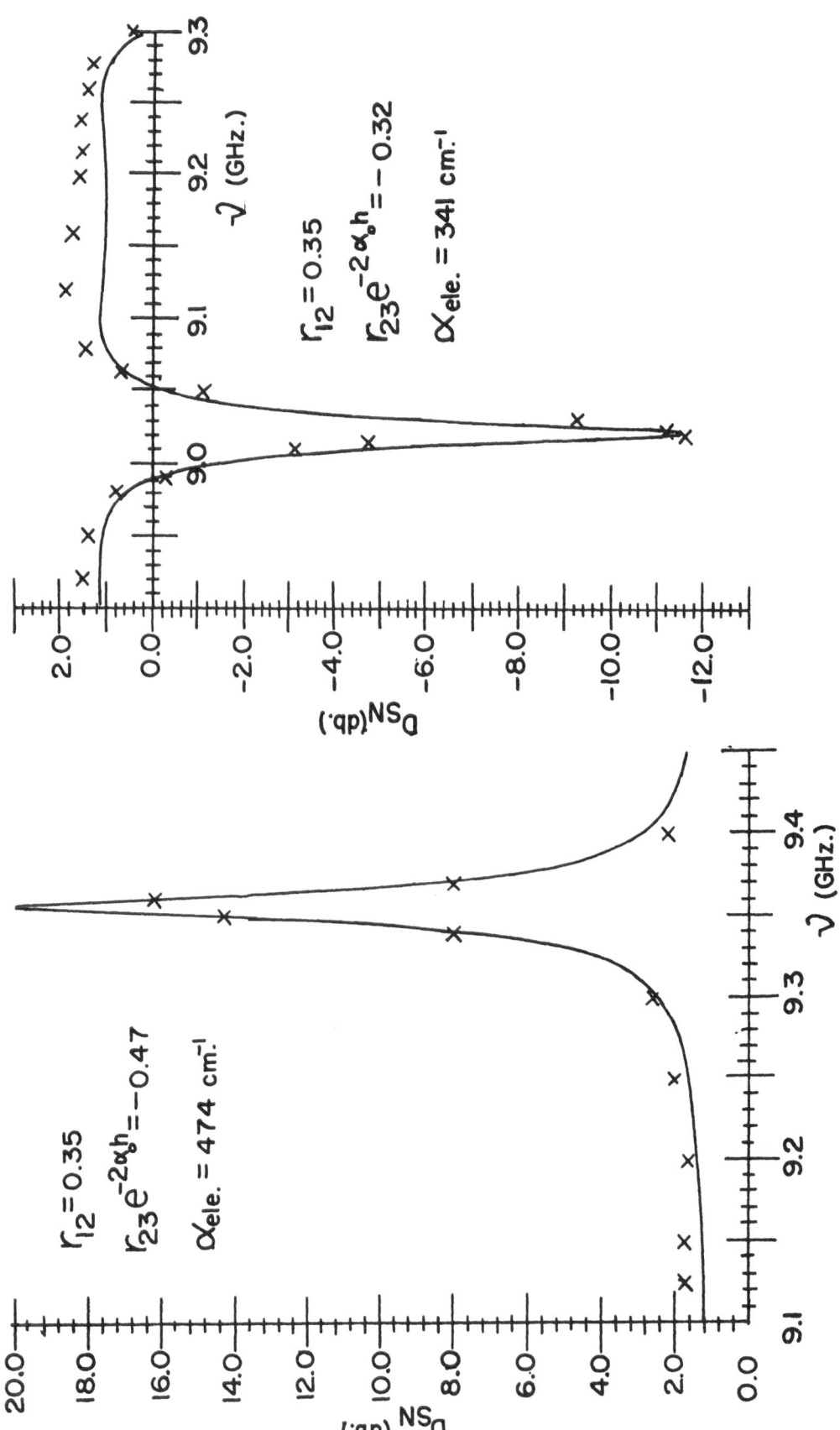

Figure 5. Data for a film in the under coupled region.

Figure 4. Data for a film in the over coupled region.

114

The parameters in Equation 6 were varied using a computer program based on the method of multiple regression to obtain the "best" fit to the data. The "best" fit was assumed to be the minimum value for the root mean squared deviation between the observed values and the calculated values. We found that the data could be fit, to within the experimental error, for a range of values of the parameters.

Some method was needed to fix one of the parameters. Thus, measurements were made on thick films in order to determine the magnitude of the first surface reflection coefficient, r_{12} . The films were so thick that the transmitted wave was completely absorbed in the film. The first surface reflection coefficient was found to be 0.35 ± 0.15. This should be compared with 0.22 for the theoretical amplitude reflection coefficient calculated from Equation 3 (3). We see that the theoretical value falls within the experimental error.

Assuming the value of r_{12} was 0.35 ± 0.15 we calculated the other parameters, $r_{23} \, e^{-2\alpha_o h}$ and $e^{-2\alpha_{ele} h}$. The electronic attenuation was found to vary from film to film between $300 \pm 50 \; cm^{-1}$ and $600 \pm 50 \; cm^{-1}$. This is consistent with the theoretical expression for the electronic attenuation, if the electron mean free path is less than the ultrasonic wavelength. Pippard derived a theoretical expression for the electronic attenuation (4). He found the electronic attenuation was a function of the mean free path divided by the ultrasonic wavelength. For the case where the mean free path of the electron

was small compared with the ultrasonic wavelength, he found
that the electronic attenuation depended linearly on the mean
free path and on the frequency squared. Thus, if we assume
the grain size is small compared to the wavelength the electronic attenuation will depend on the grain size. We will
see below, that it is necessary to assume the grain size is
less than the ultrasonic wavelength in order to explain the
data.

The parameter $r_{23} e^{-2\alpha_0 h}$ was also calculated. From Equation 6,
we see that r_{23}, the magnitude of the back surface reflection
coefficient, and the residual attenuation coefficient, α_0, can
not be separated. Thus, if one wishes to calculate r_{23} or α_0
he must make an independent measurement of one of the parameters
or assume a value for one of the parameters. If we assume a
realistic value of -0.8 for r_{23} we find that α_0 is between
500 cm^{-1} and 2,000 cm^{-1}. Thus, the residual attenuation is
between one and five times the electronic attenuation.

The residual attenuation was also found to be frequency
dependent. In Figure 5 one can note a small downward slope
to the data between 9.1 GHz. and 9.27 GHz., whereas our curve
for $D_{S,N}$ does not have this slope. One way to account for this
downward trend is to assume the attenuation is frequency dependent. If the residual attenuation is due to grain boundary
scattering, the attenuation coefficient may depend on the
frequency. When the grain size is small compared to the ultrasonic wavelength, we get the normal Rayleigh scattering with

116

a frequency dependence of the fourth power. We were able to account for the slope in the data by assuming the residual attenuation depended on the frequency to the fourth power. This is also consistent with the results for the electronic attenuation.

It will also be noted that there is a small dip in the top of the curve when $D_{S,N}$ is positive (between 9.1 GHz. and 9.3 GHz.) in Figure 5. The depth of the dip depends on the valve of the parameters. By varying the first surface reflection coefficient, r_{12} , we found the set of parameters that best fit the dip. We found that r_{12} was 0.4 ± 0.1 in agreement with the data from the thick films.

It has been found that the parameters which characterize a film vary from film to film depending on how the film was prepared. In general, we found that the area not in contact was between zero and 60 per cent. The first surface reflection coefficient was found to be larger than the theoretical value of 0.22. While the second surface reflection coefficient could not be separated from the residual attenuation, we found that the residual attenuation was one to five times the electronic attenuation, if we assumed r_{23} was -0.8. We also found the electronic attenuation to be between 300 cm^{-1} and 600 cm^{-1} in agreement with the results of Pippard.

We have presented a brief survey of the study of ultrasonic waves reflected from a metallic film. A more extensive report is expected in the future.

References

(1) E.H. Jacobsen, J. Acoust. Soc. Amer. $\underline{32}$, 949 (1960).

(2) J.M. Andrews, Q.P.R. No. 83, Research Laboratories of Electronics, M.I.T.

(3) In calculating the theoretical reflection coefficient we used the bulk density of tin 7.3 gm/cm^3 and 3.3 x 10^5 cm/sec for the longitudinal ultrasonic velocity in polycrystalline tin.

(4) A.B. Pippard, Phil. Mag. $\underline{46}$, 1104 (1955).

THE ELECTRONIC ATTENUATION OF ULTRASOUND IN THE INTERMEDIATE STATE
OF ALUMINIUM AND ZINC

M.J. Lea, K.R. Lyall, D.J. Meredith and E. Read.
Department of Physics, University of Lancaster, Lancaster LA1 4YB, England.

Introduction

The intermediate state of a type I superconductor (1) consists of
alternate laminae of superconducting and normal material, and is produced by
applying a magnetic field $\underset{\sim}{B}_a$ to a sample such that $B_p = (1 - D)B_c < B_a < B_c$
where B_c is the critical field, B_p is the penetration field and D is the
demagnetising factor for the sample. Ultrasound can be used to study the
intermediate state (2) since the electronic attenuation α_s in a superconductor
decreases rapidly below the transition temperature T_c and is zero at OK, while
the normal state attenuation α_n is usually independent of temperature in this
range. The attenuation in the intermediate state α_i depends on the relative
orientation of the sound wavevector $\underset{\sim}{q}$ and the phase boundaries which are
parallel to $\underset{\sim}{B}_a$. If $\underset{\sim}{B}_a$ is parallel to $\underset{\sim}{q}$ then the intensities of the sound
travelling in the two phases must be added (ignoring any effects specific to
the phase boundaries) to give

$$\exp(-\alpha_i) \quad = \quad \eta \exp(-\alpha_n) \quad + \quad (1 - \eta) \exp(-\alpha_s) \tag{1}$$

where η is the fraction of normal material in the sample. If the laminae are
perpendicular to $\underset{\sim}{q}$ then each part of the sound beam traverses equal fractions
of normal material and hence

$$\alpha_i \quad = \quad \eta \, \alpha_n \quad + \quad (1 - \eta)\alpha_s \quad . \tag{2}$$

We present here some ultrasonic experiments on aluminium and zinc (see also
reference 3) in the parallel field geometry.

Aluminium

Measurements of the ultrasonic attenuation in the intermediate state of
aluminium (T_c = 1.178K) were made in single crystal disks at frequencies near
9 GHz. Because of the extremely large normal state attenuation ($\approx$0.34 dB μm^{-1})
at these frequencies, the 10mm diameter samples were lapped to a thickness, L,
of less than 100 μm. For $\underset{\sim}{B}_a$ perpendicular to the disk (parallel to $\underset{\sim}{q}$), D is

then close to unity and these thin microsonic samples are in the intermediate state from almost zero applied field to the critical field. The repeat distance, a, of the intermediate state structure depends on L and B_a/B_c (4) and we estimate a to be $\sim$ 60 μm.

Figure 1 shows the measured attenuation of 8.73 GHz longitudinal phonons (wavelength λ = 0.74 μm) propagating along the ⟨100⟩ direction in a thin sample of aluminium (residual resistivity ratio R = ρ(300K)/ρ(0K) = 150) at 1.11K as a function of B_a in the parallel field geometry. The fraction of normal material is given by $\eta = (B_a - B_p)/(B_c - B_p)$ assuming that the field in the normal regions is the critical field. The solid lines are the theoretical field dependences of the attenuation using equations (1) and (2). The agreement of the data with equation (1) is excellent except at very low fields. In this experiment the electron mean free path $\ell \simeq 2.4$ μm is much smaller than a.

Figure 2 shows the results of a similar experiment at 9.3 GHz using a much purer sample of aluminium (R = 2,000) at 1.02K. The field dependence of the attenuation is now best described by equation (2) rather than equation (1). We postulate that this is due to the non-local nature of the attenuation process when ℓ ($\simeq$ 40 μm for this pure sample) is of the order of a and greater than the width of the individual laminae. Since qℓ is large for both samples ($\simeq$ 400 for the pure sample), the acoustic phonons interact only with the 'surf-riding' electrons moving perpendicular to $\underset{\sim}{q}$. Faber (5) has shown that the intermediate state structure in a disk is irregular with the normal regions linked together. Hence if ℓ is large, the effective electrons will sample both normal and superconducting regions and equation (2) should then be valid even for $\underset{\sim}{B}_a$ parallel to $\underset{\sim}{q}$, as shown in figure 2.

Zinc

Measurements on the intermediate state of zinc were made by trapping magnetic flux in the cylindrical ultrasonic samples. This was done by quickly removing a parallel magnetic field greater than B_c at temperatures well below T_c (0.84K). Figure 3 shows the electronic attenuation of 75 MHz longitudinal ultrasound in a sample of pure zinc (thickness 3 mm) as the temperature was increased. In curve I, the sample was completely superconducting while, in curve II, it contains some magnetic flux which was trapped below 0.2K. α_n is independent of temperature below 1K and hence from $\alpha_i(T)$ and $\alpha_s(T)$ we can compute the temperature dependence of η in the sound beam. If the average trapped flux density is $\bar{B}$ then

$$\eta = \bar{B}/B_c(T) \quad . \tag{4}$$

Hence if $\bar{B}$ remains constant as T is varied we expect $1/\eta$ to be proportional to $B_c(T)$ except close to T_c where the trapped flux will escape. Figure 4 shows $1/\eta$ versus $B_c(T)$ derived from the data in figure 3 using equation (1)(line A)

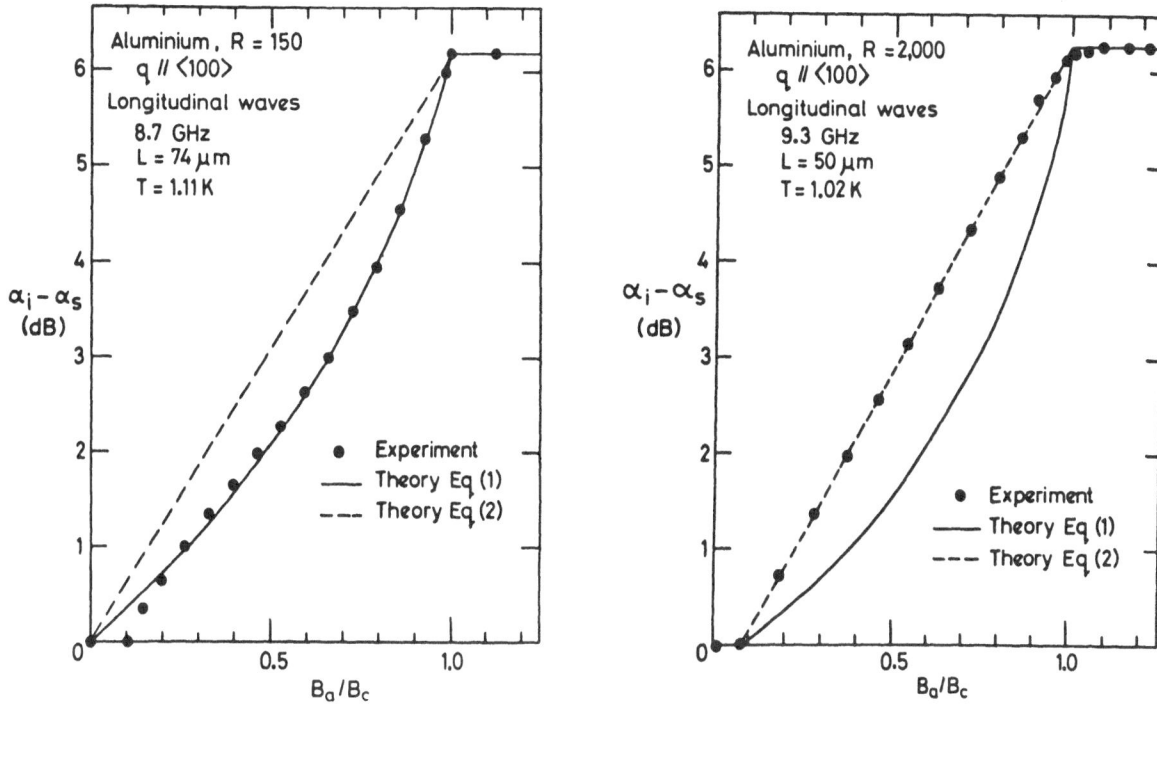

FIG. 1   FIG. 2

Ultrasonic attenuation in the intermediate state of aluminium

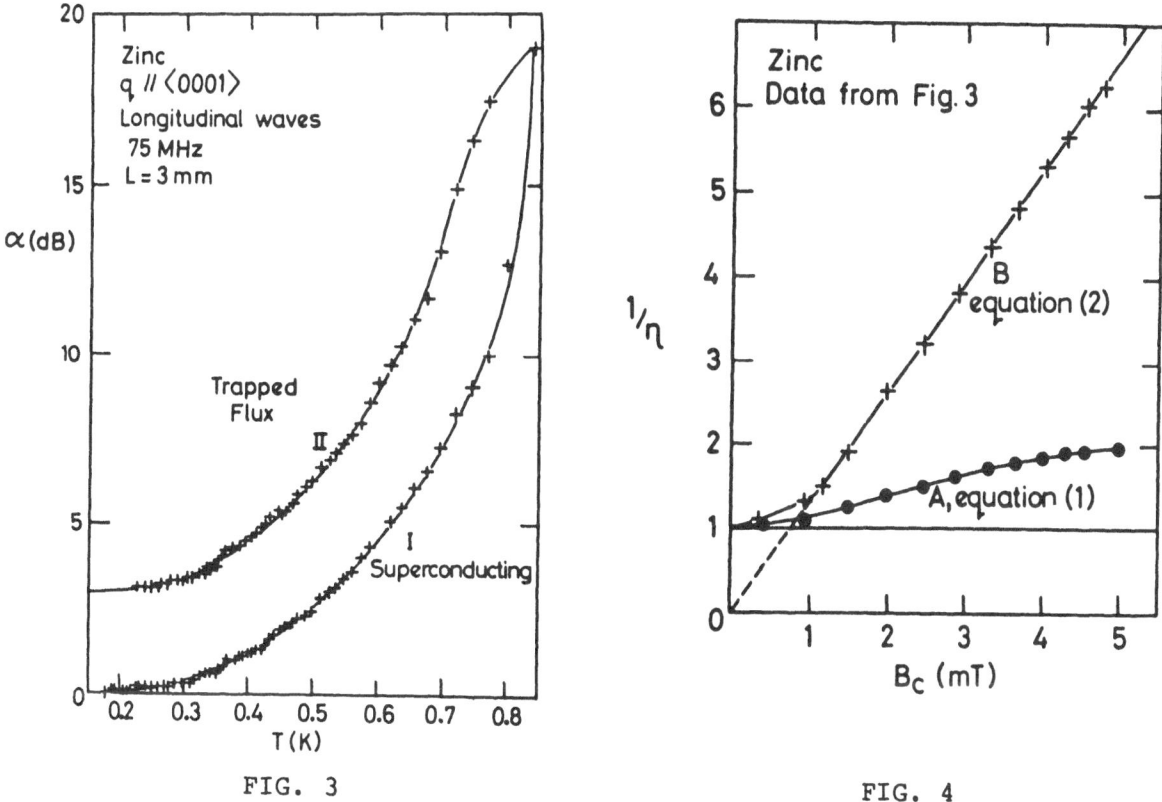

FIG. 3   FIG. 4

Ultrasonic attenuation in the intermediate state of zinc

and equation (2) (line B). The agreement with equation (4) is excellent if we assume that equation (2) is valid even though the trapped field is nominally parallel to q. We estimate a for this zinc sample to be 300 μm whereas ℓ is only $\simeq$ 35 μm. Thus non-local effects will be much less important than for our aluminium samples. However, we would also expect equation (2) to be valid if the laminae are tilted with respect to q by more than some critical angle $\theta_c \sim 5^\circ$ in this case. Because of the method used to produce the intermediate state such misorientation is quite likely and this is the most probable reason for the validity of equation (2) in this case.

Conclusion

In several experiments in the parallel field geometry, we have found equation (2) to be valid rather than equation (1) and two reasons are suggested. Firstly, when the electron mean free path exceeds the laminae thickness, the attenuation mechanism becomes non-local in nature. Secondly, with thick samples a slight misorientation of the phase boundaries and the sound wave-vector means that each part of the sound beam samples the same fraction of normal material.

References

1. J.D. Livingston and W. DeSorbo, Superconductivity, Ed. R.D. Parks, Vol.2, p. 1235. Dekker, New York (1969).

2. M. Gottlieb, M. Garbuny and C.K. Jones, Physical Acoustics, Ed. W.P. Mason, Vol. VII, p. 1. Academic Press, New York (1970).

3. M.J. Lea and E. Read, J. Phys. F, in press (1973).

4. P.G. de Gennes, Superconductivity of Metals and Alloys, p. 40 . Benjamin, New York (1966).

5. T.E. Faber, Proc. Roy. Soc. A248, 460 (1958).

ULTRASONIC ABSORPTION IN KONDO SUPERCONDUCTORS

P.A. Hilton and D.J. Meredith
Department of Physics, University of Lancaster, Lancaster LA1 4YB, U.K.

Introduction

The absorption of very high frequency phonons in superconductors provides a convenient measurement of the temperature dependence and anisotropy of the energy gap. Although the same information can often be acquired from the low frequency result of B.C.S.

$$\alpha_s/\alpha_n = 2/\left[1 + \exp(\omega_g(T)/kT)\right]$$

where $2\omega_g(T)$ is the energy gap parameter, the sharp absorption edge that occurs for longitudinal sound waves, when the phonon energy is sufficient to destroy a Cooper pair, i.e. $\hbar\omega = 2\omega_g(T)$, gives a precise and more direct measure of $\omega_g(T)$. Measurements in aluminium at 9 GHz [1], when the edge occurs only 2 mK below the transition temperature, is not infinitely narrow but has a finite width of $\sim$ 1 mK. If the broadening is associated with gap anisotropy, the addition of non-magnetic impurities to the superconductor should result in a smearing of the anisotropy, and a sharp discontinuity corresponding to a single gap would be expected when the Anderson criterion is satisfied, i.e. when the coherence length ξ and the electron mean free paths are equal. The addition of magnetic impurites leads to far more dramatic changes in the superconducting properties. A rapid decrease in the transition temperature and the order parameter (or pair potential) is accompanied by an intermediate superconducting phase where pairs of electrons form bound states and yet there is no gap in the quasiparticle spectrum. Magnetic superconductors are thus characterised by two parameters; $\Delta(T, n_i)$ a temperature and impurity dependent order parameter, which may be finite while the energy gap $\omega_g(T, n_i)$ is zero. In a BCS superconductor these two quantities are equal. For concentrations of impurity n_i greater than a critical value n_{cr}, the alloy has no superconducting phase.

Two types of magnetic impurities, rare earth and transition metal have been investigated experimentally and theoretically. For the former the initial rate of decrease of the transition temperature $dT_c/dn_i \sim 3.5$K/at.%, and their behaviour is moderately well described by the theory of Abrikosov and Gor'kov [2]. Larger depressions of T_c are found in Kondo superconductors, e.g. for Mn in Zn, $dT_c/dn_i \simeq 400K/at.\%$, and the inclusion of resonant scattering of conduction electrons by the impurities (see Müller-Hartmann and Zittartz (3)) leads to different predictions for the concentration dependence of T_c and for the formation of bound states within the energy gap, particularly when the transition temperature for the pure metal $T_{cp} \simeq T_K$, the Kondo temperature. Here the resonant scattering most strongly affects the superconducting transition as the spin flip probability reaches a maximum at T_K. The decrease in T_c is thus greatest for $T_K \simeq T_{cp}$, and Zittartz shows that for this condition, a narrow energy band which is formed at the middle of the gap, should be distinguishable from the pure superconductor excited energy states. When $T_{cp} \gg$ or $\ll T_K$, impurity scattering has a reduced effect and the formation of impurity bands close to the gap edge is similar to the broadening of the density of states predicted by Abrikosov and Gor'kov.

Theory

The absorption of high frequency phonons in pure superconductors to include pair breaking effects has been described by Bobetic (4), and ultrasonic attenuation at low frequencies in magnetic superconductors by Kadanoff and Falko (5). Snow (6) has extended the latter work to include both quasiparticle scattering and pair breaking contributions to the total attenuation. A Green's function formulation is used to express the response of the system to a longitudinal sound wave and effects of impurities are included in additional scattering and also through the A.G. parameters for the excitation energies and the quasiparticle density of states. We would not expect these results to be correct for the strong Kondo superconductors, but in the absence of any other calculations, it is useful to note the predicted changes for these more weakly coupled magnetic superconductors. The functions for α_s/α_n cannot be expressed in analytic form for either the pure or the doped superconductor when pair breaking by phonons occurs. For pure zinc with a T_c of 0.87K, the pair breaking contribution to the total attenuation of 9 GHz sound is only effective in a temperature range 5 mK below T_c. As the sharp discontinuity of Bobetic results from the confluence of two sharp density of states factors in the transition probability for phonon scattering, the addition of a small number of impurity atoms, should be to smear the absorption edge (which will now occur at a lower temperature) if their main effect on the attenuation is through the broadened density of states at the reduced gap edge. At low phonon frequencies pair breaking effects will be negligible except at high impurity concentrations when the material is gapless for much of its superconducting temperature range. With increasing phonon frequency, the absorption edge will only be observed at low impurity levels, and a gradually rising curve replaces the edge as the impurity concentration is increased. A finite attenuation at zero temperature is predicted for $n_i > 0.4n_{cr}$ when the phonon energy is 40% of the zero

temperature gap for the pure superconductor. These experimental conditions
would be satisfied at 9 GHz in the zinc alloys that we are studying.

Experimental

Measurements have been made on zinc samples containing 6.5 and 10 ppm of
manganese. The initial decrease of transition temperature was found to be
400K/at.%. In table 1, we summarise these results and, on the basis of the
Abrikosov and Gor'kov theory, indicate the probable range of gapless superconductivity. The temperature variation of the normalized attenuation α_s/α_n
for the more dilute alloy is shown in figure 1. Longitudinal sound waves were
propagated at a frequency of 118 MHz along the $(10\bar{1}0)$ direction, and the solid
curve is calculated from the BCS expression with the appropriate temperature
dependent energy gap for the host matrix. These preliminary measurements
indicate that the experimental points lie above this line, as predicted by
Kadanoff and Falko. We are currently making measurements at 9 GHz on a range
of alloy samples to determine how pair breaking is influenced by magnetic
impurities and to compare the total attenuation changes with the predictions of
the available theories.

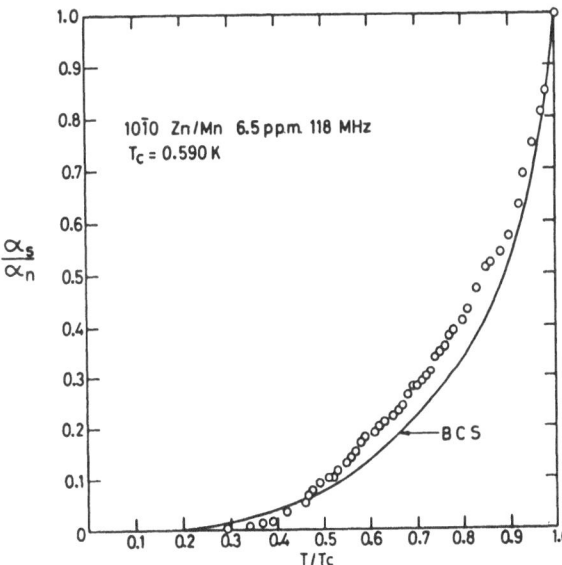

FIG. 1
Ultrasonic Attenuation in a Zn/Mn sample

TABLE 1

	T_{cp}	dT_c/dn_i	n_{cr}	$0.9n_{cr}$
Zn/Mn	0.87K	400K/at.%	15 ppm	14 ppm

Example   For T_c/T_{cp} = 0.5, T_c = 0.43K, n_i = 10 ppm, n_i/n_{cr} = 0.65

Gapless region occurs from T_c = 0.43K to T = 0.41K.

References

1. E. Hughes, D.J. Meredith and E.R. Dobbs, Phys. Lett. <u>38A</u>, 325 (1972).

2. A.A. Abrikosov and L.P. Gor'kov, Soviet Phys. J.E.P.T. <u>12</u>, 1243 (1961).

3. E. Müller-Hartmann and J. Zittartz, Z. Physik <u>232</u>, 11 (1970).
                                        "   " <u>234</u>, 58 (1970).

4. V.M. Bobetic, Phys. Rev. <u>136</u>, 1535 (1964).

5. L.P. Kadanoff and I.I. Falko, Phys. Rev. <u>136</u>, A1170 (1964).

6. J.A. Snow, Phys. Rev. <u>172</u>, 455 (1968).

ULTRASONIC ATTENUATION AND THE ELECTRICAL TRANSITION IN Ti$_2$O$_3$.

T. C. Chi and R. J. Sladek

Department of Physics, Purdue University, West Lafayette, Indiana, 47907 USA

ABSTRACT Measurements of the attenuation, α, of ultrasonic waves in single crystal samples of Ti$_2$O$_3$ have been made at various frequencies, f, between T = 298K and 525K. The α versus f data at 298K revealed that phonon viscosity losses were mainly responsible for the intrinsic attenuation. The attenuation of longitudinal waves propagating along the a-axis or close to the c-axis and of shear waves propagating along the a-axis goes through a maximum in a temperature range close to that where the gradual electrical transition occurs (400K to 500K). The peak in α for a-axis longitudinal waves has been correlated with that calculated for the $\langle \gamma^2 \rangle$ parameter in the phonon viscosity attenuation using thermal expansion and elastic constant data. The maximization of the ratio of anharmonic to harmonic lattice forces which this implies is attributed to the change in electronic screening of important interionic interactions by Ti 3d electrons as the overlap and relative populations of the a$_{1g}$ and e$_\pi$ subbands progress through the electrical transition. The screening seems to be most effective where the lattice parameters are most temperature dependent. Alternative attenuation mechanisms are considered but fail to explain the attenuation maxima satisfactorily.

Introduction

Ti$_2$O$_3$ exhibits a gradual semiconductor-to-semimetal transition (1,2,3) as the temperature is increased from about 400K to 500K, which is accompanied (4,5) by an expansion of the c-axis and a contraction of a-axis of the crystal which however maintains (6) its R$\bar{3}$c rhombohedral, α-corundum type structure. Anomalies in the specific heat (7,8), the Raman frequencies (9,10) and elastic constants (11) also occur. Based on the idea that increasing temperature causes Ti 3d a$_{1g}$ and e$_\pi$ - like Ti 3d subbands to overlap (12), quantitative interpretations of the anomalies in the elastic constants (11) and in the specific heat (8) were made by employing rigid subbands whose separation, or overlap, depends on temperature and, for the case of the elastic constants, on stress also. According to a free energy model of the transition, (13,14) which includes band electron entropy, elastic and electron-electron energies, the band overlap is driven mainly by competition between electron-electron interaction energy and band electron entropy.

Cooperative interaction between the lattice modes and the electronic states (15) has been invoked to reconcile Raman data on Ti$_2$O$_3$ with the way the electrical transition is affected by replacing Ti with V to form (Ti$_{1-x}$V$_x$)$_2$O$_3$ with 0<x<0.1. That the effect produced by doping with V is relevant to an understanding of the transition in undoped Ti$_2$O$_3$ is consistent with the fact that the spacings between the metal ions are changed in similar ways when the V content (16) or the temperature (17) is increased.

In the hope of clarifying the roles played by the electrons, the lattice and their interactions in Ti_2O_3 we began an investigation of the ultrasonic attenuation in this material. It was expected, for example, that in the vicinity of the electrical transition there might be evidence for the interband electron transfer which accounted for the anomalies in the elastic constants (11). Furthermore, ultrasonic attenuation might reveal information about other phenomena which may be connected with the electrical transition. Among these phenomena are changes in the lattice forces which are involved in the so called phonon viscosity attenuation (18, 19), rapid but continuous change in the degree of order near a critical temperature (20), and fluctuations in order which cause attenuation by scattering the ultrasonic waves (20, 21).

Below we shall report and discuss our investigation of the ultrasonic attenuation in Ti_2O_3. Our main purpose was to study the temperature dependence of the attenuation. However in the hope of obtaining a better understanding the mechanism(s) responsible for it, the attenuation was also measured as a function of frequency at a few temperatures as well.

Experimental

Measurements of the attenuation were made using an echo-voltage-ratio attenuation measuring system comprised of commercial electronic components as described previously (22). The "home-made" sample holder and furnace, the commercial temperature controller, and the temperature measuring and recording systems (thermocouples and potentiometer and strip chart recorder, respectively) were the same as used in an investigation of the elastic constants of Ti_2O_3 (11) and have been described elsewhere. (23) Quartz transducers were bonded to the large single crystal samples of Ti_2O_3 at room temperature by means of Dow Corning 710 silicone fluid. Bond failure at elevated temperatures was frequent and has severely restricted the amount of data we have obtained so far.

Results and Discussion

Attenuation versus frequency data at room temperature are presented in Fig. 1 for longitudinal waves traveling parallel to the a-axis of the crystal. From Fig. 1 it can be seen that the three term expression shown can yield a calculated curve which fits the data quite well. In view of previous work on other materials we identify the frequency independent term with bond, transducer, and other extraneous losses, and the $1/f$ term with diffraction effects. (24) The f^2 term is intrinsic to the sample, and, is, we believe, due mostly to the phonon viscosity loss mechanism for the case $2\pi f \tau_p \ll 1$ where τ_p is the thermal phonon relaxation time. This belief is based on theoretical predictions for the behavior of phonon viscosity attenuation (19). An approximate expression for such attenuation has been used with success to account for attenuation in dielectric solids (25). An equivalent version (19) of that expression is

$$\alpha_p = \langle \gamma^2 \rangle \, C_v T \, \omega^2 \, \tau_p / 3 \bar{\rho} \bar{v}^3 \tag{1}$$

where $\langle \gamma^2 \rangle$ is the average of the square of a Grüneisen-like parameter, C_v the heat capacity, T is the absolute temperature, $\omega = 2\pi f$, τ_p is taken to be $3K/C_v\bar{v}^2$ where K is the lattice thermal conductivity, $\bar{v}$ is an average velocity calculated from the Debye θ and the atomic volume (25). For Ti$_2$O$_3$ at room temperature ρ = 4.58 gm/cm^3, $\bar{v}$ = 5.1 x 10^5 cm/sec and K = 4.4 x 10^5 ergs cm^{-1} deg^{-1} sec^{-1}. As is usually the case, obtaining a value for $\langle \gamma^2 \rangle$ is a problem. (25). In order for Eq. (1) to account for the intrinsic attenuation (f^2 term) deduced for our sample, $\langle \gamma^2 \rangle$ would have to be equal to 7.8. Since this value is unusually large, we wondered if perhaps our method of deducing the intrinsic attenuation (f^2 term) was somewhat faulty. To check this possibility we tried fitting the α vs f data by means of expressions which are somewhat different from the one given in Fig. 1.

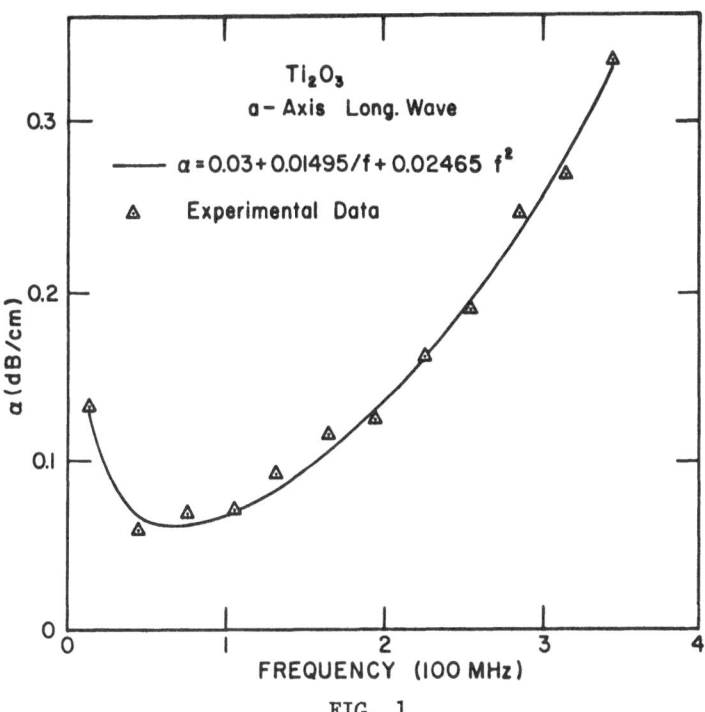

FIG. 1

Ultrasonic attenuation versus frequency at room temperature. The numerical coefficients were chosen to yield the best fit of the calculated curve to the data.

We found that about as good a fit was obtainable using an expression which had a term linear in f rather than a frequency independent term but still had 1/f and f^2 terms, although the coefficients of these terms had somewhat different values than those in the expression in Fig. 1. (The term proportional to f presumably still represents extraneous, non-diffraction losses in view of the fact there seems to be some precedent for residual attenuation to increase with increasing f. (26).) Using the new f^2 term and Equation (1) we again obtain a rather large value for $\langle \gamma^2 \rangle$, namely, 5.3. To test the reasonableness of these large values for $\langle \gamma^2 \rangle$ we shall calculate a value for $\langle \gamma^2 \rangle$ from measured quantities assuming that $\langle \gamma^2 \rangle$ is given by

$$\langle \gamma^2 \rangle = (2\gamma_a^2 + \gamma_c^2)/3 \qquad (2)$$

where γ_a and γ_c are the thermal Grüneisen parameters for the a and c crystallographic directions. In order to connect γ_a and γ_c to measured quantities we shall use expressions appropriate for hexagonal crystals. (27) These expressions should be adequate since C$_{14}$ is extremely small in Ti$_2$O$_3$ so that this material is similar elastically to a hexagonal crystal for which $C_{14} = 0.$ (28)   The expressions are

$$\gamma_a = (2 C_{13} \beta_a + C_{33} \beta_c)/C_p \qquad\qquad\qquad (3)$$

$$\gamma_c = \{(C_{11} + C_{12})\beta_a + C_{13}\beta_c)\}/C_p \qquad\qquad (4)$$

where β_a and β_c are linear thermal expansion coefficients, the C_{ij} are elastic constants, and C_p is the heat capacity   per unit volume at constant pressure. Recent references provide thermal expansion and density data (23) and give values for the C_{ij} (11) and the heat capacity (8). Employing this information in Eq. (2), (3), and (4) we obtain a value of 1.5 for $\langle\gamma^2\rangle$. It should also be noted that values of $\langle\gamma^2\rangle$ about twice as large as the 1.44 usually found for dielectric crystals (25) are sometimes needed to account for attenuation which definitely seems to be due to phonon viscosity effects. These values for $\langle\gamma^2\rangle$ thus approach those needed for the phonon viscosity to be responsible for the f^2 term in the attenuation. This fact coupled with the uncertainty about the actual size of the intrinsic attenuation and the approximate nature of Equations (1) and (2) lead us to conclude that phonon viscosity losses are a major cause of the intrinsic attenuation of longitudinal ultrasonic waves traveling in the a-direction in Ti_2O_3. Since, in some substances, thermoelastic losses (29) produce an appreciable amount of attenuation which is proportional to f^2, we have calculated a value for this type of attenuation in Ti_2O_3 and have found it to be about 10% of the phonon viscosity attenuation. Thus inclusion of thermoelastic losses would result in the calculated attenuation being only slightly closer to the intrinsic attenuation.

Attenuation versus temperature data are shown in Fig. 2 for longitudinal waves traveling parallel to the a-axis of the sample.

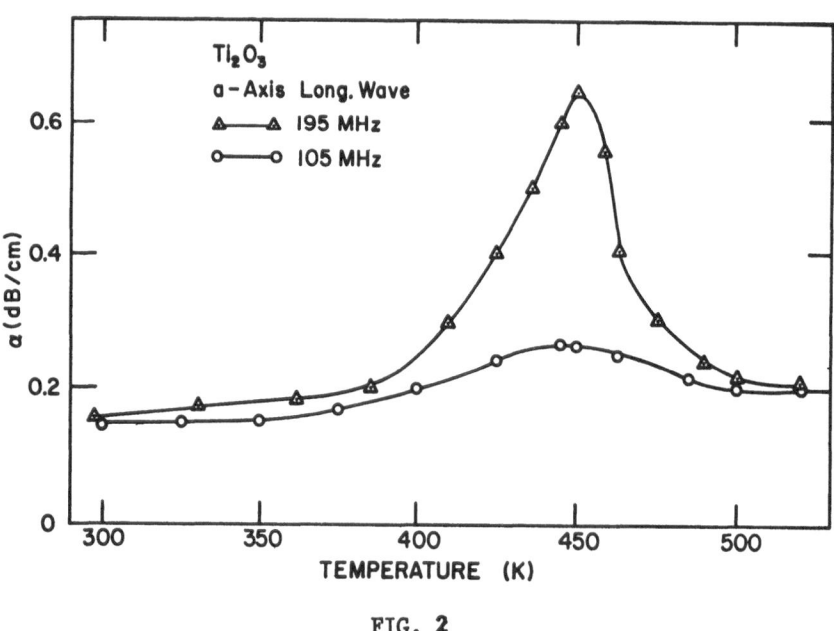

FIG. 2

Experimental ultrasonic attenuation versus temperature.

From Fig. 2 it can be seen that the attenuation goes through a maximum whose size depends strongly on the frequency of measurement. The maxima cover about the same temperature range as do the electrical transition (3) and the anomalies in the elastic constants (11). However, the attenuation reaches its highest value at a somewhat higher temperature for 195MHz than for 105MHz. If the nonintrinsic part of the attenuation is the same as it is at room temperature, then the intrinsic component of the peak is somewhat more than 4 times larger at 195MHz than at 105MHz.

We have also made some measurements of attenuation versus temperature employing 150MHz fast shear waves propagating in the direction of the a-axis. It was found that the attenuation went through a broad maximum in the temperature range of the electrical transition even though the velocity of these waves shows no minimum. (23) The significance of this will be discussed later. We have not presented a figure showing the shear attenuation data because the echo train was not quite exponential at some temperatures. Improved data will have to be obtained before drawing any quantitative conclusions about the particular size and location of the shear wave maximum.

Because of its availability, a sample, whose long direction was 9 deg. off the c-axis, was employed in measurements of the (apparent) attenuation of compressional waves as a function of frequency and temperature. Part of the observed attenuation may not be real because the group and phase velocities of the waves may not be parallel and leakage of energy into other modes might occur. This limits the significance of the particular magnitude and frequency dependence which the observed attenuation exhibits. However, we believe that the temperature dependence of the observed attenuation probably approximates that of pure longitudinal waves propagating in the direction of the c-axis and therefore we feel justified in presenting Fig. 3.

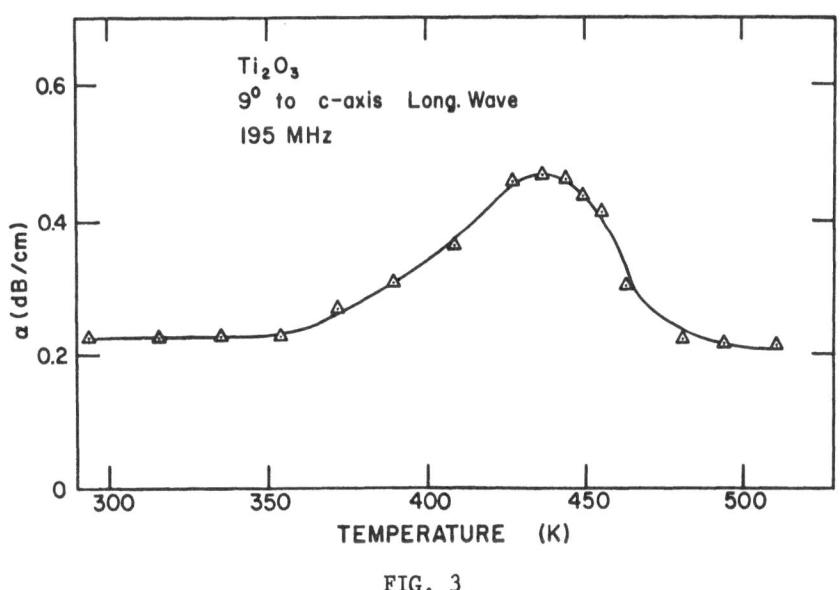

FIG. 3

Apparent ultrasonic attenuation versus temperature.

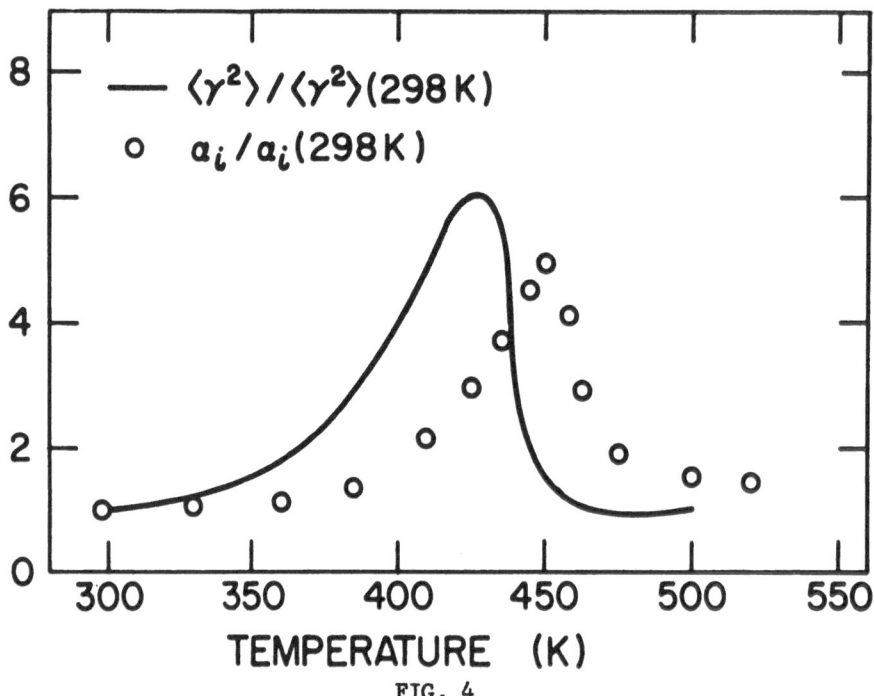

FIG. 4

Curve showing the temperature dependence calculated for $\langle \gamma^2 \rangle$ as indicated in the text and points showing the temperature dependence of the intrinsic attenuation deduced from our ultrasonic attenuation measurements at 195MHz using a-axis longitudinal waves in Ti_2O_3.

It might be asked how this conclusion is to be reconciled with the electronic interpretation (11) of the elastic constant anomalies which accompany the transition. Our answer is that the important interionic forces depend on the screening provided by Ti 3d electrons in Ti_2O_3. The screening depends on the degree to which the electrons have been transfered from a_{1g} to e_π subbands at each stage of the electrical transition and seems most effective when the lattice parameters have their maximum temperature dependences. Our interpretation of the attenuation maxima is closely related to the suggestion that there is a cooperative interaction between lattice modes and electronic configurations. (15)

A more direct way in which Ti 3d electrons might cause attenuation is via the relaxation of the stress induced transfer of electrons between the a_{1g} and e_π subbands. The relaxation time for restoring electronic equilibrium would depend on the spatial diffusion and interband scattering times. This idea for such attenuation has of course, been adapted from work on degenerate n-type semiconductors (31) in which intervalley, rather than interband, scattering occurs. This type of attenuation should be describable by a relation of the form

$$\alpha_{et} = -(C_{ij})_e \, \omega^2 \tau_{et} / 2\rho v^3 \tag{5}$$

where $(C_{ij})_e$ is the electronic contribution to the elastic constant and τ_{et} is the interband electron scattering time, which, in view of the values of spatial diffusion (31) and inter-valley scattering times (31, 32) in type IV semiconductors, is most likely to be much smaller than the spatial diffusion time and to satisfy the condition $2\pi f \tau_{et} \ll 1$. Indeed, if we

From Fig. 3 we see that the apparent attenuation has a maximum in the region of the electrical transition which is downshifted in temperature relative to the a-axis attenuation peak. (Compare Fig. 2), whereas the minimum in C_{33} occurs at a _higher_ temperature than does the minimum in C_{11}. (11)

It is now time to try to interpret the attenuation maxima which we have obtained. First let us consider if a relaxation process, characterized by a temperature dependent relaxation time τ, could be responsible for them. The most common type of maximum associated with relaxation processes is proportional to the function $\omega^2\tau/(1+\omega^2\tau^2)$. This function reaches its highest value when $\omega\tau = 1$ at a temperature which depends on frequency. The height of the maximum is proportional to the measuring frequency. Although the experimental attenuation peaks shown in Fig. 2 do seem to occur at somewhat different temperatures, as expected from the relaxation function, the ratio of the heights of their intrinsic components would require a quadratic, or even stronger frequency dependence, rather than the linear one given by the relaxation function. Therefore, we must seek another explanation for the experimental attenuation peaks.

Since the phonon viscosity mechanism accounted for much of the intrinsic attenuation deduced from the room temperature data in Fig. 1, we shall try to determine whether this mechanism could also produce the peaks observed in the attenuation. We again make use of Equation (1) since we expect the condition $\omega\tau_p \ll 1$ to be fulfilled even better at high temperatures than it is at room temperature. This expectation is based on the behavior of the lattice thermal conductivity of solids (30) which indicates that τ_p decreases strongly with increasing temperature in the temperature range of interest. The temperature dependences of all the remaining quantities in Eq. (1) except $\langle \gamma^2 \rangle$ are known and cannot cause large maxima in α like those which we observe. Thus only if $\langle \gamma^2 \rangle$ exhibits an appropriate maximum as a function of temperature could phonon viscosity losses be responsible for our experimental peaks. In order to see if $\langle \gamma^2 \rangle$ is likely to have such a behavior, we have used Equations (2), (3), and (4) to calculate $\langle \gamma^2 \rangle$ at various temperatures using appropriate values for the thermal expansion (23), elastic constants (11), heat capacity (18) and density. (23) Fig. 4 shows the temperature dependences of $\langle \gamma^2 \rangle$ and of the intrinsic attenuation.

From Fig. 4 it can be seen that $\langle \gamma^2 \rangle$ exhibits a maximum which is large enough to account for the temperature dependence of the intrinsic attenuation. The fact that $\langle \gamma^2 \rangle$ reaches its highest value at a lower temperature than the attenuation does is probably due mainly to systematic error in determining the temperature during sample length measurements but might also arise partly from quantitative deficiencies in Equations (1) to (4) and in the thermal expansion values we used. The general similarity of the behavior of $\langle \gamma^2 \rangle$ and the intrinsic attenuation indicates, we believe, that the observed attenuation peaks are explainable as phonon viscosity losses, whose temperature dependence is due mainly to the fact that the ratio of anharmonic to harmonic lattice forces goes through a maximum within the same temperature interval where the electrical transition occurs.

equate the right side of Eq. (5) to the peak values of the intrinsic attenuation, we obtain a value of about 2×10^{-13} sec for τ_{et}, which is reasonable since it is greater than the mobility relaxation time of 10^{-15} sec at 450K. For Ti_2O_3 $(-C_{11})_e$ goes through a maximum (11) and the velocity of a-axis longitudinal waves goes through a minimum (23) in about the same temperature range as where the measured attenuation has its maximum. More detailed analysis reveals that in order for Eq. (5) to fit the intrinsic attenuation peak quantiatively, τ_{et} would have to be much larger where the attenuation reaches its highest value that it is at both lower and higher temperatures. Since it seems impossible to account for the decrease in τ_{et} with decreasing temperatures at temperatures below the maximum, we believe that relaxation attenuation due to interband electron transfer cannot describe our experimental results. (For comparison we note that the intervalley scattering time in degenerate Ge decreases monotonically with increasing temperature at ordinary temperatures. (32)).

Before closing we would like to indicate briefly why we believe our attenuation peaks are not manifestations of relaxation or scattering effects of the type associated with phase transitions in which there is a continuous but rapid change in some order parameter. The attenuation peaks which occur in such cases do in fact have a quadratic frequency dependence (20) and in this respect are not unlike ours. However the peaks associated with order parameter effects always occur at a definite critical temperature, T_c, regardless of measuring frequency and, near T_c, have a shape which depends inversely on the difference between the temperature and T_c. Analysis shows that our experimental peaks do not have such a character. In addition we observed an attenuation maximum for shear waves which would not be expected when there is no change in crystal structure at the transition. Therefore we conclude that our peaks are not order parameter effects. Support for this conclusion is provided by the fact that order parameter effects are accompanied by sharper changes in elastic moduli than those implied by the anomalies observed in the elastic constants of Ti_2O_3. (11) The order parameter effect to be looked for in connection with the electrical transition in Ti_2O_3 would presumably have been connected with ordering of the Ti 3d electron system due to band motion.

Conclusions

Ultrasonic attenuation in Ti_2O_3 is due mainly to the phonon viscosity process in which the ultrasonically produced changes in the frequencies and populations of thermal phonon modes are relaxed via phonon-phonon scattering. Attenuation versus temperature data indicate that the ratio of anharmonic to harmonic lattice forces goes through a maximum within the temperature range of the electrical transition. From these results, and the behavior of the elastic constants and thermal expansion during the transition, it is concluded that the lattice forces are greatly influenced by electronic screening provided by Ti 3d electrons when the lattice parameters have their largest temperature dependences. Attenuation arising from interband electron transfer directly and from ordering, or fluctuations in ordering, were also considered and shown to be unable to account for the experimental attenuation maxima.

Acknowledgments

This work was supported by NSF Grant #GH33383 and benefited from NSF-MRL grant #GH33574. The large single crystals were provided by the Purdue Central Crystal Growth Facility supported by the latter grant.

References

1. M. Foex and J. Loriers, CR Acad. Sci. (Paris), 222, 901 (1948).

2. F. J. Morin, Phys. Rev. Letters, 3, 34 (1959).

3. J. M. Honig and T. B. Reed, Phys. Rev., 174, 1020 (1968).

4. R. E. Newnham and Y. M. DeHaan, Quart. Prog. Rept. No. XXVI, p. 10, Laboratory for Insulation Research, MIT, Cambridge, Mass., (1960).

5. C. N. R. Rao, R. E. Loehman and J. M. Honig, Phys. Letters, 27A, 271 (1968).

6. A. D. Pearson, J. Phys. Chem. Solids, 5, 316 (1958).

7. S. Nomura, T. Kawakubo and T. Yanagi, J. Phys. Soc. Japan, 16, 706 (1961).

8. H. L. Barros, G. V. Chandrashekhar, T. C. Chi, J. M. Honig, and R. J. Sladek, Phys. Rev. B, 7, 5147 (1973).

9. A. Mooradian and P. M. Raccah, Phys. Rev. B, 3, 4253 (1971).

10. S. H. Shin, R. Aggarwal, B. Lax and J. M. Honig, to be published.

11. T. C. Chi and R. J. Sladek, Phys. Rev. B, 7, 5080 (1973).

12. L. L. VanZandt, J. M. Honig, and J. B. Goodenough, J. Appl. Physics, 39, 594 (1968).

13. H. J. Zeiger, T. A. Kaplan, and P. M. Raccah, Phys. Rev. Letters, 26, 1328 (1971).

14. H. J. Zeiger, Bull. Am. Phys. Soc. 18, 399 (1973).

15. G. V. Chandrashekhar, Q. Won Choi, J. Moyo and J. M. Honig, Materials Res. Bull., 5, 999 (1970).

16. W. R. Robinson, submitted for publication.

17. P. M. Raccah, Bull. Am. Phys. Soc. 18, 339 (1973).

18. A. Akhieser, J. Phys. (USSR), 1, 277 (1939).

19. T. O. Woodruff and H. Ehrenreich, Phys. Rev. 123, 1553 (1961).

20. A. S. Nowick, and B. S. Berry, Anelastic Relaxation in Crystalline Solids, p. 463 ff, Academic Press, N. Y. (1970).

21. C. W. Garland, in Phys. Acoust. Vol. 7, p. 51, edited by W. P. Mason and R. N. Thurston, Academic Press, N. Y. (1970).

22. W. F. Boyle, Ph.D. Thesis, (Purdue University, 1973), (unpublished).

23. T. C. Chi, Ph.D. Thesis, (Purdue University, 1972), (unpublished).

24. R. Truell, C. Elbaum, and B. B. Chick, Ultrasonic Methods in Solid State Physics, p. 87ff, Academic Press, N. Y. (1969).

25. D. W. Oliver and G. A. Slack, J. Appl. Phys., 37, 1542 (1966).

26. M. J. Keck and R. J. Sladek, Phys. Rev. B, 2, 3135 (1970).

27. J. G. Collins, J. A. Cowan, and G. K. White, Cryogenics 7, 219 (1967).

28. H. B. Huntington, in Solid State Physics, Vol. 7, p. 214, edited by F. Seitz and D. Turnbull, Academic Press, N. Y. (1958).

29. W. P. Mason, in Physical Acoustics, Vol. III B, p. 255, edited by W. P. Mason, Academic Press, N. Y. (1965).

30. P. G. Klemens, in Solid State Phys., Vol. 7, p. 1, edited by F. Seitz and D. Turnbull Academic Press, N. Y. (1958).

31. M. Pomerantz, R. W. Keyes, and P. E. Seiden, Phys. Rev. Letters, 9, 312 (1962).

32. W. P. Mason and T. B. Bateman, Phys. Rev., 134, A1387 (1964).

On Mechanism of Attenuation Sound Oscillations
in Films at Superconduction Transition

V.S. Postnikov, I.V. Zolotukhin, V.E. Miloshenko, G.E. Shunin
The Polytechnical Institute Plekhanov Street 84 Voronezh USSR

The problem of attenuation of ultrasonic oscillations in a superconductor at superconduction transition has been studied by a number of workers.(see [I-7]). Only several works have been devoted to the investigation of sound oscillations ($\omega < 10^5$ c/s) [8-I2] . Specifically it was shown in works [I0-II] that at n-s transition a jump in the internal friction background (Q^{-I}) was observed, i.e. the background of the $Q^{-1}(\tau)$ superconducting state was lower than that of the normal state. In our further investigations we discovered a rather narrow internal friction peak in the region of n-s transition [I3-I6]. In the present paper we make an attempt to explain its nature.

Experimental Procedure

The investigation of internal friction of superconductors was performed at frequencies 50-6000 cps using the technique of flexural oscillations of a sample cantilever-mounted to the apparatus [I7] shown in Fig. I . Sample I having the shape of a plate I-200 μk in thickness was mounted in transducer 2 and excited electrostatically at the natural frequency. For this purpose a special transducer was designed which was included into the electronic circuit. A high-frequency generator signal modulated by the sample low frequency oscillations was received by a frequency deviator.

Intensified low-frequency signal component was transferred through an amplitude discriminator to a pulse meter. Data registration was carried out by a digital recorder. The transducer with the sample was placed into thermocontrolled bulk **3** of cryostat 4.

The experimental error of the internal friction measurements in helium temperature range was 0.5 %.

Experimental Results

Temperature dependence of internal friction of vanadium specimens is presented in Fig.2. It is seen that the internal friction peak is observed at n-s transition. The peak width for a deformed polycrystalline sample (curve I) reaches the magnitude of the order of $\mathcal{E} = \frac{T - T_c}{T_c} \sim 10^{-2}$; $\mathcal{E}$ for a deformed single crystal (curve 2) is smaller. The peak is not observed for a perfect single crystal. Below the temperature of the transition into the superconducting state a series of subsidiary peaks were observed (not shown in Fig.) , which as preliminary studies show, are of relaxation nature.

The experiments were made on polycrystalline vanadium and tantalum specimens for frequencies ranging from I to 6 kcps to determine sample oscillation frequency effects on the position and magnitude of the main peak $Q^{-1}(T)$. Fig.3 and 4 show the curves for frequencies I kcps (curve I) and 6 kcps (curve 2). It is seen that the peak width is apparently smaller with increasing frequency while its temperature position does not change.

It is to be noted that the internal friction background increases with frequency.

It can also be seen from Fig.2 and 3 that at the superconducting-normal transition of the sample the jump in the internal friction background is observed. Its magnitude is larger for polycrystalline specimens compared to that of single crystal ones.

Applied magnetic field with the magnitude of $H < H_c$ or H_{c2} to a superconductor leads to the peak shear into the lower temperature range. At $H > H_c$ (or H_{c2}) the jump and the internal friction peak dissapear.

Experimentally it is convenient to observe the superconduction transition by varying the external magnetic field since thermally activated processes leading to the increase of the background with temperature are excluded. Therefore we have investigated the external magnetic field dependence of internal friction for vanadium and tantalum specimens. The experimental results are given in Fig.5 and 6. They show that small internal friction peaks $Q^{-1}(H)$ are observed in small magnetic field range. The peaks $Q^{-1}(H)$ are also observed in the field region H_{c2} for vanadium (Fig.5) and H_c for tantalum (Fig.6). The peak width decreases with the sample deformation degree. The $Q^{-1}(H)$ peak is not observed for rather perfect single crystals in the region H_{c2} (or H_c) only the jump in the internal friction background is observed.

In conclusion it should be noted that for Lead-Indium alloy specimens (Pb + 4 at % In) in the third critical field region we observed a small internal friction peak exceeding the background by a factor I.5 .

Discussion

According to modern conceptions of superconduction theory [I8-20] at temperatures near critical it is necessary to take into account the influence of the order parameter fluctuations on superconductor kinetic properties. Let us consider fluctuation effects on sound attenuation in a superconductor at $T \to T_c$ (above). In this case the fluctuation field quantum propagation the so called "suprons" [2I] is described by the Bosonlike Green's function in the form [22]

$$\mathcal{D}(\omega_n, \bar{q}) = N^{-1}\left(\varepsilon + \frac{\pi}{8 T_c}|\omega_n| + \eta q^2\right)^{-1}, \quad (I)$$

where $\varepsilon = \frac{T - T_c}{T_c}$, $\eta = \frac{\pi \ell \upsilon_F}{24 T_c}$, $N = \frac{m^2 \upsilon_F}{2 \pi^2}$ − is the density of states on the Fermi surface; m − is the electron mass; υ_F − is the electron velocity on the Fermi surface; ℓ −is the electron mean free path; $\omega_n = 2\pi n T$; $n = 0, \pm I , \pm 2, \ldots$.

It is known that sound attenuation in metals is determined by the imaginary part of the polarization operator. In the first approximation fluctuation corrections are plotted in Fig.7 [23].

The work [22] shows that the first three expansion members do not lead to the increase of sound attenuation coefficient. Therefore we shall consider the contribution to the polarization operator given by the last expansion member, the so called

Aslamazov-Larkin diagram [23] which describes the process of supron scattering on sound phonons. This correction to the polarization operator may be analytically written in the form

$$\Pi_f(\omega_n, \bar{k}) = 4g^2 T \sum_{\omega'_n, q} \mathcal{D}(\omega'_n - \omega_n, \bar{q} - \bar{k}) \mathcal{D}(\omega'_n, \bar{q})$$

(2)

$$\times \left[T \sum_{\varepsilon_n, P} G(\varepsilon_n + \omega'_n - \omega_n, \bar{p} + \bar{q} - \bar{k}) G(\varepsilon_n + \omega'_n, \bar{p} + \bar{q}) G(-\varepsilon_n, -\bar{p}) \right],$$

where K is the sound wave vector; $G(\varepsilon_n, \bar{p}) = \dfrac{1}{i\varepsilon_n - \xi(\bar{p})}$, $\varepsilon_n = \pi n T$, $n = \pm 1, \pm 3, \pm 5, \dots$.

Since $K, q \ll P$, the summation over (ω'_n, q) and (ε_n, P) may be done separately. As a result one obtains the following expression:

$$\Pi_f(\omega_n, \bar{k}) = 4g^2 C^2 T \sum_{\omega'_n, q} \mathcal{D}(\omega'_n - \omega_n, \bar{q} - \bar{k}) \mathcal{D}(\omega'_n, \bar{q}),$$ (3)

where [6] $\quad C = T \sum_{\varepsilon_n, P} G^2(\varepsilon_n, \bar{p}) G(-\varepsilon_n, -\bar{p}) = \dfrac{m}{(2\pi)^2 v_F} ln\,\omega_{D}/2\pi T$,

ω_D is the Debye phonon frequency.

Now let us do summation in the expression (3) over discrete frequencies and analytical extension over ω_n using the technique given in [24] . As a result one obtains

$$\Pi_f(\omega, \bar{k}) = -\frac{4g^2 C^2}{N^2 \lambda_o^2} \sum_q \frac{n_-(-i\zeta_2) - n_-(-i\zeta_1)}{\omega + i\zeta_1 - i\zeta_2} ,$$ (4)

where the following notations are introduced:

$$\zeta_1 = \frac{\varepsilon + \eta \bar{q}^2}{\lambda_o} , \quad \zeta_2 = \frac{\varepsilon + \eta(\bar{q} - \bar{k})^2}{\lambda_o},$$

$$n_-(x) = \left(e^{\frac{x}{T}} - 1 \right)^{-1}, \quad \lambda_c = \frac{\pi}{8 T_c} .$$

141

Since $\zeta_1, \zeta_2 \ll T$, one gets approximately from (44)

$$\Pi_f(\omega, \bar{k}) \simeq - \frac{4g^2 C^2}{N^2 \lambda_0^2} iT \sum_q \frac{\zeta_1 - \zeta_2}{\zeta_1 \zeta_2 (\omega + i\zeta_2 - i\zeta_1)}. \quad (5)$$

The imaginary part of the polarization operator (5) has the form

$$\mathrm{Im}\, \Pi_f(\omega, \bar{k}) = -\omega g^2 \pi \frac{\ell_n^2 \omega_D / 2\pi T}{m^2 \upsilon_F^4} \sum_q (2\eta k q \cos\theta$$

$$- \eta k^2) \frac{1}{[(2\eta k q \cos\theta - \eta k^2)^2 + \frac{\pi^2}{8^2 T_c^2} \omega^2]} \quad (6)$$

$$\times \frac{1}{(\varepsilon + \eta q^2 + \eta k^2 - 2\eta k q \cos\theta)(\varepsilon + \eta q^2)} \quad .$$

Making in (6) the summation over q, in linear over ω approximation one gets the following expression:

$$\mathrm{Im}\, \Pi_f(\omega, \bar{k}) \sim -\omega g^2 \frac{\ell_n^2 \omega_D / 2\pi T}{m^2 \upsilon_F^4 \eta^{3/2} \varepsilon^{3/2}} \quad . \quad (7)$$

This expression coincides with that obtained for the imaginary part of the polarization operator in the [23]. Thus we may conclude that the static fluctuations give the main contribution to the longwave sound attenuation for superconductors near T_c. Comparing (4) with the expression $-\omega g^2 m^2 \ell \pi^{-2}$ for the imaginary part of the polarization operator of the normal state [25], one gets the expression for the temperature range in which fluctuation effects on sound attenuation are essential:

$$\frac{T - T_c}{T_c} \sim \frac{\ell_n^{4/3} \omega_D / 2\pi T_c}{m^{8/3} \upsilon_F^{8/3} \eta \, \ell^{2/3}} \quad (8)$$

From (8) one concludes that even for the case of a very dirty metal fluctuation attenuation becomes essential in the range $\varepsilon \sim 10^{-5}$ which is accessible for experimental observations. However in real superconductors there are inhomogeneities due to crystal defects. They are dislocations, crystallites, etc. All these inhomogeneities lead to the fact that the effective parameters specifying superconductor properties appear to be the coordinate functions. Near T_c even weak inhomogeneities can lead to the widening of the phase transition. The inhomogeneities causing the effective electron interaction prove to be more essential.

Let us consider dislocation effects on the fluctuation character of the order parameter near T_c. We shall consider dislocations with length ($L \gg \xi_0$) much larger than the superconductor coherence length, while the dislocation diameter is much less the coherence length ($d \ll \xi_0$). Since the presence of dislocations leads to the lattice anharmonic distorsion in the region of dislocation cores, we may suppose that the magnitude of the effective electron interaction is constant along the dislocation line and changes abruptly on going from the dislocation core. Then the suprons formed near the dislocation will possess one degree of freedom along the dislocation line. In the case when the distance between dislocations is larger than the coherence length we may suppose that the superconductor behaves as an array of one-dimensional filaments. Let us calcu - late sound attenuation in this superconductor model at $T \gtrsim T_c$.

Taking into account one dimensional character of suprons near the dislocation line, we obtain after integrating in (6)

$$\mathcal{I}m \prod_{1f}(\omega, \bar{k}) \sim -\omega g^2 \frac{\ell n^2 \omega_D/2\pi T}{m^2 v_F^4 d^2 \eta K \varepsilon^2} \qquad (9)$$

Then for the given model (with the dislocation density parallel to the sound propagation $-N_d$) we get the following expression for the imaginary part of the polarization operator

$$\mathcal{I}m \prod_{3f}(\omega, \bar{k}) \sim -\omega g^2 N_d \frac{\ell n^2 \omega_D/2\pi T}{m^2 v_F^4 \eta K \varepsilon^2} \cdot \qquad (10)$$

Comparing (10) with the expression for the imaginary part of the polarization operator of a normal metal, we obtain the following expression for the temperature range in which the fluctuation attenuation is essential :

$$\frac{T - T_c}{T_c} \sim N_d^{1/2} \frac{\ell n \, \omega_D/2\pi T_c}{m^2 v_F^2 \eta^{1/2} k^{1/2} \ell^{1/2}} \cdot \qquad (11)$$

In the case of average dislocation densities in the superconductor $N_d \sim 10^{10}$, $v_F \sim 10^8$, $\eta \sim 10^{-10}$, $\ell \sim 10^{-6}$ at sound frequency $\omega = 1000$ cps, one finds $\varepsilon \sim 10^{-1}$.

The estimate value ε coincides in the order of magnitude with experimentally observed peak width . .

It is also seen from (11) that ε decreases with the increase of the wave vector of sample sound oscillations and this is qualitatively consistent with the experimental results.

At magnetic fields near the critical field H_c (or H_{c2}) fluctuation electron pairing in the superconductor becomes essential. In the case of a small critical field $(H \ll 5.000$ Oe$)$ we may find the fluctuation correction to sound attenuation, taking into account in (II) H dependence of T_c [23]. For the field region in which fluctuation sound attenuation is essential we obtain the following expression

$$\frac{H - H_c}{H_c} \sim N_d^{1/2} \frac{\ell n \, \omega_D / 2 \pi T}{m^2 \upsilon_F^2 \, \rho^{1/2} \kappa^{1/2} \ell^{1/2}} \left(\frac{T^2}{T_c^2 - T^2} \right)^{1/2}.$$

Substituting numerical values of its magnitudes into this expression, we obtain for samples with average dislocation density the following value $\dfrac{H - H_c}{H_c} \sim 10^{-1}$, which coincides in the order of magnitude with the experimentally observed peak width in the critical field range.

In conclusion it should be noted that a small peak $Q^{-1}(H)$ in small magnetic region is apparently associated with dissipation processes which are due to nonequilibrium in conduction electron system of a superconductor. This nonequilibrium occurs in electron system at formation of mixed (intermediate) state in a superconductor.

References

1. H. Bömmel, Phys. Rev., 96, 220, (1954).

2. R. Norse, H. Bohm, Phys. Rev., I08, II75, (1957).

3. J. Rayne, C.Jones, "Phys. Acous.", v.7, New-York-London, I49, (I970).

4. M. Gottlieb, M. Larbuny, "Phys. Acous.", v. 7, New-York-London, I, (I970).

5. Б.Т. Гейликман, В.З. Кресин, Кинетические и нестационарные явления в сверхпроводниках, изд. "Наука", (I972) .

6. L. Kadanoff, A. Pippard, Proc. Royal Society, 292, 299, (I966).

7. R. Sandström, Ann. Phys. 70, 5I6, (I972).

8. B. Welber, S. Quimby, Acta met. 6, 35I, (I958).

9. R. Chambers, J. Schultz, Acta met., 8, 585, (I960).

I0. E. Kramer, C. Bauer , Phys. Rev., I63, 407, (I967).

II. Э.Л. Андроникашвили, С.Н. Ашимов, Дж. С. Цакадзе, Дж.Г. Чигвинадзе, ХЭТФ, 55 , 775, (I968).

I2. Дж.Г. Чигвинадзе. ХЭТФ , 63, 2I44, (I972) .

I3. V.S. Postnikov, Internationales, Symposium, May, I970, Dresden, Akademie Verlag, I62, (I972).

I4. В.С. Постников, И.В. Золотухин, В.Е. Милошенко. Письма в ХЭТФ, I3 , I0 , (I97I).

I5. В.С. Постников, В.Е. Милошенко, И.В. Золотухин, Г.Е. Шунин, Е.П. Шухалов, ФТТ, I4, 3447 , (I972).

I6. В.С. Постников, И.В. Золотухин, В.Е. Милошенко. ФТТ, I4, 940, (I972).

17. В.Е. Милошенко, И.В. Золотухин, В.С. Постников. ПТЭ,№ I,
    218 (1972).

18. П. Хоэнберг. УФН, 102, 239 , (1970).

19. А. Свидзинский. Теор. и матем. физ., 9, 273, (1971).

20. J. Hurault, K. Maki, Phys. Rev., 82, 2560, (1970).

21. P. Fulde, K. Maki, Phys. Kondens. Mat., 8, 371, (1969).

22. K. Maki, Progr. Theor. Phys., 40, 193, (1968).

23. Л.И. Асламазов, А.И. Ларкин, ФТТ, 10, 1104, (1968).

24. С. Малеев. Теор. и матем. физ., 4, 86, (1970).

25. T. Tsuneto, Phys. Rev., 121, 402, (1961).

Figure Captions

Fig.I   Principal diagram of experimental arrangement for
             internal friction investigations at helium temperatures.

Fig.2   Temperature dependence of internal friction of a
             deformed vanadium polycrystal (curve I); a single
             crystal (curve 2); annealed single crystal (curve 3)
             (ω $\simeq$ I kcps).

Fig.3   Temperature dependence of internal friction for
             polycrystalline vanadium at frequencies I kcps
             (curve I), and 6 kcps (curve 2).

Fig.4   Temperature dependence for polycrystalline tantalum
             at frequencies I kcps (curve I) and 6 kcps (curve 2).

Fig.5.   Magnetic field dependence of internal friction for a
             polycrystalline (curve I) ; single crystal (curve 2);
             and annealed single crystal vanadium (curve 3).

Fig.6   Magnetic field dependence of internal friction for
             polycrystalline (curve 2) and single crystal tan -
             talum (curve I).

Fig.7   Fluctuation corrections to the polarization operator.
             Solid lines show electron Green's functions $G(\mathcal{E}_n, \bar{P})$
             $= \dfrac{1}{i\mathcal{E}_n - \xi(\bar{P})}$, curly - supron Green's functions,
             dotted- renormalized constant of electron-phonon
             interaction.

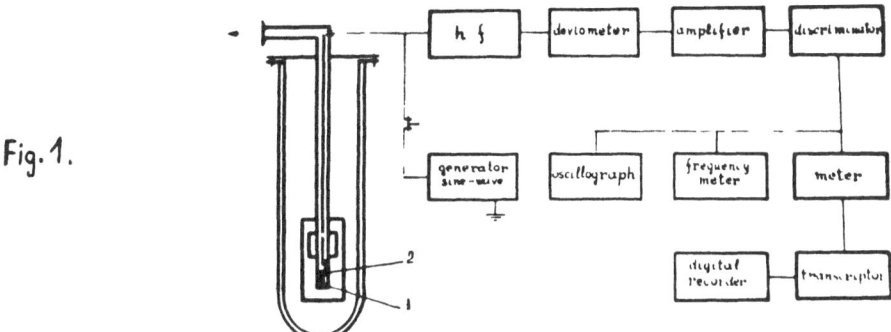

Fig. 1.

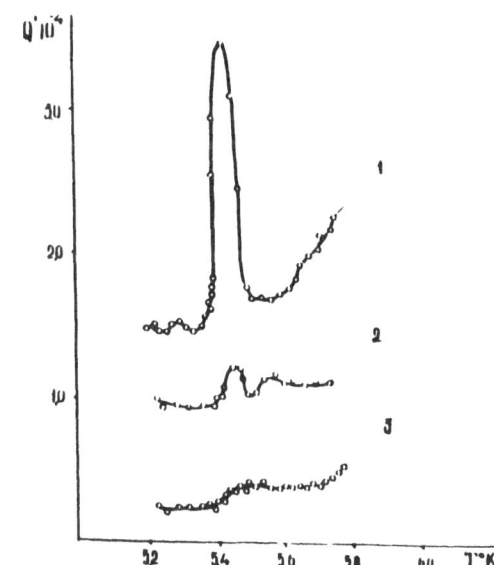

Fig. 2.

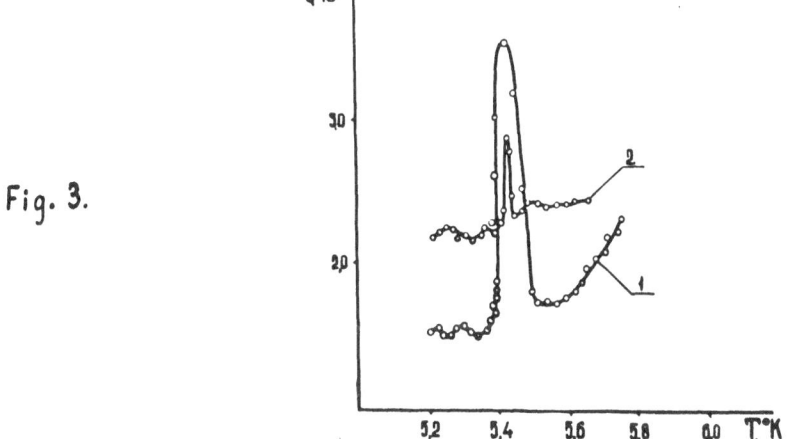

Fig. 3.

Fig. 4

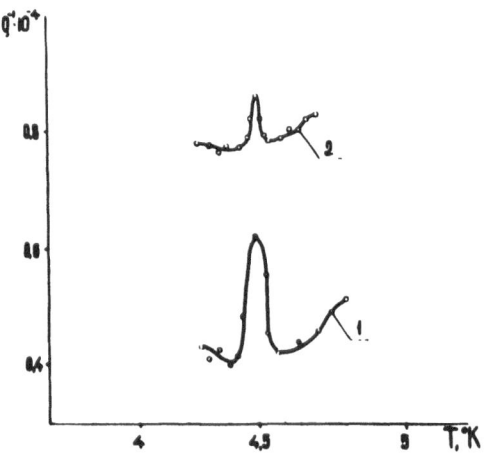

Fig. 5

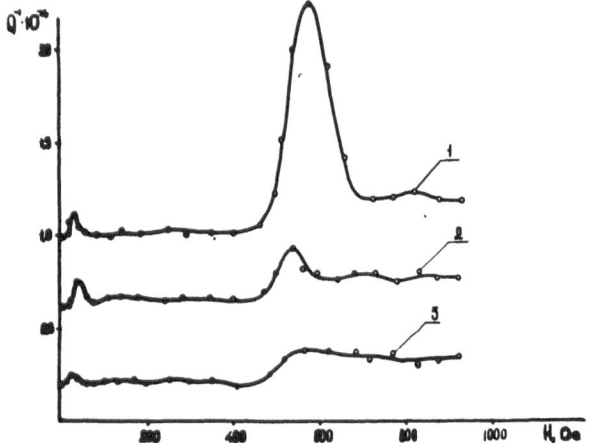

Fig. 6

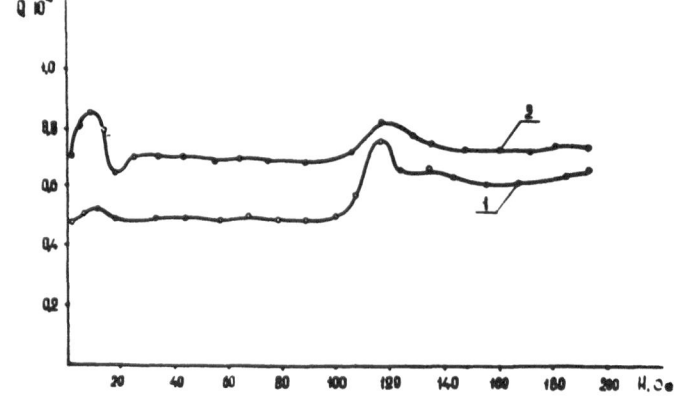

Fig. 7

$$\prod_{\xi}(\vec{k},\omega) =$$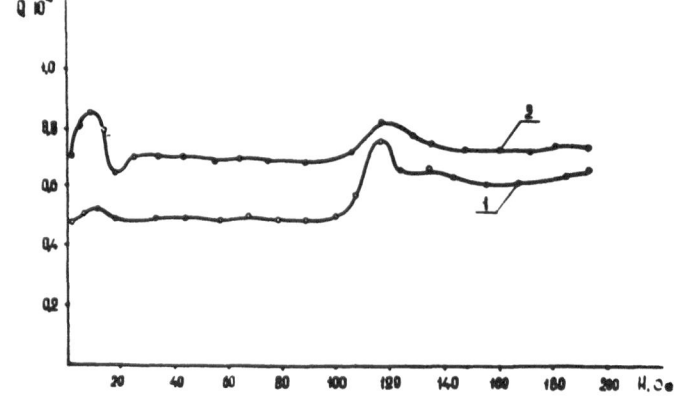

Phonon Mechanism of Internal Friction
in Polydomain Ferroelectrical Crystals

V.S. Postnikov, S.K. Turkov, B.M. Darinsky, A.V. Parshin

The Polytechnical Institute Plekhanov Street 84 Voronezh USSR

In experimental investigations of domain wall motion, dielectric properties and internal friction of ferroelectrics the ideas of viscous wall motion were often used.

The present paper gives theoretical investigation of phonon mechanism of 90°-domain wall drag in ferroelectrics with the perovskite structure. Analogous mechanism was investigated for mobile dislocations in crystalline materials [I-7] , while analogous studies for motion of flat defects of a domain wall type are absent.

The characteristic of a drag force is the phonon viscosity coefficient which will be found from consideration of refraction and reflection of elastic waves on domain walls with constant velocity v .

General problem of reflection from domain walls of elastic waves of any polarization propagating in arbitrary direction is rather difficult. Therefore we in analog to the work [8] shall consider only the simplest wave - transverse propagating in the plane of elastic symmetry with displacement vector perpendicular to this plane.

The problem will be treated in the coordinate system in which the plane XOZ is the plane of wave propagation and domain wall is in the plane YOZ , and it is assumed to move in the positive direction of the x-axis. Spontaneous polarization vector on both sides from a domain wall $\vec{P}'$ and $\vec{P}''$ lies in propagation plane and forms with the z-axis the angles α and $\pi-\alpha$ respectively.

Let us consider a flat monochromatic incident wave

$$U_y = U_0 e^{i(K_x x + K_z \cdot z - \omega_0 t)} \qquad (1)$$

In longwave approximation ($K\delta < 1$) the use of boundary conditions such as the equality of components of a displacement vector and the equality of components of an elastic force on the wall between two domains gives the relations

$$U_0 + A = B, \quad X_0 U_0 + X_2 A = X_1 B \qquad (2)$$

$$K_z = \varkappa_z = q_z; \quad \omega_0 - K_x v = \omega_2 - \varkappa_x v = \omega_1 - q_x v \qquad , \quad (3)$$

where the following notations are introduced :

δ is the domain wall thickness,

U_0, A, B are the amplitudes of incident, reflected and refracted waves respectively,

$\omega_0, \omega_2, \omega_1$ and $\vec{K}, \vec{\varkappa}, \vec{q}$ are frequencies and wave vectors of incident, reflected and refracted wave, $X_0 \equiv ctg\varphi_0 + a'$,

$X_2 \equiv ctg\varphi_2 + a'; X_1 \equiv ctg\varphi_1 + a''$. Here φ_ℓ are the angles of incidence, reflection and refraction measuring from the x-axis (index ℓ =0,I,2 will correspond to the magnitudes of incident, refracted and reflected waves respectively).

$a' = \dfrac{C'_{46}}{C'_{66}} \equiv a$, $a'' = \dfrac{C''_{46}}{C''_{66}}$, where C'_{iK} and C''_{iK} are the components of elastic moduli in the matrix form on both sides from the wall in the coordinate system connected with a domain wall. They can be expressed in the terms of components of elastic modulus tensor C_{iK} in crystallophysic coordinate system using the relations

$$C'_{44} = C''_{44} = C_{66} \sin^2 \alpha + C_{44} \cos^2 \alpha$$
$$C'_{66} = C''_{66} = C_{66} \cos^2 \alpha + C_{44} \sin^2 \alpha \qquad (4)$$
$$C'_{46} = -C''_{46} = (C_{66} - C_{44}) \cos \alpha \sin \alpha$$

Using the equation of motion for the waves understudy we obtain the dispersion relation

$$\omega_\ell = c \cdot signK_z \sqrt{X_\ell^2 + b^2} \cdot K_z \qquad (5)$$

Here $c = \sqrt{\dfrac{C'_{66}}{\rho}}$; $b^2 = \dfrac{C'_{44} C'_{66} - C'^2_{46}}{C'^2_{66}}$; ρ is the density of a crystal. For a refracted wave ($\ell = I$) it is necessary to replace in (5) C'_{iK} by C''_{iK} .

Using the expressions (2), (3), (4) and (5) let's define the following relations between wave amplitudes

$$A = \dfrac{X_1 - X_0}{X_2 - X_1} U_0, \quad B = \dfrac{X_2 - X_0}{X_2 - X_1} U_0 \qquad (6)$$

$$X_1 = \dfrac{1}{(1+\beta^2)} \Big[(2a - X_0)\beta^2 + \beta\, signK_z \sqrt{X_0^2 + b^2} + \qquad (7)$$
$$+ signK_z \sqrt{X_0^2 + \beta(X_0^2 + b^2) + 2\beta(2a - X_0)signK_z \sqrt{X_0^2 + b^2}}$$

$$X_2 = \dfrac{1}{(1-\beta^2)} \Big[-X_0(1+\beta^2) + 2\beta\, signK_z \sqrt{X_0^2 + b^2} \Big] \qquad (8)$$

153

where $\quad \beta = \dfrac{\vartheta}{C}, \quad sign\, K_z = \dfrac{K_z}{|K_z|}$

As a result of interaction of elastic waves and a mobile domain wall, the wall will be affected by a drag force acting opposite the direction of motion of the wall. The force acting on the unit of wall surface can be defined as a pulse which an elastic wave transfers to a mobile wall per the unit of time. The incident wave acts from the left on wall surface unit with the force which can be determined in the following way :

$$F_o = n(S_o - \vartheta)\hbar K_x \qquad (9)$$

where $\quad n = \dfrac{\mathcal{E}}{\hbar \omega_o}\quad$ is the density of phonons, i.e. their quantity in unit crystal volume which is determined from the density of elastic energy of an incident wave $\mathcal{E} = \dfrac{1}{2}\rho\,\omega_o^2\,u_o^2$

Taking into account both reflected and refracted waves one obtains the following expression for the force in the form

$$F = \frac{1}{2}\rho\,u_o^2\omega_o^2\left[(S_o - \vartheta)\frac{K_x}{\omega_o} + (S_2 + \vartheta)\frac{\varkappa_x}{\omega_2}\left(\frac{X_1 - X_0}{X_2 - X_1}\right)^2(1 + \vartheta\gamma_2)^2 - \right.$$

$$\left. - (S_1 - \vartheta)\frac{q_x}{\omega_1}\left(\frac{X_2 - X_0}{X_2 - X_1}\right)^2(1 + \vartheta\gamma_1)^2 \right] \equiv \mathcal{E}\cdot f(X_0, \vartheta) \qquad (10)$$

$$\widetilde{F} = \frac{1}{2}\rho\,u_o^2\omega_o^2\left[(S_o - \vartheta)\frac{K_x}{\omega_o} + S\frac{\varkappa_x}{\omega_2}(1 + \vartheta\gamma_2)^2\right] \equiv \mathcal{E}\cdot\widetilde{f}(X_0, \vartheta) \qquad (11)$$

Here S_0, S_2, S_1 are the projections of group velocities on the x-axis.

$$\gamma_1 = \frac{1}{c} \frac{X_1 - X_0 + 2a}{sign K_2 \sqrt{X_0^2 + b^2}} \; ; \; \gamma_2 = \frac{1}{c} \frac{X_2 - X_0}{sign K_2 \sqrt{X_0^2 + b^2}} \quad (12)$$

Using the dispersion relation (5) one can find

$$S_\ell = \frac{c X_\ell}{sign K_2 \sqrt{X_\ell^2 + b^2}} \quad (13)$$

Introducing of two different expressions for the force affecting the wall is caused by the onset of the effect of total reflection of incident waves in the range of sufficiently small angles between a domain wall plane and a vector of group velocity. Formula for the force (11) corresponds to the presence of total internal reflection, while (10) corresponds to the absence of it.

It is to be noted that the presence of a reflected wave and the effect of total internal reflection for a transverse wave propagating in the plane of elastic symmetry is only characteristic of a mobile wall.

Integrating over all possible frequencies of incident waves and averaging over angles of incidence one gets the following expression for the drag force acting on the unit of wall surface due to thermal lattice vibrations:

$$F = E(T) J, \quad (14)$$

where $E(T)$ is the density of internal crystal energy due to thermal vibrations of lattice atoms, while dimensionless magnitude J neglecting the members of higher order of smallness over β than the first order of smallness is given by the following expression

$$J = \frac{1}{\pi} \left\{ \iint_{2\sqrt{\beta b a}}^{\infty} [H^{\dot{+}}(X_0, v) - H^{\dot{+}}(X_0, -v) + H^{\bar{-}}(-X_0, v) - H^{\bar{-}}(-X_0, -v)] dX_0 + \right.$$ (15)

$$\left. + \int_{\beta b}^{2\sqrt{\beta b a}} [H^{\dot{+}}(X_0, v) - H^{\bar{-}}(-X_0, -v) + \widetilde{H}^{\bar{-}}(-X_0, v) - \widetilde{H}^{\dot{+}}(X_0, -v)] dX_0 \right.$$

Here $H(X, v) \equiv -\dfrac{1}{1 + (X_0 - a^2)} \cdot f(X, v); \quad \widetilde{H}(X, v) \equiv -\dfrac{1}{1 + (X_0 - a)^2} \widetilde{f}(X_0, v),$

the sign " + " and " - " over H points to the choice of the sign for the function $sign K_z$. Using the relations (13), (12), (11), (10), (8), (7), (5) from (15) we obtain

$$J = \frac{\beta}{\pi} \iint_{2\sqrt{\beta b a}}^{\infty} \left\{ 8a^2 x (1+x^2+a^2)/(x^2+b^2)^{3/2} [(1+x^2+a^2)^2 - 4x^2 a^2] + \right.$$

$$+ 16 a^2 x^3/(x^2+b^2)^{3/2} [(1+x^2+a^2)^2 - 4x^2 a^2] - 256 a^4 b^3 \beta^2 x^2 \sqrt{x^2+b^2} [1+$$

$$(x-a)^2] [x^2 + 2\beta(a-x)\sqrt{x^2+b^2} - \beta\sqrt{x^2+b^2} g(\beta, x) + x g(\beta, x)][x^2 -$$

$$- 2\beta(a-x)\sqrt{x^2+b^2} + \beta\sqrt{x^2+b^2} g(-\beta, x) + x g(-\beta, x)][x^2 +$$

$$+ g(-\beta, x) g(\beta, x)] \cdot [g(-\beta, x) + g(\beta, x)] - 256 a^4 b^3 \beta^2 x^2 \sqrt{x^2+b^2} [1+$$

$$+ (x+a)^2][x^2 - 2\beta(a+x)\sqrt{x^2+b^2} - \beta\sqrt{x^2+b^2} g(-\beta, -x) +$$ (16)

$$+ x g(-\beta, -x)][x^2 + 2\beta(a+x)\sqrt{x^2+b^2} + \beta\sqrt{x^2+b^2} \cdot g(\beta, -x) +$$

$$+ x g(\beta, -x)][x^2 + g(-\beta, -x) g(\beta, -x)][g(-\beta, -x) + g(\beta, -x)] \right\} dx$$

$$- \frac{4a\beta}{\pi(1+a^2)} \iint_{\beta b}^{2\sqrt{\beta b a}} x^2 g(\beta, x)/\beta(x^2+b^2)[x^2 + 2\beta(a-x)\sqrt{x^2+b^2} -$$

$$- \beta\sqrt{x^2+b^2} g(\beta, x) + x g(\beta, x)] + x^2 g(\beta, -x)/\beta(x^2+b^2)[x^2 +$$

$$+ 2\beta(a+x)\sqrt{x^2+b^2} + \beta\sqrt{x^2+b^2} g(\beta, -x) + x g(\beta, -x)] \right\} dx,$$

where $g(\beta,x)=\sqrt{x^2+2\beta(2a-x)\sqrt{x^2+b^2}}$.

After calculation of these integrals neglecting the degrees a higher than the second one we find

$$J=1,07a^2\beta \qquad (17)$$

Writing the density of thermal energy in the Debye approximation we can find the finite expression for phonon viscosity of the unit of surface of a domain wall, using (17), (14).

$$\eta=-\frac{F}{v}=5,42\cdot10^{-2}a^2\frac{kT\omega_{max}^3}{c_0^4}\mathcal{D}\left(\frac{\hbar\omega_{max}}{kT}\right)$$

Here $\mathcal{D}\left(\frac{\hbar\omega_{max}}{kT}\right)$ is the Debye function for threedimensional lattice, kT is the temperature in energy units, c_0 is the average sound velocity, ω_{max} is maximum phonon frequency of a crystal to the value of which the integration was made in(14) in the process of calculation of the drag force.

For the temperatures $T<\frac{\hbar\omega_{max}}{4k}$ formula (18) is accurate if ω_{max} being the magnitude of $\omega_{max}=\frac{2\pi c_0}{\delta}$.
At higher temperatures being important for the comparison with the experiment the formula gives the reduced value because of absence of the contribution of shortwave phonons. Rough calculation of the effect of shortwave phonons may be done if ω_{max} in(18) is assumed to be equal to the Debye frequency of a crystal. In these two approximations the numerical estimations of the magnitudes of viscosity coefficients for 90°-domain wall in $BaTiO_3$ at T = 300°K give η = 4$\frac{gm}{cm^2\ sec}$ and

η = 5·10^2 $\frac{gm}{cm^2\ sec}$ respectively.

The experimental value η equals $3 \cdot 10^2 \dfrac{gm}{cm^2 \ sec}$ [9],

that points to the qualitative agreement with theoretical results if viscous forces acting in experimental conditions 9 are assumed to be of phonon nature.

In order to find the magnitude of internal friction let us write the formula of pressure acting on the wall due to external stresses σ.

$$P = B \varepsilon \sigma \qquad (19)$$

where ε is the spontaneous strain of a ferroelectric,
B is the orientation factor.

The magnitude of average wall displacement from equilibrium position is determined by the equation of motion

$$\mu \ddot{\xi} + \eta \dot{\xi} + \varkappa \xi = P \qquad (20)$$

Here μ is the effective surface density of a domain wall, η is the viscosity coefficient of unit surface, $\varkappa$ is the quasi-elastic coefficient determining bending stiffness of a wall.

From (19) and (20) we find the following expression for internal friction [10]:

$$Q^{-1} = B^2 \varepsilon^2 \frac{G}{L\varkappa} \frac{\Omega/R}{(1-\Omega^2)^2 + \Omega^2/R^2} \qquad (21)$$

In formula (21) G is the shear modulus, L is the average domain size, $\Omega = \nu/\nu_0$, $\nu_0 = \sqrt{\dfrac{\varkappa}{\mu}}$ is the natural oscillation frequency of the principal tone of a domain wall, $R = \dfrac{\sqrt{\varkappa\mu}}{\eta}$.

In order to evaluate the height of internal friction peak experimentally observed it is necessary to take into account

158

not only the above-considered viscosity mechanism but some other mechanisms. They are reradiation, combination scattering due to domain motion in periodic potential lattice field etc. Therefore we do not evaluate the peak height numerically.

References

I. G. Leibfried. Zs. f. Phys. $\underline{127}$, 344, 1950.

2. W.P. Mason. J. Ac. Soc. Am., $\underline{32}$, 458, 1960.

3. A. Seeger, H. Engelke. Dislocation Dynamics.
   Ed. A.R. Rosenfield et. al. McCraw-Hill Publ.Comp.
   N.Y., 1968.

4. J. Lothe. Journ. Appl. Phys. $\underline{33}$, 2II6, 1962.

5. J.D. Eshelby. Proc. Roy. Soc. Lond. $\underline{AI97}$, 396, 1957.

6. A.D. Brailsford. Journ. Appl. Phys. $\underline{43}$, I380, 1972.

7. В.И. Альшиц, А.Г. Мальшуков. "ЖЭТФ", $\underline{63}$, 1849, 1972.

8. Г.Г. Кесенних, Д.Г. Санников, Л.А. Шувалов.
   "Кристаллография", $\underline{15}$, I022 , I970.

9. J. Fousek, B. Brezina. J. Phys. Soc. Japan, $\underline{19}$, 830, 1969.

IO. A. Granato, K. Lücke. Journ. Appl. Phys. $\underline{27}$, 583, 1956.

ULTRASONIC MEASUREMENTS IN SINGLE CRYSTAL COBALT NEAR 250°C

W. D. Wallace
Department of Physics, Oakland University
Rochester, Michigan

ABSTRACT   The attenuation and ultrasonic velocity in cobalt have been measured at 5 MHz between 25°C and 260°C for each of the 5 plane wave modes. These measurements show that anomalies of magnetic origin result from the onset of the easy axis of magnitization change at 245°C.

Introduction

In a previous study of ultrasonic attenuation in ferromagnetic metals, Taborov and Tarasov (1) reported that as the temperature was increased the attenuation of 5MHz longitudinal waves went through a well-defined maximum at the temperature where the easy axis of magnitization in cobalt begins to change. From their graphs this temperature appears to be about 230°C. It was suggested by these authors that the attenuation maximum might be a manifestation of a resonant interation between elastic and spin waves resulting from a lowering of the spin wave frequencies as the easy axis begins to change. In order to explore this point further, an investigation of the temperature and magnetic field dependence of the elastic properties of cobalt single crystals was begun and our initial results are presented here.

The elastic constants of cobalt have been measured by McSkimin (2) in the vicinity of 25°C and by Fisher and Dever (3) from -269°C to 275°C, with measurements of c_{33} extended to 438°C. Fisher and Dever report that the ultrasonic attenuation above 275°C was too great (at their 50 MHz frequency) to allow measurement of any of the elastic constants except c_{33}. Furthermore, a direct measurement of c_{44} was not possible above 25°C (at 50 MHz) because the corresponding shear modes were too highly attenuated. Instead, c_{44} was computed indirectly from measurements made by propagating sound at some angle other than parallel or perpendicular to the hexagonal axis. The authors noted that the large attenuation preventing measurements above 275°C seemed to be associated with the easy axis change, and therefore perhaps was caused by domain-wall motion which is known to be a source of attenuation in ferromagnetic metals (4). The origin of the high attenuation of the c_{44} shear waves above 25°C was unknown.

Easy Axis Directions in Cobalt

As is well known, cobalt is a ferromagnetic metal whose low temperature crystalline structure is hexagonal close-packed. Upon heating, cobalt undergoes a martinsitic transformation at about 440°C, becoming face-centered cubic, the stable high temperature phase with a Curie temperature of 1120°C.

The anisotropy energy, E_A, of hexagonal cobalt can be represented by an expression of the form

$$E_A = K_1 \sin^2 \theta + K_2 \sin^4 \theta$$

where θ is the angle the magnitization makes with the hexagonal axis. K_1 and K_2 are the first and second anisotropy constants and are both functions of temperature. Below 245°C, K_1 and K_2 are positive and the easy axis of magnitization is parallel to the hexagonal axis. At 245°C, K_1 changes sign (5) and becomes negative. Between 245°C and 340°C the easy axis changes gradually from the hexagonal axis to the basal plane and for temperatures greater than 340°C is perpendicular to the hexagonal axis (6).

Experimental Procedures

Measurements have been made of the attenuation and ultrasonic velocity of each of the five possible plane wave modes using two differently oriented single crystals cut from the same original boule. It is convenient to identify each of these modes by the appropriate elastic constants: for sound propagating <u>parallel</u> to the hexagonal axis, c_{33} is the longitudinal wave elastic constant and $(c_{44})_\parallel$ is the shear wave elastic constant; with sound propagating <u>perpendicular</u> to the hexagonal axis the elastic constants are c_{11} for the longitudinal wave, $(c_{44})_\perp$ for the shear wave polarized along the hexagonal axis and $\frac{1}{2}(c_{11}-c_{12})$ for the shear wave polarized perpendicular to the hexagonal axis. The two c_{44} shear wave elastic constants are not equal in cobalt because of the presence of ferromagnetism. One of the crystals was approximately a cube with opposite surfaces polished flat and parallel to (001) planes perpendicular to the hexagonal axis, and the other was of rectangular cross-section with opposite surfaces polished flat and parallel to (100) planes parallel to the hexagonal axis.

A single quartz transducer was used to both transmit and receive ultrasonic signals. The conventional pulse-echo technique employing a calibrated attenuator was used to measure attenuation. A phase comparison method similar to that described by Williamson (7) was used to measure changes of ultrasonic velocity with temperature and magnetic field. As the velocity changed a cw signal coherent with the ultrasonic driving pulse was maintained in quadrature with one of the amplified echo signals by adjusting the frequency. Under these conditions fractional changes in ultrasonic velocity are equal to fractional changes in the frequency provided phase changes due to variation in sample length, transducer bonds, etc., can be neglected. A thermistor controlled temperature regulator maintained sample temperatures constant to better than 0.25°C. The sample temperature was measured by placing a copper-constantan thermocouple in direct contact with the sample using a small amount of transistor heat-sink grease.

A major experimental problem to be overcome in ultrasonic measurements above room temperature is that of finding a suitable agent to acoustically bond the transducers to the sample. The problem is most severe in the case of shear waves since silicone fluids can be used for longitudinal waves up to temperatures in excess of 300°C. We have found that sodium nitrite ($NaNO_2$) makes an excellent shear wave bond on cobalt which is usable up to at least 260°C. Sodium nitrite melts at 271°C but in some instances appeared to soften somewhat below this temperature. These bonds could be repeatedly cycled between 25°C and 260°C, but are slightly hydroscopic and needed to be kept in a desicator if not in use. At the beginning of this work it was thought that measurements up to 260°C might be adequate. In view of the results reported here, it now appears that it is important to extend these measurements to higher temperatures.

Results

The attenuation of the c_{44} shear modes is so large, and increases with increasing frequency, that it was decided to work near 5 MHz, at which frequency we could still expect to produce approximately plane acoustic waves with 0.953 cm diameter transducers. Since we always initially tuned our equipment for the best appearing echo pattern, our measurements vary between about 4.5 MHz and 5.5 MHz. We omit specification of frequencies in what follows in the interest of clarity.

Zero Field Attenuation

The temperature dependence of attenuation in zero applied magnetic field is shown in Fig. 1. No attempt has been made to correct for bond losses since these appear small compared to the changes in sample attenuation. Where possible, we measured the pulse-height of several echoes to determine the attenuation. However, we were left with only one echo above about 200°C for the c_{44} shear modes and could only measure changes in the single echo pulse-height. We find about a 10 per cent variation in attenuation values under these conditions in repeated measurements. The most notable feature of these data is the large increase of attenuation of the c_{44} shear waves above 200°C.

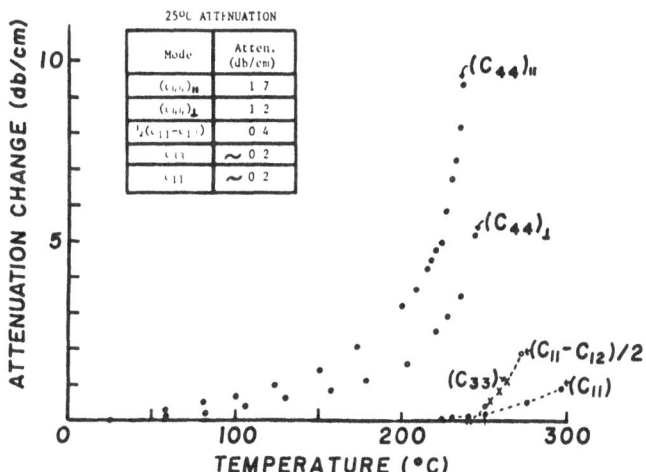

FIG. 1

Temperature dependence of attenuation of the five plane wave modes in cobalt in zero applied field.

In repeated measurements, no echo signal has been observed between about 240°C and 265°C for these modes. The attenuation is still increasing when the signal disappears into the amplifier noise. The attenuation of the c_{44} shear modes is large at 25°C (see insert in Fig. 1) and increases steadily with temperature from there.

In contrast, the $\frac{1}{2}(c_{11}-c_{12})$ shear wave and the c_{11} and c_{33} longitudinal waves show no measurable attenuation change until temperatures above 200°C are reached. The attenuation of these modes begins to rise rapidly between 230°C and 250°C. These results are in agreement with the observations of Fisher and Dever (3), but we have not found a maximum in the attenuation similar to that reported by Taborov and Tarasov (1).

<u>Attenuation in Magnetic Fields</u>

The temperature increases in attenuation are greatly reduced for <u>every</u> acoustic mode by the application of an external magnetic field of 10 kOe applied parallel to the hexagonal axis. Magnetic fields applied perpendicular to the hexagonal axis give rise to complicated results which have not been fully studied as yet. Technical saturation appears to be reached in the rectangular shaped (100) sample above about 5 kOe and the temperature dependent attenuation of the c_{11}, $\frac{1}{2}(c_{11}-c_{12})$ and the $(c_{44})_\perp$ modes appear to be completely removed by application of a saturating magnetic field. Technical saturation does not appear to be reached in the cube shaped (001) sample, and the attenuation is still decreasing with magnetic field, at any fixed temperature (up to 260°), at 13.5 kOe, our maximum available field. Figure 2 shows the temperature variation of attenuation of the $(c_{44})_\parallel$ shear wave in the (001) sample both in zero field and in a field of 10 kOe applied parallel to the hexagonal axis.

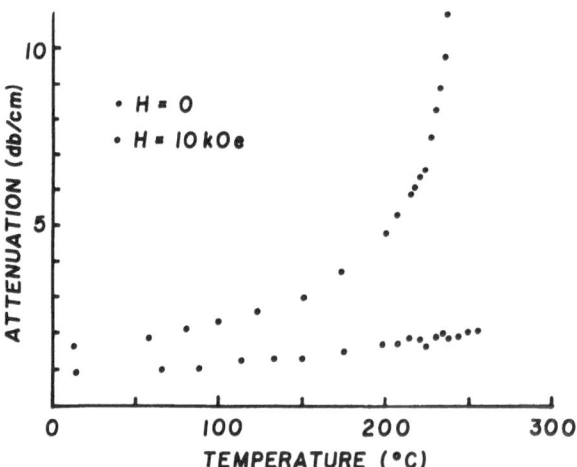

FIG. 2

Temperature dependence of attenuation of the $(c_{44})_\parallel$ shear wave
in zero field and with a field of 10 kOe applied parallel to
the hexagonal axis.

164

Temperature Variation of Ultrasonic Velocity in Zero Field

As expected, the elastic constants show anomalous temperature variations at temperatures where the attenuation changes rapidly. Figure 3 shows the fractional changes in velocity (which are equal to half the fractional changes in elastic constants neglecting sample length changes) as the temperature is increased. For each of these modes the velocity begins to decrease more rapidly with temperatures between $230^\circ C$ and $240^\circ C$ than at lower temperatures. Fisher and Dever (3) have also noted similar behavior in c_{11} and c_{33}. Our measurements of $\Delta v/v$ are generally within a few per cent of corresponding velocity changes computed from the elastic constants given by Fisher and Dever, although we find consistantly smaller values of $\Delta v/v$ for c_{11} and $\frac{1}{2}(c_{11}-c_{12})$, (e.g., 10 per cent smaller at $200^\circ C$), and the values of c_{44} computed by these authors do not show the tendency to decrease more rapidly above $230^\circ C$ that we find.

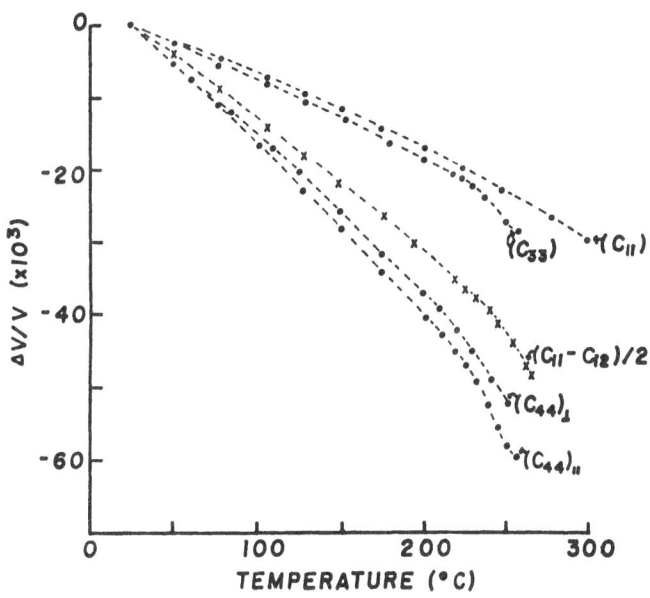

FIG. 3

Fractional velocity changes of the five plane wave modes in cobalt in zero applied magnetic field.

Temperature Variation of Ultrasonic Velocity in Applied Magnetic Fields

The temperature dependence of ultrasonic velocity in the presence of an applied magnetic field has only been measured for the c_{44} shear wave modes. The results obtained for the $(c_{44})_{\parallel}$ shear wave are shown in Figure 4, where it is seen that the onset of the more rapid decrease near $240^\circ C$ is removed by a field of 10 kOe applied parallel to the hexagonal axis. Similar results were obtained for the $(c_{44})_{\perp}$ shear wave. Velocity measurements for the other modes have not been carried out to date, but it is expected that the effect of an applied field will be similar to that found for the c_{44} modes.

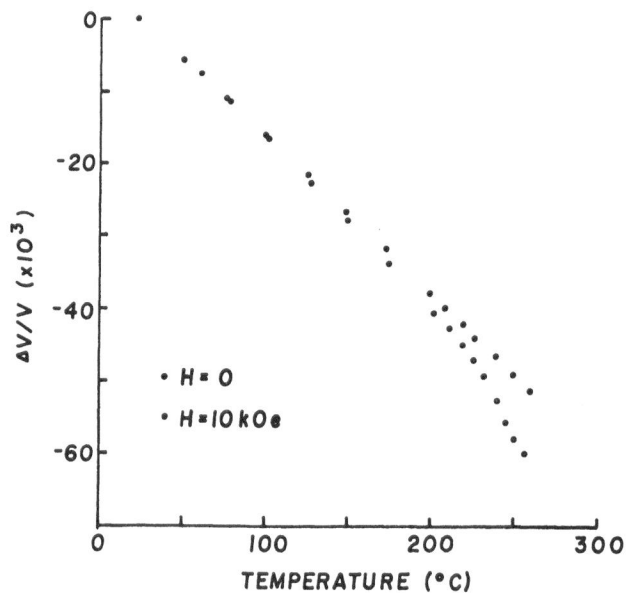

FIG. 4

Effect of a 10 kOe magnetic field applied parallel to the
hexagonal axis on the temperature variation of ultrasonic
velocity for the $(c_{44})_{\parallel}$ shear wave.

Discussion

The results presented here clearly show the existance of elastic anomalies of magnetic origin
in cobalt associated with the onset of spin reorientation near 240°C. This is not an unexpected
occurance since the existance of such anomalies in magnetic systems is well documented in the
literature (8). The origin of these anomalies has been attributed in many instances to the
softening of spin-wave modes as suggested by Taborov and Tarasov and is discussed in some detail
in reference (8). The elastic-spin wave interactions that occur at temperatures where the easy
axis is changing are usually observed as fairly sharply peaked changes in attenuation and velocity. The fact that we have not observed a maximum in the attenuation at temperatures up to
260°C may mean that in cobalt the c_{44} attenuation remains quite large until the easy axis
reaches the basal plane at 340°C. Domain-wall interactions with ultrasonic waves in metals
also give rise to magnetic field dependent attenuation and velocity changes, and these interactions disappear in saturating magnetic fields. The very rapid increase in attenuation on the
low temperature side of the easy axis change is suggestive of a spin-wave interaction, but
until we have completed measurements at temperatures above 260°C we lack enough information to
provide a detailed description of elastic effects in cobalt arising from the easy axis change.

Acknowledgements

The author would like to thank Dr. R. L. Melcher for the loan of the cobalt samples used in this work and acknowledge the assistance of Mr. P. Merlo and Mr. G. Persha during the initial stages of the cobalt measurements and the work of Mr. R. Stanton in designing and building the temperature regulator. He also wishes to thank Dr. E. S. Fisher for kindly sending a reprint of his work with Dr. D. Dever which the author was unaware of at the start of this work. Finally, it is acknowledged that this work was supported in part by a Cottrell Research Grant from Research Corporation.

References

1. V. F. Taborov and V. F. Tarasov, IEEE Trans. on Sonics and Ultrasonics SU-14, 1 (1967).

2. H. J. McSkimin, J. Appl. Phys. 26, 406 (1955).

3. E. S. Fisher and D. Dever, Trans. of the Metallurgical Society of AIME 239, 48 (1967).

4. R. Truell, C. Elbaum, B. B. Chick, Ultrasonic Methods in Solid State Physics, P. 263. Academic Press, New York (1969).

5. Y. Barnier, R. Pauthenet and G. Rimet, Comptes Rendus 252, 2839 (1961).

6. O. Boser, H. Kronmuller, A. Seeger and H. Trauble, Proceedings of the International Conference on Magnetism, p. 720. The Physical Society, London (1964).

7. R. C. Williamson, J. Acoust. Soc. America 45, 1251 (1968).

8. G. Gorodetsky and B. Luthi, Phys. Rev. B2, 3688 (1970).

ELECTROMAGNETIC GENERATION IN FERROMAGNETIC CRYSTALS

M.J.W. Povey and E.R. Dobbs
Department of Physics, University of Lancaster, Lancaster,LA1 4YB, U.K.

We have investigated the response of a ferromagnetic metal to an r.f. magnetic field of frequency ω incident at its surface in terms of a model of spin wave - sound wave - e-m wave interactions in the presence of magneto-elasticity. We show that the magnetoelastic coupling is important at low (MHz) frequencies and that it provides an additional mechanism for generating ultrasound in ferromagnets under certain conditions.

We write down the total energy of the ferromagnet and then derive the equations of motion of the lattice and the magnetisation. These equations are linearised and we look for solutions simultaneously with Maxwell's equations for the metal, in the form of plane waves whose eigenfrequencies and dispersion relations are obtained from the vanishing of the determinant of the system of plane wave amplitudes. The total energy of a ferromagnetic crystal may be written as

$$F = \underbrace{F_{Ko} + F_{mel}}_{\text{ANISOTROPY ENERGY}} + \underset{\text{ELASTIC}}{F_{el}} + \underset{\text{ZEEMAN}}{F_z} + \underset{\text{EXCHANGE}}{F_{ex}} \qquad (1)$$

The equations of motion for the lattice and the magnetisation may be written as

$$\rho \ddot{\xi}_i = (\partial/\partial x_k)(\partial F/\partial \epsilon_{ik}) \qquad (2)$$

and

$$-\underline{\dot{M}}/\gamma = \underline{M} \wedge (\partial F/\partial \underline{M}) + (\lambda/|M|^2)\underline{M} \wedge \underline{M} \wedge (\partial F/\partial \underline{M}) \qquad (3)$$

where

and ξ_i - displacement of lattice from equilibrium, M - magnetisation vector, γ - magnetomechanical ratio, λ - relaxation frequency, ρ - density and ϵ_{ik} - strain components.

Equation (2) is an expression of Newton's law for a deformable solid and equation (3) may be regarded as the equation of motion for a damped classical top applied to the magnetisation vector. The dispersion relations for this coupled sound wave/spin wave/electromagnetic wave system are plotted in Fig.1 as Re q = f(ω) on a log-log plot. They have been plotted for the case of a nickel single crystal disc with a static magnetic field B_O normal to the plane of the disc and just large enough to saturate the disc. We have plotted only right circularly polarised modes and have considered transverse sound waves propagating normally to the surface so that $B_O // q$. Longitudinal waves are uncoupled in this geometry. We have in Fig.1 ω_q = 2 x 10^8 s^{-1}, uncoupled spin wave frequency; Δ = 1.7 x 10^8 s^{-1}, the magnetoelastic coupling term; V_t = 3.5 x 10^5 $cm.s^{-1}$, shear wave velocity; $\delta^2 = \frac{10^4}{\omega}$ cm^2, electromagnetic skin depth; q = 2π/λ, wave vector; ω is the frequency of the incident r.f. magnetic field; ω_m = 4πγM_s = 1.2 x 10^{11} s^{-1}; γ, magnetomechanical ratio.

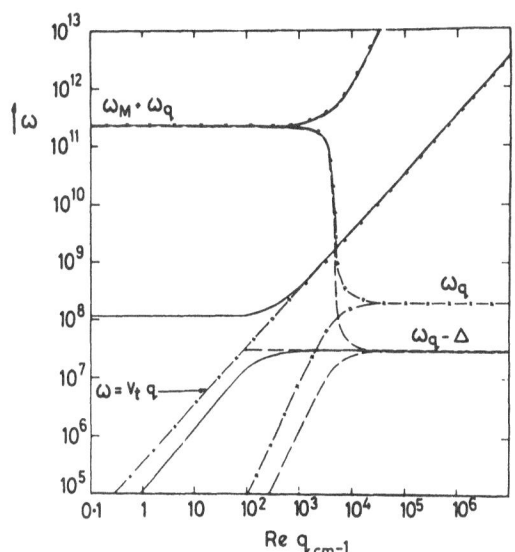

FIG. 1

Dispersion relations for a magnetoelastically coupled, sound wave/spin wave/
electromagnetic wave system.
-x-, Dispersion Relations in the absence of magnetoelasticity;
---, Electromagnetic/spin wave branch; ——, sound wave/spin wave branch.

Referring to Fig.1, we can see the anti-resonance condition for the r.f. magnetic field at ω_+ = ω_q + ω_m. Here the metal becomes transparent, in the absence of spin wave relaxation, to an r.f. magnetic field of this frequency and the correct circular polarisation. This occurs because the r.f. field, instead of exciting r.f. electric currents which are resistively damped, excites a spin wave and propagates through the metal as a spin wave. The spin wave in this case is a free precession of the magnetisation having infinite wavelength. This possibility, that magnetisation current may be excited rather than electron current, has important consequences for electromagnetic generation at these frequencies. In particular, the Lorentz force type of generation discussed by Dobbs and Povey (1), cannot occur at this frequency since it requires an electron current to flow producing a field acting on the lattice. We can show that the normal generation term, ξ_B, is modified by the magnetisation current term to $\xi_B \emptyset_{\pm}$ where

$$\emptyset_{\pm} = 1 \mp \frac{\omega_M (1 \pm i a)}{\omega \mp \omega_q - 1/\tau_s} \qquad (4)$$

where $\omega_m = 4\pi\gamma M_s$, $a = (\tau_s \omega_q)^{-1}$; $\tau_s = \gamma M_s / \lambda \omega_q$; λ, spin wave relaxation frequency; γ, magneto-mechanical ratio.

The ferromagnetic resonance condition, $\omega = \omega_q$ is modified by the magneto-elasticity and becomes $\omega = \omega_q - \Delta$, the modes in this region being mixed. The sound wave, spin wave and electromagnetic wave are strongly coupled in this region and an r.f. magnetic field at the surface will excite a spin wave which in turn excites a sound wave via the magnetoelastic interaction. From equation (4) we see that the Lorentz force mechanism is enhanced at this frequency by a factor of 3.5 if $a = 2.5 \times 10^{-2}$ (2). However, calculations show that the magneto-elastic term is much larger than the Lorentz force term when the condition $\omega = \omega_q - \Delta$ in Fig.1 is satisfied despite the enhancement of the Lorentz force mechanism. We have calculated that the conversion efficiency of r.f. power at the metal surface to sound wave power, η_F (1,3), is of the order of 0.1 for the magnetoelastic process under these conditions. This is in contrast to the conclusion of Kobayashi et al (2) who show that the ferromagnetic resonance line width in thin films is strongly influenced by magnetoelastic interactions when the acoustic plate resonance coincides with the spin wave resonance frequency. Our calculation of η_F shows that, for MHz frequencies, the magnetoelastic interaction is important for $\omega = \omega_q - \Delta$, plate resonance being unnecessary. The calculation uses a value for a quoted by Kobayashi et al (2), measured at GHz frequencies, a value which may be incorrect at our frequencies but for which we have no experimental data. In particular, the assumption that $\tau_s \propto (\omega_q)^{-1}$, implicit in the Landau-Lifschitz damping formula (equation 3), would appear dubious.

We are currently verifying these calculations experimentally and have typically obtained $\eta \sim 10^3$ for 15 MHz slow shear wave generation along [110]. In Fig.2 we present a plot of generation efficiency (in dB's and corrected for field dependent attenuation) vs log B, for a single crystal nickel sphere at room temperature. In this plot any normal (Lorentz type) generation would appear as a straight line of slope +20. We can obtain a value for the efficiency η of the generation peak in Fig.2 by assuming that the normal

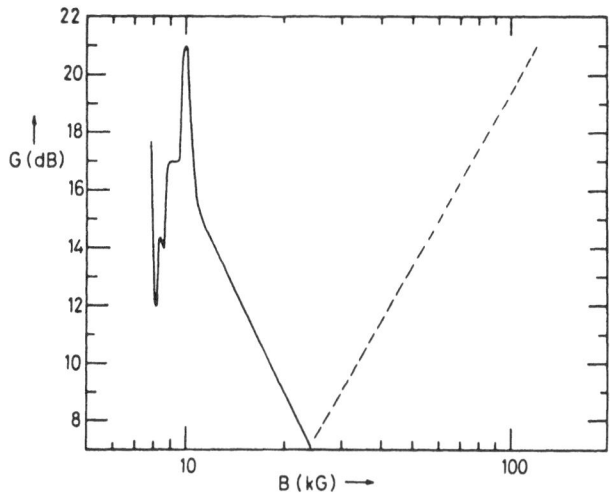

FIG.2

Field Dependence of Slow Shear Wave Generation in $N_i[110]$

generation process becomes significant at the highest fields reached in this experiment. By extrapolating a line of slope +20 to a field value at which the line has the same y axis value as the generation peak we obtain the field value at which the normal generation would have the same magnitude as the low field peak. Putting this field value into the formula for the normal generation conversion efficiency η_e (1), we obtain a lower limit for the peak generation efficiency in this experiment. The extrapolated field value is 12 Tesla and this gives an η_B of 10^{-3}. This compares well with the calculation for η_F of 0.1 and we feel that this is strong evidence that the generation peak of Fig.2 is caused by the magnetoelastic interaction enhanced by ferromagnetic resonance. Further evidence supporting this conclusion is provided by M. Gitis (4) who has demonstrated that the electromagnetic generation efficiency in Nickel drops sharply at the Curie Temperature.

To summarise, we have shown that there are two distinct processes whereby an r.f. magnetic field at the surface of a ferromagnetic metal can excite sound waves. Firstly, a Lorentz force mechanism, modified by the magnetisation current and the dynamics of the magnetisation, and secondly, a magnetoelastic process enhanced by ferromagnetic resonance. In Nickel the magneto-elastic process is calculated to be much more efficient than the Lorentz force mechanism at the ferromagnetic resonance frequency. The Lorentz force mechanism should become predominant at high fields. In a magnetic insulator the Lorentz force process will be absent but the magnetoelastic process should still be important.

References

1. E.R. Dobbs and M.J.W. Povey, Following Paper.

2. T. Kobayashi, R.C. Barker, J.L. Bleustein and A. Yelon, Phys. Rev. B7, 3273, 1973.

3. M.J.W. Povey, E.R. Dobbs and D.J. Meredith, J. Phys. F (to be published).

4. M.B. Gitis, Fizika Tverdoga Tela 14, 3563, 1972 (English Trans: Sov. Phys. Solid State 14, 2992, 1973).

EFFICIENCY OF ELECTROMAGNETIC GENERATION OF SOUND IN SOLIDS

E.R. Dobbs and M.J. Povey
Department of Physics, University of Lancaster, Lancaster, LA1 4YB, U.K.

Normal Generation

At radio frequencies ($\sim$ MHz) eddy currents induced in the plane surface of a metal in the presence of a static magnetic field B_o generate either longitudinal or transverse acoustic waves when B_o is either in, or normal to, the surface, as shown in FIG 1. The currents j are due to the incident r.f. magnetic field
$B_x(z) = B \exp\{-(1+i)z/\delta\}$, where $B = B_x(0)$,
and accompanied by an electric field $E_y(z)$
$= j_y(z)/\sigma$. The mean electromagnetic power
Q entering the surface per unit area is
$Q = \frac{c}{4\pi} \oint \frac{<Re\underset{\sim}{E}xRe\underset{\sim}{B}>.dA}{\oint dA}$, where the integral
is over the metal surface. Hence (1) we
have $Q = \frac{\omega\delta B^2}{16\pi}$, where the normal skin depth
$\delta = c(2\pi\sigma\omega)^{-\frac{1}{2}}$, σ is the conductivity (in
e.s.u.), B is in e.m.u., ω is the angular,
radio frequency and c is the speed of light.

The mean acoustic power P generated
in the metal per unit area is $P = \frac{1}{2}ds\omega^2|\xi|^2$,
where $|\xi|$ is the acoustic amplitude. We
define the conversion efficiency $\eta = P/Q$
for this generation process. Hence
$\eta = \frac{8\pi ds\omega|\xi|^2}{\delta B^2}$.

Under classical conditions, that is,
in the local limit of conduction when
$q\ell \ll 1$, where ℓ is electron mean free path,
the acoustic amplitude for longitudinal
waves (1) and for shear waves (2,3) is
$|\xi_B| \doteq \frac{B B_o}{4\pi ds\omega} \left(\frac{1}{1+\beta^2}\right)^{\frac{1}{2}}$, where the skin
depth parameter $\beta = \frac{1}{2}q^2\delta^2$, $q = \omega/s$ and s
is the relevant acoustic speed. This
type of 'normal' generation has a
conversion efficiency: $\eta_B = B_o^2/2\pi ds\omega\delta(1+\beta^2)$.

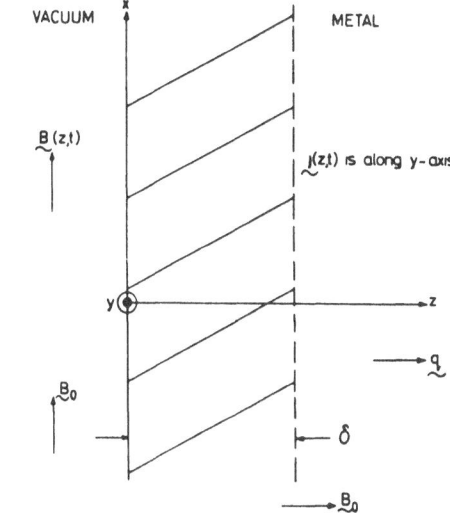

Longitudinal Waves or Transverse Waves

FIG 1

An r.f. field $\underset{\sim}{B}(z,t)$ induces an eddy current
$\underset{\sim}{j}(z,t)$ which decays within the skin depth δ.
Ultrasonic waves with wave vector $\underset{\sim}{q}$ along Oz
are polarized (a) longitudinally when $\underset{\sim}{B_o}$ // Ox
and (b) transversely when $\underset{\sim}{B_o}$ // Oz.

In the local limit acoustic generation only takes place when B_o is finite and increases linearly with B_o. Under non-local conditions, that is when $q\ell \gtrsim 1$, acoustic generation can take place in the absence of a static field, i.e. when $B_o = 0$. We consider the efficiency calculated for specular reflection of the electrons under normal skin effect and anomalous skin effect conditions and the results for a model calculation of diffuse reflection.

Normal Skin Effect

In the non-local limit, Southgate (4) showed that acoustic generation was possible when $B_o = 0$ and obtained the amplitude ξ_E (see FIG 2). For low frequencies and so $b = q\ell \simeq 1$, this amplitude becomes

$$|\xi_E| = \frac{cBm}{4\pi ds\omega e\tau} \frac{b}{5(1+\beta^2)^{\frac{1}{2}}} \quad ,$$

since the conductivity $\sigma_{xx} = \sigma_o(1-\tfrac{1}{5}b^2)$ in this limit. The corresponding efficiency is

$$\eta_E = \frac{c^2m^2b^4}{50\pi ds\omega\delta e^2\tau^2(1+\beta^2)}$$

and the relative conversion efficiency in the absence (η_E) and presence (η_B) of a static field B_o is

$$\frac{\eta_E}{\eta_B} = \frac{c^2}{25}\left(\frac{m}{e\tau}\right)^2 \frac{b^4}{B_o^2}$$

In these equations m, $-e$ and τ are the mass, charge and relaxation time of the conduction electrons and the d.c. conductivity $\sigma_o = n_o e^2\tau/m$.

Typically, for 10 MHz generation of shear waves in an aluminium crystal at 4 K with residual resistance ratio of 1000 (τ = 12 psec, b = 0.27), η_E/η_B = 48 when

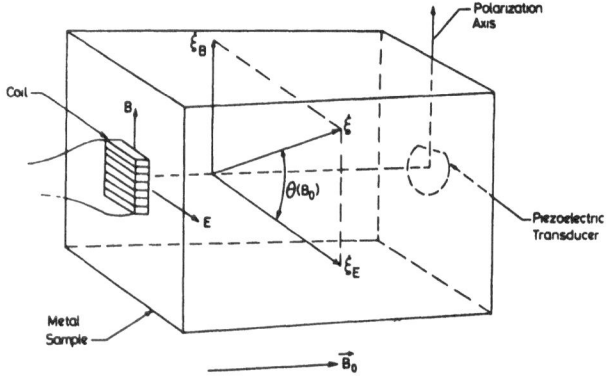

FIG 2

The r.f. fields B, E generate ξ_E when $B_o = 0$ and ξ_B when B_o is finite. Their sum ξ is a plane polarized wave making angle $\theta(B_o)$ with ξ_E). [After Wallace (5)].

$B_o = 10G$, but η_E/η_B = 0.5 when B_o = 100 G. Normal generation at 300 K with B_o = 4 kG produces $\eta_B = 10^{-5}$ for longitudinal waves and $\eta_B = 2 \times 10^{-5}$ for shear waves. At these low frequencies $\eta_B \sim B_o^2$ and normal generation is more efficient at high fields for both polarizations and at all temperatures.

Anomalous Skin Effect

Under normal skin effect conditions, η_E increases as $\omega^3\ell^2$, but for $b \gg 1$ and anomalous conditions ($\ell \gg \delta$), the acoustic amplitude becomes $|\xi_E| = \frac{B n_o e}{cdsq^2}$, the conductivity being given by: $\sigma_{xx} = 3\pi\sigma_o/4b$.   Hence the conversion efficiency $\eta_E^a = \dfrac{c^2 m^2}{2\pi ds\omega\delta e^2\tau^2} \dfrac{4}{q^4\delta^4}$ in the anomalous limit and then η_E decreases as ω^{-3} and is independent of ℓ (e.g. $\nu > 10$ GHz in Aℓ with $\ell > 1$ μm).

At these microwave frequencies, the acoustic signals are generated by microwave radiation from a cavity and we define a radiation efficiency,

$$H = \frac{\text{Generated Acoustic Power P}}{\text{Incident Electromagnetic Power W}}$$

It is easily shown that under normal skin effect conditions, $H = (2\omega\delta/c)\eta$ and in general $H/\eta = (2\omega/\pi\sigma)^{\frac{1}{2}}$.   Mertsching (6) has derived

$$H = \frac{64cn_o m\omega^2 V_F^2}{9\pi^2 d\omega^2 s^3 \rho} \cdot \frac{1}{1+(q\delta_A)^6}$$

for generation when $b \gg 1$ and $\ell > \delta$, where δ_A is the Pippard skin depth $= \frac{1}{2}(\ell\delta^2)^{\frac{1}{3}}$.   His expression agrees with η_E^a within a factor of 2, but shows additionally that between the normal and anomalous limits there is a maximum efficiency, under specular conditions, when $q\delta_A = 1$, which occurs at about 1 GHz (FIG 3).   Typically for Aℓ at 4K with $\ell = 16$ μm, at maximum $\eta_E^a \simeq 10^{-1}$ and $H \simeq 10^{-7}$, but this has not been achieved experimentally.

Diffuse Reflection

Southgate (4) has developed a useful model of isotropic electron motion, which enabled him to calculate η_E for diffuse reflection under both normal and anomalous skin effect conditions.   He finds that for $\ell > \delta$, η_E becomes independent of frequency, so that, for example, at 20 GHz in Aℓ with $\ell > 1$ μm, $\eta_E \sim 10^{-3}$ for diffuse reflection, much larger than for specular reflection (about 10^{-5}). Unlike the specular case, there is no matching condition for diffuse reflection.   Goldstein and Zemel (7) have measured $H = 5 \times 10^{-5}$ at 9 GHz in an In film at 4K and attribute this high value to diffuse reflection.

Ferromagnetic-Acoustic Resonance

We have recently shown that a ferromagnetic-acoustic resonance occurs in a magnetostrictively coupled spin wave-sound wave system in a magnetically saturated metal, in which an r.f. magnetic field excites an acoustic signal of

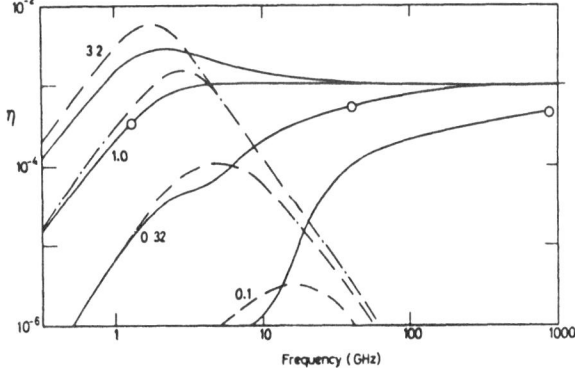

FIG 3

Efficiency η_E of generation calculated for an isotropic metal (Aℓ) for various free paths (μm); —— diffuse, — · — · specular reflection, 0 frequencies at which skin effect becomes anomalous [after Southgate (4)].

amplitude $|\xi_F| = \dfrac{\gamma B_2 B \tau_s}{ds^2 |(q_1 + q_2)|}$, where γ is the gyromagnetic ratio of the electrons, B_2 is the second magnetoelastic constant, τ_s is the spin wave relaxation time, q_1 is the wave vector of the electromagnetic branch and q_2 the wave vector of the sound wave branch. The conversion efficiency is therefore

$$\eta_F = \frac{8\pi\omega\gamma^2 B_2^2 \tau_s^2}{ds^3 \delta |(q_1 + q_2)|^2}$$

and for Nickel (110) with B_o = 10 kG, ν = 15 MHz, T = 300K we find $\eta_F \simeq 10^{-1}$ on the improbable assumption that we can extrapolate τ_s from microwave measurements using $\tau_s \sim \omega^{-1}$. This is much greater than the efficiency of normal generation, $\eta_B \sim 10^{-5}$, and experimentally we have found $\eta \sim 10^{-3}$ for shear wave generation, about 100 times greater than expected for normal generation.

A fuller account of electromagnetic generation in metals and semimetals (8) and in ferromagnetic crystals (9) will be given elsewhere.

References

1. H.L. Grubin, IEEE Trans. Sonics Ultrason. 16, 27 (1969).

2. M.R. Gaerttner, W.D. Wallace and B.W. Maxfield, Phys. Rev. 184, 702 (1969).

3. D.J. Meredith, R.J. Watts-Tobin and E.R. Dobbs, J. Acoust. Soc. Amer. 45, 1393 (1969).

4. P.D. Southgate, J. Appl. Phys. 40, 22 (1969).

5. W.D. Wallace, Int. J. Nondestructive Test. 2, 309 (1971).

6. J. Mertsching, Phys. Status Solidi, 37, 465 (1970).

7. Y. Goldstein and A. Zemel, Phys. Rev. Lett. 28, 147 (1972).

8. E.R. Dobbs, Physical Acoustics, Volume 10, Chap. 3, Academic Press, New York (1973).

9. M.J.W. Povey, E.R. Dobbs and D.J. Meredith, J. Phys. F. (to be published).

AN AUTOMATED TORSION PENDULUM WITH INTERFEROMETRIC ANGULAR POSITION TRANSDUCER

J.B.Kuipers and A.W.Sleeswyk

Laboratorium voor Fysische Metaalkunde, University of Groningen,
Nijenborgh 18, Groningen, Netherlands.

ABSTRACT

A two-beam Michelson interferometer is described, designed specifically for measuring the angular position of an automated inverted torsion pendulum. Angles up to 0.1 rad from the centre position can be measured at speeds up to 1 rad/sec. The long-term accuracy is better than 10^{-6} rad, corresponding to one count of the electronic fringe counter. The optical geometry is such that the interferometer is insensitive to parasitic rotational and translational oscillations of the pendulum, e.g. due to foundation movements. A decrement-meter, which measures the amplitude of each oscillation and calculates the logarithmic decrement is described in some detail.

Introduction

When the need was felt at our laboratory for a torsion pendulum to complement other methods already in use for investigating dislocation dynamics, a search in the literature revealed a large number of descriptions of such instruments incorporating various optical and electrical transducers, none of which however could completely satisfy our requirements. Our design aim was an instrument capable of measuring torsion angles up to 10^{-1} rad to within 10^{-6} rad at pendulum frequencies from zero (elastic after-effect) up to 10 sec^{-1}, while being sufficiently reliable and insensitive to parasitic vibrations to permit continuous automatic operation with on-line processing to determine the internal friction.

Failing a suitable model we investigated the feasability of a transducer making use of the advantages offered by interferometric techniques. An optical geometry was found which solves to a large extent all the problems peculiar to the torsion pendulum, and bi-directional fringe-counting circuits were developed with the emphasis on reliable automatic operation.

Interferometer

Basic Design: The interferometer comprises a 1 mW He-Ne laser, a specially made beam-splitting prism, a plane mirror and a detector containing two light-sensitive diodes. Fig.1 shows the set-up viewed from above, i.e. along the rotation axis. The prism is attached to the pendulum inertia member whereas the other components are fixed. The beam 1 from the laser is split into two beams 2 and 3 of approximately equal intensity at the coated mating surface of the

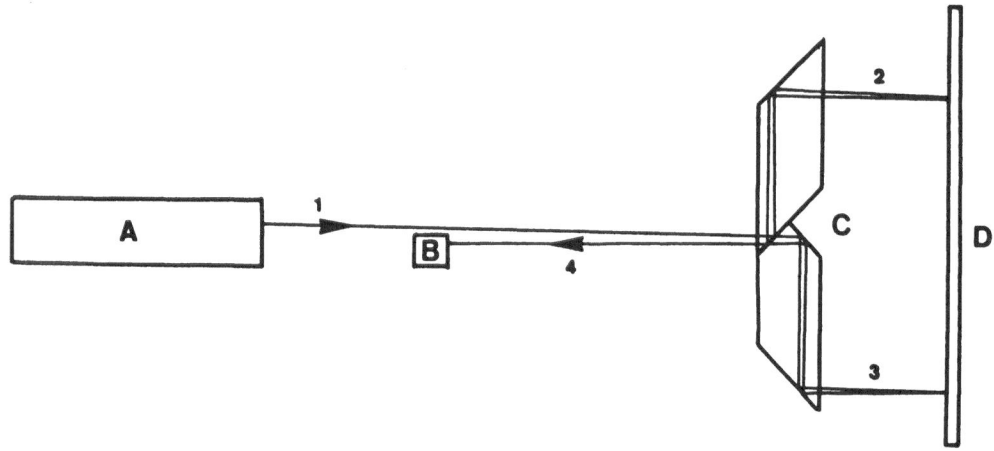

FIG.1.

Interferometer set-up. A. Laser; B. Detector; C. Rotating Prism; D. Reference mirror.
The numbers of the beams are referred to in the text.

prism halves. The two beams exit parallel to beam 1 but displaced by equal amounts in
opposite directions. They are reflected by the mirror, traverse the prism in the opposite
direction and are re-united, producing an interference pattern in the detector. Providing
all the surfaces involved are sufficiently flat, the result is a grid of illuminated parallel
fringes spaced at λ/δ, where δ is the small angle between the interfering beams.

Rotation of the pendulum causes the interference pattern to move across the detector
while beam 4 remains stationary. The resulting variations in the currents generated in the
diodes contain the information necessary for determining the magnitude and sense of the
rotation. The need for devices to separate beams 1 and 4 is obviated by setting the mirror's
normal at a small angle γ to the incoming beams.

Geometrical considerations: Providing the reflecting surfaces A and A' (Fig.2) are rigidly
connected and parallel, translatory movement of the pair AA' does not affect the optical
path-length of beam 2 (Fig.1). The same holds for the pair BB' and beam 3. Pairs AA' and BB'
must be symmetrically opposed to obtain a first-order linear relationship between the pendulum
torsion angle ϕ and the associated change in optical path-length difference Δn. A detailed
analysis of the relationship between Δn and rotations in the various degrees of freedom has
not been attempted, however the rotations ψ and θ around axes normal to the pendulum axis
certainly do not enter the expression in terms of lower than third order, so that:

$$\Delta n = d_0 \sin \phi + \text{(terms of 3rd order and higher in } \phi, \psi, \theta), \qquad (1)$$

where d_0 is the distance separating beams 2 and 3, in our case 60 mm when the prism is in its
central position.

Prism Construction: Although the rotating member could conceivably be built up with discrete
components, a prism seems the best choice in the interest of rigidity and low inertia. The
partially-transmitting coating must be deposited on the prism half which transmits the

178

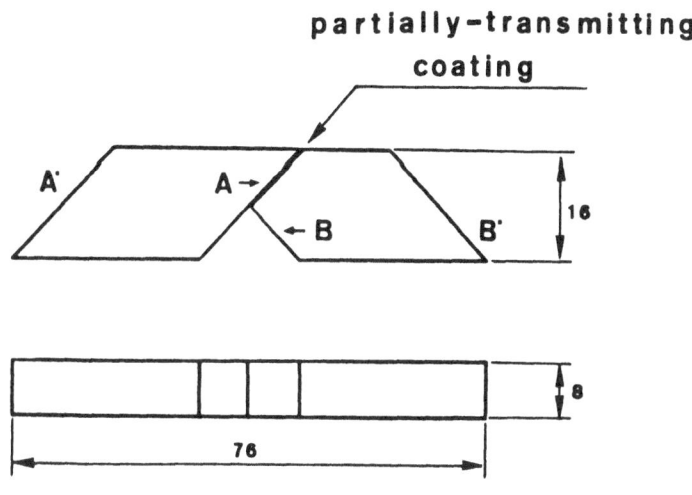

FIG.2.

Dimensions of prism, in mm.

reflected beam. Deposition on the other half would make the alignment of the halves un-
necessarily critical. A surface flatness of 0.1λ and parallelism of corresponding surfaces
to within 10 arcsec proved sufficient; the remaining angular inaccuracies are adjusted for
during the glueing procedure. This was carried out in the same jig which was used afterwards
as prism mounting on the pendulum. Alignment was checked with the aid of the laser and
mirror in a set-up identical to that of the interferometer. In our case the fringe spacing
turned out to be around 2 mm, varying slightly as the 1 mm laser beam was moved over the
8×8 mm prism aperture. From the description of the detector in the following section it will
be apparent that large variations in fringe spacing can be tolerated, so that in general no
corrective measures are called for.

Phase-sensitive detector: The detector consists of two silicon photo-diodes (BPY 10) each
having a light-sensitive area of 1.6×1.6 mm, mounted side-by-side so that the interference
fringes move from one diode to the other (Fig.3). The phase angle between the sinusoidal
currents generated in the diodes as the pendulum is rotated depends on the ratio of diode
spacing to fringe spacing, and is set at approximately $\pi/2$. The actual amount is not critical,

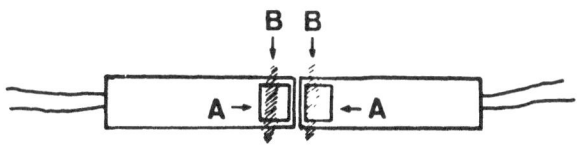

FIG.3.

Arrangement of photo-diodes relative to fringes in beam.
A. Photodiode (light-sensitive areas); B. Interference fringes in beam.

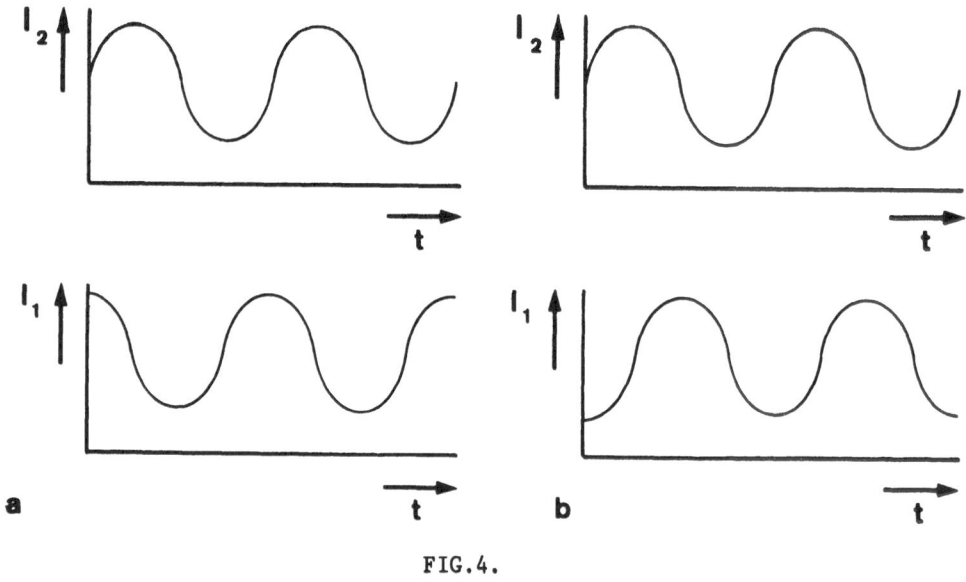

FIG.4.

Phase relation between diode outputs I_1 and I_2.
a. Pendulum rotates clockwise; b. Pendulum rotates anti-clockwise.

the important point being that the phase relationship is inverted as the prism is rotated in the opposite direction (Fig.4).

With d_0 = 60 mm and λ = 6328 Å, and at the maximum pendulum speeds envisaged the fringe counting speed approaches 1 Mhz. In view of the diode capacitance this restricts the diode load resistance to 1 kΩ, so that the amplifier input signals resulting from μA-currents lie in the mV-range. Stability problems arising from the large gain-bandwidth product required are avoided by emplying AC-coupled amplifiers. This also eliminates problems associated with variations in the total transmitted intensity over the prism aperture, caused by inhomogeneity of the partially transmitting layer or the canada balsam glue. The amplifier outputs are clipped to produce 5 V rectangular signals.

Vibrating Mirror: The use of AC-coupled amplifiers does not preclude the measurement of very slow pendulum movements. The interference pattern is permanently kept in motion by introducing an auxiliary vibration of fixed amplitude, which is later subtracted in a filtering circuit. We found that this is conveniently done by mounting a small loudspeaker on the mirror, powered by a sine-wave generator. We chose 500 Hz; the mirror amplitude of 5.10^{-6} rad was determined as the smallest value for which the electronics would function reliably. An auxiliary vibration was employed by de Lang et al. (1) for measuring displacements much smaller than the interferometer wavelength. The same means could be employed in our apparatus to resolve prism rotations of about 10^{-8} rad.

Signal Processing

Decoder and Filter: The decoder possesses two inputs and two outputs. When the logical value of one of the two inputs changes, a 1 μsec rectangular pulse is generated in a one-shot multivibrator, appearing at the "up" or "down" output depending on the logical value of the other input at that instance. Thus displacement of the interference pattern by one fringe spacing produces four pulses at one of the outputs.

The decoder outputs can be fed to a filter containing a 3-bit up-down counter which accumulates the pulses until it over- or underflows. On overflow, "up" pulses can no longer enter the counter but are passed on to the output whereas "down" pulses can only enter the counter. The inverse is true when underflow occurs. Consequently two maxima in the pendulum movement are only distinguishable at the filter output if they are separated by a minimum lower than both by more than 8 pulses, and vice versa. Although the prime objective of the filter is to subtract electronically the auxiliary vibration from the pendulum motion signal, it serves equally well in removing any other high frequency vibrations registered by the interferometer. Interference of this sort (Fig.5a) would otherwise upset the amplitude and oscillation counting in the decrement-meter.

Registration of elastic after-effects: When studying elastic after-effects, the motion of the pendulum in time is registered with an autographic recorder. In this case the filter output is fed into an up-down counter of sufficient (16-bit) capacity which in turn is connected to a digital-to-analogue converter.

Automatic decrement measurement: Many examples of automated torsion pendulums can be found in the literature (2,3), so the description of the present system (Fig.6) will be limited to the operation of the calculating circuits, which are designed to make full use of the digital

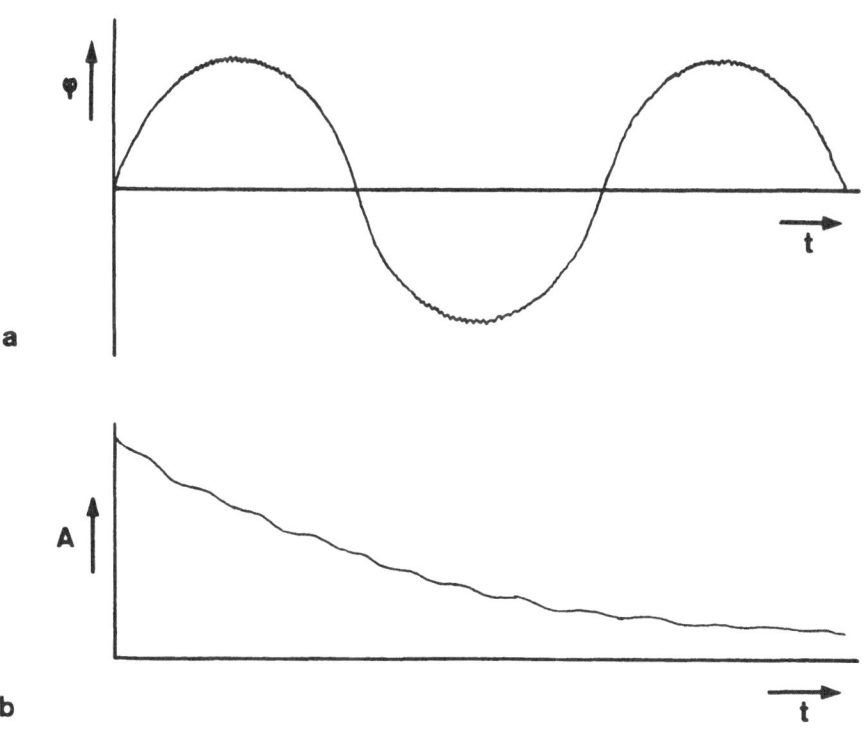

FIG.5.

Illustration of the effect of parasitic vibrations, of
a: high frequency (in comparison to period of pendulum oscillation)
or b: low frequency.

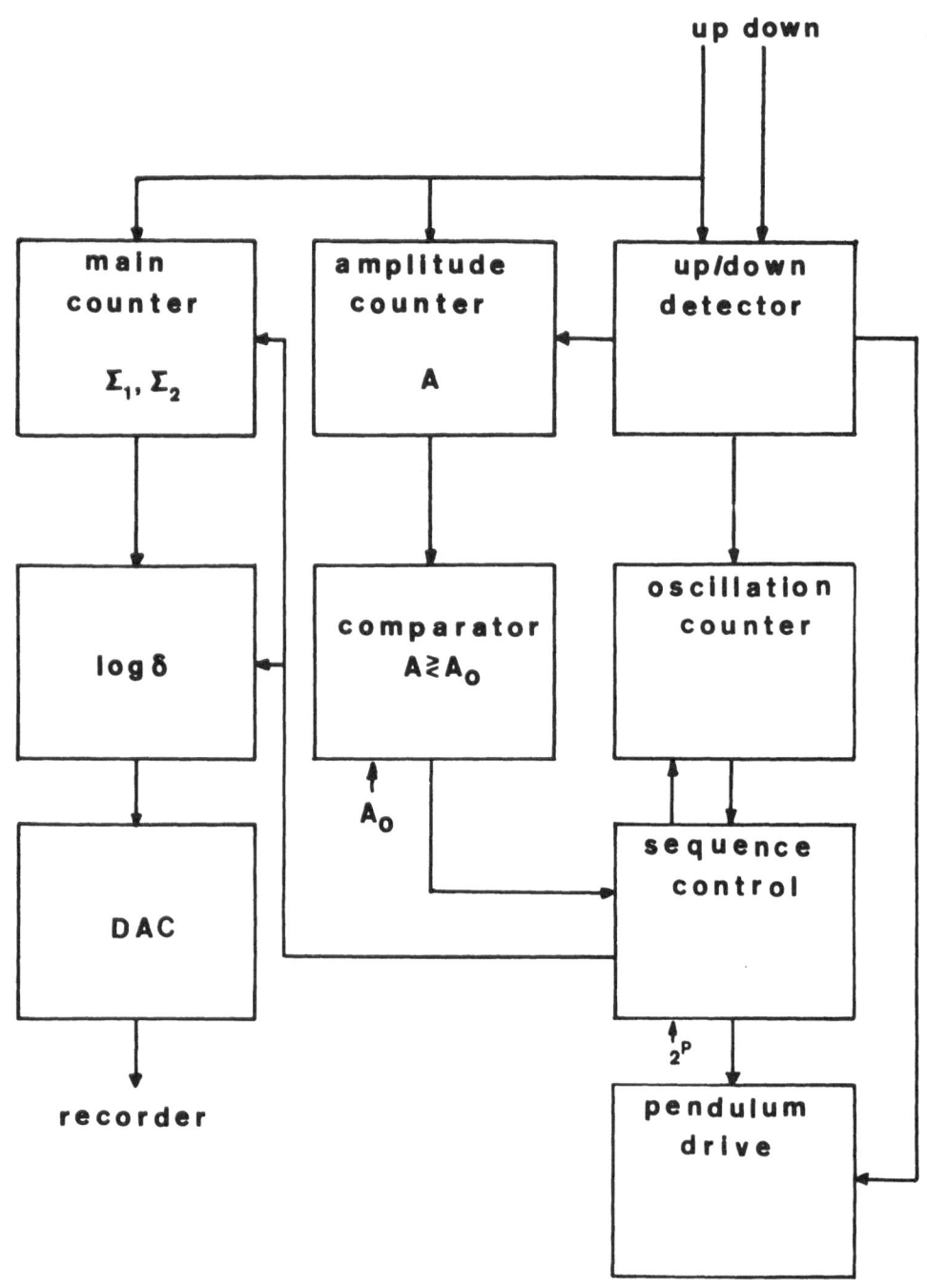

FIG.6.

Block diagram of signal flow in decrementmeter.

nature of the interferometer output. The filter ensures that the decrementmeter receives an uninterrupted flow of "up" pulses during the whole clockwise swing of the pendulum followed by an uninterrupted flow of "down" pulses during the return swing. These are entered in the inputs of the up-down detector, a bistable element of which the logical condition consequently corresponds to the pendulum rotation sense. On starting up the apparatus the up-down detector turns on the pendulum drive current during each clockwise swing until the amplitude has reached a preset value A_0. The "up" pulses on each swing are accumulated in the amplitude counter and at the first "down" pulse the total A is compared with A_0, after which it is reset at 8 to compensate for the filtering operation.

As soon as $A < A_0$ the actual measurement commences. All the "up" pulses during the first 2^{p-1} oscillations are accumulated in the decrement counter and the total $\Sigma_1 = \sum_{n=1}^{2^{p-1}} A_n$ is temporarily stored. After each oscillation one count is added in the oscillation counter. This is repeated to obtain $\Sigma_2 = \sum_{n=2^{p-1}+1}^{2^p} A_n$. The final operation is the calculation of the logarithmic decrement which is approximated by:

$$\log \delta = \frac{\Sigma_1 - \Sigma_2}{(\Sigma_1 + \Sigma_2)2^{p-2}} \ . \tag{2}$$

This approximation is correct to within 0.1 % if $A_{2^p} > 0.75\ A_1$, and the preset number p is chosen accordingly taking into account the maximum decrement of the specimen in the temperature range under investigation.

System Performance: The interferometer has proved capable of measuring rotations down to the limit of its resolving power, 10^{-6} rad in the presence of normal building vibrations. Test runs with dummy specimens show no measurable drift over a period of 24 hours. It is convenient in operation apart from the fact that the prism must not be subjected to high temperatures. Development of the electronic circuits is still under way, having been concentrated on the decoding and filtering circuits. The important property of the decoder is that it operates symmetrically, in the sense that the positive and the negative flanks of the amplifier output are used. Although decrement measurements are normally conducted without the auxiliary vibration, the filter has proved indispensable for reliable operation of the amplitude and oscillation counters.

During such measurements the pendulum amplitude is sometimes found to decrease in an oscillating manner as depicted in Fig.5b. This is attributed to the interaction of the various degrees of oscillatory freedom caused by slightly bent specimens. To prevent systematic calculating errors arising from this behaviour the averaging procedure described in the previous section was developed.

Acknowledgements

We are grateful to L.G.N. van den Houten for assistance in designing the electronic circuits. The prism was manufactured by the Technisch Physische Dienst of the T.N.O.-Institute. The work was supported by the Foundation for the Investigation of Matter (F.O.M.) and the Metaalinstituut T.N.O. at Delft, the Netherlands.

References

1. H.de Lange et al.: "Nauwkeurige digitale verplaatsingsmetingen met optische middelen", Philips Techn.T. 30, 153 (1969).

2. D.L.Smith, A.L.Winiecki and R.H.Lee: "An automated system for internal friction measurements", J.Phys.E 3, 715 (1970).

3. D.E.Barrow and Z.C.Szkopiak: "Automatic apparatus for high resolution internal friction measurement", J.Phys.E 5, 915 (1972).

POINT DEFECTS

THERMODYNAMIC TREATMENT OF THE RELAXATION AND RESONANCE PROCESSES IN ANELASTICITY

B. Vittoz

Laboratoire de Génie Atomique de l'Ecole Polytechnique Fédérale de Lausanne
(Switzerland)

INTRODUCTION

Many anelastic properties are explained by the relaxation behaviour governed by the thermodynamics of irreversible processes [1], where the deviations from a reference state, in which the system is at thermodynamic equilibrium, play the roles of thermodynamic forces, whereas the rate of return to the equilibrium state of the deviations are the thermodynamic fluxes. In the special case where the external excitation is constant, the phenomenological relations between fluxes and forces lead to the well-known exponential dependence of the measured quantity such as the amplitude of deformation.

Well-known experimental results show that on to the relaxation there are superposed inertial phenomena. In order to include this phenomena in a general theory of irreversible processes, we must take into account the relative, non-thermal velocities of the components of the system considered, with respect to the local mass centre, when forming balance equations to describe local behaviour. The inertial forces will then appear in the thermodynamic forces and the phenomenological equations lead to a behaviour in which relaxation and resonance are superposed. We recall first the essential results of the theory of irreversible processes, then we give the applications of them to pinned dislocations. The theory is based essentially on the work of de Groot and Mazur [2] where in chapter III the inertial forces appear. The works of Meixner and Reik [3], and of Glansdorff and Prigogine [4] have also been used.

THE ENTROPY BALANCE EQUATION

Let us consider a system consisting of n components (index k) amongst which r chemical reactions (index p) are possible. The general form of the entropy balance equation is :

$$\rho \, \frac{ds}{dt} \; = \; -\text{div} \, \vec{J}_s \, + \, \sigma \tag{1}$$

where ρ is the density, d/dt the (barycentric) substantial time derivative, s the specific entropy (entropy per unit mass), $\vec{J}_s$ the entropy flux (across a surface moving with the local mass centres (barycentric motion)), and σ the entropy source strength, or entropy production per unit volume and unit time.

From the second law of thermodynamics :

$$\sigma \geqslant 0 \qquad (2)$$

For the definition of the entropy s we use the Gibbs equation :

$$T\,ds = du^* - \sum_{ij} \theta_{ij}\, de_{ij} - \sum_{k} \mu_k^*\, dc_k \qquad (3)$$

where u^* is the specific internal energy (energy per unit mass) after allowing for the kinetic energy of the n components with respect to the barycentric motion, θ_{ij} (i, j = 1 to 3) is the stress tensor σ_{ij} divided by ρ, e_{ij} is the strain tensor, μ_k^* the chemical potential of the component k excluding the kinetic energy with respect to the barycentric motion and c_k the mass fraction of the component k :

$$c_k = \rho_k / \rho \qquad (4)$$

where ρ_k is the density of component k (mass of k per unit volume of the system). The quantities u^* and μ_k^* are related to the usual values u and μ_k by

$$u^* = u - \sum_{k} \frac{1}{2}\, c_k\, \Delta_k^2 \quad ; \qquad \mu_k^* = \mu_k - \frac{1}{2}\, \Delta_k^2 \qquad (5)$$

where $\vec{\Delta}_k$ is the local velocity of the component k defined with respect to the barycentric motion :

$$\vec{\Delta}_k = \vec{v}_k - \vec{v} \qquad (6)$$

$\vec{v}_k$ and $\vec{v}$ being the absolute velocities of component k, and the local mass centre.

The energy balance equation is :

$$\rho \frac{du}{dt} = -\text{div}\, \vec{J}_q + \sum_{k} \vec{J}_k \cdot \vec{f}_k + \sum_{ij} \sigma_{ij} \frac{de_{ij}}{dt} \qquad (7)$$

where $\vec{J}_q$ is the heat flux, $\vec{f}_k$ the external force per unit mass exerted on the component k, and $\vec{J}_k$ the diffusion flow of component k defined with respect to the barycentric motion :

$$\vec{J}_k = \rho_k \vec{\Delta}_k \underset{(6)}{=} \rho_k (\vec{v}_k - \vec{v}) \qquad (8)$$

The density ρ is the sum of the partial ρ_k :

$$\sum_{k} \rho_k = \rho \qquad \text{or} \qquad \sum_{k} c_k = 1 \qquad (9)$$

the linear momentum (per unit volume) is the sum of the partial linear momenta :

$$\rho \vec{v} = \sum_{k} \rho_k \vec{v}_k \qquad (10)$$

thus we have

$$\sum_{k} \vec{J}_k = \sum_{k} \rho_k \vec{\Delta}_k = 0 \qquad (11)$$

With equations (3), (5) and (7), the entropy flux $\vec{J}_s$, and the entropy source strength σ are:

$$\vec{J}_s = \frac{1}{T} (\vec{J}_q - \sum_k \mu_k \vec{J}_k) \tag{12}$$

$$T\sigma = \sum_{k=1}^{n} \vec{J}_k \cdot [\vec{f}_k - \vec{\Delta}_k - \overrightarrow{\text{grad}\,\mu_k}] - \vec{J}_s \cdot \overrightarrow{\text{grad}\,T} - \sum_{p=1}^{r} J_p A_p \tag{13}$$

where J_ρ is the chemical reaction rate (moles per unit volume and unit time) of the reaction p, and A_p the chemical affinity.

$$A_p = \sum_k \mu_k \nu_{k,p} \tag{14}$$

$\nu_{k,p} =$ molecular mass of component k which appears ($\nu_{k,p} > 0$) or disappears ($\nu_{k,p} < 0$) in the reaction p ; $\sum_k \nu_{k,p} = 0$.

These terms come from the last term of equation (3) and from the conservation of mass :

$$\rho \frac{dc_k}{dt} = -\text{div}\,\vec{J}_k + \sum_p \nu_{k,p}\, J_p \tag{15}$$

According to (13), and the second law of the thermodynamics (2), the entropy production can be expressed in a very general form

$$T\sigma = \sum_\alpha J_\alpha X_\alpha \geqslant 0 \tag{16}$$

where J_α and X_α are the independent thermodynamic fluxes and forces. The latter are the cause of irreversible processes and should all disappear in thermodynamic equilibrium, whereas the fluxes J_α are the effects of irreversible processes, but in equation (13) the n diffusion flows $\vec{J}_k$ are not independent. With equation (11) :

$$\vec{J}_n = - \sum_{k=1}^{n-1} \vec{J}_k \tag{17}$$

Substitution of (17) into (13) yields :

$$T\sigma = \sum_{k=1}^{n-1} \vec{J}_k \cdot [\vec{f}'_k - \vec{\Delta}'_k - \overrightarrow{\text{grad}\,\mu'_k}] - \vec{J}_s \cdot \overrightarrow{\text{grad}\,T} - \sum_p J_p A_p \tag{18}$$

where $\qquad \vec{f}'_k = \vec{f}_k - \vec{f}_n \;\; ; \;\; \vec{\Delta}'_k = \vec{\Delta}_k - \vec{\Delta}_n \;\; ; \;\; \mu'_k = \mu_k - \mu_n \tag{19}$

The thermodynamic fluxes can therefore be considered to be J_p, $\vec{J}_s$ (with 3 scalar components), and the (n-1) diffusion flows $\vec{J}_k$ (each with 3 scalar components). While the associated or conjugated forces are :

$$\vec{X}_k = \vec{f}_k' - \dot{\vec{\Delta}}_k' - \overrightarrow{\text{grad}}\mu_k' = \text{mechanical (and chemical) diffusion force}$$
$$(k = 1 \ldots n - 1)$$

$$\vec{X}_s = -\overrightarrow{\text{grad}}\, T = \text{thermal diffusion force}$$

$$X_p = -A_p = \text{chemical reaction force } (p = 1 \ldots r)$$

In the mechanical diffusion forces we see the inertial forces (per unit mass) appearing:

$$\dot{\vec{\Delta}}_k' = \dot{\vec{\Delta}}_k - \dot{\vec{\Delta}}_n \tag{20}$$

If the deviation from equilibrium is small enough, one may postulate the existence of linear relations, between the fluxes and the forces, which give the phenomenological equations:

$$J_\alpha = \sum_\beta L_{\alpha\beta} X_\beta \tag{21}$$

where the phenomenological coefficients $L_{\alpha\beta}$ may depend on the local state variables (temperature, composition, strain, etc.), but not on the thermodynamic forces themselves. The substitution of (21) in (16) gives the quadratic definite positive form

$$T\sigma = \sum_{\alpha\beta} L_{\alpha\beta} X_\alpha X_\beta \geqslant 0 \tag{22}$$

which implies

$$L_{\alpha\alpha} > 0 \; ; \; (L_{\alpha\beta} + L_{\beta\alpha})^2 \leq 4 L_{\alpha\alpha} L_{\beta\beta} \tag{23}$$

If there is no magnetic or spin effect, the Onsager relations give :

$$L_{\alpha\beta} = L_{\beta\alpha} \tag{24}$$

ENERGY DISSIPATION

Let us consider a system subjected to a periodic excitation of frequency $\omega/2\pi$. The amount of energy Δw dissipated per unit volume over a period $2\pi/\omega$, is equal to :

$$\Delta w = \int_{2\pi/\omega} \rho\, \frac{du}{dt}\, dt \tag{25}$$

whereas the local internal friction is given by :

$$Q^{-1} = \frac{1}{2\pi} \frac{\Delta w}{w} \tag{26}$$

where w is the maximum energy per unit volume stored during a period.

APPLICATIONS

Pinned dislocations (the Koehler, Granato-Lücke model)

If the above theory is applied to the behaviour of a crystal having pinned dislocations which are characterized by a mass m per unit length, by a line tension γ, and on which acts the Peach-Koehler force τb, one obtains the equation of the vibrating string model suggested by Koehler [5] and developed by Granato and Lücke [6]. In this equation, the inertial term and the damping term will appear from the direct application of the theory of irreversible processes.

Consider, as shown in figure 1, a dislocation pinned in the crystal at O and A, points fixed with respect to the region of good crystal which constitutes the relative frame of reference R.

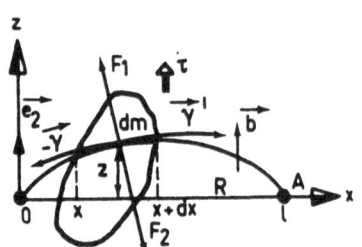

Let us take an element dm (volume dV) of the crystal, containing a segment of dislocation of equilibrium length dx ; the rest of dm is considered as a region of good crystal. The dislocation is the component k = 1, and the rest the component k = 2. Suppose that the displacement z of the dislocation with respect to its equilibrium position is small. From the expression for the entropy source strength (13) we can calculate a partial source strength σ_1 :

Fig. 1

$$T\sigma_1 = \sum_{k=1}^{2} \vec{J}_k \cdot [\vec{f}_k - \dot{\vec{\Delta}}_k] = \sum_{k=1}^{2} \rho_k \vec{\Delta}_k \cdot [\vec{f}_k - \dot{\vec{\Delta}}_k] \qquad (27)$$

$\vec{f}_k$ being the external force per unit mass exerted on component k. The segment dx of dislocation is subjected to external forces F_1 (Peach-Koehler force), and to the resultant F_2 of the line tensions $-\gamma$ and $\vec{\gamma'} = \vec{\gamma}(x + dx)$. The external force $f_1 \, dm_i = f_1 \, \rho_1 \, dV$ then takes the value :

$$\vec{f}_1 \rho_1 \, dV = \vec{e}_2 \, (\tau b \, dx + \gamma \frac{\partial^2 z}{\partial x^2} dx) \qquad (28)$$

where τ is the resolved shear stress and $\vec{b}$ the Burgers vector. In the volume dV there is an average length dx of dislocation equal to $\Lambda \, dV$, where Λ is the total length per unit volume of the dislocations ; thus after substitution for dV equation (28) becomes :

$$\vec{f}_1 \rho_1 = \vec{e}_2 \, \Lambda \, (\tau b + \gamma \frac{\partial^2 z}{\partial x^2}) \qquad (29)$$

For the velocities $\vec{\Delta}_k$ we have from previous equations :

$$\vec{\Delta}_k = \vec{v}_k - \vec{v} \quad (6) ; \qquad \rho \vec{v} = \rho_1 \vec{v}_1 + \rho_2 \vec{v}_2 \quad (10) ; \qquad \rho_k = \rho \, c_k \quad (4)$$

Let us define $\vec{w} = \vec{e}_2 \dfrac{\partial z}{\partial t} = \vec{e}_2 \dot{z} =$ the velocity of the segment of dislocation, relative to R. Since R has a translational motion (z is small, thus a rotation has no effect) with the velocity $\vec{v}_2$:

$$\vec{v}_1 = \vec{w} + \vec{v}_2 \tag{30}$$

and the absolute velocity of the mass centre is given by (10) :

$$\vec{v} = c_1 \vec{v}_1 + c_2 \vec{v}_2 = c_1 (\vec{w} + \vec{v}_2) + (1 - c_1) \vec{v}_2 , \quad \text{that is}$$

$$\vec{v} = c_1 \vec{w} + \vec{v}_2 \tag{31}$$

c_1 being the mass fraction of the dislocations :

$$c_1 = \rho_1 / \rho = m \Lambda / \rho ; \quad c_2 = 1 - c_1 \tag{32}$$

and the velocities relative to the mass centre :

$$\vec{\Delta}_1 = \vec{e}_2 (1 - c_1) \dot{z} ; \quad \vec{\Delta}_2 = -\vec{e}_2 \, c_1 \, \dot{z} \tag{33}$$

By considering that the region of good crystal is not subjected to any external force per unit mass, $f_2 = 0$, the entropy source strength given by (27) becomes :

$$T\sigma_1 = \Lambda (1 - c_1) \dot{z} \left[\tau b + \gamma \frac{\partial^2 z}{\partial x^2} - m\ddot{z} \right] \tag{34}$$

According to the theory of irreversible processes we may write :

$$\tau b + \gamma \frac{\partial^2 z}{\partial x^2} - m \ddot{z} = X_z = \text{thermodynamic force} \tag{35}$$

$$\Lambda (1 - c_1) \dot{z} = J_z = \text{thermodynamic flux} \tag{36}$$

Supposing that the other thermodynamic forces have no effect[*], the phenomenological equations (21) become :

$$\Lambda (1 - c_1) \dot{z} = L_{zz} \left[\tau b + \gamma \frac{\partial^2 z}{\partial x^2} - m \ddot{z} \right]$$

which may be written :

$$m \ddot{z} + B \dot{z} - \gamma \frac{\partial^2 z}{\partial x^2} = \tau b \tag{37}$$

and one obtains formally the equation of the vibrating string model, with its inertial term $m\ddot{z}$, and its damping term $B\dot{z}$, where $B = L_{zz}^{-1} \Lambda (1 - c_1) > 0$ as the diagonal coefficients $L_{\alpha\alpha}$ are essentially positive (24).

---

[*] We have no place here to consider the cross-effects such as the influence of grad T on the motion of the dislocation

Let ξ be the average value of displacement of the pinned dislocation

$$\xi = \langle z \rangle_x = \frac{1}{l} \int_l z \, dx \qquad (38)$$

Equation (37) gives for ξ :

$$m\ddot{\xi} + B\dot{\xi} + k\xi = \tau b \qquad (39)$$

where k is the average elastic constant of the dislocation ($\cong 12 \, \gamma/l^2$).

A knowledge of ξ permits the determination of the anelastic component of the strain tensor from which the internal friction and the modulus defect can then be deduced. Since the work of Granato and Lücke [6] this calculation has become a classic one ; nevertheless it is interesting to present a more general formalism, thermodynamicaly based, which uses the idea of internal variables (see Nowick and Berry [1] - chapter 5).

Internal variables

Consider the crystal containing pinned dislocations in equilibrium ; let us apply a small disturbance to the system which takes it into a neighbouring state of equilibrium. During this process the increase of entropy ds_1 due to the source σ_1 is, according to equation (34), equal to:

$$T \, ds_1 = \rho^{-1} T\sigma_1 \, dt = \rho^{-1} \Lambda \, (1 - c_1) \left[\tau b + \gamma \frac{\partial^2 z}{\partial x^2} \right] dz$$

Using the average value of z this gives :

$$T \, ds_1 = \rho^{-1} \Lambda \, (1 - c_1) \, [\tau b - k\xi] \, d\xi$$

In mechanical equilibrium $\dot{\xi}$ and $\ddot{\xi}$ are zero and equation (39) gives $k\bar{\xi} = \tau b$ where $\bar{\xi}$ is the value of ξ in mechanical equilibrium ; finally we have :

$$T \, ds_1 = K \, (\bar{\xi} - \xi) \, d\xi \qquad (40)$$

where

$$K = \rho^{-1} \Lambda \, (1 - c_1) \, k > 0$$

If we put

$$A = K \, (\bar{\xi} - \xi) \qquad (41)$$

equation (40) becomes :

$$T \, ds_1 = A \, d\xi \qquad (42)$$

To the variation of entropy given by the Gibbs equation (3), we add the variation given by (42) to give :

$$T \, ds = du - \sum_{ij} \theta_{ij} \, de_{ij} + A \, d\xi \qquad (43)$$

where $u = u^*$ (in mechanical equilibrium) and the mass fractions c_k have been supposed constant. Expression (43) can be considered as valid for all systems characterized by the variables u and e_{ij} and by an internal variable ξ which describes the internal structure of the

194

system (c_k can also be considered as an internal variable). The quantity A, which is called the associated **affinity** is the thermodynamic force conjugate to the flux $\dot{\xi}$, but only in the quasi-static case where we may neglect the inertial forces ; the real thermodynamic force X conjugate to the flux $\dot{\xi}$ is :

$$X = A - M \ddot{\xi} \qquad (44)$$

where M is the inertial coefficient of ξ.

The phenomenological equations, in the case where we neglect the cross-effects, give

$$\dot{\xi} = L(A - M \ddot{\xi}) \qquad (45)$$

which is the **evolution equation** for the internal variable ξ :

$$M \ddot{\xi} + L^{-1} \dot{\xi} - A = 0 \qquad (46)$$

We have obtained the same formalism as that developed by Nowick and Berry [1] - chapter 5. The affinity A may be expanded as a function of the variables, T, θ_{ij} and ξ for example. In the isothermal case we will again obtain formula (41) ; but **the evolution equation** for ξ now contains **the inertial term**. By similar limited expansions we can obtain expressions for e_{ij} and s, and hence the value of du by the Gibbs equation, and deduce the **internal friction** by using formulae (26) and (25) in **which the inertial term will appear.**

The above calculations can also be generalized to cover the case of several internal variables.

Point defect dragging by the dislocations

For a dislocation of length l, the equation of motion (39) when multiplied throughout by the length l can be written :

$$m l \ddot{\xi} = F \qquad (47)$$

where

$$F = l(\tau b - B_L \dot{\xi} - k \xi) \qquad (48)$$

Here B becomes B_L in order to indicate that it is the damping coefficient due to the proper line damping effect (dislocation -lattice interaction). As ξ represents the average value of the dislocation displacement z from equilibrium (38) over the length l, the quantity $l\xi$ is in fact an integral

$$l\xi = \int_l z \, dx \qquad (49)$$

Thus the force F is the sum of all the forces (continuous or otherwise) exerted on the dislocation ; the term $(-kl\xi)$ for instance is the sum of the forces exerted on the dislocation by the two extreme strong pinning points (O and A in figure 1).

Now consider that the dislocation is pinned by n mobile point defects between O and A ; we will assume that there is no interaction between these point defects, and that the interaction between each point defect and the dislocation may be represented by a short range force F_{pd} (force exerted by the point defect on the dislocation). With these n additional forces, the total force F_t acting on the dislocation is :

$$F_t = F + n F_{pd} \qquad (50)$$

By the principle of equality of action and reaction, the force F_{dp} exerted by the dislocation on a point defect is :

$$F_{dp} = -F_{pd} \qquad (51)$$

If we neglect the inertia of the point defects, and also assume that they are always pinning the dislocation and have no motion along the dislocation, then their average velocity v is equal to $\dot{\xi}$; and the phenomenological equation for each point defect is :

$$\dot{\xi} = v = M F_{dp} \qquad (52)$$

where the phenomenological coefficient M is the mobility, and is related to the diffusion coefficient D by :

$$M = \frac{D}{k_0 T} = \frac{D_0 \, e^{-E_M/k_0 T}}{k_0 T} \qquad (53)$$

where k_0 is the Boltzmann constant, and E_M is the activation energy for diffusion.

Substituting F_t for F in equation (47) and using equations (50) to (53) we obtain :

$$m \ddot{\xi} + (B_L + \frac{n \, k_0 \, T}{1 D}) \, \dot{\xi} + k\xi = \tau b \qquad (54)$$

As well as the usual damping B_L, we have a damping effect due to the point defect dragging by the dislocation ; this was noted by Simpson, Sosin [7] and by Winkler, Schindlmayr, Schlipf [8]. Equation (54) is used by O. Mercier et al. [9] for the interpretation of the Hasiguti peaks.

REFERENCES

[1] A. S. Nowick and B. S. Berry, Anelastic Relaxation in Crystalline Solids, Acad. Press, New York (1972)

[2] R. S. de Groot and P. Mazur, Non-Equilibrium Thermodynamics, North-Holland, Amsterdam (1962)

[3] J-Meixner and H. G. Reik, Thermodynamik der irreversiblen Prozesse, in Encyclopedia of Physics vol. III/2 (Springer, Berlin 1959)

[4] P. Glansdorff and I. Prigogine, Structure, Stabilité et Fluctuations, Masson, Paris (1971)

[5] J. S. Koehler, Imperfections in Nearly Perfect Crystals (W. Shockley et al. eds), chap. 7, Wiley, New York (1952)

[6] A. Granato and K. Lücke, J. Appl. Phys. 27, 583 (1956)

[7] H. M. Simpson and A. Sosin, Phys. Rev. B 5, 1393 (1972)

[8] Mrs. W. Winkler, R. Schindlmayr and J. Schlipf, Fifth Int. Conf. on Internal Friction and Ultrasonic Attenuation in Crystalline Solids, Aachen, 1973

[9] O. Mercier, A. Isoré and W. Benoit, Fifth Int. Conf. on Internal Friction and Ultrasonic Attenuation in Crystalline Solids, Aachen, 1973

GAS INTERSTITIALS IN B.C.C. METALS

E.J. Savino[+] and V.K. Tewary[++]
Theoretical Physics Division,
U.K.A.E.A. Research Group,
A.E.R.E., Harwell, England.

ABSTRACT

An "effective" gas-host atom interaction potential, obtained from the measured values of the dipole tensor, is used to calculate the static lattice distortion, activation energy and migration volume of nitrogen and oxygen interstitials in niobium, tantalum and vanadium, assuming both octahedral and tetrahedral occupancy. The Green function method for lattice statics has been used throughout the work and the dynamical model is based on the Flynn and Stoneham theory of light interstitials migration. The results are compared with experimental measurements where available.

Introduction

Though relaxation measurements do not distinguish between octahedral and tetrahedral occupancy in a b.c.c. structure, there is a general belief that interstitial gas atoms enter the octahedral site in the group VA metals (1). This assessment is based on the high value of the measured shape factor $|\lambda_1 - \lambda_2|$ of the strain ellipsoid (2) and the similarity of this value with that corresponding to C and N in α-Fe (3). In these latter systems a comparison of relaxation results and measurements of the tetrahedral martensite lattice parameter proves the octahedral occupancy (4). Previous theoretical calculations gave contradictory results; while Johnson et al (5) favour the octahedral site for N in V and C in α-Fe, Beshers (6) obtained the opposite conclusion for N and C in V and O, N in Ta and Nb.

To check the possibility of deducing the occupancy of an interstitial site from the measured parameters of the dipole tensor it is necessary to use an accurate lattice model which reproduces continuum and discrete properties of the material. The Green function

[+]On leave from Comision Nacional de Energia Atomica, Dep. Metalurgia, Argentina, with a Fellowship from the Consejo Nacional de Investigaciones Cientificas y Tecnicas, Argentina.
[++]On attachment from the University of Surrey.

technique developed by Tewary (7) and Bullough and Tewary (8) fulfills these conditions. We have extended it to compute the migration energy and activation volume of O and N in Nb, Ta and V in the framework of Flynn and Stoneham (9) diffusion theory; these parameters are compared with experimental results.

The following scheme has been followed throughout the work. The dipole tensor deduced from relaxation and volume expansion measurements (10) has been used to calculate the lattice distortion and relaxation energy for the interstitial and "effective" gas-host atom interaction potentials have been fitted assuming both interstitial sites (11). Then, these potentials have been used to calculate the migration energy and activation volume. A comparison of these quantities with experimental values gives support for the octahedral occupancy by the N interstitial.

Lattice distortion for the octahedral and tetrahedral site

The displacement $\underline{U}(\ell)$ of the ℓ atom of a lattice composed of N host atoms and an impurity can be obtained by solving the system of 3N non linear equations (7),(11):

$$\underline{U}(\ell) = \underline{\underline{G}}(\ell,\ell') \ \underline{F}(\ell') \qquad\qquad [1]$$

where $\underline{U}$ and $\underline{F}$, the force calculated at the relaxed lattice site, are column matrices of dimension 3N and $\underline{\underline{G}}$, the perfect lattice Green function is a square matrix of order 3N. The discrete forces $\underline{F}(\ell)$ completely describe the configuration of a crystal containing a point defect. The corresponding representation in an elastic continuum may be obtained through a singularity in the body force on the medium of the form:

$$-f_i = - \sum_j \mathcal{D}_{ij} \frac{\partial}{\partial x_j} \delta(\underline{r}) \quad . \qquad\qquad [2]$$

Hardy (12) has shown that the discrete and continuum approach are related through the equality

$$\mathcal{D}_{ij} = \sum_\ell F_i(\ell) \ R_j(\ell) \qquad\qquad [3]$$

$$R_j(\ell) : \text{j coordinate of the atom } \ell.$$

An interstitial in either the octahedral or the tetrahedral site produces a tetragonal lattice distortion. In the principal axes directions, $\mathcal{D}$ is diagonal with two independent elements (A,B). The trace of $\mathcal{D}(A+2B)$ can be obtained from volume expansion measurements and the magnitude of the difference between its independent elements through the Snoek relaxation. These values for O and N in Nb, Ta and V(10) are summarised in Table I. From these, using the equality [3] and fully exploiting the defect lattice symmetry, the forces $\underline{F}$ at the displaced lattice positions were obtained (further details of this calculation are given in (11)). The main hypothesis made is that these forces are sufficiently short range to be expressed in terms of two independent parameters. The corresponding displacements shown in Table II were obtained by solving [1].

TABLE I

Dipole tensor deduced from experimental

values, taken from (10)

System	A+2B [eV]	$\lvert A-B \rvert$ [eV]	System	A+2B [eV]	$\lvert A-B \rvert$ [eV]
Nb–N	21.6	8.43	Nb–O	21.6	6.92
Ta–N	30.6	9.3	Ta–O	22.8	6.55
V–N	19.2	6.81	V–O	19.2	12.8

TABLE II

Lattice distortion due to nitrogen and oxygen impurities at octahedral and

tetrahedral sites in niobium, tantalum and vanadium

Units: U_i are in units of a, F_i are in units of eV/a and E_f (formation energy),

A and B are in eV.

Impurity Site		Niobium		Tantalum		Vanadium	
		N	O	N	O	N	O
Octahedral	U_1	0.3143	0.2823	0.3294	0.2384	0.3183	0.4829[+]
	U_2	0.0019	0.0127	0.0141	0.0130	−0.0077	−0.0682[+]
	F_1	6.4100	5.9067	8.2000	5.9833	5.4700	7.4667[+]
	F_2	1.0975	1.2233	1.7750	1.3542	1.0325	0.5333[+]
	E_f	2.02	1.73	2.80	1.50	1.71	3.46[+]
	A	12.82	11.81	16.40	11.97	10.94	14.93
	B	4.39	4.89	7.10	5.42	4.13	2.13
Tetrahedral (Approximation –A)	U_1	0.2381	0.2170	0.2493	0.1818	0.2327	0.3333
	U_2	0.0654	0.0641	0.0537	0.0428	0.0535	0.0455
	U_3	−0.0485	−0.0390	−0.0391	−0.0276	−0.0539	−0.1037
	U_4	−0.0843	−0.0664	−0.0620	−0.0433	−0.0914	−0.1814
	F_1	5.3782	4.9801	6.8842	5.0478	4.5880	6.1447
	F_2	2.4560	2.3082	3.0512	2.3185	2.0600	2.5142
	F_3	−0.3732	−0.2268	−0.2342	−0.1562	−0.2530	−0.8113
	F_4	−0.5553	−0.3383	−0.3504	−0.2339	−0.3767	−1.1936
	E_f	3.01	2.52	3.82	2.06	2.45	4.93
	A	1.58	2.59	4.00	3.23	1.86	−2.13
	B	10.01	9.51	13.30	9.78	8.67	10.67
Tetrahedral (Approximation –B)	U_1	0.2252	0.2092	0.2433	0.1778	0.2246	0.3070
	U_2	−0.0050	0.0212	0.0210	0.0209	−0.0044	−0.1376
	U_3	−0.0351	−0.0309	−0.0330	−0.0235	−0.0466	−0.0797
	U_4	−0.0568	−0.0496	−0.0527	−0.0371	−0.0670	−0.1039
	F_1	5.0050	4.7533	6.6500	4.8917	4.3350	5.3333
	F_2	0.7900	1.2933	2.0000	1.6167	0.9300	−1.0667
	E_f	2.25	2.04	3.32	1.81	1.94	3.57

[+]not reliable

The defect space in the octahedral site was considered to be composed of the two nearest neighbours of the impurity, at the relative positions $\pm(1,0,0)$ and the four next nearest neighbours $(0,\pm1,\pm1)$. Symmetry considerations show that the displacements and forces at these atoms are $\pm(U_1,0,0)$, $\pm(F_1,0,0)$ at $\pm(1,0,0)$ and $(0,\pm U_2,\pm U_2)$, $(0,\pm F_2,\pm F_2)$ at $(0,\pm1,\pm1)$ and the dipole tensor elements $A = 2F_1a$, $B = 4F_2a$. No further assumptions were made to obtain the data shown in Table II.

The four nearest neighbours of the impurity in the tetrahedral site are $(\pm1,0,\pm\frac{1}{2})$ and the four next nearest neighbours $(\pm1,0,\pm^3/2)$. The corresponding displacements, forces and dipole tensor elements are $(\pm U_1,0,\pm U_2)$, $(\pm U_3,0,\pm U_4)$, $(\pm F_1,0,\pm F_2)$, $(\pm F_3,0,\pm F_4)$, $A = 2(F_2+3F_4)a$, $B = 2(F_1+F_3)a$. In this case the defect space is of dimension 4 (11). So, a further approximation has to be made to obtain the displacements from the date of Table I. Two possibilities were considered: i) the interaction is given by a central potential (approximation A), ii) the forces at the next nearest atoms are zero (approximation B).

From Table II it may be seen that the calculated values of the atomic displacements, forces, dipole tensor and relaxation energy are <u>physically plausible in all systems</u> (except the system V-O, where only the tetrahedral site gave a reasonable configuration[+], see (11)). This result proves it is not possible to distinguish between octahedral and tetrahedral occupancy by the magnitude of the shape factor. Our conclusion is that in the systems studied the measured relaxation values could correspond to tetrahedral or to octahedral occupancy. Any contrary assumption, at this stage of knowledge of interatomic potentials, is extremely model dependent.

The interaction forces between the impurity and host atoms were used to fit an "effective" interaction potential. A Born-Mayer function was chosen for the octahedral configuration and a 6-12 power Lennard-Jones form for the tetrahedral, where a purely repulsive potential cannot be fitted (11). These potentials were used for the calculations described in the next sections.

Activation energy

To calculate the migration energy (E) of an interstitial from an equilibrium site to an equivalent one through classical rate theory (13) it is necessary to compute the relaxation energy at an intermediate saddle point configuration. As our interaction potentials depend on the lattice configuration at the equilibrium site, to allow this calculation the validity of those at the saddle point must be assumed. This assumption is doubtful and can lead to erroneous results. The energy values calculated using this procedure were an order of magnitude smaller than those measured, in agreement with Ferro's findings on his elastic approach to the same problem (14). Then, a completely different approach was adopted by making the calculation in the framework of the Flynn and Stoneham (FS) quantum theory of light interstitials migration (9), which avoids any explicit evaluation of the saddle point

---

[+] In this system at the octahedral site the first neighbours of the impurity in the unperturbed lattice become its second neighbours in the distorted lattice and vice versa.

configuration; since our potentials are fitted to the equilibrium configurations this framework is clearly more appropriate. This theory assumes that the interstitial follows the motion of the host nuclei adiabatically and can be partly justified by the fact that the migrating particle is much lighter than the host atoms. Though the systems treated by us should be considered at the limit of validity of this assumption the agreement between calculated and experimental migration energy values is extremely good.

In FS theory $\underline{U}_p$ and $\underline{U}_q$, the initial and final displacements of the host lattice atoms when the interstitial migrates from site p to q, are decomposed into $\underline{U}_s$, the symmetric part, and $\underline{U}_a$, the antisymmetric

$$\underline{U}_p = \underline{U}_s - \underline{U}_a \qquad\qquad [4]$$

$$\underline{U}_q = \underline{U}_s + \underline{U}_a \quad . \qquad\qquad [5]$$

At the relevant temperatures the diffusivity follows an exponential dependence with a migration energy given mainly by the antisymmetric displacements:

$$E = \tfrac{1}{2} \sum_n \omega_n^2 \, \theta_{an}^2 = \tfrac{1}{2} \, \tilde{\underline{U}}_a \, \underline{\underline{\phi}}^0 \, \underline{U}_a \quad . \qquad\qquad [6]$$

Where
n : normal mode
θ_a : normal coordinate corresponding to $\underline{U}_a$
ϕ^0 : perfect lattice force constant matrix
ω^2 : normal mode frequencies, eigenvalues of $\underline{\underline{\phi}}^0$.

E is a lattice activation energy. The barrier to motion is provided by the self trapping distortion. There are cases where the transition matrix element will also be sensitive to symmetric motion, this is not expected to be the case for the first neighbour octahedral transition (see (9) for a full discussion). Then [6] must be considered as a lower limit to the measured migration energy.

Expression [6] has only a limited validity because it is based on a linear coupling (LC) model which neglect the non linear terms in the gas-host atom interaction (see, however, (15)). We have, therefore, developed a simple model which enables us to include approximately the non linear contributions to E (a more detailed discussion will be published elsewhere (16)).

The effect of higher order terms in the defect host lattice coupling is to change ω^2 which will now be the eigenvalues of $\underline{\phi}$, the perturbed force constant matrix. Then, from [4],[5] and [6]

$$E = \tfrac{1}{2} \, (\tilde{\underline{U}}_s - \tilde{\underline{U}}_p) \, \underline{\underline{\phi}} (\underline{U}_s - \underline{U}_p) \quad . \qquad\qquad [7]$$

This equation can be expressed in terms of $\underline{G}$ and $\underline{F}$, the forces calculated at the relaxed atomic sites but applied at the original lattice site (7,11), then:

$$E = \tfrac{1}{2} \, (\tilde{\underline{F}}_s - \tilde{\underline{F}}_p) \, \underline{\underline{G}} (\underline{F}_s - \underline{F}_p) \quad . \qquad\qquad [8]$$

Where $\underline{F}_p$ are the forces at the equilibrium site p and may be taken from Table II.

In the LC approximation $\underline{F}_s$ is given by:

$$\underline{F}_s = \underline{F}_s^O = (\underline{F}_p + \underline{F}_q)/2 \ . \tag{9}$$

To include non-linear effects it is necessary to calculate $\underline{F}_s$ consistently. Then, we first note that an "average" state of the crystal with the defect at a certain site can be represented by the atomic displacements or forces due to it. From [9] the state "s", defined by the force $\underline{F}_s$, can be regarded as an equally weighted superposition of the states p and q and visualized as a state in which the defect has equal probability of being present at either of the sites p or q. Then, in our model (16), $\underline{F}_s$ and $\underline{U}_s$ are given by the system of non-linear equations:

$$F_{s\alpha} = -\tfrac{1}{2} \left(\frac{\partial V_p(r)}{\partial r_\alpha} \right)_{\underline{r}=\underline{r}_p(\ell)+\underline{U}_s(\ell)} - \tfrac{1}{2} \left(\frac{\partial V_q(r)}{\partial r_\alpha} \right)_{\underline{r}=\underline{r}_q(\ell)+\underline{U}_s(\ell)} \tag{10}$$

and

$$\underline{U}_s = \underline{\underline{G}} \cdot \underline{F}_s \ . \tag{11}$$

When $V_x(\underline{r})$ is the interaction potential of the interstitial at site x with a host lattice atom at $\underline{r}$, it is obvious that if the derivatives in [10] are evaluated at the un-relaxed positions it leads to the value $\underline{F}_s = \underline{F}_s^O$ as given in the LC approximation. The state "s" can be said to correspond to a lattice saddle point configuration.

The values of E obtained through this model for the first and second neighbour octahedral and tetrahedral jump are summarised in Table III. We compare these values with those of the migration energy obtained through internal friction measurements. Evidently as the initial and final position have the same axes orientation the second neighbour tetrahedral jump cannot contribute to the relaxation. In the three metals studied the best agreement was obtained between the experimental values of E and those calculated for the first neighbour octahedral transition of N. This is consistent with the belief that this interstitial occupies an octahedral site in these crystals. However, if we consider the assumptions of the model to obtain E, the simplicity of the arguments leading to the interaction potential and the considerable divergence among the measured values of E in the same system this conclusion must be considered only as tentative.

Activation volume

While the formation volume of the defect, defined as:

$$V_f = \partial E_f / \partial P \tag{12}$$

with P the external pressure, was considered explicitly in deducing the interaction potential between the interstitial and host atoms through the approximate relation [3], the activation

TABLE III

Activation energies of N and O interstitials in Nb, Ta and V

Units: activation energy in eV and site coordinates in a

(2a = lattice constant)

System	Average experimental value[+]	O-O migration		T-T migration	
		$(1,0,0) \rightarrow (1,0,1)$	$(1,0,0) \rightarrow (0,1,0)$	$(1,\frac{1}{2},0) \rightarrow (1,0,\frac{1}{2})$	$(1,\frac{1}{2},0) \rightarrow (1,-\frac{1}{2},0)$
Nb-N	1.52	1.42	1.53	1.13	1.85
Ta-N	1.69	1.69	1.98	1.27	2.02
V-N	1.50	1.30	1.28	0.92	1.49
Nb-O	1.16	0.93	1.15	0.87	1.44
Ta-O	1.12	0.67	0.89	0.65	1.04
U-O	1.26	-	-	2.45	3.65

[+]Taken from Szkopiak (17).

or migration volume:

$$V_m = \partial E / \partial P \qquad [13]$$

is an independent parameter than can be calculated from our model.

If we assume that the response of the perfect lattice (Green function) remains unchanged with external pressure and define a strain tensor $\underline{q}$, independent of ℓ if the dilation is uniform, such that:

$$P = K^{-1} Tr(\underline{q}) \qquad [14]$$

$$q_{\alpha\beta} = q \, \delta_{\alpha\beta} \qquad [15]$$

$$\Delta V/V = 3\Delta a/a = 3 \, q \qquad [16]$$

with K^{-1} the bulk modulus, we can either solve [12] and [13] numerically through finite variations in the lattice parameter or obtain an analytic approximation to [13]. To be consistent with [3] we will assume that the change in forces and displacements due to the external pressure are linearly coupled. Then, the displacement field due to the external pressure is expressed as:

$$U_\alpha(\ell) = q_{\alpha\beta} \cdot R_\beta(\ell) = q \cdot R_\alpha(\ell) \qquad [17]$$

and from [8],[13],[14] and [17]:

$$V_m = \partial E_\alpha / \partial P = \left[\sum_\ell (F_s - F_p)_\alpha \, (\ell) \, R_\alpha(\ell) \right] / (3K^{-1}) \qquad [18]$$

this can be expressed as:

$$V_m = V_s - V_p \qquad [19]$$

which means that the motion volume is given by the difference between the volume expansion produced by the states "s" and "p"; this reinforces the visualization of "s" as a lattice saddle point. It is important to point out that in the LC approximation [19] is zero, this is obvious replacing $\underline{F}_s$ by $\underline{F}_s^0 = (\underline{F}_p + \underline{F}_q)/2$ in [19] and using the equivalence between "p" and "q" sites.

The migration volume values for the first neighbour octahedral and tetrahedral transition for N in Nb, Ta and V and O in Nb and Ta calculated through [19] are summarised in Table IV.

As a last point we should like to mention that the numerical solution of [13] produced values similar to those reported in Table IV with the same systematic difference as that obtained between a numerical evaluation of [12] and the Tr $\mathcal{D}_{ij}$ from [3].

TABLE IV

Migration volume in units of Ω
(volume of unit cell)

System	O-O transition	T-T transition	Experimental
Nb-N	.090	.072	
Nb-O	.058	.064	
Ta-N	.092	.073	
Ta-O	.037	.050	
V-N	.118	.091	.135 ± .02 (18)

Conclusions

Two main conclusions may be derived from this work:

i) The adiabatic quantum theory of light interstitial migration due to Flynn and Stoneham provides an adequate description of the dynamics of migration of O and N in V, Ta and Nb if the non-linear interaction between the gas interstitial and host lattice is adequately modelled.

ii) A comparison between the calculated and measured values of activation energy and motion volume favours the occupancy of the octahedral site by N in these systems. However the values obtained for the tetrahedral site are not unreasonable. For the V-O system the measured values of the dipole tensor could only be fitted assuming a tetrahedral site. But the calculated value of the activation energy in this case do not agree with experiment, so no definite conclusion can be reached concerning the site occupancy.

Acknowledgements

The authors are grateful to Dr. A.M. Stoneham for several useful discussions and Dr. R. Bullough for critically reading the manuscript.

205

References

1. A.S. Nowick and B.S.Berry, Anelastic Relaxation in Crystalline Solids. Academic Press, New York, London (1972).

2. L.S. Dijkstra, Philips Res. Rep. 2, 357 (1947).

3. R.W. Powers and M.V. Doyle, J. Appl. Phys. 30, 514 (1959).

4. J.C. Swartz, J.W. Schilling and A.J. Schwoeble, Acta. Met. 16, 1359 (1968).

5. R.A. Johnson, G.J. Dienes and A.C. Damask, Acta. Met. 12, 1215 (1964).

6. D.N. Beshers, J. Appl. Phys. 36, 290 (1965).

7. V.K. Tewary, A.E.R.E. Tech. Report, T.P. 388 (1969).

8. R. Bullough and V.K. Tewary, Interatomic Potentials and Simulation of Lattice Defects. Ed. P.C. Gehlen, J.R. Beeler Jnr. and R.I. Jaffee (Plenum Press, N.Y.), 155 (1972).

9. C.P. Flynn and A.M. Stoneham, Phys. Rev. B1, 3966 (1970).

10. J. Buchholz, Proc. Int. Meeting on H in Metals, Jülich, Vol.II, 544 (1972).

11. V.K. Tewary, J. Phys. F. (Metal Phys), in press.

12. J.R. Hardy, J. Phys. Chem. Solids, 29, 2009 (1968).

13. G.H. Vineyard, J. Phys. Chem. Solids, 3, 121 (1957).

14. A. Ferro, J. Appl. Phys. 28, 895 (1957).

15. M.J. Norgett and A.M. Stoneham, J. Phys. C. (Solid State Phys.), 6, 238 (1973).

16. E.J. Savino and V.K. Tewary, to be published J. Phys. F. (Metal Phys.).

17. Z.C. Szkopiak, J. de Phys. 32, Suppl. 7, Colloque C2, 1 (1971).

18. G.W. Tichelaar, R.V. Coleman and D. Lazarus, Phys. Rev. 121, 748 (1961).

SYMMETRY LOWERING OF LATTICE DEFECTS
DUE TO ANHARMONIC INTERATOMIC FORCES

J. Amran Sussmann[:]
Centre d'Etudes Nucléaires de Grenoble
Département de Recherche Fondamentale
Section de Physique du Solide
BP 85, Centre de Tri, 38041 Grenoble Cedex (France)

Prof. J. Amran Sussmann died on the 22nd of June 1973.
During his stay in Grenoble in Dautreppe's laboratories, he
began theoretical work on the symmetry lowering of lattice
defects due to anharmonic interatomic forces. The purpose
of this work was to explain the very large decrease of the
elastic modulus observed in several irradiated metals (1)
and attributed to the deformability of the interstitial (2).

This paper is a compilation of notes, documents and
ideas collected by those who lived with Prof. Amran Sussmann
during these last months (P. Moser, R. Pichon, J. Hillairet,
P. Pouilley). It is their hope that this summary of Prof.
Amran Sussmann's efforts at Grenoble inspires other high
level theoreticians to continue his work.

Amran Sussmann's ideas were expressed in a note to be
published in Acta Physica Austriaca and are reproduced here in a
completed form. We give in Fig.1 the principle : if three
atoms of a linear chain are in an extended position and
subjected to anharmonic forces, the equilibrium configuration could correspond to an off center situation for the
central atom. Fig.2 gives the published results corresponding to a first approach using a 12-6 Lennard-Jones potential.

[:]On leave of absence from Soreq Nuclear Research Center, Yavne, Israel.

These calculations were recently extended using a Johnson
potential (Fig.3) and the results of this second approach
appear in Fig.4.

___

Introduction

Physical systems containing forces, which are significantly anharmonic, very often present equilibrium configurations having a lower
symmetry than that of the system's Hamiltonian. Well known examples are
the hydrogen bond, the off-center substitutional impurities in alkali
halides and Jahn-Teller systems. Instersitials in metals may behave similarly.

Interatomic distances in the neighbourhood of lattice defects are
strongly affected. A small increase in interatomic distance may cause
a large increase in the anharmonic lowering (and it will be indicated
how this behaviour can explain the observed decrease of elastic constants
of metals due to interstitials).

Using a one dimensional model, we shall show how, and under which
conditions, the anharmonicity may render the highest symmetry configurations unstable relative to a lower symmetry one. Then, we shall show that
interstitials may satisfy the conditions obtained, whereas vacancies
do not.

The observed decrease of elastic constants of metals induced by fast
particle irradiation could be explained if the self interstitials would
behave in the above described manner.

A one dimensional model

Let us consider a linear system of three atoms, where the distance
d between the two extreme ones is a fixed parameter and where r designates the distance of the central atom to one of the extreme ones. If
$V(r)$ is the interatomic potential, the potential acting on the central
atom will be :

$$V_s(r, d) = V(r) + V(d - r) \qquad (1)$$

The choice of the type of interatomic potential will be critical.
A harmonic interatomic potential will lead to a central equilibrium
configuration, with a potential minimum at $r = d/2$. The same will happen
if we take a purely repulsive potential, say of the Born Mayer type.
This result is not meaningful as a realistic potential must have an
attractive part for large interatomic separations. Such a realistic

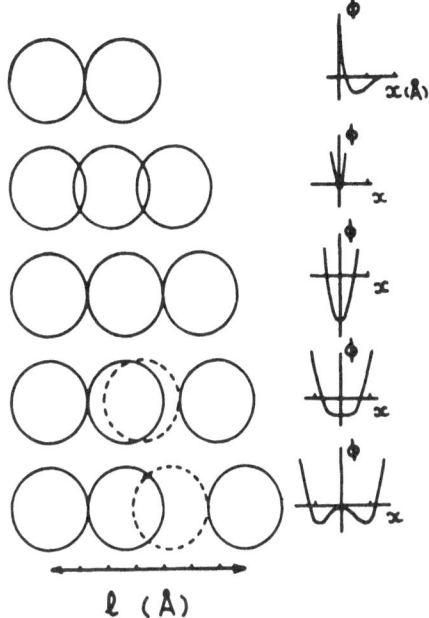

ℓ (Å)

FIG. 1

Schematic representation of the potential
well for the central atom in a linear chain
of three atoms. Above a critical distance of
the extreme atoms, the equilibrium position
of the central atom is off center

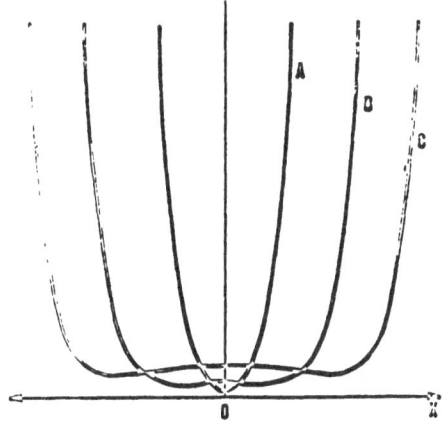

FIG. 2

Results of a first calculation
using a 12-6 Lennard-Jones
potential and giving the values
of the potential well for the
central atom for three distances
between the extreme atoms

A)   d = 1,9 r_o

B)   d = 2,3 r_o

C)   d = 2,6 r_o

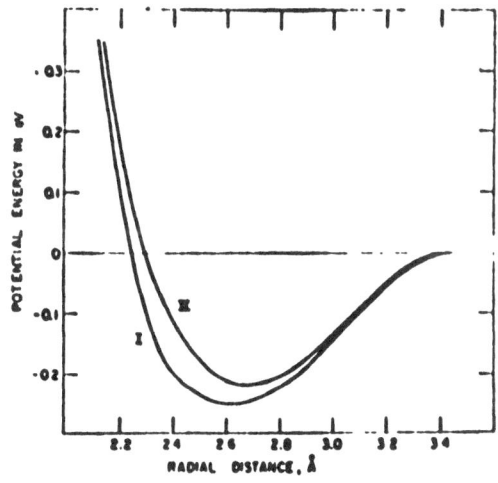

FIG. 3

Johnson's interatomic interactions used in the
second calculation. The potential has the
analytic form

Range (Å)	Potential (eV)
R<2.40	$-2.195976(R-3.097910)^3 + 2.704060R - 7.436448$
2.40<R<3.00	$-0.639230(R-3.115829)^3 + 0.477871R - 1.581570$
3.00<R<3.44	$-1.115035(R-3.066403)^3 + 0.466892R - 1.547967$

potential V (r) presenting both a repulsive and an attractive part (which
may be examplified by a Lennard Jones potential), will have a minimum at
r_o and an inflexion point at $r_i > r_o$

$$\left\{ \frac{d^2 \ V \ (r)}{dr^2} \right\}_{r \ = \ r_i} = 0 \qquad\qquad (2)$$

as necessarily V (r)$_{r \to \infty}$ must be finite.

The symmetrical configuration of the system located at r = d/2 will
be stable only if d < 2r_i, as

$$\left\{ \frac{d^2 V_s}{dr^2} \right\}_{r \ = \ d/2} = \begin{array}{ll} > 0 & \text{for} \quad d < 2r_i \\ \\ < 0 & \text{for} \quad d > 2r_i \end{array} \qquad (3)$$

For d > 2r_i, two degenerate equilibrium configurations appear, which
correspond to off-center localization of the central atom. In the figure
at A, B and C we present the behaviour of the potential felt by the
central atom, for different values of the extreme atom separation d,
for a 12-6 Lennard-Jones potential.

Symmetry lowering of equilibrium configurations

The one dimensional model allows us to anticipate the trends of
behaviour of more complex systems. We expect thus that whenever the distances from a given atom to its first neighbours are smaller than a
critical value, this atom will locate itself centrally. If, however, the
distances from that atom to its nearest neighbours exceed a critical value,
the atom will locate itself in one of several off-center positions. In a
different context, Turnbull and Cohen (3) have shown that the glass-liquid
transition is related to the increase of the interatomic distance beyond
a critical value, leading to the formation of easily redistributable free
volume. The mechanism involved is essentially the same as the one considered here.

The interatomic separations near lattice defects

We shall be concerned with vacancies and self interstitials, and we
shall consider the atoms neighbouring either the vacant site or the interstitial atom.

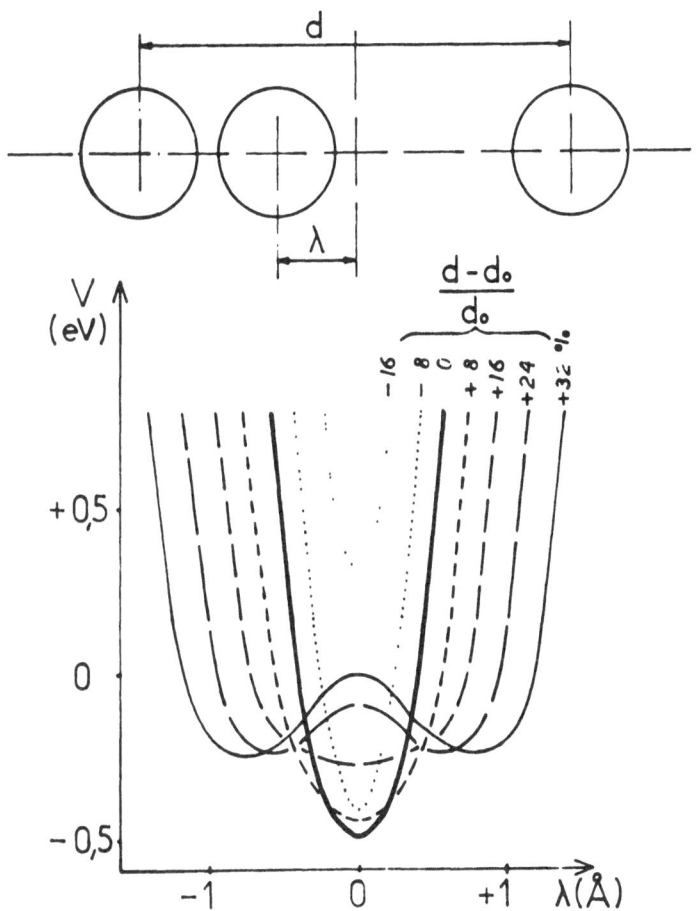

FIG. 4 Potential well for the central atom for different
distances between the extreme atoms calculated
with the Johnson potential.

Let us consider first the neighbours to a vacancy. In comparison
with the perfect lattice, they have lost a first neighbour in the radial
direction. Tangentially the interatomic distances are somewhat reduced,
due to the relaxation of the lattice. Applying the results of the one
dimensional model, we expect a tendency to asymmetric configuration in the
radial direction and no such effect in the tangential direction. The
tendency to radial asymmetry is a further contribution, but not the main
one, for the instability of a hypothetical split vacancy.

Around an interstitial the situation is quite different. Again, we
consider the atoms neighbouring the defect, in this case the added atom.
Radially they have a new neighbour at a very short distance. According
to our model, this will increase the stability of the symmetric configuration. Tangentially, however, the distances to the neighbours are
increased. If this increase is large enough, an instability of the
symmetric configuration will result and the system will take up a configuration with a lower symmetry.

In order to predict, for a given system, whether a lowering of symmetry
does occur, and what is the geometry of the resulting equilibrium configurations, numerical computation should be performed. Previous calculations
necessarily missed such distortions because, in order to reduce computing
time, the expected symmetry of the system had been introduced from the
start. Computation becomes much more involved if the symmetry of the
equilibrium configuration is not known. A much more fundamental difficulty
is connected with the choice of the interatomic potential. It has been
pointed out by Vitek, Perrin and Bowen (4) that, even if the analytic form
of the potential (which is an arbitrary choice in itself) is conserved,
the fitting to the known experimental parameters allows for very different
potential shapes. It is easy to see that the effect we are discussing is
very sensitive to the detailed characteristics the potential. Thus, actual
computations, starting from our present knowledge of interatomic potentials,
are not only very difficult, but also not very meaningful.

The influence on the elastic constants

Most metals do present upon irradiation, a very large decrease of their
elastic moduli (see Hillairet, Bonjour and Poirier (5), Wenzl et al. (1)
where further references may be found).

This effect is thought to be mainly due to the interstitials produced
by the irradiation. It has also been suggested by Melngailis (2) that the

sign and magnitude of the effect could be explained, if the defect was deformable. The effect described in this work gives an atomic explanation of the deformability of the interstitial. The tangential character of the deformability leads us to predict that self-interstitials affect mainly the shear elastic constant.

Acknowledgements

The author thanks Dr. Hillairet for calling his attention to the problem and for many fruitful discussions. He thanks most heartily the "Département de Recherche Fondamentale du Centre d'Etudes Nucléaires de Grenoble", for its exceptionally friendly and helpful hospitality.

Prof. J. Amran Sussmann wished to extend these calculations to three dimensions. The expected results were to show the possibility of non localized positions of the neighbouring atoms around an interstitial atom. Consequently, he predicted, on one hand for the self interstitial, an increased probability of existence of intermediate configurations described by Johnson (6), and on the other hand, for the hetero-interstitial, the possibility of an off-center position as observed by Turner for gold in lead (7).

One of us (P. Moser) will send with pleasure any complementary information to those interested in Prof. Amran Sussmann's ideas.

References

1. H. Wenzl, F. Kerscher, V. Fischer, K. Ehrensperger and K. Papathanassopoulos, Z. Naturforschung, 26a, 489 (1966)

2. J. Melngailis, Phys. Stat. Sol. 16, 247 (1966)

3. D. Turnbull and M.H. Cohen, J. Chem. Phys. 34, 120 (1961)

4. B.V. Vitek, R.C. Perrin and D.K. Bowen, Phil. Mag. 21, 1049 (1970)

5. J. Hillairet, E. Bonjour and J.P. Poirier, Journal de Physique, 32, C2-31 (1971)

6. R.A. Johnson, Phys. Rev. 145, 423 (1966)

7. T.J. Turner, S. Painter and C.H. Nielsen, Sol. State Comm. 11, 577 (1972)

EFFECT OF ULTRA HIGH VACUUM DEGASSING ON THE

INTERNAL FRICTION OF TANTALUM AND NIOBIUM

F. M. Mazzolai

C.N.R - Istituto di Acustica O.M. Corbino, Via Cassia 1216, Rome

Introduction

It is well known that gaseous impurities are involved in a number of anelastic effects appearing at low temperatures both in annealed and cold-worked bcc metals (1-18). A special role is played by hydrogen, which is the mobile impurity in the reorientation processes and which forms atmospheres around dislocation lines even at low temperatures. To a great extent an understanding of low-temperature effects in bcc metals is related to the possibility of controlling the content of selected interstitial impurities. Ultra high vacuum (UHV) - degassing has proved to be a successful technique for the elimination of gaseous impurities from refractory metals. The present work reports internal friction measurements made in niobium and tantalum after annealing in UHV, hydrogen loading and cold-working. Preliminary results of this investigation have already been published[19].

Experimental

Circular plates of niobium and tantalum have been prepared from 99.9% purity material, the diameters were 25 and 30mm, the thicknesses 4.5 and 6.5mm. After machining the specimens were annealed in an ultra high vacuum apparatus (Varian ASS.) at temperatures vary-ing from 2020 to 2620 K; pressures were in the range 10^{-7} - 10^{-8} torr. The samples were furnace cooled at a mean speed of 0.5 K/sec. The oxygen content of the annealed specimens was estimated by measuring the low-temperature side of Snoek peak. The temperature of the measurements was kept below 600 K, in order to avoid oxygen and nitrogen contamination. The Snoek peak was assumed to be a relaxation effect with a single relaxation time. In view of the relatively small amounts of oxygen involved here, the assumption appears to be a reasonable one. Hydrogen was introduced at room temperature by a cathodic charging technique; a deuterium loading treatment was performed at 540 K and 0.8 torr. Deformation of the specimens was carried out by axial compression at room temperature. Internal friction measurements were made as a function of temperature (from 65 to 600 K) in a vacuum of 10^{-6} at frequencies ranging from 24.0 to 158.0 kHz, the strain amplitude was of the order of 10^{-7}. The experimental technique was an improved version of that introduced by Bordoni[20] and Nuovo[21].

Results

Tantalum.   The internal friction of a tantalum specimen annealed in UHV is shown in Fig.1, curve 1. A small peak is present near 172 K (in the following labelled P_3).

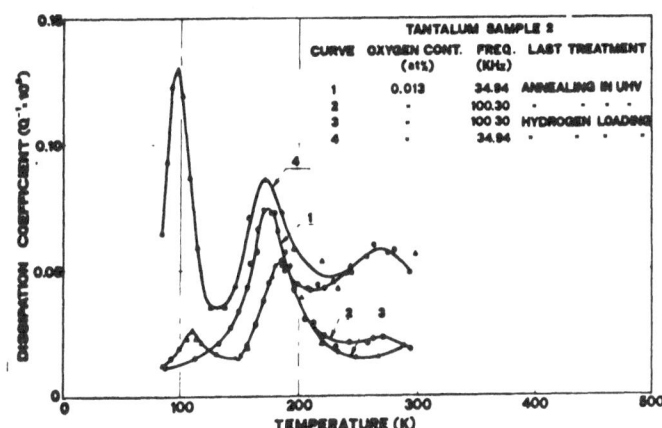

FIG. 1.   Internal friction of a tantalum specimen annealed

at 2470 K and 8 x 10^{-7} torr for 3 hours.

Hydrogen impurities give rise to (curves 2 and 3) the peak already reported by Cannelli and Verdini[1] and produce a hardly appreciable effect on P_3. This is more clearly demonstrated by measurements at higher hydrogen contents (not reported in Fig. 1), where the hydrogen peak increases to a saturation value of about 1.3 x 10^{-6}, while the dissipation coefficient in the temperature range of P_3 remains of the order of 10^{-7}. The activation energy and the limiting relaxation time are 0.26 ev and 2 x 10^{-13} sec., respectively[19]. It is to be noticed that peaks of the order of 1 x 10⁻⁷ can be resolved at low temperature. No peak was clearly

evident in the internal friction curve of a second sample annealed at 2470 and 2600 K for

3 hours; the oxygen contents after these treatments were 0.07 and 0.001 at.%, respectively.

Niobium.   The dissipation coefficient of annealed niobium sample 2 is plotted as a

function of temperature in Fig. 2.

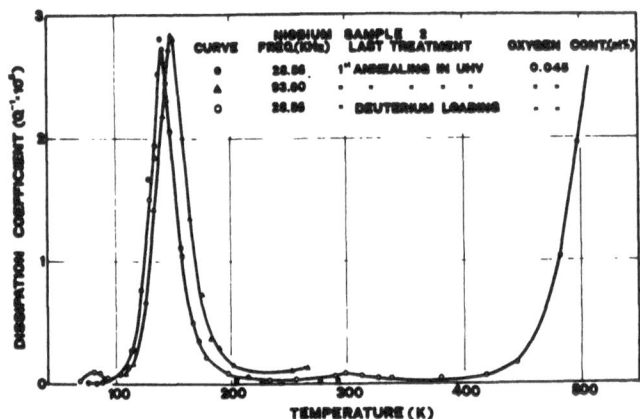

FIG. 2.   Dissipation coefficient of niobium sample 2 annealed at

2220 K and 5 x 10⁻⁷ torr for 90 minutes and subsequently

deuterium loaded.

A thermally activated relaxation peak (again labelled P_3) is exhibited at low temperature.

Its activation energy and its relaxation time are 0.20 ev and 2 x 10⁻¹³ sec, respectively[19].

In the figure are also reported measurements made at two frequencies after doping with

deuterium. The doping treatment did not significantly affect P_3, which introduced a small

peak at a lower temperature. It is worth noting that P_3 seems to be greater at higher

frequencies and consequently at higher temperatures.

The effect of plastic deformations on the internal friction of deuterium doped sample 2 is

shown in Fig. 3. Cold-working brings about an additional peak (α peak), which at first

increases with plastic strain then decreases and finally disappears at a strain of 10.1%;

the maximum value of Q^{-1} (9.5 x 10⁻⁵) was found for a plastic deformation of about 2.7%.

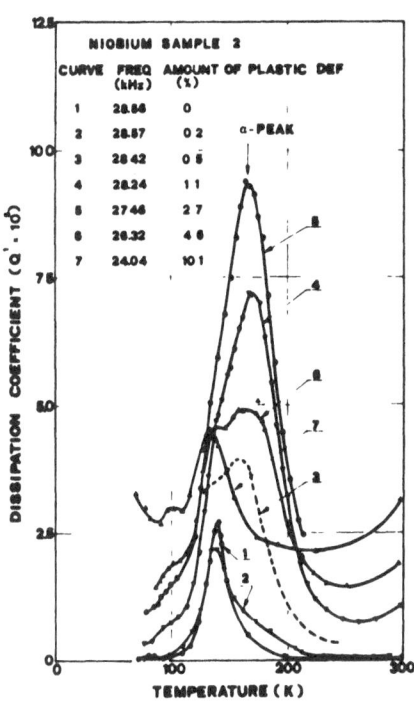

FIG. 3.   Effect of plastic deformations on the internal
friction of deuterium doped niobium sample 2.

It can be seen that the temperature of P_3 is reduced and its width increased (compare curves
1 and 7) by plastic deformation. After a strain of 10.1% the height of P_3 appears almost
the same as that of the annealed material. After the sequence of plastic deformations
reported in Fig. 3, sample 2 was annealed again. The internal friction measured after this
thermal treatment is plotted in Fig. 4, curve 2. The figure also shows measurements at two
frequencies in the same specimen aged at room temperature. Ageing induces a decrease in the
peak, which again appears to have a somewhat greater value at higher frequencies. For
comparative purposes the curves of Fig. 2 are superimposed. A hydrogen loading treatment
of the aged sample (200 mA and 20 minutes) introduced the peak of Cannelli and Verdini near
100 K and reduced P_3. A comparison between the theoretical Debye curve and the experimental
ones shows that P_3 in niobium has a single relaxation time.

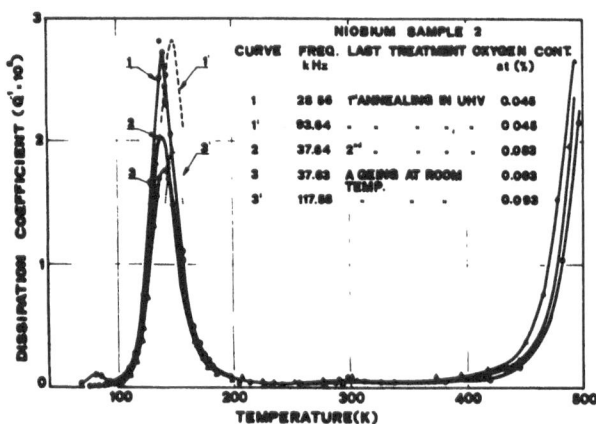

FIG. 4.   Dissipation-coefficient of niobium sample 2 annealed

at 2020 K and 2.0 x 10^{-7} for 90 minutes and then aged

at room temperature for about two years.

Sample 1 was plastically deformed without any previous deuterium or hydrogen loading. No
additional peak is brought about by plastic deformations lower than 2.5% (Fig. 5).

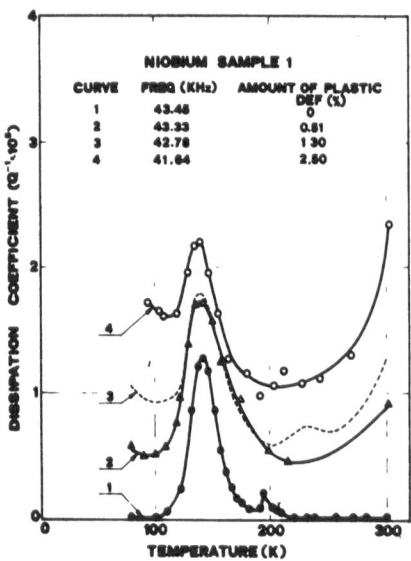

FIG. 5.   Dissipation-coefficient of niobium sample 1 annealed at

2220 K and 2.5 x 10^{-7} for 90 minutes (curve 1) and then

plastically deformed (curves 2-4).

Although the oxygen content of the sample was not measured it should presumably be the same as that of sample 2 (0.045 at.%) after their simultaneous thermal treatment. Peak P_3 is shifted to lower temperatures and broadened by plastic deformations, however the height of the peak is not significantly changed. All these effects had already been observed on sample 2. After the plastic deformations reported in Fig. 5 sample 1 was annealed (11 hours at 2520 K and 5×10^{-8} torr.), repeatedly hydrogen loaded (50 mA for 1, 3 and 20 minutes) then annealed again (11 hours at 2620 K and 8×10^{-8} torr.) and finally it was oxygen loaded (0.55 at.%). The internal friction measured after each of these treatments did not exhibit any peak P_3. A small hydrogen peak (1.2×10^{-6}) was found near 110 K following electrolytic hydrogen charges. The frequency of annealed sample 2 decreases with increasing temperature over the range 65-340 K; at temperatures greater than 340 K the frequency increases. A similar behaviour was found in sample 1 after its first annealing treatment, the minimum was in this case hardly appreciable, and occurred at a much higher temperature (480 K). After the subsequent annealing treatments sample 1 showed a monotonic decrease of the frequency over the entire temperature range. The frequency curves were retraced during cooling runs.

Discussion

Comparatively less extensive work has been done on tantalum than on niobium, however the available results allow a significant comparison between the features of relaxation effects P_3 in the two metals. Both peaks appear at about $0.05\ T_m$ (T_m is the melting temperature of tantalum or niobium), have similar relaxation parameters, need high temperature anneals for their appearance and finally they are not very sensitive to hydrogen loading. In view of these similarities we assume they are expressions of the same microscopic processes. The low value of the activation energy of P_3 rules out the possibility that the reorientation of bound vacancies and divacancies or impurities other than hydrogen may be responsible for the peak. Furthermore a mechanism involving intrinsic point defects would require a non-equilibrium concentration, which cannot be quenched-in at cooling rates as low as 0.5 K/sec.

The possibility that hydrogen may play a role in the appearance of P_3 must be carefully considered, because appreciable quantities of hydrogen may be absorbed by thin samples of tantalum[22] and niobium[23] following UHV degassing. The experiments on tantalum show that the peak due to hydrogen (at 97 K in Fig. 1) gets a saturation value of 1.3×10^{-6} after heavy electrolytic hydrogen loading, while no peak higher than 1×10^{-7} appeared in the annealed material. The hydrogen concentration C_H at saturation cannot presumably exceed the solubility limit, which at 97 K is about 3×10^{-3} at.%. Assuming a linear dependence of the Cannelli-Verdini peak on the hydrogen content, a value less than 0.20 at. ppm is estimated for C_H in the annealed material. Similar considerations may apply to the case of niobium, where a peak of 1×10^{-7} was found in the hydrogen loaded sample 2, while no trace of the peak was evident after annealing. A protective monolayer of oxide may have been formed before appreciable quantities of hydrogen were picked up. Apart from the low expected value of C_H in the annealed samples, an O-H (or N-H) pair reorientation, as well as a Snoek mechanism, should predict an increase in the peak with the hydrogen content. The increase was not observed even at hydrogen levels far below the saturation conditions for the Cannelli-Verdini peak. Furthermore a decrease of the relaxation strength with frequency should be expected from a reorientation model; this is in contrast with the experimental results. The above considerations also rule out dislocation-hydrogen complexes as possible sources of the effect. In view of the stability of the peak following ageing at temperatures as high as 600 K, P_3 does not appear to be due to fresh dislocations which might have been introduced by handling the samples.

It is the feeling of the author that some kind of characteristic dislocation array, built up by not too prolonged anneals at not too high temperatures (peak P_3 was cancelled in niobium sample 2 by an annealing of 11 hours at 2520 K), may be responsible for the observed anelasticity. These structures should not be affected by plastic deformation at room temperature. It is of some interest to note here that a very small peak, having a single relaxation time has also been found near 70 K in platinum annealed at high temperatures [24]. Even keeping in mind the different crystal structure of platinum (fcc) with respect to niobium and tantalum (bcc), it seems that a sort of relationship may exist between the two phenomena.

Lastly we call attention to the fact that plastic deformations lower than 2.5% did not introduce an α-peak in the outgassed material even at relatively small oxygen and nitrogen contents[15]. This again[4] raises doubts about the existence of a peak related to an intrinsic process involving only dislocations[25-28] in the range 150-200 K.

References

1. G. Cannelli and L. Verdini, La Ricerca Scient. 36, 98 (1966).

2. G. Carnelli and L. Verdini, La Ricerca Scient. 36, 246 (1966).

3. W. R. Heller, Acta Met., 9,600 (1961)

4. F. M. Mazzolai and M. Nuovo, Solid State Comm., 7, 103 (1969).

5. P. Shiller and A. Schneiders, Int. Conf. on Vacancies and Interstitials in Metals, Julich, 2, 871 (1968).

6. C. Baker and H. K. Birnbaum, Acta. Met., in Press.

7. J. J. Au and H. K. Birnbaum, Scripta Met., 7, 595 (1973).

8. G. Cannelli and F. M. Mazzolai, Nuovo Cim., 64B, 171 (1969).

9. G. Cannelli and F. M. Mazzolai, J. Phys. Chem. Solids 31, 1913 (1970).

10. G. Cannelli and F. M. Mazzolai, Applied Physics 1, 111 (1973).

11. C. A. Wert, D. O. Thompson and O. Buck, J. Phys. Chem. Solids, 31, 793 (1970).

12. O. Buck, D. O. Thompson and C. A. Wert, J. Phys. Chem. Solids, 32, 2331 (1971).

13. G. Baker and H. K. Birnbaum, Scripta Met., 6, 851 (1972).

14. O. Buck, D. O. Thompson and C. A. Wert, J. Phys. Chem. 34, 591 (1973).

15. H. Y. Chang and C. A. Wert, Int. Met. on Hydrogen in Metals, Julich, 2, 558 (1972).

16. R. Gibala, Scripta Met., 4, 77 (1970).

17. G. Schaumann, J. Volkl and G. Alefeld, Phys. Rev., 21, 981 (1968).

18. R. Cantelli, F. M. Mazzolai and M. Nuovo, Applied Physics, 1, 27 (1973).

19. F. M. Mazzolai, J. De Physique Suppl. C 6,. 163 (1973)

20. P. G. Bordoni, Nuovo Cimento 4, 177 (1947).

21. M. Nuovo, La Ricerca Scient. 31, 212 (1961).

22. R. Hanada, T. Suganuma and H. Kimura, Scripta Met., 6, 483 (1972).

23. K. Faber and H. Schultz, Scripta Met., 6, 1065 (1972).

24. J. Coremberg and F. M. Mazzolai, Solid State Comm., 6, 1 (1968).

25. R. H. Chambers, Physical Acoustics Vol. IIIA. Chapt.4. Academic Press, New York and London (1966).

26. G. Rieu and J. de Fouquet, J. de Physique Suppl. C2, 32, 221 (1971).

27. B. Escaig, Scripta Met., 5, 199 (1971).

28. A. Seeger and B. Sesták, Scripta Met., 5, 681 (1971).

INTERNAL FRICTION AND MODULUS DEFECT DUE TO THE MOVEMENT OF
POINT DEFECTS IN INHOMOGENEOUS STRESS FIELDS

A.Seeger and W.Hornung

Max-Planck-Institut für Metallforschung, Institut für Physik, and Institut für Theoretische
und Angewandte Physik der Universität Stuttgart, Stuttgart, Germany.

Abstract. Using frequency- and wavenumber-dependent compliances a general method for calculating the energy loss and the modulus defect associated with the movement of point defects with lower than lattice symmetry in inhomogeneous fields is given. In the case of small interaction energies with the field ($U_i \ll k_B T$) the basic equations may be linearized in $U_i/k_B T$. An inhomogeneous system of linear equations for the Fourier transforms of the "partial concentrations" (describing the distribution of the defects over the energetically distinguished defect orientations) is obtained. The general method is illustrated by considering defects with <100>-axis in cubic crystals, which are capable of reorientation with or without migration, and their interaction with straight edge dislocations.

1. Introduction

Point defects may contribute to the internal friction and the modulus defect of crystals through several mechanisms. Particularly simple examples are the Snoek effect [1,2], associated with the redistribution of point-defect orientations in a (quasi-) homogeneous stress field, and the Gorski effect [3-5], associated with the long-range migration of point defects in a stressfield that is inhomogeneous in macroscopic but not in atomic dimensions. The theoretical treatment of these effects is comparatively simple and well established [6-8]. More difficult theoretical problems arise when the energy dissipation associated with the motion of point defects in time-dependent inhomogeneous fields such as those arising from dislocations or from ferromagnetic domain walls are considered. An example is the "dislocation-enhanced Snoek effect" [9,10], which results from the amplification of the time-dependent interaction of an applied stress with point defects by the motion of dislocations under the applied stress.

It is desirable to have a theoretical method of handling, by a unified approach, both the "point-defect damping" and the energy dissipation arising from the interaction of point defects with dislocations, domain walls etc. The present paper develops such an approach under fairly general conditions though with some limitations regarding the admissable strength of the interaction between the point defects and the inhomogeneities. In Sect. 4-6 examples for the application of the general approach to specific problems will be given; it is intended to present further applications elsewhere.

The desired generality is achieved by making extensive use of Fourier transform techniques not only with regard to time t but also in space. We consider a "partial" concentration $C_j(\underline{r},t)$ of point defects with orientation j at a position $\underline{r}$. It is assumed that at a given spatial position the defects are capable of taking up n_E orientations which in a general external field are energetically different. The total defect concentration is thus given by

$$C(\underline{r},t) = \sum_{j=1}^{j=n_E} C_j(\underline{r},t) \quad .$$

(1.1)

The Fourier transform is defined by

$$\tilde{C}_j(\underline{r},\omega) = \frac{1}{(2\pi)^2} \iint C_j(\underline{r},t)\, e^{i(\underline{k}\cdot\underline{r} - \omega t)}\, d^3\underline{r}\, dt \tag{1.2a}$$

$$C_j(\underline{r},t) = \frac{1}{(2\pi)^2} \iint \tilde{C}_j(\underline{k},\omega)\, e^{-i(k\cdot r - \omega t)}\, d^3\underline{k}\, d \tag{1.2b}$$

The interaction energy (potential energy) of a point defect with orientation j at position $\underline{r}$ with the inhomogeneous field is denoted by $U_j(\underline{r},t)$ with Fourier transform $\tilde{U}_j(\underline{k},\omega)$. In the following we shall obtain, in terms of given $U_j(k,\omega)$, the partial concentrations C_j from a set of generalized diffusion equations. These equations are derived from rate equations describing the change of the partial defect concentrations at each lattice site due to jumps of the point defects. Macroscopically measurable quantities such as the energy ΔW dissipated per cycle of theoscillating field, or the time dependence of elastic moduli, magnetic or electric suscepti- bility etc. may be obtained from the partial concentrations C_j or their Fourier transforms by quadratures.

2. Derivation of Generalized Diffusion Equations

We consider point defects that may carry out three different types of jumps:

a) Change of defect orientation without displacement of the "centre of gravity" of the defect.
b) Jumps from position $\underline{r}$ to position $\underline{r}+\underline{\Delta r}$ without simultaneous change of the defect orientation.
c) Jumps from position $\underline{r}$ to position $\underline{r}+\underline{\Delta r}$ with simultaneous change of the defect orientation.

The rate of reorientation jumps without displacement of the centre of gravity from orientation i into orientation j is denoted by Γ_{ij}^R, that jump from orientation i at lattice site $\underline{r}$ to orientation j at site $\underline{r}+\underline{\Delta r}_{ij}$ by $\Gamma_{ij}^M(\underline{r},\underline{\Delta r})$.

The principle of detailed balancing requires

$$\Gamma_{ij}^R \exp\left\{-\frac{U_i(\underline{r})-U_j(\underline{r})}{k_B T}\right\} = \Gamma_{ij}^R \tag{2.1a}$$

$$\Gamma_{ij}^M(\underline{r},\underline{\Delta r})\exp\left\{-\frac{U_i(\underline{r})-U_j(\underline{r}+\underline{\Delta r}_{ij})}{k_B T}\right\} = \Gamma_{ij}^M(\underline{r}+\underline{\Delta r},-\underline{\Delta r}) \tag{2.1b}$$

where k_B denotes Boltzmann's constant and T the absolute temperature. Equations (2.1) are satis- fied if throughout the crystal we have

$$\Gamma_{ij}^R = \Gamma^R \exp\left\{[U_i(\underline{r})-U_j(\underline{r})]/2k_B T\right\} \tag{2.2a}$$

$$\Gamma_{ij}^M = \Gamma^M \exp\left\{[U_i(\underline{r})-U_j(\underline{r}+\underline{\Delta r}_{ij})]/2k_B T\right\} \quad i \neq j \tag{2.2b}$$

$$\Gamma_{jj}^M = \Gamma^D \exp\left\{[U_j(\underline{r})-U_j(\underline{r}+\underline{\Delta r}_{ii})]/2k_B T\right\} \tag{2.2c}$$

Here Γ^R is the rate of rotational jumps in the absence of external fields, with analogous meanings of Γ^M (migration with change of orientation) and Γ^D (migration without change of orientation).

The rate of change of $C_i(\underline{r},t)$ due to the processes a), b), and c) is given by

$$
C_i(\underline{r},t) = \left\{
\begin{array}{l}
\sum\limits_{j=1}^{n_E} [C_j(\underline{r},t)\ \Gamma_{ji}^R - C_i(\underline{r},t)\ \Gamma_{ij}^R] \\[2mm]
\sum\limits_{j=1}^{n_E} \sum\limits_{\underline{\Delta r}_{ji}} [C_j(\underline{r}+\underline{\Delta r}_{ji},t)\ \Gamma_{ji}^M - C_i(\underline{r},t)\ \Gamma_{ij}^M(\underline{r},\underline{\Delta r})]
\end{array}
\right\}
\tag{2.3}
$$

For a very wide range of applications the interaction energies U_i are small compared with the thermal energy $k_B T$. This suggests linearization of the system (2.3) of 10^{23} difference equations with respect to $\frac{U_i}{k_B T}$. A first step makes use of

$$
\exp\{\frac{U_i - U_j}{2k_B T}\} \approx 1 + \frac{U_i - U_j}{2k_B T}
\tag{2.4a}
$$

As a second step, again with an error of the order of $(\frac{U_i}{k_B T})^2$ or smaller, we write

$$
C_i[1 + \frac{U_i - U_j}{2k_B T}] \approx C_i + \frac{C_o}{n_E} \cdot \frac{U_i - U_j}{2k_B T} \quad,
\tag{2.4b}
$$

The total defect concentration C_o is taken as constant. Approximation (2.4b) allows a straight-forward solution of the resultant system of rate equations. Without this approximation we would be left with a system of integro-differential equations, which could be solved by an iteration method, of which the first step would be identical to approximation (2.4b).

The linearized system reads

$$
C_i = \left\{
\begin{array}{l}
\Gamma^R \sum\limits_{j=1}^{n_E} [X_j(\underline{r},t) - X_i(\underline{r},t)] \\[2mm]
+\ \Gamma^M \sum\limits_{\substack{j=1 \\ j \neq i}}^{n_E} \sum\limits_{\underline{\Delta r}_{ij}} [X_j(\underline{r}+\underline{\Delta r},t) - X_i(\underline{r},t)] \\[2mm]
+\ \Gamma^D \sum\limits_{\underline{\Delta r}_{ii}} [X_i(\underline{r}+\underline{\Delta r},t) - X_i(\underline{r},t)]
\end{array}
\right\} \quad,
\tag{2.5}
$$

where

$$
X_i(\underline{r},t) \equiv C_i(\underline{r},t) + \frac{C_o}{n_E k_B T} U_i(\underline{r},t) \quad.
\tag{2.5a}
$$

Fourier transformation with regard to space and time transforms this system of $\sim 10^{23}$ difference equations into the following inhomogeneous system of n_E linear equations as defined:

$$
i\omega\, \tilde{C}_i(k,\omega) = \left\{
\begin{array}{l}
\Gamma^R \sum\limits_{j=1}^{n_E} [\tilde{X}_j(\underline{k},\omega) - \tilde{X}_i(\underline{k},\omega)] \\[2mm]
+\ \Gamma^M \sum\limits_{\substack{j=1 \\ j \neq i}}^{n_E} \sum\limits_{\underline{\Delta r}_{ij}} [\tilde{X}_j(\underline{k},\omega)\cos(\underline{\Delta r}\cdot\underline{k}) - \tilde{X}_i(\underline{k},\omega)] \\[2mm]
+\ \Gamma^D \sum\limits_{\Delta r_{ij}} [\tilde{X}_i(\underline{k},\omega)\cos(\underline{\Delta r}\cdot\underline{k}) - \tilde{X}_i(\underline{k},\omega)]
\end{array}
\right\}
\tag{2.6}
$$

In (2.6) we have made use of $\sum \underline{\Delta r} = o$, a condition which is satisfied for crystal structures

224

of sufficiently high symmetry. Restricting ourselves to next nearest neighbour jumps and re-arranging (2.6) we obtain

$$i\omega \ \tilde{C}_i \ (\underline{k},\omega) = \left\{ \begin{array}{l} \sum\limits_{j=1}^{n_E}(\Gamma^R + p_{ji} \ \Gamma^M)[\tilde{C}_j(\underline{k},\omega) - \tilde{C}_i(\underline{k},\omega) + \dfrac{C_o}{n_E k_B T} \ (\tilde{U}_j - \tilde{U}_i)] \\[4mm] - \sum\limits_{j=1}^{n_E} \tilde{C}_j(\underline{k},\omega)k^2 D_{ji}(\underline{k}) \end{array} \right\} \qquad (2.7)$$

In (2.7) p_{ji} denotes the number of nearest neighbours sites that can be reached with simultaneous reorientation from j to i.

$$k^2 D_{ji}(\underline{k}) \equiv 2\Gamma^M \ \sum \ \sin^2 \ (\tfrac{1}{2} \ \underline{\Delta r}_{ji} \cdot \underline{k}) \quad j \neq i \qquad (2.8a)$$

$$k^2 D_{ii}(\underline{k}) \equiv 2\Gamma^D \ \sum \ \sin^2 \ (\tfrac{1}{2} \ \Delta r_{ii} \cdot k) \qquad (2.8b)$$

are functions which are periodic in k-space. In (2.8a) the summation extends over all nearest neighbour sites that can be reached with simultaneous reorientation from j to i, and in (2.8b) over all nearest neighbour sites than can be reached without simultaneous reorientation.

The solutions of (2.7) are of the form

$$\tilde{C}_i(\underline{k},\omega) \qquad \sum\limits_{l=1}^{n_E} \bar{Q}_i^l(\underline{k},\omega) \ \tilde{U}_l(\underline{k},\omega) \qquad (2.9)$$

where the $\bar{Q}_i^l(\underline{k},\omega)$ are rational functions of ω and of $k^2 D_{ij}$. These solutions comprise a number of special cases, e.g., that treated in [11], where k - periodic functions similar to the $k^2 D_{ij}(\underline{k})$ have been employed.

In many other cases, however, the macroscopically measured quantities may be obtained from simplified solutions of (2.3). This simplification is achieved by inserting Taylor's expansions at site $\underline{r}$ into equations (2.3), e.g.,

$$C_j(\underline{r}+\underline{\Delta r}) = C_j(\underline{r}) + \underline{\Delta r}.\nabla \underline{C}_j + \underline{\Delta r} \ \nabla..\nabla \underline{C} \ \underline{\Delta r} +.., \qquad (2.10)$$

and neglecting all derivatives higher than the second. This corresponds to a continuum approach to diffusion. We arrive at the following system of n_E linear differential equations:

$$C_i = \left\{ \begin{array}{l} \sum\limits_{j=1}^{n_E}(\Gamma^R + p_{ij} \ \Gamma^M) \ \{[C_j(\underline{r},t)-C_i(r,t)] + \dfrac{C_o}{n_E k_B T} \ [U_j(\underline{r},t)-U_i(\underline{r},t)]\} \\[4mm] + \ \tfrac{1}{2} \ \Gamma^M \ \sum\limits_{\substack{j=1 \\ j \neq i}}^{n_E} \ \sum\limits_{\underline{\Delta r}_{ji}} \ \{\underline{\Delta r} \ \underline{\nabla}..\underline{\nabla}C_j \ \underline{\Delta r} + \dfrac{C_o}{n_E k_B T} \ \underline{\Delta r} \ \underline{\nabla}..\underline{\nabla}U_j \ \underline{\Delta r}\} \\[4mm] + \ \tfrac{1}{2} \ \Gamma^D \ \sum\limits_{\Delta r_{ii}} \ \{\underline{\Delta r} \ \underline{\nabla}..\underline{\nabla}C_i \ \underline{\Delta r} + \dfrac{C_o}{n_E k_B T} \cdot \underline{\Delta r} \ \underline{\nabla}..\underline{\nabla}U_i \ \underline{\Delta r}\} \end{array} \right\} \qquad (2.11)$$

Again we have made use of $\sum \underline{\Delta r} = o$.

We now define partial diffusion coefficients

225

$$\underset{=}{D}^{ij} \equiv \frac{\Gamma^M}{2} \sum_{\underline{\Delta r}_{ij}} \underline{\Delta r} \, \underline{\Delta r} \quad , \quad i \neq j \tag{2.9a}$$

where the summation extends over all sites with j-orientation (j≠i) that can be reached from a defect with i-orientation at site $\underline{r}$ by one jump, and

$$\underset{=}{D}^{ii} = \frac{\Gamma^D}{2} \sum_{\underline{\Delta r}_{ii}} \underline{\Delta r} \, \underline{\Delta r} \tag{2.9b}$$

where the summation extends over all sites with i-orientation that can be reached in one jump.

Inserting equations (2.9a) and (2.9b) into (2.8) gives us

$$\dot{C}_i = \left\{ \begin{array}{l} \sum_{j=1}^{n_E} (\Gamma^R + p_{ij} \, \Gamma^M)[C_j(\underline{r},t) - C_i(\underline{r},t) + \dfrac{C_o}{n_E k_B T} \, U_j(\underline{r},t) - U_i(\underline{r},t)\}] \\[4mm] \sum_{j=1}^{n_E} \underset{=}{D}^{ij} \, \Delta C_j + \sum_{j=1}^{n_E} \underset{=}{D}^{ij} \, \Delta U_j \end{array} \right\} \tag{2.10}$$

This is a system of n_E differential equations which may be called generalized diffusion equations with drift terms. Again this can be solved by a Fourier transformation, whereby we obtain

$$i\omega \, \tilde{C}_i(\underline{k},\omega) = \left\{ \begin{array}{l} \sum_{j=1}^{n_E} (\Gamma^R + p_{ij} \, \Gamma^M + \underset{=}{D}^{ij} \cdot \cdot \underline{k} \, \underline{k})(\tilde{C}_j + \dfrac{C_o}{n_E k_B T} \, \tilde{U}_j) \\[4mm] - (n_E \, \Gamma^R + \Gamma^M \sum_{j=1}^{n_E} p_{ij})(\tilde{C}_i + \dfrac{C_o}{n_E k_B T} \, \tilde{U}_i) \; . \end{array} \right\} \tag{2.11}$$

The solutions of this system of linear equations are of the form

$$\tilde{C}_i(\underline{k},\omega) = \sum_{l=1}^{n_E} Q_i^l(\underline{k},\omega) \, \tilde{U}_l(\underline{k},\omega) \tag{2.12}$$

where the Q_i^l are rational functions of $\underline{k}$ and ω. Equations (2.7) and (2.12) are identical to the order of $\underline{k}^2$, e.g., when a potential expansion at $\underline{k}=o$ is inserted into (2.7):

$$\sum_{\underline{\Delta r}_{ij}} 2 \sin^2 \frac{1}{2} \underline{\Delta r} \cdot \underline{k} \approx \sum_{\underline{\Delta r}_{ij}} \frac{1}{2} \underline{\Delta r} \, \underline{\Delta r} \cdot \cdot \underline{k} \, \underline{k} \equiv \underset{=}{D}^{ij} \cdot \cdot \underline{k} \, \underline{k} \; , \tag{2.13}$$

3. Application to Anelastic Behaviour of Crystals Due to Point Defects

In order to show how to proceed further when the point defects move in a given field of force, we consider in the following the example of elastic interactions. Each of the possible orientations i of the defects may be characterized by an elastic dipole tensor P_{lm}^i, or by the tensor

$$\Pi_{pq}^i = \sum_{l,m} s_{lmpq} \, P_{lm}^i \tag{3.1}$$

measuring the strains caused by a defect. As usual s_{lmpq} denotes the fourth-rank tensor of elastic compliances.

From the dipole strain field tensors Π^i we obtain the macroscopic strain ε^{PD} produced by the defects according to

$$\varepsilon^{PD}_{pq} (\underline{r},t) = g \sum_{i=1}^{n_E} C_i(\underline{r},t) \, \Pi^i_{pq} \quad , \tag{3.2}$$

where the summation extends over all possible orientations and g is a geometrical factor relating the atomic partial concentrations C_i to the partial local defect density. (For example, in fcc crystals g equals $\frac{4}{a^3}$, where a is the lattice parameter.)

If the point defects are moving in the stress field $\sigma(\underline{r},t)$, the interaction energies U_i are given by

$$U_i(\underline{r},t) = - \Pi^i \,..\, \sigma(\underline{r},t) \quad . \tag{3.3}$$

Since according to Sect.2 the partial concentrations are linear functions of the interaction energies, we obtain a linear relationship between the anelastic strain caused by the defects and the stress. The total strain, i.e., the sum of the elastic strain ε^{el} and the anelastic strain ε^{an}, may be written as

$$\tilde{\varepsilon}_{pq} \equiv \tilde{\varepsilon}^{el}_{pq} + \tilde{\varepsilon}^{an}_{pq} = s_{pqrs} \, \tilde{\sigma}_{rs} + \sum_{i=1}^{n_E} \Pi^i_{pq} \, C_i(\underline{k},\omega)$$

$$\equiv s_{pqrs} \, \tilde{\sigma}_{rs} + \sum_{i=1}^{n_E} \Pi^i_{pq} \sum_{l=1}^{n_E} Q^l_i \, [\Pi^l \,..\, \tilde{\sigma}(\underline{k},\omega)] \tag{3.4}$$

or

$$\tilde{\varepsilon}_{pq} = S_{pqrs} \, (\underline{k},\omega) \, \tilde{\sigma}_{rs} \quad , \tag{3.5}$$

where we have introduced the fourth-order tensor S_{pqrs} of the complex compliances. For spatially inhomogeneous fields S_{pqrs} depends on the wave vector $\underline{k}$ if long-range migration processes are possible. Experimentally one measures a mean bulk compliance or modulus. These macroscopic quantities are related to the energy ΔW dissipated per cycle

$$\Delta W = \int d^3\underline{r} \int_{t-\frac{\pi}{\omega_o}}^{t+\frac{\pi}{\omega_o}} \sigma(\underline{r},t') \,..\, \varepsilon^{an}(\underline{r},t') dt' \quad , \tag{3.6}$$

where ω_o denotes the angular frequency of the oscillating stress field.

Parseval's theorem provides other forms of ΔW

$$\Delta W = \int d^3k \int_{t-\frac{\pi}{\omega_o}}^{t+\frac{\pi}{\omega_o}} \tilde{\sigma}(\underline{k},t') \,..\, \tilde{\varepsilon}(\underline{k},t') \, dt' \tag{3.7}$$

$$-2\pi\cdot\Delta W = \int d^3k \int_{t-\frac{\pi}{\omega_o}}^{t+\frac{\pi}{\omega_o}} dt' \iint_{-\infty}^{\infty} i\omega' \tilde{\sigma}(\underline{k},\omega) \,..\, \tilde{\varepsilon}(\underline{k},\omega') e^{-1(\omega+\omega')t'} \, d\omega \, d\omega' \tag{3.8}$$

where $\tilde{\sigma}(k, t')$ denotes the spatial Fourier transform of $\sigma(\underline{r}, t')$.

If σ and ε are dependent on time according to

$$\sigma = \sigma_0(\underline{r}) \cos \omega_0 t \ , \ \varepsilon = \varepsilon_0(\underline{r}, t) \cos \omega_0 t \tag{3.9}$$

we have

$$\Delta W = - i \frac{\pi}{2} \int \tilde{\sigma}_0(\underline{k}) .. \{\tilde{\varepsilon}_0^{an}(\underline{k}, \omega_0) - \tilde{\varepsilon}_0^{an}(\underline{k}, -\omega_0)\} d^3\underline{k} \ . \tag{3.10}$$

The energy loss ΔW is closely related to other quantities, such as the modulus defect as function of frequency or, by an inverse Fourier transformation, as function of time. With the notation and the respective formulae given by Nowick and Berry [7] we have

$$\Delta W = \pi M_2 \ \varepsilon_a^2 \ = \ \Delta W = \pi J_2 \ \sigma_a^2 \ , \tag{3.11}$$

where M_2 is the imaginary part of the dynamic modulus

$$M^*_{(a)} = M_1(\omega) + i \ M_2(\omega) \tag{3.12a}$$

and J_2 the imaginary part of the dynamic compliance

$$J^*(\omega) = J_1(\omega) - i \ J_2(\omega) \ . \tag{3.12b}$$

Since M^* and J^* are analytical functions of ω their real parts can be evaluated from their imaginary parts by means of Kramers-Kronig relations.

Very often the stress relaxation function $M(t)$ or the creep function $J(t)$ are of interest. These quasi-static properties are connected to the dynamic properties by Fourier transformations.

All the above-mentioned quantities may be obtained from (3.10):

$$M_2(\omega) = - \frac{i}{2\varepsilon_a^2} \ \int d^3k \ \tilde{\sigma}_0(k)\{\varepsilon_0(\underline{k}, \omega) - \varepsilon_0(\underline{k}, -\omega)\} \tag{3.13}$$

$$M_1(\omega) = M_R - \frac{2\omega^2}{\pi} \ \frac{i}{2\varepsilon_a^2} \ \int d^3k \ \tilde{\sigma}_0(k) \int_0^\infty \frac{1}{\alpha} \frac{d\alpha}{\omega^2 - \alpha^2} \{\varepsilon_0(\underline{k}, \alpha) - \varepsilon_0(\underline{k}, -\alpha)\} \tag{3.14}$$

$$J_2(\omega) = \frac{-i}{2\sigma_a^2} \int d^3k \ \tilde{\sigma}_0(k)\{\varepsilon_0(\underline{k}, \omega) - \varepsilon_0(\underline{k}, -\omega)\} \tag{3.15}$$

$$J_1(\omega) = J_u - \frac{i}{\pi \ \sigma_a^2} \int d^3k \ \tilde{\sigma}_0(k) \int_0^\infty \{\varepsilon_0(\underline{k}, \alpha) - \varepsilon_0(\underline{k}, -\alpha)\} \frac{\alpha \ d\alpha}{\omega^2 - \alpha^2} \tag{3.16}$$

$$M(t) = M_R - \frac{i}{4\pi\varepsilon_a^2} \int d^3k \ \tilde{\sigma}_0(k) \int_0^\infty \{\varepsilon_0(\underline{k}, \omega) - \varepsilon_0(\underline{k}, -\omega)\} \cos \omega t \ \frac{d\omega}{\omega} \tag{3.17}$$

$$J(t) = J_u + \frac{i}{4\pi\sigma_a^2} \int d^3k \; \tilde{\sigma}_o(k) \int_o^\infty \{\epsilon_o(\underline{k},\omega) - \epsilon_o(\underline{k}.-\omega)\} \; \cos \; t \frac{d\omega}{\omega} \tag{3.18}$$

Thus the macroscopically measurable quantities may be calculated within the framework of the present method apart from a constant describing the mean displacement of the atoms in the crystal under the influence of the inhomogeneous stress field.

In (3.14) and (3.16-18) we have interchanged the integration over $\underline{k}$ with the Kramers-Kronig integration or the inverse Fourier transformation. This is expedient since in general it is not possible to evaluate all the integrals in terms of known functions. Good approximations, however, can often be found if the integration over $\underline{k}$ is carried out last.

4. The Complex Compliance of FCC-Metals Containing Dumbbell Self-Interstitials

According to [12] and [13] the stable configuration of the intrinsic interstitial in fcc metals is the socalled dumbbell configuration with tetragonal symmetry and a main axis in <100>-direction. This defect may take $n_E=3$ crystallographically different orientations at each lattice site. It may reorient over two different types of saddle points and it can exert migrational jumps in the {100}-plane of its main axis with simultaneous reorientation. We have

$$\Gamma^R = 2\nu^R, \quad \Gamma^M = \nu^M, \quad \Gamma^D = o \;. \tag{4.1}$$

The partial concentrations in the three crystallographically different orientations are given by (with cyclic permutation)

$$\tilde{C}_1(\underline{k},\omega) = \frac{1}{D(\underline{k},\omega)} \{ \sum_{l=1}^{3} Q_1^{(1)} \; \tilde{U}_1(\underline{k},\omega)\} \tag{4.2}$$

where $D(\underline{k},\omega)$ denotes the following third rank determinant

$$D(k,\omega) \equiv \begin{vmatrix} [-i\omega-4(\nu^R+2\nu^M)] & 2[\nu^R+\nu^M\{\cos \underline{k}.\Delta r_1 + \cos \underline{k}.\Delta r_2\}] & 2[\nu^R+\nu^M \dots\}] \\ 2[\nu^R+\nu^M \cos \underline{k}.\Delta r_1 + \cos \underline{k}.\Delta r_2\}] & [-i\omega-4(\nu^R+2\nu^M)] & 2[\nu^R+\nu^M \dots\}] \\ 2[\nu^R+\nu^M\{\cos \underline{k}.\Delta r_3 + \cos \underline{k}.\Delta r_4\}] & 2[\nu^R+\nu^M\{\cos \underline{k}.\Delta r_5 + \cos \underline{k}.\Delta r_6\}] & [-i\omega - 4(\nu^R+2\nu^M)] \end{vmatrix} \tag{4.3}$$

with

$$\Delta r_1 = \frac{a}{2}\begin{pmatrix}1\\1\\o\end{pmatrix}, \quad \Delta r_2 = \frac{a}{2}\begin{pmatrix}1\\-1\\o\end{pmatrix}, \quad \Delta r_3 = \frac{a}{2}\begin{pmatrix}1\\o\\1\end{pmatrix}, \quad \Delta r_4 = \frac{a}{2}\begin{pmatrix}1\\o\\-1\end{pmatrix}, \quad \Delta r_5 = \frac{a}{2}\begin{pmatrix}o\\1\\1\end{pmatrix}, \quad \Delta r_6 = \frac{a}{2}\begin{pmatrix}o\\1\\-1\end{pmatrix}$$

$$\tag{4.3a}$$

As a first example we suppose the stress to oscillate harmonically and to be homogeneous throughout the crystal. Because of the tetragonal symmetry of the defects their interaction energies U_i with the stress field are given by

$$\tilde{U}_1 = 2\pi^2 \begin{pmatrix} \lambda_1 & o & o \\ o & \lambda_2 & o \\ o & o & \lambda_o \end{pmatrix} .. \begin{pmatrix} \sigma_{11} & \sigma_{12} & \sigma_{13} \\ \sigma_{12} & \sigma_{22} & \sigma_{23} \\ \sigma_{13} & \sigma_{23} & \sigma_{33} \end{pmatrix} \delta(\underline{k})[\delta(\omega-\omega_o) + \delta(\omega+\omega_o)] \tag{4.4}$$

In this case the partial concentrations read ($\nu_u = 2(\nu^P + 2\nu^M)$)

$$\tilde{C}_1(\underline{k},\omega) = 2\pi^2 \frac{C_o}{3k_BT} (\lambda_1 - \lambda_2)(\sigma_{22} + \sigma_{33} - 2\sigma_{11})\delta(k) \left[\frac{\nu_u}{(i\omega_o - 3\nu_u)} + \text{c.c.}\right] \qquad (4.5)$$

with the abbreviation

$$\nu_u = 2(\nu^R + 2\nu^M) . \qquad (4.5a)$$

Inserting (4.5) into (3.4) we obtain the tensor of the complex compliance

$$S_{lmpq} = s_{lmpq} + \frac{2\pi^2}{3k_BT} C_o(\lambda_1 - \lambda_2)^2 \frac{\nu_u}{i\omega - 3\nu_u} \delta_{lm}\delta_{pq} [1 - 3\delta_{lp}\delta_{mq}] \qquad (4.6)$$

where δ_{uv} is Kronecker's symbol.

Eq.(4.6) describes the Snoek effect of the point defect. As expected S_{lmpq} is independent of the wave vector $\underline{k}$.

5. Relaxation Times for Quasi-Static Experiments in the Presence of Straight Edge or Screw Dislocations

The zeros of the determinant $D(\underline{k},\omega)$ in (4.3) in the complex ω-plane give the reciprocal relaxation times. In order to obtain the relaxation times measured macroscopically in quasi-static experiments it is necessary to integrate expressions of the form $\sum \tilde{C}_i(\underline{k})\tilde{U}_i(\underline{k})$ over the entire $\underline{k}$-space. As may be seen from (4.2) because of this integration the Fourier transforms of the stress field pick out the relaxation times measured macroscopically.

For a screw dislocation in fcc metals it has been shown that three macroscopic relaxation times

$$\tau_1 = \frac{1}{6(\nu^R + 2\nu^M)} , \quad \tau_2 = \frac{1}{6(\nu^R + 2\nu^M)} , \quad \tau_3 = \frac{1}{7\nu^M} \qquad (5.1)$$

result. In order to demonstrate how to proceed in the general case, we consider the stress field of an infinitely long straight edge dislocation with $[1\bar{1}0]$ - Burgers vector along the $[\bar{1}\bar{1}2]$ - direction in an elastically isotropic fcc crystal. We retain the periodicity of $D(\underline{k},\omega)$ but insist on the correct $\underline{k}$-dependence only up to the second power in $\underline{k}$. This latter approximation corresponds to the description in terms of a macroscopic second-order diffusion equation. Then the solutions of $D(\underline{k},\omega) = o$ lead to

$$\tau_1^{-1} = 6(\nu^R + 2\nu^M) , \quad \tau_2^{-1} = 6(\nu^R + 2\nu^M)$$
$$\tau_3^{-1} = 8\nu^M(1 - \frac{1}{3} [\cos \frac{a}{2} k_x \cos \frac{a}{2} k_y + \cos \frac{a}{2} k_x \cos \frac{a}{2} k_z + \cos \frac{a}{2} k_y \cos \frac{a}{2} k_z]). \qquad (5.2)$$

The two relaxation times close to the Snoek relaxation times τ_1 and τ_2 are approximately independent of $\underline{k}$; thus the stress field determines only the relaxation strength. When τ_3 is inserted into the equations for the creep function or the relaxation function integrals of the form

230

$$\int e^{-\frac{t}{\tau_3}} \tilde{U}_i^2(\underline{k})\, d^3k \tag{5.3}$$

are obtained. Since the $\tilde{U}_i$ do not vary in $[\bar{1}\bar{1}2]$ - direction we have

$$\tilde{U}_i \sim \delta(k_x + k_y - 2k_z) \ . \tag{5.4}$$

From (5.2-4) it may be seen that the macroscopically measured τ_3 is equal to

$$\tau_3^{edge} = \frac{1}{8\nu^M} \tag{5.5}$$

which is the smallest possible value for τ_3. For a screw dislocation along $[1\bar{1}0]$ - direction

$$\tilde{U}_i \sim \delta(\underline{k}_x - k_y) \tag{5.6}$$

which leads to

$$\tau_3^{screw} = \frac{3}{20\nu^M} \approx \frac{1}{7\nu^M} \ \ . \tag{5.7}$$

6. The Energy Dissipated in the Field of an Oscillating Edge Dislocation

Inserting (4.2) into the (3.13 - 3.18) quantities leads to complicated expressions which will be considered in more detail elsewhere. Some informations, however, may already be obtained if the usual continuum approach to diffusion, i.e., the approximation

$$\cos \frac{a}{2} k_x \approx 1 - \frac{a^2}{8} k_x^2 \tag{6.1}$$

is used. In order to calculate the internal friction due to the reorientation and diffusion of the point defects within the frame-work of (6.1) we proceed as follows: We assume that the dislocation oscillates in a definite pattern, e.g., we prescribe the space and time dependence of the dislocation density tensor. By inserting the corresponding stress field tensor into equations (3.4) and (3.6) we obtain the energy ΔW dissipated per cycle.

We treat the edge dislocation as a rigid rod without any nodal or pinning points that oscillates around x=o with the amplitude A . Ascribing the total variation of the stress field to the lowest harmonic we then have for instance

$$\tilde{\sigma}_{11}(\underline{k},t) = -2\Omega \sin k_x A \frac{k_y^3}{(k_x^2 + k_y^2)^2} \tag{6.2}$$

where $\Omega = \frac{G}{2\pi(1-\mu)}$, G = elastic shear modulus, and μ = Poisson's constant.

Inserting (6.2) into equations (3.4) and (3.6) gives us several contributions to the dissipated energy of the following type

$$\Delta W_1(\omega) = -\frac{i\pi}{24} (\lambda_1 - \lambda_2)^2 b^2 L \frac{G}{1-\mu} \frac{C_0}{3k_BT} \int_0^{2\pi} \sin^6 \psi d\psi \int_0^\infty \frac{\sin^2(Ak\rho\cos\psi)}{k\rho} \left[\frac{\omega^2}{N} - \frac{\omega^2}{N*}\right] dk\rho \tag{6.3}$$

with $N = i\omega(i\omega - 3\nu_u) + 4\nu_u\nu_M k_\rho^2$, where $k\rho$ is measured in units of $\frac{1}{a}$. These integrals can be evaluated analytically in terms of combinations of logarithmic, rational and Bessel functions. Let us calculate the relaxation function $J(t) \equiv \frac{\varepsilon(t)}{\sigma_0}$ ($t \geq o$), which is the inverse Fourier transform of $\frac{\Delta W(\omega)}{\omega}$. If we interchange this transformation with the k-space integration in equation (6.3), we obtain in a very good approximation

$$J(t) - J_u = \text{const } e^{-6\nu_u t} E_i(6\nu_u t) + \text{const } e^{-3\nu_u t} E_i(-3\nu_u t) \qquad (6.4)$$

Eq.(6.4) shows that for small times the relaxation function varies exponentially, thus giving the Snoek-effect time dependence, whereas for $\nu_u t \gg 1$ it is proportional to $\frac{1}{t}$ characteristic of long-range diffusion. For a physical discussion of this transition see the accompanying paper [10].

Acknowledgment

The authors gratefully acknowledge fruitful discussions with Dr. W. Frank, Dr. F.J.Wagner and Dipl.-Phys. D.Werneth.

References

[1]   J.Snoek, Physica 8, 711 (1941)

[2]   D.Polder, Philips Res.Rep. 1, 1 (1945)

[3]   W.S.Gorsky, Phys.Z.SU 8, 457 (1935)

[4]   G.Schaumann, J.Völkl, and G.Alefeld, Phys.Rev.Letters 21, 891 (1968)

[5]   G.Alefeld, in: Vacancies and Interstitials in Metals (Eds. A.Seeger, D.Schumacher, W.Schilling, and J.Diehl), p.959 , North-Holland, Amsterdam 1970

[6]   R.de Batist, "Internal Friction of Structural Defects in Crystalline Solids", North-Holland, Amsterdam 1972

[7]   A.S.Nowick, B.S.Berry: "Anelastic Relaxation in Crystalline Solids", Academic Press

[8]   G.Alefeld, J.Völkl, and G.Schaumann, phys.stat.sol. 37, 337 (1970)

[9]   A.Seeger and F.J.Wagner, phys.stat.sol. 9, 583 (1965)

[10]   A.Seeger, this conference

[11]   W.Hornung, phys.stat.sol. 54, 441 (1972)

[12]   J.B.Gibson, A.N.Goland, M.Milgram, and G.H.Vineyard, Phys.Rev. 120, 1229 (1960)

[13]   A.Seeger, E.Mann, and R.von Jan, J.Phys.Chem.Solids 23, 639 (1962)

ANISOTROPY OF DIA-ELASTIC MODULUS CHANGE
IN ALUMINUM SINGLE CRYSTALS AFTER ELECTRON
IRRADIATION AT 4.2°K.

K.-H. Robrock and W. Schilling
Institut für Festkörperforschung der
Kernforschungsanlage Jülich, Jülich, Germany

The introduction of interstitials and vacancies in metals changes the coupling
constants between the atoms of their neighbourhood. This change of coupling
constants changes also the response to an external stress such that an extra
strain field is set up around the defect in addition to the homogeneous strain
in the ideal crystal. This means, one gets a change of the elastic moduli,
which for small defect concentrations can be described by

$$\Delta C^{diel.} = c \cdot \alpha$$

where ΔC is the modulus change, c the defect concentration and α a quantity,
which describes the dia-elastic polarizability of the defect by an external
strain field.

This dia-elastic effect should strongly be distinguished from the modulus
change caused by orientational relaxation. This relaxation effect is observed
at temperatures high enough such that the occupation numbers for the different
possible orientations can come into thermal equilibrium within times small
compared to the period of an oscillatory external stress. The reorientation of
the defects (which act as elastic dipoles) causes a para-elastic modulus change
with a temperature dependence:

$$\Delta C^{para} \sim \frac{c}{T}$$

In contrast to ΔC^{para} the modulus change ΔC^{dia} associated with defect polarisation is independent of temperature and may be measured even at the lowest
temperatures.

Our Measurements were done with an inverted torsion pendulum. The samples were
thin walled Al-single crystal tubes. By measuring the oscillation frequency
the change in the torsional modulus of the Al-samples has been determined
during electron irradiation at 4.2°K as a function of irradiation dose.

The tube-like sample shape was choosen to obtain a high cooling efficiency during irradiation by running liquid He through the tube.

We made measurements on samples with two different orientations that is with sample-axis parallel to a 100-cubic axis, and with sample axis parallel to a 111-cubic axis. From these one can find the change of the two cubic moduli C_{44} and $\frac{C_{11}-C_{12}}{2}$ separately.

Fig. 1 shows the result obtained on the <100>-sample. We have plotted the relative change of the appropriate modulus, which in this case is C_{44}, versus the concentration of the defects. The latter was determined from two simultaneously irradiated resistivity samples using a value of 4 $\mu\Omega$cm per atomic percent for converting the observed resistivity changes into defect concentrations. Above $5 \cdot 10^{-5}$ defect density a linear decrease of the elasticity modulus is observed. This is caused by the dia-elastic polarisation of the defects. At the beginning of the irradiation one finds a non-linear increase in modulus. This is a well known effect and due to dislocation pinning. Fortunately, this pinning process saturates at relatively low doses and the dia-elastic modulus change can be observed. The main features of this DEM are: It is linear with defect density, negative and large! For instance, a defect density of 10^{-4} changes the modulus by an relative amount of 0.7 percent!

Fig. 2 shows the corresponding measurements on a <111>-oriented sample. Again one finds the general behaviour as in the foregoing picture. Curve 1 shows the result for a first irradiation. There is a relative high amount of dislocation pinning, followed by the linear decrease. After this first irradiation the sample was annealed at room temperature. By this procedure all the radiation induced defects as observed by resistivity measurements are removed. There upon a second irradiation was started. One now finds, that the magnitude of dislocation effect is strongly reduced, but that the slope of the linear part is the same. Repeating this procedure for a third time reveals the same. Here one finds for a density of 10^{-4} defects a relative modulus change of about 0.3 percent. Thus, the modulus change per unit concentration is a factor of about 2.3 higher in the 100-sample than it is in the 111-sample. That means, DEM shows a strong anisotropy. From the data one gets for the cubic moduli the following numbers for their relative change per unit concentration of Frenkel defects

$$\frac{1}{c} \frac{\Delta C_{44}}{C_{44}} = -72.3 \quad \text{and} \quad \frac{1}{c} \frac{\Delta (C_{11}-C_{12})}{C_{11}-C_{12}} = -17.8$$

That means: A defect density of 1 % would decrease the modulus C_{44} by 72 % but "only" 17.8 % for $(C_{11}-C_{12})$.

From the foregoing numbers one can get an estimate for the relative change of
the shear modulus $\Delta G/G$ of a polycrystalline sample. Averaging the cubic moduli
with Reuss' formula one finds

$$\frac{1}{c} \frac{\Delta G}{G} \left|_{\substack{\text{Poly} \\ \text{Calc}}} \right. = -47.7$$

Some years ago Wenzl et al. (1) have measured the change in the shear modulus
of polycrystalline wires after neutron irradiation. They found

$$\frac{1}{c} \frac{\Delta G}{G} \left|_{\substack{\text{poly} \\ \text{meas}}} \right. = -47$$

which is in best agreement to our result.

In order to see whether the observed effects are really dia-elastic in nature
and not due to a defect relaxation at temperatures below 4.5°K, we investigated the temperature dependence of elastic modulus change between 6°K and
20°K, that is in a temperature range, where the interstitials are immobile.
The results are shown in Fig. 3. There is plotted the relative modulus change
after irradiation versus measuring temperature. The modulus change is constant
and does not show a 1/T dependence (dotted line) as one would expect in the
case of thermally activated defect orientation. These data therefore demonstrate, that our measured values are really due to dia-elastic polarisation
of the defects.

Now I come to the question how one can understand this large, anisotropic
softening of the lattice. - Firstly, from measurements on quenched samples
one knows, that the change of elastic moduli by vacancies is one order of magnitude smaller than the effects observed in this experiment. From that it
follows, that the main effect is caused by the interstitials. By X-ray measurements it has become clear in the last years that the structure of the
interstitial in Al is the wellknown 100 dumbbell-configuration, as shown in
Fig. 4. An extra atom has been put into the region of a face centre, forming
a dumbbell-like configuration with the other atom. This introduction of an
extra atom causes the springs, by which the atoms can be thought to be connected, to be highly compressed. This compression is especially large for the
springs between the dumbbell atoms themselves and their neighbour atoms. When
now applying a 100 stress to the crystal the atoms are displaced as shown by
the arrows in the picture. Such a displacement causes the spring between the
dumbbell atoms to expand and because of its compression it now works as leaf
spring with a negative force constant. This negative spring now compensates

partly the restoring forces of the surrounding other springs and thus causes large additional displacements or a high polarizability of the defect configuration under a 100-shear. When applying a 110-shear however, the coupling of the negative leaf springs to the external strain field is weaker and thus the polarizability is smaller. This model has been treated analytically by P.H. Dederichs (2) for Cu and his theoretical results, expressed in terms of relative modulus change per unit concentration

$$\frac{1}{c} \frac{\Delta C_{44}}{C_{44}} = -60 \; ; \qquad \frac{1}{c} \frac{\Delta(C_{11}-C_{12})}{C_{11}-C_{12}} = -12$$

reveal surprising agreement with our experimental values.

Although of course the details of the interatomic potentials used in the calculations may be different for Cu and Al, nevertheless the following relation should hold for the dumbbell interstitial:

$$- \Delta C_{44} \gg - \Delta \frac{C_{11}-C_{12}}{2}$$

This relation must not be true for other defect structures. The calculations of Dederichs show, that in the case of the crowdion for instance, because of its 110 symmetry the relation is just opposite:

$$- \Delta C_{44} < - \Delta \frac{C_{11}-C_{12}}{2}$$

But this result is in contradiction to our experimental values and thus can be used to exclude the crowdion-configuration.

Coming to the end, I want to summarize our results and conclusions.

1) Introduction of interstitials into a crystal gives a strong decrease of shear moduli. Qualitatively this means, that the atomic configuration around the interstitial is highly polarizable under shear. This polarizability is anisotropic. It is much larger for a 100-shear than for a 110-shear for instance.

2) The high polarizability can be understood by considering that the interstitial has a 100 dumbbell configuration and that the springs around the defect are highly compressed. That is they act like leaf springs with negative restoring forces for displacements perpendicularly to the spring axis. These negative leaf springs nearly cancel the effect of the other positive springs

around the defect, making the lattice very soft for certain shear deformations. The anisotropy of this polarizability is connected directly with the symmetry of the defect. The comparison between the theoretical and the experimental result suggests, that the interstitials must have the 100-dumbbell configuration. This is in accordance with recent X-ray studies.

3) One further conclusion is the following: The small restoring forces for a 100-shear of the interstitial should result also in very low eigenfrequencies if the corresponding modes are excited dynamically. Such low lying resonance modes have indeed been observed recently by Scholz and Lehmann (3, 4) in computer simulation studies of interstitial vibrations. Since low frequency modes can be excited to rather large amplitudes already at low temperatures, the existence of such low frequencies explains in a natural way why interstitials in f.c.c. metals become mobile already at low temperatures.

References

1. H. Wenzl, F. Kerscher, V. Fischer, K. Ehrensberger, and K. Papathanassopoulos, Z. f. Naturforsch., 26a, 489 (1971)
2. P.H. Dederichs, to be published
3. A. Scholz, Ch. Lehmann, Phys. Rev. B, 6, 813 (1972)
4. A. Scholz, Ch. Lehmann, to be published

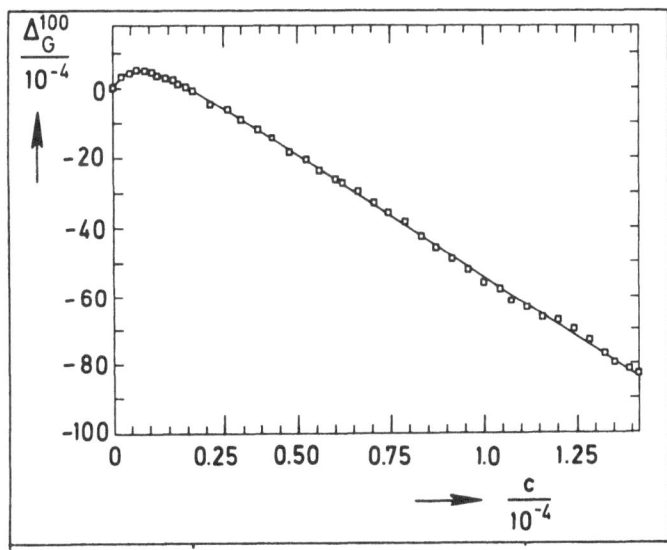

FIG. 1

Dose dependence of torsional modulus
Crystal:
Al 99.999
<100> orientation
3 MeV-electron irradiation.

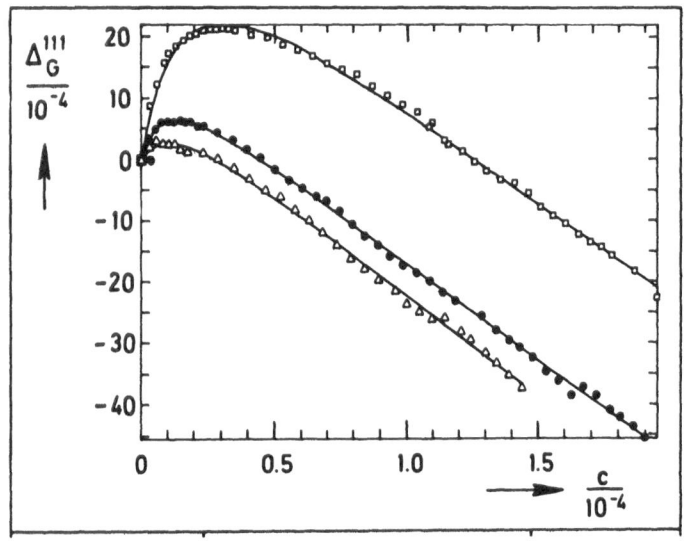

FIG. 2

Dose dependence of torsional modulus
Crystal:
Al 99.999
<111> orientation
3 MeV-electron irradiation.

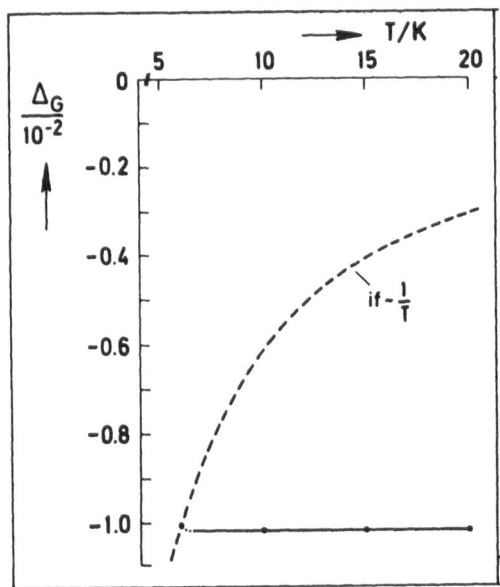

FIG. 3

Temperature dependence of modulus change. Measured data points (full line) compared with theoretical law for orientational relaxation (dotted line).

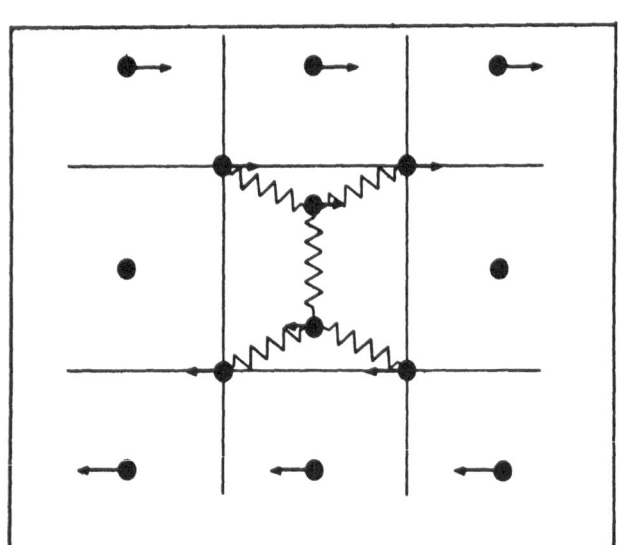

FIG. 4

<100>-shear mode of dumbbell-interstitial.

SELF INTERSTITIAL RELAXATION IN HEAVILY IRRADIATED
bcc, hcp and fcc METALS

P. Moser, J. Verdone, W. Chambron, V. Hivert and R. Pichon

Centre d'Etudes Nucléaires de Grenoble

Département de Recherche Fondamentale

Section de Physique du Solide

BP 85, Centre de Tri, 38041 Grenoble Cedex (France)

Introduction

Several peaks of internal friction were observed in heavily irradiated metals. For bcc and hcp metals, our laboratory pointed out that some of them could be attributed to the relaxation of the free interstitial migrating three dimensionally at the end of the stage I of the resistivity recovery (1-5). For fcc metals the situation is not so clear : such an interstitial was observed by magnetic after effect (6-7) but its existence was not confirmed by internal friction (8) probably because the strain ellipsoïd created by the self interstitial is practically a sphere (9-10).

The purpose of this paper is to interpret the results obtained for five metals in terms of the Nowick's theories (11-13) and to show, for nickel, that the presence of an impurity in the neighbourhood of the self interstitial can increase the anisotropy of the strain ellipsoïd and give rise to internal friction peaks.

I.bcc metals

Molybdenum

After a 28K neutron irradiation (5×10^{17}nvt>1MeV) of polycrystalline wires, two internal friction peaks were found at 31K and 40K (Fig. 1). The amplitudes of these peaks were so correlated that they appeared like "brother peaks" (4.5). They disappeared simultaneously at 43K. An important resistivity stage occured at this temperature.

The proposed mechanism is presented on Fig. 2. It corresponds to "frozen free split" phenomena described by Nowick (10). The corresponding defect, attributed to the <110> free interstitial (3) can explore the six possible orientations by either 1→2 jumps (see Fig. 2) or 1→3 jumps. From this hypothesis, we deduce that :

$$\nu_{1\to2} \Big/ \nu_{1\to3} = 2000$$

and that the number of ν_{13} jumps made by the defect during its half life is about 100.

These values mean probably that $\nu_{1\to2}$ jumps correspond to a simple (100) coplanar reorientation of the <110> defect while complex reorientation and migration are involved in ν_{13} jumps.

Iron

Pure <100> and <110> single crystals were kindly given to us by Prof Funakubo. We thank him gratefully. These crystals were tested in torsion after 28K, 5×10^{18} nvt>1MeV neutron irradiation and the obtained results confirm the previous determinations (2). The 125K peak (Fig. 3) which corresponds to a magnetic after effect zone (14-15) was attributed to the three dimensional migration of the free interstitial (16). This peak cannot correspond to a <100> oriented defect because it appears for both <100> and <110> single crystals (2).

These results are coherent with the theoretical Johnson's analysis in which the self interstitial of iron is <110> oriented and its reorientation and migration energies have the same value (17).

Consequently $\tau^{-1}_{(S_{44})}$ is practically equal to $\tau^{-1}_{(S_{11}-S_{12})}$ and both peaks of the "frozen free split" occur at the same temperature and cannot be separated.

The <110> symmetry was not confirmed by magnetic determinations. By magnetic after effect a strong <100> character was observed (16) which was recently confirmed with torque measurements under saturating magnetic field, using the same single crystals.

This discrepancy can be understood in the following way (18) :

- the elastic and magnetic techniques do not detect the symmetry of the defect but the symmetry of the perturbated volume ;

- the size and the shape of this volume depend on the range of the interaction, which differs strongly for magnetic and elastic perturbation respectively;

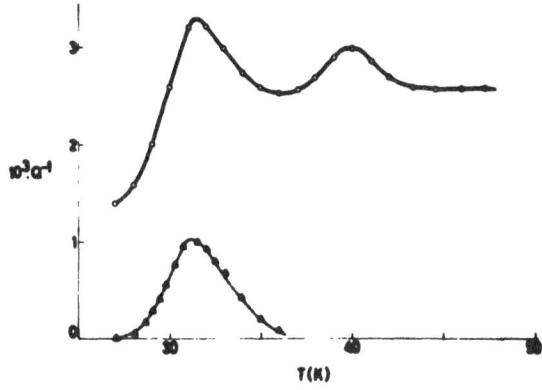

FIG. 1.

Internal friction peaks due to self interstitial relaxation in neutron irradiated molybdenum (Tirr 28K ; 5 x 10^{17} nvt > 1 MeV ; f = 0,5 Hz) (3)

FIG. 2.

Reorientation of a <110> split interstitial in a bcc metal.

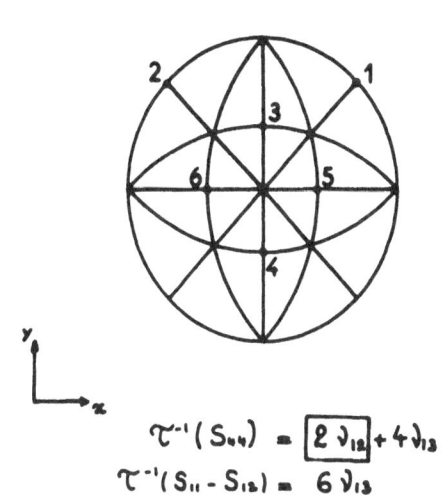

⟨110⟩ ORTHO

$$\tau^{-1}(S_{44}) = \boxed{2\,\vartheta_{12}} + 4\,\vartheta_{13}$$
$$\tau^{-1}(S_{11} - S_{12}) = 6\,\vartheta_{13}$$

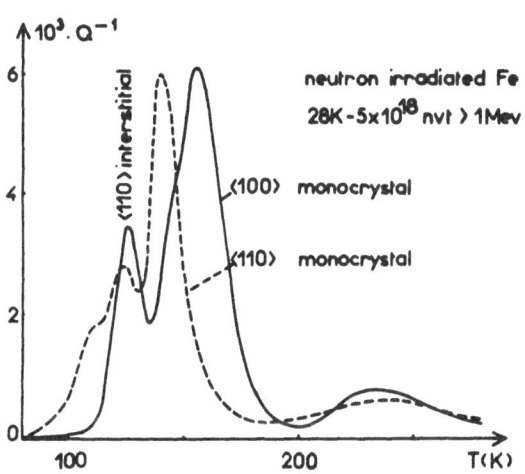

neutron irradiated Fe
28K - 5x10^{18} nvt > 1Mev

⟨100⟩ monocrystal

⟨110⟩ monocrystal

⟨110⟩ interstitial

FIG. 3.

Internal friction peaks of <100> and <110> neutron irradiated iron single crystal (f = 1 Hz).

- supposing a <110> orthorhombic symmetry for the defect, we call $\lambda_1, \lambda_2, \lambda_3$ the principal axes of the strain ellipsoïd with, for example λ_1 parallel to <110>, λ_2 to <1$\bar{1}$0> and λ_3 to <001>. We obtain $\lambda_1 \neq \lambda_2 \neq \lambda_3$ (λ_1 = 0,68 ; λ_2 = 0,47 ; λ_3 = 0,45 (2));

- by analogy, the magnetic perturbation is represented by an ellipsoïd of principal axes $\epsilon_1, \epsilon_2, \epsilon_3$ parallel to $\lambda_1, \lambda_2, \lambda_3$. The perturbated volume will present a <100> anisotropy if $\epsilon_1 = \epsilon_2$, which is not surprising for a magnetic interaction.

In other words, the <110> defect seems to be <100> oriented when magnetic techniques are used because they observe a more diffuse ellipsoïd than the strain ellipsoïd and they cannot resolve $\nu_{1 \to 2}$ jumps.

Fig 3 presents another example of "frozen free split" due to a single type of defect with the pair of peaks at 140K - 153K. The corresponding defect, probably a complex of interstitials is characterized by :

$$E_{140} = 0,30 \pm 0,03 \text{ eV and } E_{153} = 0,33 \pm 0,03 \text{ eV}$$

After electron irradiation, only one peak is found at 125K corresponding to a <110> defect.[*]

II.Hexagonal metals

Zirconium

"Frozen free split" phenomena were observed in irradiated zirconium (25-26). On Fig. 4 the 75K and 115k peaks were attributed to the relaxation of a free migrating interstitial, because their amplitudes vary simultaneously during thermal treatment (a second order law was found at the end of the annihilation corresponding to a resistivity recovery stage) (27).

[*] Before concluding about iron, we must mention the stage I_D paradox, which can be observed on Fig. 3. Iron presents an important stage I_D (80 %) of resistivity recovery (19-20). A corresponding stored energy stage was found (21). No relaxation effect (neither magnetic nor elastic) can be correlated with the defect annealing at this stage. The mechanical properties of iron, strongly affected by 77K irradiation are not concerned by this defect (22). From this basis, any interpretation such as Corbett (23) or Granato (24) models is excluded. This means that, perhaps the stage I_D does not correspond to the annihilation of defects but rather to a rearrangement such as a microrecrystallization in the perturbed volume.

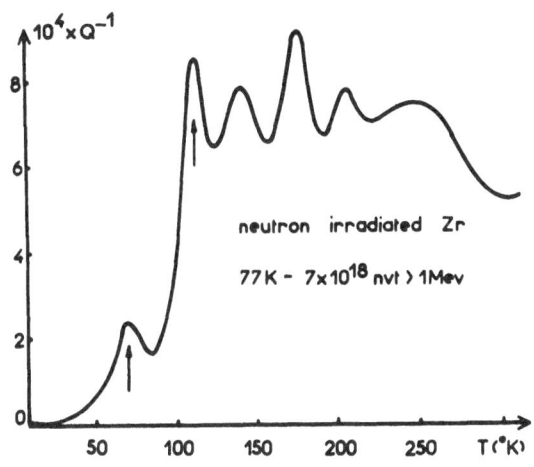

FIG. 4.

Internal friction peaks of neutron irradiated zirconium (f = 1 Hz). The peaks attributed to self interstitial relaxation are indicated by the arrows (25).

FIG. 5.

Interpretation of the relaxation of the split interstitial in irradiated zirconium or titanium.

<100> MONO

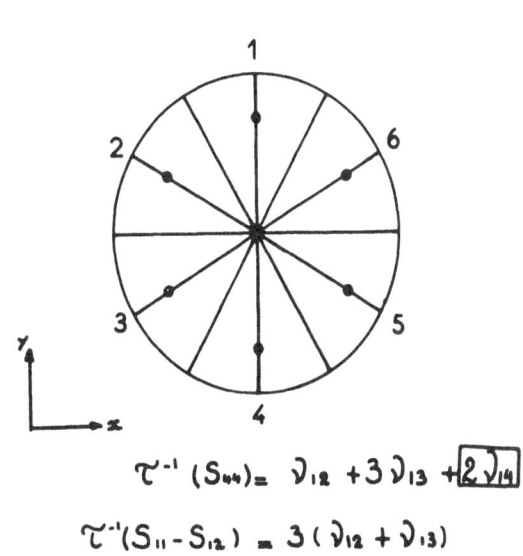

$$\tau^{-1}(S_{44}) = \nu_{12} + 3\nu_{13} + \boxed{2\nu_{14}}$$

$$\tau^{-1}(S_{11} - S_{12}) = 3(\nu_{12} + \nu_{13})$$

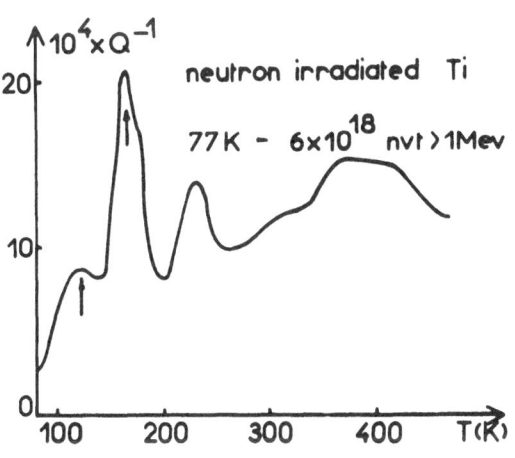

FIG. 6.

Internal friction peaks of neutron irradiated titanium (f = 1Hz) (25).

Considering the Nowick   stable (10), the simplest model attributes these peaks to the relaxation of a monoclinic <100> defect for which 1→4 jumps are easier than 1→2 or 1→3 jumps (Fig. 5). Experimental data give :

$$E_{1\to 4} = 0.17 \pm 0.02 \text{ eV}; \quad \frac{\nu_{1\cdot 4}}{\nu_{1\cdot 2} + \nu_{1\cdot 3}} = 4 \times 10^4$$

As for molybdenum, the high value of this ratio means probably that $\nu_{1\cdot 4}$ jumps correspond to coplanar reorientations of the defect and that $\nu_{1\cdot 2}$ and $\nu_{1\cdot 3}$ jumps need migrations and reorientations.

Titanium

The "frozen free split" concerns the 115K peak and one of the three components of the 162K peak (Fig. 6) (25). The model proposed for zirconium is convenient for titanium and we determined :

$$E \quad = 0.32 \pm 0.03 \text{ eV} \qquad \frac{\nu_{1\cdot 4}}{\nu_{1\cdot 2} + \nu_{1\cdot 3}} = 1.8 \times 10^4$$

III· fcc metals

Nickel

The visualization of the free migrating split interstitial in nickel by means of internal friction techniques remains an open question. Wagner observed a peak of reduced amplitude near room temperature corresponding to a free migration at stage III (28). Peretto, using magnetic after effect, found a magnetic after effect zone at 55K due to a three dimensionally migrating defect creating a strong <100> magnetic perturbation (6-7). De Keating Hart could not find a corresponding internal friction peak (8). This negative result was predicted by Huntington and Johnson (9) and Nowick (10) because the anisotropy of the strain ellipsoïd created by the <100> split interstitial is too low ($\lambda_1 - \lambda_2 = 0.025$).

Nickel doped with impurity

The anisotropy of this ellipsoïd can be increased by an impurity such as Mn, as shown in the next experiment implying magnetic after effect and internal friction measurements.

- A sample of nickel doped with 0.1 % Mn is electron irradiated at 20K. Magnetic after effect is observed during linear increase of the temperature. Fig. 7 presents the results (first run up to 70K, second run up to 110K). A nice magnetic after effect zone appears at 55K ($\frac{\delta\mu}{\mu}$ = 65 %). It is identical to that observed previously in pure nickel (7) and attributed to the three dimensionnally migrating interstitial. After nine jumps of the corresponding defect, this zone is converted into another zone similar in amplitude and spectrum width, but occuring at 65K. This last zone disappears slowly by further annealing following a first order kinetics.*

This phenomenon is interpreted in terms of trapping of the self interstitial.

The free interstitial migration is first observed, giving rise to the 55K zone. After some jumps, the interstitial is trapped, giving rise to the second zone. Fig. 8 presents an interpretation : the ellipsoïd of the magnetic perturbation due to the split interstitial is slightly altered by the presence of the impurity.

This phenomenon is found for doping by other impurities (Fig 9) and also for neutron irradiation . The nature of the impurity does not influence very much the position and the width of the second zone. On the contrary it strongly influences its stability (Temp. of half annealing : for Mn, $T_{1/2}$ = 130K, for Si, $T_{1/2}$ = 60 K, for C, $T_{1/2}$ = 70K)

_ Fig. 10 presents the results of the corresponding internal friction experiment. A sample of the same Ni - 0.1 % Mn alloy is 77K neutron irradiated and studied from 4°K. A high peak appears at 70K. This peak corresponds to the relaxation of a defect in a strain field and does not disappear under saturating magnetic field. The characteristics determined for this defect are similar to those observed with the magnetic after effect, and, for this reason, can be attributed to the self interstitial relaxation around the Mn atom.

An interpretation is given on Fig. 8 : the self interstitial (<100> split interstitial) moving at the end of stage I cannot be visualized by internal friction. The presence of an impurity in the neighbourhood gives rise to a nice relaxation peak by increasing the anisotropy of the strain ellipsoïd (for the free interstitial $\lambda_1 - \lambda_2$ = 0.025 (10) and for the trapped interstitial we determine $\lambda_1 - \lambda_2 = 0.3$).

As presented on Fig. 10, similar peaks seem to appear for other impurities. Interstitial trapping by Silicium cannot be observed by our technique because this complex is broken at 77K, as shown by magnetic after effect. The peaks at 100K, 35K and 22K, which are half annealed respectively at 120K, 150K and 200K, can be attributed to the relaxation of more and more important clusters of nickel interstitials and silicium atoms.

* A similar behaviour was observed in irradiated doped iron (29-30).

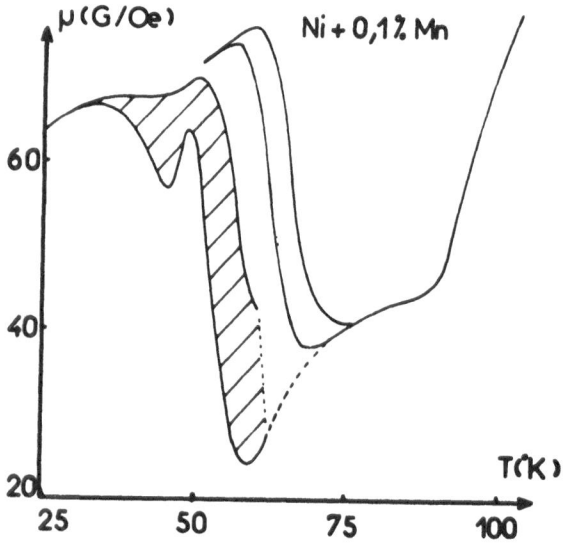

FIG. 7.

Magnetic after effect zones of 20K electron irradiated Ni-0.1 % Mn alloy appearing during linear increase of temperature. The first run (→70K) shows the magnetic after effect zone due to the free migrating split interstitial. The second run (→110K) shows the second zone due to the reorientation of the split interstitial arround a Mn atom.

FIG. 8.

Interpretation of inelastic and magnetic results of irradiated Ni-0.1 % Mn alloy. The ellipsoïd describing the magnetic perturbation is slightly influenced by the impurity. On the contrary, the strain ellipsoïd which is practically a sphere for the split interstitial becomes strongly anisotropic due to the presence of an impurity.

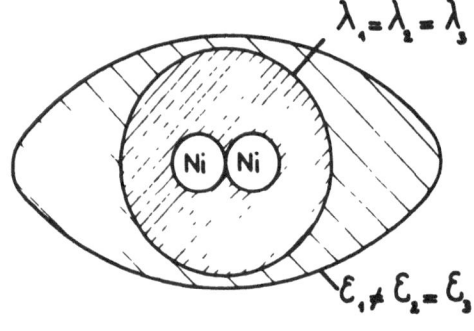

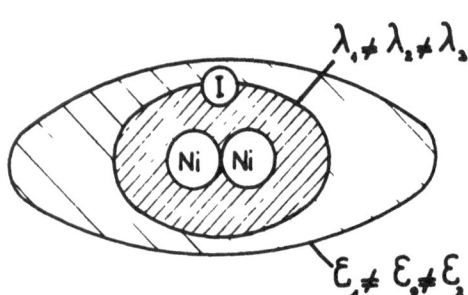

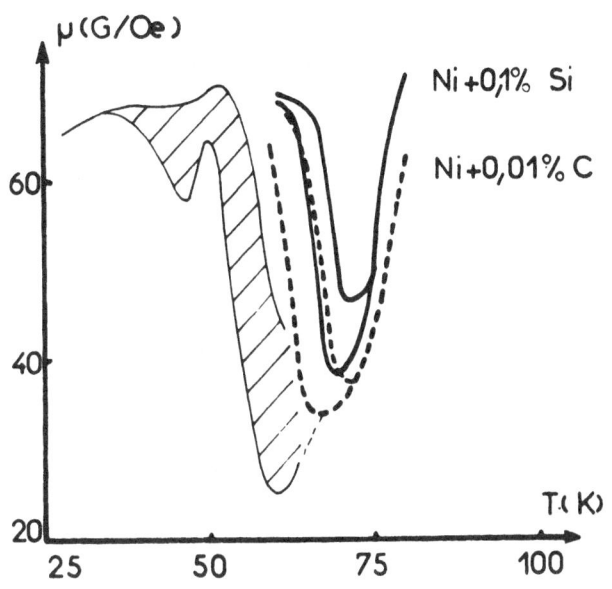

FIG. 9.

Magnetic after effect zones
of 20K electron irradiated
Ni-Si and Ni-C alloys.

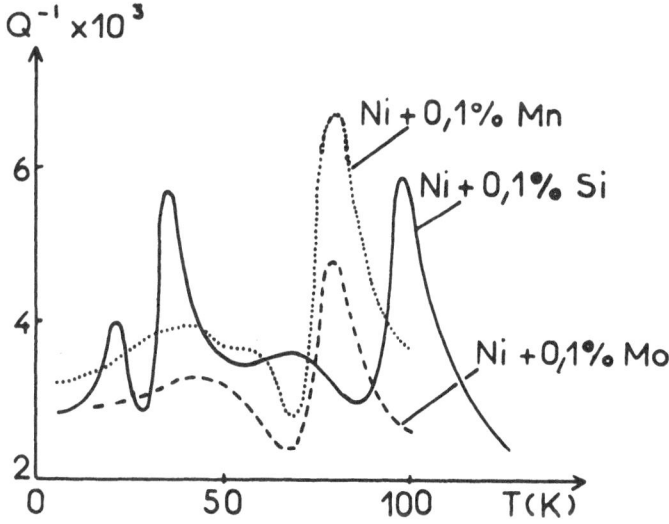

FIG. 10.

Internal friction peaks due
to trapped interstitials in
irradiated nickel alloys.

Conclusion

As predicted by Nowick (10) the number of internal friction peaks corresponding to the relaxation of a defect in a strain field could be zero, one, two... Among the reported examples, the split-interstitial of nickel does not give rise to any peak, while, if one impurity is in the neighbourhood, one peak appears. In the case of bcc and hcp metals, five examples of "frozen free split" are given.

Furthermore, Nowick pointed out that the symmetry deduced from experiments using single crystals can be either the correct or a higher symmetry (18). Examples are given for iron and nickel.

Acknowledgements

We gratefully thank E. Bisogni who initiated our hcp researches and P. Remy, Cl. Martin and Z. Kabsch for their kind experimental assistance.

References

1. Dautreppe D., Hivert V., Moser P. and Salvi A., C. R. Acad. Sci. 258, 4539 (1964)

2. Hivert V., Pichon R., Bilger H., Bichon P., Verdone J., Dautreppe D., and Moser P., J. Phys. Chem. Solids 31, 1843 (1970)

3. Pichon R., Bichon P. and Moser P., J. Phys., Paris 32, C2, 39 (1971)

4. Pichon R., Bisogni E. A. and Moser P., Defect in Refractory Metals, Mol p. 39 (1972)

5. Moser P. and Pichon R., J. Phys. F. 3, P. 363 (1973)

6. Peretto P., Thèse Grenoble CEA -R- 3772 -1969)

7. Peretto P., Moser P. and Dautreppe D., Phy. Stat. Sol. 13, 325 (1966)

8. De Keating Hart G., Thèse, Grenoble CEA-R- 3757 (1968)

9. Huntington H. B. and Johnson R. A., Acta Met. 10, 281 (1962)

10. Nowick A. S. and Berry B. S., Anelastic Relaxation in Crystalline Solids, Acad. Press, New York (1972)

11. Nowick A. S. and Heller W. R., Adv. Phys. 12, 251 (1963)

12. Nowick A. S. and Heller W. R., Adv. Phys. 12, 251 (1965)

13. Nowick A. S., Adv. Phys. 16, 1 (1967)

14. Moser P., Dautreppe D. and Brissonneau P., C. R. Acad. Sci. 250, 3963 (1960)

15. Moser P. and Dautreppe D., J. Phys., Paris, 24, 516 (1963)

16. Moser P., thèse, Mem. Sci. Rev. Metall. 63, 343 and 431 (1966)

17. Johnson R. A., Phys. Rev. 134, A, 1329 (1964)

18. Nowick A. S., Scripta Met. 1973, 7, 289 (1973)

19. Leveque J. L., Thèse, Grenoble, CEA-R- 3857 (1969)

20. Minier C., Thèse, Grenoble, CEA-R-2905 (1966)

21. Bonjour E., Moser P., C. R. Acad. Sc. 257, 1256 (1963)

22. Groh P., Vanoni F. and Moser P., "Defects in BCC Metals and their Alloys" Gaithers-burg, Maryland, USA (1973)

23. Corbett J. W., Smith R. B. and Walker R. M., Phys. Rev. 114, 1460 (1959)

24. Granato A. V. and Nilan T. H., Phys. Rev. 137, 1250 (1965)

25. Pichon R., thèse, Grenoble,(1973)

26. Pichon R., Bisogni E., and Moser P., to be published in Rad. Eff.

27. Neely H. H., Rad. Eff. 3, 189 (1970)

28. Wagner F. J., Phys. Stat. Sol. 51 b, 589 (1972)

29. Vigier P., Thèse, Grenoble, CEA-R-3280 (1968)

30. Vigier P. and Moser P., Mem. Sci. Rev. Metall. 65, p. 189 (1968)

A STUDY OF THE EFFECTS OF DEUTERIDE PRECIPITATION IN TANTALUM BY FREQUENCY AND INTERNAL FRICTION MEASUREMENTS.

G. CANNELLI and R. CANTELLI[(+)]

Consiglio Nazionale delle Ricerche - Istituto di Acustica "O. M. Corbino" - Via Cassia, 1216 Rome, Italy.

ABSTRACT - The effects of deuteride precipitation from Ta-D solid solution was investigated by vibration frequency and internal friction measurements in the temperature range (80-300) K. Initiation of deuteride precipitation is indicated by the inflexion of frequency curve, whilst the onset of the dislocation loops formation around the deuteride particles is correlated with the internal friction growth (precipitation peak). The dislocation interstitial drag model which seems to better account for the precipitation peak, is discussed. The phase diagram of the Ta-D system was determined in the range (0.2 - 7.5) at%.

1. Introduction

In these last years a relaxation effect in the binary alloys Ta-H [1] , V-H [2, 3] Nb-H [4, 5] , and V-D [6] has been extensively studied. This effect called "precipitation peak", appears when hydrogen (deuterium) precipitation from the solid solution occurs ($\alpha \to \beta$ transformation) and, if generated, persists also at temperatures higher than the transformation temperatures. So far it has not been satisfactorily clarified which physical mechanism causes the precipitation peak. Previous papers [3-5] have proposed as more reliable loss mechanisms: -i) the hydrogen (deuterium) long range diffusion through precipitates; ii) the interaction between interstitial hydrogen and dislocations generated by the precipitates (hydrogen cold-work peak); iii) the thermal kink generation on dislocation produced during precipitation (α-peak).

Besides, since experimental evidence shows that ductile-brittle transition is correlated to hydride formation [7-9] , this relaxation effect has become of great interest also in the metallurgical field. However it is not yet known whether the embrittlement does or does not coincides with the initiation of hydride precipitation.

Because the internal friction mechanism is associated with hydride formation, it has turned out to be a valid tool for determining the phase diagram of transition metals-hydrogen (deuterium) systems. The phase diagram of the Ta-D system was determined by equilibrium pressure and resistivity measurements in the high concentration range [10] , but at present no data are available at low concentration.

The purpose of present work was to acquire further information on the precipitation peak by internal friction and elastic modulus measurements. In addition the phase diagram of the tantalum-deuterium system was obtained in the range where there are no experimental data.

2. Experimental procedure

The samples used in this work were cut from 99.9% pure tantalum plate supplied by Haynes. The shape was that of circular plates of 36 mm diameter and 3 mm thickness. After machining, the specimens were annealed in an ultra high vacuum furnace at 2100°C

(+) Present adress: University of Surrey, Department of Metallurgy and Materials Technology, Guildford, England.

and 5×10^{-7} torr for 3 h. The oxygen and nitrogen content of the samples, measured after the thermal treatment by the height of the Snoek peak [11] , were always less than 350 at. ppm and 25 at. ppm respectively.

In order to dope the samples with deuterium, they were first chemically polished to remove surface oxide and then heated to 550°C for 15 h in an atmosphere of 99.5% pure deuterium at pressures varying from 0.3 to 600 torr. Finally the system was rapidly cooled to room temperature to maintain the conditions of equilibrium concentration and homogenization reached. The deuterium contents, ranging from 0.2 to 8 at % were estimated from the reduction in sample weight after the final outgassing treatment. The internal friction and vibration frequencies were measured from 70-350 K at a cooling rate of 0.2 K/min in a vacuum apparatus. The vibration frequencies were included between 14 KHz and 81 KHz and the accuracy was always better than 5×10^{-6}.

3. Results

The 1st cooling and 1st heating curves of internal friction Q^{-1} and vibration frequency f of the sample with 7.5 at % deuterium, are shown in FIG. 1 together with the Q^{-1} curve of outgassed material.

After charging, the dissipation coefficient of the cooling curve is low and comparable with the background until deuterium is in solid solution and exhibits a sharp change at a temperature T_t. Below T_t the curve reveals an increasing precipitation peak and a shallow peak (at about 130 K) due to deuterium in solid solution [12] . The Q^{-1} heating curve displays a more developed precipitation peak whose intensity remains high also above T_t.

The frequency cooling curve exhibits a linear behaviour above T_t and at this temperature undergoes an inflexion. The heating curve goes below the cooling curve values but reassumes its linear behaviour and the same slope at a temperature higher than T_t with an hysteresis of about 20 K which reduces to few degrees after 9 thermal cyclings.

A small temperature increase of the sample at T_t, was also noted, during the 1st cooling, as the slopes of Q^{-1} and f curves reveal.

FIG. 2 illustrates a sequence of cooling curves for a sample with concentration 7.5 at %. After each cooling, the peak of internal friction curves increases and, at the same time, the temperature of sharp change shifts towards higher values. After the sample was cycled 9 times, the damping reached a value close to saturation and no sharp change was then observed in the Q^{-1} curve.

FIG. 3 gives the relative difference of the vibration frequency $\Delta f/f$ of the Q^{-1} curves of FIG. 2 as a function of maximum damping Q_m^{-1}. The relative difference $\Delta f/f$ was evaluated by the frequency value at the temperature T_t and that obtained at the same temperature by a linear extrapolation from the low temperature. In the same figure the theoretical curve $\Delta f/f = 1/\gamma \cdot Q_m^{-1}$ is plotted for $\gamma = 1$ (single time relaxation process) and for $\gamma = 0.5$, where γ is the spectrum width parameter introduced by Fouss and Kirkwood [13] .

In FIG. 4 the relative difference of frequency $\Delta f/f$ shows a linear dependence upon the deuterium content. The specific relative variation of frequency is $\frac{\Delta f/f}{D/Ta} = 3.7 \times 10^{-4}/1$ at %.

The curves of the 2nd and 3rd cooling of FIG. 2 are shown in FIG. 5 together with their respective vibration frequencies. The value of frequency departs from the linear behaviour at a temperature higher than that of the corresponding internal friction sharp change. This fact was observed for all cooling curves subsequent to the 1st one.

FIG. 6 shows the 1st cooling curve of Q^{-1} and vibration frequency of a sample containing 1.2 at % deuterium. The Q^{-1} sharp change which presents at 155 K is hardly appreciable, whilst the frequency inflexion is well marked. The peak at 133 K is due to deuterium in solid solution [12] .

FIG. 7 shows the dependence of the vibration frequency upon the precipitation peak temperature T_m for the Ta-D (1 at.% deuterium content) and the Ta-H [1] systems. The activation energy and the attempt frequency of the Ta-D system are $W_D = (0.36 \pm 0.09)$ eV and

$f_{oD} = 3.2 \times 10^{12}$ Hz respectively. In the same figure the quantity $T_m Q_m^{-1}$ is plotted as a function of the reciprocal temperature of precipitation peaks relative to the 1st heating curves, in samples with the same vibration frequency and with deuterium concentrations ranging from 0.2 to 7.5 at.%. All the data of FIG. 7 refer to samples with about the same oxygen and nitrogen content.

In FIG. 8 is represented the effect of annealing on the temperature and intensity of the precipitation peak in a sample with 0.74 at.% deuterium content. That the concentration remained constant was checked by weighting the specimen before and after thermal treatment.

The temperatures at which the 1st cooling curves of internal friction and frequency sharply increase and the corresponding values of deuterium contents permit the coexistence line to be drawn between the α and β phases (FIGS. 9 and 10). The slope of the curve shown in FIG. 10 gives the molar heat of formation for the deuteride in tantalum.

4. Discussion

In earlier investigations [1-6] the sharp change of damping of transition metal - hydrogen (deuterium) alloys during the 1st cooling, was attributed to the $\alpha \rightarrow \beta$ transformation. This statement was in agreement with the results obtained by the heat capacity [14], metallographic experiments [8], resistivity [15] and neutron scattering [16]. So the temperature T_t where the 1st cooling curve of damping and the vibration frequency curve show the sharp change (FIG. 1), can be interpreted as the temperature for initiation of deuterium precipitation from solid solution, even if a supersaturation effect in the virgin crystal cannot be excluded.

The 1st heating curve of frequency becomes again a straight line above T_t. This linear behaviour is typical of the system in α phase, as we observed down to 130 K, and indicates that the inverse $\beta \rightarrow \alpha$ transformation has completely occurred above this temperature during heating. This hypothesis is also supported by the subsequent Q^{-1} sharp changes of the 2nd and 3rd cooling curves of FIG. 2, which reveal that precipitation reoccurs. The above interpretation is in agreement with resistivity measurements in the Nb-H system made by Westlake [9]. This author asserts that complete dissolution of hydride precipitates has occurred during heating when the slope of resistivity heating curve has recovered that of the cooling curve in α phase. Also Sherman et al. [8] found by metallographic observations in the V-H system, that all hydrides revert by slowly heating the alloy at temperatures a little above the transformation temperature.

The lowering of the frequency heating curve (FIG. 1) in the solid solution temperature range, is a consequence of plastic strains introduced by precipitation [17, 18] and is characteristic of several cold-worked metals [19].

The relaxation process cannot account for the observed values of the relative difference of frequency (FIG. 3) because $\Delta f/f$: - i) approaches a value different from zero when the intensity of the precipitation peak approaches zero ($[\Delta f/f]_{Q_m^{-1} \rightarrow o} = 270 \times 10^{-5}$); ii) is much higher than the theoretical curve for a single relaxation time $\gamma = 1$. This discrepancy persists even if one takes $\gamma = 0.5$, which is the mean spectrum width parameter of our experimental Q^{-1} curves. In addition $\Delta f/f$ is a quantity directly proportional to deuterium concentration (FIG. 4). Thus is must be inferred that deuteride precipitation is the only one responsible for the relative change of frequency $[\Delta f/f]_{Q_m^{-1} \rightarrow 0}$. The slope of the observed curve $\Delta f/f$ is the same as that of the theoretical curve with $\gamma = 1$, within experimental error. This fact indicates that though the precipitation peak is a complex relaxation process, it does not contribute to the modulus defect as an effect with a spectrum width parameter $\gamma = 0.5$. At present no satisfactory explanation can be given to this behaviour in the light of present formal theory of relaxation processes. However one is induced to believe that the relaxation times τ_i of contributing processes in which the precipitation peak can be resolved do not follow the Arrhenius relation or, if they do, they do not compose linearly (superposition principle).

The departure of frequency from linear behaviour of the curves of FIG. 5 begins at temperatures where the Q^{-1} curve does not yet reveal the sharp change. Again the relaxation process cannot account for this departure. For instance, for curve 1 at the temperature of T_t = 248 K, where Q^{-1} begins to increase abruptly from 3.6 x 10^{-5}, the frequency has already departed 3 Hz from the linear behaviour, while the frequency increase corresponding to a relaxation maximum with the same intensity is only 0.6 Hz. Even if one takes the value of the peak (9.5 x 10^{-5}) and considers that the corresponding total frequency increase completely occurs at the temperature T_t, Δf should be 1.5 Hz. Thus we suggest that the frequency inflexion starts at the temperature of initiation of precipitate particles formation, whereas the Q^{-1} sharp whange occurs at a lower temperature, when the deuteride particles have reached a certain critical size necessary to generate dislocation loops [20]. In fact, because the vibration frequency depends in a sensitive way, through the elastic constants, on lattice simmetry, it is to be expected that different symmetry particles which separate in the crystal matrix modify the vibration frequency. Also the decrease of the amount of deuterium in solid solution during the deuteride enucleation might contribute to the frequency inflexion. But this second effect, which is strongly dependent on the lattice orientation [21], should give rise in a polycrystalline sample to a frequency deviation opposite to that observed. Thus, it seems that of the two opposing effects, the former is predominant. The fact that, only in the 1st cooling curves (FIG. 1) the temperatures of the sharp increase of Q^{-1} and f coincide, can be explained by supposing that the critical size of the misfitting particle is very low in the virgin crystal and increases with increasing crystal damage.

Lastly it should be emphasised that the frequency is also sensitive to precipitation when internal friction does not show any sharp change, either because it has reached a saturation state (9th cooling FIG. 2) or because deuterium concentration in the sample is too low (FIG. 6). In addition it must be noted that whilst Q_m^{-1}, for a given concentration, strongly depends on thermal hystory of the sample, $\Delta f/f$ does not change appreciably with thermal cycling.

The present experimental results indicate that among the different mechanisms proposed for the precipitation peak, the model of the long range diffusion through the precipitates is not reliable, the internal friction peak persisting also in the temperature range where deuterium precipitates must be dissolved. Rather, the precipitation peak is directly related to dislocations which are generated by precipitation. This is proved from the fact that after each cooling: - i) Q_m^{-1} increases (up to a saturation value) because the dislocation density increases after every precipitation; ii) the temperature of Q^{-1} sharp change and of frequency inflexion; shift towards higher values, since generated dislocations constitute centers upon which deuteride nucleation occurs more readily than it does in the virgin or less damaged matrix. A similar shift of the transition temperature during subsequent coolings was observed by resistivity measurements in the Nb-H alloys [9].

Between the two dissipation mechanisms involving dislocations, the model based upon thermal generation of kinks on punched-out dislocations (α-peak) does not seem to explain sufficiently the precipitation peak, because it is in contrast with the following experimental observations: - i) the activation energy of the precipitation peak deduced from the shift of T_m with frequency (FIG. 7), is 0.36 eV. This value is higher than that of the α-peak in tantalum (W_α = 0.25 eV) [22]. The same fact was also observed by Chang and Wert in the V-H system [3]; ii) the plots of vibration frequency versus T_m^{-1} of the Ta-D and Ta-H systems (FIG. 7) do not coincide. This is not explicable if only the dislocations are responsible for this relaxation process; iii) cold worked polycrystalline tantalum without hydrogen and deuterium, with a content of oxygen and nitrogen comparable with our own samples, exhibits peaks placed at a lower temperature range for the same vibration frequency [23].

Schoeck's theory [24] gives for the maximum damping Q_m^{-1}, the mean relaxation time τ and the effective diffusion coefficient D of dragged impurity, the following relations :-

$$ Q_m^{-1} \propto \rho \, l_o^2 \quad , \quad \tau \propto \frac{T \, C_d \, l_o^2}{D\,(T)} \tag{1} $$

$$D = D_o \exp \left(- \frac{E_b + E_d}{k\,T} \right) \tag{2}$$

where ρ is the dislocation density, l_o the mean length of dislocation segments, C_d the concentration of impurity along dislocations, E_b the binding energy of interstitial to dislocation and E_d the activation energy for diffusion of impurity in a perfect crystal.

Using relations (1) and (2) the following expression is obtained:

$$T_m \cdot Q_m^{-1} \propto \frac{\rho}{C_d} \backsim \exp \left(- \frac{E_b + E_d}{k \cdot T_m} \right) \quad . \tag{3}$$

Since, in the concentration range of the present work (0.2 - 7.5 at.%), both ρ and C_d increase with the total concentration it may be assumed that ρ/C_d is a constant. Thus relation (3) gives, for samples vibrating at the same frequency, the activation energy of precipitation peak $W_D = (0.43 \pm 0.08)$ eV. This value, obtained from the plot of FIG. 7, is the same within experimental error as that deduced by the plot f versus T_m^{-1}. In addition relation (3) also explains the observed concomitant increase of T_m and Q_m^{-1}. The non-coincidence in the plots f versus T_m^{-1} of the Ta-D and Ta-H systems (FIG. 7), can be explained in Schoeck's model, by the different values of the diffusion coefficient of impurities dragged by dislocations.

Lastly the decrease of T_m and Q_m^{-1} with annealing in a sample with constant deuterium concentration (FIG. 8) can find its explanation in relations (1) where the mean length of dislocation segments l_o decreases as a consequence of oxygen and nitrogen migration to dislocations.

In the present work the coexistence line between the α and β phases of the Ta-D system was determined in the low concentration range. Pryde et al. [10] reported the phase boundary curve obtained for the same system at higher deuterium concentrations by measurements of equilibrium pressure and electrical resistance (FIG. 9). These authors found that the coexistence line of the Ta-D system coincided with that of the Ta-H system previously investigated by them. This circumstance is not observed at lower concentration, as it can be seen in FIG. 10, where are reported the Ta-D and Ta-H [1] solvus curves obtained from samples having about the same oxygen and nitrogen content.

5. Acknowledgements
The authors wish to express their gratitude to Prof. A. Seeger for valuable remarks. They also thank Prof. M. Nuovo for helpful discussions, S. D'Angelo and P. Rossi for their technical assistance.

6. References
1. G. CANNELLI, F. M. MAZZOLAI : Nuovo Cimento 64 B, 171 (1969)
2. G. CANNELLI, F. M. MAZZOLAI : J. Phys. Chem. Solids 31; 1913 (1970)
3. H. Y. CHANG, C. A. WERT : Int. Conf. on Hydrogen in Metals, Julich 2, 558 (1972)
4. O. BUCK, D. O. THOMPSON, C. A. WERT : J. Phys. Chem. Solids 32, 2331 (1971)
5. O. BUCK, D. O. THOMPSON, C. A. WERT : J. Phys. Chem. Solids 34, 591 (1973)
6. G. CANNELLI, F. M. MAZZOLAI : Appl. Phys. 1, 111 (1973)
7. B. LONGSON : The Hydrogen Embrittlement of Niobium, TRG Report Nr. 1035c (1966)
8. D. H. SHERMAN, C. V. OWEN, T. E. SCOTT : Trans. TMS-AIME 242, 1775 (1968)
9. D. G. WESTLAKE : Trans. TMS-AIME 245, 287 (1969)
10. J. A. PRYDE, I. S. T. TSONG : Trans. Faraday Soc. 67, 297 (1971)
11. R. W. POWERS, M. V. DOYLE : J. Appl. Phys. 30, 514 (1959)
12. G. CANNELLI, L. VERDINI : Ric. Sci. 36, 246 (1966)
13. R. M. FUOSS, J. G. KIRKWOOD : J. Chem. Phys. 63, 385 (1941)

14. T. R. WAITE, W. E. WALLACE, R. S. CRAIG : J. Chem. Phys. Solids $\underline{24}$, 634 (1956)
15. D. G. WESTLAKE : Trans. TMS-AIME $\underline{239}$, 1341 (1967)
16. R. L. ZANOWICK, W. E. WALLACE : J. Chem. Phys. $\underline{32}$, 2059 (1962)
17. N. DAHLSTROM, C. C. DOLLINS, C. WERT : Acta Met. $\underline{19}$, 955 (1971)
18. N. E. PATON, B. S. HICKMAN, D. H. LESLIE : Met. Trans. $\underline{2}$, 2791 (1971)
19. P. G. BORDONI, M. NUOVO, L. VERDINI : Nuovo Cimento 1st Suppl. $\underline{18}$, 55 (1960)
20. M. F. ASHBY, L. JOHNSON : Phil. Mag. $\underline{20}$, 1009 (1969)
21. J. BUCHHOLZ : Doktorarbeit, Technische Hochschule Aachen (1973)
22. R. H. CHAMBERS, J. SCHULTZ : Acta Met. $\underline{10}$, 466 (1962)
23. G. KNOBLAUCH : Diplomarbeit, Universität Stuttgart, Institut für theoretische und angewandte Physik (1972)
24. G. SCHOECK : Acta Met. $\underline{11}$, 617 (1963)

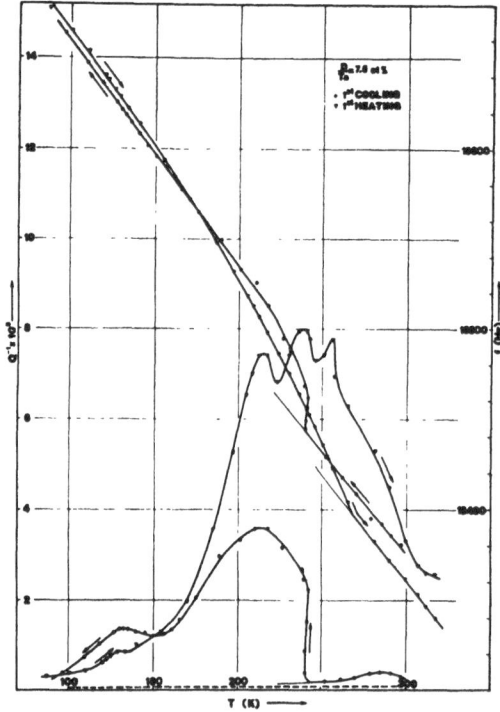

FIG. 1. Internal friction and vibration frequency of the sample with 7.5 at. % deuterium content during the 1st thermal cycling (cooling-heating). Dotted line refers to outgassed material.

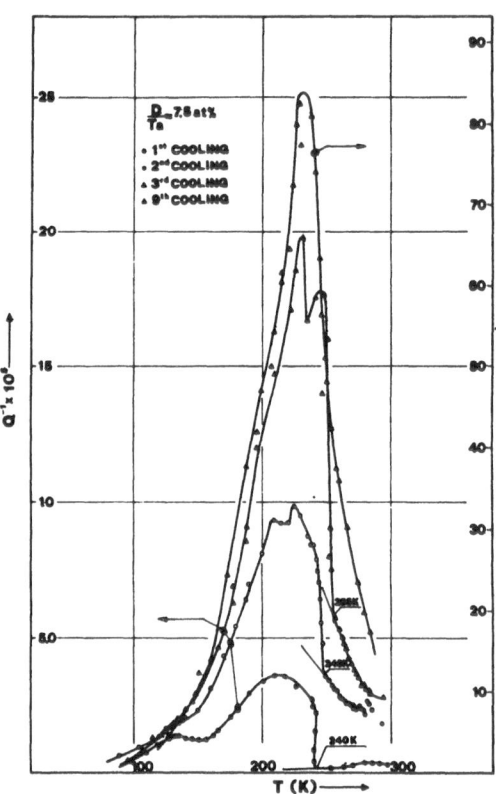

FIG. 2. Variation of peak height and shift of Q^{-1} sharp change of precipitation peak during subsequent coo - lings.

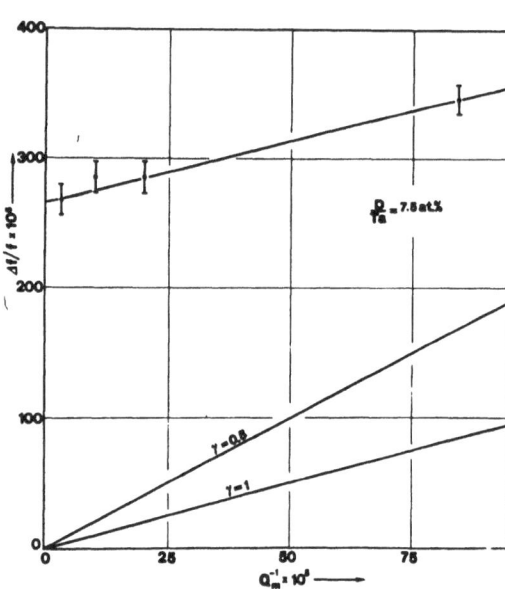

FIG. 3. Dependence of relative frequency difference on height of precipitation peak. Data refer to the curves of FIG. 2. The theoretical curve $\gamma \cdot \Delta f/f = Q_m^{-1}$ is plotted for $\gamma = 0.5$, $\gamma = 1$.

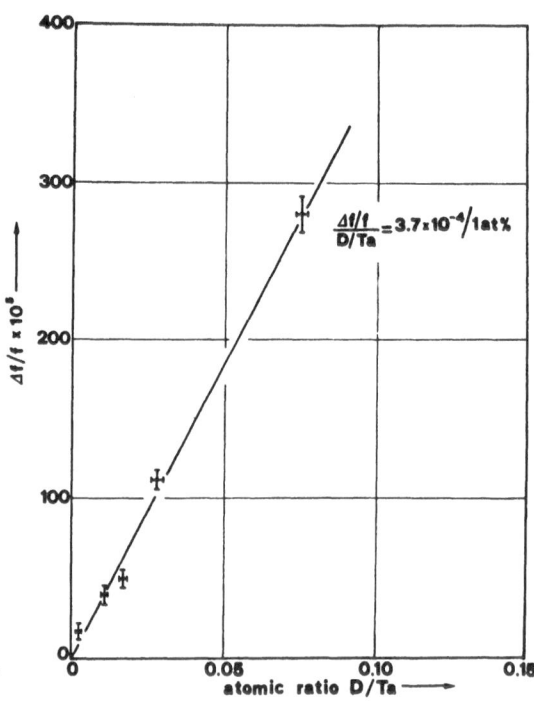

FIG. 4. Dependence of relative frequency difference on deuterium concentration.

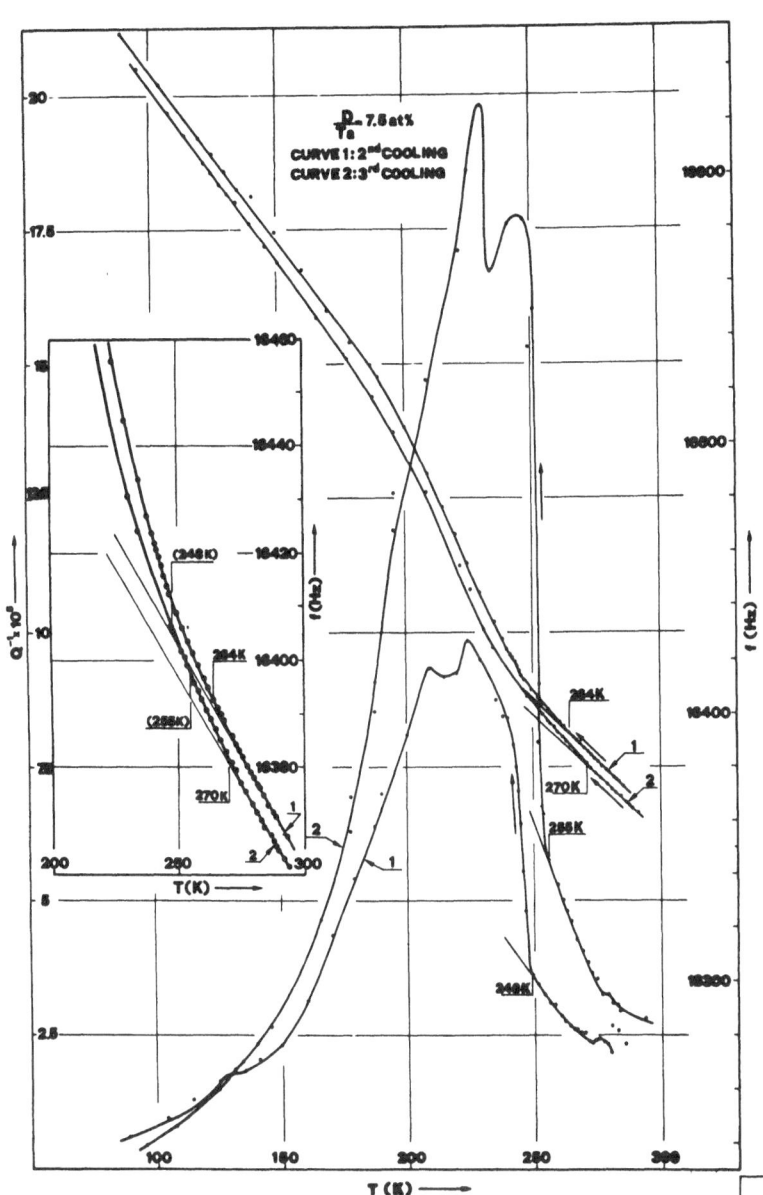

FIG. 5. Internal friction and
vibration frequency of the
2nd and 3rd cooling curves
of FIG. 2.

FIG. 6. Internal friction and vibration frequency of the sample with
1. 2at. % deuterium.

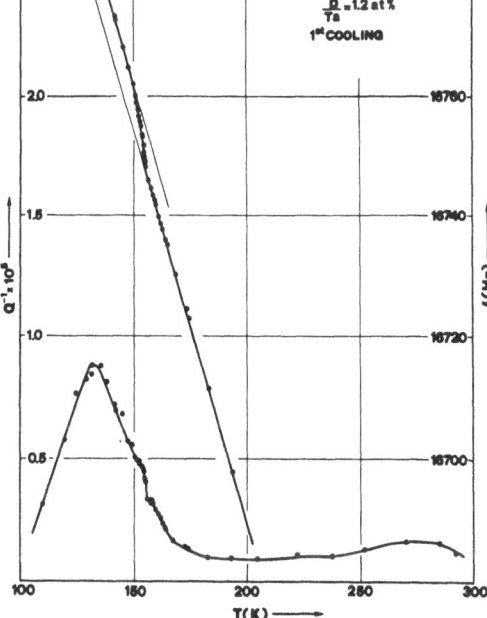

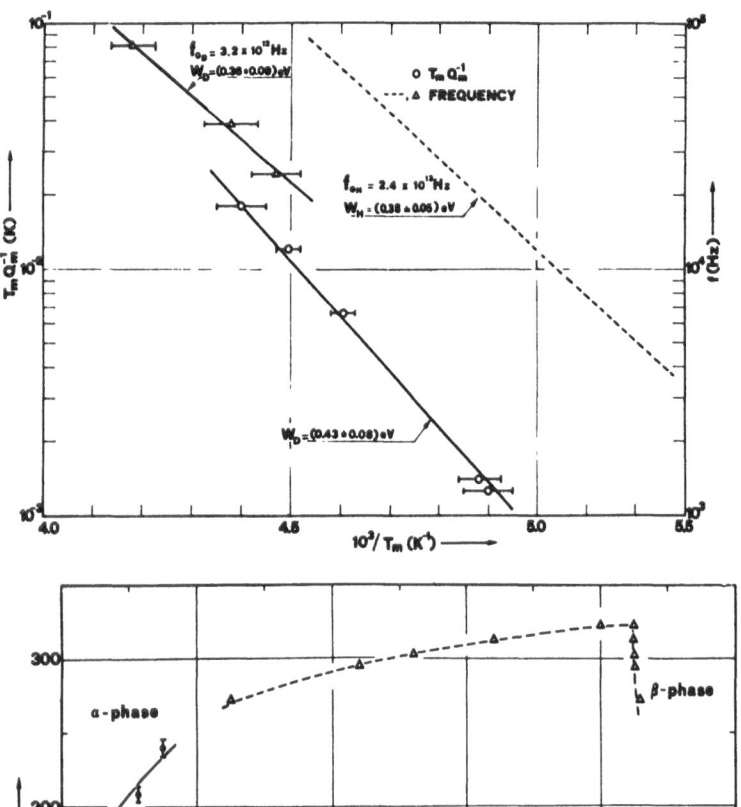

FIG. 7. Dependence of vibration frequency on reciprocal precipitation peak temperature for the Ta-D(△) and Ta-H[1](dotted line)systems. Plot of the quantity $T_m \cdot Q_m^{-1}$ as a function of reciprocal precipitation peak temperature(O) of the 1st heating curves .

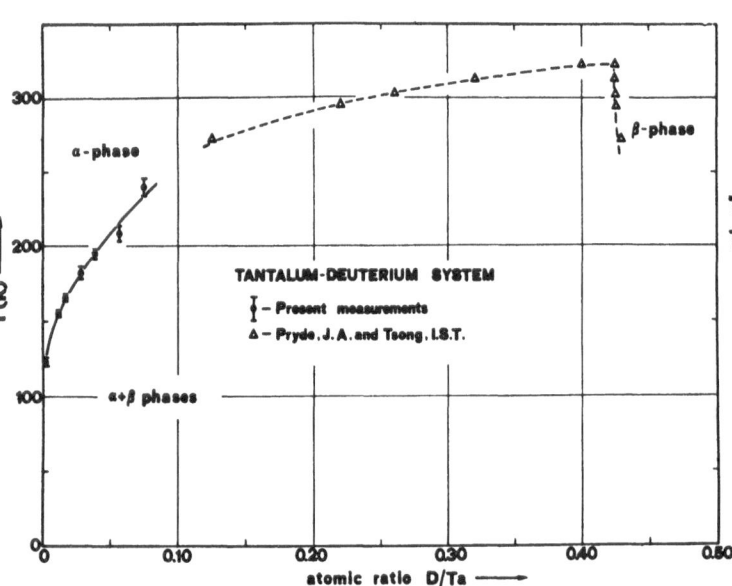

FIG. 9. Ta-D phase diagram at low (present work) and high [10] concentration.

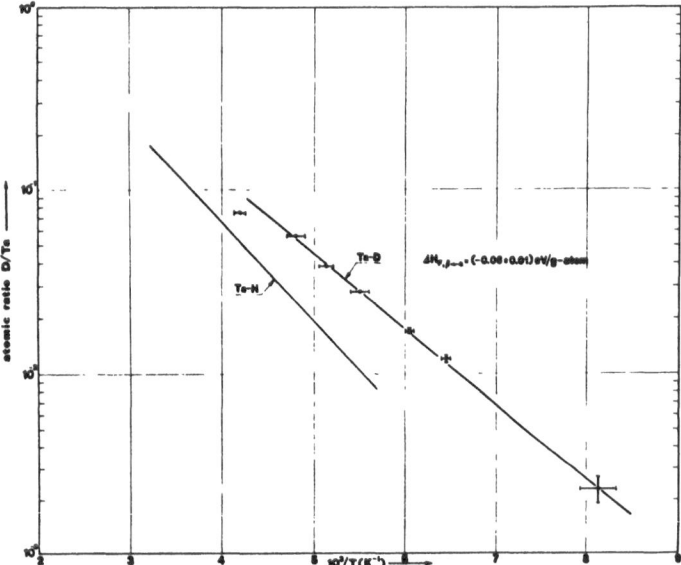

FIG. 8. Effect of annealing on the precipitation peak. Dotted line represents outgassed material.

FIG. 10. Solvus curves of Ta-D (present work) and Ta-H [1] systems.

THE α-PEAK IN VANADIUM

H. Y. Chang* and C. A. Wert
Department of Metallurgy and Mining Engineering
and
Materials Research Laboratory
University of Illinois at Urbana-Champaign
Urbana, Illinois 61801

Introduction

A model for the α-peak in the bcc metals has been extremely elusive. This peak, observed over a wide range of conditions, has been assumed by most authors to have its origin in dislocation damping, but many associated details are not well understood. Such factors as amount of deformation, the role of interstitials, irradiation effects, and effects of annealing have been difficult to evaluate.

The α-peak has been reported for several of the bcc metals, among them Nb, Ta, Mo, V and W (1-6). The damping peak, seen between 100°K and 200°K for usual frequencies of observation, initially increases in height with degree of deformation, but decreases again after a certain point. This behavior has usually been interpreted in terms both of the density of dislocations and their loop length. Most writers have, in fact, interpreted their results in terms of the Bordoni mechanism; a concise review has been given by Nowick and Berry (7). Their review does not consider impurity effects, however, and does not attempt an explanation of annealing effects.

The role of hydrogen seems important since it diffuses rapidly in the metals at usual temperatures of observation. Bruner indeed attributed the peak as being caused by an interaction of hydrogen atoms with dislocations (8). Later work supported that view (9), Cannelli and Mazzolai proposed that the peak in Ta, Nb, and V was the Snoek-Köster peak due to hydrogen (10). Other authors have doubted this view (11,12).

The present study had three goals:
1. To deduce the effect of interstitials on the α-peak.
2. To deduce the effects of annealing on the α-peak.
3. To relate a peak caused by the precipitation of hydrides to the α-peak.

*Department of Materials Science, University of Kentucky, Lexington, Kentucky

Material Used

The vanadium used in the work was wire drawn from rod of initial purity about 99.9%. It was annealed and outgassed by a high temperature vacuum treatment. The details of the experimental method are outlined in an earlier publication (13).

The α-Peak

The α-peak in the annealed and outgassed vanadium shows the characteristic behavior with deformation. Data for four levels of deformation are shown in Fig. 1. Three features are apparent: (1) The peak initially increases with deformation, then levels off. (2) The peak temperature shifts appreciably with increase in deformation. (3) The peak broadens markedly at high deformation. The first two of these features show the peak to have two regions of behavior -- one below 20% deformation, the other above, see Fig. 2.

Measurements of activation energies emphasize the difference between the regions of low and high deformation. For the low region the activation energy is about 3400 cal/mole, for the high, 6750. Thus the region of high deformation has an activation energy a little higher than those reported for other bcc metals, the low region a little lower, see Table I. For both peaks in V, the frequency factor is much lower than for Nb, Ta, Mo and W (7,14).

Effect of Hydrogen on the α-Peak

Hydrogen additions modify the α-peak. Data are shown in Fig. 3 for a specimen deformed 14% and subsequently charged with hydrogen to various levels. Clearly the relaxation spectrum is altered appreciably -- the peak shifts to higher temperatures and increases in height; both effects saturate at about 1000 at. ppm of hydrogen. We label this altered peak the Type I

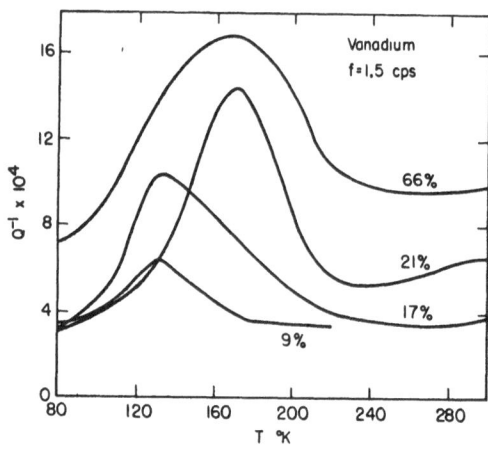

FIG. 1
The α-peak in vanadium.

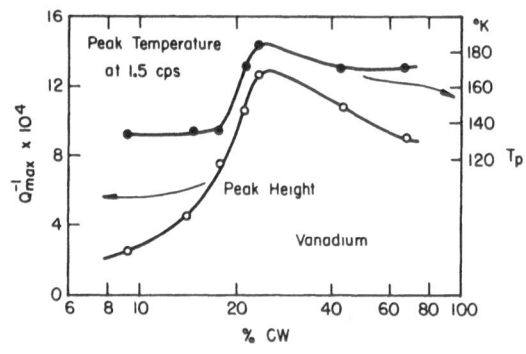

FIG. 2
Peak heights and position for α-peak as a function of deformation.

TABLE I
Parameters for the α-Peak in the bcc Metals.
$f = f_o e^{-Q/RT}$. First Four Entries Are from
Reviews of Nowick and Berry (7) and of
Chambers (14).

Metal	f_o (sec^{-1})	Q (ev)
Nb	10^{12}	0.25
Ta	10^{12}	0.25
Mo	10^{11}	0.18
W	10^{10}	0.21
V (14%CW)	10^6	0.15
V (23%CW)	10^8	0.29

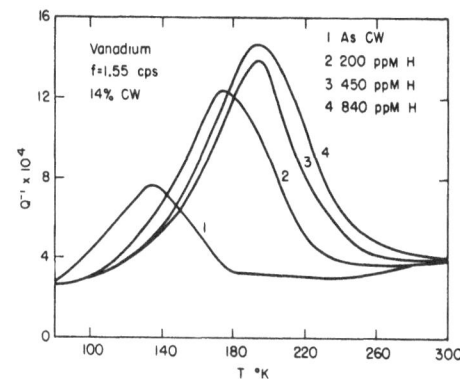

FIG. 3
Effect of hydrogen
on α-peak.

Snoek-Köster peak. It has the relaxation equation (for curve 3 at 450 at.
ppm H)

$$f = 1.3 \times 10^{10} \exp (0.38 \text{ ev}/RT).$$

Effect of Interstitial Oxygen and Nitrogen

The α-peak for both the low and high regions is suppressed by the presence of oxygen and nitrogen in solution. The effect for oxygen is shown in
Fig. 4. (The symbols VXO or VXH signify vanadium with X000 at. ppm of oxygen or hydrogen.) Clearly the α-peak is nearly suppressed by the presence
of 12,000 at. ppm of oxygen; nitrogen in the amount of 8000 at. ppm also
suppresses the peak at this level of deformation. Data for the suppression
of the peak for 14%CW may be seen in earlier publications (13,15).

Materials for which the α-peak is suppressed by the presence of oxygen
and nitrogen display a peak if hydrogen is added. An example is shown in
Fig. 5 for a specimen containing 5000 at. ppm of oxygen and cold worked 14%.
The initial measurement with no hydrogen addition is shown by curve 1; the
peak is suppressed. Hydrogen addition causes a peak to appear, the height
of which saturates at about 2000 at. ppm of hydrogen (curves 2-5). We
designate this restored peak as the Type II Snoek-Köster hydrogen peak.
Measurement of the frequency dependence of this peak yields the parameters
(for 520 at. ppm, i.e., curve 3) $f_o = 3 \times 10^{16}$ and Q = 0.55 ev. This is an
extremely large frequency factor, but not unique.

An effect similar to this has been reported by Cannelli and Mazzolai
(10). They observed no α-peak after deformation but a peak developed after
hydrogen addition. Their values of f_o and Q were 3×10^{17} per sec and 0.52 ev,
values fairly close to ours. They called this peak the "hydrogen cold work
peak" and concluded that no α-peak exists after deformation unless hydrogen
is present. We conclude that their specimen contained enough dissolved

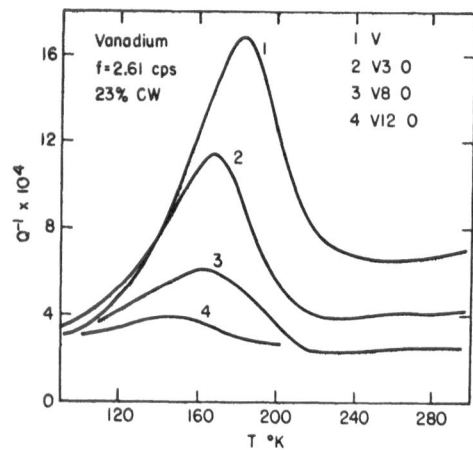

FIG. 4
Suppression of α-peak by oxygen.

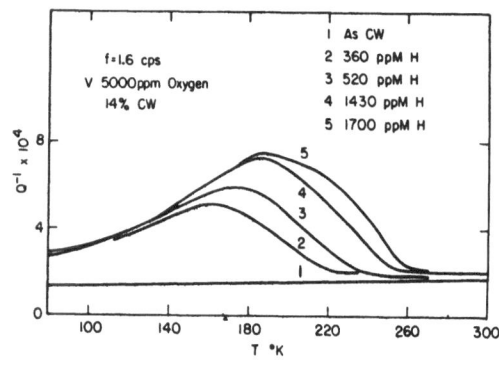

FIG. 5
Growth of a Snoek-Köster peak
by addition of hydrogen.

oxygen and nitrogen to suppress the α-peak initially (16) and that hydrogen addition produced the same peak we show in Fig. 5, a Snoek-Köster peak due to hydrogen.

Annealing Away of the α-Peak

The annealing of deformed vanadium in the temperature range 200°C to 300°C causes the α-peak to disappear. An example is shown in Fig. 6 for a specimen containing a small amount of oxygen. We believe that annealing in this temperature range causes segregation of oxygen to the dislocations, pinning them firmly. Bruner showed this also for the annealing of deformed niobium (8). We find, furthermore, that hydrogen additions do not, in this instance, cause development of a Type II Snoek-Köster peak.

Peak Caused by Precipitation of Hydride

A peak caused by the precipitation of hydrides has been observed for V (as well as for Nb). It is thought to have its origin in the dissipation of energy by dislocations generated in the metal during formation of inco- herent hydrides. Thus the peak should have the same features as the Type I Snoek-Köster peak.

A set of measurements showing the peak for different hydrogen levels is shown in Fig. 7. These are all cooling measurements; the abrupt in- crease in damping shows the temperature of the solvus line at that compo- sition. The height increases with hydrogen content, but levels off at the higher hydrogen contents. Activation energy measurements as have been at- tempted by us, by Owen and Scott (17) and by Butera and Kofstad (18) yield values around 0.4 ev. This value is about the same as that found by us for the Snoek-Köster peak of Type I.

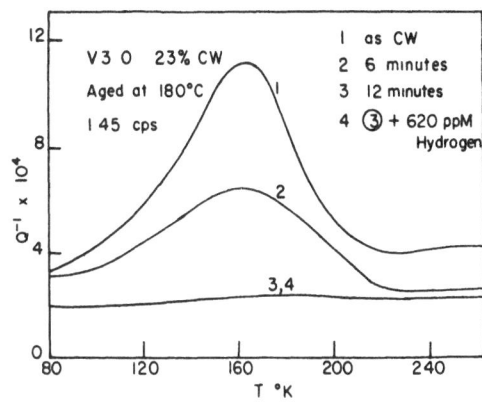

FIG. 6
Effect of annealing on α-peak.

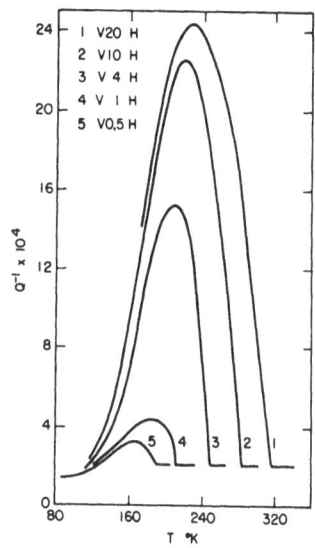

FIG. 7
Precipitation peak.

We have found differences in the solvus temperatures determined by us, by Westlake (19) by Scott and colleagues (17,20) and by Cannelli and Mazzolai (10). We attribute these differences to variation in dissolved oxygen and nitrogen impurities between the several studies; a paper describing our interpretation is in press (21).

Summary and Discussion

We believe that the α-peak (as described in many earlier papers) is not a single, unique peak. Instead it may vary with the presence of hydrogen, oxygen and nitrogen. We see from the preceding observations and references to the literature that many of the previous discrepancies can be reconciled if careful control is established of the impurity levels.

The peak in deformed vanadium of the highest purity we could achieve we call the α-peak. Two regions of deformation seem to exist for this peak. Frequency factors and activation energies for these two regions are shown by the first two lines of Table II.

The presence of immobile interstitials (oxygen and nitrogen) in solid solution suppresses the α-peak, apparently by locking the dislocations. Only a small concentration of randomly dispersed interstitial is required -- a thousand atomic ppm. A damping peak can be restored by low temperature hydrogen addition, a peak whose damping mechanism seems likely to be the dragging of a hydrogen atmosphere in the small motion left to the pinned dislocations. This damping peak -- termed the Snoek-Köster peak of Type II -- has a much higher frequency factor and a large activation energy, see lines 4 and 5 of Table II.

Annealing of deformed specimens in the temperature range 200°C to 300°C permits large scale diffusion of interstitial oxygen or nitrogen to the dislocations. These interstitials pin the dislocations so firmly at low temperature that they not only suppress the α-peak but also inhibit the formation of a Type II Snoek-Köster peak when hydrogen is added. Thus the anomalous annealing behavior referred to by earlier authors is explained.

The precipitation peak can now readily be explained. The formation of incoherent hydrides causes the generation of many dislocations. These are, of course, in the atmosphere of hydrogen in thermodynamic equilibrium with the hydride at the peak temperature. Consequently the damping peak originating from these dislocations should be the Snoek-Köster peak Type I. Indeed it seems to be; the rate parameters listed in Table II lines 6 and 7 agree well with those of the line 3 of Table II.

Finally, we caution against annealing of the bcc metals in the temperature range 1000°C to 1200°C. Even in the best vacuums available, oxygen and nitrogen pick up is inevitable. If these interstitials at levels of a few thousand ppm influence the property to be observed, the result of the annealing treatment may be to complicate interpretation of the observations.

TABLE II
Parameters for Various Peaks Observed in Vanadium

	Peak Designation	f_o (sec^{-1})	Q (ev)	
1)	α-peak 14% CW	10^6	0.15	
2)	α-peak 23% CW	10^8	0.29	
3)	Type I S-K peak	10^{10}	0.38	
4)	Type II S-K peak	10^{16}	0.55	
5)	Type II S-K peak	10^{17}	0.52	Ref. 17
6)	Precipitation peak	10^9	0.36	
7)	Precipitation peak	10^{10}	0.41	Ref. 10

Acknowledgment

The authors wish to acknowledge that this research was supported in part by the U. S. Atomic Energy Commission Contract AT(11-1)-1198.

References

1. R. H. Chambers and J. Schultz, Acta Met. 8, 585 (1960).

2. R. DeBatist, phys. sta. sol. 2, 661 (1962).

3. R. H. Chambers and J. Schultz, Acta Met. 10, 466 (1962).

4. M. W. Stanley and Z. C. Szkopiak, J. Nucl. Mat. 23, 163 (1967).

5. D. R. Muss and J. R. Townsend, J. App. Phys. 33, 1804 (1962).

6. R. DeBatist, phys. sta. sol. 3, 1475 (1963).

7. A. Nowick and B. Berry, Anelastic Relaxation in Crystalline Solids, Academic Press, New York (1972).

8. L. J. Bruner, Phys. Rev. 118, 399 (1960).

9. F. M. Mazzolai and M. Nuovo, Solid State Comm. 7, 103 (1970).

10. G. Cannelli and F. M. Mazzolai, J. Phys. Chem. Solids 31, 1913 (1970).

11. P. G. Bordoni, M. Nuovo and L. Verdini, Phys. Rev. 123, 1204 (1961).

12. A. Seeger and B. Sestak, Scripta Met. 5, 681 (1971).

13. H. Y. Chang and C. A. Wert, Ber. Bunsen-Gesell. fur Phys. Chemie 77, 47 (1973).

14. R. H. Chambers, in Physical Acoustics (Ed.: W. P. Mason) 3A, Chapter 4, Academic Press, New York (1966).

15. H. Y. Chang, Ph. D. Thesis, University of Illinois at Urbana-Champaign (1972).

16. R. Cantelli, F. Mazzolai and M. Nuovo, J. Phys. Chem. Solids 31, 1811 (1970).

17. C. V. Owen and T. E. Scott, Met. Trans. 3, 1682 (1972).

18. R. A. Butera and P. Kofstad, J. App. Phys. 34, 2172.

19. D. Westlake, Trans. AIME 239, 1341 (1967).

20. D. Shuman, C. Owen and T. Scott, Trans. AIME 242, 1775 (1968).

21. H. Y. Chang and C. A. Wert, Acta Met. In press.

INTERNAL FRICTION ASSOCIATED WITH SUBSTITUTIONAL-INTERSTITIAL
SOLUTE-ATOM CLUSTERS IN IRON-VANADIUM-NITROGEN ALLOYS AND
OTHER TERNARY NITROGEN-FERRITES

M. POPE[x], D.M. JONES AND K.H. JACK
CRYSTALLOGRAPHY LABORATORY, THE UNIVERSITY, NEWCASTLE UPON TYNE, ENGLAND

ABSTRACT   When iron containing strong nitride formers such as V, Ti, Mo and Cr is nitrided
under suitable conditions, mixed substitutional-interstitial solute-atom clusters
are produced as discs on the ferrite cube planes.   Although the internal friction
Snoek peak is then extremely small, and interaction peaks (due to stress-induced
diffusion of N between interstices in the vicinity of substitutional atoms) are
absent, relatively strong damping is observed at 150-220°C.   The peak temperature
depends upon the nitriding treatment and it is suggested that the effects arise
from the movement of the less firmly held N atoms at the edges of the clusters.

I.   Introduction

In ternary nitrogen-ferrites in which the substitutional and interstitial solutes
remain in "random" solid solution, damping peaks arising from the stress-induced diffusion
of nitrogen atoms in the vicinity of substitutional alloying atoms are observed in
addition to the well-known Snoek peak.   Fast & Meijering (1) and Dijkstra & Sladek (2)
observed the interaction peak in Fe-0.5a/o V at 88°C and 0.77 c.p.s.;   similar peaks were
also observed (2) in Fe-0.5a/o Mo and Fe-0.5a/o Cr with maxima at 75 and 47°C respectively.
These peaks can not be interpreted in terms of a single relaxation time and consequently a
number of relaxation processes are assumed to contribute to the overall damping.   Perry,
Malone & Boon (3) suggest that the peak in Fe-V-N alloys has three components while Ritchie
& Rawlings (4) analyse the equivalent peak in Fe-Cr-N into either five or six components
depending upon the Cr concentration.

In previous studies, "random" substitutional-interstitial solid solutions were produced
either by nitriding alloy-austenites at 950°C in $N_2:H_2$ mixtures and then quenching (5), or
by solution-treating at 950°C after a prior lower temperature nitriding in $NH_3:H_2$ (6).   By
nitriding at low temperatures (400-650°C) in controlled $NH_3:H_2$ gas mixtures, as in the
present work, alloy nitride formation is preceded by mixed substitutional-interstitial
solute-atom clustering.   The clusters, which contain iron as well as alloying element and
nitrogen, form as discs about 100Å in diameter and 10Å thick on $\{100\}_\alpha$ planes.   Overageing
occurs by the gradual rejection of iron from the cluster and by ordering of the solute atoms
to give the equilibrium nitride, e.g. CrN or VN.   In Fe-Mo-N the continuous ordering
transformation includes the successive formation of two intermediate metastable precipitates

[x] Now at Metallurgy Department, University of Strathclyde

$$\text{Fe-Mo-N cluster} \rightarrow \alpha'' \text{-(Fe,Mo)}_{16}N_2 \rightarrow \gamma'\text{-(Mo,Fe)}_2N$$

after which the equilibrium δ-MoN forms by discontinuous precipitation at grain boundaries. The internal friction characteristics associated with these previously unrecognised stages of homogeneous precipitation are described in the present paper for several ternary Fe-X-N systems.

II. Experimental

High-purity alloys were received as hot-rolled 12 mm diameter bar and subsequently drawn to 0.55 mm diameter wire.

Damping was measured on 200 mm long specimens in a torsional pendulum assembly similar to that described by Williams & Leak (7) using shear strains $\geqslant 2 \times 10^{-5}$. All specimens were first annealed in pure hydrogen for 4h at 850°C, furnace cooled to the nitriding temperature, and then nitrided at 400-650°C in purified $NH_3:H_2$ mixtures under precisely controlled conditions to give the required nitrogen concentration in solid solution without forming iron nitrides. After nitriding, specimens were quenched into cooled oil and further cooled to -60°C until required for damping measurements.

Accurate unit-cell dimensions of the ferrite matrix, essential for differentiating between clusters and nitride precipitates, were obtained by a Nelson-Riley extrapolation (8) from data obtained with CoKα X-radiation and either 114.8 mm diameter Philips or 90 and 190 mm diameter Unicam powder cameras.

III. Results

III.1 Microstructure

Nitrided alloys containing small concentrations of substitutional solutes such as Mo (9), Cr (10) or V (11) show large increases in hardness and nitrogen concentrations which often exceed those corresponding to the complete formation of the equilibrium alloy nitride; see Table 1. Electron and field-ion micrographs show extremely thin disc-shaped structures on ferrite cube planes which give continuous diffraction streaking in $\langle 100 \rangle_\alpha$ directions; see Figure 1. The observations are indistinguishable from those associated with GP zone formation in Al:4w/o Cu and other f.c.c. alloys and recent work at Newcastle involving determination of the ferrite matrix unit-cell dimensions before and after nitriding (12) (see Table 1) shows conclusively that mixed substitutional-interstitial solute-atom clusters are present. The increases in unit-cell dimensions on nitriding are much greater than those expected by assuming alloy nitride precipitation together with equilibration between the ferrite matrix and the nitriding atmosphere but they are accounted for if almost all the nitrogen remains in solid solution and occupies the octahedral interstices of the ferrite lattice.

The unit-cell edge length of a substituted nitrogen-ferrite (Fe-X-N) is given by:

$$a_o = a_{Fe} + a_X + a_N \qquad \qquad \dots \ (1)$$

where a_{Fe} is the unit-cell dimension of pure α-iron (2.8664Å) and a_N and a_X are the increases caused by solution of nitrogen and the substitutional solute respectively; a_x is known from the dimensions of the annealed un-nitrided alloys compared with a_{Fe}. For calculation of a_N, extrapolation of data for N-ferrite (13) to the nitrogen concentrations encountered in the present work is of doubtful accuracy and more reliable values, based on the structures of α'-N-martensite containing up to 2.3w/o N and of α''-$Fe_{16}N_2$ (14, 15), are:

$$a_N = 0.007 \pm 0.0003\text{Å per N atom per 100 metal atoms}$$
$$= 0.0285 \pm 0.0015\text{Å per w/o N in iron and dilute iron alloys}$$

which, within experimental error, agree with Wriedt & Zwell (13). This dilation of ferrite due to nitrogen in solid solution has been used in equation (1) to calculate the lattice parameters of the nitrided alloys given in Table 1 and the agreement of the latter with observed values shows that most of the solute atoms must exist in solid solution (as clusters) and not as alloy nitride precipitates.

Fe-Cr alloys nitrided at 575°C are the exception; the observed lattice parameters are lower than those calculated for complete solid solution and indicate that nitride precipitation has occurred. Indeed the data for 1.3a/o Cr nitrided in $10NH_3$:$90H_2$ at 575°C for 26h can be accounted for by complete precipitation of CrN together with the nitrogen in solution expected for pure α-iron in equilibrium with the nitriding atmosphere. However, the unit-cell dimension observed for Fe-1.3a/o Cr nitrided in $19NH_3$:$81H_2$ at the lower temperature of 450°C for 245h is in excellent agreement with that calculated for Cr and N retained in solid solution as mixed substitutional-interstitial solute-atom clusters and confirms the suggestion (16) that clusters in Fe-Cr-N alloys are less stable than in Fe-Mo-N, Fe-V-N and Fe-Ti-N.

It should be emphasised that the unit-cell dimensions of a solid solution are essentially invariant with respect to the distribution of its solute atoms. A matrix containing clusters or GP zones is a solid solution in which non-random local atomic arrangements exist but its dimensions are the same as for a random solid solution having the same solute concentration. When precipitation occurs, the lattice parameter changes to that of the depleted solid solution. Thus, in the Al:Cu system the sequence of transformations on quench-ageing is:

$$\text{supersaturated solid solution} \rightarrow GP1 \rightarrow GP2(\theta'') \rightarrow \theta' \rightarrow \theta(CuAl_2)$$

and at the GP1 stage the lattice parameter a_o is the same as that of the initial random solid solution although the material properties are quite different. With the formation of θ'' and θ', a_o increases indicating depletion of Cu atoms ($r_{Cu} < r_{Al}$) showing that these phases are intermediate precipitates and not zones. For substitutional solutes the change in unit-cell dimensions is usually small but for N and C the increase in a_o is, as shown above, about 0.007Å per a/o of interstitial solute in b.c.c. iron; for f.c.c. iron it is 0.008Å. Unit-cell dimensions of the matrix, even when measured to only 1 in 5000, therefore provide a very sensitive method of determining whether nitrogen is in solution (randomly or as zones) or whether it is precipitated in a second phase; see Figure 2. On nitriding each of the Fe-V, Fe-Mo, Fe-Cr and Fe-Ti alloys the unit-cell dimensions increase by about the amount expected if all the absorbed nitrogen goes into solid solution, yet from the electron micrographs and from the hardness of the material the solute-atom distribution can not be random.   By contrast, on nitriding Fe:W alloys there is a rapid unit-cell dimensional decrease because the first-formed clusters transform almost immediately to a metastable α''-type Fe-W-N intermediate precipitate (16).

III.2 Internal friction of Fe-V-N alloys

The two significant features shown in Figure 3 of the damping characteristics of Fe:1.08a/o V alloys nitrided at 580°C in different $NH_3:H_2$ mixtures are:

    (i)   the absence of a significant Snoek peak at about 20°C;

and (ii)   the occurrence of high-temperature damping at 150-200°C.

The N concentration after nitriding in $10NH_3:90H_2$ is approximately 1.6a/o (see Table 1) the majority of which, as shown by lattice parameter measurements, is in solid solution. The low level of damping in this material therefore indicates that only a small proportion of the dissolved nitrogen diffuses under the applied torsional stress.   Reduction in the nitriding potential, and hence the concentration of dissolved nitrogen, leads to a smaller high-temperature peak and also to an increase in the peak temperature;   compare Figures 3a and 3b.   At the same time there are small microstructural changes;   clusters are clearly visible by transmission electron microscopy after 48h in $1NH_3:99H_2$ at 580°C whereas the clusters produced with $10NH_3:90H_2$ at the same temperature are too small to be resolved even after 256h.

At higher temperatures the effect of changes in the nitriding potentials on the damping is particularly marked.   At 640°C in $6NH_3:94H_2$ a high-temperature peak is observed at 200°C together with an appreciable Snoek peak (Figure 4a) and although the nitrogen content is not inconsistent with the formation of VN precipitates, the unit-cell dimension of 2.873Å shows that some solute-atom clusters are present;   from electron micrographs the clusters are 200-250Å in diameter.   The observation of a Snoek peak and the absence of a V-N interaction peak at 80-90°C is significant because it indicates that the concentration in "random" solid solution is extremely small.   On the other hand, when the same material is nitrided under conditions which precipitate only part of the V as coarse VN particles, e.g. $1NH_3:99H_2$ at 650°C for 44h (11) the V-N interaction peak is observed;   see Figure 4b.

III.3 Internal friction of Fe-Mo-N alloys

Hardness measurements and lattice parameter determinations (Table 1) confirm the presence of mixed solute-atom clusters in nitrided Fe-2.92a/o Mo alloys.   The damping characteristics of such alloys containing a fully developed GP zone structure (Figure 5a) are very similar to those of iron-vanadium-nitrogen alloys.   A high-temperature damping peak is observed at approximately 170°C accompanied by a Snoek peak which, although small in comparison to the total nitrogen concentration, is higher than in iron-vanadium alloys nitrided at 580°C.   Differences in Snoek peak height could arise from the different inter-cluster spacing of nitrided Fe-Mo and Fe-V alloys;   certainly Snoek damping in Fe-V-N alloys is a function of inter-cluster spacing.

The internal friction of partially nitrided material in which the GP zone structure is not fully developed is particularly complex. Damping is appreciable over the whole range 50-280°C suggesting the involvement of nitrogen environments intermediate between those responsible for the molybdenum-nitrogen interaction peak and those responsible for the high-temperature peak.

III.4 Internal friction of Fe-Cr-N alloys

The internal friction characteristics of Fe-1.3 a/o Cr and Fe-2.0a/o Cr alloys have been observed at three stages of the precipitation:

(i)   after partial nitriding and prior to the complete development of the zone structure;

(ii)   at the zone stage;

and (iii) in overaged material containing nitride precipitates.

Stage (i) is shown by Figure 6a; damping is complex, as in partially nitrided Fe-2.9a/o Mo alloys and the composite Snoek and chromium-nitrogen interaction peak is broadened on the high temperature side. In Fe-2.0a/o Cr nitrided in $13NH_3:87H_2$ at 500°C for 17h (see Figure 6b), damping is similar to that observed in Fe-V-N and Fe-Mo-N at the GP zone stage with the high-temperature peak at approximately 170°C.

Overageing occurs rapidly at 575°C and is accompanied by a coarsening of the microstructure although the nitride precipitates retain the disc-shaped morphology of clusters. High-temperature damping is still observed although the peak occurs at a higher temperature, 210-230°C depending upon the nitriding time; see Figures 6c and 6d. Such data suggest that the nitrogen atom environment responsible for high-temperature damping is present together with both solute-atom clusters and fine nitride precipitates.

III.5 Internal friction of Fe-Ti-N alloys

Cluster formation in Fe-Ti-N inferred from abnormally high N contents and hardnesses (17) has been confirmed (18) by lattice parameter measurements; see Table 1. The damping of Fe-0.69a/o Ti nitrided in $5NH_3:95H_2$ at 575°C for 24h to form clusters (see Figure 5b) is very similar to that in Fe-V-N, Fe-Mo-N and Fe-Cr-N at similar stages of the precipitation sequence. High-temperature damping occurs at approximately 150°C together with a small Snoek peak.

III.6 Hydrogen reduction

Heating previously nitrided alloys in hydrogen gives some indication of the state of combination of the nitrogen. Thus, nitrogen in clusters is removed more quickly than nitrogen more strongly combined in alloy nitride precipitates, and in Fe-Mo-N the nitrogen can be removed completely leaving only Fe-Mo clusters in the ferrite matrix (19).

In Fe-V-N, the clusters are not completely reduced by such treatment at 575°C but nitrogen depletion occurs until the N:V ratio falls to about 1:1 after 40h. The ferrite lattice parameter also decreases during the early stages of this treatment but even after no further N is lost the cell dimension is much higher than the value of 2.866Å expected for the complete precipitation of VN. Re-nitriding again increases the nitrogen concentration, the final value of which is inversely related to the prior hydrogen-reduction time (11).

The internal friction of Fe-1.08a/o V was observed at three stages in the nitriding/ hydrogen reduction/re-nitriding cycle at 580°C. The damping peaks present after the initial nitriding treatment disappeared completely after hydrogen reduction for 130h suggesting that the more strongly bonded N atoms in the clusters do not contribute to this damping. On subsequent re-nitriding the high-temperature peak is again observed, but it occurs at a higher temperature (190°C) and is accompanied by an increased Snoek peak. Growth of the clusters to about 150Å diameter during the hydrogen reduction and re-nitriding cycle probably modifies the N-atom environment responsible for the high-temperature damping and leads to this increase in peak temperature.

The very small Snoek damping in alloys containing clusters is thought to be due to the lattice distortion produced by strain fields; see Figure 1. The applied torsional stress is insufficient to overcome the internal strain and produce the periodic variations in the size of the octahedral interstices necessary for a stress-induced redistribution of the interstitial atoms at the Snoek peak temperature. A similar explanation has been proposed (20) to account for the absence of Snoek damping in carbon martensite. As the inter-cluster or inter-precipitate spacing increases, the volume fraction of strain-free ferrite increases and the stress-induced diffusion of nitrogen atoms in these regions is responsible for the increase in Snoek damping.

IV. Discussion

High temperature damping was first observed (21) at approximately 200°C in cold-worked iron containing interstitials and it was subsequently shown (22) that the peak varied from 160 to 200°C depending upon the degree of cold-work. High-temperature damping in carbon-martensite has also been demonstrated (20, 23). It is generally agreed that the necessary anelastic strain for these high temperature peaks arises from the motion of dislocations under stress in the presence of carbide or nitride precipitates or of interstitial atmospheres. However, this explanation seems unlikely to account for the high-temperature damping observed in the annealed specimens used in the present work; the observed presence of mixed substitutional-interstitial solute-atom clusters allows an alternative explanation.

It is suggested that the anelastic strain responsible for high-temperature peaks in fully nitrided alloys containing clusters arises from the motion of the least firmly held nitrogen atoms at the periphery of the clusters. Such a process is likely to occur with a range of relaxation times depending upon the exact location of the contributing nitrogen in or near the cluster. Similar peaks observed in alloys containing fine nitride precipitates arise from the stress-induced diffusion of nitrogen in precipitate strain fields and should also occur with a range of relaxation times. Further data about N-atom environments are required before the dependence of the peak temperature upon nitriding conditions can be understood but the model explains the very low level of damping in comparison with the total nitrogen concentration and also the peak shape. The ideal damping curve of a process occurring with a single relaxation time is given by:

$$Q^{-1} = Q_{max}^{-1} \cdot \text{sech} \left[\frac{\Delta H}{R} \left(\frac{1}{T} - \frac{1}{T_p} \right) \right] \qquad \dots \dots (2)$$

where ΔH is the activation energy of the process and T_p is the peak temperature.

Comparison of half-peak widths calculated from equation (2) with the corresponding experimental values indicates whether or not damping occurs with a single relaxation time. ΔH can be measured by observing the shift in peak-temperature with frequency or can be calculated using a method by Wert & Marx (24). Data from both techniques are given in Table 2. The peaks are seen to be considerably broader than predicted from equation (2) and so suggest that high-temperature damping occurs with multiple relaxation times.

Damping very similar to that observed in the present investigation has previously been reported without explanation by Köster & Horn (25) in nitrided Fe-0.75 w/o V. Their nitriding method, 20h in N_2:NH_3, followed by annealing for 20h in vacuum at 590°C, would certainly produce clusters and so their observations are explicable by the present proposals. It also seems probable that the peaks interpreted as Ti-N interaction peaks (26) in brittle Fe-Ti-N alloys are characteristic of material containing a high volume fraction of clusters.

The complex damping curves observed in partially nitrided material can not be accounted for solely by Snoek peaks, metal-nitrogen interaction peaks and high-temperature damping peaks all of which are found in fully nitrided alloys. The additional damping is presumably associated with groupings of substitutional and interstitial atoms intermediate between those of the fully nitrided clustered material and those of the "random" solid solution.

V.   Conclusions

(1) Measurements of ferrite unit-cell dimensions used in conjunction with electron microscopy or observations of properties provide a means of identifying mixed substitutional-interstitial solute-atom clusters in ternary N-ferrites.

(2) Complete characterization of the state of the solutes either as precipitates, or as clusters, or existing in random solid solution is a prerequisite to the interpretation of the internal friction observed in alloy ferrites nitrided at 400-650°C in NH_3:H_2 mixtures.

(3) Damping at 160-240°C in fully nitrided Fe-Mo, Fe-Cr, Fe-V and Fe-Ti alloy arises from the stress-induced diffusion of nitrogen atoms at the periphery of solute-atom clusters or in the strained matrix adjacent to fine nitride precipitates in overaged material. These peaks are different to the cold-work peak previously observed in the same temperature range in binary Fe-N and Fe-C alloys.

(4) Internal friction of partially nitrided material is particularly complex. Snoek and substitutional-interstitial interaction peaks are observed together with the characteristic high-temperature peak associated with clusters but additional damping, thought to arise from atomic arrangements intermediate between "random" solid solution and clustering, is also present.

TABLE 1
Preparation and Properties of Fe-X-N Alloys

alloy composition		nitriding conditions					hardness increase	unit-cell dimension, Å	
X	a/o	temp °C	ratio $NH_3:H_2$	time h	N a/o	N/X	kg mm^{-2}	calc	obs
V	0.52	570	10:90	41	1.01	1.94	450		
	1.08	570	10:90	41	1.65	1.53	650		
	2.26	570	10:90	41	3.59	1.58	900		
	1.08	640	6:94	48	1.32	1.23	460	2.876	2.873
	0.52	570	3:97	94	0.70	1.35	400	2.872	2.870
	1.08	570	3:97	94	1.16	1.07	600	2.875	2.874
	2.26 (i)	570	3:97	94	2.70	1.20	850	2.887	2.880
	x (ii)	570	0:100	181	2.26	1.00	460	2.883	2.876
Ti	0.69	580	8:92	48	1.20	1.74	450	2.875	2.875
	1.11	580	8:92	48	2.41	2.17	600	2.884	2.884
	2.42	580	8:92	48	4.60	1.90	825	2.895	2.896
	0.69	580	4:96	24	0.92	1.33	400	2.873	2.873
	1.11	580	4:96	24	2.19	1.97	600	2.883	2.881
Mo	2.92	450	25:75	66	3.16	1.08	850	2.899	2.895
	2.92	580	7:93	30	2.48	0.85	780	2.891	2.890
Cr	1.30	457	19:81	240	1.18	0.91	450	2.875	2.874
	1.30	575	10:90	35	1.46	1.12	400	2.877	2.868
	1.30	575	10:90	26	1.36	1.05		2.876	2.869
	2.54	575	6:94	18	2.51	0.97	600	2.884	2.871
x nitrided, then reduced in hydrogen									

TABLE 2
Internal Friction of Fe-X-N Alloys: High-Temperature Peaks

alloy composition		nitriding conditions			ΔH, cal mol^{-1}		half peak width $^{\circ}K^{-1} \times 10^4$	
X	a/o	temp °C	ratio $NH_3:H_2$	time h	calc from (24)	obs	calc	obs
V	1.08	580	10:90	67	27,880	27,990	1.88	3.10
	1.08	580	2:98	67	29,010	-	1.80	3.15
Ti	0.69	575	5:95	24	27,367	-	1.91	3.90
Mo	2.92	580	6:94	23	27,898	28,070	1.88	3.40
Cr	1.30	575	6:94	24	31,119	-	1.68	3.20
	2.00	580	10:90	21	31,081	30,800	1.68	3.83
	2.00	580	10:90	70	32,913	-	1.63	3.45
	2.00	500	13:87	17	29,250	-	1.79	3.80

References

(1) Fast, J.D. and Meijering, J.L., Philips Res. Rep. **8**, 1, 1953.

(2) Dijkstra, L.J. and Sladek, R.J., Trans. A.I.M.E., **197**, 69, 1953.

(3) Perry, A.J., Malone, M. and Boon, M.H., J. Appl. Phys., **37**, 4705, 1966.

(4) Ritchie, I.G. and Rawlings, R., Acta Met., **15**, 491, 1967.

(5) Fast, J.D., Met. Corr. Ind., **36**, 447, 1961.

(6) Jamieson, R.M. and Kennedy, R., J.I.S.I., **204**, 1208, 1966.

(7) Williams, T.M. and Leak, G.M., Acta Met. **15**, 1111, 1967.

(8) Nelson, J.D. and Riley, D.P., Proc. Roy. Soc., **57**, 1960, 1945.

(9) Speirs, D.L., Roberts, W., Grieveson, P. and Jack, K.H., Proc. of 2nd Int. Conf. on the Strength of Metals & Alloys, **2**, 601, 1971.

(10) Mortimer, B., Grieveson, P. and Jack, K.H., Scand. J. Met., **1**, 203, 1972.

(11) Pope, M., Grieveson, P. and Jack, K.H., Scand. J. Met. **2**, 29, 1973.

(12) Krawitz, A. and Sinclair, R. to be published.

(13) Wriedt, H.A. and Zwell, L., Trans. A.I.M.E., **224**, 1242, 1962.

(14) Jack, K.H., Proc. Roy. Soc., **A208**, 200, 1951.

(15) Jack, K.H., Proc. Roy. Soc., **A208**, 216, 1951.

(16) Stephenson, A., Grieveson, P. and Jack, K.H., Scand. J. Met. **2**, 39, 1973.

(17) Jack, D.H., University of Newcastle upon Tyne, Unpublished work, 1970.

(18) Henderson, S., University of Newcastle upon Tyne, Unpublished work, 1973.

(19) Driver, J.H., Unthank, D.C. and Jack, K.H., Phil. Mag. **26**, 1227, 1972.

(20) Stark, P., Averbach, B.L. and Cohen, M., Acta Met. **4**, 91, 1956.

(21) Snoek, J.L., Physica **8**, 711, 1941.

(22) Petarra, D.P. & Beshers, D.N., Acta Met. **15**, 791, 1967.

(23) Mura, T., Tamura, I. and Brittain, J.O., J. Appl. Phys. **32**, 92, 1961.

(24) Wert, C., and Marx, J., Acta Met. **1**, 113, 1953.

(25) Köster, W. and Horn, W., Arch. Eisenhutten, **37**, 245, 1966.

(26) Szabó-Miszenti, G., Acta Met. **18**, 477, 1970.

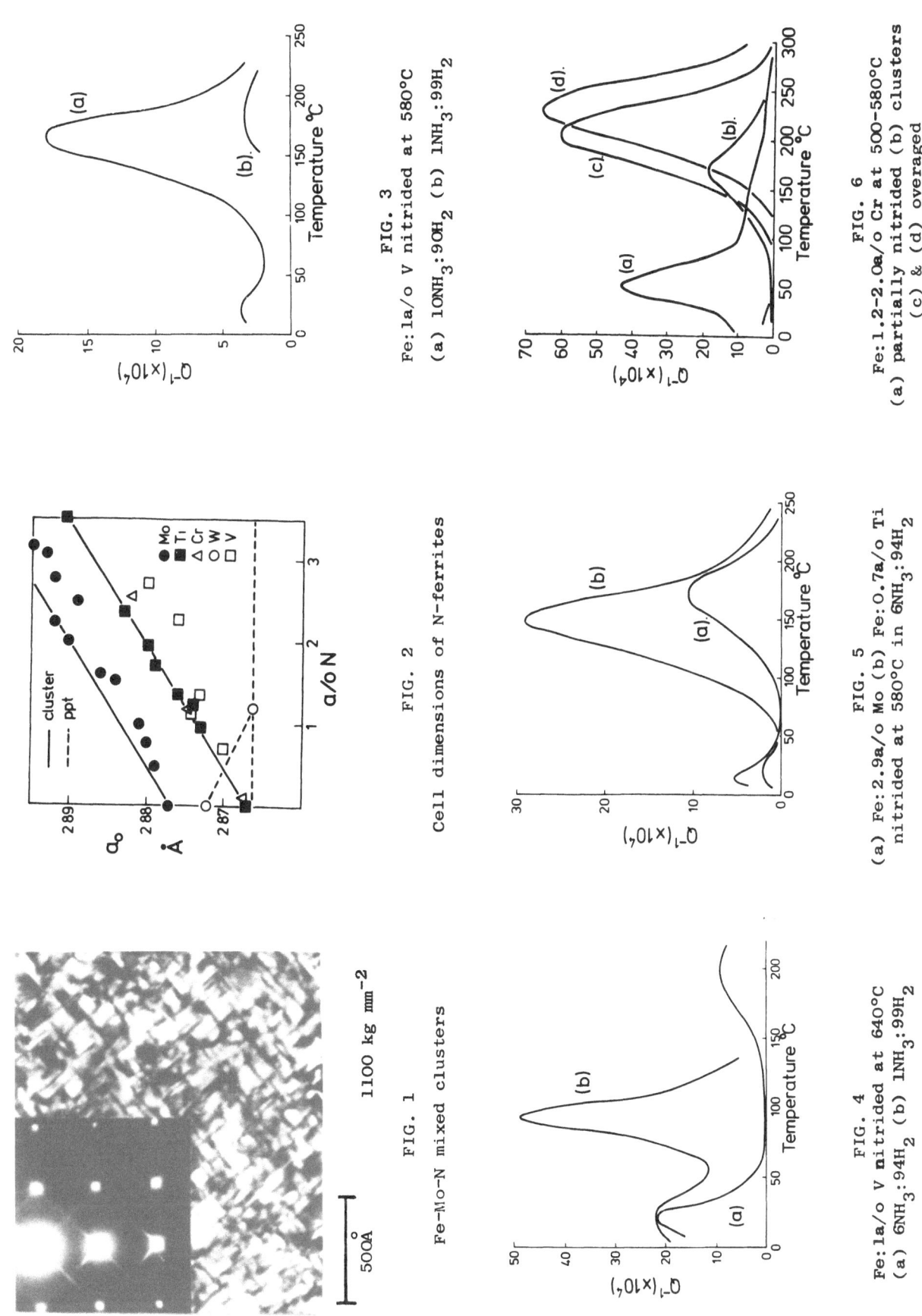

1100 kg mm^{-2}

FIG. 1

Fe-Mo-N mixed clusters

FIG. 2

Cell dimensions of N-ferrites

FIG. 3

Fe:1a/o V nitrided at 580°C

(a) 10NH$_3$:90H$_2$ (b) 1NH$_3$:99H$_2$

FIG. 4

Fe:1a/o V nitrided at 640°C

(a) 6NH$_3$:94H$_2$ (b) 1NH$_3$:99H$_2$

FIG. 5

(a) Fe:2.9a/o Mo (b) Fe:0.7a/o Ti

nitrided at 580°C in 6NH$_3$:94H$_2$

FIG. 6

Fe:1.2-2.0a/o Cr at 500-580°C

(a) partially nitrided (b) clusters

(c) & (d) overaged

RELAXATION PHENOMENA DUE TO SEMI-COHERENT PRECIPITATES

M. KOHEN [°], G. FANTOZZI, F. FOUQUET, J. MERLIN, J. PEREZ, P.F. GOBIN
Laboratoire de Physique des Matériaux - Bâtiment 502
I.N.S.A. de Lyon - 20, avenue Albert Einstein - 69621 VILLEURBANNE
FRANCE

Introduction

Over the past few years, the elastic and anelastic behavior of alloys containing precipitates has been studied by several authors (1-11). In particular, KRIVOGLAZ (1-2) has put forward a theory applicable to a biphased system, based on local variations in composition due to the passage of an elastic compression wave. These variations provoke the dispersion of the propagation speed of the longitudinal waves, and induce anelastic phenomena. In this theory, only the spherical part of the strain tensor is considered. Under those conditions, this theory cannot be applied to the case of pure torsion experiments.

For his part, SCHOECK (9) has approached the problem in the case of torsion and reached that only precipitates with incoherent interfaces, (or at very most semicoherent) can give rise to an internal friction peak. This author considers the interfaces surrounding the precipitates, as viscous boundaries. This analysis cannot be generalized, but it seems hold well for certain particular cases (for instance, Al-Ag, Cu-Si, Cu-Co alloys (8-10-11)).

In this work we propose a generalization of the problem, based on the thermodynamical properties of the system, and capable of being applied to all types of solicitations; more particulary, to the case of torsion.

Problem presentation

We shall consider an elementary cell, composed by the whole precipitate and surrounding matrix. The applicationof external stress provokes, as well as the elastic strain, a modification of the system which brings about an extra strain and an interaction energy associated to this strain.

[°]Laboratorio de Vibraciones y Ultrasonido- Facultad de Ingeniería - Universidad de Buenos Aires

If the speeds of movement of all the precipitate-matrix interfaces are equal (the study of the growth process allows us to establish this fact (12)),a volume change takes place only in the case where σ_{ii} is not zero,while when σ_{ii} equals zero there is only a shape change. Inthis last case, which corresponds to pure torsion,the interaction energy is very low (9).

On the other hand, if the speeds of movement of the different interfaces are not the same, i.e. if there is a direction of rapid growth, even in the case of torsion, a volume variation which is diffusion controlled could take place.

Let's consider a specimen in the axis of Z' which contains plate shaped precipitates subjected to a pure torsion following this axis (figure 1). At one point, we choose the axes X' and Y' in order that only the $\sigma_{Y'Z'}$ shear component is not zero. The corresponding elastic strain is

$$\varepsilon_{Y'Z'} = \frac{\sigma_{Y'Z'}}{2\mu}$$

μ being shear modulus.

We shall carry out a rotation of the axes about X', in such a way to make Z' coincide with the normal Z at the broad face of the precipitate being considered:we obtain the state of stresses shown by figure 1.

The anelastic strain , can be expressed as a function of additional parameters λ_L as follows (13),

$$\varepsilon_{ij}^{an} = \sum_L \frac{\partial \varepsilon_{ij}}{\partial \lambda_L} \Delta \lambda_L \qquad (1)$$

the parameters λ_L being a function of the time, since changes of system states are not instantaneous but diffusion controlled. Returning to the initial position of the axes, we determine

$$\varepsilon_{Y'Z'}^{an} = \sum_L \frac{\partial \varepsilon_{Y'Z'}}{\partial \lambda_L} \left\{ \frac{\partial \lambda_L}{\partial \sigma_{YZ}} \cos 2\alpha + \left(\frac{\partial \lambda_L}{\partial \sigma_{YY}} - \frac{\partial \lambda_L}{\partial \sigma_{ZZ}} \right) \sin 2\alpha \right\} \Delta \sigma_{Y'Z'} \qquad (2)$$

i) In the case of a precipitate whose lateral and broad faces have the same structure -therefore the same displacement rate in the direction of the interface's plane normal, we have

$$\frac{\partial \lambda_1}{\partial \sigma_{YY}} = \frac{\partial \lambda_1}{\partial \sigma_{ZZ}} \qquad (3)$$

In addition, if the interface defects are sessile, as in the case of θ' precipitate in Al-Cu (12), the σ_{YZ} component does not cause any anelastic strain. On the other hand, if the interface is glissible, the term $\frac{\partial \lambda_1}{\partial \sigma_{YZ}}$ becomes preponderant.

ii)In the opposite case, where one of the faces has a speed of movement sufficiently greater, the two terms of the relation (3) are no longer equal.As a result,no matter what the nature of the interface is, the anelastic strain is different from zero and there is interaction with the applied stress.

Thus, the semicoherent precipitates, taking in consideration the structure of different interfaces, give arise to anelastic phenomena which can be seen in torsion. On the other hand, the fully incoherent precipitates only give rise to this phenomena (of a considerable size) if the interfaces have a glissible structure.

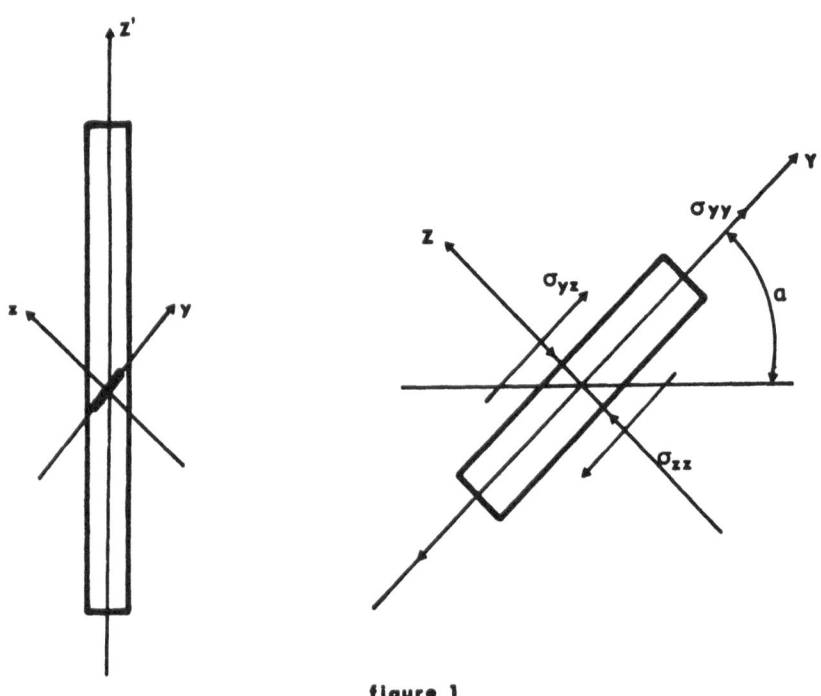

figure 1

Relaxation Model

In order to analyse the response of the system to an external solicitation, we shall use as a starting point the thermodynamical method described by KRIVOGLAZ (1) but taking into account the presence of the interfaces and their structure.

Let's consider an alloy containing plate shaped precipitates.The thermodynamical state of quasi equilibrium of the alloy is modified by the application of stress. Limiting ourselves to isothermal conditions,-i.e. the temperature equilibrium in the system is established in a period of time considerably less than the oscilation cycle -, the thermodynamical variables reduce to three: for example, σ_{ij} , C_M (local solute concentration in the matrix), and d (plate diameter).

The applied stress induces across the interface a difference of thermodynamical potential equal to

$$\Delta \mu_b = \frac{\partial \mu_b}{\partial C_{Mb}} \Delta C_{Mb} + \frac{\partial \mu_b}{\partial d} \Delta d + \frac{\partial \mu_b}{\partial \sigma_{ij}} \Delta \sigma_{ij} \qquad (4)$$

the index b showing that we are referring to the property taken at the edge of the precipitate.

Taking into account the continuity of the flux across the interface, we have

$$\Delta \mu_b = \frac{D.a}{M} \nabla C_{Mb} \qquad \text{and} \qquad V_b = \dot{\Delta} d = \frac{D}{\Delta C_I} \nabla C_{Mb} \qquad (5)$$

D: solute diffusion coefficient

a: lattice parameter

M: interface mobility

$\Delta C_I = C_I - C_M$: Difference of solute concentration between the precipitate and the matrix.

The final resolution of the problem implies the choice of a solution of diffusion equation which satisfies the physical behavior of the system. For sufficiently low temperatures, in order that the effective length of volume diffusion

$$\Lambda_{vol} = \sqrt{\frac{D \, vol}{\omega}}$$

(ω being the pulsation) can be negligeable with relation to the distance l between precipitates, we can state :

$$\frac{\dot{\Delta} C_{Mb}}{\Delta C_{Mb}} = D k^2 \qquad (6)$$

where k is an integration constant. In addition , $\dot{\Delta} C_{Mb}$ is equal to $j\omega \Delta C_{Mb}$ ($\Delta \sigma_{ij}$ being in the form $\sigma_0 e^{j\omega t}$), so that Dk^2 is equal to $j\omega$, and we can write

$$k = \sqrt{\frac{j\omega}{D}} = \sqrt{\frac{\omega}{2D}} (1+j) = \sqrt{\frac{1}{2 \Lambda^2}} (1+j) \qquad (7)$$

Taking (5) into (4) and using the relation (6) , we find the relation between the variation in concentration at the edge of the precipitate and the applied stress:

$$\Delta C_{Mb} = R_{ij} \cdot \Delta \sigma_{ij} \qquad (8)$$

The Rij expression depends on each specific case. Let's take as an example an onedimensional solution for low temperatures ($\frac{\Lambda}{l} \ll 1$)

$$R_{ij} = j\omega \frac{\partial \mu_b}{\partial \sigma_{ij}} \left\{ \frac{kD}{\Delta C_I} \cdot \frac{\partial \mu_b}{\partial d} - \left(\frac{1}{ka} + \frac{D}{M\frac{\partial \mu_b}{\partial C_{Mb}}} \right) \frac{\partial \mu_b}{\partial C_{Mb}} Dk^3 a \right\}^{-1}$$

In addition, we can state that

$$\Delta \varepsilon_{mn} = \frac{\partial \varepsilon_{mn}}{\partial \sigma_{ij}} \Delta \sigma_{ij} + \frac{\partial \varepsilon_{mn}}{\partial C_M} \Delta C_M + \frac{\partial \varepsilon_{mn}}{\partial d} \Delta d \qquad (9)$$

Thus, we obtain by replacing in the equation (9), the imaginary part of the strain and therefore the internal friction

$$\delta = W \cdot \frac{\sqrt{\frac{\omega \tau}{2}} \left(1 + Q \sqrt{\frac{\omega \tau}{2}} \right)}{\frac{\omega \tau}{2} \left(1 + Q \sqrt{\frac{\omega \tau}{2}} \right)^2 + \left(1 - \sqrt{\frac{\omega \tau}{2}} \right)^2} \qquad (10)$$

where,

$$W = \frac{\dfrac{\partial \mu b}{\partial \sigma_{ij}} \cdot \dfrac{\partial \varepsilon_{mn}}{\partial d}}{S_{ijmn} \dfrac{\partial \mu b}{\partial d}} \cdot \left(1 + \frac{2 \Delta C_I}{1} \cdot \frac{\dfrac{\partial \varepsilon_{mn}}{\partial C_M}}{\dfrac{\partial \varepsilon_{mn}}{\partial d}} \right)$$

S_{ijmn} : compliance

$$Q = \frac{2a}{\Delta C_I} \cdot \frac{\partial \mu b}{\partial d} \cdot \frac{D}{M \left(\dfrac{\partial \mu b}{\partial C_{Mb}} \right)^2}$$

$$\tau = \frac{(\Delta C_I)^2}{D} \cdot \left[\frac{\dfrac{\partial \mu b}{\partial C_{Mb}}}{\dfrac{\partial \mu b}{\partial d}} \right]^2$$

Let's examine the relation (10) in more detail. If the interface mobility is very great $Q \sqrt{\dfrac{\omega \tau}{2}}$ can be ignored, and the relation (10) can thus, be written as:

$$\delta = W \cdot \frac{\sqrt{\dfrac{\omega \tau}{2}}}{\dfrac{\omega \tau}{2} + \left(1 - \sqrt{\dfrac{\omega \tau}{2}} \right)^2} \qquad (11)$$

The internal friction as a function of the temperature, therefore shows a peak (fig.2) whose half-width is 13% greater than that which corresponds to the Debye peak, and an activation energy equal to U (U being the diffusion activation energy, that can correspond either to volume diffusion or short circuit diffusion). A thermodynamical study of W, allows us to state that the relaxation intensity is proportional to the precipitated fraction and to ΔV (ΔV being the volume variation when a solute atom is transferred from the precipitate to the matrix).

If, however the interface mobility is low, we can no longer ignore $Q \sqrt{\dfrac{\omega \tau}{2}}$. This term is governed by an activation energy equal to $(\tilde{U} - \dfrac{\dot{U}}{2})$, $\tilde{U}$ being the energy for the interface mobility. It modifies the curve shape, diminishing its height, moving its maximum to the high temperature region, and widening its half-width.

For lower mobilities, the maximum even disapears, and a monotonous curve remains.

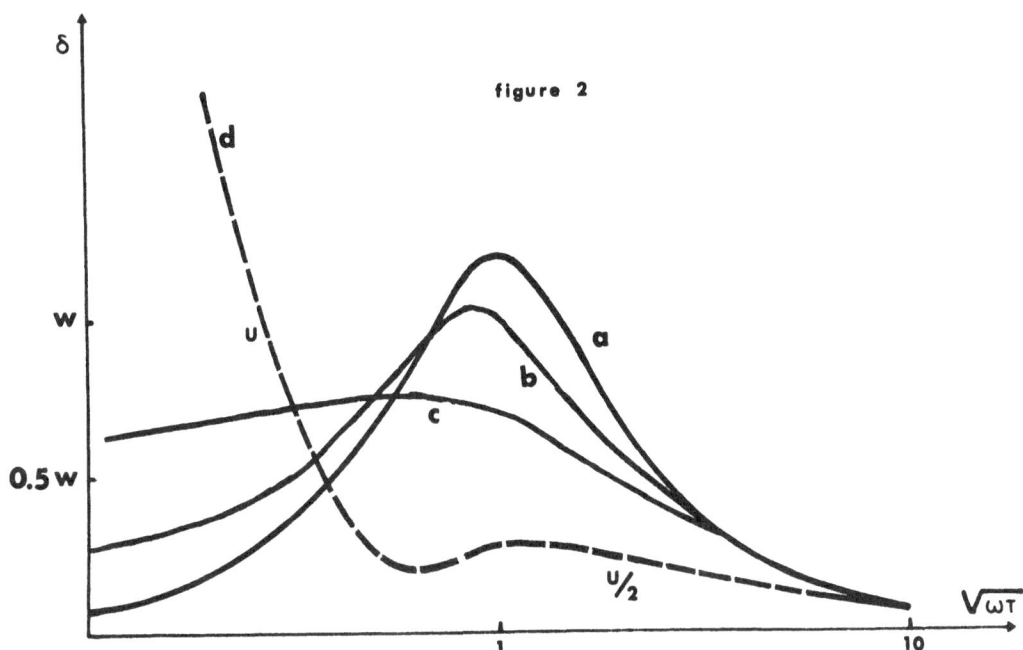

Figure 2: $\delta = f(\sqrt{\omega \tau})$

a: $Q\frac{\omega \delta}{2}=0$
b: $Q\frac{\omega \delta}{2}=0,2$; — — $\Lambda \simeq 1$ d
c: $Q\frac{\omega \delta}{2}=1$

—— $\Lambda \ll 1$

Conclusion

Thus, the presence of precipitates gives rise to relatively complex anelastic phenomena. These phenomena are noticeable only in two cases :

- when the interfaces are glissile
- when the speeds of movement of the interfaces are different (which is known to be the case for semi-coherent precipitates).

We have determined the internal friction associated with the displacement of precipitate interfaces and we have shown that the anelastic behavior depends on the interface mobility.

For example, taking the case of the θ' semi-coherent precipitate in Al-Cu. The θ' lateral interface shows sessile dislocation loops and its movement is controlled on one hand by the solute diffusion and on the other hand by dislocation climb (14-15). The broad face is coherent and its mobility is very much reduced (15-18). The activation energy U for θ' is the short circuit diffusion energy: it is of the order of 22 kcal/mole (15-17), while the activation energy of the interface mobility

$\tilde{U}$ is of the order of 10 kcal/mole (15). The corresponding shape of the curve (for a solution of the diffusion equation valid also at high temperatures) (20) is outlined in figure 2d. Since the volume variation ΔV is large (4,3%) (19), the amplitude of phenomena is noticeable. The internal friction spectrum associated with θ' shows clearly that there is a phenomenon observed in torsion which is situated around 150°C (at a frequency of the order of one hertz) (3-19). This phenomenon has the charasteristics indicated above.

Thus, the model under consideration seems to describe correctly the anelastic phenomena due to precipitates. In particular, this analysis might therefore be capable of being applied to the numerous results presented by several authors (8-10-11).

References

1. M.A. KRIVOGLAZ, Fiz. Metal. Metalloved,10, N°4, 1 (1960)

2. M.A. KRIVOGLAZ, Fiz. Metal. Metalloved,12, N°3, 31 (1961)

3. B.S. BERRY,A.S. NOWICK,Technical Note 4225, NACA, (1958)

4. A.C. DAMASK,A.S. NOWICK,J.Appl.Phys.,26,1165,(1955)

5. S.Ya.GEGUZINA, Fiz. Metal Metalloved,18,N°5,17 (1964)

6. S.Ya.GEGUZINA,M.A.KRIVOGLAZ, Fiz. Metal. Metalloved,21,N°2, 6 (1966)

7. B.M. DARINSKIY,Yu.N. LEVIN, Fiz. Metal. Metalloved,26, N°6, 98 (1968)

8. G. SCHOECK,E. BISOGNI,Phys. Stat. Sol.,32,31 (1969)

9. G. SCHOECK, Phys. Stat. Sol.,32,651 (1969)

10. M. MONDINO,G. SCHOECK,Phys.Stat. Sol.(a),6,665 (1971)

11. M. MONDINO,C. LARDONE,Scripta Met.,6,1109,(1972)

12. H.I. AARONSON,C. LAIRD, K.R. KINSMAN,Phase Transformations,p.313,ASM (1970)

13. V.T. SHMATOV,Fiz. Metal. Metalloved,6,984 (1958) and 7,321 (1959)

14. C. LAIRD,H.I. AARONSON,Trans. Met. soc. AIME,242,1393 (1968)

15. H.I. AARONSON,C. LAIRD,Trans. Met. Soc. AIME,242,1437 (1968)

16. G.C. WEATHERLY,R.B. NICHOLSON,Phil. Mag.,17,801,(1968)

17. J.D. BOYD,R.B. NICHOLSON,Acta Met.,19,1101 (1971) and 19, 1379 (1971)

18. G.C. WEATHERLY, Acta Met.,19,181 (1971)

19. C. HANAUER, Thesis, Université de Lyon,(1973)

20. M. KOHEN et al to be published

INTERNAL FRICTION ASSOCIATED WITH INTERSTITIAL DISTRIBUTIONS IN FE-N AUSTENITE AND MARTENSITE

J. Foct

Laboratoire de Métallurgie associé au C.N.R.S. n° 159 - E.N.S.M.I.M. Parc de Saurupt
54000 NANCY - FRANCE

Introduction

It is known that anelastic phenomena in ferrous austenites and martensites containing carbon interstitials strongly depend on these interstitials. The purpose of this work is to see if by replacing carbon by nitrogen, internal friction measurements are quantitatively or qualitatively modified.

The Fe-C and Fe-N austenites are both face centered cubic. The Fe-C and Fe-N martensites are both body centered tetragonal. The size of carbon atoms and of nitrogen atoms being almost the same (1), similar interactions with dislocations may be expected with carbon and nitrogen. Obviously other properties like precipitation and ordering differ. Tempering of Fe-N martensite does not give rise to the hexagonal iron nitride isomorphous with ε carbide but it precipitates the α" phase, $Fe_{16}N_2$, whose tetragonal structure has been discovered by K.H. Jack (2). At higher temperatures (> 200°C), γ' Fe_4N nitride appears in Fe-N austenite and in Fe-N martensite.

By Mössbauer spectrometry we have shown that the distributions of interstitials and their evolution in metastable conditions are different for carbon and for nitrogen. In Fe-C martensite the carbon atoms tend to cluster near room temperature (3) but in Fe-N martensite the nitrogen atoms ordering occurs indicating long range order of α" $Fe_{16}N_2$ (4). In both Fe-C and Fe-N austenites interaction between interstitials is repulsive. However the repulsion in Fe-N austenite is not strong enough to avoid formation of interstitial pairs (5).

Presently, only the study of binary Fe-N alloys is undertaken. It seems interesting to see if the different ordering phenomena observed by Mössbauer spectra are dctctable by internal friction.

Experimental Procedure

The anelastic measurements were made in a vacuum inverted pendulum over the temperature range of 80°K to 650°K at a frequency of $\sim$ 1 H_z. The heating and cooling rate was 120°C/h. The maximum temperature gradient along the 8 cm specimen was 2°C in the worst conditions. The description of measurements apparatus is given elsewhere (6).

Hydrogen-ammonia mixtures (7) were used for nitriding. Zone-melted iron wires (1,7 mm diameter) were kept at 800°C in the gaz mixture and water quenched. The ammonia concentration determines the nitrogen concentration in the solid phase (typically 2 wt %). A nitrided coat of $\sim$ 300μm is obtained after 10 hours of treatment. In this coating the greater part of the mechanical energy is dissipated during the internal friction measurements. The core of the specimen remains ferritic.

Results

The internal friction (Q^{-1}) and the oscillation period (P) are measured for austenitic samples between 20°C and 350°C (FIG. 1). At higher temperatures nitrogen losses occur and corresponding experiments must be avoided.

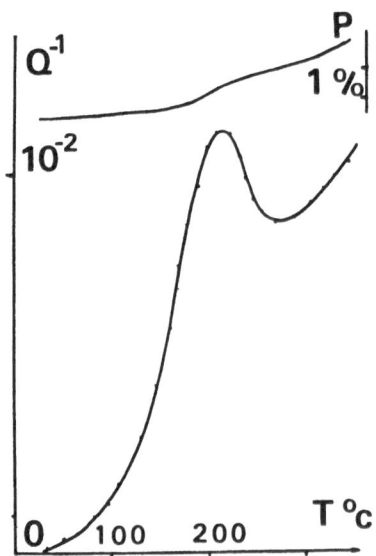

FIG. 1

Internal Friction in Fe-N Austenite

At a frequency of 1,3 H$_z$ a peak isobserved at 210°C. The activation energy measured for different oscillation frequencies is 30 kilocalories/mole. This value is much higher than that which is given by the peak width ($\sim$ 20 kilocalories/mole).

When the deformation amplitude increases from 10^{-4} to 4.10^{-4} a slight increase of Q^{-1} is observed but seems to be due to the high back ground level because Q^{-1} increases more at 300°C that at the temperature of the peak.

During three heating and cooling measurements the austenite peak exhibits a remarkable stability ; just a slight decrease of back ground can be noticed between the first heating and cooling. By cooling (200°C/h) the austenitic specimen in the pendulum to liquid nitrogen temperature a rough and discontinuous internal friction variation is observed. This phenomenon during which Q^{-1} reaches 10^{-2} is clearly related to the martensitic transformation and may be similar to what is observed in stainless steel (8).

The internal friction is measured immediately after the quench at 77°K on specimens containing virgin martensite ($\sim$ 60 %) and retained austenite ($\sim$ 40 %). The Q^{-1}(T) curve which is observed may be caracterized by three domains : A from - 100°C to 50°C, B from 50°C to 150°C and C from 150°C to 300°C (FIG.2).

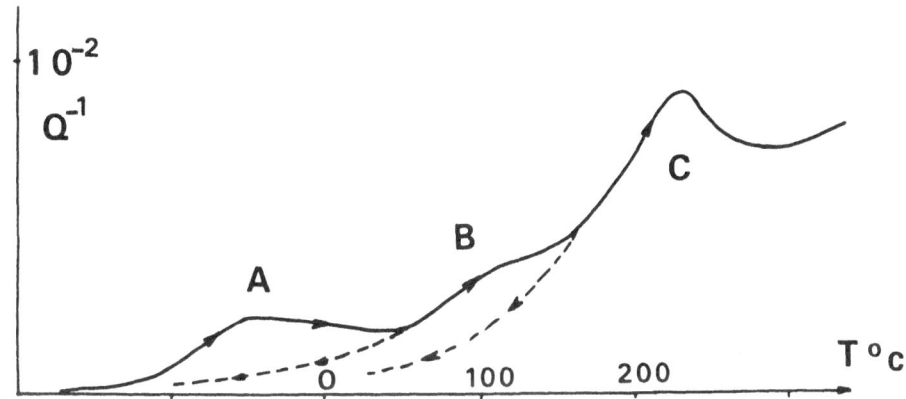

FIG. 2

Internal friction in Fe, N martensite on heating from - 196°C to 350°C. Dotted lines indicate cooling measurements in A and B domains.

Domain C is clearly caracterized by a 240°C peak which is superposed to a high temperature increasing back ground. By measuring internal friction on cooling from 350°C, the 240°C peak disappears and is replaced by a 210°C peak which should be attributed to retained austenite (FIG. 3).

As other experiments show, irreversibility on cooling in A and B domains appear at much lower temperatures : by stopping initial heating Q^{-1} measurements and cooling down from 70°C and from 200°C one observes the disappearance of an important port of Q^{-1}(FIG.2).

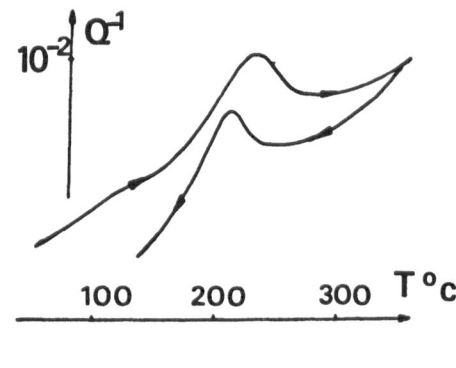

FIG. 3

Heating and cooling measurements in C domain

If the difference between the heating and the cooling Q^{-1} curves is traced very broad peaks appear near - 50°C an + 80°C.

Discussion

The observed phenomenon corresponding to 240°C peak in Fe-N austenite is due to interstitial relaxations, as indicated by : a) The peak stability on heating and cooling b) the observed period variation - c) the good correlation of peak activation energy with high temperature activation energy (9), and - d) other results quoted by Wert (10) in cubic face centered alloys. With the exception of the work of Mah and Wert on Co-C (11) previous studies on face centered alloys were done in ternary systems. Frequently, the interpretations are based on substitutionnal-interstitial pairs. In the present case the interstitial site symmetry is lowered only in the vicinity of vacancies or of other interstitials. The bounded pairs (i-v) and (i-i) may be formed.

Arguments which are used in carbon alloys to propose a (i-v) pair built by a vacancy between two interstitials along <100> may be correct in the Fe-N austenite. If one compares the relaxation intensities obtained by Ulitchny and Gibala on quenched Fe-Ni-C austenites (12) to those measured here one notes that the values are clearly much higher in Fe-N case in spite of the lower austenitizing temperature. Obviously (i-v) formation energies are not identical in the two cases but they may be comparable. The observed discrepancy should be due to the relaxation of the interstitial pairs NN $\frac{1}{2}$<110> and N FeN<100> which are detected by Mössbauer spectra. The simple reorientation of NN $\frac{1}{2}$<110> and N FeN<100> pairs is an insufficient explanation since the N-N interaction is repulsive. Destruction and recombination of (i-i) pairs must be considered as an equilibrium reaction which may be affected by stress. A more quantitative study of the relaxation mechanism would require the unknown values of strains associated to defects and a decomposition of the peak.

As far as internal friction in carbon martensite is understood our results in Fe-N martensite may be comparable to previous studies. Similarities with carbon martensites in A and C domains would confirm that the corresponding phenomena are essentially related to dislocation-interstitial interactions.

In spite of the differences in ordering for carbon and for nitrogen in the martensites, no clear difference has been detected by internal friction near and below room temperature. This may be a provisional result since Q^{-1} measurements on binary Fe-C and Fe-N martensites without tempering are difficult and rare. Preliminary results about the influence of shear strain amplitude (ε) on Q^{-1} clearly indicate a positive value of $(\frac{\delta Q^{-1}}{\delta \varepsilon})_T$ at room temperature. More accurate measurements must be compared to recent results on Fe-C martensites (13).

The C domain similarity between carbon and nitrogen martensites is in both cases related to the shear mecanisms of the martensitic transformation and to the importance of glide or of twinning. If the arguments which explain the 220°C carbon martensite peak as a Snoek-Köster (14) peak in ferrite are correct, they should equally apply to Fe-N martensite. The observed 240°C peak is to be compared to the 240°C peak studied by Ino and Sugeno (15) in cold worked Fe-N ferrite.

The internal friction which is measured in the B domain cannot be interpreted by coherency losses of precipitates as Ke proposed for Fe-C martensites in which ε carbide forms. The precipitation of α'' nitride below 200°C is coherent (16). An attempt to relate isothermal Q^{-1} variation in the B domain to nitrogen ordering phenomena detected by Mössbauer spectrometry failed on kinetic considerations. However the Mössbauer spectra reveal that some nitrogen clusters present in freshly quenched martensite are dissolved between 50°C and 150°C (16). Q^{-1} in the B domain could be due to stress ordering of jumping nitrogen atoms liberated from unstable interstitial configurations before getting trapped by stable α'' precipitates. Obviously the activation energies are different from that of Snoek peak in nitrogen ferrite. The energy difference between the x and z octahedral sites in martensite roughly varies between 0 and 0.2 eV depending upon interstitial concentration. Even for high interstitial concentrations ($\sim$ 10N/100 Fe) Zener's order parameter may, during tempering, be different from the theoretical values. This explains an interstitial flow between z and x sites which is at the origin of relaxation intensity. The broadening of the peak reveals a spectrum of relaxation times which correspond to environments of different defects such as dislocations, twins, ordered zones and precipitates.

In conclusion the relaxation peak in Fe-N austenite is related to the (i-i) pairs which are revealed by Mössbauer spectrometry and to the quenched (i-v) pairs.

The well established differences in low temperature ordering of Fe-C and Fe-N martensites are not clearly detected. Analogous behaviour near room temperature and near 200°C in the two systems may be due to similarities of interstitial-dislocation interactions and of similar martensitic transformations with carbon or nitrogen. Near 100°C internal friction in Fe-N martensite is attributed to the nitrogen atoms which are jumping before complete α'' formation and $\dot{\gamma}$ precipitation.

Aknowledgments

The author withes to thank Professeur R. Faivre for his help and his interest in this work, and Professeurs J. de Fouquet and P. Gobin for fruitful discussions.

References

1. K.H. Jack, Proc. Roy. Soc. 208, 200 (1951).
2. K.H. Jack, Proc. Roy. Soc. 208, 216 (1951).
3. J.M. Genin and P.A. Flinn, Trans. AIME. 242, 1419 (1968).
4. J. Foct and J.M. Genin, Compt. rend. 270, 1563 (1970).
5. J. Foct, Compt. rend. 276, 1159 (1973).
6. J. Perez, P. Peguin and P. Gobin, J. Sci. Instr. 42, 65 (1966).
7. D. Atkinson, T. Bell and D. Brough, JISI. 196, 836 (1965).
8. J.F. Delorme, R. Schmid, M. Robin and P. Gobin, J. Phys.,Paris 32 C2-101 (1971).
9. K. Bohnenkamp, Arch. Eisenhüttenwes. 38, 229 (1967).
10. C.A. Wert, Physical Acoustics, W.P. Mason, ed., vol. III A, p. 44, Acad? Press (1966).
11. G. Mah and C. Wert, Trans. AIME 242, 1211 (1968).
12. M.G. Ulitchny and R. Gibala, Met. Trans. 4, 497 (1973).
13. Y. Bertin, J. Parisot and J. de Fouquet, Scripta Met. 7, 769 (1973).
14. A.S. Nowick and B.S. Berry, Anelastic Relaxation in Crystalline Solids, Acad. Press (1972).
15. H. Ino and T. Sugeno, Acta Met. 15, 1197 (1967)
16. J. Foct, Mem. Sc. Rev. Met. to be published.

CRYSTALLOGRAPHIC ORIENTATION DEPENDENCE OF INTERSTITIAL ANELASTICITY IN Ta-Re-N AND Ta-Re-O ALLOYS

A.A. Sagues[x] and R. Gibala[xx]
Department of Metallurgy and Materials Science
Case Western Reserve University
Cleveland, Ohio 44106

Introduction

There have been many extensive anelastic measurements on the influence that substitutional solutes have on Snoek anelasticity in body centered cubic metals (1-3). These investigations, which usually show that the Snoek spectrum is broadened into several component relaxations by alloying, have given important information on the nature of substitutional (s) - interstitial (i) solute interactions. However, few studies have examined the dependence of various component relaxations in s-i anelastic spectra on crystallographic orientation of single crystals. This type of study is capable of giving structural information on s-i defects which cause relaxations (4). Only the investigations of Hashizume and Sugeno on Fe-Si-N alloys (5) and of Miner et al. (6) on Nb-Zr-O alloys have dealt with this problem.

In the present investigation, we have examined the orientation dependence of interstitial anelasticity in Ta-Re-N and Ta-Re-O alloys. Other complementary investigations (7, 8) have disclosed broadening of Snoek peaks by Re additions in these alloy systems. The spectra can be interpreted consistently in terms of relaxations (in addition to the standard Snoek peaks) of ReN and Re_2N clusters in Ta-Re-N alloys and of a wide spectrum of Re_xO type clusters in Ta-Re-O alloys (9).

---

[x] Now at : Henry Krumb School of Mines, Columbia University, New York, N.Y.
[xx] On leave at : C.E.N.G., Département de Recherche Fondamentale, Section de Physique du Solide, Grenoble, France.

Procedure

Single crystals of binary Ta-1.3 to 6.0 at.% Re alloys were prepared by electron beam zone melting of 99.99 % Ta rods wrapped in appropriate amounts of 99.97 % Re wires. Seed crystals were employed to obtain orientations very near <100> <110> and <111>. Wires ∿ 1.4 mm diameter and rods ∿3mm diameter were obtained by careful centerless grinding of the as grown crystals. The crystals then were outgassed near the melting point in ultra high vacuum (∿10⁻¹⁰ torr) by direct resistance heating (wires) or by high frequency induction heating (rods). Subsequent interstitial additions were made by Sievert's additions (nitrogen) or by diffusion annealing of anodic films (oxygen). Internal friction measurements were made on wire specimens in torsion at ∿ 1 Hz and on rod specimens excited longitudinally at ∿50 kHz. Details concerning all experimental techniques are given elsewhere (8, 9).

Results and Discussion

Typical results obtained for Ta-Re-N alloys are given in Fig.1 for a Ta-3.8 at.% Re - 0.06 at.% N alloy excited in torsion at 1.6 Hz. The damping spectrum, broadened by Re additions in the same manner as disclosed elsewhere (7-9), is maximum for the <111> orientation and minimum for <100>. Fig.2 shows equivalent experiments performed on Ta-Re-O alloys. An alloy of Ta-3.8 at.% Re - 0.04 at.% O excited longitudinally at 50 kHz exhibits a broad Snoek spectrum which has maximum damping for the <100> orientation and near zero damping for <111>.

To within all experimental uncertainties, the entire damping spectra in Figs.1 and 2 have the same orientation dependence. This is demonstrated in Fig.3, where we have plotted Q^{-1}<100> and Q^{-1}<110> versus Q^{-1}<111> from Fig.1 and Q^{-1}<111> and Q^{-1}<110> versus Q^{-1}<100> from Fig.2 at various temperatures in these spectra. The simple linear relationships obtained show that the respective ratios Q^{-1}<xyz> / Q^{-1}<111> and Q^{-1}<xyz> / Q^{-1}<100> are constants and that the shapes of the spectra are not functions of orientation.

Also to within all experimental uncertainties, the entire spectra in Figs. 1 and 2 have the same orientation dependence as the Snoek relaxation. The relaxation strength of the Snoek peak (and other defects with tetragonal or <100> orthorhombic symmetry) varies with orientation as (4, 10)

$$\frac{\Delta_G \ (\Gamma)}{\Delta_G \ (\Gamma = \frac{1}{3})} = \frac{S_{44} + \frac{4}{3} \ (\ S_{11} - S_{12} - \frac{S_{44}}{2} \ )}{S_{44} + 4 \ (\ S_{11} - S_{12} - \frac{S_{44}}{2} \ )\Gamma} \ 3\Gamma \qquad (1)$$

for torsional excitation and as

$$\frac{\Delta_E \ (\Gamma)}{\Delta_E \ (\Gamma = 0)} = \frac{S_{11} \ (1 - 3\Gamma)}{S_{11} - 2 \ (S_{11} - S_{12} - \frac{S_{44}}{2} \ )\Gamma} \qquad (2)$$

for longitudinal excitation. Here Δ_G and Δ_E are the relaxation strengths in torsional and longitudinal excitation, respectively, S_{ij} are the elastic compliances, and $\Gamma = \overset{3}{\underset{i \neq j}{\Sigma}} \gamma_i^2 \gamma_j^2$ is the orientation parameter, where γ_i and γ_j are the direction cosines relative to <100> directions. In Fig.4, the results for Ta-Re-N and Ta-Re-O alloys from Figs.1-3 are compared with Eqs. 1 and 2. The parameter Γ was computed from X-rays measurements, the elastic constants were those obtained by Raffo (11), and the ratios

$$Q_G^{-1}(\Gamma) \ / \ Q_G^{-1}(\Gamma = \frac{1}{3}) \quad \text{and} \quad Q_E^{-1}(\Gamma) \ / \ Q_E^{-1}(\Gamma = 0)$$

were adjusted by a constant amount to achieve the best fit with Eqs.1 and 2 respectively. The agreement is very good for both alloy systems. For the Ta-Re-N alloys, the low frequency results also suggest that there is negligible contribution from torsion-flexure coupling (4).

Because discrete relaxations caused by single N atoms, ReN pairs and Re_2N triplets have been identified and characterized thermodynamically in the Ta-Re-N alloys (7-9), the selection rules of anelasticity (4) can be applied to determine possible configurations of the Re-N defects. Fig.5 shows configurations that are possible if, as expected (12), octahedral sites are occupied. If the suggested behavior of Fig.4a is absolutely obeyed, the selection rules dictate that only tetragonal and <100> orthorhombic defect symmetries are possible. Therefore, only configurations of the type DA or DC in Fig.5a and EBF or ECF in Fig.5b are possible ; configurations such as the <110> orthorhombic pair DB or the lower symmetry DBF triplet (proposed by Mosher et al. (13)) should be ruled out.

Unfortunately, this type of analysis can not be applied rigorously. As pointed out by Nowick (14), the tetragonal distortion of the interstitial within defects of the type DB can be large enough to mask the effect of the distortion in the direction DB in Fig.5 for virtually all values of Γ. The net result is an orientation dependence almost identical to that observed for a tetragonal defect. Thus, unless very accurate measurements of internal friction are carried out on exactly oriented <100> and <111> single crystals, data on orientation dependence per se are not conclusive. The same argument applies to the Re_2N defect and to various possible Re_xO defects. This limitation on the interpretation is further substantiated when one considers the very different types of s-i defects for which there are data now available : ReN, Re_2N and Re_xO (with $x \geq 1$) in Ta ; ZrO, ZrO_2 and ZrN in Nb ; and SiN plus higher order defects in Fe. It seems unlikely, with the widely different s-i binding energies involved, that all of the defects containing single interstitial atoms have one of two common symmetries, which in turn involve only <100> configurations.

Various kinetic and thermodynamic criteria (4) were examined for additional clues concerning the proper s-i defect configurations. When an interstitial in the DA configuration jumps from A to C, it passes through B. The time of stay at B will determine whether the si defect DA or DB is predominant. Unfortunately, recent theoretical work by Nowick (15) shows that for either DA or DB predominant, the si relaxation peak will appear at higher temperatures than the Snoek peak and with a relaxation strength proportional to the concentrations of s and i. Thus to distinguish between the two configurations based perhaps on only small relative differences in expected relaxation times and strengths is very difficult.

We have examined the detailed temperature dependence of the relaxation strengths for other possible information on the correct configurations. In our other papers (8, 9), we have determined that the binding enthalpy of the ReN relaxation is positive and is approximately 3000 cal/mole ; we have obtained therefrom the concentrations of ReN defects causing the relaxation. In Fig.6 we have plotted these concentrations C_{ReN}, normalized by the concentration of unassociated interstitials C_N obtained from the Snoek peak , at ~ 360 C as a function of Re concentration. Also included are the ratios expected for random solid solutions if either of the defects DA or DB is predominant. In principal, one can make the same type of measurements at high temperatures and determine if C_{ReN} approaches the limit given by DA or DB in the figure. One preliminary experiment on a Ta-2.1 at. Re-0.06 at.% N alloy at 50 kHz disclosed no measurable ReN relaxation at the expected temperature of ∿750 C. This result, given in Figure 6, is consistent with the small magnitude of the binding enthalpy given above and is more consistent with the occurrence of DA defects than DB defects.

Summary

The crystallographic orientation dependence of substitutional-interstitial (s-i) relaxations in Ta-Re-N and Ta-Re-O alloys is found to be consistent with that predicted by the selection rules of anelasticity for relaxations caused by defects with tetragonal or <100> orthorhombic symmetry. However, the large tetragonal distortion of single interstitials within $s_x i$ type defects makes this conclusion uncertain. In general, extensive experiments on the orientation, concentration and temperature dependences done with high accuracy and sensitivity are required to obtain more definite hints on the specific s-i defect configurations causing relaxation.

Acknowledgements

We are indebted to Professor A.S. Nowick for making available the results of his recent investigations on s-i relaxations prior to publication. This work was sponsored by the United States Atomic Energy Commission. The manuscript was prepared with the assistance of facilities in the Section de Physique du Solide at C.E.N.G., Grenoble.

References

1. C.A. Wert, in Physical Acoustics, Vol.3A, ed. W.P. Mason, Academic Press, 1966.

2. P. Moser and R. Pichon, J. Phys. F : Metal Phys. 3, 363 (1973).

3. D.F. Hasson and R.J. Arsenault, in Treatise on Materials Science and Technology, Vol. 1, ed. H. Herman, Academic Press, New York, 1972.

4. A.S. Nowick and B.S. Berry, Anelastic Relaxation in Crystalline Solids, Academic Press, New York, 1972

5. H. Hashizume and T. Sugeno, Jap. J. Appl. Phys. 6, 567 (1967).

6. R.E. Miner, D.F. Gibbons and R. Gibala, Acta Met. 18, 419 (1970)

7. A.A. Sagues and R. Gibala, Scripta Met. 5, 689 (1971).

8. A.A. Sagues and R. Gibala, to be submitted to Acta Met.

9. A.A. Sagues, Ph.D. Thesis, Case Western Reserve University, 1972.

10. H. Ino, S. Takagi and T. Sugeno, Acta Met. 15, 29 (1967).

11. P.L. Raffo, Ph.D. Thesis, Case Western Reserve University, 1967.

12. D.N. Beshers, J. Appl. Physics $\underline{36}$, 290 (1965)

13. D. Mosher, C. Dollins and C. Wert, Acta Met. $\underline{18}$, 797 (1970)

14. A.S. Nowick, Scripta Met. $\underline{7}$, 289 (1973).

15. A.S. Nowick, to be published.

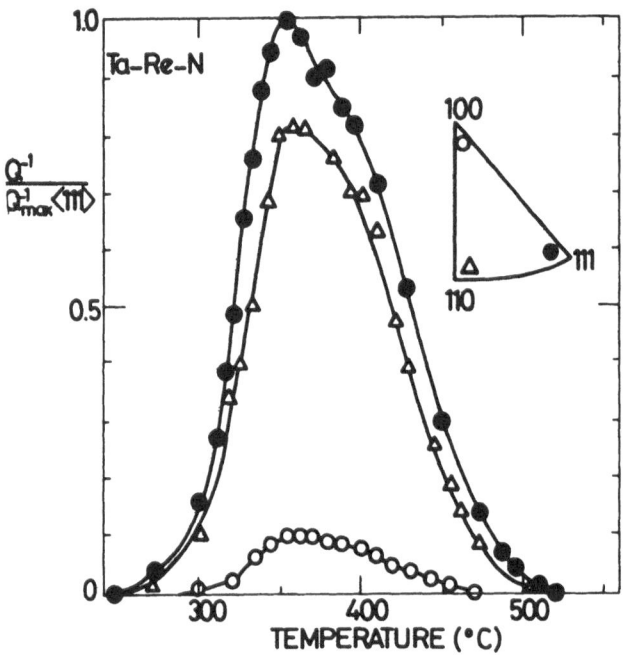

FIG. 1

Orientation dependence of the Snoek spectrum for
Ta-3.8 at.% Re - 0.06 at.% single crystals.
Frequency = 1.6 Hz.

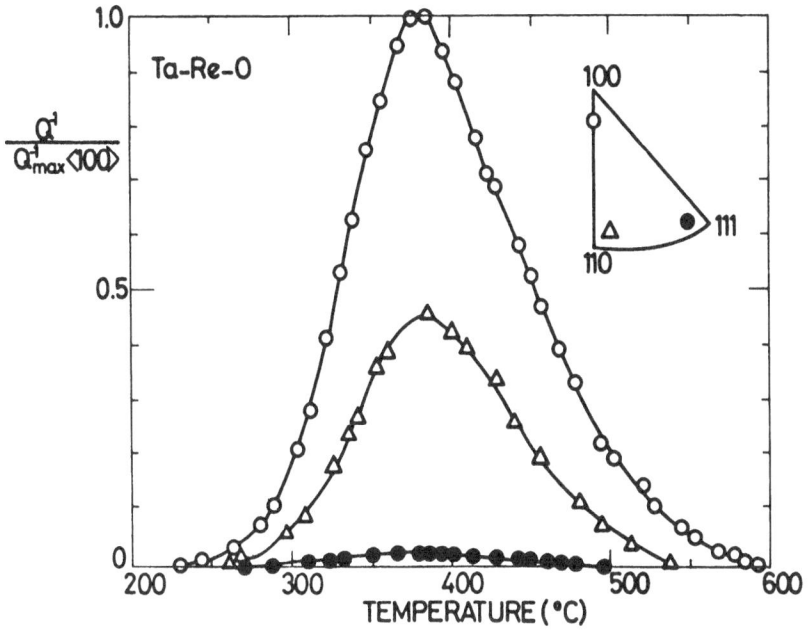

FIG. 2

Orientation dependence of the Snoek spectrum for
Ta-3.8 at.% Re - 0.04 at.% O single crystals.
Frequency = 50 kHz.

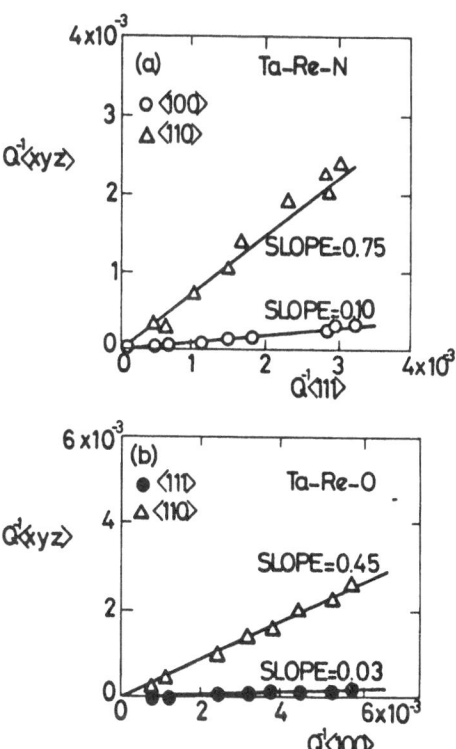

FIG. 3

Graphical analyses of Figs. 1 and 2
showing the composite peak shape does
not change with orientation.

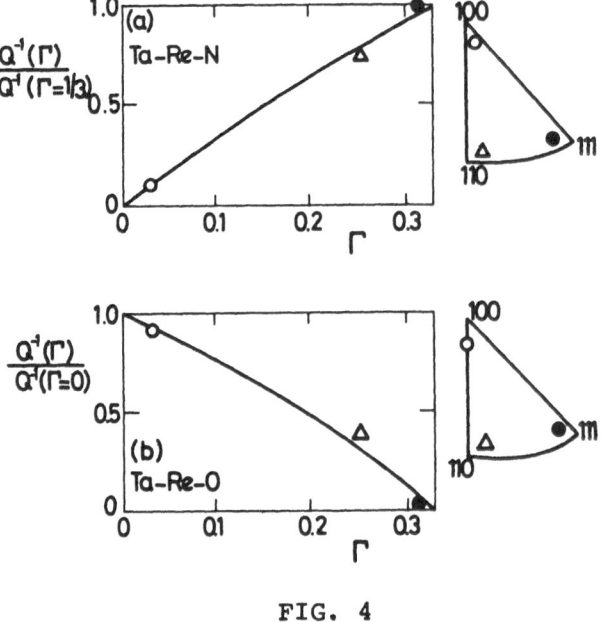

FIG. 4

Comparison between experimental data for (a)
Ta-Re-N alloys and (b) Ta-Re-O alloys and
the predicted values for ideal Snoek peaks.

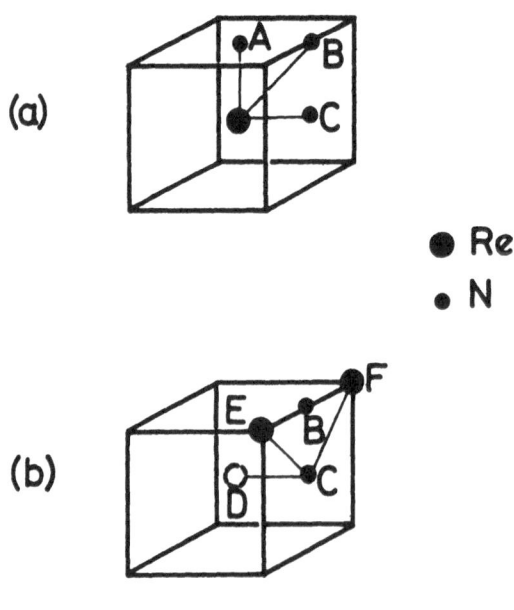

FIG. 5

Models for substitutional-interstitial defects
(a) ReN defects, (b) Re_2N defects.

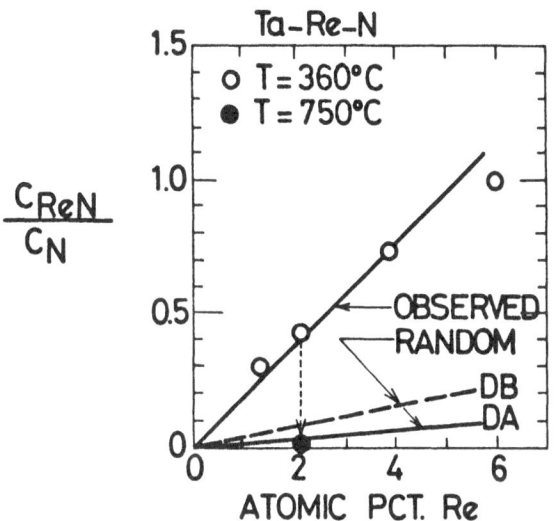

FIG. 6

The ratio C_{ReN}/C_N observed experimentally (data points)
and calculated for random solid solutions containing
DA or DB defects.

ON THE EFFECT OF GRAIN SIZE ON THE INTERNAL FRICTION PEAK DUE TO INTERSTITIALS

A. Ferro-Milone and F. Mezzetti
Istituto Elettrotecnico Nazionale Galileo Ferraris, GNSM - CNR, Torino
Istituto di Fisica dell'Università, GNSM - CNR, Ferrara

ABSTRACT

It is shown that the grain size effect on the strength of the after-effect due to interstitial atoms or any other source of directional ordering is caused, to a large extent, by the different mode of stress transmission between the grains: at constant stress for grains of the order of the cross section of the specimen, and intermediate between constant stress and constant strain for small grain sizes. Typically for an interstitial in octahedral position in α iron such as carbon, the conversion factor between the concentration c in weight and the internal friction Q^{-1} at 1 Hz in torsion is respectively 1.21 for specimens with small grain size and 0.82 for large grain size. The same factors for nitrogen are 1.4 and 0.95, with good agreement with experimental results mainly for nitrogen. In order to explain some observed discrepancies a figure of the interstitial fraction anchored at grain boundaries, impurities and lattice defects was obtained by comparing the internal friction obtained by quenching from γ phase and α phase respectively. Some experimental results for iron of different purity are given.

Introduction

The presence of grain size effect on the relaxation strength due to interstitials in polycrystalline materials has been recognized since long time. This effect is rather large (1÷11) and is important particularly when the internal friction method is used for dosing the dissolved content of carbon or nitrogen in iron.

To some extent the effect has been considered as an evidence of the segregation of these elements at grain boundaries or at small amounts of present impurities. According to some authors (4,5) for very small grain sizes the effect is possibly also due to some inevitable grain boundary precipitation during the quench so that it is impossible to reach, at room temperature, the impurity concentration values corresponding to the true equilibrium at high temperature.

Recently Swartz (6) has shown that the effect is in part due to the fact that different grain sizes are associated, as a rule, with different crystalline textures. However, the problem cannot be considered completely solved.

The fact that for single crystals there is very good agreement between theory and experiments (12,13) suggests that, to a large extent, the effect may be simply connected with the problem of finding the proper procedure for averaging the internal friction values over the different crystals in a randomly oriented polycrystal.

In earlier works this was done simply by taking the average values of direction cosines (12,14). More recently the problem was first considered carefully by Smit and Van Bueren (15) using a general function deduced experimentally which relates the ratio between the elastic constants of the polycrystal and those of the single crystal to the anisotropy of these constants.

More recently Ino et al. (16) made the theoretical calculation of the relaxation strength of a polycrystal both in bending and in torsion in the two extreme situations of uniform stress and uniform strain distribution. The authors conclude that the uniform stress calculation agrees with the experimental results on coarse grained polycrystals (i.e. when the crystal size is of the order of the cross section of the specimen) while the calculation at constant strain does not correspond to any actual physical situation. Moreover the averaging procedure used in this second case does not seem to correspond to the uniform strain approximation and does not agree with the results of Smit and Van Bueren.

Calculations

The problem was therefore considered again, starting from the results of Smit and Van Bueren. For grain sizes larger or of the same order of the cross section of the specimen, it is clear that all crystals are submitted to the same stress and therefore the relaxation average over the different orientations must be performed at constant stress. When the grain size is small with respect to the cross section of the specimen the relaxation occurs as the average between constant stress and constant strain: in the first situation the crystals can be considered in series to a given crystal and in the second one in parallel.

The expressions of the relaxation strength, Δ_E and Δ_G, for a polycrystalline specimen are then immediately obtained using the mean elastic moduli of a collection of monocrystals in the same state of stress and in the same state of strain, given respectively by Reuss and Voigt (17,18):

$$(1) \quad \left\langle \frac{1}{E} \right\rangle_\sigma = \frac{3S_{11} + 2S_{12} + S_{44}}{5} \quad , \quad \left\langle \frac{1}{G} \right\rangle_\sigma = \frac{4S_{11} - 4S_{12} - 3S_{44}}{5}$$

at uniform stress and

$$(2) \quad \left\langle \frac{1}{E} \right\rangle_\varepsilon = \frac{2C_{11} + 3C_{12} + C_{44}}{(C_{11}-C_{12}+3C_{44})(C_{11}+2C_{12})} \quad , \quad \left\langle \frac{1}{G} \right\rangle_\varepsilon = \frac{5}{C_{11} - C_{12} + 3C_{44}}$$

at uniform strain. The symbol $\langle \ \rangle$ denotes the average on all directions.

On the other hand the presence in the crystal lattice of impurity atoms gives rise to changes in the elastic moduli ΔC_{ij} and elastic constants ΔS_{ij}. As known

$$(3) \quad \begin{array}{ll} \Delta S_{11} = -2\,\Delta S_{12} & \Delta S_{44} = 0 \\[2mm] \Delta C_{11} = -2\,\Delta C_{12} & \Delta C_{44} = 0 \end{array}$$

where the ΔC_{ij} are related to the ΔS_{ij} by the equation

$$(4) \quad \Delta C_{12} = -(C_{11} - C_{12})^2 \, \Delta S_{12}$$

Therefore, by logarithmic differentiation of Eqs. (1) and (2), one obtains respectively

$$(5) \quad \begin{aligned} (\Delta_E)_\sigma &= \frac{\langle \Delta\,1/E \rangle_\sigma}{\langle 1/E \rangle_\sigma} = \frac{-4\,\Delta S_{12}}{3S_{11} + 2S_{12} + S_{44}} \\[4mm] (\Delta_G)_\sigma &= \frac{\langle \Delta\,1/G \rangle_\sigma}{\langle 1/G \rangle_\sigma} = \frac{-12\,\Delta S_{12}}{3S_{44} + 4S_{11} - 4S_{12}} \end{aligned}$$

$$(6) \quad \begin{aligned} (\Delta_E)_\varepsilon &= \frac{\langle \Delta\,1/E \rangle_\varepsilon}{\langle 1/E \rangle_\varepsilon} = \frac{5(C_{11}+2C_{12})\,\Delta C_{12}}{(2C_{11}+3C_{12}+C_{44})(C_{11}-C_{12}+3C_{44})} \\[4mm] (\Delta_G)_\varepsilon &= \frac{\langle \Delta\,1/G \rangle_\varepsilon}{\langle 1/G \rangle_\varepsilon} = \frac{3\,\Delta C_{12}}{C_{11} - C_{12} + 3C_{44}} \end{aligned}$$

Substituting Δs_{12} to Δc_{12} into Eq. (6) the relaxation strengths at constant strain become

$$(\Delta_E)_\varepsilon = \frac{-5(C_{11}+2C_{12})(C_{11}-C_{12})^2 \Delta s_{12}}{(2C_{11}+3C_{12}+C_{44})(C_{11}-C_{12}+3C_{44})}$$

(7)

$$(\Delta_G)_\varepsilon = \frac{-3(C_{11}-C_{12})^2 \Delta s_{12}}{C_{11}-C_{12}+3C_{44}}$$

Using the known values of the elastic constants for α iron and the values of Δs_{12} obtained from the tetragonal deformation of martensite (12,15) the corresponding values of Q^{-1} and of the conversion factor $K = c/Q^{-1}$ in different conditions are easily obtained, where the values for the polycrystal are the average of the two extreme ones. The results obtained are reported in Table I.

TABLE I

		Uniform Stress K_σ	Uniform Strain K_ε	Average (Polycrystal) K_p
a) Carbon Torsion	Q^{-1}	$3.82 \times 10^2 c_C/T_C$	$1.34 \times 10^2 c_C/T_C$	$2.58 \times 10^2 c_C/T_C$
	K	0.82	2.33	1.21
Bending	Q^{-1}	$3.34 \times 10^2 c_C/T_C$	$1.14 \times 10^2 c_C/T_C$	$2.24 \times 10^2 c_C/T_C$
	K	0.93	2.74	1.4
b) Nitrogen Torsion	Q^{-1}	$3.11 \times 10^2 c_N/T_N$	$1.09 \times 10^2 c_N/T_N$	$2.10 \times 10^2 c_N/T_N$
	K	0.95	2.71	1.4
Bending	Q^{-1}	$2.72 \times 10^2 c_N/T_N$	$0.93 \times 10^2 c_N/T_N$	$1.83 \times 10^2 c_N/T_N$
	K	1.08	3.17	1.61

The results at constant stress reproduce the ones of Ino et al. (16), while those for small grain size are the theoretical correspondent of the ones of Smit and Van Bueren and are actually very close to their values.

As seen the change of the factor K from small to large grains is large, and this physically corresponds to the fact that the constrictions due to the surrounding grains strongly reduce the relaxation strength occurring in the crystallographic direction in which the elastic modulus for iron has a minimum.

Once the role of the grain size has been evaluated in these two extreme conditions a figure for intermediate cases may be obtained to a first approximation by considering that the crystals on the surface are almost free, while the crystals in the interior are constricted by the surrounding ones.

Each crystal when surrounded by other crystals with different orientations can be considered as a stress center whose diameter is equal to its size. As is known the stresses around a stress center decrease with the third power of the distance. When the center of the grain is on the boundary, the constriction due to the surrounding crystals can be considered as completely relaxed, and when a crystal has its boundary at about one crystal diameter from the surface, the state of the stress produced by these constrictions is almost completely built-up. Therefore the internal friction Q^{-1} of the specimen can be considered as the average between the crystals on the surface which relax roughly at constant stress and the crystals in the interior which, as seen, relax half-way between constant stress and strain.

Then, if d is the crystal size and D the size of the specimen, the parameter K for specimens of any grain size is given by

$$(8) \qquad K = \frac{K_p}{(K_p/K_\sigma)(1 - (1 - 2\,d/D)^2) + (1 - 2\,d/D)^2}$$

where K_p is the conversion factor of polycrystalline specimens with small grain size and K_σ corresponds to the case of large crystals.

Experimental

The experimental values of K for C and N in α iron obtained by the different authors are given in Table II and in Fig. 1 a) b) in comparison with the theoretical curves given by Eq. (8) and the values of Table I. As is seen a very good agreement with the theoretical values is obtained for Nitrogen. For Carbon the situation is not so good, probably because the solubility

TABLE II   * estimated

Interstitial Content (Chemical) weight %	Quenching Temperature °C	Grain Size microns	Specimen Size microns	Internal Friction Q^{-1}	Observed Conversion Factor Q^{-1}	Theoretical Conversion Factor	Possible Fraction of Blocked Impurity weight %	Reference
a) Carbon								
0.023	710	500-2000	700	0.026	0.88	~0.82	0.002	1
0.013	630	"	700	0.013	1.0	~0.82	0.002	"
0.031	710	15-50	700	0.016	1.9	~1.11	0.013	"
0.023	630	"	700	0.0078	2.9	~1.11	0.014	"
0.0197	710	-	2500	0.0116	1.7	-	-	19
-	-	70-140	(500-1000)*	-	1.3	~0.95	-	2-3
0.02	700	70-200	"	-	1.45	~0.90	0.008	"
0.02	700	15-40	"	-	2.0	~1.11	0.010	"
b) Nitrogen								
0.032	570	400	(500-1000)*	0.032	1.0	~0.95	negligible	4
0.032	"	150	"	0.029	1.1	~1.05		"
0.032	"	50	"	0.025	1.28	~1.24		"
0.017	"	700	"	0.019	0.89	~0.95		"
0.017	"	280	"	0.016	1.06	~0.95		"
0.017	"	70	"	0.014	1.21	~1.29		"
0.006-0.045	710-450	500-2000	700	0.007-0.042	~0.95	~0.95		1
0.006-0.045	"	15-25	"	0.005-0.030	~1.35	~1.33		"

303

of C is much lower and therefore the error introduced by the interstitial fraction trapped at impurities, grain boundaries and lattice defects is much higher.

However, also in this case good agreement is obtained for large grain sizes and in any case all experimental points lie above the theoretical curve, as expected. The results for Carbon emphasize the very large difference which may occur between the actual number of interstitial atoms in solid solution and the effective interstitial concentration. The effect is clearly due to the fact that at the quenching temperature the interstitial atoms anchored at lattice defects, grain boundaries or impurities are not yet released.

In Table II the C fractions of blocked impurities which would permit the agreement between the theoretical value of K and the measured value of internal friction are given. Also an effect of the specimen texture, however, may be present.

Evidence that some interstitial fraction is blocked is for instance provided by the fact that linear extrapolation to zero concentration of some internal friction values gives a definite intercept (1). For less pure iron some fraction may be trapped to strong carbide or nitride forming impurities while the other can be trapped essentially at grain boundaries. Also vacancies and dislocations can give some contribution (19).

Typically for grain size of 30 μ and 10 interstitial atoms blocked per area element equal to the atomic diameter, the fraction trapped at grain boundaries may be of the order of 50 ppm in weight, which may be of the correct order of magnitude to explain the internal friction values observed for very small grain sizes.

To better understand this effect, some experiments were performed on iron specimens of different purity and very low interstitial content by quenching from very high temperatures in γ phase, in order to bring in solubility also the anchored interstitial fraction. In fact the solubility of interstitials in γ phase is about 1 order of magnitude higher than in α phase and therefore the free energy difference to free a blocked interstitial atom decreases accordingly. As will be seen from the experimental results, no inconveniences are caused by the phase transition which may only slightly increase the background damping due to the introduced strains. The experiments were performed with an inverted torsion pendulum on strips 0.3 x 200 x 8 mm at the frequency of about 1 Hz.

Two typical results are reported in Fig. 2: a) pure electrolitic iron and b) iron of technical purity (99.98). The materials are polycrystalline with almost isotropic crystalline distribution and grain size of about 150 μ. Several other results on other specimens with somewhat different purity are analogous.

As can be noticed a marked increase of the free interstitial content is found abruptly when quenching from γ phase. By quenching from still higher temperatures only minor changes were observed. The effect is largely reversible (Fig. 2 a)). In the two cases considered the pinned interstitial fraction results therefore about 8 ppm in weight for the electrolitic iron and about 20 ppm for the iron of technical purity, which are of a reasonable order of magnitude. The pinning centers may be, to a major extent, identified with grain boundaries, at least for the high purity iron. However, also other defects may contribute. The binding energy of interstitials to pinning points estimated from the observed dissolved quantity is of the order of 0.8 eV.

In conclusion quenching experiments from high temperatures in γ phase give the direct evidence of the interstitial fraction trapped on the defective regions of the lattice and on strong carbide or nitride forming impurities. The model proposed clears in general the grain size effect on elastic relaxation due to ordering.

References

1. G. Langersberg and A. Josefson, Acta Met. 3, 236 (1955).
2. H.J. Seeman and W. Dickenscheid, Rev. Met. 49, 376 (1952).
3. H.J. Seeman and W. Dickenscheid, Acta Met. 6, 62 (1958).
4. J.D. Fast, Le Frottement Intérieur des Métaux, p. 9, IRSID (1960).
5. P. Stark, B.L. Averbach and M. Cohen, Acta Met. 6, 149 (1958).
6. J.C. Swartz, J.W. Shilling and A.J. Schwoeble, Acta Met. 16, 1359 (1968).
7. K. Aoki, S. Sekino and T. Fijishima, J. Japan Inst. Metals 26, 47 (1962).
8. M.I. Bayazitov and Y.V. Piguzov, Fiz. Metal. Metalloved 20, 632 (1965).
9. J.D. Fast and M.B. Verrijp, J. Iron Steel Inst. 180, 337 (1955).
10. P.M. Strocchi, Energia Nucleare 13, 564 (1966).
11. J.C. Swartz, Acta Met. 17, 1511 (1969).
12. L.J. Dijkstra, Philips Res. Rpt. 2, 357 (1947).
13. A.S. Nowick and W.R. Heller, Adv. Phys. 12, 251 (1963).

14. R. Rawlings and D. Tambini, J. Iron Steel Inst. 184, 302 (1959).

15. J. Smit and H.G. Van Bueren, Philips Res. Rpt. 9, 460 (1954).

16. H. Ino, S. Takagi and T. Sugeno, Acta Met. 15, 29 (1967).

17. W. Voigt, Lehrbuch der Krystalphysik, G.B. Teubner Ed., Leipzig (1910).

18. A. Reuss, Z. Angew. Math. Mech. 9, 49 (1929).

19. E. Lindstrand, Acta Met. 3, 431 (1955).

20. M. Wuttig, J.T. Stanley and H.K. Birnbaum, Phys. Stat. Sol. 27, 701 (1968).

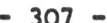

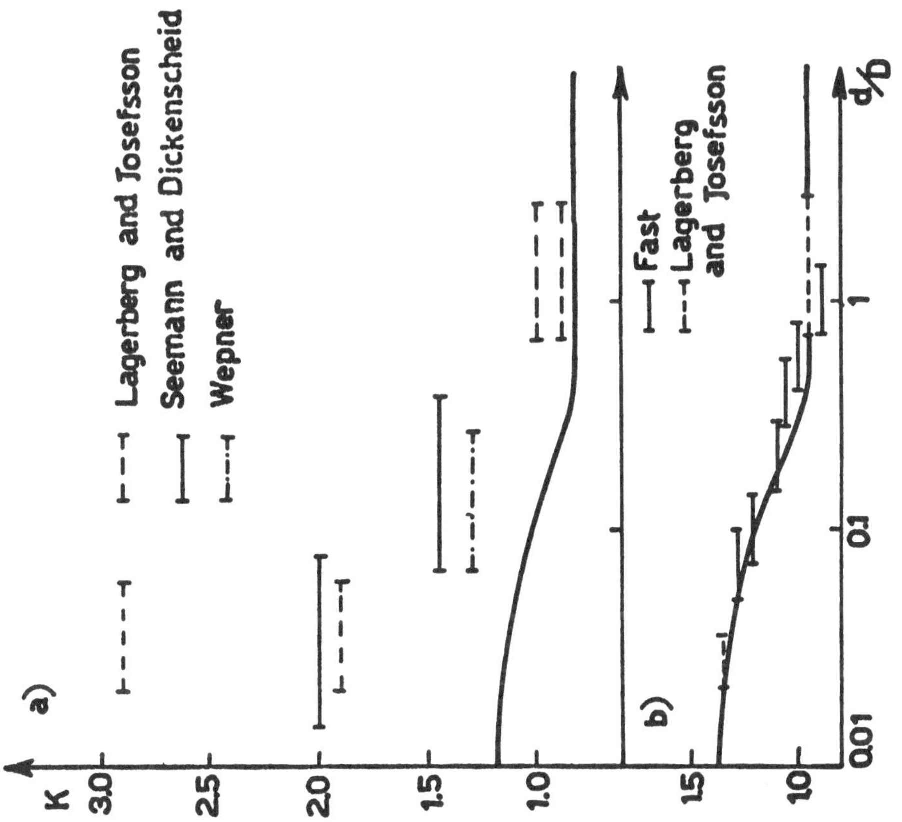
K
3.0
2.5
2.0
1.5
1.0
a)
b)
Lagerberg and Josefsson
Seemann and Dickenscheid
Wepner
Fast
Lagerberg and Josefsson
0.01
0.1
1
d/D

Fig. 1

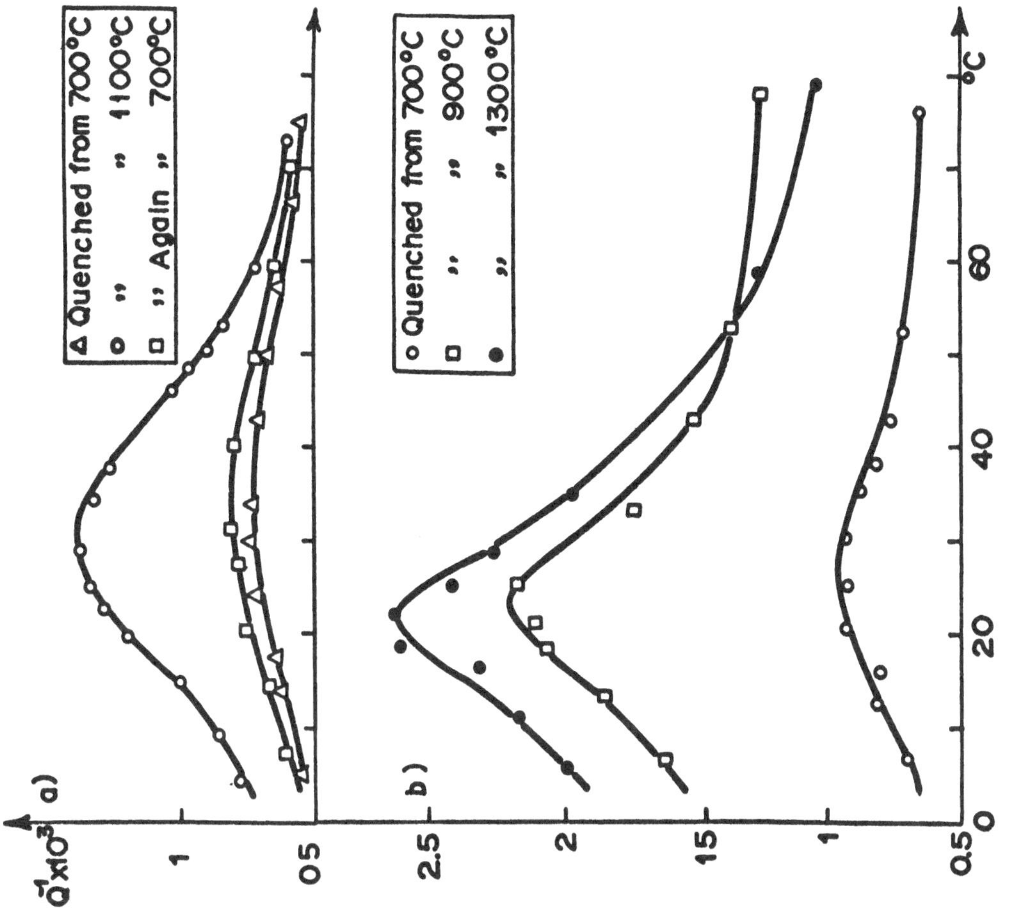
Q⁻¹x10³
a)
b)
Quenched from 700°C
" " 1100°C
" Again " 700°C
Quenched from 700°C
" " 900°C
" " 1300°C
0.5
1
1.5
2
2.5
0
20
40
60
°C

Fig. 2

EFFECTS OF GAMMA-IRRADIATION ON THE ULTRASONIC ATTENUATION
IN QUARTZ AND TOURMALINE AT LOW TEMPERATURES

M. S. Thuraisingham[1] and R. W. B. Stephens[2]

1. Imperial College, London and Department of Physics, University
of Ceylon, Katubedde Campus, Sri Lanka. (Present Address)
2. Chelsea College, London.

Introduction

The work to be described deals with part of a detailed investigation (1) that was made in order to study the effects of gamma-irradiation and annealing on the attenuation of hypersonic compressional waves in quartz and tourmaline crystals. The defect concentration in the crystals was altered by successive doses of irradiation and subsequent annealing, and the effects of these changes on the ultrasonic attenuation studied. The data relating to overall attenuation behaviour is correlated with the attenuation theory of Silverman (2).

In computing the ultrasonic attenuation in insulating crystals two regions of temperature are generally recognised. In the high temperature, or 'classical', region where the impressed sound wavelength is greater than the mean free path of the thermal phonon modes, and defined by the condition $\omega\tau < 1$, where ω is the sound frequency and τ the thermal phonon relaxation time, the interaction is treated macroscopically. The calculation is based on the mechanism originally proposed by Akhieser (3) in which the sound wave is regarded as a macroscopic driving force that, owing to anharmonic interaction, modulates the frequencies of the thermal phonons. The perturbed phonon system being out of thermal equilibrium relaxes through collisions among phonons in the different modes towards the distribution which the phonons would have if they were locally in equilibrium with the lattice. This delayed re-establishment of equilibrium in the phonon gas results in an increase in the entropy of the system and leads to the attenuation of the sound wave.

In the treatment of the low temperature, or 'quantum', region where the impressed sound wavelength is less than the mean free path of the thermal phonon modes, and governed by the condition $\omega\tau > 1$, the sound wave is considered to interact with individual phonon modes in a microscopic manner, the ultrasonic attenuation being given by the rate of energy loss in the direct process of interaction between the acoustic and thermal phonons. Here the lifetime of the thermal phonons is sufficiently long to survive many periods of the sound wave so that conditions are appropriate for such direct interaction.

Shear wave attenuation in the low temperature region is well accounted for by the quantum theory of Landau and Rumer (4) which gives the attenuation as proportional to T^4, where T is the absolute temperature. There are difficulties however when considering compressional waves, which cannot interact directly with thermal waves and therefore their attenuation should be much less than for shear waves, but the measured compressional wave attenuation in quartz does not bear this out. Various theories have been proposed to explain this discrepancy with varying degrees of success, but an explanation of promise is the recent generalised theory of Silverman (2) which has also the advantage of being applicable to attenuation in imperfect insulating crystals.

Silverman's Theory

Silverman assumes that the introduction of imperfections into the crystal has the effect of modifying the frequency and lifetime of the thermal phonons which are coupled to the sound wave via anharmonic interactions. He supposes the imperfect crystal to comprise a specified number of random defect sites which are coupled to the remaining host atoms. In calculating the consequent change in the potential energy of the imperfect from the perfect crystal the host-host as well as the host-defect interactions are taken into account, but the coupling between the defect sites themselves is neglected on the assumption that the defect density is small.

In recent experiments reported by Nava and Rodriguez (5) for neutron irradiation of quartz they explain their attenuation observations at low doses by assuming that defects are readily created in the quartz, and furthermore the observed frequency independent absorption below 15 K, for both irradiated and annealed samples, support the concept of a small number of lattice point defects being involved.

The expression obtained by Silverman for the attenuation per unit length α, which is valid for all values of $\omega\tau$, is

$$\alpha = \frac{8.68\,\gamma^2\omega CT}{8\rho V^3(1-\lambda)}\;arc\,tan\left[2(1-\lambda)(\omega\tau)\Big/\left\{1+\lambda(2-\lambda)(\omega\tau)^2\right\}\right] \qquad (1)$$
$$(1a)$$

where $\lambda = 0.32\left(a\,kT/\hbar V\right)^2$, (a) being the atomic lattice dimension, V the acoustic velocity, γ the volume Gruneisen constant, ρ the density and C the specific heat at constant volume, $\hbar$ is Planck's constant and k is Boltzmann's constant.

The above expression reduces in the high temperature region to the expression of Akhieser (3) :-

$$\alpha = \frac{8.68\,\gamma^2\omega^2 CT\tau}{4\rho V^3} \qquad (2)$$

and in the low temperature region to the expression of Maris (6) for compressional waves :-

$$\alpha = \frac{8.68\,\gamma^2\omega CT}{8\rho V^3}\left\{\frac{\pi}{2} - arc\,tan(\lambda\omega\tau)\right\} \qquad (3)$$

It is to be noted that the lowering of the effective thermal phonon lifetime due to the introduction of imperfection scattering as a result of irradiation affects the ultrasonic attenuation in the low and high temperature regions in opposite ways. At high temperatures the decrease in lifetime results, on the basis of the Akhieser damping mechanism, in a more rapid relaxation towards the instantaneous equilibrium phonon distribution and the attenuation is reduced. At low temperatures a reduction in the lifetime leads to the relaxation of the energy conservation restriction and the attenuation is increased. These considerations point to the existence of a cross-over point for the perfect and imperfect attenuation curves. Silverman has shown that such cross-over may be expected to occur in the vicinity of $\omega\tau = 1$.

Experimental

The attenuation measurements were carried out at Imperial College and were made at the frequency of 9.57 GHz. The microwave frequency ultrasonic phonons used in the measurements were generated by the surface excitation technique originated by Baranskii (7) and employed by a succession of workers in the field. The experimental system used does not require detailed or special comment as it comprises the standard set-up for surface excitation of a piezoelectric crystal at microwave frequencies. A schematic diagram of the system is given in Fig. 1 while Fig. 2 shows the construction of the cryostat. Attenuation measurements were made at liquid helium temperature (4.2 K) and thereafter from time to time during the warm-up of the sample, using a Sanders precision rotary vane attenuator. Temperature measurements were made to $\pm$ 0.2 K using copper-constantan thermocouples for the high and chromel-atomic iron in gold for the low temperatures. The attenuation in quartz was measured over the temperature range 4.2 K to 38 K and up to 150 K in tourmaline. The results have been in all cases normalized to zero attenuation at 4.2 K, i.e. the value of the attenuation at 4.2 K has been subtracted from the data in order to eliminate the additive temperature independent contribution to the attenuation.

The X-cut quartz specimen used was a cylindrical rod 0.3 cm. diameter, 2.50 cm. long with the end faces polished flat to a tenth of an optical wavelength and parallel to 3 sec. of arc. The crystal was orientated to within $\pm$ 0.5°. The absorbed doses of gamma-radiation for the quartz specimen were 0.76, 1.50, 3.81, 6.84 Mrad. After the final dosage and attenuation measurement the crystal was subjected to isothermal annealing at 500°C for successive periods of ten hours each making a total of 30 hours. The attenuation was after each period of annealing measured.

The c-axis tourmaline specimen used, 0.89 cm. long and 0.3 cm. in diameter, was polished to similar close tolerances as for quartz. It was irradiated with doses 1.49 and 5.96 Mrad. and was subsequently annealed at 500°C for 30 hours.

Results

Quartz. The attenuation of compressional waves in quartz may be expressed by the power law $\alpha = AT^n$, where α is the attenuation in dB/cm., T the absolute temperature, A a constant and n the power index. For the un-irradiated specimen the value of 4.2 obtained for the power index differed from the value of 5.7 obtained theoretically by Maris (8) and experimentally by Patterson (9) for his pure specimen. A logical inference is that the author's specimen was initially imperfect. Fig. 3 is a curve of attenuation against temperature for the smallest irradiation dose used and clearly indicates the opposing effects at low and high temperatures giving rise to the cross-over point P. Cross-over points obtained from graphs such as Fig. 4 gave for the un-irradiated, dose 1 (0.76 Mrad.) and dose 2 (1.50 Mrad.) the cross-over temperatures as 50 K, 45 K and 40.5 K respectively, the cross-over occurring at lower temperatures with increased dosage. It should be noted that in each case the cross-over point has been located relative to Patterson's curve for his pure specimen. The observed downward shift of the cross-over point with dosage appears to confirm Silverman's prediction of an extension of the Akhieser region with a corresponding diminution of the Landau-Rumer region.

Since Silverman's expression is valid for all values of $\omega\tau$ a cross-over is obtained by setting the attenuation in the irradiated crystal equal to that in the un-irradiated crystal. This leads to the expression :-

$$\omega\tau_c = \left[1/(\omega\tau_i) + \left\{ 1/(\omega\tau_i)^2 + 4\lambda(2-\lambda) \right\}^{1/2} \right] / 2\lambda(2-\lambda) \qquad (4)$$

Here ω is the excitation frequency, τ_c the thermal phonon relaxation time at the cross-over temperature for the perfect crystal, τ_i being the contribution to the effective thermal phonon relaxation time τ in the irradiated crystal due to the presence of imperfections and given by the relation :-

$$\tau^{-1} = \tau_c^{-1} + \tau_i^{-1} \qquad (5)$$

In analysing the cross-over data the value of τ_c for Patterson's specimen from the phonon relaxation time – temperature curve of Maris (8) is used. Corresponding to the cross-over temperatures of 50 K, 45 K and 40.5 K the respective values of τ_c are 7.0×10^{-11} sec., 10.1×10^{-11} sec. and 14.5×10^{-11} sec. From relations (1a) and (4) the quartz values of λ and $\tau_i^{-1} = \nu_i$ (the imperfection contribution to the relaxation frequency) may be respectively calculated. The results of the calculation show that ν_i increased by 1.4×10^{10} sec.$^{-1}$ in the author's specimen after irradiation dose 1 and by 2.7×10^{10} sec.$^{-1}$ after dose 2.

Tourmaline. Similar results were obtained for tourmaline, whose attenuation – temperature curve comparison with the quartz specimen is shown in Fig. 5, while Fig. 6 shows the existence of the cross-over point.

Conclusions

The overall attenuation behaviour for both quartz and tourmaline has been found to follow closely the predictions of Silverman's theory. The results appear to confirm his basic postulate that the effect of irradiation is to introduce scattering centres into the crystal and lower the thermal phonon relaxation time, thereby influencing the ultrasonic attenuation.

Furthermore, the power law relating attenuation and temperature has been shown by the author (1) to be influenced by irradiation. The associated velocity changes that were observed appear to suggest that the attenuation is influenced via the elastic properties of the crystal.

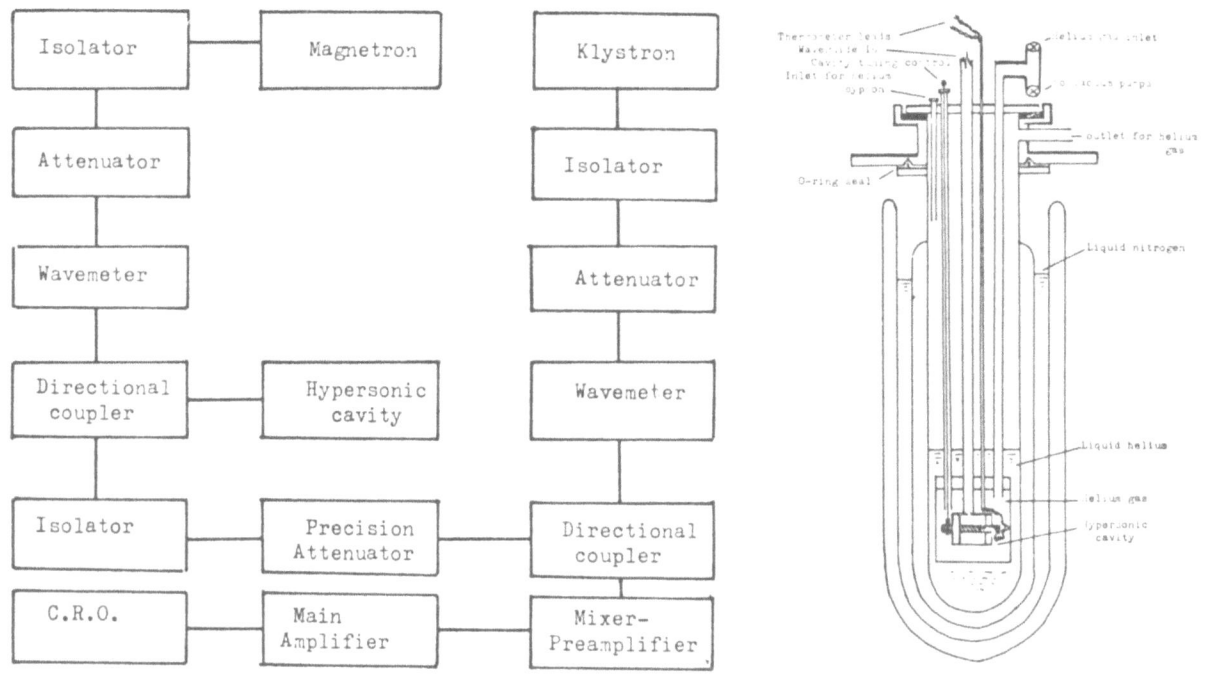

Fig. 1

Fig. 2

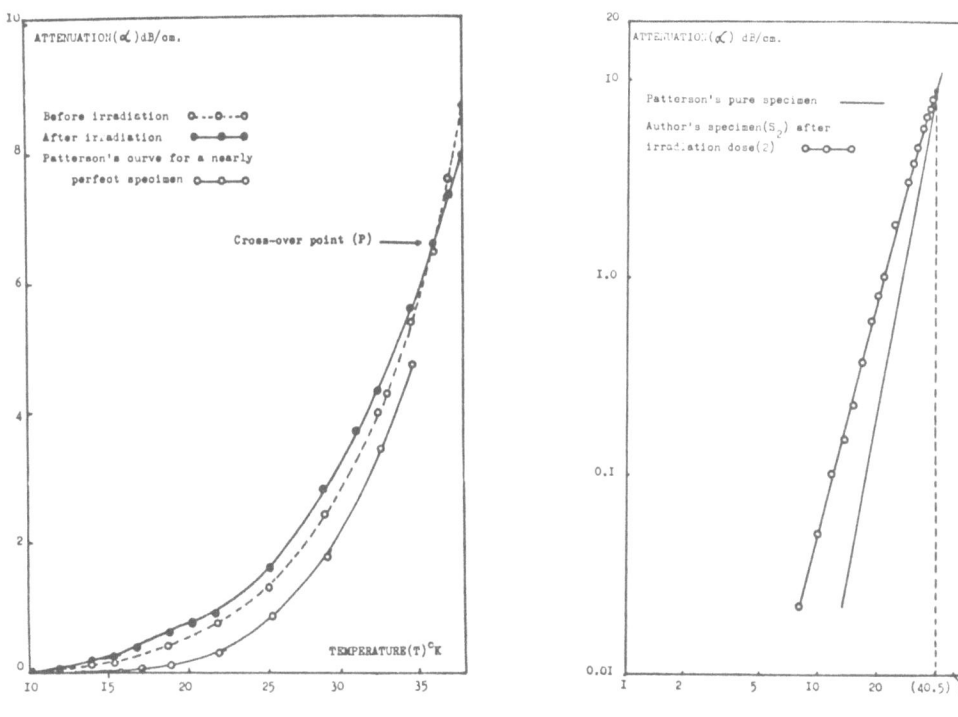

Fig. 3

Fig. 4

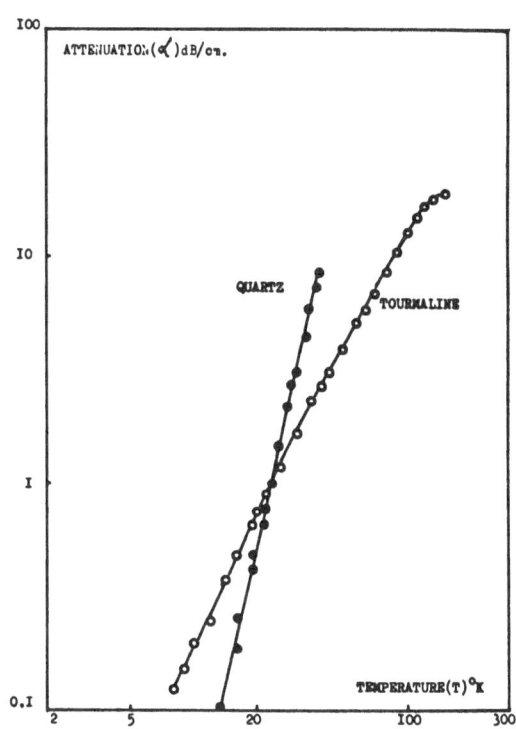

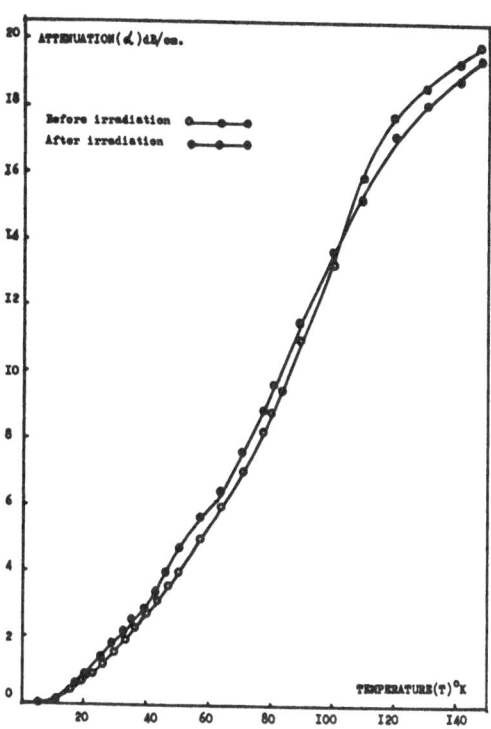

Fig. 5   Fig. 6

References

1. M. S. Thuraisingham, Ph.D. Thesis, University of London (1973).

2. B. D. Silverman, Prog. Theor. Phys. 39, 245 (1968).

3. A. Akhieser, J. Phys. (USSR) 1, 277 (1939).

4. L. Landau and G. Rumer, Physik. Z. Sowjet-Union 11, 18 (1937).

5. R. Nava and M. Rodriguez, Phonon Scattering Conference, Paris (1972).

6. H. J. Maris, Ph.D. Thesis, University of London (1963).

7. K. N. Baranskii, Soviet Phys. Doklady 2, 237 (1957).

8. H. J. Maris, Phil. Mag. 9, 901 (1964).

9. E. Patterson, Ph.D. Thesis, University of London (1968).

Relaxation Processes in Boron

F. N. Tavadze, G.V. Tsagareishvili, G.Sh. Darsavelidze

G.F. Tavadze, R.A. Khachapuridze

Institute of Metallurgy Academy of Sciences
of Georgian SSR, Tibilisi, USSR

The given work deals with the results of investigations of relaxation processes in β-rhombohedral boron by internal friction method. The experiments were carried out on films, plates, and cylindrical samples of zone melted boron with purity not less than 99,9%.

Cylindrical samples of 0.8 mm in diameter and of 80 to 120 mm length were investigated in a low-frequency vacuum relaxation apparatus of the inverted torsional pendulum type of construction described in reference 1. The measurements were carried out at a medium deformation amplitude of 5×10^{-6} in the frequency range of 0.5 to 3.5 cps at temperatures from $196^\circ C$ to $+ 800^\circ C$.

Boron samples of 1.2 mm diameter were thinned by mechanical polishing on boron carbide powder and also by chemical etching in boiling nitric acid.

The temperature dependence of the internal friction in plates and films of β-rhombohedral boron from room temperature to $+ 800^\circ C$ was investigated by a vibrational method with electrostatic exitation. The average amplitude of relative deformation was $\sim 10^{-7}$. Boron plates of $2 \times 20 \times 0.35$ mm^3 size were obtained by mechanical cutting of massive single-crystal boron samples with a diamond disc. Boron films 0.1 mm thickness were obtained by quenching from the liquid state. The falling drop was clasped by massive copper /2/. X-ray investigations of the films showed their particularly fine grained structure.

Earlier investigations /3,4/ of the temperature dependence of internal friction in cylindrical samples of β-boron at 1 cps frequency have shown four relaxation maxima at the temperatures -120°C, -20°C, +70°C, +260°C, with activation energies 0.13; 0.53; 0.76; 1.3 eV, and a sharp peak at 300°C, of non-relaxation nature.

It was established that the height of the peak at -120°C is amplitude indepedent. The temperature of the maximum is not influenced by impurities, deformation amplitude and thermal treatment; annealing lowers the peak; quenching increases its height and impurities strongly reduce the peak height. Thus, the peak has all characteristics of the stable low temperature maxima observed in metals after plastic deformation, where the peak is due to dislocation movement in the stress field.

The small relaxation peaks observed at -20 and +70°C, must be associated with redistribution of impurities; the anomalous high relaxation peak at 260°C probably is connected with twin boundary movement in the stress field.

The nature of the nonrelaxation peak observed at 300°C is more complex. We suppose that it can be associated with interaction processes of lattice defects and small complexes with short range order. On the curve of temperature dependence of internal friction of boron platelets 3 well defined maxima were found at 170°C, 300°C and 420°C (Fig. 1. curve I). The internal friction maximum at 300°C is accompanied by a sensible decrease of the modulus. In spite of several orders of frequency change, a temperature shift of the maximum was not observed, which proved once again the non-relaxational nature of the peak in accordance with the results of measurements at 1 cps. The maximum at 260°C found during measurements 1 cps with activation energy of 1.3 eV was expected at 400°C when the frequency was increased up to 4 kcps and this was confirmed experimentally. The height of the maximum increases after annealing below 300°C and decreases after heat treatment above 600°C. The maximum at 70°C (with activation energy 0.76 eV) shifts to 150 - 170°C after an increase of measurement frequency up to ~4kcps. The peak height was unaffected by low temperature annealing which indicates that this maximum can be caused by stress induced impurity movement for which high temperatures are needed.

Similar results were obtained on films prepared by the liquid drop quenching technique. The non-relaxational maximum at 300°C and the relaxational maximum at 420°C were observed (Fig. 3, curve I). If measurements were carried during cooling of the sample, the maximum at 300°C strongly decreased (Fig. 3, curve II).

It can be concluded that all internal friction maximums observed in different samples of β-rhombohedral boron (rods, plates, films) prepared by different techniques have the same nature.

References

1. U.V. Piguzov, Equipment for the Investigation of Physico-mechanical Properties of Materials, Nr. II-51-516, 1956.

2. G.F. Tavadze, G.V. Khantadze, S.G. Nakaidze, C.B. Tsagareishvili, F.N. Tavadze, Equipment and Technique of Experiment, Moscow, Nr. 2, 1973.

3. F.N. Tavadze, G.Sh. Darsavelidze, G.V. Tsagareishvili; "Internal Friction Mechanisms in Semiconductors and Metals", "Nauka", Moscow, 1972.

4. G.V. Tsagareishvili, G.Sh. Darsavelidze, F.N. Tavadze; Internal Friction Mechanisms in Semiconductors and Metals", "Nauka", Moscow, 1972.

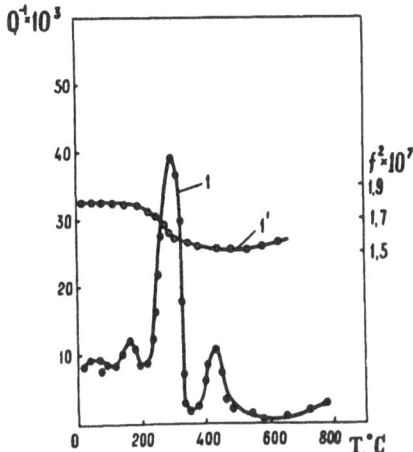

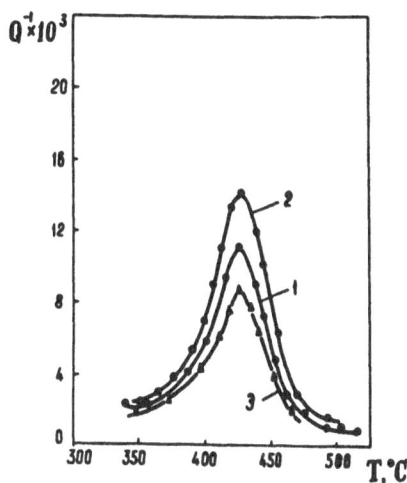

Fig. 1: Temperature dependence of internal friction and of frequency squared for boron plates.

Fig. 2: The influence of heat treatment on internal friction maximum at 420°C in boron plates

    curve I   Initial state

    curve 2   Annealing at 280°C for 4 hours

    curve 3   Annealing at 600°C for 4 hours

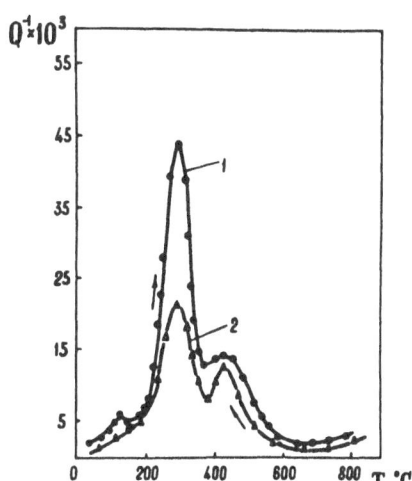

Fig. 3: Temperature dependence of internal friction in boron films.

THEORY OF THE ZENER-RELAXATION STRENGTH
(Abstract)

C.M. van Baal

Laboratorium voor Metaalkunde

Delft, The Netherlands

A straightforward method, used previously [1], here is applied to a generalized Ising model of a f.c.c. alloy in the Kikuchi approximation [2] [3]. The generalized Ising model takes into account not only the usual pair interactions, but also all 3- and 4-body interactions between nearest neighbours. The final result can be written as:

$$\Delta f = -\frac{1}{kT} \left| (F_1 a_0^2 + F_2 a_1^2 + F_3 d_2^2 + F_4 a_0 a_1 + \right.$$

$$F_5 a_0 d_2 + F_6 a_1 d_2 + \frac{1}{3} F_7 b_2^2) (\varepsilon_{xx} + \varepsilon_{yy} + \varepsilon_{zz})^2$$

$$- F_7 b_2^2 (\varepsilon_{xx}^2 + \varepsilon_{yy}^2 + \varepsilon_{zz}^2)$$

$$\left. + (F_8 b_1^2 + F_9 c_2^2 + F_{10} b_1 c_2)(\varepsilon_{xy}^2 + \varepsilon_{yz}^2 + \varepsilon_{zx}^2 \right|$$

In this Δf is the difference between the free energy in the unrelaxed and the relaxed states of the strained crystal and it gives directly the differences between the elastic constants c_{11}, c_{12}, c_{44} in these states by means of the usual formulae.

F_1 through F_{10} are known functions of the equilibrium concentrations t_0, t_1 and t_2 of tetrahedral clusters with 0, 1 and 2 B-atoms in the alloy lattice and so can be calculated as functions of concentration and temperature.

a_0, a_1, d_2, b_2, b_1 and c_2 are essentially the derivatives with respect to the strains of the three interaction energies in the generalized Ising model and so are constants for the alloy system. The first three of them can be found, if the concentration dependence of the lattice parameter and the bulk modulus are known. Assuming the other three constants to be equal to the mean value of the former a rough estimate for the torsional relaxation strength Δ_G for several alloys can be gained. It is compared with experiment in the following table.

	$\Delta_G \times 10^3$	
	exp.	estim.
Ag-Au (25)	2,8	1,8
Cu-Ni (32)	1,5	4,5
Cu-Zn (15)	5,8	4,2
Ag-Zn (24)	77	65
Au-Cu (25)	115	300

Some other results are:

1. For low concentrations the relaxation of the bulk modulus is small compared to that of the other constants.

2. The relaxation strength of all elastic constants varies for low concentrations as c^2, but for higher concentrations terms with c^3, c^4 etc. become influential in varying degrees for the various constants.

3. The temperature behaviour of the relaxation strength is mainly as $\frac{1}{T}$, but usually a "critical" temperature is present. This temperature is different for the different elastic constants and can be positive or negative for both types of alloy systems (ordering or decomposing) dependent on the constants a_0 etc. and composition.

A detailed account of this theory will appear in Physica (1974).

References: [1] C.M. van Baal, Physica 52 (1971) 410
[2] R. Kikuchi, Phys. Rev. 81 (1951) 988
[3] C.M. van Baal, Physica 64 (1973) 571.

ZENER RELAXATION ASSOCIATED WITH VACANCY
SUPERSATURATIONS UNDER A FAST PARTICLE FLUX

M. Halbwachs, J. Hillairet and E.A. Bisogni[x]
Centre d'Etudes Nucléaires de Grenoble
Département de Recherche Fondamentale
Section de Physique du Solide
BP 85, Centre de Tri, 38041 Grenoble Cedex (France)

Our general prospect is the determination of the vacancy concentration which exists in dynamic equilibrium in alloys exposed to a fast particle flux, in the high temperature range, i.e. 0.3 - 0.5 T_m.

Under these conditions, vacancies are produced in two ways. There are thermal vacancies, which are in thermodynamical equilibrium with the lattice, and radiation induced vacancies, which are in excess of this equilibrium concentration. The resulting supersaturation is of importance in many respects. Namely, it is the driving force for the enhancement under flux of many processes, such as local ordering, diffusion, precipitation, recrystallization, etc.. In particular it is one of the most important controlling parameters for the nucleation and growth of voids.

On a more fundamental level, the detailed study of the vacancy supersaturations associated with various conditions of flux, energy and type of particle, purity and metallurgical state of specimens... should bring basic information concerning the annihilation mechanisms of defects, the efficiency of sinks and the role of traps.

Zener relaxation measurements appear to be the most convenient method for this determination, at least in appropriate binary alloys. Indeed such a type of measurement has already been used with success for the study of the formation and migration characteristics of thermal vacancies in the quenched state (1, 2, 3, 4). We will see how we can

---

Present address : Departamento de Metalurgia, Comission Nacional d'Energia Atomica, Buenos Aires, Argentina.

extend its application to the case of the <u>dynamic equilibrium</u> under consideration.

Theoretical evaluations

The supersaturation state we are interested in is not a frozen state, which is the usual situation for most defect studies. In our specific case, the defects are generated at a constant rate during the course of the measurement and they are mobile enough, at the temperature of examination, to migrate and recombine or be annihilated at sinks. These opposing processes result in a steady-state concentration which is classically described by the following expressions (5)

$$\frac{dc_v}{dt} = P - \alpha_v D_v c_v - Z \nu_i c_i (c_v + c_{v,th}) = 0$$

$$\frac{dc_i}{dt} = P - \alpha_i D_i c_i - Z \nu_i c_i (c_v + c_{v,th}) = 0$$

where P is the rate of defect production by radiation, and the two other terms represent respectively the defect losses on the fixed sinks, and by mutual recombination. c_v and c_i are the atomic fractions of vacancies and interstitials in excess of thermodynamic concentrations, and $c_{v,th}$, the equilibrium vacancy concentration. D_v and D_i are the diffusion coefficients. ν_i is the mobility of interstitials and Z the recombination volume. α is a parameter which depends on the density and efficiency of sinks. If we assume that the sinks are essentially the preexisting dislocations, α is given by the diffusion equation (6, 7)

$$\alpha = \frac{2\pi\rho}{\text{Log} \frac{r_S}{r_c}}$$

with $r_S = \frac{1}{\sqrt{\pi\rho}}$ for a regular dislocation network. ρ is the dislocation density and r_c, the capture radius.

For the sake of simplicity, we shall consider only the two cases where : i) defects get annihilated at fixed sinks, ii) defects are removed both at fixed sinks and by pair recombination.

In the first case, the resolution of the above equations leads readily to a vacancy concentration given by

$$c_v = \frac{P}{\alpha_v D_v} = \frac{P}{\alpha_v a^2 \nu_0} \exp\frac{H_v^M}{kT}$$

where H_v^M is the enthalpy of motion of vacancies, ν_0 the usual vibration and a, ^{v} the lattice parameter.

In the more general case, the result is

$$c_L = -\frac{1}{2}\left(\frac{\alpha_i a^2}{Z} + c_{v,th}\right) + \frac{1}{2}\sqrt{\left(\frac{\alpha_i a^2}{Z} + c_{v,th}\right)^2 + \frac{4P}{Z}\frac{\alpha_i}{\alpha_v}\frac{1}{\nu_v}}$$

Fig. 1 is given to illustrate the corresponding features of the vacancy supersaturation. It is an Arrhenius plot showing the respective concentrations of both species : the thermal vacancy concentration follows a Boltzmann expression, leading to a slope equal to the formation enthalpy of vacancies. The radiation induced vacancy concentration decreases with increasing temperature, as the production rate is constant and the elimination rate increases with temperature. Then, there is a critical temperature T^* for which the supersaturation, which is the ratio of the two species, equals unity. Below T^* the supersaturation can reach high values. According to the prevailing elimination process, the slope is $- H_v^M$ for sinks only and $- H_v^M / 2$ when recombination and elimination at sinks operate together.

If we turn now to the way this supersaturation influences the internal friction spectrum, the qualitative picture is the following : on the high temperature side of the Zener peak, where the supersaturation is generally low, no effect is to be expected. On the low temperature side (with the not unrealistic assumption that the supersaturation becomes effectively appreciable at temperatures not too far below the peak temperature) the damping, which is directly proportional to τ^{-1}, thus to c_{v_R}, should be markedly increased for reasonably high fluxes. It is the region of interest between 0.3 and 0.5 T_M.

To show the orders of magnitude involved, we will consider a Ag 27 at % Zn alloy, which is the material we have retained for the first experimental study. This choice has been dictated by the amount of information we possess on the Zener characteristics and vacancy properties in this alloy (1). We will treat the case of electron bombardment, for which the production rate and distribution of defects are better defined than for neutrons. The flux chosen for this calculation is $3.10^{14} e^-/cm^2.S$; it is close to the maximum utilisable intensity, the limitation being the strong heating of the specimen by the slowing down of electrons.

To evaluate the production rate of defects, we have taken a total cross section of 100 barns (8) and applied a correction factor of 0.5 to take into account the instantaneous recombination of close pairs. The other numerical values are the data of Berry (1) for the vacancies : H_v^F = 0,89 eV and H_v^M = 0.55 eV. Also ν_0 = 10^{13}, S_v^F = k, Z = 12 and r_c = 40 Å.

We have varied the dislocation density in a wide range between 10^7 and 10^{10} cm^{-2} (2). The corresponding vacancy concentrations in the low temperature range, say between room temperature and 200°C lie between 10^{-7} and 10^{-10}. The calculated damping and elastic after effects associated with the above mentioned supersaturation are reported in Figs. 3 and 4.

The parametric study as a function of the dislocation density shows that the level and shape of both elastic after effect and internal friction curves are highly sensitive to the concentration of sinks. From the shape of the curve, we can obtain information not only about the supersaturation level, but also on the elimination mechanisms involved.

The c_v values reported in Fig. 3 refer to the total vacancy concentration at the indicated temperature. It is to be noted that the sensitivity of the method is remarkably good : at low frequency, concentrations as low as 10^{-9} are measurable.

As far as the elastic after effect is concerned, its sensitivity is quite high too. An interesting feature is that the sensitivity remains high to extremely low supersaturation levels. For smaller supersaturations, the curves are only shifted towards lower temperatures.

Experimental

Due to the short life time of defects in our experimental conditions, all the measurements must be performed directly under flux. Up to now, only in-pile internal friction experiments have been conducted. We will describe the apparatus and present the preliminary results.

The major problem for the in situ work arises from the importance of the mechanical vibrations of the reactor. To escape this difficulty, we use a torsional pendulum working in the medium frequency range. It is an inverted pendulum (Fig.5) with electromagnetic excitation and detection (9). An electronic device maintains the vibration amplitude at a constant level, and delivers a direct record of the damping versus temperature curve. The specimen is a cylinder 3 mm in diameter and 4 cm in length. The frequency is about 140 Hz.

Another problem we face is the nuclear heating, which leads rapidly to prohibitively high temperatures. To lower the thermal barriers and improve the cooling, the furnace is mounted compactly directly on the tube that surrounds the specimen. With use of low density materials and of a small gap between specimen and furnace, and also employing helium gas under 3 bars as a cooling agent, the temperature can be maintained below 200°C for fluxes up to 2.10^{13} fast neutrons/cm^2.s.

The first results are shown in Figs. 6 and 7. They have been obtained on two Ag-Zn specimens of similar compositions, but prepared differently.

Fig. 5 shows the results of the first in situ measurements for a Ag. 30 at % Zn alloy exposed to a flux of 2.10^{12} fast neutrons. The thermal cycling, in this preliminary experiment, was obtained by controlling the gas pressure ; the corresponding heating and cooling rates were rather fast, about 7 degrees per minute. The lower curve.has been traced in the first minutes of the reactor divergence and is practically identical to the curve in zero flux. In particular no deviation occurs in the low temperature range.

The two other curves are characteristic of the permanent regime : they reveal a significant increase of the damping on the low temperature side of the Zener peak. If this increase is caused by the enhancement of the Zener relaxation, this would lead to a vacancy supersaturation as high as 10^4 at 400K (0.4 T_m). This value is one order of magnitude higher than the estimated level, taking a dislocation density of 10^8 cm^{-2}. On the other hand, the absence of effect in the transient condition could be explained by the existence of a rather long build-up time to reach the dynamic equilibrium (5).

Fig. 7 shows the same curves for an Ag 27 at % Zn alloy without irradiation and under various fluxes. Although the maximum flux is higher than in the first case, and the conditions of measurement improved (better temperature control, higher sensitivity), no systematic deviation from the original curve could be detected.

These contradictory results can tentatively be explained in two ways :
- a difference in the sinks density, which could have been induced by a difference in the specimens preparation and anneal. We are checking this point. Berry (1) has already pointed out that the life time of vacancies

could be very sensitive to the presence of impurities.

- an alternative explanation is suggested by the much bigger amplitude of the grain boundary peak in the first sample. It is reasonable to assume that this type of relaxation is enhanced too under flux. This could bring a significant contribution to the low temperature internal friction.

Further experiments are in progress to identify the exact mechanism responsible for the observed effect. In addition, we have started a series of experiments under conditions where the necessary vacancy supersaturation should be more accessible by experiments namely in-pile elastic after effect and low frequency internal friction with employing an accelerator.

References

1. B.S. Berry, J.L. Orehotsky, Acta Met. 16, 683 (1968) and 16, 697 (1968)
2. J.R. Cost, Acta Met. 11, 1313 (1963)
3. A.S. Nowick, R.J. Sladek, Acta Met. 1, 131 (1953)
4. A.E. Roswell, A.S. Nowick, J. Metals, N.Y., 5, 1259 (1953)
5. J.V. Sharp, AERE-R-6267 (1969)
6. F.S. Ham, J. Appl. Phys. 30, 915 (1959)
7. M. Gomolinski, G. Brebec, J. Nucl. Mat. 43, 59 (1972)
8. O.S. Oen, ORNL-3813 (1965)
9. G. Dedianne, D. Lemercier, P. Lemercier, Revue de Phys. Appl. 4, 423 (1969).

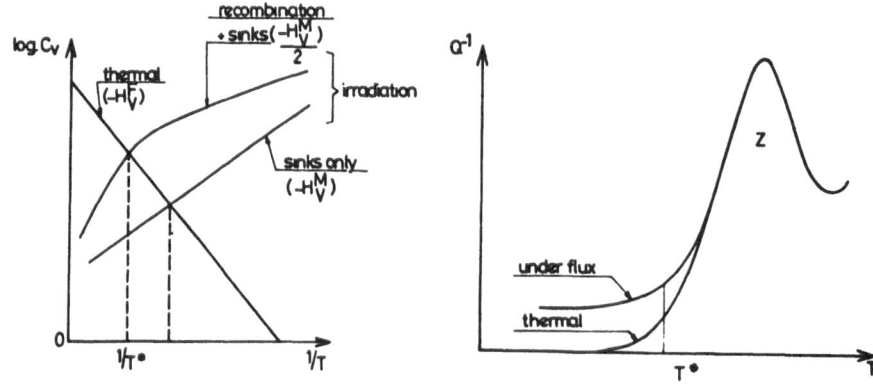

FIG. 1

Vacancy concentrations and internal friction
under a fast particle flux

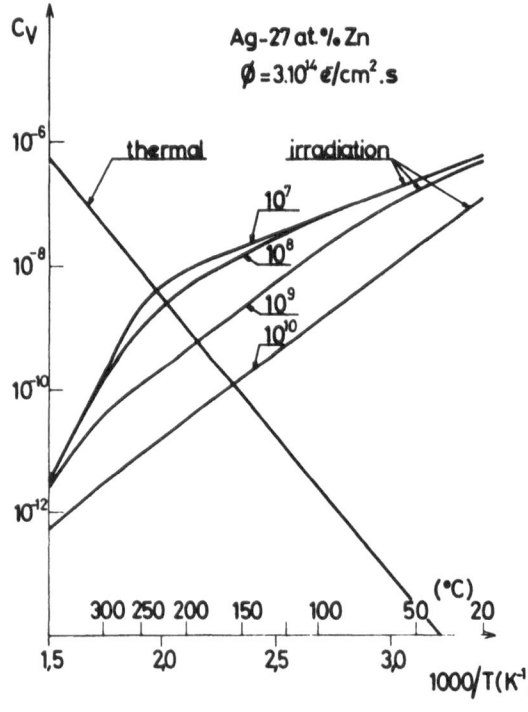

FIG. 2

Vacancy concentrations in a Ag - 27 at % Zn
alloy exposed to an electron flux of 3.10^{14}
$e^-/cm^2.s$. The numbers indicated on the figure
refer to the dislocation density in cm/cm^3.

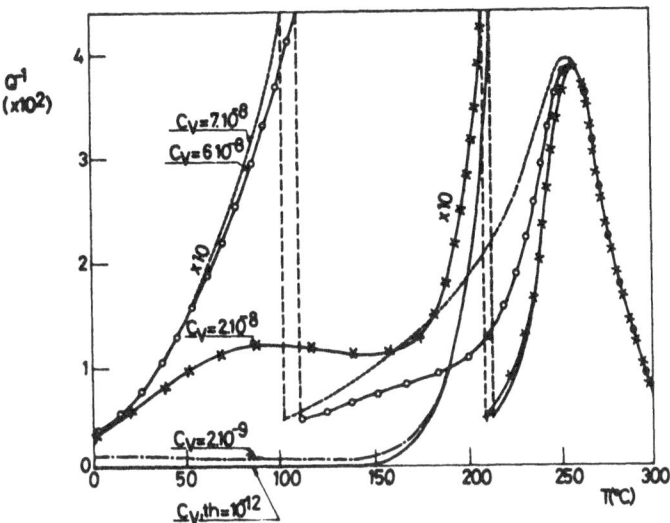

FIG. 3

Estimated Zener damping under irradiation in a Ag 27 at %
Zn alloy : influence of the sink density

$\emptyset = 0$ ⸻⸻⸻

$\emptyset = 3.10^{14}$ e⁻/cm².s dislocation density 10^{10} cm⁻² ⸺·⸺·⸺

$\emptyset =$ "   "   "   " 10^9 ✗✗✗✗

$\emptyset =$ "   "   "   " 10^8 ●—●—●

$\emptyset =$ "   "   "   " 10^7 ‐‐‐‐‐‐

c_v is the total vacancy concentration calculated for T = 360 K.
The vibration frequency is 1 c.p.s.

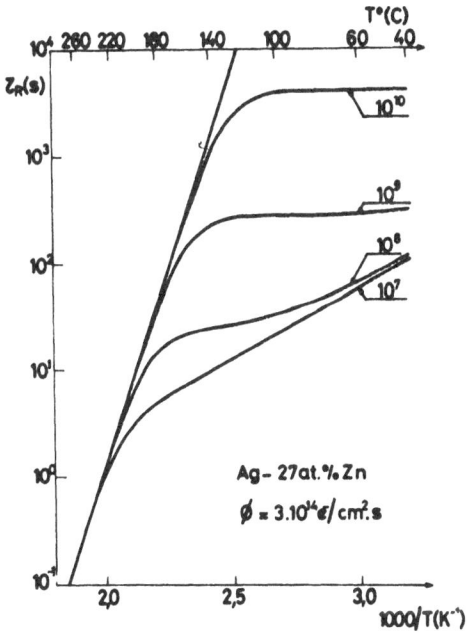

FIG. 4

Estimated Zener elastic after effect for various dislocations
densities. The numbers on the figure refer to this density
in cm/cm³.

327

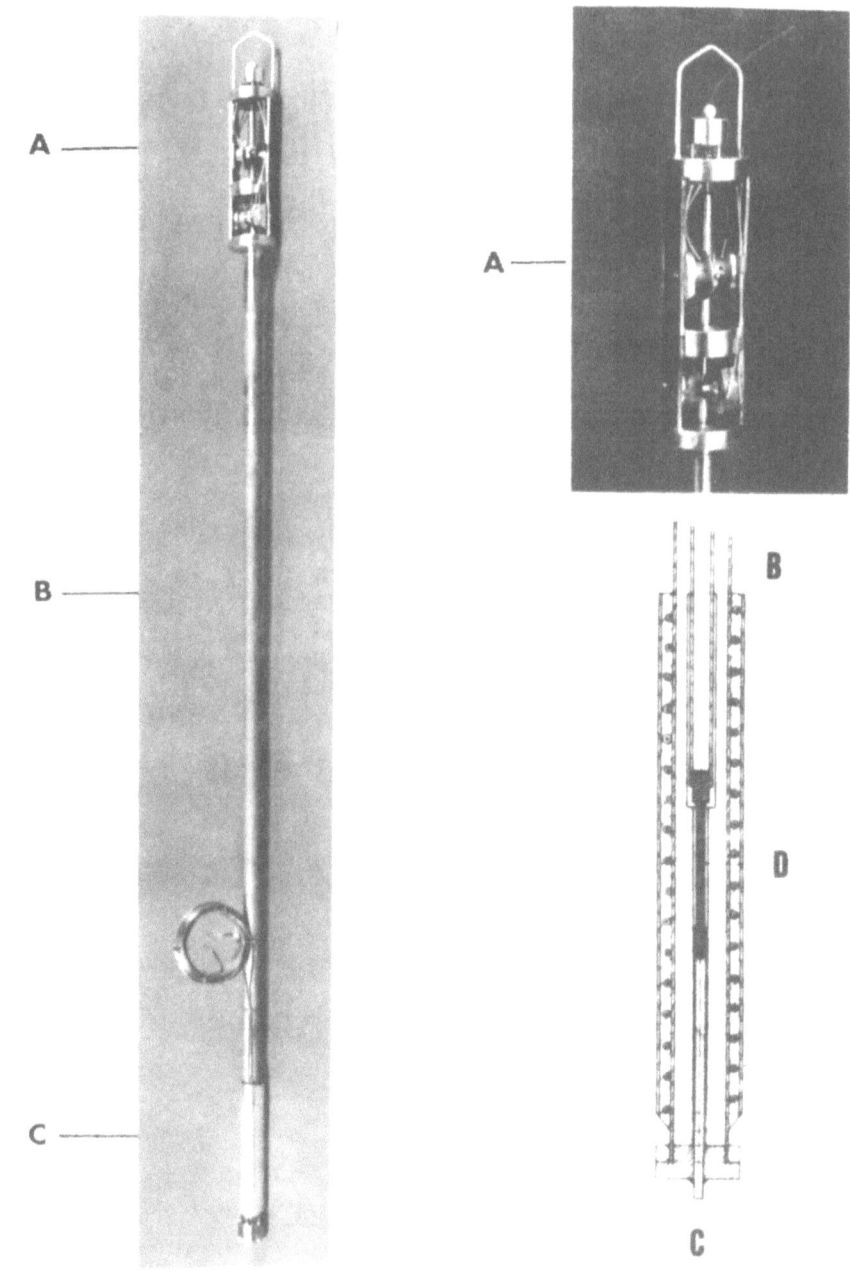

FIG. 5

Experimental apparatus

A - Drive and detection system
B - Stainless steel tube
C - Furnace
D - Specimen

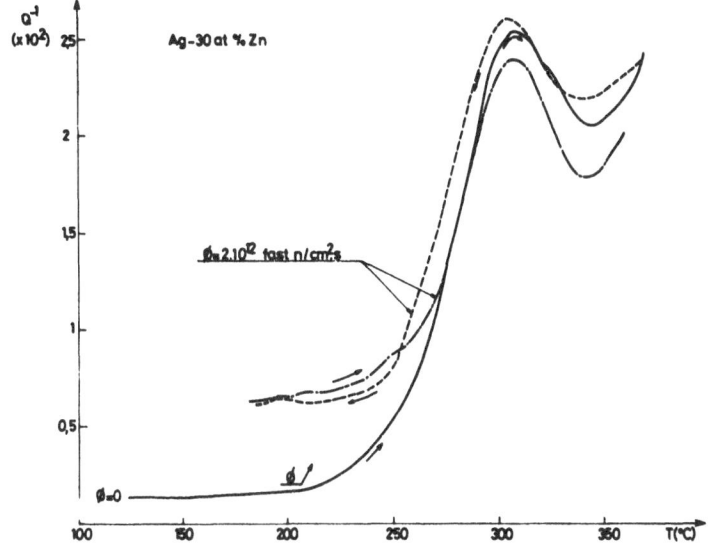

FIG. 6

Experimental internal friction curves for a
Ag 30 at % Zn alloys obtained under irradiation
in a fast neutron flux of 2.10^{12} n/cm^2.s.
The measurement frequency is about 140 cps.

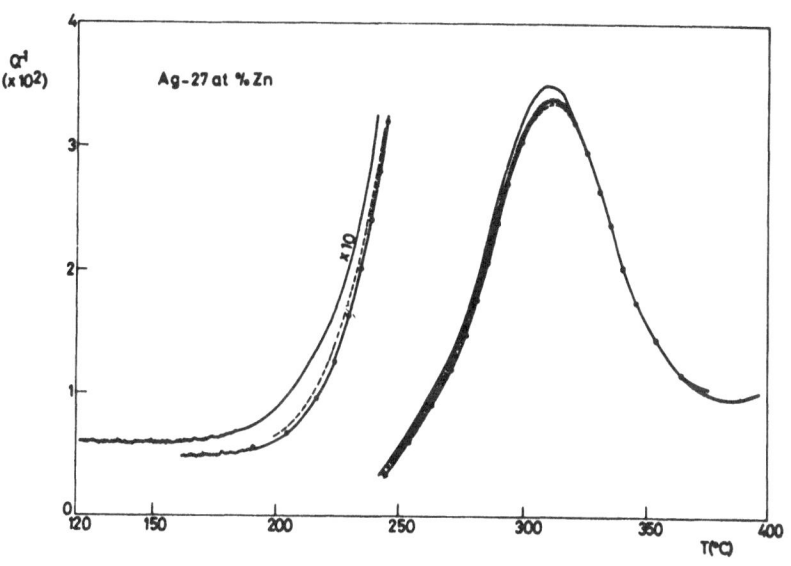

FIG. 7

Experimental internal friction curves for a
Ag 27 at % Zn alloy exposed to various fast neutron
fluxes :

$\emptyset = 0$
$\emptyset = 1,4.10^{13}$ fast neutrons/cm^2.s
$\emptyset = 2.10^{13}$ "
$\emptyset = 2,8.10^{13}$ "

The initial part of the damping curve has been
magnified by a factor of ten.

329

INTERSTITIAL RELAXATIONS DUE TO

HYDROSTATIC STRESS IN NIOBIUM-OXYGEN ALLOYS

S. N. Tewari and J. R. Cost
School of Materials Engineering
Purdue University
West Lafayette, Indiana 47907

ABSTRACT Isothermal measurements at 81.6°C of an anelastic relaxation induced by hydrostatic stresses in the range from ambient to 80,000 psi (5.52 Kbar) have been made for a Nb-2.3 at% O alloy. Analysis of the relaxation is made by both considering the relaxation spectrum to be composed of three discrete relaxation times and to be a lognormal distribution about some mean relaxation time. The relaxation which occurs during pressurization is considered to be due to the reaction of single interstitial oxygen atoms to form oxygen doublets (or higher order multiplets). The reverse reaction is considered to occur during pressure release. The relaxation strength and rate are found to be in good agreement with rough predictions based upon the above model for the reaction. The pressure dependence of the relaxation rate gives an apparent activation volume of 4 cm^3/mole.

Introduction

Studies of anelastic relaxations have become a powerful tool for investigation of the mobility of both interstitial and substitutional solutes in solids, particularly at low to moderate temperatures where the average time for solute atom jumps is in the range from 10^4 to 10^{-6} sec. For the case in which the applied stress is non-hydrostatic, the Snoek and Zener relaxations have received extensive study in interstitial and substitutional alloys respectively. However, the other case, in which the applied stress is pure hydrostatic, has never been directly investigated, mostly for reasons of experimental difficulty. Although relaxations under pure hydrostatic stress have never been directly measured, they have been carefully discussed, primarily by Nowick (1, 2) and also by Koiwa (3). In addition, Seraphim and Nowick (4) were able in an indirect manner to extract the relaxation strength in compressibility for a Li-57 at% Mg alloy by measuring both Young's modulus and the rigidity modulus relaxations as a function of crystal orientation.

Relaxations induced by hydrostatic-pressure have the advantage of allowing researchers to examine a completely new kind of relaxation in the total relaxation spectrum (1). This kind of relaxation involves reactions between defect species as opposed to the reorientations of a particular species which are induced by non-hydrostatic states of stress. This capability for studying defect reactions can be expected to provide kinds of information concerning the nature and behavior of point defects in solids which is not available from usual mechanical relaxation studies. The hydrostatic pressure counterpart to the Snoek relaxation is of particular interest, and this paper, our initial study of relaxations due to hydrostatic stress, is concerned with relaxations in B.C.C. interstitial alloy systems. The niobium-oxygen system has been chosen because it is presently well characterized and because calculations indicate that a measurable effect can be observed.

The point defects present in solid solutions of oxygen in niobium have been deduced by careful analysis of the Snoek relaxation to be single oxygen atoms, oxygen pairs, triplets and possibly quadruplets (5-7). The phenomenon of interstitial clustering which accounts for the above multiple defects has been relatively well established. Its presence has also been shown in the Ta-O (8) and Fe-N (9) systems. For cubic crystals, anelastic relaxations in the compressibility will take place if a reaction involving changes in the concentrations of the various defect species is induced by the application of a hydrostatic stress. Such a reaction will tend to occur if there is a change in volume when the reaction takes place. A calculation of the predicted magnitude of this effect for the reaction in niobium of forming an oxygen pair from two single oxygen atoms is presented in the following section.

Predicted Rate and Magnitude of the Relaxation

We use the development due to Nowick (1) for the reaction

$$2\alpha \rightarrow \eta \tag{1}$$

where the α species is single oxygen atoms and η is oxygen doublets each being in solution in the niobium. Also we ignore the reactions involving triplets and higher order multiplets.

Relaxation Rate:

Without doing an exact calculation, a simple estimate can be obtained by considering the kinetics of the reverse reaction in which a doublet which has a binding enthalpy H_b breaks up into two singlets. This event may be considered to occur by a single jump of one of the atoms in the doublet away from the other atom, and thus to have an activation barrier $H_m + H_b$, where H_m is the activation enthalpy for motion of the singlet. Ignoring entropy, the rate of the dissociation reaction can be written as

$$\tau^{-1} = \nu e^{-(H_m + H_b)/kT} = \Gamma_\alpha e^{-H_b/kT} \tag{2}$$

where ν is the atomic vibration frequency and Γ_α is the atomic jump frequency of the α species. Gibala and Wert (6) give the result $H_b = 0.07eV$ and also report Γ_α as a function of temperature. For $80^\circ C$ the result $\tau^{-1} \simeq 10^{-1} \Gamma_\alpha \simeq 2 \times 10^{-3}$ sec^{-1} is obtained. From this result we can predict that the relaxation rate should be favorable for measurement by elastic after-effect methods at temperatures roughly in the range of $100^\circ C$ and below.

Relaxation Strength:

The hydrostatic relaxation strength $\Delta v_{an}/\Delta v_{el}$, the ratio of the anelastic to the elastic volume change induced by a given hydrostatic stress, is given by Nowick (1) as

$$\Delta = \frac{\Delta v_{an}}{\Delta v_{el}} = \frac{1}{K}\frac{v_o}{kT}\frac{c_\alpha^o\, c_\eta^o}{2c_\eta^o + c_\alpha^o}(2\lambda_\alpha - \lambda_\eta)^2. \tag{3}$$

Here K is the niobium compressibility, v_o is the molecular volume, and c_α^o and c_η^o represent the equilibrium atom fractions of species α and η at the given temperature in the absence of an applied field. The factor in parenthesis is the difference in strain between the two species per unit concentration of defect undergoing the reaction.

Examination of the concentration factor indicates that the condition for a large relaxation strength is $c_\alpha^o \simeq c_\eta^o$ and that if either concentration is significantly less than the other, the magnitude of the factor will roughly be that of the low concentration species. By making use of the results which Gibala and Wert obtained for the concentrations of singlet and doublet oxygens (6), we obtain for a 2 at% O alloy in the temperature range of this study roughly $c_\alpha^o = c_\eta^o = .005$, which clearly provides a very favorable concentration factor.

To estimate the strain factor in Eq. 3, we make use of the calculations by Fisher (10) of the volume change associated with various configurations of paired carbon interstitials in B.C.C. iron and also the fractional change in molar volume of niobium due to addition of a unit concentration of oxygen as obtained from lattice parameter measurements (11). We obtain for the strain factor $(2\lambda_\alpha - \lambda_\eta) \cong 5 \times 10^{-2}$, which corresponds to a decrease in volume when the doublet is formed.

Using these results in Eq. 3 gives the strength of the relaxation as $\Delta \simeq 10^{-3}$. Thus for elastic volume strains of $\sim 10^{-2}$ which can be obtained with laboratory pressures, anelastic volume strains due to this relaxation should be roughly 10^{-5}. Strains of this magnitude can be measured without undue difficulty thus indicating that the relaxation should be measurable.

Experimental Procedure

Anelastic volume changes in a niobium specimen containing oxygen were measured by monitoring length changes of pure niobium using a three terminal capacitance technique (12). Differential length measurements were made as a function of time both after the application of pressure and after its subsequent release.

The geometry of the sample assembly and the two plates, of the capacitor is shown in Fig. 1. A niobium[*] specimen with 2.3 at% O in solution in the form of a 1/4-in. (6.35 mm) diameter

---

[*] Materials Research Corporation, VP Grade, 99.9+ at%.

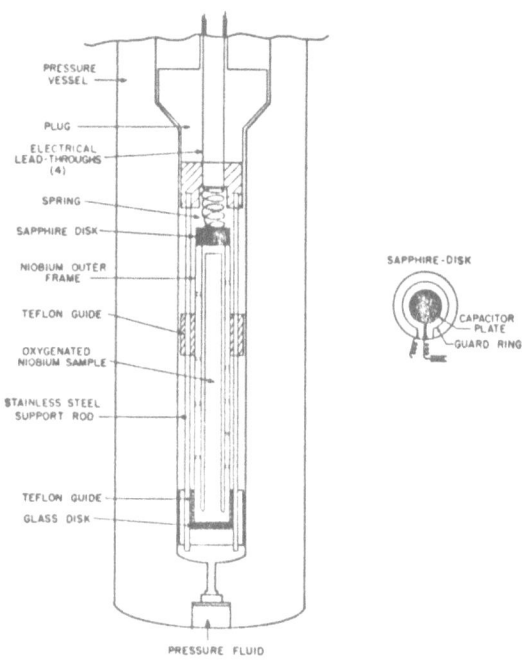

FIG. 1
Sample assembly and pressure vessel.

rod is in the center and is joined by electron beam welding at the bottom to a hollow 1/2-in.
(12.7 mm) outer diameter cylinder of pure niobium. The length of this assembly is roughly
4 in. (~100 mm) with the inner rod being approximately 4×10^{-4} in. (~10^{-2} mm) shorter in
order to form the gap of the capacitor. The two plates of the capacitor are the top surface
of the inner rod and the metal plated bottom surface of a sapphire disk which bears against
the top surface of the outer niobium sample. The geometry of the metal plated surface of the
sapphire disk is shown in the inset of Fig. 1. The outer ring is grounded and acts as the
guard-ring for the three-terminal capacitance measurement. All of these surfaces are lapped
to an optical flatness of 1/2 wave-length. Several such samples were prepared and one
prepared without oxygen did not show an anelastic relaxation. As shown in Fig. 1, the
sapphire disk is maintained in a steady position against the non-oxygenated portion of the
sample surface by a spring compressed against a supporting frame.

The sample assembly with appropriate electrical leads fits into the pressure vessel, a
1 in. inner diameter, 4 in. outer diameter, cylinder with a 6 in. long inner working space.
During measurements, the pressure vessel rests on a support frame in a constant temperature
oil bath which can be regulated to ± .01°C. Pentane is used as the pressurization medium.
Pressure is varied in the range from 2.5 Ksi (0.17 Kbar) to 80 Ksi (5.52 Kbar) and is
measured with a calibrated manganin cell and a Bourdon gauge. The low pressure limit of 2.5
Ksi was chosen because it resulted in good mechanical stability to the sample assembly.
During all constant pressure measurements, the pressure was constant to ± 30 psi.

Capacitance measurements are made using a capacitance bridge and a phase sensitive
detector which generates a voltage signal proportional to the deviation of the capacitance
from the bridge null. This voltage signal is fed to a data acquisition system and recorded
at desired intervals on teletype and paper tape. Length change measurements can be made
accurate to 1A. In operation the experimental arrangement is limited by the temperature change
in the pressure vessel which results from changes in pressure. Typically 15 minutes is
required for the temperature to stabilize following such a transient, and no data are taken
during this time.

Method of Analysis

It is obviously desirable to find an analytical fit to the data which is definitive in
that only one method of analysis provides a good fit, particularly since each method of
analysis may involve a definite physical model for the mechanism of the relaxation. The
experimental determination of one particular model over others is a goal of our research. The
basic problem is that of characterizing the unknown relaxation time spectrum from the experimental anelastic response curve. The relaxation time spectrum, of course, may be composed
of a set of discrete relaxation times or it may be a continuous spectrum which can be
characterized by a particular distribution function. An excellent discussion of various
relaxation spectra and their resulting response curves is given by Nowick and Berry (13). In
their discussion of methods for determination of spectra, they emphasize that since the
response curves for a single relaxation are spread out over roughly a decade in time, the
superimposed response curves of several discrete relaxations which differ by less than an
order of magnitude or of a continuous relaxation spectrum will have a large amount of overlap
so that the total response curve will be extremely difficult to analyze and decompose into the
relaxation spectrum. Also, it is quite apparent that such an analysis requires excellent data
taken over several decades of time.

The relaxation spectra may be approximated by a number of different distribution
functions. For the most part, however, studies requiring analysis of response curves have
made use of either of two different methods of analysis based upon two kinds of relaxation
spectra. One, the summed exponential, assumes that a relaxation which is not simple
exponential with a single relaxation time is due to a sum of two or more exponential
relaxations each with a single relaxation time and a unique relaxation strength. This is the
model which was used by Gibala and Wert in their analysis of the Snoek relaxation in the
niobium-oxygen system (6). The other method of analysis, the lognormal, considers that a
relaxation which is other than exponential is due to a continuous relaxation spectrum which
has a strength varying about an average relaxation time according to a normal or Gaussian
distribution in the logarithm of time. The use of this distribution has been discussed most
thoroughly by Nowick and Berry (13), and has been particularly successful in analyzing the
Zener relaxation (14). Both of these methods of analysis are based upon particular physical

models for the relaxation. In order to avoid the a priori choice of such a model and instead choose according to the best fit of the data, it becomes important to examine the details of these two methods of analysis, particularly with regard to the ability to discriminate whether data are from the discrete or the continuous relaxation spectrum. In this section some of the results of such an examination are presented.

A regression analysis which minimized the squared residuals was used to fit data by computer. For the summed exponential decay analysis, a fit was made to the predicted response curve

$$\chi(t) = \sum_i A_i \, e^{-t/\tau_i} \tag{4}$$

for $i = 2$ and 3 where $\chi(t)$ is the fractional anelastic response. This fit provided values of the relaxation strengths A_i and the relaxation times τ_i. For the lognormal analysis, three parameters were obtained by the fit, A the total relaxation strength, τ_m the mean relaxation time, and β the half width in the distribution at $1/e$ of its maximum expressed in natural logarithm of time. Computer fits using both methods of analysis were made to actual data as well as to data generated according to the discrete relaxation time and the lognormal distribution functions

Figs. 2 and 3 show the results of a summed exponential decay with $i = 2$ and $i = 3$ respectively computer fit to anelastic response data points from a lognormal relaxation with $\tau_m = 10$ minutes and $\beta = 3$. Examination indicates two significant facts. First, a good fit using the sum of two exponentials can not be made, especially for times of the order of the mean relaxation time and longer. Second, an excellent three exponential fit can be made to the data taken up to $10 \, \tau_m$. This is not unexpected since a total of six adjustable parameters are available for the $i = 3$ fit. It was also found that this fit improved as β was decreased. Similar good fits were obtained when a lognormal fit was forced upon exponential data. Using this fitting method, it would appear to be very difficult to distinguish whether a given set of data was from a lognormal or a discrete value type of distribution, although some significant differences appear to be showing up at long times.

Further examination indicated that the rate of the response is more effective than the residuals in discriminating between these two relaxation spectra. For instance, when the three exponential fit results of Fig. 3 are plotted on a logarithmic time scale, the curve distinctly shows five inflection points, whereas the lognormal curve only shows one. It was observed that summed exponential data show extra inflection points provided that the relaxation times are in ratio of approximately eight or more and the relaxation strengths are roughly equal.

A not unexpected correspondence was found between β, the width of the lognormal

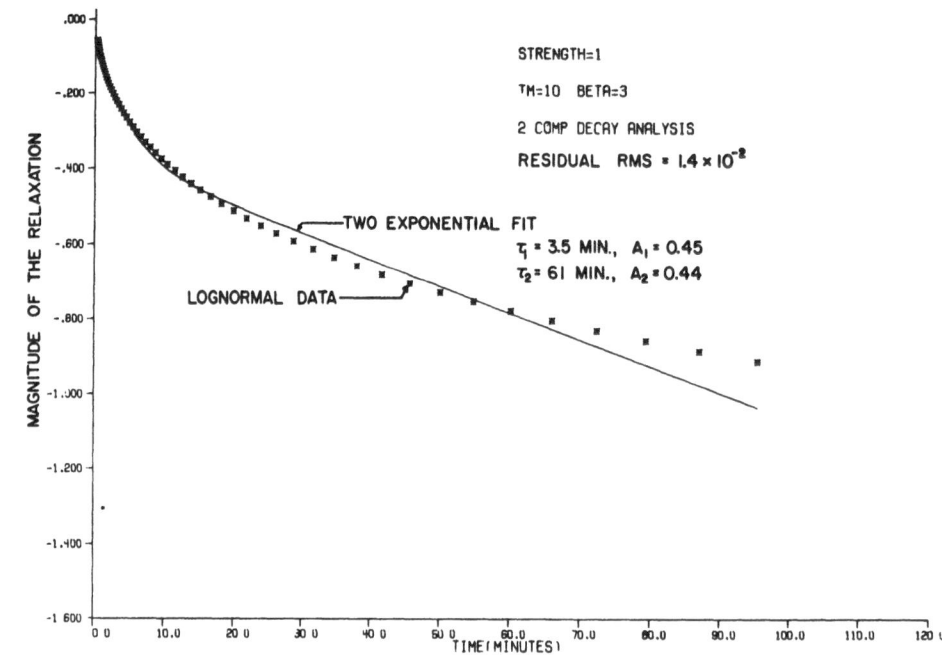

FIG. 2
Two exponential fit to elastic after-effect data based
upon a lognormal distribution of relaxation times.

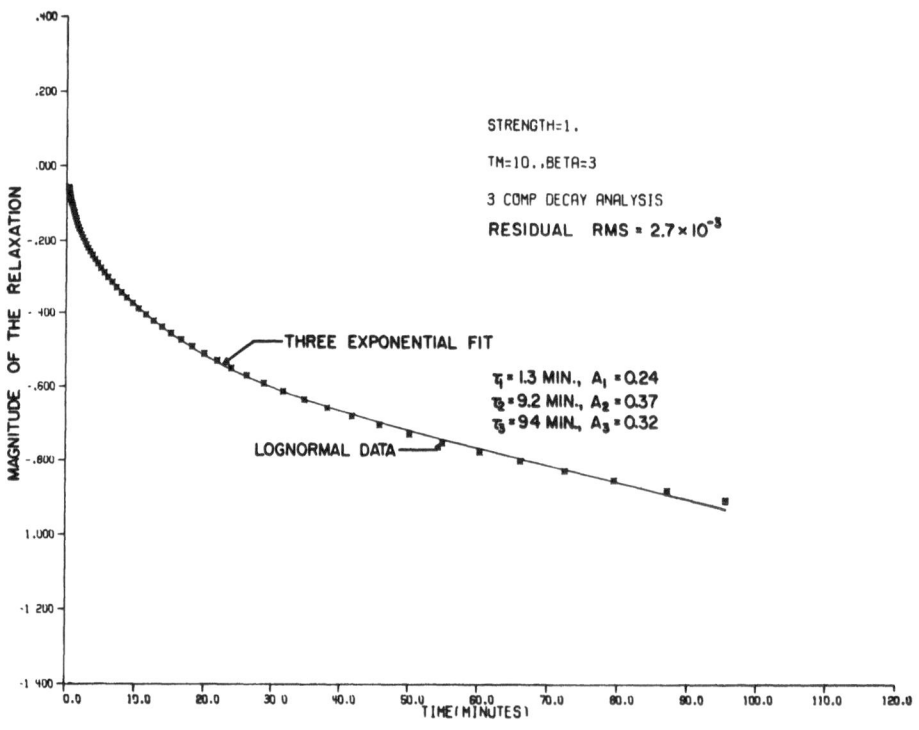

FIG. 3
Three exponential fit to lognormal data.

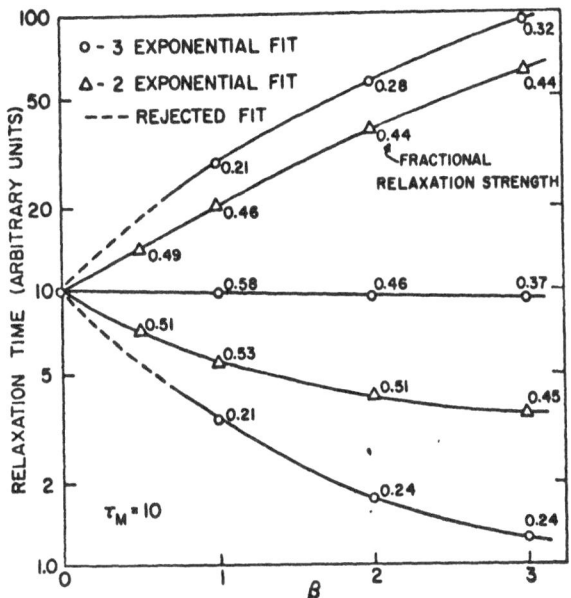

FIG. 4
Discrete relaxation times obtained for summed exponential analysis of lognormal
data as a function of β the width parameter for the lognormal distribution.

distribution, and the ratios of the discrete relaxation times obtained for a summed
exponential fit to the lognormal data. This result is shown in Fig. 4 for both i = 2 and
i = 3 fits. From this plot it is seen that for i = 3 the ratio of eight or more in relaxation times which will produce extra inflection points on a logarithmic time plot requires
that the lognormal data have roughly β ≥ 2.7. Thus experimental data from lognormal distributions with β ≥ 2.7 can be expected to be discriminated from those involving discrete relaxation times. Further study of the differences between the response curves for these two
distributions is required to allow this discrimination to be made for lower values of β or
for more closely spaced discrete relaxation times.

Results and Discussion

Our preliminary experiments indicated that a small but reproducible anelastic relaxation
induced by changes in hydrostatic pressure could be measured in the temperature range from
room temperature to 100°C. The relaxation rate at a given temperature was close to the rate
$\tau^{-1} \simeq 10^{-1} \Gamma_\alpha$ predicted earlier in this paper. The results of an initial series of experiments designed to study the pressure dependence of the relaxation at a constant temperature
are herein reported. Subsequent series of experiments will study the temperature and oxygen
concentration dependence of this relaxation. All of the results reported were obtained from
a single specimen with an oxygen concentration of 2.3 at%. The temperature chosen was
81.6°C, one at which the complete relaxation could be measured in less than two days.

The results for one of the typical hydrostatic relaxation experiments are shown in Figs. 5 and 6. Fig. 5 is for pressurization from 2.5 Ksi (0.17 Kbar) to 55.4 Ksi (3.82 Kbar), and Fig. 6 is for the pressure release to 2.5 Ksi. The capacitance change, which is the deviation of the capacitance from the bridge null, is indicated on the ordinate so that a decrease as in Fig. 5 represents a decrease in volume of the specimen. The magnitude of this decrease in terms of change in length (distance across the gap of the capacitor) is indicated by the 10A marker on the figure. Analysis of the time dependence of the relaxation indicates that it cannot be fit by a single relaxation time. It may be observed that part of the relaxation occurs very rapidly so that an appreciable portion has taken place during the initial 15 minute period of waiting for the temperature to stabilize following the pressure change. Also it may be seen that the relaxation is still continuing after one day. The release experiment shown in Fig. 6 shows essentially the same results in reverse, an anelastic increase in volume of roughly the same magnitude and a relaxation rate which is not greatly different.

The data for each experimental anelastic response curve was computer analyzed and fit by both the summed exponential with $i = 3$ and the lognormal methods. Fits with $i = 2$ were also tried but were rejected by the computer. The fits obtained for the pressurization data shown in Fig. 5 are presented in Figs. 7 and 8, where the solid line through the data points is the fitted curve. Both methods of analysis may be observed to show good fits to the data. The root mean square of the residuals is roughly the same; however, for most of the runs it tends to be less for the summed exponential. At present we are attempting to improve the accuracy of the data in order to better distinguish between the two relaxation spectra. Until a better distinction can be made, results of both methods of analysis will be presented.

It is of interest to compare values of the parameters obtained by the two methods. The three discrete relaxation times obtained for pressurization to 55.4 Ksi are roughly different by a factor of six. Also they have magnitudes which are in the same range as those observed by Gibala and Wert at this temperature. For the fit shown in Fig. 8, the mean relaxation time for the lognormal distribution is between that for the first and second discrete relaxation times (see Fig. 7); for most measurements it was found to be closer to that for the second time. In addition, values of τ_m appeared to show more scatter between different runs than did the discrete relaxation times. The total strength of the relaxation is usually found to be slightly larger when analyzed by the summed exponential method; it is roughly 15% larger for the results presented in Figs. 7 and 8. For the lognormal analysis, the values of β which were obtained varied in the range from 1.5 to 3.0, and appeared to be very sensitive to the data taken at long times.

The relaxation was investigated at pressures up to 80 Ksi (5.52 Kbar). The magnitude of the anelastic response both for pressurization and release experiments was found to be proportional, within the experimental error, to the pressure change. A relaxation strength $\Delta = 4 \times 10^{-4}$ was obtained. This is in reasonable agreement with the predicted strength of

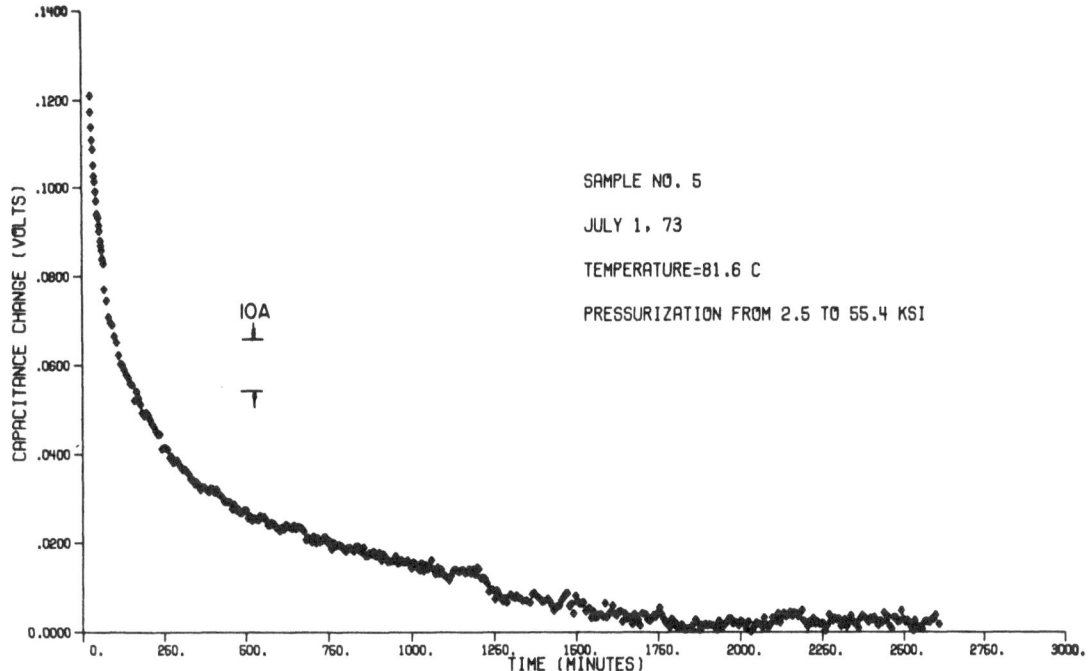

FIG. 5
Decrease in length with time (measured by change in capacitance)
due to pressurization of a Nb-2.3 at% O sample to 55.4 Ksi.

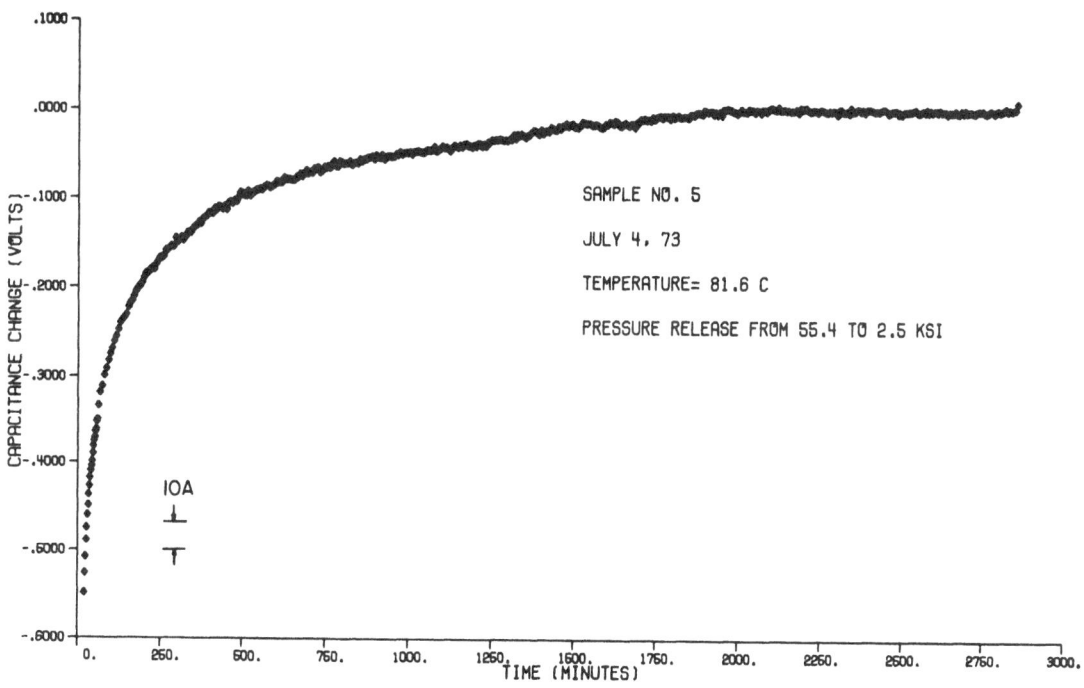

FIG. 6
Increase in length with time due to release of pressure from 55.4 Ksi to 2.5 Ksi. These
Measurements were made on the same sample two days after the pressurization shown in Fig. 5.

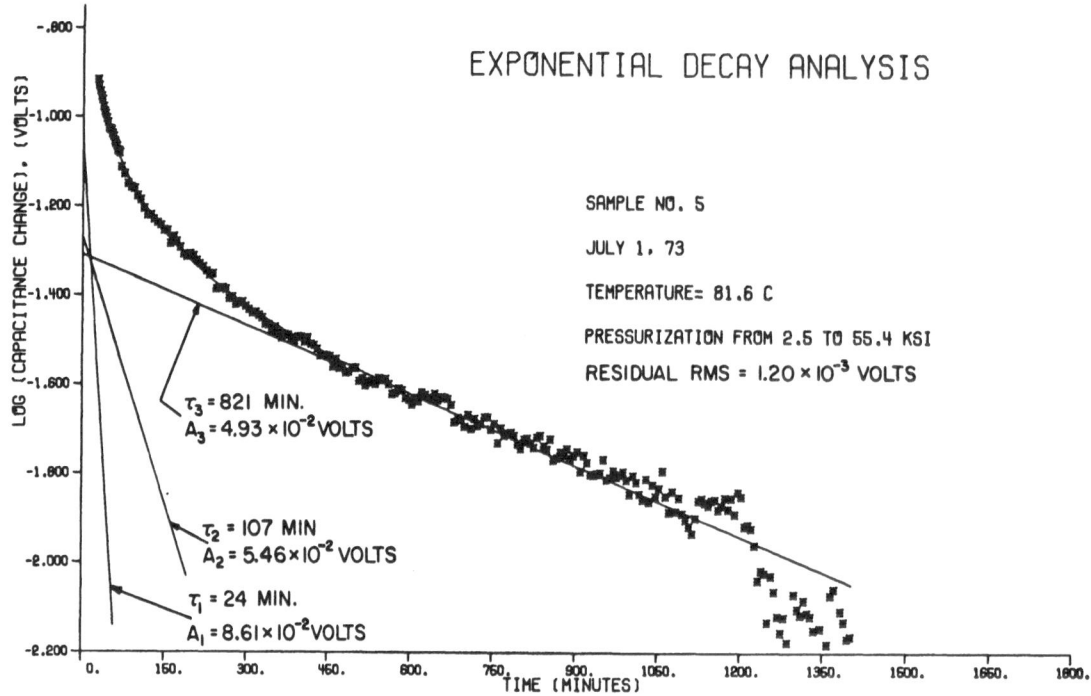

FIG. 7

Exponential decay analysis with three discrete relaxation times of the data for pressurization shown in Fig. 5. The length change is taken as proportional to the capacitance change which is the difference from the final capacitance value.

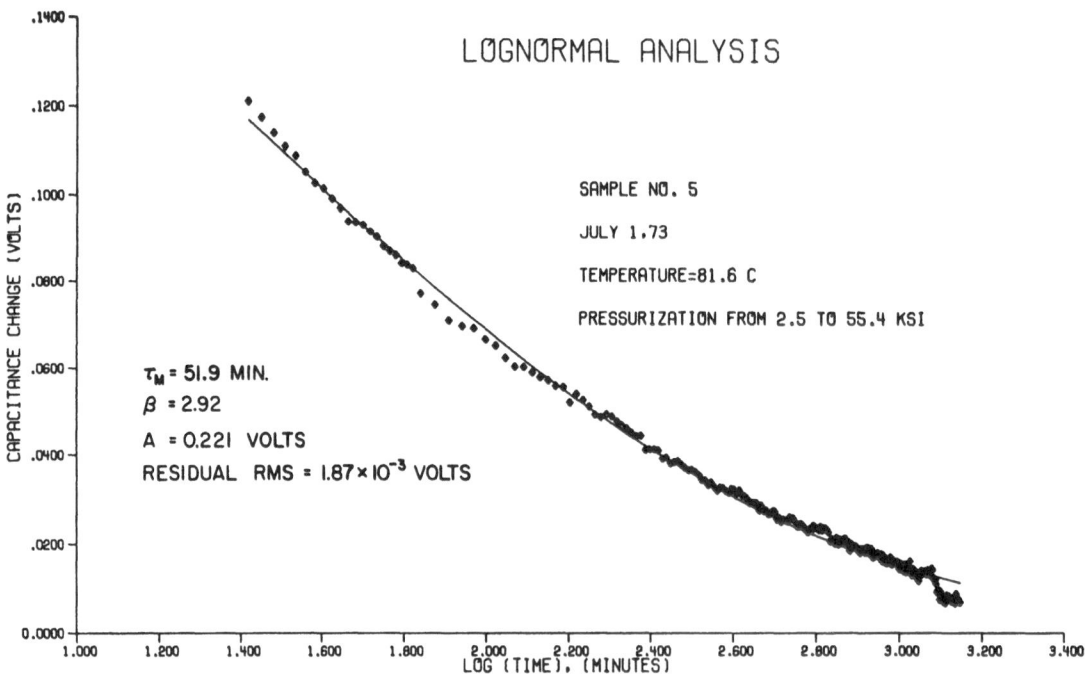

FIG. 8

Lognormal analysis of the data for pressurization shown in Fig. 5.

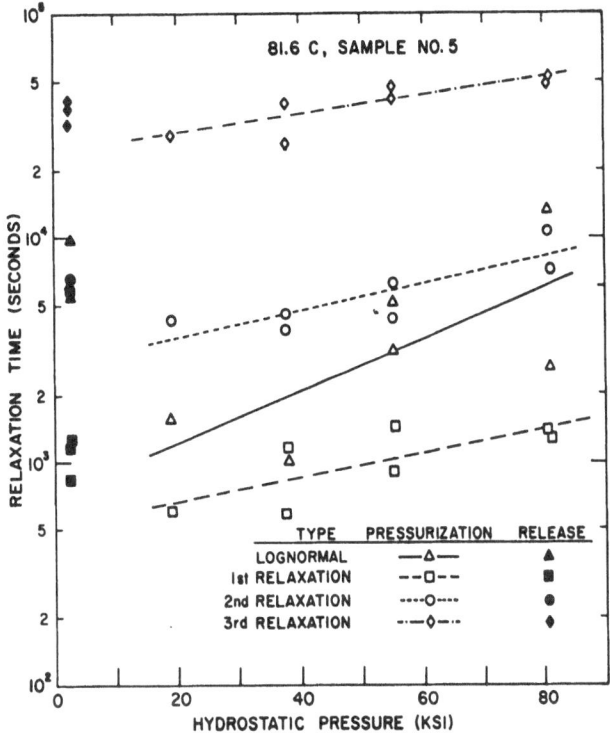

FIG. 9

Pressure dependence of the relaxation time for oxygen in niobium at 81.6°C as determined by the summed exponential decay and the lognormal methods of analysis.

$\Delta \simeq 10^{-3}$, suggesting that the estimate of the volume change due to pair formation is a good one.

The results for the effect of pressure upon the relaxation time are shown in Fig. 9 for pressurization experiments (open symbols) and for pressure release experiments (closed symbols). The latter were all measured at 2.5 Ksi ± 0.2 Ksi. The data analyzed by the summed exponential method in all cases showed the relaxation times to be in ratio of roughly 6. The set of data for the longest relaxation time appear to show the least scatter, probably because these data are obtained from response curves which have not had an exponential tail subtracted. It may be noted that the relaxation times increase with pressure. This is consistent with an expected decrease in atomic mobility with increased pressure. It is however, not clear that the increase in the logarithm of the relaxation time is linear with pressure as suggested by the straight lines fitted to the data. The slope of such a straight line, however, may be taken to obtain an apparent activation volume for the relaxation. The best data, those for the longest relaxation, give an apparent activation volume of approximately 4 cm^3/mole, roughly 40% of the atomic volume of niobium. Further experiments which are in progress are expected to provide more details of the pressure dependence. It is noted that this apparent activation volume appears to be large compared to the values of 1.7 and 1.1 cm^3/mole obtained for oxygen and nitrogen respectively diffusing in vanadium (15).

In Fig. 9 the relaxation times measured for pressure release are roughly a factor of two larger than those for the pressurization experiments extrapolated to 2.5 Ksi. This finding requires confirmation by pressure release experiments to pressures above 2.5 Ksi. It might well be expected, however, that the reaction to form doublets would have a different rate constant than the reverse dissociation reaction. Subsequent experiments on specimens with lower oxygen concentrations can be expected to provide an additional test for this since if doublets or other interstitial clusters are being formed, the rate of formation should be less when single oxygen atoms are more widely separated.

Acknowledgements

The authors wish to express their thanks to Professor P. Winchell for interesting them in the problem and for providing both stimulating discussions and continuous encouragement. Thanks are also due to Professor R. Sladek for making high pressure facilities available. Appreciation is expressed for financial support from the Advanced Research Projects Agency and the National Science Foundation.

References

1. A. S. Nowick, J. Chem. Phys., 53, 2066, (1970).

2. A. S. Nowick and W. R. Heller, Adv. Phys., 12, 251, (1963).

3. M. Koiwa, Phil. Mag., 24, 539, (1971).

4. D. P. Seraphim and A. S. Nowick, Acta Met., 9, 85, (1961).

5. R. Gibala and C. Wert, Acta Met., 14, 1095, (1966).

6. R. Gibala and C. Wert, Acta Met., 14, 1105, (1966).

7. M. S. Ahmad and Z. C. Szkopiak, J. Phys. Chem. Solids, 31, 1799, (1970).

8. R. W. Powers and M. V. Doyle, Acta Met., 4, 233, (1956).

9. D. Keefer and C. Wert, Acta Met., 11, 489, (1963).

10. J. C. Fisher, Acta Met., 6, 13, (1958).

11. E. Gebhardt and R. Rothenbacher, Z. Metallk., 54, 443, (1963).

12. A. M. Thompson, I.R.E. Trans. Instr., I-7, 245, (1958).

13. A. S. Nowick and B. S. Berry, Anelastic Relaxation in Crystalline Solids, p. 91, Academic Press, N. Y. (1972).

14. B. S. Berry and J. L. Orehotsky, Acta Met., 16, 683, (1968).

15. G. W. Tichelaar, R. V. Coleman and D. Lazarus, Phys. Rev., 121, 748, (1961).

INTERNAL FRICTION DUE TO OXYGEN
IN ULTRA PURE ZIRCONIUM

J.L. GACOUGNOLLE, S. SARRAZIN, J. de FOUQUET
Laboratoire de Mécanique et de Physique des Matériaux
E.R.A. au C.N.R.S. n° 123
E.N.S.M.A., rue Guillaume VII - 86034 - POITIERS

INTRODUCTION

It is well known that titanium, zirconium and hafnium are characterized by a high solubility in oxygen atoms which occupy the octahedral interstitial sites of the h.c.p. lattice. The octahedral sites possess trigonal symmetry and it was shown by Povolo and Bisogni (1) that the oxygen single atoms may cause a shear relaxation only if they are paired either to a substitutional atom (O-S) or to another oxygen atom (O-O). In titanium-oxygen alloys the O-S pairs relaxation has been investigated by Gupta and Weinig (2) and Miller and Browne (3) ; a O-O peak has been also studied by Miller et al. (3). In zirconium-oxygen alloys the O-S peak has been identified by Gacougnolle et al.(4), Browne (5), and Mishra et al. (6) ; but the existence of O-O peak is still being discussed (11). The purpose of this paper is to present results obtained in ultra pure zirconium-oxygen alloys, showing the existence of a peak which appears only when the oxygen concentration becomes sufficiently high.

EXPERIMENTAL PROCEDURE

The basic material used was an unique sample (70x7x0.8) of zone refined zirconium, supplied by the Centre National de Recherches Métallurgiques, and containing less than 2 ppm of metallic impurities (7). An oxydation-diffusion treatment was carried out to introduce oxygen step by step into this sample. After oxydation at 650°C in an oxygen atmosphere the specimen was annealed in the $\alpha+\beta$ phase at 1200°C for 24 h and then in the α phase at 870°C for 24 h in a 4.10^{-7} Torr vacuum. The oxygen concentration was evaluated from weight increasing during oxydation and then verified by electrical resistivity and microhardness. Microhardness measurements proved that the oxygen concentration was homogeneous through the whole sample section ; for 8.8 at.% little cracks appeared excluding any test at 5.3 Hz and further increases in oxygen concentration.

Internal friction measurements were carried out on an inverted pendulum enclosed in 10^{-5} torr vacuum (8). Measurements were taken on heating at a rate of about 20°C/hr. An anneal in situ at 700°C for 30 mn was given to specimen prior testing.

EXPERIMENTAL RESULTS

Figure 1 shows the internal friction versus temperature curves obtained as a function of the oxygen content. To allow a further peak analysis, it was assumed that the internal friction spectrum in pure zirconium is made of two superposed phenomena (9) as depicted in Figure 2 : i) a high temperature background fitting the equation (10)

$$\Delta = \frac{A}{T} \exp - \frac{22000}{RT} \qquad (1)$$

the activation energy of 22 kcal/mole corresponding with self-diffusion energy seems to be insensitive to oxygen concentration, ii) a grain boundary peak located at about 700°C. The influence of this grain boundary peak may be neglected at 400°C ; therefore only the background given by equation (1) was substracted from experimental curves. The peaks for 4.8 at.% - 6.6 at.% - 7.8 at.% and 8.8 at.% oxygen corrected for background are shown in Figure 3. The characterizing parameters are listed in table 1.

TABLE I
Oxygen peak parameters

Oxygen Contents at.%	Position for 3,6s 1000/T	Peak Height $Q^{-1}.10^4$		Activation Energies : Kcal/mole		
				Temperature shift	Width at half-height	
		3,6s	0,2s		3,6s	0,2s
4.8	1.488	1.5	1.6	52.3	45.5	46.5
6.6	1.494	2.4	3	50.3	45.	41.7
7.8	1.493	4.5	5.4	49.5	43.5	36
8.8	1.493	6.3			36.3	

Several important pieces of information are contained in this table :

1/ High oxygen concentrations are necessary to give rise at this peak ; for 3 at. % oxygen the peak is very small and its parameters are not measurable. The variation of the peak height versus oxygen concentration, depicted in

Figure 4, does not agree with a parabolic function of interstitial content for the high concentrations ; a cubic function of the form

$$Q_M^{-1} = 0.026 \ c^2 + 0.6 \ c^3$$

is necessary to take into account all the experimental points. Thus it must be assumed that clusters of more than two oxygen atoms are involved in the relaxation process for concentrations higher than 6.6 at. % .

The peak height variation in titanium-oxygen after Browne (12) is reported in Figure 4 ; it may be seen that for a same oxygen concentration, the relaxation strength is about ten times lower in zirconium. For 4.8 at.% the peaks are indentical to the two frequencies, Figure 5 shows that the height of peak obtained for 7.8 at.% oxygen at approximatively 5 Hz is about 20% higher compared with that observed at 0.3 Hz ; the corresponding temperature shift being close to 45°C.

2/ Activation energy obtained from this shift is 50 ± 2 Kcal/mole in good agreement with values of 48 ± 4 - 52.5 - 50 ± 5 Kcal/mole previously reported for the O-S peak (4,5,6). However the peak is broader than a Debye peak and the width increases for oxygen concentration higher than 6.6 at.%. Moreover Figure 5 shows that for 7.8 at.% oxygen the peak at 5 Hz is broader than at 0.3 Hz.

3/ For a same frequency, the temperature of peak is about 17°C below that of the O-S peak detected by Browne (5). With a activation energy of 50 Kcal/mole this discrepancy corresponds to a frequency ratio of 4 ; thus, the reorientation of oxygen clusters must be regarded as 4 times faster than the reorientation of O-S pairs.

DISCUSSION

From the above results, there appears to be little doubt that the internal friction peak observed at about 400°C in this pure zirconium containing various amount of oxygen originates from a relaxation mechanism involving pairs of intertitial atoms or higher order clusters :

i) The substitutional atoms concentration is too low to give rise to a seizable relaxation effect. In Figure 6 is reported the variation with oxygen concentration of the peak height obtained in technical zirconium containing 400 ppm of metallic impurities ; it may be seen that the concentration dependence is vey different from that for purer metal.

ii) The activation energy found for the peak is very close to the of 50.8 Kcal/mole corresponding to the bulk diffusion of oxygen in zirconium, at the same temperature (17).

iii) As already mentioned the height of peak at a fixed frequency, as a function of the oxygen concentration,follows a cubic law.

However two points must be discussed : the very low observed peak height, and, the variation of the relaxation intensity with frequency and temperature.

As established by Nowick and Heller (13) the quantitative expression for the peak height due to a reordering effect is always of the form

$$Q_M^{-1} = (\alpha C_o v_o / KT) \cdot (\delta\lambda)^2$$

where α is a numerical constant, C_o the total concentration·of defects, v_o the atomic volume, and $\delta\lambda$ an appropiate difference among components of the λ-tensor which measures the distortion of the lattice per unit concentration of defects in a given orientation.For pairs having the same configuration it can be expected that higher distortion will be associated with the atom, providing, when isolated, the greater size factor. The relative volume variation associated with oxygen in titanium and zirconium are respectively $\frac{\Delta V}{V} = 11.6\%$ (14) and $\frac{\Delta V}{V} = 6.2\%$ (15). Thus, if a sample pair model can be considered, the relaxation strength in titanium-oxygen alloys and in zirconium-oxygen alloys must differ in a ratio of approximatively four. The obtained experimental ratio, of about ten, suggests that the concentration C_o and the pair configurations are different in the two metals.

On the other hand, studies of surstructure in titanium-oxygen and zirconium-oxygen solid solutions show that structures of type Ti_6O and $Zr_3 O_{1-x}$ are observed by ordering arrangement of the oxygen atoms (16). The available sites for oxygen in the two lattices are illustrated in Figure 7. Moreover, in contrast with titanium, a zirconium atom with two nearest-neighbour oxygen atom is attracted by them; such a situation implies partially ionic bonds ; and never in the two metals, two superposed octahedral sites along the c axis are together occupied. Accordingly, it seems that the probability for oxygen atoms to occupy nearest-neighbour sites, allowing the existence of O-O pairs, is higher in titanium than in zirconium. Thus, the magnitude of C_o and $\delta\lambda$ associated with oxygen pairs in Zr must be low.

Further more, the existence of no nearest-neighbour pairs in zirconium suggests that, even at low concentration, higher-order clusters involving three or four oxygen atoms are predominently responsible for the relaxation;that could explain the cubic law observed for the variation of the peak height with the concentration. When temperature increases, the increasing of the peak height would indicate an augmentation of concentration pairs consecutive to the desordering-effect.

In short it can be said that oxygen gives rise to a relaxation peak in pure zirconium-oxygen alloys for concentration sufficiently high. In addition to the size factor effect,the fact that only a small peak is observed seems related to the partially ionic bond which characterize these solide solutions : in contrast with pure titanium-oxygen alloys where the dominant relaxation is produced by O-O pairs (3), more complex clusters would be here responsible of the relaxation.

ACKNOWLEDGEMENT

The authors are grateful to Dr. J.P. Langeron for preparing the high purity zirconium specimen and valuable discussions.

REFERENCES

(1) F. Povolo and E.A. Bisogni, Acta Met 14, 711 (1966).

(2) D. Gupta and S. Weinig, Acta Met. 10, 292 (1962).

(3) D.R. Miller and K.M. Browne, "The Science, Technology and Application of Titanium" p. 401.

(4) J.L. Gacougnolle, S. Sarrazin and J. de Fouquet, J. de Phys. 32, C2-21, (1971).

(5) K.M. Browne, Scripta Met. 5, 519 (1971).

(6) S. Mishra and M.K. Asundi, Trans. Indian Inst. Metals 73 (Sept. 1970).

(7) J.P. Langeron, Thesis Paris, 1964.

(8) J.L. Gacougnolle and J. Duvaud (to be published).

(9) J.L. Gacougnolle and. J.P. Langeron (to be published).

(10) J. Woirgard and J. de Fouquet (in this issue).

(11) S. Mishra and M.K. Asundi, Scripta Met. 5, 973 (1971).

(12) K.M. Browne, Acta Met 20, 507 (1972).

(13) A.S. Nowick and W.R. Heller, Adv. Phys. 14, 101 (1965).

(14) E. Garcia, J. Com-Nougue, D. David, P. Boisot and G. Béranger, Journées d'Automne de la Métallurgie, Paris, 1972.

(15) P. Boisot and G. Béranger, C.R. Acad. Sci. 296, Série C, 587 (1969).

(16) A. Dubertret, Thesis Paris 1970.

(17) J.P. Pemsler, J. Electrochem. Soc. 105, 315 (1958).

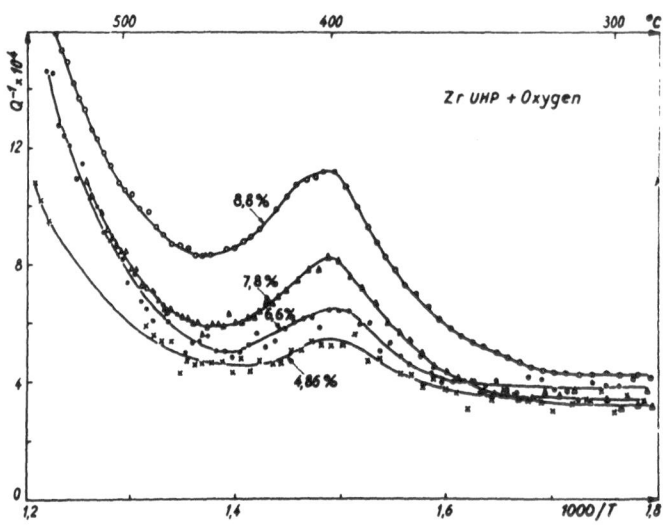

FIG. 1

Experimental internal friction curves of high purity
zirconium alloyed with different oxygen concentrations.

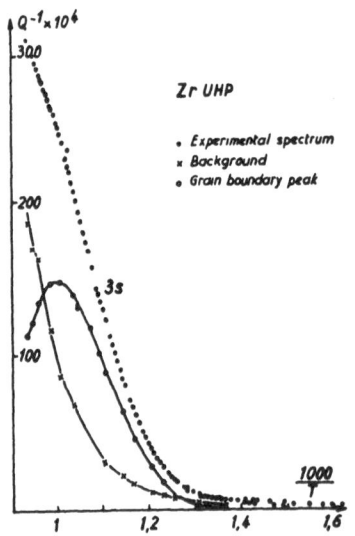

FIG. 2

Decomposition of experimental
spectrum of pure zirconium as
constitued by a high temperature background and a grain
boundary peak.

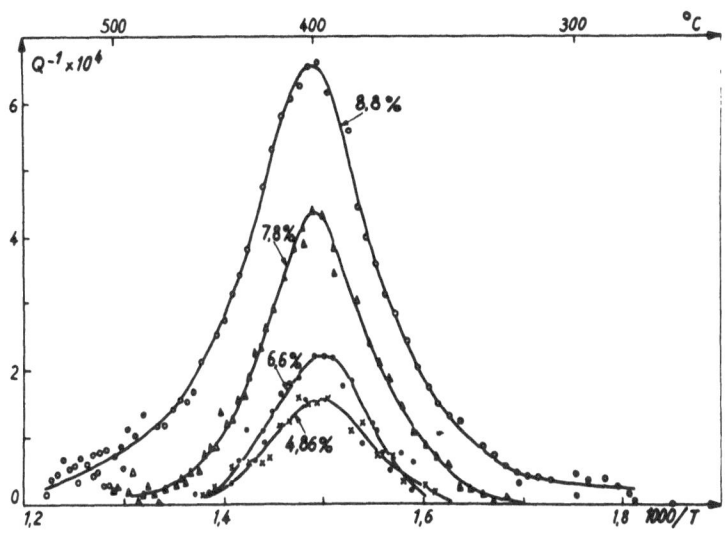

FIG. 3

Oxygen peak corrected for background in high purity zirconium

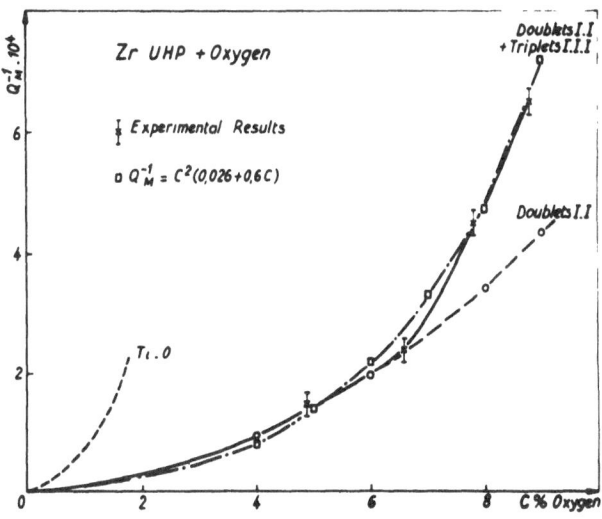

FIG. 4

Oxygen peak height variation versus oxygen concentration.
Ti-O curve after Browne (12).

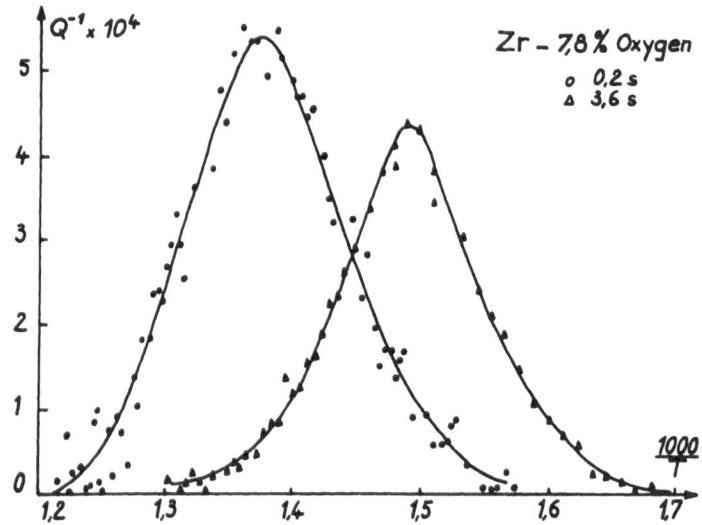

FIG. 5

Oxygen internal friction peak shift in temperature
with frequency at 7.8 at.%

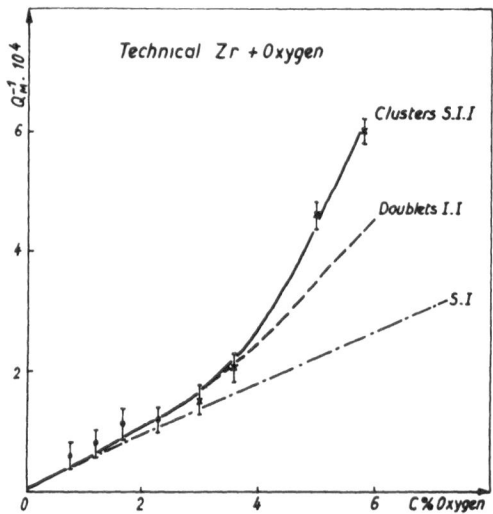

FIG. 6

Height of the oxygen internal friction peak versus oxygen content
in technical zirconium with about 400 ppm metallic impurities.

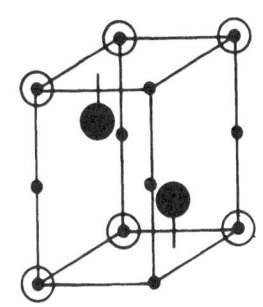 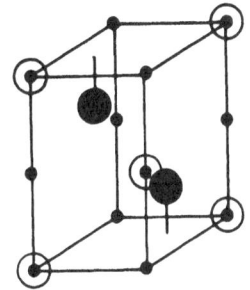

Titanium - Oxygen   Zirconium - Oxygen

● Metal atoms
○ Oxygen atoms
• Octahedral interstices

FIG. 7

Structures of Ti_6O and $Zr_3 O_{1-x}$ solid solutions after Dubertret (16).

INFLUENCE OF INTERSTITIAL IMPURITY ATOMS ON POINT DEFECT RELAXATION IN NEUTRON IRRADIATED IRON

M. Weller and J. Diehl

Max-Planck-Institut für Metallforschung, Stuttgart, Germany

Introduction

The internal friction of iron after neutron irradiation has been studied by several research groups under different conditions as regards purity of the samples (or impurity content) on the one hand and irradiation temperature and neutron dose on the other hand. Rather pure iron was studied quite extensively by the Grenoble group (1,2) after irradiations at temperatures $\leq$ 77 K. During warming up of the samples four internal friction peaks were observed between 110 and 245 K (1 cps) which were labeled I_D^*, I_E^*, II^*, III^*, apparently originating from intrinsic lattice defects. One of the conclusions derived from these studies and correlated investigations of the recovery behaviour was that peak I_E is caused by the relaxation of a $\langle 110 \rangle$-split interstitial which is able to reorient and begins to migrate freely through the lattice at about 120 K.

After irradiation at a somewhat higher temperature (140 K) Wagenblast and Damask (3) did not observe any internal friction peaks in pure iron. If, however, the iron contained 500 at.ppm carbon, internal friction peaks were observed (4) between 100 and 400 K after irradiation at the same temperature and annealing between 37 and 50°C. This points to the fact well known from different types of investigations that in b.c.c. metals intrinsic lattice defects as produced by the irradiation and interstitial impurity atoms can interact strongly.

The aim of the present investigation was to study the combined influence of neutron irradiation and interstitial impurities on the above mentioned low temperature internal friction peaks as well as on those appearing at higher temperatures after annealing, in a more systematic manner, using irradiations at $<$ 77 K and introducing alternatively carbon and nitrogen as interstitial impurities. The main results of these investigations are reported here.

Experimental

Polycrystalline iron wires, 100 mm in length, were used as samples. They were prepared from material of two different suppliers and are marked in the following accordingly: MRC (Materials Research Corp.; 0,75 mm ∅) and VAC (Vacuumschmelze Hanau; 1 mm ∅), respectively. The latter material contained a higher amount of metallic impurities, mainly nickel (300 ppm).

By annealing in dry hydrogen (70 h, 750°C) carbon and nitrogen contents were decreased according to Snoek peak heights below 1 wt.ppm; the resistivity ratios (ρ(300)/ρ(4.2)) were 1600 and 70, respectively. A part of the samples was afterwards doped with C or N up to 300 at.ppm (for details see (5)). The neutron irradiation was performed in the Karlsruhe reactor FR 2 at about 20 K with $3 \cdot 10^{17}$ n/cm^2 (E > 0.1 MeV). After irradiation, the samples were transferred under liquid nitrogen in an inverted torsion pendulum without intermediate warming up. Internal friction was measured at 1 cps. To avoid internal friction effects due to Bloch wall motion a magnetic field of 200 Oe was applied.

Results

Low Temperature Internal Friction Peaks

The internal friction after irradiation was measured at temperatures increasing in steps (5 K/5 min), starting at 77 K.

Fig. 1 shows the internal friction of purified iron. There arise, in accordance with (1,2) four peaks centred at 103 K (I_D^*), 123 K (I_E^*), 147 K (II*) and 232 K (III*). On account of the rather low neutron dose the peaks are relatively small as compared to those obtained by the Grenoble group after $2 \cdot 10^{18}$ n/cm^2.

In Fig. 2 two examples are presented for doped samples (Fig. 2a: Fe + 170 at. ppm carbon; Fig. 2b: Fe + 300 at.ppm nitrogen). A comparison with purified iron shows that there appear similar internal friction peaks. Peak III* is slightly altered in shape and splits up into two peaks III_1^* and III_2^*. The more predominant one, III_2^*, seems to be identical with peak III* in pure iron. Peak II* shows a narrowing by the presence of impurities (especially due to carbon).

TABLE 1

Low Temperature Relaxation Peaks in Neutron Irradiated Iron
($\emptyset$ t = $3 \cdot 10^{17}$ n/cm^2, T_{irr} = 20 K)

Peak No.	Peak Temp. (K)	Activation Energy (eV)[+]	Peak Temp. after ref. (1,2,8) (K)
I_D^*	103	0.28	108 – 110
I_E^*	118 – 123	0.32	124 – 126
II*	146 – 152	0.4	146 – 151
III* (III_2^*)	225 – 232	0.6	233 – 245
III_1^*	194 – 208	0.55	–

[+] Calculated from peak temperatures with the relations given by Wert, Marx (6) and Stephenson (7).

353

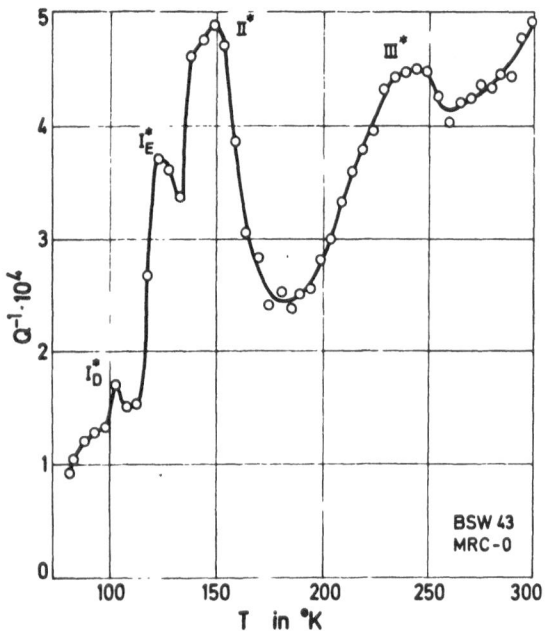

FIG. 1:
Internal friction of purified iron (MRC) after low temperature neutron irradiation $(3 \cdot 10^{17}$ n/cm$^2)$.

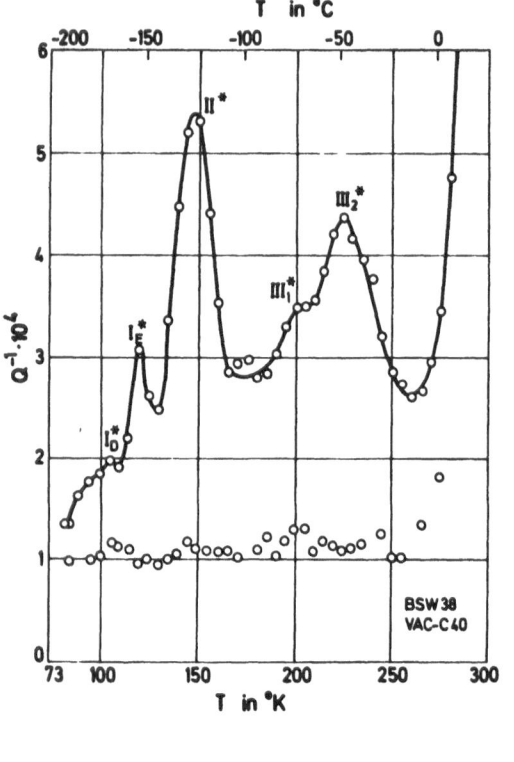

a

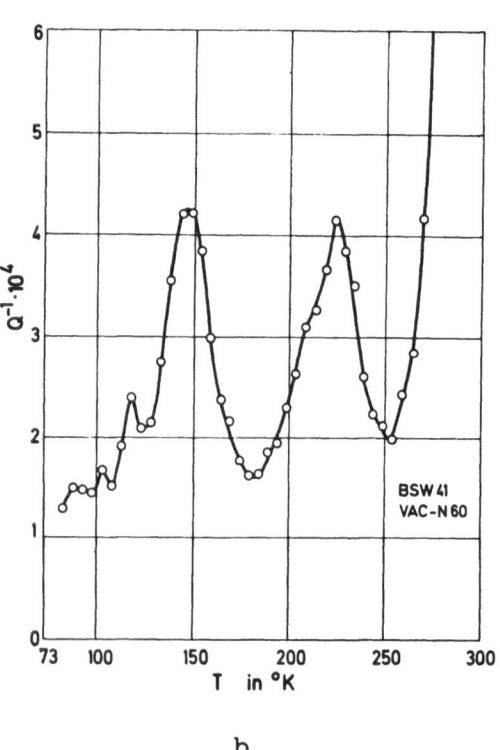

b

FIG. 2: Internal friction of iron (VAC) with 170 at.ppm carbon (a) and 300 at.ppm nitrogen (b) after low temperature neutron irradiation $(3 \cdot 10^{17}$ n/cm$^2)$ (lower points in (a) after annealing at 300 K).

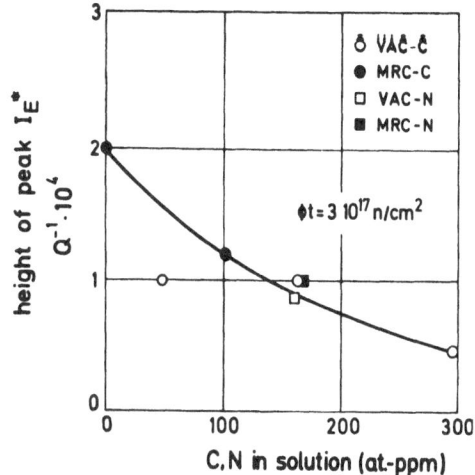

FIG. 3:

Dependence of the height of peak I_E^* on carbon or nitrogen concentration.

Table 1 summarizes the ranges of peak temperatures for all investigated puri-fied and doped iron samples and furthermore estimates of the corresponding activation energies of relaxation. Results on pure iron from ref. (1,2,8) are included.

Whereas the heights of peaks II^* and III^* remain nearly unchanged there exists a strong influence of carbon and nitrogen on the height of peak I_E^*. This can be seen in Figs. 1 and 2, but is more clearly demonstrated in Fig. 3, which com-prises various peak I_E^* measurements as a function of the concentration of car-bon or nitrogen (in solution). A steady decrease of peak I_E^* with increasing impurity contents can be seen.

Internal Friction Peaks after Annealing above 300 K

After heating to about 300 K peaks I_D^*, I_E^*, II^* and III^* have disappeared in purified as well as in doped iron. One example of a measurement after warming up to 300 K can be seen in Fig. 2a (lower data points).

Further annealing to higher temperatures did not reveal any additional re-laxation peaks in purified iron.

In iron containing carbon or nitrogen the well known Snoek peaks appear at 315 K and 300 K, respectively. Parts of the low temperature shoulders of these peaks can be seen in Figs. 2a, b near 300 K. The Snoek peak heights, measured immediately after warming up, are the same as before irradiation.

Further annealing at or above the Snoek peak temperatures influences, however, the internal friction spectrum drastically. Internal friction curves after dif-ferent isochronal annealing steps ($\Delta T_a = 40^\circ C$, $\Delta t = 20$ min) up to $360^\circ C$ are shown in Figs. 4 and 5 for iron with carbon and nitrogen, respectively. The carbon Snoek peak (Fig. 4a) decreases by annealing between $T_a = 40$ and $160^\circ C$

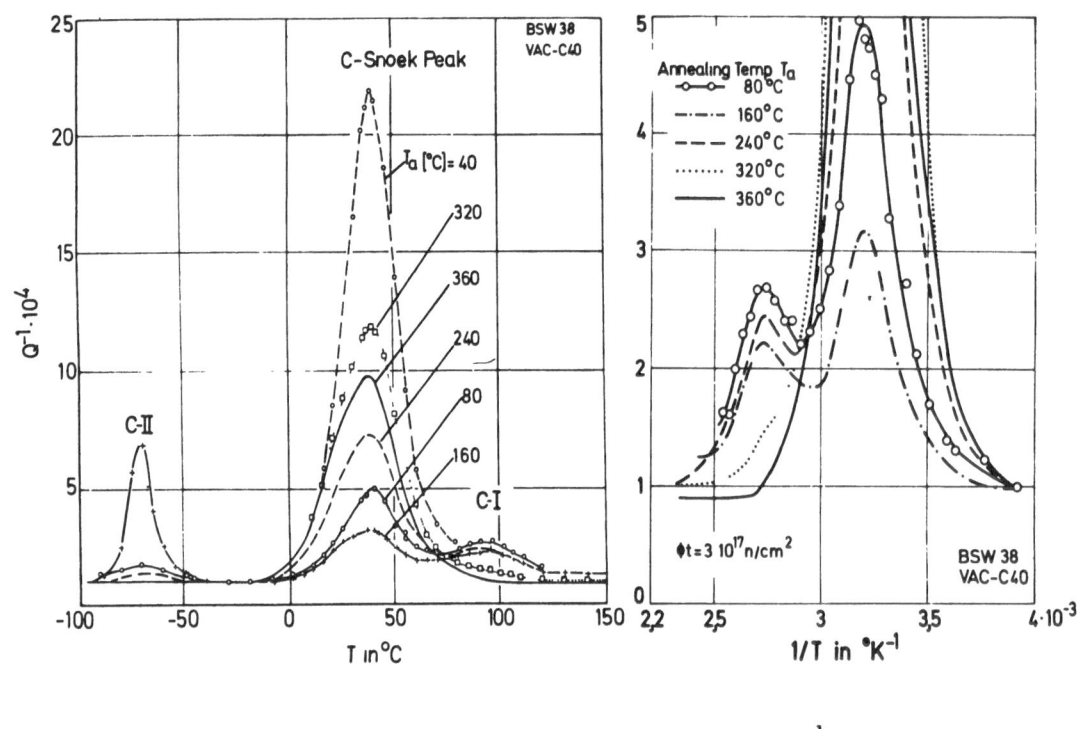

a   b

FIG. 4: Internal friction of neutron irradiated iron (VAC) with 170 at.ppm carbon after isochronal annealing treatments (ΔT_a = 40°C, Δt = 20 min).

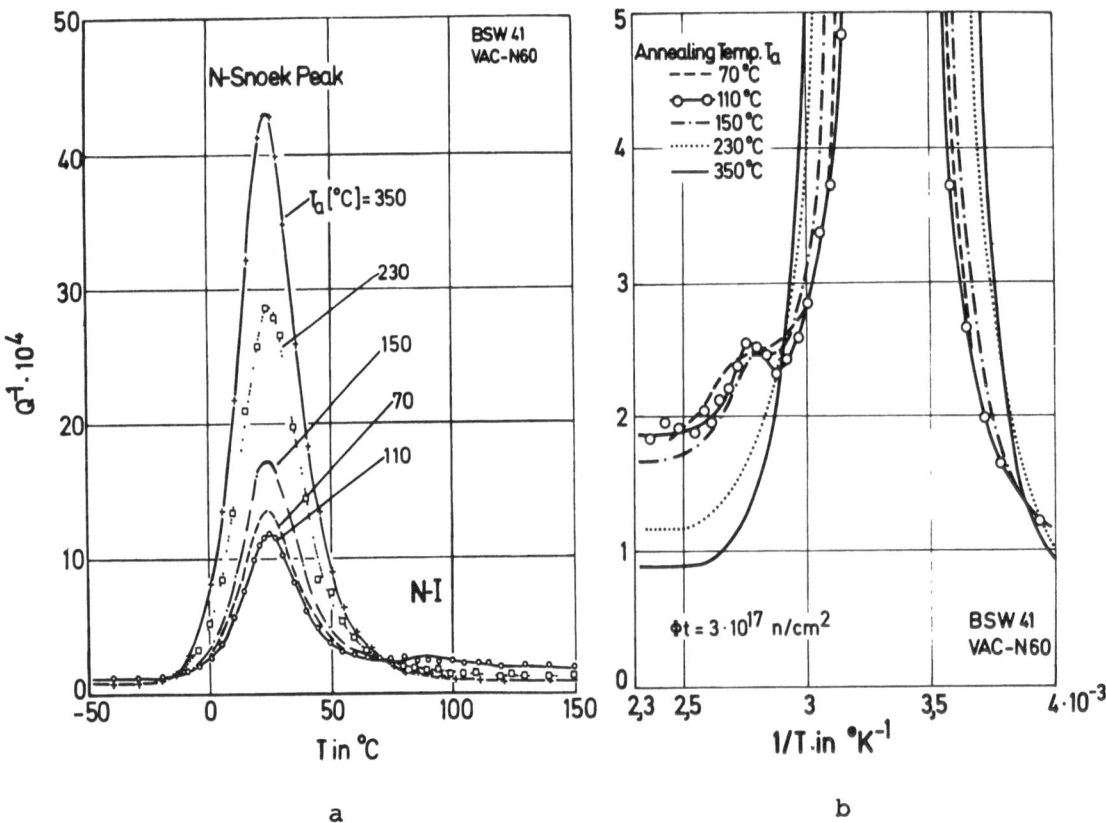

a   b

FIG. 5: Internal friction of neutron irradiated iron (VAC) with 300 at.ppm nitrogen after isochronal annealing treatments (ΔT_a=40°C, Δt=20 min).

from a height corresponding to 170 at.ppm to one corresponding to 10 at.ppm. Simultaneously a new peak appears at 93°C, called by us C-I. This peak is stable up to T_a = 240°C. Annealing at higher temperatures is combined with a reappearance of the Snoek peak. In Fig. 4b internal friction in the range of the C-I peak and the Snoek peak is plotted vs. 1/T, in a way which gives a clearer resolution of peak C-I. For one curve individual data points are plotted in order to give an impression of the precision of the measurements.

In iron with nitrogen the Snoek peak decreases and increases by annealing in a similar way as for iron with carbon (Figs. 5a, b). A new peak, N-I, corresponding to C-I appears at 88°C. The N-I peak begins to anneal already at T_a = 150°C.

An additional, thermally rather unstable internal friction peak C-II (Fig. 4a) was found at 194 K but only in iron with carbon, not with nitrogen. The internal friction curves in Figs. 4 and 5 indicate a close correlation between the decrease of the Snoek peak and the increase of the C-I or the N-I damping peaks and vice versa.

Especially for peak C-I this has been verified quantitatively. Fig. 6 shows measurements on samples with different carbon contents. In addition two N-I measurements are plotted in this diagram. Peak C-I increases almost linearly with the concentration of trapped carbon (or nitrogen), derived from the decrease of the Snoek peak.

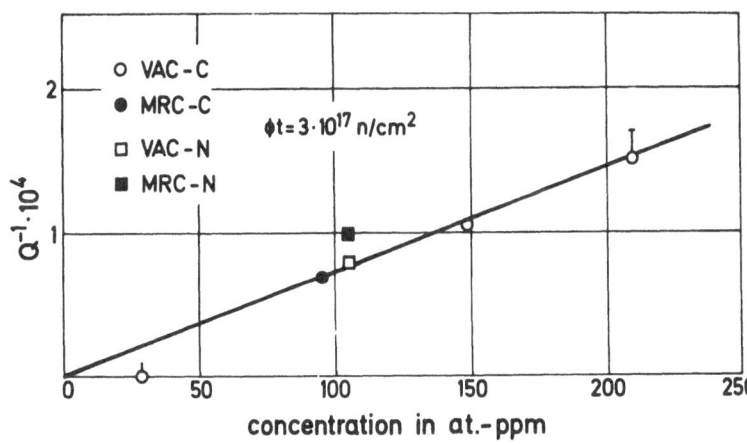

FIG. 6:
Dependence of peak C-I and N-I on the concentration of trapped carbon or nitrogen.

Discussion

With respect to the influence of carbon or nitrogen we have to distinguish two temperature ranges according to the mobility of these impurity atoms. At lower temperatures ($\lesssim$ 300 K) they are immobile, whereas for annealing treatments at higher temperatures it has to be taken into account that they can migrate by thermal activation. Accordingly, the discussion of our results will be divided into two parts.

Low Temperature Internal Friction

A survey of the low temperature internal friction peaks, as given in Tab. 1, shows that essentially the same peaks appear in purified and doped samples. The peak temperatures don't seem to depend strongly on impurity content or neutron dose (comparison with the results from Grenoble). This indicates that the relaxation peaks in our experiments on iron containing nitrogen or carbon are determined by the same point defect configurations as in purified samples irradiated with the same or a higher dose. They were discussed in detail by the Grenoble group (1,2). We, therefore, can base our further considerations on the interpretations given in ref. (1,2).

Among the low temperature peaks, peak I_E^* assigned to the reorientation of a free <110> self-interstitial is most strongly influenced by the presence of carbon or nitrogen. At first sight it seems that the rapid decrease of the height of this peak with increasing interstitial impurity content could be easily explained by trapping of self-interstitials at carbon or nitrogen due to their ability to migrate freely with the same elementary step (and activation energy) with which they reorient. For the given dose and range of impurity concentrations the average number of jumps necessary to find an impurity atom and being trapped there can be estimated to be 10^3 to 10^4 (larger or of the same order of magnitude as for annihilation at vacancies). At the temperature of peak I_E^* the self-interstitials can, however, perform only a few jumps per second. Even if the temperature steps before reaching the peak temperature and the time of measurement are taken into account, the total number of jumps carried out up to the peak I_E^* turns out to be at least one order of magnitude smaller than the above mentioned number necessary for trapping. This makes trapping due to thermal migration of the self-interstitial rather unlikely to cause the suppression of peak I_E^*.

We, therefore, favour another explanation: The interstitial atoms are generated by replacement collision chains, which can propagate over large distances. At the end of their dynamical ranges they convert into the I_E^* defect. If interstitial impurities are present some of the replacement collision chains will be captured at impurity atoms, thereby reducing the number of "free" (uncaptured) self-interstitials and consequently peak I_E^*. Using a simple model and assuming a capture radius of 2 to 4 atomic distances for carbon or nitrogen in iron we deduced from the decrease of peak I_E^* (Fig. 3) a range of 40 to 160 atomic distances for the replacement collision chains.

It should be mentioned that according to (11,12) metallic impurities at even higher concentrations (at.% of Ni, Si) do apparently not strongly affect the formation of free self-interstitials, but accelerate the recovery of them. This seems to indicate that in contrast to N and C the metallic impurities are less effective in capturing replacement collision chains but act as trapping sites during thermal migration of the interstitials.

Among the peaks which are only weakly or not at all influenced by carbon or nitrogen, peak I_D^* has been assigned to the relaxation of a close Frenkel-pair configuration in good agreement with its insensitivity towards interstitial impurities reported here. Peak II^* which in pure iron consists of two overlapping neighbouring peaks (1) was interpreted by small agglomerates of di-interstitials and di-vacancies. If this interpretation is correct, the narrowing of this peak due to interstitial impurities might be caused by the suppression of the relaxation of one of these configurations by the presence of impurities.

The relatively broad peak III^* seems not to originate from the relaxation of a simple point defect. As it was pointed out in (2) dislocations play a rôle. The splitting up into two peaks (III_1^* and III_2^*) by impurities seems to indicate that in addition to dislocations and point defects also impurities are involved in a relatively complicated manner. A comparison with cold worked iron, which also contains these types of lattice defects, seems to be adequate. Swartz (9) observed after cold work two peaks at 155 and 220 K, one of which possibly corresponds to III^* or III_2^*, the other one to II^*. From measurements of Hasiguti et al. (10) at 10^3 cps it can be estimated that for 1 cps two peaks obtained by these authors would appear at 170 and 210 K, also within the range of the peaks III^*. Therefore, we tentatively assign the damping peaks III^*, III_1^*, III_2^* to peaks of the Hasiguti-type.

There seems to be a difference between the I_E^*-defect becoming mobile only during post irradiation annealing or being mobile already during irradiation, especially with respect to the appearance of the peaks II^* and III^*. This can be deduced from the absence of any annealing peaks after 140 K irradiation reported in (3). For a more detailed interpretation of these relaxation peaks it has to be considered that they seem to be suppressed by immediate annihilation or agglomeration of the self-interstitials during irradiation.

High Temperature Internal Friction

It follows from the above mentioned 140 K irradiations and from our results, that in pure iron after low temperature irradiation and annealing above 300 K no internal friction peaks appear, indicating that no intrinsic lattice defects which are able to reorient and to cause relaxation remain in the lattice.

The internal friction experiments in iron with carbon or nitrogen point, however, to the fact well known from other measurements that intrinsic defects are retained in the lattice also above 300 K. The decrease of the Snoek peak and the appearance of the peaks C-I, N-I and C-II (see Figs. 4 and 5) can only be interpreted by carbon or nitrogen atoms being attached to lattice defects. Thereby quite clearly complexes are formed which are able to reorient and to give rise to internal friction. Wagenblast and Swartz (4) were the first who reported about complexes, which cause relaxation in neutron irradiated iron with carbon. One of their peaks, centred at 400 K (at 30 cps) is identical with our peak C-I.

Due to the similarities between the peaks C-I and N-I the assumption seems to be justified that N-I originates from the same kind of complex but formed with N instead of C.

A more detailed analysis of the peaks C-I and N-I, based on Figs. 4b and 5b and taking into account the background damping, leads to the following results: The activation energies of relaxation estimated from the peak temperatures (6,7) turns out to be ≈1.0 eV. Assuming a simple Debye relaxation curve, an activation energy of 0.95 is obtained from the half widths of the measured curves. The agreement between these two values seems to justify the conclusion that C-I and N-I originate from simple relaxation processes.

Investigations on the complex formation (Snoek peak decrease) and annealing (14) indicate, that the C-I and N-I complexes consist of pairs of carbon or nitrogen, respectively, with single lattice defects. From the concentration dependence of the relaxation strength in Fig. 6, a dipole strength (per impurity atom bound in a complex) of $\delta\lambda \approx 0.1$ can be derived.

Considering all these results it seems to us most likely that the relaxation peaks C-I and N-I are caused by carbon-vacancy and nitrogen-vacancy complexes. This agrees with what was first proposed by Wagenblast and Swartz for irradiated iron with carbon (4). Complexes with interstital atoms can be ruled out because of the low value of $\delta\lambda$. According to theoretical calculations of Johnson and Damask (15) the formation of metastable complexes of vacancies with ex-centric carbon atoms should be possible. We assume that this is true also for complexes of nitrogen with vacancies.

Peak C-II, which only arises in iron with carbon, can tentatively be interpreted by a di-carbon-vacancy complex.

Summary and Concluding Remarks

The present investigations demonstrate that interstitial impurities influence the internal friction of iron after low temperature neutron irradiation in temperature ranges in which only intrinsic lattice defects are able to cause relaxation or to migrate and the interstitial impurities are immobile as well as in those in which in addition to the intrinsic defects the interstitial impurities can migrate through the lattice. In the first case, as indicated by the decrease of peak I_E^* with increasing impurity content an athermal (dynamic) process, the trapping of replacement collision chains at impurity atoms seems to be responsible for the most striking effect of impurities. In the other case in

which the interstitial impurities are mobile, they react with remaining intrinsic defects, leading to a decrease of the Snoek peak and to reaction products (complexes) which are themselves able to reorient, thereby giving rise
to additional internal friction peaks.

As already mentioned in the introduction also other properties sensitive to
lattice defects reveal interactions and reactions between intrinsic point defects and interstitial impurity atoms. This is especially true for the yield
stress, which exhibits a strong influence of impurities on irradiation effects
after low temperature irradiation as well as after irradiation at elevated
temperatures (see e.g. (16-20)). Most of these effects have also been accounted for by the formation of complexes between intrinsic defects and impurity
atoms. It should be pointed out, however, that depending on the experimental
parameters, the formation of a greater variety of configurations of such "complexes" seems to be possible (20), which is indicated e.g. by the observation
that not all of them cause internal friction peaks, like the peaks C-I, C-II
and N-I, described here. Therefore, internal friction is a tool to sort out
those complexes, which have a low symmetry stress field and the ability to
reorient. Apparently these complexes are preferably formed during annealing
after low temperature irradiation, as described here, and there exists a rather
close correspondence between internal friction and hardening measurements (radiation anneal hardening (19,20)) for this kind of irradiation and annealing
treatment.

Acknowledgements

We are indebted to the staff of the reactor operations division at the FR 2
in Karlsruhe for their assistance during the low temperature irradiations.
We furthermore express our thanks to many colleagues in our institute for
much help and valuable discussions, especially to those working at the low
temperature irradiation facility of the FR 2.

References

1. H. Bilger, J. Verdone, J.L. Leveque, J.C. Soulie, Internat. Conf.
    Vacancies and Interstitials in Metals, Jül.-Conf.-2 (Vol. II),
    1968, p. 751.

2. V. Hivert, P. Pichon, H. Bilger, P. Bichon, J. Verdone, D. Dautreppe,
    P. Moser, J. Phys. Chem. Solids 31, 1843 (1970).

3. H. Wagenblast, A.C. Damask, Acta Met. 10, 333 (1962).

4. H. Wagenblast, J.C. Swartz, Acta Met. 13, 42 (1965).

5. M. Weller, Dr. rer. nat. - Thesis, University Stuttgart, 1972.

6. C. Wert, J. Marx, Acta Met. 1, 113 (1953).

7. E.T. Stephenson, Trans. AIME 233, 1183 (1965).

8. D. Dautreppe, V. Hivert, P. Moser, A. Salvi, C.R. Acad. Sci. 258,
    4539 (1964).

9.   J.C. Swartz, Acta Met. <u>10</u>, 406 (1962).

10. R.R. Hasiguti, N. Igata, G. Kamoshita, Acta Met. <u>10</u>, 442 (1962).

11. P. Vigier, V. Hivert, P. Moser, E. Bonjour, C.R. Acad. Sci. <u>260</u>, 3359 (1965).

12. P. Vigier, C. Minier-Cassayre, P. Moser, V. Hivert, phys. stat. sol. <u>17</u>, 317 (1966).

13. F.E. Fujita, A.C. Damask, Acta Met. <u>12</u>, 331 (1964).

14. M. Weller, to be published.

15. R.A. Johnson, A.C. Damask, Acta Met. <u>12</u>, 443 (1964).

16. G.P. Seidel, phys. stat. sol. <u>25</u>, 175 (1968).

17. J. Diehl, G.P. Seidel, M. Weller, Trans. Jap. Inst. Met. <u>9</u>, Suppl. 219 (1968).

18. J. Diehl, G.P. Seidel, in "Radiation Damage in Reactor Materials", Vol. 1, IAEA, Vienna, 1969, p. 187.

19. S.M. Ohr, M.S. Wechsler, C.W. Chen, N. Hinkle, Proc. 2nd Intern. Conf. Strength of Metals and Alloys, 1970, p. 742.

20. J. Diehl, U. Merbold, O. Reimold, M. Weller, in "Defects and Defect Clusters in B.C.C. Metals and Their Alloys", Nuclear Metallurgy, Vol. 18, R.A. Arsenault (edt.), 1973, p. 69.

ORDERING AND INTERNAL FRICTION IN ALPHA-BRASSES

C.J. Spears

Middlesex Polytechnic, Enfield, England

ABSTRACT   The internal friction of copper and alpha-brasses of constant grain size containing up to 30 atomic percent zinc was measured between 120 and 300 K at 12.5 kHz at a strain amplitude of 10^{-7}, as a function of tensile pre-strain. A maximum, Q_{MAX}^{-1}, of the internal friction occurs, for any given composition, at strains of 1.5-2.0% corresponding approximately to the end of the linear hardening stage. Q_{MAX}^{-1} decreases monotonously with zinc content, except close to the "ordering" composition Cu_3Zn, at which a local maximum occurs in the Q_{MAX}^{-1}/zinc content relation. The effect is ascribed to the amelioration of internal stress fields at dislocations facilitated by local order.

Introduction

Recent work on the background internal-friction in the kHz range in alpha-brasses (1), at strain amplitudes of the order of 10^{-7}, show that it is strongly influenced by the intra-crystalline micro-stress fields and by the alloy content. Q^{-1} decreased steadily with zinc content (1), but an anomaly in this uniformity appeared to occur at concentrations corresponding to 25 atomic percent zinc, i.e. to Cu_3Zn. The deviations, which were not studied in detail in (1), were indicative of ordering effects. The object of the present work was to examine the internal friction within a narrow composition range centred on 25 atomic percent zinc and to check the occurrence of such ordering and, if present, to assess its significance.

Experimental

The experiments, which were an extension of earlier work (1) with brasses containing 10, 20 and 30% Zn were again carried out on polycrystals of 45μm grain size with zinc contents of 15, 17.5, 22.5, 25 and 27.5 atomic percent, i.e. at concentrations within a range where ordering effects were expected to be observable. The main impurities were, as before, traces of iron (<10 ppm), tin and bismuth. Rod-shaped specimens were oscillated longitudinally in vacuo at 12.5 kHz at a strain amplitude of about 10^{-7} and the decay of free oscillations was monitored to determine Q^{-1}. It was assumed that, as in the previous work, Q^{-1}, was amplitude independent. Within the temperature range examined (120-300 K), Q^{-1} varied uniformly with temperature as is apparent from Fig. 1. The diagram also shows the effect of pre-strain on Q^{-1} in the case of brass containing 25 atomic percent zinc.

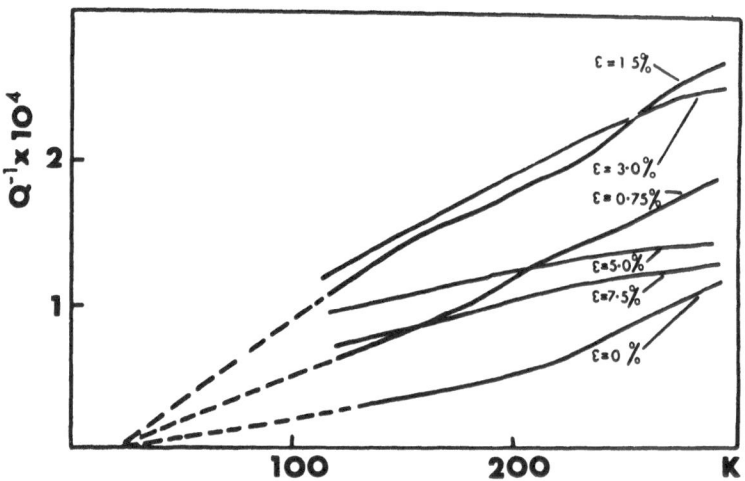

Fig. 1 – Dependence of internal friction of 75/25 alpha-brass of
grain size 4.5 x 10^{-3} cm on temperature and tensile pre-strain

Isotherms for annealed and pre-strained specimens of the same brass, derived from
Fig. 1, which are also characteristic of similar results for brasses of the other compositions,
are shown in Fig. 2.

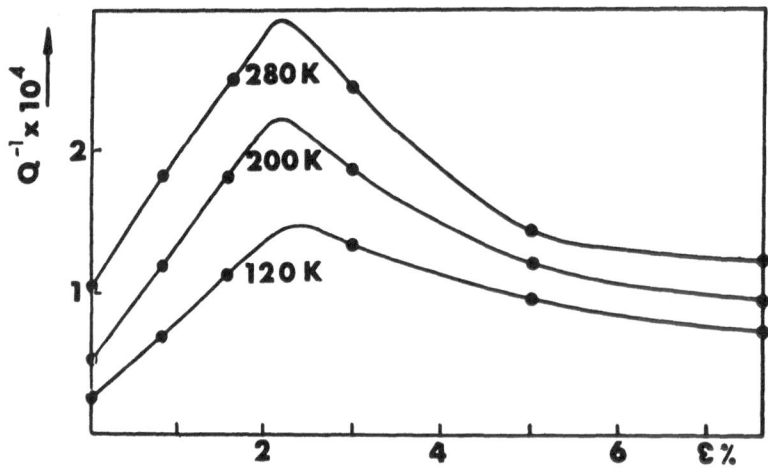

Fig. 2 – Internal friction/strain isotherms of 75/25 alpha-brass of
4.5 x 10^{-3} cm grain size as function of tensile strain, ε
at three temperatures.

The maximum value of internal friction, i.e. Q_{MAX}^{-1}, obtained from Fig. 2 and similar sets of curves for the other compositions, are shown in Fig. 3 for T = 280 K. Corresponding sets of curves for other temperatures have similar characteristics.

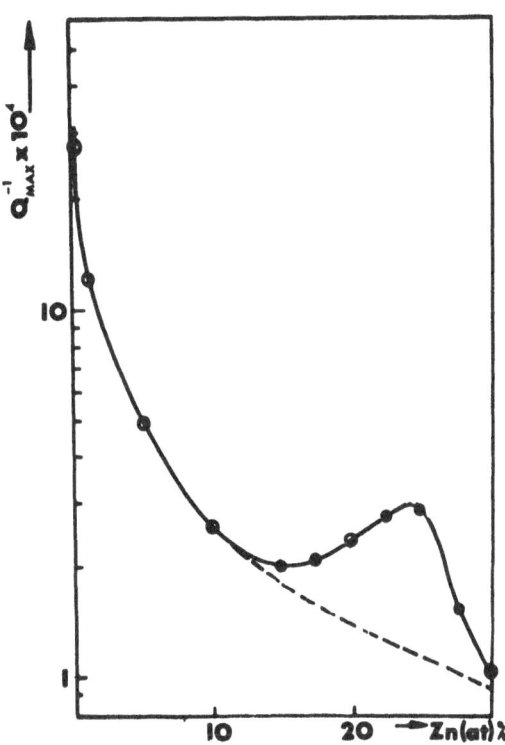

Fig. 3 - Dependence of Q_{MAX}^{-1} on zinc content, empty circles from references (1) and (2).

Discussion

A tentative explanation of the "peak" in Q_{MAX}^{-1} (Fig. 3) close to the composition Cu_3Zn is proposed on the basis of the equation previously advanced to account for the amplitude – independent internal friction in the alpha-brasses (2).

In that equation, i.e.

$$Q^{-1} = \frac{\Omega N_{max} L_{max} G b^3}{2\lambda\sigma_{\infty}} \; (kT/H_{max}) \; \ln(\omega_d/\omega), \qquad (1)$$

Ω is an orietation factor, λ the mean spacing between adjacent geometric kinks on dislocations in the absence of an applied stress, G the shear modulus, σ_{∞} a constant equal to about $10^{-6}G$, ω_d a "Debye" frequency and ω the vibrational frequency of the specimen. In the model dislocation segments of length $L_1 \leq L \leq L_{max}$ are visualised held up at a distribution of internal energy –

barriers $H_1 \leqslant H \leqslant H_{max}$. The distribution of segment lengths and activation energies is approximated by the relation H/L = constant, which is based on the details of the "kink-kinetics" assumed in the model. Similarly the product N(L).L is taken to be constant, i.e. the density of segments per unit volume increases with decreasing length. In (2) the effect of alloying on Q^{-1} was tentatively expressed by the semi-empirical relation.

$$H_{max}(C) = H_{max}(0) \left[1 + \alpha C \right] \qquad (2)$$

where C is expressed as atomic fraction, and α is a constant; it implies an increase in H_{max} with alloy content, resulting from the enhancement of the internal stress field by an increase in the strength of locking of dislocations by the higher concentration of zinc atoms. The present observations suggest that the linear relation (2) would not hold at compositions close to Cu_3Zn. Assuming validity of equation (2), local order appears to facilitate a reduction in the locking effectiveness of the alloy for vibrations of small amplitude, at least statistically. Equation (2) would therefore have to be modified. A semi-empirical form, which would describe the effect of alloying in the range here investigated is:-

$$H_{max}(C) = H_{max}(0) \left[1 + \alpha C - e^{-\beta(C - C_0)^2} \right],$$

where C_0 corresponds to that composition giving greatest deviation of Q_{MAX}^{-1} i.e. from the "background" indicated by a broken line in Fig. 3, it is approximately equal to 25 atomic percent zinc. Short and long range order is well known to occur in alpha-brasses (3); however the precise mechanism responsible for the observed non-uniformity of the internal friction, which appears to arise from $L1_2$ type of order, requires further elucidation.

References

1. C.J. Spears, J. de Physique 32. C2 - 183 (1971)
2. C.J. Spears and P. Feltham, Jl. Mat. Sci. 7. 969 (1972)
3. L.M. Clarebrough and M.H. Loretto, Proc. Roy. Sco. A.257. 326 (1960)

GRAINBOUNDARIES

AND

PHASETRANSFORMATIONS

GRAIN BOUNDARY RELAXATIONS - AN ASSESSMENT

G. Roberts[*] and G. M. Leak

Department of Metallurgy, University of Manchester

INTRODUCTION

The presence of grain boundaries in a metal has been known to produce
internal friction peaks since the work first carried out by Kê (1947).
Since then low temperature grain boundary peaks have been observed in
num erous materials, though extensive investigations of any one material
have been rare. This lack of systematic study, coupled with an uncertainty
of the structure of medium and high angle grain boundaries has hindered the
development of quantitative models for the low temperature peak. Attempts
have been made to link grain boundary damping parameters with dislocation
properties without much success. This note considers some features
showing that dislocation movements, although essential, are not necessarily
directly linked to the measured parameters.

EXPERIMENTAL DATA

The two main parameters obtained from grain boundary relaxation studies
are the activation energy which is found from the frequency sensitivity of
the peak temperature, and the peak height which is a measure of the
relaxation strength. Contrary to the recent association of these two
parameters with stacking fault energy (Roberts and Barrand 1968a) it is
considered that they can be controlled by entirely different processes. The
activation energy will govern the kinetics of relaxation, and hence the
temperature at which relaxation takes place but other conditions prevailing
at that temperature will govern the amount of grain boundary movement and
therefore the relaxation strength. In general, the low temperature peaks
are much wider than an ideal peak width predicted from the standard linear

---

[*] Now at C.E.G.B., Berkeley, Research Laboratories

solid model. The relaxation strength may be taken directly as the peak height, or, following an analysis by Nowick and Berry (1961), the peak height may be converted to that of an equivalent Debye peak using a distribution parameter.

Data found for the low temperature peak in various materials is shown in Table I.

It is apparent that both the activation energy and relaxation strength can vary markedly between materials, though reasonable consistency is obtained from different investigations of the same material. However, some of the marked differences in measurements, for example, on Zn and on Mg are known to be dependent on grain size.

THEORIES

Zener (1941) and Raj and Ashby (1971) have carried out elasticity analyses of the stress build-up which opposes grain boundary sliding, and which thus governs the ratio of relaxed to unrelaxed modulus M_R/M_U. The results they obtained differ, presumably because of the approximations which are inherent or necessary in their approaches. Both results can be expressed in terms of Poisson's ratio ν,

$$\text{Zener} : \quad \frac{M_R}{M_U} = \frac{2(7 + 5\nu)}{5(7 - 4\nu)} \qquad \ldots\ldots (1)$$

$$\text{Raj and Ashby} : \quad \frac{M_R}{M_U} = \frac{1}{0.57(1 - \nu) + 1} \qquad \ldots\ldots (2)$$

These expressions have never been applied to experimental grain boundary damping data. To do so it is necessary to consider the damping equation derived from the standard linear solid model (Zener, 1947):

$$\frac{\Delta}{\pi} = \frac{M_U - M_R}{(M_U M_R)^{\frac{1}{2}}} \quad \frac{W \tau}{1 + W^2 \tau^2} \qquad \ldots\ldots (3)$$

This equation relates the logarithmic decrement Δ to oscillatory frequency W and relaxation time τ, and has a maximum value of:

$$\Delta_M = \frac{\pi}{2} \frac{M_U - M_R}{(M_U M_R)^{\frac{1}{2}}} \qquad \dots\dots (4)$$

when $W\tau = 1$.

Equation (4) may be re-written as:

$$\Delta_M = \frac{\pi}{2} \left\{ \frac{M_U}{M_R} + \frac{M_R}{M_U} - 2 \right\}^{\frac{1}{2}} \qquad \dots\dots (5)$$

which allows an estimation of Δ_M using equations (1) and (2) for various values of Poisson's Ratio, ν. The comparison of theory and experiment is shown in figure (1), for measured peak height and for distribution-corrected peak height. Whilst the theories predict the correct trend of variation of relaxation strength with Poisson's ratio there is a discrepancy in the absolute magnitude. This discrepancy amounts to an over-estimation of the extent of grain boundary sliding by up to a factor of four. Despite this difference the correlation of relaxation strength with Poisson's ratio indicates that the elasticity approach is a promising method of analysis. One outstanding problem is that both theories predict that the relaxation strength is independent of grain size, which does not agree with experimental findings (Leak 1961, Roberts and Barrand 1968b).

An alternative mechanism of grain boundary migration rather than sliding under cyclic stress, was proposed by Leak (1961). The suggestion has been quantified, to some extent, by Roberts and Leak (1973a), assuming that the relaxed modulus is limited by the increased grain boundary area (and therefore energy) on bowing out. The modular ratio is given by:-

$$\frac{M_R}{M_U} = 1 + \frac{Ao.\mu}{\gamma} \qquad \dots\dots (6)$$

where γ is the grain boundary energy. Ao is a constant which depends on grain boundary mis-orientation, ϕ the orientation between the grain boundary and the stress axis, θ, the shear modulus μ for the metal arises from the applied stress acting on the boundary.

The experimental variation of relaxation strength with grain boundary energy is compared with theoretical estimates in figure 2 for the distribution corrected and uncorrected peak height data, using several values of the constant, Ao. Values of grain boundary free energy, (Hondros 1969) have been corrected to the grain boundary peak temperature using:

$$\gamma_T = \gamma_{OK} - T.\Delta S$$

taking the entropy term ΔS as 0.3 erg. cm^{-2} (degree K)$^{-1}$. Values of μ have been taken from Hirth and Lothe (1968). Correlation for an assumed value of Ao is not too unreasonable. Dependence on grain size, however, is still indeterminate. The constant Ao, as well as containing orientation terms also contains a measure of grain size 2ℓ, Ao = $\theta.\phi.\ell$. At first sight this interpretation suggests that the peak height should increase with increasing grain size. However, the orientation terms θ and ϕ will also depend on grain size, in particular upon how mis-orientations vary as grain growth occurs.

ACTIVATION ENERGY

The activation energy governing the relaxation process will necessarily reflect the intrinsic grain boundary properties and so will not be comparable to larger scale deformation mechanisms such as diffusion creep. Qualitative suggestions have accounted for the activation energy being equal to that for self-diffusion (Roberts and Barrand 1968a) or grain boundary migration (Leak 1961). Extensions of either theory to explain the whole spectrum of activation energies observed have proved complex (Roberts 1972). Roberts and Leak (1973b) have recently re-examined the activation energy data in terms of the Mott island model, (1948) prompted by two factors. As the activation energy distribution is the main contributor to the peak width (Cordea and Spretnak 1965), then even in materials having a mean activation energy equal

to that of volume self-diffusion some boundaries will be relaxing with values
considerably in excess of the mean. This indicates that apparent equivalence
of the measured value of activation energy with that of volume self-diffusion
can have no particular significance. Furthermore, work on zinc and on
magnesium has shown the effect of grain size on measured activation energies.
Fine grained material ($\sim 10^{-5}$m) generated at a low temperature peak with an
activation energy equal to half that for volume self-diffusion (Wegria et al.
1970 for Zn, Smith 1972 for Mg). At a coarser grain size ($> 10^{-4}$m) the data
supported the activation energy's being equal to that of volume self-diffusion
(Roberts et al. 1969, for Zn, Smith 1972 for Mg), figure 3.

The island model when applied to the data allows an estimation of the
good-fit island size. Following a suggestion by Gifkins (1967) that the
disordered regions between the islands could be considered as channels of
relaxed vacancies, it is possible to derive grain boundary energies. The
values found are given in Table II and compared with the experimental
results taken from the review by Hondros (1969).

The picture here is of grain boundary sliding taking place by disordering
and re-ordering of atoms and relaxed vacancies at the edges of the islands.
The grain boundary energy is given by :

$$\frac{2}{a^2 n} \cdot Q_v$$

where a is the interatomic spacing, in the lattice, n the width of an island
in number of atoms (assumed square for ease of calculation) and Q_v the energy
of formation of a relaxed vacancy (assumed to be half the energy of self-
diffusion for purposes of calculation). The Mott island model gives the
value of grain boundary damping activation energy Q_b as $\frac{nL}{k}$ where L is the
latent heat of melting and k is Boltzmanns constant. Table II has then
been constructed from values of n estimated from damping data.

Reasonable agreement is obtained which lends support to this view of grain boundary structure.

The migration theory is not so easy to fit into the island model. Thus on balance a model based on grain boundary sliding can account for most of the results. It is apparent that dislocation movements, whilst essential to allow sliding to occur are not going to be rate-controlling and thus correlation with stacking fault energies must be fortuitous.

TABLE I

INVESTIGATOR	MATERIAL	Δ_M (Uncorr.)	Δ_M (Corr.)	T_P/T_M	Q_M, kJ mole^{-1}	Q_M/Q_V
Cordea and Spretnak (1965)	Al	0.253	0.407	0.614	144	1.02
Williams and Leak (1967)	"	0.140	–	0.609	159	1.12
Kê (1947)	"	0.252	–	0.598	134	0.94
Cordea and Spretnak (1965)	Cu	0.040	0.125	0.349	132	0.67
Williams and Leak (1967)	"	0.047	–	0.386	151	0.76
De Morton and Leak (1966)	"	0.050	–	0.400	129	0.66
Peters et al. (1964)	"	0.064	~ 0.14	0.408	157	0.80
Weinig and Machlin (1957)	"	0.100	–	0.467	167	0.85
Kê (1949)	"	0.100	–	0.459	–	–
Rotherham and Pearson (1956)	"	0.064	–	0.423	138	0.70
Marsh (1954)	"	0.080	–	0.415	155	0.79
Cordea and Spretnak (1965)	Ni	0.106	0.274	0.406	308	1.05
Roberts and Barrand (1968b)	"	0.100	–	0.397	293	1.00
Postnikov (1958)	"	0.099	–	0.384	264	0.90
De Morton and Leak (1966)	Au	0.03	–	0.340	95	0.55
Marsh and Hall (1953)	"	0.03	–	0.399	144	0.83
Cordea and Spretnak (1965)	Ag	0.012	~ 0.03	0.353	174	0.94
Roberts (1968)	"	0.011	–	0.319	92	0.50
Pearson and Rotherham (1956)	"	0.025	–	0.367	92	0.50
Rotherham et al. (1951)	Sn	0.04	–	0.679	107	0.74
Leak (1961)	α-Fe	0.028	–	0.44	268	0.72
Roberts et al. (1969)	Zn	0.30	0.44	0.55	95	1
Wegria et al. (1970)	"	~ 0.06	~ 0.148	0.50	60	0.62
Delaplace et al. (1970)	Mg	0.1	–	0.58	133	1*
Smith (1972)	"	.10	.36	0.59	107	0.8+
		.11	.55	0.59	159	1.2†
Berlec (1970)	W	–	–	~0.6	505	0.8

* (Grain Size < 0.1mm.)   + (Grain Size 0.08 mm.)   † (Grain Size 0.15 mm.)

TABLE II

Boundary Energy $J.m^{-2} \times 10^3$

	Island Size n	Derived	Experimental
Al	13	200	625
Cu	9	500	534
Ni	16	440	770
Ag	8	415	399
Au	7	435	390
Zn	14	150	1118
Fe	18	475	795
Sn	15	200	160
W	13	950	1070

REFERENCES

Berlec I., (1970), Met. Trans., $\underline{1}$, 2677.

Cordea J. N. and Spretnak J. W., (1965) Trans. Met. Soc. A.I.M.E., $\underline{233}$, 1685.

Delaplace J., Nicoud J. C. and Trabut L., (1970), J. Nucl. Mats. $\underline{35}$, 167.

Gifkins R. C., (1967), Mat. Sci. Eng., $\underline{2}$, 181.

Hirth, J. P. and Lothe J., (1968), "Theory of Dislocations" (McGraw-Hill, New York).

Hondros E. D., (1969), "Interfaces" ed. Gifkins, R. C. (Butterworths).

Kê T. S., (1947), Phys. Rev., $\underline{71}$, 533; $\underline{72}$ 41

Kê T. S., (1949), J. Appl. Phys., $\underline{20}$, 274.

Leak G. M., (1961), Proc. Phys. Soc. Lond., $\underline{78}$, 1520.

Mash D. R. and Hall, L. D., (1953), Trans. Met. Soc. A.I.M.E., $\underline{197}$, 937.

de Morton M. and Leak G. M., (1966), Acta Met., $\underline{14}$, 1140.

Mott N. F., (1948), Proc. Phys. Soc., $\underline{60}$, 391.

Nowick A. S. and Berry B. S., (1961), I.B.M., J. Res. and Devel., $\underline{5}$, 297, 312.

Pearson S. and Rotherham L., (1956), J. of Metals, 206, 895.

Peters D. T.,,Bisseliches J. C. and Spretnak J. W., (1964), Trans. Met. Soc. A.I.M.E., $\underline{230}$, 530.

Postnikov V. S., (1958), Sov. Phys. Uspecki, $\underline{66}$, 1.

Raj R. and Ashby M. F., (1971), Met. Trans., $\underline{2}$, 1113.

Roberts G., (1972), Ph.D. Thesis, University of Manchester.

Roberts G., Barrand P. and Leak G. M., (1969), Scripta Met. $\underline{3}$, 409.

Roberts G. and Leak G. M., (1973a). To be published.

Roberts G. and Leak G. M., (1973b). To be published.

Roberts J. T. A., (1968), Ph.D. Thesis, University of Manchester.

Roberts J. T. A., and Barrand P., (1968a), Trans. Met. Soc. A.I.M.E., 242, 2299.

Roberts J. T. A. and Barrand P., (1968b), J. Inst. Mets., $\underline{96}$, 172.

Rotherham L. and Pearson S., (1956), J. of Metals, $\underline{206}$, 881.

Smith C. C., (1972), Ph.D. Thesis, University of Manchester.

Wegria J., Gouzou J. and Habraken L., (1970), C.N.R.M., 25, 45.

Weinig S. and Machlin E.S., (1957), Trans. Met. Soc., A.I.M.E., 209, 32.

Williams T. M. and Leak G. M., (1967), Acta Met., 15, 1111.

Zener C., (1941), Phys. Rev., 60, 906.

Zener C., (1947), Elasticity and Anelasticity of Metals" Chicago Univ. Press.

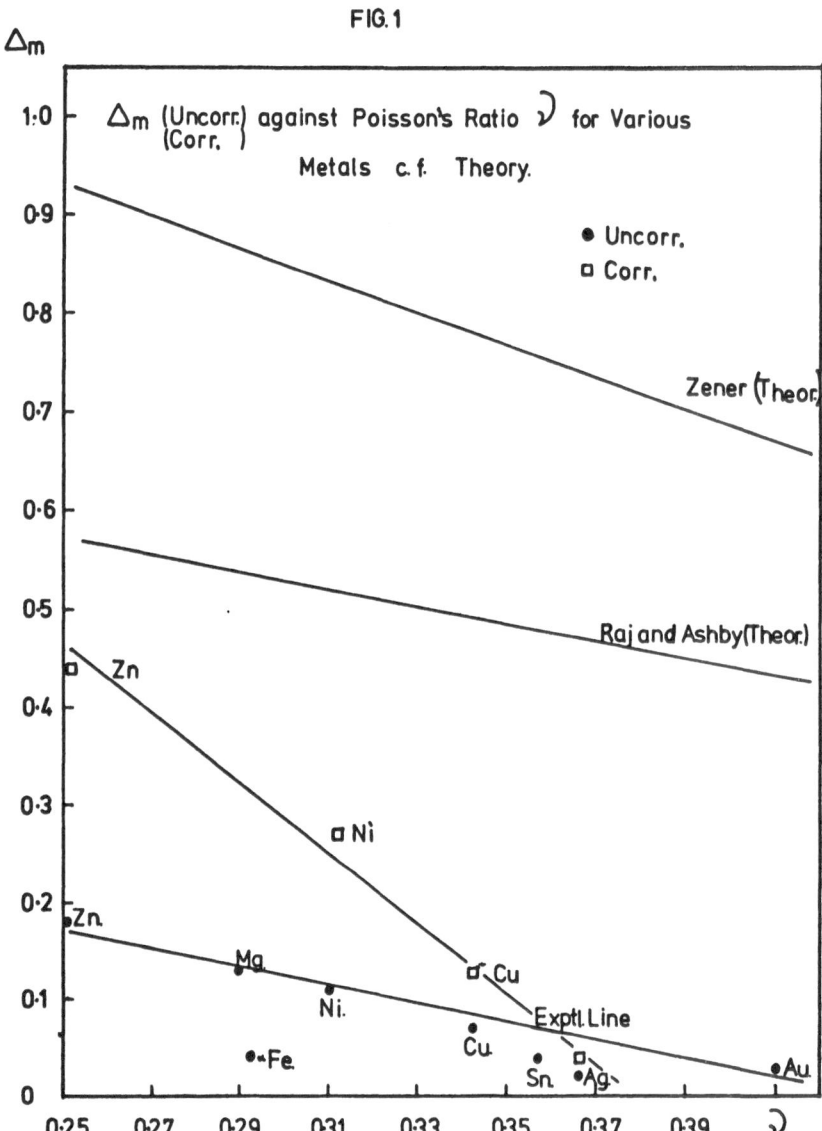

FIG. 1

Δ_m (Uncorr.) against Poisson's Ratio ν for Various (Corr.) Metals c.f. Theory.

• Uncorr.
□ Corr.

Zener (Theor.)

Raj and Ashby (Theor.)

Zn

□ Ni

Zn.

Ma

Ni.

□ Cu

Exptl. Line

Cu.

Fe.

Sn. □ Ag.

Au.

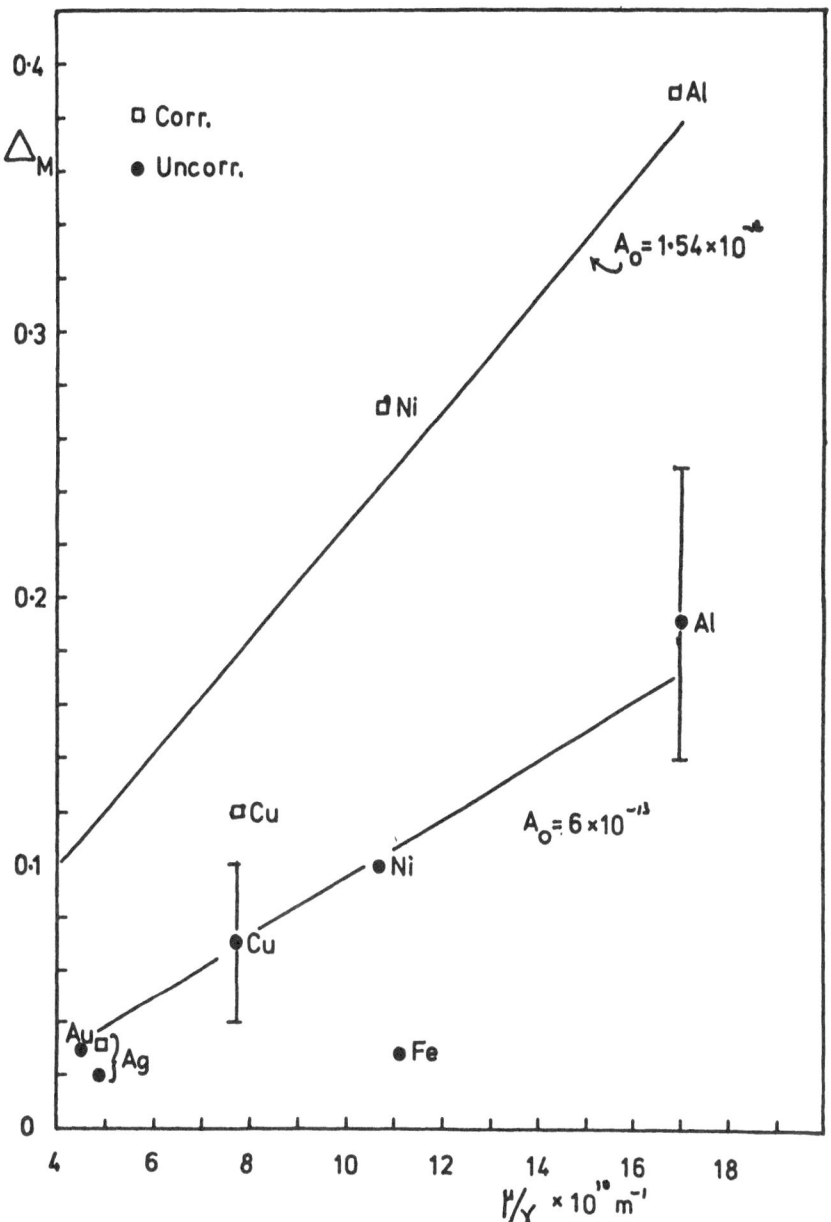

FIG. 2.

$\Delta_{M \text{($\beta$ Corr.)}}^{\text{(β Uncorr.)}}$ against μ/γ for Roberts - Leak theory.

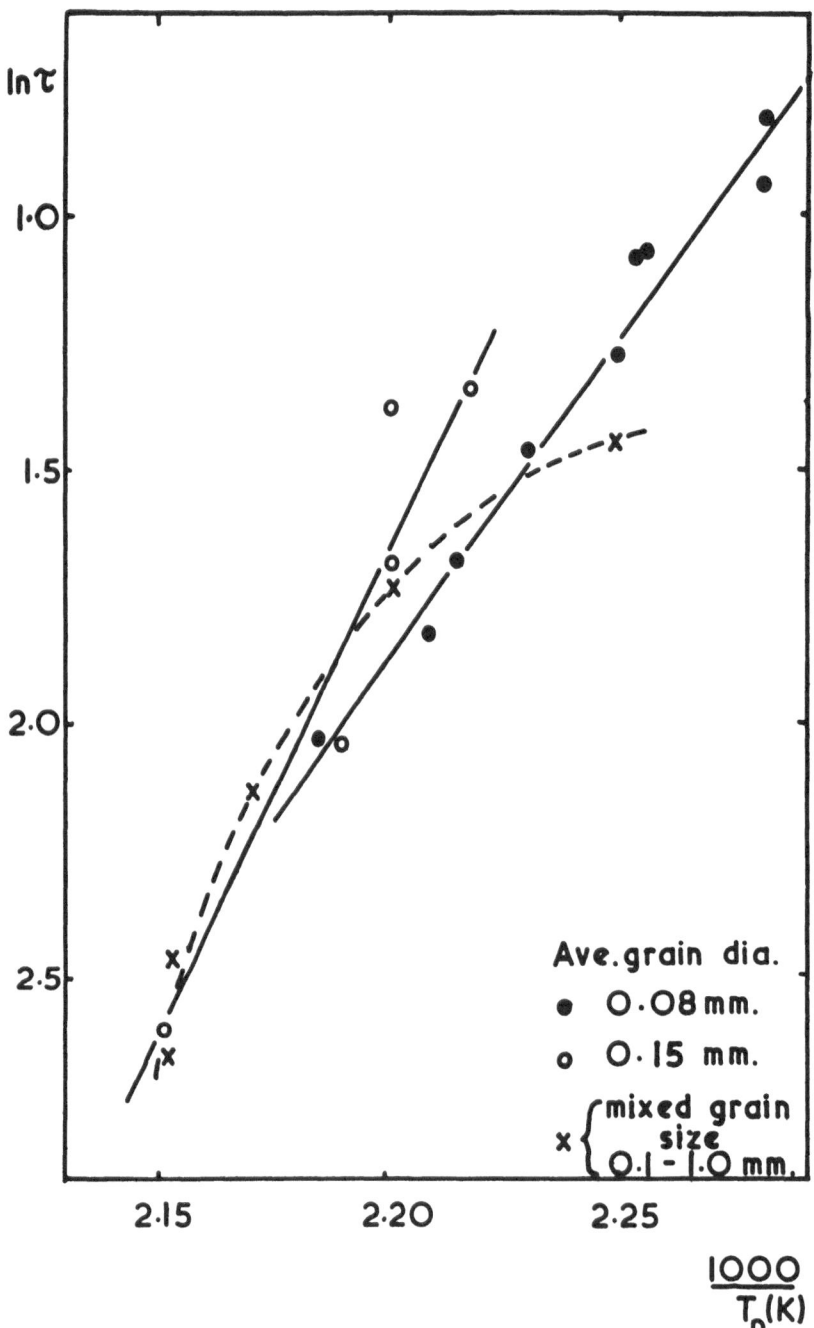

Amplitude Dependence Associated With Grain Boundary Damping.

C.C.Smith and G.M.Leak.
Department of Metallurgy,
University of Manchester,
Manchester. England.

Introduction.

Theories proposed for the orthodox or low temperature damping peak in pure metals (L.T.P.) have been based on a comparison of the observed activation energy with those of either lattice self diffusion or grain boundary self diffusion. The trend has been to accept that the former indicates a grain boundary sliding mechanism (Kê,ref 1) and the latter a grain boundary migration mechanism (Leak,ref 2). To account for the experimental observation that, in some cases, the damping activation energy is intermediate between those for lattice diffusion and grain boundary diffusion, Roberts and Barrand (3,4) proposed a qualitative model based on empirical correlations between stacking fault energy and both activation energy and relaxation strength first reported by Cordea and Spretnak (5). This model does not purport to establish the relaxation mechanism as either sliding or migration but is concerned with the relative contributions of dislocation glide and climb.

In each of these models there are several implicit assumptions which may be open to serious objection.

(a) The peaks are assumed to be wholly anelastic in origin. Any analysis is based on the formal theory of anelasticity developed by Zener (6) in some cases with modifications due to Nowick and Berry (7) to take account of the large peak width normally observed.

(b) The mechanism of relaxation and the associated activation energy are treated as structure insensitive properties of the material. Recent evidence (8,9) suggests that the activation energy is different at different grain sizes. The work of Gleiter (10) and of Rutter and Aust (11) also indicates that the activation energy for grain boundary migration is sensitive to the fine structure of the boundary and the misorientation across it.

(c) In order to isolate the grain boundary contribution, the background damping is estimated in the region of the peak by a smooth interpolation between the high and low temperature extremes of the peak. Such a technique may be justifiable in cases where the peak and background have different origins (Snoek, Zener). If the grain boundary peaks originate in part or in whole from dislocation movement, their separation from the background damping may be more complex.

(d) Modern concepts of grain boundary structure involve the presence of dislocations in grain boundaries (12,13,14). Grain boundary damping may then arise from the movement of these dislocations. Such damping might be expected to show amplitude dependent characteristics, an aspect neglected in previous work.

The purpose of the present paper is to present the results of an investigation of the amplitude dependence of damping in polcrystalline magnesium over a range of temperature in which a grain boundary peak occurs. An attempt is made to analyse the results in terms of models previously proposed for damping due to thermally activated dislocation movement.

Experimental.

Damping measurements were carried out on polycrystalline magnesium (99.98%) of mean grain diameter 0.14 mm. at a frequency of 1 Hz over the temperature range from 400°C to 50°C. A grain boundary peak occurs at this frequency at 185°C. The damping was measured as a function of strain amplitude over the range $10^{-6} \leqslant \epsilon \leqslant 1.5 \times 10^{-4}$ in an inverted torsional pendulum evacuated to 10^{-5} torr (15).

Results.

Figs 1-3 show the relationship between logarithmic decrement and strain amplitude for a selection of temperatures. The principal points of interest are that the damping is strongly amplitude dependent and that the form of the amplitude dependence is itself temperature dependent. Fig 4 shows the relationship between logarithmic decrement and temperature as a function of strain amplitude. The strains used in this investigation were not sufficiently low to enable an amplitude independent region to be established. However, Roberts (16) has shown that at strains $\sim 10^{-8}$ an amplitude independent

region exists for aluminium-zinc alloys over the temperature range of a
grain boundary peak. The amplitude independent peak (at zero strain) was
estimated from the present results by extrapolation of δ vs ϵ curves to
zero strain. This was then subracted from the total damping to yield the
amplitude dependent peaks shown in Fig 5. The amplitude dependent peaks
are seen to be ill-defined at low strains but increasingly well defined
as the strain amplitude increases.

The amplitude dependent background damping was estimated by interpolation
in the region of the peak in order to separate background and peak contributions.
Figs 6-8 show the relationship between logarithmic decrement and strain
amplitude for the background (Fig 6) and grain boundary (Figs 7,8) damping.

Discussion.

The discussion of these results will be confined to a consideration
of the amplitude dependent damping.

(a) Background damping.

Models for amplitude dependent damping due to thermally activated
dislocation movement have been proposed by Friedel (17) and by Peguin et al
(18). The Friedel analysis predicts a linear relationship between ln δ_H and ϵ
from the slope of which the mean spacing (L_c) between obstacles on dislocations
may be estimated. Fig 9 shows this relationship for the present results.
The linear relationship is obeyed over most of the strain range. The
deviation at low strains is due to the nature of the plot (as $\delta_H \rightarrow 0$, ln $\delta_H \rightarrow -\infty$)

The Peguin analysis results in a similar equation to that of
Friedel but predicts a linear relationship between ln($\delta_H \epsilon$) and ϵ from the
slope of which an apparent activation volume may be estimated. Fig 10 shows
that this relationship is also obeyed by the present results, again with
the expected deviation at low strains.

The values of L_c and activation volume estimated from Figs 9 and
10 are listed in Table 1. L_c is $\sim 10^{-5}$ cm. and the activation volume for
the high temperature background is 2.5×10^{-20} cm^3. The interpretation of
activation volume for dislocation movement leads to its definition as

$$V = L_c b^2$$

where b is the Burgers vector.

Values of L_c derived from this equation are $\sim 10^{-5}$ cm. so that the two sets of results are compatible.

The implications of these results are therefore that the background damping arises from the thermally activated movement of dislocations whose motion is impeded by obstacles spaced $\sim 10^{-5}$ cm. apart along the dislocation lines. It is not possible to identify the nature of the obstacles with certainty until a more detailed analysis, at present being carried out, is completed. However the spacing of 10^{-5} cm. seems rather large for solutes segregated at dislocation lines and rather small for a dislocation network spacing. It might however be the spacing between jogs on dislocation lines. On this basis, the damping would be due to the thermally activated motion of edge-type jogs on screw dislocations.

(b) Grain boundary damping.

The form of the δ_H vs ϵ relationship shown in Figs 7 & 8 suggests that the grain boundary contribution may consist of two components. One component rises steeply with amplitude at low strains and then saturates at $\epsilon \sim 4 \times 10^{-5}$. The second component rises steeply with increasing strain, showing similar characteristics to the background damping. The Friedel and Peguin analyses have been applied to the high $(\epsilon \geqslant 4 \times 10^{-5})$ strain part of the grain boundary results for two temperatures, one on each side of the peak temperature. The results are shown in Figs 11 & 12. The expected linear relationships are obeyed and from the slopes values of activation volume and L_c have been estimated (Table 1). The values are seen to agree well with those for the background damping.

On the basis of these results it is suggested that the grain boundary damping peaks consist of three components:

An amplitude independent component at low $(\epsilon < 10^{-6})$ strains, possibly anelastic in origin. The mechanism giving rise to this component is not clear. Further work to establish the frequency dependence is required.

(b) An amplitude dependent component which saturates at $\epsilon \sim 4 \times 10^{-5}$. The most probable explanation of this is that it is due to small reversible

movements of grain boundary dislocations. These are constrained to move
only in the boundary and are therefore limited in their ability to contribute
to damping. The amplitude dependence of this component then arises from
the activation of an increasing number of boundary dislocations. The number
present and the total number able to contribute to the damping is determined
by the fine structure of the boundary. The effect reaches saturation when
all available sources are operating.

(c)An amplitude dependent component which shows the same amplitude dependence
as the background damping. The proposed mechanism for this is that as the
strain amplitude increases, the stress on the boundary regions, which has
not been relaxed by the anelastic and grain boundary dislocation contributions,
is relaxed by the nucleation of dislocations at such sources as grain
boundary ledges. These dislocations, which are not required to remain at the
boundary to maintain misorientation, are able to move into the grains and
give rise to a contribution to the total damping which has the same characteristics
as the background damping. Mechanisms for the nucleation of dislocations by
shear of grain boundary ledges have been proposed by Li (19) and by Price
and Hirth (20).

In view of the present results and the interpretation of them outlined
above, it is perhaps not surprising that no satisfactory treatment of grain
boundary internal friction has yet emerged in spite of a wealth of published
results. If the amplitude dependence observed in the present work is a
general effect, redetermination of much of the data will be necessary in
order to apply anelasticity theory only to data which is amplitude independent.
The production of dislocations at grain boundary sources and their subsequent
motion in the grains probably accounts for the apparent correlation of
relaxation parameters with stacking fault energy. But this is only true for
one component of a complex sequence of relaxation processes..

References.

1. T.S.Kê, Phys. Rev., v 71, 533, (1947).

2. G.M.Leak, Proc. Phys. Soc., v 78, 1520, (1961).

3. J.T.A.Roberts, and P.Barrand, Trans. A.I.M.E., v 242, 2299, (1968).

4. J.T.A.Roberts, J.Inst. Met., v 98, 381, (1970).

5. J.N.Cordea and J.W.Spretnak, Trans. A.I.M.E., v 236, 1685, (1966).

6. C.Zener, Elasticity and Anelasticity of Metals, Chicago Univ. Press., (1947).

7. A.S.Nowick and B.S.Berry, I.B.M. J.Res. and Dev., v 5, 297 & 312, (1961).

8. J.Wegria, J.Gouzou and L.Habraken, C.N.R.M. Report No. 25 (1970).

9. C.C.Smith, Ph.D. Thesis, University of Manchester, (1972).

10. H.Gleiter, Acta. Met., v 17, 853, (1969).

11. J.Rutter and K.Aust, ibid, v 13 181, (1965).

12. D.G.Brandon, Acta Met., v 14, 1479, (1966).

13. D.H.Warrington and W.Bollmann, Phil. Mag., v 25, 1195, (1972).

14. J.P.Hirth and R.W.Balluffi, Acta Met., v 21, 929, (1973).

15. M.E. de Morton and G.M.Leak, Acta Met., v 14, 1140, (1966).

16. G.Roberts, Ph.D. Thesis, Manchester University, (1972).

17. J.Friedel, Dislocations, Pergamon Press, (1964).

18. P.Peguin, J.Perez and P.Gobin, Trans. Met. Soc. A.I.M.E., v239, 438, (1967).

19. J.C.M.Li, Trans. A.I.M.E., v 227, 239, (1963).

20. C.W.Price and J.P.Hirth, Mater. Sci. Eng., v 9, 15, (1972).

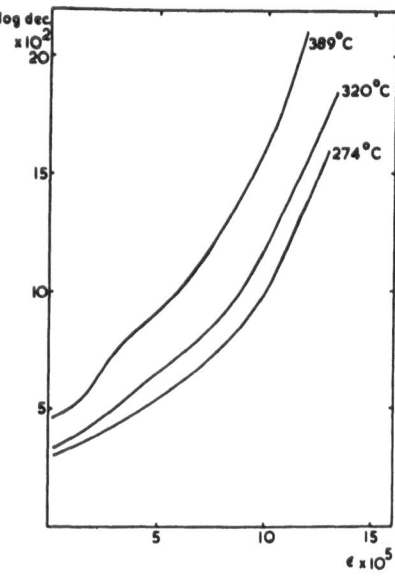

FIG. 1

Relationship between logarithmic decrement
and strain amplitude, 399°C to 274°C.

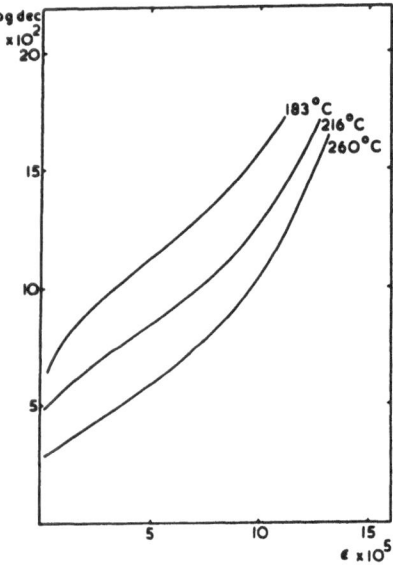

FIG. 4

Effect of strain amplitude on damping peaks.

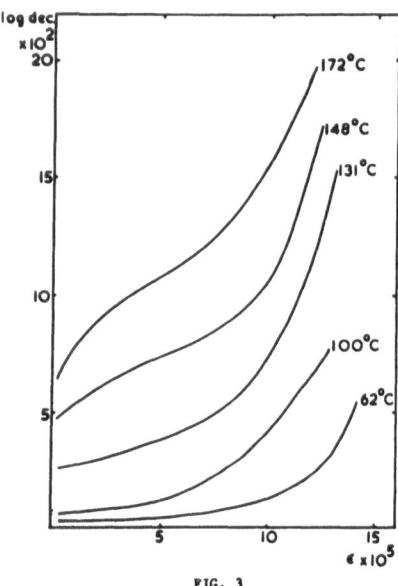

FIG. 2

Relationship between logarithmic decrement
and strain amplitude, 260°C to 183°C.

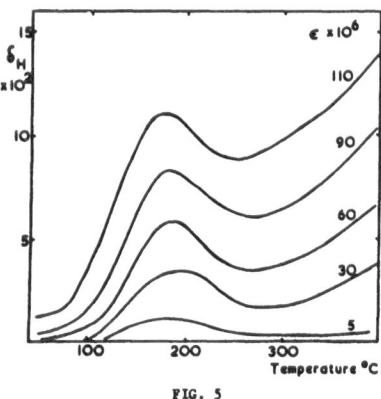

FIG. 3

Relationship between logarithmic decrement
and strain amplitude, 172°C to 62°C.

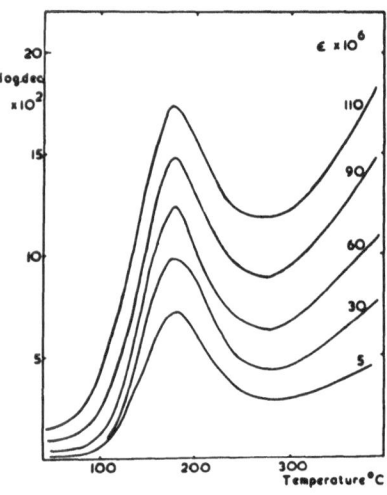

FIG. 5

Effect of strain amplitude on amplitude
dependent damping peaks.

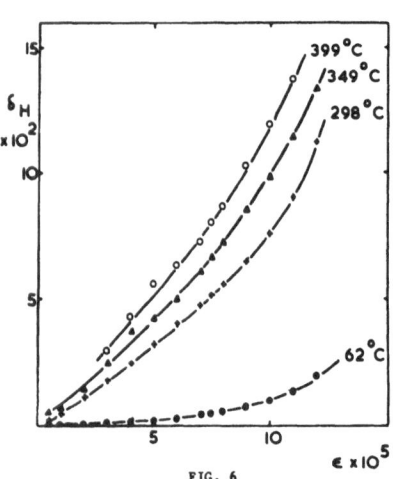

FIG. 6

Amplitude dependent damping vs amplitude -
background damping.

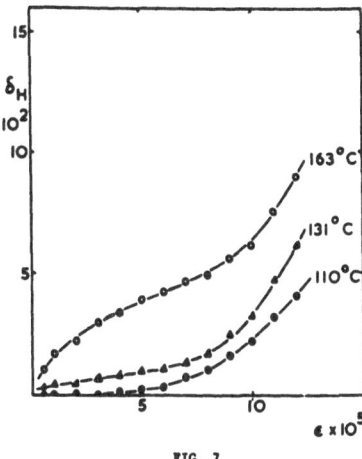

FIG. 7

Amplitude dependent damping vs amplitude –
grain boundary damping, T < T_p.

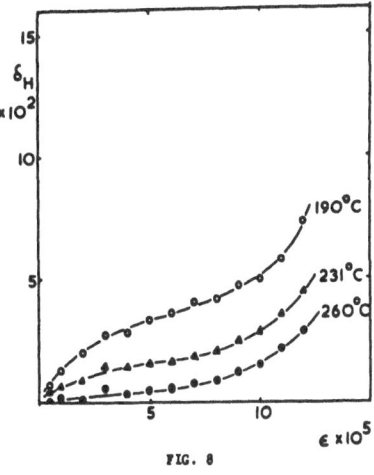

FIG. 8

Amplitude dependent damping vs amplitude –
grain boundary damping, T > T_p.

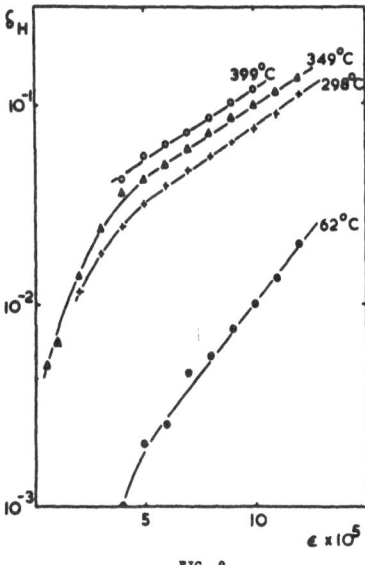

FIG. 9

Ln δ_H vs ϵ – background damping.

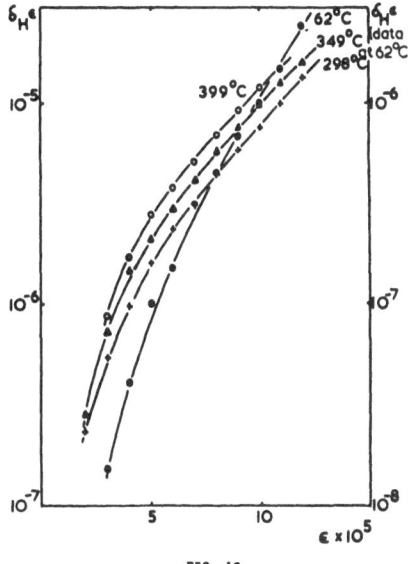

FIG. 10

Ln ($\delta_H\epsilon$) vs ϵ – background damping.

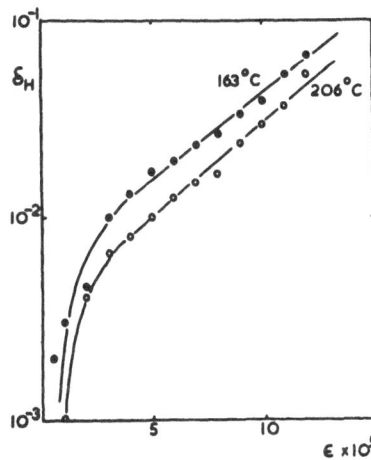

FIG. 11

Ln δ_H vs ϵ – grain boundary damping.

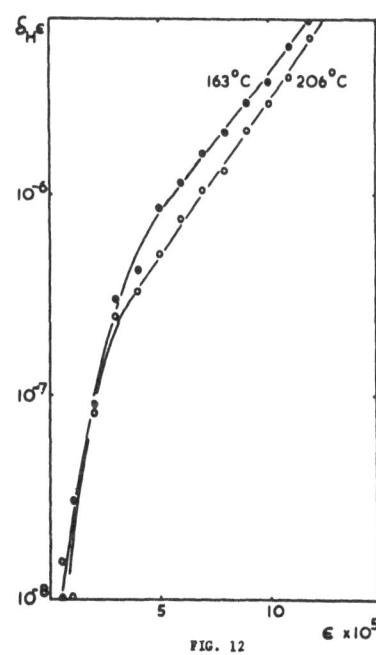

FIG. 12

Ln ($\delta_H\epsilon$) vs ϵ – grain boundary damping.

391

A DISLOCATION MODEL FOR GRAIN BOUNDARY PEAKS
IN PURE FACE CENTERED CUBIC METALS

J. WOIRGARD, J.P. AMIRAULT, J. de FOUQUET
Laboratoire de Mécanique et de Physique des Matériaux
E.R.A. au C.N.R.S. n° 123
Université de Poitiers
E.N.S.M.A., rue Guillaume VII - 86034 - POITIERS

INTRODUCTION

Following Kê's fondamental work (1) it is usually considered that the
internal friction peaks observed in pure polycrystalline metals between 0.3 and
$0.5\ T_m$ (T_m = absolute melting temperature) are essentially due to a viscous behaviour of the grain boundaries. In support of this assumption it is claimed
that the effects are absent in single crystals, and become very weak when the
grain size is large compared with the specimen diameter (1) to (9). However it
must be noted that annealings at high temperature producing large grain sizes,
also provide low dislocation densities, and then it is not possible from the
experimental data available in the literature to exclude an eventual influence
of lattice dislocations.

The following results have been obtained in single crystals, bicrystals
and polycrystals of pure F.C.C. metals. A quantitative model, involving dislocation climb by vacancy diffusion, is then proposed, for the different peaks
and for the high-temperature background observed in single crystals and in polycrystals.

EXPERIMENTAL METHODS

Internal friction measurements were carried out in an apparatus described in detail elsewhere(10), and allowing flexural tests at low frequencies,
between 20°C and 1100°C, on flat bars of 50 x 6 x 2 mm. Single crystals, bicrystals and polycrystals where obtained from the same metals containing less
than 10 p.p.m. of metallic impurities. A possible influence of the misorientation

has been investigated on bicrystal specimens, produced from two seed crystals rotated with respect to each other by 50° and 53° about the common <100> axis, lying along the specimen axis ; the boundary plane, close to the surface, was submitted to an alternative tensile stress as depicted in Fig. 1 ; the selected misorientations by 50° and 53°, correspond respectively to a low and a high coincidence lattice density.

EXPERIMENTAL RESULTS

a - COPPER.   The results obtained·in a copper polycrystal (1 mm grain diameter), and in copper single crystals slightly strained and annealed at 600° and 900°C, are shown in Fig. 2. A notable relaxation effect is present in mono-crystal in the same temperature range as the grain boundary relaxation peaks in polycrystal.

Figure 3 shows the relaxation spectra obtained in copper single crystals and bicrystals (53° about <100>) : no detectable difference between the two spectra may be found. In the same way, and in spite of the fact that bi-crystals with 50° and 53° misorientations about a <100> axis have a quite different behaviour in creep tests (11), internal friction curves did not reveal any difference between them.

It may be noted that single crystals slightly strained by bending exhibit a high low temperature damping (fig. 4) which rapidly decreases during further annealings ; subsequent annealings at higher temperatures increase the high temperature side of the spectra (in the 0.5-0.6 T_m range), while the low temperature one nearly vanishes (fig. 5 and 6).

In order to obtain the constituent peaks of the spectra, some consistent with experimental data assumptions have been made :

1) In the case of copper, it has been assumed that all spectra may be analysed in 4 elementary peaks, with fixed widths and temperatures at constant frequency ; the relaxation strength will then be the only parameter allowed to change ; the validity of this assumption has been checked on a great number of experimental results.

2) As it will be shown, the high temperature background may be considered as a nearly exponential function of 1/T, where T is the absolute temperature, according to the relation :

$$\Delta_F = \frac{A_o}{T} \exp \left(- \frac{H_c}{kT} \right) \qquad (1)$$

where H_c will be shown to be very close to the self diffusion energy along dislocation cores, i.e. 0.5 H_v, H_v being the bulk self diffusion energy.

Only the extreme peaks P_1 and P_4, which are determined with a good accuracy, will be considered here.

Table 1 provides a summary of the data on the kinetic aspects of the relaxation peaks, for copper monocrystals and polycrystals : H is the activation energy deduced from the shift of the peak temperature with frequency,

τ_0 is the limit relaxation time, β the lognormal distribution parameter and Δj the individual relaxation strength.

It may be seen that the only significant difference between single crystals and polycrystals is relaxation strengths much higher in the last case.

TABLE 1

Relaxation parameters for P_1 and P_4 peaks in
copper single crystals and polycrystals

	$H_{kcal/mole}$	τ_{os}	β	$\Delta_j \times 10^4$	
single crystals	25.1	1.1×10^{-11}	1.5	92	P_1
	53	3.4×10^{-14}	2.5	169	P_4
polycrystals	26.9	2.6×10^{-12}	3.5	1087	P_1 -
	48.2	1.2×10^{-12}	2.5	755	P_4

b - ALUMINIUM. Similar tests have been performed in polycrystalline bicrystalline and monocrystalline aluminium specimens; in the case of aluminium it may be seen that the spectra can always be analysed in three symmetric peaks. Figure 7 shows the internal friction curve for an aluminium bicrystal corresponding to a 53° misorientation about a <100> axis, and spectra obtained in a polycrystal and an unstrained single crystal are shown in figure 8 ; the weakness of the relaxation effect in the unstrained single crystal possibly explains that this one had not been previously reported.

In addition to the above tests, experiments were performed on Silver, Lead, Nickel, polycristalline samples :

In table 2 are listed the apparent activation energies obtained from the shift of the peaks with frequency, for the extreme peaks, compared with the self diffusion activation energy H_v ; the results in Au are due to De Morton and Leak (12). In view of the fact that the discrete spectra are not the same for all these metals, the lowest temperature peak is labelled P and the highest temperature one P'.

TABLE II

Relaxation parameters for the extreme peaks

P and P' in some F.C.C. metals

	H_P	$0.5\ H_V$	$H_{P'}$	H_V
Ag	20.3	22.2	46.2	44.3
Au	(11) 22.8	20.9	$40^{(11)}$	41.7
Cu	26.1	23.6	48.2	47.1
Al	25.4	17	29.6	34
Pb	17	12.1	23	24.2
Ni	58.2	32.5	66.5	65

It may be seen that the activation energy associated with P peak is close to $0.5\ H_V$, the self diffusion energy along dislocation cores, except for the high stacking fault energy metals : Al, Ni and Pb ; and, in every case, the measured activation energy for P' is very close to H_V.

DISCUSSION

The above results show that seizable relaxation effects are observed in slightly strained monocrystals and bicrystals of copper and aluminium in the same temperature range that the relaxation grain-boundary peaks in poly-crystals ; the discrete spectra obtained in monocrystals and bicrystals agree with those obtained in polycrystals ; in every case it has been verified that no recristallization effect was interfered during experiments. On the other hand, a good correlation is observed for all the studied metals between the activation energy for the peaks at lowest and highest temperatures and the dis-location-core self-diffusion and self-diffusion energies.

These observations strongly suggest in every case, a relaxation process based on climb of dislocations.It has been shown elsewhere (13) that a relaxation peak can be obtained when the dislocations, bowing between pinning points, climb by vacancy diffusion.

The corresponding relaxation strength is given by :

$$\Delta j = \frac{\rho \Lambda^2}{3\beta} \quad (2)$$

and the relaxation time by :

$$\tau = \frac{kT\Lambda^3}{3\beta G\Omega^2 A} \quad (3)$$

where : ρ = dislocation density.

Λ = half mean spacing between pinning points.

G = shear modulus.

b = Burgers vector.

k = Boltzmann constant.

T = absolute temperature.

Ω = vacancy volume.

A = factor related to the exact diffusion path.

β = geometric constant such that $T_\ell = \beta Gb^2$ (T_ℓ being the line tension).

Different diffusion paths are possible corresponding to the different peaks observed.

DIFFUSION ALONG DISLOCATION CORES. It is assumed that vacancies diffuse from one dislocation to an other one along dislocation lines with a diffusion coefficient D_c much higher than the bulk diffusion coefficient D_v. The sink-source distance will be the mean spacing between jogs exchanging vacancies.

One obtains then :

$$\tau = \frac{kT}{6\pi\beta G\Omega} \cdot \frac{\Lambda^3 \ell}{b^2 D_c} \qquad (4)$$

ℓ being the sink-source distance.

This effect would be associated with the P(or P_1) peak observed at lowest temperature.

VOLUME DIFFUSION. It is assumed that vacancies travel through the crystals from a dislocation acting as a source to an another dislocation acting as a sink, this mechanism would correspond to the highest temperature peak with :

$$\tau = \frac{kT}{6\pi\beta G\Omega} \cdot \frac{\Lambda^2}{D_v} \text{ Log } \frac{R}{\sqrt{\ell b}} \qquad (5)$$

where : R = cut-off parameter.

ℓ = mean spacing between dislocations.

DIFFUSION ALONG GRAIN BOUNDARIES. In polycrystals one may expect that dislocations responsible for the relaxation are grain boundary dislocations allowed to move in the boundary plane by glide and climb, and vacancies travel in grain boundaries with a diffusion coefficient D_g. The case of a non symmetrical tilt boundary, making an average angle $\pi/4$ with the applied stress, as depicted Fig. 9, has been treated elsewhere (13): Such a boundary is constituted of two different types of dislocations with perpendicular Burgers vectors. Φ being the angle between the Burgers vectors of type 1 dislocations and the boundary plane (see fig. 8) the relaxation magnitude is given by :

$$\Delta j = \frac{\rho\Lambda^2}{3\beta} \cdot \sin^2 \Phi\cos\Phi \qquad (6)$$

and the relaxation time is :

$$\tau = \frac{kT}{3\beta G\Omega} \cdot \frac{\Lambda^2 \ell}{D_b \delta} \cdot \sin^2\Phi\cos^2\Phi \qquad (7)$$

where : δ = boundary width.

ℓ = mean spacing between grain boundary dislocations of different type.

396

HIGH TEMPERATURE BACKGROUND.   As shown in figures 10 and 11, a very good agreement is obtained between the experimental results and the equation (1). Moreover in table 3 it may be seen that the relation holds for all the F.C.C. metals studied.

TABLE III

Activation energies obtained from the high temperature
background damping in some F.C.C. metals

Metal	H_f	$0.5\ H_v$
Ag	22.8	22.2
Cu	24.6	23.6
Pb	13.6	12.6
Al	14.7	17

If one supposes that at high temperatures some pinning points become mobile, and that consequently Λ gets large in equation (1), the corresponding background internal friction would then be given by :

$$\Delta_F = \frac{2\pi G\Omega}{\omega kT} \cdot \frac{\rho b^2 D_c}{\Lambda \ell} \qquad (8)$$

where ω is the angular vibration frequency. It seems then reasonnable to admit that the high temperature background is associated with highly unpinned dislocations, and the observed activation energy suggests that vacancy diffusion occurs along dislocation cores.

The above expressions for the relaxation strength and the relaxation times quite well agree with the experimental results in every case. For example, for the bulk diffusion (see equation 5) if it is assumed that :

$\beta = 1$
$\rho = 3 \times 10^8 \ cm/cm^{-3}$
$\Lambda = 10^{-5} \ cm$
$\ell = 10^{-4} \ cm$
$D_v = 1 \ cm^2.s^{-1}$
$R = 10^4 b$

one obtains :

$\Delta j = 10^{-2}$
$\tau_0 = 2.10^{-12} s$

values which are in reasonable agreement with the experimental ones.

CONCLUSION

While the above model provides a satisfactory description of the observed effects in polycristals and single crystals, it should be noted that this over simplified model does not take into account the shear stress and length loop distributions which must enlarge the relaxation peaks in accordance with the experimental observations, and also ignores the interaction between dislocations which constitute boundaries. Nevertheless, at present time it seems clearly established that at least in pure metals, the "grain boundary peaks", must be explained, by mechanisms which are not specific of grain boundary itself, but which more generally involve the climb and glide of lattice dislocations , and of grain-boundary dislocations in the polycristal case.

REFERENCES

1) - T.S. KE (1947), Phys. Rev. , 71, 533.

2) - W. KOSTER, L. BANGERT and W. LANG (1955), Z. Metallk, 48, 84.

3) - J.T.A. ROBERTS and P. BARRAND (1968), Jour. Inst. Metals, 96, 172.

4) - G.M. LEAK (1961), Proc. Phys. Soc. , 78, 1520.

5) - D. SIDDEL and Z.C. SZKOPIAK (1972), Met. Trans., 3, 1907.

6) - J.N. CORDEA and J.W. SPRETNAK (1966), Trans. A.I.M.E., 12, 1685.

7) - D.T. PETERS, J.C. BISSELICHES and J.W. SPRETNAK (1964), Trans. A.I.M.E. , 4, 530.

8) - T.M. WILLIAMS and G.M. LEAK (1967), Acta Met, 15, 1111.

9) - A.S. NOWICK, B.S. BERRY, Anelastic Relaxation in Crystalline Solid Acad. Press. (1972).

10) - J. WOIRGARD, J.P. AMIRAULT, H. CHAUMET and J. de FOUQUET, Rev. de Phys. Appl. 6 (1971) p. 355.

11) - M. BISCONDI, (1971), Thèse, Paris.

12) - M.E. DE MORTON and G.M. LEAK, (1966), Acta Met. 14, 1140.

13) - J. WOIRGARD, to be published.

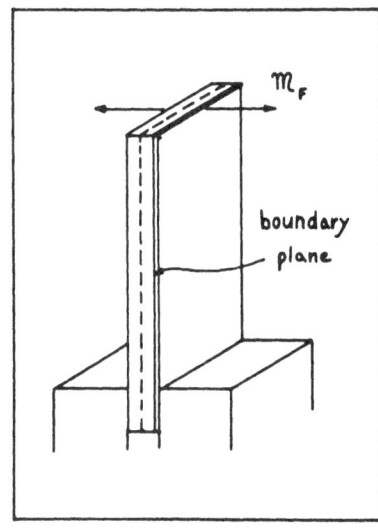

FIG. 1

Bicrystalline sample with
the boundary plane close
to the surface.

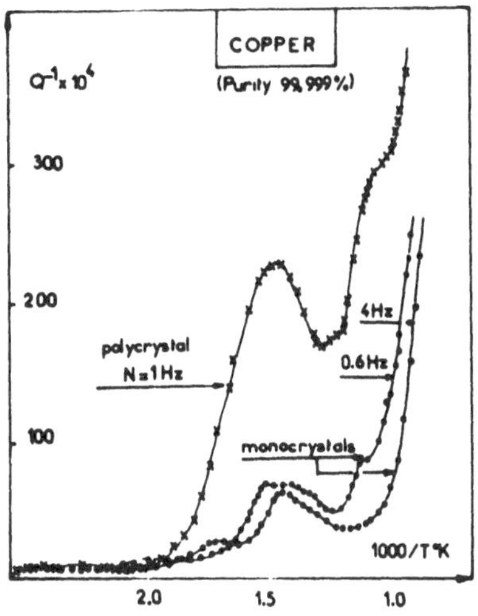

FIG. 2

Spectra obtained in a copper
polycrystal and copper single
crystals slightly strained by
bending.

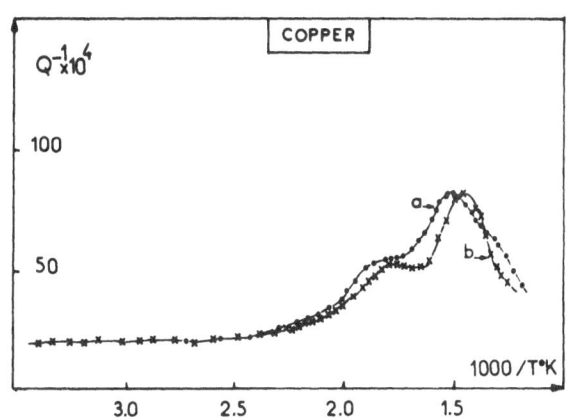

FIG. 3

Damping spectra obtained in :

a) Copper single crystal.

   N = 0.67 Hz

b) Copper bicrystal misoriented of 53°
   about an <100> axis.

   N = 4.4 Hz

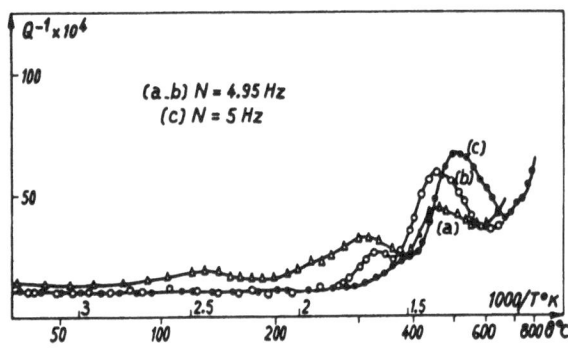

FIG. 4

Influence of successive annealings in
a slighlty bent copper single crystal.
a) 1ᵗʰ heating
b) 2ⁿᵈ heating
c) 3ʳᵈ heating

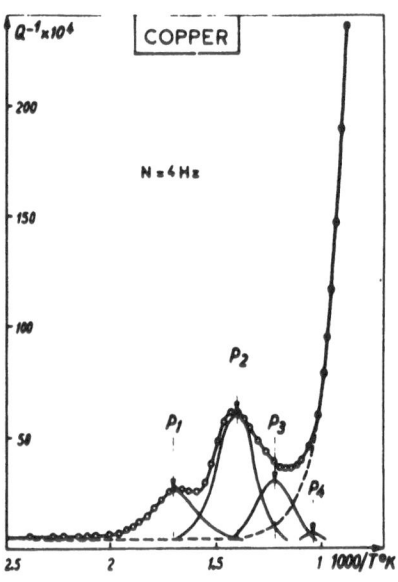

FIG. 5

Damping spectrum obtained in
a slightly strained copper
single crystal annealed at
600°C.

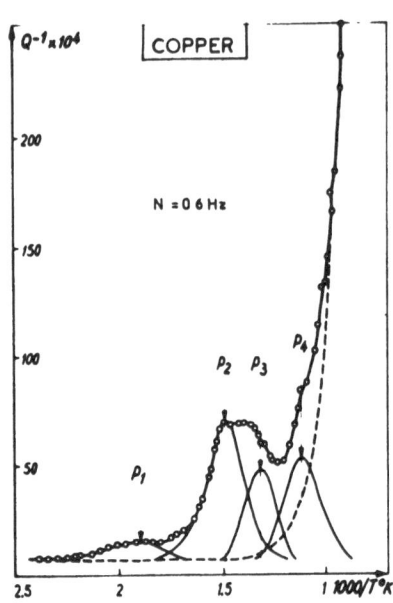

FIG. 6

Damping spectrum obtained in
a slightly strained copper
single crystal annealed at
900°C.

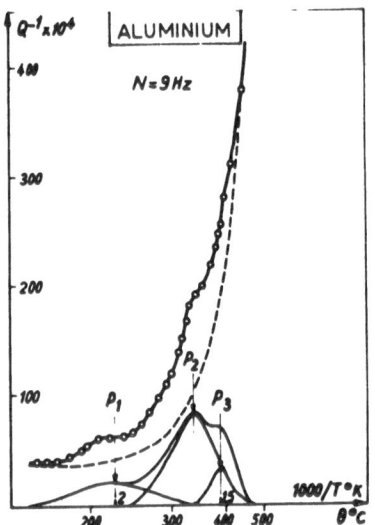

FIG. 7

Damping spectrum in an aluminium
bicrystal with a 53° misorientation about a <100> axis.

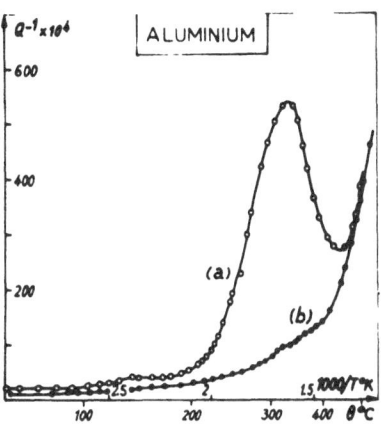

FIG. 8

Damping curves in an aluminium
polycrystal and an unstrained
single crystal. N = 0.85 Hz.

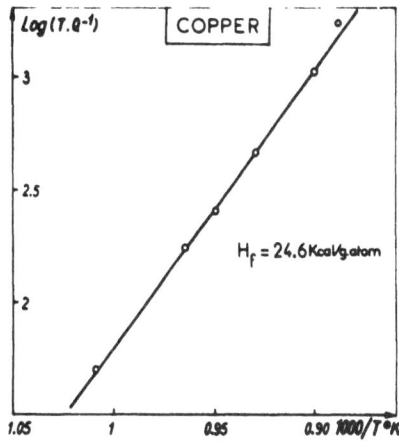

FIG. 9

Apparent activation energy
obtained from the high temperature background in Copper
single crystals.

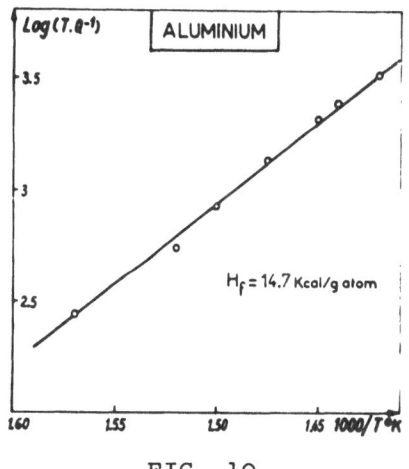

FIG. 10

Apparent activation energy
obtained from the high temperature background in Aluminium single crystals.

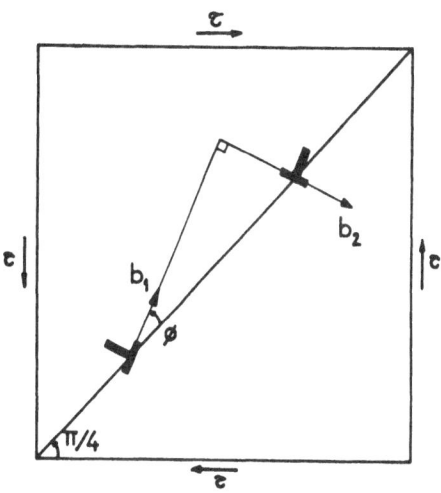

FIG. 11

Non symmetrical tilt boundary made
of two types of edge grain boundary
dislocations.

INTERNAL FRICTION STUDIES OF THE RECRYSTALLIZATION OF SILVER [*]

A. Isoré, O. Mercier and W. Benoit[1]

Laboratoire de Génie Atomique de l'Ecole Polytechnique Fédérale de Lausanne

The study of the evolution of internal friction and modulus defect in cold worked metals during stage V annealing has received considerable attention (1-9). It has been shown that both properties are strongly influenced by the recrystallization process.

During recrystallization the metal may be considered as a two phase mixture consisting of deformed regions containing a high density of defects (point defects and dislocations), and of recrystallized regions with only a very low defect concentration.

In order to explain observations made during recrystallization it must therefore be decided which phase (deformed or recrystallized) is producing the observed variations in internal friction and modulus defect. The aim of this paper is to present clear evidence that, for silver, the evolution of the abovementioned properties is related to the fraction of metal recrystallized, and to show this relationship for different conditions of recrystallization (different nucleations and driving forces) ; thus confirming the conclusions of earlier authors (6-7).

I. EXPERIMENTAL DETAILS

The material investigated was 99,995 wt % purity polycrystalline silver, provided by Métaux Précieux S.A. (Switzerland), with a grain size $\approx$ 0,3 mm.

The measurements of internal friction and modulus defect were conducted at room-temperature on bulk samples (50x5x3mm), excited in the transversal mode at a frequency of about 3 kHz, in an apparatus allowing "in situ" annealing in a helium atmosphere (10^{-2} Torr), at an amplitude of vibration of 10^{-8} [**]. The recrystallization range was determined by means of microhardness measurements and optical microscopy observations.

---

[*] This work was supported by the "Fonds National Suisse de la Recherche Scientifique", subsidy no.: 2.776.72

[**] This is in the amplitude independent range of Q^{-1} and $\Delta E/E$.

[1] Now Visiting Professor at the "Institut für Theoretische und Angewandte Physik, Universität Stuttgart"

We define the modulus defect $\Delta E/E$ as follows :

$$\Delta E/E = \frac{f^2 - f_0^2}{f_0^2}$$

f_0 being the frequency after annealing for one hour at 400°C.

II. RESULTS AND DISCUSSION

II.1 Presentation of problem

Fig. 1 shows, for a specimen cold rolled at 20°C and then annealed for periods of one hour at increasing temperatures T_R, the variation of internal friction Q^{-1} and modulus defect $\Delta E/E$, measured at 20°C. In the range of annealing temperature between 20 and 350°C, three separate stages of annealing can be distinguished (5) :

stage 1 : "recovery", from 20 to 195°C

stage 2 : "recrystallization", between 195 and 235°C

stage 3 : "post-recrystallization", above 235°C.

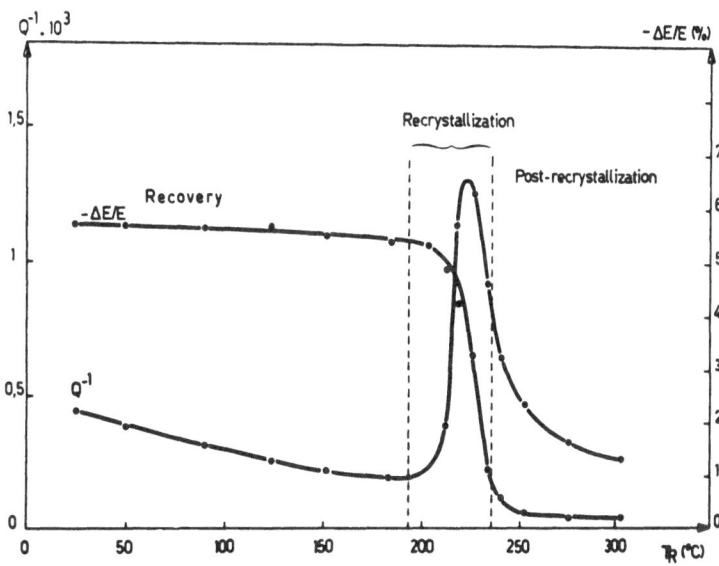

Fig. 1 Internal friction and modulus defect versus temperature of isochronal annealings (1 hour) measured at 20°C on silver polycrystal rolled 7% at room temperature

The first stage, "recovery", corresponds to a pinning of dislocations and has been attributed to vacancy migration, since many experimental observations have proved that, for silver, the dislocation network does not evolve perceptibly before the start of recrystallization (9).

The next stage, "recrystallization", is characterized by a peak in the internal friction and a drop in the modulus defect.

Finally, after the primary recrystallization is complete, and with no grain growth, one observes a decrease in the internal friction, and a simultaneous but very small evolution in the modulus defect during the "post-recrystallization" stage.

The background internal friction observed may be assumed to be essentially due to dislocations ; however, the increase at a temperature of about 200°C could be interpreted in several ways, even if its beginning were to coincide with that of the primary recrystallization. For instance, the increase in Q^{-1} might be caused by the evolution of a dislocation network in the deformed phase (e. g. a depinning of dislocations), or by the appearance of a recrystallized phase with a high internal friction. In order to clarify this point we considered that the simultaneous measurements of internal friction, modulus defect and recrystallized-phase fraction would be of considerable importance.

II. 2   Recrystallized-phase fraction f , internal friction Q^{-1}
and modulus defect $\Delta E/E$

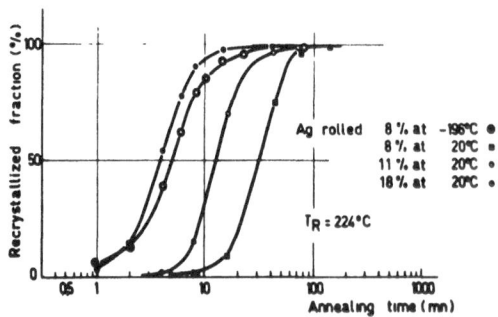

Fig. 2   Evolution of the volume fraction of recrystallized material f versus logarithm of time, measured on polycrystalline silver samples rolled at 20 and -196°C (8, 11 and 18% at 20°C, and 8% at -196°C ; annealing temperature T_{R} = 224°C)

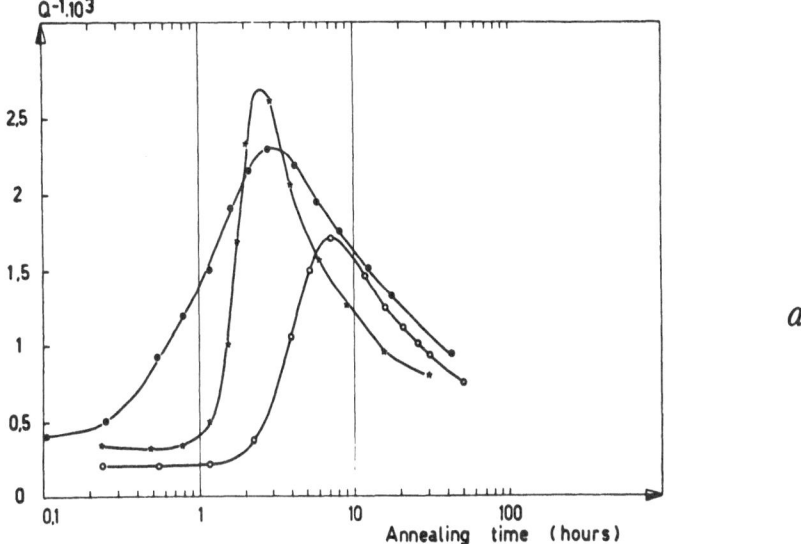

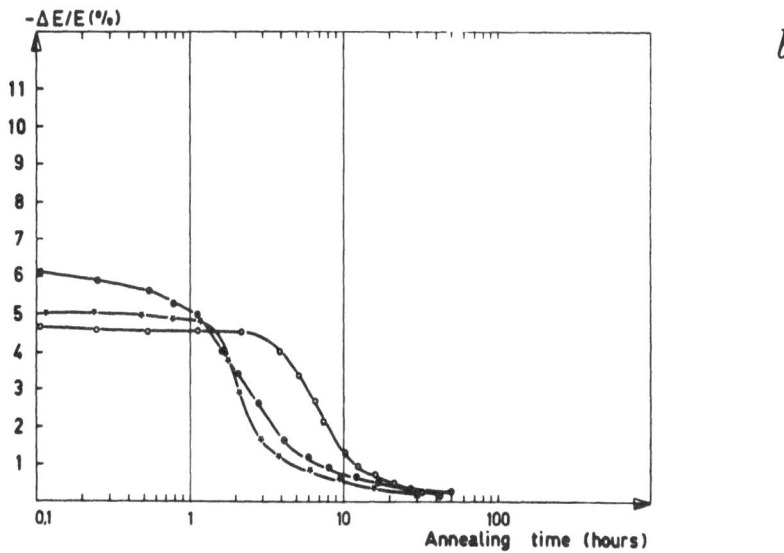

Fig. 3   a)   Evolution of internal friction Q^{-1} versus logarithm of time (isothermal
annealing at 200°C) measured at 20°C on rolled silver samples :

      ● rolled   7% at -196°C
      o rolled   7% at room temperature
      ✹ rolled 12% at room temperature

    b)   Evolution of modulus defect $\Delta E/E$ under same conditions as in a)

The evolution of the volume fraction of metal in the recrystallized phase f *, during an isothermal anneal at 224°C, for samples having undergone different mechanical treatments, is shown in fig. 2 plotted as a logarithmic function of time**. Several features should be noticed :

    i)   At a constant temperature of deformation (20°C) the recrystallization rate is higher for a larger deformation ε_p ; $(df/dt)_{max}$ increases with ε_p . This behaviour is well known.

    ii)   For the sample rolled 8% at -196°C the recrystallization starts as for the sample rolled 18% at 20°C, but the rate of evolution is lower. This is a quite general result and is independent of the purity of the sample (7-9).

           Since we are interested only in the general features of f , we will not enter into the details of this phenomenon (see(9)).

    iii)   For two samples deformed the same amount at temperatures of 20 and -196°C, the rate of recrystallization is greater for the sample deformed at the lower temperature.

Figs. 3a and 3b show, for samples rolled at 20°C and -196°C and then annealed at 200°C, the variation of internal friction and modulus defect (measured at 20°C) with the annealing time at 200°C. The results may be summarized as follows :

    i)   At a constant temperature of deformation of 20°C, the internal friction begins to rise earlier for greater amounts of deformation ; the rate of increase of Q^{-1}, and the maximum value Q^{-1}_{max} attained, are also higher for a larger deformation.

    ii)   For the sample deformed 7% at -196°C, the increase of internal friction begins as for a sample deformed more at a higher temperature (20°C), but the subsequent rate of evolution and the maximum value of Q^{-1} attained are lower.

    iii)   As figure 3b shows, the above comments are also valid for the drop in modulus defect.

---

* f was determined by "point counting" (10) from the microhardness measurements

**   This graph, plotted using a linear time scale shows that the recrystallization starts at t = 0

II.3 Discussion

As we can see the same features are to be noticed for the evolution of the re-crystallized fraction f, internal friction Q^{-1} and modulus defect $\Delta E/E$. The characteristics of the internal friction increase and maximum, and those of the modulus defect drop, follow exactly those of the recrystallized phase.

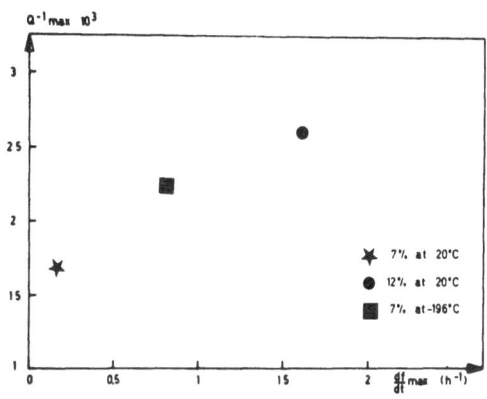

It clearly appears that the internal friction evolves the most rapidly and obtains its highest maximum value for the fastest recrystallization rate. The relation between Q^{-1}_{max} and $\mathrm{d}f/\mathrm{dt}|_{max}$ for samples with different deformation conditions is shown in fig. 4: one observes that Q^{-1} increases with $\mathrm{d}f/\mathrm{dt}|_{max}$.

This fact can be explained if one considers that the recrystallized grains cause an internal friction higher than that caused by the deformed grains. In this case Q^{-1} will increase accompanying the recrystallization process, which means it will follow the changes in the recrystallized fraction f. The subse-

Fig. 4   Internal friction maximum Q^{-1}_{max} versus $\mathrm{d}f/\mathrm{dt}|_{max}$ in silver samples under different conditions of deformation, 7 and 12% at 20°C and 7% at -196°C (annealing temperature T_R = 200°C)

quent drop in the measured Q^{-1} should be due to a simultaneous decrease in the recrystallized phase internal friction (5-7). Therefore, for a given rate of internal friction evolution in the recrystallized phase (at constant temperature), the higher the recrystallized rate, the larger the amount of material having a high internal friction at a given moment, and consequently the higher the observed value of Q^{-1}_{max}.

If the large internal friction values were due to an evolution of the dislocation network in the deformed phase, the subsequent drop in the measured internal friction should be due to the shrinking of this phase by recrystallization (at least in part). So the Q^{-1}_{max} height should decrease with increasing recrystallization rate, and the increase in the internal friction should be independent of the recrystallization evolution. The experimental results reveal a very different behaviour.

The internal friction maximum is therefore due to the appearance and further evolution of the recrystallized phase, and not to an evolution of the deformed phase.

On the other hand it seems that the observed modulus defect variations are only influenced by the disappearance of the deformed phase. This behaviour seems to be different from that of the internal friction. Because the measured values for Q^{-1} and $\Delta E/E$ are the addition of two terms there is only an apparent difference ; it can be written (9) :

$$Q^{-1} \text{ measured} = Q_d^{-1}(1 - f) + Q_r^{-1} \cdot f$$

and
$$\Delta E/E \text{ measured} = \Delta E/E_d (1 - f) + \Delta E/E_r \cdot f$$

The subscripts "d" and "r" being related respectively to the deformed and recrystallized phases. We must therefore consider that $Q_d^{-1} \ll Q_r^{-1}$ (the measured internal friction follows the evolution of the recrystallized phase internal friction) and $\Delta E/E_d \gg \Delta E/E_r$ (the measured modulus defect is mainly due to the deformed phase). It must be noticed that sometimes (for instance in higher purity samples), $\Delta E/E_r$ is not negligible therefore the modulus defect evolution shows a maximum during the recrystallization stage (5-7).

The change in $\Delta E/E$ during recrystallization corresponds to a Young's modulus increase. This evolution could be very well explained by the annealing-out of defects such as dislocations. However, for silver which is elastically anisotropic an appearance or evolution of texture could perfectly explain the modulus changes (9). In fact this variation in texture by recrystallization has been observed by several authors in the case of heavily cold-worked metals (8), but studies carried out by the Schulz goniometric method (11) failed to reveal any texture in our samples.

CONCLUSION

During primary recrystallization of deformed silver samples there exists a strong relation between the evolutions of the measured internal friction and modulus defect on the one hand, and the measured recrystallized fraction on the other hand ; the internal friction evolution exhibits a maximum which can be explained by the appearance and internal evolution of the recrystallized phase. The simultaneous modulus defect variations are mainly due to the shrinking of the deformed phase.

REFERENCES

(1)   M. E. de Morton, Trans. Met. Soc. AIME, 221, 395 (1961)

(2)   R. Kamel and E. A. Attia, Phil. Mag. 4, 644 (1959)
                    Acta Met. 9, 1047 (1961)

(3)   B. Dubois and G. Bouquet, J. de Phys. 32 (7), C2-201 (1971)

(4)   L. M. Robinson and P. N. Richards, Phil. Mag. 11, 407 (1965)

(5)   A. Isoré and W. Benoit, Mém. Sc. Rev. Mét. 69, 223 (1972)

(6)   A. Isoré, W. Benoit and O. Mercier, Scripta Met. 6, 933 (1972)

(7)   A. Isoré, O. Mercier and W. Benoit, Mém. Sc. Rev. Mét. 70, (1973)

(8)   E. Kovacs-Csetényi and B. Sas, Phys. Stat. Sol. (a) 15, 687 (1973)

(9)   A. Isoré, Thèse EPF-Lausanne (1973)

(10) J. E. Hilliard in "Recrystallization, Grain Growth and Textures",
      Am. Soc. Met. , Metals Park Ohio, p. 267 (1966)

(11) L. G. Schulz, J. Appl. Phys. 20, 1030 (1949)

INTERNAL FRICTION IN COBALT AND COBALT-IRON ALLOY.
INTERACTION BETWEEN RECRYSTALLISATION AND PHASE TRANSFORMATION.

G. Bouquet et B. Dubois

Ecole Nationale Supérieure de Chimie

Laboratoire de Métallurgie et Matériaux.

11 rue Pierre et Marie Curie

75230 Paris Cedex 05.

The evolution of cold rolled cobalt as a function of annealing temperature was followed by internal friction measurements ; measurements were performed at room temperature on a torsion pendulum at 0,5Hz. Isochronal annealings of three hours were done under vacuum (10^{-5}Torr). Electrolytic cobalt (99,7% pure) has a resistivity ratio $R_{20,3K}/R_{294K} = 110.10^{-4}$.

A previous study (1, 2) showed that the internal friction - annealing temperature spectrum $Q^{-1} = f(T_R)$ presents an important maximum at 430-440°C (fig. 1). This maximum is preceded by another small peak or shoulder, the position of which shifts with the initial rate of cold working of the material. In a 40% cold worked sample this shoulder appears at 350°C (fig. 1).

We have paid particular attention to the high temperature maximum. The explanation of this maximum was difficult : we have to know if the damping decrease which occurred beyond 440°C was related to the end of the recrystallisation which is observed at this temperature, or to the occurrence of allotropic transformation of cobalt, also seen at this temperature.

To resolve this problem, we have tried to separate the end of recrystallisation from the allotropic transformation. We have changed the mechanical treatments of our specimens. Fig. 2 shows the spectra $Q^{-1} = f(T_R)$ of 20% and 30% cold rolled cobalt. The start of recrystallisation in the two samples differs by 40 degrees. If this difference is the same for the end of recrystallisation, we should observe an internal friction maximum at different temperatures for different samples, if the maximum is really associated with the end of recrystallisation. However we have always found the maximum at 440°C. It is tempting to implicate the phase change in the damping decrease beyond 440°C. Nevertheless if at 440°C, the recrystallisation rates are very different for each sample, we know (2) that recrystallisation is ended when internal friction begins to decrease.

Therefore there is a close interaction between the end of recrystallisation and allotropic transformation in cobalt, but this experiment is not a complete explanation of the problem.

To obtain complete recrystallisation before phase change, we have modified the thermal conditions of our samples. We have followed the isothermal recrystallisation at 405°C by measuring internal friction after short annealing treatments. The results are plotted in fig. 3 : at 405°C, internal friction increases rapidly during the early period in relation to the rapid evolution of recrystallisation. Then the curve inflects and we have a large maximum indicating complete recrystallisation. Optical micrography has shown that all recrystallised grains were in contact each other after 100h. at 405°C (2). Further annealings give a slight increase of the damping values. At 440°C after more marked increase, we can see the first drop of damping. If we go on to increase annealing temperature, the internal friction decrease is amplified. This experiment supports the hypothesis that the allotropic transformation alone is responsible for the damping drop observed after 440°C in $Q^{-1} = f(T_R)$ spectrum, if we consider that the end of recrystallisation is the coming into contact of all recrystallised grains. Nevertheless we have seen that phase change accelerates the completion of recrystallisation, at the migration boundaries scale. It was necessary to be sure that it had not influenced the distribution dislocations at the end of recrystallisation ; in so far as the transformation process is based on dislocations reactions (3, 4). Also an electron microscopic study was carried out.

In order to obtain complete recrystallisation in the hexagonal phase, a cobalt sample, 40% cold rolled, was annealed for
120h at 420°C. Fig. 4a shows the numerous stacking faults limited
by $\frac{a}{3}\langle\bar{1}100\rangle$ extended dislocations. After additionnal annealing at
460°C for 3h (in the f.c.c. phase), the same sample exhibits a diffe-
rent aspect (fig. 4b). We can see f.c.c. grains, maintained at room
temperature and showing stacking faults in the $\{111\}$ plans. These
stacking faults cross over the whole grains or are stopped by other
stacking faults. Faulted volumes, that are in the hexagonal phase,
coming from the f.c.c. phase, do not contain extended dislocations
(5). The remaining f.c.c. matrix must contain few perfect dislocations
because we have not seen any. So the h.c. ⟶ f.c.c. ⟶ h.c. cycle
involves a decrease in the dislocation density. We think that this
phenomenon is responsible for the damping drop beyond 440°C.

The same internal friction study was carried out with 99,99%
cobalt ($R_{20,3K}/R_{294K} = 89.10^{-4}$). Fig. 5 shows the spectrum $Q^{-1} = f(T_R)$.
In addition to some unexplained anomalies between 100 and 300°C, we
notice a maximum at 440°C followed by a marked drop of the internal
friction. Besides this, a less marked maximum appears at 380°C. Optical microscopy indicates that this maximum is not related to the end
of recrystallisation but rather to a slackening of this phenomenon.
It seems that this maximum is similar to those observed at 350°C
(fig. 1), 350 and 375°C (fig. 2). The internal friction increase after
these phenomena may be related with an evolution of the cobalt before
its phase change. This would be supported by the damping increase at
440°C observed on a recrystallised cobalt (fig. 3).

The specific contribution to the 440°C maximum of the allotropic transformation can be illustrated by the next experiment ; annealing cycles were carried out on the previous sample on the vicinity
of the phase change (dotted curves in fig. 5). An internal friction
maximum appears during every cycle : this maximum is related to the
allotropic transformation. The internal friction increase is difficult

to explain but the decrease may be explained by the following observations : the hexagonal phase produced during primary recrystallisation below 440°C is not completely transformed in f.c.c. phase by
annealing at 500°C. Annealing cycles in the vicinity of the phase
change would allow the hexagonal phase rate to increase, producing
f.c.c. phase during heating. Accordingly, during cooling, the hexagonal phase rate produced from f.c.c. phase would increase corresponding by. There are less extended dislocations in this hexagonal
phase than in the primary one ; so that every cycle in volves a
dislocation density decrease and an internal friction drop.

   We have also attempted to discover the influence of the phase
change on the beginning of recrystallisation. Under the same conditions
we have worked out and Co-5% Fe alloy and studied its internal friction. The spectrum $Q^{-1} = f(T_R)$ after 45% cold rolling can be seen
in fig. 6. Dilatometric experiments on castalloy indicate an allotropic transformation at 260°C. We notice no anomaly at this temperature in the internal friction-annealing temperature spectrum, even
after an annealing cycle. Electron microscopy demonstrates that
recrystallisation starts at 380°C. As in pure cobalt, cold working
induces hexagonal structure (6). This point was confirmed by X ray
diffraction. Up to 380°C we have recovery of the alloy in hexagonal
phase. At 380°C recrystallisation appears with the f.c.c. phase and
we can consider the following transformation :

   cold worked h.c. $\longrightarrow$ recrystallised f.c.c.
Dilatometric study of cold worked alloy gives an anomaly at about
380°C. In the hexagonal phase recovered at elevated temperature
(for instance 300°C), the density of free dislocations is high and
drops sharply at the transformation (fig. 7). This study is a proof
of the very low specific internal friction of the f.c.c. phase ; it
equally supports the assumption that the internal friction drop in
cobalt, beyond 440°C, is closely related to the phase change.

REFERENCES

1. B. Dubois and G. Bouquet, Jnal de Phys. 32, C2, 201 (1971).

2. G. Bouquet and B. Dubois, C.R.Acad. Sc. Paris, 274, 1031 (1972).

3. A. Seeger, Z. Metallk. 44, 247 (1953).

4. J.M. Drapier, E. Votava and L. Habraken, Journées Int. App. Cobalt Bruxelles (1964).

5. E. Votava, Jnal Inst. Metals 90, 129 (1961).

6. A.G. Stacey and E.R. Petty, Cobalt 53, 206 (1971).

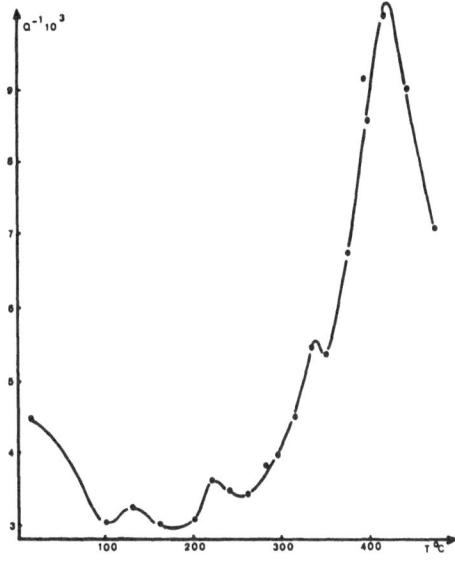

FIG. 1

Internal friction versus annealing temperature
$Q^{-1}=f(T_R)$ of a 40% cold rolled cobalt (99,7% pure).

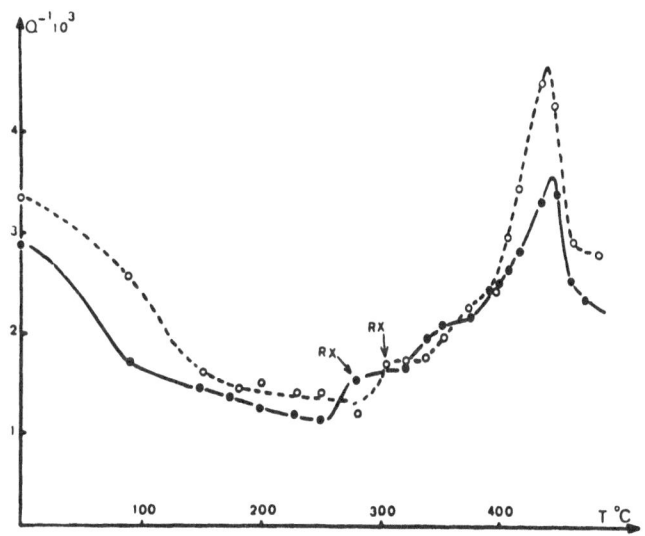

FIG. 2

$Q^{-1}=f(T_R)$ of cobalt (99,7%)
30% cold worked
20% cold worked
The recrystallisation start is detected by
X ray diffraction.

415

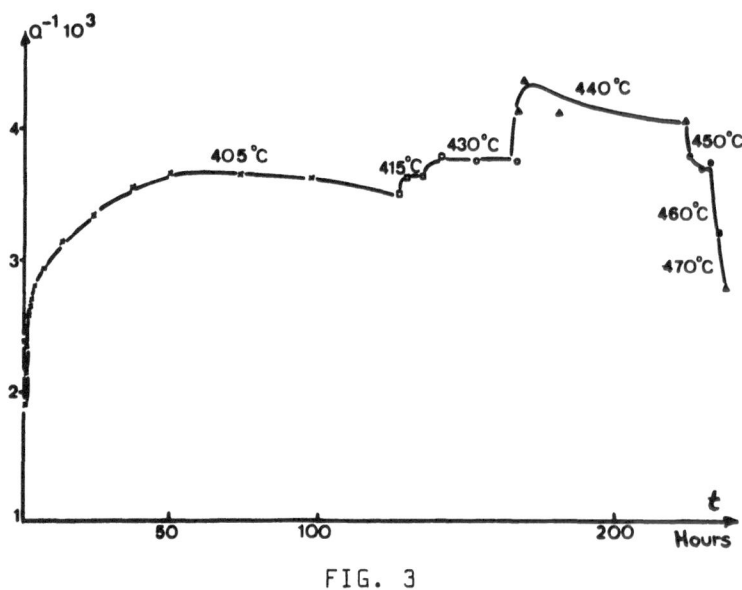

FIG. 3

Annealing time dependence of the internal
friction of a 40% cold rolled cobalt.

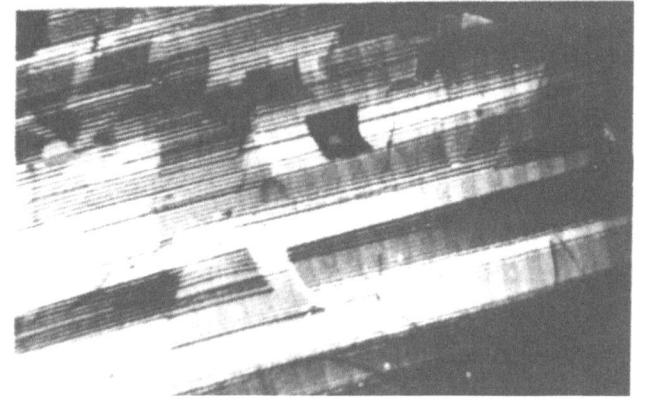

FIG 4a

40% cold rolled cobalt after
120h at 420°C. The extended
dislocations limit the
stacking faults. G × 13000

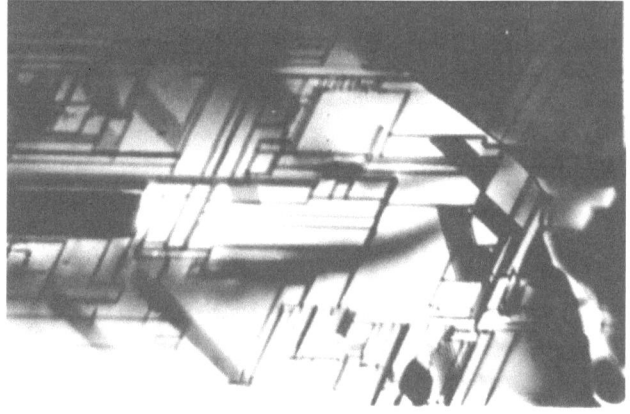

FIG 4b

40% cold rolled cobalt after
120h at 420°C + 3h at 460°C.
We notice the stacking faults
blocking up. G × 12000.

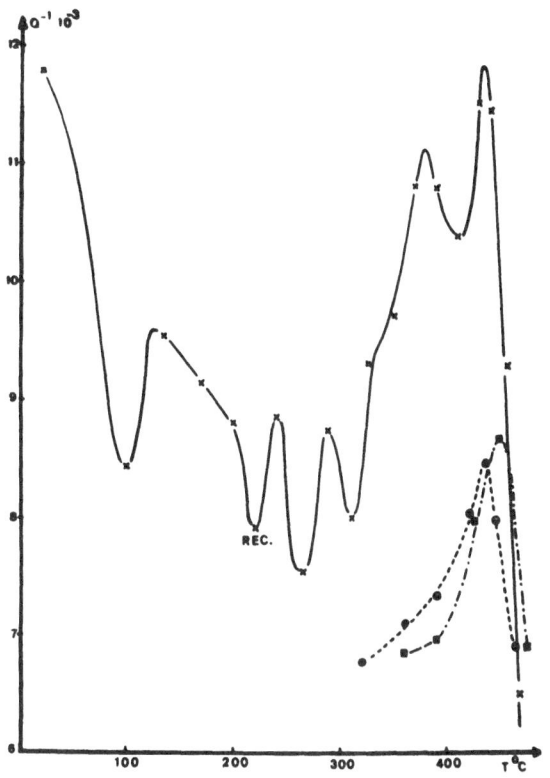

FIG. 5

$Q^{-1}f(T_R)$ of a 99,99% cobalt.
Recrystallisation starts at
REC. as seen by optical
microscopy.

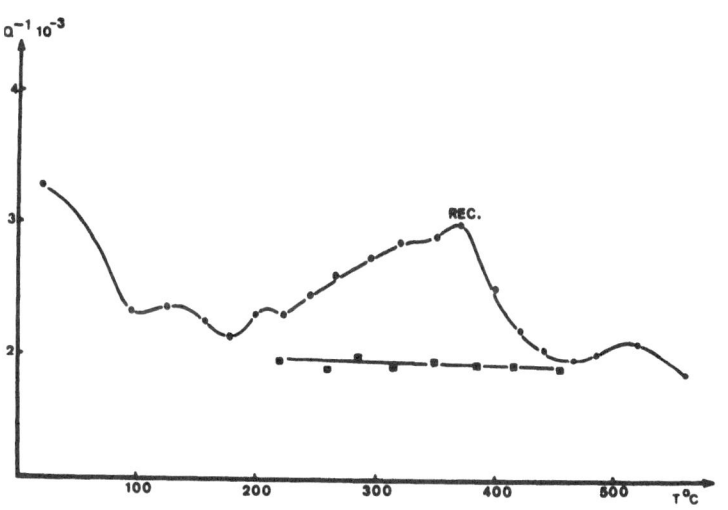

FIG. 6

$Q^{-1}f(T_R)$ of a Co-5% Fe alloy.
Recrystallisation starts at
REC. as seen by electron
microscopy.

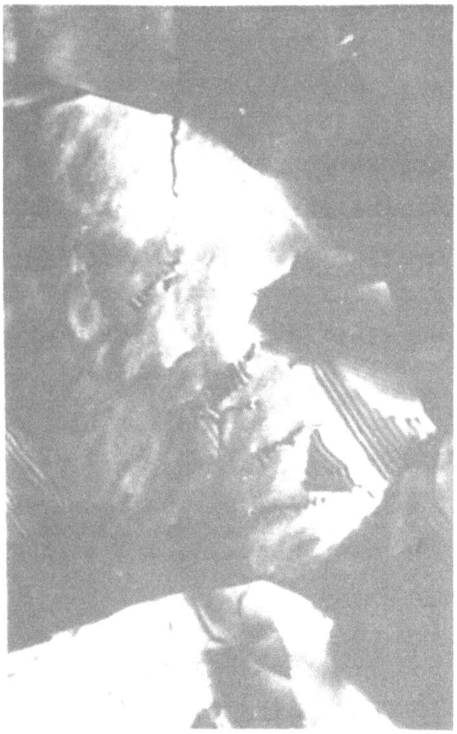

FIG. 7

Alloy structure after internal
friction measurement. G × 36000

INTERNAL FRICTION PEAK ASSOCIATED WITH PHASE TRANSFORMATION
IN Mn - Cu ALLOYS

K. Sugimoto and T. Mori

Institute of Scientific and Industrial Research, Osaka University
Osaka, Japan

Introduction

Relaxation near a Lambda transition has been extensively studied for the last few years in ferroelectric, ferromagnetic and antiferromagnetic materials and a very sharp internal friction peak has been observed in the vicinity of the transition temperature (1). Very little is known, however, about the behavior of internal friction associated with phase transformations in metals and alloys. Among a few examples sharp internal friction peaks are reported on the order - disorder transition in β -brass (2) and the martensitic (diffusionless and shear-like) transformation in TiNi (3).

It is well-known that the physical properties of Mn - Cu alloys are very peculiar. Internal friction of these alloys were first measured by Worrell (4). He found a well-defined internal friction peak around 0°C. This was explained by Zener (5) as a stress relaxation phenomenon across the twin boundaries in the fct phase. Aoyagi et al (6) showed that the relaxation peak may be explained by an elementary process involved in the movement of twin boundaries, i.e., the formation of step-rings on surface dislocations.

The present authors have been studying the internal friction in this alloy from the viewpoint of its high damping capacity for engineering use and reported the experimental results on the variation in internal friction and Young's modulus with increasing temperature in several Mn - Cu alloys containing 11.74 - 26.08 wt% Cu (7). They confirmed that a 'main peak' appear around 3°C in all specimens containing less than 21 wt% Cu. They also found a new 'sub-peak' at the temperature of the reverse martensitic transformation in each specimen. The nature of the 'sub-peak', however, was not clarified in the previous paper. The purpose of the present paper is to report the behavior of such 'sub-peaks' and to expain the nature from the mechanism of the phase transformation.

Methods

Table 1 shows the chemical composition of alloy samples used. They were polycrystalline specimens and were solution-treated at 850 or 900°C (sample A), followed by rapid quenching into 10% KOH solution. Specimens from sample C were then aged for 21 h at 375°C.

Table 1 Chemical Composition of Samples (wt%)

sample	Mn	Cu	Fe	Co	Si
A	88.26	11.74	-	-	-
B	84.57	15.18	0.01	0.01	0.23
C	55.42	44.58	-	-	-

Internal friction and elastic modulus were measured by a flexural vibration method in a frequency range 300 - 900 Hz. The strain-amplitude was about 6×10^{-6}. Measurements were made both on heating and cooling with a constant rate 0.03 - 1.5°C/min by means of a PID-controller, the temperature fluctuation being less than 0.2°C.

Results and Discussion

Figs. 1 - 3 show the typical temperature dependence of internal friction and Young's modulus in three alloys. A very pronounced 'main peak' and a small 'sub-peak' can be seen in samples A (Fig. 1) and B (Fig. 2). The 'main peak' in sample A had an activation energy of 5.15×10^4 J/g atom (12.3 kcal/mol) and a frequency factor $1/\mathcal{T}_o = 2.4 \times 10^{13}$/s (7). It was, therefore, concluded that the peak is due to a stress relaxation associated with the movement of {101}-twin boundaries in the fct phase. This conclusion is in accordance with those reported by previous workers (4,5,6). The 'main peak' was not so clearly observed in sample C as in other two samples (Fig.3) but a broad peak was observed at temperatures higher than 3°C. The broad peak will be discussed in the following sections.

Behaviors of the 'sub-peak' and the broad peak

As can be seen in Figs. 1 and 2 the temperatures of the 'sub-peaks' coincide with those of the elastic anomalies and the peak temperatures of the specific heat. These temperatures correspond to the temperature of the martensitic transformation from the fcc to the fct structure, according to our previous work (7). It is supposed, therefore, that this 'sub-peak' may be a phase transformation peak. But in sample C the 'main peak' and the 'sub-peak' are not clearly distinguishable and an asymmetrical broad peak is visible at a temperature slightly below that of the elastic anomaly (Fig.3). Such a different behavior of the internal friction in the sample would be due to the effect of compositional fluctuation which was caused by the formation of Mn- or Cu-clusters during ageing at 375°C (8).

The 'sub-peaks' in samples A and B were first found by the present authors, while a similar broad peak as that of sample C was reported by Schwaneke et al (9). The behavior of these peaks are quite unusual; i.e., they do not exhibit a normal peak shift with change in the frequency of vibration but always appear around the temperature of the elastic anomaly. In addition they are considerably sharp and are clearly visible both on heating and cooling. All these features seem to be typical of the internal friction peaks due to phase transformations.

Fig. 4 shows an example of such a sharp internal friction peak observed in sample A on heating. It may probably be due to the effect of compositional fluctuation and the nature of polycrystalline aggregates that the peak is not so sharp as in the case of relaxation near a Lambda transition in ferroelectric materials etc. The effect of the compositional fluctuation was most pronounced in the case of sample C as shown in Fig. 5. The compositional fluctuation was so large that the internal friction peak was observed over a wide range of temperature (about 40°C). On the other hand in the case of sample A the internal friction peak almost completed in a narrow range of temperature (about 10°C) as shown in Fig. 4.

Thermal Hysteresis and the Nature of Phase Transformation

The height of the 'sub-peak' and the broad peak in sample C seemed to be sensitive to the thermal history of the specimen. Fig. 6(A) shows an example of such an effect. The height of the internal friction peak Q_{max}^{-1} varried from 7.5 $\times 10^{-3}$ to 3.0×10^{-3} for the 33 runs made succesively on heating and cooling at various rates from 0.03 to 1.5°C/min. In some cases a subsidiary peak appeared at the lower temperature side of the internal friction peak and the height of the original peak was considerably reduced. The reason why the height and the shape of the internal friction peak were influenced by the thermal history was not clear.

Both the temperature of the elastic anomaly T_t and of the internal friction peak T_p are observed to be higher on heating than on cooling. This is shown in Figs. 6(B) and (C). In addition, both T_t and T_p are dependent on the rate of heating or cooling. The difference between the two temperatures (T_t on heating and T_t on cooling, etc), i.e., the thermal hysteresis of the phase transformation ΔT, increases with increasing rate of heating or cooling. The extrapolation of the observed temperatures to $v = 0$ shows that T_t does not coincide exactly with T_p. T_t is always lower than T_p by 2 or 3°C in the case of sample A. In sample C, on the contrary, the situation is somewhat different, i.e., T_t is always higher than T_p by about 10°C as shown in Fig. 5. The reason for such different behaviors of the internal friction peak in aged samples is not clear at present but they might be connected with the compositional fluctuation. Further work is now underway for aged samples.

Table 2 summarizes the observed values of the thermal hysteresis ΔT in the three Mn - Cu alloys, along with that of ΔT in a Cu - Al - Ni single crystal (10). ΔT is much smaller in Mn - Cu alloys than in the Cu - Al - Ni single crystal which is known to exhibit a thermoelastic martensite transformation (first order transformation) (11). This might suggest that the phase transformation in Mn - Cu alloys may have a characteristics of the second order transition rather than the first order.

Table 2 Thermal Hysteresis of Transformation Peak
(v = 1.05°C/min)

sample	thermal hysteresis ΔT(°C)
A	8 - 12
B	10 - 11
C	7
Cu-14.5%Al-4.98%Ni (single crystal)	38

The change in the unit cell volume of a 87 at% Mn - Cu alloy has been reported by Makuhrane (12). The volume change occurs quite continuously around the temperature of the phase transformation. This is a very important evidence for the second order transition in this alloy. On the other hand Bazinski et al (13) and Sugimoto (14) observed such a change in microstructure during the phase transformation. The martensite phase, which contained fine {101}-twins, coexisted with the parent phase. The twinned region grew, when the temperature was lowered. Such a result of the microscopic examination might suggest that the phase transformation were of the first order. However, the phase transformation in Mn - Cu alloys should be understood as a second order transformation which accompanies the formation of {101}-twin boundaries as a result of magnetic energy. The {101}-twin boundaries are, therefore, believed to be the antiferromagnetic domain boundaries, as suggest by Hedley (8).

Origin of the Internal Friction Peak

The 'sub-peaks' in samples A and B and the broad peak in sample C were observed near the temperature of the elastic anomaly, which was previously explained by the strain-dependent exchange model (7,12). The phase transformation was confirmed to be of the second order. Although the peaks were not so sharp as in the case of ferromagnetic materials etc, they had a very similar characteristics to those near a Lambda transition. Hence, it is concluded that the internal friction peak associated with the phase transformation in Mn - Cu alloys is a kind of Lambda transition peak. As for the internal variable the degree of the antiferromagnetic ordering will be the most important.

References

1. A. S. Nowick and B. S. Berry, Anelastic Relaxation in Crystalline Solids, p.463. Academic Press, New York (1972).

2. W. Köster, Z. Metallk. 32, 145 and 151 (1940).

3. R. R. Hashiguti and K. Iwasaki, J. Appl. Phys. 39, 2182 (1968).

4. F. T. Worrell, J. Appl. Phys. 19, 929 (1948).

5. C. Zener, Elasticity and Anelasticity of Metals, p.159. Univ. Chicago Press (1948).

6. T. Aoyagi and K. Sumino, Physica Status Solidi. 33, 317 (1969).

7. K. Sugimoto, T. Mori and S. Shiode, Met. Sci. J. (to be published).

8. J. A. Hedley, Met. Sci. J. 2, 129 (1968).

9. A. E. Schwaneke and J. W. Jensen, J. Appl. Phys. Suppl. to 33, 1350 (1962).

10. K. Otsuka and K. Shimizu, unpublished.

11. K. Otsuka and K. Shimizu, Jap. J. Appl. Phys. 8, 1196 (1969).

12. P. Makuhrane and P. Gaunt, J. Phys. C (Solid St. Phys.). Ser. 2, 2, 959 (1969).

13. Z. S. Bazinski and J. W. Christian, J. Inst. Met. 80, 659 (1951-52).

14. K. Sugimoto, Nippon Kinzoku Gakkai Kaiho. 10, 44 (1971), in Japanese.

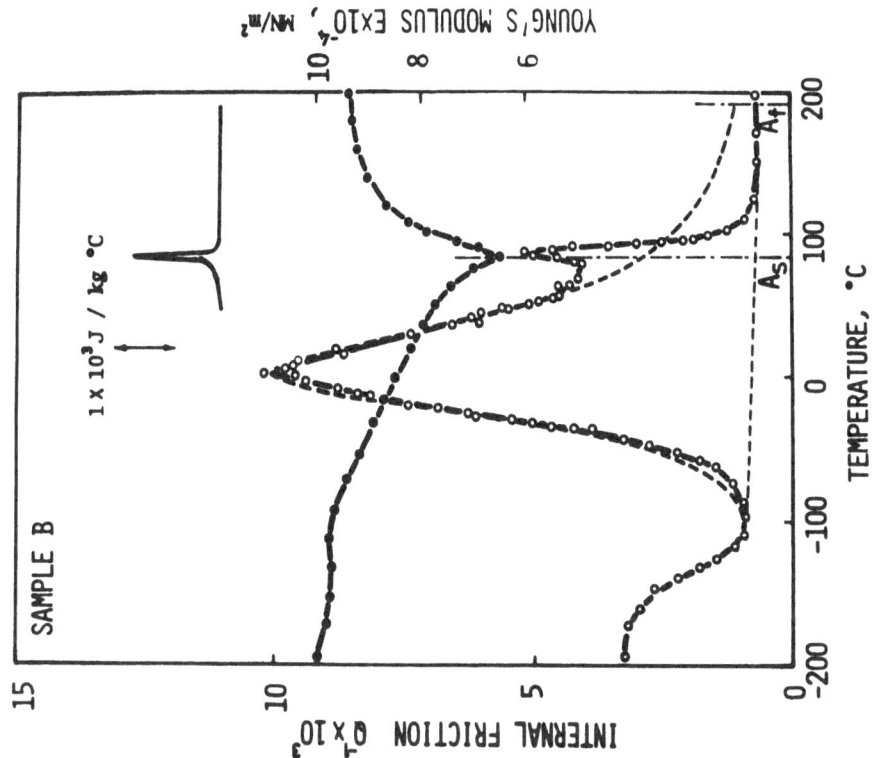
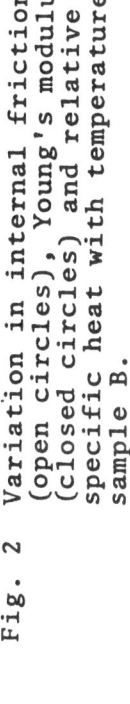

Fig. 1 Variation in internal friction
(open circles), Young's modulus
(closed circles) and relative
specific heat with temperature
in sample A.

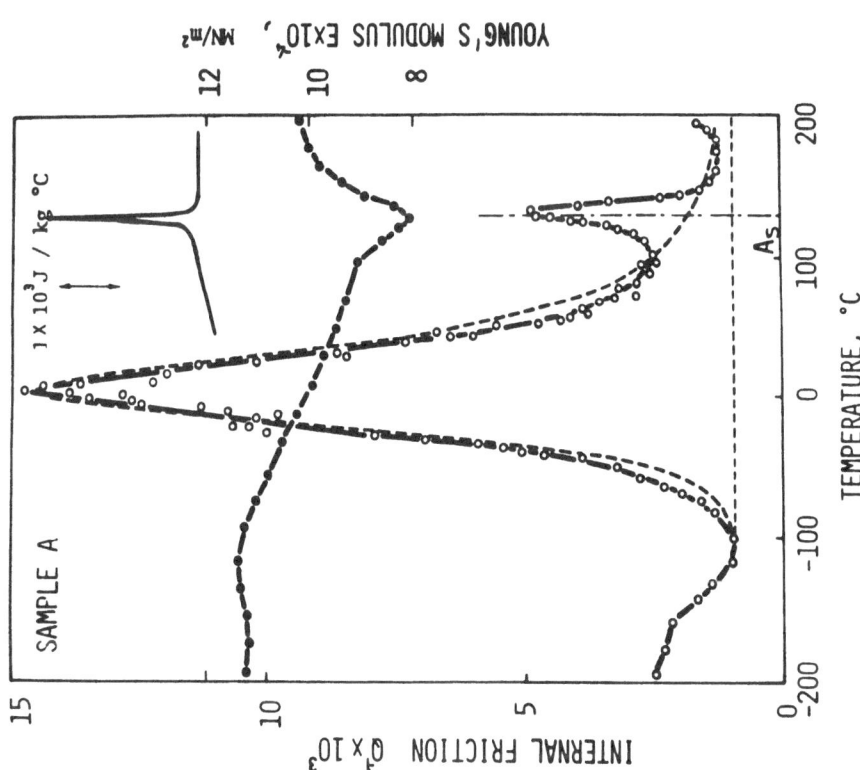

Fig. 2 Variation in internal friction
(open circles), Young's modulus
(closed circles) and relative
specific heat with temperature in
sample B.

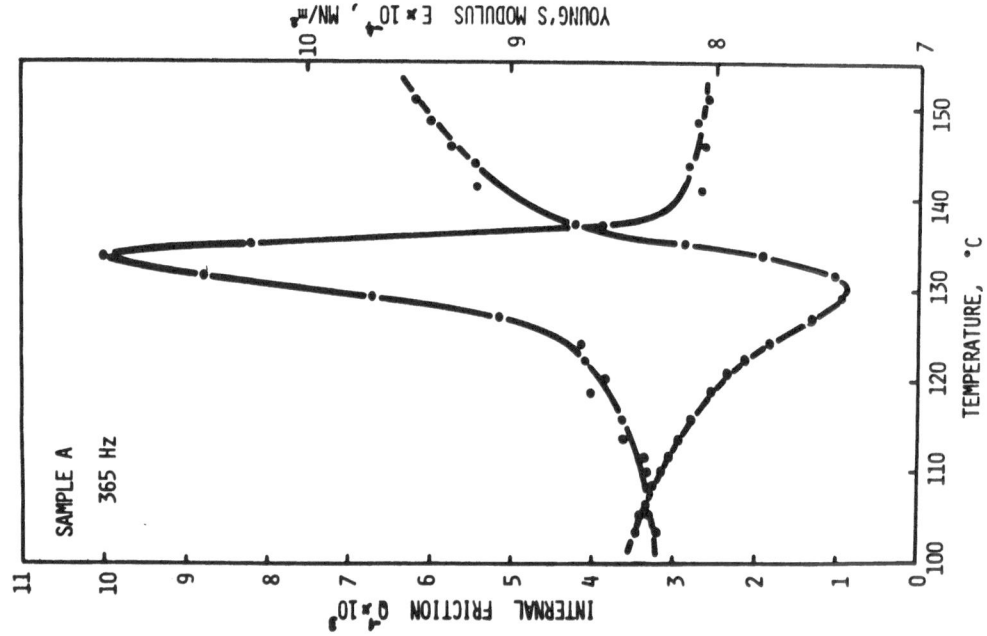

Fig. 4 Internal friction peak (open
circles) and elastic anomaly
(closed circles) near the phase
transformation temperature in
sample A.

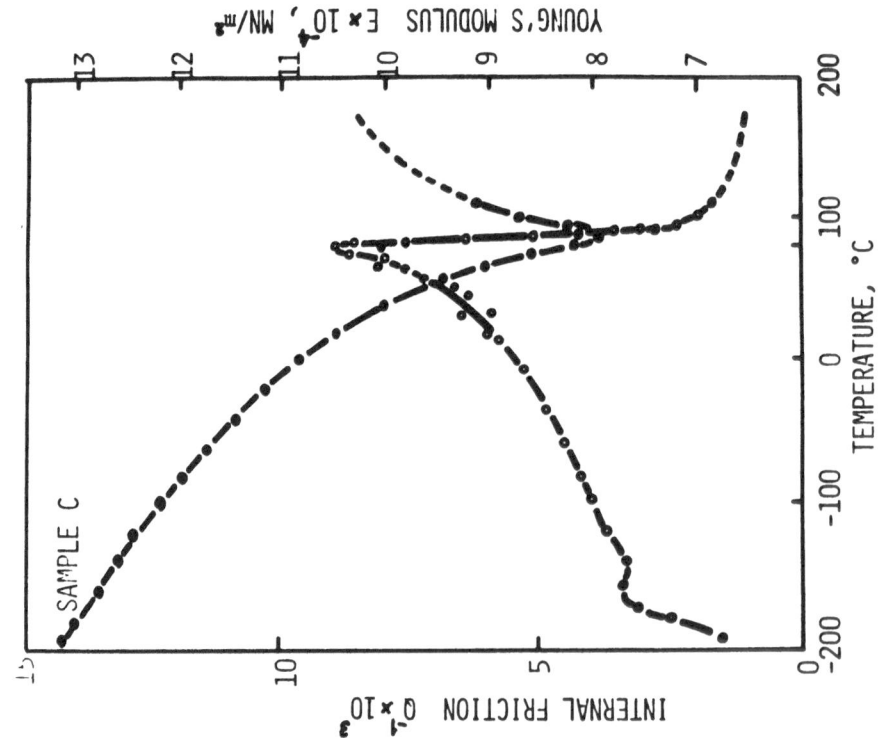

Fig. 3 Variation in internal friction
(open circles) and Young's modulus
(closed circles) with temperature
in sample C aged for 21 h at 375°C.

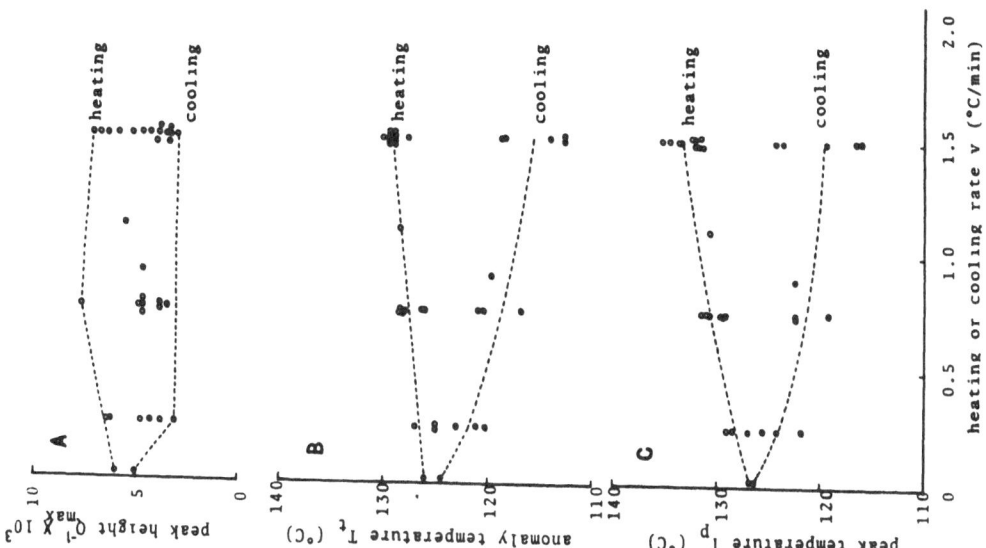

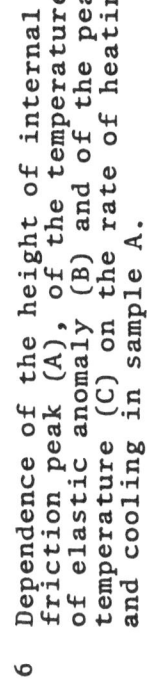

Fig. 6 Dependence of the height of internal friction peak (A), of the temperatures of elastic anomaly (B) and of the peak temperature (C) on the rate of heating and cooling in sample A.

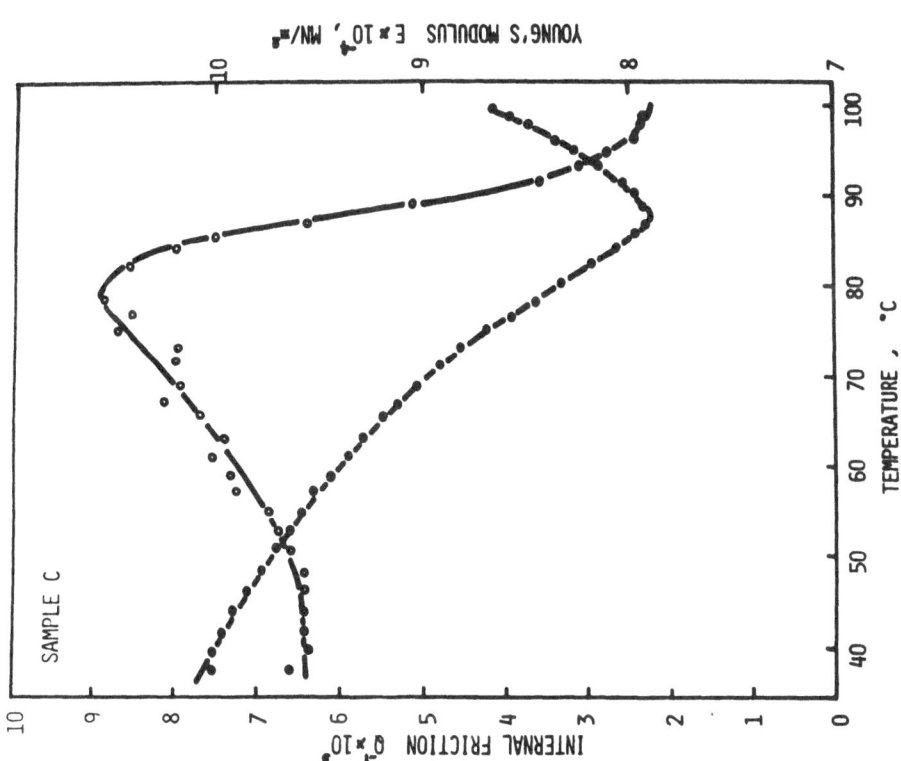

Fig. 5 Internal friction peak (open circles) and elastic anomaly (closed circles) near the phase transformation temperature in sample C aged for 21 h at 375°C.

INTERNAL FRICTION IN FERROUS MARTENSITES

G. J. Klems, R. E. Miner and F. A. Hultgren
Republic Steel Research Center
Cleveland, Ohio 44131

and

R. Gibala
Case Western Reserve University
Cleveland, Ohio 44106

INTRODUCTION

A great deal of internal friction data have been accumulated in the literature on the behavior of carbon in ferrous martensites and cold worked ferrites (1-5). In the case of cold worked ferrites, such investigations have led to an accurate theoretical description of the "cold-work" peak in iron (6,7) and to extensive use of the peak for investigation of dislocation-interstitial solute interactions in many alloy systems (8). Examination of this literature reveals that, by comparison, the internal friction of ferrous martensites is poorly understood. The source of this problem appears to be that, in the interpretation of results, little advantage has been taken of the abundant information available on the microstructural differences and similarities (viz. martensite structure and morphology and carbide precipitation) which exist between the various kinds of martensites in ferrous materials. In the present investigation, we show how these factors can be used to analyze our own data on martensites in Fe-Ni-C alloys (9) and on all other data in the literature on ferrous martensites. Moreover, we demonstrate that this metallurgical information is vital to the understanding of the internal friction mechanisms involved.

PROCEDURES

The experimental procedures in our experiments (9) and in all work reported in the literature involve measurements of internal friction at ~ 1 Hz in the temperature range from near room temperature to 300–400 C with a torsional pendulum. The materials examined can be classified into four categories according to the types of martensitic substructure expected (10):

(A) martensites with subzero M_s temperatures and which exhibit a twinned morphology (10);

(B) martensites with M_s temperatures well above room temperature, which exhibit a heavily dislocated substructure and which are auto-tempered during quenching and holding at room temperature prior to measurements (10–12);

(C) martensites with high M_s temperatures and predominantly dislocated substructures, but rapidly (e.g., brine) quenched (to avoid auto-tempering) and refrigerated at 77 K prior to measurements (also avoiding tempering but in addition causing retained austenite to transform to twinned martensite) (10–12);

(D) martensites with high M_s temperatures and auto-tempered on quenching, but with ϵ carbide as the sole tempering product (10–12).

Besides martensite substructure and carbide precipitation, the metallurgical variables investigated include effects of carbon content, tempering time, tempering temperature, and cold work (1–3,9). For brevity, we limit discussion mainly to the last two of these variables in this paper.

RESULTS

Fig. 1 is a composite figure which illustrates the types of internal friction spectra that are observed in ferrous martensites and cold worked ferrites at a frequency of ~ 1 Hz. In general, one observes (i) one relaxation peak at ~ 160 C, as in A in the figure, in twinned martensites with low M_s temperatures (9,13); (ii) one relaxation peak at ~ 250 C, as in B and D in the figure, in dislocated and auto-tempered martensites with high M_s temperatures and in cold worked ferrites, irrespective of the types of carbides present, i.e., ϵ carbide and/or cementite (2,13,14); and (iii) two relaxation peaks in the vicinity of 160 C and 250 C, as in C in the figure, in dislocated + partially twinned martensites with high M_s temperatures (13). Hereafter, for convenience, we refer to these peaks simply as the 160° peak and the 250° peak.

It is easy to show that the 160° peak and the 250° peak are truly different relaxation peaks because of several phenomenological differences in their behaviors. Some of these differences are summarized in Table 1 and Fig. 2.

Table 1 illustrates in summary form that the 160° peak and 250° peak have fundamentally different responses to tempering above room temperature and to cold work. The 160° peak is first reduced and then eliminated totally by tempering at temperatures as low as 200–250 C. The 250° peak, much like the cold work peak in ferrite (15), is resistant to tempering below 250 C and is not eliminated until temperatures well above 400 C are reached. The difference in the influence of cold work on the two peaks is more dramatic: the 160° peak is reduced by cold working, whereas the 250° peak is enhanced in the same manner as the cold work peak (4).

We can also note that the relaxation rate parameters for the 160° peak and the 250° peak are significantly different. Fig. 2 shows that the relaxation time $\tau = \tau_o$ exp (Q/RT) for the 160° peak in twinned Fe-Ni-C martensites is characterized by values of τ_o of approximately 10^{-12} sec and activation energies Q of the order of 20–22 kcal/ mole. On the other hand, the 250° peak and the cold work peak in ferrite are characterized for the most part by τ_o's $\sim 10^{-16} - 10^{-18}$ sec and Q's $\sim$ 29–45 kcal/mole, giving τ's indicated by the band of data in Fig. 2. It should be noted that the activation energy of the 160° peak is close to that of the carbon Snoek peak, also given in Fig. 2 (16), although the τ_o's are different by almost three orders of magnitude.

Finally, we can correlate the occurrence of the 160° peak and the 250° peak with the presence of pertinent microstructural features of the various martensites. Such a correlation is summarized in Table 2 and Fig. 3: (A) All of the Fe-Ni-C martensites with low M_s temperatures have $\{259\}_\gamma$ – $\{3\ 10\ 15\}_\gamma$ habits, have a twinned substructure, always precipitate only ϵ up to ~ 200-250 C, and exhibit only the 160° peak. (B) The Fe-C martensites usually have either a $\{111\}_\gamma$ or $\{225\}_\gamma$ habit, have a dislocated substructure, precipitate either ϵ or Fe_3C (depending on the extent of auto-tempering), and exhibit only the 250° peak. (C) Similar alloys that are rapidly quenched and refrigerated at 77 K tend toward $\{111\}_\gamma$ to $\{225\}_\gamma$ habits, but with some $\{259\}_\gamma$ habits from the formation of martensite at 77 K. They also have a mixed dislocated + twinned substructure, precipitate ϵ or Fe_3C, and exhibit both peaks. (D) The Fe-Si-C martensite has a $\{225\}_\gamma$ habit and a dislocated substructure, but the first precipitate is ϵ, which is very stable and persists to very high temperatures. This alloy exhibits only the 250° peak.

Examination of the data in Table 2 discloses that neither peak can be correlated with any of the carbide precipitation products. In particular, it is clear that the 160° peak cannot be associated with ϵ carbide because of its absence in the silicon steels. However, there is a clear correlation between the existence of these peaks and the martensite habits and the as-quenched substructures. When _twinned_ $\{259\}_\gamma$ or $\{3\ 10\ 15\}_\gamma$ martensites are present, the _160°_ peak occurs; when _dislocated_ $\{111\}_\gamma$ $\{225\}_\gamma$ martensites are present, _the 250° peak_ occurs; and if both habit/substructure classes of martensite are present, both peaks are present also. Fig. 3 summarizes this correlation by indicating the occurrence of the 160° peak and 250° peak along with the regions of the reported martensite habit planes on a unit stereographic triangle. As the habit plane moves toward $\{259\}_\gamma$ or $\{3\ 10\ 15\}_\gamma$, the 160° peak is observed at the expense of the 250° peak.

DISCUSSION

From the results presented and from other supporting evidence (9) that we will publish in greater detail later, we have concluded that the relaxation mechanism of the 160° peak must be associated with twins and that the mechanism of the 250° peak is associated with dislocations. Of course, carbon atoms play a vital role in both mechanisms since neither peak occurs in carbon-free materials.

Because of its detailed similarity to the cold work peak in ferrites, the 250° peak is best interpreted in terms of models devised to explain the cold work peak. In particular, the model of Schoeck (6), depicted schematically in Fig. 4, is most appropriate. This theory attributes the peak to the motion of dislocations under the influence of the viscous drag caused by their attendant interstitial atmospheres.

We have devised a similar model, but involving twin boundaries rather than dislocations to explain the 160° peak (9). In this model, also given in Fig. 4, the application of a stress on the twin causes the twin interface to move (by translation of _partial_ dislocations); an internal friction peak occurs from the relocation of carbon atoms in the vicinity of the twin to new boundary sites as the twin moves. The process is similar to the Schoeck mechanism. It is also akin to reorientation of carbon atoms by magnetic domain wall motion to produce a magnetic Snoek effect, except that the interstitial sites at a twin interface are somewhat distorted compared to normal interstitial sites away from the twin. From such a model, one expects an activation energy approaching that of the Snoek relaxation but a very different (smaller) effective attempt frequency, as observed in Fig. 2. We have developed this simple model quantitatively (9), and it does predict qualitatively all of the results that are presently available on 160° peaks. However, additional data on these peaks are needed before more refined versions of the models are developed.

SUMMARY

(1) We have shown that two distinct internal friction peaks can occur in

ferrous martensites which contain carbon. (2) One relaxation peak at ~ 160 C at ~ 1 Hz is associated with twin boundary-carbon interaction; the other is a cold work type peak at ~ 250 C at ~ 1 Hz and is associated with dislocation-carbon interaction. (3) Identification of the appropriate mechanisms of these peaks has been accomplished by detailed correlation of the peak characteristics with martensite structure, martensite morphology, and carbon precipitation behavior.

ACKNOWLEDGMENTS

A portion of the work conducted at Case Western Reserve University was sponsored by the U. S. Atomic Energy Commission. The financial assistance provided for G. J. Klems by Republic Steel Corporation is gratefully acknowledged.

REFERENCES

1. R. Ward and J. M. Capus, J. Iron Steel Inst. 201, 1038 (1963).

2. T. Gladman and F. B. Pickering, ibid. 204, 112 (1966).

3. J. N. McGrath and R. Rawlings, Metal Sci. J. 2, 37 (1968).

4. D. P. Peterra and D. N. Beshers, Acta Met. 15, 791 (1967).

5. H. Ino and T. Sugeno, ibid. 15, 1197 (1967).

6. G. Schoeck, ibid. 11, 617 (1963).

7. R. E. Miner, F. A. Hultgren and R. Gibala, to be published.

8. A. S. Nowick and B. S. Berry, Anelastic Relaxation in Crystalline Solids, p. 401, Academic Press, New York, 1972.

9. G. J. Klems, Ph.D. Thesis, Case Western Reserve University, 1971.

10. C. M. Wayman, in Advances in Materials Science, vol. 3, ed. H. Herman, p. 147, Interscience, New York, 1968.

11. S. Murphy and J. A. Whiteman, Met. Trans. 1, 843 (1970).

12. G. R. Speich, Trans. Met. Soc. AIME 245, 2553 (1969).

13. T. S. Kê and Y. L. Ma, Scientia Sinica 6, 81 (1957); ibid. 5, 19 (1956).

14. I. Tamura, T. Mura, and J. O. Brittain, Trans. Met. Soc. AIME 221, 1158 (1961).

15. W. Köster, L. Bangert, and R. Hahn, Arch. Eisenhütten. 25, 569 (1954).

16. A. E. Lord and D. N. Beshers, Acta Met. 14, 1959 (1966).

TABLE 1

Summary of the Effect of Tempering Temperature and Cold Work on the Internal Friction Peaks Observed in Ferrous Martensites

EFFECT OF TEMPERING TEMPERATURE	160° PEAK	250° PEAK
UP TO ~250°C	CONTINUOUS DECREASE IN PEAK HEIGHT	NO CHANGE
~250°C TO~400°C	ELIMINATED	CONTINUOUS DECREASE IN PEAK HEIGHT
OVER ~400°C	ELIMINATED	NEARLY ELIMINATED
COLD WORK	CONTINUOUS DECREASE IN PEAK HEIGHT	INCREASE IN PEAK HEIGHT; REACHES SATURATION VALUE

TABLE 2

Summary of Martensite Habits, Structures, Precipitates and Internal Friction on Peaks Observed in Ferrous Martensites

MARTENSITE ALLOY	MARTENSITE HABIT	BASIC SUBSTRUCTURE	1st TEMPER PRECIPITATE	160°C PEAK	250°C PEAK
Fe-Ni-C	$\{259\}_\gamma$ $\{3\ 10\ 15\}_\gamma$	TWINNED	ϵ	YES	NO
Fe-C (REFRIGERATED)	$\{225\}_\gamma$ TO $\{259\}_\gamma$	DISLOCATED AND TWINNED	ϵ OR Fe_3C	YES	YES
Fe-C	$\{111\}_\gamma$ TO $\{225\}_\gamma$	DISLOCATED	ϵ OR Fe_3C	NO	YES
Fe-Si-C	$\{225\}_\gamma$	DISLOCATED	ϵ	NO	YES

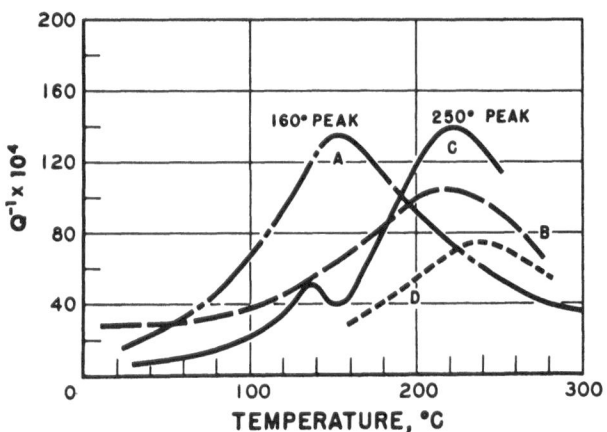

FIG. 1

Composite of Reported Internal Friction Peaks Observed in Ferrous Martensites.
A) Fe–25.5% Ni – 0.32% C; B) Fe–0.4% C; C) Fe–0.8% C; D) Fe–2.4% Si – 0.39% C.

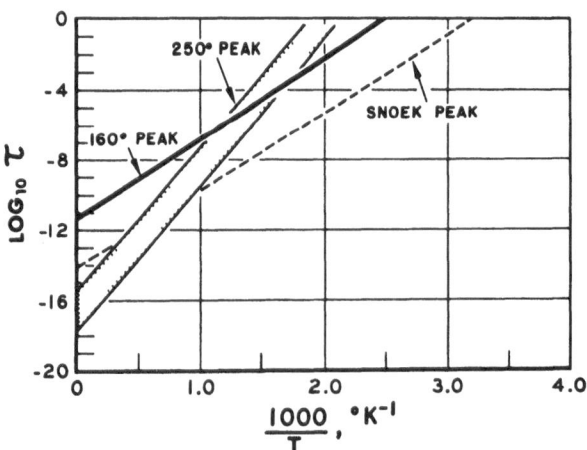

FIG. 2

Temperature Dependence of Relaxation Parameter τ for the 160° Peak, 250° Peak and
Snoek Peak.

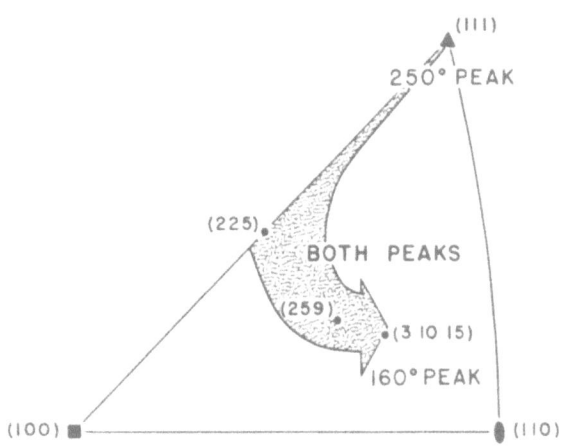

FIG. 3

Martensite Habit Planes and the Occurrence of Internal Friction Peaks Overlayed on a Unit Stereographic Triangle.

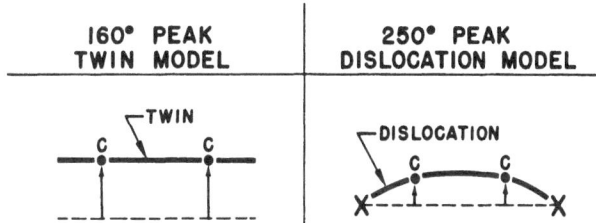

FIG. 4

Schematic Representations of the Twin Model for the 160° Peak and the Dislocation Model for the 250° Peak.

STUDY OF RELAXATION PEAKS IN AUSTENITIC STEELS.

F. Mezzetti and L. Passari
Istituto di Fisica dell'Universita' and Gruppo G.N.S.M.- Ferrara.
D. Nobili
Laboratori LAMEL, C.N.R.- Bologna

1. Introduction.

It has been known for a long time that in f.c.c. alloys a low frequency
relaxation peak (at about 1 cps) appears in the range between $250^{o}C$
and $350^{o}C$, depending on the type of material.
First Finkelstein and Rozin (1) described an I.F. peak at 300ºC in an austenitic steel containing 25% of Cr and 20% of Ni. They gave no theoretical
explanation of this peak, but suggested it would be probably due to the stress induced diffusion of the carbon atoms present in the material.
Later on, analogous peaks were observed in some other f.c.c. alloys.
Kè and Tsien (2) found in a manganese steel alloy a linear relationship
between dissolved carbon and height of the I.F. peak. To explain this,
they supposed that the relaxation peak would arise from the movement of
pairs of an interstitial carbon atom linked to a substitutional impurity
atom, which strain the original f.c.c. cells. A similar model, with a second carbon atom in place of the substitutional impurity, was proposed,
from a suggestion of Cheng and Chang (3), by Wu and Wang (4) to take into
account a parabolic and then linear relation between the height of the I.F.
peak and the quantity of carbon in solid solution, found in a Fe-Ni alloy.
Afterward, the last model was assumed, without any essential changes, by
Verner (5) to interpret data about measurements on some alloys of steel
with Ni, Cr, Mn and Co.
More recently Golovin and Belkin (6) have shown in a 18/8 steel, a large
dependance of the height of the I.F. peak on the percent of cold- work. As
this treatment induces the formation of a large number of dislocations in

the material, the authors have suggested either a direct interaction of the Carbon atoms (interstitial) with the dislocations produced in the original solid solution or a migration of Carbon atoms in the austenitic phase. The effective I.F. mechanism is not yet understood and it needs further informations. Especially, since during cold work an appreciable martensitic trasformation is induced, the possible influence of this new phase will be taken into account in this paper.

2. Experimental

The low frequency I.F. measurements have been carried out by an inverted torsional pendulum in which the free oscillation decrement has been recordered by a mechanico-optical system (spot follower). For the high temperature measurements an automatically controlled furnace has been used.
The materials studied were austenitic steels of different kind and after different treatments. The AISI designations and the compositions of the steels are reported on Table I. The technological treatments performed on the specimens are those of Tables II, III and IV.

TABLE I

Composition of austenitic steels (in percent*).

Type of Steel	C	Si	Mn	Cr	Ni	S	P	Mo
301	0.094	0.46	1.24	16.60	8.20	0.010	0.022	0.18
302	0.080	0.35	1.47	16.75	7.96	0.019	0.024	0.18
304	0.055	0.55	1.57	17.54	9.29	0.020	0.030	0.12
304 L	0.018	0.53	1.44	17.94	9.93	0.009	0.023	-

Before the measurements all the specimens have been vacuum heated at 1050°C for 30 minutes then quenched in water.
The degree of cold-work was evaluated from the area reduction.
All the samples have been examined by X-rays (Debye-Scherrer method) in

* In weight

order to detect the structural changes and the changes in the amount of the various phases. On Table II,III, and IV the qualitative informations so deduced are reported.

3. Experimental results and discussion.

In fig. 1 the results of the measurements of I.F. versus temperature for the 302 steel, as quenched and after different amounts of cold-work, are plotted. All measurements are made at a frequency of 1 cps. The 320°C peak clearly increases with increasing cold-work.

This behaviour is better emphasized by fig. 2 where the height of the peak versus the degree of cold-work is reported. Here the peak height (Q_m^{-1}) increases until the degree of cold-work reachs a value of about 30%, after this it seems to have a little decrease. (The values of Q_m^{-1} are always given after the background subtraction).

In fig.1 it is possible to note, in addition to the 320°C peak, a second peak centered at about 100°C. This peak also depend on cold-work, but it disappears very rapidly, together with the background, during annealings which have no effect on the 320°C peak, as shown by curve 3 of fig.1 . We do not investigate the nature of this peak.

The effects of the annealing on the 320°C peak may be deduced from table IV. The peak height decréases when the annealing temperature increases and nearly disappears at a temperature of about 500 °C.

In fig.3 the height of the 320°C peak as a function of the carbon content is plotted. One can observe that Q_m^{-1} increases nearly quadratically with the carbon concentration. This trend is the same for all the three curves 1, 2 and 3 of fig.3, which respectively refers to as quenched, 10% and 30% cold-worked specimens.

The mean activation energy for the 320°C peak has been obtained from the Marx-Wert plot given in fig.4 and results of about 1 eV, both for quenched and col-worked specimens.

The X-rays diffraction data about the effects of the various treatments on the phases in the alloys shows that the content of α-phase* in 301 steel,

---

* Actually this α-phase is a martensitic body centered tetragonal phase.

already present, in little amount **in** the γ matrix, after quenching, increases with the deformation. When the cold-work is over 30 % the phase is only α and the γ content is no more appreciable. On the countrary, in 304 L steel, the α-phase is always absent, except for traces in an heavily cold worked (36%) specimen.

The annealing at temperature below 600°C does not change, in an appreciable way, the ratio between the different phases present in the samples.

All these effects are summarized in Table II, III and IV.

TABLE II

Effects of quenching

Steel	Phases	Q_m^{-1}x 10^4
301	γ+(α)	3.5
302	–	2.3
304	–	1.2
304 L	γ	0

TABLE III

Effects of cold-work

Steel	Cold-work %	Phases	Q_m^{-1}x 10^4	Steel	Cold-work %	Phases	Q_m^{-1}x 10^4
301	5	γ+(α)	7.5	302	9	–	.8
	10	"	11.5		10	–	9
	17	–	14.5		30	–	13
	20	α+γ	17		70	–	12.5
	30	–	18	304	10	–	4.5
	35	α	18	304 L	10	γ	0
	50	α	16.5		36	γ+((α))	2.5

TABLE IV

Effects of annealing

Steel	Previous coldwork %	Annealing treatments			Phases	$Q_m^{-1} \times 10^4$
301	5	3^h	at	325 °C	$\gamma + (\alpha)$	3.5
	10	10'		400	–	5
	17	"		500	–	3
	"	30'		"	–	2
	20	10'		450	$\alpha + \gamma$	4.5
	35	"		"	α	5
	50	15'		325	α	11
302	9	10'		450	–	3.5
	10	"		400	–	4
	40	30'		200	–	12.5
	"	10'		365	–	9
	70	"		450	–	5
304	10	"		480	–	0
304 L	36	"		"	–	0

From these experimental data we can make the following remarks:

- the 320 °C peak is always related to the presence of carbon atoms in the solid solution and increases nearly quadratically with the carbon content,

- the peak height increases with deformation, reaching a saturation level at about 30% of area reduction (for 301 and 302 alloys) after which it decreases slowly,

- during the cold-work, besides the dislocations, a lot of α-phase (martensite) is produced in the austenitic steels,

- the annealing treatments above 300 °C decrease the Q_m^{-1}, but do not change the phase concentrations in the matrix,

- the activation energy value is similar to that found by several authors
  for interstitial (carbon) atom diffusion.

The models proposed for the explication of the 320°C peak are not completely in agreement with this results.

The theory of Kè and Tsien involves a linear relation between the Q_m^{-1} and
the carbon content and a critical value of this, which are not confirmed
by our measurements.

From the theory of the carbon atom pair riorientation in f.c.c. lattice, the
approximately quadratic low for the peak height versus carbon content can
be expected, and also the activation energy is of the right order of magnitude, but this theory cannot explain the cold-work effects.

Concerning the theory of the interaction of carbon atoms with dislocations
in the austenite, as proposed by Golovin and Belkin, it is in agreement
with the experimental results for the cold-work effects, but the activation energy is not the right one and the possible effects of the α-phase
has not been sufficently considered.

It is well strengthened that with the 320°C peak three elements are always
present: carbon, α-phase and dislocations. The mean difficulty to understand the mechanism of dissipation is to separate the effects of α-phase and
dislocations. In the course of the cold-work treatment they are produced
together, so it is not clear if the peak is directly related to the presence of α-phase or rather due to the high density of dislocations introduced during the formation of martensite.

The hypothesis we may suggest is a diffusion of interstitial pairs of carbon atoms in the b.c.t. martensite, induced by the interaction with the
field of the moving dislocations. The annealing effects are probably caused by a decrease of carbon content in the solid solution due to the formation of carbides and a partial elimination of the moving dislocations followed by a blocking due to the carbide precipitation. The saturation of
the cold-work effects may be the result of the complete trasformation of
the γ in α phase, while the further increase of dislocations, which tangle
each other, could better explain the decrease of Q_m^{-1} for high values of deformation. The temperature of the peak is probably related to the high

degree of distorsion of the lattice in the α-phase.

References.

1. B.N. Finkelstein and K.M. Rozin, Dok. Ak. Nauk S.S.S.R. 91, 811 (1953)

2. T.S. Kè and C.T. Tsien, Scientia Sinica, 5, 625 (1956)

3. K.C. Cheng and S.K. Chang, Acta Physica Sinica, 14, 71 (1958)

4. T.L. Wu and C.M. Wang, Scientia Sinica, 7, 1029 (1958)

5. V.D. Verner, Soviet Phys. Solid State, 7, 1870 (1964)

6. S.A. Golovin and K.N. Belkin, Fiz. Met. Metalloved., 20, 763 (1965)

Figures.

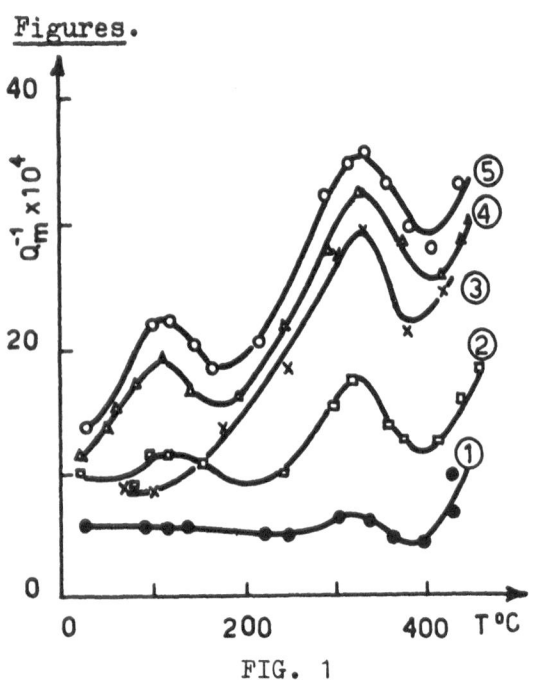

FIG. 1

Effects of cold-work on the temperature
dependence of the I.F. of the 302 steel.
1- quenched; 2- 9% cold-worked;
3- 40% cold-worked; 4- 30% cold-worked;
5- 70% cold-worked.

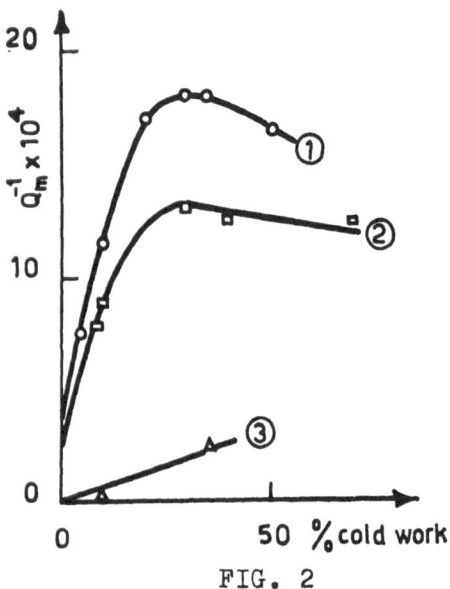

FIG. 2

Q_m^{-1} versus degree of cold-work.
1- 301 steel; 2- 302 steel;
3- 304 L steel.

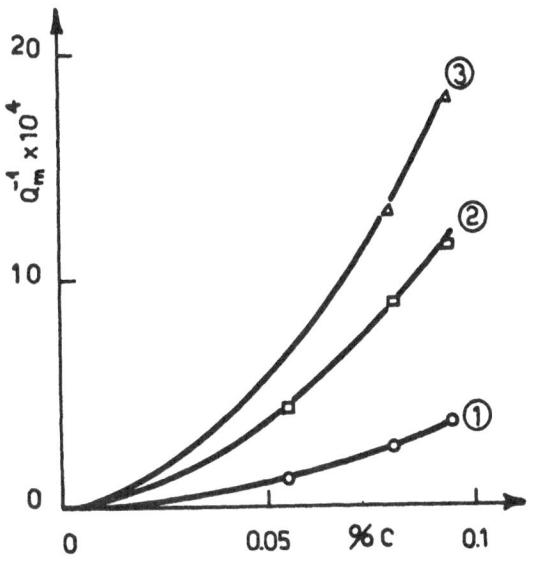

FIG. 3

Q_m^{-1} versus carbon content.

1- quenched; 2- 10% cold-worked
3- 30% cold-worked.

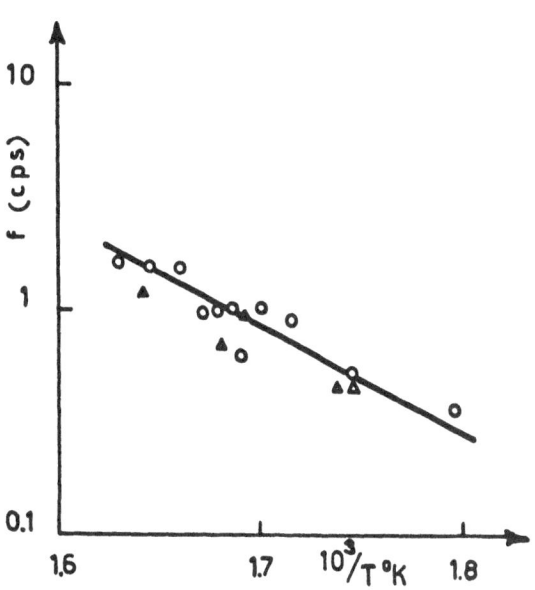

FIG. 4

Marx-Wert plot for 320 °C peak.
△ quenched; ○ cold-worked.

INTERNAL FRICTION ANOMALIES IN Ni_3Fe
NEAR THE ORDER-DISORDER TRANSITION

M. Lebienvenu and B. Dubois

Ecole Nationale Supérieure de Chimie

Laboratoire de Métallurgie et Matériaux

11 Rue, Pierre et Marie Curie

75230 Paris Cedex 05

Ni_3Fe alloy is well known for its magnetic properties, but it is also interesting for an order-disorder transition around 500-520°C. The constitution diagram was recently reviewed by Calvayrac and Fayard (1) and is presented in figure 1. We considered that it would be valuable to attempt some internal friction measurements in this temperatures range.

Alloy was worked out from ex-carbonyl nickel and electrolytic iron . The composition was about $Ni_{75}Fe_{25}$. Measurements were performed on an inverted torsion pendulum oscillating at 0,67 Hz, the samples were strips 70 x 8 x 0,45 mm^3 and the amplitude deformation was 2.10^{-6}.

1. INTERNAL FRICTION OF COLD WORKED ALLOY.

Preliminary measurements dealt with magnetomechanical effect. After annealing the sample at 650°C for half an hour, the effect of the intensity of a direct magnetic field on the internal friction was studied. The results were similar to those obtained with nickel (2). A maximum was observed at 10 oersteds and magnetic saturation appeared at about 200 oersteds. The maximum is not easy to explain because magnetostriction in Ni_3Fe is not well understood.

After a strong cold rolling (90% reduction), the internal friction of the alloy was measured at room temperature under a 250 oersteds direct

magnetic field as a function of annealing temperature under vacuum
between 20 and 900°C; the results are shown in figure 2 : it can
be seen that recovery is like that obtained for pure metals, for
instance nickel and cobalt (3). After a decrease of internal
friction from room temperature to 200°C, there is an unexplained
maximum at 275°C and another at 520°C related to recrystallisation,
as indicated by X ray diffraction and microhardness tests (4). More
interesting is the maximum observed at 460°C which we think is
related to ordering in the alloy. We tried to obtain more information
by isothermal annealings at 480°C, but measurements at room temperature
failed. We therefore decided to measure the variation in internal
friction with temperature.

2. <u>INTERNAL FRICTION OF QUENCHED ALLOY AS A FUNCTION OF TEMPERATURE.</u>

The alloy was quenched from 650°C by retiring furnace; X ray
diffraction detects no short range order, the reason being the slowness
of diffusion below this temperature.

After mounting it in the pendulum, the sample was heated at
5°C/mn to 540°C under an argon atmosphere. We can see in figure 3 an
internal friction peak at 470°C. After cooling, this peak is described
again on reheating; it becomes clear that this peak is due to stress-
induced ordering. The change of frequency shows that we have a thermally
activated short range ordering under the effect of ctress : when the
frequency is changed from 0,67 Hz to 0,44 Hz, the peak becomes visible
at 455°C; if we suppose a single relaxation time, we can estimate an
activation energy of 55 kcal/mole; this value is slighty smaller than
self-diffusion energies and may be characteristic of a limited number
of jumps (5) or a different weighting of atomic jumps in the environment(6).

Further evidence that this is a stress induced ordering Zener
peak is its disappearance with subsequent annealing out of the

pendulum (fig.4) : the height of the peak decreases after annealing for
6 h. at 495°C and vanishes completely after 48 h. at this temperature.
Calvayrac and Fayard (7) have shown that 8 h.at 495°C is the minimum time
required to detect long range order by X ray diffraction, but after
48 h. at 495°C, they have found an ordered domain size of 80 Å and a
L.R.O. parameter equal to 0,7. Therefore we can say that the disappearance
of the peak attributed to short range order is related to the development
of long range order.

Nevertheless it was interesting to follow internal friction
variation in isothermal conditions below and above critical temperatures;
from figure 1 we chose 480, 495°C and 525°C.

Before this study, we had noticed that the height of the peak
was a function of quench temperature : on quenching from 850°C it is
decreased and it completely vanishes after quenching from 1150°C (fig.5).
It is possible that quenched-in vacancies accelerate the appearance of
long range order, exactly ordered domains (8, 9).

3. ISOTHERMAL STUDIES AFTER RECRYSTALLISATION AND QUENCH FROM 650°C.

We thought that internal friction might be used to follow the
appearance of long range order during isothermal conditions at 480° and
495°C, because this phenomenon is sufficiently slow in nickel-iron.

After quenching from 650°C, we measured the internal friction
of the sample kept at 480°C (fig.6). Three maxima become visible :
M_1 at 3 h 30 mn after the beginning of heating, M_2 after 10 h. and
M_3 after 66 h. In a new experiment performed at 495°C, we found again
three maxima at shorter times, 2 h 30 mn, 8 h. and 55 h. respectively.
We avoided studies between A_1 and A_3 (fig.1) and the last experiment
was done at 525°C. There is a suggestion of the first maximum, but
the others have vanished.

These results can be explained by the following assumptions :

- M_1 is related to short range order; from calorimetric study of the isothermal ordering of Ni_3Fe, Iida has concluded that short range order forms before long range order manifests itself (10); Davies and Stoloff pointed out a parameter variation after an hour at 480°C corresponding to the etablishment of short range order (11); at 525°C, there seems to be an internal friction change (fig.6).

- M_2 is attributed to the nucleation of ordered domains : long range order develops from nuclei which grow at the expense of the matrix until they come into contact in M_3; this point of view is supported by Calvayrac's results which show the contact beginning after 70 h. at 480°C (7). If an isothermal study is made at 495°C (fig.6) M_2 and M_3 emerge at earlier times; the difference between M_2 and M_3 is 47 h. at 495°C and 56 h. at 480°C : nuclei grow more quickly at 495°C.

The internal friction background grows slowly after M_3 at 480 and 495°C; if we suppose that this represents the growth of ordered domains, an activation energy can be obtained. The experiment was performed in the following way : after quench from 650°C, samples were annealed for 100 h. at different temperatures below 500°C : ordered domains come into contact at 470°, 485°C and 495°C. From the same manner that Kamel (12) we have drawn $a = Q^{-1}_{100h+t}/Q^{-1}_{100h}$ as a function of t. From these curves we deduce $Log\ t = f\ (1/T)$ for a given value of a. We obtain parallel lines corresponding to $t.\ exp\ (-Q/RT) = cte$ from the classical law $D^2 = kt$ with $k = A\ exp\ (-Q/RT)$. The value of the activation energy is 68 kcal/mole, which is near to the 73 kcal/mole found by Calvayrac. These values are larger than that obtained from the Zener peak.

4. VARIATION OF QUENCHING TEMPERATURE.

Other experiments were performed after quenching from 850 and 1150°C. As we have already said, a sample quenched from 1150°C does not present a short range order peak (fig.5).

Isothermal internal friction was measured at 490° on a sample quenched from 850°C. Only two maxima M_2 and M_3 appeared at 4 h. and 40 h. (fig.7), nothing emerged at 530°C. In our opinion, the existence of more vacancies permits this speedier etablishment of short and long range order.

The last experiment concerns Ni_3Fe quenched from 1150° and heated at increasing temperatures for two hours. The results are plotted in fig.8; only at 470°C does the internal friction increases and then decreases as this temperature is maintained; after quenching from 1150°C, we know that short range order does not appear and we think that this phenomenon is related to nucleation of ordered domains at 470°C.

ACKNOWLEDGEMENTS

The authors are particularly indebted to Y. Calvayrac and M. Fayard for many helpful discussions.

REFERENCES

1. Y. Calvayrac and M. Fayard. Mat. Res. Bull. 7, 891 (1972).

2. B. Dubois. Mem. Sci. Rev. Met. 63, 409 (1966).

3. B. Dubois. and G. Bouquet. Jnal de Phys. 32, C-2, 201 (1971).

4. M. Lebienvenu. Thèse 3ème cycle. Paris VI (1973).

5. D. Leclaire. Phil. Mag. 7, 141, (1962).

6. D.P. Séraphim, A.S. Nowick and B.S. Berry. Act. Met. 12, 891 (1964).

7. Y. Calvayrac and M. Fayard. Phys. Stat. Sol. 17, 407 (1973).

8. A. Ferro and G. Griffa. Jnal Phys. Chem. Sol. 31, 2789 (1970).

9. D. Gratias, M. Condat and M. Fayard. Phys. Stat. Sol. 14, 123 (1972).

10. S. Iida. Jnal Phys. Soc. Jap. 7, 373 (1952); 9, 346 (1954).

11. R.G. Davies and N.S. Stoloff. Act. Met. 11, 1347 (1963).

12. R. Kamel. Jnal Inst. Met. 84, 55 (1955-56).

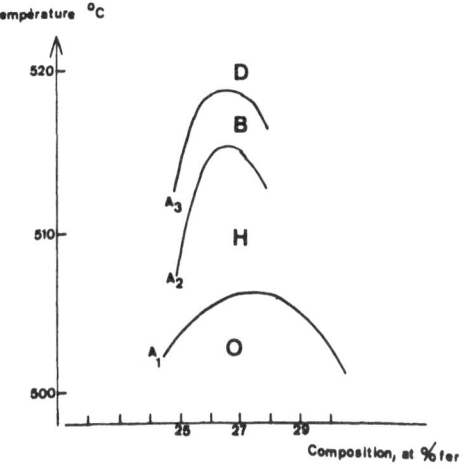

FIG. 1

Structure of Ni-Fe alloys
from CALVAYRAC (1).
D - Disorder.
B - Two phases domain.
H - Hysterisistic single domain.
O - Long range order.

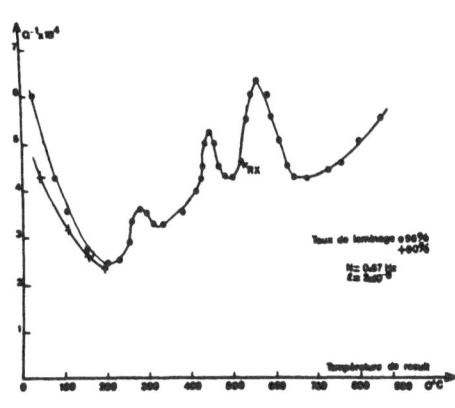

FIG. 2

Internal friction of cold worked
Ni_3Fe with annealing temperature.

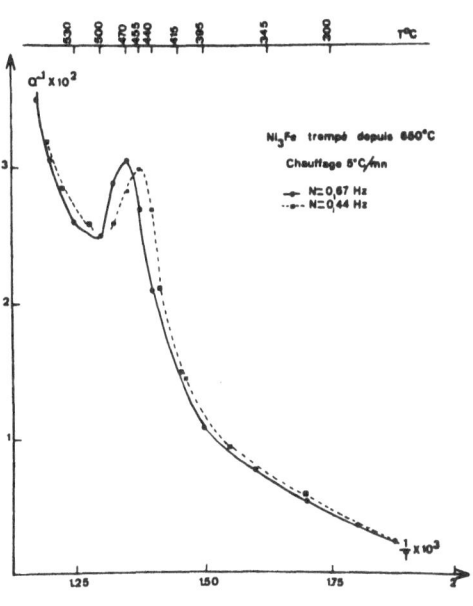

FIG. 3

Variation of Q^{-1} with temperature
for Ni_3Fe quenched from 650°C.
Effect of a frequency change.

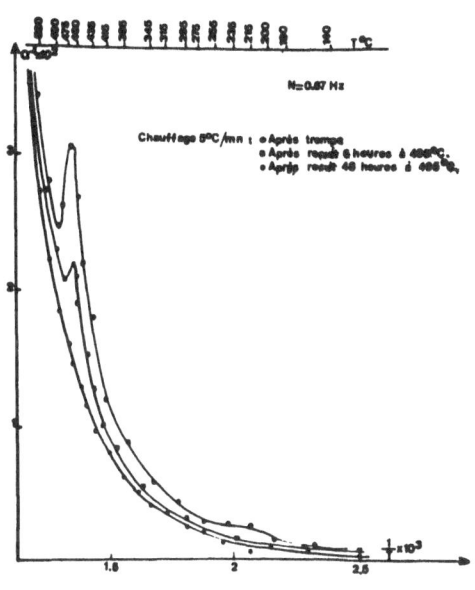

FIG. 4

Damping curves of Ni_3Fe quenched
from 650°C and annealed at different
temperatures.

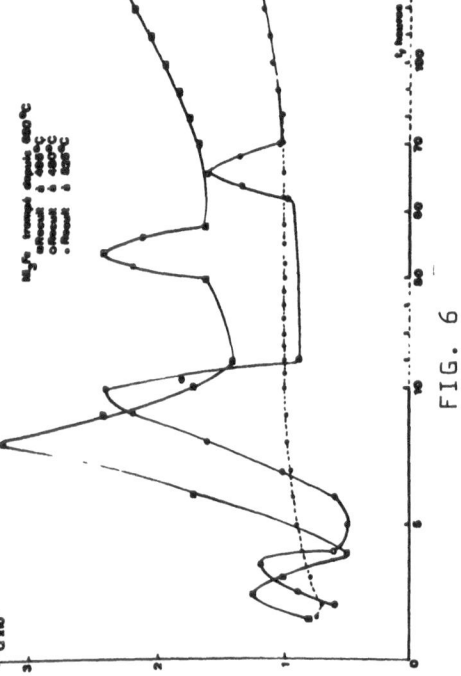

FIG. 5

Influence of quenching
temperature on the
damping – temperature
curves of Ni₃Fe.

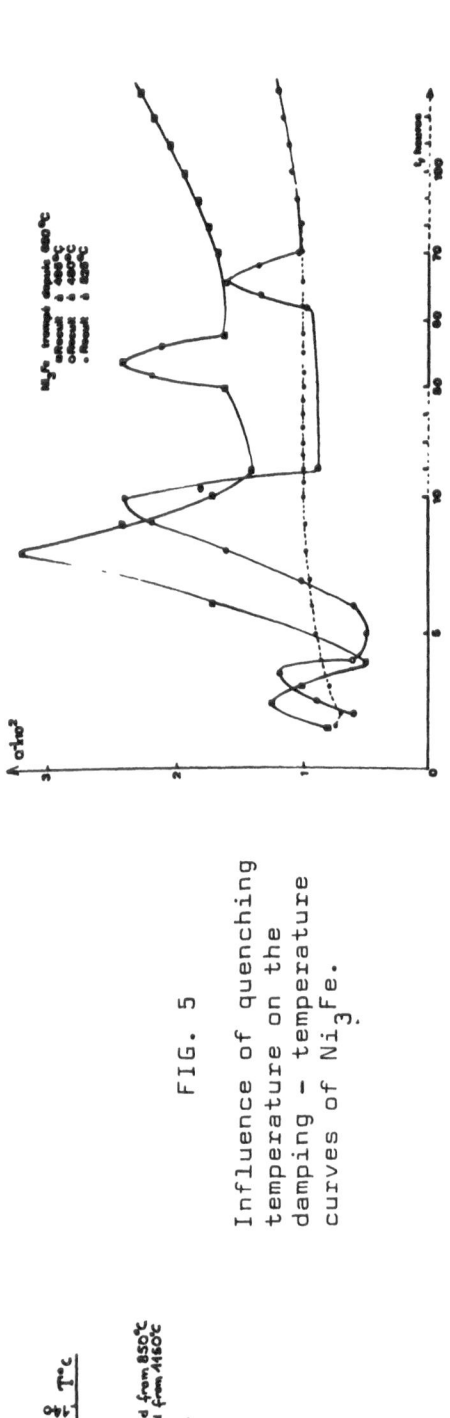

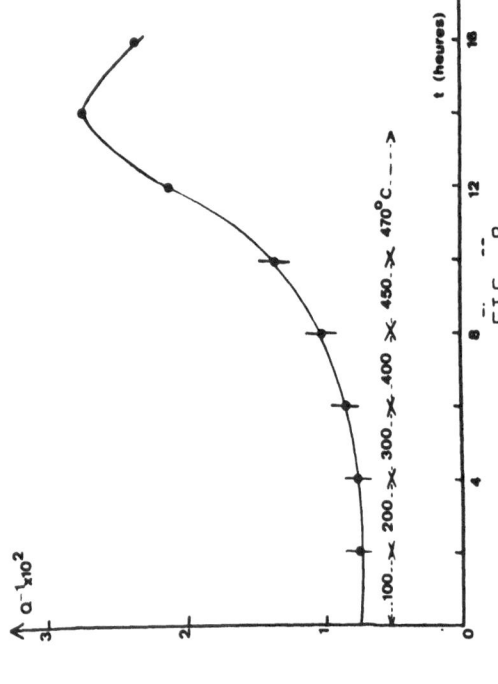

FIG. 6

Isothermal internal friction curves
after quench from 650°C.

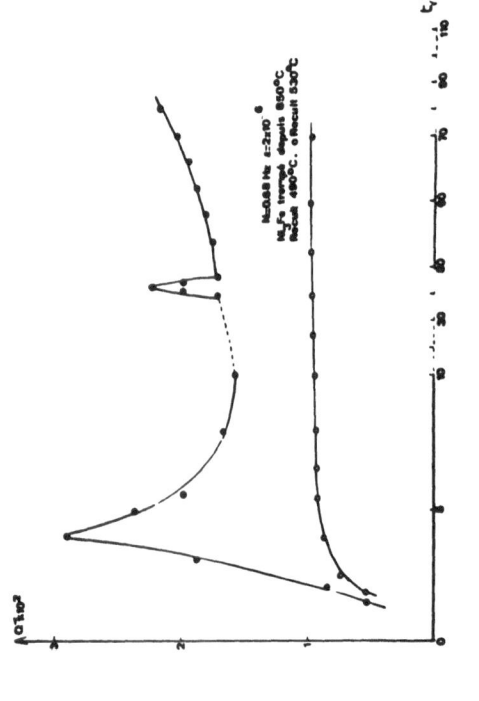

FIG. 8

Internal friction as a function
of time at different temperatures
(Ni₃ Fe quenched from 1150°C).

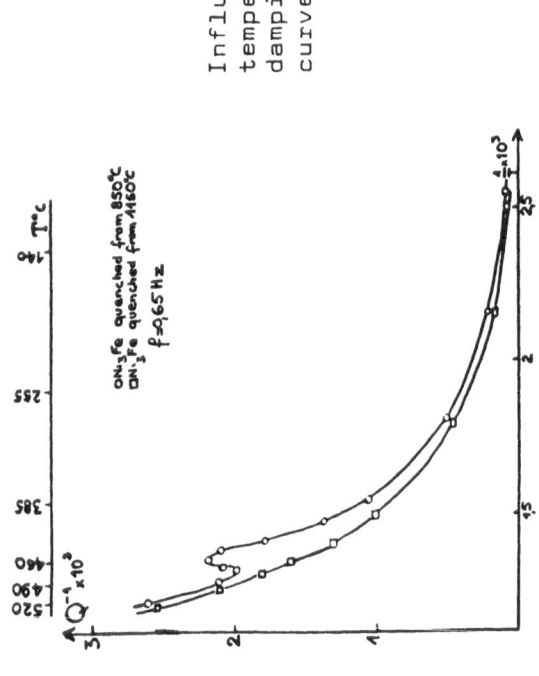

FIG. 7

Isothermal internal friction
curves after quench from 850°C.

THE RELATION BETWEEN INTERNAL FRICTION SPECTRA AND THE ATHERMAL
β ⇄ ω TRANSFORMATION IN Ti-V, Ti-V-O and Ti-V-H ALLOYS

O. Buck, D. O. Thompson, N. E. Paton, and J. C. Williams
Science Center, Rockwell International
Thousand Oaks, California 91360

Abstract

Unexpectedly large effects of oxygen and hydrogen on the low temperature internal friction peaks associated with athermal β ⇄ ω transformation in Ti-V alloys have been observed using combined damping and modulus measurements and cold stage electron microscopy. The measurements show that O suppresses the omega start temperature, whereas H raises it. It is speculated that not only the size of the dopants but also their electronic state contribute to changes occurring in metastable β alloys.

Introduction

In alloys containing IV-B elements, Ti, Zr, Hf and V-B or VI-B elements, such as V, Nb, or Mo in suitable amounts, the high temperature bcc allotropic form can be retained in a metastable state by quenching to room temperature from the single phase bcc field [1]. The resulting bcc phase can undergo a series of decomposition reactions depending on the alloy composition and reaction temperature. Such reactions lead to the formation of two types of martensite and several metastable phases [2]. The decomposition of metastable bcc Ti alloys has been discussed in several recent papers [3-7] both from an experimental and a theoretical point of view with particular emphasis on the formation of the athermal β ⇄ ω transformation. It has been postulated [5] that this athermal β ⇄ ω transformation can be explained by a two-dimensional ordering of linear one-dimensional defects with each defect consisting of a 1/3 vacancy and a 1/3 interstitial in a vernier-like fashion [8]. The possibility that such defects not only form the nuclei of the phase transformation but are also responsible for electron diffraction [3,9] and neutron diffraction effects well above the transformation temperature [10] as well as internal friction [6,11] and anomalous diffusion [12] in such alloys has been pointed out.

Recently, using cold stage electron microscopy and selected area electron diffraction, Paton and Williams [13] have shown that an increase in oxygen content markedly reduces the β ⇄ ω transformation temperature. These authors have suggested that this observation is consistent with an interaction of interstitial oxygen atoms with the linear defects mentioned above.

The internal friction spectrum of high purity metastable β phase alloys was studied before by Nelson et al. [15], Doherty and Gibbons [11], and more recently by Sommer et al. [6]. In general, two very strong internal friction peaks are observed in these alloys, both of which can be attributed to thermally-activated processes. Experiments performed at 20 KHz [6] show that the low-temperature peak (P_L) occurs at about 30°K (or below) while the high-temperature peak (P_H) occurs at about 120-140°K. The process responsible for P_L has not been identified in all details thus far. The only explanation is due to Nelson et al. [15] who speculate that possibly a Jahn-Teller-type distortion [16] could be responsible for this relaxation process. P_H is probably caused by a relaxation process in the β phase; this process has been connected with the $\beta \rightleftarrows \omega$ transformation [6,11,15]. The height of P_H passes through a maximum with increasing solute concentration with the maximum occurring at an alloy composition whose ω_s, the start temperature for ω formation, corresponds to a temperature at which ^sP_H occurs. In an earlier paper, De Fontaine and Buck [5] have discussed the connection between the linear one-dimensional defects, mentioned above, and the $\beta \rightleftarrows \omega$ transformation. They have also suggested that these defects would act as an elastic dipole oriented along <111> which can give rise to an internal friction peak although the exact mechanism is unspecified at the present time [5].

The purpose of the present paper is to investigate the influence of systematic variations in substitutional solute concentration on the internal friction spectrum of Ti-V alloys and then to examine the effect of oxygen and hydrogen content since at least the former of these has been shown to alter independently the $\beta \rightleftarrows \omega$ transformation temperature [13]. Cold stage electron microscopy and electron diffraction studies have also been conducted to support the internal friction results.

Experimental Procedures

The internal friction and modulus measurements were performed on titanium-vanadium alloys in the form of cylindrical bars 12.5 cm long and 6 mm in diameter. The bars were mounted with three set screws at their center of gravity with both ends free; their resonant frequency in the fundamental mode was about 20 KHz and the maximum strain amplitude about 2×10^{-7}. The apparatus used has been described in detail earlier [14]. Measurements were made over a temperature range from about 20 to 300°K. Hydrogenation of a Ti-30V specimen and oxygenation of a Ti-20V specimen were accomplished by heating them up to about 900°C in a micro-Sieverts apparatus containing a known quantity of oxygen or hydrogen. Following charging, the samples were rapidly cooled by withdrawing the furnace from the glass Sieverts apparatus tube and cooling the tube containing the samples to room temperature by a cold air stream. The cooling rate obtained by this procedure was found to be sufficiently rapid to retain the β-phase in Ti-30V but the Ti-20V specimen had to be homogenized at 900°C in argon for about 24 hours and then quenched into ice water.

All measurements discussed in the following were taken during warm-up at a rate of about 0.3°K/min. Measurements were taken during cool-down also but they are less complete than the warm-up measurements since the internal friction was so large that the power input into the drive coil exceeded the heat extraction capability of the heat exchanger, resulting in cessation of cooling at a temperature just above the peak temperature. In those cases the measurements were interrupted until a temperature of 20°K was reached.

Thin foils of several high-purity and oxygenated or hydrogenated alloys were prepared using standard techniques. They were examined in the cooling stage of a Phillips 300 electron microscope at temperatures from room temperature down to about 100°K.

Results

Typical results of internal friction and Young's modulus measurement in high-purity Ti-20V and Ti-30V alloys are shown in Fig. 1. P_H dominates the internal friction spectrum of Ti-20V, whereas P_L dominates the spectrum of Ti-30V. This figure shows that a large modulus defect accompanies the damping peaks in both alloys. The peak height δ_{max} of P_L and P_H as a function of V concentration over

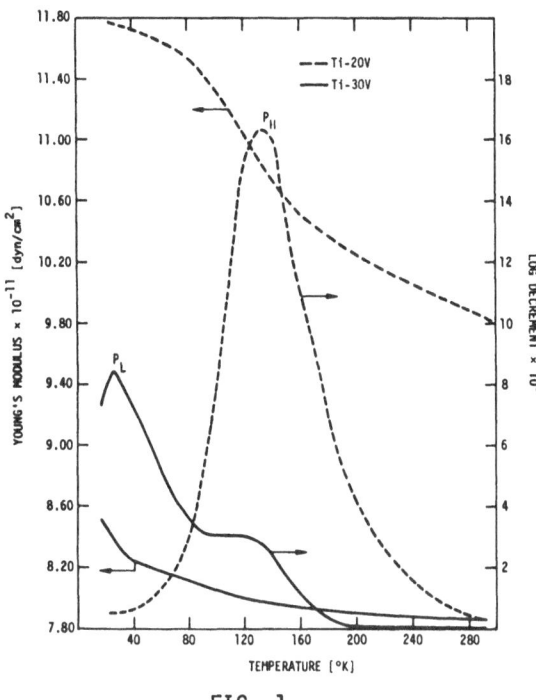

FIG. 1

Young's modulus and logarithmic decrement of Ti-20V and Ti-30V

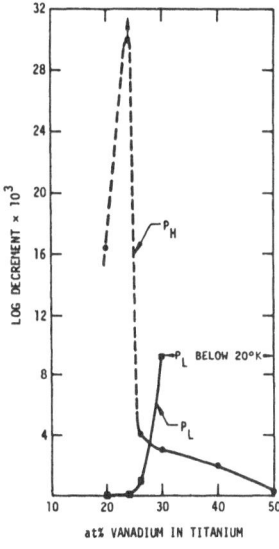

FIG. 2

The heights of the internal friction peaks P_L and P_H, respectively, as a function of alloy composition

the range 20V to 50V are exhibited in Fig. 2 from which it can be seen that the maximum damping is associated with P_H and occurs in the vicinity of 24V; a value of 3×10^{-2} has been established as a lower limit for this maximum. Figure 2 also shows that P_L starts to grow as soon as P_H starts to disappear. Above 30V, P_L drops below 20°K which is out of the temperature range of the present experiment.

The effects of oxygen concentration on the internal friction of a Ti-20V alloy are shown in Fig. 3(a). The starting material had a residual concentration of about 0.08 at% O ($\simeq 0.027$ wt% O). The internal friction measurements are consistent with the results shown in Fig. 1, with the addition of a shoulder in the damping curve at the high temperature side of P_H. Additionally, the low temperature modulus (Fig. 3(b)) is somewhat smaller than shown in Fig. 1. This can be ascribed to a slight variation in V concentration between the two alloys since the modulus is a very sensitive function of V over this concentration range. Increasing the oxygen concentration over the base line level results in a pronounced decrease in the height of P_H (Fig. 3(a)). Concomitant with the decrease in P_H, P_L becomes detectable, although it is not possible at the present time to give a functional relation of its height with oxygen concentration (see also Fig. 3(a)). The modulus measurements (Fig. 3(b)) reflect the damping measurements in that both P_H and P_L are accompanied by modulus defects; quantitative statements cannot be made at present regarding the relation between modulus and oxygen concentration.

Very pronounced changes in the internal friction spectrum accompany hydrogen charging Ti-30V (residual hydrogen content in these alloys is about 0.03 at% H = 6 wt ppm H). These results are illustrated in Figs. 4(a) and (b): Fig. 4(a) shows that the height of P_H increases with increasing hydrogen concentration. Simultaneously, the height of P_L

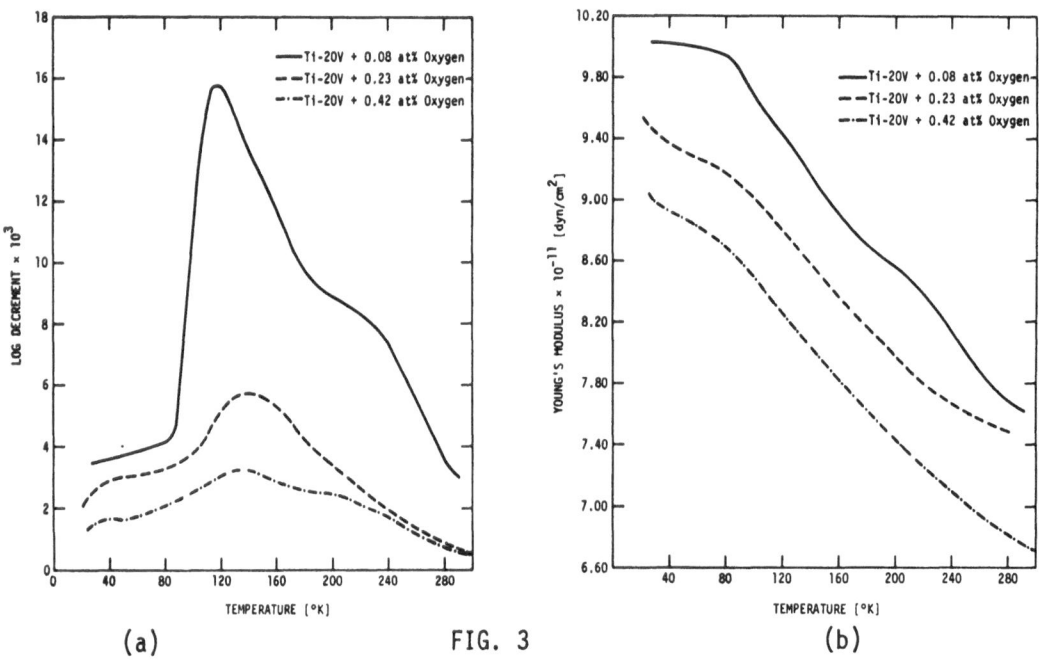

FIG. 3

(a) The logarithmic decrement, and (b) Young's modulus of Ti-20V as a function of temperature and oxygen content

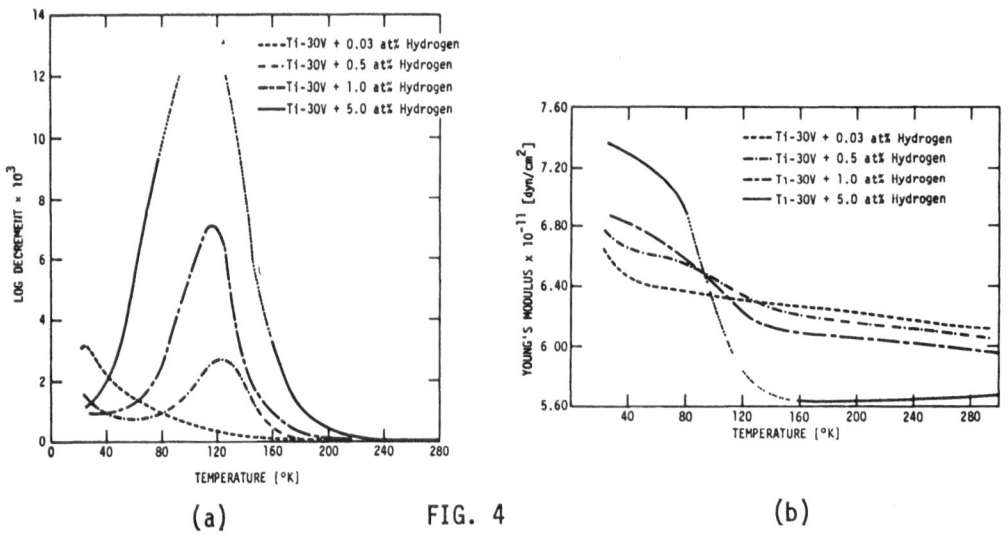

FIG. 4

(a) The logarithmic decrement, and (b) Young's modulus of Ti-30V as a function of temperature and hydrogen content

454

decreases and disappears completely at hydrogen concentrations above 0.5 at% H. The absolute height of P_H for 5 at% H cannot be given because the power inputs required to measure exceed the machine capacity but it can be estimated from the modulus defect at 120°K (Fig. 4(b)). Such an estimate leads to a value for 5 at% H about five times larger than δ_{max} for 1 at% H or a value of about 3.5×10^{-2}. The modulus defect due to P_L disappears with increasing H concentration, whereas that due to P_H increases. Additionally, the temperature coefficient of the modulus above P_H becomes negative in the sample containing 5 at% H. The connection between the magnitude of P_H and the omega start temperature, ω_s, is best illustrated with the aid of Fig. 2, where the height of P_H for the binary Ti-V alloys is a maximum when ω_s is close to the temperature at which P_H occurs (140°K at 20 KHz). If ω_s is above or below 140°K then the magnitude of P_H decreases. Thus the effect of adding O or H on ω_s in Ti-V alloys can be determined by measuring P_H. In the Ti-20V, adding oxygen lowers P_H and from this it can be inferred that ω_s was lowered. On the other hand, adding hydrogen to the Ti-30V raised P_H and ω_s.

The correctness of these deductions was checked using cold stage transmission electron microscopy. The microscopy and diffraction studies showed that the addition of oxygen depresses ω_s. No ω phase was observable at temperatures as low as 100°K for the alloy Ti-20V + 0.42 at% O. This is in contrast to observations on the starting material which contained 0.08 at% O and had an ω_s just below room temperature. Limited microscopy results showed that addition of hydrogen raises ω_s of Ti-30V.

Discussion and Summary

The results obtained in this work generally verify earlier observations on the internal friction spectra of high-purity Ti-V alloys and earlier electron diffraction studies of structural changes [13] in oxygenated Ti-V alloys. No systematic studies on the influence of hydrogen on either the internal friction spectrum or structure of Ti-V alloys have been reported previously; therefore, no comparison with earlier work is possible.

Four specific observations have been made:

1. The internal friction measurements indicate that the omega start temperature, ω_s, is strongly affected by the presence of oxygen and hydrogen; oxygen suppresses ω_s in Ti-20V whereas hydrogen raises ω_s in Ti-30V. This is schematically indicated in Fig. 5 which shows the manner in which O and H shifts ω_s in Ti-V alloys.

2. The low temperature peak P_L seems to grow only when the high temperature peak disappears. This has been found in high-purity Ti-V alloys of variable V content (see Fig. 2), as well as in hydrogenated Ti-30V (see Fig. 6) and in oxygenated Ti-20V (as indicated by the results presented in Fig. 3).

3. The internal friction results on peak P_H and limited electron microscopy results are in mutual agreement and, therefore, support the suggestion that the same mechanism which causes streaking in the diffraction patterns also causes the peak P_H.

4. The effects of oxygen and hydrogen on Young's modulus are unexpectedly large.

As has been pointed out before [15] the high temperature peak P_H can be explained by assuming that the compressional part of the linear displacement defect forms an elastic dipole in a <111> close-packed direction. It is suggested that an

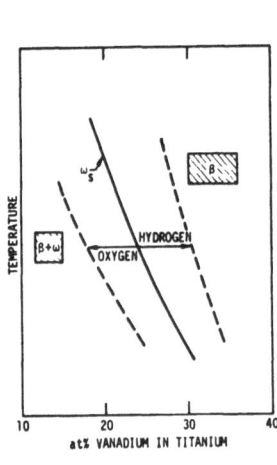

FIG. 5

The omega start temperature as a function of alloy composition (schematically)

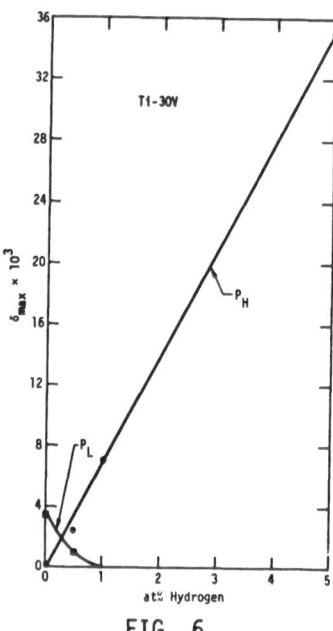

FIG. 6

The heights of the internal friction peaks P_L and P_H, repsectively, for Ti-30V as a function of hydrogen content.

external stress would modify the population of <111> defects in preferred directions thereby giving rise to energy dissipation. At the same time this defect can be locked into ω positions by the cooperative motion of similar defects on neighboring <111> rows thereby forming the nuclei for the athermal ω phase. Combined electron transmission and internal friction experiments indicate that the dissipation process operates to the maximum extent at a temperature just above that where the ω-phase actually appears since such a temperature corresponds to the greatest population of linear displacement defects [6].

A complete explanation of the doping experiments presented above is not possible at present due to the complexity of the subject. However, it should be recalled that earlier studies [17] have shown that the β-phase becomes more stable as the electron:atom ratio increases. Such increases in stability are accompanied by a reduction in the ω_s temperature. The present results suggest that size and/or electronic effects of oxygen and hydrogen can contribute to the changes occuring in these alloys. For the concentrations studied here, oxygen and hydrogen exist in the bcc lattice as an interstitial solid solution but the two atom types probably occupy different sites in the Ti-V lattice.

The effects of hydrogen on the internal friction spectrum of Ti-30V could then be rationalized by assuming that the hydrogen becomes a negatively charged ion (acceptor) or possibly forms a negatively-charged VH complex. Thus the addition of H would have the same effects as a reduction in V concentration. Indeed, the results in Figs. 2 and 6 show that the height of P_H in Ti-30V + 5 at% H compares favorably to the height of P_H in a high-purity alloy of composition close to Ti-24V. The size of the H ion should be small so that it fits well into an interstitial site without generating a stress field since H does not increase the yield stress σ_y signficantly [18].

If the above line of reasoning is applied to rationalize the effect of oxygen on the internal friction spectrum of Ti-V alloys, it is obvious, as will be shown below, that electronic effect alone cannot account for the observation that oxygen suppresses the $\beta \rightleftarrows \omega$ transformation [13]. It is proposed that the observed effects of oxygen are the combined result of an electronic and an elastic interaction of oxygen with the bcc Ti-V matrix. Assuming that the oxygen forms a positively-charged ion (donor), the addition of 0 would have the same effect as an increase in the V concentration. In light of the results presented in Figs. 2 and 3, the height of P_H in Ti-20V + 0.42 at% O compares favorably to the height of P_H in high-purity Ti-30V. It is hard to imagine, however, that such a small O addition changes the e/a ratio as much as is expected from the change in V content. The observation [18] that small additions of O increase σ_y significantly, suggests a large misfit of the O ion in the Ti-V lattice. Its associated stress field interacts with the stress field of the linear displacement defect, thus suppressing the formation of the ω phase. However, the presence of electronic effects should not be neglected since they are supported by the earlier work of Sass [19] who found that the meta-stable bcc phase could only be retained on quenching a Ti-75% Zr alloy if it was contaminated by oxygen. This suggests that oxygen acts like a donor thus increasing the electron concentration in Ti-Zr alloys in a manner similar to the addition of V, Mo, Cr, and other known β-stabilizing elements.

It should be mentioned that the present authors are aware that the assumption of a positively charged O ion is in contradiction to arguments which predict a negatively charged ion based upon the electronegativity difference [20] of O with respect to Ti or V. As has been pointed out elsewhere [21] such arguments do not hold in very dilute alloys, and should therefore not be used in the present context.

An interpretation of the low-temperature peak P_L cannot be given at the present time. Considering the height of P_L in the Ti-30V (see Fig. 1), it is suggested, however, that it is due to an alternate, low-temperature atomic re-arrangement which is competitive with the linear displacement defect and which is suppressed in alloys which form the ω phase. Low-temperature electron microscopy (in the vicinity of the liquid helium temperature) is necessary to yield a definite answer.

The present results may have raised more questions than have been answered. Further experiments are necessary to obtain a clearer picture not only on the internal friction measurements, but also (and perhaps more important) on the mechanical properties of high-strength titanium alloys.

Acknowledgements

Many helpful discussions with Drs. H. L. Marcus, T. Wolfram, C. S. Burton, and F. J. Morin are acknowledged. The experimental assistance of M. W. Mahoney, P. Q. Sauers, R. A. Spurling, and R. V. Inman also are gratefully acknowledged.

References

1. R. I. Jaffee in "Progress in Metal Physics," Vol. VII, Pergamon Press, London, 1958, p. 65.

2. J. C. Williams in "Titanium Science and Technology," Vol. 3, R. I, Jaffee and H. M. Burte, Eds., Plenum Press, 1973, p. 1433.

3. D. De Fontaine, N. E. Paton and J. D. Williams, Acta Met. 19, 1153 (1971).

4. N. E. Paton and J. C. Williams, Proc. II Int. Conf. on Strength of Metals and Alloys, American Soc. for Metals, 1970, p. 108.

5. D. De Fontaine and O. Buck, Phil. Mag. 27, 967 (1973).

6. A. W. Sommer, S. Motokura, K. Ono, and O. Buck, Acta Met. 21, 489 (1973).

7. J. C. Williams, D. De Fontaine, and N. E. Paton, Met. Trans. (in press).

8. H. R. Paneth, Phys. Rev. 80, 708 (1950).

9. S. L. Sass, "Proceedings of the Conference on the Local Structural Order and Decomposition of Titanium, Uranium, and Zirconium-Base B.C.C. Solid Solutions," Organized by S. L. Sass, Cornell University, New York, 1972 (unpublished).

10. D. T. Keating, J. D. Axe, and S. C. Moss, "Proceedings of the Conference on the Local Structural Order and Decomposition of Titanium, Uranium and Zirconium-Base B.C.C. Solid Solutions," Organized by S. L. Sass, Cornell University, New York, 1972 (unpublished).

11. J. E. Doherty and D. F. Gibbons, Acta Met. 19, 275 (1971).

12. A. D. LeClaire in "Diffusion in Body-Centered Cubic Metals," American Soc. for Metals, Metals Park, Ohio, 1965, p. 3.

13. N. E. Paton and J. C. Williams, Scripta Met. 7, 647 (1973).

14. D. O. Thompson and F. M. Glass, Rev. Scient. Instr. 29, 1034 (1958).

15. C. W. Nelson, D. G. Gibbons, and R. F. Heheman, J. Appl. Phys. 37, 4677 (1966)

16. L. E. Orgel, "Introduction to Transition-Metal Chemistry: Ligand-Field Theory," John Wiley & Sons, New York, 1960.

17. C. A. Luke, R. Taggart, and D. H. Polonis, Trans. Am. Soc. Metals 57, 143 (1964).

18. J. C. Williams and N. E. Paton, unpublished results.

19. S. L. Sass, Acta Met. 17, 813 (1969).

20. L. Pauling, "Nature of the Chemical Bond," Cornell Univ. Press, Ithaca, New York, 1959.

21. G. G. Libowitz and T. R. P. Gibb, J. Phys. Chem. 60, 510 (1956).